MANUFACTURING

"DPC Corporation"
Problem: How to anticipate customer complaints
Type of analysis: Sampling, confidence intervals

Where Found: Part I: page 28
              Part II: page 72
              Part III: page 365

"Strangways Textiles, Inc."
Problem: Evaluation of flame-retardant qualities of additives used in connection with textiles manufacturing
Type of analysis: Analysis of variance

Where found: page 566

MARKETING

"Glamour Burger Systems"
Problem: To determine how the company can improve its competitive position vis-à-vis Burger King and McDonald's
Type of analysis: Survey sampling; cross-classification analysis; hypothesis testing involving two population proportions; chi-square analysis

Where found: Part I: page 27
             Part II: page 125
             Part III: page 464
             Part IV: page 620

"The Duplikwik Corporation"
Problem: Determining the optimum method of allocating a salesperson's time
Type of analysis: Multiple regression and correlation analysis

Where found: page 742

RESEARCH

The "Universal Mutual Life Insurance Company"
Problem: To determine the public's attitudes toward life insurance and evaluate the impact of the company's television commercials
Type of analysis: Survey sampling; interpretation of summary tables

Where found: Part I: page 70
             Part II: page 120

"A.F.R. Consulting"
Problem: To evaluate the feasibility of a proposed new product
Type of analysis: Confidence intervals

Where found: page 369

SERVICES

The U.S. Postal Service
Problem: How to reduce mistakes in mail handling
Type of analysis: Sampling

Where found: page 73

"Dixon and Doxey Engineering, Inc."
Problem: To determine whether power plant construction-time delays are greater for pressurized-water-reactor plants or for boiling-water-reactor plants
Type of analysis: Hypothesis testing for means of two populations; analysis of variance

Where found: Part I: page 436
             Part II: page 503

Applied Business Statistics

Applied Business Statistics

Text, Problems, and Cases

STEPHEN K. CAMPBELL
University of Denver

1817

HARPER & ROW, PUBLISHERS, New York
Cambridge, Philadelphia, San Francisco, Washington,
London, Mexico City, São Paulo, Singapore, Sydney

To Scott and Melanie, the Campbell kids

Sponsoring Editor: Peter Coveney
Project Editor: Susan Goldfarb
Text Design: Leon Bolognese
Cover Design: Jack Ribik
Text Art: Vantage Art, Inc.
Production Manager: Jeanie Berke
Compositor: York Graphic Services, Inc.
Printer and Binder: R. R. Donnelley & Sons Company

Grin and Bear It by Fred Wagner © by and permission of News America Syndicate. Cartoons appear on the following pages: 9, 34, 35, 141, 202, 290, 385, 571, 661, 778, 884, 902, 925

Applied Business Statistics: Text, Problems, and Cases

Copyright © 1987 by Harper & Row, Publishers, Inc.

Library of Congress Cataloging in Publication Data

Campbell, Stephen K. (Stephen Kent)
    Applied business statistics.

    Includes index.
      1. Commercial statistics. 2. Commercial statistics—Case studies. I. Title.
HF1017.C36   1987   519.5′024658   86-9833
ISBN 0-06-141167-8

86 87 88 89 9 8 7 6 5 4 3 2 1

Brief Contents

Detailed Contents

Note: An asterisk to the left of an entry signifies that the topic is optional.

9 INTRODUCTION TO STATISTICAL HYPOTHESIS TESTING 373

16   INTRODUCTION TO TIME SERIES ANALYSIS 747

17 SHORT-TERM FORECASTING WITH AUTOPROJECTIVE METHODS 803

Optional Sections in *Applied Business Statistics*

To the Student

For many years astrologers, tarot card interpreters, crystal ball gazers, phrenologists, iridologists, spiritualists, faith healers, prophets, character analysts, tea leaf readers, and other members of the fortune-telling fraternity have made profitable use of a complicated bit of trickery known as the "cold reading." Here is how a cold reading works: A client with a problem seeks out the fortune-teller for advice. If the client harbors any initial skepticism, it is quickly dispelled as the fortune-teller proceeds to demonstrate that he or she knows more about the client's personality and problems than any ordinary mortal could possibly know. As a layman, I don't pretend to know very much about how this aspect of fortune-telling works. I do know that sometimes a deft reading of the client's clothing, jewelry, and mannerisms will serve to make the fortune-teller appear remarkably insightful. I also know that when the financial stakes are sufficiently high—as they often are with spiritualism and faith healing—the fortune-teller will engage in prior research regarding the client's life, health, and relationships. Whatever the specific techniques used, the client is encouraged to believe that such marvelous revelations could only have been supplied by a supernatural source. Naturally, the client is going to place a high value on—and be willing to pay dearly for—advice from someone having a direct personal pipeline to the supernatural.

What does all this have to do with statistics? Nothing, really. In fact, statistics and fortune-telling are about as far apart as two activities can be. I resort to this rather far-out introduction because somehow I feel impelled to attempt a cold reading of you. Even though we have never met, my psychic vibrations tell me that

1. You are taking a course in statistics because it is required and not because of any long-standing thirst for knowledge about the subject.
2. You are a little vague on what statistics is all about, anyway.
3. You are quite skeptical about the subject's usefulness to you in your future career endeavors.
4. You are at least a little nervous about taking the course because other students have said such scary things about it.

Does this cold reading fit you pretty well? If not, I shall divert attention away from my embarrassment by inviting you to proceed immediately to Chapter 1. The remainder of this preface is aimed at the reader who is fairly accurately described by my cold reading.

How did I achieve such uncanny accuracy sight unseen? Certainly not by supernatural means. I merely drew upon my experiences as a conscripted student of statistics back in my undergraduate days and as a teacher of the subject now. I have found that the majority of students in introductory business statistics courses do indeed fit the above profile. Not only that, but so do I—or, at least, so *did* I. I recall very vividly my first course in statistics. To say that I was frightened would be an understatement. Although I had been a reasonably good student of mathematics, I felt no particular fondness for the subject. Moreover, it seemed as if a greater number of frightening rumors circulated about the statistics course than about any other required course in the college of business. Imagine my surprise when I actually found myself enjoying the subject! Imagine my surprise when I discovered that the techniques I was learning about could indeed be applied to practical business problems. This reluctant romance of mine with the subject of statistics led to a dramatic and permanent change in my career plans—a change I have never regretted.

Granted, your experiences with the subject are not likely to be identical to my own. I draw upon my experience to emphasize that my cold reading once applied to me just as aptly as it now applies to you. While writing this textbook, I have taken pains never to forget that fact. I have tried to write a book that is "user-friendly," as some of my computer-oriented colleagues might put it.

What are the characteristics of a user-friendly statistics textbook? As I see it, such a book should

- Be written in a friendly, relaxed style and with sufficient clarity that students can learn some of the material by themselves, if necessary.
- Provide intuitively satisfying explanations of why the various statistical methods work the way they do.
- Emphasize especially important formulas, definitions, assumptions, and precautions.
- Make ample use of plausible anecdotal material to show how statistical techniques can be applied to business problems.
- Draw upon cases to demonstrate how statistical procedures have helped real managers address and deal with actual business challenges.
- Treat the subject in such a way that only a modest amount of mathematical knowledge is required of students.
- Provide a large selection of practice problems.
- Point out at the beginning of each chapter which earlier material students might need to review before proceeding and just what they are expected to learn from the chapter.

I have tried to write a book having all of these characteristics. You, of course, will be the only meaningful judge of how well I have succeeded.

In addition to meeting the above criteria, a user-friendly statistics textbook should have some well-designed supplemental learning aids. For this reason, Per Olsson and I have prepared a workbook to accompany this text. The workbook contains (1) some general instructions on using available computer packages; (2) additional worked problems, (3) additional practice problems, and (4) self-help quizzes.

Notice that I have nowhere said that a user-friendly statistics textbook has to be easy. I have tried to make this one as rigorous as necessary—but only as rigorous—to convey the concepts and techniques correctly. Moreover, I have not shrunk from treating some fairly advanced topics. Several optional sections of this text, especially those in the later chapters, deal with some statistical topics in more depth and detail than is true of most introductory textbooks in the field. My goal has not been to avoid difficult topics; it has been to make certain that those topics don't appear any more difficult than they actually are.

It is my sincere wish that you and this textbook will become—if not fast friends—at least amiable companions.

Stephen K. Campbell

To the Instructor

Applied Business Statistics: Text, Problems, and Cases has been designed with two-semester (or two-quarter) graduate or undergraduate introductory business statistics courses in mind.* The only mathematical prerequisite is high school algebra. Actually, unless your teaching pace is a good deal more brisk than my own, you will have trouble covering all of the material, even in two semesters. I like to think that this is more of an advantage than a disadvantage. Perhaps I have been unduly influenced by my fondness for do-it-yourself banana splits—the kind where *you* choose the flavors of ice cream and the toppings and decide whether *you* want whipped cream, nuts, and cherries. In any event, I have tried to write a textbook that provides the instructor with some of this same kind of discretionary power. Granted, certain subjects simply have to be covered in a basic statistics course. Still, between "Cases for Analysis and Discussion" and the optional sections appearing in most chapters (both of which are described more completely below), the teacher of a two-semester course should enjoy a somewhat greater sense of freedom to emphasize certain topics and de-emphasize others than most other introductory business statistics textbooks permit conveniently.

Even though this text is quite orthodox in many ways, there are some unorthodoxies and special features. These are discussed below under the headings "Organization," "Breadth of Coverage," "Questions and Problems," "Cases for Analysis and Discussion," and "Special Pedagogical Aids."

Organization

I have not departed very much from the kind of organizational structure used in most introductory business statistics textbooks. The following remarks, therefore, are limited to the differences which *are* found here.

1. Material on sampling, usually split between the beginning of a chapter on sampling distributions and a chapter near the end of the book, is treated in Chapter 2 of this text. Although I understand the reasons behind the traditional manner of positioning this material, I have never agreed with it. Most of the chapters in any statistics textbook are dependent upon the assumption that sampling has been carried out. In my view, the subject should be dealt with right away to get the student thinking about sampling—why it is used and how it is done.
2. The subject of index numbers is treated in Chapter 5—much earlier than in most basic textbooks. My experience has suggested that the subject works better in close proximity to the topic of weighted averages than it does when combined with, or

*This book could be used in a one-semester (or one-quarter) course by covering Chapters 1 through 4, 6 through 10, and, if time permits, choosing one or more chapters from among Chapters 5, 11, 14, 16, 18, and 19.

placed close to, the subject of time series analysis. However, this chapter is quite self-contained and can be taught just about anywhere you prefer.

3. The subjects of sampling distributions and confidence intervals are introduced together in Chapter 8. In this chapter I adopt a question-and-answer format.

4. There are two chapters on analysis of variance. Chapter 11 is introductory in nature and deals with the subject adequately enough to permit your skipping over Chapter 12, if that is your wish. Chapter 12 is a fairly rigorous chapter concentrating on the application of analysis-of-variance methods to business-oriented experiments. Between the two chapters, analysis of variance receives a broader and deeper treatment than it enjoys in most competitive textbooks.

5. Chi-square analysis and other nonparametric methods are presented in Chapter 13, a rather comprehensive chapter on the subject.

6. There are two chapters on time series analysis. Chapter 16 deals with the classical approach. Chapter 17 concentrates on autoprojective methods; that is, no-change models, moving averages, exponential smoothing, and autoregressive models. A substantial part of this chapter is labeled optional and deals with Box-Jenkins methods.

Breadth of Coverage

This textbook boasts a greater breadth of material than that offered by most competitive texts. I realize that this is a claim easily and often made and one difficult to quantify. Nevertheless, the list of optional sections on p. xx should lend some credence to this claim. It will also serve to amplify the point made above regarding the flexibility this text affords.

Questions and Problems

This text contains approximately 700 "Questions and Problems." These problems have about four parts each, on average. The "Questions and Problems" are quite varied in nature. For example, all chapters have problems of the following kinds:

1. *"Define the following terms."* These problems force the student to review and rephrase definitions of key terms introduced within the chapter.

2. *"Distinguish between. . . ."* These problems get the student to focus on the sometimes subtle differences among similar-appearing terms, symbols, formulas, and theorems.

3. *"Indicate which of the following statements you agree with and which you disagree with, and defend your opinions."* These problems also require the student to pay close attention to subtleties.

4. *Problems to solve.* These are primarily skill-building problems. Some require only application of the appropriate statistical techniques. Most, however, require the student to grasp a described situation, recognize which statistical procedure is called for, and apply the selected procedure. The great majority of these problems ask the student to do more than just come up with an answer. Usually, the student is encouraged to interpret the results and tell how the results might be put to practical use.

5. *"Take charge."* These are relatively unstructured problems requiring the student to design and carry out a very small-scale research project.

A number of chapters contain at least one "Computer Exercise" involving the analysis of large data sets. Some chapters contain several.

A few chapters contain what I call "Special-Challenge Problems." This simply means that use of techniques described in the chapter *will not quite lead to the correct solution.* These techniques must be extended or modified in some way. The student must determine what kind of modification is called for before proceeding with the computational steps.

Cases for Analysis and Discussion

Each chapter of this text boasts from two to five "Cases for Analysis and Discussion." There are over 50 such cases altogether. I am quite proud of these, even though I had relatively little to do with their preparation. Indeed, my pride in them stems largely from the fact that I *didn't* have much to do with their preparation. With only a few exceptions, these cases were provided by middle managers of quite a wide variety of businesses and governmental organizations who have graduated from the University of Denver's Executive MBA program. I have honored the wishes of these donors to withhold their names and the names of their companies (pseudonyms are signified by quotes). I have also taken the liberty of reducing the length of some of the narratives. Still, the problems addressed by these problems and the statistical methods used in connection with them are very real. Unlike the "cases" in some recently published textbooks, the ones appearing here have not been made up by the author, nor have they been drawn from readily accessible academic journals. In my opinion, a contrived case is seldom better than no case at all. Moreover, cases appearing in academic journals, valuable as they are for some purposes, suffer from a "basic research feel" as distinguished from a practical problem-solving feel.

All cases presented here are accompanied by one or more questions for the student to answer or (since many do not have single, pat answers), at least, ponder. Most lend themselves to in-class argumentation if you choose to use them in that way.

Where it has been meaningful to do so, both the problems and the cases have been coded according to their areas of applicability—marketing, management, and so forth. The following emblems are employed for this purpose:

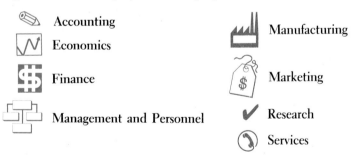

Accounting

Economics

Finance

Management and Personnel

Manufacturing

Marketing

Research

Services

Special Pedagogical Aids

The following special features have been incorporated into this text in the hope of making the study of statistics more enjoyable and/or easier to grasp than it might otherwise be:

- Each chapter begins with a section titled "What You Should Learn from This Chapter." This section lists the concepts and skills the student should seek to master while proceeding through the chapter.
- Supplementing the list of topics under "What You Should Learn from This Chapter" is a summary of subjects from previous chapters the student might want to review. This feature is apparently unique to this text.
- Each chapter ends with a summary; a list of terms introduced in the chapter, with a reference to the appropriate page number; and a list of formulas introduced in the chapter.
- Especially important definitions, formulas, assumptions, warnings, and procedural notes are highlighted in boxes throughout the text.
- In an effort to convince students that statistics can actually be put to practical use, I have included much anecdotal material. The illustrative cases within the chapters, most of the questions and problems, and the cases at the end of each chapter are all anecdotal in nature.
- There are five large data sets presented in the text to be used in connection with the computer exercises.
- A number of "Grin and Bear It" cartoons have been included just for the fun of it.
- An Appendix gathers together the traditional statistics tables for easy reference.

Supplementing the text are (1) the student workbook mentioned in the preface to the student; (2) an instructor's resource manual including thoughts on using the textbook, suggestions for teaching the individual chapters, computer instructions, and transparency masters; (3) an instructor's solutions manual containing answers to the questions and problems as well as notes on the cases for analysis and discussion; and (4) a test bank.

I would like to thank the following reviewers, who provided invaluable suggestions and comments at various stages of the writing and revising process:

Paul Baroutsis, Slippery Rock State College
Robert F. Brooker, Gannon University
Daniel G. Brooks, Arizona State University
William E. Burrell, Wayne State University
P. O. Claypool, Oklahoma State University
Dinesh S. Dave, Marshall University
Ann Maruchek, University of North Carolina
John Nash, University of Ottawa
Leonard Presby, William Paterson College
Carol Stamm, Western Michigan University
Donald R. Williams, North Texas State University
William Ziemba, University of British Columbia

Should the gods decree that there is to be a second edition of this text, your suggestions for improving upon it so that it will suit your purposes more effectively will be warmly welcomed.

Stephen K. Campbell

Applied Business Statistics

1 Statistics and Modern Business

> The difference between well-managed companies and not-so-well-managed companies is the degree of attention they pay to numbers, the thermometer chart of their business.
> —Harold S. Geneen

> The new tools of applied mathematics offer insights and rationality of extraordinary power and value. No manager of the future will be able to ignore them, since the days of the purely intuitive manager are numbered. —George S. Odiorne

What You Should Learn from This Chapter

The primary goal of this chapter is to show you how statistical analysis can aid management in the ongoing process of business decision making. The chapter also introduces some important terminology. When you have finished studying Chapter 1, you should be able to

1. Define *statistics* in both its narrower and broader meanings.
2. Define *elementary unit*, *variable*, and *observation*.
3. Identify variables that are (*a*) qualitative, (*b*) quantitative, (*c*) discrete, and (*d*) continuous.
4. Distinguish between a population and a sample.
5. Distinguish between descriptive statistics and inductive statistics.
6. Distinguish between deductive inference and inductive inference.
7. Distinguish between a survey and a designed experiment.
8. Describe the process involved in carrying out a statistical analysis.
9. Describe how statistical analysis may be used within the managerial decision process.

1.1 Introduction

You are about to begin studying a subject that will require diligent effort to master.

That's the bad news.

The good news is that the results will almost certainly be worth the effort. Quite aside from the stimulating intellectual challenge offered by the subject, there are two very good practical reasons for you to take the study of statistics seriously.

First, as you pursue your business degree, whether in finance, marketing, management, accounting, economics, or something else, you will find statistics and other quantitative subjects to be an integral part of your course work. As recently as 20 years ago, this was not the case; at most universities, courses in these disciplines had little, if any, statistical content. Today, the student unskilled in statistics toils under a serious competitive disadvantage.

Second, the past few decades have witnessed a rapidly growing emphasis on the use of statistical techniques in connection with business decision making. Let us consider some fairly representative examples:

ILLUSTRATIVE CASE 1.1

Management of a division of Standard Brands, Inc., a major food-processing company, was considering whether to introduce a new brand of margarine, one made entirely from corn oil rather than from the usual cottonseed oil. The question of whether to produce and market the new margarine on a large scale was doubly difficult in that almost nothing was known about consumer interest in a corn oil margarine and the new margarine would necessarily be more expensive than competing brands.

To reduce their uncertainty, management decided to conduct a small-scale exploratory study involving personal and mail interviews with a cross section of consumers. Finding that considerable interest in a corn oil margarine *did* exist, they proceeded with a market test on a larger scale. The market test confirmed the results of the exploratory analysis. The story has a happy ending: The margarine we have been talking about is Fleischmann's, for many years one of the leading brands in margarine sales.[1]

ILLUSTRATIVE CASE 1.2

The Pillsbury Company at one time was faced with the question of whether to switch from a box to a bag to package a certain grocery product. The sales manager favored retaining the box on the grounds of greater consumer appeal. But the brand manager preferred the bag because of its lower cost; he supported his position with a statistical decision analysis (the subject of the final two chapters of this book) based on his best assessment of the bag's acceptance. Even when the sales manager's more conservative assessment of the bag's acceptance was substituted in the relevant equations, the bag still appeared potentially more profitable. During the period of debate, some Pillsbury executives urged that the bag be test-marketed before a final decision was made. The statistical decision analysis, however, showed a nine-in-ten chance that the bag would be profitable. A simple supplementary analysis showed that the value of any information obtained from a marketing test could not conceivably offset the cost of acquiring it. Management's confidence in this statistical decision analysis was vindicated when the bag proved a conspicuous success.[2]

ILLUSTRATIVE CASE 1.3

Business economists at a major brokerage firm sought to determine what factors cause airline revenues to fluctuate. A search for helpful explanatory variables disclosed that next year's revenues are rather strongly dependent on this year's aggregate corporate profits. Evidently, the realization of high profits during one year encourages corporate leaders to be lenient about granting approval for employee air travel the following year. This finding was helpful because it enabled the economists to use a known corporate profits figure to estimate year-ahead revenues for airlines and pass the resulting estimate along to the firm's security analysts who specialize in this important industry. The security analysts used this information to determine whether the industry as a whole would be healthy or sluggish during the upcoming year. The information also helped them to determine which airlines would probably do best, because some airlines are more sensitive than others to fluctuations in overall corporate profits.

ILLUSTRATIVE CASE 1.4

Although many critics disliked it, the movie *The Omen* was a great box office success. The following article from *Forbes Magazine*[3] summarizes the steps involved in planning the picture and points out the central role that sampling played.

"OMEN TORCHES," said *Variety*, the show biz weekly. Translation: Twentieth Century Fox's low budget movie *The Omen*—it cost a mere $2.8 million to make—has caught fire. It has already grossed $80 million in North America, and probably will double that with foreign exhibition.

Marketing movies has been very largely a matter of gut feel, but this campaign was planned as carefully as one for a new toothpaste or a new brand of cigarettes. The marketer: Brooklyn-born Allan Freeman, 39. Freeman, Fox's marketing research director, realized he had a problem with this story of a monstrous child, literally the anti-Christ. After the success of *The Exorcist,* so many occult-type cheapies flooded the market that the whole genre got a bad name. How to market another satanic script? A former marketing analyst at Bristol-Myers and General Foods, Freeman decided to take several market surveys. Using a randomly selected sample of 6000 moviegoers, he discovered that women were turned off by the violence of previous occult films; seeing one was enough. That was half the potential audience gone. And older folks were not attracted. Only young urban males said they would see another horror occult film. From this survey came the name Omen— denoting mystery and the unknown, but not violence—"ominous but not bloodchilling" in Freeman's words. He conceived an ad which was all text—unusual for a movie—with a black background. In a subliminal way the ad denoted quality—but scariness. The unknown, yes. Blood and gore, no.

It worked.[*]

An even more global picture of the broad use of quantitative techniques in business is painted by the following tables excerpted from a 1984 report by Forgionne.[4] The report was based on a questionnaire survey of a random sample of corporate executives. Each executive represented a different firm among the 1500 largest American-operated corporations listed in the Economic Information Systems Directory. Table 1.1 shows the results of a somewhat general question, and Table 1.2 the results of a question pertaining to marketing in particular. (Of the analytical methods listed in the tables, regression analysis, statistical decision analysis, forecasting, market share analysis, surveys and experiments, and time series analysis receive substantial emphasis in this book.) Notice in particular the "Percent Using Methodology" columns.

Table 1.1   Extent of Use: Mathematical and Statistical Methodologies

Tool	Frequency of Use			Percent Using Methodology	Rank by Usage
	Never	Moderate	Frequent		
	(Percent of Respondents)				
Regression analysis	13.8	33.8	52.4	86.2	1
Statistical decision analysis	23.1	32.3	44.6	76.9	2
Linear programming	27.7	49.2	23.1	72.3	3
Calculus	35.4	40.0	24.6	64.6	4

Source: From "Economic Tools Used by Management in Large American Operated Corporations" by Guisseppi A. Forgionne. *Business Economics* (April 1984), Tables 1 and 2, p. 6. Reprinted by permission.

[*]Reprinted by permission of Forbes, Inc.

Table 1.2 Extent of Use: Analytical Tools Used in Marketing Analysis

Tool	Frequency of Use			Percent Using Methodology	Rank by Usage
	Never	Moderate	Frequent		
	(Percent of Respondents)				
Forecasting	6.2	18.5	75.4	93.8	1
Market share analysis	10.8	20.0	69.2	89.2	2
Surveys, experiments	16.9	43.1	40.0	83.1	3
Time series analysis	21.5	29.2	49.2	78.5	4
Demand curves	30.8	32.3	36.9	69.2	5
Elasticity	33.8	32.3	33.8	66.2	6
Input-output analysis	44.6	38.5	16.9	55.4	7

Source: From "Economic Tools Used by Management in Large American Operated Corporations" by Guisseppi A. Forgionne. *Business Economics* (April 1984), Tables 1 and 2, p. 6. Reprinted by permission.

1.2 What Is Statistics?

The word *statistics* has two interrelated meanings.[5]

Statistics, in its most common usage, is simply a synonym for *figures* or *data*. Statistics, then, in its less glamorous sense, refers to sets of numerical facts. The numerical facts may be in the form of measurements, counts, or ranks; they may even be summary measures, such as totals, averages, or percentages of several such measurements, counts, or ranks.

No one knows when statistics in this familiar sense first appeared, but it must have been a very long time ago because the earliest written records contain statistical information. The Bible, for example, contains references to data on taxes, wars, agriculture, and even athletic events. Fragmentary evidence suggests that the earliest attempts at census-taking may have been made around 4500 B.C. in Babylonia. However, it was much later than this when the words *statistic* and *statistician* came to be used. The term *statistik* was coined by the German philosopher Gottfried Achenwall around the middle of the eighteenth century.

In the past, the principal job of statisticians, or whatever they were called before Achenwall, was that of collecting and organizing numerical facts and presenting them in digestable forms. Small wonder that statistics, as a livelihood, came to suggest in the minds of many a dull, repetitious labor.

It is true, of course, that even today statistical analysis is concerned with the collection, organization, and presentation of numerical facts. But that is only part of the story. Nowadays the word *statistics* is also used in a second, more comprehensive, and more exciting sense:

> *Statistics*, in this broader sense, refers to a body of methods for obtaining, organizing, summarizing, presenting, interpreting, analyzing, and acting upon numerical facts related to an activity of interest.

Usually, the numerical facts utilized in a formal statistical analysis represent partial, rather than complete, knowledge about a situation, as is the case when a sample is used in lieu of a complete census, or when past data for a specific measure of business activity are used to obtain projections of future values.

Numerical facts are usually subjected to statistical analysis with a view to helping someone make wise decisions in the face of uncertainty. Therefore, it is not surprising that the most distinctive characteristic of the recent increase in the use of statistical methods has been a shift in emphasis from *descriptive statistics* to *inductive statistics*.

The distinction between descriptive statistics and inductive statistics is one of the most important points you should glean from this chapter. However, you will find the distinction more meaningful if some basic statistical terms are already part of your vocabulary. Accordingly, the next two sections are devoted to terminology.

1.3 Elementary Unit, Variable, and Observation

Statistical studies examine the characteristics of people, businesses, products, accounts, and so forth.

An *elementary unit* is a specific person, business, product, account, and so on, with some characteristic to be measured or categorized.

In any kind of real-world study, the factor under investigation can take on different possible values or outcomes. Such a factor is called a *variable*. A variable differs from a *constant* in that the latter term implies that the values or outcomes are always the same.

A variable may be either quantitative or qualitative.

A variable is *quantitative* if it is expressed numerically in terms of measurements. A variable is *qualitative* if it is described in terms of a set of categories.

Examples of quantitative variables are height, weight, income, age, number of units of product produced, and so forth. Examples of qualitative variables are sex, job classification, type of business, color of clothing worn, quality category of a manufactured product, and so forth.

We may also subdivide quantitative variables into the two broad categories *discrete* and *continuous*.

> A *discrete variable* is one that may assume only certain numerical values, with no intermediate values.

The number of persons in the public relations departments of the nation's 500 largest corporations would be an example of a discrete variable. The number of persons could be 1, 2, 3, and so forth, but it could not be 5.8 or 14.7 or some other number suggesting a fraction of a person.

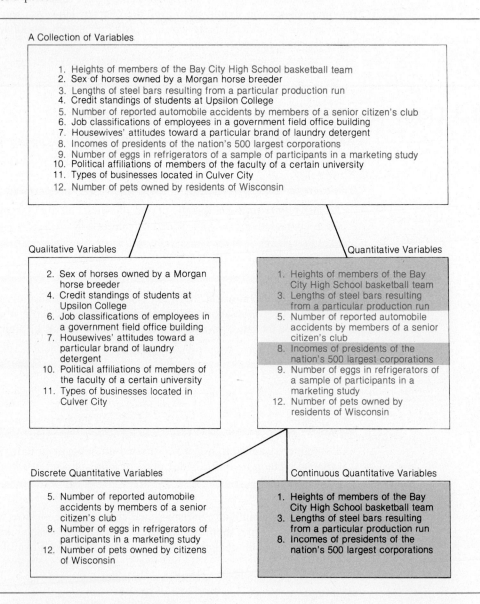

A Collection of Variables

1. Heights of members of the Bay City High School basketball team
2. Sex of horses owned by a Morgan horse breeder
3. Lengths of steel bars resulting from a particular production run
4. Credit standings of students at Upsilon College
5. Number of reported automobile accidents by members of a senior citizen's club
6. Job classifications of employees in a government field office building
7. Housewives' attitudes toward a particular brand of laundry detergent
8. Incomes of presidents of the nation's 500 largest corporations
9. Number of eggs in refrigerators of a sample of participants in a marketing study
10. Political affiliations of members of the faculty of a certain university
11. Types of businesses located in Culver City
12. Number of pets owned by residents of Wisconsin

Qualitative Variables

2. Sex of horses owned by a Morgan horse breeder
4. Credit standings of students at Upsilon College
6. Job classifications of employees in a government field office building
7. Housewives' attitudes toward a particular brand of laundry detergent
10. Political affiliations of members of the faculty of a certain university
11. Types of businesses located in Culver City

Quantitative Variables

1. Heights of members of the Bay City High School basketball team
3. Lengths of steel bars resulting from a particular production run
5. Number of reported automobile accidents by members of a senior citizen's club
8. Incomes of presidents of the nation's 500 largest corporations
9. Number of eggs in refrigerators of a sample of participants in a marketing study
12. Number of pets owned by residents of Wisconsin

Discrete Quantitative Variables

5. Number of reported automobile accidents by members of a senior citizen's club
9. Number of eggs in refrigerators of participants in a marketing study
12. Number of pets owned by citizens of Wisconsin

Continuous Quantitative Variables

1. Heights of members of the Bay City High School basketball team
3. Lengths of steel bars resulting from a particular production run
8. Incomes of presidents of the nation's 500 largest corporations

Figure 1.1. Partitioning of variables.

> A *continuous variable* is one that can, in theory, assume any numerical value over some range of values.

The lengths of steel bars produced by a certain manufacturing process might be measured to the nearest inch, or eighth of an inch, or hundredth or thousandth of an inch. In short, there is no limit, in theory, to the degree of accuracy that might be employed in measuring the length of a steel bar. Figure 1.1 illustrates this partitioning of variables.

Worth noting is the fact that a statistical study is concerned with analyzing information about characteristics of individual persons or items, not with the persons or items themselves. The *measurements* associated with some specific *quantitative characteristic* of the elementary units, or the *categories* associated with some specific *qualitative characteristic* of the elementary units, are the raw material of a statistical investigation.

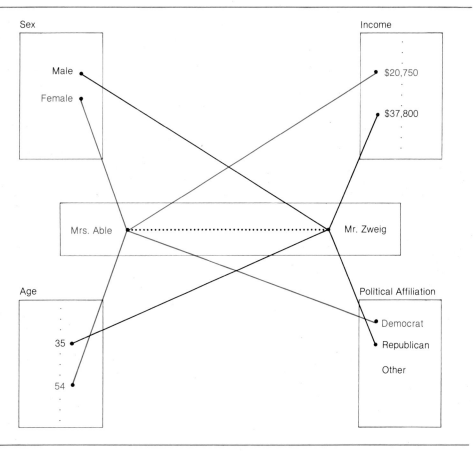

Figure 1.2. How a single set of elementary units can be a source of observations for many different variables.

Table 1.3 The Meaning of "Observation" When (I) One Variable is Studied and (II) Two or More Variables Are Studied Simultaneously

Part I: One Variable

Person (Elementary Unit)	Income (Variable)	
Mrs. Able	$ 20,750	
Mr. Bond	18,430	
Ms. Fenwick	27,000	
Mr. Watkins	150,500	
Mr. Zwieg	(37,800)	—One observation

Part II: Two or More Variables Analyzed Together

Person (Elementary Unit)	Income	Age	Number of Children	
		(Variables)		
Mrs. Able	$ 20,750	54	0	
Mr. Bond	18,430	19	2	
Ms. Fenwick	27,000	24	0	
Mr. Watkins	150,500	59	6	
Mr. Zwieg	(37,800	35	5)	—One observation

> When we study one variable at a time, an individual measurement or an indication of the category associated with a characteristic of an individual elementary unit is an *observation*.

Clearly, an unlimited variety of sets of observational data on different variables can be obtained from the same elementary units. This point is illustrated in Figure 1.2 (p. 7), in which a group of people, typified by a Mrs. Able and a Mr. Zwieg, is categorized according to two quantitative variables, age and income, and two qualitative variables, sex and political affiliation.

> If several variables are examined together, an *observation* covers the entire set of measurements or outcomes associated with a specific elementary unit. (See Table 1.3.)

1.4 Population and Sample

> The largest collection of observations on a variable in which there is an interest constitutes the *population* of such observations. A *sample* is a smaller collection, that is, a

subset, of such observations. (Important to note is the fact that a population exists whether all observations are actually known [a complete enumeration], only some observations are known [a sample], or no observations are actually known.)

If test scores achieved by entering freshmen at Clearwater College make up the largest collection of observations in which there is an interest, those test scores, whether they are known or unknown to someone contemplating a statistical analysis, constitute the population. The scores achieved by freshman students Kathy Smith, Tim Jones, Fred Naylor, and Lynn Hathaway, on the other hand, would constitute a sample.

Of course, if the scores obtained by Kathy Smith, Tim Jones, Fred Naylor, and Lynn Hathaway constituted the largest set of scores of interest, those scores would be the population. The scores achieved by, say, Kathy Smith and Fred Naylor would be a sample.

In this example, whether the population of interest is the larger one (the scores for the entire entering freshman class) or the smaller one (the scores of only four students), it is a *finite population*.

A *finite population* is one that is not indefinitely large. The population observations are countable—at least in theory.

In some studies, especially those involving experimentation, the population will be *infinite*.

GRIN & BEAR IT   **BY WAGNER**

News America Syndicate
© News Group Chicago, Inc., 1985

1-4   Wagner

"I counted all the sheep."

An *infinite population* is one that is indefinitely large. The population observations cannot be counted even in theory. (In practice, a finite population considerably larger than the sample size is often treated as if it were infinite.)

For example, if weight data were obtained for a sample of Canadian rats fed a special diet for a period of time—a diet rats outside the laboratory environment would not get—the relevant population would be weight measures for all the Canadian rats that ever could have been given, are now being given, or ever will be given the special diet.

It will sometimes be convenient to speak as if the population of interest were a collection of elementary units (i.e., people, rats, businesses, etc.). Still, our ultimate interest is in the observations and not in the elementary units themselves.

1.5 Descriptive Statistics and Inductive Statistics

The distinction between descriptive statistics and inductive statistics can be readily demonstrated through the use of an example. Let us consider a problem faced by Universal Foods, Inc., a fictitious company.

ILLUSTRATIVE CASE 1.5

Universal Foods, Inc., developed a new soybean cereal intended for distribution through grocery stores because management had noted a trend in recent years toward a health-food consciousness on the part of the grocery-buying public. Management felt that, if properly marketed, a soybean cereal might prove profitable. One aspect of the marketing mix had to do with the kind of container to use—a box or a jar.

After much discussion, the concerned members of management decided to resort to a test-marketing analysis. Twenty grocery stores, located in different geographical areas, were selected and cooperation was secured from the stores' managers to stock the new cereal for a period of ten weeks. Ten randomly assigned stores carried the cereal in clear jars having only a label that indicated the brand name and ingredients. The other ten stores carried the cereal in boxes having the same quantity of cereal as the jars. The price charged was the same for jars and boxes.

A convenient way of comparing results might be to determine the total number of boxes sold and the total number of jars sold, and see which sum is larger. For boxes, let us say, the total was 144 dozen; for jars, 149 dozen.

How might management describe these findings? Two quite different reactions are possible:

First reaction: For the specific stores and the specific ten weeks included in the experiment, total sales of jars were 149 dozen, whereas total sales of boxes amounted to only 144 dozen.

Second reaction: Jars are more effective than boxes in generating sales of the cereal.

What can be said about the first reaction? To begin with, it is certainly accurate. The statement has to do with observed fact and observed fact only. Alas, such bald accuracy, while safe, is not very helpful. The experiment was conducted with a view to determining whether a difference in effectiveness between boxes and jars actually exists.

And what about the second reaction? If the statement that jars are superior to boxes is true, important information has been gained. Such a statement goes well beyond the observed data; management would be using observed facts about a limited situation to draw a conclusion related to a much broader situation. But is the statement true? That depends on several things. For example: (1) Were the stores used in the experiment representative of all stores where the new cereal might be sold? (2) Were the geographical areas representative of all areas where the cereal might be sold? (3) Were the decisions regarding which stores would display boxes and which jars made in such a way that other influencing factors can be assumed to have been distributed about evenly between the two sets of stores? (4) Do the two totals differ by an amount great enough that this difference, five dozen units, cannot properly be interpreted as nothing more than a sampling fluke?

So we see that the second reaction, though potentially much more useful than the first reaction, is not necessarily true. It is a riskier conclusion, because it involves the use of factual information from *samples* to draw conclusions pertaining to the corresponding infinite *populations*, which brings us to the formal distinction between descriptive statistics and inductive statistics.

> *Descriptive statistics* is the branch of statistical analysis concerned with describing certain characteristics of a set of observed data (usually a sample)—that is, what it is shaped like, what number the values tend to cluster around, how much variation is present in the data, and so forth. Viewed in the strict sense, in descriptive statistics the characteristics are observed or measured because they are of interest in their own right and not because they are to be used as guides in determining the nature of a larger body of data (the population).

> *Inductive statistics*, on the other hand, is concerned with the process of drawing inferences about specific characteristics of a population based on information obtained from sample data.

A closely related distinction between deductive inference and inductive inference must also be made at this juncture.

1.6 Description, Deductive Inference, and Inductive Inference

Let us suppose that you awaken one morning, rub your eyes, yawn, push back the bedroom curtains, and look out the window. You discover that snow is falling in the backyard. Just then a family member calls to you and asks what the weather looks like. Let us consider three of several possible answers you might give:

First possible answer: "Snow is falling in the backyard."

Second possible answer: "Snow is falling in the backyard and, although I can't see it from this angle, I presume that snow is also falling on the doghouse in the backyard."

Third possible answer: "Snow is falling in the backyard and therefore is probably falling all over the city."

The first possible answer is strictly descriptive. It is a statement of observed fact with no inference of any kind being drawn.

The second possible answer includes a *deductive inference*. You *deduce* that, because snow is falling on a fairly large area (the backyard), it must be falling on a small area contained within the larger area. This deductive inference is probably correct. A snowstorm would have to be pretty unusual to fall in the backyard and yet miss the doghouse. However, unless there is something very special about the doghouse or its occupant—in which case you would presumably have safeguarded them against snow a long time ago—your deductive inference is unlikely to prove a useful guide for action.

The third possible answer includes an *inductive inference*. The conclusion has to do with a geographical area much larger than the area actually observed. This inference is also probably correct. However, snow has been known to fall on small, isolated areas before; this might be one of those times. In any event, your inductive inference might well prove to be a useful guide for action. For example, you might decide to wear your overcoat and galoshes and to allow more time than usual for traveling to school or work. If your inductive inference is correct, you will be well protected against the discomforts that snowstorms can bring. If your inductive inference is wrong, little harm is done in this case.

> In statistics, a *deductive inference* is a conclusion drawn about a sample based on known information about the population from which the sample came. An *inductive inference* is a conclusion drawn about a population based on information obtained from a sample selected from that population.

Let us assume that a sample is drawn from a population, as illustrated in part (*a*) of Figure 1.3. Depending on the nature of the descriptive information available, we may draw either a deductive or an inductive inference. If we have descriptive information about the population, we may wish to use it to infer (deductively) something about the characteristics of the sample, as in part (*b*) of Figure 1.3. For example, we might know the population average and wish to estimate the sample average. Which brings us to two more important terms, *parameter* and *statistic*.

> A *parameter* is a descriptive measure (such as the average) of *population* data. A *statistic* is a descriptive measure (again, such as the average) of *sample* data.[6]

Obviously, a deductive inference is seldom very useful because, if the population data are available, the sample observations will also be available. Therefore, if we wish to know the sample average, we can calculate it rather than having to estimate it.

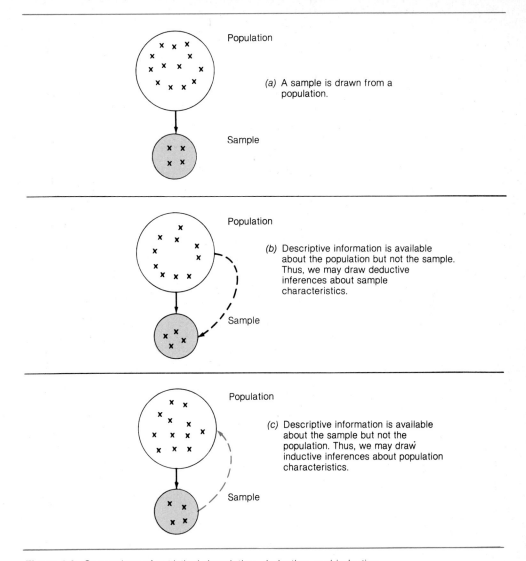

Figure 1.3. Comparison of statistical description, deduction, and induction.

If, as is more likely in practice, we have descriptive information about a sample, we may use it to infer (inductively) something about the characteristics of the population, as in part (c) of Figure 1.3. This can be quite a fruitful endeavor because the population observations will be mostly unknown. For example, we may wish to use the sample average (the statistic) to estimate the unknown value of the population average (the parameter), or to test a hypothesis about the value of the population average. (Although these topics will not be pursued in any depth at this time, you should know that in statistics induction consists of (1) estimating unknown values of population parameters and (2) testing hypotheses about population characteristics. See Figure 1.4 for this partitioning of statistical analysis.)

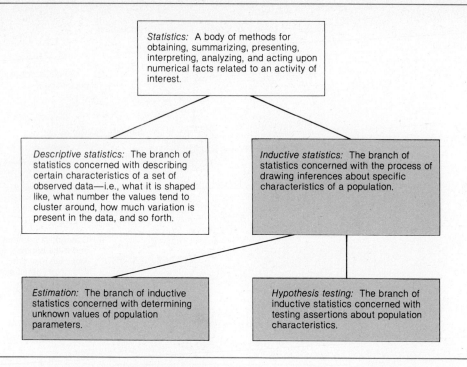

Figure 1.4. Partitioning of statistical analysis.

This book is primarily concerned with induction because that is the payoff area of statistical analysis as it is applied to business decision making. However, description and deduction are also discussed. Actually, these three divisions of statistical analysis are tightly intertwined: Descriptive measures must be used in conjunction with inductive procedures. Moreover, justification for our inductive procedures rests upon some general facts and theorems discovered by others through research of a primarily deductive nature.

1.7 General Outline of a Statistical Problem

Figure 1.5 shows in flowchart form the steps involved in a typical statistics problem.

The first step entails formulating the problem in such a way that it can be analyzed through use of an appropriate statistical technique. This step will sometimes require that the managerial statement of the problem be reworded. Needless to say, any such rewording should not destroy management's original intent. (A reasonable rewording of a management problem is illustrated in the next section.)

The second step is to identify clearly the population to which conclusions derived from the statistical analysis will apply. This step, like the preceding one, calls for a clear understanding of the managerial problem requiring solution.

The third step consists of determining the appropriate survey-sampling plan or the type of experimental design to be used. The word *survey* may conjure up images of someone inter-

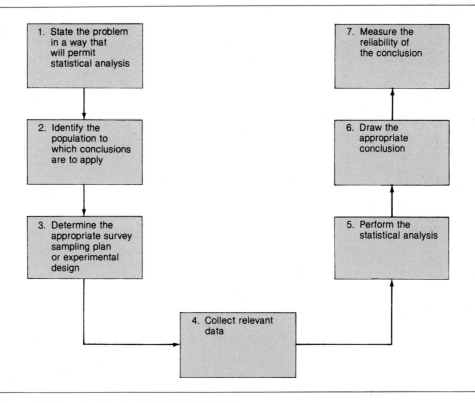

Figure 1.5. Flowchart of a typical statistics problem.

viewing other people and extracting opinions or intentions. In this text *survey* is used more broadly than that.

Henceforth, the word *survey* will be used to refer to situations in which someone attempts to determine characteristics of a population without affecting the elementary units involved.

Thus, a quality-control inspector conducts a survey when he examines a batch of gate valves even though, one hopes, he makes no attempt to converse with them.

A survey can be distinguished from a designed experiment.

In a *designed experiment* certain variables are varied systematically with the possible effect of altering the observations taken on the elementary units.

For example, in a study undertaken to determine what kind of point-of-purchase advertising is most effective in generating sales of a product, a designed experiment would almost certainly be used. Different point-of-purchase ads would be used in different stores in the hope that one such kind of ad would prove more effective than the others.

> In *survey sampling*, the analyst is an onlooker; in designed experiments, the analyst is an influencer.

Survey sampling is the subject of the next chapter; the subject of designed experiments is treated in Chapter 12.

The fourth step in a statistical analysis consists of collecting the relevant data.

The fifth step involves application of the formal statistical analysis to the data collected. The kind(s) of statistical analysis used will depend greatly on the problem statement and the nature of the data gathered.

Drawing appropriate conclusions about the population under study makes up the sixth step.

Finally, the seventh step entails passing along to management not only the results of the statistical analysis but an indication of the reliability of the results as well.

> By *reliability* we mean the probability that a correct conclusion has been drawn.

1.8 Statistical Analysis as Part of the Decision Process

Figure 1.5 is limited to the statistical analyst's responsibilities. By placing this entire flowchart within a more broadly conceived managerial flowchart, we can see the relationship between statistical analysis and the managerial decision process. Figure 1.6 shows such a flowchart.

First, there is a real-world problem or potential problem.

Second, management recognizes the problem or potential problem and frames it in the form of one or more operational questions.

> An *operational question* is a question expressed in such a way that, when answered, it will point management toward the proper action to take.

Let us pause at this point to consider why it matters just how a managerial problem is stated.

ILLUSTRATIVE CASE 1.6

The main office of a large temporary help company had been located in a busy downtown location for several years, but now a move to a suburban location was being contemplated. A

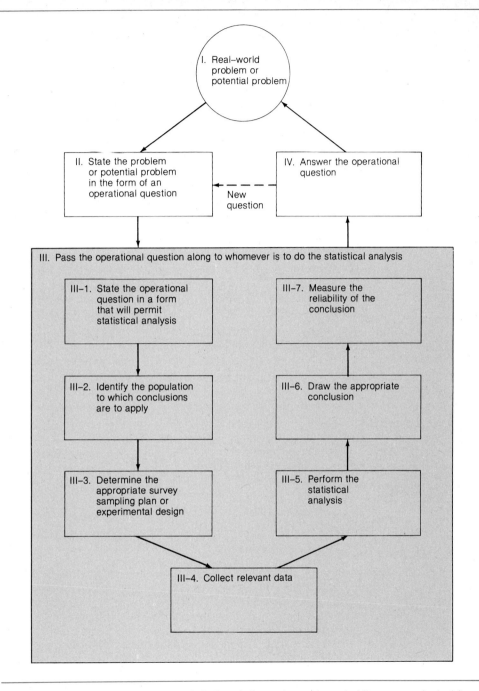

Figure 1.6. Flowchart showing how statistical analysis can be made part of the manager's decision process.

staff employee suggested that the problem definition be: "What factors affect the relative profitability of service companies located in downtown areas as opposed to suburban areas?" The president of the company vetoed this way of stating the problem, preferring instead his own operational question: "Will a move to the proposed suburban location increase the company's profits?"

Clearly, if the problem were framed as suggested by the staff employee, there would be no end to the quantity of data that could be collected and analyzed. Worse yet, there would be no assurance that the results would point unambiguously to a specific action. The president's statement pins down precisely what he wishes to know. It has the virtue not only of inviting an answer that would point toward a specific action (presumably the right one if the research is conducted properly), but it also limits and circumscribes the quantity and nature of the data to be collected.

In a similar vein, a question such as "Why does the demand for the company's products tend to decline during the summer months?" is in many ways more complex and yet less potentially useful than one worded like this: "What actions can be taken to bolster demand for the company's products during the normally slow summer months?"—even though some of the same kinds of information would have to be gathered with both forms of the question. The question "What is there about the ABC Company that enables it to submit lower bids on most projects than we are able to do?" will probably require more research and still produce less useful results than the question "What changes can we make to enable us to compete more effectively, yet profitably, on bids against the ABC Company?"—even though, again, some of the data requirements would be the same.

Third in the list of steps involved in the decision process is the performance of the statistical analysis. This step encompasses all steps listed in section 1.5.

Fourth, an answer is provided to the operational question.

Fifth, one of two things happens: (1) The answer to the operational question is capable of solving the real-world problem or potential problem, or (2) the answer points toward a new operational question.

Let us see how this entire process was handled by the "Royal Oil Company," a real company appearing here under an alias.*

ILLUSTRATIVE CASE 1.7

In recent years oil companies have been able to offer some of the most comprehensive employee benefit packages of all the major industries. "Royal Oil" has one of the best such packages among the larger oil companies.

As part of the benefit package, Royal offers what is referred to as the Thrift Plan. Under this plan, the company matches employee contributions up to and including 4% of the employee's annual salary. The combined funds can be used to purchase company common stock or U.S. Series E bonds. Or the funds can simply be held in the form of cash. Although Royal encourages all eligible employees to participate in the plan, many do not.

The personnel in the Financial Planning Department have gained the impression over the years that female, Hispanic, and black employees are less likely to participate in the plan than male Caucasians are.

*Quotes at the first occurrence of a name signify that a pseudonym has been substituted for the name of a real company.

The circle labeled I in Figure 1.6 shows the existence of a real-world problem or a potential problem. The designation "potential problem" appears to be the proper one in this case: The fact that many employees do not elect to participate in the plan is not in itself a problem. Participation is a matter of free choice. What makes this a *potential problem* is that some employees are less likely to participate.

This is still not necessarily even a potential problem. First, it is not certain that the perceptions of the financial planning people are correct. Second, if they are correct but if female, Hispanic, and black employees decline to participate for reasons quite apart from management policy or practice, then no change need be made in the structure or methods of communicating the plan. However, if there is something subtly discriminatory in the terms of the plan, or the way it is presented to employees, or something else about the plan, then the company could conceivably be inviting future adverse publicity and maybe even costly litigation.

Let us call the real-world *potential* problem the fact that many employees do not participate in the plan. (See the circle labeled I in Figure 1.7.)

The box labeled II in Figure 1.6 says "State the problem or potential problem in the form of an operational question." In this case, *several operational questions* might be posed. Some possibilities are: (1) Should something about the plan be changed to make it *generally* more attractive? (2) Should the methods used for making *all* employees aware of the plan be altered? (3) Should the plan be changed in some respect(s) so that it appeals more to female employees? (4) To Hispanic employees? (5) To black employees? And so forth.

For simplicity, we will concentrate only on the third operational question, the one pertaining to female employees. This is shown in the box labeled II in Figure 1.7. The arrows emanating from the circle labeled I but not leading to the box labeled II in this figure are included as a reminder that this case probably has more than one operational question associated with it.

ILLUSTRATIVE CASE 1.7 (CONTINUED)

A member of the Legal Department, trained in statistics, conducted a statistical study. [See the box labeled III in Figures 1.6 and 1.7.] She reworded the operational question somewhat to put it into the form of a statistical hypothesis: "Sex and participation in the Thrift Plan are independent." [See the box labeled III-1 in Figures 1.6 and 1.7. At this stage she was merely trying to determine whether female employees *really do* have a tendency to reject the plan.]

She viewed the relevant population to be all full-time employees of the Royal Oil Company. [See the box labeled III-2 in Figures 1.6 and 1.7.]

She selected random samples from each district office and combined the results. The final sample consisted of 50 employees. [See the box labeled III-3 in Figures 1.6 and 1.7.] The information used was obtained from permanent personnel records for each member of the sample. [See the box labeled III-4 in Figures 1.6 and 1.7.]

The analyst tested the hypothesis of independence statistically and concluded that there is some kind of dependency between sex and participation in the Thrift Plan. [See the boxes labeled III-5 and III-6 in Figures 1.6 and 1.7.] Close examination of the data revealed that women do in fact participate in the plan to a lesser extent than would be expected if the hypothesis of independence were true.

The analyst in this case rejected the hypothesis of independence at the .05 level of significance. Briefly stated, this means that, if the hypothesis of independence were really

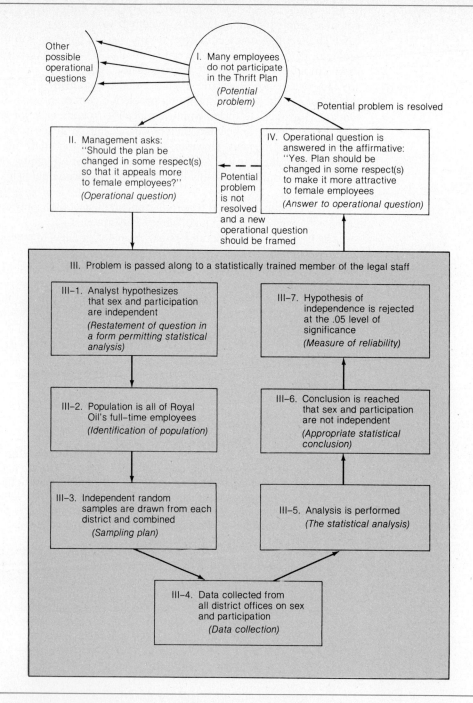

Figure 1.7. The Royal Oil Company problem shown in the same flowchart form as Figure 1.6.

true, in only five times out of 100, on the average, with repeated sampling from this same population would the hypothesis be rejected incorrectly. Details aside (hypothesis testing is treated in Chapters 9 through 13; don't worry about the techniques involved right now), these are not bad betting odds under the circumstances. If you could buy a state lottery ticket that was guaranteed to have only a .05 chance of losing, you would probably be tempted to risk a dollar. This is the sort of thing that is meant when we speak of measuring the reliability of the results of the statistical analysis. (See the box labeled III-7 in Figures 1.6 and 1.7.)

The answer to the operational question, then, is evidently "Yes, the plan should be changed in some respect(s) to make it more attractive to female employees." The word "evidently" is used in the preceding sentence to convey the idea that this answer to the operational question is more tentative than is often the case. After all, women may fail to participate in the plan for reasons having nothing whatever to do with inadvertent discriminatory practices on the part of management. Even if it is correct to say that the plan should be changed, this answer to the operational question does not constitute a solution to the potential real-world problem. Instead, it would lead to a change in the operational question (dashed arrow pointing to the box labeled II in Figures 1.6 and 1.7), a change related to determining *why* women fail to participate in the plan to the extent expected.

Let us assume, contrary to fact, that the hypothesis of independence had been accepted. If management were concerned only with possible, unintentional discrimination against women, the answer to the operational question, "No change is necessary," would have taken care of the real-world problem (solid arrow pointing to the circle labeled I in Figures 1.6 and 1.7).

1.9 Summary

Briefly stated, the statistical approach to problem solving usually involves (1) collecting and organizing relevant data, (2) computing essential descriptive measures, and (3) deciding on the basis of a small amount of data what would probably have been found if it had been possible and feasible to collect all relevant data. This book deals with all these tasks but places primary emphasis on the third.

The study of statistics will require you to learn many new terms and to develop a number of technical skills that may seem difficult at first. But the payoff will almost certainly be worth the trouble. More and more, business decisions are being made with the help of statistical and other quantitative techniques. Judgmental decision making will never become extinct; there will always be questions that mathematical methods, however sophisticated, simply cannot supply complete answers to. The decision maker requires a great many pieces of input data—data on business conditions, the political situation, competitors' activities, the weather, the firm's image, and on and on—to analyze complicated problems properly. Although statistical analysis is not a panacea, it *can* provide *many* of these pieces of input data in a convenient, easy-to-understand form. For this reason, it is safe to say that statistical analysis will be used in business to an ever-increasing extent and in an ever-growing number of ways in the future to supplement and sharpen human judgment. The decision maker familiar with and skilled in the use of the many quantitative techniques available will enjoy an advantage over his or her less well-equipped colleague or competitor.

Notes

1. Adapted from remarks made by Caleb C. Whitaker, Vice-President, Planters Division, Standard Brands, Inc. Cited in *Market Testing of Consumer Products, Experiences in Marketing Management, No. 12* (New York: The National Industrial Conference Board, 1967), p. 51.
2. Cited in Rex V. Brown, "Do Managers Find Decision Theory Useful?" *Harvard Business Review* (May–June 1970), pp. 78–89.
3. "Omen Torches," *Forbes Magazine* (December 15, 1976), p. 73. Reprinted by permission.
4. Guisseppi A. Forgionne, "Economic Tools Used by Management in Large American Operated Corporations," *Business Economics* (April 1984), pp. 5–17.
5. A third meaning is presented in section 1.6.
6. It is assumed that you are already familiar with the common meaning of the word "average." It is what statisticians call the arithmetic mean and is obtained by summing all values of interest and dividing by the number of those values. Much more is said about this and related descriptive measures in Chapter 4.

Terms Introduced in This Chapter

descriptive statistics (p. 11)
designed experiment (p. 15)
elementary unit (p. 5)
estimation (p. 14)
hypothesis testing (p. 14)
inductive statistics (p. 11)
inference (p. 12)
   deductive inference (p. 12)
   inductive inference (p. 12)

observation (p. 5)
operational question (p. 16)
population (p. 8)
   finite population (p. 9)
   infinite population (p. 10)
population parameter (p. 12)
sample (p. 8)
sample statistic (p. 12)
statistics (broader sense) (p. 4)

statistics (narrow sense) (p. 4)
survey (p. 15)
variable (p. 5)
   qualitative variable (p. 5)
   quantitative variable (p. 5)
      continuous quantitative variable (p. 7)
      discrete quantitative variable (p. 6)

Questions and Problems

1.1 Explain the meaning of each of the following terms:
   a. Population
   b. Descriptive statistics
   c. Inductive inference
   d. Operational question
   e. Reliability of a statistical conclusion
   f. Parameter
   g. Survey

1.2 Explain the meaning of each of the following terms:
   a. Elementary unit
   b. Quantitative variable
   c. Discrete variable
   d. Sample
   e. Finite population
   f. Estimation in inductive statistics
   g. Designed experiment

1.3 Distinguish between:
   a. Population and sample
   b. Descriptive statistics and inductive statistics
   c. Estimation and hypothesis testing in inductive statistics
   d. Deductive inference and inductive inference

1.4 Distinguish between:
  a. Description and deductive inference
  b. Parameter and statistic
  c. *Statistics* in the narrower and in the broader sense

1.5 Distinguish between:
  a. Elementary unit and observation
  b. Figures 1.5 and 1.6 in the chapter
  c. Quantitative variable and qualitative variable

1.6 Distinguish between:
  a. Discrete variable and continuous variable
  b. Finite and infinite population
  c. Survey and designed experiment
  d. A statistical conclusion and the reliability of the statistical conclusion

1.7 Indicate which of the following statements you agree with and which you disagree with, and defend your opinions:
  a. A knowledge of statistics is needed only by people who plan to become professional statisticians.
  b. The subject of statistics is concerned with the collection, organization, and presentation of numerical facts.
  c. The subject of statistics is concerned *only with* the collection, organization, and presentation of numerical facts.
  d. Statistical techniques supplement and sharpen human judgment but do not eliminate the need for human judgment in business decision making.

1.8 Indicate which of the following statements you agree with and which you disagree with, and defend your opinions:
  a. Income measurements on a roomful of people attending a self-help workshop would constitute a population.
  b. Income measurements on a roomful of people attending a self-help workshop would constitute a sample.
  c. Income measurements on a roomful of people attending a self-help workshop could conceivably be a population one minute and a sample the next.

1.9 Indicate which of the following statements you agree with and which you disagree with, and defend your opinions:
  a. Information about the incomes of a group of people, whether the income data are viewed as a population or as a sample, need not necessarily lead to any kind of inference.
  b. A statistical conclusion and the related measure of reliability will always resolve the real-world problem or potential problem.
  c. Stating a managerial problem in the form of an operational question, rather than in nonoperational form, will usually reduce the quantity of data that must be collected and analyzed.
  d. A single set of elementary units can give rise to an unlimited number of sets of observational data on different variables.

1.10 Indicate which of the following statements you agree with and which you disagree with, and defend your opinions:
  a. A finite population is a population of small size.
  b. If observations on a population's elementary units are not known, then the population does not really exist.
  c. An average is properly regarded as a parameter if and only if it pertains to population data.
  d. Estimation and hypothesis testing are both subcategories of inductive statistics.

1.11 Cite a realistic hypothetical problem in the areas of (1) marketing, (2) management, (3) finance, and (4) accounting to illustrate the distinction between descriptive statistics and inductive statistics.

1.12 Tell which of the following variables are qualitative and which are quantitative. If a variable could be either, explain under what circumstances it could be qualitative and under what circumstances it could be quantitative.
   a. Number of guns owned by members of the local chapter of the National Rifle Association
   b. Sales achieved on a given day by members of the sales staff of the ABC Company
   c. IQ scores of students at Ranum High School
   d. Lengths of time nonmanagerial employees of the Compu-Aid Company have been engaged in their present jobs
   e. Weights of apples in a basket
   f. Political party affiliations
   g. Output per day from a certain manufacturing process over a month's time
   h. Marksmanship designations bestowed on members of a shooting team

✔ 1.13 Tell which of the following studies, utilizing formal statistical methods, sound as if they might have involved survey sampling and which sound as if they might have involved designed experimentation. Justify your choices.
   a. Lengths of metal chain are subjected to destructive testing to make certain the population conforms to specified stress standards.
   b. Grapes are displayed in paper bags in some grocery stores and in transparent bags in others; loose displays are used in still other grocery stores. The investigators hope to determine whether a particular way of displaying grapes is superior in generating sales compared to other ways.
   c. People are called on the telephone and asked what television show they are watching at the time.
   d. An auditor wishes to estimate the average size of accounts receivable for a population containing 15,000 such accounts.
   e. One group of Canadian rats is force-fed large quantities of saccharine-intensive liquid. Another group of Canadian rats is allowed to drink water at whatever times and in whatever quantities they wish. The object of the study is to determine whether saccharine "causes" cancer.

 1.14 Ten percent of a population of rivets are known to be defective. A foreman estimates that, if a sample of 100 rivets were drawn from this population, somewhere between 8% and 12% of the rivets would be defective.
   a. Tell which kind of inference, deductive or inductive, is being drawn by the foreman.
   b. Could the foreman's conclusion be wrong? Explain.

 1.15 A shipment of guitar picks is received from a new supplier. A random sample of 30 picks is drawn and 3 of these (10%) are found to have flaws in them. The manager of the music store receiving the shipment believes that at least 10% of the picks in the entire shipment are probably defective.
   a. Tell which kind of inference, deductive or inductive, is being drawn by this manager.
   b. Could the manager's conclusion be wrong? Explain.

✔ 1.16 A consumer rating service using a complex driving simulation machine to test two brands of automobile fan belts for length of useful life performed the following experiment: Each belt in a sample of 100 Ajax Brand fan belts and each belt in a sample of 100 Butler Brand fan belts was subjected to continuous use until it snapped. The number of miles of (simulated) driving for each fan belt used in the experiment was recorded and the average number of miles of use was computed for each brand. The average number of miles of use associated with the Ajax Brand was found to be 50,722 and for the Butler Brand, 61,542. Two newspapers reported the results of the experiment quite differently. The two versions are summarized here.

1. "In an experiment conducted by the Consumer Products Testing Institute of America, the average number of miles of useful life for 100 Butler Brand fan belts was 61,542, whereas the average for 100 Ajax Brand fan belts was only 50,722."
2. "Results of an experiment conducted by the Consumer Products Testing Institute of America indicate that Butler Brand fan belts last longer, on the average, than Ajax Brand fan belts."

a. Explain why sampling was absolutely necessary in this experiment.
b. Write an evaluation of the two newspaper versions of the results of this experiment. In your evaluation, indicate the principal strengths and weaknesses of each version.

 1.17 Two appraisers of homes were tested to determine how similar their estimates tended to be. A random sample of 50 homes was selected, and each appraiser, working independently, assigned values to each home. Appraiser A came up with an average appraised value of $95,000; Appraiser B came up with an average appraised value of $97,500. A report on these results submitted to the real estate firm employing the two appraisers said: "Appraiser A had an average value of $95,000; Appraiser B, an average value of $97,500." Upon examining the report, one of the managers of the company said to another, "Appraiser A is the more conservative appraiser."

a. Explain why a sample of homes was used in this experiment.
b. Why do you think the appraisers were asked to work with the same sample of homes rather than with two separate samples?
c. Write an evaluation of the way the information was stated in the report, and then evaluate the manager's comment on the report. In your evaluation, indicate the principal strengths and weaknesses of each version.

 1.18 One year after the Traffic Department of Midway City converted several major streets from two-way to one-way, a study indicated that the accident rate had fallen by 5% from the year before. The study stated only that a decline had occurred; it made no effort to explain why. A local newspaper, *The Midway Times*, ran the headline: ONE-WAY STREETS PROVED SAFER THAN TWO-WAY.

Write an evaluation of the way the results were stated in the study, on the one hand, and by *The Midway Times* on the other. In your evaluation, indicate the principal strengths and weaknesses of each version.

 1.19 An analyst for an airline studied reservation data for a sample of 30 days to estimate the percent of people who failed to keep their reservations. The analysis revealed the "no-show" rate to be 3%. This information was passed along to management without comment. Thereafter, management assumed a 3% "no-show" rate whenever discussions of overbooking policy came up.

Is any kind of inductive inference evident in this problem? Explain.

 1.20 When testing the effectiveness of a new drug, two patient groups are generally used. One group is given the actual drug (though they may not know that it is the drug and, usually, neither does the administering physician) and the other group is given a placebo, an inert substance having no real medical value.

a. What do you suppose is the purpose of the placebo?
b. Is the group receiving the drug a sample or a population? Explain.
c. Is the group receiving the placebo a sample or a population? Explain.
d. Would the relevant populations be finite or infinite? Explain.

TAKE CHARGE

1.21 This assignment presupposes that you have chosen a field of concentration in a college of business program. If you have not yet chosen a field of concentration, pick one you *think* you might like to pursue.
  a. Speak to a professor associated with the field of concentration you have chosen and ask him or her to describe at least three important applications of statistical analysis in that field.
  b. Arrange an appointment with a manager or staff employee working in your field of concentration in a business in your area. Ask this person whether he or she makes any use, directly or indirectly, of statistical analysis. If the answer is Yes, find out in what specific ways statistical analysis is used and how strategic it is considered to be.
  c. Write a brief summary of what you learned from parts *a* and *b* of this assignment.

Cases for Analysis and Discussion

1.1 THE "DELTA COFFEE COMPANY": PART I*

Very much aware of a decrease in market share in the early months of 1960, the product manager of the "Delta Coffee Company" in November of that year faced several decisions about how to improve the attractiveness of the company's product. One such decision had to do with whether Delta should follow the lead of a competitor and change the manner of packaging the firm's coffee. More specifically, it was his responsibility to determine whether to (1) convert the packaging of Delta Coffee to a new quick-strip can immediately, (2) initiate research into the effects on demand of a change to a quick-strip can, or (3) take a "wait and see" attitude regarding forthcoming actions of the company's major competitors. The product manager had at his disposal various kinds of information describing the effect on demand of one competitor's change a year earlier from the traditional key can to the no-key can, as well as information on sales trends and market shares for almost all of the major coffee companies.

1. What are some apparent advantages and disadvantages of the following general approaches to addressing the problem: (*a*) a purely intuitive approach, (*b*) a purely statistical approach, and (*c*) an approach combining intuition and statistical analysis?

2. In its statistical aspects, would this problem be more properly categorized as a problem in descriptive statistics or as a problem in inductive statistics? Give reasons for your choice.

This case is continued at the end of Chapter 18.

1.2 THE "LAMPSON COMPANY": PART I

The "Lampson Company," manufacturer and distributor of travel trailers, employs approximately 700 people in activities directly related to the manufacturing process. One problem the company has always faced is that of dealing with employee turnover, which has been about 50% per year, on the average. This rate of turnover means, in theory, that there is a completely new production force every two years; 350 people must be hired and trained each year in order to keep 700 workers in the shop.

Lampson's problem is compounded by the fact that it is somewhat of a custom manufacturing company and, as such, pays close attention to the needs and wishes of each customer.

*See footnote on p. 18.

Such personal attention means that many of the tasks to be performed require skilled labor, particularly skilled welders, urethane machine operators, and riveters. People with these skills are usually relatively scarce.

Lampson's management is willing to give newly hired employees the training needed to qualify them as skilled workers. However, such training requires a commitment by the company of both time and money. This commitment represents an investment that, for Lampson, is no small gamble in view of the high employee turnover rate.

The question facing the personnel manager in late 1981 was: How does one go about determining what kinds of people will stay with the company for a period of at least one year (a period sufficiently long to permit the company to break even on its investment)? The personnel manager decided to analyze data on past employees. His approach involved categorizing past employees according to the following five characteristics: (1) employee had or had not lived in the state for at least one year prior to the time he was employed; (2) employee was or was not married at the time he was hired; (3) employee had or had not completed high school prior to the time he was hired; (4) employee was or was not under 27 years of age at the time he was hired; (5) employee had or had not held fewer than three jobs during the two years prior to the time he was hired.

The personnel manager then proceeded to analyze employee data to determine what the best combination of employee characteristics is in terms of ensuring a high probability of the company's recovering, on each newly hired employee, its investment in personnel training.

1. Would this problem more properly be described as a problem in descriptive statistics or as a problem in inductive statistics? Give reasons for your choice.

2. How would you state the real-world problem?

3. How would you state the operational question?

4. Do you believe the personnel manager would be wise to limit his analysis to employee characteristics, or are there aspects of this case suggesting that management itself should be subjected to careful scrutiny? Elaborate.

This case is continued at the end of Chapter 6.

1.3 "GLAMOUR BURGER SYSTEMS": PART I

After acquiring "Glamour Burger Systems" in early 1968, "Twentieth Century Foods, Inc." embarked on an ambitious nationwide investment program in new restaurants. Between 1968 and 1970 the acquiring company invested over $200 million in 500 new restaurants, either owned by the company itself or leased for a 20-year period to individual owners. The 500 new restaurants brought the total number of restaurants in the Glamour Burger Systems to 1300.

Unfortunately, sales failed to meet planned objectives and over 200 restaurants were either closed or forced to operate at a loss. It was evident at this time to the new management of Twentieth Century Foods that they lacked adequate understanding of the fast-food industry and, more specifically, the factors motivating consumers to choose a specific kind of restaurant to patronize.

Although Glamour Burger was one of the largest national fast-food chains at that time in terms of number of units, it was running a poor third in sales to its major competitors, McDonald's and Burger King. Management of Twentieth Century Foods recognized that they must develop new policies and objectives based on reliable research data from the marketplace. Accordingly, they authorized two major attitude and usage surveys to obtain necessary data from which decisions could be made regarding future objectives and plans.

1. How would you state the real-world problem or potential problem?

2. How would you state the operational question(s)?

3. In its statistical aspects, is this problem more properly describable as a problem in descriptive statistics or as a problem in inductive statistics? Give reasons for your choice.

This case is continued at the end of Chapter 3.

1.4 "DIGITAL COMPU-AID INC."

"Digital Compu-Aid (DC)" is a small corporation whose primary function is that of storing and servicing digital data storage equipment used by computer systems. At present, DC leases two magnetic computer tape-cleaning machines, which are operated primarily by Mr. White, the company's operations manager. When Mr. White is on customer call, the other operating personnel run the machines. During its short existence, the firm's business has grown rapidly to the point where it now cleans more than 100 tapes per day on the vast majority of working days. As a result, the task of overseeing and assisting with the tape-cleaning activities has come to consume inordinate amounts of time for Mr. Brown, the company's vice-president, and Mr. Black, one of the company's operations officers, both of whom could be better employed in marketing tasks and in servicing new accounts. The increased business has led the management of Digital Compu-Aid to the conclusion that an appropriate number of new tape-cleaning machines must be purchased and new personnel hired.

1. How would you state the real-world problem or potential problem?

2. How would you state the operational question(s)?

3. What are some apparent advantages and disadvantages to the following general approaches to dealing with this problem: (1) a purely intuitive approach, (2) a purely statistical approach, or (3) an approach that combines intuition and statistical analysis?

1.5 "DPC CORPORATION": PART I

"DPC Corporation" is a manufacturer of office machines that are generally rented or leased by customers; maintenance service is included in the lease or rental price. DPC is very sensitive to the reliability of its products for two major reasons: First, reliability is a key element in customer satisfaction with the machines. In a rental/lease environment, customers will not continue their contract unless they are satisfied with the product. Second, since the cost of maintenance is a variable to DPC and its price is fixed to the customer, reliability is an important determinant of the profitability of DPC's products.

When DPC introduced a new model called DE-1, the company was using its traditional methods of quality control. In addition to inspection and testing of purchased parts, they performed quality checks at interim points in the assembly process and a final inspection prior to shipment to the customer. The final inspection consisted of running the machines and checking the operation and quality of output. Adjustments were made as required, and a record of faults was kept and fed back into the manufacturing operation for control purposes. Few statistics were employed, and no attempts were made to predict reliability in a customer environment even though reliability criteria were established.

DPC decided as part of an overall examination of its quality and reliability procedures to institute a sampling test of the DE-1, whose purpose was to predict the number of service calls that a customer might require during the first 3 months of operation. They believed that this information would be more useful than their old method because the prediction of service calls is a prediction of customer satisfaction and of service costs, as well as a means of providing feedback to manufacturing regarding numbers and types of failures. These predictions could be useful in shaping the activities of the marketing and service organizations. They would also provide early data to product management as to how DE-1 was tracking according to its reliability criteria during each month of manufacture. (Typical of any new product, the DE-1 was expected to improve in reliability during the first 18 to 24 months of manufacture.) Knowing how it was tracking 3 to 5 months earlier than when the comparable field data became available would permit earlier response to any problems. This, in turn, would affect customer satisfaction and service costs.

The procedure as currently performed calls for 2% of the machines produced in any 1 week to be run almost continuously until 3 months of average customer operation has been simulated. This process requires 3 weeks to perform. During this time, a record is kept of the number of service calls required, their nature, and the number of problems found per call. The number of calls is an indication of customer satisfaction, and when combined with the number of problems per call, it relates to service cost.

The resulting data can be portrayed in various ways to make them more meaningful to specific, interested departments. For example, if the average number of calls per 3 months of average use is 2, then by rearrangement, it can be stated that, for a typical user, the average time between service calls is 1.5 months. (This assumes, according to observed fact, that service calls are about evenly distributed over the 3-month period. The figure 1.5 is obtained by dividing 3 [months elapsed] by 2 [calls], yielding 1.5 months elapsed per call.) This might be a more useful way for the marketing department to view matters as it attempts to evaluate customer satisfaction level as a function of frequency of failure.

Figure 1.8 shows the pattern of average number of service calls per 3 weeks of simulated operation (or 3 months of on-the-job use). Notice the slight decline in these averages as time passes. For a specific 3 weeks of simulated use, the average number of service calls was found to be 1.84 for the sample of machines. This corresponds to a .95 confidence interval of 1.66 to 2.02. What this interval means is that 95 times out of 100, with repeated samples of the same size from the same week's output, the confidence interval obtained will include the true but unknown population average of service calls per 3-week period of simulated activity. (This subject is treated in Chapter 8; don't worry about the details right now.) Thus, the .95 figure is the *measure of reliability* of the statistical conclusion in this case. The range of values representing lengths of time between service calls runs from 1.49 to 1.81 months, with .95 confidence.

1. The advantages of the simulation method used by DPC in connection with the DE-1 model are many relative to the old method. What is the most obvious relative disadvantage?

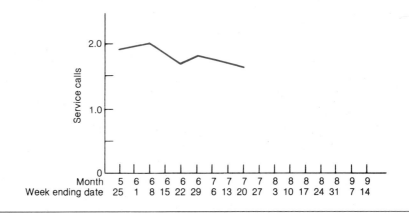

Figure 1.8. Plant test service calls per average 3 weeks of simulated operation.

2. In the simulation procedure, the physical environment—that is, temperature, humidity, air quality, and so forth—was controllable and could be made very similar to the physical environments in user plants. Can you think of any relevant factors in the real environments that probably could not be so well simulated within DPC's controlled environment? If so, name some.

This case is continued at the end of Chapter 2.

Essentials of Survey Sampling

Everyone who has poured a highball into the nearest potted plant after taking one sip has had some experience in sampling.—Morris James Slonim

You don't have to eat the whole ox to know the meat is tough.—Samuel Johnson

What You Should Learn from This Chapter

By its very nature, inductive statistics is dependent on sampling. However, unless the sampling process is carried out properly, the resulting sample may be invalid and the inductive conclusions incorrect. In this chapter we consider a variety of topics related to the how's and why's of sampling. When you have finished studying this chapter, you should be able to

1. Defend the use of sampling.
2. Distinguish between sampling errors and nonsampling errors.
3. Describe the qualities of a good sample.
4. Distinguish between probability sampling and nonprobability sampling, and cite the major advantages of probability sampling.
5. Identify the following methods: (1) simple random sampling, (2) systematic random sampling,* (3) stratified random sampling,* (4) random cluster sampling,* and (5) multiple random sampling.*
6. Distinguish between target population and sampled population, and tell why the distinction is important.
7. Cite several considerations related to the determination of sample size.

Before proceeding, you should be sure you are clear on the distinction between parameter and statistic (section 1.6 of Chapter 1) and the distinction between survey sampling and designed experiment (section 1.7); this chapter deals exclusively with survey sampling.

2.1 Introduction

- A woman dips her hand into a tub of bathwater to make certain the temperature is right.
- A quality-control inspector examines a batch of engine cylinder liners and concludes that the production process is in control.
- A grocery shopper selects a cup of strawberries based on the appearance of the berries on top; while in the produce section, he also squeezes a couple of nectarines to see how ripe they are and, hence, how ripe the unsqueezed ones *probably* are.
- A holder of political office initiates a public opinion poll trying to determine her chances for reelection.

*These topics are optional.

- A doctor extracts a small amount of blood from a patient and concludes that his blood-sugar level is normal.
- An applied research organization polls homeowners, asking whether they intend to buy a new refrigerator during the next six months.
- A dog loses interest in his dinner after eating one mouthful.
- A television station charges viewers 50 cents apiece for the privilege of phoning in their opinions on genetic engineering.
- The Internal Revenue Service selects the names of several citizens to undergo tax audits.
- A young man who has been told on two separate occasions by two different girlfriends with red hair that they never want to see him again decides to stop dating women with red hair altogether.

The point, of course, is that sampling plays a role in many facets of our lives. We all encounter sampling several times each day from the moment we take our first cautious sip of hot coffee at the breakfast table to the time we fiddle with the television knobs at night to see whether anything appears worth staying up to watch. Even animals engage in sampling, as the Morris the Cat commercials have reminded us in a humorous way.

Admittedly, there are all kinds of sampling procedures, with varying characteristics, ranging from the very casual to the very formal, and the preceding list includes examples that run the gamut. Some investigations can be carried out with flawless safety despite a sampling procedure that may seem recklessly sloppy. For example, a doctor examining a patient's blood-sugar level need not extract a huge quantity of blood nor even stick the needle into a randomly selected assortment of places on the patient's body. A small amount of blood taken from a single location is adequate because the chemical composition of one's blood is effectively uniform. On the other hand, the grocery shopper who judges a cup of strawberries by the appearance of those on top may be taking a chance; perhaps the berries have been "stacked," like a deck of cards is sometimes stacked, so that the least attractive strawberries are hidden from view. Similarly, in view of the great diversity of human characteristics, tastes, responses, and interactions, the young man who swears off women with red hair because of bad luck with two of them is making a gigantic inductive leap, not at all justified by his meager sample data.

Even though the act of sampling undoubtedly predates recorded history, only in recent years have systematic efforts been made to develop sampling procedures having specific desirable properties. In this chapter we consider a number of topics concerned with sampling in situations where formal procedures are called for.

2.2 Why Sample?

Sampling plays a most important role in statistical analysis. Often it is simply not possible, not feasible, or not desirable to examine the entire population of interest. Here are some of the most important reasons that samples are used:

- Sample data are usually more up to date than data obtained from an entire population study, called a *census*, because samples require less time for data collection and processing. This is no small consideration to the decision maker who would like to have had the data yesterday. A manager can seldom wait, as many a pure scientist has had to wait, for data requiring several months or even years to collect.

- Because of its smaller size, a sample study can provide useful information at a much lower cost than a census. Consider, for example, a study concerned with determining consumer preferences in package design for a widely used product: If all the people who buy the product with some regularity had to be contacted, the cost of gathering the preference data might far exceed the increase in profits resulting from use of the preferred package.
- Paradoxically, a sample study will often provide data that are more accurate than those which could have been obtained from a complete census. Since sampling is an undertaking on a smaller scale, it is easier to select and supervise interviewers and investigators. Moreover, errors resulting from various kinds of routine activities—tabulating data, transcribing responses, and so on—are often kept lower when a sample, rather than a census, is used. Of course, such efficiencies from sampling are not inevitable, but they are often realizable because of the greater control that is possible.
- Some investigations, such as certain quality-control inspections, require the destruction of the product. Only by destroying an elevator cable can one be sure of the amount of pressure it can withstand; only by burning a light bulb until it ceases to work can one be sure of its length of useful life. Clearly, if all the elevator cables or light bulbs (or clotheslines, firecrackers, etc.) produced were subjected to destructive testing, there would be none left to sell to customers. Consequently, sampling is absolutely necessary in such situations.
- Some populations of interest are infinite in size, thereby ruling out even the possibility of a complete census. Such populations are usually associated with designed experiments, such as the one described in Illustrative Case 1.5, where jars and boxes were used to package a new soybean cereal.

Some people are highly suspicious of sampling for understandable, though not necessarily valid, reasons. Their arguments tend to run something like, "I have never had a polling organization ask me who I planned to vote for; I have never had anyone phone to ask what television show I was watching; no marketing research company has ever sought my opinion about a new product. How can samples tell the truth when *my* opinions are so blatantly and consistently ignored?" The late humorist Fred Allen undoubtedly spoke for many when he castigated the whole idea of sampling through his attack on the Hooperatings (a radio listenership survey service), with obvious bitterness, after learning that the ratings of his own radio show had been slipping: "The Hooperatings is a so-called service that allegedly tells you approximately how many listeners the average radio show theoretically has. It's like taking a bite out of a roll and telling you how many poppy seeds there are in the entire country."[1]

Adding to the bad press that sampling has received is the fact that many sample studies are poorly conducted—sometimes unbelievably so. For example, shortly before the presidential election of 1968 a widely read news magazine reported, with utmost seriousness, that a recent poll conducted by a Wisconsin editor had turned up some puzzling results.[2] It then elaborated on the "puzzling" findings. And just how was the poll conducted? It seems that the Wisconsin editor had questioned eight men in a bar! The only thing puzzling about the results is that they were considered worthy of space in a national magazine. Other examples of faulty sampling would not be hard to find.[3]

Despite the many criticisms of sampling, both just and unjust, the myriad uses to which the method is put attests to its value. Today, sampling is used extensively in

- Quality-control inspections within manufacturing companies and firms buying goods from manufacturing companies

GRIN AND BEAR IT   by Lichty & Wagner

© Field Enterprises, Inc., 1983

"The only thing I have opinions on is sports."

- Public opinion polling on political and social issues
- Television viewership surveys
- Testing for pollutants in the environment
- Regular collection of data related to the state of the general economy
- Accounting and auditing
- Testing for mineral quality and quantity in mining activities
- A great many different kinds of marketing research investigations
- Election-night projections of winners of political offices

and many other activities.

2.3 Errors in Statistical Data

Data obtained from a sample are subject to two broad categories of error—sampling and nonsampling.

> A *sampling error* is the difference between the result obtained from a sample study and the result that would have been obtained from an *equal complete coverage*—that is, from an investigation of the entire population conducted in exactly the same manner as in the sample study.

Consider, for example, a survey conducted to determine the average amount of alcoholic liquid consumed by respondents in a "typical" week. The equal complete coverage in this case is the average quantity of alcoholic liquid consumed according to the results of a study of *all* persons of interest using the same survey methods as in the sample study.

Unfortunately, this equal complete coverage may not be the quantity really sought if, for example, people knowingly understate the extent of their drinking. Or overstate it. Or do not know what is meant by a "typical" week. Or do not make a habit of counting drinks and are unsure themselves of the actual quantity. Under such circumstances, equal complete coverage will itself be in error even though it contains only nonsampling errors.

> *Nonsampling errors* are those errors that can arise even in an equal complete coverage.

Nonsampling errors can enter into a statistical investigation because (1) questions were not worded properly or clearly, (2) biases or mistakes on the part of the interviewers are present, (3) respondents do not furnish accurate information, as suggested in the example about alcoholic beverage consumption, (4) some sample units are not available to be interviewed or examined, and so forth. Still other nonsampling errors can occur in the editing, tabulating, and calculating steps.

"Oh, I'm so glad you're here! I answered those questions yesterday, but I've changed my mind!"

Because the final accuracy of a statistical study depends on the adequacy with which both sampling and nonsampling errors have been controlled, every effort compatible with available resources of money, time, and manpower should be made to minimize these errors. When a valid random sampling procedure is already being used, sampling errors can be reduced only by increasing the size of the sample. A list of ways to control nonsampling errors could be quite long. Such precautions as thorough training of interviewers, pretesting of questionnaires, and followups and callbacks on nonrespondents should certainly be taken insofar as financial and other resources permit.

2.4 Characteristics of a Good Sample

The goodness of a sample is determined by how well it represents the population it is presumed to represent. The more closely it resembles the population, the greater its *accuracy* is said to be. Accuracy can be evaluated according to the two criteria of *unbiasedness* and *precision*.

Unbiasedness

An *unbiased sample* is one in which the sample observations exhibit no overall systematic tendency to underestimate or overestimate the population observations.

Put another way, the "center of gravity" of the sample is in about the same location as that of the population; sample observations below the "center of gravity" of the population are roughly balanced by sample observations above the population's "center of gravity."

Precision

Precision refers to the adequacy with which sampling information can be marshalled to provide estimates of population parameters. Estimates close to the true parameter values are said to be precise.

A sample statistic can be expected to differ from the corresponding population parameter to some extent because of random fluctuations in the sampling process. This is referred to as sampling error, a subject introduced in the preceding section. However, other things being equal, the smaller the sampling error the better we like it; a small sampling error often augurs for population estimates that differ little from the true values. A large sampling error may imply less precise estimates.

The difference between unbiasedness and precision can be clarified through use of an analogy: Suppose that a physical education teacher is asked to pick his best archer to compete

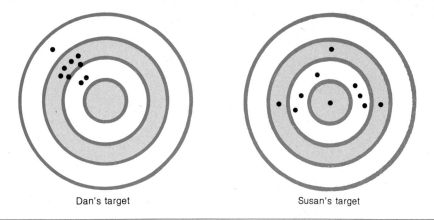

Dan's target   Susan's target

Figure 2.1. Results of ten shots at a target by archers Dan and Susan.

in a prestigious international tournament. Some preliminary screening narrows the choices down to Dan and Susan. Dan and Susan are asked to shoot ten arrows at separate targets. The results are shown in Figure 2.1.

We see that Dan's results are more tightly bunched than Susan's. However, he shows a bias for one area of the target. Susan, on the other hand, is apparently not biased, but she is not particularly precise either, as indicated by the scattering of her hits. On the basis of this trial, the teacher is inclined to pick Susan for the big competition on the grounds that she, at least, tends to hit in and around the bull's-eye.

Before breaking the bad news to Dan, the teacher works with him for a short time hoping to improve Dan's aiming ability. Specifically, he has Dan concentrate on a point both a little lower and further to the right than the boy had been using. The two archers are then told to fire ten more arrows at their respective targets. Figure 2.2 shows the results. With his inher-

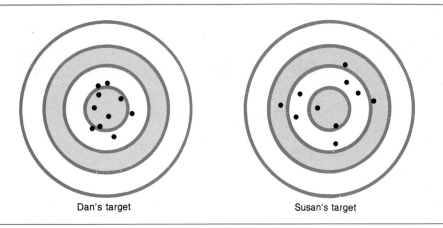

Dan's target   Susan's target

Figure 2.2. Results of ten shots at a target by archers Dan and Susan after Dan is given some special instruction on aiming.

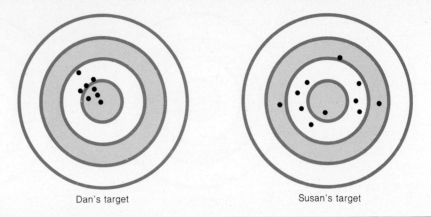

Dan's target   Susan's target

Figure 2.3. The archery example again, showing that, if precision is good and bias is small, the presence of bias may not be too serious a concern.

ent tendency to bunch his shots more tightly and his newfound unbiasedness, Dan is seen to be more precise as well and proves to be the more promising archer.

Actually, Dan's bias, had it been much smaller, might not have interfered with his being judged the better archer in the first place. Suppose, for example, that the first set of ten shots by each archer resembled the results of Figure 2.3 (that is, his shots are more precise). Despite the presence of bias, Dan is seen to have hit the bull's-eye more often than Susan. Similarly, in statistics a procedure need not be rejected solely because of the presence of bias. In fact, a biased procedure with a small sampling error will often be preferred to an unbiased procedure with a large sampling error.

2.5 Overview of Approaches to Survey Sampling

Elementary units may be selected from a population in a variety of ways. The kind of sampling plan chosen will depend on the objectives of the investigation and on such mundane matters as time and financial constraints. A most important distinction is that between probability and nonprobability sampling plans.

> In *probability sampling* each elementary population unit has a known probability of being included in the sample. With *nonprobability sampling* personal judgment is exercised in determining which elementary population units are to make up the sample. Samples having a nonprobability basis are also called *judgment samples*.

Thus, a nonprobability, or judgment, sample is one selected on the basis of convenience, availability of expert opinion, or something else having nothing to do with formal probability considerations. It may be the method of choice under circumstances where the sample must

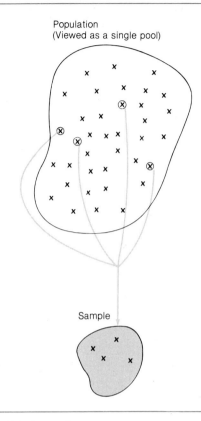

Population
(Viewed as a single pool)

Sample

Figure 2.4. Example of unrestricted sampling.

be small, because of limited funds or other constraints, and measures of reliability and precision are not required. Judgment samples have a place in business investigations but will be largely ignored in this book for reasons to be noted shortly.

A distinction can also be made between unrestricted sampling and restricted sampling.

> If the sampling plan calls for elementary units to be selected from a population viewed as a single pool, as in Figure 2.4, it is called *unrestricted sampling*. If the population has to be divided into subpopulations (as in Figure 2.9) or if clusters of elementary population units are selected just as if they were single population units (Figure 2.11) or if some other departure from the single pool view of the population is used, it is called *restricted sampling*.

Table 2.1 lists several sampling plans according to whether they are probability or non-probability types, on the one hand, and unrestricted or restricted on the other. In the following five sections, we will consider the *probability sampling plans* appearing in this table. The

Table 2.1 Sampling Procedures Classified According to Whether They Are (1) Probability or Nonprobability and (2) Unrestricted or Restricted

	Probability	Nonprobability
Unrestricted	Simple random Systematic random	Convenience
Restricted	Stratified random Random cluster Multiple random	Purposive Expert choice Quota

nonprobability sampling plans will not be described. Probability methods are emphasized throughout this text because of their clear-cut superiority. This is not to suggest that use of judgment sampling is necessarily bad. A particular judgment sample may portray excellently the characteristics of the population. On the other hand, it could be very poor. Everything depends on the skill, wisdom, and, perhaps, luck of the investigator. The unavoidable problem with judgment sampling is that no objective method exists for measuring the precision or reliability of estimates made from the sample. With probability samples, the precision with which estimates of parameter values can be made is measurable from the sample itself—an important advantage, as will be demonstrated in later chapters.

2.6 Simple Random Sampling

The term *simple random sample* is often misunderstood and, thus, must be defined with care.

> With *simple random sampling* the sample is selected in such a way that each possible different combination of a specified number of elementary population units has the same probability of being selected.

For example, if we call N the number of elementary population units and n the desired sample size, the number of different samples of size n (denoted N^*) that could be drawn from this population is equal to

$$\frac{N!}{n!(N-n)!}$$

The ! sign means *factorial*. The factorial of a number is the product of a series of factors that change by decrements of 1 until the figure 1 has been reached. That is, N! indicates that the product of $N(N-1)(N-2)(N-3) \ldots 1$ is to be obtained.

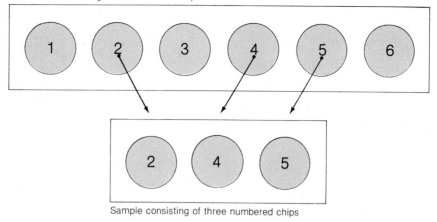

Population consisting of six numbered chips

Sample consisting of three numbered chips

Figure 2.5. Illustration of one possible sample of size 3 which could be selected from a population of size 6. This specific sample is only 1 of 20 obtainable from the population of 6 chips. (See Table 2.2 for an exhaustive list of sample possibilities.)

Let us designate a set of elementary population units with numbers running from 1 to 6 and assume that three of these units are selected purely at random to make up our sample, as shown in Figure 2.5. In this figure, population units 2, 4, and 5 are assumed to be the ones selected. What do we mean when we say that these elementary units were selected purely at random? We mean that they were selected by means of a *random mechanism* rather than by personal judgment. We could, for example, place six numbered chips—identical in size, shape, and texture—into a hat, shake the hat vigorously to ensure that the chips are well mixed, and, without looking into the hat, select three chips to constitute the sample. One such exercise in sample selection produced the results shown in Figure 2.5, in which chips numbered 2, 4, and 5 were the ones selected. This sample is only one among many—20 to be exact—that could have been selected. That is,

$$N^* = \frac{N!}{n!(N-n)!} = \frac{6!}{3!3!} = \frac{(6)(5)(4)(3)(2)(1)}{[(3)(2)(1)][(3)(2)(1)]} = 20$$

The entire list of possibilities is shown in Table 2.2. The expression $N!/[n!(N-n)!]$ provides us with a convenient method for determining the number of possible sample combinations without our having to enumerate them.

To return to our definition of simple random sampling: Since there are 20 different sample combinations shown in Table 2.2 and since any one of these is just as likely to have been selected as any other, we could have said prior to selection that the probability of getting any specific one of these sample combinations is 1/20 or, more generally,[4]

$$1/\frac{N!}{n!(N-n)!}$$

Table 2.2 All Possible Samples of Size 3 That Could Have Been Selected from a Population of 6 Numbered Chips

1, 2, 3	2, 3, 4
1, 2, 4	2, 3, 5
1, 2, 5	2, 3, 6
1, 2, 6	2, 4, 5 ←The specific sample we assume
1, 3, 4	2, 4, 6 was selected
1, 3, 5	2, 5, 6
1, 3, 6	3, 4, 5
1, 4, 5	3, 4, 6
1, 4, 6	3, 5, 6
1, 5, 6	4, 5, 6

Selecting a Simple Random Sample from a Finite Population

We spoke above of listing the different samples of a given size n that could be selected from a specific population of size N. In practice, of course, we usually work with a single sample—never with all possible samples. Let us now consider some how-to aspects of simple random sample selection.

ILLUSTRATIVE CASE 2.1

A certified public accountant called in to audit the fiscal year records of the Femme Fatale Fashion Shoppe wishes, among other things, to obtain an independent check on the store's claimed accounts receivable in order to estimate (1) the average size of receivables and, in turn, the total amount of money owed the store by its charge customers and (2) the average of and the amount of variation in the ages of these accounts to determine whether the company's allowance for probable bad debts is realistic. Because the population of accounts receivable includes 794 such accounts and because the auditor considers it impractical to examine each account individually, she decides to select a simple random sample for close examination.

How might this auditor go about picking her sample? Let us assume, for simplicity, that she wishes to work with a sample of only ten accounts receivable. She would probably proceed as follows:

1. Prepare a list of customers' names associated with each of the 794 accounts receivable that make up the population. Such a list is called a *sampling frame*.
2. Assign a number to each account on the list. The first account would be assigned the number 001—a three-digit number because the highest number on the list, 794, is a three-digit number. This step is illustrated in Table 2.3.
3. Determine what random mechanism will be used to select the ten sample units from this population. The auditor could proceed in the manner illustrated above, that is, by placing numbered chips in a hat, shaking the hat, and removing ten chips without first looking into the hat. Fortunately, there is an easier and surer way—namely, that of making use of a *table of random digits*. Published tables of random digits present the integers 0, 1, 2, . . . , 9 in such a way that knowledge of one such integer gives no guide whatever to the integer following it, preceding it, above it, beneath it, or in any other specified location. Table 2.4 will be used to illustrate how such a table of random digits might be employed.[5]

Table 2.3 Data for 794 Charge Customers of the Femme Fatale Fashion Shoppe

Number Assigned	Account Name	Size of Account	Age of Account
001	Aarmond, Betty	$ 300.00	1 month
002	Abel, Marsha	73.75	3 months
003	Abelson, Burton	10,465.23	14 months
.	.	.	.
.	.	.	.
.	.	.	.
794	Zwindell, Tess	235.95	1 month

The auditor in our example may make use of the table of random digits in any manner she chooses provided that she (1) develops some self-imposed rules prior to consulting the table and (2) does not deviate from those rules when using the table. She might, for example, decide to begin with the left-most column of three-digit numbers, recording the fifth such number down from the top and then every third three-digit number thereafter. When she arrives at the bottom of the left-most column, she decides to move to the next column to the right, still keeping an interval of three between selected numbers. To illustrate: The fifth three-digit number down from the top in the left-most column is 540. Therefore, the account assigned the number 540 in the list of population units would be the first to be included in the sample. Skipping three lines, we find the number 538. So account number 538 is included in the sample. Next come 775, 514, and 852. Since the population identification numbers only go as high as 794, the number 852 is not used. Then comes 651. Then 971; this one is also ignored. Then 076, 108, 385, 272, and 489. The sample of $n = 10$ items, therefore, consists of accounts numbered

540   538   775   514   651   076   108   385   272   489

You will notice that no number appears more than once in this sample. In this specific case, the absence of duplicate numbers was a matter of luck. In general, when duplications occur, they should be ignored the second and subsequent times because all the information one can extract from a specific elementary unit is extracted on the first examination of it.

The sample obtained in the manner just described was chosen to give every possible sample combination of ten elementary population units an equal opportunity of being selected; hence, it is, by definition, a simple random sample.

Table 2.4 Random Digits

091	775	380	595	128	328	868	817	755	650
900	195	374	385	217	519	699	013	237	388
731	218	971	306	195	476	932	870	946	651
757	514	218	654	672	209	681	471	182	355
540	995	731	272	605	662	624	959	131	075
083	337	076	913	538	785	935	030	190	360
283	852	605	684	246	813	113	079	844	743
538	841	835	489	830	815	440	069	547	389
917	567	108	069	164	616	173	995	426	361
894	651	398	104	607	003	876	436	142	571

Target Population Versus Sampled Population

The sampling-frame idea is conceptually quite simple. But putting it into practice is often more difficult than Illustrative Case 2.1 might lead one to believe. The problems arise from the fact that some populations simply do not lend themselves readily to the development of an operationally feasible frame. Let us consider a seemingly innocuous hypothetical example.

ILLUSTRATIVE CASE 2.2

Tyrone Tyson of Tyler City is thinking about opening a shop devoted exclusively to the sale of men's ties. He would like to get some idea of the potential for such a specialty shop and feels that a sensible way to begin would be to talk to knowledgeable managers of existing retail outlets selling men's ties. Having taken a statistics course, he is sensitive to the possibility of permitting a bias to enter into the selection of people to interview. Therefore, he decides to obtain a list of retail stores in Tyler City in which men's ties are sold and select a simple random sample from the list.

Unfortunately, it is not long before he encounters unforeseen problems: First, there is no such ready-made list of relevant retail stores available. So he decides to develop his own list.

But then he finds he has difficulty identifying and defining the population units. For example, it seems obvious that all, or almost all, department stores and men's clothing stores should be included in the sampling frame. But what about the occasional variety store that sells men's ties? The individual who makes ties as a hobby and manages to sell a few? Mail-order specialty-tie operations headquartered elsewhere but making some sales within Tyler City? Western-wear stores? He also wonders whether tie rental services, including those renting ties with tuxedos, theatrical costumes, and so on, would be at all relevant to his search for information.

Granted, most of Tyson's concerns could be resolved adequately by his asking himself some searching questions about what kinds of retail outlets are likely to have managers with meaningful opinions about his likelihood of success. That is, he could presumably come up with a satisfactory "working definition" to serve as a basis for his sampling frame. Nevertheless, the point is: A population that seemed clear-cut to begin with is now seen as something of a challenge to identify properly. Nor is this necessarily the last of Tyson's problems: It is possible that many managers of existing stores will refuse to grant him interviews once they learn that he has intentions of becoming a future competitor. Thus, even after he has identified and defined the relevant population, he may find that his *sampled population* is not exactly the same as his *target population*.

> A *target population* is the population one wishes to sample. A *sampled population* is the one that actually is sampled.

Under ideal circumstances, the target population and the sampled population are the same. However, there are times when finding or developing an operationally feasible sampling frame is so difficult that the target and sampled populations differ in important respects. This is no small consideration, since the success or failure of a statistical survey often hinges on how closely the population which can be sampled resembles the population one would like to sample. Other factors besides sampling-frame problems can also lead to a disparity between sampled and target populations. For example, people who cannot be contacted even

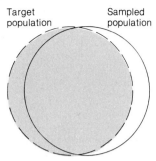

Target population contains some elementary units not included in the sampled population, and the sampled population contains some elementary units not included in the target population

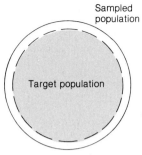

Sampled population somewhat more inclusive than target population

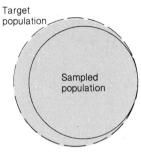

Target population somewhat more inclusive than sampled population

Figure 2.6. Some ways that target and sampled populations can differ slightly. If the sampled population can be judged to contain important and relevant information, the investigation can be conducted provided that those involved realize that any inductive inferences will pertain to the sampled population and not necessarily the target population.

after several callbacks may be quite different in important respects from others in the sampling frame who are contacted with ease; people who refuse to be interviewed may be quite different from others in the sampling frame who are more obliging; and so forth. If the target and sampled populations differ somewhat but the latter can be judged to contain important and relevant information, the investigation can be conducted as long as those involved realize that any subsequent inductive inferences will pertain to the sampled population and not necessarily the target population (Figure 2.6). If the sampled and target populations differ markedly and nothing can be done to improve the correspondence, the investigation should be dropped (Figure 2.7).

Limitations of Simple Random Sampling

Frequently, a sampling plan other than simple random sampling will be used in practice either because (1) simple random sampling is impractical to employ—perhaps because population lists are not available or because sample members are separated by great distances—or (2) simple random sampling does not promise sufficient precision in the estimation of population parameters without the use of a prohibitively large sample. Let us consider some alternative probability sampling plans.

2.7 Systematic Random Sampling

When conditions are right for the use of *systematic random sampling*, the equivalent of a simple random sample can be obtained with greater ease this way than with a table of

*A large asterisk to the left of a heading or item means that the topic is optional.

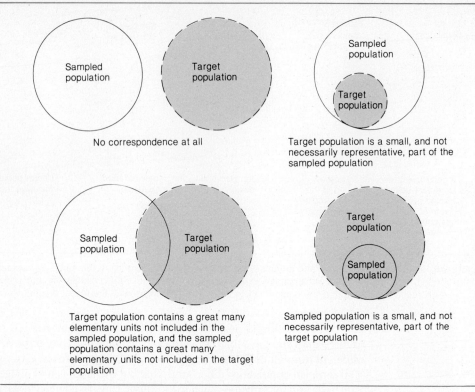

Figure 2.7. Some ways that target and sampled populations can differ significantly. If nothing can be done to improve the correspondence, the investigation should be dropped.

random digits. The conditions necessary for systematic random sampling are: (1) The elementary population units are spatially ordered or listed, as in a file drawer or in computer storage, and (2) this spatial ordering or listing is random.

> In *systematic random sampling* one of the first *g* elementary population units is selected using a random start. Then, it and every *g*th unit thereafter are included in the sample.

For example, the auditor in the Femme Fatale Fashion Shoppe problem (Illustrative Case 2.1) could have employed systematic random sampling because (1) she had available a list of charge customers comprising the entire population of interest and (2) the fact that the names of the charge customers were arranged alphabetically suggests a random ordering of size and age-of-account data. She could have proceeded in the following manner, as illustrated in Figure 2.8.

1. Determine the *sampling fraction f* by dividing the population size N into the sample size n. Recall that the population size in this illustrative case was $N = 794$. For

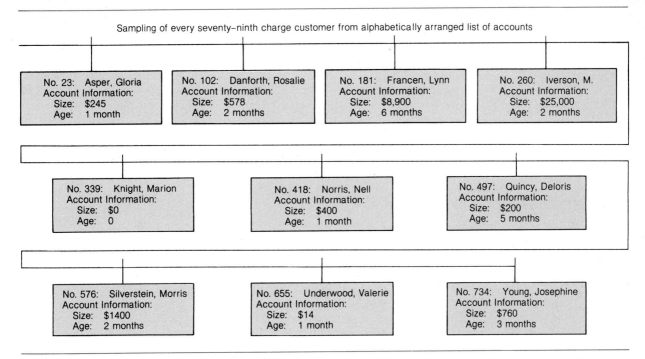

Figure 2.8. Illustration of sample selection by systematic random sampling. The sample consists of every seventy-ninth account, starting with Account 23.

simplicity, we declared the sample size to be $n = 10$. Therefore, the sampling fraction is $f = n/N = .0126$.

2. Convert the sampling fraction into the corresponding *sampling interval g* by dividing the sampling fraction into 1. That is, $1/.0126 = 79.37$, or about 79. This means that every 79th charge customer on the list would be included in the sample.

3. Use a table of random digits to determine which one of the first 79 charge customers will be the first one included in the sample. Let us say that such a random procedure produced the number 23. Thus, the account positioned 23 down from the beginning of the list would be one of the sample units.

4. Add 79 to 23, to get 102; add 79 to 102, to get 181; and so forth until $n = 10$ such numbers have been determined. These numbers will correspond to positions on the list of elementary population units. In this case, the ten elementary units making up the sample would be the 23rd, the 102nd, the 181st, the 260th, the 339th, the 418th, the 497th, the 576th, the 655th, and the 734th.

The simplicity of this method makes its use highly desirable under some circumstances. However, anyone using it without heeding the requirement that the elementary population units be ordered randomly to begin with runs the risk of getting a deceptively unrepresentative sample. Recall that with simple random sampling every combination of n elementary units within the population has the same probability of being picked for the sample. This condition is not met when systematic random sampling is used, regardless of the arrangement of the elementary population units. The reason is that systematic sampling rules out combinations having two elementary units closer to each other than g. When the elementary population units are arranged randomly, as can usually be assumed when alphabetization is used or when the investigation is concerned with fabricated items emerging from a production process, a systematic random sample can be expected to be as representative of the population as a simple random sample of the same size. However, if periodicities exist in the population observations, a systematic sample would likely be less representative than its simple random counterpart. Generally speaking, it is best to limit the use of this method to situations where one can reasonably assume a random ordering of populations units.

---

*2.8 Stratified Random Sampling

In simple random sampling, the population is treated as an undifferentiated whole—or, to use the terminology employed earlier, as a "single pool"—from which the sample units are selected. Often it is both possible and desirable to break the population up into mutually exclusive (that is, nonoverlapping) *subpopulations* and obtain a simple random sample from each such subpopulation. This process is known as *stratification*. When the separate simple random samples are combined, the result is a particular kind of probability sample called a *stratified random sample*.

Let us suppose that we wish to interview a sample of people to determine their attitudes toward buying fresh fruit in specialty stores rather than in grocery stores. We fear, however, that simple random sampling might result in too many (or too few) housewives. The population could be broken down into two subpopulations—housewives and others—assuming, of course, that relatively recent data are available to make such a breakdown feasible. One simple random sample would then be selected from the subpopulation "housewives" in the same manner as in our Femme Fatale Fashion Shoppe example (Illustrative Case 2.1) and another from subpopulation of "others." The process is shown in Figure 2.9, where unrealistically small subpopulation sizes have been assumed for ease of illustration.

A common procedure is to select the same proportion from each subpopulation. That is, select sample units in such a way that $n_1/N_1 = n_2/N_2$, where N_1 is the number of persons in the subpopulation consisting exclusively of housewives, N_2 is the number of persons in the subpopulation consisting of people with occupations other than housewife, n_1 is the number of persons in the sample from the housewife subpopulation, and n_2 is the number of persons in the sample from the other-than-housewife subpopulation. (More generally, $n_1/N_1 = n_2/N_2 = n_3/N_3 = \cdots = n_k/N_k$.) This practice is called *proportional allocation*, and it is appropriate when (1) the cost of sampling is about the same for all subpopulations and (2) the amount of variation is about the same within all subpopulations. If the cost of sampling differs among subpopulations, that fact should be taken into consideration because the usual goal of sampling is to achieve maximum precision in parameter estimation at minimum cost. However, if the cost of sampling is the same for all subpopulations, the object is to secure maximum precision with the smallest sample. In this case, the important criterion in determining the number of elementary units to draw from each subpopulation is the amount of variation within the subpopulations. If one subpopulation has much more variation than the other(s), more observations should be taken from it.

The main purpose of stratification is to force a certain amount of representativeness upon the overall sample. With a properly selected simple random sample, we might reasonably

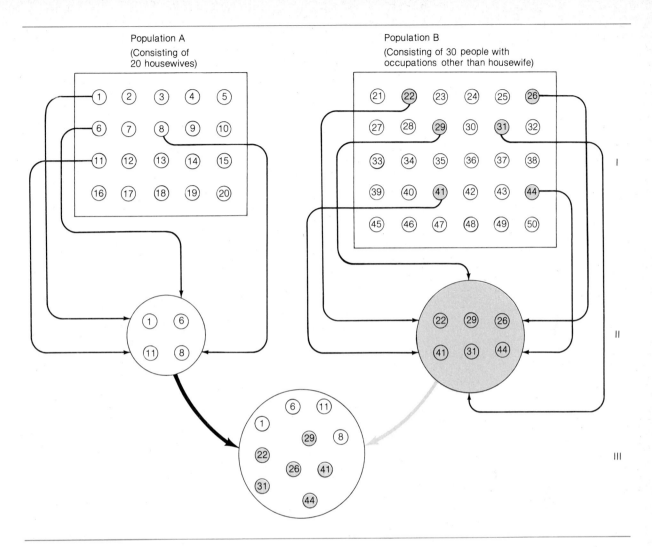

Figure 2.9. Illustration of stratified random sampling with proportional allocation. On Line I, two separate populations have been distinguished—a population consisting entirely of housewives (Population A) and a population consisting entirely of people with occupations other than housewife. Line II depicts the results of two separate simple random samplings. Four housewives were selected at random from the 20 housewives making up Population A, and 6 persons were selected at random from the 30 persons making up Population B. (Note: 4/20 = 6/30; hence, we see that proportional allocation was used.) In Line III, the separate simple random samples are combined to make a stratified random sample of 10 persons.

expect the sample to resemble the population in general configuration, but that is by no means inevitable. In any given case, a simple random sample can be quite unrepresentative of the population. With stratified random sampling, the opportunity for a sample to differ markedly from the population is more restricted, provided, of course, that the criterion used to subdivide the population is a valid one. For example, if attitudes toward specialty fruit stores are affected by whether the respondent is or is not a housewife, then stratification on

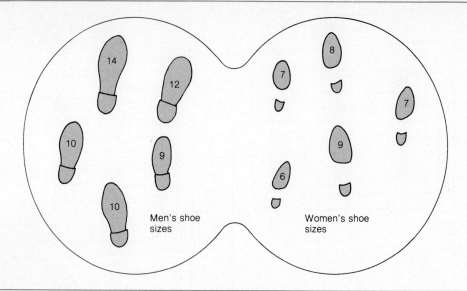

Figure 2.10. Illustration of how stratified random samples can lead to more precise estimates of population parameters than simple random samples of the same size. The population of interest is ten shoe sizes (lengths only). The population total of the shoe sizes is 92. Now let us suppose that two shoe sizes are to be selected from this population and used, by multiplying the sample total by 5, to estimate the population total. With simple random sampling the two largest shoe sizes could occur together in the same sample. That is, (14 + 12)5 = 130. Or the two smallest shoe sizes could occur together in the same sample: (6 + 7)5 = 65. Both estimates, 130 and 65, are markedly different from the true population value of 92. However, if we stratify according to whether the shoe sizes are men's or women's, thereby guaranteeing that we have one man's shoe size and one woman's shoe size in the sample, the most extreme possible estimates are (14 + 9)5 = 115 and (9 + 6)5 = 75. With both kinds of sampling plans, less extreme samples would provide even closer estimates. However, with simple random sampling, we will find more fairly extreme samples than we will with stratified random sampling.

this basis will result in sample units that are relatively homogeneous within subpopulations and relatively heterogeneous between subpopulations. Such a situation helps to ensure that the sample is reasonably representative of the population, as illustrated in Figure 2.10. Usually, a stratified random sample will deliver more precision in parameter estimation than a simple random sample of the same size.

*2.9 Random Cluster Sampling

The term *primary sampling unit* is essential to the understanding of how random cluster sampling works.

Primary sampling units may be elementary population units or they may be *groups* or *clusters* of elementary population units.

For example, the elementary population units of interest may be people within a certain geographical area old enough to have a meaningful opinion on a political issue. The primary sampling units in such a situation might be households within the geographical area. In a different context, the elementary population units might be students and the primary sampling units, schools.

> If it is elementary units whose characteristics are to be studied, but primary sampling units in the form of clusters of such elementary population units are selected at random in the initial stage of sampling, the sampling method is called *random cluster sampling*.

If all elementary units comprising the clusters are to be examined, the method is called *single-stage sampling*. But if, as is more common, the clusters are to be subsampled so that the elementary units are selected by a second random sampling process, the method is called *two-stage sampling*. Of course, there could be more than two stages. For example, the first stage might consist of manufacturing businesses; the second stage, separate and identifiable production facilities within the manufacturing businesses; and the third stage, production workers at these facilities.

The most important form of cluster sampling in business research is *area sampling*. In area sampling, the clusters usually consist of individuals or households living within compact geographical areas such as city blocks. Use of this sampling plan allows us to sidestep one of the practical problems of simple random sampling: It is often easy to obtain a list of areas even when it is impossible to obtain a list of individuals residing in the areas. All one needs to launch a sample survey is a map permitting identification of nonoverlapping areas. With a list of areas, one can readily select a simple random sample of the areas. Having selected the areas—let us say city blocks—making up the simple random sample, the investigators may either proceed to interview someone in each of the households in those blocks or they may decide to use a multistage approach and select individual households at random from each of the randomly selected city blocks (Figure 2.11).

Unlike stratified random sampling, where increasing the precision of estimates is often a prime goal, area samples (or other cluster samples) are not ordinarily used to increase precision in estimation. Indeed, they usually produce less precise estimates than simple random samples of comparable size. The chief advantage of area sampling is lower cost. Savings can be achieved with area sampling through (1) the ready availability of lists of primary sampling units and (2) the fact that many of the elementary sample units will ordinarily be in close proximity to one another.

*2.10 Multiple Sampling

The probability sampling plans described above all involve the prior determination of a sample size. Once the sample size has been determined, a sample of that specified size is selected by the appropriate method and the entire sample is inspected.

Sometimes, important economies can be realized by making use of less-than-complete samples. Let's look at an example.

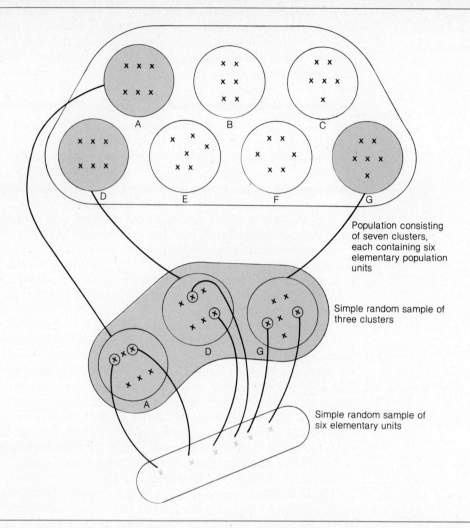

Population consisting of seven clusters, each containing six elementary population units

Simple random sample of three clusters

Simple random sample of six elementary units

Figure 2.11. Illustration of a two-stage random cluster sampling process. The population consists of seven primary sampling units, say, city blocks A, B, C, D, E, F, and G. A simple random sampling of primary sampling units results in the selection of blocks A, D, and G. Simple random sampling within blocks results in the elementary units, say, households indicated at the bottom.

ILLUSTRATIVE CASE 2.3

The Booth and Hobbs Drilling Company purchases many heavy-duty drill bits from several suppliers. In the past, the Receiving Department used the following procedure for determining whether a particular shipment should be accepted or rejected:

1. Select a simple random sample of 60 bits.
2. Subject the sample of bits to intensive simulated drilling activity.
3. Reject the shipment if 6 or more bits develop blunt, chipped, or cracked points within three hours of simulated activity.

4. Accept the shipment if 5 or fewer bits develop blunt, chipped, or cracked points within three hours of simulated activity.

The head of the Receiving Department has noted, however, that at times 6 or more defective bits have been found among the first 20 or 30 inspected. At such times, it has been possible to reject the shipment without examining all the bits in the sample. He wondered whether a modified procedure, involving a smaller sample size to begin with, would be better. He reasoned that because (1) the bits are expensive, (2) destructive testing must be employed to be certain of the quality of the bits, and (3) the testing process itself is rather time-consuming and costly, use of smaller samples in those situations where the evidence, good or bad, was dramatic could mean a much less expensive inspection process.

Consequently, the process was modified. The new steps in the process were as follows:

1. Select a simple random sample of 30 drill bits.
2. Subject the sample bits to intensive simulated drilling activity.
3. Reject the shipment if 4 or more bits develop blunt, chipped, or cracked points within three hours of simulated activity.
4. Accept the shipment if only 1 bit (or none) develops a blunt, chipped, or cracked point within three hours of simulated activity.
5. Select another simple random sample of size 30 if either 2 or 3 bits develop blunt, chipped, or cracked points within three hours of simulated activity.

This modified procedure is called *double sampling* and, as already mentioned, it is more economical than single sampling. By the same token, *triple sampling*, where the procedure involves one more step, is better still. Perhaps 20 drill bits could be selected at first and the accept/reject/continue sampling decision made on the basis of them. If the decision were to continue sampling, 20 more bits could be examined and the accept/reject/continue sampling decision made on the basis of the total of 40 bits. If the decision were, again, to continue sampling, 20 more bits would be inspected and the accept/reject decision made on the basis of all 60 bits (Figure 2.12).

The logical extreme of this kind of process is called *sequential sampling*. In sequential sampling, one item is inspected at a time and the inspector makes an accept/reject/continue sampling decision after each inspection.

2.11 The Problem of Sample Size

"The population I plan to study consists of 10,000 people. How large a sample do I need?" People regularly engaged in statistical analysis encounter this kind of question often. You may be surprised to learn that the question cannot be answered definitively. However, we *can* discuss several considerations that bear upon the question.

First, population size has very little to do with the matter. A commonly held belief is that, unless a sample is large, it cannot possibly be representative of the population. One sometimes hears it said that a sample is too small if it doesn't consist of at least 10% of the population. A little reflection will reveal why this is not necessarily true: Suppose the population observations all had exactly the same value. In such a case, a sample of size $n = 1$ would contain all the information to be had from this population and it would not matter whether the population contained 10, 10,000, or 1 billion elementary units. Such a situation is admittedly farfetched, but the point is that population size matters much less than certain other factors. We can briefly consider five factors here.

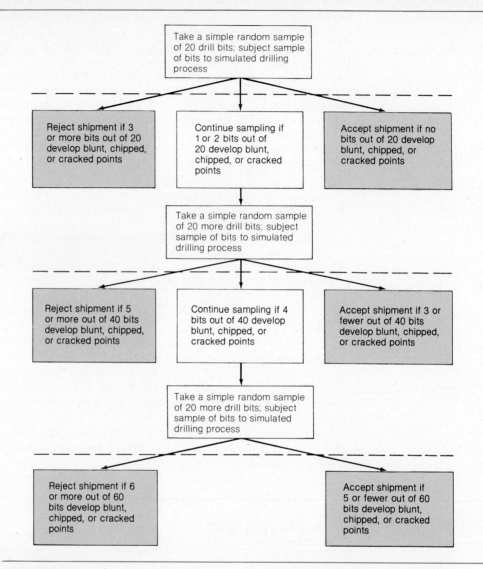

Figure 2.12. Flowchart showing how *triple sampling* might be used in connection with Illustrative Case 2.3.

The Uniformity of the Population

An important consideration in sample size determination is whether there is little or much variation among the population observations. The greater the variation, the larger the sample size must be to deliver a specified degree of precision in estimating parameter values. In practice, the actual variation in the population will usually be unknown, but can, perhaps, be judged from a small exploratory sample, past experience with similar sampling problems, or expert opinion. In any event, the fact that this information is both important and usually unknown is what makes sample size determination partly an art.

The Kind of Sampling Plan Used

We have seen that certain kinds of sampling plans are capable of achieving greater precision in parameter estimation for a given sample size than other plans. Ordinarily, if one employs stratified random sampling, a smaller sample size will be required than when, say, simple random sampling is used.

The Precision Requirements of Management

Other things being equal, the greater the precision required in connection with parameter estimation, the larger the sample size must be. (The relationship between sample size and precision will be spelled out in Chapter 8.)

The Rarity of the Event

If the study involves attribute data—that is, information about what proportion of elementary units possesses a certain characteristic and what proportion does not possess that characteristic—the required sample size depends greatly on the frequency with which the characteristic occurs. For example, in a study concerned with determining whether a certain drug "causes" multiple births, a fairly large sample of mothers-to-be would be required since, presumably, multiple births will be fairly rare with or without the drug. In a small sampling, maybe no multiple births at all would be found.

Cost Constraints

Unfortunately for those of an idealistic bent, cost is often a most important consideration in sample size determination. Almost all statistical investigations are conducted under some kind of budgetary constraint. This will not only influence decisions about sample size but about the type of sampling plan to use and methods of extracting information as well. It may even dictate the use of nonprobability sampling, since probability sampling costs tend to be higher. Probability sampling may entail payment of fees to statistical consultants, costs associated with developing a sampling frame which truly captures the target population, costs of repeated visits to sample members not at home during the first call, and other costs of little or no consequence when more casual sampling methods are used.

One benefit of a budgetary constraint is that it can be used quite explicitly as a guide to sample size determination. Let us assume, for simplicity, that simple random sampling is to be used in connection with a sample survey and that the survey is to cost no more than $50,000. We will call this amount C.

C consists of

C_f = fixed costs: Those costs such as statistical consultant fees, clerical work, and so on, which are not affected to any meaningful degree by the number of interviews conducted.

C_v = variable costs: Those costs, such as communication and transportation costs, directly related to the number of interviews conducted.

The cost equation would be

$$C = C_f + C_v n$$

We can rearrange this equation so that maximum n can be determined:

$$n = \frac{C - C_f}{C_v}$$

If fixed costs are estimated to be $20,000 and variable costs $12.00 per interview, the cost equation for this specific case would be $C = \$20,000 + \$12n$, and n would be ($50,000 − $20,000)/$12 = 2500.

Ways of determining sample size when cost is not the sole consideration are discussed in Chapters 8 and 9.

2.12 Procedural Note on the Sampling Method Assumed Hereafter

Simple random sampling, although often less practical to use than some available alternative plans, still forms the basis for statistical sampling theory. Moreover, mathematical methods used in sampling problems are easiest for beginning students to understand when the sampling procedure is assumed to be simple random.

> Consequently, in the remainder of this book, whenever mention is made of random sampling in connection with examples and demonstrations, you are asked to assume that simple random sampling was the specific approach used.

2.13 Summary

This chapter has attempted to convey three principal messages: (1) The use of sampling is often desirable—and, sometimes, absolutely essential—in many kinds of business research problems, (2) samples can be selected in many different ways, and (3) probability sampling is usually to be preferred to nonprobability sampling because the former permits the development of reliability measures whereas the latter does not.

Frequently used probability sampling methods include simple random sampling (the approach assumed throughout the remainder of this text), systematic random sampling, stratified random sampling, random cluster sampling, and multiple sampling.

Whatever the sampling approach used, efforts must be made to ensure that the sampled population and the target population are as nearly identical as possible, and that both sampling and nonsampling errors are kept as small as possible.

The adequacy of a sample can be judged by how well it represents the population it is presumed to represent. Two criteria by which to judge a sample are unbiasedness and precision.

Notes

1. Quoted in *Newsweek*, January 17, 1949, p. 48.
2. *U.S. News and World Report*, October 21, 1968, p. 28.
3. For more on misuses of sampling, see Stephen K. Campbell, *Flaws and Fallacies in Statistical Thinking* (New Jersey: Prentice-Hall, 1974), Chapter 12, and Darrell Huff, *How to Lie with Statistics* (New York: Norton, 1954), Chapter 1.
4. We are alluding here to the a priori concept of probability. According to this approach to determining the probability of a specified event, one finds the desired probability through use of the ratio $n(S)/n(I)$, where $n(I)$ is the number of equally likely outcomes that could occur and $n(S)$ is the number of such outcomes that are viewed as favorable. More will be said about this in Chapter 6 (Section 6.2).
5. This particular table is hypothetical. However, entire books of random digits have been published; a widely used book of random digits is the Rand Corporation's *A Million Random Digits with 100,000 Normal Deviates* (Glencoe, Illinois: The Free Press, 1955). Appendix Table 12 presents a table of random digits sufficiently extensive to accommodate problems at the end of this chapter.

Terms Introduced in This Chapter

area sampling (p. 51)
double sampling (p. 53)
equal complete coverage (p. 35)
judgment sampling (p. 38)
multiple sampling (p. 51)
nonprobability sampling (p. 38)
nonsampling errors (p. 35)
precision (p. 36)
primary sampling unit (p. 50)
probability sampling (p. 38)

proportional allocation (p. 48)
random cluster sampling (p. 51)
restricted sampling (p. 39)
sampled population (p. 44)
sampling error (p. 35)
sampling fraction (p. 46)
sampling frame (p. 42)
sampling interval (p. 47)
simple random sampling (p. 40)

single-stage sampling (p. 51)
stratified random sampling (p. 48)
systematic random sampling (p. 46)
table of random digits (p. 42)
target population (p. 44)
triple sampling (p. 53)
two-stage sampling (p. 51)
unbiased sample (p. 36)
unrestricted sampling (p. 39)

Formulas Introduced in This Chapter

The formula for determining the number of different samples of size n contained within a population of size N:

$$N^* = \frac{N!}{n!(N-n)!}$$

The formula for determining the cost of a survey based on simple random sampling, where C represents total cost, C_f represents fixed costs—those associated with statistical consultant fees, clerical work, and so forth, which are not affected to any meaningful degree by the number of interviews conducted, and C_v stands for variable costs—such as communication and transportation costs directly related to the number of interviews conducted:

$$C = C_f + C_v n$$

This formula can be rearranged in such a way that the maximum sample size can be determined:

$$n = \frac{C - C_f}{C_v}$$

Questions and Problems

2.1 Explain the meaning of each of the following terms:
   a. Census
   b. Destructive testing
   c. Simple random sampling
   d. Equal complete coverage
   e. Sampling frame
   f. Sample combination

2.2 Explain the meaning of each of the following terms:
   a. Sampling error
   b. Probability sampling
   c. Unrestricted sampling
   d. Target population
   e. Sampling fraction
   f. Subpopulation

2.3 Explain the meaning of each of the following terms:
   a. Nonsampling errors
   b. Judgment sampling
   c. Primary sampling unit
   d. Area sampling
   e. Single-stage sampling
   f. Random cluster sampling

2.4 Explain the meaning of each of the following terms:
   a. Unbiased sample
   b. Restricted sampling
   c. Sampled population
   d. Systematic random sampling
   e. Sampling interval in systematic random sampling
   f. Two-stage sampling
   g. Multiple sampling

2.5 Distinguish between:
   a. n and N
   b. Target population and sampled population
   c. Sampling error and nonsampling error
   d. Probability sampling and nonprobability sampling
   e. Unrestricted sampling and restricted sampling

2.6 Distinguish between:
   a. Unbiasedness and precision
   b. Simple random sampling and systematic random sampling
   c. Two-stage sampling and double sampling
   d. Stratified random sampling and random cluster sampling

2.7 Indicate which of the following statements you agree with and which you disagree with, and defend your opinions:
   a. Data obtained from a sample must inevitably contain both sampling and nonsampling errors.
   b. Data obtained from a sample study, provided that probability sampling is used, will be the same in all important particulars as data that could have been obtained from the entire population.
   c. Since data obtained from a census will be free of sampling errors, one should, whenever possible, conduct population studies in preference to sample studies.
   d. Simple random sampling is a particular kind of probability sampling plan.
   e. Information obtained from a sample is always inferior to information obtained from a complete census.

2.8 Indicate which of the following statements you agree with and which you disagree with, and defend your opinions:
   a. An unbiased sample will invariably permit more precise estimates of parameter values than a biased sample of the same size.
   b. The principal advantage of probability sampling over nonprobability sampling is ease of sample selection.

c. Statistical investigations can sometimes be safely completed even though the sampled population and the target population are not exactly the same.

d. If the sampled population differs markedly from the target population and nothing can be done to reduce the disparity, the statistical investigation should be abandoned.

2.9 Indicate which of the following statements you agree with and which you disagree with, and defend your opinions:

a. A systematic random sample will resemble a simple random sample of the same size obtained from the same population regardless of the arrangement of the elementary population units.

b. A simple random sample will usually permit more precise estimates of parameter values than a stratified random sample of the same size.

c. A simple random sample will usually permit more precise estimates of parameter values than a random cluster sample of the same size.

d. If a person wishes to investigate a population consisting of 1 million elementary population units, that person must plan on using a sample of size $n = 100,000$ or larger.

2.10 Indicate which of the following statements you agree with and which you disagree with, and defend your opinions:

a. Other things being equal, the larger the population size, the larger the sample size must be.

b. Other things being equal, the greater the variation among the population observations, the larger the sample size must be.

c. Other things being equal, the rarer an event of interest, the larger the sample size must be.

d. Other things being equal, the more precise one wishes estimates of parameter values to be, the larger the sample size must be.

 2.11 Each week the Department of Labor Statistics releases figures on initial claims for unemployment insurance. Do you think these figures are more likely obtained from sample studies or population studies? Defend your choice.

2.12 Cite a realistic hypothetical problem in the areas of (1) marketing, (2) finance, (3) accounting, and (4) management in which a sample study might be preferred to a complete census. Indicate clearly the advantage(s) of sampling in each of the situations you describe.

2.13 Cite a realistic hypothetical problem in the areas of (1) marketing, (2) finance, (3) accounting, and (4) management in which stratified random sampling might be used. Indicate clearly why stratified random sampling would be the sampling plan preferred.

2.14 Cite a realistic hypothetical problem in the areas of (1) marketing, (2) finance, (3) accounting, and (4) management in which random cluster sampling might be used. Indicate clearly why random cluster sampling would be the sampling plan preferred.

2.15 Under what circumstances, if any, might a complete census be preferred to a sample study? Illustrate with a hypothetical example.

2.16 a. Is it ever possible for a sample designed as a probability sample to turn out to be a judgment sample in practice? Explain.

b. Would such a result be desirable or undesirable? Explain using a hypothetical example.

c. Using the same hypothetical example as in part *b*, indicate some precautions that could be taken to ensure that a sample designed as a probability sample will *not* turn out to be a judgment sample in practice.

2.17 Consider the following target populations:
  a. All white-collar workers in your city
  b. All residents of your city who voted in the last presidential election
  c. All candy bars sold in your city last week
  d. All organized crime members in your city
Describe some possible problems that might arise in the course of conducting a sample survey which could make the sampled population differ from the target population.

2.18 Reflect on the following statements about simple random sampling:
  1. A simple random sample is a sample selected in such a way that each elementary population unit has an equal chance of being included in the sample.
  2. A simple random sample is a sample selected in such a way that each sample combination of a specified size n has an equal chance of being selected.
  a. Explain the difference between the two statements.
  b. For finite populations, the first statement is regarded as a necessary, but not a sufficient, condition for simple random sampling. The second statement, on the other hand, is regarded as both a necessary and a sufficient condition. Explain why the second statement is the more precise one. (*Hint:* You may wish to employ a hypothetical example to show how the first condition could conceivably be met even though a sampling plan other than simple random sampling was used.)

2.19 For which of the following situations do the most sample combinations exist: (1) finite population size of $N = 8$ and sample size of $n = 3$, or (2) finite population size of $N = 12$ and sample size of $n = 7$? Demonstrate clearly how you arrived at your answer.

2.20 Determine the number of sample combinations associated with each of the following pairs of population sizes and sample sizes:

	Population Size, N	Sample Size, n
a.	9	3
b.	9	4
c.	11	7
d.	11	5
e.	15	2

2.21 A population consists of $N = 5$ people named Adams, Burford, Carlton, Dillman, and Ephriam.
  a. How many different sample combinations of size $n = 2$ could be drawn from this population?
  b. List all of the possible sample combinations of size $n = 2$.

2.22 A population of $N = 7$ automobile engine parts consists of parts A, B, C, D, E, F, and G.
  a. How many different sample combinations of size $n = 3$ could be drawn from this population?
  b. List all of the possible sample combinations of size $n = 3$.

 2.23 The following companies have agreed to participate in a study to determine whether (1) recipients of MBA degrees from certain universities advance more rapidly in a company than those who received their MBA degrees from other universities and, in turn, whether (2) MBA degree holders, in general, advance more rapidly than employees not holding this degree:

Baker and Wilhelm, Inc.	Drover and Smythe, Inc.	Synergistix Machinery
Calvin Industries	Jaeger Tool Company	Company
C-T-Z, Inc.	Leaseback, Inc.	Zippy Services, Inc.
Draper Company	Neptune Boats, Inc.	

Use Appendix Table 12 to select a simple random sample of four companies. Which four did you select? Describe and defend the procedures you used when employing Appendix Table 12.

2.24 Refer to problem 2.23.

a. What would be the total cost C of the study if fixed costs were estimated to be $3000 and variable costs $20 per interview.

b. How large a sample size n could have been used if total cost C was $4240, fixed costs C_f were $4000, and variable costs C_v were $40 per interview? Assume that cost is the only consideration in the determination of sample size.

2.25 The following people constitute the production crew of the Clark Cable Company:

Adams	Hamilton	Porter
Belvedere	Jenkins	Quenton
Benton	Kane	Rasmussen
Chapman	Lambert	Ross
Clinton	Lucas	Steinberg
Davis	Lyle	Stevens
Ferguson	Martin	Tidwell
Fye	Melville	Thompson
Graham	Ortega	Williams
Grummon	Partridge	Young

Suppose that nine of these employees are to be selected to determine worker attitudes toward the possibility of using robots for a routine welding task.

Use Appendix Table 12 to select a simple random sample of nine production workers. Which nine did you select? Describe and defend the procedures you used when employing Appendix Table 12.

2.26 Refer to problem 2.25.

a. What would be the total cost C of the study if fixed costs were estimated to be $750 and variable costs $10 per interview?

b. How large a sample size n could have been used if total cost C was $1000, fixed costs C_f were $850, and variable costs C_v were $15 per interview? Assume that cost is the only consideration in the determination of sample size.

2.27 Refer to problem 2.23. Suppose that the following companies are primarily involved in manufacturing activities, whereas the other companies are engaged primarily in marketing activities:

Baker and Wilhelm, Inc.   Jaeger Tool Company
C-T-Z, Inc.   Synergistix Machinery Company
Draper Company

a. Use Appendix Table 12 to select a stratified random sample of size $n = 4$ from the population of $N = 10$ companies. Do so by (1) selecting a simple random sample of two companies from the primarily manufacturing group and a simple random sample of two companies from the primarily marketing group and (2) combining the two resulting samples of size 2.

b. Tell which four companies you selected. Describe the procedure you used when employing Appendix Table 12.

c. If you worked problem 2.23, indicate how many primarily manufacturing firms were included in your simple random sample. Would you say that your simple random sample was (1) on a par with, (2) better than, or (3) worse than your stratified random sample in terms of "representing" the entire population? Assume that the manufacturing-versus-marketing distinction is an important basis for stratification.

2.28 Refer to problem 2.25. Suppose that the following workers had been with the company for five years or longer, whereas the others had been with the company for less than five years:

Adams	Graham	Lyle
Benton	Jenkins	Rasmussen
Davis	Kane	Steinberg
Fye		

   a. If the overall sample size is to be $n = 9$ and proportional allocation were to be used in selecting a stratified random sample, how many people from the above list should be chosen? How many people from among the remaining 20 production workers should be chosen?

   b. Proceeding in a manner consistent with your answer to part *a*, use Appendix Table 12 to select a stratified random sample of size $n = 9$ from this population of $N = 30$.

   c. Tell which 9 employees you selected. Describe the procedure you used when employing Appendix Table 12.

   d. If you worked problem 2.25, indicate how many workers in your simple random sample have been with the company for five years or longer. Would you say that your simple random sample was (1) on a par with, (2) better than, or (3) worse than your stratified random sample in terms of "representing" the entire population? Assume that number of years with the company is an important basis for stratification.

2.29 A railroad car contains 1000 wooden crates. Each such crate contains 24 cardboard boxes. Inside each cardboard box there are 48 shelf brackets. Suppose that you must examine a sample of 5000 shelf brackets in order to determine whether the shipment should be accepted. Would you be most likely to use simple random sampling, systematic random sampling, stratified random sampling, or some kind of cluster sampling? Defend your choice. Then describe the steps you might employ to carry out the sampling procedure.

2.30 A population consists of 20 statisticians, 100 accountants, and 300 production workers employed by the Cummings Manufacturing Company. A stratified random sample of size 84 is to be selected, using type of worker as the basis for stratification.

   a. If proportional allocation is used, how many statisticians, how many accountants, and how many production workers will there be in the overall sample?

   b. If simple random sampling with $n = 84$ had been used, what would be the maximum number of accountants that could be included in the sample? Explain.

   c. If (1) the workers were identified by name on a list showing the 20 statisticians first, the 100 accountants second, and the 300 production workers third, (2) a systematic random sample of size 84 were selected by the methods described in section 2.7, and (3) statistician No. 5 was selected using a random start, how many statisticians, how many accountants, and how many production workers would there be in the overall sample?

2.31 Using your own hypothetical example, explain how stratification can increase the precision with which parameter values can be estimated.

2.32 A television commercial which you may have seen shows a man holding a microphone in what looks like the theater district of a large city on a warm summer evening. As cars wait for a red light to turn green, he thrusts his head and the microphone into the open windows of accessible cars and asks the driver whether he or she uses a certain dental product. The driver answers either "yes" or "no." In the commercial, two out of three say they use the product. Is this process of sample selection a random one? Tell why you believe it is or is not.

Questions 2.33 to 2.38 pertain to the following situation:

The city of Del Norte is made up of only six city blocks which we will refer to as Block I, II, . . . , VI. Block II has an apartment building on it which houses ten households the majority of which are single-person households. Blocks I, II, and III are situated in the poorer part of town, whereas Blocks

IV, V, and VI are in the nicer part of town. Residents of Block VI are all exceptionally well-to-do. A survey is to be conducted to determine what proportion of households own at least one personal computer. The purpose of the study is to determine the feasibility of opening a personal computer service and accessory store in Del Norte.

M and F next to surnames indicate whether the head of the household is male or female. E, J, H, and C next to surnames indicate whether the highest level of formal education attained by the head of the household is elementary school graduate—E, junior high school graduate—J, high school graduate—H, or college graduate—C. For example, at the beginning of the list of residents of Block I, we see Whitney M-J. This means that (1) the household surname is Whitney, (2) the head of the household is male, and (3) the head of the household graduated from junior high school only.

Block I			Block III			Block V	
Whitney	M-J		Kane	M-J		Kaufmann	F-C
Soss	F-E		Baker	M-E		Gervin	F-H
List	M-H		Anderson	M-H		Bennett	M-H
Bruner	M-E		Crean	F-E		Dreshfield	M-C
Lustig	M-J		Bombel	F-C		Cone	F-C
McCulloush	F-E		Landeck	M-C		Lawson	M-C
Pike	F-J		Richardson	M-H		Wilcox	M-C
Rivera	M-H		Semba	F-H		VanDrehle	M-C
Stein	M-J		Walters	M-H		Zulkefier	F-C

Block II			Block IV			Block VI	
Martin	M-H		Frisch	M-H		Barnard	M-C
Neighbors	M-J		Conn	F-H		Melvin	M-C
Hoffman	F-H		Lind	M-E		Billars	M-C
Fleming	F-C		Bueli	F-C		Morgan	M-C
Arbrough	M-E	apartment	Sidwell	F-H			
Acker	F-H		Spilke	M-C			
Owen	M-H		Reyerson	M-C			
Saladino	M-J		Hanson	M-H			
Mundt	F-E		Mann	M-C			
Harri	M-E						
Milewski	F-E						
Kidner	M-H						
Hagen	M-H						
Oser	M-J						
Wingle	M-C						
McGowan	M-J						
Fernandez	M-H						
Elzahr	F-H						
Burger	F-C						
Evans	M-J						

✔ **2.33 a.** Use Appendix Table 12 to select a simple random sample of $n = 12$ households from the city of Del Norte. Describe and defend the procedures you used when employing Appendix Table 12.

   b. What proportion of your sample of households are headed by females? How does this compare with the proportion for the total population?

   c. What proportion of your sample of households have heads who graduated from college? How does this compare with the proportion for the total population?

   d. Do you believe that simple random sampling was a good procedure to use in this situation? Tell why or why not.

✔ **2.34 a.** Use Appendix Table 12 to pick one of the first five households in Block I to serve as a random start for a systematic random sampling procedure. Then choose every fifth household to be included in the sample.

b. What proportion of your sample of households are headed by females? How does this compare with the proportion for the total population?

c. What proportion of your sample of households graduated from high school only? How does this compare with the proportion for the total population?

d. Since personal computer ownership is probably at least partly a function of household income, do you believe it was a good idea or a bad idea to use systematic random sampling in this situation? Explain fully.

2.35 a. Stratify the population according to sex of head of household. How many male heads of household are there in the population? How many female heads of household?

b. Use Appendix Table 12 to select a stratified random sample of size 12 with proportional allocation. Tell exactly how you used Appendix Table 12 to obtain your sample.

c. What proportion of your sample of households are located in Block II? How does this compare with the proportion for the entire population?

d. What proportion of your sample of households have heads who graduated from college? How does this compare with the proportion for the total population?

e. Do you believe that stratification on the basis of sex of head of household helped you achieve a more representative sample than if simple random sampling had been used? Tell why or why not.

2.36 a. Stratify the population according to educational attainment of head of household. How many heads of households in the population graduated from elementary school only? How many graduated from junior high school only? How many graduated from high school only? How many graduated from college?

b. Use Appendix Table 12 to select a stratified random sample of size 12 with proportional allocation. Tell exactly how you used Appendix Table 12 to obtain your sample.

c. What proportion of your sample of households are headed by females? How does this compare with the proportion for the total population?

d. What proportion of your sample of households are located in Block II? How does this compare with the proportion for the entire population?

e. Do you believe that stratification on the basis of educational level attained by heads of households helped you achieve a more representative sample than if simple random sampling had been used? Tell why or why not.

2.37 a. Use Appendix Table 12 to select at random two city blocks from among the six blocks in the city. Tell exactly how you used Appendix Table 12 to obtain your sample.

b. What two blocks did you choose? Is the number of households the same for both blocks?

c. What proportion of households within the two blocks are headed by females? How does this compare with the proportion for the entire population?

d. What proportion of households within the two blocks are headed by college graduates? How does this compare with the proportion for the total population?

e. Use Appendix Table 12 to select a simple random sample of four residents from each of the two blocks. Tell exactly how you used Appendix Table 12 to obtain your sample. Also tell which households you picked.

f. In your particular case, was it a good idea or a bad idea to pick the same number of households from each of the two blocks? Explain.

g. What are some advantages and disadvantages of using area sampling in this case?

2.38 Suppose that the survey of Del Norte residents is being conducted on behalf of a person hoping to open a personal computer service and accessory shop in the town. He estimates that if at least 30% of Del Norte households own at least one personal computer, the shop can be profitable; if fewer than 30% own personal computers, he will abandon his plan. Can you think of any way that multiple sampling could be used to advantage in this situation? Explain fully. (Assume that the cost of interviewing residents of this city is sufficiently high to provide an incentive for using multiple sampling.)

TAKE CHARGE

2.39 a. Search newspapers, popular magazines, and other publications for two articles containing information obtained from survey sampling procedures. One article should be based on a properly drawn sample and the other on an improperly drawn sample.

b. Tell why you believe the one sampling procedure was "proper" and the other "improper."

c. Cite some ways in which the article based on the improperly drawn sample could be misleading to readers.

Computer Exercise

2.40 Data in Table 2.5 represent number of people (in thousands) employed by the largest 500 industrial corporations outside the United States.

a. Have the computer select a simple random sample of size $n = 40$. List the identification numbers of the 40 sample observations.

b. The letters P, M, and O in Table 2.5 stand for "Petroleum," "Motor Vehicles," and "Other," respectively. Have the computer select a simple random sample of $n_1 = 6$ petroleum companies, $n_2 = 4$ motor vehicle companies, and $n_3 = 30$ other companies. Then combine the results to get a single stratified random sample of size $n = 40$. List the identification numbers of the 40 observations selected.

c. Do you have any reason to think the stratified random sample selected in connection with part *b* is any more or less representative of the population of $N = 500$ observations than the simple random sample selected in connection with part *a*? Explain.

Table 2.5 Employee Data for Problem 2.40

I.D. No.	No. of Employees	Industry Category	I.D. No.	No. of Employees	Industry Category
001	156.0	P	025	117.9	O
002	131.6	P	026	187.2	O
003	134.0	P	027	96.6	M
004	515.9	O	028	74.2	O
005	267.0	O	029	137.5	O
006	57.8	M	030	203.0	M
007	46.8	P	031	43.4	M
008	77.6	P	032	7.0	P
009	124.8	O	033	14.2	P
010	56.8	P	034	121.1	O
011	343.0	O	035	103.0	O
012	157.0	P	036	6.6	P
013	155.6	O	037	146.0	O
014	313.0	O	038	27.9	P
015	108.1	M	039	46.2	M
016	231.7	M	040	2.1	O
017	184.9	M	041	268.0	O
018	174.8	O	042	148.7	O
019	179.8	O	043	178.0	O
020	210.0	M	044	133.6	O
021	243.8	M	045	23.2	P
022	140.4	O	046	73.8	O
023	114.1	O	047	14.7	P
024	76.2	M	048	130.0	O

(continued)

Table 2.5 *(Continued)*

I.D. No.	No. of Employees	Industry Category	I.D. No.	No. of Employees	Industry Category
049	102.3	O	100	7.7	P
050	53.0	P	101	9.8	O
051	79.2	O	102	54.0	O
052	72.8	O	103	66.4	O
053	33.5	M	104	15.0	O
054	27.9	O	105	7.2	P
055	32.8	P	106	59.3	O
056	69.3	O	107	70.0	O
057	139.3	O	108	88.8	O
058	3.6	P	109	9.7	P
059	72.3	O	110	27.3	O
060	110.0	O	111	18.8	O
061	4.2	P	112	56.7	O
062	15.0	O	113	22.1	O
063	38.6	O	114	42.6	O
064	19.0	P	115	44.2	O
065	64.4	O	116	1.8	P
066	40.0	M	117	97.5	O
067	136.3	O	118	15.4	P
068	44.5	P	119	52.6	O
069	59.4	M	120	49.2	O
070	73.1	O	121	51.5	O
071	81.0	O	122	41.0	O
072	68.9	O	123	68.0	O
073	109.7	M	124	74.2	O
074	8.4	P	125	2.5	P
075	69.0	O	126	21.0	O
076	104.8	O	127	3.5	P
077	35.7	O	128	45.9	O
078	63.9	M	129	2.6	P
079	81.1	O	130	21.1	O
080	102.0	O	131	4.3	P
081	19.5	P	132	12.0	M
082	66.3	O	133	78.0	O
083	3.9	P	134	34.4	O
084	48.0	M	135	24.7	O
085	101.3	M	136	70.8	O
086	90.6	O	137	1.4	P
087	2.0	P	138	38.0	O
088	62.7	O	139	14.7	P
089	3.4	P	140	28.4	O
090	32.8	O	141	6.6	P
091	22.8	O	142	32.3	O
092	43.2	M	143	73.0	O
093	3.9	P	144	23.8	O
094	21.2	O	145	35.5	O
095	29.3	O	146	3.7	P
096	42.7	O	147	38.1	O
097	91.5	O	148	18.6	O
098	76.6	O	149	8.6	P
099	21.4	O	150	14.9	O

(continued)

Table 2.5 *(Continued)*

I.D. No.	No. of Employees	Industry Category	I.D. No.	No. of Employees	Industry Category
151	37.9	O	202	38.8	O
152	21.0	M	203	26.0	O
153	10.3	O	204	47.1	O
154	4.4	P	205	50.0	O
155	57.8	O	206	11.7	O
156	5.2	P	207	24.4	O
157	54.0	M	208	36.8	O
158	62.0	O	209	71.2	O
159	34.6	M	210	22.0	O
160	4.5	P	211	13.7	M
161	28.4	M	212	14.6	O
162	205.6	O	213	15.4	O
163	34.2	O	214	5.7	O
164	34.3	O	215	21.6	O
165	49.8	O	216	43.3	O
166	35.4	O	217	20.3	O
167	24.8	O	218	7.1	P
168	16.7	O	219	11.2	O
169	41.9	O	220	45.0	O
170	35.1	O	221	40.9	O
171	27.3	O	222	14.6	O
172	39.1	M	223	3.8	O
173	39.3	O	224	14.6	O
174	39.1	O	225	42.7	O
175	7.6	O	226	28.5	O
176	42.6	O	227	6.3	O
177	32.0	P	228	20.4	O
178	37.1	O	229	6.7	P
179	14.2	O	230	27.8	O
180	16.3	O	231	57.0	O
181	16.0	O	232	183.0	O
182	30.9	O	233	23.1	O
183	25.9	O	234	124.2	O
184	12.9	O	235	43.0	O
185	10.2	O	236	11.4	O
186	30.5	O	237	16.0	O
187	14.7	O	238	10.9	O
188	47.8	O	239	42.3	O
189	1.9	P	240	32.2	O
190	21.7	O	241	13.4	O
191	33.0	P	242	6.5	O
192	1.4	P	243	10.0	M
193	53.0	O	244	10.0	O
194	15.0	O	245	64.6	M
195	15.5	P	246	30.2	O
196	18.5	O	247	13.6	O
197	11.6	P	248	15.0	O
198	23.0	P	249	31.2	O
199	24.6	O	250	10.6	O
200	31.8	O	251	3.3	P
201	17.7	M	252	10.7	M

(continued)

Table 2.5 *(Continued)*

I.D. No.	No. of Employees	Industry Category	I.D. No.	No. of Employees	Industry Category
253	1.3	P	304	7.4	O
254	32.4	O	305	0.8	P
255	7.9	P	306	34.3	O
256	15.8	O	307	1.9	O
257	62.6	O	308	15.2	O
258	2.7	P	309	25.7	O
259	26.1	O	310	15.2	O
260	6.7	O	311	63.5	O
261	25.5	O	312	9.5	O
262	5.8	O	313	13.1	O
263	40.9	O	314	13.5	O
264	1.6	P	315	26.5	O
265	31.7	O	316	31.2	O
266	25.7	O	317	3.6	O
267	27.8	O	318	17.1	O
268	18.2	O	319	20.3	O
269	23.1	O	320	8.4	O
270	8.7	O	321	11.5	O
271	22.8	O	322	6.3	O
272	11.7	O	323	10.1	O
273	16.7	O	324	28.2	O
274	50.0	O	325	8.1	O
275	38.1	O	326	8.5	O
276	2.3	P	327	3.5	P
277	2.4	O	328	9.5	O
278	59.4	O	329	29.0	O
279	30.7	O	330	20.5	O
280	6.9	M	331	18.2	O
281	22.8	O	332	7.4	O
282	16.8	O	333	11.0	O
283	9.3	O	334	14.7	O
284	5.7	O	335	30.8	O
285	11.5	M	336	15.0	O
286	27.8	O	337	43.5	O
287	15.5	O	338	9.6	O
288	19.5	O	339	6.1	O
289	19.3	O	340	0.7	P
290	19.0	O	341	11.5	O
291	16.2	O	342	5.1	O
292	8.2	M	343	26.7	O
293	19.0	O	344	32.1	O
294	7.2	O	345	22.6	O
295	22.6	O	346	46.7	O
296	9.0	O	347	24.2	O
297	7.9	O	348	21.2	M
298	26.0	O	349	12.0	P
299	26.0	O	350	17.7	O
300	10.4	O	351	11.4	O
301	13.0	O	352	37.0	O
302	18.4	O	353	4.8	O
303	13.1	O	354	19.0	O

(continued)

Table 2.5 *(Continued)*

I.D. No.	No. of Employees	Industry Category	I.D. No.	No. of Employees	Industry Category
355	15.2	O	406	18.1	O
356	11.7	O	407	17.0	O
357	18.5	O	408	6.2	O
358	8.8	P	409	3.9	O
359	9.7	M	410	14.4	O
360	3.8	O	411	4.6	O
361	21.3	O	412	11.3	O
362	28.9	O	413	15.1	O
363	23.7	O	414	0.9	P
364	5.1	O	415	18.8	M
365	18.4	O	416	27.3	O
366	11.6	O	417	15.2	O
367	4.7	O	418	4.8	O
368	16.0	O	419	15.4	O
369	21.3	O	420	3.7	O
370	12.2	O	421	13.4	O
371	13.7	O	422	21.6	O
372	27.6	M	423	22.1	O
373	43.9	O	424	10.4	O
374	27.6	O	425	3.3	O
375	17.3	O	426	22.4	O
376	8.2	O	427	6.6	O
377	11.0	O	428	31.7	O
378	27.0	O	429	0.3	P
379	5.6	O	430	22.2	O
380	4.6	O	431	14.3	O
381	11.0	O	432	17.6	O
382	N.A.	P	433	25.2	M
383	5.4	P	434	6.5	M
384	7.9	O	435	12.1	O
385	13.5	O	436	5.7	O
386	14.2	O	437	17.3	O
387	6.3	O	438	9.4	M
388	5.7	P	439	24.9	O
389	14.7	O	440	6.7	O
390	15.7	O	441	27.9	O
391	8.7	O	442	16.0	O
392	4.6	O	443	6.0	M
393	5.4	O	444	3.9	O
394	3.6	O	445	12.3	M
395	13.0	O	446	11.4	O
396	27.6	O	447	8.5	O
397	18.4	O	448	1.1	P
398	20.4	M	449	23.1	O
399	10.8	O	450	9.8	O
400	12.5	O	451	5.0	O
401	1.0	P	452	13.7	O
402	5.4	O	453	6.5	O
403	7.0	M	454	6.0	O
404	18.6	O	455	5.3	O
405	17.5	O	456	9.4	O

(continued)

Table 2.5 *(Concluded)*

I.D. No.	No. of Employees	Industry Category	I.D. No.	No. of Employees	Industry Category
457	7.8	O	479	15.9	O
458	15.1	O	480	8.3	O
459	0.8	P	481	6.0	O
460	9.5	O	482	3.8	O
461	8.2	O	483	14.0	O .
462	4.3	O	484	13.3	O
463	1.9	O	485	23.4	O
464	14.0	O	486	7.6	O
465	7.4	M	487	2.0	O
466	22.3	O	488	6.9	O
467	5.9	M	489	14.7	O
468	6.3	M	490	37.0	O
469	3.5	P	491	14.1	O
470	9.9	O	492	10.9	O
471	4.8	O	493	18.3	O
472	N.A.	O	494	16.2	O
473	15.5	M	495	17.8	O
474	0.1	O	496	7.2	O
475	16.1	O	497	22.0	O
476	21.2	O	498	12.4	O
477	0.8	P	499	4.0	O
478	6.7	O	500	9.4	O

N.A.: Not available.
P: Petroleum.
M: Motor vehicles.
O: Other.

Source: From *Fortune* (August 20, 1984), pp. 202–211. Copyright © 1984 by Time, Inc. All rights reserved. Reprinted by permission.

Cases for Analysis and Discussion

2.1 THE "UNIVERSAL MUTUAL LIFE INSURANCE COMPANY": PART I

The "Universal Mutual Life Insurance Company" is the sixth largest company of its kind in the United States. The company bills itself as "The Peace-of-Mind Company" and has a history of very conservative business practices which include refusal to write any group-type insurance, either life or disability. Its close to $7 billion in assets has been accumulated largely by the sales force selling one policy at a time since 1857.

A few years ago Universal began a $1.5 million advertising campaign which included the sponsoring of several athletic events and a weekly news program on television. Management felt that this advertising effort would help agents make contact more easily with prospective clients.

A local agency of Universal wished to determine how effective this advertising had been in its community and, at the same time, attempt to determine people's attitudes toward life insurance. To meet these objectives, management of this local agency decided to survey potential clients at random to determine answers to the specific questions:

a. Do people recognize Universal as "The Peace-of-Mind Company?"
b. What are respondents' attitudes regarding the ownership of life insurance?

An undergraduate student at a nearby university was hired to develop a questionnaire and to design a sampling procedure. The instructions given interviewers are presented in Figure 2.13.

The sampling procedure used was as follows. Two young girls were each given a supply of survey forms, two clipboards, and a copy of the instructions shown in Figure 2.13. Both girls were placed on a street corner in a busy downtown part of the city. Their instructions were to stop relatively successful looking men and women and to ask them to complete the form. The assumptions implicit in the use of this procedure, as spelled out by a member of management, were

a. The population about which the company was most interested (18 years of age and over, business-type individuals, who could afford life insurance) would be fairly represented in the downtown area of the city where the two interviewers were stationed.
b. The screening device (observation) would provide the type of respondent in whom the company was most interested.
c. The respondents would be more cooperative with female interviewers than they would be if men did the interviewing.

Eighty-six usable questionnaires were completed—36 by men and 50 by women.

```
INSTRUCTIONS FOR THE ADMINISTERING OF QUESTIONNAIRE SURVEYS

A.   Demographics:
     1.    Have each respondent check M____or F____.
     2.    Income:  This concerns only the respondent's income
           (annual); not spouses. Explain this to respondent
           before he or she attempts to check the appropriate
           income level.
     3.    Age:  Each category is nonconclusive.  For example,
           if the respondent is 22 years of age, check box 2.
           If there is any question as to the correct age level,
           have respondent record his or her exact age.

EACH SURVEY WILL TAKE APPROXIMATELY 20 SECONDS.  CONVEY THIS
TO THE PROSPECTIVE RESPONDENT BEFORE YOU ATTEMPT TO ADMINISTER
IT.

B. Objections:
     1.    If a prospective respondent demands to know the nature
           of the survey, tell him or her it is designed to help
           improve relations between a corporation and the
           general public.
     2.    If no questions are asked, do not offer any information
           as to the purpose of the survey.

Two clipboards will be furnished to you, which will be given to
the respondents so they may record their answers.  Also, two
boards will allow you to have two respondents going at once.
```

Figure 2.13. Instructions to interviewers participating in the "Universal Mutual Life Insurance Company" survey study.

1. Was the sampling procedure used really a random one? Explain why or why not.

2. How might the sampling procedure have been improved?

This case is continued at the end of Chapter 3.

 ## 2.2 "DPC CORPORATION": PART II

Read Part I of the "DPC Corporation" case (case 1.5 at the end of Chapter 1).

1. The concerned employees decided that simple random sampling would be adequate for their purposes. Do you agree? Tell why or why not.

2. What, if anything, would have been wrong with their simply picking some machines conveniently at hand for their simulation studies?

This case is continued at the end of Chapter 8.

 ## 2.3 "ETRA AEROSPACE, INC."

With the winding down of the Vietnam War, many changes in the aerospace industry took place. For example, management of "Etra Aerospace, Inc.," one of the larger companies in the industry, discovered that the business data processing procedures used to support a number of functional departments within the organization were in need of major rework and redesign. The principal changes in the company's data processing needs had occurred as a result of a shift in emphasis away from activities requiring repetitive manufacture of antiballistic missiles and launch vehicles and into activities necessitating the engineering, designing, and manufacture of one-of-a-kind type of hardware required by the company's entry into the nation's space program.

Consequently, many of the computer systems, particularly those supporting the engineering, configuration, and contracts departments, were no longer effective in meeting the users' data processing requirements. Moreover, the increasingly competitive nature and cost consciousness of the aerospace industry and its customers (the Department of Defense and NASA) made better cost controls imperative. The computer procedures then in use, many of which were designed originally for second-generation equipment six to eight years earlier, did not take advantage of the current state-of-the-art in computer hardware and software. Because of these recognized failings in the present computer methods, the data processing department initiated a study to determine what improvements might be effected. The study's objectives and the methods used to obtain the data are described below.

The general objective of the study was to provide a flexible, responsive package of application systems that met current and anticipated needs of the users at substantially reduced costs. To accomplish this objective, three study phases were conducted each of which was concerned with determining—in increasingly greater detail—answers to these questions:

a. What were the significant problems confronting management that could be alleviated through system modification and replacement?
b. What systems required revision?
c. What systems should be replaced?
d. What new systems were required?
e. What were the various cost trade-offs associated with revision or replacement?
f. What implementation schedules were realistic?

Only the initial phase of the study is described here. The initial survey was primarily keyed to the following system aspects:

a. Identification of significant problem areas.
b. Degree of concern expressed by various system users.

2 Essentials of Survey Sampling

c. Importance of systems covered in the study to new or anticipated programs or projects.
d. Ability to implement within an effective time frame.
e. Potential for reduced cost or increased efficiency through system merger or interrelationships.

A questionnaire was prepared and sent to the 106 users of the various systems involved in the company's operations. The 106 questionnaire recipients were viewed as a complete population of users. These users ranged from clerical to management personnel, and covered a wide range of individual projects and departments. The questionnaire consisted of two basic parts: Part One requested the user to rank, on a scale of 0 to 5, each system with which he was familiar on a variety of topics. Topics included were (1) degree of familiarity with the system, (2) utility of the system to the individual and his or her department, (3) degree of cost effectiveness, (4) ease with which data could be input, (5) adequacy of reporting, and (6) degree of relevance of the system data content. Part Two of the questionnaire was more general in nature. It contained nine subjective questions, requesting the users to state their views of the business systems with which they were in some way involved.

Responses were received from 52 recipients of the questionnaire—slightly under 50%. According to one member of management, a man directly involved in conducting the study:

> The results, therefore, constitute a sample rather than a complete census of the population of users. In addition, it is probable that some bias was introduced since respondents did not likely represent a true statistical cross section of the surveyed population. Nevertheless, it is believed that the results represented a reasonably good sampling of the opinions of the various users.

Certain exclusion criteria were applied to the data prior to tabulation. These included limiting the tabulation to those with at least five responses—a condition thought to suggest average or above-average familiarity with the system in question. In addition, certain data were excluded where "it was well established that it had arisen from faulty respondent premises."

The results of the survey were then analyzed according to the priorities and criteria cited above. Particularly heavy weight was given to those responses from primary-system users and those with a high degree of departmental importance.

1. In view of the exclusion criteria and the different weights assigned to respondents' opinions, what is *your* definition of the target population? Do you think that your definition of the target population coincides with the one used by the member of management quoted above? Explain.

2. Discuss the propriety of starting out to conduct a population study, or census, and then calling the result a sample study after it is found that many recipients of the questionnaire did not return it.

3. Judging from the case description, it seems reasonable to assume that at least two or three sources of disparity exist between the target population and the sampled population. Tell what they are. Also tell how the procedure could have been carried out differently to ensure a closer correspondence between target and sampled populations.

ⓒ 2.4 THE U.S. POSTAL SERVICE

The U.S. Postal Service has developed an elaborate sampling procedure aimed at reducing the quantity of mishandled mail. The sampling process is applied to the Multiposition Letter Sorting Machine (MPLSM) operations. Figures 2.14 and 2.15 show one such machine from the console side and from the sweep side, respectively. Figure 2.16 shows a closeup view of the operator console assembly.

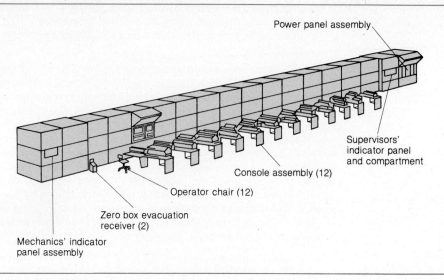

Figure 2.14. Multiposition letter sorting machine, console side.

In brief summary, the physical process involved works as follows. Letters are placed manually on edge in the letter feeder tray of the operator console (Figure 2.16) and are advanced automatically to the pickoff position. Letters are then removed individually by the vacuum pickoff head and dropped into the console inserter tube. The letters continue to move automatically at machine-paced intervals to the operator's viewing position above the console keyboard. At this position each letter is stopped momentarily to allow the operator to read the address. As the letter starts to move from the viewing position into the machine, the operator enters a scheme code number associated with the address by manually depressing appropriate keys on the console keyboard.

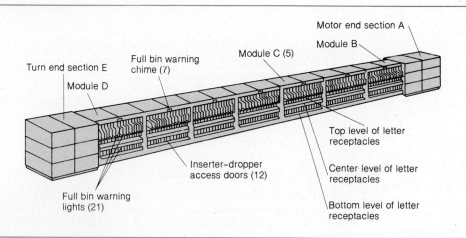

Figure 2.15. Multiposition letter sorting machine, sweep side.

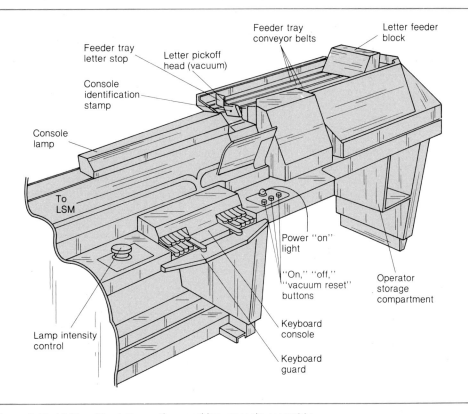

Figure 2.16. Multiposition letter sorting machine, console assembly.

After each letter leaves the viewing position, it is advanced into an inserter-dropper, which projects the letter downward into the operator-coded compartment of a traveling letter cart. The letter carts travel within the machine on an endless set of conveyor chains. When the coded-letter compartment passes over an identically coded-letter receptacle, a door in the bottom of the compartment is opened and the letter falls into the proper bin. A MPLSM has 277 letter bins. If something has gone amiss in the process of keying in the scheme code numbers, the letter ends up in a zero or a 400 bin.

The MPLSM Sort Quality System (SQS) was developed to identify the mishandled mail rate and, in addition, identify the various contributors to mishandling. Sample mishandling rates are determined to estimate the population percentage of mail processed that must be rehandled as a result of an MPLSM failure, regardless of whether the failure resulted from human error or from an equipment malfunction.

After the mishandling rate is determined, reasons for mishandling can be identified through diagnostic tests and necessary corrective action can be taken. The ultimate purpose of the SQS test is to reduce costs and improve service by providing accurate information on the size and scope of the mishandling problem. A mishandling is defined as a piece of mail improperly oriented in a bin or in the wrong bin. And here is where the sampling comes in.

Sampling for the SQS test is on live mail during normal operations. The total sample size used is a minimum of 5000 pieces of mail, a figure thought to provide adequate confidence at reasonable cost. A computerized system allocates the total sample size based on a specified density and transfers it to the sweepside bins or to the zero or 400 bins. For a total sample of 5000 pieces, a bin must have a density of at least 0.01% of capacity before it is sampled.

When the SQS test is used to establish the mishandling rate for an entire facility, random selection is used in connection with

a. MPLSMs
b. Bins within a given MPLSM
c. Order in which bins within a given MPLSM are inspected
d. Letters within selected bins

The bin-machine sampling order is determined by use of a table of random digits. The first digit of a two-digit random number determines the bin (eight bins are selected from each randomly selected MPLSM), and the second digit determines the specific MPLSM. As already noted, the sample of 5000 pieces of mail is allocated among the selected bins on a basis of volume densities of the bins. When the process has been completed, the ten bins in the entire facility with the largest number of mishandlings are identified for management action.

1. In your opinion, does the system provide a true probability sample? Explain.

2. If the system does provide a true probability sample, is it best described as simple random, systematic random, stratified random, random cluster, or multiple random? Or is it a combination of two or more of these methods? Justify your choice.

3 Data Organization and Exploratory Data Analysis

Good order is the foundation of all good things.—Edmund Burke

Everything should be made as simple as possible, but not simpler.—Albert Einstein

What You Should Learn from This Chapter

This is a menu chapter, and section 3.2 is the main course. This section contains material that will be referred to in Chapter 4 and in several later chapters. The other sections in this chapter may be viewed as side dishes to be ordered at your instructor's discretion. However, since in statistics the name of the game is to learn what the data are trying to say, it is suggested that you give the entire chapter at least a once-over-lightly. Although the data tend to speak most loudly when subjected to appropriate formal statistical analysis, the roles of data organization and exploratory data analysis should not be underestimated.

When you have finished this chapter, you should be able to

1. Construct a standard frequency distribution.
2. Demonstrate an awareness of common frequency distribution shapes.
* 3. Construct a stem-and-leaf display. *
* 4. Construct a cumulative frequency distribution.
* 5. Construct and interpret a box-and-whiskers plot.
* 6. Construct and interpret bivariate and trivariate relative frequency distributions.

All of these tools are designed to give the analyst a valid "feel" for the data with only a modest investment in data handling and computational effort.

Before proceeding, you should be sure you are clear on the distinctions between (1) quantitative variable and qualitative variable, (2) discrete quantitative variable and continuous quantitative variable (both in section 1.3 of Chapter 1), and (3) sample and population (section 1.4 of Chapter 1). Sample data are assumed throughout this chapter. However, the methods described could be applied without modification to population data as well.

3.1 Introduction

The task of arriving at business decisions systematically often entails the handling of large quantities of data. Unless the data are organized and simplified, they defy comprehension— and, if they cannot be comprehended, they certainly cannot be acted upon intelligently. In this chapter we consider some helpful ways of organizing and presenting statistical data so that they can be readily comprehended.

*See footnote on p. 45.

3.2 Introduction to Frequency Distributions

A data-organizing device of long standing is the *frequency distribution*, the nature of which can be conveyed most readily through an example.

ILLUSTRATIVE CASE 3.1

Management of the Gotham City Hardware Company, a large wholesaler of hardware products, wished to assess the adequacy of the company's order-filling system and, at the same time, evaluate the capabilities of individual order fillers. They gave each of 75 order fillers a standardized list of 45 hardware items and instructed them to fill the order correctly as quickly as possible. The results are shown in Figure 3.1.

Figure 3.1 displays what would technically be called a *univariate quantitative frequency distribution*—"univariate" because only one variable, speed of order fillers, is being consid-

Time (in Minutes)	Number of Employees (Frequency)	
10 and under 20	5	
20 and under 30	15	Data for illustrative Case 3.1
30 and under 40	20	shown in tabular form
40 and under 50	14	
50 and under 60	11	
60 and under 70	7	
70 and under 80	3	
Total	75	

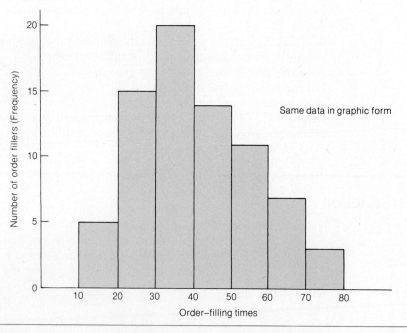

Same data in graphic form

Figure 3.1. Tabular and graphic presentations of data for Illustrative Case 3.1.

ered, and "quantitative" because the single variable is quantitative rather than qualitative. However, because this particular tool of data organization is so commonly used, we will refer to it hereafter simply as *frequency distribution* and reserve qualifying adjectives for similar but less frequently used methods.

> A *frequency distribution* is a tabular or graphic device for displaying the data of interest for a *single quantitative variable* grouped into several classes along with the number of observations, called the *class frequency*, associated with each indicated class.

We will refer to Illustrative Case 3.1 later in this chapter and in the next chapter. In the meantime, let us assume that someone has an unorganized mass of data which she wishes to capture in frequency distribution form.

ILLUSTRATIVE CASE 3.2

A portfolio manager keeps a close watch on price-earnings ratios (defined as current market price divided by earnings for the most recent four quarters) for common stocks. She reasons that, when the majority of stocks in a representative sample have low price-earnings (P-E) ratios by historical standards, it is time for her to become an aggressive buyer. Low P-E's may mean that investors in general are unrealistically pessimistic. Moreover, stocks with low P-E's can benefit in a twofold way when earnings increase: (1) higher earnings multiplied by a constant P-E ratio means a higher market price and (2) rising earnings are usually accompanied by rising P-E ratios.

Naturally, if the majority of stocks in a representative sample sport high P-E ratios by historical standards, she will begin to sell aggressively because the process just described can also work in reverse.

At periodic intervals this portfolio manager selects a simple random sample of stocks from among those listed on the New York Stock Exchange and scrutinizes their P-E ratios. The results of one such study involving a sample of 200 stocks are presented in Table 3.1.

Steps in Constructing a Frequency Distribution

The analyst constructing a frequency distribution must be flexible. In the final analysis, common sense, not a rigid set of rules, must prevail. Still, we will present here a series of steps that have been found to work quite well when applied to many different kinds of data. In sections 3.3 and 3.4 we will take a cook's tour of some alternative methods sometimes preferable in view of the peculiarities of the data or the informational requirements of the analyst.

The standard steps in the construction of a frequency distribution are as follows:

1. Organize the values of the variable into an array.
2. Determine the appropriate number of classes to use.
3. Determine the appropriate class interval to use.
4. Set up the classes.
5. Count and record the number of observations associated with each class.
6. Present the resulting frequency distribution as a histogram or frequency polygon.

Let us consider each of these steps in turn.

1. *Organize the values of the variable into an array.*

An *array* is simply an ordering of the data from smallest to largest, or from largest to smallest.

Although this step is optional, use of an array is usually advisable since it helps the analyst to minimize the number of errors in subsequent steps—no small consideration when the number of observations is very large. The price-earnings data introduced in Table 3.1 appear in array form, ordered from smallest to largest, in Table 3.2.

2. *Determine the appropriate number of classes to use.* No inalterable rule exists that is capable of eliminating the need for a pinch of judgment in determining the number of classes to use. The main point to remember is that the observations should not be assigned to only two or three very lumpy classes (as in Figure 3.2) nor spread too thinly over a great many classes (as in Figure 3.3). Table 3.3 has been prepared with a view to providing the beginner with a rough guide to determining an adequate number of classes.[1]

The suggested number of classes for $n = 200$ is about 9.

Table 3.1   Price-Earnings Ratios for 200 Common Stocks

11.1	12.6	26.7	5.2	8.3	5.5	6.8	7.6
7.3	18.1	14.6	10.9	7.2	9.5	9.2	11.8
12.0	16.9	10.1	14.6	5.2	7.5	11.1	19.9
14.9	7.4	6.0	39.9	29.3	35.1	6.8	39.0
6.1	6.2	26.8	33.7	9.6	16.6	10.9	11.2
22.6	46.0	7.3	29.7	10.3	6.4	9.6	7.6
10.3	5.0	14.4	11.6	8.3	7.9	17.8	7.5
7.8	7.3	6.0	20.2	5.6	8.3	7.7	10.7
8.6	14.5	6.0	5.4	12.9	14.8	9.2	14.1
15.7	10.4	7.0	11.0	6.3	8.7	7.6	16.9
7.9	8.3	13.1	9.8	8.2	18.0	26.6	7.8
4.1	10.6	15.3	7.2	35.5	6.1	10.2	6.1
7.8	8.1	30.0	15.0	6.1	15.4	10.1	9.6
6.8	4.4	6.8	9.1	16.3	5.4	5.9	6.5
7.9	44.9	13.8	12.3	10.9	9.3	11.9	10.0
7.6	17.9	7.1	8.4	35.5	7.4	7.7	8.3
15.8	8.3	23.1	8.4	12.4	7.8	8.2	9.8
13.7	15.8	4.7	7.9	26.4	6.2	11.4	13.2
8.6	11.7	8.6	13.7	9.3	16.6	8.7	39.7
14.0	9.1	7.1	10.9	23.4	13.3	10.9	24.0
11.9	8.7	15.6	27.7	10.4	16.9	6.9	5.5
22.8	8.5	22.2	5.8	14.7	8.0	7.5	10.5
4.4	7.1	63.8	12.5	13.3	10.5	5.5	16.0
53.1	7.4	24.1	15.3	29.1	11.0	9.9	36.3
9.6	6.6	5.1	7.8	8.4	38.3	20.4	9.1

Table 3.2 Array of Price-Earnings Ratios for 200 Common Stocks

4.1	6.3	7.6	8.4	10.1	11.9	15.3	23.4
4.4	6.4	7.6	8.4	10.1	11.9	15.3	24.0
4.4	6.5	7.6	8.5	10.2	12.0	15.4	24.1
4.7	6.6	7.7	8.6	10.3	12.3	15.6	26.4
5.0	6.8	7.7	8.6	10.3	12.4	15.7	26.6
5.1	6.8	7.8	8.6	10.4	12.5	15.8	26.7
5.2	6.8	7.8	8.7	10.4	12.6	15.8	26.8
5.2	6.8	7.8	8.7	10.5	12.9	16.0	27.7
5.4	6.9	7.8	8.7	10.5	13.1	16.3	29.1
5.4	7.0	7.8	9.1	10.6	13.2	16.6	29.3
5.5	7.1	7.9	9.1	10.7	13.3	16.6	29.7
5.5	7.1	7.9	9.1	10.9	13.3	16.9	30.0
5.5	7.1	7.9	9.2	10.9	13.7	16.9	33.7
5.6	7.2	7.9	9.2	10.9	13.7	16.9	35.1
5.8	7.2	8.0	9.3	10.9	13.8	17.8	35.5
5.9	7.3	8.1	9.3	10.9	14.0	17.9	35.5
6.0	7.3	8.2	9.5	11.0	14.1	18.0	36.3
6.0	7.3	8.2	9.6	11.0	14.4	18.1	38.3
6.0	7.4	8.3	9.6	11.1	14.5	19.9	39.0
6.1	7.4	8.3	9.6	11.1	14.6	20.2	39.7
6.1	7.4	8.3	9.6	11.2	14.6	20.4	39.9
6.1	7.5	8.3	9.8	11.4	14.7	22.2	44.9
6.1	7.5	8.3	9.8	11.6	14.8	22.6	46.0
6.2	7.5	8.3	9.9	11.7	14.9	22.8	53.1
6.2	7.6	8.4	10.0	11.8	15.0	23.1	63.8

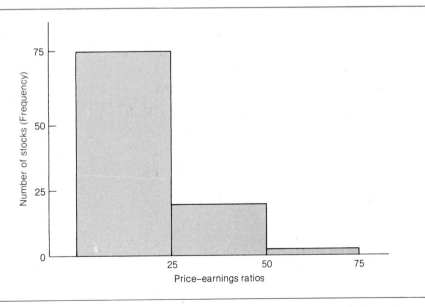

Figure 3.2 A frequency distribution that is unsatisfactory because of an excessively large class interval.

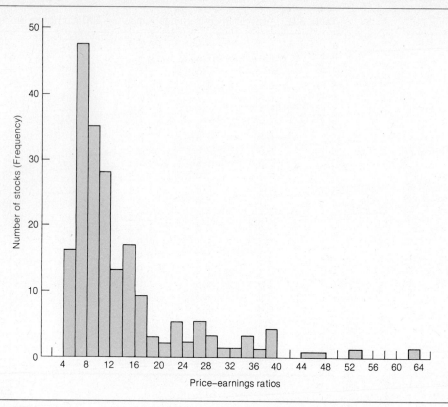

Figure 3.3. A frequency distribution whose usefulness is questionable because of too small a class interval.

Table 3.3 Guide to Determining the Number of Classes to Use[1]

Number of Observations, n	Suggested Number of Classes
10	4
20	5
50	7
100	8
200	9
500	10
1,000	11
2,000	12
5,000	13
10,000	14
100,000	18

3. *Determine the appropriate class interval to use.* The smallest and largest values that can fall into a given class of a frequency distribution are called the *class limits*. The *class interval* is the difference between the upper limit of a given class and the lower limit of the same class. For example, the frequency distribution in Figure 3.1 has as its first class "10 and under 20." Therefore, the class interval is 20 (ignoring the "and under" qualification, which is simply a device serving to make absolutely clear where an observation with a value of 20 would be placed) minus 10. Thus, $20 - 10 = 10$.

A convenient way of determining the appropriate class interval is to (1) find the smallest value in the array and subtract it from the largest value, and (2) divide the resulting difference by the number of classes previously decided upon (Step 2). For example, referring to Table 3.2, we find that the largest P-E ratio in the list is 63.8, and the smallest is 4.1. The difference of 59.7 is, therefore, divided by 9 and the resulting quotient of 6.62 is obtained. Because it is inconvenient to work with a class interval that is not a whole integer, 7, rather than 6 plus a fraction, is used in this case for the class interval. (For that matter, little harm would result from use of a class interval of, say, 5, or even 10, instead of 7. The procedures described here are merely intended to put you into the right "ballpark.")

4. *Set up the classes.* Having determined the size of the class interval, the analyst would next select a value for the lower class limit of the beginning class (following the conventional procedure of arranging the data from small to large) and proceed to set up nine classes. Any value such that, if 7 is added to it, we obtain a range of values that includes 4.1, the smallest P-E ratio in the array, will serve adequately as the lower class limit of the beginning class. Let us assume that our analyst chooses—perhaps quite arbitrarily—the value 3. Classes would then be determined by her adding the constant class interval of 7 to 3, giving a class running from 3 to 10; adding 7 to 10, giving a class running from 10 to 17; and so forth. This procedure would be repeated until a class has been obtained whose limits surround the value 63.8, the largest P-E ratio. The resulting classes, with the "and under" clarifying device being used, are shown in column (1) of Table 3.4.

Table 3.4 Frequency Distribution for 200 Price-Earnings Ratios

(1) Classes of P-E Ratios	(2) Frequency
3 and under 10	99
10 and under 17	65
17 and under 24	12
24 and under 31	11
31 and under 38	5
38 and under 45	5
45 and under 52	1
52 and under 59	1
59 and under 66	1
Total	200

5. *Count and record the number of observations associated with each class.* The analyst will now examine the array as presented in Table 3.3 and count the number of observations having values of 3 or more but under 10 and place the result of the count alongside the class labeled "3 and under 10." That number in the present case is 99. This same procedure is then carried out with each of the remaining classes. When all such counts have been made and recorded, as in column (2) of Table 3.4, the construction of the frequency distribution is complete.

6. *Present the resulting frequency distribution in the form of a* histogram or frequency polygon. This step is optional. However, a primary reason for constructing the frequency distribution in the first place was to make the jumble of data more comprehensible. Therefore, one more step, that of presenting the frequency distribution in graphic form, is usually advisable. One generally begins, and often ends, with a *histogram.*

A *histogram* is a particular kind of bar chart. It shows the list of class limits along the horizontal axis and frequencies along the vertical axis. The number of observations associated with the classes, called class frequencies, are presented as bars of varying heights, as shown in Figure 3.4.

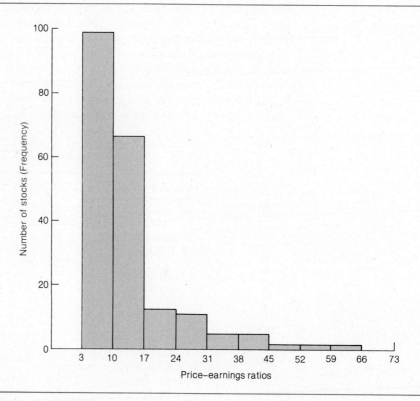

Figure 3.4. Histogram of price-earnings ratios for 200 common stocks. (See Table 3.4 for tabular presentation of these same data.)

3 Data Organization and Exploratory Data Analysis

Since, according to Table 3.4, 99 observations are associated with the class labeled "3 and under 10," the height of the leftmost bar is placed at 99 on the vertical scale in Figure 3.4. The class labeled "10 and under 17" has 65 observations associated with it; hence, a bar of height 65; and so forth. A convention widely followed when constructing histograms is to leave an empty class on each end of the set of vertical bars, the reason being that, if the histogram is converted into a frequency polygon, a somewhat nicer graphic presentation results. (Because such a disproportionate number of observations is associated with low P-E values, this practice has been followed on the right-hand end of the histogram in Figure 3.4 but not on the left-hand end.)

A histogram can be converted into a *frequency polygon* by (1) placing a dot at the midpoint of the top of each bar in a histogram, (2) connecting the dots by straight lines, and

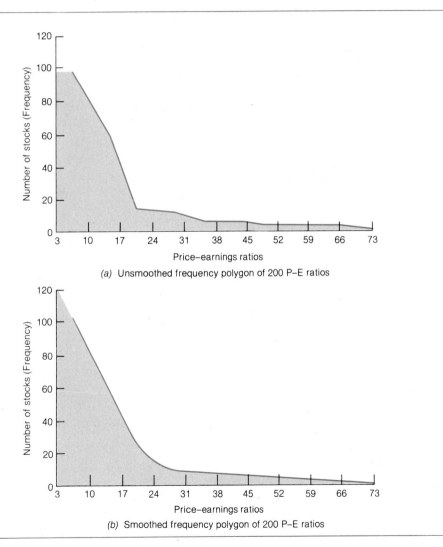

(a) Unsmoothed frequency polygon of 200 P–E ratios

(b) Smoothed frequency polygon of 200 P–E ratios

Figure 3.5. Unsmoothed and smoothed versions of the frequency polygon corresponding to the histogram presented in Figure 3.4.

(3) erasing all other lines. A midpoint is defined as the lower class limit of a specific class plus the upper class limit of the same class, divided by 2. Referring to the histogram in Figure 3.4, for example, we find that the lower class limit of the leftmost class is 3 and the upper class limit is 10. Using the formula $(LCL + UCL)/2$, where LCL and UCL represent lower class limit and upper class limit, respectively, we find that the midpoint of this class is $(3 + 10)/2 = 6.5$. Therefore, a dot is placed at the top of the leftmost bar directly above the point where 6.5 is found on the horizontal axis of the histogram. This procedure is repeated for all bars and for the empty class on the right-hand end of the set of bars (at both ends when the data are more definitely clustered around some central value). The dots are then connected by straight lines and the bars erased. The frequency polygon corresponding to the histogram in Figure 3.4 appears in part *(a)* of Figure 3.5 (p. 85).

The histogram and frequency polygon are simply alternative ways of displaying frequency distributions graphically. When only one set of data is being considered, use of a histogram is usually preferred for reasons illustrated in Figure 3.6. However, if a graphic comparison among two or more similar variables is desired, frequency polygons are usually better because they present fewer confusing lines. For example, if our portfolio manager wished to compare her collection of 200 P-E ratios obtained recently with a like number obtained, say, one year ago and two years ago, such a visual comparison is easily achieved through use of frequency polygons, as shown in Figure 3.7. By the way, Figure 3.7 suggests that the stock

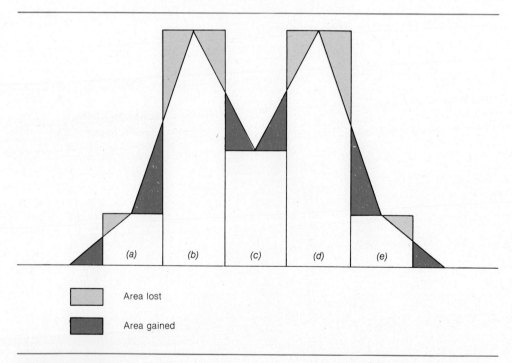

Area lost

Area gained

Figure 3.6. Demonstration of why a histogram is usually preferred over a frequency polygon when only one variable is presented. A frequency polygon can distort the appearance of the data in important ways. For example, notice how much area is lost from classes *(b)* and *(d)* and how much is gained by classes *(a)*, *(c)*, and *(e)*.

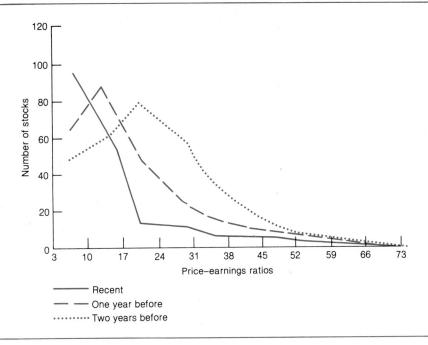

Figure 3.7. Comparison of frequency polygons for samples of 200 stocks at a recent time, one year earlier, and two years earlier.

market might be at or near a buying mode, according to this portfolio manager's buying and selling strategy as spelled out in Illustrative Case 3.2.

When frequency polygons are used, a sometimes useful procedure is to present them in smoothed form. That is, for a very large number of observations having the same general pattern as the frequency polygon in part (*a*) of Figure 3.5, the act of progressively increasing the number of classes and reducing the size of the class interval would presumably yield a frequency polygon approaching the shape of the smoothed curve shown in the (*b*) part of this same figure. At various points in the following pages, such smoothed frequency polygons will be used for convenience and for clarity of explanation. Figure 3.8 shows, in the form of smoothed frequency polygons, several other shapes that frequency distributions sometimes take.

✳3.3 Other Ways of Constructing Frequency Distributions

To this point, we have considered a somewhat pat set of procedures for constructing frequency distributions. Occasionally, the analyst will wish to depart from this suggested methodology in order better to provide an overall impression of the data than the standard method might convey. Therefore, some variations on the standard method of frequency distribution construction are presented in this section.

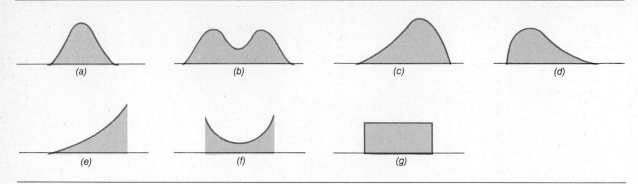

Figure 3.8. A few of the many shapes that smoothed frequency distributions may display: *(a)* bell-shaped or normal, *(b)* bi-modal, *(c)* skewed left, *(d)* skewed right, *(e)* J-shaped, *(f)* U-shaped, *(g)* uniform.

Unequal Class Intervals

When data depart conspicuously from a symmetrical shape, sometimes a frequency distribution having unequal class intervals will be used. One such frequency distribution is shown in Figure 3.9, where the histogram of the 200 P-E ratios is shown with a class interval of 3 throughout the part of the distribution where observations are heavily concentrated and a class interval of 10 toward the end where observations are more sparse. The better definition on the left-hand end suggests that this modified method suits our P-E data somewhat better than the standard method.

Open-Ended Distributions

An open-ended frequency distribution is one with an open-ended class—that is, a class with only one limit—at either or both ends. An example is shown in Figure 3.10 where the P-E data are presented with a constant class interval of 3 up to a P-E value of 30, and then all ratios greater than 30 are bunched into a single open-ended class.

In general, this method has little to recommend it. A disadvantage of open-ended classes is that the arithmetic mean and standard deviation—important descriptive measures described in the next chapter—cannot be calculated from the frequency distribution.

Relative Frequency Distributions

The number of observations associated with any specific class can easily be converted into a *relative frequency* by dividing by the total number of observations altogether. This is illustrated in Figure 3.11 where, for example, the frequency value 99 associated with the "3 and under 10" class is converted into a relative frequency by dividing by the sample size of 200. That is, $99/200 = .495$. Relative frequencies for all other classes were calculated in the same way.

Price–Earnings Ratios	Number of Stocks (Frequency)
3 and under 6	16
6 and under 9	68
9 and under 12	43
12 and under 15	22
15 and under 18	17
18 and under 28	17
28 and under 38	9
38 and under 48	6
48 and under 58	1
58 and under 68	1
Total	200

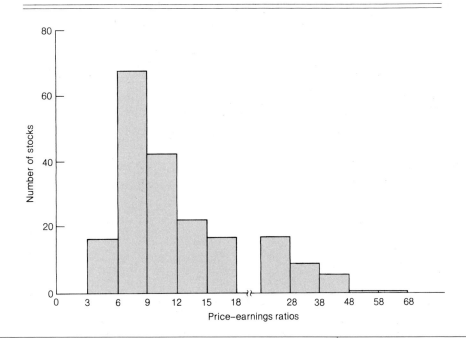

Figure 3.9. Tabular and graphic presentations of the price-earnings data. Unequal class intervals are used.

The analysis of relative frequencies can sometimes be more helpful than analysis of absolute frequencies. Suppose, for example, that our portfolio manager wished to make a comparison like that shown in Figure 3.7 but had used a sample of size $n = 1000$ two years ago and a sample of size $n = 500$ one year ago. Such a pattern might reflect some experimentation on her part; perhaps she had found, through trial and error, that a sample of size 200 conveys as much information for her purposes as a sample of size 500 or even 1000. For purposes of graphic comparison, as in Figure 3.7, she would be well advised to convert all frequencies to relative frequencies.

Relative frequency distributions also play a fundamental role in probability theory, as we will see in Chapters 6 and 7.

Price–Earnings Ratios	Number of Stocks (Frequency)
3 and under 6	16
6 and under 9	68
9 and under 12	43
12 and under 15	22
15 and under 18	17
18 and under 21	5
21 and under 24	5
24 and under 27	6
27 and under 30	4
30 and over	14
Total	200

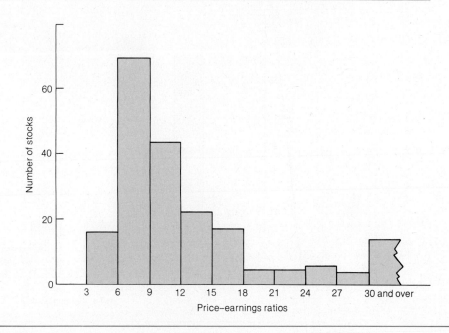

Figure 3.10. An open-ended frequency distribution in tabular and graphic form.

Frequency Distributions for Discrete Variables

When a variable of interest is definitely discrete, it will be graphed somewhat differently than a continuous one: Instead of bars, vertical lines with spaces between them will be used to remind the viewer that interpolation should not be attempted. For example, let us suppose that a marketing research study is conducted to determine how many working radios are owned by a sample of 1000 households in a certain large city so that the degree of market saturation can be judged. The results, let us say, are as shown in Figure 3.12.

(1) Price–Earnings Ratios	(2) Frequency	(3) Relative Frequency
3 and under 10	99	.495
10 and under 17	65	.325
17 and under 24	12	.060
24 and under 31	11	.055
31 and under 38	5	.025
38 and under 45	5	.025
45 and under 52	1	.005
52 and under 59	1	.005
59 and under 66	1	.005
Total	200	1.000

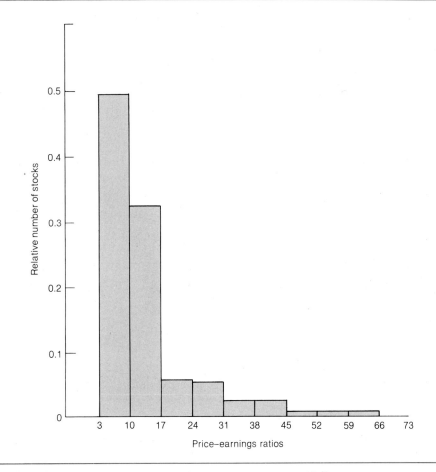

Figure 3.11. A relative frequency distribution in tabular and graphic form.

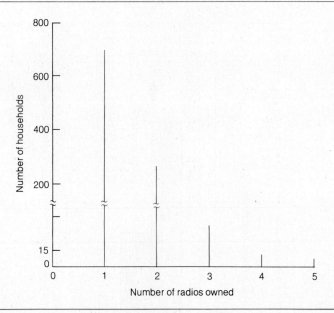

Number of Radios Owned	Number of Households (Frequency)
0	5
1	700
2	275
3	13
4	4
5	3
Total	1000

Figure 3.12. Frequency distribution for a discrete variable in tabular and graphic form.

The Stem-and-Leaf Display

A particularly useful variation on standard frequency distribution construction is the *stem-and-leaf display*. The method will be illustrated using the Gotham City Hardware example (Illustrative Case 3.1). Recall that 75 employees were told to fill an order consisting of 45 hardware items as fast as possible. The frequency distribution was presented in both tabular and graphic forms in Figure 3.1. But let us start at the beginning and organize the data by quite a different means. Table 3.5 shows the original raw data. All order-filling times have been rounded to the nearest whole minute for convenience of illustration.

We will begin with the stem part of the display. In this case, all of our numbers have two digits; we will call the first digit the stem and the second the leaf. Part (*a*) of Figure 3.13 shows the stem values only. A vertical line is placed at the right of these stem values. Part (*b*) of this figure shows how the first three leaf values are placed in the emerging chart. The first three 2-digit numbers in Table 3.5 are 25, 41, and 75. Therefore, 5 is placed to the right of

Table 3.5 Times Required for 75 Employees to Fill a Standardized Order (Rounded to Nearest Whole Minute)

25	41	37	67	36	35	57	38
41	40	65	18	34	55	45	31
75	55	19	54	25	20	24	32
31	49	41	65	42	21	21	22
32	32	55	67	46	49	28	37
61	42	39	76	23	52	32	
14	68	50	57	24	43	37	
52	13	34	23	31	33	36	
54	61	24	43	49	36	31	
41	14	72	24	24	24	57	

```
1                1                1 | 4 3 4 9 8                                      (5)
2                2 | 5            2 | 5 4 3 4 5 3 4 4 0 1 4 4 1 8 2                    (15)
3                3                3 | 1 2 2 7 9 4 6 4 1 5 3 6 2 7 6 1 8 1 2 7        (20)
4                4 | 1            4 | 1 1 1 0 9 2 1 3 2 6 9 9 3 5                      (14)
5                5                5 | 2 4 5 5 0 4 7 5 2 7 7                           (11)
6                6                6 | 1 8 1 5 7 5 7                                   (7)
7                7 | 5            7 | 5 2 6                                           (3)

(a) Stem         (b) Stem values  (c) Completed
    values           and first        stem–and–
    only             three leaf       leaf display
                     values
```

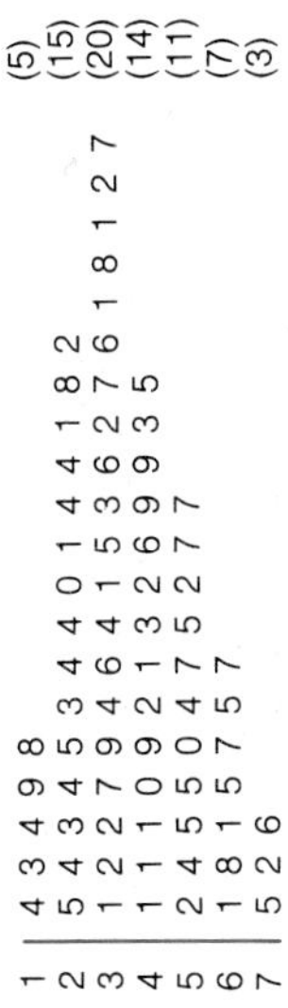

(d) Stem–and–leaf display in part (c) sitting on its stem part

Figure 3.13. Construction of a stem-and-leaf display for data in Table 3.5.

the vertical line in alignment with the stem value of 2. Then, 1 and the other 5 are placed to the right of the vertical line in alignment with 4 and 7, respectively. Part (*c*) of Figure 3.13 shows the completed stem-and-leaf display.

The principal advantage of the stem-and-leaf display over the standard frequency-distribution methodology is the ease with which it can be constructed. Notice that it is not necessary to array the data before beginning to develop the graph. Notice also what happens when we give part (*c*) of Figure 3.13 a counterclockwise twist, so that it ends up sitting on its stem part, as in part (*d*). We see that the stem-and-leaf display becomes an instant histogram. Clearly, this method permits some consolidation of steps and a noticeable saving of time.

Qualitative Frequency Distributions

A qualitative frequency distribution is a special kind of discrete distribution. An example of a qualitative frequency distribution is shown in Figure 3.14 where data on stage of completion of new houses in a particular city are displayed. Stage of completion in this case is conveyed through the use of three descriptive, as distinguished from quantitative, categories. Figure 3.14 follows the convention of placing frequencies on the horizontal, rather than the vertical, axis when the variable of interest is qualitative.

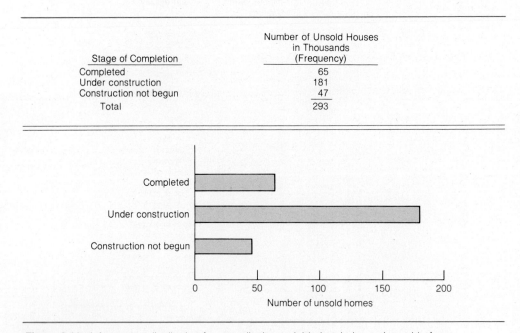

Stage of Completion	Number of Unsold Houses in Thousands (Frequency)
Completed	65
Under construction	181
Construction not begun	47
Total	293

Figure 3.14. A frequency distribution for a qualitative variable in tabular and graphic form.

*3.4 Cumulative Frequency Distributions

Sometimes an analyst of statistical data will wish to know how many observations are above a certain value or below a certain value. Such information can be obtained from a regular frequency distribution, but not so conveniently as when cumulative frequencies are employed. The cumulative frequencies for a set of data may be determined by adding the number of observations associated with a specific class to the sum of the frequencies for all preceding classes. This point will be illustrated using the P-E data from Table 3.4. The cumulative frequency for P-E ratios of less than 17, for example, is found by adding the number of observations associated with the class "3 and under 10" to the number of observations associated with the class "10 and under 17": 99 + 65 = 164. Similarly, the cumulative frequency for P-E ratios of less than 24 is found by adding the number of observations associated with the class "17 and under 24" to 164. And so it goes. This procedure is demonstrated in Figure 3.15. This kind of cumulative frequency distribution is intended to provide a quick answer to the question: "How many (or what percent of the) observations have a value of less than _____ [some specified value]?" Cumulative distributions can convey such information with exactness only when it is a class limit that is specified, though approximate answers for other values can also be obtained.

Actually, there are two kinds of cumulative frequency distributions. The kind we have been discussing, probably the more frequently used, is called a *"less than" distribution* for obvious reasons. If we wanted just the opposite kind of information—that is, the number of observations having values of a specified size or more—we would use a process of successive subtraction, rather than successive addition, thereby obtaining what is called an *"or more" distribution*. The construction and graphic presentation of this kind of cumulative distribution are shown in Figure 3.16.

You will notice that in both Figures 3.15 and 3.16, the points connected by straight lines are points associated with class limits—not the midpoints as in the construction of a standard frequency polygon.

*3.5 Box-and-Whiskers Plots

A relative newcomer to the realm of exploratory data analysis is the box-and-whiskers chart.[2] A box-and-whiskers chart is based on only five numbers obtained from a set of arrayed data:

1. The smallest value
2. The value 25% of the way into the array
3. The value 50% of the way into the array
4. The value 75% of the way into the array
5. The largest value

The method will be illustrated using, first, the P-E data appearing in Table 3.2 and then the data for order-filling times shown in part (c) of Figure 3.13.

Looking at Figure 3.2, we quickly note that the smallest P-E ratio is 4.1 and the largest is 63.8. That is all there is to items 1 and 5 on the above list.

Price–Earnings Ratios	Frequency	Class Limit	Number of Observations Having a Value Less Than the Specified Class Limit	Relative Frequency (If Desired)
3 and under 10	99	3	— — = 0	.000
10 and under 17	65	10	0 + 99 = 99	.495
17 and under 24	12	17	99 + 65 = 164	.820
24 and under 31	11	24	164 + 12 = 176	.880
31 and under 38	5	31	176 + 11 = 187	.935
38 and under 45	5	38	187 + 5 = 192	.960
45 and under 52	1	45	192 + 5 = 197	.985
52 and under 59	1	52	197 + 1 = 198	.990
59 and under 66	1	59	198 + 1 = 199	.995
	200	66	199 + 1 = 200	1.000

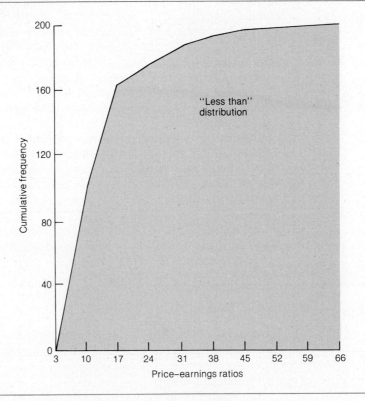

Figure 3.15. Construction of a "less than" cumulative frequency distribution.

Items 2, 3, and 4 are more difficult, but not much. The numbers we seek, by the way, are known, respectively, as the *first quartile*, Q_1, the *second quartile* (or *median*, a subject treated in greater detail in Chapter 4), Q_2, and the *third quartile*, Q_3. They are called quartiles because they divide the arrayed data into four equal parts. The difference between Q_3 and Q_1 is known as the *interquartile range*. Fifty percent of the observations on any variable will be between Q_1 and Q_3.

Price–Earnings Ratios	Frequency	Class Limit	Number of Observations Having a Value as High as the Specified Class Limit or More	Relative Frequency (If Desired)
3 and under 10	99	3	— — = 200	1.000
10 and under 17	65	10	200 − 99 = 101	.505
17 and under 24	12	17	101 − 65 = 36	.180
24 and under 31	11	24	36 − 12 = 24	.120
31 and under 38	5	31	24 − 11 = 13	.065
38 and under 45	5	38	13 − 5 = 8	.040
45 and under 52	1	45	8 − 5 = 3	.015
52 and under 59	1	52	3 − 1 = 2	.010
59 and under 66	1	59	2 − 1 = 1	.005
	200	66	1 − 1 = 0	.000

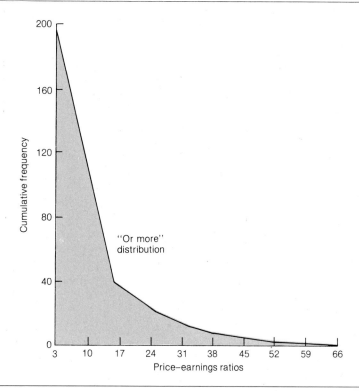

Figure 3.16. Construction of an "or more" cumulative frequency distribution.

Q_1, Q_2, and Q_3 can be determined most readily through use of the following position-location formulas:

$$Q_1 = \text{Value associated with the } \frac{n+1}{4} \text{ ordered observation}$$

$$Q_2 = \text{Value associated with the } \frac{n+1}{2} \text{ ordered observation}$$

$$Q_3 = \text{Value associated with the } \frac{3(n+1)}{4} \text{ ordered observation}$$

Since there are 200 P-E ratios displayed in Table 3.2, we calculate that $(n+1)/4 = 201/4 = 50.25$, or about 50. This tells us that the first quartile, Q_1, is the 50th value in the array, counting from the top of the leftmost column in Figure 3.2. The 50th value is found to be 7.6. Then, we find that $(n+1)/2 = 201/2 = 100.5$. When the result has a .5 on the end, it is customary to average the competing values. In this case, we would split the difference between the 100th observation and the 101st observation. The 100th observation is found to be 10; the 101st, 10.1. Therefore, Q_2 is 10.05. Finally, we calculate that $[3(n+1)]/4 = [3(201)]/4 = 150.75$, or about 151. Thus, Q_3 is the 151st observation in the array, namely, 15.3.

We now know the five numbers needed for a box-and-whiskers chart. We know that

The smallest value is 4.10.

The first quartile, Q_1, is 7.60.

The second quartile, Q_2, is 10.05.

The third quartile, Q_3, is 15.30.

The largest value is 63.80.

The box-and-whiskers chart for these data is shown in Figure 3.17. The leftmost dot represents the smallest value, the leftmost vertical line represents Q_1, the middle vertical line represents Q_2, the rightmost vertical line represents Q_3, and the rightmost dot represents the largest value. The extreme lack of symmetry in this distribution is reflected here in the facts that (1) the middle vertical line is noticeably closer to the leftmost than to the rightmost vertical line and (2) the length of the right-hand whisker is considerably greater than that of the left-hand whisker.

To apply this procedure to the order-filling data, we will present once again part (c) of Figure 3.13, shown below as part (a) of Figure 3.18, and then show it in slightly rearranged form as part (b) of this same figure—namely, with the leaf values in arrayed order. We quickly see that the smallest value is 13. Next we compute

$$\frac{n+1}{4} = \frac{76}{4} = 19$$

Thus, the first quartile is the 19th value, namely, 25 minutes.

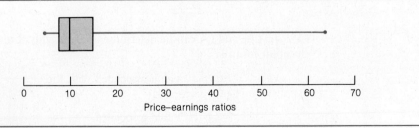

Figure 3.17. Box-and-whiskers chart of the price-earnings data presented in Table 3.2.

(a) Stem–and–leaf display of the data for order–filling times (originally presented as (c) of Figure 3.13)

```
1 | 4 3 4 9 8                                      (5)
2 | 5 4 3 4 5 3 4 4 0 1 4 4 1 8 2                  (15)
3 | 1 2 2 7 9 4 6 4 1 5 3 6 2 7 6 1 8 1 2 7        (20)
4 | 1 1 1 0 9 2 1 3 2 6 9 9 3 5                    (14)
5 | 2 4 5 5 0 4 7 5 2 7 7                          (11)
6 | 1 8 1 5 7 5 7                                  (7)
7 | 5 2 6                                          (3)
```

(b) Data in (a) in array form

```
1 | 3 4 4 8 9                                      (5)
2 | 0 1 1 2 3 3 4 4 4 4 4 4 5 5 8                  (15)
3 | 1 1 1 1 2 2 2 2 3 4 4 5 6 6 6 7 7 7 8 9        (20)
4 | 0 1 1 1 1 2 2 3 3 5 6 9 9 9                    (14)
5 | 0 2 2 4 4 5 5 5 7 7 7                          (11)
6 | 1 1 5 5 7 7 8                                  (7)
7 | 2 5 6                                          (3)
```

Figure 3.18. Stem-and-leaf displays for data on order-filling times presented in original and array form.

We then calculate that

$$\frac{n+1}{2} = \frac{76}{2} = 38$$

So the second quartile is the 38th number—in this case, 37 minutes.
Then we see that

$$\frac{3(n+1)}{4} = \frac{3(76)}{4} = 57$$

The 57th observation is found to be 52 minutes. Finally, the largest value is 76.

The resulting box-and-whiskers chart is shown in Figure 3.19. It resembles the one for the P-E ratios somewhat in that the middle vertical line favors the lower end of the box and the right-hand whisker is longer than the left-hand whisker. However, the departure from symmetry of this distribution is much less extreme than that for the P-E ratios.

The shape of a frequency distribution is partially reflected in the corresponding box-and-whiskers chart. Figure 3.20 shows the box-and-whiskers counterparts of the smoothed frequency distributions presented earlier in Figure 3.8.

*3.6 Bivariate and Trivariate Frequency Distributions

A two-way, or bivariate, frequency distribution is useful when one desires an overview of where the frequencies tend to cluster when two separate variables are considered. Certainly, if the analyst in our example about the price-earnings ratios (Illustrative Case 3.2) were interested in determining what factors influence a stock's price-earnings ratio, she would not get very far by limiting her analysis to a univariate frequency distribution of P-E ratios.

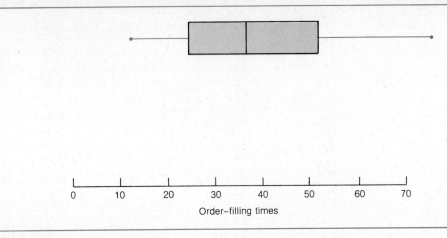

Figure 3.19. Box-and-whisker chart of the order-filling data as presented in part *(b)* of Figure 3.18.

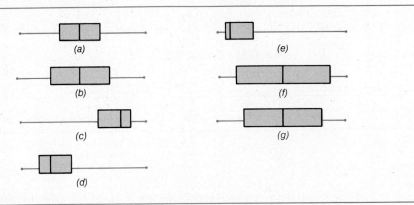

Figure 3.20. Box-and-whisker charts corresponding to frequency distribution shapes shown in Figure 3.8: *(a)* bell-shaped or normal, *(b)* bimodal, *(c)* skewed left, *(d)* skewed right, *(e)* J-shaped, *(f)* U-shaped, and *(g)* uniform.

Suppose that our portfolio manager wonders whether high P-E ratios tend to be associated with high earnings. A sensible procedure would be for her to collect frequency distribution data not only for her 200 price-earnings ratios but for the corresponding earnings per share as well, as displayed in Table 3.6. Notice that the individual frequency distributions are open-ended, despite our earlier disparaging advice regarding this kind of distribution. Also, a class interval of 3, rather than 7, has been used in those parts of the distribution where the P-E ratios tend to be tightly bunched. These decisions, admittedly arbitrary, were motivated by the fact that, in a bivariate distribution, the number of observations associated with *one class* of a specific univariate distribution will be spread over *all classes* of the other univariate distribution. Unless suitable precautions are taken, the frequencies will be spread too thin.

Let us begin our discussion of how to construct a bivariate frequency distribution by examining the partially completed one in Table 3.7. One of the first things you should notice is that in the right-most column labeled "Total" are listed the number of P-E ratios that fall

Table 3.6 Two Univariate Frequency Distributions to be Combined into a Single Bivariate Frequency Distribution

Part I		Part II	
P-E Ratios	Frequency	Earnings per Share (in Dollars)	Frequency
3 and under 6	16	0.00 and under 1.00	31
6 and under 9	68	1.00 and under 2.00	73
9 and under 12	43	2.00 and under 3.00	57
12 and under 15	22	3.00 and under 4.00	23
15 and over	51	4.00 and over	16
Total	200	Total	200

within the range of "3 and under 6," then within the range of "6 and under 9," and so forth. This is simply the list of frequencies from the univariate distribution labeled Part I in Table 3.6.

Similarly, in the row labeled "Total" at the bottom of Table 3.7, the 200 observations are distributed according to the earnings-per-share variable, as presented in Part II of Table 3.6.

Now let us consider the 25 spaces, or *cells*, as they are called, in the body of the table. How should we determine the number of observations belonging to each cell? We proceed by asking ourselves: How many of the 16 observations associated with P-E ratios of 3 and under 6 are also associated with earnings of $0 and under $1.00? Then by asking: How many of the 16 observations associated with P-E ratios of 3 and under 6 are also associated with earnings of $1.00 and under $2.00? And so forth. To answer such questions, we must return to the original data, since definite answers cannot be determined from the two univariate frequency distributions alone.

Table 3.7 Partially Completed Bivariate Frequency Distribution of Price-Earnings Ratios and Per-Share Earnings for 200 Common Stocks

Price-Earnings Ratios	Earnings per Share (in Dollars)					Column ↓
	0 and Under 1.00	1.00 and Under 2.00	2.00 and Under 3.00	3.00 and Under 4.00	4.00 and Under	Total
3 and under 6						16
6 and under 9						68
9 and under 12						43
Row→ 12 and under 15						22
15 and over						51
Total	31	73	57	23	16	200

Cell

Table 3.8 Completed Bivariate Frequency Distribution of Price-Earnings Ratios and Per-Share Earnings for 200 Common Stocks

Price-Earnings Ratios	Earnings per Share (in Dollars)					
	0 and Under 1.00	1.00 and Under 2.00	2.00 and Under 3.00	3.00 and Under 4.00	4.00 and Under	Total
3 and under 6	1	8	4	2	1	16
6 and under 9	5	19	28	7	9	68
9 and under 12	10	13	13	6	1	43
12 and under 15	4	9	4	3	2	22
15 and over	11	24	8	5	3	51
Total	31	73	57	23	16	200

Table 3.8 shows the completed bivariate frequency distribution. The process of classifying elementary units jointly on two or more variables is called *cross classification*. Cross classification can be used to determine whether two or more variables are related, as will be demonstrated shortly.

Relative Bivariate Frequency Distributions

The usefulness of a bivariate distribution can be increased if the actual cell frequencies are converted to relative frequencies.

> An important general rule when a visual impression is desired to determine whether a relationship exists between two variables is: *Compute percentages across classes of the dependent variable.*

> The *dependent variable*, as we use the term, is simply the variable of fundamental interest in a statistical investigation. Other variables in the same investigation are called *independent variables*.

Table 3.9 shows relative frequencies obtained by following the above rule. Because the correct interpretation of the rule is crucial, let us dwell on it a little. Looking again at Table 3.8, we see that the total number of observations associated with the column headed

Table 3.9 A Relative Bivariate Frequency Distribution (Percentages Based on Column Totals)

Price-Earnings Ratios	Earnings per Share (in Dollars)					
	0 and Under 1.00	1.00 and Under 2.00	2.00 and Under 3.00	3.00 and Under 4.00	4.00 and Over	Total
2 and under 6	3.2	11.0	7.0	8.7	6.2	8.0
6 and under 9	16.1	26.0	49.1	30.4	56.2	34.0
9 and under 12	32.3	17.8	22.8	26.1	6.2	21.5
12 and under 15	12.9	12.3	7.0	13.0	12.5	11.0
15 and over	35.5	32.9	14.0	21.7	18.8	25.5
Total	100.0 (31)	100.0 (73)	99.9 (57)	99.9 (23)	99.9 (16)	100.0 (200)

"$0 and Under $1.00" is 31. Since the P-E ratios are viewed here as the variable of fundamental interest—that is, as the dependent variable—the value 31 is divided into the values above it in the same column, namely, 1, 5, 10, 4, and 11. The resulting quotients are, respectively, .032, .161, .323, .129, and .355. These quotients, presented in percentage form, are shown in the column "$0 and Under $1.00" of Table 3.9. Similarly, in Table 3.8 the column total for the "$1.00 and Under $2.00" "Earnings per Share" data is 73; 73 is divided into the frequency values above it in the same column—namely, 8, 19, 13, 9, and 24. The resulting quotients are .110, .260, .178, .123, and .329, respectively. The same procedure is repeated for the "$2.00 and Under $3.00," "$3.00 and Under $4.00," and "$4.00 and Over" columns.

We find in Table 3.9 no evidence of a relationship between price-earnings ratios and earnings per share. If these variables were really related in the direct manner that one might reasonably assume, the relative frequencies in the lower left-hand corner would tend to be small and those in the lower right-hand corner large. Figure 3.21 elaborates on this point.

Trivariate Frequency Distributions

The general principles for presenting and interpreting bivariate distributions apply also to frequency distributions with three or more variables represented. For example, the portfolio manager in our example might wish to categorize common stocks according to P-E ratios, earnings per share, and whether or not a dividend is paid. Such a trivariate relative frequency distribution, as it is called when three variables are considered, is presented in Table 3.10. Notice that the independent variable, dividend versus no dividend, is "nested" under class headings for the other independent variable, earnings per share. Again, percents are computed across classes of the dependent variable, the P-E ratios. Dividends do appear to affect P-E ratios to some degree. Table 3.10 indicates that, for any given class of earnings, P-E ratios associated with dividend-paying stocks tend to be somewhat greater than for stocks

Earnings per Share (in Dollars)

Price–Earnings Ratios	0 and under 1.00	1.00 and under 2.00	2.00 and under 3.00	3.00 and under 4.00	4.00 and over
3 and under 6	H	L	L	L	L
6 and under 9	L	H	H	L	L
9 and under 12	L	H	H	H	L
12 and under 15	L	L	H	H	L
15 and over	L	L	L	L	H
Total	100.0	100.0	100.0	100.0	100.0

(a) Example of a relative bivariate frequency distribution when the variables are *directly* related. High percentages (H) form a pattern descending toward the right.

Earnings per Share (in Dollars)

Price–Earnings Ratios	0 and under 1.00	1.00 and under 2.00	2.00 and under 3.00	3.00 and under 4.00	4.00 and over
3 and under 6	L	L	H	L	H
6 and under 9	H	H	H	L	H
9 and under 12	L	L	L	H	L
12 and under 15	L	H	L	L	L
15 and over	H	H	L	L	H
Total	100.0	100.0	100.0	100.0	100.0

(b) Example of a relative bivariate frequency distribution when the variables are unrelated. The high (H) and low (L) percentages form no clear pattern.

Earnings per Share (in Dollars)

Price–Earnings Ratios	0 and under 1.00	1.00 and under 2.00	2.00 and under 3.00	3.00 and under 4.00	4.00 and over
3 and under 6	L	L	L	L	H
6 and under 9	L	L	H	H	H
9 and under 12	L	H	H	H	L
12 and under 15	H	H	L	L	L
15 and over	H	H	L	L	L
Total	100.0	100.0	100.0	100.0	100.0

(c) Example of a relative bivariate frequency distribution when the variables are inversely related (i.e., as one variable goes up the other tends to go down, and vice versa). High percentages (H) form a pattern ascending toward the right.

Figure 3.21. How a relative bivariate frequency distribution would be expected to look if (a) the variables are directly related, (b) the variables are unrelated, and (c) the variables are inversely related.

Table 3.10 Trivariate Relative Frequency Distribution Showing the Effect of (1) Per-Share Earnings and (2) Dividend Status on Price-Earnings Ratios for 200 Common Stocks

Price Earnings Ratios	*Earnings per Share (in Dollars) and Dividend Status*									
	0 and Under 1.00		*1.00 and Under 2.00*		*2.00 and Under 3.00*		*3.00 and Under 4.00*		*4.00 and Over*	
	Divi-dend	No Divi-dend	Divi-dend	No Divi-dend	Divi-dend	No Divi-dend	Divi-dend	No Divi-dend	Divi-dend	No Divi-dend
3 and under 6	0.0	7.1	5.6	16.2	0.0	11.4	0.0	18.2	0.0	10.0
6 and under 9	11.8	21.4	19.4	32.4	36.4	57.1	16.7	45.5	16.7	80.0
9 and under 12	35.3	28.6	19.4	16.2	27.3	20.0	16.7	36.4	0.0	10.0
12 and under 15	11.8	14.3	13.9	10.8	13.6	2.9	25.0	0.0	33.3	0.0
15 and over	41.2	28.6	41.7	24.3	22.7	8.6	41.7	0.0	50.0	0.0
Total	100.1	100.0	100.0	99.9	100.0	100.0	100.1	100.1	100.0	100.0

paying no dividend. Consider, for example, the earnings class "$0 and Under $1.00." Over 88% of the dividend-paying stocks have P-E ratios of 9 or greater, whereas only 71.5% of the stocks paying no dividend have comparably high price-earnings ratios. A similar pattern can be seen in the other columns as well.

Seldom does it pay to push tabular presentation too far when attempting to determine whether variables are related. Much more precise methods are available for realizing this goal; some of these are presented later in this book. Generally speaking, a trivariate table is about as complicated a tabular device as one will ever want to use.

*3.7 Computer-Assisted Data Organization

Most of the data-organizing techniques described in this chapter can be employed with surprising ease through use of one of several available computer software packages. The following illustrative case provides a glimpse of one organization's use of such a package.

ILLUSTRATIVE CASE 3.3

The gasoline shortage of the mid-1970s prompted the marketing department of a certain university to conduct a study designed to determine whether driving habits had been altered as a result of the shortage. The researchers asked questions about the influence of the shortage on (1) driving to banks, (2) driving to convenience stores, (3) driving to department stores, (4) driving for drugs or prescriptions, (5) driving to gasoline stations, (6) driving to grocery stores, (7) driving for pleasure only, and (8) driving to service-connected businesses such as laundries, beauty parlors, and so forth. The computer output shown in Figure 3.22 pertains to number (7) on this list—driving for pleasure only.

The parts of the questionnaire for which the computer output provides information were as follows:

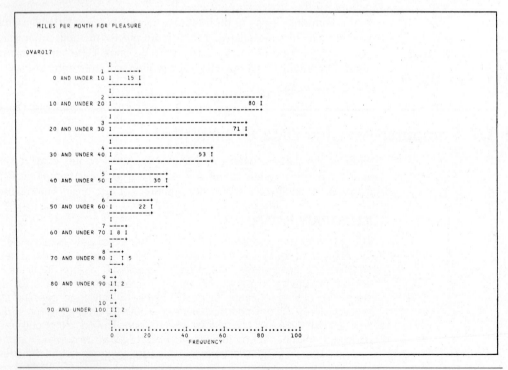

```
      MILES PER MONTH FOR PLEASURE

0- - - - - - - - - - - - - - - - - - - C R O S S T A B U L A T I O N   O F  - - - - - - - - - - - - - - - -
      VAR004                                          BY  VAR016
0 - - - - - - - - - - - - - - - - - - - - - - - - - - - - - - - - - - - - - - - - - - PAGE 1 OF 1
0                    VAR016
           COUNT  :
           ROW PCT :G.I.    S.I.    R.S    S.D.    G.D.    N.A.     ROW
           COL PCT :                                               TOTAL
           TOT PCT :   1:     2:     3:     4:     5:     9:
      VAR004  -------:------+------+------+------+------+------+
              1 :      5 :    3 :   97 :   18 :   17 :    3 :    143
      MALE      :    3.5 :  2.1 : 67.8 : 12.6 : 11.9 :  2.1 :   49.7
                :   50.0 : 75.0 : 55.7 : 32.7 : 48.6 : 30.0 :
                :    1.7 :  1.0 : 33.7 :  6.3 :  5.9 :  1.0 :
                -------:------+------+------+------+------+------+
              2 :      3 :    1 :   58 :   25 :   12 :    6 :    105
      FEMALE    :    2.9 :  1.0 : 55.2 : 23.8 : 11.4 :  5.7 :   36.5
                :   30.0 : 25.0 : 33.3 : 45.5 : 34.3 : 60.0 :
                :    1.0 :   .3 : 20.1 :  8.7 :  4.2 :  2.1 :
                -------:------+------+------+------+------+------+
              9 :      2 :      :   19 :   12 :    6 :    1 :     40
      N.A.      :    5.0 :      : 47.5 : 30.0 : 15.0 :  2.5 :   13.9
                :   20.0 :      : 10.9 : 21.8 : 17.1 : 10.0 :
                :     .7 :      :  6.6 :  4.2 :  2.1 :   .3 :
                -------:------+------+------+------+------+------+
           COLUMN     10      4    174     55     35     10     288
           TOTAL     3.5    1.4   60.4   19.1   12.2    3.5   100.0


      0VAR017
      0                                           CUMULATIVE
           VALUE LABEL          VALUE  FREQUENCY  PERCENT  PERCENT

      0 AND UNDER 10              1       15       5.2     5.2
      10 AND UNDER 20             2       80      27.8    33.0
      20 AND UNDER 30             3       71      24.7    57.6
      30 AND UNDER 40             4       53      18.4    76.0
      40 AND UNDER 50             5       30      10.4    86.5
      50 AND UNDER 60             6       22       7.6    94.1
      60 AND UNDER 70             7        8       2.8    96.9
      70 AND UNDER 80             8        5       1.7    98.6
      80 AND UNDER 90             9        2        .7    99.3
      90 AND UNDER 100           10        2        .7   100.0
                                       -------   -------
                            TOTAL        288     100.0
```

Figure 3.22. Sample computer output related to Illustrative Case 3.3.

Approximately how many miles per month do you now travel for pleasure only? _____

We would now like to find out about any change that may have occurred since the beginning of the gasoline shortage. Would you say that the distance traveled per month for pleasure only has: (Check one)

Greatly increased?   _____
Slightly increased?   _____
Remained the same? _____
Slightly decreased?   _____
Greatly decreased?   _____

Are you male or female?   (Check one)   Male_____   Female_____

3.8 Summary

Remarkable though it is in so many ways, the human mind is limited in its ability to extract meaningful messages from large sets of unorganized data. It fares much better with data that have been organized and simplified.

A first step toward organizing and simplifying a set of data is the array. In an array the values are arranged in order of size—largest to smallest or smallest to largest.

Even more useful is a data condensation method such as a frequency distribution, a box-and-whiskers plot, or (in the case of two variables being considered together) bivariate cross classification. These techniques give the analyst a quick, easy-to-grasp overview of the data of interest.

A standard frequency distribution displays classes of the (quantitative) variable and the corresponding class frequencies. It is constructed by: (1) arraying the data, (2) determining the appropriate number of classes to use, (3) determining the appropriate class interval, (4) setting up the classes, (5) counting and recording the number of observations associated with each class, and (6) presenting the results graphically in the form of a histogram or a frequency polygon. Frequency distributions can display the class frequencies in their original form (i.e., straight counts), in terms of relative frequencies, or in terms of cumulative frequencies.

A box-and-whiskers plot makes use of only five values from a set of arrayed data: the lowest value, the first-quartile value (the value one-fourth the way into the array), the second-quartile value (the value one-half the way into the array), the third-quartile value (the value three-fourths the way into the array), and the highest value. A box-and-whiskers plot plays a less fundamental and less ubiquitous role in the broad discipline of statistical analysis than the frequency distribution. However, it does represent a convenient way of describing a set of data briefly and meaningfully.

When relationships among variables are of interest, cross classification can be most helpful. In bivariate cross classification, for example, the number of observations associated with one class of a specific univariate distribution is spread over all classes of the other univariate distribution. Bivariate cross-classification frequency distributions may be either absolute or relative. Usually, a relative bivariate frequency distribution, where relative frequencies are obtained by dividing across classes of the dependent variable, is the most useful choice from the standpoint of revealing any relationship that may exist between the variables of interest.

Many computer software packages are available that provide a wealth of information about one's data very quickly.

Notes

1. The suggested number of classes in this table came from application of a formula known as Sturges' rule, which states

$$\text{Number of classes} = 1 + (3.3)(\log n)$$

   Results have been rounded to the nearest whole number.
2. For more on box-and-whiskers charts, stem-and-leaf displays, and numerous other innovative aids to exploratory data analysis, see John W. Tukey, *Exploratory Data Analysis* (Reading, Mass.: Addison-Wesley, 1977.)

Terms Introduced in This Chapter

array (p. 80)
bell-shaped frequency distribution (p. 87)
bimodal frequency distribution (p. 87)
bivariate frequency distribution (p. 100)
box-and-whiskers plot (p. 95)
class frequency (p. 79)
class interval (p. 83)
class limits (p. 83)
class midpoint (p. 86)
cross classification (p. 102)
cumulative frequency distribution (p. 95)

dependent variable (p. 102)
frequency distribution (p. 79)
frequency polygon (p. 85)
histogram (p. 84)
independent variable (p. 102)
interquartile range (p. 96)
J-shaped frequency distribution (p. 87)
left-skewed frequency distribution (p. 87)
normal frequency distribution (p. 87)
open-ended frequency distribution (p. 88)
qualitative frequency distribution (p. 94)

quartiles (p. 96)
first quartile (p. 95)
second quartile (p. 95)
third quartile (p. 95)
relative frequency distribution (p. 88)
right-skewed frequency distribution (p. 87)
stem-and-leaf display (p. 92)
trivariate frequency distribution (p. 103)
Sturges' rule (p. 108)
uniform frequency distribution (p. 87)
U-shaped frequency distribution (p. 87)

Formulas Introduced in This Chapter

Where *LCL* represents the lower class limit and *UCL* the true upper class limit of a given class in a frequency distribution, the class midpoint is

$$\frac{LCL + UCL}{2}$$

Position-location formulas for determining the first, second, and third quartiles (Q_1, Q_2, and Q_3, respectively):

$$Q_1 = \text{Value associated with the } \frac{n + 1}{4} \text{ ordered observation}$$

$$Q_2 = \text{Value associated with the } \frac{n + 1}{2} \text{ ordered observation}$$

$$Q_3 = \text{Value associated with the } \frac{3(n + 1)}{4} \text{ ordered observation}$$

where *n* represents the number of observations making up the data set.

Sturges' rule for determining the number of classes to use when constructing a frequency distribution, where *log* refers to the common logarithm and n is the number of observations making up the data set:

Number of classes = $1 + (3.3)(\log n)$

Questions and Problems

3.1 Explain the meaning of each of the following:
  a. Frequency distribution
  b. Univariate
  c. Class frequency
  d. Array
  e. Class interval
  f. Sturges' rule
  g. Histogram

3.2 Explain the meaning of each of the following:
  a. Frequency polygon
  b. Open-ended frequency distribution
  c. Relative frequency distribution
  d. Stem-and-leaf display
  e. Qualitative frequency distribution
  f. Cumulative frequency distribution
  g. Box-and-whiskers plot

3.3 Explain the meaning of each of the following:
  a. "Less than" distribution
  b. Quartile
  c. Interquartile range
  d. Bivariate frequency distribution
  e. Trivariate frequency distribution
  f. Cross classification
  g. Dependent variable
  h. Independent variable

3.4 Distinguish between:
  a. Class interval and class limit
  b. Constant class interval and unequal class intervals
  c. Histogram and frequency polygon
  d. Bell-shaped frequency distribution and skewed frequency distribution
  e. Relative frequency distribution and qualitative frequency distribution

3.5 Distinguish between:
  a. The method for showing a continuous quantitative variable and the method for showing a discrete quantitative variable in chart form
  b. The method for showing a quantitative variable and the method for showing a qualitative variable in chart form
  c. First quartile, Q_1, and third quartile, Q_3
  d. Relative univariate frequency distribution and relative bivariate frequency distribution
  e. Stem-and-leaf display and standard frequency distribution (as described in section 3.2)

3.6 Distinguish between:
  a. "Less than" distribution and "or more" distribution
  b. Box-and-whiskers plot and standard frequency distribution (as described in section 3.2)
  c. Median and interquartile range
  d. Bivariate and trivariate frequency distributions
  e. Dependent variable and independent variable

3.7 Indicate which of the following statements you agree with and which you disagree with, and defend your opinions:
  a. Some information about a set of data is lost when use is made of a frequency distribution.

b. A histogram is always preferable to a frequency polygon when frequency-distribution data are to be presented in graphic form.

c. Use of a bivariate frequency distribution is limited to quantitative variables.

d. The shape of a smoothed frequency distribution is reflected partially in a box-and-whiskers plot.

e. Identification of the first, second, and third quartiles serves to divide a set of arrayed data into five equal parts.

3.8 Indicate whether each of the distributions in Table 3.11 is a univariate quantitative frequency distribution and defend your conclusions.

Problem 1.16 of Chapter 1 alluded to two sets of fan belt longevity data. One of these sets of data is displayed in Table 3.12. The next four problems pertain to these data.

✔ 3.9 a. Organize the data in Table 3.12 into array form with numbers running from small to large. Show your array.

b. Use Table 3.3 to determine the number of classes to be used in the construction of a standard frequency distribution.

Table 3.11   Three Kinds of Frequency Distributions

Part I: Geographical Distribution of Nonresident Skiers at a Colorado Ski Resort During a Specific Week

Area of U.S. Where Skier Resides	Number of Skiers
Southwest-Pacific	384
South Central	684
Midwest	1536
Atlantic/South	276
Other	1016
Total	3896

Part II: Age Distribution of Shoppers at the Fixit-City Hardware Store During a Specific Week

Age of Shopper	Number of Shoppers
5 and under 15	6
15 and under 25	22
25 and under 35	769
35 and under 45	1043
45 and under 55	1286
55 and under 65	522
65 and under 75	47
Total	3695

Part III: Distribution of Families According to the Number of Tubes of Super-Dazzle Toothpaste in Their Homes at the Time of a Spot Check

Number of Tubes of Super-Dazzle Toothpaste in Home	Number of Homes
0	5997
1	426
2	53
3	4
4 or more	2
Total	6482

c. Subtract the smallest value in the array from the largest value and divide the result by the result of part *b* to determine the (constant) class interval. (Round to the nearest whole number.)

d. Using 35.0 as the lower class limit of the beginning class, present the data as a frequency distribution in tabular form.

e. Present the frequency distribution obtained in connection with part *d* in the form of a histogram.

f. Indicate with an $\overline{X}$ about where on the horizontal axis of your histogram the average value of 50,722 miles would be found.

g. Determine the first-, second-, and third-quartile values, and show about where each would be found on the horizontal axis of your histogram. Use the symbols Q_1, Q_2, and Q_3.

h. Would you say that the data are distributed in an approximately symmetrical or in a skewed manner?

✔ **3.10** Refer to problem 3.9. (Problem 3.9 must have been worked before you can work this one.) Convert the frequency distribution you obtained in connection with part *d* of problem 3.9 into a relative frequency distribution and tell under what circumstances this form of data organization might be a more useful way of presenting the fan-belt longevity data than the form obtained in connection with part *d* of problem 3.9.

✔ **3.11** Refer to problem 3.9. (Problem 3.9 must have been worked before you can work this one.)

a. Convert the frequency distribution you obtained in connection with part *d* of problem 3.9 into a "less than" cumulative frequency distribution. Present your results in both tabular and graphic forms.

b. Convert the frequency distribution you obtained in connection with part *d* of problem 3.9 into an "or more" cumulative frequency distribution. Present your results in both tabular and graphic forms.

c. Tell under what circumstances one might prefer having the data in the form obtained in connection with part *a* of this problem over an ordinary frequency distribution. Then do the same for the form obtained in connection with part *b* of this problem.

✔ **3.12** a. Prepare a box-and-whiskers chart of the data in Table 3.12. Comment on any evidence of symmetry, or lack of it, shown by your chart.

b. Prepare a stem-and-leaf display of the data in Table 3.12. (For simplicity, you may round to the nearest whole number.)

The next four problems all pertain to the adjusted gross income data displayed in Table 3.13.

Table 3.12 Number of Miles of Continuous (Simulated) Driving Before Fan Belt Snapped, Ajax Brand Belts (in Thousands)

48.4	37.8	55.4	52.8	41.6	65.4	48.4	37.4	55.8	39.2
41.6	44.8	65.4	44.2	61.0	38.8	55.8	45.2	41.0	63.6
55.8	59.6	37.4	39.2	41.0	47.0	39.2	44.8	60.6	48.6
61.0	55.8	38.8	39.6	55.4	49.0	40.6	59.4	44.8	61.0
39.2	58.8	45.2	47.4	65.4	55.4	60.6	52.8	55.8	65.4
41.0	48.6	47.0	63.6	44.8	44.2	37.8	39.2	51.4	58.8
40.6	51.4	44.8	55.2	57.8	39.6	59.6	47.4	55.4	65.4
55.4	55.2	49.0	53.2	58.8	63.6	55.8	55.2	38.8	49.0
60.6	58.8	59.4	48.6	51.4	53.2	48.6	48.6	44.8	39.6
65.4	57.8	55.4	39.8	58.8	39.8	55.2	55.4	55.4	39.8

Table 3.13 Adjusted Monthly Gross Income, Merwin Mortuary
Company (in Dollars)

16,200	14,600	19,400	14,800	16,600
9,500	14,600	16,350	26,100	17,700
8,550	15,150	19,050	18,900	19,850
8,750	15,300	14,000	17,700	20,600
9,100	20,550	18,950	23,300	25,050
6,150	10,300	17,500	11,950	22,850
12,650	13,950	18,750	20,550	25,650
11,150	11,750	24,400	15,700	18,500
8,300	20,350	15,200	21,400	16,600
18,250	16,100	13,600	22,400	19,500

3.13 a. Organize the data in Table 3.13 into array form with numbers running from small to large. Show your array.
b. Use Table 3.3 to determine the number of classes to use in the construction of a standard frequency distribution.
c. Subtract the smallest value in the array from the largest value and divide the result by the result of part *b* to determine the (constant) class interval. (Round to the nearest whole number.)
d. Using $6000 as the lower class limit of the beginning class, present the data as a frequency distribution in tabular form.
e. Present the frequency distribution obtained in connection with part *d* in the form of a histogram.
f. Indicate with an $\overline{X}$ about where on the horizontal axis of your histogram the average value of $16,683 would be found.
g. Determine the first-, second-, and third-quartile values, and show about where each would be found on the horizontal axis of your histogram. Use the symbols Q_1, Q_2, and Q_3.
h. Would you say that the data are distributed in an approximately symmetrical or in a skewed manner?

3.14 Refer to problem 3.13. (Problem 3.13 must have been worked before you can work this one.) Convert the frequency distribution you obtained in connection with part *d* of problem 3.13 into a relative frequency distribution and tell under what circumstances this form of data organization might be a more useful way of presenting the adjusted gross income data than the form obtained in connection with part *d* of problem 3.13.

3.15 Refer to problem 3.13. (Problem 3.13 must have been worked before you can work this one.)
a. Convert the frequency distribution you obtained in connection with part *d* of problem 3.13 into a "less than" cumulative frequency distribution. Present your results in both tabular and graphic forms.
b. Convert the frequency distribution you obtained in connection with part *d* of problem 3.13 into an "or more" cumulative frequency distribution. Present your results in both tabular and graphic forms.
c. Tell under what circumstances one might prefer having the data in the form obtained in connection with part *a* of this problem over an ordinary frequency distribution. Then do the same for the form obtained in connection with part *b* of this problem.

3.16 a. Prepare a box-and-whiskers chart of the data in Table 3.13. Comment on any evidence of symmetry, or lack of it, shown by your chart.
b. Prepare a stem-and-leaf display of the data in Table 3.13. (For simplicity, you may round to the nearest thousand dollars. That is, the stem values would be just 0, 1, and 2.

Table 3.14 Average Weekly Catsup Sales (Number of Bottles)

45	30	41	34	52
51	47	35	38	40
44	38	30	32	28
71	32	21	27	40
54				

The next four problems pertain to the catsup sales data shown in Table 3.14.

3.17 a. Organize the data in Table 3.14 into array form with numbers running from small to large.
  b. Use Table 3.3 to determine the number of classes to use in the construction of a standard frequency distribution.
  c. Subtract the smallest value in the array from the largest value and divide the result by the result of part *b* to determine the (constant) class interval. (Round to the nearest whole number.)
  d. Using 20 as the lower class limit of the beginning class, present the data as a frequency distribution in tabular form. Was it necessary to use the "and under" device when setting up the class limits? Explain.
  e. Present the frequency distribution obtained in connection with part *d* in the form of a histogram.
  f. Indicate with an $\overline{X}$ about where on the horizontal axis of your histogram the average of 39.5 bottles would be found. Does it bother you that this is an average of several averages? Discuss.
  g. Determine the first-, second-, and third-quartile values and show about where on the horizontal axis of your histogram each would be found. Use the symbols Q_1, Q_2, and Q_3.
  h. Would you say that the data are distributed in an approximately symmetrical or in a skewed manner?

3.18 Refer to problem 3.17. (Problem 3.17 must have been worked before you can work this one.) Convert the frequency distribution you obtained in connection with part *d* of problem 3.17 into a relative frequency distribution and tell under what circumstances this form of data organization might be a more useful way of presenting the catsup sales data than the form obtained in connection with part *d* of problem 3.17.

3.19 Refer to problem 3.17. (Problem 3.17 must have been worked before you can work this one.)
  a. Convert the frequency distribution you obtained in connection with part *d* of problem 3.17 into a "less than" cumulative frequency distribution. Present your results in both tabular and graphic forms.
  b. Convert the frequency distribution you obtained in connection with part *d* of problem 3.17 into an "or more" cumulative frequency distribution. Present your results in both tabular and graphic forms.
  c. Tell under what circumstances one might prefer having the data in the form obtained in connection with part *a* of this problem over an ordinary frequency distribution. Then do the same for the form obtained in connection with part *b* of this problem.

3.20 a. Prepare a box-and-whiskers chart of the data in Table 3.14. Comment on any evidence of symmetry, or lack of it, shown by your chart.
  b. Prepare a stem-and-leaf display of the data in Table 3.14.

3.21 Figure 3.23 shows some data on delinquent accounts in histogram form. Can you think of a more useful way of displaying these data graphically? Explain.

 3.22 Figure 3.24 shows some duration-of-employment data in histogram form. Can you think of a more useful way of displaying these data graphically? Explain.

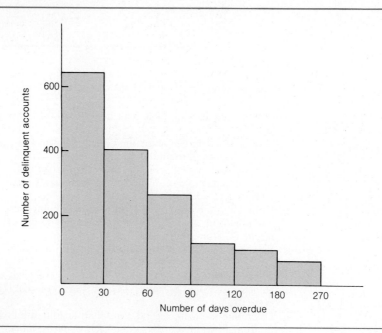

Figure 3.23. Histogram of number of days that 1655 delinquent accounts are overdue in Day Department Stores.

Figure 3.24. Histograms showing duration of employment of 65 marketing employees (solid line) and 95 manufacturing employees (dashed line) of the Drexel Business Machines Company.

3 Data Organization and Exploratory Data Analysis

Table 3.15 Bonuses Paid Salespersons with Territories in Eastern States Versus Bonuses Paid Salespersons with Territories in Western States, ABC Corporation

Size of Bonus (in Dollars)	Number of Salespersons Western United States	Number of Salespersons Eastern United States
100 and under 200	2	4
200 and under 300	3	7
300 and under 400	15	28
400 and under 500	20	20
500 and under 600	32	40
600 and under 700	10	85
700 and under 800	7	43
800 and under 900	5	30
900 and under 1000	1	15
1000 and under 1100	0	8
Total	95	280

The next three problems all pertain to the sales-bonus data shown in Table 3.15.

3.23 a. Plot frequency polygons for both sets of sales-bonus data using the same graph.
  b. Describe in words the main differences and similarities between the two frequency polygons.

3.24 a. Convert both frequency distributions shown in Table 3.15 into relative frequency distributions.
  b. Plot relative-frequency polygons for both sets of sales-bonus data using the same graph.
  c. Describe in words the main differences and similarities between the two relative-frequency polygons.
  d. Do you prefer the use of absolute frequencies or relative frequencies for purposes of comparing these two sets of data graphically? Give reasons for your choice.

3.25 a. Convert both frequency distributions shown in Table 3.15 into "less than" cumulative frequency distributions.
  b. Show both "less than" frequency distributions graphically using the same graph.
  c. Describe in words the main differences and similarities between the two "less than" cumulative distributions.
  d. Would your graphic comparison have been more meaningful if you had used *relative* "less than" cumulative distributions? Tell why or why not.

The next three problems all pertain to the financial data shown in Table 3.16.

3.26 a. Plot histograms for each set of data using a single graph.
  b. Plot frequency polygons for each of the same three sets of data using a single graph.
  c. Compare the merits and shortcomings of the frequency polygon approach for presenting these data.
  d. Using the frequency polygons obtained in connection with part *b*, describe in words the main differences and similarities of the three distributions.

3.27 a. Convert all three frequency distributions shown in Table 3.16 into relative frequency distributions.

Table 3.16 Net Income as a Percent of Stockholders' Equity for 40 Large Industrial Companies, 40 Large Commercial Banks, and 40 Large Retailers

Net Income as a Percent of Stockholders' Equity	Industrial	Commercial Banking	Retailing
6 and under 9	0	0	10
9 and under 12	0	17	12
12 and under 15	0	21	10
15 and under 18	7	2	5
18 and under 21	19	0	3
21 and under 24	5	0	0
24 and under 27	5	0	0
27 and under 30	1	0	0
30 and under 33	1	0	0
33 and under 36	1	0	0
36 and under 39	1	0	0
Total	40	40	40

b. Plot relative frequency polygons for all three sets of data using the same graph.

c. Describe in words the main differences and similarities of the three relative frequency polygons.

3.28 a. Convert all three frequency distributions in Table 3.16 into "or more" cumulative frequency distributions.

b. Show all three "or more" distributions graphically using the same graph.

c. Describe in words the main differences and similarities of the three "or more" cumulative distributions.

d. Would your graphic comparison have been more meaningful if you had used *relative* "or more" cumulative distributions? Tell why or why not.

3.29 A marketing research study conducted for the purpose of determining whether men or women tend to watch television more frequently produced the results shown in Table 3.17.

a. Convert frequencies in Table 3.17 into relative frequencies in such a way that you can determine whether frequency of television watching is related to sex of viewer. (That is, construct a relative bivariate frequency distribution.

b. What denominator(s) did you use when calculating the percents obtained in connection with part *a*? What was the reason for your choice?

c. Is there any evidence of a statistical relationship between the two variables under study? Justify your answer.

3.30 Refer to problem 3.29. When a third variable, marital status, was added to the analysis, the results shown in Table 3.18 were obtained.

a. Present these data in the form of a *relative* trivariate frequency distribution.

b. What denominator(s) did you use when calculating the percents obtained in connection with part *a*? What was the reason for your choice?

c. Judging from the table you constructed in connection with part *a*, does there appear to be a statistical relationship between viewer category and sex? Between viewer category and marital status? Justify your answers.

Table 3.17 Sex and Viewer Category Data for a Sample of Television Viewers

Part I: Television Viewers Classified as "Frequent Watchers"

Sex	Number of Viewers
Male	235
Female	345
Total	580

Part II: Television Viewers Classified as "Moderate Watchers"

Sex	Number of Viewers
Male	420
Female	410
Total	830

Part III: Television Viewers Classified as "Infrequent Watchers"

Sex	Number of Viewers
Male	300
Female	200
Total	500

Table 3.18 Viewer-Category, Sex, and Marital Status Data for a Sample of Television Viewers

Description	Number of Viewers
Frequent watchers, male, married	135
Frequent watchers, male, single	100
Frequent watchers, female, married	135
Frequent watchers, female, single	210
Moderate watchers, male, married	250
Moderate watchers, male, single	170
Moderate watchers, female, married	200
Moderate watchers, female, single	210
Infrequent watchers, male, married	200
Infrequent watchers, male, single	100
Infrequent watchers, female, married	150
Infrequent watchers, female, single	50
Total	1910

Table 3.19 Quality-Perception and Place-of-Residence Data
for a Sample of Respondents

Description	Number of Respondents
Perceive products as high quality, eastern U.S.	710
Perceive products as high quality, western U.S.	1005
Perceive products as medium quality, eastern U.S.	2175
Perceive products as medium quality, western U.S.	2200
Perceive products as low quality, eastern U.S.	1010
Perceive products as low quality, western U.S.	525
Total	7625

3.31 A manufacturer of home appliances conducted a study to determine how consumers perceive the quality of the company's products. The breakdown for the quality-perception variable consisted of (1) high quality, (2) medium quality, and (3) low quality. In addition, management desired to know whether the geographical location of the respondent's place of residence was in any way related to quality perception. The results obtained are shown in Table 3.19.

a. Present these data in the form of a relative bivariate frequency distribution by converting frequencies into percents and presenting the results in a cross-classification table.

b. What denominator(s) did you use when calculating the percents obtained in connection with part *a*? What was the reason for your choice?

c. Is there any evidence of a statistical relationship between the two variables under study? Justify your answer.

Table 3.20 Quality-Perception, Sex, and Place-of-Residence Data
for a Sample of Respondents

Description	Number of Respondents
Perceive products as high quality, eastern U.S., male	200
Perceive products as high quality, eastern U.S., female	510
Perceive products as high quality, western U.S., male	300
Perceive products as high quality, western U.S., female	705
Perceive products as medium quality, eastern U.S., male	1000
Perceive products as medium quality, eastern U.S., female	1175
Perceive products as medium quality, western U.S., male	1100
Perceive products as medium quality, western U.S., female	1100
Perceive products as low quality, eastern U.S., male	750
Perceive products as low quality, eastern U.S., female	260
Perceive products as low quality, western U.S., male	400
Perceive products as low quality, western U.S., female	125
Total	7625

3.32 Refer to problem 3.31. When a third variable, sex of respondent, was added to the analysis, the results shown in Table 3.20 were obtained.
  a. Present these data in the form of a relative trivariate frequency distribution by converting frequencies into percents and presenting the results in a cross-classification table.
  b. What denominator(s) did you use when calculating the percents obtained in connection with part *a*? What was the reason for your choice?
  c. Judging from the table you constructed in connection with part *a*, does there appear to be a statistical relationship between quality perception and sex? Between quality perception and place of residence? Justify your answers.

✔ **3.33** Examine the top part of Figure 3.22 (p. 106), the part showing the cross-classification data. Each cell in this table contains three percent values. Which of the percents should an analyst pay most attention to if he wishes to answer the question: Does sex of respondent have a bearing on change in driving habits for pleasure since the gas shortage? Defend your choice.

3.34 A sample of 500 families was categorized according to (1) political party affiliation (head of household), (2) family income, and (3) number of automobiles owned. The results are shown in Table 3.21. Construct a separate univariate frequency distribution for each of these categories.

TAKE CHARGE

3.35 Find a business- or economics-related data set of interest to you. The data set should be made up of at least 100 quantitative observations.
  a. Using a constant class interval, show the data in the form of a standard frequency distribution.
  b. Show the data in the form of a box-and-whiskers plot.
  c. Using information obtained from parts *a* and *b*, briefly describe the appearance of your data.
Caution: Keep a copy of your data set; it will be used in connection with a later "Take Charge" assignment.

Table 3.21 Data Obtained from a Sample of 500 Households

Political Party Affiliation	Income Class	Number of Automobiles Owned	Number of Families
Democrat	$30,000 and under $40,000	2	105
Democrat	$40,000 and under $50,000	1	40
Libertarian	$20,000 and under $30,000	2	3
Republican	$40,000 and under $50,000	3	115
Democrat	$20,000 and under $30,000	2	95
Democrat	$30,000 and under $40,000	2	75
Republican	$100,000 and under $110,000	2	7
Republican	$70,000 and under $80,000	3	45
Independent	$20,000 and under $30,000	1	5
Not sure	$30,000 and under $40,000	2	10
Total			500

Computer Exercise

3.36 Refer to Table 2.5.
  a. Have the computer arrange the first 200 observations into frequency distribution form.
  b. Have the computer arrange the first 200 observations into bivariate frequency distribution form with categories of numbers of employees as the dependent variable and industry classes P, M, and O as the independent variable.

Cases for Analysis and Discussion

3.1 THE "UNIVERSAL MUTUAL LIFE INSURANCE COMPANY": PART II

Reread Part I of this case (case 2.1 at the end of Chapter 2). The results of "Universal's" survey are presented in Table 3.22.

1. Do you believe that any benefits resulted from the conversion of actual frequencies into relative frequencies? Why or why not?

2. Is there a better way of presenting the survey results? Explain.

3. Review the instructions to interviewers (Figure 2.18) and the frequency distributions in Table 3.22. How well do you think the instructions were followed by the two interviewers? Explain.

4. Review the stated goals of this research undertaking, and then project yourself into the role of a member of management in this company. Tell how these research results might be useful to you in making future company decisions. If you believe they would not be helpful, tell why.

Table 3.22 Survey Results for the "Universal Mutual Life Insurance Company" Case

Part I: Sex of Respondent		
Sex	Frequency	Relative Frequency
Male	36	.419
Female	50	.581
Total	86	1.000

Part II: Income		
Income ($000)	Frequency	Relative Frequency
0 and under 3	17	.205
3 and under 5	12	.145
5 and under 8	16	.193
8 and under 11	12	.145
11 and under 15	11	.133
15 and under 25	11	.048
25 and over	4	.048
Total	83	1.002

(continued)

Table 3.22 (*Continued*)

Part III: Age

Age	Frequency	Relative Frequency
Under 18	5	.058
18 and under 22	21	.244
22 and under 26	18	.209
26 and under 33	24	.279
33 and under 45	11	.128
Over 45	7	.081
Total	86	.999

Part IV: Question Number 1: "Do you feel that life insurance is something that everyone with an adequate income and dependents should have?"

Response	Frequency	Relative Frequency
Yes	75	.872
No	11	.128
Don't know	0	.000
Total	86	1.000

Part V: Question Number 2: "Do you feel that the image of the life insurance industry is _____ ?"

Response	Frequency	Relative Frequency
Poor	12	.140
Fair	24	.279
Getting better	15	.174
Good	22	.256
Excellent	3	.035
No feeling	10	.116
Total	86	1.000

Part VI: Question Number 3: "Do you plan on purchasing more (or some) life insurance in the future?"

Response	Frequency	Relative Frequency
Yes	40	.465
No	37	.430
Don't know	9	.105
Total	86	1.000

(*continued*)

Table 3.22 (*Concluded*)

Part VII: Question Number 4: "Have you ever heard of the 'Universal Mutual Life Insurance Company'?"

Response	Frequency	Relative Frequency
Yes	52	.605
No	34	.395
Total	86	1.000

Part VIII: Question Number 5: "Did you watch [a major televised sports broadcast]?"

Response	Frequency	Relative Frequency
Yes	78	.918
No	7	.082
Total	85	1.000

Part IX: Question Number 6: "Do you recall an advertisement for the 'Peace-of-Mind Company' during this sports broadcast?"

Response	Frequency	Relative Frequency
Yes	28	.333
No	56	.667
Total	84	1.000

Part X: Question Number 7: "What was the name of the company?"

Response	Frequency	Relative Frequency
Universal Mutual Life	6	.286
Metropolitan Life	1	.048
Northwestern National	1	.048
Don't know	13	.619
Total	21	1.001

3.2 THE "COLUMBIA TELEPHONE SYSTEM"

The policy of "Columbia Telephone System" regarding absenteeism has two major aspects:

    a. A global one spelled out in the various company rules and applied evenly to all employees.

    b. A local one, generally less formal, in which management is concerned about individual attendance.

A question such as how many paid days off per year an employee should be allowed for unexpected and unavoidable absences is often a subject for union negotiations and is a good example of what is meant by management's global policy. On the other hand, the need to decide whether an individual telephone operator has exhibited satisfactory attendance arises when the operator is considered for transfer or promotion. This is a good example of local policy.

Within the Columbia Telephone System, attendance has been evaluated in local terms (i.e., by comparison with the attendance of other operators in the same office). This has been done in both a subjective and informal way in the past. Although the informality is an advantage for both management and employees, the subjectivity of the evaluations is clearly a disadvantage.

To implement a "bias free" approach (one that is both objective and consistent), Columbia recently developed a procedure for evaluating attendance that removes much of the subjectivity normally associated with such evaluations. The procedure is based on a mathematical model fashioned on the results of a study conducted in 1984. We will concern ourselves with a rather small aspect of the overall model-building task.

The development of the model began with an awareness that there are two kinds of absences:

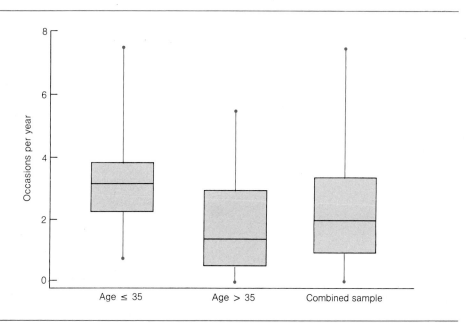

Figure 3.25. Box-and-whiskers plots for the frequency of incidental absences (IAs).

a. Incidental absences (IA), which are usually short, fairly frequent, and to some extent controllable.

b. Disabled absence (DA), which are usually long, less frequent, and uncontrollable.

In formal terms, a DA is any absence that lasts six or more days and that results from an illness (an exception is an on-the-job accident in which case the DA period can be shorter than six days). Any other absence is defined as an IA. This part of the case is concerned only with the IAs.

The procedure referred to above began in September of 1984 and involved examination of the attendance records of 112 New England telephone operators. Out of 112 records, 6 covered approximately one year, 63 approximately two years, and 43 approximately three years. Altogether 560 IAs were identified.

One aspect of the study had to do with the question of whether younger or older operators tend to have the most IAs. As part of their search for an answer, the investigators developed the box-and-whiskers charts shown in Figure 3.25 (p. 123).

1. Explain in words what Figure 3.25 appears to be saying about whether younger or older operators have the most IAs.

2. Why do you suppose box-and-whiskers plots were used in preference to, say, histograms or frequency polygons?

3.3 "GLAMOUR BURGER SYSTEMS": PART II

Reread Part I of this case (case 1.3 at the end of Chapter 1). The first survey was carried out during August of 1970. Some resulting data are presented in Table 3.23.

The second survey, conducted in September of 1971, was a comparison shopper study utilizing professional shoppers assigned the task of evaluating various aspects of the "Glamour Burger," McDonald's, and Burger King operations. The research was designed with a view to developing ratings for each chain based on the analysis of various factors thought to influence a representative customer's choice of establishment. The restaurants included in the study were drawn from all parts of the country and consisted of both company-owned and (where relevant) licensed units. These results are summarized in Table 3.24.

Table 3.23 Bivariate Frequency Distribution Data for the "Glamour Burger" Case

Part I: Who Patronizes the Three Fast-Food Chains?

Kind of Group Respondent Was in When Visiting the Restaurant	Type of Restaurant			
	"Glamour Burger"	McDonald's	Burger King	Total
Mother with children	242	330	222	794
Mother and father with children	283	371	323	977
Father with children	40	41	61	142
Adult—not part of family group	950	928	990	2868
Teenager—not part of family group	485	330	404	1219
Total	2000	2000	2000	6000

(continued)

Table 3.23 (*Continued*)

Part II: Where Were Customers Before They Came to the Restaurant?

| Location of Respondent When Decision Was Made to Go to Restaurant | Type of Restaurant | | | |
	"Glamour Burger"	McDonald's	Burger King	Total
Home	977	1178	1252	3407
Work	116	89	132	337
Shopping	744	600	484	1828
Place of entertainment	163	133	132	428
Total	2000	2000	2000	6000

Part III: Which Member of the Family Made the Decision to Eat Out?

| Family Member | Type of Restaurant | | | |
	"Glamour Burger"	McDonald's	Burger King	Total
Mother	291	328	350	969
Father	172	284	175	631
Child under 13 years	74	94	70	238
Child 13 years or more	28	36	11	75
Total	565	742	606	1913

Part IV: Which Member of the Family Made the Decision to Patronize the Particular Restaurant?

| Family Member | Type of Restaurant | | | |
	"Glamour Burger"	McDonald's	Burger King	Total
Mother	225	260	279	764
Father	225	337	235	797
Child under 13 years	74	99	74	247
Child 13 years or more	41	46	18	105
Total	565	742	606	1913

Part V: What Was the Principal Reason for Selecting the Particular Restaurant?

| Principal Reason Given | Type of Restaurant | | | |
	"Glamour Burger"	McDonald's	Burger King	Total
Good food	696	786	992	2474
Convenience	740	480	504	1724
Fast service	312	360	296	968
Price	252	374	208	834
Total	2000	2000	2000	6000

1. Apply appropriate cross-classification analyses on data in Table 3.23, making clear at the outset exactly what questions you are attempting to answer. State your conclusions clearly. (*Note:* Type of restaurant is the dependent variable.)

2. Utilizing the data in Table 3.24 and the results of your cross-classification analyses, tell what you believe is (are) the principal competitive advantage(s) of Glamour Burger in relation to McDonald's. In relation to Burger King. Tell what you believe is (are) the principal disadvan-

Table 3.24 Results of "Glamour Burger" 1971 Comparison Shopper Study

	"Glamour Burger"	McDonald's	Burger King
Service			
Time in line (1 minute or less)	73%	76%	64%
Time from order (1 minute or less)	17	36	32
Personnel			
Politeness	58%	82%	48%
Friendliness	52	67	42
Grooming	53	91	58
Cleanliness	77	91	64
Cleanliness of uniform	75	91	68
Suggestive selling	12	15	3
Appearance/Net Impression			
Cleanliness of parking lot	55%	84%	77%
Appearance of restaurant	42	70	58
Cleanliness of washroom	40	70	74
Inside restaurant	65	85	74
Outside restaurant	60	97	74
Employees	58	88	64
Food			
Super Burger—Whopper	39%	—	61%
Super Burger—Big Mac	50	52%	—
French fries	52	55	52%
Chocolate shake	35	48	42

Note about ratings: All attributes were rated on a 1 to 7 scale, where 1 is unsatisfactory and 7 is excellent. Results are reported in terms of the percent of very good or excellent scores achieved except for (1) service and suggestive selling, which is actual, and (2) appearance of restaurant and washroom, which is the percent of *very clean* ratings.

tage(s) of Glamour Burger in relation to McDonald's. In relation to Burger King. (The confusing "note about ratings" is quoted almost verbatim from an actual report.)

3. Project yourself into the role of a top manager of Twentieth Century Foods and write a report (about 300 to 500 words) spelling out some of the reasons that Glamour Burger had been running a poor third in sales to McDonald's and Burger King. Suggest ways in which Glamour Burger could change in order to (1) capitalize on its relative strengths, if any, and (2) overcome its relative weaknesses.

This case is continued at the end of Chapter 10.

4 Measures of Central Tendency and Variation

When you can measure what you are speaking about and express it in numbers you know something about it; but when you cannot express it in numbers, your knowledge is of a meagre and unsatisfactory kind.—Lord Kelvin

When I was a young man practicing at the bar, I lost a great many cases I should have won. As I got along, I won a great many cases I ought to have lost; so on the whole justice was done.
—Lord Justice Matthews

What You Should Learn from This Chapter

This is the second of three chapters concerned with statistical description. Chapter 3 showed you how to take a large, unruly mass of data and organize it in a way that would allow you a quick, global grasp of things. This chapter will show you how to obtain several kinds of descriptive measures so that data may be conveniently described via a few relatively simple summary numbers. After studying this chapter, you should be able to

1. Distinguish between measures of central tendency and measures of variation.
2. Distinguish among arithmetic mean, median, and mode.
3. Compute the arithmetic mean and the median from (1) data not arranged in frequency distribution form and (2) data arranged in frequency distribution form.
4. Identify the range, the mean deviation, the standard deviation, the variance, and the coefficient of variation.
5. Compute the standard deviation from (1) data not arranged in frequency distribution form and (2) data arranged in frequency distribution form.

Before proceeding, you should make certain you are clear on the distinction between descriptive statistics and inductive statistics (section 1.5 of Chapter 1). With respect to the material in the present chapter having to do with the calculation of descriptive measures from data in frequency distribution form, a review of section 3.2 of Chapter 3 is advised.

Except where otherwise indicated, sample data are assumed throughout this chapter.

4.1 Introduction

A police department rap sheet might describe a wanted criminal in these terms: Eyes: brown; Hair: black; Age: 45; Height: 6 feet; Weight: 225 pounds; Distinguishing features: Gray around the temples, slight limp, tattoo of a ship on left forearm. Such a description provides limited but easily assimilated and communicated information about the person sought.

Similarly, users of statistical information can describe data in terms of easily understood *descriptive measures*. Of course, in doing so, they will ignore a great many details, but, in exchange, they will receive a clearer overview, a better "feel," for the mass of numbers on which a decision will eventually be based.

In this chapter we consider two categories of descriptive statistical measures—namely, *measures of central tendency* and *measures of variation*.

4.2 Measures of Central Tendency

When analyzing business data, the manager or other user of statistical information often finds it helpful to have the data described in terms of a single value which in its own unique way *represents all the numbers under study*. Such a "representative" value is called a *measure of central tendency*. The term *central tendency* is based on the fact that a great many frequency distributions are dome shaped—that is, the frequencies tend to be high in the middle of the distribution and lower on the ends. Thus, a measure of central tendency is a value describing about where the observations are most concentrated.[1]

Three different measures of central tendency—the arithmetic mean, the median, and the mode—are discussed here. When treating these measures, we will first consider the procedures involved in determining their values from *ungrouped data* and then how the procedures would be modified if the data were *grouped*.

Ungrouped data are raw data, either in jumbled or arrayed form, but *not* organized in standard frequency-distribution form. By *grouped data* we mean data presented in frequency-distribution form.

4.3 The Arithmetic Mean

Chances are you are already an expert of sorts on the arithmetic mean. It is simply the measure you have known since grade school days as the "average." Your author has even taken the liberty of using the term "average" or "center of gravity" on occasion in previous chapters on the assumption that this is not your first exposure to the measure. In any event:

The *arithmetic mean* of a set of measurement data is the sum of all the values of interest divided by the number of the values summed.

Ungrouped Data

Let us begin our discussion of the mean from ungrouped data by assuming that the observations under study represent a sample.

The Arithmetic Mean for Sample Data. Since the arithmetic mean is a measure that can be obtained from any set of numbers, we will find it convenient to have a general formula for it. To arrive at such a formula, we begin by designating a set of values by some symbol such as X or Y. Taking, say, X as the symbol for a given set of data, we then proceed to designate the first value in the set (such as the value of 4.1 in Table 3.2, for example) as X_1, the second X_2, and so forth. The last value of the set may be designated X_n. Therefore, a generally applicable expression for the arithmetic mean of ungrouped (sample) data is

$$\overline{X} = \frac{X_1 + X_2 + X_3 + \cdots + X_n}{n}.$$

A more compact way of expressing the same procedure is

$$\overline{X} = \frac{\Sigma\, X}{n} \qquad (4.3.1)$$

where $\overline{X}$ (read "X bar") is the arithmetic mean of a set of sample data
  n is the number of observations, that is, the number of X values in the
  set
  Σ means "summation" ($\Sigma\, X$, therefore, is read "sum all of the X values")

For the 200 price-earnings ratios presented in Table 3.2, for example, we have

$$\overline{X} = \frac{\Sigma\, X}{n} = \frac{4.1 + 4.4 + 4.4 + \cdots + 63.8}{200} = 13.32$$

The value 13.32 is a "representative" number in the sense that the actual numerical value of each of the 200 P-E ratios contributed to the final result.

The Arithmetic Mean for Population Data.   Usually, in practice, the procedures for determining the arithmetic mean of a set of data will be applied to sample, rather than population, data for reasons cited in section 2.2 of Chapter 2. Nevertheless, we will have a number of occasions in the chapters concerned with probability and inductive statistics to refer to the arithmetic mean of a population. When the data of interest constitute a population, we will compute, or, if more relevant, think in terms of computing

$$\mu = \frac{\Sigma\, X}{N} \qquad (4.3.2)$$

where μ (pronounced "mew" or "moo") is the population arithmetic mean
  N is the number of observations making up the population

The population mean μ is called a *parameter* which, as noted earlier, means that it is a particular kind of descriptive measure of *population data*. A sample mean $\overline{X}$ is known as a *statistic*, that is, a descriptive measure of *sample* data. The advantage of employing different notations and terms to population data, on the one hand, and sample data, on the other, will become apparent in Chapter 8 and subsequent chapters.

The Weighted Mean. A sometimes useful variation on the above procedure, and one which will help you to understand the methodology recommended for grouped data, involves assigning different weights to the different observations. For example, let us say that a shopper buys bread at $1.00 a loaf, peanuts at $2.25 a jar, and canned shrimp at $5.25 a can. Her average expenditure would be

$$\mu = \frac{\Sigma X}{N} = \frac{\$1.00 + \$2.25 + \$5.25}{3} = \$2.83$$

if and only if she bought only one of each of these grocery items. But what if she bought four loaves of bread, two jars of peanuts, and one can of shrimp? In this case the average cost per item would more properly be computed by the formula

$$\mu_w = \frac{\Sigma wX}{\Sigma w} \tag{4.3.3}$$

where μ_w stands for the weighted mean
X is the price per unit of a specified item
w stands for the weight assigned a particular item

(*Note:* Population notations are used here because we are considering *all* items purchased by this shopper during this particular trip to the grocery store. If the items of interest constituted a sample, the appropriate notation for the weighted mean would be $\overline{X}_w$.)

For our problem: $\mu_w = [(4)(\$1.00) + (2)(\$2.25) + (1)(\$5.25)]/7 = \$1.96$.

As Figure 4.1 illustrates, an arithmetic mean, whether unweighted or weighted, can be thought of as a fulcrum. If you think of the line in this figure as a seesaw of sorts resting on a triangular base (representing the mean μ), the seesaw will be in balance if the numbers are placed along it according to how far each is away from the mean of $2.83 (part (*a*) of Figure 4.1). In the case of the weighted mean, the same idea holds except that now it is as if there were more numbers to be positioned along the seesaw (part (*b*) of Figure 4.1) and the fulcrum (i.e., the mean of $1.96) is appropriately repositioned to keep the seesaw in balance.

Grouped Data

The arithmetic mean of data in frequency-distribution form is computed in much the same way as described above except that the midpoint of each class is used as a representative value of all the X values in the class. Then, the midpoint is multiplied by the corresponding class frequency. This procedure is repeated for all classes of the frequency distribution. The resulting products are then summed; the sum so obtained is then divided by the total number of observations. In symbols (assuming sample data), this procedure may be described as follows:

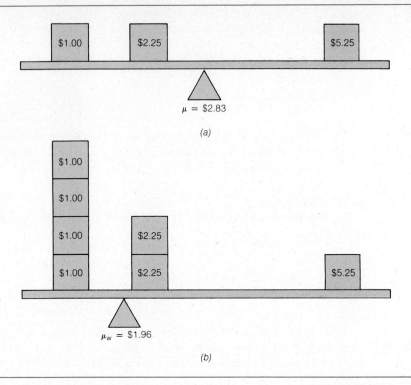

Figure 4.1 *(a)* The arithmetic mean as a center of gravity, and *(b)* the weighted mean as a center of gravity.

$$\overline{X} \cong \frac{\Sigma fX_m}{n} \qquad\qquad (4.3.4)$$

where $\cong$ means "is approximately equal to"
f is the frequency for a specified class
X_m is the midpoint of a specified class
$n = \Sigma f$ and represents the number of observations in the entire sample

Application of this formula using the frequency distribution shown in Table 3.4 is demonstrated in Table 4.1.

The value obtained through this procedure, 13.08, is slightly lower than the 13.32 obtained from the same data in ungrouped form. The reason is that the arithmetic mean obtained from grouped data is merely an approximation of the true arithmetic mean. In our example—as in most others—the difference is small.

The procedure just described was illustrated using a frequency distribution with a constant class interval. This is not a necessary condition for its use. For example, the price-earnings data as presented in Figure 3.9, where the interval is 3 for some classes and 10 for others,

Table 4.1 Method of Computing the Arithmetic Mean from a Frequency Distribution, 200 Price-Earnings Ratios

(1) Price-Earnings Ratios	(2) Class Midpoint X_m	(3) Class Frequency f	(4) [(2) × (3)] fX_m
3 and under 10	6.5	99	643.5
10 and under 17	13.5	65	877.5
17 and under 24	20.5	12	246.0
24 and under 31	27.5	11	302.5
31 and under 38	34.5	5	172.5
38 and under 45	41.5	5	207.5
45 and under 52	48.5	1	48.5
52 and under 59	55.5	1	55.5
59 and under 66	62.5	1	62.5
Total		200	2616.0

$$\overline{X} \cong \frac{\Sigma fX}{n} = \frac{2616.0}{200} = 13.08$$

were subjected to Equation (4.3.4) with a resulting sample mean of 13.36. Notice how the approximation was even improved in this case when unequal class intervals were used. This procedure is valid regardless of whether it is sample or population data being analyzed.

The arithmetic mean cannot be determined for frequency distributions having open-ended classes since the midpoints of the open-ended classes can only be guessed. If use of such a distribution cannot be avoided, the median will usually serve as a more appropriate measure of central tendency.

4.4 The Median

Let us now consider a second measure of central tendency, one that is preferable to the arithmetic mean under some circumstances. This measure is the median.

> The *median* (which is exactly the same thing as the second-quartile, Q_2, measure touched on in Chapter 3) is the middle value when numerical observations have been arranged in order of size.

Ungrouped Data

When the data are not in frequency-distribution form, the median is obtained by (1) organizing the ungrouped data into an array and (2) picking the middle value in the array if n is an odd number, or averaging the middle two values if n is an even number. For example, in section 3.5 of Chapter 3, we obtained the median for the 200 price-earnings ratios by noting that two such ratios, the 100th (with a value of 10) and the 101st (with a value of 10.1) were in contention for the honor of being picked. So we split the difference and obtained the value of 10.05 for the median. On the other hand, the order-filling time data in part (b) of Figure 3.18 had an odd number, 75, for n. Hence, the 38th value in the array, 37 minutes, was identified as the median. In both cases, the position-location formula of $(n + 1)/2$ helped us to identify the median.

Grouped Data

When determining the median from ungrouped data, we merely counted from one end of a set of ordered observations until we found the observation exactly in the middle. With grouped data we do much the same thing but with two modifications: (1) Instead of counting in a literal sense, we cumulate class frequencies until we reach the class containing the median—hereafter referred to as the *median class*—and (2) instead of counting into the median class to find the median value—an impossible task when data are in grouped form—we estimate what percentage of the way into the median class we must go to locate the median value. The formula for the median from grouped data merely expresses this procedure symbolically. It is

$$\text{Median} \cong LCL + c\left(\frac{n/2 - cf}{f_{\text{med}}}\right) \qquad (4.4.1)$$

where LCL is the lower class limit of the median class
c is the class interval of the median class
cf is the cumulated frequency for all classes prior to (i.e., positioned above) the median class
f_{med} is the number of observations associated with the median class

This formula is demonstrated in Table 4.2 using the price-earnings data from Table 3.4 and the order-filling time data from Figure 3.1.

The first step in applying Equation (4.4.1) is to determine the subscript of the observation located exactly in the middle of the array (the frequency distribution being a kind of an array) by computing $n/2$. This is a slightly different position-location formula than the $(n + 1)/2$ used in connection with ungrouped data. With ungrouped data, we sought the $(n + 1)/2$ observation. When working with grouped data, we seek a number that divides the total area of the histogram into two equal parts with each part representing $n/2$ observations. In our example involving the price-earnings ratios, we have $n/2 = 200/2 = 100$. When we cumulate the entries in the f column, the successive subtotals are found to be 99, 164, and so forth. The first subtotal of 99 is slightly less than the $n/2 = 100$ target value, so we continue to cumulate frequencies. The second subtotal is 164, which, of course, is much higher than the target value of 100. We know, therefore, that the median class is the "10 and under 17" class. We put 10 into the formula as the LCL (lower class limit) value. Next, we determine how far into the median class to proceed.

Table 4.2 Method of Calculating the Median from Grouped Data

Part I: Method Applied to Price-Earnings Data in Table 3.4 of Chapter 3

(1) Price- Earnings Ratios	(2) Frequency f	(3) Cumulative Frequency cf
3 and under 10	99	99
10 and under 17[a]	65	164
17 and under 24	12	176
24 and under 31	11	187
31 and under 38	5	192
38 and under 45	5	197
45 and under 52	1	198
52 and under 59	1	199
59 and under 66	1	200
Total	200	

$$\text{Median} \cong LCL + c\left(\frac{n/2 - cf}{f_{\text{med}}}\right) = 10 + 7\left(\frac{100 - 99}{65}\right) = 10.11$$

Part II: Method Applied to Order-Filling Time Data in Figure 3.1 of Chapter 3

(1) Time (in Minutes)	(2) Frequency f	(3) Cumulative Frequency cf
10 and under 20	5	5
20 and under 30	15	20
30 and under 40[a]	20	40
40 and under 50	14	54
50 and under 60	11	65
60 and under 70	7	72
70 and under 80	3	75
Total	75	

$$\text{Median} \cong LCL + c\left(\frac{n/2 - cf}{f_{\text{med}}}\right) = 30 + 10\left(\frac{17.5}{20}\right) = 38.75$$

[a] Indicates the median class.

> When determining how far into the median class to go to locate the median value, we *assume* that the observations are distributed evenly throughout the class.

The proportion of the way into the median class we must go to find the actual median is determined by the number of observations we have left to count to get to the $n/2$ observation, expressed as a ratio of the f_{med} observations in that class. The cumulative frequency in our example was 99 by the time we reached the beginning of the median class. Our target was the 100th observation. Thus, we had only one observation to go. Since there are 65 observations associated with the median class, we determined that we must proceed 1/65th of the way into that class. The part of the formula that reads $(n/2 - cf)/f_{med}$ expresses this idea. That is, $(200/2 - 99)/65 = 1/65$. When 1/65 is multiplied by the class interval of 7 and the resulting product added to the lower class limit of 10, we obtain the value of 10.11 for the median. Clearly, the value of 10.11 so obtained is only an approximation of the true median.

In the order-filling time problem, the target value was the $n/2 = 75/2 = 37.5$th value. The cumulative frequency by the end of the beginning class was 5; by the end of the second class, 20; by the end of the third class, 40. The value 40 is higher than the target of 37.5; thus, we conclude that the median class is the third one, the class of "30 and under 40." The quantity $n/2 - cf$, by the end of the second class, is $37.5 - 20 = 17.5$. Putting this value over the number of observations associated with the median class, $f_{med} = 20$, we get the fraction 17.5/20, which means that we must proceed 17.5/20 of the way into the median

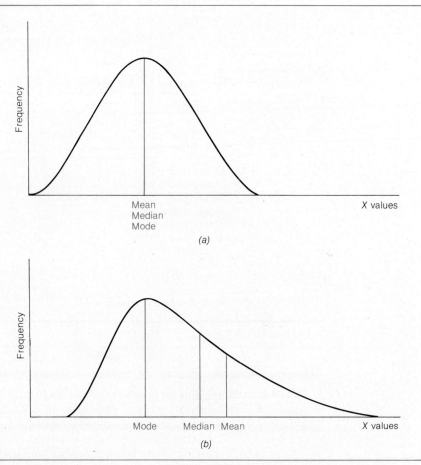

Figure 4.2. *(a)* Symmetrical distribution with location of mean, median, and mode identified. *(b)* Skewed distribution with locations of mean, median, and mode identified.

class to find the median value. In this example, the class interval is 10. Therefore,

$$\text{Median} \cong 30 + 10\left(\frac{17.5}{20}\right) = 38.75$$

somewhat higher than the true median of 37.

The median is affected by the presence of, but not the actual magnitudes of, extreme observations; hence, it can be determined even for open-ended distributions. Moreover, no special difficulties arise when the class interval is not the same for all classes, since the c symbol in the formula refers to the class interval of the median class only.

4.5 The Mode

One final measure of central tendency that is conceptually useful but not always possible to compute is the *mode*.

> As its name suggests, the *mode* is the "most popular," that is, the most frequently occurring, value.

Because the mode cannot always be determined mathematically (and because some distributions possess more than one mode), the modern tendency is to forego actual calculation of the modal value. Still, the concept itself is sometimes quite useful. In a unimodal distribution (such as the ones illustrated in both parts of Figure 4.2), the mode may be thought of as the value on the horizontal axis lying below the highest point on the curve. The mode in this graphic sense indicates where a maximum of clustering occurs and is often used as a standard against which the adequacy of the arithmetic mean or the median as "representative" values is measured.

4.6 Characteristics of the Mean, Median, and Mode

When the frequency distribution under study is symmetrical and unimodal, the mean, median, and mode are identical, as illustrated in part (*a*) of Figure 4.2, and are, therefore, equally good representative measures. On the other hand, when the distribution is not symmetrical, the three measures may differ markedly because of differences in relative sensitivity to extreme values. The arithmetic mean is the most sensitive to extreme values and, as a result, is the measure of central tendency pulled farthest in the direction of the long tail of the distribution; the mode is the least sensitive of the three measures. This point is illustrated in part (*b*) of Figure 4.2.

Because much business data are distributed in ways that are definitely not symmetrical, the arithmetic mean is often a poor representative value *when used for descriptive purposes only*. For this reason, the analyst in our example about the price-earnings ratios would be well advised to view the median of 10.11 (from grouped data) as a better representative value than the mean of 13.08 (also from grouped data) or maybe even consider using the mode.

On the other hand, we will see that the arithmetic mean is an extremely important measure in many realms of *inductive statistics*. A quality of the arithmetic mean, admittedly one difficult to appreciate at this time but most useful in connection with the inductive tasks of estimation and hypothesis testing, is its *stability*. That is, if we were to take a great many samples of the same size n from the same population and compute the mean, median, and mode for each such sample, we would find that the $\overline{X}$ values vary less from sample to sample than do values for either the median or the mode. This quality, plus the fact that the mean lends itself well to mathematical and theoretical treatment, makes it the most desirable of measures of central tendency when formal inductive inferences are to be drawn from sample data.

4.7 Introduction to Measures of Variation

You may have heard the old story of the man who was told that a certain river was only two feet deep on the average. This news led him to try crossing it on foot. Unfortunately, he drowned almost immediately. The point, of course, is that the arithmetic mean, or any other measure of central tendency, is not necessarily all the descriptive information one needs to make a sensible decision. It was the variation in the depth, not the central tendency, that led to the death of the unfortunate man in this apocryphal tale.

Variation is found in every facet of life: No two items coming off an assembly line are ever exactly alike, however well controlled the manufacturing process may be. Sales of virtually any product will vary according to the day of the week, the season of the year, the stage of the business cycle, or the degree of market saturation. The amount of accounts receivable, bad debts, and discounts for prompt payment will all vary from one company to another and from one accounting period to another. The rewards associated with a specific kind of work vary from one part of the country to another—and so does the cost of living. Consequently, the business decision maker must act within an environment in which variation, rather than uniformity, is the natural state of things.

In the next several sections, we consider some helpful ways of expressing the rather abstract notion of variation in quantitative terms. Each such quantitative measure tells in its own way how much variation, or dispersion, or "spread" (the terms are used synonymously) there is in a set of data. By way of introduction, let us contemplate the following illustrative case.

ILLUSTRATIVE CASE 4.1

Management of the Gotham City Hardware Company, the large wholesaler of hardware products referred to in Chapter 3, wished to know (1) whether the average length of time required for the company's order fillers to complete a standardized order is dependent on length of employment with the company and (2) whether the amount of variation in the time required to fill the standardized order is affected by the length of employment with the company. Toward these ends, management selected 10 order fillers who had been with the company for less than one year and 10 who had been with the company one year or longer to participate in an experiment. Each of the 20 employees was given a list of 15 identical items and instructed to fill the order correctly as quickly as possible. The results are shown in Table 4.3 and Figure 4.3.

Let us address the central tendency part of the task first. According to the sample data (Table 4.3), the average length of time required to fill the order declined from 12.2 to 10.5 minutes as length of employment increased. (However, as will be shown in Chapter 10, these results do not necessarily mean that such a decline has also occurred on the population level. We are looking right now at *sample descriptive results* only.) We will now refer to this illustrative case in connection with three available measures of variation—the range, the mean deviation, and the standard deviation.

Table 4.3 Time (in Minutes) Required for 20 Employees to Fill a Standardized Order for 15 Hardware Items

(1) Times for 10 Employees with Company Less Than One Year	(2) Times for 10 Employees with Company One Year or Longer
X	Y
5	5
7	10
9	10
10	10
11	10
12	10
14	10
16	10
18	10
20	20
122	105

$$\overline{X} = \frac{\Sigma X}{n} = \frac{122}{10} = 12.2$$

$$\overline{Y} = \frac{\Sigma Y}{n} = \frac{105}{10} = 10.5$$

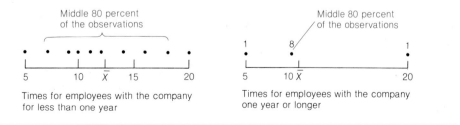

Times for employees with the company for less than one year

Times for employees with the company one year or longer

Figure 4.3. Data from Table 4.3 in diagram form.

4.8 The Range

Many people have a "feel" for some of our principal measures of central tendency—the mean, the median, and the mode—long before they study them in a formal way in statistics courses. Perhaps the desire to express disparate things in a simplified, summary manner is a deep-seated human inclination. However, the task of measuring—or even thinking in terms of—variation is something else again. The idea of measuring the extent to which several numbers differ from one another is, for many, unpleasantly foreign. That being so, you might find it helpful to project yourself into the role of a person living in a world where no formal measures of variation exist. Suppose, in fact, that a highly placed political figure singles you out and charges you with the responsibility of "inventing" a workable measure of variation having as few disadvantages as possible. This fantasy has two virtues: (1) It will direct your attention to the underlying logic of the measures rather than to the formulas per se, and (2) it will serve to emphasize the fact that, whereas variation is inherent in life, methods of measuring variation are necessarily man-made.

Chances are, the first measure of variation you would come up with would be the *range*.

> The *range* of a set of numbers is simply the difference between the largest value and the smallest value.

The most obvious advantage of this "invention," the range, is its easy calculation—and that says it all, as you will quickly discover. Its limitations, you find, are great because of its sole dependence on the two extreme values in the array. In some situations the extreme values are very extreme indeed. Moreover, this measure is limited in its ability to discriminate among different sets of data with respect to variability. Applying your new tool to the two sets of numbers in Table 4.3, for example, you find: For employees with the company for less than one year: Range $(X) = 20 - 5 = 15$; for employees with the company for one year or longer: Range $(Y) = 20 - 5 = 15$. Thus, the range is the same in both cases despite the fact that the two sets of numbers differ considerably except at the ends of the arrays. (Figure 4.3 emphasizes this point.) You decide, let us say, to continue your quest for an acceptable measure of variation.

4.9 The Mean Deviation

Since the principal shortcoming of the range is its failure to take account of any observations other than the most extreme ones, it occurs to you, we will suppose, that a preferable measure of variation would be one that takes all observations in a set into account. This realization gives you a start—but only a start. Now you must ponder more deeply what exactly one means when one says he wishes to measure "variation." You might come to reason as follows: Since the amount of variation is small when the numbers are very close to one another and large when they differ a great deal, a sensible way of measuring variation would be to calculate the average difference among all the observations. Better yet, you could select a representative value—say the arithmetic mean—and determine the extent to which the individual observations differ from it. That is, you could determine by how much observation X_1 differs from the arithmetic mean, $\overline{X}$. Then you could do the same for X_2, X_3, and so forth, until all values, including X_n, have been taken into account. To this point you would have listed what are called *deviations from the mean*: $X_1 - \overline{X}$, $X_2 - \overline{X}$, $X_3 - \overline{X}$, . . . , $X_n - \overline{X}$. Having made a list of these deviations, it would then be a simple matter to average them. Behold: When you try to calculate the average of this collection of deviations, you get $\Sigma\, (X - \overline{X})/n = 0$. This, you find after some experimentation, is always the case. (For example, column (3) in both parts of Table 4.4 sums to 0 despite the obviously different natures of the two sets of data.) Here's why: Since $\Sigma\, (X - \overline{X})$ can be written $\Sigma\, X - \Sigma\, \overline{X}$, we can place the sum of the X's over n and $\Sigma\, \overline{X}$ over n separately, giving $(\Sigma\, X/n) - (\Sigma\, \overline{X}/n)$. But $\Sigma\, \overline{X}$ is the same thing as $n\overline{X}$. Therefore, $\Sigma\, X/n - n\overline{X}/n$, which is the same as $\Sigma\, X/n - \overline{X}$. But we know that $\Sigma\, \overline{X}/n = \overline{X}$. Thus, we have $\overline{X} - \overline{X} = 0$.

Alas, the average deviation from the mean, obtained in this intuitively appealing way, measures nothing at all. Still, you decide, the basic idea of averaging deviations from the mean has some merit. The challenge is to find some acceptable way of dealing with the

GRIN AND BEAR IT   by Lichty & Wagner

"Everything came out even . . . For every man with two left feet, we had one with two right feet."

negative deviations, which, collectively, have the disconcerting habit of offsetting *exactly* the positive deviations overall.

One possibility is to express all deviations in terms of *absolute* values, that is, treat all deviations as if they were positive. In symbols,

$$MD = \frac{\Sigma |X - \overline{X}|}{n}$$

where *MD* stands for mean deviation and the vertical bars are used to indicate absolute values of deviations. If you were to apply this measure to the hardware company data, you would find that $MD(X) = \Sigma |X - \overline{X}|/n = 38.4/10 = 3.84$ minutes, and $MD(Y) = \Sigma |Y - \overline{Y}|/n = 19.0/10 = 1.90$ minutes. (Numerators for these measures were obtained from column (3) of Parts I and II of Table 4.4, with negative signs ignored.)

According to these *purely descriptive* sample results, the variability in the time required to fill the order declines as length of service increases. (Though at this time we are unprepared to conclude that the same is true for the populations. We address this kind of question in Chapter 10.)

The mean deviation possesses much intuitive appeal and has found some useful applications in descriptive statistical analysis. On the other hand, it is difficult to calculate when the number of observations is large; moreover, the absolute deviations make theoretical developments difficult. For most purposes, a close cousin of the mean deviation is preferred—namely, the standard deviation (or its square, the variance).

4.10 The Standard Deviation and the Variance

Because you are interested only in the *magnitudes* of the deviations from the arithmetic mean and not in their *signs*, you soon find that you can accomplish much the same thing as was accomplished with absolute deviations if you resort to (1) squaring the deviations from the mean, thereby eliminating negative signs, and then (2) averaging the *squared deviations*. The resulting measure, if you were to stop at this point, is the *variance*. The variance is an important measure of variation in its own right, and one which we will make much use of in several later chapters. For now, we will emphasize its square root, the *standard deviation*. The standard deviation is a somewhat "friendlier" measure in that it is expressed in terms of the same units as the original observations—that is, minutes, dollars, pounds, and so on—whereas the variance is in terms of "square units," so to speak. The standard deviation expressed in formula form, according to the above description, would be

$$\sqrt{\frac{\Sigma (X - \overline{X})^2}{n}}$$

Having come this far, let us drop the fantasy of your having been charged with inventing a satisfactory measure of variation. I hope this little imaginary exercise has given you some insight into the kind of reasoning that went into the eventual development of the much-used standard deviation and variance.

As a purely descriptive measure of variation for a set of sample observations, the standard deviation obtained from the above formula would be most adequate. However, because we ordinarily use the sample standard deviation as a means of estimating the true, but unknown, population standard deviation σ, we modify the formula slightly and have it read

$$s = \sqrt{\frac{\Sigma (X - \overline{X})^2}{n - 1}} \qquad (4.10.1)$$

The $n - 1$ denominator in Equation (4.10.1) reflects a peculiarity of the sample standard deviation and variance. Let us focus briefly on the variance, s^2, computed by $\Sigma (X - \overline{X})^2/n - 1$.

If (1) all different samples of size n contained within a population of interest were identified, (2) s^2 were obtained using a denominator of n for each such sample, and (3) the resulting s^2's were used individually to estimate the true population variance σ^2, more such s^2's would be too low than too high. Or, to state the matter more briefly, *the sample variance is a biased estimator of the population variance.* Therefore, we follow the practice of inflating the sample variance somewhat by using an $n - 1$ denominator.

You can get a pretty good intuitive feel for the source of the underestimation bias from the following argument: According to a rule of summation notation, $\Sigma (X - \mu)^2/n = \Sigma (X - \overline{X})^2/n + (\overline{X} - \mu)^2$, where μ is the population mean. Except in the rare case where $\overline{X} = \mu$, $\Sigma (X - \overline{X})^2/n$ must be less than $\Sigma (X - \mu)^2/n$.

You should be aware that use of $n - 1$ eliminates the underestimation bias in the sample *variance*—not in the sample standard deviation. Still, its use in connection with the sample standard deviation makes this statistic a more nearly unbiased estimator of the population standard deviation.

The $n - 1$ in the sample variance and the standard deviation formulas is referred to as *degrees of freedom.* The term *degrees of freedom*, discussed more completely in section 8.22 of Chapter 8, will be used frequently throughout this text.

A general definition of *degrees of freedom*, which will serve us well in many cases, is: $n - \lambda$, where n is the sample size and λ stands for the number of population parameters being estimated by the corresponding sample statistic(s). In the case of the ordinary variance or standard deviation, $\overline{X}$ is used to estimate the true μ value. Hence, we say that one degree of freedom is lost, or that the number of degrees of freedom is $n - 1$. Later you will see that λ can have a value greater than 1.

Let us now return to the hardware company problem. The application of Equation (4.10.1) to this problem is demonstrated in Table 4.4. The standard deviations in this problem, declining as they do when experience increases—from 4.80 to 3.69 minutes—tell the same kind of story that the mean deviations told about the two samples. You should note, however, that the standard deviations are higher in both cases than the corresponding mean deviations. The reasons are two: (1) The mean deviations were computed using an n denominator, whereas the standard deviations were computed using an $n - 1$ denominator; and (2) with the standard deviation, the taking of the square root at the end of the calculations compensates only partially for the fact that the deviations from the mean were squared. Thus, we see that the standard deviation is a little more difficult to interpret than the mean deviation. Whereas the mean deviation is defined as the average absolute deviation from the arithmetic mean, the best we can say about the standard deviation is that it is *a kind of*

Table 4.4 Calculation of Standard Deviations for the Gotham City Hardware Company Case

Part I: Employees with Company for Less Than One Year

(1) X	(2) $\overline{X}$	(3) $X - \overline{X}$	(4) $(X - \overline{X})^2$
5	12.2	-7.2	51.84
7	12.2	-5.2	27.04
9	12.2	-3.2	10.24
10	12.2	-2.2	4.84
11	12.2	-1.2	1.44
12	12.2	$-.2$	.04
14	12.2	1.8	3.24
16	12.2	3.8	14.44
18	12.2	5.8	33.64
20	12.2	7.8	60.84
Σ 122		0.0	207.60

$$s_X = \sqrt{\frac{\Sigma\,(X - \overline{X})^2}{n - 1}} = \sqrt{\frac{207.60}{9}} = 4.80 \text{ minutes}$$

Part II: Employees with Company for One Year or Longer

(1) Y	(2) $\overline{Y}$	(3) $Y - \overline{Y}$	(4) $(Y - \overline{Y})^2$
5	10.5	-5.5	30.25
10	10.5	$-.5$	.25
10	10.5	$-.5$	.25
10	10.5	$-.5$	.25
10	10.5	$-.5$	.25
10	10.5	$-.5$	.25
10	10.5	$-.5$	.25
10	10.5	$-.5$	.25
10	10.5	$-.5$	.25
20	10.5	9.5	90.25
Σ 105		0.0	122.50

$$s_Y = \sqrt{\frac{\Sigma\,(Y - \overline{Y})^2}{n - 1}} = \sqrt{\frac{122.50}{9}} = 3.69 \text{ minutes}$$

average deviation from the arithmetic mean. (For a geometric interpretation of the standard deviation and variance, see Figure 4.4, p. 148.) Fortunately, what the standard deviation lacks in definitional clarity, it more than makes up for through its various roles in inductive statistical analysis.

Although Equation (4.10.1) is correct and has the advantage of keeping the logic underlying the standard deviation in plain view, it is not the simplest standard deviation formula in the statistician's tool kit to use, especially when n is large. An alternative formula delivering the same result with less computational time and effort is

$$s = \sqrt{\frac{\Sigma X^2 - \dfrac{(\Sigma X)^2}{n}}{n - 1}} \qquad (4.10.2)$$

Use of this formula is demonstrated in Table 4.5.

We will not, as a rule, derive formulas in this text. However, Equation (4.10.2) and variations on it will come into play on many future occasions. Therefore, you really should know how to get from $\Sigma (X - \overline{X})^2$ to $\Sigma X^2 - (\Sigma X)^2/n$.

Let us begin by taking only one deviation from the mean and squaring it:

$$(X - \overline{X})^2 = X^2 - 2\overline{X}X + (\overline{X})^2$$

Next, we proceed as if the same squaring process had been applied to all n deviations.

Table 4.5   Calculation of Standard Deviations for the Gotham City Hardware Company Case by the Alternative Ungrouped-Data Formula

Part I: Employees with Company for Less Than One Year

(1) X	(2) X^2
5	25
7	49
9	81
10	100
11	121
14	196
16	256
18	324
20	400
Σ 122	1696

$$s_X = \sqrt{\frac{\Sigma X^2 - \dfrac{(\Sigma X)^2}{n}}{n - 1}} = \sqrt{\frac{1696 - \dfrac{(122)^2}{10}}{9}}$$
$$= 4.80 \text{ minutes}$$

Part II: Employees with Company for One Year or Longer

(1) Y	(2) Y^2
5	25
10	100
10	100
10	100
10	100
10	100
10	100
10	100
10	100
20	400
Σ 105	1225

$$s_Y = \sqrt{\frac{\Sigma Y^2 - \dfrac{(\Sigma Y)^2}{n}}{n - 1}} = \sqrt{\frac{1225 - \dfrac{(105)^2}{10}}{9}}$$
$$= 3.69 \text{ minutes}$$

This we do by utilizing appropriate summation notation. Hence:

$$\Sigma X^2 - 2\overline{X}\Sigma X + n(\overline{X})^2$$

We know that $\overline{X} = \Sigma X/n$; therefore, we may substitute $\Sigma X/n$ for $\overline{X}$ in the above expression, giving $\Sigma X^2 - 2\Sigma X/n\Sigma X + n(\Sigma X/n)^2$. This expression is readily rearranged to give $\Sigma X^2 - (\Sigma X)^2/n$.

Grouped Data

When sample data are in frequency-distribution form, we may determine the standard deviation in much the same way as the arithmetic mean from grouped data—that is, by using class midpoints to represent all observations in the related classes and weighting by class frequencies. The basic formula is

$$s \cong \sqrt{\frac{\Sigma f(X_m - \overline{X})^2}{n-1}} \qquad (4.10.3)$$

where f stands for class frequency
X_m is the midpoint of a specified class

Application of this formula is demonstrated in Table 4.6, using the price-earnings data as presented in Table 3.4.

Table 4.6   Calculation of Standard Deviation for Grouped Data (200 Price-Earnings Ratios)

(1) Price-Earnings Ratios	(2) Midpoint X_m	(3) Frequency f	(4) $\overline{X}$	(5) $X_m - \overline{X}$	(6) $(X_m - \overline{X})^2$	(7) $f(X_m - \overline{X})^2$
3 and under 10	6.5	99	13.08	− 6.58	43.2964	4286.3436
10 and under 17	13.5	65	13.08	.42	.1764	11.4660
17 and under 24	20.5	12	13.08	7.42	55.0564	660.6768
24 and under 31	27.5	11	13.08	14.42	207.9364	2287.3004
31 and under 38	34.5	5	13.08	21.42	458.8164	2294.0820
38 and under 45	41.5	5	13.08	28.42	807.6964	4038.4820
45 and under 52	48.5	1	13.08	35.42	1254.5764	1254.5764
52 and under 59	55.5	1	13.08	42.42	1799.4564	1799.4564
59 and under 66	62.5	1	13.08	49.42	2442.3364	2442.3364
Total		200				19074.7200

$$s_X' = \sqrt{\frac{\Sigma f(X_m - \overline{X})^2}{n-1}} = \sqrt{\frac{19074.7200}{199}} = 9.79$$

An alternative formula, which gets the job done noticeably faster when n is large, is the following:

$$s \cong \sqrt{\frac{\Sigma\, f(X_m)^2 - \dfrac{(\Sigma\, fX_m)^2}{n}}{n - 1}}$$

(4.10.41)

Application of this formula is presented in Table 4.7.

Population Data. We have thus far implicitly assumed that the data whose variation is to be measured represent a sample rather than a population. When the data constitute a population, the basic formula used to obtain the standard deviation is

$$\sigma = \sqrt{\frac{\Sigma\, (X - \mu)^2}{N}}$$

(4.10.5)

where σ is the standard deviation for population data
μ is the arithmetic mean for population data
N is the number of observations making up the population

The operations are obviously the same as with sample data—see Equation (4.10.1)—except that the denominator N need not be modified to help offset an underestimation bias. Since with population data there is no estimation, no such bias is possible.

Table 4.7 Calculation of Standard Deviation Using the Alternative Grouped-Data Formula

(1) Price- Earnings Ratios	(2) Midpoint X_m	(3) Frequency f	(4) $[(2) \times (3)]$ fX_m	(5) $(X_m)^2$	(6) $[(3) \times (5)]$ $f(X_m)^2$
3 and under 10	6.5	99	643.5	42.25	4182.75
10 and under 17	13.5	65	877.5	182.25	11846.25
17 and under 24	20.5	12	246.0	420.25	5043.00
24 and under 31	27.5	11	302.5	756?25	8318.75
30 and under 38	34.5	5	172.5	1190.25	5951.25
38 and under 45	41.5	5	207.5	1722.25	8611.25
45 and under 52	48.5	1	48.5	2352.25	2352.25
52 and under 59	55.5	1	55.5	3080.25	3080.25
59 and under 66	62.5	1	62.5	3906.25	3906.25
Total		200	2616.0		53292.00

$$s \cong \sqrt{\frac{\Sigma\, f(X_m)^2 - \dfrac{(\Sigma\, fX_m)^2}{n}}{n - 1}} = \sqrt{\frac{53{,}292.00 - \dfrac{(2616.0)^2}{200}}{199}} = 9.79$$

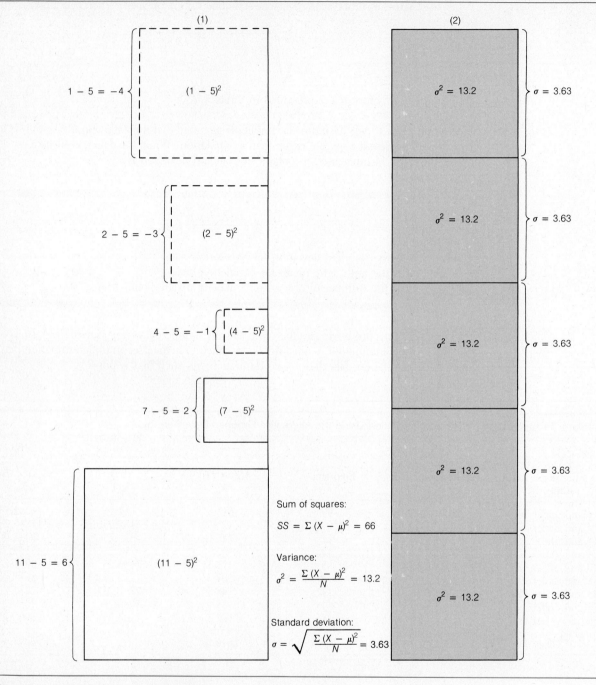

Figure 4.4. A geometric interpretation of the standard deviation and variance of a population.

4   Measures of Central Tendency and Variation

The geometric interpretation of the standard deviation, referred to earlier, is presented in Figure 4.4. In this figure, five simple population values are used: 1, 2, 4, 7, and 11. (We could have assumed sample data about as easily, except that the $n - 1$ denominator complicates matters somewhat.) These population values have an arithmetic mean, μ, of 5. Therefore, the deviations from the mean are $1 - 5 = -4$, $2 - 5 = -3$, $4 - 5 = -1$, $7 - 5 = 2$, $11 - 5 = 6$. The corresponding squared deviations are, respectively, 16, 9, 1, 4, and 36. In column (1) of the figure, a regular deviation from the mean is represented by any arbitrarily selected edge of the relevant left-hand square. (Squares associated with negative deviations are drawn with dashed lines, and those associated with positive deviations are drawn with solid lines.) The corresponding squared deviation is represented by the entire square. The sum of all the squares, called exactly that—*the sum of squares*, when divided by $N = 5$ gives the variance. The variance is represented by any one of the squares of uniform size in column (2) of Figure 4.4. The standard deviation is represented by any edge of any of these uniform squares.

4.11 The Coefficient of Variation

In connection with Illustrative Case 4.1, we compared without apology two sample standard deviations. Although this kind of comparison is not misleading in this particular case, there are times when it will be: If two sets of data are expressed in different units of measurement (say, one in pounds and the other in inches), the comparison will be meaningless. If the units of measurement are the same but the arithmetic means are markedly different, the comparison can convey a completely faulty message. Let's look at another example.

ILLUSTRATIVE CASE 4.2

The Gemini Athletic Equipment Company manufactures several qualities of weight-lifting equipment. The company's top-of-the-line equipment is used in such important events as the Olympics and some of the more prestigious professional power-lifting competitions. Consequently, it is very important that if a barbell plate is stamped, say, "50 pounds" it weigh very close to 50 pounds. In addition, a barbell plate must have a hole just slightly larger than 1 inch in diameter so that it will slip onto the 1-inch diameter bar easily but fit snuggly when it is in place. A recent sampling of 10-pound and 50-pound barbell plates revealed the following information:

1. Weights of 10-pound plates had an arithmetic mean $\overline{X}_1$ of 10.013 pounds and a standard deviation s_{X1} of .124 pounds.
2. Weights of 50-pound plates had a mean $\overline{X}_2$ of 50.032 pounds and a standard deviation s_{X2} of .465 pounds.
3. Diameters of holes in the 10-pound plates had a mean $\overline{Y}_1$ of 1.22 inches and a standard deviation s_{Y1} of .187 inches.
4. Diameters of holes in the 50-pound plates had a mean $\overline{Y}_2$ of 1.20 inches and a standard deviation s_{Y2} of .183 inches.

One question to be answered from this study was whether the production process associated with one size of barbell plate produced more variable results than the production process associated with the other size.

Considering the weight data first, we find that the standard deviation for the sample of 50-pound plates (.465 pounds) is almost four times as great as the standard deviation of the sample of 10-pound plates (.124 pounds). Does this mean that the production process associated with the heavier plates is inferior to the other? Not necessarily. A standard deviation will reflect the scale of the measurements as well as the true variation. Because 50-pound plates are, on the average, five times as heavy as 10-pound plates, it is not surprising that the standard deviation of the weights of the heavier plates is greater than that of the lighter plates. A sometimes better way of comparing variation between two or more sets of data is by using a measure of *relative variation*, a frequently used one being the *coefficient of variation*.

The coefficient of variation is defined as

$$CV(X) = \frac{s}{\overline{X}}(100) \qquad (4.11.1)$$

where CV(X) stands for the coefficient of variation of a variable designated "X"

Applying Equation (4.11.1) to the barbell-plate weight data, we find

$$CV(X_1) = \frac{s_{X1}}{\overline{X}_1}(100) = \frac{.124}{10.013}(100) = 1.24\%$$

$$CV(X_2) = \frac{s_{X2}}{\overline{X}_2}(100) = \frac{.465}{50.032}(100) = .93\%$$

Viewed in this light, the production process associated with the 50-pound plates may actually be superior with respect to variability.[3]

Turning to the hole-diameter data, we find

$$CV(Y_1) = \frac{s_{Y1}}{\overline{Y}_1}(100) = \frac{.187}{1.22}(100) = 15.33\%$$

$$CV(Y_2) = \frac{s_{Y2}}{\overline{Y}_2}(100) = \frac{.183}{1.20}(100) = 15.25\%$$

Hence, the two production processes turn out virtually identical results with respect to the variation in diameters of holes that the bar must pass through.

If there were any reason to do so, one could even compare the variation in diameters (in inches) to the variation in weights (in pounds). For example, to say that s_{X1} of .124 pounds is less than s_{Y1} of .183 inches is nonsense. But to say that $CV(X_1)$ of 1.24% is less than $CV(Y_1)$ of 15.33%, or $CV(Y_2)$ of 15.25%, is valid. The act of dividing a standard deviation expressed in some unit of measurement by the corresponding arithmetic mean, expressed in the same units, produces a quotient stripped of the original units of measurement. It is merely a percentage of the mean.

Although this example has been useful in illustrating the nature and possible advantages of the coefficient of variation, it has not pretended to solve all of Gemini's quality-control problems. For example, it might not be sufficient that $CV(X_1)$ be less than $CV(X_2)$; management might require that it be considerably less. Even though it might be permissible for a 10-pound plate to vary between, say, 9.9 and 10.1 pounds, it does not necessarily follow that it is permissible for a 50-pound plate to vary by approximately five times as much—that is, between about 49.5 and 50.5 pounds. The overall quality-control task may have more dimensions than our illustrative case has suggested. Still, as a general rule, when descriptive comparisons of variation are desired, one is usually well-advised to consider use of the coefficient of variation.

4.12 Summary

You have now been introduced to three measures of central tendency—the *arithmetic mean, the median*, and *the mode*. For many data sets these measures convey, in their own individual way, approximately where the observations tend to be most concentrated.

The preferred measure of central tendency in a given situation will greatly depend on the way the data are distributed and whether the analysis is descriptive or inductive. For descriptive purposes only, the three measures will be nearly the same and probably of equal usefulness when the data are distributed in an approximately symmetrical manner. If the data are highly skewed, the *median* or even the *mode* might be preferred to the *mean*, because the mean is most sensitive to extreme values. For inductive purposes the *mean* is preeminent. The *arithmetic mean* possesses more stability from sample to sample than either the *median* or the *mode*. It is also an unbiased estimator of its population counterpart for all values of n. These traits are regarded as desirable in statistical induction; for that reason, much more use will be made of the *arithmetic mean* throughout this text than either of the other measures of central tendency.

This chapter also introduced you to several measures of variation. These are measures that show us by how much, overall, the observations under study differ from one another. The *range* has little to recommend it except simplicity. For descriptive purposes only, the *mean deviation*, defined as the average absolute deviation from the arithmetic mean, has much in its favor—including ease of interpretation.

In formal statistical induction, the *variance* and the *standard deviation* (the choice depends on the kind of analysis being conducted) are used vastly more often than other measures of variation.

When descriptive comparisons of variation are to be made between two or more sets of data, the comparison may be, and often should be, carried out using *coefficients of variation*. The *coefficient of variation* expresses the standard deviation of a data set as a percent of the mean of the same data set. Consequently, differences in units of measurement or gross differences in scale are not allowed to dominate the comparison.

Note

1. As Figure 3.8 suggests, this term is a misnomer when applied to the relatively rare frequency distribution that reveals no tendency for the individual observations to cluster around some more-or-less central value. However, it is an apt term for the great majority of frequency distributions.

Terms Introduced in This Chapter

Formulas Introduced in This Chapter

The arithmetic mean of a set of (ungrouped) sample data (n is the number of observations making up the sample, Σ means summation):

$$\overline{X} = \frac{\Sigma\,X}{n}$$

The arithmetic mean of a set of (ungrouped) population data (N is the number of observations making up the population):

$$\mu = \frac{\Sigma\,X}{N}$$

The weighted arithmetic mean of a set of (ungrouped) population data (w is the weight assigned a specified X value):

$$\mu_w = \frac{\Sigma\,wX}{\Sigma\,w}$$

The arithmetic mean of a set of (grouped) sample data ($\cong$ means "is approximately equal to," f is the frequency of a specified class, X_m is a midpoint of a specified class):

$$\overline{X} \cong \frac{\Sigma\,fX_m}{n}$$

The median of a set of (grouped) data (LCL is the lower class limit of the class containing the median, c is the class interval of the class containing the median, cf is the cumulated frequency for the classes positioned above the class containing the median, f_{med} is the number of observations associated with the class containing the median):

$$\text{Median} \cong LCL + c\left(\frac{n/2 - cf}{f_{\text{med}}}\right)$$

The mean deviation for a set of (ungrouped) sample data (| | indicates that absolute values are used):

$$MD = \frac{\Sigma\,|X - \overline{X}|}{n}$$

The variance of a set of (ungrouped) sample data:

$$s^2 = \frac{\Sigma\,(X - \overline{X})^2}{n - 1}$$

The standard deviation of a set of (ungrouped) sample data:

$$s = \sqrt{\frac{\Sigma\,(X - \overline{X})^2}{n - 1}}$$

Shortcut formula for obtaining the standard deviation for a set of (ungrouped) sample data:

$$s = \sqrt{\frac{\Sigma X^2 - \frac{(\Sigma X)^2}{n}}{n-1}}$$

Formula for obtaining the standard deviation for a set of (grouped) sample data:

$$s \cong \sqrt{\frac{\Sigma f(X_m - \overline{X})^2}{n-1}}$$

Shortcut formula for the standard deviation of a set of (grouped) sample data:

$$s \cong \sqrt{\frac{\Sigma f(X_m)^2 - \frac{(\Sigma fX_m)^2}{n}}{n-1}}$$

The variance of a set of (ungrouped) population data:

$$\sigma^2 = \frac{\Sigma (X - \mu)^2}{N}$$

The standard deviation of a set of (ungrouped) population data:

$$\sigma = \sqrt{\frac{\Sigma (X - \mu)^2}{N}}$$

The coefficient of variation for a variable designated X:

$$CV(X) = \frac{s}{\overline{X}} (100)$$

Questions and Problems

4.1 Explain the meaning of each of the following:
  a. Measure of central tendency   e. Range
  b. Grouped data   f. Mean deviation
  c. Σ g. Variance
  d. Arithmetic mean   h. Coefficient of variation

4.2 Explain the meaning of each of the following:
  a. Weighted mean   e. Standard deviation
  b. $\cong$ f. Mode
  c. Median   g. Sum of squares
  d. Measure of variation

4.3 Distinguish between:
  a. Measure of central tendency and measure of variation
  b. Arithmetic mean and median
  c. Arithmetic mean and weighted mean
  d. Ungrouped data and grouped data
  e. $\overline{X}$ and μ

4.4 Distinguish between:
   a. Median and mode
   b. Mean deviation and standard deviation
   c. s and σ
   d. N and n
   e. Sum of squares and variance

4.5 Indicate which of the following statements you agree with and which you disagree with, and defend your opinions:
   a. Use of the term *measure of central tendency* reflects the empirical fact that all frequency distributions exhibit a tendency for the observations to cluster around some central value.
   b. The arithmetic mean obtained from ungrouped data is usually somewhat more accurate than the arithmetic mean for the same data in grouped form.
   c. The arithmetic mean is more sensitive to extreme observations than either the median or the mode.
   d. The standard deviation obtained from grouped data is usually somewhat more accurate than the standard deviation for the same data in ungrouped form.

4.6 Indicate which of the following statements you agree with and which you disagree with, and defend your opinions:
   a. The median is always the middle value in a set of numerical data.
   b. The sum of the deviations from the arithmetic mean is always equal to 0.
   c. The sum of the deviations from a number not equal to the arithmetic mean can never equal 0.
   d. When computing the standard deviation of a set of data, the number of observations to be divided into the sum of squares—that is, $\Sigma (X - \overline{X})^2$ or $\Sigma (X - \mu)^2$—should be reduced by 1 in order to reduce an underestimation bias regardless of whether the data under study represent a sample or a population.

4.7 Indicate which of the following statements you agree with and which you disagree with, and defend your opinions:
   a. If two different sets of data are expressed in different units of measurement, it is not possible to compare them to see which set is the more variable.
   b. The median obtained from ungrouped data is usually somewhat more accurate than the median obtained from the same data in grouped form.
   c. The reason for using an $n - 1$ denominator when computing the standard deviation of a set of sample data is to convert the ordinarily biased sample standard deviation into an unbiased estimator of the population standard deviation.
   d. The mean deviation is easier to interpret than the standard deviation obtained from the same set of data.

4.8 Cite a realistic hypothetical problem in each of the business specialties listed below in which (1) a suitable measure of central tendency would provide all the descriptive information needed and (2) a suitable measure of central tendency, though perhaps helpful, would not provide all the descriptive information needed and would necessarily have to be accompanied by a suitable measure of variation. Be as specific as possible.
   a. Management
   b. Marketing
   c. Finance

4.9 The following numbers were obtained from a table of random digits: 55, 63, 18, 52, 56, 32.
   a. Compute the arithmetic mean of these numbers.
   b. Demonstrate that the sum of deviations from the mean obtained in connection with part *a* is 0.
   c. Demonstrate that the sum of the deviations from the number 48 does not equal 0.

    d. Demonstrate that the sum of the deviations from the number 55 does not equal 0.

    e. Does the size of the difference between the arithmetic mean and another specified number have any bearing on how closely the resulting sum of deviations approximates 0? Elaborate.

4.10 Refer to problem 4.9.

    a. Compute the sum of the squared deviations from the arithmetic mean.

    b. Compute the sum of the squared deviations from 48.

    c. Compute the sum of the squared deviations from 55.

    d. Does the size of the difference between the arithmetic mean and another specified number have any bearing on the size of the sum of squared deviations? Elaborate.

4.11 The following data represent sales achieved by the Thorndike Company over a ten-year period (in millions of dollars): 10, 12, 15, 16, 20, 21, 23, 25, 28, 29.

    a. Compute the arithmetic mean of these numbers.

    b. Demonstrate that the sum of the deviations from the mean obtained in connection with part *a* is 0.

    c. Demonstrate that the sum of the deviations from the number 14 does not equal 0.

    d. Demonstrate that the sum of the deviations from the number 10 does not equal 0.

    e. Does the size of the difference between the arithmetic mean and another specified number have any bearing on how closely the resulting sum of deviations approximates 0? Elaborate.

4.12 Refer to problem 4.11.

    a. Compute the sum of the squared deviations from the arithmetic mean.

    b. Compute the sum of the squared deviations from 14.

    c. Compute the sum of the squared deviations from 10.

    d. Does the size of the difference between the arithmetic mean and another specified number have any bearing on the size of the sum of the squared deviations? Elaborate.

4.13 Given these *population* values—5, 7, 2, 4, 6:

    a. Compute the arithmetic mean of these data.

    b. Compute the standard deviation of these data.

    c. Now add 4 to each of the original values and compute the arithmetic mean and standard deviation of the altered numbers.

    d. What effect did the addition of a constant 4 to the original values have on the arithmetic mean? What is the explanation for this effect?

    e. What effect did the addition of a constant 4 to the original values have on the standard deviation? What is the explanation for this effect?

4.14 Given these population values—2, 1, 2, 3, 5, 1, 4:

    a. Compute the arithmetic mean of these data.

    b. Compute the standard deviation of these data.

    c. Now multiply each of the original values by 3 and compute the arithmetic mean and the standard deviation of the altered numbers.

    d. What effect did multiplication of the original values by a constant 3 have on the arithmetic mean? What is the explanation for this effect?

    e. What effect did multiplication of the original values by a constant 3 have on the standard deviation? What is the explanation for this effect?

4.15 In a unimodal frequency distribution skewed to the left (that is, having a long tail on the left-hand side of the histogram or frequency polygon), which measure of central tendency—the arithmetic mean, the median, or the mode—will have the lowest value? Why? Which will have the highest value? Why?

4.16 *Real Class*, a men's magazine, in an attempt to lure new advertisers runs a full-page ad in which it claims that subscribers of the magazine spend an average of over $1000 a year on gifts for others. Suppose, for simplicity, that the magazine has only five subscribers: Al, who spends $100 on gifts each year; Bill, who spends $300; Chuck, who spends $300; Dan, who spends $400; and Ed, who spends $600.

  a. Compute the arithmetic mean. Is the magazine's claim correct?
  b. Compute the median.
  c. Compute the mode (defined as the most frequently occurring value).
    Now, let us change our assumption about the number of subscribers. Let us say that the magazine has six subscribers made up of (1) the five subscribers listed above and (2) Fred, who spends $1500 a year on gifts.
  d. Compute the arithmetic mean for the six subscribers. Is the magazine's claim correct?
  e. Compute the median for the six subscribers.
  f. Compute the mode for the six subscribers.
    Now, let us change one more time our assumption about the number of subscribers. Let us say that the magazine has seven subscribers made up of (1) the original five subscribers; (2) Fred, who spends $1500 a year on gifts; and (3) Greg, who spends $4000 a year on gifts.
  g. Compute the arithmetic mean for the seven subscribers. Is the magazine's claim correct?
  h. Compute the median for the seven subscribers.
  i. Compute the mode for the seven subscribers.
  j. Describe what happened to the arithmetic mean as you moved from part *a* to part *d* to part *g* of this problem.
  k. Describe what happened to the median as you moved from part *b* to part *e* to part *h* of this problem.
  l. Describe what happened to the mode as you moved from part *c* to part *f* to part *i* of this problem.
  m. What do your results suggest about the (descriptive) relative stability of the arithmetic mean? Of the median? Of the mode?
  n. How do you feel about the ethics of the magazine's advertising that claims the average subscriber spends over $1000 a year on gifts for others?

 4.17 Refer to the gift-expenditure data given in problem 4.16.

  a. Compute the range associated with the gift expenditures of the original five subscribers. Of the six subscribers. Of the seven subscribers.
  b. Compute the mean deviation associated with the gift expenditures of the five subscribers. Of the six subscribers. Of the seven subscribers.
  c. Compute the standard deviation (using the population formula) associated with the gift expenditures of the five subscribers. Of the six subscribers. Of the seven subscribers.
  d. Which of these three measures of variation is most sensitive to the introduction of extreme values? Elaborate. Which is least sensitive to the introduction of extreme values? Elaborate.

 4.18 The College of Business Administration of Gruwell University states in some of its promotional material that the average graduate of the college earns over $200,000 a year. Assume, for simplicity, that only four people have graduated to date: Sam, who earns $25,000 a year; Tom, who earns $35,000; Ula, who earns $35,000; and Vivian, who earns $60,000.

  a. Compute the arithmetic mean. Is the college's claim correct?
  b. Compute the median.
  c. Compute the mode (defined as the most frequently occurring value).
    Now, let us change our assumption about the number of graduates. Let us say that five people have graduated. They consist of (1) the four listed above and (2) Walt, who earns $150,000.
  d. Compute the arithmetic mean for the five graduates. Is the college's claim correct?
  e. Compute the median for the five graduates.
  f. Compute the mode for the five graduates.

Now, let us change one more time our assumption about the number of graduates. Let us say that six people have graduated. They consist of (1) the four original ones; (2) Walt, who earns $150,000 a year; and (3) Xavier, who earns $1 million a year.

   g. Compute the arithmetic mean for the six graduates. Is the college's claim correct?

   h. Compute the median for the six graduates.

   i. Compute the mode for the six graduates.

   j. Describe what happened to the arithmetic mean as you moved from part *a* to part *d* to part *g* of this problem.

   k. Describe what happened to the median as you moved from part *b* to part *e* to part *h* of this problem.

   l. Describe what happened to the mode as you moved from part *c* to part *f* to part *i* of this problem.

   m. What do your results suggest about the (descriptive) relative stability of the arithmetic mean? Of the median? Of the mode?

   n. How do you feel about the ethics of this college in claiming that the average graduate earns over $200,000 a year?

 4.19 Refer to the income data given in problem 4.18.

   a. Compute the range associated with the incomes of the four graduates. Of the five graduates. Of the six graduates.

   b. Compute the mean deviation associated with the incomes of the four graduates. Of the five graduates. Of the six graduates.

   c. Compute the standard deviation (using the population formula) associated with the incomes of the four graduates. The five graduates. The six graduates.

   d. Which of these three measures of variation is most sensitive to the introduction of extreme values? Elaborate. Which is least sensitive to the introduction of extreme values? Elaborate.

 4.20 Refer to Table 3.12. Use only the 20 observations in the left-most column.

   a. Calculate the mean lifetime of Ajax Brand fan belts in this sample of 20.

   b. Calculate the median lifetime of Ajax Brand fan belts in this sample of 20.

   c. What does the comparison of the mean and median suggest about the symmetry of the distribution of lifetimes of the 20 Ajax Brand fan belts? Explain.

   d. Can the median value obtained in connection with part *b* above be interpreted as indicating that one-half of the fan belts in this sample of 20 had lifetimes of less than this value? Explain.

 4.21 Refer to Table 3.12. Use only the 20 observations in the left-most column.

   a. Compute the sum of the deviations from the arithmetic mean. What is your result?

   b. Compute the sum of the deviations from the median. What is your result?

   c. Select an observation value arbitrarily and compute the sum of the deviations from that value. What is your result?

 4.22 Refer to Table 3.12. Use only the 20 observations in the left-most column.

   a. Calculate the range of lifetimes of Ajax Brand fan belts for this sample of 20.

   b. Calculate the mean deviation of this same sample of 20 observations. In what units is the mean deviation expressed?

   c. Calculate the standard deviation of this same sample of 20 observations. In what units is the standard deviation expressed?

   d. Since the range, the mean deviation, and the standard deviation are all measures of variation, should they all have about the same value when obtained from the same set of data? Why or why not?

   e. If all 100 observations in Table 3.12 had been used, would you expect the range to be greater or smaller than the value found in connection with part *a* above? Explain. Would you expect the standard deviation to be greater or smaller than the value found in connection with part *c* above? Explain.

4.23 Refer to Table 3.13. Use only the 20 observations in the *two* left-most columns.
   a. Compute the mean monthly adjusted gross income for this sample of 20 observations.
   b. Compute the median monthly adjusted gross income for this sample of 20 observations.
   c. What does the comparison of the mean and median suggest about the symmetry of the distribution of the 20 monthly adjusted gross income figures? Explain.
   d. Can the median value obtained in connection with part *b* above be interpreted as indicating that one-half of the monthly adjusted gross income values in this sample of 20 were above this value? Explain.

 4.24 Refer to Table 3.13. Use only the 20 observations in the *two* left-most columns.
   a. Compute the sum of the deviations from the arithmetic mean. What is your result?
   b. Compute the sum of the deviations from the median. What is your result?
   c. Select an observation value arbitrarily and compute the sum of the deviations from that value. What is your result?

 4.25 Refer to Table 3.13. Use only the 20 observations in the *two* left-most columns.
   a. Calculate the range of the monthly adjusted gross income values in this sample of 20 observations.
   b. Calculate the mean deviation of this same sample of 20 observations. In what units is the mean deviation expressed?
   c. Calculate the standard deviation of this same sample of 20 observations. In what units is the standard deviation expressed?
   d. Since the range, the mean deviation, and the standard deviation are all measures of variation, should they all have about the same value because they are obtained from this same set of 20 observations? Explain.
   e. If all 50 observations in Table 3.13 had been used, would you expect the range to be greater or smaller than the value found in connection with part *c* above? Explain.

 4.26 Refer to Table 3.11, Part II.
   a. Calculate the mean age of the shoppers. If you had to make any assumptions when computing the mean, state the assumption(s).
   b. Calculate the median age of the shoppers. If you had to make any assumptions when computing the median, state the assumption(s).
   c. Would you expect the mean age obtained from these same data in ungrouped form to be the same as the mean value obtained in connection with part *a* above? Explain.
   d. Would you expect the median obtained from these same data in ungrouped form to be the same as the median obtained in connection with part *b* above? Explain.
   e. Calculate the standard deviation for ages of shoppers (using the sample formula). If you had to make any assumptions, state the assumption(s).
   f. Would you expect the standard deviation obtained from these same data in ungrouped form to be the same as the standard deviation value obtained in connection with part *e* above? Explain.

4.27 Refer to Table 3.11, Part I.
   a. Can the arithmetic mean be determined from this frequency distribution? Why or why not?
   b. Can the median be obtained from this frequency distribution? Why or why not?
   c. Can the standard deviation be obtained from this frequency distribution? Why or why not?

4.28 Refer to Table 3.11, Part III.
   a. Can the arithmetic mean be determined from this frequency distribution? Why or why not?
   b. Can the median be obtained from this frequency distribution? Why or why not?
   c. Can the standard deviation be obtained from this frequency distribution? Why or why not?

4.29 Refer to Figure 3.1.
  a. Calculate the mean time required to fill the standardized order. If you had to make any assumptions when computing the mean, state the assumption(s).
  b. Calculate the median time required to fill the standardized order. If you had to make any assumptions, state the assumption(s).
  c. Would you expect the mean time obtained from these same data in ungrouped form to be the same as the mean value you obtained in connection with part *a* above? Explain.
  d. Would you expect the median time obtained from these same data in ungrouped form to be the same as the median value you obtained in connection with part *b* above? Explain.
  e. Calculate the standard deviation (using one of the formulas for sample data) of times required to fill the standardized order. If you had to make any assumptions, state the assumption(s).
  f. Would you expect the standard deviation obtained from these same data in ungrouped form to be the same as the standard deviation value you obtained in connection with part *e* above? Explain.

4.30 Refer to Table 3.15.
  a. Using data in this table, compute the mean size of bonuses for salespersons in the (1) western United States and (2) eastern United States. Do salespersons in the West or in the East earn larger bonuses, on the average?
  b. Compute the standard deviation of bonuses for salespersons in the (1) western United States and (2) eastern United States.
  c. According to the results you obtained in connection with part *b* above, are bonuses more variable among salespersons in the western United States or salespersons in the eastern United States?
  d. Compute the coefficient of variation for salespersons in the (1) western United States and (2) eastern United States.
  e. What advantages, if any, might use of coefficients of variation have over use of the standard deviations in a comparison of the amount of variation in the two sets of data?
  f. Do your findings from part *d* above lead you to the same conclusion about the amount of variation in the two sets of bonus data as did your findings from part *b*? Elaborate.

4.31 Refer to Table 3.16.
  a. Compute the mean of net income as a percent of stockholders' equity for (1) industrial companies, (2) commercial banks, and (3) retailing companies. Indicate which kind of company has the largest mean. The smallest mean.
  b. Compute the standard deviation of net income as a percent of stockholders' equity for (1) industrial companies, (2) commercial banks, and (3) retailing companies.
  c. According to the results you obtained in connection with part *b* above, for which category of companies are values for net income as a percent of stockholders' equity the most variable? The least variable?
  d. Compute the coefficient of variation for (1) industrial companies, (2) commercial banks, and (3) retailing companies.
  e. What advantages, if any, might use of coefficients of variation have over the use of standard deviations in a comparison of the amount of variation in the three sets of data?
  f. Do your findings from part *d* above lead you to the same conclusion about the amount of variation in the three sets of data as did your findings in connection with part *b*? Elaborate.

The next six problems are quite theoretical and relatively difficult and tedious. However, if you can work them correctly and with understanding, the exercise should help to clarify some points made only briefly in the text thus far as well as prepare you for a successful assault on Chapter 8, one of the most strategic chapters in the book, when the times comes.

4.32 A small population of accounts receivable consists of the following $N = 5$ observations (in hundreds of dollars): 1, 2, 3, 3, 4. A sample of size $n = 3$ is to be selected from this population. The sample selected could be any one of the following:

1, 2, 3	1, 3, 4
1, 2, 3	2, 3, 3
1, 2, 4	2, 3, 4
1, 3, 3	2, 3, 4
1, 3, 4	3, 3, 4

a. Compute the population mean μ.
b. Compute the population median.
c. Compute the arithmetic mean $\overline{X}$ of each of the ten possible samples.
d. Compute the median of each of the ten possible samples.
e. Compute the arithmetic mean of the ten sample means you obtained in connection with part c. How does this mean of the means compare with the population mean you obtained in connection with part a?
f. Compute the arithmetic mean of the ten medians you obtained in connection with part d. How does the mean of the medians compare with the population median you obtained in connection with part b?
g. Which sample measure of central tendency—the arithmetic mean or the median, if either—is an unbiased estimator of the corresponding population parameter? Defend your choice.

 4.33 Refer to problem 4.32. (You must have worked problem 4.32 in order to work this one.)
a. Compute the standard deviation of the ten sample means you obtained in connection with part c of problem 4.32. (Use a denominator of 10.)
b. Compute the standard deviation of the ten sample medians you obtained in connection with part d of problem 4.32. (User a denominator of 10.)
c. Which of the standard deviations is smaller, the one associated with the ten sample means or the one associated with the ten sample medians?
d. What does your answer to part c of this problem suggest about the relative (sample-to-sample) stability of the arithmetic mean? Of the median?

 4.34 Refer to the data in problem 4.32.
a. Compute the standard deviation and the variance of the population data.
b. Compute the standard deviation and variance of each of the ten possible samples. (Use $n - 1$ denominators.)
c. Compute the arithmetic mean of the ten standard deviations you obtained in connection with part b. How does your result compare with the true population standard deviation?
d. Compute the arithmetic mean of the ten variances you obtained in connection with part b. How does your result compare with the true population variance?
e. What sample measure of variation—the standard deviation or the variance, if either—is an unbiased estimator of the corresponding population parameter? Defend your choice.

✔ **4.35** A small population of households is known to own the following numbers of toothbrushes: 0, 3, 3, 4, 4, 4. A sample of size $n = 4$ is to be selected from this population to determine how much interest there is in a recently developed "lifetime" toothbrush. The sample selected could be any one of the following:

0, 3, 3, 4	0, 3, 4, 4	3, 3, 4, 4
0, 3, 3, 4	0, 3, 4, 4	3, 3, 4, 4
0, 3, 3, 4	0, 3, 4, 4	3, 3, 4, 4
0, 3, 4, 4	0, 3, 4, 4	3, 4, 4, 4
0, 3, 4, 4	0, 4, 4, 4	3, 4, 4, 4

a. Compute the population arithmetic mean.
b. Compute the population median.
c. Compute the population mode (defined as the most frequently occurring observation).
d. Compute the arithmetic mean $\overline{X}$ of each of the 15 possible samples.

e. Compute the median of each of the 15 possible samples.
f. Compute the (single) mode of as many of the 15 possible samples as you can.
g. Compute the arithmetic mean of the 15 sample means you obtained in connection with part *d*. How does the mean of the means compare with the population mean you obtained in connection with part *a*?
h. Compute the arithmetic mean of the 15 sample medians you obtained in connection with part *e*. How does the mean of the medians compare with the population median you obtained in connection with part *b*?
i. Compute the arithmetic mean of the 12 modes you obtained in connection with part *f*. How does the mean of the modes compare with the population mode you obtained in connection with part *c*?
j. Which sample measure of central tendency—the arithmetic mean, the median, the mode, if any—is an unbiased estimator of the corresponding population parameter? Defend your choice.

 4.36 Refer to problem 4.35. (You must have worked problem 4.35 in order to work this one.)
a. Compute the standard deviation of the 15 sample means you obtained in connection with part *d* of problem 4.35. (Use a denominator of 15.)
b. Compute the standard deviation of the 15 sample medians you obtained in connection with part *e* of problem 4.35. (Use a denominator of 15.)
c. Compute the standard deviation of the 12 modes you obtained in connection with part *f* of problem 4.35. (Use a denominator of 12.)
d. Which standard deviation is smallest, the one associated with the 15 sample means, the one associated with the 15 sample medians, or the one associated with the 12 sample modes?
e. What does your answer to part *d* suggest about the relative (sample-to-sample) stability of the arithmetic mean? Of the median? Of the mode?

 4.37 Refer to the data in problem 4.35.
a. Compute the standard deviation and variance of the population.
b. Compute the standard deviation and variance of each of the 15 possible samples. (Use $n - 1$ denominators.)
c. Compute the arithmetic mean of the 15 standard deviations you obtained in connection with part *b*. How do your results compare with the true population standard deviation?
d. Compute the arithmetic mean of the 15 variances you obtained in connection with part *b*. How does your result compare with the true population variance?
e. Which sample measure of variation—the standard deviation or the variance, if either—is an unbiased estimator of the corresponding population parameter? Defend your choice.

TAKE CHARGE

4.38 Refer back to problem 3.35.
a. Think of the data set you gathered in connection with problem 3.35 as a population and compute the arithmetic mean μ and standard deviation σ from the *ungrouped data*.
b. Compute the arithmetic mean μ and standard deviation σ from the *frequency distribution* you developed in connection with problem 3.35. How much difference is there between the arithmetic mean value you obtained in connection with part *a* of this problem and the one you obtained from grouped data? How much difference is there between the standard deviation value you obtained in connection with part *a* of this problem and the one you obtained from grouped data?
c. Using the results you obtained in connection with part *a* of this problem, (1) add and subtract the value of the standard deviation to/from that of the arithmetic mean, (2) multiply the standard deviation value you obtained in connection with part *a* by 2, and add

and subtract the resulting product to/from the arithmetic mean, and (3) multiply the standard deviation value you obtained in connection with part *a* by 3, and add and subtract the resulting product to/from the arithmetic mean.

d. What percent of the observations making up your data set are within the range $\mu \pm \sigma$? Within the range $\mu \pm 2\sigma$? Within the range $\mu \pm 3\sigma$?

e. It is known that, if a frequency distribution is bell-shaped or normal (see part (*a*) of Figure 3.8), the following percents will occur:

Range of Values	Percent of Values Within Range
$\mu \pm \sigma$	68.27
$\mu \pm 2\sigma$	95.44
$\mu \pm 3\sigma$	99.74

Comment on how well, or how poorly, your data conform to the bell shape.

Computer Exercise

4.39 Refer to Table 2.5. Have the computer determine the mean, median, variance, and standard deviation of the first 100 observations.

Cases for Analysis and Discussion

4.1 THE "H. STILLMAN COMPANY": PART I

The "H. Stillman Company," a highly successful brewery, has for many years maintained its own construction department which presently employs approximately 900 people. This department's sole function is to provide for internal expansion of the company through the construction and installation of new structures and equipment.

This case concerns certain problem-solving procedures within the concrete production area and has to do specifically with economically maintaining desired levels of compressive strength for psi (pounds per square inch) concrete. Of the various concrete strength levels required in industrial construction, strengths of 5000 psi and greater are the most difficult to maintain on a consistent basis, a situation necessitating frequent strength tests on the part of the quality-control staff.

The process of maintaining the desired compressive strength for 5000 psi concrete at H. Stillman's entails the use of several statistical aids, including a graphic device known as the Histogram and Coefficient of Variation Chart. At the end of each quarter, the quality-control department prepares an updated Histogram and Coefficient of Variation Chart on which is shown (1) the smooth frequency polygon of psi strengths obtained from all tests conducted during the quarter just ended, (2) the "ideal" distribution (plotted as a smooth frequency polygon), and (3) coefficients of variation.

When the engineering department issues a set of plans for a new structure, it indicates the compressive strength requirement (such as 5000 psi) and also a specification of how many strength tests it will allow to fall below the indicated strength requirement. For example, they might say that no more than 10% of the test results can be under 5000 psi.

The quarterly Histogram and Coefficient of Variation Charts thus indicate, through the compression strength values, how close to the "ideal" value compressive strength is being maintained and, through the coefficients of variation, how uniformly that control level is being maintained. From time to time, the information presented will indicate the need for a change in either the operational procedures used or the mix design. Changes in operational procedures do not ordinarily involve an extra outlay of money. Changes in mix design, on the other hand, normally involve a money outlay that is added to the average cost per cubic yard of concrete. It is known from past records that an additional $0.42 of cost will have to be added to the average cost per cubic yard to increase the average strength by 300 psi.

For a construction project begun in 1982, it was determined that to ensure a compressive strength of 5000 psi or higher in over 90% of the tests run, an "ideal" value for compressive strength would be 5740 psi. (That is, the average for all tests run during a particular period of time should be 5740 psi. The specific manner in which this was obtained need not concern us. The main point is that the actual average strength achieved must be set above 5000 psi to ensure that only a small minority of tests turn up results under 5000 psi.) Data for the third and fourth quarters of 1982 are presented in Figure 4.5.

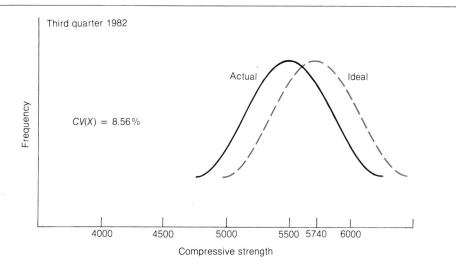

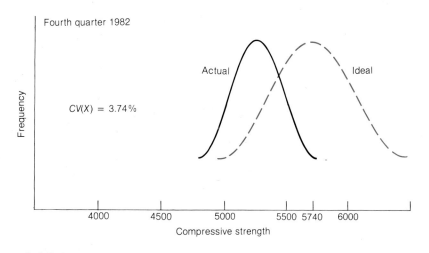

Figure 4.5. Histogram and Coefficient of Variation Charts for the third and fourth quarters of 1982 for the "H. Stillman Company."

Worth noting in connection with Figure 4.5 are the following universal standards for the coefficient of variation: $CV(X)$ less than 10, excellent; $CV(X)$ of 10 and under 15, good; $CV(X)$ of 15 and under 20, fair; and $CV(X)$ equal to or greater than 20, poor.

1. Examine Figure 4.5. How well did the average compressive strength compare with the "ideal" compressive strength during the two quarters? Elaborate.

2. How well did the coefficients of variation measure up against the universal standards for these coefficients? Elaborate.

3. At the end of the third quarter of 1982, a change was made in the operational procedures used. Did this change affect the coefficient of variation?

4. Why do you suppose a change was made in operational procedure rather than in mix design?

5. If a change were to be made in mix design, what would you estimate to be the increase in cost per cubic yard associated with an increase in average compressive strength from where it was at the end of the fourth quarter of 1982 to the "ideal" value of 5740? Show how you arrived at your answer.

This case is continued at the end of Chapter 7.

4.2 GOTHAM CITY HARDWARE COMPANY

The Gotham City Hardware Company, a large wholesaler of hardware products, determined from a careful cost analysis conducted in late 1984 that many of its sales were in small dollar amounts, a situation leading to undesirably large handling and bookkeeping costs per order. In an effort to encourage the company's customers to place few large orders rather than many small ones, management decided to grant a discount of 1% on list price to all buyers whose purchases in 1985 averaged (that is, had an arithmetic mean of) $5000 or more.

One member of management suggested that use of the median would be preferable to the arithmetic mean, because the former measure is affected less by extreme values. The prevailing view, however, was that orders of unusually large size should be encouraged rather than discouraged and that use of the arithmetic mean would serve that purpose better.

Management instituted the new discount policy on a trial basis at first with a view to making it permanent should it prove effective in increasing the dollar amount of the average order. As a means of making this determination, they tabulated dollar amounts of orders placed with the company during 1984; at the end of 1985 similar tabulations were to be made and conclusions drawn about the effectiveness of the new discount policy. Data for three regular customers of Gotham are presented in Table 4.8.

Management was also interested in determining which of its customers was most consistent in terms of size of order placed. The statistical supervisor was asked to develop for each customer a measure that would indicate the amount of variation in the dollar amounts of 1984 orders. He recognized immediately that just about any measure of variation would suffice, but decided on the much used standard deviation.

At the end of 1985, management proceeded to evaluate the discount program. For the three purchasers used here for purposes of illustration, means and standard deviations were computed, as shown in Table 4.9, which is partially completed.

1. Refer to Table 4.8, Parts I, II, and IV. If the discount had been in effect in 1984, would Chase Lumber have qualified for it? Vrontikis Brothers? Hanlin and Haycock? Justify your answers.

2. Refer to Table 4.8, Part III. According to the ungrouped data, if the discount had been in effect in 1984, would Hanlin and Haycock have qualified for it? Justify your answer.

Table 4.8 1984 Purchases of Three Customers of Gotham City Hardware Company

Part I: Chase Lumber Company (Figures in Dollars)	*Part II: Vrontikis Brothers (Figures in Dollars)*
$ 19.95 124.75 3,795.14 5,023.95 7,422.00	$ 1,875.25 3,799.95 5,024.76 7,899.60 8,755.75 10,899.93 11,002.87

Part III: Hanlin and Haycock—Ungrouped Data (Figures in Thousands of Dollars)					*Part IV: Hanlin and Haycock—Grouped Data (Figures in Thousands of Dollars)*	
1.1	3.3	4.3	5.1	6.0	**Dollar Value of Orders**	**Number of Orders f**
1.2	3.3	4.4	5.1	6.1		
1.2	4.0	4.4	5.1	6.1		
1.3	4.0	4.4	5.1	6.3		
2.0	4.0	4.4	5.2	6.6	$ 1 and under 2	4
2.0	4.0	4.4	5.2	7.2	2 and under 3	8
2.1	4.1	4.5	5.2	7.3	3 and under 4	10
2.1	4.1	4.5	5.2	7.3	4 and under 5	30
2.1	4.1	4.6	5.2	7.4	5 and under 6	28
2.2	4.2	4.7	5.3	7.4	6 and under 7	5
2.3	4.2	4.8	5.3	7.4	7 and under 8	7
2.3	4.2	4.9	5.4	7.6	8 and under 9	3
3.0	4.2	5.0	5.4	8.1	9 and under 10	0
3.0	4.2	5.0	5.4	8.2	10 and under 11	3
3.1	4.3	5.0	5.5	8.4	11 and under 12	0
3.1	4.3	5.0	5.5	10.0	12 and under 13	1
3.1	4.3	5.0	5.6	10.0	13 and under 14	1
3.2	4.3	5.0	5.7	10.0		
3.2	4.3	5.0	5.8	12.2	Total	100
3.3	4.3	5.1	5.9	13.0		

Table 4.9 Comparison of Means and Standard Deviations for Three Customers

	Purchaser	Number of Orders	Arithmetic Mean	Standard Deviation
1984	Chase Lumber	5		
	Vrontikis Brothers	7		
	Hanlin and Haycock (Ungrouped data)	100		
1985	Chase Lumber	5	2077.97	3117.42
	Vrontikis Brothers	10	8423.75	3758.41
	Hanlin and Haycock (Ungrouped data)	90	6200.50	1988.00

3. Insofar as Hanlin and Haycock is concerned, if your answers to the two previous questions were not the same, which form of the data—grouped or ungrouped—do you consider better as a basis for determining whether Hanlin and Haycock qualified for the discount? Why?

4. Refer to Table 4.8, Parts I, II, and IV. Compute the standard deviation for each company. Which company's purchases exhibit the least variation? The most variation?

5. Refer to Table 4.9. Fill in the missing information and indicate whether, if the only information you had available was that shown in this table, you would be likely to conclude in favor of implementation of the proposed discount policy. Justify your answer.

4.3 "T.H.E., INCORPORATED": PART I

"T.H.E., Incorporated" is a medium-sized manufacturing company in the fluid power industry. The fluid power industry is composed of manufacturers of products that are used in systems designed for the transmission, application, preparation, and control of pneumatic and hydraulic power.

T.H.E. manufactures more than 15,000 different products. Many of these products are similar to one another and are only slightly different configurations of the same basic product. For example, some units might have a ¼-inch hole instead of a ½-inch hole. Other units might have different colors or make use of different gasket materials, but be exactly the same in all other respects. The current manufacturing values of the products range from $4 to $100, with a weighted mean of approximately $20.

T.H.E.'s potential customer base in the United States is approximately 60% of the manufacturing establishments with more than 20 employees. Many of these plants use the products in their facilities and are called *user customers*. Other manufacturers place the product on the equipment they make and are called *OEMs* (original equipment manufacturers). The majority of the product usage is within the metalworking industry, with one-third of the usage within the machinery manufacturing industry.

Because the fluid power industry is very cyclical, accurate corporate forecasts are desired so that reliable projections of overall manpower requirements, capital expenditures, inventories, and budgets can be made. If forecasts are too high, excess inventory, capacity, and manpower levels will occur. This results in worker layoffs and necessitates other severe cost-cutting measures. If, on the other hand, forecasts are too low, production delays resulting from inadequate inventory, capacity, and manpower levels will result in increased costs, as well as declining sales and market share loss, because customers will go to the competition.

T.H.E. requires forecasts on a yearly, quarterly, and monthly basis. The needed forecasts are at the summary, or corporate, level and are not at the individual-product or product-group level. After several years of experimenting with different approaches to forecasting and checking them against the sobering standard of real-world experience, Sam Rollins, the person in charge of forecast preparation, gradually developed a procedure he considered quite

Table 4.10 Forecasting Methods and Their Weights as Used by "T.H.E., Incorporated"

Name of Method	Weight (%)
1. Short-term projections based on the Bureau of Labor Statistics Composite Index of Leading Indicators	40
2. Forecast prepared by the Institute for Trend Research	15
3. Input from T.H.E.'s own Marketing Department	15
4. Cahner's Fluid Power Market Early Warning Forecast	15
5. Ratio using company sales to consensus gross national product forecast	15

satisfactory (in early 1981, at any rate). The procedure involved a regular calculation of a weighted mean of estimates produced by five different forecasting methods. Weights were assigned to these methods according to their accuracy during the most recent ten-year period. The methods (which you are not expected to understand) and their weights are shown in Table 4.10.

1. In view of the relatively heavy weight assigned the first forecasting method listed in Table 4.10, why do you think Sam Rollins and his co-workers still prefer to combine estimates obtained from five different forecast sources?

2. Do you believe the weights were determined from a rigorous mathematical procedure, or is it more likely that judgment played an important role? Defend your choice.

This case is continued at the end of Chapter 5.

5 Index Numbers

You can't add apples and oranges.—Old Saying

The cost of living has gone up $1.00 a quart.—W. C. Fields

What You Should Learn from This Chapter

This is the third chapter on descriptive statistics; it is also the last one until we deal with time series analysis in Chapter 16. When you have completed this chapter, you should be able to

1. Tell what is meant by *index number*.
2. Distinguish between a simple relative index and a composite index.
3. Construct a simple relative index.
4. Recite and discuss some fundamental considerations in index number construction.
5. Distinguish between a Laspeyres index and a Paasche index, and recite some advantages and disadvantages of each.
6. Construct a Laspeyres index and a Paasche index.
7. Differentiate between weighted aggregate indexes and weighted relative indexes.
8. Deflate a value series using a price index.

This chapter is self-contained, though you may wish to review section 4.3 of Chapter 4, including the material on the weighted mean.

5.1 Introduction

During 1980 more than 7 million wage earners received automatic pay increases of several cents an hour. During this same year, Reginald Van Steen enjoyed a $10,000 a year increase in spending money from a trust fund established by his billionaire father. The same year Sally Roberts of Akron, Ohio, started getting $75 more each month in child support payments from her ex-husband, Gerald. These seemingly unrelated occurrences actually did have a common cause—namely, an unprecedented 13.5% increase in the Consumer Price Index (CPI).

During this year of rapid inflation, some people, like those mentioned above, enjoyed some purchasing power protection through contractual provisions tying their incomes to increases of specified size in the CPI. Many others, not covered by such accommodating contracts, suffered a temporary decrease in purchasing power but largely recouped their losses over the following year, when sizable wage increases were negotiated. Alas, still others lost purchasing power which might never be regained.

People gain purchasing power, merely hold their own, or lose purchasing power depending on whether their nominal income increases faster than, on a par with, or more slowly than the rate of increase in the prices of consumer goods. This gives you some idea of how important an index such as the CPI can be. To be sure, it is the underlying inflation, not the way it is measured, which actually wields the influence. Still, when millions of workers receive automatic pay increases each time the Consumer Price Index increases by, say, a mere 0.4%, we see that the method of measuring inflation is of importance in its own right.

Index numbers have other uses as well. For example, many a business leader finds such broad indexes as the Consumer Price Index, the Index of Industrial Production, the Index of Net Business Formation, the Index of Construction Contracts, and others invaluable as transmitters of news about diverse facets of the general economy. In addition, index numbers constructed to serve more narrowly defined purposes are also helpful to the business decision maker. For example, index numbers can (1) provide simple methods for making comparisons over time, from place to place, from product line to product line, and so forth; (2) facilitate comparisons of changes in activities measured in different units; and (3) provide a means of adjusting data expressed in current dollar terms into data expressed in terms of dollars of constant purchasing power.

5.2 What Is an Index Number?

A really complete definition of either *index* or *index number* could be rather involved because indexes and index numbers are used in a wide variety of ways in different disciplines. However, for our purposes the following definitions will suffice:

> An *index* is a device used to measure the relative change in a set of measurements over time, compared to a specific base period. The measurements of interest may represent a single variable or several variables whose values have been combined in some manner.

> An *index number* is simply one of a series of measures of relative change resulting from the application of an index formula to the relevant data.

These definitions can be readily clarified through use of the simplest kind of index, the *simple relative*.

> A *simple relative index* is one that expresses each observation making up a single variable as a percent of a base-period value.

We will soon be analyzing an illustrative case in which the costs of various grains are of considerable importance to a manufacturer of food products. Thus, it will be convenient to illustrate the concept of simple relative index by referring to the historical costs of one such grain. In Table 5.1 the average annual prices (in dollars per bushel) received by farmers for

Table 5.1 Construction of a Simple Relative Index for Wheat Prices

Year	(1) Price of Wheat (Dollars per Bushel)	(2) Simple Relative Index Number
1963	$(2.04 \div 1.37) \cdot 100 =$	148.9
1964	$(1.85 \div 1.37) \cdot 100 =$	135.0
1965	1.37 etc.	100.0
1966	1.35	98.5
1967	1.63	119.0
1968	1.39	101.5
1969	1.24	90.5
1970	1.24	90.5
1971	1.33	97.1
1972	1.31	95.6
1973	1.76	128.5
1974	3.96	289.1

wheat between 1963 and 1974 are shown in column (1); the corresponding simple price relatives, using 1965 as the base year, are shown in column (2).

We see in Table 5.1 that the 1965 price of wheat was $1.37 per bushel, whereas the price in 1966 was $1.35. Therefore, the simple price relative associated with 1966 is simply the ratio of $1.35 to $1.37 multiplied by 100. That is, $(1.35/1.37)(100) = 98.5$. (Index numbers are percentages, but usually the percent sign is omitted.) The 98.5 value indicates that the 1966 price was 98.5% of the 1965 price.

The simple price index may be expressed symbolically in terms of the base-period price p_0 and the price in some other year p_i using the formula

$$I_i = \frac{p_i}{p_0} \cdot 100 \qquad (5.2.1)$$

Sometimes clarity is increased through the use of calendar year subscripts, such as

$$I_{66} = \frac{p_{66}}{p_{65}} \cdot 100 \qquad (5.2.2)$$

Simple relative indexes have the advantage of (1) being easy to compute and comprehend and (2) requiring a bare minimum of human judgment in their development. On the other hand, the simple relatives are of limited usefulness, expressing as they do a single variable in a somewhat altered form and nothing more. More challenging to construct, more dependent on sound human judgment, but more potentially useful are indexes collectively known as *composite indexes*.

Composite indexes are indexes measuring relative changes in several variables in combination.

This definition immediately prompts a question to which much of the remainder of this chapter addresses itself: How can one best combine prices (quantities, sales volumes, etc.) of several items to obtain a kind of average, or composite, summary of them?

5.3 Some Fundamental Considerations in Composite Index Construction

The development of index numbers is seldom solely an exercise in applied mathematics. The task requires the judicious combining of mathematical methods and informed judgment. Generally speaking, the simpler the situation to be depicted by index numbers, such as a situation which can be handled adequately with a simple relative index approach, the greater the relative contribution of impersonal mathematics. Conversely, the more complex the situation, the greater the relative importance of the wisdom of the person constructing the index. Thus, before proceeding with the mathematical aspects of composite index construction, we must consider several qualitative matters, including

1. Purpose of the index
2. Items to be included
3. Nature and the quality of the data required
4. Base period to be used
5. Specific manner in which the index is to be constructed

Few inalterable rules are available regarding any of these items. The discussions that follow will consist largely of generalizations distilled from the experiences of many of those working with index numbers, but these should be adapted to suit the specific situation for which an index is to be developed. In an effort to increase the meaningfulness of several important points, we can look at a specific case example.

ILLUSTRATIVE CASE 5.1

Western Food Products, Inc., produces a wide variety of dry cereals and animal feeds as well as a few bakery items; consequently, the company makes considerable use of several kinds of grain. It purchases its grain directly from farmers at prices corresponding with those determined on the organized grain exchanges. Because of the wide price fluctuations in the grains used, the threat always exists of the company's having to adjust its own selling prices often and by substantial amounts, a situation considered undesirable for a number of reasons, most of them obvious. Fortunately, grain prices do not often fluctuate in unison. The company can, within certain well-defined limits, substitute for a grain whose price is rising rapidly grains whose prices are declining—or, at worst, rising less rapidly. Because of Western's ability to substitute one grain for another, price increases in one or a few grains will not necessarily put pressure on the company's profit margin and force an increase in selling prices. On the other hand, when most or all relevant grain prices are increasing rapidly in unison, substitution among grains can only help a little. At such times, since the company's profit margins are already relatively thin, Western Food Products faces cost-push pressures that it can cope with only by increasing its own selling prices.

Because when grain prices are rising on a broad front Western must put selling price increases into effect promptly, management in 1975 desired a method of monitoring its grain costs in an aggregative and up-to-date manner. Toward this end, they charged their chief statistician with the responsibility of constructing a grain price index that could be computed on a yearly, quarterly, monthly, weekly, and, if possible, even daily basis. Management decided that the price index should measure the combined prices of the following grains: (1) corn, (2) grain sorghums, (3) oats, (4) rice, and (5) wheat. Other grains, they reasoned, were used irregularly and in such small quantities that they need not be considered. It was thought that 1965 was a fairly typical year for grain prices and sufficiently recent (at that time) to serve as a base. Historical prices for the strategic grains were known to be readily available and comparable over time.

How should the statistician proceed to construct the composite grain price index? For simplicity of discussion, we will take a few liberties with the case and deal with only three years of *annual* data. But first, let us consider some of the more qualitative aspects of index number construction.

Purpose of the Index

The purpose for which an index is to be used should be clearly stated before any attempt is made to construct it. To be sure, indexes *do* sometimes find important uses for which they were not originally intended. Still, a clear, precise statement of the purpose of the index will serve as a helpful organizing device because it will suggest—indeed, in some cases, dictate— answers to certain other questions such as those pertaining to items to be included and the nature, quality, and availability of needed historical data.

In the Western Food Products case the statistician needs a price index that is easily kept up to date and that reflects the collective price changes for five specific grains. Therefore, management has defined the purpose of the index in relatively precise terms. Such clarity of purpose immediately renders the problem of determining items to be included nonexistent and, at the same time, suggests much about data requirements.

Items to Be Included

In the Western Food Products case, the purpose of the index dictates the items to be included. Management specified that the price index to be constructed should reflect price movements in five specific grains. In statistical terms, what management has done is declare that the items whose prices are to be reflected in the resulting index constitute, for all practical purposes, an entire population of grains of interest. The problem, then, is clearly one belonging to the realm of descriptive statistics.

If, however, there had been ten different grains of concern to the company and, for simplicity, only five had been selected for inclusion in the price index, the problem would be one belonging to the realm of inductive statistics. At the very least, the five grains would have to be, in some sense, "representative" of the ten. In such an instance, the task of constructing a price index would have been more complex because the problem of sampling would have been an integral part of the overall project.

More often than not, index number construction does involve sampling. Some reasons for this are obvious. First, sometimes the number of items of interest is so large that, for simplicity of index construction, as well as simplicity of keeping the numbers up-to-date, a few items will be selected to represent the many. Second, certain economies can be realized when sampling is employed.

Usually, when sampling is used in index number construction, it is judgment rather than random sampling. At least two reasons for this preference can be discerned. (1) In most cases, no formal techniques of statistical inference are used and thus there is no real need for random selection, and (2) such considerations as speed of calculations, currency, and quality of data are often of overriding importance—a fact sometimes rendering random sampling impracticable. Ideally, when judgment sampling is used, some serious research should be undertaken to determine the most important items for inclusion.

Another valid generalization about the selection of items, though one often impossible to comply with without loss of representativeness, is that commodities whose physical characteristics are substantially the same over time are preferable to those whose characteristics change. Here again, the statistician with Western Food Products enjoys an advantage not always available to the index-number worker. The change in the price of a particular grade of wheat, for example, can be measured more accurately than the change in the price of, say, a power lawn mower. The quality of a particular grade of wheat remains the same from one year to the next, whereas the quality of a power lawn mower may not. Suppose that a change were made in a particular brand of power mower which allowed the machine to cut closer to walls and shrubs than had previously been the case. In such an event, an increase in price would reflect, in whole or in part, a change in quality.

Nature and Quality of the Data Required

It goes without saying that one cannot make appropriate comparisons, whether through the use of index numbers or some other statistical intermediary, if the necessary data cannot be obtained. Anyone aspiring to construct a composite index, therefore, is well advised to find out which relevant data are available and which are not just as soon as the purpose of the index has been defined and some criteria for selection of items have been set up. If having the index numbers up to date is a matter of utmost importance, as it is in the Western Food Products case, one cannot very well include items for which data on prices or, if relevant, quantities do not become available until after a considerable passage of time. Moreover, historical data should be comparable over time both in definition and method of measurement.

Base Period to Be Used

Regardless of the degree of simplicity or complexity of an index, a past period must be selected whose value will serve as a standard against which values for other periods can meaningfully be compared.

In the final analysis, selection of a base period is largely arbitrary. However, there *are* two rules of base period selection that many index number workers find helpful. But you should be aware that the rules are quite general and their valid application depends greatly on the knowledge and experience of the person(s) constructing the index. The first rule is that the period used as a base should be a "typical" or "representative" period. Ideally, the base period should be neither too high nor too low, relative to the values for other periods. If the base value is too high, the resulting index numbers will appear chronically depressed, because most of them will be less than 100. If the base value is too low, distortion of the opposite kind may result. The second general rule is that the base of an index should be relatively recent because, generally speaking, one is most interested in comparing current values with values generated by a roughly similar economic environment.

Specific Manner in Which the Index is to Be Constructed

You will recall that in Chapter 4 we discussed three measures—the mean, the median, and the mode—that in their own unique ways serve as representative values for a set of data. We also touched on the weighted mean, which is computed much like the ordinary arithmetic mean, except that individual observations are assigned weights reflecting their relative importance. These measures are all "averages" even though they are computed differently and do at times differ markedly in their values. If one desires an "average," he or she must first determine what, in the situation at hand, is the most appropriate kind of average to calculate.

The same point holds for selection of a composite index formula. The person constructing a composite index has many formulas from which to choose; if he is wise, he will be familiar with as many of these as possible. An examination of several methods available for composite index construction, as well as some of the strengths and weaknesses of each, is the subject matter of most of the remainder of this chapter.

The symbols we will use are as follows:

I_i = Index number associated with a specific time period
p_i = Price of a particular commodity during a specific time period
p_0 = Price of a particular commodity during the base period
q_i = Quantity bought or sold of a particular commodity during a specific time period
q_0 = Quantity bought or sold of a particular commodity during the base period
N = Number of commodities included in the index
V = Value = pq

5.4 Aggregate Index Numbers

In Table 5.2 the prices of the five grains of principal concern to Western Food Products, Inc., are provided for the years 1965, 1971, and 1974. A price index for several grains may be determined in a manner similar to that used in section 5.2 for a single grain. This would involve the calculation of an *aggregate* of the price for each year by summing together the prices for the five grains. Such an aggregate may be either weighted or unweighted.

Unweighted Aggregate Price Index

Let us deal with unweighted aggregates first. The unweighted aggregate price index is determined by the formula

Table 5.2 Price and Quantity for Five Grains 1965, 1971, and 1974

Grain	Price (Dollars per Bushel)			Quantity Purchased (Thousands of Bushels)		
	1965	1971	1974	1965	1971	1974
Corn	1.17	1.33	2.55	9	7	4
Grain sorghums	1.07	1.16	2.16	5	4	5
Oats	0.73	0.71	1.34	3	6	5
Rice	2.21	2.33	6.21	7	7	4
Wheat	1.37	1.33	3.96	10	20	8

$$I_i = \frac{\Sigma\, p_i}{\Sigma\, p_0}\,(100) \qquad\qquad (5.4.1)$$

Using 1965 as the base year, we obtain from the data in Table 5.2 the price aggregates

$$\Sigma\, p_{65} = 1.17 + 1.07 + 0.73 + 2.21 + 1.37 = 6.55$$
$$\Sigma\, p_{71} = 1.33 + 1.16 + 0.71 + 2.33 + 1.33 = 6.86$$
$$\Sigma\, p_{74} = 2.55 + 2.16 + 1.34 + 6.21 + 3.96 = 16.22$$

Therefore, the unweighted aggregate price index values for the years 1971 and 1974 would be

$$I_{71} = \frac{\Sigma\, p_{71}}{\Sigma\, p_{65}}\,(100) = \frac{6.86}{6.55}\,(100) = 104.7$$

$$I_{74} = \frac{\Sigma\, p_{74}}{\Sigma\, p_{65}}\,(100) = \frac{16.22}{6.55}\,(100) = 247.6$$

We see, then, that the aggregate price index for the five grains of interest increased by 4.7% (104.7 − 100.0 = 4.7) from 1965 to 1971 and by a startling 147.6% from 1965 to 1974.

Such unweighted price indexes have the virtue of simplicity, but they also have a potentially important shortcoming: the resulting index numbers are affected by something other than price—namely, the units in which the prices are expressed. When we computed these index numbers, we used the per-bushel price for each of the grains. But suppose that, whereas Western Food Products customarily purchases corn, oats, and wheat in bushel units, as we have assumed, they have bought grain sorghums and rice in units of 100 pounds (approximately 1.79 and 2.22 bushels, respectively). If we had used in our calculations prices for these two grains expressed in dollars per hundredweight, we would have made use of grain sorghum cost values of $1.92, $2.08, and $3.87 and rice cost values of $4.91, $5.17, and $13.79. The resulting simple aggregates would be

$$\Sigma\, p_{65} = 1.17 + 1.92 + 0.73 + 4.91 + 1.37 = 10.10$$
$$\Sigma\, p_{71} = 1.33 + 2.08 + 0.71 + 5.17 + 1.33 = 10.62$$
$$\Sigma\, p_{74} = 2.55 + 3.87 + 1.34 + 13.97 + 3.96 = 25.69$$

The corresponding simple aggregate index numbers for 1971 and 1974 would be

$$I_{71} = \frac{\Sigma\ p_{71}}{\Sigma\ p_{65}}\ (100) = \frac{10.62}{10.10}\ (100) = 105.1$$

$$I_{74} = \frac{\Sigma\ p_{74}}{\Sigma\ p_{65}}\ (100) = \frac{25.69}{10.10}\ (100) = 254.4$$

Clearly, these results differ somewhat from those obtained when all prices were expressed in per-bushel terms. One way to avoid having units of measurement affect the index numbers is to apply weights to the prices. The weights, if wisely selected, can (1) allow any kind of units (including any mix of units) to be used and (2) provide measures of relative importance of the component commodities.

In price index construction the units normally used are quantities, denoted by q. The product of $p \times q$ represents total value and is unaffected by the units chosen since p is the price per unit and q is the number of units. By summing the $p \times q$ terms for several items, we obtain weighted aggregate values, $\Sigma\ pq$.

Laspeyres Index

Various kinds of indexes can be computed using weighted price aggregates. The differences among those presented in this text result from the choice of quantities used. We first consider the *Laspeyres index*, which is calculated from

$$I_i = \frac{\Sigma\ p_i q_0}{\Sigma\ p_0 q_0}\ (100) \tag{5.4.2}$$

Table 5.3 Data from Table 5.2 Subjected to Laspeyres Index-Number Calculations

	Price (Dollars per Bushel)			Quantity	Value $p_{65}q_{65}$	$p_{71}q_{65}$	$p_{74}q_{65}$
Grain	1965	1971	1974	1965	1965	1971	1974
Corn	1.17	1.33	2.55	9	10.53	11.97	22.95
Grain sorghums	1.07	1.16	2.16	5	5.35	5.80	10.80
Oats	0.73	0.71	1.34	3	2.19	2.13	4.02
Rice	2.21	2.33	6.21	7	15.47	16.31	43.47
Wheat	1.37	1.33	3.96	10	13.70	13.30	39.60
				Total	47.24	49.51	120.84

$$I_{71} = \frac{\Sigma\ p_{71}q_{65}}{\Sigma\ p_{65}q_{65}} \cdot 100 = \frac{49.51}{47.24} \cdot 100 = 104.8$$

$$I_{74} = \frac{\Sigma\ p_{74}q_{65}}{\Sigma\ p_{65}q_{65}} \cdot 100 = \frac{120.84}{47.24} \cdot 100 = 255.8$$

The Laspeyres index uses quantity weights q_0 from the base period so that only prices are allowed to change. The denominator $\Sigma\, p_0 q_0$ provides the total value of those commodities when they were purchased in the base period. The numerator $\Sigma\, p_i q_0$ represents what the total value of those commodities would be in the more recent time period *if exactly the same quantity of each was purchased as was purchased in the base period.*

> Therefore, we can say that a *Laspeyres index number* tells us how much a particular collection of commodities (kinds and quantities held constant) would cost during a period of interest relative to its base-period cost.

The Laspeyres index numbers for 1971 and 1974 are calculated in the lower portion of Table 5.3.

Paasche Index

Let us take a look at another kind of weighted aggregate index, the Paasche index, which is calculated using the formula

$$I_i = \frac{\Sigma\, p_i q_i}{\Sigma\, p_0 q_i}\,(100) \tag{5.4.3}$$

The Paasche index is similar to the Laspeyres except that *the quantities associated with recent periods*, rather than with the base period, are used in computing the price aggregates. If, as is likely, the quantities purchased of the various grains change over time, the two indexes will provide different percent changes.

Let us use the current-period quantity weights as presented in Table 5.2 to compute the Paasche index numbers for the years 1971 and 1974. The procedures are demonstrated in the lower portion of Table 5.4.

> The *Paasche index numbers* tell us how much a particular collection of commodities would cost, relative to its base-period cost, in a period of interest if exactly the same quantities of exactly the same commodities had been purchased during the base period as were purchased in the period of interest.

Laspeyres and Paasche Indexes Compared

According to the Paasche index results shown in Table 5.4, grain prices increased by slightly smaller percentages than indicated by the Laspeyres index approach. We see that for 1971 the Laspeyres index indicated 104.8 and the Paasche index 102.3; for 1974 the Laspeyres

Table 5.4 Data from Table 5.2 Subjected to Paasche Index-Number Calculations

Grain	Price			Quantity		Value			
	1965	1971	1974	1971	1974	$p_{65}q_{71}$	$p_{65}q_{74}$	$p_{71}q_{71}$	$p_{74}q_{74}$
Corn	1.17	1.33	2.55	7	4	8.19	4.68	9.31	10.20
Grain sorghums	1.07	1.16	2.16	4	5	4.28	5.35	4.64	10.80
Oats	0.73	0.71	1.34	6	5	4.38	3.65	4.26	6.70
Rice	2.21	2.33	6.21	7	4	15.47	8.84	16.31	24.84
Wheat	1.37	1.33	3.96	20	8	27.40	10.96	26.60	31.68
					Total	59.72	33.48	61.12	84.22

$$I_{71} = \frac{\Sigma\, p_{71}q_{71}}{\Sigma\, p_{65}q_{71}} \cdot 100 = \frac{61.12}{59.72} \cdot 100 = 102.3$$

$$I_{74} = \frac{\Sigma\, p_{74}q_{74}}{\Sigma\, p_{65}q_{74}} \cdot 100 = \frac{84.22}{33.48} \cdot 100 = 251.6$$

index said 255.8 whereas the Paasche index indicated 251.6. As suggested by these results, a general rule is that the size of the discrepancies between Laspeyres and Paasche index numbers can be expected to increase as the time differential between the base period and the more recent period becomes greater. There are two principal reasons for this: One is the changing tastes of consumers which, in the Western Food Products example, will wield an effect upon the quantities of the specific grains purchased by the company. The Laspeyres index would make use of the earlier quantities for all grains, giving, perhaps, unrealistically large weights to some grains purchased in smaller amounts in the more recent period than during the base period.

A second explanation for the growing discrepancies between Laspeyres and Paasche index numbers is that in a period of inflation of some prices, there may be a downward shift in quantities bought of the relatively higher-priced commodities. As mentioned in Illustrative Case 5.1, the substitution of cheaper grains for dearer ones, within specified limits, is an avowed policy of Western Food Product's management.

For Western's purposes, then, use of the Paasche index approach might be preferable to the Laspeyres index approach because changes in both tastes and relative prices—as reflected in revised quantity weights—are taken into account.

On the other hand, the Paasche index numbers are more difficult to compute. From the standpoint of sheer practicality, the use of base-period quantities, as in the Laspeyres index, is preferable to the current quantity values of the Paasche index. One reason is that collecting data to serve as current quantity weights is sometimes a very expensive and time-consuming pursuit, so that to do so every year (or quarter, month, week, or day, as Western Food Products would like) would be extremely impractical. Moreover, even assuming that this company has an advantage not enjoyed by many users of index numbers—namely, that of knowing instantly when a shipment of grain arrives (or an order is placed)—the daily and weekly, maybe even monthly or quarterly, index numbers would seemingly have to be computed using 0 weights for some of the grains.

5.5 Composite Price-Relative Indexes

In section 5.2 we considered the simplest kind of price index available, the simple relative index. The index numbers, you will recall, were determined by expressing a recent price p_i for a single commodity as a ratio to the base price p_0 for the same commodity and multiplying the resulting quotient by 100. In this section we examine ways of combining such simple relative indexes to make composite relative indexes.

Unweighted Price-Relative Index

The simplest kind of index to compute in this broad category is the *unweighted* price-relative index. The index numbers are obtained by

$$I_i = \frac{\sum\left[\left(\dfrac{p_i}{p_0}\right)(100)\right]}{N} \qquad (5.5.1)$$

Table 5.5 demonstrates how the individual index numbers are computed for the Western Food Products problem. Results for the years 1971 and 1974 are shown at the bottom of this table.

Table 5.5 Data from Table 5.2 Subjected to Unweighted Price-Relative Index Calculations

| | | | | | Price Relative | |
| | | Price | | | $(p_{71}/p_{65}) \cdot 100$ | $(p_{74}/p_{65}) \cdot 100$ |
Grain	1965	1971	1974		1971	1974
Corn	1.17	1.33	2.55		113.7	217.9
Grain sorghums	1.07	1.16	2.16		108.4	201.9
Oats	0.73	0.71	1.34		97.3	183.6
Rice	2.21	2.33	6.21		105.4	281.0
Wheat	1.37	1.33	3.96		97.1	289.1
				Total	521.9	1173.5

$$I_{71} = \frac{\sum\left[\left(\dfrac{p_{71}}{p_{65}}\right)(100)\right]}{N} = \frac{521.9}{5} = 104.4$$

$$I_{74} = \frac{\sum\left[\left(\dfrac{p_{74}}{p_{65}}\right)(100)\right]}{N} = \frac{1173.5}{5} = 234.7$$

This kind of index has an important advantage over and above mere simplicity: The resulting index numbers are independent of the units in which commodity prices are expressed. For example, the same index values would be obtained whether grain sorghums and rice were priced in terms of bushels or in terms of hundredweights. The principal disadvantage of this kind of index is one it shares with the unweighted price aggregate index—namely, all items are treated as equally important. As we have seen, this feature can make an index less meaningful as a measure of how a company such as Western Food Products is affected by general increases in costs.

Weighted Price-Relative Index

Recall that when we discussed aggregative price indexes we saw that quantities may often serve as appropriate weights for assigning different degrees of importance to prices of the various commodities. For price relatives, value weights will accomplish the same end. For a grain price index, a specific value weight would be the total amount paid for the corresponding grain. A weighted price-relative index would then be expressed by the weighted average

$$I_i = \frac{\Sigma\left[\left(\dfrac{p_i}{p_0} \cdot 100\right) \cdot V\right]}{\Sigma V} \qquad (5.5.2)$$

When $V = p_0 q_0$, the index is obtained by

$$I_i = \frac{\Sigma\left[\left(\dfrac{p_i}{p_0} \cdot 100\right) \cdot p_0 q_0\right]}{\Sigma p_0 q_0} \qquad (5.5.3)$$

Since the p_0's in the numerator cancel, we see that this expression is the algebraic equivalent of that for the Laspeyres index. The computational procedures are demonstrated in Table 5.6. The index numbers shown in the lower portion of this table, 104.8 for 1971 and 255.8 for 1974, are exactly those obtained in Table 5.3 for the Laspeyres index.

When $V = p_0 q_i$, the following index formula results:

$$I_i = \frac{\Sigma\left[\left(\dfrac{p_i}{p_0} \cdot 100\right) \cdot p_0 q_i\right]}{\Sigma p_0 q_i} \qquad (5.5.4)$$

In this formula, the p_0's again cancel, leaving us with the Paasche formula.

Table 5.6 Data from Table 5.2 Subjected to Value-Weighted Price-Relative Index Calculations

		Price		Price Relative $\left(\dfrac{p_{71}}{p_{65}}\right)\left(\dfrac{p_{74}}{p_{65}}\right)\cdot 100$		Quantity	Value	Value-Weighted Price Relative $v\left(\dfrac{p_{71}}{p_{65}}\right)\cdot 100 \quad v\left(\dfrac{p_{74}}{p_{65}}\right)\cdot 100$	
Grain	1965	1971	1974	1971	1974	1965	$V = p_{65}q_{65}$		
Corn	1.17	1.33	2.55	113.7	217.9	9	10.53	1197.26	2294.49
Grain sorghums	1.07	1.16	2.16	108.4	201.9	5	5.35	579.94	1080.16
Oats	0.73	0.71	1.34	97.3	183.6	3	2.19	213.09	402.08
Rice	2.21	2.33	6.21	105.4	281.0	7	15.47	1630.54	4347.07
Wheat	1.37	1.33	3.96	97.1	289.1	10	13.70	1330.27	3960.67
		Total		521.9	1173.5		47.24	4951.10	12084.47

$$I_{71} = \frac{\sum\left[\left(\dfrac{p_{71}}{p_{65}}\cdot 100\right)\cdot p_{65}q_{65}\right]}{\sum p_{65}q_{65}} = \frac{4951.10}{47.24} = 104.8 \qquad \frac{\sum\left[\left(\dfrac{p_{74}}{p_{65}}\cdot 100\right)p_{65}q_{65}\right]}{\sum p_{65}q_{65}} = \frac{12084.47}{47.24} = 255.8$$

Whether to use price relatives instead of aggregates is usually a matter of data availability. The statistician for Western Food Products, for example, would have no compelling reason to employ weighted price relatives because price data for the relevant grains are readily available. Moreover, even if current-period quantity weights were used, data on current purchases of the grains would presumably be at his disposal. On the other hand, if the index was concerned with measuring price movements in manufactured steel products, members of the industry might have an easier time supplying value weights in the form of *value added in manufacturing* (sales minus cost of raw materials) than information on the number of bars of this and the number of sheets of that produced during the period of interest. If so, use of a weighted price-relative index would be appropriate.

5.6 Deflating a Value Series Using a Price Index

Many important measures of business activity, whether for the general economy or for an industry or a firm, express the current market value of a collection of commodities. Whenever use is made of such data, a problem of interpretation is potentially ever present. Changes in the dollar volume of a company's sales, for example, reflect not only changes in the physical volume of goods sold, but changes in selling prices as well. Therefore, dollar sales volume cannot be used as a measure of physical volume of goods sold—unless one can find a way of removing the effect of changes in selling prices. Illustrative Case 5.2 illustrates this point.

ILLUSTRATIVE CASE 5.2

The sales manager of Harth and Holmes Appliances, a manufacturer of an extensive line of home appliances, was severely criticized in a board of directors' meeting for the failure of sales to increase adequately during his nine years as a member of top management. He countered with the argument that sales volume had increased by fully 54.5% during his years

Table 5.7 Use of a Price Index to Convert Current Dollar Figures into Constant Dollar Figures

Year	(1) Company Sales (Millions of Current Dollars)	(2) Index of Producers' Prices for Finished Consumer Goods (1967 = 100)	(3) Company Sales (Millions of 1967 Dollars)
1972	$39.6	116.6	$33.9
1973	38.0	129.2	29.4
1974	41.0	149.3	27.5
1975	43.6	163.6	26.7
1976	43.0	169.0	25.4
1977	48.9	178.9	27.3
1978	51.9	192.6	26.9
1979	59.8	215.7	27.7
1980	61.2	246.8	24.8

Source: Bureau of Labor Statistics.

as sales manager, a perfectly respectable rate of increase. Mr. Holmes, the president of the company, reminded him that these were highly inflationary years; when sales volume is measured free of the effects of price inflation, the growth was actually somewhat negative—not a very good showing for almost a decade of effort. Mr. Holmes reinforced his argument through use of the information in Table 5.7.

In column (1) we see the sales volume data that the sales manager used to determine his value for percent increase in dollar sales volume. His reasoning ran as follows: In 1972, the year just prior to his advancement to the sales manager position, sales were $39.6 million; in 1980, the recent full year, sales were $61.2 million. Therefore, $[(61.2 - 39.6)/39.6] \cdot 100 = 54.5\%$.

However, these data represent not only the increase in the real volume of sales but the increase in prices of the goods produced as well. Mr. Holmes dramatized this unfortunate fact by deflating the current dollar sales-volume figures of column (1) in Table 5.7 by using the index of Producers' Prices for Finished Consumer Goods, listed in column (2). The resulting price-deflated sales-volume data are shown in column (3) of the table. Mr. Holmes calculated the percent change as $[(24.8 - 33.9)/33.9] \cdot 100 = -2.7$.

Whether the specific index used in this illustrative case was the best possible deflator for these sales data is a question on which opinions would undoubtedly differ. Nevertheless, the president's position was valid enough: When current dollar data are stripped of the effects of price inflation, the true rate of growth has been lackluster at best.

5.7 Summary

Index numbers are used in a wide variety of ways: Some, like the Consumer Price Index (CPI), serve as the basis for automatic wage increases for many millions of workers. Several important indicators of the health of the general economy are in index-number form. Special-purpose indexes facilitate making comparisons over time, from place to place, from product line to product line, and so forth, even when the constituent variables are expressed in different units. Finally, price indexes are used to adjust data expressed in current dollar terms into data expressed in terms of dollars of constant purchasing power.

Index construction can vary in complexity over a considerable range. At one extreme is the simple relative index, which merely expresses each observation for a single variable as a percent of a base-period value. At the opposite extreme we find weighted composite indexes which may be either in the form of a weighted aggregate index or a weighted relative index. These may involve use of a great many variables—variables of interest in their own right or of interest because they are considered representative of a still larger set of variables—and the manner of assigning weights may be quite complex.

The two most well-known kinds of weighted aggregate *price indexes* are the Laspeyres, which utilizes base-period quantity weights, and the Paasche, which utilizes current-period quantity weights. A weighted relative price index may be the algebraic equivalent of a Laspeyres or Paasche index. Often the decision to use a weighted aggregate index or a weighted relative index is based on the nature of the available data.

When composite indexes are to be constructed, serious attention must be paid such considerations as (1) the purpose of the index, (2) the items to be included, (3) the nature and quality of the data required, (4) the base period to be used, and (5) the specific manner in which the component variables are to be combined.

Terms Introduced in This Chapter

aggregate index (p. 174)
   unweighted aggregate index (p. 175)
   weighted aggregate index (p. 176)
base period (p. 169)
composite index (p. 170)

index (p. 169)
index number (p. 169)
Laspeyres index (p. 177)
Paasche index (p. 177)

relative index (p. 179)
   simple relative index (p. 169)
   unweighted relative index (p. 179)
   weighted relative index (p. 180)
value series (p. 181)

Formulas Introduced in This Chapter

Glossary of Symbols:

I_i = Index number associated with a specific time period
p_i = Price of a particular commodity during a specific time period
p_0 = Price of a particular commodity during the base period
q_i = Quantity bought or sold of a particular commodity during a specific time period
q_0 = Quantity bought or sold of a particular commodity during the base period
N = Number of commodities included in the index
V = Value = pq

Simple relative price index number for time period i:

$$I_i = \frac{p_i}{p_0} (100)$$

Unweighted aggregate price index number for time period i:

$$I_i = \frac{\Sigma \, p_i}{\Sigma \, p_0} (100)$$

Laspeyres index number for time period i:

$$I_i = \frac{\Sigma \, p_i q_0}{\Sigma \, p_0 q_0} (100)$$

Paasche index number for time period i:

$$I_i = \frac{\Sigma \, p_i q_i}{\Sigma \, p_0 q_i} (100)$$

Unweighted price-relative index number for time period i:

$$I_i = \frac{\Sigma \left[\left(\dfrac{p_i}{p_0} \right)(100) \right]}{N}$$

General formula for a weighted relative price index for time period i:

$$I_i = \frac{\Sigma \left[\left(\dfrac{p_i}{p_0} \cdot 100 \right) V \right]}{\Sigma \, V}$$

A weighted relative price index number for time period i. Gives results equivalent to a Laspeyres index:

$$I_i = \frac{\Sigma \left[\left(\dfrac{p_i}{p_0} \cdot 100 \right) p_0 q_0 \right]}{\Sigma \, p_0 q_0}$$

A weighted relative price index number for time period i. Gives results equivalent to a Paasche index:

$$I_i = \frac{\Sigma \left[\left(\dfrac{p_i}{p_0} \cdot 100 \right) p_0 q_i \right]}{\Sigma \, p_0 q_i}$$

Questions and Problems

5.1 Explain the meaning of each of the following:
    a. Index number   **c.** Weighted price-relative index
    b. Simple relative index   **d.** Price-deflated value series

5.2 Distinguish between:
    a. Simple relative index and composite index
    b. Unweighted aggregate price index and weighted aggregate price index
    c. The Laspeyres index and the Paasche index
    d. Unweighted price-relative index and weighted price-relative index
    e. Weighted price-relative index and Laspeyres price index

5.3 Indicate which of the following statements you agree with and which you disagree with, and defend your opinions:

  a. One shortcoming of an unweighted aggregate index is that the index numbers may be affected by the units in which the constituent items are measured.

  b. The Paasche-type index utilizes the total value that a certain collection of items would have in the more recent period if exactly the same quantity of each was purchased as was purchased in the base period.

  c. Generally speaking, a Paasche-type aggregate weighted index is more difficult to keep up to date than is a Laspeyres-type aggregate weighted index.

5.4 Cite a realistic hypothetical situation in which the development of an index would be potentially useful, and then indicate why the answer to each of the following questions would be important in determining the specific manner of constructing the index:

  a. What is the purpose of the index?

  b. What items are to be included in the index?

  c. What kinds of data are required?

  d. What quality of data is required?

  e. What base period should be used?

  f. What weights should be assigned to items included in the index?

5.5 A random sample of 10,000 households was selected during each of three years to develop a price index for fractional-horsepower household appliances. The average price paid for each item was determined as well as the number of households purchasing the items during the years under consideration. The data are shown in Table 5.8.

Using 1978 as the base year, compute the unweighted aggregate price index numbers for 1978, 1979, and 1980. What was the percent increase in prices between 1978 and 1980?

5.6 Refer to problem 5.5. Using the 1978 quantity weights, compute the Laspeyres price index numbers for 1978, 1979, and 1980. What was the percent increase in prices between 1978 and 1980?

5.7 Refer to problem 5.5. Using the quantities for the current year (as distinguished from the base year) as weights, compute the Paasche price index numbers for 1978, 1979, and 1980. What was the percent increase in prices between 1978 and 1980?

5.8 Refer to problem 5.5. Compute the unweighted price-relative index numbers for 1978, 1979, and 1980. What was the percent increase in prices between 1978 and 1980?

5.9 Refer to problem 5.5. Using 1978 value weights, compute the weighted price-relative index numbers for 1978, 1979, and 1980. What was the percent increase in prices between 1978 and 1980?

Table 5.8 Price and Quantity Data for Constructing a Price Index of Fractional-Horsepower Household Products

Products	Prices (Dollars per Unit)			Quantities (Number of Units Purchased by the Sample Households)		
	1978	1979	1980	1978	1979	1980
Electric toothbrushes	6.00	6.00	6.50	500	550	550
Electric steak knives	8.00	8.50	9.00	500	650	800
Foot vibrator	9.50	11.00	10.00	200	300	350
Electric scissors	5.00	5.50	6.00	400	400	400

Table 5.9 Price and Quantity Data for Constructing a Cost Index for Repair Work Ordered by the Titan Machinery Company

Type of Worker	Prices (Dollars per Hour)				Quantities (Number of Hours During Year)			
	1977	1978	1979	1980	1977	1978	1979	1980
Carpenter	20	22	25	30	10	25	50	80
Electrician	15	17	20	25	5	7	25	35
Plumber	25	30	35	40	3	5	8	12
General repair	10	13	18	22	300	400	400	600

 5.10 From time to time, the Titan Machinery Company must hire skilled workers to tend to repair jobs as they come up. The company pays these workers by the hour, and so it now wishes to construct a cost index for such repair work. Data for cost per hour for four kinds of workers and the number of hours each kind of worker was employed, in each of four years, are shown in Table 5.9.

Using 1977 as the base year, compute the unweighted aggregate cost index numbers for 1977, 1978, 1979, and 1980. What was the percent increase in costs between 1977 and 1980?

5.11 Refer to problem 5.10. Using the 1977 quantity weights, compute the Laspeyres-type cost index numbers for 1977, 1978, 1979, and 1980. What was the percent increase in costs between 1977 and 1980?

5.12 Refer to problem 5.10. Using the quantities for the current year (as distinguished from the base year) as weights, compute the Paasche-type cost index numbers for 1977, 1978, 1979, and 1980. What was the percent increase in costs between 1977 and 1980?

5.13 Refer to problem 5.10. Compute the unweighted cost-relative index numbers for 1977, 1978, 1979, and 1980. What was the percent increase in costs between 1977 and 1980?

5.14 Refer to problem 5.10. Using 1977 value weights, compute the weighted cost-relative index numbers for 1977, 1978, 1979, and 1980. What was the percent increase in costs between 1977 and 1980?

5.15 The Stewart family owns a show-quality Afghan hound which they take very good care of. The dog is now six years old and in top contest-winning form. The Stewarts wish to determine by how much the cost of raising their pet has increased during the prior five-year period. Data are shown in Table 5.10.

Using 1978 as the base year, compute the unweighted cost-relative index numbers for 1978, 1979, 1980, 1981, and 1982. What was the percent change in costs between 1978 and 1982?

5.16 Refer to problem 5.15. Using the 1978 quantity weights, compute the Laspeyres-type index numbers for the years 1978 through 1982. What was the percent change in cost during this time?

5.17 Refer to problem 5.15. Using the quantities for the current year (as distinguished from the base year) as weights, compute the Paasche-type cost index numbers for the years 1978 through 1982. What was the percent change in costs during this time?

Table 5.10 Cost Data for the Stewart Family's Afghan Hound

	Price per Unit					Number of Units				
Nature of Expense	1978	1979	1980	1981	1982	1978	1979	1980	1981	1982
Dog food	7.00	8.00	8.50	9.50	10.00	13	11	12	12	13
Veterinarian	15.00	20.00	27.00	30.00	35.00	6	2	5	3	3
Supplemental vitamins	6.00	6.50	8.00	8.00	9.00	15	16	15	16	16
Boarding	25.00	30.00	35.00	35.00	40.00	14	14	14	14	14
Obedience training	150.00	175.00	175.00	200.00	250.00	20	10	5	5	5

5.18 Refer to problem 5.15. Compute the unweighted cost-relative index numbers for the years 1978 through 1982. What was the percent change in costs during this time?

5.19 Refer to problem 5.15. Using 1978 value weights, compute the weighted cost-relative index numbers for the years 1978 through 1982. What was the percent change during this time?

5.20 Current dollar values for the part of the nation's gross national product concerned with housing, called Residential Fixed Investment, are shown in column (1) of Table 5.11. Values of a special price index for this kind of spending are presented in column (2) of this same table.
   Deflate the current dollar values using the corresponding index numbers and comment on the extent to which use of the price index in this manner alters the configuration of the Residential Fixed Investment data for the ten years under study.

5.21 Current dollar values for consumer expenditures for nondurable goods are shown in column (1) of Table 5.12. Constant dollar values are presented in column (2) of this same table.

Table 5.11 Ten Years of Expenditures Data for Residential Structures and Corresponding Price Index Numbers

Year	(1) Residential Structures (Billions of Current Dollars)	(2) Price Index (1972 = 100)
1967	28.6	77.0
1968	34.5	80.7
1969	37.9	87.7
1970	36.6	90.6
1971	49.6	94.9
1972	62.0	100.0
1973	66.1	110.6
1974	55.1	122.1
1975	51.2	133.2
1976	67.7	143.9

a. Comment on the manner in which use of the price index has altered the configuration of the nondurables' expenditure data.

b. Derive the index numbers that were used to deflate the current dollar data.

Table 5.12 Ten Years of Current and Constant Dollar Data for Personal Nondurables' Consumption Expenditures

Year	(1) Current Dollars (Billions)	(2) Constant Dollars (Billions)
1967	$212.6	$259.5
1968	230.4	270.2
1969	247.0	276.4
1970	264.7	282.7
1971	277.7	287.5
1972	299.3	299.3
1973	333.8	309.3
1974	376.2	303.5
1975	409.1	306.1
1976	440.4	319.2

Table 5.13 Price and Quantity Data for Seven Farm Products, 1965–1983

	Corn		Grain Sorghums		Oats	
Year	Price (Cents per Bushel)	Domestic Disappearance (Millions of Bushels)	Price (Cents per 100 Pounds)	Domestic Disappearance (Hundredweight in Thousands)	Price (Cents per Bushel)	Domestic Disappearance (Millions of Bushels)
1965	117	332.5	208	237.8	73.2	886.0
1966	116	372.2	198	582.0	66.4	856.0
1967	124	369.8	211	614.0	72.5	833.0
1968	103	388.5	196	545.0	70.2	784.0
1969	108	396.6	197	626.0	64.1	840.0
1970	116	418.9	207	647.0	63.2	843.0
1971	133	397.7	232	694.0	70.9	883.0
1972	108	438.7	205	701.0	66.0	837.0
1973	157	473.3	324	666.0	80.0	804.0
1974	255	465.3	464	708.0	130.0	759.0
1975	303	367.7	501	443.0	168.0	662.0
1976	254	409.3	446	514.0	166.0	647.0
1977	215	412.1	349	425.2	174.0	572.7
1978	202	433.4	354	467.9	127.0	593.9
1979	225	494.4	400	555.7	143.0	602.8
1980	252	519.4	465	496.7	157.0	567.0
1981	311	487.4	536	312.5	209.0	505.8
1982	250	498.4	429	442.1	214.0	529.1
1983	268	520.0	496	524.6	165.0	543.2

(continued)

TAKE CHARGE

5.22 One of the oldest continually published price indexes available to students of business conditions is the Bureau of Labor Statistics' Producer Price Index (PPI—originally called the BLS Wholesale Price Index). The PPI is a Laspeyres-type weighted aggregate index utilizing, at this writing, a 1967 base period and covering some 2800 commodities.

Journalists and business commentators in general often examine the Producer Price Index for clues about future movements in the Consumer Price Index. The rationale for this is that changes in the inflation climate will show up at the producer wholesale level right away. Producers will attempt to adjust their prices appropriately; however, it will take a few months for such price changes to become felt at the consumer level.

a. Gather ten years of monthly data for both the Consumer Price Index and the Producer Price Index.

b. Plot both sets of price data on the same sheet of graph paper.

c. Evaluate the claim that marked changes in consumer prices can be foreseen by examining the Producer Price Index.

Computer Exercise

5.23 Table 5.13 shows 19 years of price data (average price received by farmers) and quantity data (measured by domestic disappearance, that is, quantity used up in the United States) for seven farm products.

Using a base period of 1977 and domestic disappearance values as weights, have the computer determine Laspeyres and Paasche price index values for the 19 years represented.

Table 5.13 *(Continued)*

	Rice		Wheat		Barley		Soybeans	
Year	Price (Cents per 100 Pounds)	Domestic Disappearance (000 cwt)	Price (Cents per Bushel)	Domestic Disappearance (Millions of Bushels)	Price (Cents per Bushel)	Domestic Disappearance (Millions of Bushels)	Price (Cents per Bushel)	Domestic Disappearance (Millions of Bushels)
1965	490	31.1	137	643.6	95	369.9	262	526.3
1966	493	30.8	135	731.2	102	318.7	254	589.1
1967	495	31.9	163	673.1	105	339.3	275	612.4
1968	497	33.7	139	633.3	101	335.0	249	633.6
1969	500	35.6	124	739.7	92	359.0	243	659.6
1970	495	33.1	124	763.9	89	388.0	235	797.5
1971	517	34.4	133	772.0	97	427.0	285	824.4
1972	534	35.4	131	849.3	104	408.0	303	786.1
1973	673	35.7	176	798.8	117	383.0	437	803.5
1974	1380	37.8	396	753.5	203	336.0	565	897.3
1975	1120	41.0	404	671.9	258	339.0	664	781.2
1976	835	40.3	355	725.8	238	334.8	492	935.5
1977	702	42.7	273	754.4	235	329.6	681	865.5
1978	949	37.7	233	859.0	168	333.3	588	1003.2
1979	816	49.2	298	837.0	180	384.7	666	1103.0
1980	1050	49.2	382	783.1	216	375.7	628	1208.0
1981	1280	54.5	396	776.2	260	349.3	761	1109.0
1982	905	59.4	368	855.7	221	376.4	605	1129.0
1983	811	61.0	352	934.5	171	412.8	588	1203.0

Source: Commodity Year Book, various years.

Cases for Analysis and Discussion

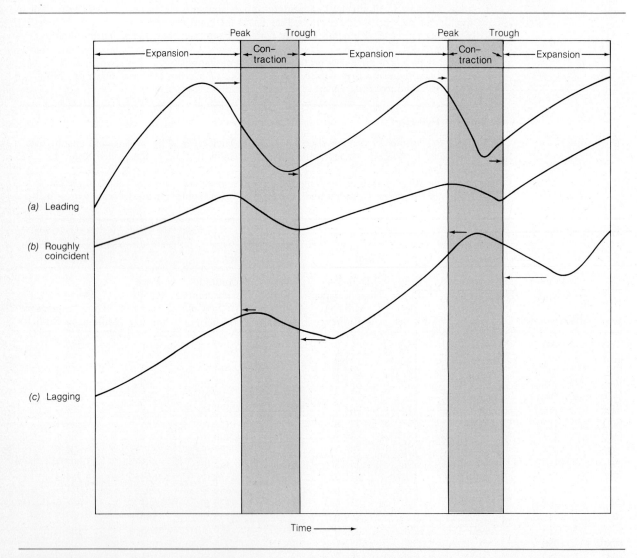

5.1 NBER COMPOSITE INDEXES OF BUSINESS ACTIVITY

The National Bureau of Economic Research (NBER), a respected nonprofit economic research organization, has determined that some economic measures tend to move through periods of cyclical expansion and contraction in a manner exhibiting a more or less consistent timing relationship with corresponding cyclical phases in the general economy. These eco-

Figure 5.1. Illustration of hypothetical leading, roughly coincident, and lagging indicators. Sections labeled "Expansion" and "Contraction" represent periods of sustained upward and downward movements, respectively, in the *general economy*. "Peak" represents the point in time when the general economy ceased its upward movement and began a downward movement. "Trough" represents the point in time when the general economy ceased its downward movement and began an upward movement.

nomic measures undergo changes in cyclical direction at about the same time as the general economy (and, thus, have come to be labeled "roughly coincident indicators"); others tend to change cyclical direction somewhat earlier ("leading indicators") or somewhat later ("lagging indicators"). Figure 5.1 illustrates this point in a somewhat idealized way. A partial list of such indicators (those making up the composite indexes we will soon discuss) is presented in Table 5.14.

Another achievement of National Bureau economists is that they have pinpointed as closely as possible the specific months in which expansions and contractions in the *general economy* have come to an end. Their chronology of business cycle turning points reaches clear back to 1854. This turning-point information for the most recent business cycles is shown in Figure 5.2; the shaded columns in this figure represent periods of contraction (or recession).

The successful identification of leading, roughly coincident, and lagging indicators has led to an interest in combining groups of indicators in order to average out the sometimes large irregular movements in individual indicators. Consequently, the National Bureau of Economic Research, in cooperation with the U.S. Department of Commerce, has combined (1) several leading indicators, (2) several roughly coincident indicators, and (3) several lagging indicators into three separate composite indexes. Current and historical values for these indexes are

---

Table 5.14 Partial List of the NBER Leading, Roughly Coincident, and Lagging Indicators

I. Leading Indicators

1. Average workweek of production workers, manufacturing
2. Average weekly initial claims, state unemployment insurance
3. New orders for consumer goods and materials, 1972 dollars
4. Vendor performance, percent of companies receiving slower deliveries
5. Index of net business formation (1967 = 100)
6. Contracts and orders for new plant and equipment, 1972 dollars
7. New building permits, private housing units index (1967 = 100)
8. Net change in inventories on hand and on order, 1972 dollars
9. Change in sensitive crude materials prices
10. Change in total liquid assets
11. Index of Stock Prices, 500 common stocks (1941–1943 = 10)
12. Money Supply—M_2, 1972 dollars

II. Roughly Coincident Indicators

1. Employees on nonagricultural payrolls
2. Personal income less transfer payments, 1972 dollars
3. Index of industrial production (1967 = 100)
4. Manufacturing and trade sales, 1972 dollars

III. Lagging Indicators

1. Average duration of unemployment
2. Manufacturing and trade inventories, 1972 dollars
3. Index of labor cost per unit of output, manufacturing (1967 = 100)
4. Average prime rate charged by banks
5. Commercial and industrial loans outstanding, large commercial banks
6. Ratio, consumer installment credit to personal income

Source: Business Conditions Digest.

---

published each month in *Business Conditions Digest.** The methodology used to combine these indicators is too complex for treatment here. Suffice to say that each composite index is a weighted index, wherein each component variable is weighted according to how well it performs as judged by a set of criteria having to do with such things as (1) dependability of its timing relationship to the general economy, (2) speed of availability of data, (3) smoothness, and so forth.† Each month *Business Conditions Digest* also presents values for the ratio of the composite index of coincident indicators to the composite index of lagging indicators. This ratio tends to change from expansion to contraction even earlier than the composite index of leading indicators. Data for the three composite indexes and this ratio are presented graphically in Figure 5.2.

1. Refer to Table 5.14. Several indicators on this list are expressed in terms of 1972 dollars. Why do you think they are not expressed in terms of current dollars?

2. In your opinion, should 1966 have been officially declared a year of recession? Tell why or why not.

3. Using only the information contained within Figure 5.1, would you say that the U.S. economy was about to emerge from a recession at the end of 1982? Justify your answer.

 ## 5.2 "T.H.E., INCORPORATED": PART II

Reread Part I of this case (Case 4.3 at the end of Chapter 4) and Case 5.1.

The following account, in Sam Rollins' own words, describes what he perceived to be the uncertainties he faced in June of 1981:

> The April detail of the Composite Index of Leading Indicators was reviewed. Seven of the ten available components [two become available on a delayed basis] were positive in April [a fact not known for sure before late May]; one indicator was flat and the other two were down.
>
> On June 26, I gave my department's fairly upbeat forecast [based on the weighted average method referred to in connection with Case 4.3] to T.H.E.'s president for comments. After review, the president accepted the forecast.
>
> The Composite Index of Leading Indicators for May 1981 was scheduled for release on Tuesday June 30, 1981. A few minutes after 10:00 A.M. the president called me and mentioned he had heard that all of the leading indicators except one were down, and the drop in the overall index was significant. He asked me to check this out, review my forecast, and suggest changes if appropriate.
>
> It was true. Nine of the ten components of the May leading indicators were down, and the drop in the overall composite index was 1.8%. [Later revisions resulted in a smaller percentage decline.]
>
> Was my forecast wrong? What was the proper thing for me to do? To make a long story short, after much anxious review of the efforts of my co-workers and myself, I told the president: "Make no change at this time."

*U.S. Department of Commerce, Bureau of Economic Analysis.
†For the details, see "Composite Indexes: A Brief Explanation and the Method of Construction," *Handbook of Cyclical Indicators: A Supplement to the Business Conditions Digest* (Washington, D.C.: U.S. Department of Commerce, Bureau of Economic Analysis, 1977), p. 73. Also helpful is this article which appeared in the same publication: Victor Zarnowitz and Charlotte Boschan, "Cyclical Indicators: An Evaluation and New Leading Indexes."

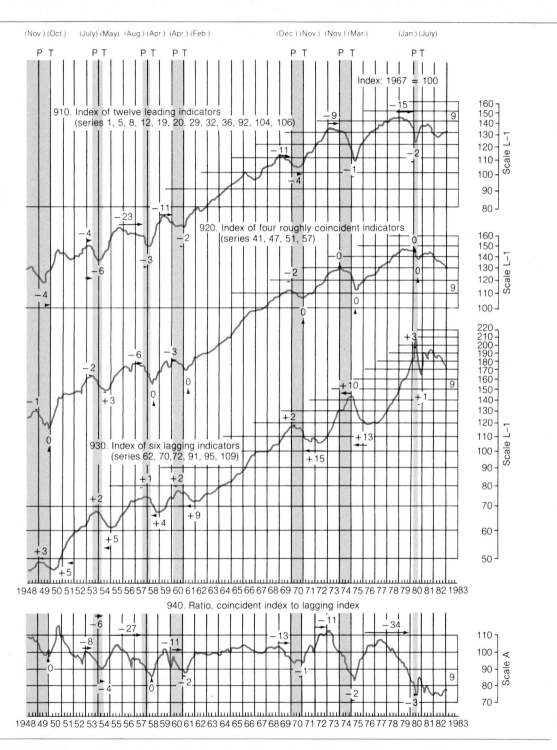

Figure 5.2. NBER indicators. Numbers entered on the chart indicate length of leads (−) and lags (+) in months from reference turning dates. (*Source: Business Conditions Digest,* Bureau of Economic Analysis, U.S. Department of Commerce)

Cases for Analysis and Discussion

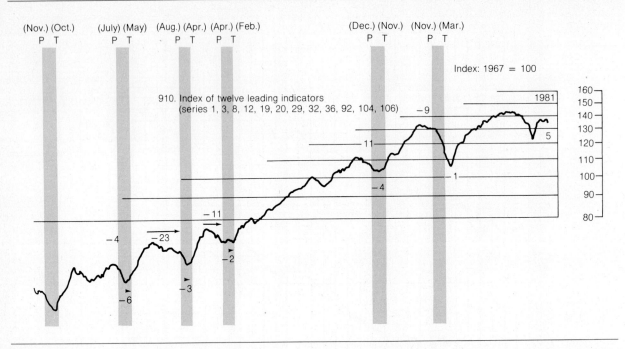

Figure 5.3. Composite index of leading indicators. (*Source: Business Conditions Digest,* Bureau of Economic Analysis, U.S. Department of Commerce)

Figure 5.3 shows the Composite Index of Leading Indicators as it would have appeared in graphic form if the most recent news could have been charted instantly and shown alongside past data.

1. Judging only by what Figure 5.3 shows, did Sam Rollins make the right decision? Putting it another way: Is it the decision you would have made had you been in his place? Tell why or why not.

2. Examine the top part of Figure 5.2, which shows the Composite Index of Leading Indicators in a somewhat more up-to-date form. Does this chart change your opinion as expressed in connection with Part I of this case? Explain.

3. What must the other forecasting methods have been suggesting to Sam Rollins when he said, "Make no change at this time"?

 ## 5.3 THE CONSUMER PRICE INDEX (CPI)

The stated intent of the Bureau of Labor Statistics—the builder of the famous, some would say infamous, Consumer Price Index—has always been more limited than widely realized: The CPI was designed to measure change in the average price of a representative sample of goods and services purchased by typical wage earners and clerical workers in urban areas of the United States. This concept is still employed. However, with the 1978 updating and revisions, two Consumer Price Indexes have come into being. One is an updated version of the index of long standing, the Consumer Price Index for Urban Wage Earners and Clerical Workers. The other one is a new Consumer Price Index for All Urban Consumers. This new index reflects the spending patterns of, in addition to wage earners and clerical workers, previously neglected groups such as professionals, managers, technical workers, short-term workers, the self-employed, and others. Thus, the breadth of coverage of the new index is about twice that of the revised version of the old Consumer Price Index. Nevertheless, the old version is still being used because organized labor has felt that it better reflects the spending patterns of blue-collar workers.

Both versions of the CPI attempt to measure changes in total dollar cost of a specific

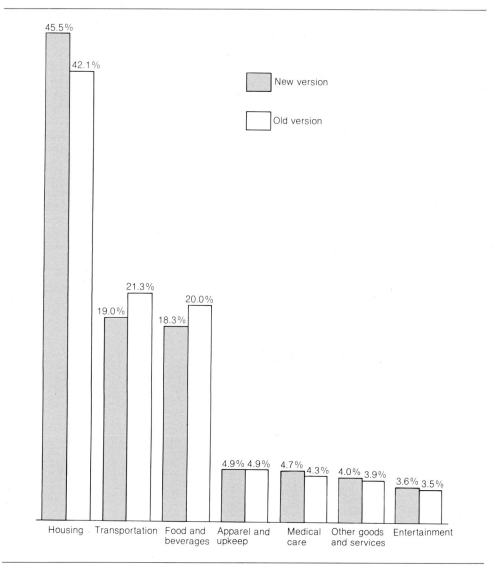

Figure 5.4. Relative weights used to compute current values of the two versions of the Consumer Price Index.

combination of goods and services. Current index values are computed by use of the formula

$$CPI_t = \frac{\Sigma\, p_0 q_0 \left(\dfrac{p_t}{p_0}\right)}{\Sigma\, p_0 q_0}$$

where CPI_t is the value of the index in the current period, p_t values are the various component prices in the current period, p_0 values are the component prices in the base period, and q_0 values are the component quantities in the base period. Figure 5.4 (p. 195) shows the relative importance of major categories (covering roughly 400 individual items) of goods and services whose prices are reflected in current values of both versions of the CPI. These quantity weights are revised from time to time as a result of shifts noted in consumer spending patterns as revealed by periodic surveys conducted by the Bureau of Labor Statistics (BLS). These surveys have been conducted in 1917–1919, 1934–1936, 1950–1951, 1960–1961, and 1972–1973. Some revision of the weights was made in 1978.

One of the most controversial components of the Consumer Price Index (either version) is housing. During the inflationary period of the late 1960s, the 1970s, and early 1980s, many economists came to take a dim view of the way the BLS measured housing costs. The Bureau's methods pertained almost entirely to the cost of buying a home, including interest costs. This approach, some believe, overstates the rate of inflation—that is, as inflation increases and mortgage rates are forced upward, this measure exerts a disproportionately large upward movement in the CPI. Add this to the fact that the typical urban resident does not buy a house in any given month and that the great majority of homeowners are really paying interest rates agreed upon prior to the relatively recent emergence of high rates of inflation, and you have a serious upward bias in the inflation numbers. As this is being written, BLS is in the process of altering the way the housing component is treated so that the cost of renting is given more consideration.

1. The Consumer Price Index is commonly referred to as a *modified* Laspeyres index. In what sense is it a Laspeyres index? In what sense is it modified?

2. Judging from the weights shown in Figure 5.4, do you think that organized labor is being sensible, or just obstinate, in insisting that the old index better reflects the spending behavior of blue-collar workers?

3. Which of the two ways referred to above for treating housing expenditures makes the most sense to you? Why?

6 Fundamentals of Probability Theory

The race is not always to the swift nor the battle to the strong—but that's the way to bet.
—Damon Runyon

It is remarkable that a science which began with the consideration of games of chance should have become the most important object of human knowledge.—Pierre Simon Laplace

What You Should Learn from This Chapter

This is the first of three chapters concerned with probability theory. In this chapter you are introduced to some fundamental terminology and principles pertaining to this important subject. When you have finished studying Chapter 6, you should be able to

1. Define what is meant by *random experiment* and give examples of random experiments.
2. Distinguish among the (1) a priori, (2) relative-frequency, and (3) subjective approaches to determining probabilities.
3. Know what is meant by (1) sample space, (2) basic outcome, (3) event, and (4) probability distribution.
4. Know what is meant by (1) mutually exclusive, (2) independent, (3) complementary, and (4) collectively exhaustive events.
5. Know the difference between conditional and unconditional probability.
6. Be able to work problems involving the addition theorem of probability—both the general case and the special case.
7. Be able to work problems involving the multiplication theorem of probability—both the general case and the special case.
8. Be able to work problems involving the use of the addition theorem of probability and the multiplication theorem of probability in combination.
9. Know what kinds of problems can be worked using Bayes' theorem and be able to apply the theorem.

This chapter is self-contained and *requires* no review of material presented earlier. However, your study of topics in this chapter related to bivariate probability distributions will probably benefit from a reading of the earlier optional material on bivariate frequency distributions (section 3.6 of Chapter 3).

6.1 Introduction

Have you ever decided against studying certain course material because you were pretty sure you wouldn't be tested on it? Tried out a new hair style just to see how your friends would react? Asked a total stranger for a date? Drawn to an inside straight in poker? Headed for the golf course despite threatening clouds overhead? Bet on the outcome of a football game?

Bought a lottery ticket? If your answer to any of the above is Yes, then you already know something about probability because you have engaged in what statisticians call a *random experiment*.

> A *random experiment* is a well-defined course of action that may, at least in theory, be repeated a great many times under similar conditions and which must during a single *trial* result in one of two or more possible outcomes. Moreover, because of the presence of chance factors, the outcome of a specific trial of the experiment cannot be predicted with certainty.
>
> A *trial* is a single repetition of the random experiment, just as, say, one push-up represents one repetition within a set of push-ups.

A random experiment can be simple or complex. For example, any time a gambler carries out the simple act of letting a pair of dice roll from his moist palm onto the crap table, he is generating a trial of a random experiment:

- He is engaging in a well-defined course of action (rolling two dice).
- He can continue to roll the dice as long as he is interested in doing so or until his money runs out (possibility of a great many trials of the experiment).
- His action may result in his (1) making his point and winning money, (2) getting craps and losing his stake, or (3) getting a result which is neither point nor craps and which will permit him another roll (two or more possible outcomes).
- Because a great many little things play a role in determining the outcome of the single trial under consideration—the positions of the dice in his hand, the force of his roll, the thickness of the felt on the table, any inherent bias in the dice themselves, and so forth—he cannot know with certainty in advance how the roll will turn out (uncertainty is present).

Near the opposite end of the complexity spectrum is the business manager introducing a new product into a company's line. The action is well defined. In theory, the action can be repeated a great many times under similar conditions (repetition of the random experiment). The action will result in a success or a failure, or in some degree of success or some degree of failure (at least two possible outcomes). Because of the fickleness of consumers' tastes, countervailing actions on the part of competitors, and a great many other factors which perhaps can only be guessed at, the manager cannot predict with certainty what the outcome of her action will be (uncertainty is present).

Because business leaders must act under conditions of uncertainty, they will make use of probability theory—in some manner, formally or informally—in lieu of the certainty they would like to have but which is never really attainable.

6.2 Ways of Looking at Probability

It would be advantageous indeed if *probability* could be defined in a short, simple, and absolutely clear sentence. Unfortunately, the word is not nearly so accommodating. Therefore, we will have to content ourselves in this section with attempting to distill a working definition by looking at the idea of probability from a variety of vantage points.

The Probability Scale

In theory, all of life's activities can be placed along a scale like that shown in Figure 6.1, where 0 appears at one end and 1 at the other end, and where the placement of each activity is determined by its probability of occurrence.

The probability-scale idea gets us off to a good start, but leaves a most important question unanswered: How does one go about determining the numerical values we call probabilities? There are three basic ways: (1) the a priori approach, (2) the relative-frequency approach, and (3) the subjective approach.

A Priori Probability

You might be surprised to learn that probability theory as a distinct branch of mathematics began with attempts to answer certain questions arising in connection with gambling games. It seems that a seventeenth-century French nobleman and playboy, Antoine Gombauld, known as the Chevalier de Méré, thought he had hit upon a sure-fire strategy for winning with dice. He was right—until he tried to push the strategy a little too far.

De Méré prospered for a time by betting that he could get at least one 6 in four rolls of a single die. He was sure that this bet must win much more often than it loses. His conclusion was correct, but for the wrong reason. He reasoned that, because he had an even chance of rolling any one of the six numbers on the die the first roll, the likelihood of a 6 is the same as for any of the other numbers—namely, ⅙ (correct). For four rolls, he thought—and this is the point where his reasoning becomes faulty, as will be demonstrated in section 6.8—the chances would be four times as good. The probability of getting at least one 6 on the four rolls, therefore, should be at least ⁴⁄₆, or ⅔. This approach to probability determination led de Méré to conclude that he would, in the long run, win two wagers for every one he would lose. The fact that his reasoning was faulty failed to bother de Méré because he was making

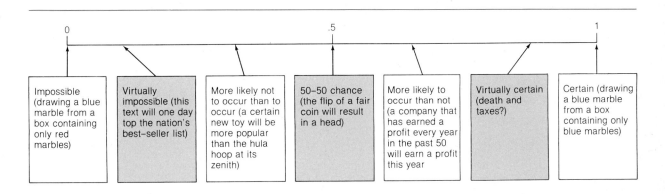

Figure 6.1 The probability scale.

money from the bet anyway. (The correct probability is about .52, as will be demonstrated shortly.)

The wager that gave formal probability theory its start was a variation on the winning one: Apparently his victims got wise to the fact that if they continued to accept de Méré's wager they would eventually all end up paupers. Whatever the reason, de Méré switched to betting that, within a sequence of 24 rolls of two dice, he could get at least one 12. Using the same line of reasoning as before, he decided that, because the probability of getting 12 on a single role is $\frac{1}{36}$ (correct), in 24 rolls there must be a probability of $\frac{24}{36}$ of getting at least one 12. (Incorrect: The true probability is just slightly over .49.) De Méré's fortunes began to dwindle slowly but surely, a setback which led him to try to discover the flaw in his reasoning. Toward this end, he wrote a letter to the mathematician and philosopher Blaise Pascal. In finding an answer to de Méré's puzzle, Pascal, in conjunction with Pierre de Fermat, launched an inquiry into the theory of probability. From this "improbable" beginning there emerged an elegant and useful science.

The work of Pascal and de Fermat led to the development of what is today called *a priori probability*—so named because probabilities can be determined *prior to* (or, really, in lieu of) actual experimentation. The a priori approach to probability determination defines the probability of a specified outcome of a trial of a random experiment in the following way:

If there are $n(I)$ *equally likely* possibilities, of which one must occur, and $n(S)$ such possibilities are regarded as favorable, that is, constituting what we refer to as a "success," then the probability of a success in a single trial of a random experiment is given by

$$\frac{n(S)}{n(I)}$$

For example, we know that the probability of getting a head on a single flip of a fair coin is .5 because (1) the coin has two sides, $n(I) = 2$; (2) only one of the sides is the head side, $n(S) = 1$; (3) either head or tail must occur on a single flip (the possibility of its landing on its side being disregarded on the grounds of extreme unlikelihood); and (4) on a "fair" coin the occurrence of head is no more nor no less likely than the occurrence of tail.

We also know that the probability of getting a 6 on a single roll of a fair die is $\frac{1}{6}$ because (1) the die has six sides, $n(I) = 6$; (2) only one of the sides has six dots, $n(S) = 1$; (3) one of the six sides has to be the one showing on top; and (4) the die is described as "fair."

By the same token, if a population of interest is made up of, say, 2000 elementary units one of which is to be selected at random for close examination, the probability that population unit number 594 will be the one selected is

$$\frac{n(S)}{n(I)} = \frac{1}{2000} = .0005$$

Helpful though the a priori view of probability is in connection with games of chance and appropriately designed sampling problems, its usefulness in most other situations is greatly restricted by the necessity of assuming equally likely outcomes—a qualification which is itself undefined and considered an intuitively obvious foundation concept. Suppose, for example, that an oil company desires to know the probability of striking oil with the next exploratory hole it drills. The possible outcomes are not difficult to determine: Either oil will be found or it will not be. But can this be construed to mean that the company has a probability of .5 of striking oil with its next exploratory drilling? Obviously not. The likelihood of not finding oil is much greater than that of finding oil. Hence, the a priori approach is inappropriate here. Similarly, the next item turned out by a certain production process will be either defective or it will be acceptable, but, one hopes, the likelihood of its being acceptable greatly surpasses that of its being defective. In both cases, then, use of the a priori approach would have to be passed over in favor of an alternative approach tied to empirical observation.

Relative-Frequency Probability

A probability in the relative-frequency sense may be determined or, better, approximated by calculating the proportion of times a success occurs in the long run under a constant cause system. That is, if we now define S to be the number of successful outcomes actually observed and n to be the number of actual independent trials of a random experiment, the ratio S/n is approximately equal to the true probability of a success in a single trial.

Naturally, the greater the number of trials, the more confidence we have in S/n as an approximation of the true probability.

Returning to our example of the oil company, what the relative-frequency view of probability translates into is: Compute the (approximate) probability of striking oil in the next exploratory drilling by determining from historical data what proportion of exploratory holes results in the discovery of oil—that is,

$$\frac{\text{Number of oil-producing exploratory drillings}}{\text{Number of exploratory holes drilled}}$$

As noted, the more extensive the historical data, the more reliable is the ratio as an approximation of the true probability—assuming a *constant cause system*.

The term *constant cause system* means that the collection of influencing factors, whatever these may be and however many there may be, affecting the outcomes of repeated trials of a random experiment remain the same from trial to trial.

How well the assumption of a constant cause system holds up in a specific set of trials of the random experiment must often be a matter of judgment. For instance, the management of the oil company in our example would be well advised to consider whether new discoveries are about as easy or about as difficult to come by today as they were over the span of time used to determine the relative-frequency probability. Perhaps better technology for discover-

GRIN AND BEAR IT   by Lichty & Wagner

© Field Enterprises, Inc., 1983

"It looks like a 30 percent chance of rain or shine."

ing promising oil sites exists today than was available in the past. Or perhaps the sites having the greatest potential for oil have already been exploited, leaving only less promising sites for new drilling.

Although the relative-frequency approach to calculating probabilities possesses some weaknesses, as the above example illustrates, for many purposes it is a vast improvement over the a priori approach. There are many business problems for which this approach is really the only one available with a rigorously objective justification. A conspicuous example of an industry very dependent on the relative-frequency view of probability is insurance. Judging what proportion of people in specified age categories, geographical locations, occupational designations, and so on, will die, have automobile accidents, or have debilitating illnesses during a given year is in essence an exercise in determining relative-frequency probabilities.

Subjective Probability

The roots of the subjective approach to determining probabilities almost certainly goes back to the very dawn of human (or perhaps even nonhuman) reasoning. For example, Ugh the

Caveman probably quickly learned that it was safer to venture out of his cave at certain times of the day than at other times. One could even interpret certain animal behavior in this same light: When a dog shows disdain for his dinner after eating only one or two bites, it is almost as if he were saying to himself: "The part I tried was terrible; therefore, there is a high probability that the part I haven't tried is also terrible."

Curiously, the "formal" application of subjective probability to statistical problems has been a relatively recent phenomenon.

> According to this view, the probability associated with a specific outcome of a trial of a random experiment is the degree of belief or degree of confidence placed in that outcome by a particular individual based on the evidence available to him.

The evidence available to the individual might consist of relative-frequency data, or other quantitative, or even qualitative, information. For example, the management of the oil company in our example might observe from historical data that over the most recent several years the ratio of strikes to exploratory drillings has been declining gradually. Such a situation would (or should) discourage them from relying blindly on a single past ratio. They would be better off taking the gradual decline into account and arriving at the desired probability in at least a semijudgmental manner.

In some situations, there may be no relevant quantitative data available, or else the available data might be highly untrustworthy. In such cases, the decision maker may have to make use of nonquantitative information such as his or her own past experience, opinions of knowledgeable advisors, and so forth.

> The manner in which a specific subjective probability is arrived at is as follows: If an individual believes it unlikely that a particular outcome will result from a trial of the random experiment, he assigns its occurrence a probability close to 0; if the individual believes it is very likely that a specific outcome will occur, he assigns it a probability close to 1. Sometimes the individual will have no convictions whatever and no reliable information on which to base a judgment. In such a dismal situation—and they are rarer than you might suppose—one may elect to assign equal probabilities to all of the possible outcomes.

Mention was made earlier of a company bringing a new product onto the market. Presumably, in doing so, the people making this decision are not acting blindly. Maybe they have information about the structural superiority of the new product over existing products of a similar kind. Or perhaps they know that the success rate of new products of this general type has been high in the past. Perhaps they have marketing research information suggesting that the new product will be successful. If any or a combination of these advantages is present in the decision environment, the decision makers would assign a rather high probability to a successful outcome.

As you might suspect, the encroachment of subjective probability into statistical analysis has generated much heated controversy among statisticians. Adherents to the classical view have argued, in substance, that what statistical analysis has always offered the decision maker is an objective basis for arriving at conclusions, a basis unaffected by the prejudices, whims, moods, aspirations, and other idiosyncrasies of fallible mortals. They have also pointed out, quite correctly, that different people will assign different probabilities to the same outcome because of differences in experience, attitudes, accessibility to relevant data, ways of interpreting the same data, and so forth.

Despite these clearly valid arguments, the fact remains that the majority of real-world business decisions are such that the decision maker derives little or no aid from either the a priori or the relative-frequency approach. The subjective approach is the only one of the three flexible enough to permit business leaders to avail themselves fully of the advantages probability theory has to offer. In this text we look on the use of the subjective approach to probability determination as a respectable and most reasonable one in connection with business decision making.

Relative Frequency as an Integrative Concept

Often in this text we will concern ourselves little with the specific manner in which a probability was determined. We will assume, for simplicity of discussion, that the probability associated with a specified outcome of a trial of a random experiment was determined by the most appropriate means available under the circumstances. However, because of its intuitive appeal, it will often be convenient to speak as if a probability had been determined by the relative-frequency approach.

In a way, the concept of probability as a relative frequency is implied in both the a priori and judgmental approaches. When we assert, for example, that in the flip of a fair coin, the probability of getting a head is .5 because there are two sides to the coin only one of which is a head, we are really saying that, if we were to flip that coin a great many times, the ratio of number of heads to number of flips will tend toward .5. Similarly, when a business manager assigns a subjective probability of, say, .7 to the idea that a newly introduced product will be a success, he or she is saying, in effect, "If I were to introduce similar products onto the market under similar conditions a great many times, I would expect a successful outcome about 70% of the time and an unsuccessful outcome about 30% of the time." Clearly, then, the relative-frequency concept of probability may be viewed as a kind of common theoretical element in each of the three alternative approaches discussed above.

6.3 The Language of Probability

In this section we define and illustrate several terms that one must understand in order to make skillful use of probability theory. It will be convenient to relate these terms to a specific illustrative case.

ILLUSTRATIVE CASE 6.1

The sales manager of the Exmont Company wished to determine what factors are related to success or failure of the company's sales staff. After a painstaking review of the sales quota assigned each salesperson last year, she concluded that the quotas were sufficiently realistic to serve as a viable basis for separating salespeople into categories indicating the relative Sales Success Status (SSS) of each. Accordingly, she determined that any salesperson who surpassed his or her quota by more than 15% during the previous 12-month period would be rated *highly successful; moderately successful* salespeople would be those whose sales during that period were within the range of 10% below to 15% above quota; finally, a salesperson whose performance had been more than 10% below quota would be designated *unsuccessful*. The sales force consisted of 200 people of whom 60 were rated highly successful, 120 moderately successful, and 20 unsuccessful, according to her criteria. These results are shown in the form of a univariate frequency distribution in Part I of Table 6.1.

In an effort to identify factors that might be related to Sales Success Status, the sales manager attempted numerous cross classifications relating SSS to a great many independent variables. Part II of Table 6.1 shows the results of one such cross classification. The independent variable in this example is age, the categories being *under 45* and *45 and over*. (Another related cross classification appears in Table 6.4.)

We noted in section 3.3 of Chapter 3 that actual frequencies listed in a frequency distribution can be converted into relative frequencies by dividing by the number of observations altogether. This has been done for both sets of data in Table 6.1; the results are shown in Table 6.2.

By now, you may be wondering what all this has to do with probability. The situation just described merely entails the presentation of observed frequency data which, as always, can be converted into relative-frequency data if one so desires. Both frequency and the relative-frequency data constitute descriptive information having no uncertainty associated with them. But the very word "probability" implies some uncertainty. We will introduce the element of uncertainty by assuming that the sales manager plans to select a single salesperson at random with a view to answering such questions as: "What is the probability that a salesperson so selected will be one rated *highly successful?*" "What is the probability that the

Table 6.1 Sales Success Status Information for 200 Salespeople Employed by the Exmont Company

Part I: Univariate Frequency Distribution for Sales Success Status

Sales Success Status	Number of Salespeople
Highly successful (A_1)	60
Moderately successful (A_2)	120
Unsuccessful (A_3)	20
Total	200

Part II: Bivariate Frequency Distribution for Sales Success Status and Age

Sales Success Status	Age of Salesperson		Total
	Under 45 (B_1)	45 and Over (B_2)	
Highly successful (A_1)	15	45	60
Moderately successful (A_2)	50	70	120
Unsuccessful (A_3)	15	5	20
Total	80	120	200

Table 6.2 Two Probability Distributions Associated with
Illustrative Case 6.1

*Part I: Univariate Relative Frequency (Probability) Distribution
for Sales Success Status*

Sales Success Status	Relative Frequency (Probability)
Highly successful (A_1)	.300
Moderately successful (A_2)	.600
Unsuccessful (A_3)	.100
Total	1.000

*Part II: Bivariate Relative Frequency (Probability) Distribution for the
Variables Sales Success Status and Age*

Sales Success Status	Age of Salesperson		Total
	Under 45 (B_1)	45 and Over (B_2)	
Highly successful (A_1)	.075	.225	.300
Moderately successful (A_2)	.250	.350	.600
Unsuccessful (A_3)	.075	.025	.100
Total	.400	.600	1.000

salesperson selected will be *highly successful* and *under 45* years of age?" And so forth. Viewed in this way, the relative frequencies combined with the list of SSS categories would constitute what we call a *probability distribution*. Similarly, the relative frequencies coupled with the list of SSS categories and a list of age categories would constitute a *probability distribution*, though of a different kind.

A *probability distribution* is a tabular or graphic method of displaying all of the possible outcomes associated with a trial of a random experiment and the probability of occurrence of each such possible outcome.

The random experiment in our example is, of course, the random selection of a single salesperson from among the 200 employed by the Exmont Company.

We refer to the different possible results of a random experiment as *basic outcomes*.

For example, in Part I of both Table 6.1 and 6.2, we observe that there are three basic outcomes: highly successful, moderately successful, and unsuccessful. In Part II of the same table, we find six basic outcomes: (1) highly successful and under 45, (2) moderately success-ful and under 45, (3) unsuccessful and under 45, (4) highly successful and 45 and over, (5) moderately successful and 45 and over, and (6) unsuccessful and 45 and over.

> The set of all possible basic outcomes of a trial of a random experiment constitutes the *sample space* of the experiment.
>   A sample space in which the collection of basic outcomes refers to a single charac-teristic (such as Sales Success Status in our example) is called a *univariate sample space*. One in which the collection of basic outcomes refers to two characteristics (such as SSS and age) is called a *bivariate sample space*.

The idea of a sample space can be shown symbolically, as in Figure 6.2. In part I of this figure, the part concerned only with Sales Success Status, a univariate sample space is presented. The only basic outcomes possible are *highly successful*, denoted O_1; *moderately successful*, O_2; and *unsuccessful*, O_3. This part of the figure is simply Part I of Table 6.1 (or 6.2) presented in a general symbolic way.

Part II of Figure 6.2 shows the symbolic counterpart of Part II of Table 6.1 (or 6.2). This is a bivariate sample space having six possible basic outcomes. The symbol O_{11} is used to denote the occurrence of *highly successful* and *under 45*—or putting it another way, it represents the basic outcome associated with row 1 and column 1. By the same token, O_{21} denotes *moderately successful* and *under 45* (i.e., the basic outcome associated with row 2 and column 1), and so forth.

The nonoverlapping rectangles surrounding the O's in Figure 6.2 are used to convey the idea that the basic outcomes themselves are nonoverlapping—that is, they are *mutually exclusive*.

I. A univariate sample space

Row Number	Sales Success Status	
1	Highly successful (A_1)	O_1
2	Moderately successful (A_2)	O_2
3	Unsuccessful (A_3)	O_3

II. A bivariate sample space

Row Number	Sales Success Status	Column 1: Under 45 (B_1)	Column 2: 45 and Over (B_2)
1	Highly successful (A_1)	O_{11}	O_{12}
2	Moderately successful (A_2)	O_{21}	O_{22}
3	Unsuccessful (A_3)	O_{31}	O_{32}

Figure 6.2 Symbolic presentation of a univariate sample space and a bivariate sample space.

Two or more basic outcomes are, by definition, *mutually exclusive*. What this means is that, if in a single trial of a random experiment a particular basic outcome results, no other basic outcome can result from the same trial.

In Part I of Table 6.1 (or 6.2), for example, we see that a salesperson is either highly successful, moderately successful, or unsuccessful. He or she cannot be both, say, unsuccessful and moderately successful. The same is true of the age categories. Clearly, a particular salesperson is either under 45, or 45 and over; he or she cannot belong to both age categories. With a bivariate sample space, if the categories associated with the two characteristics of interest are all mutually exclusive, then the basic outcomes themselves are necessarily mutually exclusive.

A sample space consists of *collectively exhaustive* basic outcomes—that is, all possible basic outcomes are included and the corresponding probabilities sum to 1. This point is emphasized in Table 6.2 by the fact that the relative frequencies (viewed as probabilities) sum to 1 in both parts of the table.

Once a random experiment and the basic outcomes constituting the sample space have been defined, it often happens that interest will center on results that may be broader than an individual basic outcome. The term *event* will be used to allow for this possibility.

An *event* consists of one or more basic outcomes making up the sample space.
Procedural Note: Hereafter, *specific events* will usually be denoted by capital letters such as the A and B labels in Tables 6.1 and 6.2 and Figure 6.2, but excluding the letters O and E. The letter O will be reserved for basic outcomes which *may* correspond with an event of interest but often will not. The letter E will be used to denote events, but only when a very general point (such as a general theorem) is to be made.

We can distinguish between basic outcome and event, where they are indeed distinguishable, by referring once again to Figure 6.2. In part I of the figure, A_1, A_2, and A_3 have been used to denote the events *highly successful, moderately successful*, and *unsuccessful*, respectively. However, we see in this part of the table that there is *not necessarily* any distinction between event and basic outcome. For example, the event *highly successful* is the same thing as the basic outcome *highly successful*. In probability symbols: $P(A_1) = P(O_1)$. However, if the sales manager were interested in determining whether a randomly selected salesperson would be designated *successful*, where *successful* is defined as either *highly successful* or *moderately successful*, the event, call it A_4 for now, would be equal to the sum of $P(O_1)$ and $P(O_2)$ and, of course, would be more encompassing than either of the individual basic outcomes. In symbols: $P(A_4) = P(O_1) + P(O_2)$.

As we just observed, the *probability of an event* is equal to the sum of the separate probabilities of the basic outcomes comprising the event.

Unlike two or more basic outcomes, which must be mutually exclusive, two or more events may or may not be mutually exclusive. They are mutually exclusive if they do not contain common basic outcomes; they are not mutually exclusive if they do contain common basic outcomes.

Looking at part II of Figure 6.2, for example, we note that the event A_1, that of selecting a salesperson who is *highly successful*, and the event A_2, selecting one who is *moderately successful*, do not have any basic outcomes in common and are, therefore, mutually exclusive. On the other hand, event A_1, selecting a salesperson who has been designated *highly successful*, and event B_1, selecting a salesperson who is *under 45*, do have common basic outcomes because some people—15 out of the 200 to be exact (Table 6.1)—are both *highly successful* and *under 45*. These points are emphasized within the diagrams in Figure 6.3.

Sometimes in probability calculations it is convenient to make use of *complementary events*.

I. Mutually exclusive events

Screened area represents the event "Successful," where "Successful" means either "Highly successful" or "Moderately successful." The constituent events, A_1 and A_2, are seen to have no basic outcomes in common.

Row Number	Sales Success Status	Column 1: Under 45 (B_1)	Column 2: 45 and Over (B_2)	
1	Highly successful (A_1)	O_{11}	O_{12}	Event (A_1)
2	Moderately successful (A_2)	O_{21}	O_{22}	Event (A_2)
3	Unsuccessful (A_3)	O_{31}	O_{32}	

II. Events not mutually exclusive

The lightly screened areas represent the two relevant events, A_1 and B_1, whereas the darkly screened area represents the overlapping of the two events, indicating that the events are not mutually exclusive. We see, then, that the probability of selecting a salesperson who is either "Highly successful" or "Under 45" cannot be determined accurately merely by obtaining the sum $P(A_1) + P(B_1)$.

Row Number	Sales Success Status	Column 1: Under 45 (B_1)	Column 2: 45 and Over (B_2)	
1	Highly successful (A_1)	O_{11}	O_{12}	Event (A_1)
2	Moderately successful (A_2)	O_{21}	O_{22}	
3	Unsuccessful (A_3)	O_{31}	O_{32}	

Event (B_1)

Figure 6.3 Symbolic presentation of two events which are (1) mutually exclusive and (2) not mutually exclusive.

The *complementary event*, Not-E, to the event E consists of all of the basic outcomes in the sample space that are not contained in E. In probability symbols: $P(\text{Not-}E) = 1 - P(E)$.

For example, from Table 6.2, we note that the probability $P(\text{Not-}A_1) = 1 - P(A_1)$. Since we know $P(A_1)$ to be .3, $P(\text{Not-}A_1)$ must be $1 - .3 = .7$. Similarly, $P[\text{Not-}(A_1 \text{ and } B_1)] =$

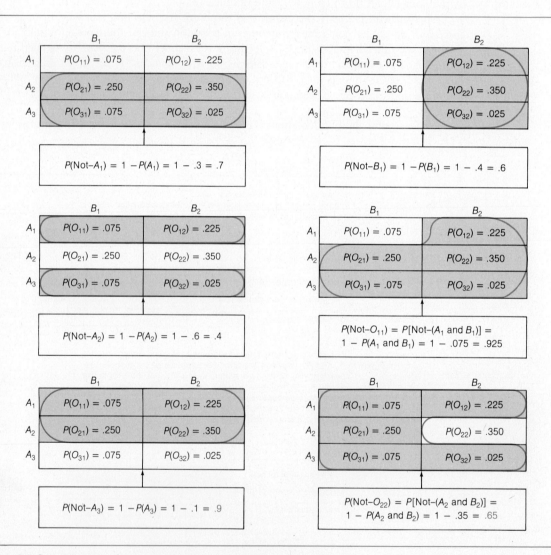

Figure 6.4 Calculation of several complementary-event probabilities. (Circled areas represent complementary-event probabilities. The basic outcome probabilities used here are from Part II of Table 6.2.)

$1 - P(A_1 \text{ and } B_1) = 1 - .075 = .925$. (More will be said in section 6.5 about the implications of the term *and* in some probability formulas.)

Several examples of complementary-event probabilities are presented in Figure 6.4.

A convenient alternative approach to illustrating the idea of complementarity, as well as several other probability concepts to be introduced shortly, is through the use of Venn diagrams. To construct a Venn diagram, we simply present a closed shape representing the sample space. We need not identify in the Venn diagram all population observations. However, we do have to imagine each such observation as represented by a point situated within the enclosed space. Although these points need not be distributed uniformly throughout the enclosed shape, certain pedagogical advantages will result from our assuming that they are. In Figure 6.5, the rectangular area labeled I represents the entire population and the circle labeled A_1 encloses the number of points representing "success." (In our example, the rectangle would enclose 200 uniformly distributed points and the circle would enclose 60—which is to say .3, or 30%—of these points.) All other points within the rectangle (140, or 70%, in our example) constitute the event Not-A_1.

A sometimes useful feature of the complementary-event concept is that it permits use of a time-saving, indirect approach to solving certain kinds of probability problems. This point will be developed in section 6.8.

There are times when it is important to determine a probability of the following kind: "*Given that* a randomly selected salesperson is *under 45* years of age, what is the probability that he or she is classified as *highly successful?*" Such a probability is called a *conditional probability*. The probability of interest relates to that category of salespeople—and to that category only—*under 45*. In our example, there are 80 such people (Part II of Table 6.1). We state the conditional probability in this case as $P(A_1|B_1) = {}^{15}\!/_{80} = .188$, where $P(A_1|B_1)$ is read: "The probability of getting A_1 given the occurrence of B_1." The vertical line between A_1 and B_1 is read "given" or "given that." Anything to the right of the vertical bar should be interpreted as hard fact—that is, *as a condition having absolutely no uncertainty associated with it*. (See Figure 6.6.)

In the example just cited we could have determined the desired probability of .188 readily from Part II of Table 6.1. However, there are times when the available information is in such a form that a more roundabout attack becomes necessary. It is at times like this that an awareness of the following rule can be most helpful:

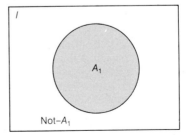

Figure 6.5 Venn diagram illustrating the idea of a complementary event: $P(A_1) + P(\text{Not-}A_1) = 1$.

When we wish to determine $P(A_1|B_1)$, the B_1 column is the only one considered. The (unscreened) B_2 column is ignored; the (lightly screened) event B_1 probability is divided into (the darkly screened) event (A_1 and B_1) probability.

When we wish to determine $P(B_1|A_1)$, the A_1 row is the only one considered. The (unscreened) A_2 and A_3 rows are ignored; the (lightly screened) event A_1 probability is divided into the (darkly screened) event (A_1 and B_1) probability.

Figure 6.6 Symbolic presentation of reasoning behind the determination of conditional probabilities.

If we have two events, A_i and B_j, based on a bivariate sample space, the conditional probability of A_i given B_j is defined as

$$P(A_i|B_j) = \frac{P(A_i \text{ and } B_j)}{P(B_j)}$$

where it is assumed that B_j does not equal 0. Similarly, the conditional probability of B_j given A_i is defined as

$$P(B_j|A_i) = \frac{P(A_i \text{ and } B_j)}{P(A_i)}$$

where it is assumed that $P(A_i)$ does not equal 0.

Note: $P(A_i|B_j)$ and $P(B_j|A_i)$ will seldom have the same value, since the denominators, $P(A_i)$ in the one case and $P(B_j)$ in the other, will usually be different.

Notice that in the example we have been addressing, $P(A_1) = .3$ and $P(A_1|B_1) = .188$. That is, $P(A_1) \neq P(A_1|B_1)$. This inequality suggests a *dependency* between events A_1 and B_1. Put another way: There is something about the age of a salesperson that has an effect on sales performance. Since $P(A_1|B_2) = [P(A_1 \text{ and } B_2)]/P(B_2) = {}^{45}/_{120} = .375$ is twice as large as the $P(A_1|B_1) = .188$ value obtained above, older age, presumably related to more sales experience, is advantageous.

Dependencies like this are common in business:

- The probability that this year's sales will exceed a specified amount might be dependent on last year's sales or on the size of the sales force.
- The probability of a new movie's being a "success" might be dependent on the time of the year the film is released.
- The probability that a manufactured item will be of acceptable quality might be dependent on the regularity with which certain maintenance operations are performed on the equipment.

6 Fundamentals of Probability Theory

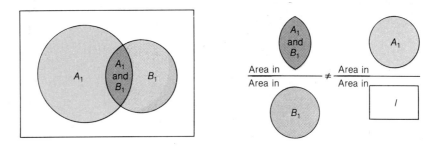

Figure 6.7 Venn diagram illustrating a dependency between events A_1 and B_1.

And so forth.

This idea of dependency is illustrated in the Venn diagram appearing in Figure 6.7. This diagram is drawn in such a way that the area in A_1 *and* B_1 (the area of overlap) divided by the area in B_1 is *not* equal to the area in A_1 divided by the area in I. Had these areas been equal, we would say that events A_1 and B_1 were *independent*. More will be said about independence and its implications in section 6.5.

Now that you are familiar with some basic probability terminology, we are in a position to treat two extremely important probability theorems having to do with how two or more event probabilities can be combined. These theorems are the addition theorem, discussed in section 6.4, and the multiplication theorem, the subject of section 6.5.

6.4 The Addition Theorem of Probability

> The addition theorem of probability equips us to answer questions like "What is the probability that *either* _____ [some specified event] *or* _____ [some alternative specified event] will occur in a single trial of a random experiment?"
>
> Notice the emphasis attached to the words *either* and *or*. A success results from a trial of a random experiment if one or the other of two (or, when relevant, more than two) specified events occurs.

General Case

Let us refer again to Part II of Table 6.1 and assume that we wish to know the probability that a randomly selected salesperson will be *either highly successful* (A_1) or *under 45* (B_1). We count as a success the selection of a salesperson who is *either* one *or* the other (or, of course, both—see Figure 6.8). The general case of the addition theorem is applicable to this problem:

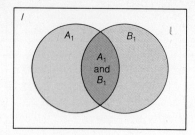

Figure 6.8 Venn diagram illustrating the general case of the addition theorem of probability when the question asked is: What is the probability that a salesperson selected at random will be either "highly successful" or "under 45"? Circle A_1 represents the "highly successful" salespeople; circle B_1 represents the salespeople who are "under 45"; and the intersection represents the salespeople who are both "highly successful" and "under 45." Since people represented by the intersection are counted in connection with "highly successful" as well as in connection with "under 45," their numbers must be subtracted out one time.

> The *general case of the addition theorem of probability* states that for any two events, E_1 and E_2, defined in a sample space:
>
> $$P(E_1 \text{ or } E_2) = P(E_1) + P(E_2) - P(E_1 \text{ and } E_2)$$

For the problem posed above, we would have $P(A_1 \text{ or } B_1) = P(A_1) + P(B_1) - P(A_1 \text{ and } B_1) = {}^{60}\!/_{200} + {}^{80}\!/_{200} - {}^{15}\!/_{200} = .300 + .400 - .075 = .625$.

As noted in a different connection, when we add $P(A_1)$ and $P(B_1)$ we count twice 15 (out of 200 = .075) salespeople who are both *highly successful* and *under 45*; hence, the probability .075 must be subtracted out one time. The source of the double counting was shown in graphic form in part II of Figure 6.3. The same idea is presented in Venn diagram form in Figure 6.8.

Here are some additional examples.

1. Table 6.3 shows the number of defective items turned out by a company's crew of 30 lathe operators on a certain day along with the number of years of experience of each worker.

   If a lathe operator is to be selected at random, what is the probability that the one selected will have (*a*) had less than 2 years experience or turned out 2 defective items? (*b*) had less than 2 years experience or turned out 0 defective items? (c) turned out either 3 or 4 defective items?

   Answers:

   (*a*) $P(A_1 \text{ or } B_3) = P(A_1) + P(B_3) - P(A_1 \text{ and } B_3) = {}^{10}\!/_{30} + {}^{4}\!/_{30} - {}^{2}\!/_{30} = .3333 + .1333 - .0667 = .4$

   (*b*) $P(A_1 \text{ or } B_1) = P(A_1) + P(B_1) - P(A_1 \text{ and } B_1) = {}^{10}\!/_{30} + {}^{10}\!/_{30} - {}^{1}\!/_{30} = .3333 + .3333 - .0333 = .6333$

Table 6.3 Years of Experience of and Number of Defective Items Turned Out by 30 Lathe Operators

Years of Experience	Number of Defective Items					
	0 (B_1)	1 (B_2)	2 (B_3)	3 (B_4)	4 (B_5)	Total
Less than 2 years (A_1)	1	3	2	1	3	10
2 to 5 years (A_2)	4	1	1	3	2	11
Over 5 years (A_3)	5	2	1	0	1	9
Total	10	6	4	4	6	30

(c) $P(B_3 \text{ or } B_4) = P(B_3) + P(B_4) - P(B_3 \text{ and } B_4) = \frac{4}{30} + \frac{4}{30} - \frac{0}{30} = .1333 + .1333 - .0000 = .2667$

Note: Since there is no double counting possible in part (c), $P(B_3$ and $B_4)$ was assigned a value of 0.

2. An investment advisory service has notified its subscribers that over the next six months Stocks Q and R will probably rise. Their subjective probability assessments are as follows: The probability that Stock Q will rise, $P(Q)$, is .85, the probability that Stock R will rise, $P(R)$, is .8, the probability that both Stock Q and Stock R will rise, $P(Q$ and $R)$, is .7. What is the probability that either Stock Q or Stock R will rise?
   Answer:

   $P(Q \text{ or } R) = P(Q) + P(R) - P(Q \text{ and } R) = .85 + .80 - .70 = .95$

3. According to the historical records of the Vickey Car Company, the probability of turning out a car with a defective steering column is $P(S) = .1$, the probability of turning out a car with defective brakes is $P(B) = .05$, and the probability of turning out a car with both a defective steering column and defective brakes is $P(S$ and $B) = .02$. If you were to buy a car made by this company, what is the probability that the car you bought would have either a defective steering column or defective brakes?
   Answer:

   $P(S \text{ or } B) = P(S) + P(B) - P(S \text{ and } B) = .10 + .05 - .02 = .13$

Special Case

Referring again to Illustrative Case 6.1, let us suppose that we were to apply the addition theorem to find the probability that a salesperson selected at random would be *either highly successful* (A_1) *or moderately successful* (A_2). As noted earlier, a salesperson must be one or the other (if either); he or she cannot be both *highly successful* and *moderately successful*. Therefore, events A_1 and A_2 have no basic outcomes in common. According to the general case of the addition theorem, we would calculate $P(A_1 \text{ or } A_2) = P(A_1) + P(A_2) - P(A_1$ and $A_2) = \frac{60}{200} + \frac{120}{200} - \frac{0}{200} = .3 + .6 + .0 = .9$, and the operations reduce to $P(A_1$ or $A_2) = P(A_1) + P(A_2)$ or, more generally:

If two events, E_1 and E_2, are mutually exclusive, the probability that *either* E_1 *or* E_2 will occur in a single trial of a random experiment is equal to the sum of the separate probabilities, as shown in Figure 6.9. That is,

$$P(E_1 \text{ or } E_2) = P(E_1) + P(E_2)$$

Here are some more examples.

1. Refer to Table 6.3. What is the probability that a lathe operator picked at random will have (*a*) turned out either 3 or 4 defective items, (*b*) had 5 years experience or less, (*c*) turned out 3 defective items or had over 5 years of experience?

   Answers:

   (*a*) $P(B_4 \text{ or } B_5) = P(B_4) + P(B_5) = \frac{1}{30} + \frac{9}{30} = .3333$
   (*b*) $P(A_1 \text{ or } A_2) = P(A_1) + P(A_2) = \frac{10}{30} + \frac{11}{30} = .7$

   Note: In this problem it was necessary to recognize that, given the experience categories used, "5 years of experience or less" includes both "2 to 5 years" and "less than 2 years."

   (*c*) $P(A_3 \text{ or } B_4) = P(A_3) + P(B_4) = \frac{9}{30} + \frac{1}{30} = .4333$

   Note: In this problem there was every reason to think that the events A_3 and B_4 were *not* mutually exclusive. However, the 0 in the A_3-B_4 cell indicates that the events are indeed empirically, if not theoretically, mutually exclusive.

2. A personnel department has three choices after interviewing a prospective employee: (*a*) hire immediately (done about half the time), (*b*) defer a decision (done about 30% of the time), and (*c*) reject immediately (done about 20% of the time). On the basis of this very limited information, determine the probability that a particular job applicant, Floyd Boyd, will either be hired immediately or have the decision deferred.

   Answer: Call the probability of his being hired immediately $P(H)$ and the probability of his having the decision deferred $P(D)$. Since one and only one of the indicated results must occur, the events are mutually exclusive. Therefore,

$$P(H \text{ or } D) = P(H) + P(D) = .5 + .3 = .8$$

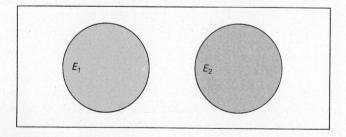

Figure 6.9 Venn diagram illustrating the special case of the addition theorem of probability. The probability sought, $P(E_1$ or $E_2)$, is represented by the sum of the two screened areas. Since there is no overlapping of the two circles, no double counting occurs, and thus nothing has to be subtracted from the sum obtained from $P(E_1) + P(E_2)$.

6 *Fundamentals of Probability Theory*

6.5 The Multiplication Theorem of Probability

> The multiplication theorem of probability equips us to answer questions like "What is the probability that *both* _____ [some specified event] *and* _____ [some alternative specified event] will occur together in a single trial of a random experiment?"

Probabilities obtained in situations where the multiplication theorem is applicable are called *joint probabilities* because two (or more) events must occur together, or "jointly," if a success is to be claimed.

General Case

> The general case of the multiplication theorem states that the probability of any two events, E_1 and E_2, occurring together can be found by
>
> $$P(E_1 \text{ and } E_2) = P(E_1) \cdot P(E_2|E_1) \quad \text{or} \quad P(E_1 \text{ and } E_2) = P(E_2) \cdot P(E_1|E_2)$$

You can see from this theorem why conditional probabilities are important; these are the probabilities that take *dependencies* between two or more events into account.

Referring again to Illustrative Case 6.1 (Tables 6.1 and 6.2), let us suppose that we wish to know the probability that a randomly selected salesperson would be *both highly successful* (A_1) and *under 45* (B_1). Looking at Part II of Table 6.1, we see that the desired probability could be obtained directly simply by placing the frequency value associated with the A_1-B_1 cell, namely 15, over 200, giving .075. This value appears in the corresponding cell of Table 6.2. We see from this that when frequencies have been converted into relative frequencies (interpreted as probabilities) *the entries in the cells of the bivariate relative-frequency distribution are all joint probabilities*.

Because the information needed to solve probability problems of this kind will not always be available in such convenient tabular form, one must know how to obtain joint probabilities through use of the multiplication theorem. It can be readily demonstrated that the multiplication theorem will also result in the answer .075: $P(A_1 \text{ and } B_1) = P(A_1) \cdot P(B_1|A_1) = (60/200)(15/60) = 15/200 = .075$. The same result can be obtained from the form $P(A_1 \text{ and } B_1) = P(B_1) \cdot P(A_1|B_1) = (80/200)(15/80) = 15/200 = .075$.

Here are some more examples to ponder.

1. Refer to Table 6.1, Part II. What is the probability that a randomly selected salesperson will be *unsuccessful* and *45 and over?*

    Answer:

    $$P(A_3 \text{ and } B_2) = P(A_3) \cdot P(B_2|A_3) = (20/200) \cdot (5/20) = 5/200 = .025$$

    or.

    $$P(A_3 \text{ and } B_2) = P(B_2) \cdot P(A_3|B_2) = (120/200) \cdot (5/120) = 5/200 = .025$$

2. A manager believes that the probability of his being promoted to a more responsible, higher-paying position before the end of the current year is .65. However, if his immediate superior will submit a favorable recommendation on his behalf, the probability will jump to .86. The probability of his being promoted if his superior does not write a favorable recommendation is only .16. The probability that his superior will write a favorable recommendation is .7. What is the probability that this manager will (a) receive a favorable recommendation and be promoted? (b) receive a favorable recommendation and not be promoted? (c) not receive a favorable recommendation and be promoted anyway?

Let $P(A)$ be the probability that the manager will be promoted, that is, "advanced." We are told that this probability is .65.

Let $P(F)$ be the probability of a favorable recommendation by the manager's immediate superior. We are told that this probability is .7.

Let $P(A|F)$ be the probability that the manager will be promoted given a favorable recommendation. We are told that this probability is .86.

Let $P(A|\text{Not-}F)$ be the probability that the manager will be promoted given an unfavorable recommendation. We are told that this probability is .16.

Assuming that the recommendation must be either favorable or unfavorable and that the manager must either be promoted or not be promoted, we may also surmise from the above information that

$$P(\text{Not-}A) = 1 - P(A) = 1 - .65 = .35$$
$$P(\text{Not-}F) = 1 - P(F) = 1 - .7 = .30$$
$$P(\text{Not-}A|F) = 1 - P(A|F) = 1 - .86 = .14$$
$$P(\text{Not-}A|\text{Not-}F) = 1 - P(A|\text{Not-}F) = 1 - .16 = .84$$

Answers:

(a) $P(F \text{ and } A) = P(F) \cdot P(A|F) = (.7)(.86) = .602$

(b) $P(F \text{ and Not-}A) = P(F) \cdot P(\text{Not-}A|F) = (.7)(.14) = .098$

(c) $P(\text{Not-}F \text{ and } A) = P(\text{Not-}F) \cdot P(A|\text{Not-}F) = (.3)(.16) = .048$

3. According to the historical records of the Vickey Car Company, the probability of turning out a car with a defective steering column is $P(S) = .10$, the probability of turning out a car with defective brakes is $P(B) = .05$, and the probability of turning out a car with a defective steering column given that it has defective brakes is $P(S|B) = .15$. If you were to buy a car made by this company, (a) what would be the probability of its having both a defective steering column and defective brakes? (b) what would be the probability of its having defective brakes given that it has a defective steering column?

Answers:

(a) $P(S \text{ and } B) = P(S) \cdot P(B|S) = (.1) \cdot P(B|S)$. We do not know $P(B|S)$ yet. Let us use the alternative procedure: $P(S \text{ and } B) = P(B) \cdot P(S|B) = (.05)(.15) = .0075$.

(b) $P(B|S)$ can be determined in more than one way. A convenient way, since we now know $P(S \text{ and } B)$ to be .0075, is to recognize that both $P(B) \cdot P(S|B)$ and $P(S) \cdot P(B|S)$ must be .0075. Thus, $P(S) \cdot P(B|S) = .0075$. $(.1)P(B|S) = .0075$. Solving for $P(B|S)$, we get $.0075/.1 = .075$.

Special Case

The special case of the multiplication theorem of probability is based on the assumption of *independent events*.

> *Independence of events* means that the occurrence or nonoccurrence of event E_1 does not affect the probability of occurrence or nonoccurrence of event E_2 and, conversely, the occurrence or nonoccurrence of event E_2 does not affect the probability of occurrence or nonoccurrence of event E_1.

In many situations the above definition is the one which must be used and its validity evaluated judgmentally. However, in view of the completeness of the data presented in connection with Illustrative Case 6.1, we can apply a more rigorous and more readily demonstrable definition of independence.

In Table 6.4 we find a bivariate frequency distribution not previously examined. In this table the characteristic of principal interest is Sales Success Status, which is thought to be affected by type of product sold. As a matter of company policy, each salesperson sells only consumer products (C_1) or industrial products (C_2). No salesperson sells both.

In Table 6.5 the frequencies appearing in Table 6.4 are shown in the form of conditional and unconditional probabilities for each of the type-of-product categories (Part I) and each of the Sales Success Status categories (Part II).

We see that the probability data in Table 6.5 are of a very special nature. For both categories of products, the conditional probabilities associated with the individual categories of Sales Success Status are the same and they, in turn, are the same as the corresponding unconditional probabilities (Part I of the table). Consequently, it makes no difference whether a salesperson sells consumer products or industrial products—the probability of being, say, *highly successful* is the same for sellers of both types of products. We instinctively feel that Sales Success Status and type of product are independent. The evidence mounts when we examine Part II of Table 6.5. This time for each level of Sales Success Status, the conditional probabilities associated with categories of product sold correspond exactly with one another and with the corresponding unconditional probabilities.

Table 6.4 Bivariate Frequency Distribution Showing the Relationship Between Sales Success Status and Type of Product Sold for 200 Salespeople Employed by the Exmont Company

	Type of Product		
Sales Success Status	Consumer (C_1)	Industrial (C_2)	Total
Highly successful (A_1)	45	15	60
Moderately successful (A_2)	90	30	120
Unsuccessful (A_3)	15	5	20
Total	150	50	200

Table 6.5 Data from Table 6.4 Converted into Conditional and
Unconditional Probabilities

Part I: Probability Table Showing $P(A_i)$ Values and $P(A_i|C_j)$ Values

| | Conditional Probability of Sales Success Status Given That Salesperson Deals in | | |
Sales Success Status	Consumer Products (C_1)	Industrial Products (C_2)	Unconditional Probability
Highly successful (A_1)	.300	.300	.300
Moderately successful (A_2)	.600	.600	.600
Unsuccessful (A_3)	.100	.100	.100
Total	1.000	1.000	1.000

Part II: Probability Table Showing $P(C_j)$ Values and $P(C_j|A_i)$ Values

| | Type of Product | | |
Conditional Probability for Type of Product Given That Salesperson Is Rated	Consumer Products (C_1)	Industrial Products (C_2)	Total
Highly successful (A_1)	.750	.250	1.0
Moderately successful (A_2)	.750	.250	1.0
Unsuccessful (A_3)	.750	.250	1.0
Unconditional probability	.750	.250	1.0

We may redefine independence.

> Two events, E_1 and E_2, are independent if the conditional probability $P(E_1|E_2)$ equals the unconditional probability $P(E_1)$ and the conditional probability $P(E_2|E_1)$ equals the unconditional probability $P(E_2)$. (See Figure 6.10.)

From Table 6.5, Part I, we note that $P(A_1|C_1)$, for example, is equal to $P(A_1)$ and, in Part II, $P(C_1|A_1) = P(C_1)$. Having defined independence, we may now state the special case of the multiplication theorem.

> If two events, E_1 and E_2, are independent, the probability that E_1 and E_2 will occur together in a single trial of a random experiment is equal to the product of their separate probabilities. That is,
>
> $$P(E_1 \text{ and } E_2) = P(E_1) \cdot P(E_2)$$

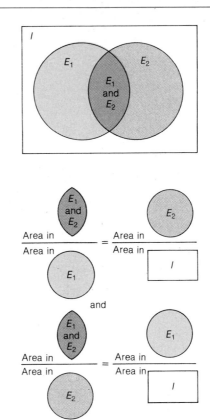

Figure 6.10 Venn diagram demonstrating independence of events.

For example, suppose we were told by the investment advisory service mentioned above that the probability that Stock A will go up during the next six months is $P(A) = .8$, the probability that Stock B will go up is $P(B) = .7$, and that their movements are independent. What is the probability that both stocks will rise during the upcoming six months?

Answer:

$$P(A \text{ and } B) = P(A) \cdot P(B) = (.8)(.7) = .56$$

6.6 More on Mutually Exclusive and Independent Events

A sharp distinction must be drawn between mutually exclusive events on the one hand and independent events on the other. Beginning students sometimes think of them as very nearly synonymous when, in fact, the existence of one as good as precludes the existence of the other. Consider the following argument.

Table 6.6 Kinds of Relationships That Can Exist Between Two Events, E_1 and E_2

	Dependent	
Independent	Mutually Exclusive	Nonmutually Exclusive
Occurrence or nonoccurrence of event E_2 does not affect the probability of occurrence of event E_1	Occurrence of event E_2 automatically causes the conditional probability of E_1 to go to zero	Occurrence or nonoccurrence of event E_2 affects the conditional probability of event E_1 but does not preclude the occurrence of E_1
$P(E_1) = P(E_1|E_2)$	$P(E_1|E_2) = 0$	$P(E_1) \neq P(E_1|E_2) \neq 0$

If, for example, events A_1 and B_1 are independent, then $P(A_1) = P(A_1|B_1)$. In a Venn diagram like that shown earlier in Figure 6.7, this means that if the area associated with event A_1 is equal to, say, 25% of the area of the entire rectangle I, then the area of overlap, "A_1 *and* B_1," must be equal to 25% of the area associated with event B_1. Consequently, the area of overlap cannot be zero.

However, if events A_1 and B_1 are mutually exclusive, they cannot occur jointly and the area "A_1 *and* B_1" must be zero. Clearly, both of these conditions cannot occur together—the sole exception being when $P(B_1) = 0$, a possibility worth noting but not one rife with practical implications.

Table 6.6 presents the various ways that two events can be related.

6.7 Combining Addition and Multiplication Methods

It is not at all uncommon to encounter probability problems calling for the use of the addition and multiplication theorems in combination. Here are two examples.

1. A gold prospector posts claims at three mining sites. He is more realistic than many gold prospectors and estimates the probability of finding gold in any meaningful quantity to be .05 at each site. Assuming independence, what is the probability of his finding gold at at least one site?

   Answer: Let $P(G_1)$ be the probability of finding gold at Site 1, $P(G_2)$ be the probability of finding gold at Site 2, and $P(G_3)$ be the probability of finding gold at Site 3. Before we try to determine the probability asked for, it will be helpful to know what the possibilities are. Table 6.7 lists all the ways that gold could be found at at least one site.

   The top probability in the right-hand column of Table 6.7, for example, corresponds to not finding gold at Site 1 *and* not finding gold at Site 2 *and* finding gold at Site 3. That is, $P(\text{Not-}G_1 \text{ and Not-}G_2 \text{ and } G_3) = (.95)(.95)(.05) = (.95)^2(.05) = .0451$. If it had not been for the "at least" in this problem, all that would be necessary at this point would be to multiply .0451 by 3 because there are three different ways that gold could be found at one site and not at the other two. As it is, however, the probability associated with gold at exactly two sites

Table 6.7 Various Ways That Gold Could Be Found at at Least One of Three Sites

Site 1	Site 2	Site 3	Probability
Not-G	Not-G	G	$(.05)(.95)^2 = .0451$
Not-G	G	Not-G	.0451
G	Not-G	Not-G	.0451
Not-G	G	G	$(.05)^2(.95) = .0024$
G	Not-G	G	.0024
G	G	Not-G	.0024
G	G	G	$(.05)^3 = .0001$
		Total	.1426

and at exactly three sites must also be determined. These are also shown in Table 6.7. The answer sought, therefore, is the sum of all the joint probabilities in the right-hand column of this table, namely .1426.

2. There are six similar-looking rubies in a jeweler's showcase. Four of these are of top quality and two are of secondary quality. A customer is interested in buying four of the rubies and is unconcerned about quality. If she were to select four rubies at random, what would be the probability of her getting at least one of secondary quality?

Let us denote top quality by T and secondary quality by S.

Answer: We must determine the probability of the customer's getting one or two secondary-quality rubies. The different ways she could get one such ruby are listed here.

$$TTTS \qquad TSTT$$
$$TTST \qquad STTT$$

This is a good start. But we must be wary. This problem is concerned with random sampling from an extremely small population, a condition ensuring that the events are not independent. Let us consider the $TTTS$ sequence with some care: The probability of her getting T with her first selection is ⁴⁄₆, or ⅔; the probability of her getting T with her second selection is ⅗ (only three top-quality rubies left out of five altogether); the probability of T on the third selection is ²⁄₄, or ½ (only two top-quality rubies left out of four altogether); and the probability of S on the fourth selection is ⅔ (two secondary-quality rubies left out of three altogether). Therefore, $P(T_1$ and T_2 and T_3 and $S_4) = (⅔)(⅗)(½)(⅔) = $¹²⁄₉₀ $= .1333$. The probability of the second arrangements of T's and S's, $TTST$, is the same—$(⅔)(⅗)(½)(⅔) = .1333$. The same probability results when a similar line of reasoning is applied to the third and fourth arrangements of three T's and one S. Therefore, $(4)(.1333) = .5332$.

The possible arrangements of two T's and two S's are as follows:

$$TTSS \qquad STTS$$
$$TSST \qquad STST$$
$$SSTT \qquad TSTS$$

Taking the first arrangement, we calculate: $(\frac{2}{3})(\frac{3}{5})(\frac{1}{2})(\frac{1}{3}) = .0667$. Since there are six such arrangements, $(6)(.0667) = .4002$. Summing .5332 and .4002, we get .9334, the answer we were seeking.

These problems have been worked using the direct approach in order to show the mechanics, and to explain the rationale behind the mechanics, of mixing addition and multiplication procedures. Actually, they could have been worked much more easily through judicious use of complementary-event probabilities. This is demonstrated in the next section.

6.8 Simplifying Probability Calculations Through the Use of Complementary Events

In section 6.3 we noted without demonstration that certain kinds of complex probability problems can be rendered quite simple by using complementary-event probabilities. In this section we demonstrate the truth of this earlier assertion.

· Recall that problem 1 in the preceding section asked: What is the probability that the prospector will find gold at at least one of his three claims? The probability of gold at any specific site was given as .05.

To answer this question, let us rephrase it temporarily and have it read: What is the probability that the prospector will *not* find gold on any of the three sites? The probability of Not-G at all three sites is $P(\text{Not-G}_1)$. $P(\text{Not-G}_2) \cdot P(\text{Not-G}_3) = (.95)(.95)(.95) = (.95)^3 = .8574$. The complement of this is $1 - .8574 = .1426$, exactly the same result we attained through the more laborious procedure described above.

Problem 2 in the preceding section asked: If a customer chooses four rubies at random from a population consisting of four top-quality and two secondary-quality rubies, what is the probability that she will get at least one ruby of secondary quality?

Again, we can rephrase the question: What is the probability that she will get *no* rubies of secondary quality? In other words, what is the probability of *TTTT*? $P(T_1 \text{ and } T_2 \text{ and } T_3 \text{ and } T_4) = (\frac{2}{3})(\frac{3}{5})(\frac{1}{2})(\frac{1}{3}) = \frac{6}{90} = .0667$. The complement is $1 - .0667 = .9333$, the same answer as before save for a .0001 error because of rounding.

Before leaving this subject, let us return to de Méré's dice wagers (section 6.2).

1. De Méré bets that he can roll at least one 6 in four rolls of a single die. He figures the probability to be $\frac{4}{6}$ (or $\frac{2}{3}$ or .6667). What is the probability of getting at least one 6 in four rolls of a die?

   As we have seen, this kind of question can be temporarily rephrased to read: What is the probability of getting *no* 6's on four rolls of the die? You now know that $P(\text{Not-}6_1 \text{ and Not-}6_2 \text{ and Not-}6_3 \text{ and Not-}6_4) = P(\text{Not-}6_1) \cdot P(\text{Not-}6_2) \cdot P(\text{Not-}6_3) \cdot P(\text{Not-}6_4) = (\frac{5}{6})(\frac{5}{6})(\frac{5}{6})(\frac{5}{6}) = (\frac{5}{6})^4 = .4823$. Thus, to find the probability of at least one 6, we compute $1 - .4823 = .5177$.

2. De Méré bets that he can roll at least one 12 in 24 rolls of a pair of dice. He figures the probability to be $\frac{24}{36}$ (or $\frac{2}{3}$ or .6667). What is the probability of getting at least one 12 on 24 rolls of a pair of dice?

   The question temporarily rephrased is: What is the probability of getting *no* 12's on 24 rolls of a pair of dice?

P(Not-12$_1$ and Not-12$_2$ and Not-12$_3$ and . . . and Not-12$_{24}$) = P(Not-12$_1$) ·
P(Not-12$_2$) · P(Not-12$_3$) . . . P(Not-12$_{24}$) = ($^{35}/_{36}$)($^{35}/_{36}$)($^{35}/_{36}$) . . . ($^{35}/_{36}$) =
($^{35}/_{36}$)^{24} = .5086. Therefore, the probability of at least one 12 is $1 - .5086 =$
.4914. Had de Méré specified 25 rolls, the probability of no 12's would have
been ($^{35}/_{36}$)^{25} = .4945; and $1 - .4945 = .5055$. It is interesting to ponder whether
we would have any formal probability theory today if de Méré had been just a
shade more cautious.

6.9 Bayes' Theorem

Just about everyone knows something about probability. However, for many this knowledge
is not in a form permitting precise mathematical manipulation. Although this widespread
shortcoming matters little in some of the more frivolous arenas of decision making, there are
many times in the world of business when one can benefit greatly by being precise and
explicit in probability assessments. We can most effectively illustrate this point by taking a
somewhat indirect approach: Suppose that someone has two bags of poker chips that appear
identical. One of the bags contains 70 blue and 30 white chips; the other, 20 blue and 80
white chips. However, this person does not know which is the predominately blue and which
is the predominately white bag. This he will try to determine by selecting a sample of chips
from one of the bags. He flips a fair coin to determine from which of the bags he will take
some chips. He then reaches into the bag determined by the coin toss, thoroughly mixes the
chips, and randomly selects one chip. He next makes a note about the color of the chosen
chip and returns that chip to the bag from which it was drawn. Reaching into the same bag
a second time, he mixes the chips and randomly selects one, records its color, and returns it
to the bag. He follows this procedure ten times. The result is 7 blues and 3 whites.

Now try answering the following question relying only on your intuition: On the basis of
the evidence, what is the probability that the bag from which the 7 blue and 3 white chips
were obtained is the predominately blue bag?

What is your guess?

Your author has asked this question of a great many classes of statistics students. The
guesses have ranged from .5 to .95. The point of maximum cluster has varied a little but has
always been within the range of .6 to .8. Guesses of .9 or above have been extremely rare.

The answer is .997—virtual certainty!

Some decision theorists who have conducted similar experiments have drawn a most
provocative conclusion from such results: Humans are ultraconservative evaluators of evidence. Somewhat like the person who throws away a tube of toothpaste after only a fraction
of its contents has been used, people tend to get out of empirical evidence only a fraction of
what they are entitled to get.

If this conclusion is correct, the implications for business decision making are profound
indeed: If we could calculate, even approximately, the extent to which the probability of a
correct decision increases when some reliable empirical evidence is brought to bear on the
problem, our decision-making ability should be conspicuously sharper than that of most
people.

Figure 6.11 illustrates some aspects of the structure of the chip-selection problem. Note
that the .5 values shown below both bars representing bags of chips are unconditional proba-

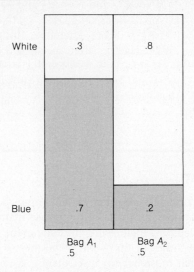

White   .3   .8

Blue   .7   .2

Bag A_1 Bag A_2
.5   .5

Figure 6.11 Unconditional and conditional probabilities for the chip-selection problem.

bilities indicating the chance of a specific bag's being selected from the coin toss. In symbols, $P(A_1) = .5$ and $P(A_2) = .5$. The probabilities inside the bars, on the other hand, are conditional probabilities. For example, the .7 in the lower part of Bar A_1 is the conditional probability that a blue chip will be selected in a single drawing of a single chip *given that* the bag selected from tossing the coin is Bag A_1. Similarly, the probability of .8 in the upper part of the bar labeled A_2 represents the probability of getting a white chip in a single drawing *given that* Bag A_2 is the one selected from the coin toss.

Clearly, with the information given, we are prepared to answer certain *already familiar* kinds of probability problems such as: *What is the probability that a chip selected from either of the two bags will be blue?*

Although the answer in this case, .45, is intuitively obvious, we might reason that the bag from which the chip is selected has a 50-50 chance of being A_1 and, if A_1 is the bag selected, the probability is .7 that the chip will be blue. In symbols, we simply make use of the general case of the multiplication theorem—that is, $P(A_1 \text{ and } B) = P(A_1) \cdot P(B|A_1) = (.5)(.7) = .35$, where B stands for the selection of a single blue chip. Of course, we are not yet finished with the problem because we have considered only one way out of two possible ways in which a blue chip might be obtained. The other possible way is to choose Bag A_2 as a result of the coin toss and, in addition, select a blue chip from it. That is, $P(A_2 \text{ and } B) = P(A_2) \cdot P(B|A_2) = (.5)(.2) = .10$. Since a blue chip can be obtained by either of the two means just described and since, if it happens one way it cannot simultaneously happen the other way, we observe that the special case of the addition theorem of probability must also be brought into the calculations. So, $P(B_1) = P(A_1) \cdot P(B|A_1) + P(A_2) \cdot P(B|A_2) = .35 + .10 = .45$.

Now, let us turn the problem around. Suppose a *single chip* is selected from among the total of 200 blue chips and we are given the *additional information* that the chip selected is blue. What assessment would we now make of the probability that the chip came from Bag A_1? That is, what is the conditional probability $P(A_1|B)$? This, of course, is the same question posed earlier except that, for simplicity, we are assuming that *one* chip was selected rather than ten. (We will return to the ten-chip version in Chapter 7 after you have had a chance to get acquainted with binomial probabilities.)

> It is important to note the nature of the probability question. The probability $P(A_1)$ may be thought of as a *prior probability* because it is determined prior to the observation of some specific kind of experimental information. The probability $P(A_1|B)$ may be thought of as a *posterior probability* or a *revised probability* because it is determined after the observation of additional information. The posterior, or revised, probability is the type computed when Bayes' theorem is employed.

Our probability assessment of the event "Bag A_1 is chosen as a result of the coin toss" is .5. How should we revise the probability if we are given the additional information that the chip selected is blue? We can reason as follows: Of all the chips in the two bags $.5 \times .7 = .35$ are in Bag A_1 and blue in color, whereas $.5 \times .2 = .10$ are in Bag A_2 and blue in color. Hence, the probability that the blue chip actually observed is from Bag A_1 is equal to $.35/(.35 + .10) = .78$. The logic behind this calculation can be expressed algebraically by Bayes' theorem, named in honor of Reverend Thomas Bayes, an eighteenth-century British scholar. His theorem is simply a formula for computing a certain kind of conditional probability. Bayes' theorem for a two-event situation is

$$P(A_1|B) = \frac{P(A_1) \cdot (B|A_1)}{P(A_1) \cdot P(B|A_1) + P(A_2) \cdot P(B|A_2)} \qquad (6.9.1)$$

A more general way of expressing Bayes' theorem—a way allowing for more than two possible events (more than two bags in our poker chip example)—is as follows:

$$P(A_1|B) = \frac{P(A_1) \cdot P(B|A_1)}{\Sigma_i[P(A_i) \cdot P(B|A_i)]} \qquad (6.9.2)$$

Either version of the Bayes' formula may look a little scary. But it is possible to show that it can be derived from some probability rules with which you are now familiar.

Let us write the joint probability $P(A_1 \text{ and } B_1)$ in two different, but equally correct, ways:

$$P(A_1 \text{ and } B) = P(A_1) \cdot P(B|A_1)$$

and

$$P(A_1 \text{ and } B) = P(B) \cdot P(A_1|B)$$

Since $P(A_1 \text{ and } B)$ can be expressed either way, it follows that

$$P(B) \cdot P(A_1|B) = P(A_1) \cdot P(B|A_1)$$

Dividing both sides of the equation by $P(B)$, we get

$$P(A_1|B) = \frac{P(A_1) \cdot P(B|A_1)}{P(B)} \qquad \text{where } P(B) \neq 0$$

We note that the A_i events are mutually exclusive. We also note that one of these events must occur if B is to occur (e.g., in order for a blue chip to be selected from a bag, a bag, either A_1 or A_2, must first be chosen). We therefore observe that

$$P(B) = P(A_1) \cdot P(B|A_1) + P(A_2) \cdot P(B|A_2) + \cdots + P(A_k) \cdot P(B|A_k) = \Sigma_i[P(A_i) \cdot P(B|A_i)]$$

Substituting $\Sigma_i[P(A_i) \cdot P(B|A_i)]$ for $P(B)$ in the equation $P(A_1|B) = [P(A_1) \cdot P(B|A_1)]/P(B)$, we arrive at

$$P(A_1|B) = \frac{P(A_1) \cdot P(B|A_1)}{\Sigma_i[P(A_i) \cdot P(B|A_i)]}$$

which is Equation (6.9.2), the general equation for Bayes' theorem.

For the problem about the chips, a problem involving only two events, we will use Equation (6.9.1) to find $P(A_1|B)$:

$$P(A_1|B) = \frac{(.5)(.7)}{(.5)(.7) + (.5)(.3)} = .78$$

We see that on the basis of a *single drawing of a blue chip*, a modest amount of information to be sure, the probability associated with the selection of Bag A_1 has increased from .5—in the absence of any experimental information—to .78. The steps involved in using the Bayes' formulas can be kept straight most easily by using a tabular setup like that shown in Table 6.8. The first column of this table lists the events of interest. Column (2) shows the prior probability assigned to each of the possible events. Column (3) shows the conditional probabilities for the empirical evidence *given* each specific event. As noted in the caption, such conditional probabilities are referred to as "likelihoods." The fourth column gives the joint probabilities. When these joint probabilities are divided by their total—in this case, by .45, the results are the revised probabilities shown in column (5). The first revised probability shown in the fifth column is the one sought in this illustrative problem.

Table 6.8 Bayes' Theorem Calculations for Illustrative Problem Involving the Selection of One Chip

(1) Events A_i	(2) Prior Probabilities $P(A_i)$	(3) Likelihoods $P(B|A_i)$	(4) Joint Probabilities $P(A_i) \cdot P(B|A_i)$	(5) Revised Probabilities $P(A_i|B)$
Bag A_1	.5	.7	.35	$.35/.45 = .78$
Bag A_2	.5	.2	.10	$.15/.45 = .22$
	1.0		.45	1.00

Note: B stands for the selection of one blue chip.

Table 6.9 Bayes' Theorem Applied to the Exmont Company Problem

(1) Events A_i	(2) Prior Probabilities $P(A_i)$	(3) Likelihoods $P(B\|A_i)$	(4) Joint Probabilities $P(A_i) \cdot P(B\|A_i)$	(5) Revised Probabilities $P(A_i\|B)$
A_1: Highly successful	.3	.95	.285	.3775
A_2: Moderately successful	.6	.75	.450	.5960
A_3: Unsuccessful	.1	.20	.020	.0265
Total	1.0		.755	1.0000

Note: B stands for achievement of a passing score on the selling aptitude test.

We may also use the Exmont Company data (Table 6.1) to illustrate the application of Bayes' theorem. Let us recall that the Exmont sales force consists of 200 salespeople of whom 30% were rated highly successful, 60% moderately successful, and 10% unsuccessful.

We will suppose that among the information gathered for the sales staff were scores on a selling aptitude test and that, of the salespeople classified as highly successful, 95% had achieved a passing score. Of those classified as moderately successful, 75% had achieved a passing score. Finally, of those classified as unsuccessful, 20% had achieved a passing score. We now ask: What is the probability that a randomly selected salesperson would be rated (1) highly successful given a passing score, (2) moderately successful given a passing score, and (3) unsuccessful given a passing score? The steps involved in answering these questions are shown in Table 6.9.

We see that the aptitude test does appear to have some power to discriminate between potentially successful and potentially unsuccessful salespeople, though not so impressively as might be hoped. The revised, or posterior, probability associated with *highly successful* is .3775, up somewhat from the prior probability of .3; the revised probability for *moderately successful* is down slightly from the prior probability of .6; and somewhat encouraging is the fact that the revised probability associated with *unsuccessful* is only .0251, down from the prior probability of .1.

We will make further use of Bayes' theorem in Chapters 7 and 19. This brief introduction has been intended merely to demonstrate how probability theory used in conjunction with observed information can serve to reduce uncertainty and increase one's chances of making successful decisions.

6.10 Summary

The goal of this chapter has been to introduce you to several basic probability concepts and operations. Because probability theory is the foundation stone of all inductive procedures to be described in later chapters, it is important that you acquire a mastery of this subject. Table 6.10 presents most of the probability principles introduced in this chapter.

A topic touched on in this chapter—that of *probability distributions*—will be pursued at greater length in the next chapter.

Table 6.10 Probability Principles Introduced in This Chapter

Principle	Interpretation
$P(O_1) + P(O_2) + P(O_3) + \cdots + P(O_k) = 1$	A sample space consists of *collectively exhaustive* basic outcomes. Hence, the probabilities of the basic outcomes will sum to 1.
$P(O_1) + P(O_2) = P(E)$	The probability of an event is equal to the sum of the separate probabilities of the basic outcomes comprising the event.
$P(\text{Not-}E) = 1 - P(E)$	The probability of the complementary event Not-E is equal to the probability of event E subtracted from 1.
$P(A_i\|B_j) = \dfrac{P(A_i \text{ and } B_j)}{P(B_j)}$	The conditional probability of event A_i given event B_j is equal to the joint probability of the two events divided by the probability of B_j (where $B_j \neq 0$).
$P(E_1 \text{ or } E_2) = P(E_1) + P(E_2) - P(E_1 \text{ and } E_2)$	When two events, E_1 and E_2, are not mutually exclusive, the probability that *either* E_1 *or* E_2 will occur in a single trial of a random experiment is obtained by summing the separate event probabilities and subtracting the probability that the two events will occur together.
$P(E_1 \text{ or } E_2) = P(E_1) + P(E_2)$	When two events, E_1 and E_2, are mutually exclusive, the probability that *either* E_1 *or* E_2 will occur in a single trial of a random experiment is equal to the sum of the separate event probabilities.
$P(E_1 \text{ and } E_2) = P(E_1) \cdot P(E_2\|E_1) = P(E_2) \cdot P(E_1\|E_2)$	When two events, E_1 and E_2, are not independent, the probability that *both* E_1 *and* E_2 will occur together in a single trial of a random experiment is equal to (1) the product of the unconditional probability of E_1 times the conditional probability of $E_2\|E_1$ or (2) the unconditional probability of E_2 times the conditional probability of $E_1\|E_2$.
$P(E_1 \text{ and } E_2) = P(E_1) \cdot P(E_2)$	When two events, E_1 and E_2, are independent, the probability that *both* E_1 *and* E_2 will occur together in a single trial of a random experiment is equal to the product of the separate probabilities.
$P(E_1) = P(E_1\|E_2)$ and $P(E_2) = P(E_2\|E_1)$	Demonstration that events E_1 and E_2 are independent.
$P(A_1\|B) = \dfrac{P(A_1) \cdot P(B\|A_1)}{P(A_1) \cdot P(B\|A_1) + P(A_2) \cdot P(B\|A_2)}$	Two-event version of Bayes' theorem. This theorem can be applied to situations where $P(A_1)$ and $P(B\|A_1)$ are known and one wishes to determine $P(A_1\|B)$.
$P(A_1\|B) = \dfrac{P(A_1) \cdot P(B\|A_1)}{\Sigma_i[P(A_i) \cdot P(B\|A_i)]}$	General version of Bayes' theorem. This version will accommodate any number of events.

Terms Introduced in This Chapter

addition theorem of probability—
    general case (p. 214)
addition theorem of probability—
    special case (p. 216)
a priori probability (p. 200)
basic outcome (p. 206)
Bayes' theorem (p. 227)
collectively exhaustive (p. 208)
complementary events (p. 210)
conditional probability (p. 212)
constant cause system (p. 201)
dependent events (p. 212)
event (p. 208)

independent events (p. 219)
joint probability (p. 217)
likelihood (p. 228)
multiplication theorem—general case
    (p. 217)
multiplication theorem—special case
    (p. 219)
mutually exclusive (p. 208)
prior probability (p. 227)
posterior probability (p. 227)
probability distribution (p. 206)
random experiment (p. 198)
relative-frequency probability (p. 201)

revised probability (p. 227)
sample space (p. 207)
    univariate sample space (p. 207)
    bivariate sample space (p. 207)
subjective probability (p. 203)
trial (p. 198)
success (p. 200)
Venn diagram (p. 211)

Formulas Introduced in This Chapter

See Table 6.10.

Questions and Problems

6.1 Explain the meaning of each of the following:
    a. Random experiment   e. Constant cause system
    b. Uncertainty   f. Probability distribution
    c. A priori probability   g. Sample space
    d. Success   h. Mutually exclusive

6.2 Explain the meaning of each of the following:
    a. Trial   e. Likelihood
    b. Probability scale   f. Conditional probability distribution
    c. Relative-frequency probability   g. Event
    d. Basic outcome   h. Addition theorem of probability

6.3 Explain the meaning of each of the following:
    a. Complementary event   e. Independence
    b. Conditional probability   f. Bayes' theorem
    c. Multiplication theorem of probability   g. Collectively exhaustive
    d. Joint probability   h. Subjective probability

6.4 Distinguish between:
    a. A priori probability and relative-frequency probability
    b. The general case of the addition theorem of probability and the special case of the same
       theorem

c. The general case of the multiplication theorem of probability and the special case of the same theorem
d. Basic outcome and event
e. Independent and dependent events
f. Prior probability and posterior, or revised, probability

6.5 Distinguish between:
a. Mutually exclusive events and independent events
b. Conditional and unconditional probability
c. Univariate sample space and bivariate sample space
d. Sample space and probability distribution
e. Random experiment and trial

6.6 Distinguish between:
a. $P(A)$ and $P(A|B)$
b. $P(A|B)$ and $P(A|B \text{ and } C)$
c. $P(A \text{ and } B)$ and $P(A \text{ or } B)$
d. $P(A) \cdot P(B|A)$ and $P(A) \cdot P(B)$
e. $P(A) + P(B) - P(A \text{ and } B)$ and $P(A) + P(B)$
f. $P(B|A_1)$ and $P(A_1|B)$ in a Bayes' theorem problem

6.7 Indicate which of the following statements you agree with and which you disagree with, and defend your opinions:
a. The term "random experiment" refers only to games of chance such as coin flipping or dice rolling.
b. In real-world decision making, probabilities must be used, in some manner, perhaps quite informally, as a substitute for certainty.
c. Using the relative-frequency approach to probability determination, one could calculate the probability that one's next purchase of a common stock will prove profitable by counting the number of one's profitable past purchases and dividing that number by the number of common stock purchases, both profitable and unprofitable, one has made altogether.
d. The use of personal judgment to determine the probability that an event will occur must be avoided at all cost when statistical methods are employed in connection with business decision making.

6.8 Indicate which of the following statements you agree with and which you disagree with, and defend your opinions:
a. A relative frequency is not necessarily a probability.
b. Two basic outcomes may or may not be mutually exclusive.
c. Two events may or may not be mutually exclusive.
d. If we have two events, A_i and B_j, based on a bivariate sample space, the conditional probability of A_i given B_j is defined as $P(A_i|B_j) = [P(A_i \text{ and } B_j)]/P(A_i)$.

6.9 Indicate which of the following statements you agree with and which you disagree with, and defend your opinions:
a. If we have two events, A_i and B_j, based on a bivariate sample space, the conditional probability of A_i given B_j is defined as $P(A_i|B_j) = [P(A_i \text{ and } B_j)]/P(B_j)$, regardless of the value of $P(B_j)$.
b. If the probability of event E_1 is .5 and the probability of event E_2 is .3, the probability that either event E_1 or event E_2 will result from a single trial of a random experiment is $.5 + .3 = .8$.
c. If the probability of event E_1 is .5 and the probability of event E_2 is .3, the probability that events E_1 and E_2 will occur together in a single trial of a random experiment is .5 times $.3 = .15$.

6.10 Indicate which of the following statements you agree with and which you disagree with, and defend your opinions:
   a. $P(E_1|E_2)$ is always equal to $P(E_1)$.
   b. $P(E_1|E_2)$ is never equal to $P(E_1)$.
   c. $P(A|B)$ is always equal to $P(B|A)$.
   d. $P(A) \cdot P(B|A)$ is always equal to $P(B) \cdot P(A|B)$.

6.11 Indicate which of the following statements you agree with and which you disagree with, and defend your opinions:
   a. $P(A)$ will always be equal to the sum of all relevant joint probabilities of the form $P(B) \cdot P(A_i|B)$.
   b. The sum of all relevant conditional probabilities of the form $P(A_i|B_1)$ will always be 1.
   c. P(at least 5 successes in 10 trials) $= 1 - P$(no successes in 10 trials).
   d. $P(A_1)$ can never be larger than $P(B|A_1)$ in a Bayes' theorem problem.

6.12 An automobile rental company has a Ford, a Chevrolet, and an Oldsmobile which have been involved in minor accidents and are in need of body and fender work. The manager sends an employee out to obtain cost estimates for each car from three body-and-fender companies: Al's, Deluxe, and Snappy. Let (1) A_1 denote the event that the Ford is examined, (2) A_2 the event that the Chevrolet is examined, (3) A_3 the event that the Oldsmobile is examined, (4) B_1 the event that Al's is the place doing the examining, (5) B_2 the event that Deluxe is the place doing the examining, and (6) B_3 the event that Snappy is the place doing the examining.
   a. Present the entire sample space, using O_{ij} symbols like those shown in the second part of Figure 6.2.
   b. List the basic outcomes making up each of the following:
   1. A_2
   2. B_1 and A_2
   3. B_1 or A_2
   4. $(B_1$ and $A_2)$ or $(B_2$ and $A_3)$
   5. B_2 and Not-A_3
   6. $B_2|A_2$

6.13 A law firm has branches in two western states, California and Nevada, and in two southern states, Florida and Louisiana. These branches specialize in divorces, bankruptcies, and tax problems. Let (1) A_1 denote the event that the state is California, (2) A_2 the event that the state is Nevada, (3) A_3 the event that the state is Florida, (4) A_4 the event that the state is Louisiana, (5) B_1 the event of a divorce case, (6) B_2 the event of a bankruptcy case, and (7) B_3 the event of a tax-problem case.
   a. Present the entire sample space, using O_{ij} symbols like those shown in the second part of Figure 6.2.
   b. List the basic outcomes making up each of the following sets:
   1. A_4
   2. A_3 and B_2
   3. A_4 and B_3
   4. A_2 or A_4
   5. $(B_1$ and $A_3)$ or $(B_2$ and $A_4)$
   6. $(B_1$ or $A_3)$ and $(B_2$ or $A_3)$
   7. B_2 and Not-A_2
   8. $(B_1$ or $B_2)|A_1$

6.14 The Sebastian Cabinet Company, manufacturer of high-priced wooden cabinets, employs 50 skilled cabinetmakers, 26 of whom have received formal trade school and apprenticeship training and 24 of whom are self-taught. The production manager categorized these craftsmen according to type of training received and according to quality of work, 1 being the

best possible rating. The bivariate frequency distribution he obtained is shown in Table 6.11. What is the probability that a cabinetmaker picked at random will
a. Be self-taught?
b. Do work classified as quality-level 1?
c. Do work classified as quality-level 1 given that he or she is self-taught?
d. Do work classified as quality-level 3 or 4 given that he or she is self-taught?
e. Be self-taught given that he or she achieved a quality-level rating of 2?
f. Be self-taught or do work classified as quality-level 3?
g. Do work classified as quality-level 1 and be self-taught?
h. Be formally trained and do work classified as either quality-level 2 or quality-level 3?
i. Do work classified as quality-level 3 given that he or she is either formally trained or self-taught?

 6.15 Refer to Table 6.11.
a. Does $P(A_1|B_1) = P(B_1|A_1)$?
b. Does $P(A_1 \text{ and } B_1)$ taken directly from the table equal $P(A_1) \cdot P(B_1)$? What does your answer convey about the independence, or lack of it, of events A_1 and B_1?
c. Are events A_1 and B_2 independent? Explain.
d. Are events A_1 and B_3 independent? Explain.
e. Are events A_2 and B_2 independent? Explain.

 6.16 Table 6.12 (containing data from Table 3.8) shows data on price-earnings ratios and earnings per share associated with 200 common stocks. What is the probability that a stock picked at random will have
a. Earnings per share in the $4 and over class?
b. A price-earnings ratio in the 15 and over class?
c. A price-earnings ratio in the 15 and over class and earnings per share in the $4 and over class?
d. A price-earnings ratio in the 15 and over class given that its earnings per share is in the $4 and over class?
e. A price-earnings ratio in the 9 and under 15 range?
f. A price-earnings ratio in the 9 and under 15 range or earnings per share in the $2 and under $4 range?
g. A price-earnings ratio in the 9 and under 15 range and earnings per share in the $2 and under $4 range?

 6.17 Refer to Table 6.12.
a. Does $P(A_3|B_1) = P(B_1|A_3)$?
b. Does $P(A_3 \text{ and } B_2)$ taken directly from the table equal $P(A_3) \cdot P(B_2)$? What does your answer convey about the independence, or lack of it, of events A_3 and B_2?

Table 6.11 50 Cabinetmakers Classified According to Type of Training and Quality of Work

Type of Training	Quality Rating of Work				Total
	1 (B_1)	2 (B_2)	3 (B_3)	4 (B_4)	
Trade school and apprenticeship (A_1)	6	11	7	2	26
Self-taught (A_2)	3	6	10	5	24
Total	9	17	17	7	50

Table 6.12 200 Common Stocks Classified According to Price-Earnings Ratio and Earnings per Share

Price-Earnings Ratio	Earnings per Share					Total
	$0 and under $1 ($B_2$)	$1 and under $2 ($B_2$)	$2 and under $3 ($B_3$)	$3 and under $4 ($B_4$)	$4 and Over ($B_5$)	
3 and under 6 (A_1)	1	8	4	2	1	16
6 and under 9 (A_2)	5	19	28	7	9	68
9 and under 12 (A_3)	10	13	13	6	1	43
12 and under 15 (A_4)	4	9	4	3	2	22
15 and over (A_5)	11	24	8	5	3	51
Total	31	73	57	23	16	200

c. Are events A_1 and B_1 independent? Explain.
d. Are events A_5 and B_5 independent? Explain.
e. Are events A_1 and B_5 mutually exclusive? Explain.

✔ 6.18 Table 6.13 shows data on television-watching habits of a group of men and women. What is the probability that a viewer picked at random will be
   a. Female?
   b. Classified as a frequent viewer?
   c. Classified as an infrequent viewer given that she is female?
   d. Classified as a frequent viewer or as a moderate viewer given that she is female?
   e. Male given that he is a frequent viewer?
   f. Female and an infrequent viewer?
   g. Male or a moderate viewer?

✔ 6.19 Refer to Table 6.13.
   a. Does $P(A_1|B_1) = P(B_1|A_1)$?
   b. Does $P(A_3$ and $B_2)$ taken directly from the table equal $P(A_3) \cdot P(B_2)$? What does your answer convey about the independence, or lack of it, of events A_3 and B_2?
   c. Are events A_1 and B_1 independent? Explain.
   d. Are events A_2 and B_2 independent? Explain.

Table 6.13 1875 Television Viewers Classified by Frequency of Viewing and Sex

Viewing Category	Sex		Total
	Male (B_1)	Female (B_2)	
Frequent (A_1)	200	300	500
Moderate (A_2)	400	600	1000
Infrequent (A_3)	150	225	375
Total	750	1125	1875

✔ 6.20 A manufacturer of home appliances conducted a study to determine how consumers perceive the quality of the company's products in general and whether geographical location of a respondent's place of residence was in any way related to product quality perception. The results are shown in Table 6.14. What is the probability that a respondent chosen at random will

a. Perceive the company's products as being of high quality?
b. Reside in the eastern United States?
c. Perceive the company's products as being of low quality given that he or she resides in the western United States?
d. Perceive the company's products as being of medium quality or reside in the eastern United States?
e. Reside in the western United States and perceive the company's products as being of either high or medium quality?

✔ 6.21 Refer to Table 6.14.
a. Does $P(A_1|B_2) = P(B_2|A_1)$?
b. Does $P(A_1) \cdot P(B_2|A_1) = P(B_2) \cdot P(A_1|B_2)$?
c. Are events A_1 and B_2 independent? Explain.
d. Are events A_3 and B_2 independent? Explain.

✔ 6.22 The manufacturer of home appliances referred to in problem 6.20 undertook a more extensive study of a similar kind. This time, the respondents were classified according to quality perception, place of residence, and sex. Table 6.15 shows the results. What is the probability that a respondent picked at random will

a. Perceive the company's products as being of high quality?
b. Reside in the western United States?
c. Be female?
d. Be female given that she resides in the eastern United States?
e. Perceive the company's products as being of high quality given that she is female?
f. Perceive the company's products as being of high quality given that she is a female residing in the western United States?
g. Be male, reside in the eastern United States, and perceive the company's products as being of low quality?
h. Be male or reside in the eastern United States or perceive the company's products as being of high quality?

Table 6.14 7615 Consumers Classified According to Their Perception of the Quality of a Company's Products and Whether They Live in the Western United States or the Eastern United States

Respondent Perceives Quality of Company's Products as	Place of Residence		
	Western United States (B_1)	Eastern United States (B_2)	Total
High (A_1)	1005	700	1705
Medium (A_2)	2200	2175	* 4375
Low (A_3)	525	1010	1535
Total	3730	3885	7615

Table 6.15 7615 Consumers Classified According to Their Perception of the Quality of a Company's Products, Place of Residence, and Sex

Respondent Perceives Quality of Company's Products as	Place of Residence				Total
	Western United States (B_1)		Eastern United States (B_2)		
	Sex		Sex		
	Male (C_1)	Female (C_2)	Male (C_1)	Female (C_2)	
High (A_1)	300	705	200	510	1715
Medium (A_2)	1100	1100	1000	1175	4375
Low (A_3)	400	125	750	260	1535
Total	1800	1930	1950	1945	7625

 6.23 Refer to Table 6.15.

    a. Does $P(A_1|C_1) = P(C_1|A_1)$?

    b. Does $P(A_1|B_1 \text{ and } C_1) = P(C_1|A_1 \text{ and } B_1)$?

    c. Does $P(A_1) \cdot P(B_1|A_1) \cdot P(C_1|A_1 \text{ and } B_1) = P(B_1) \cdot P(C_1|B_1) \cdot P(A_1|B_1 \text{ and } C_1)$?

    d. Are events A_2 and B_2 independent? Explain.

    e. Are events A_2 and C_2 independent? Explain.

    f. Are events B_2 and C_2 independent? Explain.

    g. Are events A_1 and C_1 mutually exclusive? Explain.

 6.24 The owner of a sporting goods store has, through long observation of and conversations with his customers, determined that the probability that the next person entering the store will

    1. Buy a tennis racket is $P(BTR) = .1$.

    2. Be someone who already owns a tennis racket is $P(OTR) = .4$.

    3. Buy tennis balls is $P(BTB) = .3$.

    4. Buy tennis balls given that he or she also buys a tennis racket is $P(BTB|BTR) = .4$.

    5. Buy tennis balls given that he or she already owns a tennis racket is $P(BTB|OTR) = .5$.

    6. Buy tennis balls given that he or she does not already own a tennis racket is $P(BTB|NOTR) = .167$.

    What is the probability that

    a. The next person entering this store will buy a tennis racket and tennis balls?

    b. The next person entering this store will be someone who already owns a tennis racket given that he or she buys tennis balls? Explain carefully how you arrived at your answer.

 6.25 Of 60 workers in a weaving mill, it is known that 20 are Caucasian, 40 are female, and 14 are Caucasian females. If one of the 60 workers were picked at random, what is the probability that the one selected is

    a. A non-Caucasian female?

    b. A Caucasian male?

    c. Caucasian given that she is female?

    d. Female given that she is Caucasian?

6.26 Refer to problem 6.25.
    a. Are events Caucasian and male independent? Explain.
    b. Are events Caucasian and male mutually exclusive? Explain.

6.27 Refer to problem 6.25. If three employees of this weaving mill were picked at random, what is the probability that
    a. At least one will be Caucasian?
    b. At least one will be female?

6.28 It is known that in a shipment of 1000 long-handled paint scrapers (1) 12 are of inadequate quality, (2) 850 were manufactured by the supplying company whereas the rest were purchased by the supplying company from another company so that the order could be filled completely, and (3) six of those purchased by the supplying company from another company are defective. If a paint scraper is picked at random, what is the probability that it will be
    a. Defective?
    b. Defective and manufactured by the supplying company itself?
    c. Defective given that it was not manufactured by the supplying company?
    d. Acceptable and bought by the supplying company from another company?
    e. Defective or bought by the supplying company from another company?

6.29 Refer to problem 6.28.
    a. Is the event "defective" independent of the event "manufactured by supplying company"? Explain.
    b. Are events "defective" and "manufactured by supplying company" mutually exclusive? Explain.

6.30 Refer to problem 6.28. If four paint scrapers are selected at random, what is the probability that
    a. At least one will be defective?
    b. At least one will have been manufactured by the supplying company?

6.31 It is known that of 12 members of the board of directors (1) all work for the company, (2) four aspire to become chairman of the board, (3) seven came up through the ranks within the company whereas the others were hired at the middle-management level from outside the firm, and (4) two were hired from outside and aspire to become chairman of the board. If one board member were picked at random, what is the probability that he or she
    a. Was hired at the middle-management level from outside the company?
    b. Aspires to become chairman of the board?
    c. Was hired at the middle-management level from outside or aspires to become chairman of the board?
    d. Came up through the ranks and has no desire to become chairman of the board?

6.32 Refer to problem 6.31.
    a. Are the events "hired at the middle-management level from outside" and "desires to be chairman of the board" independent? Explain.
    b. Are the events cited in part *a* mutually exclusive? Explain.

6.33 Refer to problem 6.31. If five members of the board of directors were picked at random, what is the probability that
    a. At least one aspires to become chairman of the board?
    b. At least one came up through the ranks?

6.34 A certain manufactured item is processed by two machines, A and B, in series; the item can be produced if and only if both machines are functioning. If, on any given day, the probability that Machine A will function is .96 and the probability that Machine B will function is .9, and if the machines operate independently, what is the probability that

a. The item can be produced on the day in question?
b. The item cannot be produced on the given day because either Machine A or Machine B (but not both) fails to function?
c. The item cannot be produced on the given day because either Machine A or Machine B (or both) fails to function?
d. The item cannot be produced on the given day because both machines fail to function?

 6.35 Refer to problem 6.34. Describe the sample space for this problem.

 6.36 Refer to problem 6.34. Suppose that management has developed a contingency plan to call into use if and only if the manufacturing process involving Machines A and B fails to function (either because Machine A or Machine B, or both, fail to function). The contingency plan involves utilizing older machines, C and D. The probability that Machine C will function on a given day is .8; the probability that Machine D will function on a given day is .82. Assume that (1) both Machine C and Machine D must function if the item is to be produced by this alternative pair of machines and (2) Machines C and D operate independently. What is the probability that
a. The item can be produced on a given day even if either Machine A or Machine B (or both) fails to function?
b. The item cannot be manufactured on a given day because the system consisting of Machines A and B fails to function and, in addition, either Machine C or Machine D (or both) fails to function?

 6.37 The success, S, or failure, Not-S, of a new employee incentive program depends in part on whether the Tobias Company's strongest competitor (1) leaves its present incentive system unchanged, U, or (2) copies the Tobias Company's change, Not-U. A few probability assessments have been made; these are shown here.

	S	Not-S	Total
U			
Not-U	.5		.7
Total	.7		1.0

a. Explain in symbols and in words what the probability .5 represents.
b. Complete the table by filling in the empty cells and marginal totals.
c. In the second row of the above table, the probabilities .5 and .2 are summed to get .7. What rule of probability is being employed here?
d. Demonstrate with numbers that $P(S|U)$ does not equal $1 - P(S|Not\text{-}U)$.
e. Demonstrate with numbers that $P(U) \cdot P(S|U) + P(Not\text{-}U) \cdot P(S|Not\text{-}U) = P(S)$.
f. Demonstrate with numbers that $P(S) \cdot P(U|S) + P(Not\text{-}S) \cdot P(U|Not\text{-}S) = P(U)$.

 6.38 The earnings outlook for a company is thought to be closely tied to the state of the general economy. A few probability assessments have been made and are shown here.

	Lose Money (B_1)	Break Even (B_2)	Earn Profit (B_3)	Total
Recession (A_1)				
Modest growth (A_2)		.3	.1	.5
Substantial growth (A_3)		.1	.2	.3
Total		.5	.3	1.0

a. Explain in symbols and in words what the probability .2 represents.
b. Complete the table by filling in the empty cells and marginal totals.
c. In the second row of the table, the probabilities .1, .3, and .1 are summed to get .5. What rule of probability is being employed here?
d. Demonstrate with numbers that the sum of $P(A_1|B_1) + P(A_2|B_1) + P(A_3|B_1)$ is 1.
e. Demonstrate with numbers that $P(B_1|A_1)$ does not equal $1 - [P(B_1|A_2) + P(B_1|A_3)]$.
f. Are events A_3 and B_1 mutually exclusive? Explain.
g. Demonstrate with numbers that $P(A_1) \cdot P(B_1|A_1) + P(A_2) \cdot P(B_1|A_2) + P(A_3) \cdot P(B_1|A_3) = P(B_1)$.
h. Demonstrate with numbers that $P(B_1) \cdot P(A_1|B_1) + P(B_2) \cdot P(A_1|B_2) + P(B_3) \cdot P(A_1|B_3) = P(A_1)$.

 6.39 Historically, the Belchfire Automobile Company has produced defective doors with a certain relative frequency. It has been found that the probability of a four-door model's having four defective doors is .0016. Assuming independence and a constant probability associated with a single defective door, what is the probability that a certain four-door car produced by this company would have
a. Exactly one defective door?
b. At least one defective door?

 6.40 The Streaper Construction Company must begin a major water-pipe-laying job next Monday morning. Anticipating the need for more trucks than they presently own, management ordered six new custom-made trucks, three excavators and three tractor pipelayers, from the Lampson Company. However, Streaper management has just received a note from the Lampson Company which states, "We regret to inform you that we will be able to deliver only two of the trucks you ordered on time." No high official of the Lampson Company could be contacted for further elaboration. What is the probability that both trucks delivered Monday morning will be excavators?

 6.41 Out of ten finalists for four job openings, three are women. Assume that the four applicants who will get the jobs are to be selected at random.
a. What is the probability that no women will be selected?
b. What is the probability that half of those selected will be women?

6.42 Given: $P(A) = .8$, $P(B) = .4$, $P(A \text{ and } B) = .32$.
a. Are events A and B independent? Explain.
b. Are events A and B mutually exclusive? Explain.

6.43 Given: $P(A) = .5$, $P(B) = .4$, $P(A|B) = .375$, $P(B|A) = .3$. Find $P(A \text{ and } B)$ two different ways.

6.44 Given: $P(A_1) = .3$, $P(A_2) = .3$, $P(A_3) = .2$. If there are five mutually exclusive A-type events altogether, what is $P(A_4 \text{ or } A_5)$? Can you determine $P(A_4)$ from this information? $P(A_5)$?

Special-Challenge Problem

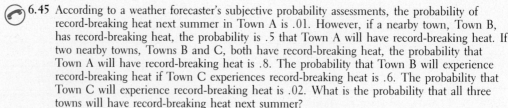 6.45 According to a weather forecaster's subjective probability assessments, the probability of record-breaking heat next summer in Town A is .01. However, if a nearby town, Town B, has record-breaking heat, the probability is .5 that Town A will have record-breaking heat. If two nearby towns, Towns B and C, both have record-breaking heat, the probability that Town A will have record-breaking heat is .8. The probability that Town B will experience record-breaking heat if Town C experiences record-breaking heat is .6. The probability that Town C will experience record-breaking heat is .02. What is the probability that all three towns will have record-breaking heat next summer?

Special-Challenge Problem

6.46 Art Rugged would like to be chief of police (an appointed position). Under the present administration, the probability of his being appointed to that position is only .1. Bill Slye would like to be district attorney (also an appointed position). Under the present administration the probability of his being appointed to that post is only .1. However, both Art and Bill are good friends of Jane Rule who will be running for mayor in an upcoming election. Art figures that if Jane wins the election, the probability of his being appointed chief of police is .7 and Bill's being appointed district attorney is .6. Moreover, Art figures that if Jane wins the election and then appoints Bill to the post of district attorney, that will indicate that Jane is not particularly concerned about accusations of cronyism; therefore, the probability that he (Art) would be appointed chief of police would increase to .8. The probability that Jane will win the election is .4. What is the probability that Jane will be mayor, Bill will be district attorney, and Art will be chief of police?

Special-Challenge Problem

6.47 A survey of 100 residents of a certain city revealed the following about the kinds of newsmagazines the residents subscribe to:

30 subscribe to *Time*
35 subscribe to *Newsweek*
25 subscribe to *U.S. News and World Report*
12 subscribe to both *Time* and *Newsweek*
14 subscribe to both *Time* and *U.S. News and World Report*
10 subscribe to both *Newsweek* and *U.S. News and World Report*
 4 subscribe to *Time, Newsweek,* and *U.S. News and World Report*

What is the probability that a resident of this city picked at random will
a. Subscribe to none of the three magazines?
b. Subscribe either to *Time* or *Newsweek?*
c. Subscribe to exactly one of the three magazines?

Special-Challenge Problem

6.48 A study of 500 business majors at a certain university revealed the following about the courses being taken this quarter:

100 are taking a management course
200 are taking an accounting course
150 are taking a finance course
 50 are taking a course in management and a course in accounting
 40 are taking a course in management and a course in finance
 20 are taking a course in accounting and a course in finance
 10 are taking a course in management, a course in accounting, and a course in finance

What is the probability that a student picked at random will be taking
a. No courses in these three subjects?
b. Either an accounting course or a finance course?
c. Either a management course or an accounting course or a finance course?

6.49 Carton A_1 contains 2 cracked eggs and 10 uncracked ones; Carton A_2 contains 4 cracked eggs and 8 uncracked ones. An egg is chosen at random from a carton chosen at random and is found to be cracked.
a. What is the probability that the egg came from Carton A_1?
b. What is the probability that the egg came from Carton A_2?

6.50 A large retailer buys assembled bicycles from three companies, A, B, and C. The retailer has bought from Company A 50% of the time, Company B 30% of the time, and Company C 20% of the time. In the past the assembly work has been done improperly at Company A 3% of the time, at Company B 5% of the time, and at Company C 2% of the time. A bicycle has just arrived and is inspected. The assembly work is found to be improper. What is the probability that the supplying company is
   a. A?
   b. B?
   c. C?

6.51 Presidential Candidate A has been emphasizing the need for reductions in personal income taxes. Experts estimate that the probability of Candidate A's being elected is .6 and that, if he is elected, the probability of a reduction in personal taxes is .8. Candidate B also favors reductions in taxes, but gives the matter a lower priority than does Candidate A. Experts estimate that the probability of Candidate B's being elected is .4 and that, if he is elected, the probability of a reduction in personal taxes is .5. Show that, before the election, the probability of a reduction in personal taxes is .68. (*Hint:* Assume that the experts are right in their probability assessments and that Candidates A and B are the only ones running.)

6.52 Refer to problem 6.51. After the election a reduction in personal taxes actually occurs. What is the probability that the person elected was
   a. Candidate A?
   b. Candidate B?

6.53 An investor recently completed a painstaking fundamental analysis of the XYZ Company. On the basis of his analysis, he purchased several thousand shares of XYZ common stock. He estimated the probability to be .7 that the market price of the stock would increase within a specified number of months. In addition to his fundamental analysis, the investor kept a chart of daily price and volume data for the stock. After the passage of several weeks he was heartened to find his chart displaying an unmistakable head-and-shoulders upside breakout. (In other words, it was saying that a definite increase in market price was underway and was likely to continue for a time.) In an effort to determine the past reliability of this technical formation, the investor consulted Gunslinger Guthrie, an investment advisor, who informed him that in the past head-and-shoulders upside breakouts had preceded sizable increases in market price 80% of the time; only 20% of the time had this chart pattern given an incorrect signal. Assuming that the 80% and 20% figures are correct, what is this investor's posterior probability of a forthcoming sizable increase in the market price of XYZ stock?

6.54 Refer to problem 6.53. Suppose that this investor happened upon a scholarly article written by Professor Sneer who had conducted a considerable amount of computer research into the predictive power of head-and-shoulders upside breakouts. The article concluded, let us say, that 50% of the time, when this technical pattern had signaled higher prices, higher prices actually followed and that the other 50% of the time the signal had proved false. Assuming that Professor Sneer's conclusions are correct (and ignoring Gunslinger Guthrie's claims), what would be this investor's posterior probability of a forthcoming sizable increase in the market price of XYZ common stock? Does your answer seem reasonable? Tell why or why not.

6.55 Five percent of the holes drilled in the frame of a company's baby cribs by an automatic drilling machine have been found to be insufficiently deep to permit the eventual buyer to assemble the crib using nothing more than a pair of pliers and a screwdriver. Since the company's advertising had emphasized how easily the crib can be assembled and since the hole is used to fasten the safety bar to the crib, management considered this kind of defect to be quite serious. Consequently, they hired an inspector to examine the depth of each freshly drilled hole. Assume that if this inspector declares a hole's depth to be adequate, the crib part is permitted to pass onto the next stage of fabrication. If he declares the depth inadequate, the crib part must be run through the drilling process a second time and then be reinspected. Let us suppose that this inspector is incorrect in his assessments 2% of the time. That is, 2% of the time when he says a hole has adequate depth, its depth is really inadequate and 2% of the time when he says a hole has inadequate depth, its depth is really adequate. What is the posterior probability that, if the inspector declares a specific hole adequately deep, it is in fact adequately deep?

6.56 Refer to problem 6.55. If the inspector declares a hole to be inadequately deep, what is the posterior probability that it is really deep enough?

6.57 Refer to problem 6.55. If the inspector declares a hole inadequately deep, has the crib part run through the drilling process a second time, and then declares the hole to be satisfactory, what is the probability that the hole is still not deep enough?

6.58 Refer to the poker chip example (section 6.9 and Table 6.8). Suppose that Bag A_1 contained 60 blue chips and 40 white chips and Bag A_2 contained 40 blue chips and 60 white chips. If the concerned person (1) chose one bag at random by flipping a coin, (2) selected one chip at random from the randomly selected bag, and (3) found that the chip selected was blue, what is the posterior probability that the bag selected was Bag A_1?

TAKE CHARGE

6.59 One of the more fascinating, albeit sometimes chilling, things that can happen to a person is to find oneself the central figure in a coincidence. Even when the coincidence is of a trivial nature, it is still sometimes difficult to shake the feeling that the experience was orchestrated by some kind of higher power.

For example, this is being written at the end of the summer quarter of 1984. At the beginning of the quarter your author passed out a syllabus to his graduate-level students in quantitative methods. This syllabus was more detailed than usual in that it indicated what material was to be covered during each class period throughout the quarter. After the quarter got underway, a student pointed out that no mention had been made of Monday, July 23, and wished to know whether the omission signaled a planned day off. This writer assured the student that the omission was merely human error and that class would be held that day as usual. However, when Monday, July 23, arrived, your author found himself with a severe case of laryngitis and was unable to meet his classes. Thus, the one day which had inadvertently been omitted from the syllabus turned out to be the only day that quarter when class was cancelled!

An admittedly rough (and, in view of arguments presented in connection with problem 6.60, maybe fallacious) calculation suggests that the probability of leaving a specific day off the syllabus *and* having laryngitis on that same day is about one in 1300:

$$\frac{1}{13} \times \frac{1}{100+} = \frac{1}{1300+}$$

(15 classes scheduled; first and last are left out of the calculations on the grounds that they would be highly unlikely to be omitted)

(Over the years, laryngitis has put this writer out of commission less often than one teaching day out of 100)

(*Note:* Independence is assumed.)

Write up a coincidence that has happened to you or to someone you know and, if possible, calculate the approximate probability associated with it.* If you cannot calculate the probability, at least describe the procedure you think should be used to arrive at it. Also state any assumptions underlying your calculations (or description).

TAKE CHARGE

6.60 Refer to problem 6.59. (You must have worked problem 6.59 in order to do this one.)

The following excerpts on the subject of coincidence are from a thought-provoking article by Ruma Falk:

It is conceivable that the element of surprise . . . stems from the fact that we attend to all the detailed components of the event as it actually occurred. The probability that an event would happen *precisely that way* is indeed minute.

When I happened to meet, while in New York, my old friend Dan from Jerusalem, on New Year's Eve and precisely at the intersection where I was staying, the amazement was overwhelming. The first question we asked each other was: "What is the probability that this would happen?" However, we did not stop to analyze what we meant by "this." What precisely was the event the probability of which we wished to ascertain?

I might have asked about the probability, while spending a whole year in New York, of meeting, at any time, in any part of the city, anyone from my large circle of friends and acquaintances. The probability of this event, the *union* of a large number of elementary events, is undoubtedly large. But instead, I tended to think of the *intersection* of all the components that converged in that meeting (the specific friend involved, the specific location, the precise time, etc.) and ended up with an event of miniscule probability. I would probably have been just as surprised had some other combination of components from that large union taken place. The number of such combinations is immeasurably large; therefore, the probability that at least one of them will actually occur is close to certainty. . . .

One's uniqueness in one's own eyes makes all the components of an event that happened to oneself seem like a singular combination. It is difficult to perceive one's own adventures as just one element in a sample space of people, meetings, places, and time. . . .

One characteristic of coincidences is that we do not set out to seek them in a predetermined time and place; they simply *happen* to us. Since they stand out in some strange combination, we single them out and observe them under a magnifying glass. A *selection fallacy* is here in action. . . . When we single out an extraordinary coincidence and claim that it is significant, we commit a logical error. Computing the probability that a surprising event of that kind would take place is not permissible since the design of the experiment was not determined in advance and the sample space was not clearly outlined. Our "experiment," in the course of which we came across our anecdote, was not preconfined in time, space, or population. It attracted our attention because of its rarity; otherwise, we would never have stopped to consider it. Having arrived at it in this way, we are not permitted to claim that it is highly unlikely. Instead of starting by drawing a random sample and then testing for the occurrence of a rare event, we select rare events that happened and find ourselves marveling at their nonrandomness. This is like the archer who first shoots an arrow and then draws the target circle around it.†

---

*Your author collects coincidences as a hobby and would enjoy receiving a copy of a description of your coincidence. Write to: Stephen K. Campbell, Department of Statistics and Operations Research, College of Business, University of Denver, 2020 South Race Street, Denver, Colorado 80210.

†From "On Coincidences" by Ruma Falk. *The Skeptical Inquirer* (Winter 1981–82), pp. 18ff. Reprinted by permission.

Do the above arguments suggest that

a. Your author should have performed the probability calculations shown in problem 6.61 in some other way? Explain why or why not.

b. You should have performed the probability calculations for the coincidence you cited in connection with problem 6.59 in some other way? If so, what changes would be in order?

Cases for Analysis and Discussion

6.1 THE "LAMPSON COMPANY": PART II

Study Part I of the "Lampson Company" case (found at the end of Chapter 1).

The following is a verbatim description of the procedures used by the personnel manager of the "Lampson Company" for determining the probability that an unskilled prospective employee will stay with the company for at least one year:

I used as my original population the unskilled workers hired during the period July 1, 1971, through June 30, 1972. After separating out the unskilled persons from all persons hired, I determined which unskilled persons were still working for Lampson one calendar year from date of hire. For those persons who were no longer with Lampson, I gathered the background data in the five categories. The progression went as follows:

Schedule A [Table 6.16] shows the more detailed results.

Here are my steps leading to Schedule B [Table 6.17].

1. Determine what percentage each subcategory is of total.
2. Multiply each subcategory percentage of total unskilled, times percentage quit for that subcategory, times the percentage which total quits represent of total unskilled workers (118/232). The result is the conditional probability of a specific unskilled worker's quitting for each subcategory.
3. Assign each subcategory a code from 0 to 9 (example: age ≥ 27 is assigned code number 2).
4. List all possible combinations of subcategories (see Schedule B [Table 6.17]). Since there are five categories with two subcategories each, the total number of possible combinations is $2^5 = 32$.
5. Assuming the subcategories are not mutually exclusive, I used the following probability equation to compute the conditional probability of each combination:

$$P(A + B + C) = P(A) + P(B) + P(C) - P(AB) - P(BC) - P(AC) + P(ABC)$$

Table 6.16 Schedule A

Characteristics of Respondent	Total Unskilled		Quit		Conditional Probability of Quitting
	Actual	Percent	Actual	Percent	
1. Lived in area					
≥ One year	150	64.7	41	34.7	.115
< One year	82	35.3	77	65.3	.118
2. Age					
≥ 27	93	40.1	30	25.4	.052
< 27	139	59.9	88	74.6	.228
3. Marital status					
Married	136	58.6	78	66.1	.197
Single or other	96	41.4	40	33.9	.071
4. Education					
≥ High school degree	124	53.4	71	60.2	.163
< High school degree	108	46.6	47	39.8	.094
5. Number jobs held last two years					
< 3	157	67.7	90	76.3	.263
≥ 3	75	32.3	28	23.7	.039

Now, if one assumes that all of the categories have a bearing on the chances that any specified unskilled prospect will quit within one year, Schedule B can be used as a guide to determining whether or not to hire him. Since 118 of the total 232 unskilled persons hired quit within one year, the overall conditional probability of quitting is .509, given that a person is unskilled. To determine if an unskilled applicant should be hired based on this statistical analysis, one would find the combination [of subcategories] that applies and see what the conditional probability of quitting is for the person based on Schedule B [Table 6.17]. If it is greater than or equal to .509, the mean, then using this method alone, one should conclude that the applicant should not be hired.

I have chosen to stop my statistical analysis here because I feel that the ultimate decision is a subjective one, based on an interview of the applicant with myself or another personnel administrator. I do want to mention, though, that this could be carried much further statistically by determining the conditional probability distribution and computing the standard deviation of this conditional probability distribution to serve as a guide to evaluating what degree of risk we might accept with respect to a specified job applicant.

1. Using your own words, summarize the procedures used to determine the numbers in Schedules A and B, Tables 6.16 and 6.17.

2. Write a detailed critical analysis of the probability methodology used.

3. If you believe that this personnel manager's probability calculations are deficient in some way or ways, explain clearly what changes in methodology you would recommend.

4. What, in general, is your feeling about the idea of using a probability approach to determining which employees are likely to stay with the company for a specified period of time and which will not?

Table 6.17 Schedule B

Combinations	Conditional Probability of Quitting
02468	.669
02469	.508
02478	.631
02479	.454
02568	.569
02569	.408
02578	.523
02579	.346
03468	.790
03469	.629
03478	.751
03479	.574
03568	.713
03569	.552
03578	.667
03579	.490
12468	.671
12469	.510
12478	.633
12479	.456
12568	.571
12569	.410
12578	.525
12579	.348
13468	.791
13469	.630
13478	.752
13479	.575
13568	.714
13569	.553
13578	.668
13579	.491

6.2 THE "CRANDALL SUPPLY COMPANY"

The "Crandall Supply Company" is a large supplier of a wide variety of products used by diverse types of industrial companies. It handles over 10,000 items, including machine tools, fasteners, mining equipment, power transportation equipment, and fire and safety items. In addition, the company has three small manufacturing divisions which produce specialty slings, gaskets, and conveyor belts.

After exhibiting virtually no real growth between 1959 and 1969, Crandall's sales exploded during the period between 1970 and 1974. More specifically, in 1974, sales were about 250%

greater than they had been in 1969. According to one member of management who will be quoted rather extensively below, the dramatic surge in sales resulted primarily from astute management decisions regarding the opening of branch supply houses and the taking on of new lucrative merchandise items. The present case has to do with a decision related to the question of whether Crandall should have taken on two new lines of merchandise.

A study conducted in early 1974 revealed that, of the products taken on by the company during the preceding 24-month period, the ones that had proved profitable had at least one characteristic in common: They were products easily sold from stock and, thus, did not require any special training of the company's sales staff. Generally speaking, these products, individually, accounted for only small fractions of Crandall's overall sales.

In late 1974, management found itself contemplating the addition of two new product lines which, if successfully marketed, had the potential of contributing a very significant proportion to total company sales. One of these was a line of crushers, screens, and material-handling equipment for the construction and sand-and-gravel industries. The other was a line of fluid power components produced by the Boston Gear Division of North American Rockwell. These product lines possessed some conspicuous similarities according to a brief analysis performed by the Crandall manager mentioned above. According to him, the similarities were: "They both are thought to have an annual $2-to-$3 million sales potential within Crandall's area of distribution; they both would be sold to relatively large, and clearly defined, users; they are both high-quality lines; and they would both require that Crandall maintain relatively large and costly inventories. Also, the two manufacturing companies both demanded two days of training for salesmen." This last similarity was regarded as quite important, despite the shortness of the training period, in that most products distributed by Crandall required no special sales training whatever.

According to this same manager:

There are basically three approaches that have been used to market goods through Crandall Supply in the past. The first involves using the standard industrial salesperson. These are excellent salespersons, but they are spread thin, both in terms of product knowledge and in terms of the large number of accounts that each must handle. The second is that of utilizing a lead salesperson to help the others in securing product knowledge. This lead salesperson has normal accounts and only calls on other accounts at the request of their assigned salespersons. Thus, Crandall Supply, when using this approach, had depended on the general salesperson to take the initiative in calling on the lead salesperson for assistance. The third approach utilizes a specialist whose primary job responsibility is that of selling a single line of goods. As a general rule, he has no specific accounts and may call upon any assigned accounts or upon any companies he considers potential buyers.

In November of 1974, a group of Crandall managers and outside advisers met with a view to settling upon the most desirable manner of selling the two new lines should they be taken on. The collective thinking of this group, expressed in conditional probability terms, was that (1) if the standard industrial salesperson were used, the probability of achieving high sales (defined as annual sales in the neighborhood of the upper limit of the $2-to-4 million range) would be .2 and the probability of achieving low sales (defined as a figure in the neighborhood of the lower end of the $2-to-4 million range) would be .5; (2) if the lead-salesperson approach were employed, the probability of achieving high sales would be .3 and the probability of achieving low sales also .3; and (3) if the specialist-salesperson approach were used, the probability of achieving high sales would be .5 and the probability of achieving low sales .2.

In describing the subsequent steps the manager quoted above asserted:

The next [type of] information needed was the past record of Crandall Supply's ability to take on a new line and make good. Thirty percent of the time, by our definition, an adequate or better job has been done. This will be referred to as "high sales." Seventy percent of the time a less than adequate job, or failure, has occurred. This will be referred to as "low sales."

The manager then compiled Table 6.18, which is partially completed, using the above probability information to arrive at his results.

As determined from this table, the most promising approach would be that of using a specialist salesperson. Keep in mind that the table pertains to products with very large potentials, which have relatively high engineering contents. If they represented all lines handled by Crandall, the company would have gone bankrupt long ago.

Table 6.18 Bayes' Theorem Calculations for the "Crandall Supply Company" Case

Part I: Regular-Salesperson Approach

(1) Events A_i	(2) Prior Probabilities $P(A_i)$	(3) Likelihoods $P(B_1\|A_i)$	(4) Joint Probabilities $P(A_i)P(B_1\|A_i)$	(5) Revised Probabilities $P(A_i\|B_1)$
High sales		.2		
Low sales		.5		
Total	1.0	.7		

Part II: Lead-Salesperson Approach

(1) Events A_i	(2) Prior Probabilities $P(A_i)$	(3) Likelihoods $P(B_2\|A_i)$	(4) Joint Probabilities $P(A_i)P(B_2\|A_i)$	(5) Revised Probabilities $P(A_i\|B_2)$
High sales		.3		
Low sales		.3		
Total	1.0	.6		

Part III: Specialist-Salesperson Approach

(1) Events A_i	(2) Prior Probabilities $P(A_i)$	(3) Likelihoods $P(B_3\|A_i)$	(4) Joint Probabilities $P(A_i)P(B_3\|A_i)$	(5) Revised Probabilities $P(A_i\|B_3)$
High sales		.5		
Low sales		.2		
Total	1.0	.7		

Note: B_1 stands for regular-salesperson approach is used.
  B_2 stands for lead-salesperson approach is used.
  B_3 stands for specialist-salesperson approach is used.

1. In view of the prior probabilities stated above, does it seem reasonable or unreasonable to think that a Bayesian approach to deciding whether to take on a new line of merchandise might be beneficial? Elaborate.

2. Complete the three parts of Table 6.18.

3. Write a carefully prepared critical analysis (approximately 200 to 500 words) of the Bayesian procedures used in this case.

4. Do you believe that the full Bayesian analysis contributed any information not contributed by the likelihoods alone? Justify your answer.

Special Probability Distributions

He couldn't read through the Binomial Theorem without tears coming to his eyes—the whole concept . . . was so shatteringly beautiful.—James Hilton, *Random Harvest*

Whenever a large sample of chaotic elements are taken in hand and marshaled in the order of their magnitude, an unsuspected and most beautiful form of regularity proves to have been latent all along. The tops of the marshaled rows form a flowing curve of invariable proportions; and each element, as it is sorted into place, finds, as it were, a preordained niche accurately adapted to fit it.—Sir Francis Galton

What You Should Learn from This Chapter

Many kinds of business and economic phenomena are distributed in ways which can be closely approximated by probability distributions of theoretical origin. In this chapter we deal with some of the most useful of such theoretical probability distributions. When you have completed this chapter, you should be able to

1. Define what is meant by a random variable.
2. Obtain the mean and standard deviation of a discrete random variable.
3. Determine which kind of theoretical probability distribution model is appropriate to use in connection with a described situation.
4. Use the binomial, hypergeometric,* normal, Poisson,* and exponential distribution* models to calculate probabilities for random variables of interest.

With respect to item 4 above, we can set some priorities. It is absolutely essential that you become familiar with the normal probability distribution (section 7.6) and skilled in using the table of areas under the normal curve. The normal curve will be either employed or referred to frequently in many of the remaining chapters of this book.

Of secondary importance is the binomial probability distribution (section 7.4). This distribution is helpful in (1) solving a large class of probability problems, (2) building a bridge between basic probability concepts introduced in the preceding chapter and inductive procedures described in later chapters, and (3) contributing to the solution of certain kinds of nonparametric statistical problems, the subject of Chapter 13.

The desirable amount of emphasis to be given the hypergeometric, the Poisson, and the exponential distributions (sections 7.5, 7.7, and 7.8) is more difficult to state in general. Much depends on the importance your instructor attaches to these subjects, whether you will be taking a required course in management science after this one, and whether you personally think you might want to pursue the study of statistics further. In any event, these topics are not referred to in later chapters.

Topics introduced in earlier chapters which you may wish to review before proceeding are discussed in the following sections: Chapter 1—1.3; Chapter 2—2.6; Chapter 4—4.3, 4.10; Chapter 6—6.1, 6.4, 6.5, 6.7, 6.9.

*These topics are optional.

7.1 Introduction

In section 6.3 of the preceding chapter the term *probability distribution* was defined. It was also illustrated, but only with qualitative variables. In this chapter we continue with the subject of probability distributions but extend the coverage to quantitative variables.

7.2 Random Variables

Let us refer once again to the Exmont Company case (Illustrative Case 6.1). We will now suppose that the sales manager, along with her other lines of investigation, examines the record of each salesperson for the previous 12-month period. For each salesperson she prepares a frequency distribution consisting of a list of number of sales, defined as the number of signed contracts attained, and the number of selling days in which a specified number of sales was made. Let us further suppose that there were, allowing for weekends, holidays, vacations, conferences, and so forth, 250 normal selling days. Such a frequency distribution for a specific salesperson, let us call him Bob Smythe, is shown in columns (1) and (2) of Table 7.1.

As you now know, frequencies may readily be changed into probabilities by converting them into relative frequencies and introducing uncertainty by assuming random selection of an elementary unit—in this case, a normal selling day. In column (3) of Table 7.1 we see the resulting probabilities. Data in columns (1) and (3) together constitute a probability distribution, but one of quite a different kind than any exhibited earlier. In the present case the events to which the probabilities are assigned are *quantitative* in nature. That is, according to the empirically determined facts, salesman Smythe may achieve 0, 1, 2, 3, or 4 sales in a normal selling day.

When dealing with probability distributions in this and later chapters we will often employ the term *random variable*.

Table 7.1 Frequency Distribution and Probability Distribution of Number of Sales per Day Achieved by Bob Smythe During a 12-Month Period

(1) Number of Sales per Day X	(2) Number of Days in Which the Indicated Number of Sales Was Achieved f	(3) Probability of Achieving the Indicated Number of Sales on Any Given Day $P(X)$
0	50	.200
1	100	.400
2	50	.200
3	30	.120
4	20	.080
5 or more	0	.000
Total	250	1.000

> A *random variable* is a variable whose value or outcome is determined from a random experiment.

The random variable of present interest, Bob Smythe's sales, is a *discrete random variable* because he can achieve only a limited number of completed sales. To use $2\frac{1}{2}$ or $3\frac{3}{4}$ for the number of signed contracts, for example, is simply not possible. If Smythe's sales had been measured in dollars and cents, we could view the sales variable as continuous because, in theory, sales could be expressed to the nearest cent or even to the nearest fraction of a cent.

The Mean, or Expected Value, of a Random Variable

Let us determine the mean number of sales per day achieved by Bob Smythe during the 12-month period under consideration. An intuitively appealing way of attacking this problem would be to begin by recognizing that, from Table 7.1, 0 sales occurred on 20% of the days, 1 sale on 40% of the days, 2 sales on 20% of the days, and so forth. Hence, we reason, a weighted average of the values 0, 1, 2, 3, and 4, with the corresponding probabilities used as weights, should give us the mean number of sales per day for this particular salesperson. In this case, unlike some, our intuition would not lead us astray, for that is precisely how the mean, or expected value, of a random variable is determined. In symbols,

$$\mu = E(X) = \frac{\Sigma\,[P(X) \cdot X]}{\Sigma\,P(X)}$$

or, because $\Sigma\,P(X)$ always equals 1, simply

> $$\mu = E(X) = \Sigma\,[P(X) \cdot X] \qquad (7.2.1)$$
>
> **where** $E(X)$ is the "expected value" of X
> X is a specific value of a random variable
> $P(X)$ is the probability associated with the occurrence of a specific X value

The term *expected value* is used here momentarily as an excuse to point out that the number obtained from use of Equation (7.2.1) extends beyond the purely descriptive interpretation usually attached to the arithmetic mean. In our example, we interpret $E(X)$ as follows: Over a very large number of selling days, the mean number of sales per day that Bob Smythe can expect to achieve is $(.20)(0) + (.40)(1) + (.20)(2) + (.12)(3) + (.08)(4) = 1.48$. In general, we interpret the expected value of X as the average value the random variable will assume in a long series of trials of a random experiment under a constant cause system.

Actually, if the $P(X)$ values constitute an accurate description of the relative frequencies for the population of data (all normal sales days for Bob Smythe, for example), then $E(X) = \mu$, the mean of the population. From now on, we will assume that such is the case and let $E(X)$ be synonymous with μ.

Table 7.2 Determination of the Standard Deviation of a Random Variable (Basic Data from Table 7.1)

(1) Sales per Day X	(2) μ	(3) $X - \mu$	(4) $(X - \mu)^2$	(5) $P(X)$	(6) (4) × (5) $P(X)(X - \mu)^2$
0	1.48	−1.48	2.1904	.200	.438080
1	1.48	−.48	.2304	.400	.092160
2	1.48	.52	.2704	.200	.054080
3	1.48	1.52	2.3104	.120	.277248
4	1.48	2.52	6.3504	.080	.508032
					1.369600

$$\sigma = \sqrt{\Sigma \ [P(X)(X - \mu)^2]} = \sqrt{1.369600} = 1.17$$

Standard Deviation of a Random Variable

In Chapter 4 we defined the standard deviation of a set of numbers as the square root of the average of the squared deviations from the arithmetic mean—or, in symbols, assuming the numbers represent observations on a finite population: $\sigma = \sqrt{\Sigma \ (X - \mu)^2/N}$. The standard deviation of a random variable is defined similarly—namely, as the square root of the *weighted average* of the squared deviations from the mean of the random variable. The weights used, of course, are the probabilities. In symbols, we define the standard deviation of a random variable as $\sigma = \sqrt{\Sigma \ [P(X)(X - \mu)^2]/\Sigma \ P(X)}$, or, because $\Sigma \ P(X)$ is always 1, simply

$$\sigma = \sqrt{\Sigma \ [P(X)(X - \mu)^2]} \qquad (7.2.2)$$

To find the standard deviation of the probability distribution for Bob Smythe's sales, we begin by recalling that the mean was 1.48 and then proceed as shown in Table 7.2.

Methods shown above for determining the mean and standard deviation of a random variable apply strictly only to *discrete* random variables. The mean and standard deviation of a continuous random variable are interpreted in the same way but cannot be computed without use of mathematics beyond the scope of this book. Fortunately, in most applications in which continuous random variables will be employed later on, actual calculation of the mean and standard deviation from the relevant probability distribution itself will not be necessary.

7.3 Introduction to Theoretical Probability Distributions

Thus far in our discussion of probability distributions, we have dealt only with cases for which the desired probabilities were determined empirically by means of a relative-frequency approach. We now turn our attention to probability distributions for which probabilities are obtained by more theoretical means.

Although a large number of theoretical probability distributions have been identified, investigators in various disciplines have found that a relatively small number of these have impressively widespread usefulness. In the remainder of this chapter we will consider five such widely used theoretical probability distributions: the binomial, the hypergeometric, the normal, the Poisson, and the exponential. Still other theoretical probability distributions will be introduced in later chapters.

7.4 The Binomial Probability Distribution

The simplest kind of theoretical probability distribution to deal with first is the kind concerned with dichotomous—or "either-or"—events. For example, the toss of a coin must result in either heads or tails. In a political poll where the preference categories are, say, "Prefers Candidate A" and "Does Not Prefer Candidate A," a voter selected at random will be assigned to one or the other category. A manufactured item turned out by a production process will either be classified as "acceptable" or as "defective." And so forth.

If we sample with replacement or if the relevant population is infinite—such as in the coin-tossing example referred to above where there is no limit to the number of times the coin could be flipped—then the probability of a success in a single trial will remain the same from trial to trial. In such a situation we can determine the probabilities associated with specified numbers of successes through the use of the binomial probability formula:

$$P(X) = \frac{n!}{X!(n-X)!} p^X q^{n-X} \qquad (7.4.1)$$

where $P(X)$ is the probability of getting exactly X successes in n independent trials of a random experiment
X is the specified number of successes
$n - X$ is the (implicitly) specified number of failures
p is the probability of a success in a single trial
$q = 1 - p$ is the probability of a failure in a single trial

Equation (7.4.1), though perhaps rather unfriendly in appearance, is nothing more nor less than the embodiment of the special cases of the addition and multiplication theorems of probability. This point will be elaborated on shortly. First, let us consider a problem for which this formula could be used to advantage.

ILLUSTRATIVE CASE 7.1

A small oil company owns four geographically separated but equally promising drilling sites. The company is presently contemplating drilling exploratory holes at Sites 1 and 2. Management knows from past experience that only about one exploratory drilling in ten will result in the discovery of oil. Since drilling for oil is an expensive undertaking, management wishes to know in advance of any drilling the probability that (1) neither drilling will result in oil, (2) one drilling will result in oil, and (3) both drillings will result in oil.

In other words, management wishes to know the probability distribution for the discrete random variable consisting of 0, 1, and 2 successes.

Before attempting to answer management's questions using Equation (7.4.1), let us assure ourselves that the conditions necessary for its proper use are present.

CONDITIONS REQUIRED FOR PROPER USE OF THE BINOMIAL FORMULA

1. There are only two possible events that can result from a single trial of a random experiment.
2. The two possible events are mutually exclusive.
3. The probability of a success in a single trial remains the same from trial to trial.
4. The results are independent—that is, the result of one trial does not affect the probability of success or failure for any other trial.

For our illustrative case, these requirements *do* appear to be met. First, there are only two results possible from a single drilling—oil or no oil—and if one result occurs, the other cannot occur. In other words, the two possible events are mutually exclusive. In the present context, requirements 3 and 4 will be met if the probability of striking oil is the same at both sites and the results of the two drillings are independent. In the description of the case, the sites are referred to as "equally promising," meaning, we will presume, that management has no reason to believe that the probability of striking oil at a given site is either greater than or less than the probability of striking oil at the other site. Moreover, the sites are described as being geographically separated; presumably, the discovery of oil at one site will not affect the probability of discovering oil at the other site.

Let us begin our analysis in a manner similar to that used in connection with several problems appearing in Chapter 6—namely, by listing the various *possible* results of two drillings. Letting S stand for "success in finding oil" and F for "failure to find oil," the possible results of the two drillings are as follows:

Site 1	Site 2
F	F
F	S
S	F
S	S

We see, then, that there are four possible joint events. It is tempting to assert that the probability associated with each is $\frac{1}{4}$. However, the temptation should be resisted because the possibilities in this case are not equally likely. Using the special case of the multiplication theorem and our knowledge that p, the historical ratio of the number of successes to number of trials, is .1, we determine that

The probability of getting F at Site 1 and F at Site 2 is $(.9)(.9) = .81$.

The probability of getting F at Site 1 and S at Site 2 is $(.9)(.1) = .09$.

The probability of getting S at Site 1 and F at Site 2 is $(.1)(.9) = .09$.

The probability of getting S at Site 1 and S at Site 2 is $(.1)(.1) = .01$.

We see that the probability of achieving a dry hole at both sites is a rather high .81, and the probability of achieving a dry hole at one or the other site (but not both sites), *the specific site being of no concern*, is $.09 + .09 = .18$. Finally, the probability of the company's finding oil at both sites is a low .01—that is, management could expect such a fortunate result only one time in 100 drillings of two holes under similar conditions. The theoretical probability distribution for this case is shown in tabular form in Part I of Table 7.3.

Thus far we have not actually used the binomial formula to determine the probability of 0, 1, and 2 successful drillings. Instead we have relied solely on the special case of the multiplication theorem and, in the part of the exercise concerned with exactly one success, the special case of the addition theorem of probability. Let us now see what the results would be if we were to apply Equation (7.4.1) to the problem under discussion.

In the case of $P(0)$, that is, no successes, we would substitute values into the binomial formula in the following manner: $P(0) = (2!/[0!2!])(.1)^0(.9)^2$. Since 0! is always equal to 1, we see that there is only one way—$2!/(0!2!) = (2)(1)/[(1)(2)(1)] = 1$—that two failures can occur—namely, by finding oil at *neither* site. Since .1 raised to the zero power is equal to 1, as is any number raised to the zero power, we find that the probability $P(0)$ is equal to $(1)(1)(.9)^2$. Clearly, this reduces to a straightforward application of the special case of the multiplication theorem; the resulting product is .81.

To determine $P(1)$, we substitute values into the binomial formula as follows: $P(1) = (2!/[1!1!])(.1)^1(.9)^1 = (2)(.1)(.9) = .18$. Finally, for $P(2)$ we have $(2!/[2!0!])(.1)^2(.9)^0 = (1)(.1)^2(.9)^0 = .01$. Therefore, through the use of the binomial formula we have been able to obtain—without the tedium of enumerating all possible arrangements of successes and failures—results that are exactly the same as those obtained from the longer alternative procedure.

From the above demonstration, we note that the $p^X q^{n-X}$ part of the binomial formula represents the probability of getting X successes and $n - X$ failures *in a specified order* in n independent trials of a random experiment. Thus, the special case of the multiplication theorem of probability is represented by $p^X q^{n-X}$. But remember that when an order is not specified—that is, when the question prompting the analysis is of the form: "What is the probability of getting X successes and $n - X$ failures in n trials of a random experiment?"—it is necessary to take into account all possible ways that X successes and $n - X$ failures could occur. Because the probability obtained from $p^X q^{n-X}$ is the same for every possible different combination of X successes and $n - X$ failures, that probability may be summed as many times as it appears or, more simply, multiplied by the number of possible combinations, $n!/[X!(n - X)!]$. (You were introduced to this subject in section 2.7 of Chapter 2.) Thus, the first part of Equation (7.4.1) brings the special case of the addition theorem into the calculations.

The fact that use of the binomial formula permits construction of a probability distribution without the necessity of enumerating all possible results of the random experiment is quite an advantage for problems in which n is larger. To demonstrate: If the oil company in our example had been interested in drilling at three of its sites and finding the corresponding probability distribution, a complete listing of possibilities would require eight lines rather than only four. Fortunately, we need not list the possible results in order to obtain the desired probability distribution. Instead, we merely substitute 3 for n in Equation (7.4.1) and let X vary from 0 to 3, inclusive, as follows:

$$P(0) = \left(\frac{3!}{0!3!}\right)(.1)^0(.9)^3 = .729$$

$$P(1) = \left(\frac{3!}{1!2!}\right)(.1)^1(.9)^2 = .243$$

$$P(2) = \left(\frac{3!}{2!1!}\right)(.1)^2(.9)^1 = .027$$

$$P(3) = \left(\frac{3!}{3!0!}\right)(.1)^3(.9)^0 = .001$$

If four drillings were planned at four equally promising sites, that is, if $n = 4$, a situation involving 16 possible arrangements of successes and failures, we would determine the desired probabilities in the following manner:

$$P(0) = \left(\frac{4!}{0!4!}\right)(.1)^0(.9)^4 = .6561$$

$$P(1) = \left(\frac{4!}{1!3!}\right)(.1)^1(.9)^3 = .2916$$

$$P(2) = \left(\frac{4!}{2!2!}\right)(.1)^2(.9)^2 = .0486$$

$$P(3) = \left(\frac{4!}{3!1!}\right)(.1)^3(.9)^1 = .0036$$

$$P(4) = \left(\frac{4!}{4!0!}\right)(.1)^4(.9)^0 = .0001$$

Binomial probabilities for two, three, and four drillings have been collected and are presented in tabular form in Table 7.3.

Table of Binomial Probabilities

For certain combinations of n and p values the computational work of finding binomial probabilities can be eliminated completely through use of Appendix Table 1, part of which is presented here as Table 7.4.

For example, suppose we wish to find the probability of obtaining exactly six successes ($X = 6$) in ten independent trials ($n = 10$) of a random experiment if the probability of a success in a single trial is .4 ($p = .4$). We would recognize that the circled part of Table 7.4 pertains to situations where $n = 10$ (indicated by the number 10 under n at the extreme left-hand side of the table). Since the specified number of successes in this example is six, we locate 6 under X toward the left-hand edge. Reading across from 6, we locate the table entry directly down from .40 under p at the top of the table. This value we find is $P(6) = .1115$.

If we wished to know the probability of, say, four, five, or six successes in ten independent trials of a random experiment when the probability of a success in a single trial is .2, we would use the table in a similar manner as described above but would have to identify $P(4)$, $P(5)$, and $P(6)$ and sum the results. That is, $P(4) = .0881$, $P(5) = .0264$, and $P(6) = .0055$. Therefore, $.0881 + .0264 + .0055 = .1200$.

Table 7.3 Binomial Probability Distributions Associated with Drillings for Oil Where $p = .1$

Part I: n = 2

Number of Oil-Producing Wells Discovered X	Probability $P(X)$
0	.81
1	.18
2	.01
Total	1.00

Part II: n = 3

Number of Oil-Producing Wells Discovered X	Probability $P(X)$
0	.729
1	.243
2	.027
3	.001
Total	1.00

Part III: n = 4

Number of Oil-Producing Wells Discovered X	Probability $P(X)$
0	.6561
1	.2916
2	.0486
3	.0036
4	.0001
Total	1.000

The Binomial Distribution Family

Although the three probability distributions associated with the oil company case bear some resemblance to one another (Table 7.3), they are certainly not identical. Close examination of the probabilities reveals that, as n increases, the degree of skewness of the distribution decreases even though p is a constant .1. Thus, we observe that what is referred to as "*the binomial distribution*" is really a large family of similarly determined distributions, the specific shape in a given case being dictated by the values of n and p.

What Happens to the Binomial Probability Distribution When p Varies and n Remains Constant? Let us focus on the $n = 4$ version of the oil company problem. But let us now

Table 7.4 Binomial Probabilities (From Appendix Table 1)

n	X	.05	.10	.15	.20	.25	p .30	.35	.40	.45	.50
9	5	.0000	.0008	.0050	.0165	.0389	.0735	.1181	.1672	.2128	.2461
	6	.0000	.0001	.0006	.0028	.0087	.0210	.0424	.0743	.1160	.1641
	7	.0000	.0000	.0000	.0003	.0012	.0039	.0098	.0212	.0407	.0703
	8	.0000	.0000	.0000	.0000	.0001	.0004	.0013	.0035	.0083	.0716
	9	.0000	.0000	.0000	.0000	.0000	.0000	.0001	.0003	.0008	.0020
10	0	.5987	.3487	.1969	.1074	.0563	.0282	.0135	.0060	.0025	.0010
	1	.3151	.3874	.3474	.2684	.1877	.1211	.0725	.0403	.0207	.0098
	2	.0746	.1937	.2759	.3020	.2816	.2335	.1757	.1209	.0763	.0439
	3	.0105	.0574	.1298	.2013	.2503	.2668	.2522	.2150	.1665	.1172
	4	.0010	.0112	.0401	.0881	.1460	.2001	.2377	.2508	.2384	.2051
	5	.0001	.0015	.0085	.0264	.0584	.1029	.1536	.2007	.2340	.2461
	6	.0000	.0001	.0012	.0055	.0162	.0368	.0689	.1115	.1596	.2051
	7	.0000	.0000	.0001	.0008	.0031	.0090	.0212	.0425	.0746	.1172
	8	.0000	.0000	.0000	.0001	.0004	.0014	.0043	.0106	.0229	.0439
	9	.0000	.0000	.0000	.0000	.0000	.0001	.0005	.0016	.0042	.0098
	10	.0000	.0000	.0000	.0000	.0000	.0000	.0000	.0001	.0003	.0010
11	0	.5688	.3138	.1673	.0859	.0422	.0198	.0088	.0036	.0014	.0005
	1	.3293	.3835	.3248	.2362	.1549	.0932	.0518	.0266	.0125	.0054
	2	.0867	.2131	.2866	.2953	.2581	.1998	.1395	.0887	.0513	.0269
	3	.0137	.0710	.1517	.2215	.2581	.2568	.2254	.1774	.1259	.0806
	4	.0014	.0158	.0536	.1107	.1721	.2201	.2428	.2365	.2060	.1611
	5	.0001	.0025	.0132	.0388	.0803	.1321	.1830	.2207	.2360	.2256
	6	.0000	.0003	.0023	.0097	.0268	.0566	.0985	.1471	.1931	.2256
	7	.0000	.0000	.0003	.0017	.0064	.0173	.0379	.0701	.1128	.1611
	8	.0000	.0000	.0000	.0002	.0011	.0037	.0102	.0234	.0462	.0806
	9	.0000	.0000	.0000	.0000	.0001	.0005	.0018	.0052	.0126	.0269
	10	.0000	.0000	.0000	.0000	.0000	.0000	.0002	.0007	.0021	.0054
	11	.0000	.0000	.0000	.0000	.0000	.0000	.0000	.0000	.0002	.0005
12	0	.5404	.2824	.1422	.0687	.0317	.0138	.0057	.0022	.0008	.0002
	1	.3413	.3766	.3012	.2062	.1267	.0712	.0368	.0174	.0075	.0029
	2	.0988	.2301	.2924	.2835	.2323	.1678	.1088	.0639	.0339	.0161
	3	.0173	.0852	.1720	.2362	.2581	.2397	.1954	.1419	.0923	.0537
	4	.0021	.0213	.0683	.1329	.1936	.2311	.2367	.2128	.1700	.1208
	5	.0002	.0038	.0193	.0532	.1032	.1585	.2039	.2270	.2225	.1934
	6	.0000	.0005	.0040	.0155	.0401	.0792	.1281	.1766	.2124	.2256
	7	.0000	.0000	.0006	.0033	.0115	.0291	.0591	.1009	.1489	.1934
	8	.0000	.0000	.0001	.0005	.0024	.0078	.0199	.0420	.0762	.1208
	9	.0000	.0000	.0000	.0001	.0004	.0015	.0048	.0125	.0277	.0537

assume that p varies, having values of .1, .3, .5, .7, and .9. The five resulting binomial distributions are shown in Figure 7.1. They are obviously five quite different probability distributions, though the ones associated with $p = .1$ and $p = .9$ are clear-cut opposites, as are those associated with $p = .3$ and $p = .7$.

What Happens to the Binomial Probability Distribution When n Varies and p Remains Constant? Now let us suppose that p is known to be .1, as was originally assumed, but that

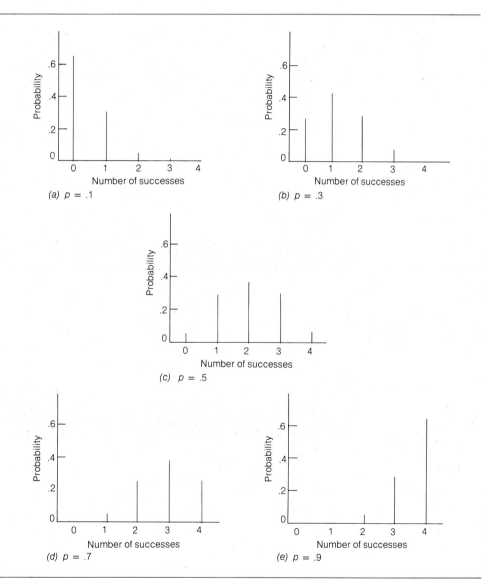

Figure 7.1 Illustration of the way the binomial distribution changes when n is a constant 4 and p varies.

the number of equally promising and geographically separated drilling sites can be increased without limit. The latter assumption is patently preposterous but will serve to aid in the illustration of another interesting feature of binomial distributions. Let us assign values of 4, 10, 20, and 50 to n and determine the $P(X)$ values. The results are shown in Figure 7.2.

When n trials of a random experiment are carried out, there will be $n + 1$ possible values of $P(X)$—namely, $P(0)$, $P(1)$, $P(2)$, . . . , $P(n)$. Many of the possible $P(X)$ values obtainable from assuming that $n = 4$, 10, 20, and 50 while p is a constant .1 cannot be shown in Figure 7.2 because they are extremely small. For present purposes this loss is of little consequence since the point we wish to emphasize is the tendency, when n is increased, for the binomial distribution to become increasingly more symmetrical. Eventually, the binomial distribution becomes not only symmetrical but bell-shaped in appearance. This property of the binomial distribution holds for any value of p, excluding 0 and 1, provided that n may be increased without limit, a fact that points us in the direction of a most important kind of

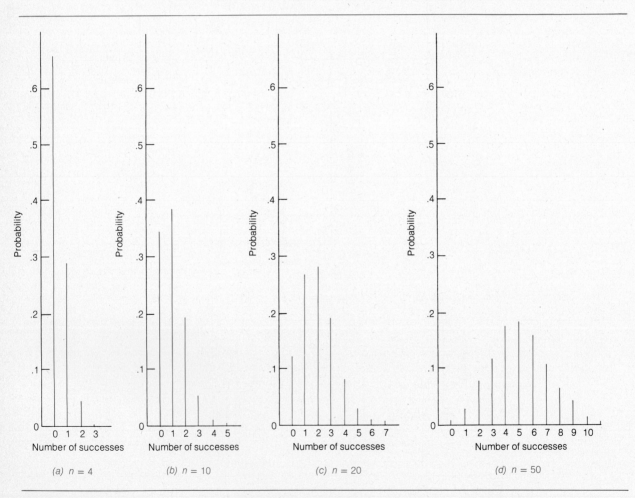

Figure 7.2 Illustration of the way the binomial distribution changes when p is a constant .1 and n increases.

7 Special Probability Distributions

theoretical probability distribution—namely, the normal distribution, the subject of section 7.6.

Mean and Standard Deviation of a Binomial Distribution

Because a binomial distribution is discrete, we could, of course, compute its mean and standard deviation by applying Equations (7.2.1) and (7.2.2), respectively; these formulas may be applied to any kind of discrete probability distribution. For example, if n and p are assumed to be 4 and .1, respectively, we can determine the mean and standard deviation of the corresponding binomial probability distribution by utilizing the $P(X)$ values from Table 7.3, Part III, in the following manner:

$$\mu = \Sigma \; [P(X)X] = (.6561)(0) + (.2916)(1) + (.0486)(2) + (.0036)(3)$$
$$+ (.0001)(4) = .4$$

and

$$\sigma^2 = \Sigma \; [P(X)(X - \mu)^2] = (.6561)(0 - .4)^2 + (.2916)(1 - .4)^2 + (.0486)(2 - .4)^2$$
$$+ (.0036)(3 - .4)^2 + (.0001)(4 - .4)^2 = .36$$

Therefore, $\sigma = \sqrt{.36} = .6$.

Fortunately, if we are certain that a discrete probability distribution is indeed a binomial distribution, much simpler formulas can be used to obtain these same results. These formulas are

$$\mu = np \qquad\qquad (7.4.2)$$
$$\sigma = \sqrt{npq} \qquad\qquad (7.4.3)$$

Therefore, $\mu = np = (4)(.1) = .4$ and $\sigma = \sqrt{npq} = \sqrt{(4)(.1)(.9)} = \sqrt{.36} = .6$.

The Binomial Formula Applied to Bayes' Problems

Now that you know how to use the binomial probability formula, we can work through the original version of the Bayes' problem introduced in section 6.9 of the preceding chapter.

Recall that the problem was as follows: A man has two bags of poker chips, Bag A_1, which is known to contain 70 blue and 30 white chips, and Bag A_2, which is known to contain 20 blue and 80 white chips. But he doesn't know which bag is A_1 and which is A_2. He decides to try to find out which is which by performing an experiment: He picks a bag by flipping a fair coin and then selects 10 chips at random, with replacement, from the chosen bag. The result is 7 blues and 3 whites. What is the probability that the bag selected is Bag A_1?

The first steps required to work this problem are exactly the same as those demonstrated in Table 6.8, where the one-chip version was worked. We begin with a list of the possible bags that could be selected and a list of the corresponding prior probabilities—.5 for each bag. These are presented in columns (1) and (2) of Table 7.5. It is in connection with column (3) that a change in procedure is called for. The B in the present version of this problem stands

Table 7.5 The Binomial Formula Used in Connection with a Bayes' Problem: The Poker-Chip Problem Assuming Ten Random Selections

(1) Events A_i	(2) Prior Probabilities $P(A_i)$	(3) Likelihoods $P(B\|A_i)$	(4) Joint Probabilities $P(A_i), P(B\|A_i)$	(5) Revised Probabilities $P(A_i\|B)$
A_1: Bag A_1	.5	.266827	.133414	.997
A_2: Bag A_2	.5	.000799	.000400	.003
Total	1.0		.133814	1.000

Note: B stands for the occurrence of seven blue chips in ten random selections.

for the occurrence of 7 blue chips in 10 random selections. Therefore, we need to know the probability of *B given that* A_1 *was the bag selected* and the probability of *B given that* A_2 *was the bag selected.* We compute: $P(B|A_1) = (10!/7!3!)(.7)^7(.3)^3 = (120)(.082354)(.027) = .266827$ and $P(B|A_2) = (10!/7!3!)(.2)^7(.8)^3 = (120)(.000013)(.512) = .000799$. We place these results in column (3) of Table 7.5. Multiplying the prior probabilities by the likelihoods we just obtained, we get .133414 and .000400, the values shown in column (4) of Table 7.5. These joint probabilities sum to .133814. Column (5) shows that $.133414/.133814 = .997$ and that $.000400/.133814 = .003$. Thus, a modest amount of information has increased the probability associated with Bag A_1 from .5, indicating the greatest possible amount of uncertainty, to .997, indicating just about no uncertainty at all.

Here is another example.

A large retailer buys assembled bicycles from three companies, X, Y, and Z. In the past this retailer has purchased 20% of its assembled bicycles from Company X, 30% from Company Y, and 50% from Company Z. Of the bicycles received from Company X in the past, 3% have been assembled improperly. For Companies Y and Z, the rates of improper assemblages have been 5% and 2%, respectively. Twelve bicycles arrived this morning, and three were assembled improperly. Unfortunately, a new employee on the loading dock has

Table 7.6 The Binomial Formula Used in Connection with a Bayes' Problem: The Bicycle Problem

(1) Events A_i	(2) Prior Probabilities $P(A_i)$	(3) Likelihoods $P(B\|A_i)$	(4) Joint Probabilities $P(A_i)P(B\|A_i)$	(5) Revised Probabilities $P(A_i\|B)$
A_1: Company X	.2	.004516	.000903	.132075
A_2: Company Y	.3	.017332	.005200	.760568
A_3: Company Z	.5	.001467	.000734	.107357
Total	1.0		.006837	1.000000

Note: B stands for the occurrence of 3 improperly assembled bicycles in a shipment of 12.

misplaced the invoice and is thus unable to identify the supplying company. What is the probability that the supplier was Company X? Company Y? Company Z?

We begin, as usual, by listing the possible events and the associated prior probabilities. The prior probabilities are listed in column (2) of Table 7.6. The likelihoods are determined in the following manner, where B stands for the occurrence of 3 improper assemblies out of 12 bikes: $P(B|\text{Company X}) = (12!/3!9!)(.03)^3(.97)^9 = (220)(.000027)(.760231) = .004516$; $P(B|\text{Company Y}) = (12!/3!9!)(.05)^3(.95)^9 = (220)(.000125)(.630249) = .017332$; and $P(B|\text{Company Z}) = (12!/3!9!)(.02)^3(.98)^9 = (220)(.000008)(.833748) = .001467$. These likelihoods are placed in column (3) of Table 7.6. The remainder of the work is shown in columns (4) and (5) of this same table. It appears likely that Company Y is the offending supplier.

✳ 7.5 The Hypergeometric Distribution

Don't forget that use of the binomial formula, described in the preceding section, is strictly appropriate only when two mutually exclusive events are being considered *and* the probability of a success in a single trial remains constant from trial to trial. This second condition usually implies that the population from which the sample is taken is infinite or, what amounts to the same thing, that sampling is being done *with replacement*. But what, you might well ask, does one do when sampling is done without replacement from a finite population? The question is pertinent because in practice one will not often sample with replacement. After all, when one has examined a sample observation he or she will have gleaned from it all the relevant information it contains. If, in addition, the population is finite, as it usually will be in survey sampling studies, the probability of a success in a single trial will not remain constant from trial to trial.

> Strictly speaking, when (1) a problem involves two mutually exclusive events, as with a binomial problem, and, in addition, (2) the probability of a success in a single trial does not remain constant from trial to trial, *hypergeometric probabilities*, rather than binomial probabilities, should be determined.

Recall that in section 6.7 of the preceding chapter we examined a problem involving some dependency between possible events resulting from successive trials of a random experiment. The problem being referred to read as follows: "There are six rubies that appear similar in a jeweler's showcase. Four of these are of top quality and two are of secondary quality. A customer is interested in buying four of the rubies and is unconcerned about quality. If she were to select four rubies at random, what would be the probability of her getting at least one of secondary quality?"

Let us broaden the question somewhat and pursue the entire probability distribution. That is, we now wish to know the probability of her getting (1) no rubies of secondary quality, (2) exactly one ruby of secondary quality, and (3) exactly two rubies of secondary quality. (Even though the customer will be selecting four rubies—$n = 4$—it would make no sense to try to determine the probability associated with three or four secondary-quality rubies because there are only two such rubies in the population.)

The formula for determining hypergeometric probabilities is

$$P(X) = \frac{\left[\dfrac{S!}{X!(S-X)!}\right] \cdot \left[\dfrac{(N-S)!}{(n-X)!(N-S-n+X)!}\right]}{\dfrac{N!}{n!(N-n)!}} \qquad \begin{array}{l} \text{for } X = 0,\ 1,\ \ldots,\ S \\ \text{or if } n < S\colon X = 0,\ 1,\ \ldots,\ n \end{array}$$

(7.5.1)

where $P(X)$ is the probability of getting exactly X successes in n nonindependent
trials of a random experiment
N is the number of items in the population
n is the number of trials, that is, the number of items in the sample
S is the number of successes in the population
N − S is the number of failures in the population
X is the specified number of successes in the sample
n − X is the (implicitly) specified number of failures in the sample

Therefore,

1. $\dfrac{N!}{n!(N-n)!}$ = Total number of equally likely different samples of n items
which could be obtained from the N items making up the population

2. $\left[\dfrac{S!}{X!(S-X)!}\right] \cdot \left[\dfrac{(N-S)!}{(N-X)!(N-S-n+X)!}\right]$ = Number of ways X successes
and n − X failures can be combined in a sample of size n

By dividing (2) above by (1), we get the *probability* of getting exactly X successes and
n − X failures in a sample of size n.
Applying Equation (7.5.1), we find

$$P(0) = \frac{\left[\dfrac{2!}{0!2!}\right]\left[\dfrac{4!}{(4-0)!(6-2-4+0)!}\right]}{\dfrac{6!}{4!2!}} = \frac{\left(\dfrac{2!}{0!2!}\right)\left(\dfrac{4!}{4!0!}\right)}{\dfrac{6!}{4!2!}}$$

$$= \frac{(1)(1)}{15} = \frac{1}{15} = .0667$$

$$P(1) = \frac{\left[\dfrac{2!}{1!1!}\right]\left[\dfrac{4!}{(4-1)!(6-2-4+1)!}\right]}{\dfrac{6!}{4!2!}} = \frac{\left(\dfrac{2!}{1!1!}\right)\left(\dfrac{4!}{3!1!}\right)}{\dfrac{6!}{4!2!}}$$

$$= \frac{(2)(4)}{15} = \frac{8}{15} = .5333$$

7 Special Probability Distributions

$$P(2) = \frac{\left[\frac{2!}{2!0!}\right]\left[\frac{4!}{(4-2)!(6-2-4+2)!}\right]}{\frac{6!}{4!2!}} = \frac{\left(\frac{2!}{2!0!}\right)\left(\frac{4!}{2!2!}\right)}{\frac{6!}{4!2!}}$$

$$= \frac{(1)(6)}{15} = \frac{6}{15} = \underline{.4000}$$

Therefore, the hypergeometric probability distribution for this problem is as follows:

Number of Rubies of Secondary Quality X	P(X)
0	.0667
1	.5333
2	.4000
Total	1.000

As a practical matter, when the population size is substantially larger than the sample size (a widely used rule-of-thumb being $N/n \geq 20$), hypergeometric and binomial probabilities will be sufficiently close to justify use of the, usually, more convenient binomial procedures. When this condition exists, the population is effectively infinite.

Mean and Standard Deviation of a Hypergeometric Distribution

The mean and standard deviation of a hypergeometric probability distribution can be obtained through the use of formulas that are very similar to their binomial counterparts. They are

$$\mu = np$$

where p is the ratio of number of successes, S, in the population to the population size, N, and

$$\sigma = \sqrt{npq\left(\frac{N-n}{N-1}\right)}$$

where q is the ratio of number of failures, $(N - S)$, in the population to the population size, N.

The $\sqrt{(N-n)/(N-1)}$ multiplier contained within the formula for the standard deviation of a hypergeometric probability distribution is called the *finite correction factor*. It shows that the standard deviation of a hypergeometric probability distribution will ordinarily be smaller than (and never larger than) that of a binomial probability distribution having the same n and p values. This is so because the ratio of $N - n$ to $N - 1$ will always be less than 1. However, if the population size, N, is at least 20 times as great as the sample size, n, the value of the finite correction factor will be very close to 1 and can be ignored. Under such circumstances, the mean and standard deviation of the hypergeometric and binomial distributions will be, for all practical purposes, identical—that is, $\mu = np$ and $\sigma = \sqrt{npq}$. More will be said about the finite correction factor in Chapter 8.

7.6 The Normal Distribution

The Nature of the Normal Distribution

The normal, or bell-shaped, probability distribution towers above all others in importance for modern statistical theory and practice. Unlike the binomial and hypergeometric distributions, this theoretical probability distribution pertains to continuous random variables. The formula for the normal distribution, like the formula for any kind of continuous curve, is in the form of $Y = f(X)$, read "Y is a function of X." This means that, as X varies, Y varies in some systematic way, the specific way being determined by the form of the functional relationship. For example, let us consider the function $Y = 9 - (X - 3)^2$. If we let X assume values of, say, 0, 1, 1.5, 2, 3, 4, 4.75, 5, and 6, the corresponding Y values will be as shown in Figure 7.3.

Notice that the Y values form a symmetrical curve having a maximum value of 9. Nine is the Y value associated with $X = 3$. Since, according to the functional expression, $X - 3$ is squared, the resulting Y will be the same regardless of whether $X - 3$ is positive or negative. The maximum ordinate (the largest Y value) is the one associated with $X = 3$ because, when $X = 3$, $(X - 3)^2 = 0$ and $9 - 0 = 9$. For all other values of X, some positive quantity is subtracted from 9.

The function $Y = 9 - (X - 3)^2$ has no special significance; however, it *does* provide us with a useful analogy for the much more complicated mathematical function used to describe the normal curve.

The formula for a normal probability distribution is

$$Y = \frac{1}{\sqrt{2\pi}\sigma} \cdot e^{\frac{-(X - \mu)^2}{2\sigma^2}}$$

(7.6.1)

where π and e are the familiar constants 3.14159 and 2.71828, respectively, and μ and σ represent the arithmetic mean and standard deviation of some set of observations designated X. Clearly, we could, if we had any reason to do so, (1) substitute values of π and e into Equation (7.6.1), (2) compute the values of μ and σ for the set of X values of interest and substitute these into Equation (7.6.1), (3) substitute the X values themselves into the equation one at a time, and (4) solve for the corresponding Y values. These Y values, when plotted against the X values, would form a set of points outlining a normal curve. The fact that we would have to know μ and σ introduces some complications. On the other hand, once we *do* know these parameter values, we know everything necessary to develop the curve for the entire set of X values. Also worth noting is that Equation (7.6.1) indicates that a normal probability distribution is symmetrical around μ just as the curve in our analogy is symmetrical around $X = 3$.

Because μ and σ are such important pieces of information for one wishing to employ Equation (7.6.1), we see that "*the* normal probability distribution" is really an entire family

X	$Y = 9 - (X - 3)^2$
0	0
1	5
1.5	6.75
2	8
3	9
4	8
4.75	5.9375
5	5
6	0

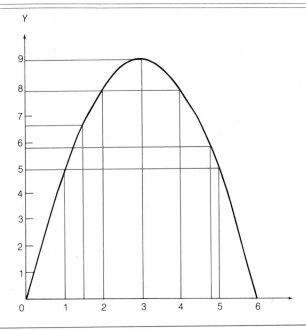

Figure 7.3 Tabular and graphic presentation of the function $Y = 9 - (X - 3)^2$.

of similarly shaped distributions. Two sets of normally distributed numbers may have different amounts of variation, since that is dependent on the value of σ, and different locations on the X axis, since that is dependent on the value of μ. Part (a) of Figure 7.5 (page 272) illustrates this point.

Areas Under the Normal Curve

When working with binomial distributions, we determined probabilities associated with specific discrete values of X. When a binomial distribution was presented in graphic form, the probabilities were indicated by the heights of vertical lines erected above the possible X values. Because the normal distribution is a continuous rather than a discrete distribution, we speak of the probability associated not with a specific X value (since, in theory, this probability is zero—see Figure 7.4 and the footnote on p. 271) but rather with a range of X values; that is, we speak of the probability that a random observation lies between two designated X values.

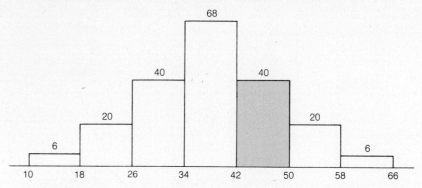

(a) Histogram of 200 observations on a continuous variable. Class interval = 8.

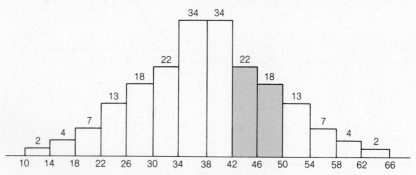

(b) Histogram of 200 observations on a continuous variable. Class interval = 4.

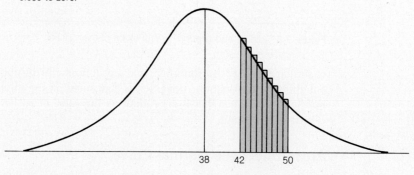

(c) Histogram of 200 observations on a continuous variable. Class interval is very close to zero.

Figure 7.4 What happens to a continuous random variable when the class interval is made progressively smaller. (See explanation in the note on page 271.)

The range of a normal curve is infinite. Thus, as we consider X values further and further removed from μ in either direction, Y approaches the X axis more and more closely but never quite reaches it. On the other hand, the approach of Y to the X axis is rather rapid. For example, if we pick a range of X values extending from $\mu - \sigma$ to $\mu + \sigma$, we find that the corresponding area represents 68.3%, or roughly two-thirds, of the entire area under the curve. Stating the same point in the language of probability: The probability is .683 that a randomly selected observation taken from a set of normally distributed X values will be within the range of the arithmetic mean plus and minus one standard deviation. Part (b) of Figure 7.5 illustrates this point. This figure also shows that the area associated with μ plus and minus 2 σ is 95.4%, and the area associated with μ plus and minus 3 σ is 99.7%, of the total area under the normal curve. *

The area under the normal curve associated with any specified range of X values can, in theory, be computed by integrating Equation (7.6.1). Fortunately, however, we will not have to work with this unwieldy formula directly because the areas are conveniently available in special tables. Appendix Table 2 at the end of this book is one such table; some of its uses will be described in detail very shortly. But first you must become familiar with the concept of *standard units* and the *standard normal curve*.

As noted above, the specific appearance of a normal probability distribution is dependent on μ and σ. As Part (a) of Figure 7.5 demonstrates, probability distributions may differ greatly in variation and in their positions on the horizontal axis and still be normal. This implies that, when finding areas under the normal curve (the probabilities of interest), the investigator must either (1) make use of Equation (7.6.1) to determine the desired area or (2) make use of a prepared normal curve table from among an infinite number of such tables, a different one for each combination of μ and σ. Clearly, neither alternative is desirable. Fortunately, we need not resort to either because a third, more attractive, alternative is also available.

> This more attractive alternative entails our transforming normally distributed X values into normally distributed *standard units*, or Z *values* as we shall frequently refer to them, where the standard unit associated with a specific X value is defined as
>
> $$Z = \frac{X - \mu}{\sigma} \qquad (7.6.2)$$

*To explain what happens to a continuous variable when the class interval is made progressively smaller, let us suppose that we gather 200 observations on a continuous random variable. We then construct a histogram of the data, using a class interval of 8, and seven mutually exclusive classes, as shown in part (a) of Figure 7.4. Here the mass (i.e., the 200 observations or the total probability of 1) is *spread over* seven classes rather than being concentrated at a few points as is the case with a discrete random variable. The probability associated with a specific class can be determined by dividing the height of the bar associated with that class by 200. For example, the probability associated with the leftmost class is 6/200 = .03.

Since the variable under scrutiny is continuous, the class interval used in histogram construction is arbitrary. Indeed, we could have chosen any value greater than zero for the class interval and still have had a nonzero probability for each class within the range of observations running from 10 to 66. Part (b) of Figure 7.4 shows the histogram for the same set of 200 observations, but this time a class interval of 4, rather than 8, was used. As the class interval is decreased in size, the number of classes increases, and hence the mass is distributed more thinly than in part (a). For example, the probability associated with the leftmost class in part (b) is 2/200 = .01.

Now suppose that the class interval were made extremely small. In such a case the number of classes would be very great and the probabilities associated with the classes very small. The smooth curve in part (c) shows what would eventually happen if the class interval were only slightly greater than zero.

As the size of the class interval approaches zero, the probability associated with a given class also approaches zero. This is why, when working with a continuous random variable, we do not speak in terms of the probability associated with a specific X value. Instead, we speak of the probability associated with a range of X values.

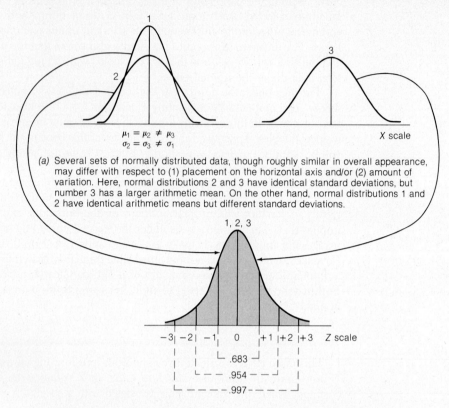

$\mu_1 = \mu_2 \neq \mu_3$
$\sigma_2 = \sigma_3 \neq \sigma_1$

X scale

(a) Several sets of normally distributed data, though roughly similar in overall appearance, may differ with respect to (1) placement on the horizontal axis and/or (2) amount of variation. Here, normal distributions 2 and 3 have identical standard deviations, but number 3 has a larger arithmetic mean. On the other hand, normal distributions 1 and 2 have identical arithmetic means but different standard deviations.

1, 2, 3

$-3 \quad -2 \quad -1 \quad 0 \quad +1 \quad +2 \quad +3$ Z scale

.683
.954
.997

(b) However, if the several sets of data are truly normally distributed, they can all be transformed into a *standardized* normal distribution having an arithmetic mean of 0 and a standard deviation of 1. Moreover, whether data have or have not been so transformed, the area under the curve associated with (1) the mean plus and minus 1 standard deviation is approximately 68.3 percent of the total, (2) the mean plus and minus 2 standard deviations is approximately 95.4 percent of the total, and (3) the mean plus or minus 3 standard deviations is approximately 99.7 percent of the total.

Figure 7.5 Illustration of the fact that different normal distributions may not have identical means or standard deviations before the X values are converted into their Z-value equivalents. However, when they have been converted to Z values, they are all *standardized,* which is to say they will all now have an arithmetic mean of 0 and a standard deviation of 1.

If such a transformation were performed on all relevant, normally distributed X values, the result would be a new normal distribution having an arithmetic mean of 0 and a standard deviation of 1. This is always so, regardless of the original μ and σ values—as long as the X values are truly normally distributed in the first place. (This point is presented pictorially in Figure 7.5.) Since the X values are symmetrical, subtracting μ from each such value—that is, performing the $X - \mu$ calculations—results in (1) a series of negative differences associated with the smaller X values, (2) a 0 value where $X = \mu$, and (3) a series of positive differences associated with the larger X values which offset exactly the negative differences referred to in connection with (1). Thus, the mean of the Z values must be zero regardless of the value of the original mean, μ. When all differences of the $X - \mu$ type are divided by σ, the result is a new set of values having a standard deviation of 1. This must be so because, for example, an X value which just happened to lie one standard deviation below μ would result in $X - \mu = -\sigma$. The $-\sigma$ divided by σ is -1. Similarly, if an X value were one standard deviation above the mean, $X - \mu = \sigma$ and σ divided by σ is 1. All $X - \mu$ values would be systematically altered, but when each is divided by σ, the result is a set of values having a standard deviation equal to 1.

We see, then, that by converting X values into their Z-value equivalents by $Z = (X - \mu)/\sigma$, we are merely performing a change of scale. In practice, even this procedure would be tedious (impossible actually) if we really did have to transform *all* X values into standard units. Fortunately, we can get by with converting only a few, usually only one or two, X values in this manner and using Appendix Table 2 *as if* all the normally distributed X values had been so converted. This point is demonstrated below in connection with several examples.

Some Practice Problems Utilizing the Table of Areas Under the Normal Curve

In our discussion of the Exmont Company case (Illustrative Case 6.1), we stated that, prior to classifying salespeople according to Sales Success Status categories, the sales manager evaluated the company's sales-quota system for realism. Let us suppose that, in the process of doing this evaluation, she collected a *sales-quota percentage*, or SQP score, for each of the 200 salespeople. Each SQP score was computed, let us say, by dividing the dollar amount of actual sales achieved by a salesperson over the previous 12-month period by that same salesperson's quota and multiplying the resulting quotient by 100. That is,

$$\left(\frac{\text{Sales achieved by Salesperson } i}{\text{Sales quota for Salesperson } i} \right) \cdot 100 = \text{SQP for Salesperson } i$$

Let us further suppose that the resulting SQP scores were found to be normally distributed around an arithmetic mean of 100 and to have a standard deviation of 12. Since the shape of the distribution is known to be normal, a considerable amount of additional information can be obtained. Let us attempt to answer eight quite different questions pertaining to these scores by utilizing Appendix Table 2 (reproduced here as Table 7.7).

1. What is the probability that a randomly selected salesperson would have achieved an SQP score between 100 and 125?

   Whenever you are faced with a problem involving the determination of an area under the normal curve—as you frequently will be as you proceed through this text—you are well advised to make a sketch so that you will be absolutely clear about the area sought. Figure 7.6 is such a sketch. You will notice that two scales are presented under the normal curve. One scale, which for convenience we call the X scale, represents the normally distributed SQP scores. The other scale, the Z scale, shows the SQP scores after each has been, in theory, run through Equation (7.6.2). In the present problem, only two X values, 100 and 125, are indicated—that is because they are the only X values that must actually be converted into standard units.

   To work this problem, we express the values 100 and 125 in standard units as follows: Since 100 coincides with μ, we recognize immediately that its Z-value

Table 7.7 Areas for the Standard Normal Probability Distribution (From Appendix Table 2)

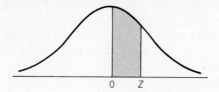

Example: For Z = 2.05, shaded area is .4798 out of the total area of 1.

Z	.00	.01	.02	.03	.04	.05	.06	.07	.08	.09
0.0	.0000	.0040	.0080	.0120	.0160	.0199	.0239	.0279	.0319	.0359
0.1	.0398	.0438	.0478	.0517	.0557	.0596	.0636	.0675	.0714	.0753
0.2	.0793	.0832	.0871	.0910	.0948	.0987	.1026	.1064	.1103	.1141
0.3	.1179	.1217	.1255	.1293	.1331	.1368	.1406	.1443	.1480	.1517
0.4	.1554	.1591	.1628	.1664	.1700	.1736	.1772	.1808	.1844	.1879
0.5	.1915	.1950	.1985	.2019	.2054	.2088	.2123	.2157	.2190	.2224
0.6	.2257	.2291	.2324	.2357	.2389	.2422	.2454	.2486	.2518	.2549
0.7	.2580	.2611	.2642	.2673	.2703	.2734	.2764	.2794	.2823	.2852
0.8	.2881	.2910	.2939	.2967	.2995	.3023	.3051	.3078	.3106	.3133
0.9	.3159	.3186	.3212	.3238	.3264	.3289	.3315	.3340	.3365	.3389
1.0	.3413	.3438	.3461	.3485	.3508	.3531	.3554	.3577	.3599	.3621
1.1	.3643	.3665	.3686	.3708	.3729	.3749	.3770	.3790	.3810	.3830
1.2	.3849	.3869	.3888	.3907	.3925	.3944	.3962	.3980	.3997	.4015
1.3	.4032	.4049	.4066	.4082	.4099	.4115	.4131	.4147	.4162	.4177
1.4	.4192	.4207	.4222	.4236	.4251	.4265	.4279	.4292	.4306	.4319
1.5	.4332	.4345	.4357	.4370	.4382	.4394	.4406	.4418	.4429	.4441
1.6	.4452	.4463	.4474	.4484	.4495	.4505	.4515	.4525	.4535	.4545
1.7	.4554	.4564	.4573	.4582	.4591	.4599	.4608	.4616	.4625	.4633
1.8	.4641	.4649	.4656	.4664	.4671	.4678	.4686	.4693	.4699	.4706
1.9	.4713	.4719	.4726	.4732	.4738	.4744	.4750	.4756	.4761	.4767
2.0	.4772	.4778	.4783	.4788	.4793	.4798	.4803	.4808	.4812	.4817
2.1	.4821	.4826	.4830	.4834	.4838	.4842	.4846	.4850	.4854	.4857
2.2	.4861	.4864	.4868	.4871	.4875	.4878	.4881	.4884	.4887	.4890
2.3	.4893	.4896	.4898	.4901	.4904	.4906	.4909	.4911	.4913	.4916
2.4	.4918	.4920	.4922	.4925	.4927	.4929	.4931	.4932	.4934	.4936
2.5	.4938	.4940	.4941	.4943	.4945	.4946	.4948	.4949	.4951	.4952
2.6	.4953	.4955	.4956	.4957	.4959	.4960	.4961	.4962	.4963	.4964
2.7	.4965	.4966	.4967	.4968	.4969	.4970	.4971	.4972	.4973	.4974
2.8	.4974	.4975	.4976	.4977	.4977	.4978	.4979	.4979	.4980	.4981
2.9	.4981	.4982	.4982	.4983	.4984	.4984	.4985	.4985	.4986	.4986
3.0	.4987	.4987	.4987	.4988	.4988	.4989	.4989	.4989	.4990	.4990
3.1	.4990	.4991	.4991	.4991	.4992	.4992	.4992	.4992	.4993	.4993
3.2	.4993	.4993	.4994	.4994	.4994	.4994	.4994	.4995	.4995	.4995
3.3	.4995	.4995	.4995	.4996	.4996	.4996	.4996	.4996	.4996	.4997
3.4	.4997	.4997	.4997	.4997	.4997	.4997	.4997	.4997	.4998	.4998
3.5	.4998	.4998	.4998	.4998	.4998	.4998	.4998	.4998	.4998	.4998
3.6	.4998	.4998	.4999	.4999	.4999	.4999	.4999	.4999	.4999	.4999
3.7	.4999	.4999	.4999	.4999	.4999	.4999	.4999	.4999	.4999	.4999

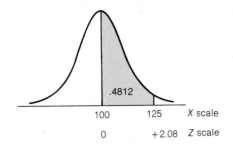

Figure 7.6 Sketch used as an aid in solving normal curve problem 1. (Shading indicates area sought.)

equivalent will be 0, indicating that the value of 100 is situated 0 standard deviations away from the mean, and we need not go through the formal calculations. The Z-value equivalent of 125, on the other hand, must be obtained through use of Equation (7.6.2). Thus, $Z = (125 - 100)/12 = 2.08$. We see that 125 is slightly over two standard deviations above the arithmetic mean of 100 on the X scale. The area sought is the shaded area shown in Figure 7.6.

Now let us look at Table 7.7. To find the area under the normal curve between 100 and 125, we begin by recognizing that the Z-value counterpart of 125 is the 2.08 value determined above. We locate 2.08 in the margins of our normal curve table. The way this is done is: Locate the first two digits of the Z value of interest (here, then, we seek 2.0) in the column headed "Z." The value 2.0 is 21 lines down from the top. We then locate the third digit of the Z value of interest in the top margin of the table. Since the third digit of 2.08 is 8, we look for .08 in the top margin and find that the value directly across from 2.0 and directly down from .08 is .4812. Thus, 48.12% of the normally distributed SQP scores are between 100 and 125.

This kind of normal curve problem is easier to solve than some because the area sought is given directly in Table 7.7. You will notice that in the sketch at the top of Table 7.7, which depicts normal curve areas, the area shaded is that between 0 and some positive value on the Z scale. Therefore, all we had to do was convert X to Z and find the corresponding area directly from the table.

2. What is the probability that a randomly selected salesperson would have achieved a score above 130 (see Figure 7.7)?

To work this problem the X value 130 must be expressed in standard units. We find that $Z = (130 - 100)/12 = 2.50$. Consulting Table 7.7, we learn that the area under the curve between a mean of 0 and a Z value of 2.50 is .4938. However, this is not the area sought; this time, we wish to know the probability measured by the area under the curve to the right of 2.50 (to the right of 130 on the X scale). Since any probability distribution has a total area of 1 and since the normal probability distribution is symmetrical (ensuring an area of .5 under each side of the curve), the table value of .4938 must be subtracted from .5000. By doing so, we get $.5000 - .4938 = .0062$, the probability sought. Therefore, in fewer than 1 time out of 100 tries would one get, from random selections, a salesperson whose SQP score topped 130.

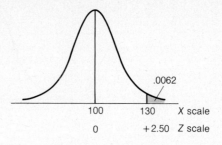

Figure 7.7 Sketch used as an aid in solving normal curve problem 2. (Shading indicates area sought.)

3. What is the probability that a randomly selected salesperson would have achieved a score below 90 (see Figure 7.8)?

   This problem is essentially the same as the preceding one except that, this time, we make use of the left-hand end of the normal curve. Converting 90 to its Z-value equivalent, Z = (90 − 100)/12, we get −0.83. The related area, according to our normal curve table, is .2967. Subtracting .2967 from .5000 to determine the area in the lower tail, we get .5000 − .2967 = .2033. Thus, we would expect to select a salesperson with an SQP score below 90 about 1 time in 5 trials of this random experiment.

4. What is the probability that a randomly selected salesperson would have achieved a score between 85 and 115 (see Figure 7.9)?

   This problem involves the addition of two areas. Converting 85 to its Z-value equivalent, we get (85 − 100)/12 = −1.25, a Z value associated with an area of .3944. A similar conversion of 115 yields +1.25. This Z value is also associated with an area of .3944. Therefore, 2(.3944) = .7888. If the stated scores had not been symmetrical around the mean of 100, areas associated with the two separate Z values would be found and summed. (We observe that one score was less than

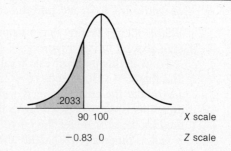

Figure 7.8 Sketch used as an aid in solving normal curve problem 3. (Shading indicates area sought.)

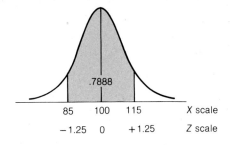

.7888

| 85 | 100 | 115 | X scale |

| −1.25 | 0 | +1.25 | Z scale |

Figure 7.9 Sketch used as an aid in solving normal curve problem 4. (Shading indicates area sought.)

100 and the other greater than 100; if both scores had been less than the mean or greater than the mean, the appropriate procedure would be as described in connection with the next question.)

5. What is the probability that a randomly selected salesperson would have achieved a score between 112 and 118 (see Figure 7.10)?

This problem also requires that two areas be determined, but, since the stated scores are both greater than the mean of 100, subtraction rather than addition must be used. We calculate

$$Z_1 = \frac{112 - 100}{12} = 1.00$$

$$Z_2 = \frac{118 - 100}{12} = 1.50$$

The area between 0 and $Z = 1.00$ is .3413, and the area between 0 and $Z = 1.50$ is .4332. The difference, $.4332 - .3413 = .0919$, is the answer sought.

6. Between what two values would 75% of the scores be found, assuming symmetry around the arithmetic mean (see Figure 7.11)?

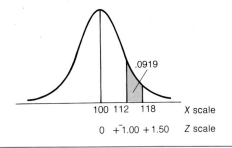

.0919

| 100 | 112 | 118 | X scale |

| 0 | +1.00 | +1.50 | Z scale |

Figure 7.10 Sketch used as an aid in solving normal curve problem 5. (Shading indicates area sought.)

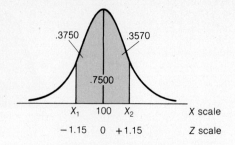

Figure 7.11 Sketch used as an aid in solving normal curve problem 6.

This problem entails using our normal curve table in reverse—that is, locating an area and then reading to the margins. We begin by recognizing that it is two unknown X values that we seek and that the area under the curve between these unknown X values is .7500. Since our normal curve table is constructed to give us directly the Z value associated with only one-half of the area, we divide .7500 by 2, getting .3750. We next try to find .3750 within the table and settle for .3749, the closest value to .3750 shown. Now, reading to the margins, we find that the Z value associated with an area of .3749 is 1.15. We reason that if .3749 is the area under the curve between 0 and +1.15, it must also be the area under the curve between 0 and −1.15. Substituting values into Equation (7.6.2), we solve the problem in the following way:

$$-1.15 = \frac{X_1 - 100}{12} \text{ and } X_1 = 86.2$$

and

$$+1.15 = \frac{X_2 - 100}{12} \text{ and } X_2 = 113.8$$

7. Above what score would only 4% of the scores be found (see Figure 7.12)?
   This problem also requires that we use the normal curve table in reverse. We wish to find the score that divides up the right-hand half of the curve in such a way that only 4% of all the scores are greater than that value and 46% of all the

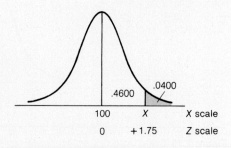

Figure 7.12 Sketch used as an aid in solving normal curve problem 7.

7 Special Probability Distributions

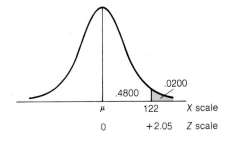

Figure 7.13 Sketch used as an aid in solving normal curve problem 8.

scores are between that value and 100. We begin by using the 46%—the .4600 value—and the Z scale. Corresponding to an area of .4600 (actually .4599) is a Z value of 1.75. Substituting this value and known values for the arithmetic mean and standard deviation into Equation (7.6.2), we obtain

$$+1.75 = \frac{X - 100}{12} \quad \text{and } X = 121$$

Therefore, the probability is only about .04 that a randomly selected salesperson would have achieved an SQP score in excess of 121.

8. Assume that the arithmetic mean of all the scores is not known but it *is known* that a salesperson with an SQP score of 122 scored above 98% of the salespeople. What must the arithmetic mean be (see Figure 7.13)?
   This is a switch on the previous kind of problem. We subtract .5000 from .9800 to get the area that will lead us to the needed Z value. The area is .4800 and the corresponding Z value is 2.05. Therefore,

$$+2.05 = \frac{122 - \mu}{12} \quad \text{and} \quad \mu = 97.4$$

WARNING

The X values can be converted to normally distributed Z values if and only if the X values themselves are normally distributed. A common misconception is that any set of numbers may be made normal through application of Equation (7.6.2). This is dead wrong. If a set of X values is distributed nonnormally, the corresponding set of Z values will also be distributed nonnormally.

Using the Normal Curve to Approximate Binomial Probabilities

Ponder the following questions and see whether you can determine what kind(s) of probability procedure(s) should be used:

1. A fair coin is to be flipped 12 times. What is the probability of getting 8, 9, or 10 heads?
2. A production process is known to turn out 1 defective item for every 5 items manufactured, on the average. If a random sample of size $n = 100$ is to be selected, what is the probability that between 15 and 30 defectives will be found?
3. If, in the long run, 5% of the trains leaving a busy station are delayed, what is the probability that during a certain week when 300 trains are scheduled to leave the station, at least 25 will be delayed?
4. In the past, 75% of the people taking a standard job aptitude test have achieved passing scores. It is known that last year 200 people took the test. What is the probability that, at most, 135 passed?

If you concluded that all of these problems could properly be worked using the binomial formula, give yourself an A. These problems have certain things in common that can tip us off to the appropriateness of the binomial approach. First, they are all concerned with two mutually exclusive events. In number 1, a flip of a coin will either produce a head or a tail; in number 2, a manufactured item will either be defective or acceptable; in number 3, a train will either be delayed or it will not be; in number 4, a person will either pass or fail the test. Second, the probabilities given, the p values, were determined by either the a priori approach or by the relative-frequency approach based on a large number of observations. Whatever its basis, the probability of a success in a single trial of the relevant random experiment does not, in these problems, vary from trial to trial. Third, all questions are of the form: "What is the probability of getting a specified number of successes (or a collection of such specified successes) in a specified number of trials of the random experiment?"

Regrettably, the problems also have one other thing in common: They would all be rather difficult to work using the binomial formula literally. The closest thing to an exception is problem 1, the one concerned with 12 flips of a coin. Because it is at least manageable, we will use this problem as a means of demonstrating the advantages of an alternative approach to be described shortly. Let us work problem 1 using the binomial formula. We know that $n = 12$, $p = .5$, $q = .5$, and wish to know $P(8) + P(9) + P(10)$. We calculate:

$$P(8) = \left(\frac{12!}{8!4!}\right)(.5)^8(.5)^4 = (495)(.003906)(.0625) = .1208$$

$$P(9) = \left(\frac{12!}{9!3!}\right)(.5)^9(.5)^3 = (220)(.001953)(.125) = .0537$$

$$P(10) = \left(\frac{12!}{10!2!}\right)(.5)^{10}(.5)^2 = (66)(.000977)(.25) = \underline{.0161}$$

Total   .1906

Thus, $P(8) + P(9) + P(10) = .1906$.

Fortunately, our newfound knowledge of the normal curve will permit us to solve this kind of problem, to a high order of approximation, in a matter of seconds. We make use of the fact, noted in section 7.4, that as n gets larger, regardless of the value of p—save for 0 and 1—the binomial distribution comes to resemble more and more closely the normal curve. Logically, we should be able to exploit this fact by using the normal curve to obtain approximations to probabilities which could be obtained exactly, though laboriously, via the binomial formula.

Let us rework problem 1 using the normal curve: We begin by rephrasing the question somewhat and having it read: "What is the probability of getting between 7.5 and 10.5 heads?" The subtraction of .5 from 8 and the addition of .5 to 10 reflects the application of what is called the *continuity correction*. This continuity correction helps to improve the accuracy of the probabilities approximated in a situation, like the present one, where a *continuous distribution is used to obtain probabilities which really should be obtained from a discrete distribution.* Figure 7.14 demonstrates the role played by the continuity correction.

RULES FOR THE APPLICATION OF THE CONTINUITY CORRECTION

1. A discrete value X becomes a range running from $X - .5$ to $X + .5$. The range associated with $X = 3$, for example, is 2.5 to 3.5.
2. A collection of consecutive discrete values requires subtraction of .5 from the smallest X value in the collection and the addition of .5 to the largest X value in the collection. For example, $X = 3$, 4, or 5 becomes a range running from 2.5 to 5.5.
3. If a problem involves an open-ended range and makes use of the words "at least," or their equivalent—for example, at least 3—the relevant range runs from 2.5 to $+\infty$.
4. In a situation like that described in point 3 but using the words "more than," or their equivalent—for example, more than 3—the relevant range runs from 3.5 (since 3 itself is excluded) to $+\infty$.
5. If a problem involves an open-ended range and makes use of the words "at most," or their equivalent—for example, at most 3—the relevant range runs from 3.5 to $-\infty$.
6. In a situation like that described in point 5 but using the words "less than," or their equivalent—for example, less than 3—the relevant range runs from 2.5 (since 3 itself is excluded) to $-\infty$.

To approach the solution of problem 1 (illustrated in Figure 7.15), we recall that the mean and standard deviation of a binomial distribution are obtained by $\mu = np$ and $\sigma = \sqrt{npq}$. For this problem: $\mu = np = (12)(.5) = 6$, and $\sigma = \sqrt{npq} = \sqrt{(12)(.5)(.5)} = 1.73$.

At this point, the problem becomes an ordinary normal curve problem like some of those worked above. We must convert 7.5 and 10.5 to their Z-value equivalents. Therefore,

$$Z_1 = \frac{7.5 - 6.0}{1.73} = 0.87$$

$$Z_2 = \frac{10.5 - 6.0}{1.73} = 2.60$$

The areas corresponding to these Z values are, from Table 7.7, .3078 and .4953, respectively. Because they are both on the same side of the normal curve, we subtract the smaller area from the larger, getting $.4953 - .3078 = .1875$; this amount differs from the .1906 obtained from using the binomial formula, but only by .0031 ($.1906 - .1875 = .0031$).

The advantage of using the normal curve approximation approach can be even more

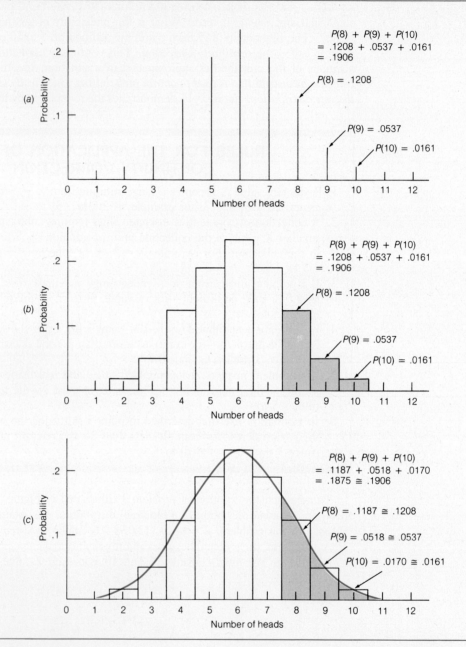

Figure 7.14 Demonstration of the normal curve approximation and the role of the continuity correction. (a) Binomial distribution for $n = 12$ and $p = .5$. Probabilities shown by heights of vertical lines. (b) Same as (a) but probabilities shown as areas. Continuity correction is used to fill in the spaces between discrete X values. (c) Same as (b) but with normal shape superimposed.

7 *Special Probability Distributions*

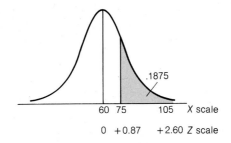

Figure 7.15 Sketch used as an aid in solving normal curve approximation problem 1. (Shading indicates area sought.)

dramatically demonstrated if, as we work through problems 2, 3, and 4, you will think about the amount of work that would be involved if we were to use the binomial formula.

> The closer p is to .5 and the larger n is, the closer the probabilities obtained from the normal curve approach will be to the corresponding true binomial probabilities. As a general rule, if the products of np and nq are both greater than 5, the normal curve approximations will be close enough to the true binomial probabilities for all practical purposes.

Now, let us apply the normal curve approximation method to the other problems.

Problem 2 (Figure 7.16) rephrased is: "If a random sample of size $n = 100$ is to be selected, what is the probability that between 14.5 and 30.5 defectives will be found?" We know that $n = 100$ and $p = .2$.

The mean of the binomial distribution is $\mu = np = (100)(.2) = 20$, and the standard deviation is $\sigma = \sqrt{npq} = \sqrt{(100)(.2)(.8)} = 4$. Therefore,

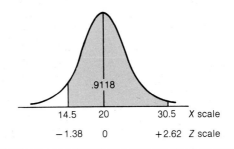

Figure 7.16 Sketch used as an aid in solving normal curve approximation problem 2. (Shading indicates area sought.)

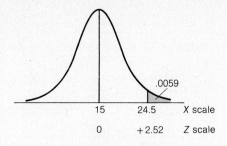

Figure 7.17 Sketch used as an aid in solving normal curve approximation problem 3. (Shading indicates area sought.)

$$Z_1 = \frac{14.5 - 20.0}{4} = -1.38$$

$$Z_2 = \frac{30.5 - 20.0}{4} = +2.62$$

The associated areas are .4162 and .4956, respectively. Since these two areas are on different sides of the mean of 20, we add them, getting .4162 + .4956 = .9118.

Problem 3 (Figure 7.17) rephrased becomes: "What is the probability that during a certain week when 300 trains are scheduled to leave the station at least 24.5 will be delayed?" We know that $n = 300$ and $p = .05$.

$$\mu = np = (300)(.05) = 15$$
$$\sigma = \sqrt{npq} = \sqrt{(300)(.05)(.95)} = 3.77$$
$$Z = \frac{24.5 - 15.0}{3.77} = 2.52$$

The associated area is .4941. But, because we seek the area under the right-hand tail, we subtract this value from .5000, getting .5000 − .4941 = .0059.

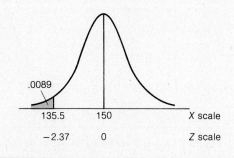

Figure 7.18 Sketch used as an aid in solving normal curve approximation problem 4. (Shading indicates area sought.)

Problem 4 (Figure 7.18) rephrased is: "What is the probability that at most 135.5 passed the test?" We know that $n = 200$ and $p = .75$.

$$\mu = np = (200)(.75) = 150$$
$$\sigma = \sqrt{npq} = \sqrt{(200)(.75)(.25)} = 6.12$$
$$Z = \frac{135.5 - 150.0}{6.12} = -2.37$$

The associated area is .4911. But this value must be subtracted from .5000 to give the area under the left-hand tail. Therefore, $.5000 - .4911 = .0089$.

✳ 7.7 The Poisson Distribution

Study the following illustrative case and see whether you can determine the appropriate probability procedure for answering the questions at the end.

ILLUSTRATIVE CASE 7.2

Jan Cole, owner and sole employee of Heavensent Computer Dating Services, receives, on the average, 18 telephone calls per each nine-hour day. (She brown-bags it at lunchtime; the hour from 12:00 to 1:00 P.M. is just as busy as any other hour.) She seldom misses any time from work. However, next Tuesday she has a dental appointment which could not be scheduled for any other time. She estimates that she will be away from the office for three hours. What is the probability that there will be (1) no incoming calls during this particular three-hour period and (2) at least 12 incoming calls during this particular three-hour period?

If you decided that these questions could be answered using the binomial formula, give yourself an F. On second thought, a gentleman's C might be fairer because the problem *does* resemble a binomial problem in one obvious respect: You are asked to determine the probability associated with a specific number, or collection of consecutive specified numbers, of successes. The probabilities sought are $P(0)$ and $P(\text{at least } 12)$.

However, this problem also differs from a binomial probability problem in an important respect:

> Whereas we can state a discrete number of successes, we cannot state a discrete number of failures nor even a discrete number of trials.

Clearly, it would make no sense to speak of the number of "non-calls." The background variable in this problem, time, is continuous even though we may concern ourselves with 0, 1, 2, or some other discrete number of incoming calls.

> We may say, therefore, that we are dealing with a *discrete variable on a continuous field*. The "field" or "background variable" is time (continuous), whereas the variable under study is number of telephone calls (discrete).

When a problem involves the determination of probabilities associated with a discrete number of successes *within a continuum of time or space*, the appropriate probability distribution to employ is the Poisson, provided that certain other conditions are also present. These other conditions are:

1. It is possible to divide a time interval (or unit of space) into many smaller intervals (or units). (For example, a 9-hour day can be divided into two 4½-hour periods, three 3-hour periods, nine 1-hour periods, 540 1-minute periods, 32,400 1-second periods, and so forth.)
2. The probability of a success remains the same throughout the time intervals (or units of space). (Jan Cole's most reasonable expectation, we will say, is for a steady stream of calls throughout the day.)
3. The probability of two or more successes during a very small subinterval (or subunit of space) is so small that it can be ignored. (It is impossible, in effect, for Jan to receive two or more calls during the same split second.)
4. Events are independent. (The probability that a certain person will phone is unaffected by the fact that some other person or persons has (have) already phoned.)

The Nature of the Poisson Distribution

When use of Poisson probabilities is appropriate, the probability of a specified number of successes, X, is determined by

$$P(X) = \frac{e^{-\mu} \cdot \mu^X}{X!} \tag{7.7.1}$$

where e is the constant 2.7183. . . .

Before attempting to solve Jan's problem, let us try to make some sense out of the Poisson probability formula itself. Since the constant e plays a critical role, we will begin by taking a close look at it.

The constant e is the sum of the numerical terms in the following series:

$$e = 1/0! + 1/1! + 1/2! + 1/3! + 1/4! + 1/5! + \cdots$$

The terms on the right-hand side go on indefinitely but become rapidly smaller. In numbers: $1 + 1 + .5 + .166667 + .041667 + .008333 + .001389 + \cdots = 2.7183$, accurate to four decimal places.

If we let e be raised to some power, call it μ, then $e^\mu = (2.7183)^\mu$. (Any power could have been used, but some advantages will accrue from assuming a power equal to μ.) This equality can be rewritten: $e^\mu = \mu^0/0! + \mu^1/1! + \mu^2/2! + \mu^3/3! + \mu^4/4! + \mu^5/5! + \cdots$.

To utilize this series as a probability distribution, we must make certain of two things: (1) The sum of the terms is 1, as with any probability distribution, and (2) each term in the series has an identifiable meaning. With respect to the first requirement, a little algebraic tinkering will suffice. A rule of algebra states that $e^{-\mu} \cdot e^{\mu} = e^0 = 1$. This suggests that the product $e^{-\mu} \cdot e^{\mu}$ might be useful when written in the form

$1 = e^{-\mu} \cdot e^{\mu} = e^{-\mu}(\mu^0/0! + \mu^1/1! + \mu^2/2! + \mu^3/3! + \mu^4/4! + \mu^5/5! + \cdots)$ And useful it is, for it satisfies both of the requirements of a probability distribution stated above. From this expansion, we can deduce that

$$\text{Probability of 0 successes is } P(0) = \frac{e^{-\mu}(1)}{1} = e^{-\mu}$$

$$\text{Probability of 1 success is } P(1) = \frac{e^{-\mu} \cdot \mu}{1} = e^{-\mu} \cdot \mu$$

$$\text{Probability of 2 successes is } P(2) = \frac{e^{-\mu}\mu^2}{2!}$$

$$\text{Probability of 3 successes is } P(3) = \frac{e^{-\mu}\mu^3}{3!}$$

And so forth. Or, in general terms, as noted above,

$$P(X) = \frac{e^{-\mu} \cdot \mu^X}{X!}$$

An interesting characteristic of the Poisson distribution is that the mean, μ, and the variance, σ^2, are identical. This identity will be demonstrated shortly.

Applying the Poisson Formula

For the Heavensent case, we know that μ for nine-hour periods is 18. For simplicity's sake and because the problem is concerned with a specific three-hour period, we will redefine μ and let it be 6 per three-hour period. (Easier yet might be 2 per one-hour period.) Conditions 1 and 2 permit us to do this. If such a change is thought improper for any reason, the appropriateness of using Poisson probabilities should be seriously reviewed.

To answer the first question, the one having to do with 0 calls during the specific three-hour period, we recognize X to be 0 and substitute it into the Poisson probability formula:

$$P(0) = \frac{(2.7183)^{-6}(6)^0}{0!} = (2.7183)^{-6} = \frac{1}{(2.7183)^6} = 1/403.4450 = .0025$$

Chances are excellent—$1 - .0025 = .9975$—that Jan will miss some calls.

Had we defined μ to be 2 per one-hour period, we would have proceeded as follows:

$$P(0) = \frac{(2.7183)^{-2}(2)^0}{0!} = (2.7183)^{-2} = \frac{1}{(2.7183)^2} = 1/7.3892 = .1353$$

The .1353 value is the probability of 0 calls during a particular one-hour period. The probability of Jan's getting 0 calls during three back-to-back one-hour periods must be determined by using the special case of the multiplication theorem. That is, $(.1353)(.1353)(.1353) = (.1353)^3 = .0025$—exactly the same probability obtained when $\mu = 6$ was used.

Using the Poisson probability formula is not especially difficult when μ is an integer, as in our example. However, since real life seems to delight in serving up μ values that are not integers, you are advised to become acquainted with Appendix Table 4. Use of this table is simplicity itself. Just find the relevant μ value at the top of a column, read down the column,

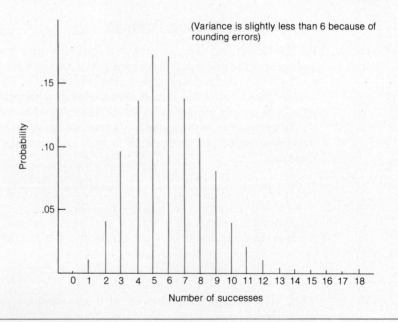

Number of Successes X	Probability When $\mu = 6$ P(X)	Variance Calculations		
		$X - \mu$	$(X - \mu)^2$	$P(X)(X - \mu)^2$
0	.0025	−6	36	.0000
1	.0149	−5	25	.3725
2	.0446	−4	16	.7136
3	.0892	−3	9	.8028
4	.1339	−2	4	.5356
5	.1606	−1	1	.1606
6	.1606	0	0	.0000
7	.1377	1	1	.1377
8	.1033	2	4	.4132
9	.0688	3	9	.6192
10	.0413	4	16	.6608
11	.0225	5	25	.5625
12	.0113 ⎫	6	36	.4068
13	.0052 ⎪	7	49	.2548
14	.0022 ⎬ Sum	8	64	.1408
15	.0009 ⎪ of .02	9	81	.0729
16	.0003 ⎪	10	100	.0300
17	.0001 ⎪	11	121	.0121
18	.0000 ⎭		Total	5.8959

(Variance is slightly less than 6 because of rounding errors)

Figure 7.19 Presentation of Poisson probability distribution in tabular and graphic form. Also calculation of the variance.

7 *Special Probability Distributions*

and extract the probabilities associated with specified numbers of successes. For example, referring once again to Illustrative Case 7.2, there is no column corresponding to $\mu = 18$ per day, but there is one for $\mu = 6$ per three-hour period (or, for that matter, 2 per one-hour period, 1 per half-hour period, and so forth). The Poisson probabilities associated with $\mu = 6$ are shown in both tabular and graphic form in Figure 7.19. We see that $P(0) = .0025$. The correspondence between μ and σ^2 is also suggested in this figure.

The values in Figure 7.19 will help us to answer the second of Jan's questions. The question was: "What is the probability that there will be at least 12 incoming calls during the three-hour period when Jan is away from the office?" In other words, we wish to know P(at least 12) $= P(12) + P(13) + P(14) + \cdots$. From Figure 7.19 (or Appendix Table 4), we find that $P(12) = .0113$, $P(13) = .0052$, $P(14) = .0022$, $P(15) = .0009$, $P(16) = .0003$, $P(17) = .0001$, $P(18) = .0000$. The sum is .02.

As mentioned above, use of Poisson probabilities is not limited to problems involving a discrete number of successes within a continuum of time. Some uses of Poisson probabilities pertain to situations where discrete numbers of successes occur within a continuum of space. For example: If, on the average, four frays are found per every 100 feet of nylon cord produced by the Jensen Company, what is the probability that Dale Vale, who buys a 20-foot length, will find at least two frays? The answer is that the μ of 4 per 100 feet is the same as μ of .8 per 20-foot length. Appendix Table 4 reveals that for $\mu = .8$, $P(2) = .1438$, $P(3) = .0383$, $P(4) = .0077$, $P(5) = .0012$, $P(6) = .0002$, and $P(7) = .0000$. The sum is .1912. We also learn from Appendix Table 4 that $P(0) = .4493$, meaning that he has less than a 50–50 chance of having no fray in the length of nylon cord he bought.

Poisson Approximations to Binomial Probabilities

Another useful feature of Poisson probabilities is that

> When n is large and p is very small, Poisson probabilities provide close approximations to binomial probabilities.*

For example, let us say that p is .1 and n is 20, so that $\mu = np = 2$. The closeness of the corresponding binomial and Poisson probabilities is evident in Table 7.8.

The advantage of using Poisson probabilities to approximate binomial probabilities can be demonstrated using a hypothetical problem: Suppose that the portion of defective tracks turned out by a company is known to be .005. The tacks are packaged 500 to a box. A customer buying one of the boxes wishes to know the probability that the box contains no more than 6 defective tacks. Assume that the 500 tacks in the box purchased amounts to a random sample of size $n = 500$ from the process population.

The binomial probability approach to solving this problem would entail

(text continues on p. 291)

---

*A frequently used but (as Table 7.8 suggests) rather conservative rule of thumb is: $p \le .05$ and $n \ge 20$.

Table 7.8 Comparison of Binomial and Poisson Probabilities for $n = 20$, $p = .1$

Number of Successes X	P(X) Values for		Number of Successes X	P(X) Values for	
	Binomial	Poisson		Binomial	Poisson
0	.1216	.1353	5	.0319	.0361
1	.2702	.2707	6	.0089	.0120
2	.2852	.2707	7	.0020	.0034
3	.1901	.1804	8	.0004	.0009
4	.0898	.0902	9	.0001	.0002
				.0000 hereafter	.0000 hereafter

GRIN and BEAR IT   by FRED WAGNER

"Hello, dear . . . This is the one in a million you married."

$$P(0) = \left(\frac{500!}{0!500!}\right)(.005)^0(.995)^{500}$$

$$+ \ P(1) = \left(\frac{500!}{1!499!}\right)(.005)^1(.995)^{499}$$

$$+ \ \cdots$$

$$+ \ P(6) = \left(\frac{500!}{6!494!}\right)(.005)^6(.995)^{494}$$

Needless to say, one could spend a very long time trying to determine the answer. The Poisson formula, on the other hand, can be used with ease and will provide results sufficiently accurate for all practical purposes.

We begin by recognizing that, for a binomial problem, $\mu = np$. In this case, $\mu = (500)(.005) = 2.5$. The Poisson probability table, Appendix Table 4, reveals that $P(0) = .0821$, $P(1) = .2052$, $P(2) = .2565$, $P(3) = .2138$, $P(4) = .1336$, $P(5) = .0668$, and $P(6) = .0278$. Summing, we get $P(\text{no more than } 6) = .9858$.

---

* 7.8 The Exponential Distribution

A most important continuous probability distribution for many kinds of operations research applications is the exponential. The theoretical requirements are the same for the exponential distribution as for the Poisson.

> If the number of successes occurring within a specified time interval is distributed in a Poisson manner, then the *exponential distribution* can be used to determine the probability associated with a specified length of time between successes.

The exponential probability distribution has been employed in connection with such concerns as time between calls at a fire department, waiting time in a line at a fast-food restaurant, length of time between emergency arrivals at a hospital, time between arrivals at the drive-in window of a bank, length of life of automobile transmissions, and so forth.

> Because both the background variable and the variable under scrutiny are time variables, we may describe an exponential probability distribution in terms somewhat similar to those used in connection with the Poisson. However, whereas with a Poisson-distributed variable we spoke of a discrete variable on a continuous field, with an exponentially distributed variable we speak of a *continuous variable on a continuous field*.

We will use the Heavensent case (Illustrative Case 7.2) for purposes of showing how this distribution can be used. We were told that Jan Cole receives an average of $\mu = 18$ telephone calls per day, or $\mu = 2$ calls per hour. The conditions, let us recall, appeared right for assuming a Poisson probability distribution of number of calls per time unit.

If we take one hour to be the time unit of interest, we may deduce that if $\mu = 2$, then the average length of time between calls must be $1/\mu = 1/2 = .5$, or half an hour. The exponential distribution associated with $\mu = 2$, and hence, $1/\mu = .5$, is shown in graphic form in Figure 7.20 and in tabular form in Table 7.9, the latter having been excerpted from Appendix Table 5. In this table, the symbol T represents elapsed time between successes. Notice that, as the time interval becomes longer and longer (i.e., T becomes larger and larger), the probability that the same amount of time or more will elapse before the next success becomes smaller and smaller. Notice also that an exponential probability has a value other than 0 only when T is equal to or greater than 0 and when μ is greater than 0.

The formula for determining exponential probabilities is

$$Y = F(T) = \mu e^{-\mu T} \qquad (7.8.1)$$

where $F(T)$ means "function of time"
T is elapsed time between successes
μ is the average number of successes during a specified period of time
e is the constant 2.7183

Let us suppose that Jan has just hung up the telephone and now wishes to know the probability of an hour's elapsing before the next incoming call. One way to answer this question would be to employ integral calculus to find the area in the right-hand tail of the relevant exponential distribution. Fortunately, this is not necessary because Table 7.9 provides us with the values of areas (that is, probabilities) under the exponential curve. Notice that this table has two columns, one of these is labeled T and the other $F(T)$. Numbers in

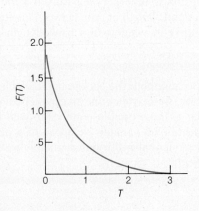

Figure 7.20 Exponential distribution for $\mu = 2$ (or $1/\mu = .5$).

7 *Special Probability Distributions*

Table 7.9 Exponential Distribution (From Appendix Table 5)

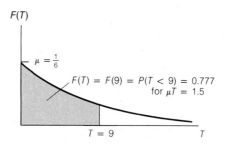

Example: If $\mu = \frac{1}{6}$, the probability of observing a value less than $T = 9$ is found by $F(T)$ for $\mu T = \frac{1}{6}(9) = 1.5$; $P(T < 9) = 0.777$.

μT	$F(T)$	μT	$F(T)$	μT	$F(T)$	μT	$F(T)$
0.0	0.000	2.5	0.918	5.0	0.9933	7.5	0.99945
0.1	0.095	2.6	0.926	5.1	0.9939	7.6	0.99950
0.2	0.181	2.7	0.933	5.2	0.9945	7.7	0.99955
0.3	0.259	2.8	0.939	5.3	0.9950	7.8	0.99959
0.4	0.330	2.9	0.945	5.4	0.9955	7.9	0.99963
0.5	0.393	3.0	0.950	5.5	0.9959	8.0	0.99966
0.6	0.451	3.1	0.955	5.6	0.9963	8.1	0.99970
0.7	0.503	3.2	0.959	5.7	0.9967	8.2	0.99972
0.8	0.551	3.3	0.963	5.8	0.9970	8.3	0.99975
0.9	0.593	3.4	0.967	5.9	0.9973	8.4	0.99978
1.0	0.632	3.5	0.970	6.0	0.9975	8.5	0.99980
1.1	0.667	3.6	0.973	6.1	0.9978	8.6	0.99982
1.2	0.699	3.7	0.975	6.2	0.9980	8.7	0.99983
1.3	0.727	3.8	0.978	6.3	0.9982	8.8	0.99985
1.4	0.753	3.9	0.980	6.4	0.9983	8.9	0.99986
1.5	0.777	4.0	0.982	6.5	0.9985	9.0	0.99989
1.6	0.798	4.1	0.983	6.6	0.9986	9.1	0.99989
1.7	0.817	4.2	0.985	6.7	0.9988	9.2	0.99990
1.8	0.835	4.3	0.986	6.8	0.9989	9.3	0.99991
1.9	0.850	4.4	0.988	6.9	0.9990	9.4	0.99992
2.0	0.865	4.5	0.989	7.0	0.9991	9.5	0.99992
2.1	0.878	4.6	0.990	7.1	0.9992	9.6	0.99993
2.2	0.889	4.7	0.991	7.2	0.9993	9.7	0.99994
2.3	0.900	4.8	0.992	7.3	0.9993	9.8	0.99994
2.4	0.909	4.9	0.993	7.4	0.9993	9.9	0.99995

Source: Adapted from Harnett and Murphy, *Statistical Analysis for Business and Economics,* © 1985, Addison-Wesley, Reading, Massachusetts, p. A54, Table IV. Reprinted with permission.

the $F(T)$ column represent areas between $T = 0$ and $T =$ specified values of interest. For the problems posed above, we multiply μ and T together, getting 2 times $1 = 2$. Therefore, we locate 2.0 in the column headed μT. Reading to the next column, we find a probability of .865. Since .865 is the area under the curve between $T = 0$ and $T = 1$, we subtract .865 from 1, getting $1.000 - .865 = .135$, a rather low probability of her having to wait an hour or longer for the next incoming call. If the T value of interest had been, say, 18 minutes, then the appropriate μ would be $2/60 = .03333$ per minute and $\mu T = (.03333)(18) = .6$. Corresponding to a μT of .6 is an $F(T)$ of .451. Thus, $1 - .451 = .549$ for the probability of a wait at least this long.

Here is another example. Trucks arrive at the loading dock of a fruit-packaging company on the average of three times per hour. What is the probability that after the departure of a truck 80 minutes or more will elapse before the arrival of another truck? The answer is that $\mu = 3$ per hour is the same as $\mu = .05$ per minute, meaning that the average time interval between arrivals is $1/.05 = 20$ minutes. The product of $\mu = .05$ and $T = 80$ is 4. Corresponding to 4.0 in Appendix Table 5 is the area .982; $1 - .982 = .018$.

7.9 Summary

This chapter has covered quite a lot of statistical ground. It began with the concept of random variable, which was defined as a quantitative variable whose values may be viewed as resulting from a random experiment. The values of the random variable and the corresponding probabilities together constitute a (quantitative) probability distribution. This idea was illustrated using a distribution whose probabilities had been determined empirically. Methodology for computing the mean, or expected value, and standard deviation of a discrete random variable was then shown.

We noted that the probabilities associated with values of a random variable need not always rest on empirical observation; if certain specified conditions are present, the probabilities sought can be determined through more convenient theoretical means.

For example, if a problem is concerned with two mutually exclusive events and if the probability of a success in a single independent trial of a random experiment is constant from trial to trial, the binomial probability formula may be used to construct the desired probability distribution. This is possible provided that the number of trials and the probability of a success in a single trial are known.

If a problem is concerned with two mutually exclusive events and if sampling without replacement from a finite population is carried out—so that the probability of a success in a single trial is not constant from trial to trial—the hypergeometric probability formula may be used to obtain the desired probability distribution. If the population size is substantially larger than the sample size—$N/n \geq 20$, as a rule of thumb—the binomial probability formula will provide close approximations to hypergeometric probabilities.

If, in a binomial probability problem, np and nq are both greater than about 5, the normal curve, a continuous bell-shaped curve, may be used to obtain close approximations to the binomial probabilities sought. The normal curve is also quite an accurate model for a great many different kinds of continuous populations.

(As if the two preceding points were not enough to ensure the normal curve's secure position in the statistical Hall of Fame, this theoretical probability distribution model also plays a most important role in sampling theory. This point will be elaborated on and illustrated in the next chapter.)

When we are faced with a situation requiring that we determine probabilities associated with a discrete variable but the background variable is continuous (such as number of successes during an interval of time or number of successes within a unit of space), the desired probabilities may be obtained through use of the Poisson probability formula as long as certain conditions are present. These conditions are: (1) It is possible to divide an interval of time or a unit of space into smaller intervals or units; (2) the probability of success remains constant throughout the time interval or unit of space; (3) the probability of two or more successes during a very small subinterval of time or subunit of space is so small that it can be ignored; and (4) the events are independent.

Finally, if the number of successes occurring within an interval of time are distributed in a Poisson manner, the exponential probability formula can be used to determine the probabilities associated with specified lengths of time between successes.

Terms Introduced in This Chapter

bell-shaped probability distribution (p. 268)
binomial approximation to hypergeometric probabilities (p. 267)
binomial probability distribution (p. 255)
continuity correction (p. 281)
expected value (p. 253)

exponential probability distribution (p. 291)
finite correction factor (p. 267)
hypergeometric probability distribution (p. 265)
normal curve approximation to binomial probabilities (p. 279)
normal probability distribution (p. 268)

Poisson approximations to binomial probabilities (p. 289)
Poisson probability distribution (p. 285)
random variable (p. 253)
standard normal probability distribution (p. 271)
Z value (p. 271)

Formulas Introduced in This Chapter

General formula for the mean, or expected value, of a discrete random variable:

$$\mu = E(X) = \Sigma\,[P(X) \cdot X]$$

General formula for the standard deviation of a discrete random variable:

$$\sigma = \sqrt{\Sigma\,[P(X)(X - \mu)^2]}$$

Formula for computing binomial probabilities:

$$P(X) = \frac{n!}{X!(n - X)!}\,p^X q^{n-X}$$

Formula for the mean of a binomial or hypergeometric probability distribution:

$$\mu = np$$

Formula for the standard deviation of a binomial probability distribution:

$$\sigma = \sqrt{npq}$$

Formula for computing hypergeometric probabilities:

$$P(X) = \frac{\left[\dfrac{S!}{X!(S-X)!}\right]\left[\dfrac{(N-S)!}{(n-X)!(N-S-n+X)!}\right]}{\dfrac{N!}{n!(N-n)!}}$$

Standard deviation of a hypergeometric probability distribution:

$$\sigma = \sqrt{npq\left(\frac{N-n}{N-1}\right)}$$

Formula for a normal probability distribution:

$$Y = \frac{1}{\sqrt{2\pi}\sigma}e^{\frac{-(X-\mu)^2}{2\sigma^2}}$$

Formula for transforming normally distributed X values into standard units:

$$Z = \frac{X-\mu}{\sigma}$$

Formula for computing Poisson probabilities:

$$P(X) = \frac{e^{-\mu}\cdot\mu^X}{X!}$$

Formula for computing exponential probabilities:

$$Y = F(T) = \mu e^{-\mu T}$$

Questions and Problems

7.1 Explain the meaning of each of the following:
a. Random variable   d. Z value
b. Expected value   e. Continuity correction
c. Standard normal curve

7.2 Distinguish between:
a. Binomial probability distribution and normal probability distribution
b. Binomial probability distribution and Poisson probability distribution
c. Poisson probability distribution and exponential probability distribution
d. Normal curve approximation of binomial probabilities and Poisson approximation of binomial probabilities
e. Random variable and probability distribution
f. Binomial and hypergeometric probability distributions

7.3 Distinguish between:
a. $P(X)$ and p
b. p and q
c. π and e, on the one hand, and μ and σ on the other (in connection with the normal curve)
d. p and $\dfrac{n!}{X!(n-X)!}$
e. $\Sigma\,[P(X)\cdot X]$ and np

7.4 Distinguish between:
  a. Method of determining the arithmetic mean of a frequency distribution and method of determining the arithmetic mean of a probability distribution
  b. Method of determining the standard deviation of a frequency distribution and method of determining the standard deviation of a probability distribution
  c. $\sqrt{\Sigma\,[P(X)(X - \mu)^2]}$ and $\sqrt{npq}$
  d. Mean of a Poisson distribution and mean of an exponential distribution
  e. Normal distribution and standard normal distribution
  f. Standard deviation of a binomial probability distribution and standard deviation of a hypergeometric probability distribution

7.5 Indicate which of the following statements you agree with and which you disagree with, and defend your opinions:
  a. A random variable is a quantitative or qualitative variable whose values are determined from a random experiment.
  b. If p is a constant .2 and n is increased from 5 to 20, the mean of the binomial probability distribution will become larger.
  c. If p is a constant .2 and n is increased from 5 to 20, the standard deviation of a binomial probability distribution will become larger.

7.6 Indicate which of the following statements you agree with and which you disagree with, and defend your opinions:
  a. If n is constant and p changes from .3 to .6, the arithmetic mean of a binomial probability distribution will become larger.
  b. If n is constant and p changes from .2 to .8, the standard deviation of a binomial probability distribution will become larger.
  c. The $p^X q^{n-X}$ part of the binomial probability formula is simply a way of expressing the special case of the multiplication theorem of probability.
  d. The $n!/[X!(n - X)!]$ part of the binomial probability formula is simply a way of expressing the special case of the addition theorem of probability.

7.7 Indicate which of the following statements you agree with and which you disagree with, and defend your opinions:
  a. If two wells each with a probability of .7 of having water are to be dug within 10 feet of each other, the binomial probability formula may be safely used to determine the probability that 0, 1, or 2 wells will have water.
  b. What we call the *binomial distribution* is really a large family of similarly constructed probability distributions.
  c. The normal and Poisson distributions are both continuous probability distributions.

7.8 Indicate which of the following statements you agree with and which you disagree with, and defend your opinions:
  a. The normal curve is symmetrical around its arithmetic mean.
  b. The normal curve is symmetrical around its median.
  c. A normal curve may be broad or skinny depending on its standard deviation.
  d. When working with the normal curve, the first thing one ordinarily does is compute the values of π and e for the data of interest.

7.9 Indicate which of the following statements you agree with and which you disagree with, and defend your opinions:
  a. What we call the *normal probability distribution*, or *normal curve*, is really a family of similarly appearing probability distributions.
  b. Any set of normally distributed data can be converted into a standard normal curve having a mean of 0 and a standard deviation of 1 by running each original X value through the Z formula.

c. It matters little whether a set of data is normally distributed to begin with because, when the X values are converted into Z values, the resulting distribution will be normal.

d. In theory, the $Y = f(X)$ values associated with a normal distribution will approach but never quite meet the horizontal axis.

7.10 Indicate which of the following statements you agree with and which you disagree with, and defend your opinions:

a. The Z formula can be used to solve for Z but never for X, μ, or σ.

b. A binomial probability problem involving a large value of n can usually be solved adequately through use of some kind of approximation.

c. As a rule of thumb, both np and nq should be smaller than 5 if the normal curve approximation approach is to be used to solve for binomial probabilities.

d. If the table of areas under the normal curve is to be used, *all* normally distributed X values must be converted into their Z-value equivalents.

7.11 Indicate which of the following statements you agree with and which you disagree with, and defend your opinions:

a. As a rule of thumb, np should be over 7 if the Poisson approximation approach is to be used to solve for binomial probabilities.

b. If the drive-in window of a bank is much busier during the hours 12:00 to 1:00 P.M. and 4:00 to 6:00 P.M. than at any other times, the Poisson probability distribution can still safely be employed to determine the probability of 0, 1, 2, and so on, customers during any arbitrarily selected half-hour period.

c. In the situation described in part *b* above, the exponential probability distribution can safely be employed to determine the probability associated with specified times between arrivals of customers.

d. Use of Poisson probabilities may be appropriate in situations where a discrete number of successes can be identified but not a discrete number of failures.

7.12 Indicate which of the following statements you agree with and which you disagree with, and defend your opinions:

a. Whereas the assumption of independent events is very important in connection with the proper use of binomial probabilities, it has no bearing upon the proper use of Poisson probabilities.

b. If the arithmetic mean associated with discrete numbers of successes within a continuum of time or space is 4, then the arithmetic mean associated with lengths of time between such occurrences will be .4.

c. An exponential probability will always be some nonzero, positive number regardless of the values of T and μ.

d. According to the exponential probability formula, as the time interval between successes becomes longer and longer, the probability that the same amount of time or more will elapse before the next success becomes larger and larger.

7.13 Indicate which of the following statements you agree with and which you disagree with, and defend your opinions:

a. If $X = 20$, $\mu = 25$, and $\sigma = 4$, then Z must be $+1.25$.

b. If $Z = -0.70$, $\mu = 50$, and $\sigma = 14$, then X must be 45.

c. If $Z = 1.50$, $X = 70$, and $\sigma = 20$, then μ must be 40.

d. If $Z = 2.20$, $X = 3$, $\mu = 2$, then σ must be .67.

7.14 Describe the conditions that must be present before each of the following theoretical probability distributions may be properly employed:

a. The binomial

b. The Poisson

c. The exponential

7.15 Determine the probability (actual or approximate) of exactly $X = 4$ successes for each of the following combinations of n and p values:
  a. $n = 5$, $p = .4$
  b. $n = 7$, $p = .9$
  c. $n = 200$, $p = .025$
  d. $n = 500$, $p = .3$

7.16 Historically, the Wheeze Motor Company has produced one defective door per every five doors manufactured. Assuming statistical independence, if you were to buy a four-door model, what would be the probability of your getting:
  a. Exactly two defective doors?
  b. At least two defective doors?
  c. At most two defective doors?
  d. Either two or three defective doors?

7.17 Refer to problem 7.16.
  a. Present the entire probability distribution in both tabular and graphic form, and briefly describe its appearance.
  b. Had the problem stated that this company produces, in the long run, one defective door per every 20 doors manufactured, in what way or ways would you expect the appearance of the probability distribution to be changed? Give reasons for your answer.
  c. Compute the arithmetic mean, or expected value, for the probability distribution obtained in connection with part *a* above.
  d. Had the problem been based on the p value implied in part *b* of the present problem and $n = 4$, what effect, if any, would use of this alternative information have on the arithmetic mean of the resulting probability distribution? State clearly the reasoning behind your answer.
  e. Compute the standard deviation for the probability distribution obtained in connection with part *a* above.
  f. Had the problem been based on the p value implied in part *b* of the present problem and $n = 4$, what effect, if any, would use of this alternative information have on the standard deviation of the resulting probability distribution? State clearly the reasoning behind your answer.

7.18 The Mega City Police Department has ten unsolved missing-children cases on its books. A man claiming to have psychic powers which enable him to divine the whereabouts of missing children 80% of the time offers his services to the police department. In the past, the whereabouts of missing children has been successfully determined by natural means only about 30% of the time.
  a. Assuming independence and assuming the alleged psychic is telling the truth, what is the probability that he can successfully divine the whereabouts of at least six of the missing children?
  b. Assuming independence and assuming that the alleged psychic has no special powers whatever, what is the probability that the whereabouts of at least six of the missing children will be discovered anyway?

7.19 A speculator in the commodities futures market is presently holding buy contracts for six different commodities whose price movements she believes to be independent of one another. She estimates the probability of her making a profit on any one of these commodities to be .3. (Since the profit could be huge, the relatively unattractive odds do not discourage her.) What is the probability that she will realize a profit from
  a. Exactly three of the commodities for which she holds futures contracts?
  b. At least three of the commodities for which she holds futures contracts?
  c. At most three commodities for which she holds futures contracts?

7.20 Refer to problem 7.19.

    a. Present the entire probability distribution in both tabular and graphic form, and briefly describe its appearance.

    b. If the problem had stated that this speculator holds buy contracts for eight, rather than six, different commodities whose prices are believed to move independently of one another and that she judged the probability of making a profit on any one of these commodities to be $p = .3$, would you expect the appearance of the probability distribution to be altered? In what way or ways?

    c. Compute the arithmetic mean for the probability distribution obtained in connection with part *a* above.

    d. If the problem had been stated as in part *b* of the present problem, what effect, if any, would use of this alternative information have on the arithmetic mean of the resulting probability distribution? State clearly the reasoning behind your answer.

    e. Compute the standard deviation for the probability distribution obtained in connection with part *a* above.

    f. If the problem had been stated as in part *b* of the present problem, what effect, if any, would use of this alternative information have on the standard deviation of the resulting probability distribution. State clearly the reasoning behind your answer.

 7.21 In a certain suburban city, 30% of the adult residents are opposed to their city's being annexed to a nearby major city. If a random sample of seven adult residences were selected, what is the probability that

    a. None would be opposed to annexation?

    b. All would be opposed to annexation?

    c. Four or five would be opposed to annexation?

    d. At least four would be opposed to annexation?

 7.22 An inspector in a semiautomated woodworking shop examines chair arms in batches of five after they emerge from a smoothing process. If one or more of the chair arms are judged still too rough, they are all sent back through the smoothing process. Find the probability that a certain batch will have to be resmoothed if the true process proportion of inadequately smoothed chair arms is

    a. $p = .1$

    b. $p = .2$

    c. $p = .5$

 7.23 The Reliable Armored Services Company owns five armored cars. The probability is .3 that at any given moment a car will be standing idle and ready for emergency service. The .3 probability applies to each of the five cars and is unaffected by what another car is doing. If the Dazzle Jewell Company needs a shipment of diamonds sent across town immediately, what is the probability that Reliable will have no cars available?

7.24 The number of trucks being loaded on a certain company's loading dock between 8:00 A.M. and 12:00 noon on Mondays has been found to have the following probability distribution:

Number of Trucks Being Loaded X	Probability (Relative Frequency) P(X)	Number of Trucks Being Loaded X	Probability (Relative Frequency) P(X)
0	.10	4	.20
1	.15	5	.07
2	.25	6	.03
3	.20		

a. Using the equation $\mu = \Sigma [P(X) \cdot X]$, find the arithmetic mean of this probability distribution.

b. Could the same result as obtained in part *a* have been obtained through use of the equation $\mu = np$? Tell why or why not.

c. Using the equation $\sigma = \sqrt{\Sigma [P(X) \cdot (X - \mu)^2]}$, find the standard deviation of this probability distribution.

d. Could the same result as obtained in part *c* have been obtained through use of the equation $\sigma = \sqrt{npq}$? Tell why or why not.

 7.25 Two different mail-order ads were run for several days on a split-run basis in a local newspaper. The purpose of the ads was to encourage people to return a coupon indicating an interest in having a salesperson call on them to tell about the advantages of buying a cemetary plot. One ad emphasized the peace of mind that comes with knowing you will not be a burden to your family. (We will call this appeal A.) The other ad emphasized the beauty and serenity of the surroundings (Appeal B). The probability distribution for each appeal follows:

Number of Responses per Day	Probability	
	Appeal A	Appeal B
0	.0	.1
1	.1	.1
2	.2	.4
3	.2	.2
4	.3	.0
5	.1	.0
6	.0	.1
7	.1	.1
Total	1.0	1.0

a. Compute the mean and standard deviation for Appeal A.

b. Compute the mean and standard deviation for Appeal B.

c. If the appeal with the higher mean is to be used hereafter, which appeal should be used?

d. If the appeal with the smaller standard deviation is to be used hereafter, which appeal should be used?

e. Could the mean and standard deviations of these probability distributions have been obtained using the formulas $\mu = np$ and $\sigma = \sqrt{npq}$? Tell why or why not.

7.26 If a certain population is known to be normally distributed with a mean of 65 and a standard deviation of 10,

a. What proportion of the values are between 65 and 80?

b. What proportion of the values are above 75?

c. What is the probability that an observation picked at random from this population is less than 60?

d. What is the probability that an observation picked at random from this population is less than 70?

e. Above what value will only 5% of the observations be?

7.27 Suppose it is known that the useful life of a certain brand of faucet washer is approximately normally distributed with a mean, μ, of 5 years and a standard deviation, σ, of 1 year.

a. What proportion of faucet washers have a useful life of over 10 years?

b. What proportion of faucet washers have a useful life of less than $4\frac{1}{2}$ years?

c. What length of useful life is exceed by 88% of the faucet washers?

7.28 The Falstaff Fragrance Company fills its cans of bayberry air freshener by machine. Each can bears a label stating that it contains 8 fluid ounces of air freshener. Suppose that the weights of the contents of the cans are known to be normally distributed with a standard deviation, σ, of 1.5 fluid ounces, a quantity over which management has no control, and a mean, μ, which can be set at whatever level management chooses.

   a. Assume that management wishes that no more than 1% of the cans of air freshener have weights of less than the 8 fluid ounces claimed. What would μ have to be in order for the company to meet this goal with no unnecessary wastage of the product?

   b. By how much could μ be reduced if management were to insist that no more than 3% of the cans have weights of less than the 8 fluid ounces claimed? How much of a *percentage reduction* in μ from that determined in connection with part *a* does your answer represent?

7.29 A large department store has 5000 accounts receivable for dollar amounts that are known to be normally distributed with a mean of $125 and a standard deviation of $30.

   a. How many of the accounts will be over $151?

   b. How many of the accounts will be between $65 and $185?

   c. Between what two dollar amounts will 4115 of the accounts lie assuming symmetry around the arithmetic mean?

   d. What will be the dollar amount of the account such that only 947 of the accounts are larger?

7.30 The average number of miles a Goodgo tire can be driven before the tread is too worn down to pass a safety inspection is 30,000, with a standard deviation of 1000 miles. The manufacturer wishes to offer a warranty for the tires, for replacing a tire if the tread wears down before a certain number of miles of travel. Assume that the number of miles of wear for this brand of tires is approximately normally distributed. For how many miles should the manufacturer warrant the tires to limit the number of replacements to no more than 4% of the tires sold?

7.31 The Laird Marketing Research Company hires part-time interviewers and compensates them according to the following incentive scheme: Interviewers are paid (1) $2 per interview for any number of interviews from 0 to 10 during a single day, (2) $3 per interview for any number of interviews between 10 and 15 in a single day, and (3) $4 per interview for any number of interviews from 15 on up. For example, if one of the workers were to complete 9 interviews, another 14 interviews, and still another 17 interviews in a single day, they would earn 9 times $2 = $18, 14 times $3 = $42, and 17 times $4 = $68, respectively. (*Note:* Although it would be a matter of practical importance whether the $3 rate started with the tenth or the eleventh interview on a given day and whether the $4 rate started with the fifteenth or the sixteenth interview on a given day, you need not concern yourself with this detail. Since you will be obtaining estimates based on use of the normal distribution, which is continuous, you should view 10 as both the upper limit of the 0-to-10 range and the lower limit of the 10-to-15 range for the number of interviews conducted; similarly, 15 should be treated as both the upper limit of the 10-to-15 range and the lower limit of the 15-and-over range for number of interviews conducted.)

   The daily number of interviews for Ms. Able and Mr. Baker for the most recent 100 working days are found to be approximately normally distributed with the following characteristics:

	Ms. Able	Mr. Baker
Mean	9	11
Standard deviation	2.5	3.5

a. On what percentage of the days did Ms. Able receive $2 per interview?
b. On what percentage of the days did Ms. Able receive $3 per interview?
c. On what percentage of the days did Mr. Able receive $4 per interview?
d. What is your estimate of the number of interviews completed by Ms. Able during the 100 working days under analysis? Explain how you arrived at your estimate.
e. What is your estimate of the amount of money earned by Ms. Able during the 100 working days under analysis? Explain how you arrived at your estimate.
f. What is your estimate of the amount of money earned by Ms. Able *per interview* during the 100 working days under analysis? Explain how you arrived at your estimate.

7.32 Refer to problem 7.31.
  a. On what percentage of the days did Mr. Baker receive $2 per interview?
  b. On what percentage of the days did Mr. Baker receive $3 per interview?
  c. On what percentage of the days did Mr. Baker receive $4 per interview?
  d. What is your estimate of the number of interviews completed by Mr. Baker during the 100 working days under analysis? Explain how you arrived at your estimate.
  e. What is your estimate of the amount of money earned by Mr. Baker during the 100 working days under analysis? Explain how you arrived at your estimate.
  f. What is your estimate of the amount of money earned by Mr. Baker *per interview* during the 100 working days under analysis? Explain how you arrived at your estimate.

7.33 Refer to problem 7.31.
  a. Comment on the fairness, or lack of it, of this system of compensation insofar as Ms. Able and Mr. Baker are concerned assuming: (1) The completion of a high average number of interviews per day is considered much more important than day-to-day consistency; (2) day-to-day consistency is considered much more important than the average number of interviews conducted per day.
  b. Which of the above considerations does the Laird Company management appear to consider more important? Explain.

7.34 The following classified ad appeared in a magazine:

Boy or girl?
What will baby be?

Famous psychic can tell from a snapshot or photo of full front view of mother-to-be. Must be after three months of pregnancy. Money-back guarantee. Send $10.00 with photo to . . .

[Name and box number given]

Assume that this "famous psychic" actually does return the $10 when he is wrong. Also assume that he has no psychic power and just guesses. Viewing net profit per guess (the possible values being $0 and $10) as the random variable of interest, determine his expected net profit per guess.

7.35 A grocery store buys eggs in about equal quantities from two suppliers, A and B. In the past, 2% of the eggs received from Supplier A and 1% of those received from supplier B have been broken. The grocery store manager has just opened a carton of eggs from an incoming shipment and found four broken eggs. Viewing the carton of eggs examined as a random sample of size $n = 12$ from the population of eggs supplied by either Supplier A or B, find the probability that the carton examined came from (1) Supplier A and (2) Supplier B. Also comment on the advisability of viewing the eggs in a single carton as a random sample from the supplier population.

7.36 Crate X contains 3 damaged and 21 acceptable oil filters, whereas Crate Y contains 5 damaged and 19 acceptable oil filters. If five random selections of a single filter are made, with replacement, from a crate picked at random and if two of the five selections are damaged, what is the probability that the crate from which the sample was drawn is
a. Crate X?
b. Crate Y?

7.37 A company sends its parcel post packages by Jiffy Express and Kwick Express. In the past, Jiffy has delivered 97% of the company's packages sent by it on time and Kwick has delivered 95% on time. On a particular day last week six packages were sent by one of these delivery services and two were late. Viewing the six packages as a random sample from the population of the company's packages sent by one or the other delivery service, determine the probability that the delivery service employed that day was
a. Jiffy
b. Kwick

7.38 The Ness Company has found from past experience that it should accept any shipment from the Elliott Company if the proportion of defectives is only .05 and reject it if the proportion of defectives is either .10 or .15. In the past, 60% of the shipments have contained 5% defective items, 30% of the shipments have contained 10% defective items, and 10% of the shipments have contained 15% defective items. Assume that the Ness Company inspectors select a random sample of ten items, with replacement, and find two defectives. What is the probability that
a. The shipment should be accepted?
b. The shipment should be rejected?

7.39 An automobile company estimates that 1% of its 1984 Panther Brand cars had a potentially serious defect in the front axle. If a random sample of 1000 Panther cars is selected, what is the probability that the defect will be found on
a. None of them?
b. Four of them?
c. At least 11 of them?
d. At most 15 of them?

7.40 An auditing firm believes that 5% of a firm's invoices contain errors. A sample of 12 invoices is selected. Assume that the auditing firm's estimate is correct and use the Poisson approximation procedure to determine the probability that
a. None contain errors.
b. All contain errors.
c. At least 8 contain errors.
d. No more than four contain errors.

7.41 Refer to problem 7.40.
a. Use a binomial formula or table to answer part *a* above.
b. Use a binomial formula or table to answer part *b* above.
c. Use a binomial formula or table to answer part *c* above.
d. Use a binomial formula or table to answer part *d* above.
e. If you worked problem 7.40, compare your two sets of results. Also, comment on the accuracy, or lack of it, of the probabilities obtained from the Poisson approximation procedure.

7.42 A hotel has 152 rooms. In the past, about 15% of the people who made reservations failed to show up. Therefore, for a particular Saturday night the hotel has made $152 + (152)(.15) \cong 175$ reservations. Find the probability that a room will be available for all who do show up.

7.43 Five percent of the squash rackets produced by the Huff Company contain at least one minor defect. If the Gale Brothers chain of sporting goods stores buys 200 rackets from this company, what is the probability that
    a. None will contain defects?
    b. At least 6 will contain defects?
    c. No more than 12 will contain defects?

 7.44 Glass mixing bowls produced by the Dyer Company are known to contain an average of one pit per every six square inches of glass. A particular mixing bowl has an estimated 24 square inches of glass. What is the probability that it will have
    a. No pits at all?
    b. No more than 5 pits?
    c. At least 10 pits?

 7.45 A plant supervisor estimates that 2.5 on-the-job accidents occur each month, on the average. What is the probability that next month
    a. There will be no accidents?
    b. There will be at least 4 accidents?
    c. There will be either 2 or 3 accidents?

 7.46 Refer to problem 7.45. Assume that an on-the-job accident has just occurred. What is the probability that
    a. At least one month will pass before the next accident?
    b. At least two months will pass before the next accident?
    c. No more than two months will pass before the next accident?

 7.47 Refer to problems 7.45 and 7.46. What assumptions must implicitly be made when working these problems?

 7.48 Airplanes arrive at an airport at the rate of 5 per hour, on the average. The conditions for use of Poisson probabilities are met.
    a. What is the probability that from 3 to 7 airplanes will arrive in a given hour?
    b. What is the probability that from 6 to 14 airplanes will arrive in a given two-hour period?

 7.49 Refer to problem 7.48. Assume that an airplane has just arrived. What is the probability that
    a. At least 12 minutes will elapse before the next airplane arrives?
    b. At least 24 minutes will elapse before the next airplane arrives?
    c. No more than 36 minutes will elapse before the next airplane arrives?

7.50 The number of cars (X) arriving per hour at the drive-up window of a fast-food restaurant is observed over a 24-hour period. The results are as shown below.

X	0	1	2	3	4	5	6	7
Frequency	0	5	7	5	3	2	2	1

    a. Assuming that the Poisson distribution is appropriate for describing these data, obtain an estimate of the parameter μ for a Poisson distribution.
    b. Why is the value you obtained in connection with part *a* of this problem only an estimate of μ?

7.51 Refer to problem 7.50. Based on your estimate of μ, what is the probability that during any given hour
    a. At least two cars will arrive at the drive-up window?
    b. No more than two cars will arrive at the drive-up window?
    c. No cars will arrive at the drive-up window?

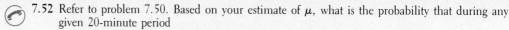 **7.52** Refer to problem 7.50. Based on your estimate of μ, what is the probability that during any given 20-minute period
   a. No cars will arrive at the drive-up window?
   b. Exactly one car will arrive at the drive-up window?
   c. No more than two cars will arrive at the drive-up window?

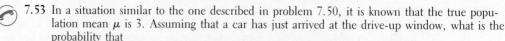 **7.53** In a situation similar to the one described in problem 7.50, it is known that the true population mean μ is 3. Assuming that a car has just arrived at the drive-up window, what is the probability that
   a. At least 20 minutes will elapse before the next car arrives?
   b. At least 40 minutes will elapse before the next car arrives?
   c. No more than 10 minutes will elapse before the next car arrives?

 7.54 A box contains eight very similar-appearing gaskets of which five are good and three are defective. Two gaskets are selected at random, without replacement, from the box. Develop the probability distribution of the number of defectives in this sample of $n = 2$.

 7.55 Five out of 12 sterling silver pitchers on sale at the Hines Restaurant Supply Company are flawless, while the remaining 7 have slight flaws. If a certain restaurant owner were to buy 4 of these pitchers, selecting them at random, what is the probability that at least 2 of those he selected would be flawless? What is the probability that at most 2 would be flawless?

 7.56 Crate X contains three damaged and seven acceptable oil filters, whereas Crate Y contains five damaged and five acceptable oil filters. If six oil filters are selected at random, without replacement, from a crate picked at random and if two of the six selections are damaged, what is the probability that the crate from which the sample was drawn is Crate X? Crate Y?

TAKE CHARGE*

7.57 Place 20 white and 5 blue poker chips in a bowl. Let the blue chips represent defective items and the white chips acceptable items in an industrial quality-control situation.

   Do the following exercises utilizing what you now know about the theoretical binomial probability distribution.
   a. Construct a theoretical binomial probability distribution for this situation if the number of defective items is the random variable of interest, assuming that $n = 5$ chips are to be selected, at random and with replacement, from the population of $N = 25$ chips. Briefly describe the shape of your theoretical binomial distribution.
   b. Compute the mean, or expected value, of your theoretical binomial probability distribution.
   c. Compute the standard deviation of your theoretical binomial probability distribution.
   d. Redo part *a* assuming that the random sample size is now $n = 10$.
   e. What has happened to the shape of the theoretical binomial probability distribution with the doubling of the sample size?
   f. Redo part *b* assuming that the random sample size is now $n = 10$.
   g. What effect, if any, did the doubling of the sample size have on the mean of your theoretical binomial probability distribution?
   h. Redo part *c* assuming that the random sample size is now $n = 10$.
   i. What effect, if any, did the doubling of the sample size have on the standard deviation of your theoretical binomial probability distribution?
   j. Redo part *a* assuming that the random sample size is now $n = 20$.
   k. What has happened to the shape of the theoretical binomial probability distribution with the quadrupling of the original sample size?

*Group participation is suggested.

l. Redo part *b* assuming that the random sample size is now $n = 20$.
m. What effect, if any, did the quadrupling of the sample size have on the mean of your theoretical binomial probability distribution?
n. Redo part *c* assuming that the random sample size is now $n = 20$.
o. What effect, if any, did the quadrupling of the sample size have on the standard deviation of your theoretical binomial probability distribution?

TAKE CHARGE*

7.58 Refer to problem 7.57. (You must have worked through problem 7.57 in order to complete this one.)

a. Select a sample of $n = 5$ chips, at random and with replacement, from your bowl containing $N = 25$ chips. Record the number of defectives (that is, blue chips) in the sample. Draw another sample of $n = 5$ chips and record the number of defectives. Repeat the same procedure until 20 random samples of 5 chips each have been selected.
b. Construct the empirical binomial probability distribution for this situation. That is, list the number of possible defective items in a sample (0, 1, 2, 3, 4, and 5) and the associated relative frequencies. Comment on how well or how poorly your empirical probability distribution compares with the theoretical one you developed in connection with part *a* of problem 7.57.
c. Compute the mean of your empirical binomial probability distribution and comment on how close it is to the theoretical mean obtained in connection with part *b* of problem 7.57.
d. Compute the standard deviation of your empirical binomial probability distribution and comment on how close it is to the theoretical standard deviation you obtained in connection with part *c* of problem 7.57.
e. For your empirical binomial probability distribution, determine what proportion of observations is associated with a range of X values between $\mu - \sigma$ and $\mu + \sigma$. Between $\mu - 2\sigma$ and $\mu + 2\sigma$. Between $\mu - 3\sigma$ and $\mu + 3\sigma$.
f. Repeat parts *a*, *b*, *c*, *d*, and *e*, using a sample size of $n = 10$.
g. In what way or ways, if any, did the doubling of the sample size affect (1) the appearance of your empirical binomial probability distribution, (2) the mean of your empirical binomial probability distribution, (3) the standard deviation of your empirical binomial probability distribution, and (4) the proportion of observations within the ranges $\mu \pm \sigma$, $\mu \pm 2\sigma$, and $\mu \pm 3\sigma$?
h. Repeat parts *a*, *b*, *c*, *d*, and *e*, using a sample of size $n = 20$.
i. In what way or ways, if any, did the quadrupling of the sample size affect (1) the appearance of your empirical binomial probability distribution, (2) the mean of your empirical binomial probability distribution, (3) the standard deviation of your empirical binomial probability distribution, and (4) the proportion of observations within the ranges $\mu \pm \sigma$, $\mu \pm 2\sigma$, $\mu \pm 3\sigma$?
j. Does the correspondence between the shape of your theoretical binomial probability distribution (constructed in connection with problem 7.57) and your empirically determined binomial probability distribution appear to improve as the sample size is increased? Elaborate.
k. Does the correspondence between the mean of your theoretical binomial probability distribution and your empirically determined binomial probability distribution appear to improve as the sample size is increased? Elaborate.
l. Does the correspondence between the standard deviation of your theoretical binomial probability distribution and the empirically determined binomial probability distribution appear to improve as the sample size is increased? Elaborate.

*Group participation is suggested.

m. Is there any evidence to suggest that your empirically determined binomial probability distribution becomes more nearly normal in shape as the sample size is increased? Elaborate.

Cases for Analysis and Discussion

7.1 THE "H. STILLMAN COMPANY": PART II

Reread Part I of this case (Case 4.1 at the end of Chapter 4).

In the beginning of the final paragraph, it is stated: "For a construction project begun in 1982, it was determined that to ensure a compressive strength of 5000 psi or higher in over 90% of the tests run, an "ideal" value for [average] compressive strength would be 5740 psi."

1. Explain how you believe the value of 5740 psi was arrived at.
2. What assumption(s), if any, underlies the 5740 psi value?

7.2 "TOMORROW AEROSPACE, INC."

The following case was written in 1983 by a manager of "Tomorrow Aerospace."

Over the past 20 years companies in the aerospace and government contracts businesses have had to increasingly bid contracts in a fixed price mode due to government pressure from previous cost overruns. The difficulty of accurately costing in such a high-technology business and subsequently bidding fixed prices was, at first glance, awesome. However, analysis of previous historical data by statistical experts showed clearly that errors tended to cancel out not only in the long run over several contracts but also in the short run by a particular division of costing and bidding which greatly increased the probability of making a profit on individual projects.

Typical cost items for projects undertaken at Tomorrow are as follows:

1. Development and engineering costs
2. Unit cost of production
3. Marketing costs
4. Sales volume

Price items are unit price and sales volume. Therefore,

Profit = Price × Volume − [Development cost + (Unit cost of production × Volume) + Marketing cost]

If the probability of each cost or price item listed above were accurately known, then a combined probability distribution could be derived for profit.

At Tomorrow, except under unusual circumstances, a probability of .166 is assigned to management's estimates for the cost and price items under the best combination of circumstances and under the worst combination of circumstances. A probability of .667 is assigned to their middle estimate. Our experts tell us that these probabilities are in rough accord with those that would be obtained if the estimates were continuous, rather than discrete, and followed a normal curve.

Assume that the distributions look like those in Exhibit A [Figure 7.21].

If all of the above are of major significance to profit *and* if they are reasonably independent of each other, then they can be combined into a profit probability curve which would

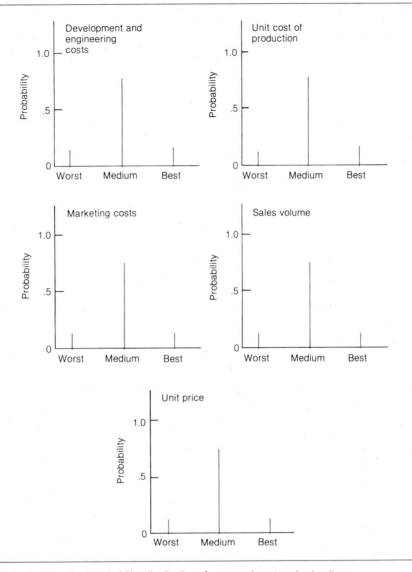

Figure 7.21 Hypothetical probability distributions for several cost and price items.

look like the very narrow one shown in Exhibit B [Figure 7.22], derived by combining all the cost and price probability distributions.

The theory is that the items in the profit equation are independent and, as a result, one is just as likely to overestimate as to underestimate any individual item. In total, there would be some compensation.

A critical feature of the model is the need to ensure that the people estimating the individual distributions do so as carefully as possible. To ensure this, some simple instructions must be followed. All estimators are asked to make explicit their scenarios for best

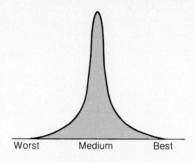

Worst Medium Best

Figure 7.22 Hypothetical probability distribution for profit obtained from combining the five probability distributions in Figure 7.21.

case, medium case, and worst case conditions. By requiring this, we hope to ensure that all the good things as well as the bad things that could happen are taken into account.

A sample of an actual case is now shown to display the technique: This case involves the development and sale of a new sophisticated component for a consumer product. Since this component is new to the marketplace, there was considerable uncertainty regarding what consumers would pay for it. Also, since it involves state-of-the-art technology there was much uncertainty over what it would cost to develop and produce it. The numbers in Exhibit C [Table 7.10] are the original ones used at the start of the project, and the numbers in boxes are the actual or currently anticipated actual ones.

Table 7.10 Data Used to Obtain Expected Profits for the "Tomorrow Aerospace" Case

Cost or Price Item	Estimate for		
	Best Conditions	Medium Conditions	Worst Conditions
Development and engineering cost (millions of dollars)	2	3.1 [4.1]	4.5
Unit cost of production (dollar cost of each)	4 [5.1]	6	10
Sales volume (millions)	4	2 [1.7]	1
Marketing cost (millions of dollars)	0.42	0.6	0.75 [0.96]
Price each (dollar cost of each)	10 [9.30]	9	7
Profit expected (millions of dollars)	21.58	2.3	(8.25 loss)
Profit actual (millions of dollars)		[2.08]	
Difference (millions of dollars)		0.22	

Table 7.11 Medium Estimate, Actual Price, and Effect of Profit for Individual Cost and Price Items

Cost or Price Item	Medium	Actual	Contribution to Profit When Other Items Are at Medium Estimate (Millions of Dollars)
Development and engineering cost (millions of dollars)	3.1	4.1	⟨1.0 loss⟩
Unit cost of production (dollars)	6.0	5.1	1.8
Sales volume (millions of units)	2	1.7	⟨1.8 loss⟩
Marketing cost (millions of dollars)	0.6	0.96	⟨0.36 loss⟩
Price (dollar cost)	9.0	9.3	0.6

An analysis of actual results shows that very few of the original estimates came in near the original medium case; however, there was sufficient compensation that the project as a whole came in very close to the original prediction for medium profit. Exhibit D [Table 7.11] shows the deviation from the medium case for each cost or price item (keeping all other items at medium).

As can be seen, each item by itself deviates substantially from medium and this deviation was a significant contributor (positive or negative) to profit.

1. Do you agree with the expectation expressed in Figure 7.22 that, if the probability of each cost and price item were accurate, the probability distribution for profit would be much more tightly clustered around the medium estimate than would be true of the individual items? If not, tell why not. If you do agree, explain why you think such a reduction in variation would occur.

2. To make certain you understand the process used, derive the expected profit of $2.3 million from the data given.

3. Write a scenario in which, with this method and a problem involving the introduction of a new product, the medium profit estimate and the actual profit eventually realized would be very far apart.

4. In general, is this a sensible procedure to use for estimating profit? Tell why you do or do not think so.

8 Sampling Distributions and Confidence Intervals

The loftier the building, the deeper must the foundation be laid.—Thomas à Kempis

THE

NORMAL

LAW OF ERROR

STANDS OUT IN THE

EXPERIENCE OF MANKIND

AS ONE OF THE BROADEST

GENERALIZATIONS OF NATURAL

PHILOSOPHY ◆ IT SERVES AS THE

GUIDING INSTRUMENT IN RESEARCHES

IN THE PHYSICAL AND SOCIAL SCIENCES AND

IN MEDICINE AGRICULTURE AND ENGINEERING ◆

IT IS AN INDISPENSABLE TOOL FOR THE ANALYSIS AND THE

INTERPRETATION OF THE BASIC DATA OBTAINED BY OBSERVATION AND EXPERIMENT

—W. J. Youden*

What You Should Learn from This Chapter

This is the last of three chapters devoted primarily to probability concepts. It is also the first chapter to introduce you to some applied inductive procedures (sections 8.10, 8.11, 8.20, and 8.21).

It is no exaggeration to state that this is the most important chapter in the book. I have found that the subject of sampling distributions, much more than any other, tends to divide students into two clearly identifiable groups. One group is made up of those who grasp, by whatever means, what the idea of sampling distribution is all about and learn well the strategic facts about sampling distributions. These students find later applied subjects relatively easy—even exciting—because they quickly recognize similarities existing among subjects which to members of the second group appear as different as night and day. The second group consists of those who never seem to get an inkling of what a sampling distribution is and why it matters; as a result, they tend to get stuck at the level of mechanically pushing numbers through formulas. As the formulas multiply, members of this group often find themselves overwhelmed. Fair warning?

Because of the special importance of this chapter, it has been organized quite differently from the others: It is made up of 22 sections. Each section pertains to a question which you must know the answer to and, where possible, the rationale behind the answer if you are to emerge from this chapter a member of the first, more enviable, group. The aim of this kind of organization is to spell out just exactly what it is about sampling distributions you need to know in order to understand and master later subjects. If you will make certain you know and, more important, understand the answers to all 22 questions before proceeding to the next chapter, you should experience few difficulties with later material.

*Published in the *American Statistician*, Vol. 4, April–May, 1950, p. 11. Reprinted by permission.

Unfortunately, the converse is also true.

When you have completed this chapter, you should be able to answer the following questions:

1. What, in general, is meant by the term *sampling distribution?*
2. How does a *sampling distribution* differ from a *sample distribution* or a *population distribution?*
3. What is a sampling distribution of $\overline{X}$?
4. How would one go about constructing a sampling distribution of $\overline{X}$?
5. What is the relationship between the mean of a sampling distribution of $\overline{X}$ and the mean of the parent population?
6. What is the relationship between the standard deviation of a sampling distribution of $\overline{X}$ and the standard deviation of the parent population?
7. What effect does sample size have on the standard deviation of a sampling distribution of $\overline{X}$?
8. What can be said about the shape of a sampling distribution of $\overline{X}$?
9. How might the important properties of a sampling distribution of $\overline{X}$ be harnessed to help us draw a deductive inference about the mean of a single random sample?
10. How might the important properties of a sampling distribution of $\overline{X}$ be harnessed to help us draw an inductive inference about the mean of the population?
11. In a problem concerned with estimating the population mean, how might the concept of sampling distribution of $\overline{X}$ help us determine the necessary sample size?
12. What is meant by the term *sampling distribution of $\overline{p}$?*
13. How would one go about constructing a sampling distribution of $\overline{p}$?
14. What is the relationship between the mean of a sampling distribution of $\overline{p}$ and the corresponding population proportion?
15. How might the standard deviation of a sampling distribution of $\overline{p}$ be obtained from knowledge of the population proportion?
16. What effect does sample size have on the standard deviation of a sampling distribution of $\overline{p}$?
17. What can be said about the shape of a sampling distribution of $\overline{p}$?
18. How might the important properties of a sampling distribution of $\overline{p}$ be harnessed to help us draw a deductive inference about the proportion for a single random sample?
19. How might the important properties of a sampling distribution of $\overline{p}$ be harnessed to help us draw an inductive inference about the proportion for the population?
20. In a problem concerned with estimating the population proportion, how might the concept of sampling distribution of $\overline{p}$ help us determine the necessary sample size?
21. May we safely conclude that the important properties of a sampling distribution of $\overline{X}$ or a sampling distribution of $\overline{p}$ also hold for the sampling distributions of all other sample statistics of possible interest?
22. In what way must our confidence interval procedures for the arithmetic mean be altered if the sample size is small?

You may wish to review topics that have already been discussed before proceeding. In particular, it may be helpful to skim section 1.6 of Chapter 1 on how to distinguish among description, deductive inference, and inductive inference; then section 3.2 of Chapter 3 on standard frequency distributions; section 4.3 of Chapter 4 on the arithmetic mean and section 4.10 of the same chapter on the standard deviation and variance; then proceed to section 7.2 of Chapter 7 on the arithmetic mean and standard deviation of random variables and section 7.6 of the same chapter, which discusses the normal distribution.

Now, on with the questions and answers.

8.1 What, in General, Is Meant by the Term *Sampling Distribution?*

In this chapter we limit our attention to sampling distributions for only two kinds of sample statistics—that is, only two kinds of descriptive measures of sample data—the arithmetic mean, $\overline{X}$, and the proportion, $\overline{p}$. However, some later material should be clearer if you will begin at once to think of the concept of *sampling distribution* as a very general one—one that can be applied to any kind of statistic of interest. Accordingly, let us designate some statistic $\overline{\theta}$ (read "theta bar") and define a sampling distribution of $\overline{\theta}$ as follows:

> A *sampling distribution of* $\overline{\theta}$ is a probability distribution made up of a listing of all possible values that $\overline{\theta}$ could have in a sampling situation (where the sample size, n, is the same for all samples) and the corresponding probabilities of occurrence.

Or, in somewhat more concrete, how-to terms:

> A sampling distribution of $\overline{\theta}$ is a probability distribution obtained by:
> 1. Listing observations comprising each different sample of size n that could be selected from the same population.
> 2. Computing the value of the statistic of interest, $\overline{\theta}$, for each of the samples listed in connection with Step 1.
> 3. Listing the various values of $\overline{\theta}$ obtained in Step 2 and the corresponding relative frequencies (to be viewed as probabilities).

This procedure is demonstrated for sample statistics $\overline{X}$ and $\overline{p}$ in sections 8.4 and 8.13, respectively.

8.2 How Does a Sampling Distribution Differ from a Sample Distribution or a Population Distribution?

It is of utmost importance to recognize that, in any situations where formal inductive inferences are to be drawn, three separate and distinct kinds of distributions exist. The three distributions are (1) the population distribution, (2) the sample distribution, and (3) the sampling distribution of a statistic of interest ($\overline{\theta}$ for now).

The Population Distribution

> The *population distribution* is a frequency distribution of population data. Like any other frequency distribution, it consists of classes into which the relevant observations have been organized along with the corresponding class frequencies.

The population distribution, if it were directly observable, would serve as a vehicle for computing descriptive measures, such as the arithmetic mean or the standard deviation. This distribution, however, is seldom directly observable for reasons cited in Chapter 2 (section 2.2).

The Sample Distribution

> The *sample distribution* is a frequency distribution of data representing *a single sample* from the population of interest.

Because a single sample is only a part of the population under study, the sample distribution will not have exactly the same shape or descriptive measurements as the population distribution. However, it is the sample distribution that we actually observe, not the population distribution. Given knowledge of a sample characteristic, it is often the task of the analyst to estimate the corresponding unknown population characteristic. This is where knowledge of the samp*ling* distribution of the relevant statistic proves useful.

The Sampling Distribution

The selection of a random sample from a population is a random experiment. The sample space for the experiment is determined by all possible samples of a given size n, which could be drawn from that population. The probability distribution associated with a specific descriptive measure of sample data $\bar{\theta}$ is the sampling distribution for that statistic. By way of elaboration, let us consider the sampling distribution of $\bar{X}$.

8.3 What Is a Sampling Distribution of $\bar{X}$?

> A *sampling distribution of* $\bar{X}$ is a sampling distribution associated with $\bar{X}$, the sample mean.

This definition is elaborated on in the next section.

8.4 How Would One Go About Constructing a Sampling Distribution of $\overline{X}$?

In practice, one wouldn't. The characteristics of sampling distributions are what is useful in real-world statistical investigations. A little reflection will reveal why it is fortunate that we can bypass the actual construction of sampling distributions in practice. One can readily conjure up situations in which so many different samples of possibly large size would have to be considered that the work of constructing a sampling distribution would require a great many years to complete. You, your children, and your grandchildren might have to devote entire lifetimes to the task.

Even though sampling distributions are not actually constructed in practice, the construction of some simple ones should be instructional if for no other reason than to give the concept a concreteness that it has not gained from our discussion up to this point.

Let us suppose, for the sake of illustration, that a certain sales manager has six salespeople under his supervision and that last year their sales were as shown in Table 8.1.

The sales data in Table 8.1 represent population values. If the population had consisted of a great many more sales values, such data would presumably be presented in frequency-distribution form. In such a form they would literally be a *population distribution*. Since for simplicity of illustration we are assuming that the population is made up of only six observations, we need not group them.

Now, let us say that a sample of two salespeople ($n = 2$) is selected at random for an in-depth analysis of sales performance. Suppose, for example, that the two sales values selected happen to be those achieved by Bardwell and Davis. In such a case, the values of 80 and 120 would be the observations comprising the single sample.

Let us suppose that the statistic of primary interest to the sales manager is the arithmetic mean $\overline{X}$, and that he wishes to construct a sampling distribution for this statistic. Clearly, the $\overline{X}$ value associated with the sample consisting of the sales achieved by Bardwell and Davis would be $\overline{X} = (80 + 120)/2 = 100$. But discovering that fact would be only a small part of the sales manager's task. To construct a sampling distribution of $\overline{X}$, he must calculate the $\overline{X}$ value associated with *every possible different sample of size 2 that could be selected from this population of six salespeople.* How many different sample combinations of size 2 are contained within a population of size 6? As we saw in Chapter 2 (section 2.7), this question can be answered readily through use of the formula for determining combinations. We hereafter denote the number of sample combinations by N^*. Thus, $N^* = N!/[n!(N - n)!] = 6!/(2!4!) = 15$. The specific pairs of salespeople whose combined sales would account for all

Table 8.1 Sales for a Population of Six Salespersons

Name of Salesperson	Last Year's Sales ($000)
Adams	60
Bardwell	80
Carpenter	100
Davis	120
Elsea	140
Farnum	160

15 possibilities are listed in Table 8.2 along with their sales values (from Table 8.1), their totals, and their arithmetic means.

If we examine the extreme right-hand column of Table 8.2, we notice that certain $\overline{X}$ values appear only once, whereas others appear two or even three times. To get a clearer impression of the relative frequency associated with each sample mean, we can collect these data and present them in the manner shown in Table 8.3.

Table 8.2 Sales Data for Fifteen Samples of Two Salespersons Which Could Be Obtained from a Specific Population of Six Salespersons

Sample Number	Salespersons Represented in Sample (for $n = 2$)	Sales ($000)	Total Sales	Sample Mean $\overline{X}$ (Total Sales ÷ 2)
1	Adams and Bardwell	60, 80	140	70
2	Adams and Carpenter	60, 100	160	80
3	Adams and Davis	60, 120	180	90
4	Adams and Elsea	60, 140	200	100
5	Adams and Farnum	60, 160	220	110
6	Bardwell and Carpenter	80, 100	180	90
7	Bardwell and Davis	80, 120	200	100
8	Bardwell and Elsea	80, 140	220	110
9	Bardwell and Farnum	80, 160	240	120
10	Carpenter and Davis	100, 120	220	110
11	Carpenter and Elsea	100, 140	240	120
12	Carpenter and Farnum	100, 160	260	130
13	Davis and Elsea	120, 140	260	130
14	Davis and Farnum	120, 160	280	140
15	Elsea and Farnum	140, 160	300	150

Table 8.3 Data from Table 8.2 Presented in Frequency-Distribution and Relative-Frequency (Probability)-Distribution Form

(1) Sample Mean $\overline{X}$	(2) Frequency f	(3) Relative Frequency (Probability) $P(\overline{X}) = f/N^*$
70	1	.0667
80	1	.0667
90	2	.1333
100	2	.1333
110	3	.2000
120	2	.1333
130	2	.1333
140	1	.0667
150	1	.0667
Total	$15 = N^*$	1.0000

Note: N^* represents the total number of different samples of size n.

The data in columns (1) and (3) of Table 8.3 constitute a sampling distribution of $\overline{X}$. (So, for that matter, do the data in columns (1) and (2) in combination. However, it will be convenient to speak henceforth of sampling distributions as if they were always probability distributions.) The same data are presented in graphic form in Figure 8.1, which dramatizes the fact that the sampling distribution of $\overline{X}$ in this case is a discrete distribution because, for such small values for N and n as used here, the number of possible different values $\overline{X}$ could have will be quite limited. Had it been possible to generate a much larger number of sample combinations, the number of different values of $\overline{X}$ would necessarily also be much greater and the sampling distribution would come to resemble more closely a continuous distribution. More is said about this feature of a sampling distribution of $\overline{X}$ in section 8.8.

Let us now consider some properties of this new kind of probability distribution with a view to determining whether any dependable generalizations can be made.

8.5 What Is the Relationship Between the Mean of a Sampling Distribution of $\overline{X}$ and the Mean of the Parent Population?

As we have seen, if the observations comprising every different sample of a given size n, which could be obtained from the same population, were listed and the mean $\overline{X}$ for every such sample computed, the $\overline{X}$ values thus determined would vary from sample to sample.

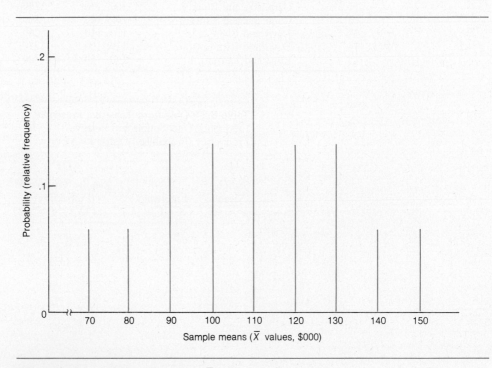

Figure 8.1 The sampling distribution of $\overline{X}$ for $N = 6$ and $n = 2$ presented in columns (1) and (3) of Table 8.3.

When we worked with X values in Chapter 4, we noted that when variation exists in a set of data, as it inevitably will, it is often helpful to determine a representative value that can give the analyst a quick overall feel for the data under analysis. We saw in Chapter 4 that one such measure is the mean obtained by $\overline{X} = \Sigma\, X/n$, if the data of interest represent a single sample, or $\mu = \Sigma\, X/N$ in the case of population data.

The mean of a sampling distribution of $\overline{X}$ is computed like any other mean—namely, by summing the observations of interest and dividing by the number of observations summed. Although the operations involved are not new, some special symbols must be introduced by which to convey the idea that a certain mean is the average of several sample means:

$$\mu_{\overline{X}} = \frac{\Sigma\, \overline{X}}{N^*} \tag{8.5.1}$$

where $\mu_{\overline{X}}$ is the mean of a sampling distribution of $\overline{X}$
$\overline{X}$ is the arithmetic mean of a single sample of size n
N^* is the number of different sample combinations of size n—and, hence, the number of different sample means altogether

Actually, in the discussion that follows we will not make direct use of Equation (8.5.1) even though this manner of expressing the procedure for determining $\mu_{\overline{X}}$ is useful because of its obvious similarity to the methods discussed in Chapter 4 for computing $\overline{X}$ and μ. Instead, we will make use of the formula for obtaining the mean, or expected value, of a probability distribution but with symbols modified appropriately to express the idea that we are determining the mean of $\overline{X}$ values rather than X values. The formula we will use is

$$\mu_{\overline{X}} = \Sigma\, [P(\overline{X})\overline{X}] \tag{8.5.2}$$

where $\overline{X}$ is the mean of a single sample of size n
$P(\overline{X})$ is the relative frequency (viewed as the probability) with which a specified value of $\overline{X}$ occurs in the entire sampling distribution of $\overline{X}$ values

We are now in a position to answer the question that introduced this section: What is the relationship between the mean of a sampling distribution of $\overline{X}$ and the mean of the parent population? The answer is

$$\mu_{\overline{X}} = \mu$$

which is read "the mean of the sampling distribution of $\overline{X}$ is equal to the mean of the population from which the various samples of size n are assumed to have been selected."

This correspondence between $\mu_{\overline{X}}$ and μ *always holds* regardless of the shape or size of the population, the value of μ, or the sample size.*

No attempt will be made here to prove this assertion. However, we can demonstrate, using data introduced earlier, that the alleged correspondence between $\mu_{\overline{X}}$ and μ does indeed hold. Let us begin by making use of the population data presented in Table 8.1 and the sampling distribution of $\overline{X}$ appearing in Table 8.3. Table 8.1 reveals that the relevant population values are 60, 80, 100, 120, 140, and 160 (in thousands of dollars). Therefore, using Equation (4.3.2), we learn that the population mean in this case is $(60 + 80 + 100 + 120 + 140 + 160)/6 = 110$.

Column (1) of Table 8.3 informs us that the $\overline{X}$ values for the 15 samples of size $n = 2$ which could be obtained from this population are (again, in thousands of dollars) 70, 80, 90, 100, 120, 130, 140, and 150, but that these $\overline{X}$ values do not occur with identical relative frequencies (column (3) of this table). Employing relative-frequency (probability) weights, we find that

$$\mu_{\overline{X}} = \Sigma\,[P(\overline{X})\overline{X}] = (.0667)(70) + (.0667)(80) + (.1333)(90) + (.1333)(100)$$
$$+ (.2000)(110) + (.1333)(120) + (.1333)(130) + (.0667)(140)$$
$$+ (.0667)(150) = 110$$

Thus, we see that both μ and $\mu_{\overline{X}}$ equal \$110.

WARNING

You should take careful note of the fact that it is the *mean of the sampling distribution of $\overline{X}$* which equals the population mean—*not the mean of a specific sample*. Occasionally, a particular $\overline{X}$ value will equal the corresponding μ value; much more often, $\overline{X} \neq \mu$ (a fact adequately demonstrated in Table 8.3). But $\mu_{\overline{X}}$ always equals μ.

8.6 What Is the Relationship Between the Standard Deviation of a Sampling Distribution of $\overline{X}$ and the Standard Deviation of the Parent Population?

As noted, in the preceding section, the $\overline{X}$ values comprising a sampling distribution of $\overline{X}$ will differ from one another, that is, they will exhibit variation. Whenever a set of numerical values—whether they are X values, $\overline{X}$ values, or something else—exhibits variation, the amount of variation can be measured by the standard deviation. We saw in Chapter 4 that the standard deviation of a finite population of X values is computed by

---

*When the mean of a sampling distribution of a statistic is equal to the corresponding population parameter for all values of n, we say that the statistic is an *unbiased estimator* of the parameter because, on the average, the statistic neither overestimates nor underestimates the true parameter value.

$$\sigma = \sqrt{\frac{\Sigma (X - \mu)^2}{N}}$$

and the standard deviation for a single sample from that population by

$$s = \sqrt{\frac{\Sigma (X - \overline{X})^2}{n - 1}}$$

It should come as no surprise that the standard deviation of a sampling distribution of $\overline{X}$—*hereafter referred to as the standard error of the mean and symbolized* $\sigma_{\overline{X}}$—can be calculated in much the same way—namely, by

$$\sigma_{\overline{X}} = \sqrt{\frac{\Sigma (\overline{X} - \mu_{\overline{X}})^2}{N^*}} \qquad (8.6.1)$$

where $\sigma_{\overline{X}}$ is the standard error of the mean, which is to say the standard deviation of the sampling distribution of $\overline{X}$
$\overline{X}$ is the arithmetic mean of a single sample of size n
$\mu_{\overline{X}}$ is the mean of the sampling distribution of $\overline{X}$
N^* is the number of different samples of size n—and, hence, the number of different sample means—altogether

Of course, since we will be viewing the sampling distribution as a probability distribution, it will be appropriate to use the following alternative formula for $\sigma_{\overline{X}}$:

$$\sigma_{\overline{X}} = \sqrt{\Sigma \left[P(\overline{X})(\overline{X} - \mu)^2 \right]} \qquad (8.6.2)$$

where $P(\overline{X})$ is the relative frequency of occurrence (or probability of occurrence) of a specified $\overline{X}$ value
(Recall that $\mu_{\overline{X}} = \mu$; therefore, one can be freely substituted for the other.)

The above is simply the formula (introduced in section 7.2) for determining the standard deviation of a discrete probability distribution; the only difference is that a few symbols have been changed to convey the idea that the values whose variation we wish to measure are $\overline{X}$ values, rather than X values. Use of Equation (8.6.2) on data presented in Table 8.3 is demonstrated in Table 8.4.

The value of $\sigma_{\overline{X}}$, which was found to be 21.61 in Table 8.4, is appreciably smaller than the standard deviation of the population of X values. The latter, σ, computed at the bottom of Table 8.5, is seen to be 34.16.

The $\sigma_{\overline{X}}$ value will always be smaller than σ whenever the sample size n exceeds 1, which, of course, it always will in a real-world sample study. However, $\sigma_{\overline{X}}$ and σ are related in a mathematically expressible manner—namely,

$$\sigma_{\overline{X}} = \frac{\sigma}{\sqrt{n}} \cdot \sqrt{\frac{N-n}{N-1}} \qquad (8.6.3)$$

We can readily demonstrate this equality by using our example of the salespeople. We begin by dividing the population standard deviation—$\sigma = 34.16$—by the square root of the sample size—$n = 2$, as called for in Equation (8.6.3). We find that $\sigma/\sqrt{n} = 34.16/\sqrt{2} = 24.15$. The value 24.15 would be the true standard error of the mean, $\sigma_{\overline{X}}$, if the population

Table 8.4 Calculation of the Standard Error of the Mean $\sigma_{\overline{X}}$

(Data from Table 8.3)

Sample Mean ($000) $\overline{X}$	Relative Frequency (Probability) $P(\overline{X})$	$\mu_{\overline{X}} = \mu$	$\overline{X} - \mu$	$(\overline{X} - \mu)^2$	$P(\overline{X})(\overline{X} - \mu)^2$
70	.0667	110	−40	1600	106.72
80	.0667	110	−30	900	60.03
90	.1333	110	−20	400	53.32
100	.1333	110	−10	100	13.33
110	.2000	110	0	0	0.00
120	.1333	110	10	100	13.33
130	.1333	110	20	400	53.32
140	.0667	110	30	900	60.03
150	.0667	110	40	1600	106.72
Total	1.0000				466.80

$$\sigma_{\overline{X}} = \sqrt{\Sigma \, [P(\overline{X})(\overline{X} - \mu)^2]} = \sqrt{466.80} = \$21.61$$

Table 8.5 Calculation of the Population Standard Deviation σ

(Data from Table 8.1)

Salespersons	Sales ($000) X	μ	$X - \mu$	$(X - \mu)^2$
Adams	60	110	−50	2500
Bardwell	80	110	−30	900
Carpenter	100	110	−10	100
Davis	120	110	10	100
Elsea	140	110	30	900
Farnum	160	110	50	2500
Total	660			7000

$$\sigma = \sqrt{\frac{\Sigma \, (X - \mu)^2}{N}} = \sqrt{\frac{\$7000}{6}} = \$34.16$$

of interest were infinitely large. Since, however, the population under study is finite and quite small at that, 24.15 is somewhat too high. But it can be reduced by the appropriate amount through multiplication by $\sqrt{(N - n)/(N - 1)}$, a device introduced in Chapter 7 and known as the *finite correction factor*. To apply the finite correction factor to the present example, we substitute values into the expression $\sqrt{(N - n)/(N - 1)}$ as follows: $\sqrt{(6 - 2)/(6 - 1)} = .894427$. Multiplying 24.15 by .894427 gives 21.60, which is, aside from a small rounding error, the same value obtained directly from the sample means themselves in Table 8.4.

Frequently the relationship between $\sigma_{\overline{X}}$ and σ will be expressed as in Equation (8.6.3), but without the finite correction factor—that is,

$$\sigma_{\overline{X}} = \frac{\sigma}{\sqrt{n}} \tag{8.6.4}$$

A little reflection will reveal the justification, under many circumstances, for this convenient simplification. The larger N becomes relative to n, the closer $\sqrt{(N - n)/(N - 1)}$ will come to equaling 1. Many statisticians use the working rule that if N/n is no less than 20, the finite correction factor may be ignored. In the present example, the ratio N/n is considerably less than 20. However, in most of the upcoming sections of this chapter and in Chapters 9 and 10, we assume for convenience of exposition—but quite realistically—that in practice the population size will be at least 20 times as great as the sample size; this assumption will permit us to ignore the finite correction factor.

8.7 What Effect Does Sample Size Have on the Standard Deviation of a Sampling Distribution of $\overline{X}$?

The relationship between sample size n and the variability of the corresponding sampling distribution of $\overline{X}$, as measured by $\sigma_{\overline{X}}$, can be shown most readily by referring to Equation (8.6.4) which states $\sigma_{\overline{X}} = \sigma/\sqrt{n}$. Clearly, the larger the sample size n, the more concentrated will be the $\overline{X}$ values around $\mu_{\overline{X}} = \mu$. This fact holds important implications for studies calling for formal inductive inferences. For example, as will be demonstrated when the subject of confidence intervals is considered in section 8.10, the use of a relatively large sample size leads to a more precise estimate of the true, but unknown, population mean than does the use of a smaller sample size.

However, Equation (8.6.4) also reveals that one will encounter some difficulty in reducing the size of $\sigma_{\overline{X}}$ by increasing the sample size, the reason being that $\sqrt{n}$, not n itself, is the denominator of $\sigma_{\overline{X}}$. Hence, a quadrupling of n results in only a 50% reduction in $\sigma_{\overline{X}}$.

Two more points concerning the size of $\sigma_{\overline{X}}$ are in order at this time: First, the amount of variation in the population, as measured by σ, also affects the size of $\sigma_{\overline{X}}$. Because $\sigma_{\overline{X}} = \sigma/\sqrt{n}$, we observe that for a given sample size n, we obtain greater precision when estimating the value of the true but unknown population mean μ, when the standard deviation of the population σ is small than when it is large.

Second, the amount of variation in the sampling distribution of $\overline{X}$, contrary to what common sense suggests, is not generally a function of population size, N. Keep in mind that for infinite populations and finite populations at least 20 times as large as the sample size, the standard error of the mean is obtained by $\sigma_{\overline{X}} = \sigma/\sqrt{n}$, which says that $\sigma_{\overline{X}}$ is determined by the standard deviation of the population and the sample size—not the population size.

8.8 What Can Be Said About the Shape of a Sampling Distribution of $\overline{X}$?

In our example about the population of six salespeople, we noted that only 15 different sample combinations of size $n = 2$ could be obtained from this rather small population. In this example, values of N and n—and, hence, the number of possible sample combinations—were kept small intentionally so that the idea of "sampling distribution of $\overline{X}$ could be illustrated simply. However, in real-world sampling problems the number of sample combinations of a given size contained within a population is usually very large. This fact motivates us to ask whether any generalizations can be made about the *appearance* of a sampling distribution of $\overline{X}$.

Without a doubt one of the most important theorems in all of statistics is the *central limit theorem*. Indeed, the central limit theorem tells us so much about the shape of a sampling distribution of $\overline{X}$ that *it permits us to bypass the actual construction of a sampling distribution of $\overline{X}$ in real-world sample investigations.*

> The *central limit theorem* states that, regardless of the shape of the population, the sampling distribution of $\overline{X}$ is approximately normal if the random sample size n is sufficiently large.

This theorem applies to infinite as well as to finite populations. It is a little unfortunate that we cannot easily state in specific terms that have general validity just how large n must be to ensure that the corresponding sampling distribution of $\overline{X}$ will be approximately normal. That depends largely on the degree of skewness of the population; the more highly skewed the population distribution, the larger the value of n must be to ensure close correspondence with the normal curve. Still, the theorem has an almost astonishing degree of applicability when we consider the vast variety of shapes that populations can have. Moreover, for many population shapes a very small value of n—not uncommonly a value as small as 4 or 5—is sufficient to produce an approximately normal sampling distribution of $\overline{X}$. This extremely important theorem is illustrated for a highly skewed population in Figure 8.2.

When the population sampled is itself normally distributed, the central limit theorem as stated above need not be employed. Instead, we may rely on a related theorem that states:

> If the population sampled is normal, the sampling distribution of $\overline{X}$ is exactly normal for any sample size.

The implications of these two theorems are profound indeed. We see that a sampling distribution of $\overline{X}$ is either exactly normal, if the population sampled is normal, or approxi-

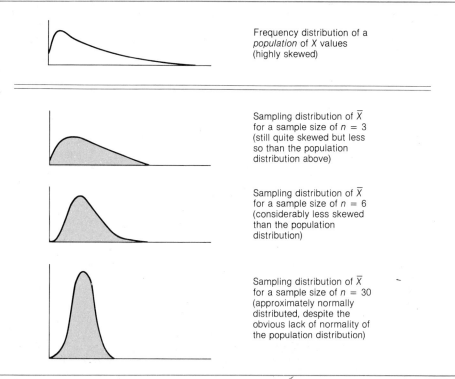

Frequency distribution of a *population* of X values (highly skewed)

Sampling distribution of $\overline{X}$ for a sample size of $n = 3$ (still quite skewed but less so than the population distribution above)

Sampling distribution of $\overline{X}$ for a sample size of $n = 6$ (considerably less skewed than the population distribution)

Sampling distribution of $\overline{X}$ for a sample size of $n = 30$ (approximately normally distributed, despite the obvious lack of normality of the population distribution)

Figure 8.2 Demonstration of the central limit theorem, showing the tendency for the sampling distribution of $\overline{X}$ to become increasingly normal in shape as the sample size n increases, despite the obvious absence of normality in the corresponding population of X values.

mately normal if the population sampled is not normal but the sample size n is reasonably large. Because we will not ordinarily know the shape of the population being sampled, the central limit theorem is especially important since it permits us to use the table of areas under the standard normal curve (Appendix Table 2) despite our ignorance of the population shape. Figure 8.3 attempts to emphasize the point that neither the population distribution nor the distribution of a specific sample need be normal to justify use of the normal distribution. *The distribution that we can usually count on to be normal is the sampling distribution of $\overline{X}$,* an entity which will not even exist in reality—only in our minds—but whose nonexistence need not disturb us because we can use it just as effectively as if it actually existed. Not many disciplines can boast such an accommodating tool.

8.9 How Might the Important Properties of a Sampling Distribution of X Be Harnessed to Help Us Draw a Deductive Inference About the Mean of a Single Random Sample?

Before proceeding to answer and illustrate this question, let us briefly summarize in tabular form (Table 8.6) what we now know about the characteristics of the three kinds of distribu-

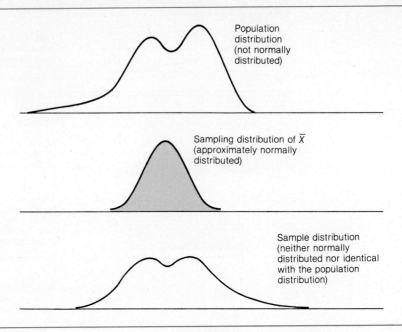

Figure 8.3 We depend on the sampling distribution of $\overline{X}$ to be approximately normally distributed—not the population distribution nor the sample distribution.

Table 8.6 Summary of Properties of Three Kinds of Distributions Used in Inductive Statistics

	Population	Sample	Sampling Distribution of $\overline{X}$
Mean	$\mu = \dfrac{\Sigma X}{N}$	$\overline{X} = \dfrac{\Sigma X}{n}$	$\mu_{\overline{X}} = \Sigma\,[P(\overline{X})\overline{X}] = \mu$
Standard deviation	$\sigma = \sqrt{\dfrac{\Sigma\,(X-\mu)^2}{N}}$	$s = \sqrt{\dfrac{\Sigma\,(X-\overline{X})^2}{n-1}}$	$\sigma_{\overline{X}} = \sqrt{\Sigma\,[P(\overline{X})(\overline{X}-\mu)^2]}$ $= \dfrac{\sigma}{\sqrt{n}}\ \text{ or }\ \dfrac{\sigma}{\sqrt{n}}\cdot\sqrt{\dfrac{N-n}{N-1}}$ If (1) population is infinite or (2) population is finite but N is at least 20 times as great as the sample size n — If population is finite and N is not at least 20 times as great as the sample size n
Type of distribution	May be of any type	May be of any type	Approximately normal when the sample size n is reasonably large

tions present in any situation calling for formal inductive inferences about a population mean—the population distribution, the sample distribution, and the sampling distribution of $\overline{X}$.

We will now illustrate by example how the important properties of a sampling distribution of $\overline{X}$ may be put to use.

ILLUSTRATIVE CASE 8.1

The Trent and Travis Company has 5000 accounts receivable averaging $350 each—that is, μ = $350—with a standard deviation σ of $200. Management is thinking of selecting a sample of n = 100 accounts for analysis and is curious to know what the probability is that the arithmetic mean $\overline{X}$ of the sample selected will be in the range of $300 to $400.

You should note that in this illustrative case we have not been told the shape of the population of accounts receivable; such information, fortunately, is not needed to solve the problem. Nor are we required to make any assumption about the shape of the specific *sample distribution*. We may work exclusively through the theoretical sampling distribution of $\overline{X}$ which, according to the central limit theorem, we may confidently assume is distributed approximately normally. The situation faced in the Trent and Travis case is illustrated in Figure 8.4, which shows the normally distributed sampling distribution of $\overline{X}$ for n = 100 accounts.

WARNING

Do not interpret the normal curve in Figure 8.4 as either the population distribution or the sample distribution, neither of which is shown in this figure. The normal curve shown here is an approximation of the sampling distribution of $\overline{X}$.

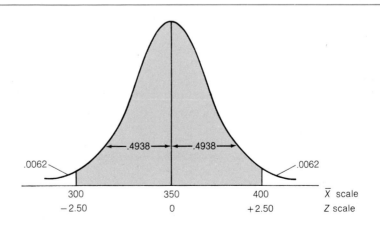

Figure 8.4 The approximately normal sampling distribution of $\overline{X}$ for a sample size of n = 100 used to solve the problem posed in the Trent and Travis case (Illustrative Case 8.1).

In Figure 8.4 two scales are shown beneath the normal curve—an $\overline{X}$ scale and a Z scale. You will recall that in Chapter 7 we defined Z to be $(X - \mu)/\sigma$. However, this way of expressing Z suggests that it is the X values that are approximately normally distributed. In our present problem, *it is the $\overline{X}$ values that are approximately normally distributed*. Therefore, we restate the Z formula and have it read $Z = (\overline{X} - \mu_{\overline{X}})/\sigma_{\overline{X}}$. Now, how do we determine $\mu_{\overline{X}}$ and $\sigma_{\overline{X}}$?

First, we know that the mean $\mu_{\overline{X}}$ of the sampling distribution of $\overline{X}$ is the same as μ, the mean of the population sampled. Hence, $\mu_{\overline{X}} = \$350$.

Second, we know that the standard deviation $\sigma_{\overline{X}}$ of the sampling distribution of $\overline{X}$, or the standard error of the mean as it is more briefly referred to, is equal to the standard deviation σ of the population divided by the square root of the sample size n. (Notice that we have dispensed with the finite correction factor because $N/n = 50$, which is substantially greater than the cutoff value of 20.) Therefore, $\sigma_{\overline{X}} = \sigma/\sqrt{n} = 200/\sqrt{100} = \$20$

We may reason that $Z = (\overline{X} - \mu_{\overline{X}})/\sigma_{\overline{X}} = (\overline{X} - \mu)/(\sigma/\sqrt{n})$ and that the Z equivalents of $\overline{X}_1 = \$300$ and $\overline{X}_2 = \$400$ are, therefore,

$$Z_1 = \frac{\overline{X}_1 - \mu}{\dfrac{\sigma}{\sqrt{n}}} = \frac{300 - 350}{20} = -2.50 \quad \text{and} \quad Z_2 = \frac{\overline{X}_2 - \mu}{\dfrac{\sigma}{\sqrt{n}}} = \frac{400 - 350}{20} = +2.50$$

Consulting Appendix Table 2, we learn that the area under the normal curve associated with the range of Z values running from 0 to $+2.50$ is .4938. This same area, of course, is associated with the range of Z values running from 0 to -2.50. Thus, .4938 times 2 = .9876 is the probability that a single random sample of size 100 selected at random from this population of 5000 accounts receivable will have a mean value $\overline{X}$ between \$300 and \$400.

8.10 How Might the Important Properties of a Sampling Distribution of $\overline{X}$ Be Harnessed to Help Us Draw an Inductive Inference About the Mean of the Population?

The problem worked in the preceding section is admittedly one having no direct practical importance, the reason being that the population parameters μ and σ were assumed to be known. In practice, they would not be known. Still, this example does illustrate something about how we might employ the concept of sampling distribution of $\overline{X}$ to take the inductive step from known sample mean to unknown population mean. After all, if the sampling distribution of $\overline{X}$ will allow us to reason from population to sample, why should it not be just as useful as an aid to reasoning from sample to population?

Let us turn the Trent and Travis case upside down and have it read as follows:

ILLUSTRATIVE CASE 8.2

An auditor engaged by the Trent and Travis Company is concerned about the number of and the average size of accounts receivable. One piece of information he seeks is an estimate of the true but unknown average amount for the *total population* of accounts receivable. Toward this end, he selects a simple random sample of size $n = 100$ and finds the sample mean $\overline{X}$ to be $320 and the sample standard deviation s to be $198. He wishes to make use of an *interval estimate* and to be *95% confident* that the interval obtained includes the true value of the population mean μ.

How might the auditor in this illustrative case proceed to estimate μ? One could argue, of course, that he already has such an estimate. He could simply assert that $\mu = \$320$—the same value as the observed sample mean $\overline{X}$. If he were to use such an approach, he would be making what statisticians call a *point estimate*.

> A *point estimate* is an estimate of an unknown population parameter in which a single number—usually the known value of the corresponding sample statistic—is asserted to be the value of the parameter.

Point estimates obtained in such a manner have two conspicuous advantages: (1) The estimate is very easily obtained, and (2) the estimate is extremely precise. The principal disadvantage, on the other hand, is a potentially devastating one: *Point estimates are usually wrong!* That is, the true population mean μ in our example is almost certainly *not* $320. To claim that it is $320 is to invite error because, as we have seen, the arithmetic mean $\overline{X}$ of a single random sample is seldom equal to the true population mean. Moreover, there is no available objective method that will provide the auditor with a value representing the probability that $320 is correct.

Fortunately, the auditor can employ an alternative approach—one which utilizes the properties of the theoretical sampling distribution of $\overline{X}$ and, as a result, permits him to assign a definite probability to his being right in his estimate. Because the auditor cannot correctly expect each sample mean to equal the true but unknown population mean, it would seem prudent for him to allow some leeway for error by viewing the population mean as having some value lying within a limited range of values. Whereas he is practically certain that the mean of the population is not $320, he might be almost certain that the population mean is some value between, say, $285 and $355. Such a range of values would constitute a *confidence interval*.

> A *confidence interval* is a range of values thought to include the true value of the parameter of interest.

An *interval estimate*, as this approach to parameter estimation is called, is obviously less precise than a point estimate but is much more likely to be correct. In addition, when certain prescribed procedures are followed, it is possible to know—even control—the probability that the limits of the confidence interval will include the true value of the parameter. This probability is called the *confidence coefficient*.

The *confidence coefficient* is the degree of certainty, expressed as a probability, that the process described below for obtaining the confidence interval will result in a pair of limits surrounding the true but unknown value of the parameter of interest.

To realize this sometimes most important advantage, we will follow a straightforward procedure.

First, in any confidence interval problem, it is wise to begin with an inventory of known facts—not only of the sample data but of the relevant sampling distribution as well. With respect to the Trent and Travis case (the version described in Illustrative Case 8.2), what facts do we know with certainty? We know:

1. The arithmetic mean $\overline{X}$ of the single random sample selected is $320.
2. The sample standard deviation s is $198.
3. The sample size n is 100.
4. Since it is the population mean μ that the auditor wishes to estimate, the theoretical sampling distribution of $\overline{X}$ is the appropriate one to employ.
5. The mean $\mu_{\overline{X}}$ of the theoretical sampling distribution of $\overline{X}$ is equal to the mean of the parent population μ.
6. The standard deviation of the theoretical sampling distribution of $\overline{X}$ is equal to the population standard deviation σ, divided by the square root of the sample size n.
7. The shape of the theoretical sampling distribution of $\overline{X}$ can be approximated by a normal curve.
8. The confidence coefficient—as set by the auditor himself—is .95.

Armed with this information, we may proceed to the second step—construction of a confidence interval. Keep in mind that the sampling distribution of $\overline{X}$ for $n = 100$ will be approximately normal. Moreover, if all the normally distributed $\overline{X}$ values were converted into their Z-value equivalents, the resulting Z values would necessarily also be distributed approximately normally. According to Appendix Table 2, the Z values associated with the central 95% of the normal curve are -1.96 and $+1.96$. We may, therefore, say that, if a specific randomly selected sample mean $\overline{X}$ were converted into its corresponding Z value, then $P(-1.96 \leq Z \leq +1.96) = .95$. We have already noted that, when working with a theoretical sampling distribution of $\overline{X}$, a value of Z is obtained by

$$Z = \frac{\overline{X} - \mu}{\dfrac{\sigma}{\sqrt{n}}}$$

Therefore, we may say

$$P\left(-1.96 \leq \frac{\overline{X} - \mu}{\dfrac{\sigma}{\sqrt{n}}} \leq +1.96\right) = .95$$

Let us now rearrange the expression within parentheses to see whether we can develop a new expression describing a procedure for obtaining a confidence interval for μ while leaving the probability of .95 undisturbed.

We begin with

$$-1.96 \leq \frac{\overline{X} - \mu}{\frac{\sigma}{\sqrt{n}}} \leq +1.96 \qquad (8.10.1)$$

Multiplying each term in this inequality by $\sigma/\sqrt{n}$, we get

$$-1.96 \frac{\sigma}{\sqrt{n}} \leq \overline{X} - \mu \leq +1.96 \frac{\sigma}{\sqrt{n}} \qquad (8.10.2)$$

Next, we subtract $\overline{X}$ from each term, multiply through by -1, do a little rearranging, and end up with the expression

$$\overline{X} - (1.96)\left(\frac{\sigma}{\sqrt{n}}\right) \leq \mu \leq \overline{X} + (1.96)\left(\frac{\sigma}{\sqrt{n}}\right) \qquad (8.10.3)$$

which says, in effect, "If we subtract the product 1.96 times the standard error of the mean, $\sigma_{\overline{X}} = \sigma/\sqrt{n}$, from the sample mean $\overline{X}$ and add the same product to $\overline{X}$, the results may be viewed as the limits of a range of values that would be expected, with a probability of .95, to include the value of the population mean μ."

Let us substitute known quantities into Equation (8.10.3) to obtain the confidence interval sought. We have

$$320 - (1.96)\left(\frac{200}{\sqrt{100}}\right) \leq \mu \leq 320 + (1.96)\left(\frac{200}{\sqrt{100}}\right)$$

$$= \$280.80 \leq \mu \leq \$359.20$$

Thus, we may say that we are 95% confident that the population mean has a value somewhere between 280.80 and 359.20. If it were not for one unfortunate feature of Equation (8.10.3), the preceding is just about all we would need to say about how to obtain a confidence interval for estimating μ. But alas, the procedure described in Equation (8.10.3) requires that we know the value of the population standard deviation σ—generally speaking, an unrealistic requirement. Clearly, if our goal is to estimate the population mean, we will seldom be in any position to know the population standard deviation. Hence, σ must be replaced with an *estimate*—namely, the standard deviation of the sample s. *If the sample size n is reasonably large*, not only will the sample standard deviation tend to approximate the population standard deviation but also any discrepancy between σ and s will be substantially stripped of practical importance by virtue of our dividing s by $\sqrt{n}$. (The ratio $s/\sqrt{n}$ will frequently be denoted $s_{\overline{X}}$ and called the *estimated standard error of the mean*.) Of course, we pay a small price for the practical advantages realized from this convenient substitution: Whereas up to this point we have spoken of being "95% confident" that our confidence interval includes the true but unknown population mean, the best we can now say is that we are "*approximately* 95% confident."

In univariate confidence interval problems, it is customary to refer to a sample of size $n > 30$ as "large" and to one of size $n \leq 30$ as "small." This cutoff value of 30 is really quite arbitrary. Still, since our present problem concerns a proposed sample of size $n = 100$, it obviously qualifies as "large"; therefore, s may be substituted for σ. What would we do with a problem like that under discussion if the sample size were small? We will address this question in section 8.22.

For the present problem, we replace σ with s and Equation (8.10.3) with

$$\overline{X} - (1.96)\left(\frac{s}{\sqrt{n}}\right) \le \mu \le \overline{X} + (1.96)\left(\frac{s}{\sqrt{n}}\right) \qquad (8.10.4)$$

Accordingly,

$$320 - (1.96)\left(\frac{198}{\sqrt{100}}\right) \le \mu \le 320 + (1.96)\left(\frac{198}{\sqrt{100}}\right)$$

$$= \$281.19 \le \mu \le \$358.81$$

Thus, he asserts, with a probability of *approximately* .95 of being correct, that this interval includes the true population mean.

Some elaboration on the meaning of the confidence coefficient is in order at this point. In any given problem, the parameter we are attempting to estimate either does or does not have a value that is between the limits of the confidence interval obtained. The probability does not attach itself to the parameter but instead to the procedure used. What we mean when we say we are "95% confident" is that, if we were to draw a large number of samples of the same size from the same population and follow the same procedures as described above each time when constructing the confidence interval, we could expect to get an interval that would include the parameter 95 times out of 100 and an interval that "misses" the true parameter the other 5 times out of 100. This interpretation is illustrated in Figure 8.5. An analogy might help. Think of the stake in a game of horseshoes as the parameter of interest. The stake does not change its location, jumping in front of the pitched horseshoe, when necessary for a ringer. It is the horseshoe which moves. Obviously some pitches will hit and some will miss. Hence, we speak of the probability that a specific thrown horseshoe will surround the stationary target—or that a specific confidence interval will surround the fixed population parameter.

Up to now we have made use of a confidence coefficient of .95. However, this degree of confidence may or may not be satisfactory in a specific case; that is a matter of personal judgment and is greatly dependent on the distastefulness of the consequences of failing to "ring" the unknown parameter. Fortunately, the confidence coefficient may be set at a more rigorous (i.e., a higher probability) value or at a less rigorous value. Let us suppose, for example, that the auditor in our illustrative case believes he should be approximately 99% confident. That is his privilege. The only change in procedure would be that of multiplying $s_{\overline{X}} = s/\sqrt{n}$ by a somewhat larger Z value. We note that one-half of .99 is .4950 and consult Appendix Table 2 to determine the Z values associated with this probability value. Reading across to the margins of this table, we find that Z could be either 2.57 or 2.58. We will arbitrarily split the difference and use 2.575. We calculate

$$320 - (2.575)\left(\frac{198}{\sqrt{100}}\right) \le \mu \le 320 + (2.575)\left(\frac{198}{\sqrt{100}}\right)$$

$$= \$269.02 \le \mu \le \$370.98$$

We may assert, with a probability of approximately .99 of being correct, that this interval includes the true population mean. Notice that the limits are now further apart than when a .95 confidence coefficient was used. A moral can be drawn from this: The surer one is, the less he has to be sure about. In other words, for a given sample size, confidence and precision of estimate work at cross purposes. If an analyst uses a relatively large confidence coefficient, he or she will get a relatively large confidence interval. The only way to achieve both a

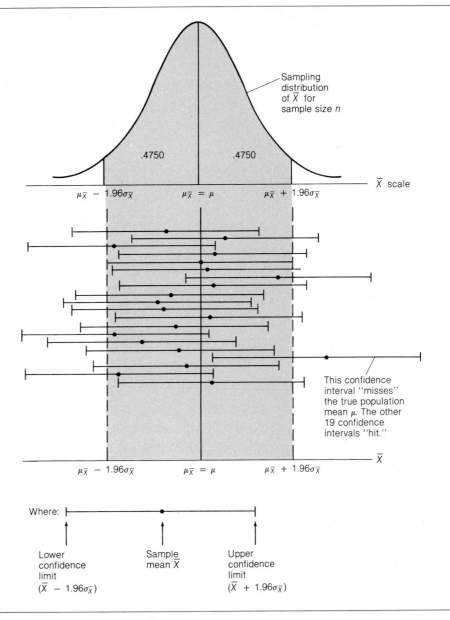

Figure 8.5 Sampling distribution of $\overline{X}$ and twenty 95 percent confidence intervals.

higher degree of confidence and a higher order of precision simultaneously is by increasing the sample size.

Before leaving the subject of how to develop a confidence interval, we should emphasize that the appropriate formula is more general than those we have used for purposes of introducing the subject: When the population standard deviation σ is unknown and the sample size n is large, we determine the confidence interval as indicated in Equation (8.10.5) which follows. The Z value in this formula is the one associated with the confidence coefficient.

$$\overline{X} - Z_{\frac{\alpha}{2}} \cdot \frac{s}{\sqrt{n}} - \leq \mu \leq \overline{X} + Z_{\frac{\alpha}{2}} \cdot \frac{s}{\sqrt{n}} \qquad (8.10.5)$$

The symbol $Z_{\alpha/2}$ requires some explanation. Let us call the confidence coefficient selected C and define α to be $1 - C$. (The symbol α will be dealt with in much greater detail in Chapter 9; for now, just viewing it as $1 - C$ will suffice.) Hence, $\alpha/2 = (1 - C)/2$. For example, if the confidence coefficient chosen were, say, $C = .92$, then $1 - C = .08$ and $\alpha/2 = .04$. The .04 value represents the area in a single tail of the standard normal curve. Therefore, we look for the area .4600 in the standard normal curve table (Appendix Table 2) and read across to the margins. The result is $Z_{\alpha/2} = 1.75$.

Similarly, if the confidence coefficient selected were .997, $\alpha = 1 - C = .003$ and $\alpha/2 = .0015$. Thus, we seek a value of .4985 in the standard normal curve table and read across to the margins. The result is $Z_{\alpha/2} = 2.96$ or 2.97.

8.11 In a Problem Concerned with Estimating the Population Mean, How Might the Concept of Sampling Distribution of $\overline{X}$ Help Us Determine the Necessary Sample Size?

We noted in Chapter 2 that sample size selection must usually be based on many considerations, not all of which are mathematical. What follows in this section should not be construed as a once-and-for-all answer to the question of how large a sample should be. However, the following methodology can be quite helpful in situations where the confidence and precision of estimates are matters of primary importance.

Early in the preceding section we observed that the expression for a .95 confidence interval for the arithmetic mean could be derived by rearranging the inequality

$$-1.96 \leq \frac{\overline{X} - \mu}{\frac{\sigma}{\sqrt{n}}} \leq +1.96.$$

In the process of moving from this inequality to the confidence interval formula, we obtained Equation (8.10.2), which is

$$-1.96 \frac{\sigma}{\sqrt{n}} \leq \overline{X} - \mu \leq +1.96 \frac{\sigma}{\sqrt{n}}$$

or, more generally,

$$-Z_{\frac{\alpha}{2}} \frac{\sigma}{\sqrt{n}} \leq \overline{X} - \mu \leq +Z_{\frac{\alpha}{2}} \frac{\sigma}{\sqrt{n}} \qquad (8.11.1)$$

Clearly, if $\overline{X}$ is to be used as a point estimate of μ, our error, which we shall denote E, is given by the difference between the value of $\overline{X}$ and the unknown value of μ. Therefore, we can rewrite Equation (8.11.1) and have it read

$$-Z_{\frac{\alpha}{2}}\frac{\sigma}{\sqrt{n}} \leq E \leq +Z_{\frac{\alpha}{2}}\frac{\sigma}{\sqrt{n}} \tag{8.11.2}$$

This formula, properly harnessed, can be used to provide the sample size n necessary for a specified confidence coefficient and a desired degree of precision. For example, let us consider the Trent and Travis case (Illustrative Case 6.2) once more, but assume that the auditor has not yet selected his sample of 100 accounts receivable. Let us further assume that he continues to use a confidence coefficient of .95 but that he now also specifies that his estimate of μ be *no more than* $25 away from the correct value. What will be the necessary sample size?

The fact that the auditor in this example wishes to be 95% confident tells us immediately that we may assert that, if $\overline{X}$ is used as an estimate of μ the error will be equal to or less than $(1.96)(\sigma/\sqrt{n})$. Since this quantity, symbolized E, is supposed to equal no more than $25, we can write

$$(1.96)\left(\frac{\sigma}{\sqrt{n}}\right) \leq \$25 \tag{8.11.3}$$

Except for the unfortunate fact that we still have two unknowns to contend with—n, the value we wish to determine, and σ, a value which will not ordinarily be known in practice—the required sample size could be determined quite easily. As it is, an assumption must be made about the value of σ. If the auditor has had experience with this company previously or with similar companies, then he may be in a position to make an informed guess of σ. Another possibility would be for him to take a very small exploratory sample and estimate σ using s. If these possibilities are not available to him for any reason, he would presumably be well advised to make a conservative estimate—that is, to assume σ to be greater than he really thinks likely. Let us suppose, for example, that he believes it highly unlikely that σ exceeds $300. Substituting $300 into Equation (8.11.3), he finds

$$(1.96)\left(\frac{300}{\sqrt{n}}\right) \leq 25$$

which is the same as

$$n \geq \left[\frac{(1.96)(300)}{25}\right]^2$$

Ignoring the inequality sign long enough to solve for n, we get $n = 533.19$, or, honoring the inequality sign, $n \geq 533.19$. This result is interpreted to mean that for the desired precision and confidence, the sample should consist of no fewer than 534 accounts.

A general formula for determining the desired sample size is

$$n \geq \left(\frac{Z_{\frac{\alpha}{2}}\sigma}{E}\right)^2 \tag{8.11.4}$$

8.12 What Is Meant by the Term *Sampling Distribution of $\bar{p}$?*

Sometimes we find it necessary to make estimates in situations where the observations of interest are attributes rather than measurements. For example, we might be interested in determining the proportion of people in a population who plan to purchase a new automobile during the upcoming 12 months. If the population of interest is quite large, a sample study rather than a complete census will be made, and a sample proportion will be used to estimate the true but unknown population proportion.

Just as we found it necessary to distinguish between the arithmetic mean of a population and the arithmetic mean of a sample from that population, so must we also distinguish between the population proportion and the proportion associated with a single sample.

> The population proportion will be denoted by p and defined as $p = X_P/N$, where X_P represents the number of elementary population units that have a specified characteristic (the number of people in the population planning to buy a new car during the next 12 months, for example) and N is the number of elementary population units altogether.
>
> The sample proportion will be denoted by $\bar{p}$ and defined as $\bar{p} = X_s/n$, where X_s is the number of elementary sample units possessing a specified characteristic and n is the number of elementary sample units altogether.

When we speak of *sampling distribution of $\bar{p}$*, we have in mind a sampling distribution associated with the statistic $\bar{p}$. This conclusion follows naturally from the general definition of *sampling distribution* presented in section 8.1. That is, the general notation $\bar{\theta}$ used in section 8.1 is now replaced by $\bar{p}$ just as it was replaced by $\bar{X}$ in section 8.3.

By way of illustration, let us consider once more the example introduced in section 8.4 concerning the population of six salespeople whose names are Adams, Bardwell, Carpenter, Davis, Elsea, and Farnum. Whereas we previously listed the sales (in thousands of dollars) achieved by these salespeople last year, we will now shift our attention to a characteristic—namely, whether a specified salesperson received a Christmas bonus. Let us suppose that the relevant bonus information is as shown in Table 8.7.

Since our interest centers on the proportion of salespeople in this small population who received a Christmas bonus, rather than on the proportion who did not receive a bonus, we determine p by counting the number of salespeople who received a bonus—namely, 4—and expressing that number as a ratio to the total number of salespeople in the population—namely, 6. Hence, the population proportion is $p = X_P/N = \frac{4}{6}$ or .6667.

8.13 How Would One Go About Constructing a Sampling Distribution of $\bar{p}$?

For reasons mentioned in section 8.4, one would not actually construct a sampling distribution of $\bar{p}$ in practice. Still, we will construct a simple one in the hope of making the concept of sampling distribution of $\bar{p}$ more concrete. We proceed by (1) specifying a sample size n, (2) listing the elementary units making up each different sample combination of the specified size which could be drawn from that population, (3) calculating the $\bar{p}$ value associated with each different sample combination, and (4) organizing the resulting $\bar{p}$ values into a probability distribution. Let us suppose, for simplicity of illustration, that the sample size selected is $n = 2$. Table 8.8 shows the possible sample combinations and the corresponding $\bar{p}$ values.

Collecting the information in Table 8.8, we get the sampling distribution shown in Table 8.9 (presented there in both frequency-distribution and probability-distribution form).

Table 8.7 Information Concerning Christmas Bonuses Distributed to a Population of Six People

Salesperson	Information Concerning Christmas Bonus
Adams	Did not receive bonus
Bardwell	Received bonus
Carpenter	Did not receive bonus
Davis	Received bonus
Elsea	Received bonus
Farnum	Received bonus

$$P = \frac{X_P}{N} = \frac{4}{6} = 0.6667$$

Table 8.8 Bonus Data for Fifteen Samples of Two Salespeople

Sample Combination	Number in Sample Who Received Bonus X_s	Proportion in Sample Who Received Bonus $\bar{p} = X_s/2$
1. Adams and Bardwell	1	0.5
2. Adams and Carpenter	0	0.0
3. Adams and Davis	1	0.5
4. Adams and Elsea	1	0.5
5. Adams and Farnum	1	0.5
6. Bardwell and Carpenter	1	0.5
7. Bardwell and Davis	2	1.0
8. Bardwell and Elsea	2	1.0
9. Bardwell and Farnum	2	1.0
10. Carpenter and Davis	1	0.5
11. Carpenter and Elsea	1	0.5
12. Carpenter and Farnum	1	0.5
13. Davis and Elsea	2	1.0
14. Davis and Farnum	2	1.0
15. Elsea and Farnum	2	1.0

Table 8.9 Data from Table 8.8 Presented in Frequency-Distribution and Relative-Frequency-Distribution Form

Sample Proportion $\bar{p}$	Frequency f	Relative Frequency (Probability) $P(\bar{p}) = \dfrac{f}{N^*}$
0.0	1	.0667
0.5	8	.5333
1.0	6	.4000
Σ	$15 = N^*$	1.0000

8.14 What Is the Relationship Between the Mean of a Sampling Distribution of $\bar{p}$ and the Corresponding Population Proportion?

> The mean of a sampling distribution of $\bar{p}$ is always equal to the population proportion p.

This property of a sampling distribution of $\bar{p}$ will be demonstrated using data from Table 8.9.
Table 8.9 reveals that (1) 1 out of 15 (or .0667) of the $\bar{p}$ values are 0, (2) 8 out of 15 (or .5333) are .5, and (3) 6 out of 15 (or .4000) are 1. Therefore, we can compute the value of $\mu_{\bar{p}}$ using the following formula:

$$\mu_{\bar{p}} = \Sigma \, [P(\bar{p}) \cdot \bar{p}] \qquad (8.14.1)$$

This formula is nothing more than the formula for obtaining the arithmetic mean, or expected value, of a discrete probability distribution—Equation (7.2.1)—but with appropriate minor changes in notations. We get $\mu_{\bar{p}} = (.0667)(0) + (.5333)(.5) + (.4000)(1) = .6667$ or $\frac{2}{3}$, the same value obtained earlier from the population data (shown under Table 8.7).

8.15 How Might the Standard Deviation of a Sampling Distribution of $\bar{p}$ Be Obtained from Knowledge of the Population Proportion?

Before seeking an answer to this question, it will be helpful to consider how the standard deviation of a sampling distribution of $\bar{p}$, hereafter referred to as the *standard error of the proportion*, would be computed directly using the various sample proportions. The calculation is straightforward because we merely make use of the formula for computing the standard deviation of a discrete probability distribution—Equation (7.2.2)—except for a few appropriate changes in the notations:

$$\sigma_{\bar{p}} = \sqrt{\Sigma\ [P(\bar{p})(\bar{p} - p)^2]} \tag{8.15.1}$$

The work is shown in Table 8.10, and the resulting $\sigma_{\bar{p}}$ value is found to be .2982.

In practice, fortunately, we need not actually develop a sampling distribution of $\bar{p}$ to measure the variation among sample $\bar{p}$ values.

Statistical theory assures us that the standard error of the proportion $\sigma_{\bar{p}}$ can be obtained by

$$\sigma_{\bar{p}} = \sqrt{\frac{pq}{n}} \tag{8.15.2}$$

where p is the population proportion and q is $1 - p$, if the population is infinite or finite but at least 20 times as great as the sample size n, or

$$\sigma_{\bar{p}} = \sqrt{\frac{pq}{n}} \cdot \sqrt{\frac{N - n}{N - 1}} \tag{8.15.3}$$

if the population is finite and not at least 20 times as great as the sample size.

Substituting values into Equation (8.15.3), we find

$$\sigma_{\bar{p}} = \sqrt{\frac{(.6667)(.3333)}{2}} \cdot \sqrt{\frac{6 - 2}{6 - 1}} = .2981$$

Table 8.10 Calculation of $\sigma_{\bar{p}}$ for Data Appearing in Table 8.9

Sample Proportion $\bar{p}$	Relative Frequency (Probability) $P(\bar{p})$	$\mu_{\bar{p}} = p$	$\bar{p} - p$	$(\bar{p} - p)^2$	$P(\bar{p})(\bar{p} - p)^2$
0.0	.0667	.6667	−.6667	.444489	.029647
0.5	.5333	.6667	−.1667	.027789	.014820
1.0	.4000	.6667	.3333	.111089	.044436
					.088903

$$\sigma_{\bar{p}} = \sqrt{\Sigma\ [P(\bar{p})(\bar{p} - p)^2]} = \sqrt{.088903} = .2982$$

Clearly, except for a minor rounding error, these results do serve to demonstrate that Equation (8.15.1) and Equation (8.15.3)—or, where appropriate, Equation (8.15.2)—are equally accurate ways of computing the standard error of the proportion $\sigma_{\bar{p}}$.

8.16 What Effect Does Sample Size Have on the Standard Deviation of a Sampling Distribution of $\bar{p}$?

For a given p value not equal to 0 or 1, the larger the sample size, the smaller will be the standard error of the proportion $\sigma_{\bar{p}}$. The implication of this fact is that, in a problem involving a statistical estimation of the population proportion p, greater precision can be attained for a specified confidence coefficient if the sample size is large. If the population is infinite in size, a quadrupling of the sample size is needed to reduce $\sigma_{\bar{p}}$ by 50%. The same is approximately true in the case of large finite populations.

8.17 What Can Be Said About the Shape of a Sampling Distribution of $\bar{p}$?

The central limit theorem discussed in section 8.8 is also applicable to a sampling distribution of $\bar{p}$.

> The *sampling distribution of $\bar{p}$* is approximately normal if the random sample size n is sufficiently large.

If the sampled population is finite, the population size N must be considerably greater than the sample size n for this theorem to apply. Fortunately, such is usually the case in practice.

Many statisticians rely on a simple rule of thumb to tell them whether they may confidently assume that the sample size is sufficiently large. The rule is

> If both np and nq are greater than 5, the sampling distribution of $\bar{p}$ will be approximately normal.

Therefore, a sample of only 16 elementary units from a population with $p = .6667$ would be sufficiently large to be approximated adequately by the normal curve because $np = (16)(.6667) = 10.67$ and $nq = (16)(.3333) = 5.33$. Clearly, for a sample size as small as $n = 2$, as was actually used in our example, we may not validly assume that the sampling distribution of $\bar{p}$ is approximately normal.

8.18 How Might the Important Properties of a Sampling Distribution of $\bar{p}$ Be Harnessed to Help Us Draw a Deductive Inference About the Proportion for a Single Random Sample?

Let us suppose, contrary to what we have hitherto assumed, that the company in our example actually has substantially more than six salespeople on the payroll; let us say that $N = 900$. Let us further suppose that it is known that two-thirds of them received a Christmas bonus last year. We will ask: What is the probability that a random sample of 25 salespeople drawn from this population will have a $\bar{p}$ value within 10% points of the population proportion p?

We know, then, that $N = 900$, $p = \frac{2}{3} = .6667$, and $n = 25$. We begin by making two quick checks: First, we determine whether the population size is at least 20 times as large as the sample size. We find that $N/n = \frac{900}{25} = 36$, a value sufficiently large to permit us to ignore use of the finite correction factor in the formula for the standard error of the proportion $\sigma_{\bar{p}}$; see Equation (8.15.3). Second, we determine whether np and nq are both greater than 5. Since $np = (25)(.6667) = 16.67$ and $nq = (25)(.3333) = 8.33$, we observe that both products are indeed greater than 5. As was pointed out in section 8.17, this knowledge permits us to use the normal curve to approximate the sampling distribution of $\bar{p}$.

Recall that the problem asks us to determine the probability that $\bar{p}$ for a particular random sample of size 25 will be within 10% points (that is, within .6667 plus and minus .1000) of $p = .6667$. To find the desired probability using areas under the standard normal curve, we must convert $\bar{p}$ values into their Z-value equivalents. Recall that we originally defined Z to be $(X - \mu)/\sigma$. However, since in the present case our random variable is a collection of $\bar{p}$ values, Z is $(\bar{p} - \mu_{\bar{p}})/\sigma_{\bar{p}}$. Moreover, because we know that $\mu_{\bar{p}} = p$ and that $\sigma_{\bar{p}}$ can be obtained by $\sigma_{\bar{p}} = \sqrt{pq/n}$, we may do a little substituting and arrive at the following new way of expressing Z:

$$Z = \frac{\bar{p} - p}{\sqrt{\dfrac{pq}{n}}}$$

We next substitute known values into this new formula and solve for Z values associated with $\bar{p}_1 = .5667$ and $\bar{p}_2 = .7667$. Therefore,

$$Z_1 = \frac{\bar{p}_1 - p}{\sqrt{\dfrac{pq}{n}}} = \frac{.5667 - .6667}{\sqrt{\dfrac{(.6667)(.3333)}{25}}} = -1.06$$

and, similarly,

$$Z_2 = \frac{\bar{p}_1 - p}{\sqrt{\dfrac{pq}{n}}} = \frac{.7667 - .6667}{\sqrt{\dfrac{(.6667)(.3333)}{25}}} = +1.06$$

We next consult Appendix Table 2 to determine the area under the normal curve associated with Z values running from 0 to $+1.06$ and find the area of interest to be .3554. Of course, this value represents only the area under the curve between $Z = 0$ and $Z = +1.06$; we must double this area to arrive at the probability sought. Thus, we may assert that the probability is

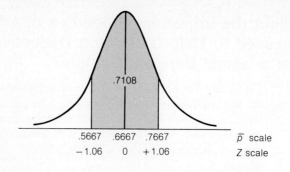

$$.7108$$

| .5667 | .6667 | .7667 | $\bar{p}$ scale |
| -1.06 | 0 | +1.06 | Z scale |

Figure 8.6 Sketch used as an aid in solving the deductive version of the Christmas bonus problem.

.7108 that the proportion of salespeople receiving a Christmas bonus in this sample of size $n = 25$, drawn at random from the population of 600 salespeople, will be between .5667 and .7667, as shown in Figure 8.6.

This problem is admittedly not a very practical one because the population proportion p had to be known—a condition almost always contrary to fact. Moreover, the conclusion is drawn about a single random sample, whereas in practice we are usually concerned about using sample information to aid us in drawing a conclusion about some characteristic of the parent population. In the next section, we will turn the procedure just described upside down so that a probabilistic conclusion about a population proportion p may be drawn from knowledge of the sample proportion $\bar{p}$.

8.19 How Might the Important Properties of a Sampling Distribution of $\bar{p}$ Be Harnessed to Help Us Draw an Inductive Inference About the Proportion for the Population?

The answer to this question is pretty much a replay of the answer to the question posed in section 8.10. By following the same procedure described in connection with this earlier question (and assuming a .95 confidence coefficient), we arrive at the expression for the desired confidence interval:

$$\bar{p} - 1.96\sqrt{\frac{pq}{n}} \leq p \leq \bar{p} + 1.96\sqrt{\frac{pq}{n}} \qquad (8.19.1)$$

which is the mathematically correct way of expressing how a .95 confidence interval may be determined. Of course, Equation (8.19.1) would, without modification, be totally useless in practice, the reason being that the true standard error of the proportion $\sigma_{\bar{p}}$ is obtained by $\sigma_{\bar{p}} = \sqrt{pq/n}$. Thus, the population p, the very thing we wish to estimate, must be known in order for one to use this formula. If it were known, the formula would not be needed anyway.

In section 8.10 we approximated the standard error of the mean $\sigma_{\bar{X}}$ by using $s_{\bar{X}} = s/\sqrt{n}$ in place of $\sigma_{\bar{X}} = \sigma/\sqrt{n}$. We may make a similar substitution in the case of the standard error of the proportion: If the sample size is large, we may substitute $\bar{p}$ for p and have the (estimated) standard error formula, which is

$$s_{\bar{p}} = \sqrt{\frac{\bar{p}\bar{q}}{n}}$$

Therefore, the confidence interval associated with a probability of (approximately) .95 of including the true but unknown population proportion is written

$$\bar{p} - 1.96\sqrt{\frac{\bar{p}\bar{q}}{n}} \leq p \leq \bar{p} + 1.96\sqrt{\frac{\bar{p}\bar{q}}{n}} \qquad (8.19.2)$$

or, more generally,

$$\bar{p} - Z_{\frac{\alpha}{2}}\sqrt{\frac{\bar{p}\bar{q}}{n}} \leq p \leq \bar{p} + Z_{\frac{\alpha}{2}}\sqrt{\frac{\bar{p}\bar{q}}{n}} \qquad (8.19.3)$$

Now, let us suppose that an investigator has selected a random sample of $n = 36$ salespeople from the population of 900 salespeople employed by the company and has determined that 28 of the sample members received a Christmas bonus; that is, $\bar{p}$ is $X_s/n = \frac{28}{36} = .7778$. She wishes to estimate the true population proportion p using a .95 confidence interval. The appropriate expression for conveying this wish would be

$$P\left(\bar{p} - Z_{\frac{\alpha}{2}}\sqrt{\frac{\bar{p}\bar{q}}{n}} \leq p \leq \bar{p} + Z_{\frac{\alpha}{2}}\sqrt{\frac{\bar{p}\bar{q}}{n}}\right) \cong .95$$

Substituting known values into the expression within parentheses, she gets $P(.7778 - 1.96\sqrt{(.7778)(.2222)/36} \leq p \leq .7778 + 1.96\sqrt{(.7778)(.2222)/36})) \cong .95$, which, when solved, is $P(.6420 \leq p \leq .9136) \cong .95$.

8.20 In a Problem Concerned with Estimating the Population Proportion, How Might the Concept of Sampling Distribution of $\bar{p}$ Help Us Determine the Necessary Sample Size?

For reasons nearly identical with those described in section 8.11, the appropriate sample size n for a specified confidence coefficient and a specified minimum error E can be determined from

$$n \geq pq\left(\frac{Z_{\frac{\alpha}{2}}}{E}\right)^2 \qquad (8.20.1)$$

Since an analyst is unlikely to know the true population proportion p (if he did, he would have no need for a confidence interval), he will necessarily have to substitute in Equation

(8.20.1) either (1) an informed estimate of p or (2) the most conservative possible estimate of p—namely, $p = .5$.

8.21 May We Safely Conclude That the Important Properties of a Sampling Distribution of $\overline{X}$ or a Sampling Distribution of $\overline{p}$ Also Hold for Sampling Distributions of All Other Sample Statistics of Possible Interest?

A thoroughgoing answer to this question could be very lengthy, but we will not strive for thoroughness. The answer is a flat no. The characteristics of a sampling distribution of a specific sample statistic will depend on what that statistic is. What has been said about the sampling distribution of $\overline{X}$ and the sampling distribution of $\overline{p}$ should be viewed as strictly appropriate to these two statistics only.

8.22 In What Way Must Our Confidence Interval Procedures for the Arithmetic Mean Be Altered If the Sample Size Is Small?

The confidence interval methodology described in section 8.10 is appropriate only when the sample size is large—a qualification usually taken to mean n in excess of about 30 or 40. The distinction between large and small samples will be of no concern if the population standard deviation σ is known. However, in practical problems we will seldom have access to such information. When we do not know σ and the sample size n is about 30 or less, the confidence interval procedures must be modified somewhat. The reason hinges on the definition of Z. We will illustrate by using the expression for the Z-value equivalent of an $\overline{X}$ value. We have seen that, courtesy of the central limit theorem, the ratio $(\overline{X} - \mu)/\sigma_{\overline{X}} = (\overline{X} - \mu)/(\sigma/\sqrt{n})$ is approximately normally distributed provided that the sample size is reasonably large. Worth noting is the fact that in $(\overline{X} - \mu)/(\sigma/\sqrt{n})$ only one measure varies from sample to sample; that measure is $\overline{X}$. The n value is assumed to be fixed and μ and σ are parameters that will be constant even though their sample counterparts, $\overline{X}$ and s, will not be.

In section 8.10, by substituting known s for unknown σ, we, in effect, assumed that $(\overline{X} - \mu)/s_{\overline{X}} = (\overline{X} - \mu)/(s/\sqrt{n})$ is also normally distributed. In doing so, however, we took pains to emphasize that substituting s for σ and using the normal curve is permissible only if the sample size is large. When n is large, the numerator $\overline{X} - \mu$ will be approximately normal thanks to the central limit theorem, and the denominator $s/\sqrt{n}$ will be relatively stable and close enough to being an unbiased estimator of $\sigma/\sqrt{n}$ for all practical purposes. But note that in $(\overline{X} - \mu)/(s/\sqrt{n})$ two measures, rather than only one, vary from sample to sample—namely, $\overline{X}$ and s. This fact makes quite a difference when the sample size is small. When n is small, we cannot count on $\overline{X} - \mu$ being distributed normally. Moreover, $s/\sqrt{n}$ will show a systematic tendency to underestimate $\sigma/\sqrt{n}$; the smaller the sample size, the greater will be this underestimation bias. For a given value of n, combining all possible values of $\overline{X} - \mu$ and $s/\sqrt{n}$, and organizing the results into probability-distribution form, produces a sampling distribution bearing a superficial resemblance to the normal curve but which is really flatter in the center and higher on the ends than the normal, as illustrated in

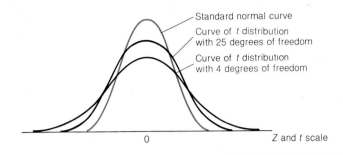

Standard normal curve

Curve of *t* distribution with 25 degrees of freedom

Curve of *t* distribution with 4 degrees of freedom

0   *Z* and *t* scale

Figure 8.7 Comparison of the standard normal curve, with curves for *t* distributions associated with (1) 4 degrees of freedom and (2) 25 degrees of freedom.

Figure 8.7. The distribution of $(\overline{X} - \mu)/(s/\sqrt{n})$ values is called the *t* distribution, and a particular $(\overline{X} - \mu)/(s/\sqrt{n})$ value is called a *t* statistic.

As Figure 8.7 suggests, the *t* distribution, unlike the normal, is really a family of similar distributions *even in standard form.* This is so because the specific shape of a standard *t* distribution is determined by the number of degrees of freedom—for present purposes $n - 1$. (Remember that the smaller the sample size, which is to say, the smaller the $n - 1$ value, the greater the underestimation bias of $s/\sqrt{n}$.) When $n - 1$ is very small, the *t* distribution will differ markedly from the normal. Jumping to the opposite extreme: It has been shown that, when $n - 1$ is infinity, the *t* and the normal distributions are identical. Not surprisingly, then, the larger the value of $n - 1$, the more closely the *t* distribution comes to resembling the normal distribution. In fact, when the sample size *n* is greater than about 30 or 40, the *t* and the normal distributions are so nearly identical that use of the Z statistic in lieu of the more theoretically correct *t* statistic is permissible. The procedure for using the table of areas for the *t* distribution (Appendix Table 3, reproduced here as Table 8.11) in a confidence interval problem is demonstrated in the following example.

Let us suppose that an efficiency expert takes observations on the typing pool of a company at random times and observes how many of the total of 75 typists are engaged in productive work. She takes a sample of six such observations and obtains the following counts: 71, 66, 75, 72, 60, and 53. She wishes to estimate, using a .95 confidence interval, the mean number of secretaries working at a given time.

> The facts that the sample size is obviously very small and σ is unknown alert us to the need to use *t* instead of Z in the confidence interval calculations.

We begin by computing the sample mean $\overline{X}$ and standard deviation *s*:

$$\overline{X} = \frac{\Sigma X}{n} = 66.17$$

$$s = \sqrt{\frac{\Sigma (X - \overline{X})^2}{n - 1}} = 8.33$$

Therefore, we may assert that

$$P\left(66.17 - t_{\frac{\alpha}{2}}\frac{8.33}{\sqrt{6}} \le \mu \le 66.17 + t_\alpha\frac{8.33}{\sqrt{6}}\right) = .95$$

where the t value in this expression is found in Appendix Table 3 and Table 8.11. Unlike the table of areas under the normal curve, this t table shows areas under the t distribution in the top margin of the table. Each such area represents the combined areas in the two tails. To obtain the needed t value from this table, we begin by subtracting the confidence coefficient from 1. In our problem, the confidence coefficient was given as .95. Therefore, $1.00 - .95 = .05$. We locate the column with the heading .05 in the table and read down that column until we reach the value across from $n - 1 = 5$ degrees of freedom. We find that the appropriate t value for our problem is 2.571, an appreciably larger multiplier than the value of 1.96 we would use if the normal curve were appropriate. We may now assert

$$P\left(66.17 - 2.571\frac{8.33}{\sqrt{6}} \le \mu \le 66.17 + 2.571\frac{8.33}{\sqrt{6}}\right) = .95$$

or

$$P(57.43 \le \mu \le 74.91) = .95$$

Before leaving the t distribution for now, we must point out that use of this distribution is strictly legitimate only if the distribution of the variables in the numerator and the denominator of the t statistic are independent. Such a condition will exist only if the *population is normally distributed*. This potentially can be a very restrictive requirement because, in many sampling situations, the shape of the parent population will be unknown. This fact is of no consequence in the case of Z because it is the sampling distribution of $\overline{X}$ that is assumed to be normal—not the population itself. But in the case of t, if nonnormality of the population were seriously to affect the distribution of the t statistic, the usefulness of the procedures described in this section would be very limited. Fortunately, it has been shown that the t statistic is relatively stable for nonnormal populations when such populations are roughly symmetrical with observations concentrated in the center and thinning out toward the ends.

Note on Degrees of Freedom

Up to now you have been asked to view "degrees of freedom" as synonymous with $n - 1$. While this view is appropriate in connection with subjects dealt with in this section and those to be dealt with in Chapter 9, it does not constitute a general definition by any means. The concept of degrees of freedom has its basis in fairly advanced linear algebra and refers to the number of linearly independent observations in a sum of squares. To develop the subject fully would take us far astray. However, an intuitive feel for the term may be achieved by considering a set of deviations from the arithmetic mean, the basis for a standard deviation or variance calculation. We know by a derivation shown in section 4.9 of Chapter 4 that $\Sigma (X - \overline{X})$ is always 0. This means that, if we know $n - 1$ such deviations, we as good as know the value of the nth deviation because it is "fixed." For example,

$$-4 + 5 + 3 - 2 + ? = 0$$

The value that must be in the position presently occupied by the question mark is -2 because that is the only value which, when added to the others, will produce a sum of 0. All

Table 8.11   *t* Distribution (From Appendix Table 3)

Areas in Both Tails Combined

Example: To find the value of *t* which corresponds to an area of .10 in both tails of the distribution combined, when there are 19 degrees of freedom, look under the .10 column, and proceed down to the 19 degrees of freedom row; the appropriate *t* value there is 1.729.

Degrees of Freedom	Area in Both Tails Combined			
	.10	.05	.02	.01
1	6.314	12.706	31.821	63.657
2	2.920	4.303	6.965	9.925
3	2.353	3.182	4.541	5.841
4	2.132	2.776	3.747	4.604
5	2.015	2.571	3.365	4.032
6	1.943	2.447	3.143	3.707
7	1.895	2.365	2.998	3.499
8	1.860	2.306	2.896	3.355
9	1.833	2.262	2.821	3.250
10	1.812	2.228	2.764	3.169
11	1.796	2.201	2.718	3.106
12	1.782	2.179	2.681	3.055
13	1.771	2.160	2.650	3.012
14	1.761	2.145	2.624	2.977
15	1.753	2.131	2.602	2.947
16	1.746	2.120	2.583	2.921
17	1.740	2.110	2.567	2.898
18	1.734	2.101	2.552	2.878
19	1.729	2.093	2.539	2.861
20	1.725	2.086	2.528	2.845
21	1.721	2.080	2.518	2.831
22	1.717	2.074	2.508	2.819
23	1.714	2.069	2.500	2.807
24	1.711	2.064	2.492	2.797
25	1.708	2.060	2.485	2.787
26	1.706	2.056	2.479	2.779
27	1.703	2.052	2.473	2.771
28	1.701	2.048	2.467	2.763
29	1.699	2.045	2.462	2.756
30	1.697	2.042	2.457	2.750
40	1.684	2.021	2.423	2.704
60	1.671	2.000	2.390	2.660
120	1.658	1.980	2.358	2.617
Normal distribution	1.645	1.960	2.326	2.576

Source: Taken from Table III of Fisher and Yates, *Statistical Tables for Biological, Agricultural and Medical Research*, published by Longman Group Ltd., London (previously published by Oliver & Boyd, Edinburgh), by permission of the authors and publishers.

of this is equivalent to saying that if we knew $n - 1$ of the X values and the arithmetic mean $\overline{X}$, we would as good as know the nth X value. To illustrate: Suppose we know the arithmetic mean $\overline{X}$ to be 5. This means that deviation -4 in the above list must arise from an X value of 1 because $1 - 5 = -4$. The other X values must be 10, 8, and 3. Thus,

$$1 + 10 + 8 + 3 + ? = ?$$

Since we know the first four X values and the arithmetic mean and since $\Sigma X = n\overline{X}$, we deduce that ΣX must be 25 and the fifth X value must be 3. Speaking very generally, and somewhat loosely, we can say that in this situation we have $n - 1$ number of X values which are free to vary independently and one which is fixed as a result of our knowing $\overline{X}$. Call $\overline{X}$ one "constraint" and we have the basis for a rather more general definition of degrees of freedom—namely, $n - \lambda$, where n is the number of sample observations and λ is the number of population parameters being estimated by their sample counterparts. This way of looking at degrees of freedom will serve us adequately throughout Chapters 9, 10, 11, and 12.

8.23 Summary

The 22 questions serving as a study guide for this chapter are listed here for one last time. They are accompanied by brief answers.

1. What, in general, is meant by the term *sampling distribution?*
   Answer: A sampling distribution of $\overline{\theta}$ (where $\overline{\theta}$ is some sample statistic) is a probability distribution showing all possible values that $\overline{\theta}$ could have in a sampling situation (where the sample size n is the same for all samples) and the corresponding probabilities of occurrence.

2. How does a *sampling distribution* differ from a *sample distribution* or a *population distribution?*
   Answer: A population distribution is a frequency distribution of population data.
   A sample distribution is a frequency distribution of data representing *a single sample* from the population of interest.
   A sampling distribution of a statistic is a probability distribution showing the various values that statistic could have in a sampling situation (assuming constant n) and the corresponding probabilities of occurrence.

3. What is a *sampling distribution of* $\overline{X}$?
   Answer: A sampling distribution where the statistic of interest is $\overline{X}$.

4. How would one go about constructing a sampling distribution of $\overline{X}$?
   Answer: In practice, one wouldn't; it is the characteristics of this sampling distribution that one makes use of. However, if one had to construct a sampling distribution of $\overline{X}$, he would (1) list observations comprising each different sample of size n which could be selected from the population, (2) compute the value of $\overline{X}$ for each such sample, and (3) organize the resulting $\overline{X}$ values into probability distribution form.

5. What is the relationship between the mean of a sampling distribution of $\overline{X}$ and the mean of the parent population?

   Answer: They are equal. That is, $\mu_{\overline{X}} = \mu$.

6. What is the relationship between the standard deviation of a sampling distribution of $\overline{X}$ and the standard deviation of the population?

   Answer: $\sigma_{\overline{X}} = \sigma/\sqrt{n}$ if the population is (1) infinite or (2) finite but at least 20 times as great as the sample size, or $\sigma_{\overline{X}} = \sigma/\sqrt{n} \cdot \sqrt{(N-n)/(N-1)}$ if the population is finite and not at least 20 times as great as the sample size.

7. What effect does sample size have on the standard deviation of a sampling distribution of $\overline{X}$?

   Answer: Since $\sigma_{\overline{X}} = \sigma/\sqrt{n}$ or $\sigma/\sqrt{n} \cdot \sqrt{(N-n)/(N-1)}$, the larger the sample size, the smaller the standard error of the mean, $\sigma_{\overline{X}}$.

8. What can be said about the shape of a sampling distribution of $\overline{X}$?

   Answer: Almost regardless of the shape of the population, the sampling distribution of $\overline{X}$ is approximately normal if the sample size n is sufficiently large. This is the central limit theorem.

9. How might the important properties of a sampling distribution of $\overline{X}$ be harnessed to help us draw a deductive inference about the mean of a single random sample?

   Answer: Like any similar normal curve problem, but we use $Z = (\overline{X} - \mu)/\sigma_{\overline{X}}$ rather than $Z = (X - \mu)/\sigma$, where $\sigma_{\overline{X}} = \sigma/\sqrt{n}$ or $\sigma/\sqrt{n} \cdot \sqrt{(N-n)/(N-1)}$.

10. How might the important properties of a sampling distribution of $\overline{X}$ be harnessed to help us draw an inductive inference about the mean of the population?

    Answer: Assuming, for simplicity, that the population size is considerably larger than the sample size, then use $\overline{X} \pm Z_{\frac{\alpha}{2}}(\sigma/\sqrt{n})$ if the population standard deviation is known or $\overline{X} \pm Z_{\frac{\alpha}{2}}(s/\sqrt{n})$ if the population standard deviation is not known and the sample size is large. If the population standard deviation is not known and the sample size is small, use the formula presented in connection with question 8.22.

11. In a problem concerned with estimating the population mean, how might the concept of sampling distribution of $\overline{X}$ help us to determine the necessary sample size?

    Answer: Through use of $n \geq (Z_{\frac{\alpha}{2}}\sigma/E)^2$, where $E = \overline{X} - \mu$ and it is understood that the population standard deviation σ would have to be estimated (1) on the basis of similar experience, (2) from a small exploratory sample, or (3) by assigning a larger value to σ than one really believes likely.

12. What is meant by the term *sampling distribution of $\overline{p}$*?

    Answer: A sampling distribution associated with the sample proportion $\overline{p}$.

13. How would one go about constructing a sampling distribution of $\overline{p}$?

    Answer: In practice, one wouldn't; it is the characteristics of this sampling distribution that one makes use of. However, if one had to construct a sampling distribution of $\overline{p}$, he would (1) identify the "either-or" type observations associated with each different sample of size n contained within the population, (2) compute $\overline{p}$ for each such sample, where $\overline{p} = X_s/n$ and X_s stands for the number of sample items having the characteristic of interest, and (3) organize the resulting $\bar{p}$ values into probability distribution form.

14. What is the relationship between the mean of a sampling distribution of $\bar{p}$ and the corresponding population proportion?

    Answer: They are equal—that is, $\mu_{\bar{p}} = p$.

15. How might the standard deviation of a sampling distribution of $\bar{p}$ be obtained from knowledge of the population proportion?

    Answer: $\sigma_{\bar{p}} = \sqrt{pq/n}$ if the population is (1) infinite or (2) finite but at least 20 times as great as the sample size or $\sigma_{\bar{p}} = \sqrt{pq/n} \cdot \sqrt{(N-n)/(N-1)}$ if the population is finite and not at least 20 times as great as the sample size.

16. What effect does sample size have on the standard deviation of a sampling distribution of $\bar{p}$?

    Answer: Since $\sigma_{\bar{p}} = \sqrt{pq/n}$ or $\sqrt{pq/n} \cdot \sqrt{(N-n)/(N-1)}$, the larger the sample size, the smaller the standard error of the proportion $\sigma_{\bar{p}}$.

17. What can be said about the shape of a sampling distribution of $\bar{p}$?

    Answer: Regardless of the population proportion ($p = 0$ or 1 excluded), the sampling distribution of $\bar{p}$ will be approximately normal if the sample size is sufficiently large. A good rule is: If np and nq are both greater than 5, a normal sampling distribution of $\bar{p}$ may be assumed.

18. How might the important properties of a sampling distribution of $\bar{p}$ be harnessed to help us draw a deductive inference about the proportion for a single random sample?

    Answer: Like any similar normal curve problem but we use $Z = (\bar{p} - p)/\sigma_{\bar{p}}$ instead of $Z = (X - \mu)/\sigma$, where $\sigma_{\bar{p}} = \sqrt{pq/n}$ or $\sqrt{pq/n} \cdot \sqrt{(N-n)/(N-1)}$.

19. How might the important properties of a sampling distribution of $\bar{p}$ be harnessed to help us draw an inductive inference about the proportion for the population?

    Answer: If the sample size is large, $\bar{p} \pm Z_{\frac{\alpha}{2}}/s_{\bar{p}}$, where $s_{\bar{p}} = \sqrt{\bar{p}\bar{q}/n}$ or $\sqrt{\bar{p}\bar{q}/n} \cdot \sqrt{(N-n)/(N-1)}$.

20. In a problem concerned with estimating the population proportion, how might the concept of sampling distribution of $\bar{p}$ help us to determine the necessary sample size?

    Answer: Through use of $n \geq pq(Z_{\frac{\alpha}{2}}/E)^2$, where $E = \bar{p} - p$ and p must be estimated by some means or assigned a conservative value like .5.

21. May we safely conclude that the important properties of a sampling distribution of $\overline{X}$ or a sampling distribution of $\bar{p}$ also hold for sampling distributions of all other statistics of possible interest?

    Answer: No.

22. In what way must our confidence interval procedures for the arithmetic mean be altered if the sample size is small?

    Answer: If the sample size is small and if, in addition, the population standard deviation σ is unknown and must be estimated using s, then instead of applying the formula $\overline{X} \pm Z_{\frac{\alpha}{2}}(s/\sqrt{n})$, use $\overline{X} \pm t_{\frac{\alpha}{2}}(s/\sqrt{n})$. When using this second formula, it is necessary to assume that the population shape is normal.

Terms Introduced in This Chapter

central limit theorem (p. 324)
confidence coefficient (p. 330)
confidence interval (p. 329)
interval estimate (p. 329)
point estimate (p. 329)
population distribution (p. 315)

proportion (p. 336)
sample distribution (p. 315)
sampling distribution (p. 314)
sampling distribution of $\bar{p}$ (p. 336)
sampling distribution of $\bar{X}$ (p. 315)
standard error of the mean (p. 321)

standard error of the proportion
    (p. 338)
t distribution (p. 345)
t statistic (p. 345)
unbiased estimator (p. 320)

Formulas Introduced in This Chapter

Formula for determining the number of different samples of size n contained within a population of size N:

$$N^* = \frac{N!}{n!(N-n)!}$$

Formulas for determining the mean of a sampling distribution of $\bar{X}$ directly from the sample means:

$$\mu_{\bar{X}} = \Sigma \; [P(\bar{X})\bar{X}] \text{ or}$$

$$\mu_{\bar{X}} = \frac{\Sigma \; \bar{X}}{N^*}$$

Formulas for determining the standard deviation of a sampling distribution of $\bar{X}$ directly from the sample means:

$$\sigma_{\bar{X}} = \sqrt{\Sigma \; [P(\bar{X})(\bar{X} - \mu)^2]}$$

or

$$\sigma_{\bar{X}} = \sqrt{\frac{\Sigma \; (\bar{X} - \mu)^2}{N^*}}$$

Formula for determining the standard deviation of a sampling distribution of $\bar{X}$ through knowledge of the population standard deviation σ. The $\sqrt{(N-n)/(N-1)}$ part implies that the population is finite and not as least 20 times as great as the sample size:

$$\sigma_{\bar{X}} = \frac{\sigma}{\sqrt{n}} \cdot \sqrt{\frac{N-n}{N-1}}$$

Formula for determining the standard deviation of a sampling distribution of $\bar{X}$ through knowledge of the population standard deviation σ, when the population is infinite or finite but at least 20 times as great as the sample size:

$$\sigma_{\bar{X}} = \frac{\sigma}{\sqrt{n}}$$

A practical substitute for the formula $\sigma_{\bar{X}} = (\sigma/\sqrt{n}) \cdot \sqrt{(N-n)/(N-n)}$ useful when the population standard deviation σ is unknown:

$$s_{\bar{X}} = \frac{s}{\sqrt{n}} \cdot \sqrt{\frac{N-n}{N-1}}$$

A practical substitute for the formula $\sigma_{\bar{X}} = (\sigma/\sqrt{n})$ useful when the standard deviation of the population σ is unknown:

$$s_{\overline{X}} = \frac{s}{\sqrt{n}}$$

Formula for converting an $\overline{X}$ value into standard units:

$$Z = \frac{\overline{X} - \mu}{\dfrac{\sigma}{\sqrt{n}}}$$

Mathematically correct formulas for obtaining a confidence interval expected to include the unknown value of the population mean:

$$\overline{X} - Z_{\frac{\alpha}{2}}\left(\frac{\sigma}{\sqrt{n}}\right) \leq \mu \leq \overline{X} + Z_{\frac{\alpha}{2}}\left(\frac{\sigma}{\sqrt{n}}\right)$$

or

$$\overline{X} \pm Z_{\frac{\alpha}{2}}\left(\frac{\sigma}{\sqrt{n}}\right)$$

Practical formulas for obtaining a confidence interval expected to include the unknown value of the population mean. Used when the standard deviation of the population σ is unknown and the sample size n is large:

$$\overline{X} - Z_{\frac{\alpha}{2}}\left(\frac{s}{\sqrt{n}}\right) \leq \mu \leq \overline{X} + Z_{\frac{\alpha}{2}}\left(\frac{s}{\sqrt{n}}\right)$$

or

$$\overline{X} \pm Z_{\frac{\alpha}{2}}\left(\frac{s}{\sqrt{n}}\right)$$

Practical formula for obtaining a confidence interval expected to include the unknown value of the population mean. Used when the standard deviation of the population σ is unknown and the sample size n is small:

$$\overline{X} - t_{\frac{\alpha}{2}}\frac{s}{\sqrt{n}} \leq \mu \leq \overline{X} + t_{\frac{\alpha}{2}}\frac{s}{\sqrt{n}}$$

or

$$\overline{X} \pm t_{\frac{\alpha}{2}}\frac{s}{\sqrt{n}}$$

Formula for determining the desired sample size when the population mean μ is to be estimated with a specified degree of confidence and a specified degree of precision:

$$n \geq \left(\frac{Z_{\frac{\alpha}{2}} \cdot \sigma}{E}\right)^2$$

Formula for computing the population proportion:

$$p = \frac{X_P}{N}$$

Formula for computing the proportion for a single sample:

$$\overline{p} = \frac{X_s}{N}$$

Formulas for determining the mean of a sampling distribution of $\bar{p}$ directly from the sample proportions:

$$\mu_{\bar{p}} = \Sigma\ [P(\bar{p})\bar{p}] \quad \text{or} \quad \mu_{\bar{p}} = \frac{\Sigma\ \bar{p}}{N^*}$$

Formulas for determining the standard deviation of a sampling distribution of $\bar{p}$ directly from the sample proportions:

$$\sigma_{\bar{p}} = \sqrt{\Sigma\ [P(\bar{p})(\bar{p} - p)^2]} \quad \text{or} \quad \sigma_{\bar{p}} = \sqrt{\frac{\Sigma\ (\bar{p} - p)^2}{N^*}}$$

Formula for determining the standard deviation of a sampling distribution of $\bar{p}$ through knowledge of the population proportion. The $\sqrt{(N - n)/(N - 1)}$ part implies that the population is finite and not at least 20 times as great as the sample size:

$$\sigma_{\bar{p}} = \sqrt{\frac{pq}{n}} \cdot \sqrt{\frac{N - n}{N - 1}}$$

Formula for determining the standard deviation of a sampling distribution of $\bar{p}$ through knowledge of the population proportion when the population is infinite or finite but at least 20 times as great as the sample size:

$$\sigma_{\bar{p}} = \sqrt{\frac{pq}{n}}$$

Formula for converting a $\bar{p}$ value into standard units:

$$Z = \frac{\bar{p} - p}{\sqrt{\dfrac{pq}{n}}}$$

Formulas for obtaining a confidence interval expected to include the unknown value of p:

$$\bar{p} - Z_{\frac{\alpha}{2}}\sqrt{\frac{\bar{p}\bar{q}}{n}} \leq p \leq \bar{p} + Z_{\frac{\alpha}{2}}\sqrt{\frac{\bar{p}\bar{q}}{n}}$$

or

$$\bar{p} \pm Z_{\frac{\alpha}{2}}\sqrt{\frac{\bar{p}\bar{q}}{n}}$$

Formula for determining the desired sample size when the population proportion p is to be estimated with a specified degree of confidence and a specified degree of precision:

$$n \geq pq\left(\frac{Z_{\frac{\alpha}{2}}}{E}\right)^2$$

Formula for converting an $\overline{X}$ value into standard units when the population standard deviation σ is unknown and the sample size n is small:

$$t = \frac{\overline{X} - \mu}{\dfrac{s}{\sqrt{n}}}$$

Questions and Problems

8.1 Explain the meaning of each of the following:
  a. Sampling distribution
  b. Sampling distribution of $\overline{X}$
  c. Sampling distribution of $\overline{p}$
  d. Sampling distribution of the mode
  e. Unbiased estimator
  f. Point estimate

8.2 Explain the meaning of each of the following:
  a. Population distribution
  b. Finite correction factor
  c. Standard error of the mean
  d. Standard error of the proportion
  e. $\sigma/\sqrt{n}$
  f. $\sqrt{pq/n}$
  g. Central limit theorem

8.3 Explain the meaning of each of the following:
  a. Sample distribution
  b. Interval estimate
  c. Confidence interval
  d. Confidence coefficient
  e. Proportion
  f. t statistic
  g. t distribution

8.4 Distinguish between:
  a. Population distribution and sample distribution
  b. Sample distribution and sampling distribution
  c. Population distribution and sampling distribution
  d. Point estimate and interval estimate

8.5 Distinguish between:
  a. Interval estimate and confidence interval
  b. Standard normal probability distribution and the t distribution
  c. μ and $\mu_{\overline{X}}$
  d. $\overline{X}$ and $\mu_{\overline{X}}$
  e. $\sigma/\sqrt{n}$ and $\sigma/\sqrt{n} \cdot \sqrt{(N-n)/(N-1)}$

8.6 Distinguish between:
  a. $\sqrt{pq/n}$ and $\sqrt{pq/n} \cdot \sqrt{(N-n)/(N-1)}$
  b. $\sigma/\sqrt{n}$ and $s/\sqrt{n}$
  c. $\sigma_{\overline{X}}$ and $s_{\overline{X}}$
  d. $\sqrt{pq/n}$ and $\sqrt{\overline{p}\overline{q}/n}$
  e. $\overline{\theta}$ and θ

8.7 Distinguish between:
  a. p and $\overline{p}$
  b. σ and $\sigma_{\overline{X}}$
  c. $\overline{X} - Z_{\frac{\alpha}{2}}(s/\sqrt{n}) \leq \mu \leq \overline{X} + Z_{\frac{\alpha}{2}}(s/\sqrt{n})$ and $\overline{X} - t_{\frac{\alpha}{2}}(s/\sqrt{n}) \leq \mu \leq \overline{X} + t_{\frac{\alpha}{2}}(s/\sqrt{n})$
  d. $\sqrt{(N-n)/(N-1)}$ and N/n
  e. $\sigma_{\overline{X}}$ and $\sigma_{\overline{p}}$

8.8 Distinguish among:
  a. $\overline{X}$, $\mu_{\overline{X}}$, μ, and $P(\overline{X})$
  b. $\overline{p}$, p, $\mu_{\overline{p}}$, and $P(\overline{p})$
  c. s, $s_{\overline{X}}$, $\sigma_{\overline{X}}$, and σ
  d. n, N, N^*, and degrees of freedom (as used in section 8.22)

8.9 Indicate which of the following statements you agree with and which you disagree with, and defend your opinions:

a. A sampling distribution is obtained by selecting a random sample from a population of interest and organizing the sample observations into a frequency or a probability distribution.

b. The mean of a sampling distribution of $\overline{X}$ will usually equal the population mean μ.

c. The mean of a sampling distribution of $\overline{p}$ can be expected to equal the population mean μ.

d. When we say that $\overline{X}$ is an unbiased estimator of μ, we are, in effect, saying that $\overline{X} = \mu$.

8.10 Indicate which of the following statements you agree with and which you disagree with, and defend your opinions:

a. When we say that $\overline{X}$ is an unbiased estimator of μ, we are, in effect, saying $\mu_{\overline{X}} = \mu$.

b. When the population size N is at least 20 times as great as the sample size n, we may dispense with the use of the finite correction factor when computing the standard error of the mean or the standard error of the proportion.

c. When deriving a confidence interval for the population mean μ, we must always use t instead of Z in the confidence interval formula if the sample size n is equal to or less than 30.

d. The principal purpose for using the finite correction factor is to ensure that the sampling distribution of $\overline{X}$ (or the sampling distribution of $\overline{p}$) is approximately normal.

8.11 Indicate which of the following statements you agree with and which you disagree with, and defend your opinions:

a. Since in practice we will ordinarily work with a single sample drawn from a population of interest, the concept of sampling distribution will be devoid of practical importance.

b. We can make use of a sampling distribution even though it does not exist in reality.

c. If a sample is selected at random from a population of interest, we can count on its being a miniature replica, both in shape and with respect to all important descriptive measurements, of the parent population.

d. $\overline{X} = \mu$.

8.12 Indicate which of the following statements you agree with and which you disagree with, and defend your opinions:

a. Because a point estimate of a population parameter is so very precise, it is to be preferred over a confidence interval for the same parameter.

b. Since the amount of variation in a sampling distribution of $\overline{X}$, as measured by $\sigma_{\overline{X}}$, is dependent on the amount of variation in the parent population, as measured by σ, there is little an investigator can do to reduce $\sigma_{\overline{X}}$.

c. If both np and nq exceed 5, we know immediately that the finite correction factor need not be used when computing the standard error of the proportion.

d. Since the mathematically correct formula for computing a confidence interval for μ, $\overline{X} - Z_{\frac{\alpha}{2}}(\sigma/\sqrt{n}) \leq \mu \leq \overline{X} + Z_{\frac{\alpha}{2}}(\sigma/\sqrt{n})$, contains Z, when using this formula we implicitly assume that the population is normally distributed.

8.13 Indicate which of the following statements you agree with and which you disagree with, and defend your opinions:

a. $\overline{p} = p$.

b. $\mu_{\overline{p}} = \mu$.

c. If the sample size n is doubled, the standard error of the mean $\sigma_{\overline{X}}$ will be halved.

d. Neither the use of $\overline{X} - Z_{\frac{\alpha}{2}}(s/\sqrt{n}) \leq \mu \leq \overline{X} + Z_{\frac{\alpha}{2}}(s/\sqrt{n})$ nor $\overline{X} - t_{\frac{\alpha}{2}}(s/\sqrt{n}) \leq \mu \leq \overline{X} + t_{\frac{\alpha}{2}}(s/\sqrt{n})$ requires that any assumption be made about the shape of the population.

8.14 Explain in words how you would construct a sampling distribution of

a. $\overline{X}$

b. $\overline{p}$

c. Median

8.15 A population consists of five company treasurers whose incomes last year (in thousands of dollars) are as follows:

Treasurer	Income
Adams	25
Billings	27
Carlton	30
DeVries	35
Emory	50

A simple random sample of two company treasurers is to be selected in connection with a salary survey conducted by a professional organization.
  a. Compute the population mean μ.
  b. Construct the sampling distribution of $\overline{X}$.
  c. What is the probability that the sample mean $\overline{X}$ will be
    1. $32,500?
    2. Less than $28,000?
    3. Over $35,000?
  d. What is the probability that the sample mean $\overline{X}$ will be the same as the population mean μ?
  e. What is the probability that the sample mean $\overline{X}$ will differ from the population mean μ by
    1. More than $5000?
    2. Not more than $10,000?
  f. Compute the mean of the sampling distribution of $\overline{X}$. Does it equal the population mean μ?

8.16 Refer to problem 8.15. (You must have worked part *b* of problem 8.15 to work this one.)
  a. Compute the population standard deviation σ.
  b. Compute the standard deviation of your sampling distribution of $\overline{X}$ directly from the sample means using Equation (8.6.2).
  c. Compute the standard deviation of the sampling distribution of $\overline{X}$ (that is, the standard error of the mean $\sigma_{\overline{X}}$) by using Equation (8.6.3). Is your answer the same as the result you obtained in connection with part *b* of this problem?

8.17 Refer to problem 8.15. (You must have worked part *b* of problem 8.15 to answer this one.) Is your sampling distribution of $\overline{X}$ approximately normal? Elaborate.

8.18 Refer to problems 8.15 and 8.16. Suppose the sample size had been $n = 3$, rather than $n = 2$.
  a. Would the mean of the sampling distribution of $\overline{X}$ be any different from the value that would be obtained from working part *f* of problem 8.15? Explain your reasoning.
  b. Would the standard error of the mean $\sigma_{\overline{X}}$ be any different from the value which would be obtained from working part *b* of problem 8.16? Explain your reasoning.

8.19 A population consists of six skilled craftsmen employed by the Tommy Wrought Corporation; the name of each population member and his or her output of a certain decorative wrought-iron item on a given day is as follows:

Craftsman	Number of Items Completed
Abbott	4
Babson	6
Carlton	6
Dodd	6
Eaton	8
Ferguson	10

A simple random sample of three of these craftsmen is to be selected for study in connection with a proposed revision in the company's incentive bonus system.

a. Compute the population mean μ.

b. Construct the sampling distribution of $\overline{X}$.

c. What is the probability that the sample mean $\overline{X}$ will be
   1. 6.67?
   2. 6 or under?
   3. Over 7?

d. What is the probability that the sample mean $\overline{X}$ will be the same as the population mean μ?

e. What is the probability that the sample mean $\overline{X}$ will differ from the population mean μ by
   1. Less than 1?
   2. More than 2?

f. Compute the mean of the sampling distribution of $\overline{X}$. Does it equal the population mean μ?

8.20 Refer to problem 8.19. (You must have worked part *b* of problem 8.19 to work this one.)

a. Compute the standard deviation of the population σ.

b. Compute the standard deviation of the sampling distribution of $\overline{X}$ directly from the sample means using Equation (8.6.2).

c. Compute the standard deviation of the sampling distribution of $\overline{X}$ (that is, the standard error of the mean $\sigma_{\overline{X}}$) by using Equation (8.6.3). Is your answer the same as the result you obtained in connection with part *b* of this problem?

8.21 Refer to problem 8.19. (You must have worked part *b* of problem 8.19 to answer this one.) Is your sampling distribution of $\overline{X}$ approximately normal? Elaborate.

8.22 Refer to problems 8.19, 8.20, and 8.21. Suppose the sample size had been $n = 4$ rather than $n = 3$.

a. Would the mean of the sampling distribution of $\overline{X}$ be any different from the value that would be obtained from working part *f* of problem 8.19? Explain your reasoning.

b. Would the standard error of the mean $\sigma_{\overline{X}}$, be any different from the value that would be obtained from working part *b* of problem 8.20? Explain your reasoning.

c. Would the shape of the sampling distribution of $\overline{X}$ be very different from the shape that would be obtained using a sample size of $n = 3$? Explain your reasoning.

8.23 A simple random sample of size $n = 400$ is to be selected from a population of $N = 10,000$ items having a mean μ of 117.5 and a standard deviation σ of 16.

a. What is the probability that the mean of the random sample $\overline{X}$ will not differ from the population mean μ by more than 1?

b. The sample mean will fall within what range about 85% of the time? Assume symmetrical limits around the population mean.

c. In working parts *a* and *b*, you were required to proceed as if the shape of the parent population were of no concern. If you had been told that the population is distributed approximately normally, would that information have led to any changes in the procedures you used? Why or why not? If you had been told that the population is distributed in a nonnormal manner, would that information have led to any changes in the procedures you used? Why or why not?

d. Did you make use of the finite correction factor? Why or why not?

e. Did you make any use of a sampling distribution of $\overline{X}$. Explain clearly the reasoning behind your answer.

8.24 Refer to problem 8.23. Redo parts *a* and *b* assuming that the population standard deviation σ is known to be 8 rather than 16, but that all other facts are the same. What effect, if any, did the smaller population standard deviation have on your answer to part *a* of problem 8.23? Part *b* or problem 8.23?

8.25 Refer to problem 8.23. Redo parts *a* and *b* assuming that the sample size is specified to be $n = 100$ rather than $n = 400$, but that all other facts are the same. What effect, if any, did the smaller sample size have on your answer to part *a* of problem 8.23? Part *b* of problem 8.23?

8.26 The Tramar Corporation employs 5000 people. Last year the average number of days lost because of illness by this population of employees was $\mu = 15$; the standard deviation was $\sigma = 6$. A simple random sample of 30 employees is to be selected in connection with a planned experiment to determine the health benefits of a company-sponsored exercise program.
   a. What is the probability that the sample mean $\overline{X}$ is at least 17?
   b. What is the probability that the sample mean $\overline{X}$ will not differ from the population mean μ by more than one day?
   c. If the sample size were to be $n = 60$ rather than $n = 30$, what would the answer to parts *a* and *b* of this problem be?
   d. If the population standard deviation σ were 4 rather than 6, what would the answers to parts *a* and *b* of this problem be (assume $n = 30$)?

8.27 A random sample of 25 common stocks is to be selected from a population of 10,000 common stocks to obtain an estimate of the population mean dividend yield. Suppose that the population mean μ is actually 8% and the population standard deviation σ is 4%.
   a. What is the probability that the mean dividend yield $\overline{X}$ for this sample of common stocks will be less than 6.75%?
   b. What is the probability that the mean dividend yield $\overline{X}$ for this sample of common stocks will be between 7% and 9%?
   c. If the sample size were to be $n = 121$ rather than $n = 25$, what would the answers to the *a* and *b* parts of this problem be?
   d. If the population standard deviation σ were 3% rather than 4%, what would the answers to parts *a* and *b* of this problem be (assume $n = 25$)?

8.28 A simple random sample of 400 manufacturing firms from a population of 15,000 such firms yielded a value of average after-tax profits $\overline{X}$ of $2 million and a standard deviations of $0.5 million.
   a. What is your point estimate of average after-tax profits for the entire population of manufacturing firms? Why is such an estimate called a point estimate?
   b. What are the principal advantages and disadvantages of using a point estimate of the true but unknown population mean μ?
   c. Develop an interval estimate of the population average after-tax profit using a confidence coefficient of .95.
   d. Explain carefully the meaning of your confidence interval. What is one 95% confident of?
   e. Why is the confidence coefficient in this problem really only *approximately* .95?
   f. If another sample of size $n = 400$ were selected from this same population of 15,000 manufacturing companies, would the same .95 confidence interval be obtained as was obtained in connection with part *c*? Why or why not?

8.29 Refer to problem 8.28. Redo part *c* using a .98 confidence coefficient. Is the resulting confidence interval wider or narrower than the one obtained in connection with part *c* of problem 8.28? Which confidence interval do you believe would be more useful? Defend your choice. Which confidence interval involves a greater risk of failing to surround the true but unknown population mean? Defend your choice.

8.30 A simple random sample of 36 mutual funds from a population of 850 such funds showed an average gain in assets per share during the most recent year of $\overline{X} = 10\%$ and a standard deviation of $s = 4\%$.

a. What is your point estimate of the population average percent gain in assets per share? Why is such an estimate called a point estimate?

b. Develop an interval estimate of the population mean using a confidence coefficient of .90.

c. Explain clearly the meaning of your confidence interval. What is one 90% confident of?

d. Why is the confidence coefficient in this problem really only *approximately* .90?

e. If another sample of size $n = 36$ were selected from this same population of mutual funds, would the same .90 confidence interval be obtained as was obtained in connection with part *b*? Why or why not?

 8.31 Refer to problem 8.30. Redo part *b* using a .95 confidence coefficient. Is the resulting confidence interval wider or narrower than the one obtained in connection with part *b* of Problem 8.30? Which confidence interval do you believe would be more useful? Defend your choice. Which confidence interval would involve the greater risk of failing to surround the true but unknown population mean? Defend your choice.

 8.32 A study conducted with a view to estimating the average number of radios per household in a certain city found the mean number of radios per household $\overline{X}$ for a sample of $n = 900$ households to be 3.4. The standard deviation s was found to be 1.2. There are 100,000 households in this city.

a. What is your point estimate of the mean number of radios owned per household for the entire population of households? Why is such an estimate called a point estimate?

b. Develop an interval estimate of the population mean number of radios per household using a confidence coefficient of .99.

c. Explain carefully the meaning of your confidence interval. What is one 99% confident of?

d. Why is the confidence coefficient in this problem only *approximately* .99?

e. If another sample of size 900 were selected from this same population of 100,000 households, would the same .99 confidence interval be obtained as was obtained in connection with part *b*? Why or why not?

f. Obtain a .99 confidence interval for the total number of radios owned by this population of households. Explain clearly how you obtained this confidence interval.

 8.33 Refer to problem 8.32.

a. Redo part *b* using a .95 confidence coefficient.

b. Is the resulting confidence interval wider or narrower than the one obtained in connection with part *b* of Problem 8.32?

c. Which confidence interval do you believe would be more useful? Defend your choice.

d. Which confidence interval involves a greater risk of failing to surround the true but unknown population mean? Defend your choice.

e. Obtain a .95 confidence interval for the total number of radios owned by this population of households. Explain clearly how you obtained this confidence interval.

 8.34 A random sample of insurance salespeople is taken from a population of 400 such salespeople and the bonuses received on sales made last week were recorded. They are (in dollars) as follows:

135	148	227	93
180	225	118	173
185	206	220	170
192	248	190	205

Assume that the population of bonuses is approximately normal.

a. Obtain a .95 confidence interval that you would expect to include the true but unknown population mean bonus.

b. Why was it necessary to assume that the population was normally distributed?

8.35 At the manufacturer's prompting, a taxicab company agreed to test a new brand of synthetic tire on a small scale for a specified length of time. If the test showed sufficient evidence that the average useful life of this new brand of tire was greater than that for the brand regularly used, the company would order a very large quantity of the new tires to use on its fleet of taxicabs. The test procedure was as follows: One tire of the new type was used in a randomly selected position on each of 6 of the company's cabs along with 3 ordinary tires. The test cars would be used under ordinary driving conditions and the tires replaced just as soon as they ceased having enough tread to meet the company's long-standing safety specifications. The number of miles each tire traveled before replacement became necessary was recorded.

Suppose that the useful lives of the 18 ordinary tires (i.e., 6 taxicabs times 3 tires of the ordinary type per cab) were found to be (in thousands of miles) as follows:

50	50	44	38	57	36	41	56	41
56	52	50	51	55	45	47	62	51

Also suppose that the useful lives of the 6 test tires were found to be (in thousands of miles) as follows:

62	68	63	62	63	72

a. Compute the mean $\overline{X}$ and the standard deviation s of the data for the ordinary tires.
b. Using a .99 confidence coefficient, construct a confidence interval which you would expect to include the true but unknown population mean number of miles of useful life for the ordinary tires.
c. Compute the mean $\overline{X}$ and the standard deviation s of the data for the new brand of tires.
d. Using a .99 confidence coefficient, construct a confidence interval which you would expect to include the true but unknown population mean number of miles of useful life for the new brand of tires.
e. Do you feel that any conclusion can safely be drawn regarding the superiority or inferiority of the new brand of tire with respect to the mean number of miles of useful life? Explain fully the reasoning behind your answer.
f. With respect to parts b and d of this problem, was it necessary to make any assumption about the shapes of the parent populations? Tell why or why not.

 8.36 A study is to be conducted in Metropolis City. The purpose of the study is to obtain a dependable estimate for the mean expenditure per household on durable goods (i.e., automobiles, home appliances, jewelry, etc.) during a recent month when several special promotions were in effect. It is important that the error of estimation not be more than $50. It is also important that a confidence coefficient of .95 be used. Informed estimates of the population standard deviation σ range between $100 and $200. If the researchers prefer to err in the direction of a larger-than-necessary sample size, what sample size n would they use?

 8.37 Refer to problem 8.36. Assume that it is important that (1) the error of estimation not be greater than $75 and that (2) the confidence coefficient be .98. Taking the population standard deviation to be $200, what must the sample size be?

 8.38 A population of eight stores is to be sampled to determine what proportion of stores sell Buck Rogers' Ray Guns for under $5. A list of this small population of stores and an indication of the ones which do and the ones which do not sell this toy for less than $5 appears in Table 8.12.

A simple random sample of two stores is to be selected from this population in connection with a study aimed at reevaluating the manufacturing company's suggested retail price.
a. What proportion of stores in the population does sell the toy gun at prices under $5?
b. Construct the sampling distribution of $\overline{p}$.

Table 8.12 Information on Stores Which Do and Do Not Sell the Buck Rogers' Ray Gun for Under $5

Store	Characteristic
A	Sells ray gun for less than $5
B	Sells ray gun for less than $5
C	Does not sell ray gun for less than $5
D	Sells ray gun for less than $5
E	Sells ray gun for less than $5
F	Sells ray gun for less than $5
G	Does not sell ray gun for less than $5
H	Sells ray gun for less than $5

c. What is the probability that the sample proportion $\bar{p}$ will be
   1. 1?
   2. .5?
   3. 0?
   4. Same as the population proportion p?
   5. Less than the population proportion p?
d. Determine the mean of your sampling distribution. Is it equal to the population proportion?
e. Determine the standard deviation of your sampling distribution (i.e., the standard error of the proportion $\sigma_{\bar{p}}$) $\sigma_{\bar{p}}$ directly from the sample $\bar{p}$ values using Equation (8.15.1).
f. Compute the standard deviation of the $\bar{p}$ values (the standard error of the proportion $\sigma_{\bar{p}}$) using Equation (8.15.3). Did you arrive at the same answer you obtained in connection with part *e*?
g. Is your sampling distribution of $\bar{p}$ approximately normal? Elaborate.

8.39 Refer to problem 8.38, parts *d*, *e*, and *g*. Suppose that the sample size had been $n = 4$ rather than $n = 2$.
   a. Would the mean of the sampling distribution of $\bar{p}$ be any different from the value you obtained in connection with part *d* of problem 8.38? Explain your reasoning?
   b. Would the standard deviation of the sampling distribution of $\bar{p}$ be any different from the value you obtained in connection with part *e* of problem 8.38? Explain your reasoning.
   c. Would the shape of the sampling distribution be any different? Explain your reasoning.

8.40 Management of the XYZ Company has been presented with a merger proposal by the ABC Company. The opinions of the five top officers are presented in Table 8.13.

Table 8.13 Opinions of XYZ's Top Officers Regarding a Proposed Merger with ABC

Officer	Opinion
President	Does not favor merger
Vice-President, Finance	Does not favor merger
Vice-President, Production	Favors merger
Vice-President, Marketing	Favors merger
Secretary-Treasurer	Does not favor merger

A simple random sample of three of these top members of management is to be selected.

    a. What is the population proportion of officers favoring the merger?

    b. Construct the sampling distribution of $\bar{p}$.

    c. What is the probability that the sample proportion $\bar{p}$ will be

        1. 1?

        2. .6667?

        3. .3333?

        4. Same as the population proportion p?

        5. No more than .3 away from the population proportion p?

    d. Determine the mean of your sampling distribution. Is it equal to the population proportion?

    e. Determine the standard deviation of your sampling distribution (i.e., the standard error of the proportion $\sigma_{\bar{p}}$) directly from the sample $\bar{p}$ values using Equation (8.15.1).

    f. Determine the standard deviation of the sampling distribution of $\bar{p}$ (the standard error of the proportion $\sigma_{\bar{p}}$) using Equation (8.15.3). Did you arrive at the same answer you obtained in connection with part *e*?

    g. Is your sampling distribution approximately normal? Elaborate.

 8.41 Refer to problem 8.40. Before arriving at a decision on whether the merger should be consumated, the president of the XYZ Company felt obliged to survey the attitudes of employees who had been with the company ten years or longer. Since there were 700 such employees, he decided to settle for a sample based on $n = 36$ personal interviews. Unknown to the president, the true population proportion of long-time employees favoring the merger was $p = .2$.

    a. What is the probability that the proportion of employees interviewed by the president who favor the merger will not differ from the true population proportion by more than .1?

    b. The sample proportion would fall within what range about 75% of the time? Assume symmetrical limits around the population proportion.

    c. Did you use the finite correction factor? Why or why not?

    d. Did you make any use of a sampling distribution of $\bar{p}$. Explain clearly the reasoning behind your answer.

 8.42 A random sample of 25 employees of a company is to be selected from the population of 700 employees to determine what proportion have set up Individual Retirement Accounts (IRAs). Assume that the true population proportion is .3.

    a. What is the probability that the proportion of employees surveyed who have set up such retirement accounts will not differ from the true population proportion by more than .1?

    b. The sample proportion would fall within what range about 50% of the time? Assume symmetrical limits around the population proportion.

    c. Did you use the finite correction factor? Why or why not?

    d. Did you make any use of a sampling distribution of $\bar{p}$? Explain clearly the reasoning behind your answer.

 8.43 A sample study was conducted to determine what proportion of television viewers in a certain large city recalled having seen any of the Oxnard Insurance Company commercials during the past week. The sample size was $n = 1000$, and 300 of the sample members indicated that they had seen at least one such commercial.

    a. What is your point estimate of the proportion of television viewers in this large population who, if questioned, would be able to recall having seen any of Oxnard's commercials? Why is such an estimate called a point estimate?

    b. Is the true population proportion of people who recall having seen any of Oxnard's commercials necessarily the same as the population proportion p of people who had in fact seen some of Oxnard's commercials during the week in question? Justify your answer. Explain, using hypothetical facts as needed, how failure to make this distinction could lead to erroneous decisions.

c. Develop an interval estimate of the true population proportion of people who, if questioned, would be expected to indicate that they had seen Oxnard Insurance commercials during the week in question. Use a .99 confidence coefficient.

d. Explain carefully the meaning of your confidence interval. What is one 99% confident of?

e. Why is the confidence coefficient in this problem really only *approximately* .99?

f. If another sample of size $n = 1000$ were selected from this same population of television viewers, would the same .99 confidence interval be obtained as was obtained in connection with part c? Why or why not?

8.44 Refer to problem 8.43.

a. Redo part c using a .95 confidence coefficient.

b. Is the resulting confidence interval wider or narrower than the one obtained in connection with part c of problem 8.43?

c. Which confidence interval do you believe would be more useful? Defend your choice.

d. Which confidence interval involves a greater risk of failing to include the true but unknown population proportion? Defend your choice.

8.45 A medical insurance company conducted a sample survey to estimate the proportion of people over age 65 who suffer from severe chronic illnesses. A random sample of 2500 people over 65 years of age was selected from among all of those in this age group in the United States. Of the sample members, 1000 were found to be suffering from severe chronic illnesses.

a. What is your point estimate of the proportion of people over 65 in the United States who suffer from severe chronic illnesses? Why is such an estimate called a point estimate?

b. Develop an interval estimate of the true population proportion of people over 65 who suffer from severe chronic illnesses. Use a .95 confidence coefficient.

c. Explain carefully the meaning of your confidence interval. What is one 95% confident of?

d. Why is the confidence coefficient in this problem really only *approximately* .95?

e. If another sample of size $n = 2500$ were selected from this same population of persons over 65, would the same .95 confidence interval be obtained as was obtained in connection with part b? Why or why not?

8.46 Refer to problem 8.45.

a. Redo part b using a .98 confidence interval.

b. Is the resulting confidence interval wider or narrower than the one obtained in connection with part b of problem 8.45?

c. Which confidence interval do you believe would be more useful? Defend your choice.

d. Which confidence interval involves a greater risk of failing to include the true but unknown population proportion? Defend your choice.

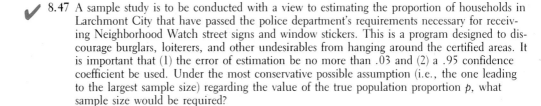

8.47 A sample study is to be conducted with a view to estimating the proportion of households in Larchmont City that have passed the police department's requirements necessary for receiving Neighborhood Watch street signs and window stickers. This is a program designed to discourage burglars, loiterers, and other undesirables from hanging around the certified areas. It is important that (1) the error of estimation be no more than .03 and (2) a .95 confidence coefficient be used. Under the most conservative possible assumption (i.e., the one leading to the largest sample size) regarding the value of the true population proportion p, what sample size would be required?

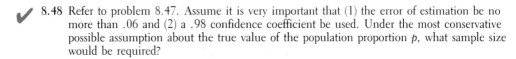

8.48 Refer to problem 8.47. Assume it is very important that (1) the error of estimation be no more than .06 and (2) a .98 confidence coefficient be used. Under the most conservative possible assumption about the true value of the population proportion p, what sample size would be required?

TAKE CHARGE*

8.49 Use the integers 1, 2, 3, 4, 5, 6, 7, 8, and 9 to construct a population of any shape you desire. The population size N should be 100. Thus, you might, for example, use 1 thirty times, 2 twenty times, 3 ten times, 4 ten times, 5 ten times, and 6, 7, 8, and 9 five times each. When developing the population, you should (1) make use of more than one of the integers listed above and (2) avoid making the population approximate a normal curve too closely. Once you have determined how many of each of the integers you are going to use, write the integers the appropriate number of times on 100 identical small squares of paper. Place the small squares of paper into a hat or bowl, and mix them thoroughly.

a. Compute the mean μ of the population you developed.

b. Compute the standard deviation σ of the population you developed.

c. Select a sample, at random and without replacement, of $n = 4$ of the small squares of paper. Compute $\overline{X}$ for the numbers shown thereon. Return the 4 squares of paper to the hat or bowl, and mix all the pieces of paper thoroughly. Select another sample, at random and without replacement, of $n = 4$ of the squares of paper. Compute $\overline{X}$ for the 4 associated integers. Return these squares of paper and mix. Repeat this procedure until 20 samples of size 4 have been selected and their means recorded.

d. Determine the mean of the 20 sample means. Is the mean of the $\overline{X}$ values close to the value of μ you calculated in connection with part a above? Elaborate. What would have to be done to ensure that the mean of the $\overline{X}$ values was exactly equal to the population mean μ?

e. Determine the standard deviation of the 20 sample means. Is this standard deviation about equal to $\sigma/\sqrt{4}$? Elaborate. What would have to be done to ensure that the standard deviation of the $\overline{X}$ values was exactly equal to $\sigma/\sqrt{4}$?

f. Organize the 20 sample means into a relative-frequency distribution. Is this distribution's shape approximately normal? Elaborate. Does it appear to be more nearly normal than the population shape?

g. Follow the instructions given in part c above, only this time select 20 samples of size $n = 12$ and compute their $\overline{X}$ values.

h. Determine the mean of the 20 sample means. Is the mean of the $\overline{X}$ values close to the value of μ you calculated in connection with part a above? Elaborate. Is the mean of the $\overline{X}$ values of the 20 samples of size 12 closer to the population mean μ than the mean of the $\overline{X}$ values of the 20 samples of size 4?

i. Determine the standard deviation of the 20 sample means. Is this standard deviation about equal to $(\sigma/\sqrt{12}) \cdot \sqrt{(100-12)/(100-1)}$? Elaborate. Is the standard deviation of the $\overline{X}$ values of the 20 samples of size 12 closer to $(\sigma/\sqrt{12}) \cdot \sqrt{(100-12)/(100-1)}$ than the standard deviation of the $\overline{X}$ values you obtained in connection with part e was to $\sigma/\sqrt{4}$?

j. Organize the 20 sample means into a relative-frequency distribution. Is this distribution's shape approximately normal? Does it appear to be more nearly normal than the relative-frequency distribution you constructed in connection with part f above?

TAKE CHARGE*

8.50 Place 100 blue and white poker chips in a hat or bowl. You may use as many blue and as many white chips as you like provided that (1) at least one chip of each color be used and (2) the number of blue chips plus the number of white chips equals 100. Think of this set of chips as a population of attribute data.

*Group participation is suggested.

a. Compute the population proportion p assuming that the proportion of blue chips is what you are interested in.
b. Select a sample, at random and without replacement, of $n = 5$ chips. Compute $\bar{p}$ for this sample. Return the 5 chips to the hat or bowl, and mix all the chips thoroughly. Select another sample, at random and without replacement, of $n = 5$ chips. Compute $\bar{p}$ for this sample. Repeat this procedure until 10 samples of size $n = 5$ have been selected and their proportions recorded.
c. Determine the mean of the 10 sample proportions. Is this mean of the $\bar{p}$ values close to the value of p you calculated in connection with part a above? Elaborate. What would have to be done to ensure that the mean of the $\bar{p}$ values was exactly equal to the population proportion p?
d. Determine the standard deviation of the 10 sample proportions. Is this standard deviation about equal to $\sqrt{pq}/5$?
e. Organize the 10 sample proportions into a relative-frequency distribution. Is this distribution's shape approximately normal? Elaborate.
f. Follow the instructions given in part b, only this time select 10 samples of size $n = 20$ and compute their $\bar{p}$ values.
g. Determine the mean of the 10 sample proportions. Is this mean of the $\bar{p}$ values close to the value of p you calculated in connection with part a above? Elaborate. Is the mean of the $\bar{p}$ values of the 10 samples of size 20 closer to the population proportion p than the mean of the $\bar{p}$ values of the 10 samples of size $n = 5$?
h. Determine the standard deviation of the 10 sample proportions. Is this standard deviation of the $\bar{p}$ values about equal to $\sqrt{pq}/20 \cdot \sqrt{(100 - 20)/(100 - 1)}$? Elaborate. Is this standard deviation of the $\bar{p}$ values of the 10 samples of size $n = 20$ closer to $\sqrt{pq}/20 \cdot \sqrt{(100 - 20)/(100 - 1)}$ than the standard deviation of the $\bar{p}$ values you obtained in connection with part d was to $\sqrt{pq}/5 \cdot \sqrt{(100 - 5)/(100 - 1)}$?
i. Organize the 10 sample proportions into a relative-frequency distribution. Is this distribution's shape approximately normal? Does it appear to be more nearly normal than the relative-frequency distribution you constructed in connection with part e above?

Cases for Analysis and Discussion

8.1 "DPC CORPORATION": PART III

Reread Part I of this case (Case 1.5 at the end of Chapter 1). The final paragraph states:

For a specific 3 weeks of simulated use, the average number of service calls was found to be 1.84 for the sample of machines. This corresponds to a .95 confidence interval of 1.66 to 2.02. . . . The range of values representing lengths of time between service calls runs from 1.49 to 1.81 months, with .95 confidence.

1. Tell in words how you think the .95 confidence interval for average number of service calls was determined.

2. Tell in words how you think the .95 confidence interval for average length of time between service calls was determined.

8.2 "MID-CITY BANK OF BIG TOWN": PART I

Prior to 1967, commercial banks in "Big Town" had not become involved in the credit-card business despite the fact that credit cards in many other parts of the United States had enjoyed wide acceptance. For a time in these other areas, there were many entries into the revolving credit-line business. As time passed, however, some of these businesses sold out to larger credit-card operations and some ceased operating altogether. Two cards eventually emerged as preeminent, BankAmericard and MasterCharge.

Although Big Town had had exposure to credit cards issued by the larger department store organizations, both national and local, there had never been an "all purpose" credit card in use. Finally, in 1967 the major banks in Big Town began to recognize the potential for the revolving credit-card business. As a result, the "Fourth National Bank" attained the franchise for BankAmericard. Within a few months the "Mid-City Bank of Big Town" and two other prominent Big Town banks formed an association that obtained the MasterCharge rights for the Rocky Mountain states and the Dakotas.

There quickly came to be considerable competition among the major Big Town banks to sign on affiliate, or agent, banks to issue a specific credit card to their most credit-worthy customers. The cardholder base grew as did outstanding balances, the balances on which interest is charged. The Mid-City Bank of Big Town, a MasterCharge bank, competed aggressively for customers and for agent banks and was very successful in its efforts.

As time passed a rather large base of credit-card customers and outstanding balances was built by Mid-City and the venture did indeed prove to be highly profitable. Management then came to be increasingly concerned with the question, "Now that the desired base of cardholders and outstanding balances has been essentially built, how can we further increase the profitability of the program without raising the interest rate charged on outstanding balances?" The interest rate charged was $1\frac{1}{2}\%$ on the unpaid balance each month (or 18% if expressed as an annual rate) and was considered sacrosanct.

The minimum payback requirement had been set at 6% of the outstanding balance or $10 per month, whichever was greater. Since these minimum payback provisions could be changed, Mid-City's management wondered whether a lowering of the monthly payback requirement would result in higher outstanding balances and, in turn, greater profits.

Table 8.14 Frequency Distributions for "Mid-City Bank of Big Town's" Sample Study Results

Part I: 12-Month Average Outstanding Balances

Size of Outstanding Balance (Dollars)	Number of Customers			
	Minimum-Payback Customers	Medial-Payback Customers	Full-Payback Customers	All Customers
0 and under 110	18	18	105	141
110 and under 220	15	15	13	43
220 and under 330	18	16	3	37
330 and under 440	7	18	3	28
440 and under 550	14	6	2	22
550 and under 660	6	3	1	10
660 and under 770	4	2	1	7
770 and under 880	4	1	0	5
880 and under 990	2	0	0	2
Total	88	79	128	295
	$n = 88$	$n = 79$	$n = 128$	$n = 295$
	$\overline{X} = \$337.50$	$\overline{X} = \$276.39$	$\overline{X} = \$95.39$	$\overline{X} = \$216.08$
	$s = \$239.26$	$s = \$181.77$	$s = \$110.32$	$s = \$206.63$
	$s_{\overline{X}} = \$25.51$	$s_{\overline{X}} = \$20.45$	$s_{\overline{X}} = \$9.75$	$s_{\overline{X}} = \$12.03$

(continued)

Mid-City Bank had 84,415 cardholders who were billed each month. Not all customers paid "a little each month," as the ads suggested. Some paid the full amount owed (a group hereafter referred to as "full-payback customers"), some paid an amount greater than the specified minimum but less than the full balance ("medial-payback customers"), and some paid only the minimum amount required ("minimum-payback customers").

The Mid-City managers presented the president of MasterCharge with the proposal that they be permitted to lower the minimum payback to 3% a month or $5, whichever was greater. The MasterCharge president expressed interest in the plan but said that he would withhold approval until Mid-City had completed a small-scale study of the effects of such a change and had reported its findings to MasterCharge.

Such a study was undertaken. Mid-City selected a random sample of 295 cardholders and informed them of the new minimum-payback terms. The sample members were not told that they were participants in an experimental study, but they *were* informed that the lower payback requirement might be rescinded—with substantial advance notice being given—at a future time. The study extended over an 18-month period; during the first 6 months, regarded as a period of adjustment to the new requirements, no attempt was made to evaluate the new program. For the remaining 12-month period, data were collected, organized, and analyzed. Parts I and II of Table 8.14 show sample distributions and relevant summary measures associated with the three categories of customers. Table 8.15 presents a summary of some of the sample study results.

1. Using the sample results presented in Tables 8.14 and 8.15 and a .99 confidence coefficient, develop interval estimates of the true mean outstanding balance per customer per month for (1) minimum-payback customers, (2) medial-payback customers, (3) full-payback customers, and (4) all customers.

Table 8.14 (*Continued*)

Part II: 12-Month Average Monthly Payback

Size of Payback (Dollars)	Number of Customers			
	Minimum-Payback Customers	Medial-Payback Customers	Full-Payback Customers	All Customers
0 and under 50	86	62	70	218
50 and under 100	2	12	30	44
100 and under 150	0	3	5	8
150 and under 200	0	2	4	6
200 and under 250	0	0	3	3
250 and under 300	0	0	3	3
300 and under 350	0	0	3	3
350 and under 400	0	0	3	3
400 and under 450	0	0	3	3
450 and under 500	0	0	2	2
500 and under 550	0	0	2	2
Total	88	79	128	295
	$n = 88$	$n = 79$	$n = 128$	$n = 295$
	$\overline{X} = \$26.14$	$\overline{X} = \$40.19$	$\overline{X} = \$95.31$	$\overline{X} = \$59.92$
	$s = \$7.49$	$s = \$33.35$	$s = \$122.71$	$s = \$88.36$
	$s_{\overline{X}} = \$0.80$	$s_{\overline{X}} = \$3.75$	$s_{\overline{X}} = \$10.85$	$s_{\overline{X}} = \$5.14$

Table 8.15 Summary Table for "Mid-City Bank's" Sample Study

Type of Customer	(1) Average Outstanding Balance per Customer per Month	(2) Average Payback per Customer per Month	(3) Number of Accounts in Sample n	(4) Number of Accounts in Population N
Minimum payback	$337.50	$26.14	88	25,181
Medial payback	276.39	40.19	79	22,606
Full payback	95.39	95.31	128	36,628
All	$216.08	$59.92	295	84,415

Type of Customer	(5) Total Outstanding Balance per Month for Sample (1) × (3)	(6) Total Payback per month for Sample (2) × (3)	(7) Total Outstanding Balance per Month after Monthly Payback (5) − (6)
Minimum payback	$29,700.00	$ 2,300.32	$27,399.68
Medial payback	21,834.81	3,175.01	18,659.80
Full payback	12,209.92	12,199.68	10.24
All	$63,744.73	$17,675.01	$46,069.72

Type of Customer	(8) Total Interest Payment per Month for Sample (7) × 0.015	(9) Average Interest Payment per Customer per Month[a] (8) ÷ (3)
Minimum payback	$411.00	$4.67 ($0.80)
Medial payback	279.90	3.54 ($1.40)
Full payback	0.1536	0.0012 ($0.20)
All	$691.0536	$2.3412 ($0.98)

[a]Sample standard deviations for this measure are shown in parentheses.

2. Using the sample results presented in Tables 8.14 and 8.15 and a .99 confidence coefficient, develop interval estimates of the true population mean payback per customer per month for (1) minimum-payback customers, (2) medial payback customers, (3) full-payback customers, and (4) all customers.

3. Can you obtain a .99 confidence interval estimate of the true population *total* outstanding balances per month for (1) minimum-payback customers, (2) medial-payback customers, (3) full-payback customers, and (4) all customers? If your answer is Yes, derive these interval estimates. If your answer is No, explain why you believe that such interval estimates cannot be obtained from the data given.

4. Can you obtain a .99 confidence interval estimate of the true population *total* payback per month for (1) minimum-payback customers, (2) medial-payback customers, (3) full-payback customers, and (4) all customers? If your answer is Yes, derive these interval estimates. If your answer is No, explain why you believe that such interval estimates cannot be obtained from the data given.

5. Using the sample results presented in Tables 8.14 and 8.15 and a .99 confidence coefficient, develop interval estimates of the true population mean interest payment per customer per month for (1) minimum-payback customers, (2) medial-payback customers, (3) full-payback customers, and (4) all customers.

6. Can you obtain a .99 confidence interval estimate of the true population *total* interest payments per month for (1) minimum-payback customers, (2) medial-payback customers, (3) full-payback customers, and (4) all customers? If your answer is Yes, derive these interval estimates. If your answer is No, explain why you believe that such interval estimates cannot be obtained from the data given.

7. Do any of the results obtained from your work done in connection with questions 1 through 6 above provide any evidence to indicate that the new payback program either was or was not successful? Justify your answer.

This case is continued at the end of Chapter 9.

8.3 "A.F.R. CONSULTING"

The owner of a small consulting firm wrote up the following case in 1985. It has to do with a project in progress at that time.

"A.F.R. Consulting," a small engineering firm, is in the business of information technology and transfer. A potential new product idea which fits general company business guidelines and which appears to have an available market segment has been suggested. This new product idea is an inexpensive, simple-to-use, automobile mileage computer. A.F.R. wishes to commercialize this product. Before this can be done, we need to know whether the product concept has market characteristics favorable to both user and producer. Specifically, before A.F.R. would undertake further product development, we require information on:

- Size of potential market
- Size of attainable market
- Price/demand curve [not treated in this case]

Information on product attributes and potential uses is shown in Exhibit A [Figure 8.8].

A nine-county area was chosen for an exploratory study. There are 1.76 million cars and light trucks registered in this area and, theoretically, a mileage computer could be sold for each one of these vehicles. It is likely, however, that the upper limit of sales would be the number of households in this nine-county market area. Therefore, this potential market is taken as the target population.

Based on two preliminary personal interview samples using an initial version of the questionnaire, it was determined that a random sample size of 100 would give the desired confidence limits on the mean price that would be paid by those willing to buy.

Frequency-distribution data, pertinent to the information needed at this stage of new product development, are as follows:

Proposed New Product

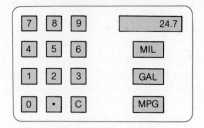

Product Attributes

1. Electronic device that calculates MPG attained between refueling
2. Battery powered with battery life greater than one year in normal use
3. Can be attached to visor or dash, carried in glove compartment, or carried in pocket or purse
4. Extremely accurate
5. Simple and easy to use
6. Reliable
7. Rugged
8. "Remembers" MPG and miles at last refueling
9. Displays MPG any time on demand
10. Turns off automatically after use to conserve battery

Product Use—to calculate new MPG

1. Turn "on" by pressing MPG key.
2. Key in the mileage reading and press the MIL key.
3. Key in the amount of fuel purchased and press the GAL key.
4. Press the MPG key.

Figure 8.8 Product information for the "A.F.R. Consulting" case.

Vehicles per Household	Frequency
0	1
1	19
2	42
3 or more	38
Total	100

Probably Would Buy At	Frequency
$ 9	21
12	12
15	21
18	10
21	2
25	5
Would not buy	29
Total	100

Two limitations of the data are immediately apparent.

1. A large proportion of households have three or more vehicles. The questionnaire should have asked specifically "how many?"
2. The prices willing to be paid by those who said they would probably buy are not normally distributed.

As mentioned, the most important data to be determined from the survey at this stage of the new product development is propensity to buy, market size, and the price/demand curve. It was felt that the confidence interval about the true mean price willing to be paid by those who said they would probably buy would be a good indicator of confidence intervals about other parameters. A confidence interval of $\pm$ $1 at the 90% level was considered adequate. On the basis of two small preliminary surveys ($n = 6$; $n = 7$), the population variance was estimated and a sample size of $n = 100$ was calculated to be sufficient to meet accuracy standards.

The mean price per se, as determined from the survey, is not particularly important but serves as a check on the assumptions used to determine sample size. As shown in the frequency distribution plot [Figure 8.9], the sample itself is not normally distributed.

Some calculations follow:

$$\overline{X} = \frac{\Sigma X}{n} = \$14.01$$

$$s = \sqrt{\frac{\Sigma (X - \overline{X})^2}{n - 1}} = \$4.56$$

$$s_{\overline{X}} = \frac{s}{\sqrt{n}} = .541$$

$$n = 71$$

$$CI = \overline{X} \pm t_{\frac{\alpha}{2}} s_{\overline{X}} = 14.01 \pm (1.67)(.541) = 14.01 \pm .90 = \$13.26 \text{ to } \$14.76$$

The error of estimation is 90 cents, which is less than the $1 required.

The survey indicated that 71 households out of 100 should buy if the price did not exceed $9. This is a conservative claim since 50 said they would buy at higher prices. The validity of this ratio, even as a conservative estimate, depends ultimately on the extent to which buyers had clearly formulated intentions and then carried them out.

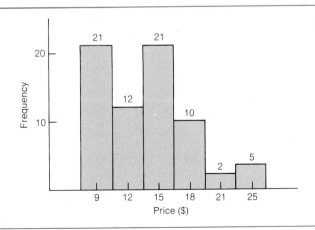

Figure 8.9 Frequency histogram of prices potential buyers said they would be willing to pay.

The standard error of the proportion is

$$s_{\bar{p}} = \sqrt{\frac{\bar{p}\bar{q}}{n}} = \sqrt{\frac{(.71)(.29)}{100}} = .0454.$$

$CI = \bar{p} \pm t_{\frac{\alpha}{2}} s_{\bar{p}} = .71 \pm (1.67)(.0454) = .71 \pm .08 = .63$ to $.79$. Thus, between 63% and 79% of the target population will buy the product at a price not exceeding $9.

If one ignores the fact that some error is introduced in the survey by not accurately counting the number of households with three or more cars, the mean number of vehicles per household is

$$\bar{X} = 2.17$$

$$s = .766$$

$$s_{\bar{X}} = .0766$$

$CI = \bar{X} \pm t_{\frac{\alpha}{2}} s_{\bar{X}} = 2.17 \pm (1.67)(.0766) = 2.17 \pm .13 = 2.04$ to 2.30

Since there are 1.76 million cars and light trucks registered within the market area, the number of households and, thus, the potential market is:

Potential market $|_{min}$ = 765,000 households

Potential market $|_{max}$ = 863,000 households

and the "attainable market" is

Attainable market $|_{min}$ = (.63)(765,000) = 482,000

Attainable market $|_{max}$ = (.79)(863,000) = 682,000

1. Although it probably served the firm's purpose adequately to define the target population to be a nine-county area, what, in your opinion, is the true target population? Explain your reasoning.

2. The writer seems somewhat worried about the nonnormal shape of the frequency distribution for prices potential buyers say they would be willing to pay. Can you offer him any words of comfort? If so, what are they?

3. According to the rules set forth in this chapter, was it necessary for the firm to use t, rather than Z, in the confidence interval calculations? Explain. Was the use of t rather than Z wrong? Explain.

4. What is your estimate of the population standard deviation that emerged from the two small preliminary surveys involving sample sizes of 6 and 7. Explain how you arrived at your result.

5. Explain in your own words how the potential market estimates of about 765,000 to 863,000 households were obtained.

6. Explain in your own words how the "attainable market" estimates of about 482,000 to 682,000 households were obtained.

7. With respect to the estimates in question 6, would it be proper to attach a .90 confidence coefficient? Tell why or why not.

8. In general, what do you think of this approach to gauging the possible market for a new product? Elaborate in about 100 to 200 words.

9 Introduction to Statistical Hypothesis Testing

There is often a sin of omission as well as commission.—Marcus Aurelius

Between two evils, I always pick the one I never tried before.—Mae West

What You Should Learn from This Chapter

This is the first of five chapters devoted exclusively to statistical tests of hypotheses, a fact which gives you some idea about the potential breadth and richness of this approach to analysis.

This chapter is introductory but will take us well into the subject. The contents of Chapter 9 should not seem startlingly new to you. If they do, that might be an indication that Chapter 8 should be reread, because again we make use of the concepts of sampling distribution of $\overline{X}$ and sampling distribution of $\overline{p}$. Moreover, we use the same measures as were used in connection with confidence intervals. The primary difference between this chapter and the applied parts of the preceding chapter is one of emphasis.

When constructing confidence intervals, we began with no preconceptions about a population parameter; we merely let $\overline{X}$ (or $\overline{p}$) be our guide to arriving at an estimate of μ (or p). Techniques described in this chapter require that we begin by making an assertion about the value of a parameter and then, using $\overline{X}$ (or $\overline{p}$) and our knowledge of the related sampling distribution, determine whether our assertion is probably correct or probably incorrect. If we conclude that our assertion—or *null hypothesis* as it is called—is probably correct, we take one course of action; if probably incorrect, we take the opposite course of action.

When you have completed this chapter, you should be able to

1. List the seven steps involved in testing just about any kind of statistical hypothesis.
2. Explain the meaning of Type I error and Type II error.
3. Identify and work two-tailed and one-tailed tests of hypotheses about the arithmetic mean of a single population.
4. Identify and work two-tailed and one-tailed tests of hypotheses about the proportion for a single population.
* 5. Compute probabilities of Type II errors.
* 6. Construct an operating characteristic curve.
* 7. Determine the necessary sample size when the probability of a Type I error and the probability of a Type II error are specified.

Before proceeding, review the following topics in Chapter 8 if you feel you do not know them well: the meaning of sampling distribution of $\overline{X}$ (section 8.3); the three important characteristics of a sampling distribution of $\overline{X}$ (sections 8.5, 8.6, and 8.8); the meaning of sampling distribution of $\overline{p}$ (section 8.12); the three important characteristics of a sampling distribution of $\overline{p}$ (sections 8.14, 8.15, and 8.17).

9.1 Introduction

Much of the rationale behind methods described in this chapter can be introduced most simply by means of a completely nonstatistical problem.

Imagine that you are a company president who has just discovered that someone has been embezzling large sums of money and concealing the fact by doctoring the accounting records. Your first inclination is to accuse the treasurer, a man with ready access to the accounting records and the technical knowledge to alter them skillfully. But, rather than accusing the treasurer right away, let us suppose you opt for a more cautious approach: You hire a private detective to keep the suspect under surveillance and to report on any suspicious behavior. Let us further suppose that the private eye soon reports that the treasurer has recently sent large amounts of money to a Swiss bank and is paying the rent on an apartment occupied by Gloria De Gare, a model. Such news would naturally heighten your suspicions even though the evidence still does not constitute incontrovertible proof of guilt.

This, admittedly fanciful, problem is similar in some basic essentials to those which will be treated statistically in this chapter. You, as the company president, begin with a hypothesis concerning the treasurer's guilt. Since you are a believer in the principle that a person is innocent until proved guilty, you state your hypothesis in the form: "The treasurer is not guilty of embezzlement."

> In statistics this kind of initial hypothesis is called the *null hypothesis* and is denoted H_0. The null hypothesis is a hypothesis that is viewed as true until persuasive empirical evidence to the contrary is obtained.

Of course, this hypothesis is not the only one you could formulate. Despite your adherence to the principle of innocent until proved guilty, you do in fact suspect the treasurer; you recognize, therefore, that the null hypothesis might be wrong. Hence, another possible hypothesis would be: "The treasurer is guilty of embezzlement."

> We will refer to this second kind of hypothesis as the *alternative hypothesis*, denoted H_a. The alternative hypothesis is a hypothesis against which the null hypothesis is tested and which is taken as true when the null hypothesis is declared false.

In some writings, the alternative hypothesis is called the *research hypothesis* or the *scientific hypothesis*, terms which tend to convey better than *alternative* the relative importance of this other hypothesis. After all, in this problem, as in most, it is the alternative hypothesis that prompted the investigation in the first place. It is this hypothesis that one ordinarily wishes to "prove" correct. A corollary of this is that the null hypothesis is the one the investigator ordinarily wishes to "prove" wrong, or to "nullify." Nevertheless, in this text we will use the conventional terminology. Appropriate ways of expressing the null and alternative hypotheses will be treated a little later. For now, keep in mind this fact:

> The null and alternative hypotheses are usually expressed in such a way that, if one is correct, the other must necessarily be incorrect.

Exceptions to this rule will be revealed and dealt with in sections 9.5 and 9.6. In the meantime, it is best not to worry about them.

You will notice that in our hypothetical problem there are two possible ways you could be right: (1) conclude that the treasurer is not guilty (do not reject H_0) if he is in fact not guilty (H_0 is true) or (2) conclude that the treasurer is guilty (reject H_0) if he is in fact guilty (H_0 is false). Unfortunately, you are in no position to know with certainty whether the suspect is really guilty or not guilty. You are dependent on your suspicions and the suggestive evidence gathered by the private detective. Therefore, just as there are two possible ways you can be right, so also are there two possible ways you can be wrong: (1) conclude that the treasurer is not guilty (do not reject H_0) even though he is in fact guilty (H_0 is false) or (2) conclude that the treasurer is guilty (reject H_0) even though he is in fact not guilty (H_0 is true). The structure of this problem is illustrated on the left-hand side of Figure 9.1. This same structure, with the parts framed in more general terms, is shown on the right-hand side of this same figure.

In view of our dependence on partial information (the private detective's findings in this example and the values of $\overline{X}$ and $\overline{p}$ in other examples in this chapter), we must recognize that, whether a decision problem is statistical or nonstatistical, our information will sometimes lead us to a correct decision and sometimes lead us astray. For example, if (1) our

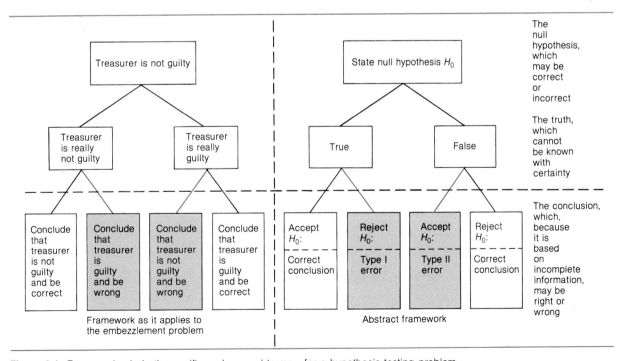

Figure 9.1. Framework—in both specific and general terms—for a hypothesis-testing problem.

observed evidence leads us to reject H_0 and (2) H_0 is in fact false, our evidence will have served us well. On the other hand, if (1) our observed evidence leads us to reject H_0 and (2) H_0 is really true, our evidence will have mislead us; our reaction to the evidence and, necessarily, our resulting action will be wrong. The mistake we make in such an instance is called the *Type I error*.

> The *Type I error* is the kind of error made when one *rejects* the null hypothesis even though it is really *true*.

Conversely, if (1) our evidence leads us to "accept" H_0 and (2) H_0 is true, we are lead to a correct conclusion. But if (1) our evidence leads us to "accept" H_0 and (2) H_0 is false, we will have committed an error—a Type II error this time.

> The *Type II error* is the kind of error made when one *accepts* the null hypothesis even though it is really *false*.
>
> *Warning:* Do not get the impression that in hypothesis testing one can only make mistakes. Remember, there is always the possibility of rejecting a false null hypothesis or of accepting a true null hypothesis.

The words *accept* and *reject* when applied to the null hypothesis are adequate with respect to suggesting what action (including inaction) should be taken once the test is completed. However, they do not convey adequately the difference in "status" of the two alternative conclusions. When we *reject* the null hypothesis, we usually do so with a high degree of conviction because we will already have set the probability of a Type I error at a low level. Hence, when we say "reject," we really mean "reject." On the other hand, when we *accept* the null hypothesis, we are not necessarily saying very much in its favor. A more technically correct, albeit more cumbersome, way of describing an "accepted" null hypothesis would be to say: "The sample evidence available is not adequate to lead to rejection of the null hypothesis." The analogy between a legal case and a test of hypothesis is still apt: When a defendant is found "guilty," it means, under our legal system, that a substantial quantity of persuasive evidence has pointed to his or her guilt. When the defendant is found "innocent" or, more to the point, "not guilty," it means that the evidence against that person was inadequate to establish his or her guilt beyond a reasonable doubt. We will continue to use the words *accept* and *reject* for the sake of convenience; however, you should be aware of the lack of symmetry between the two opposing conclusions.

9.2 Two-Tailed and One-Tailed Tests

We may view the subject of statistical hypothesis testing as involving two broad categories of problems—namely, two-tailed tests and one-tailed tests. One-tailed tests can be subdivided into lower-tail tests and upper-tail tests. When we speak of two-tailed and one-tailed tests in this chapter and the next one, the "tails" referred to are the tails of either the normal or the *t* distribution. It is these probability distributions—one or the other, whichever is more appropriate in a specific problem—which allow us to supplement our reaction to the null hypothesis with a probability statement concerning the reliability of that reaction.

Whether a hypothesis-testing problem calls for a two-tailed or a one-tailed test depends entirely on the kind of question to be answered and, in turn, the nature of the null hypothesis to be tested. For example, if one wishes to know whether a population parameter has a specific value and it is appropriate to assert in the alternative hypothesis that that specific value is not correct, the test called for would be two-tailed.

> If H_0: θ = some specific value and H_a: $\theta \neq$ the specific value stated in the null hypothesis, a *two-tailed test* is called for. The equal and not-equal signs, or words implying these signs, tell us that. (The symbol θ is used here to represent any kind of parameter of interest.)

The value used in the statement of the null hypothesis might be one obtained from a similar past investigation, or from an analysis of a similar group of elementary units, or it might merely be one which "sounds reasonable." Whether its basis is formal or informal, persuasive or "make-do," the procedures to be described shortly can be employed regardless.

> If we wish to determine (1) whether a population parameter θ has a value of *less than* some specified value or, alternatively, (2) whether a population parameter θ has a value *greater than* some specified value, we have the beginning of a *one-tailed test*.

More specifically:

> If the hypotheses are of the form
>
> $\qquad H_0$: $\theta \geq$ some specified value
> $\qquad H_a$: $\theta <$ the value specified in the null hypothesis
>
> a *lower-tail test* is called for. If they are of the form
>
> $\qquad H_0$: $\theta \leq$ some specified value
> $\qquad H_a$: $\theta >$ the value specified in the null hypothesis
>
> an *upper-tail test* is called for.

More will be said on this subject in section 9.7.

Be aware that there are times when the null and alternative hypotheses for a problem necessitating a one-tailed test should more properly be stated

$$H_0: \theta = \text{some specified value}$$
$$H_a: \theta < \text{the value specified in the null hypothesis}$$

or

$$H_0: \theta = \text{some specified value}$$
$$H_a: \theta > \text{the value specified in the null hypothesis}$$

In problems like these, the null hypothesis is referred to as a *simple hypothesis* because only one possible parameter value can be consistent with it. In one-tailed tests like those alluded to in the above box, the null hypothesis is called a *compound hypothesis* because many possible population parameter values are consistent with it.

Because (1) compound hypotheses are probably more common in one-tailed tests than simple hypotheses, (2) the decision rule will be unaffected by whether we view the null hypothesis as simple or compound, and (3) we wish to keep the number of types of hypothesis-testing problems treated here to a manageable few, we will assume in this chapter and the next that the null hypothesis for a one-tailed test is always compound.

9.3 Two-Tailed Test of the Arithmetic Mean of a Single Population When the Population Standard Deviation Is Known

Some Fundamentals

In a two-tailed test, if the relevant statistic, $\bar{\theta}$ for now, suggests that the true population parameter θ is *either* smaller than *or* greater than the specific value used in the statement of the null hypothesis, H_0 is rejected and H_a is accepted; otherwise, H_0 is accepted and H_a rejected.

This point can be clarified through use of a specific example.

ILLUSTRATIVE CASE 9.1

Business economists with the Deft Tool Manufacturing Company have learned from past experience that several kinds of data on new orders serve collectively as reliable indicators of company sales for the upcoming six-month period. Hence, they rely on such data when advising management on inventory policy for parts used in the manufacture of a wide variety of power tools. One of the new-order measures that the economists watch closely is average size (in price-deflated dollars) of new orders relative to a year ago.

Recently, random sampling was used to determine whether the average size of new orders for the first quarter of this year was approximately the same as that of the first quarter of last year, a time when the *entire population* of new orders had been studied and a mean size of new orders of $3600 was found. This year, for reasons of economy, *a random sample of n = 225 new orders* was used in lieu of a complete population study. The sample mean $\bar{X}$ was found to be $3582. The economists now wish to compare the $3582 value of this year with the $3600 of last year to determine whether a *change* has occurred.

At first, the question the economists are addressing may seem trivial—even silly. The $3582 value is obviously not the same number of dollars as $3600. Still, one should not be too quick to conclude that a real change has occurred. Keep in mind that last year population data were available for analysis, whereas this year only sample data are available. Statistically speaking, $3582 may be close enough to $3600 to be consistent with the hypothesis that no change in the average size of new orders has occurred.

Two possible explanations may be discerned for the observed discrepancy between last year's $3600 and this year's $3582: (1) The average size of new orders has indeed changed from last year's average or (2) the average size of new orders has not changed but, as a result of chance errors of sampling, the specific sample with a mean of $\overline{X}=\$3582$ is erroneously conveying the impression that a change has occurred. Which of these totally different explanations is right is something the economists will never know for sure. However, thanks to probability theory, they can marshal some rather persuasive evidence to suggest which is *more likely* correct.

In broad strokes, we may state the rationale behind the probability aspect of a statistical hypothesis-testing problem in the following way: If the true population mean in our illustrative case, for example, were really still $3600 and if we were to examine the $\overline{X}$ value for each possible sample of size $n = 225$ which could be drawn from the population of N new orders, we would expect the $\overline{X}$ values to be clustered around the population mean of $3600. Only a relatively small number of such $\overline{X}$ values would differ markedly from $3600. On the other hand, if the true population mean were some value other than $3600, then the $\overline{X}$ values would tend to cluster around that number. In such a situation, relatively few of the $\overline{X}$ values would be close enough to $3600 to suggest (incorrectly) that the true population mean is still $3600.

The implications of all this for a single sample of size $n = 225$ are that (1) if $3582 can be viewed as close enough to $3600 to be one of a large number of possible $\overline{X}$ values differing relatively little from $3600, then $\overline{X}=\$3582$ would be said to be consistent with the null hypothesis and the economists would accept H_0 (and reject H_a), or (2) if $3582 can be viewed as differing from $3600 by an amount great enough to suggest that such a discrepancy between sample mean and hypothesized population mean would occur only rarely if the population mean were really $3600, they would be led to reject H_0 (and accept H_a) (Figure 9.2). In the second case, the economists would have a choice between (1) betting that a "long shot" has occurred as a result of random sampling or (2) betting that the observed $\overline{X}$ value is

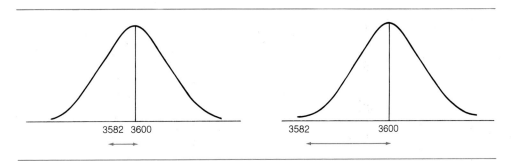

Figure 9.2. Query: At what point does a sample mean differ sufficiently from a hypothesized population mean to convince one that the difference is not just a sampling fluke? To determine this, we look at (1) the size of the difference and (2) the amount of variation $\sigma_{\overline{X}}$ in the associated sampling distribution of $\overline{X}$. If $\sigma_{\overline{X}}$ is large, then a rather large difference between $\overline{X}$ and the hypothesized population mean may be a common occurrence and not grounds for concluding that the population mean has changed. However, if $\sigma_{\overline{X}}$ is small, even a little difference between $\overline{X}$ and the hypothesized population mean may be highly improbable as a result of random sampling. The explanation *probably* lies elsewhere—that is, the null hypothesis is *probably* wrong.

attempting to tell them that the true population mean is no longer $3600. Ordinarily, the economists would be well-advised to accept (2) because, as will be shown, the probability of (1) being correct is under the control of the investigator and is usually set at a low level.

Steps Involved in Testing the Null Hypothesis

The steps involved in testing just about any kind of statistical hypothesis are as follows:

1. State the null and alternative hypotheses.
2. Establish the level of significance.
3. Determine the appropriate kind of sampling distribution to rely on.
4. Establish a decision rule.
5. Take a random sample of size n, calculate the statistic of interest, and (if appropriate) convert it into its standard value.
6. Apply the decision rule to the observed sample statistic and either accept or reject the null hypothesis.
7. Take the appropriate action.

Let us consider each of these steps as they might be employed in the Deft Tool Company problem. For simplicity of presentation we will begin with some, probably unrealistic, assumptions: First, we will assume that the population standard deviation is known to be $\sigma = \$200$; this assumption can easily be relaxed a little later. Second, we will disregard the finite correction factor when computing the standard error of the mean $\sigma_{\overline{X}}$. However, if it were known that $N < 20n$, it really should be used. Clearly, we are implicitly assuming that the population size N is pretty large.

The steps listed for testing a statistical hypothesis will serve as subsection headings.

State the Null and Alternative Hypotheses. The word *change* in the final sentence of Illustrative Case 9.1 tips us off to the need for a two-tailed test.

Words like *change, difference, departure from,* and so forth, are neutral with respect to direction. Therefore, a sample mean, $\overline{X}$, differing significantly from the hypothesized population mean will lead to rejection of the null hypothesis regardless of whether the difference results from $\overline{X}$ being lower than μ_0 or $\overline{X}$ being higher than μ_0. (The symbol μ_0 will be used to denote the true population mean *according to the null hypothesis*. Because the null hypothesis may be either true or false, μ_0 may or may not be equal to the true but unknown population mean μ.)

With this in mind, we write the hypotheses as follows:

H_0: No change from last year's $3600 has occurred in average size of new orders

In symbols, the null hypothesis would read:

$$H_0: \mu = \$3600$$

Of course, when one states a null hypothesis, an opposing hypothesis of some kind is implied. Often it is advantageous to make the other hypothesis explicit. Hence:

H_a: The true average size of new orders is no longer $3600

Or, in symbols:

$$H_a: \mu \neq \$3600$$

Establish a Level of Significance. We have spoken earlier of the desirability of supplementing a statistical conclusion with a measure of the reliability of that conclusion. In hypothesis testing such a measure of reliability is called *the level of significance of the test.*

> For convenience, we refer to the probability of committing a Type I error as α ("alpha"). As noted, this probability is also called the level of significance of a test. If α is set at, say, .05 and the null hypothesis that $\mu =$ some specified value is rejected, we say that the null hypothesis was rejected at the .05 (or 5%) level of significance. (In a similar vein, β ("beta") will be used to denote the probability of committing a Type II error.)

Some reflection will reveal a similarity between α in a hypothesis-testing problem and the confidence coefficient in an estimation problem: They are both judgment calls brought into the analysis from outside, so to speak. How does one know what value of α to choose? Unfortunately, there is no definite answer to this question. Much depends on the unpleasantness of the consequences of a Type I error, on the one hand, and the unpleasantness of the consequences of a Type II error on the other. Ideally, one should reflect carefully on this trade-off, though it is probably seldom done in practice. The term *trade-off* is apt because α and β work at cross purposes just as confidence and precision of estimation work at cross purposes in a confidence interval problem. For a given sample size n, when α is small, β will be large, and vice versa. Much more will be said about this subject. In the meantime, the point to remember is that α cannot be set at an extremely low value arbitrarily without raising the specter of an unacceptably large β.

As a general rule, the Type I error is regarded as the more serious error in business applications because it means an unnecessary (and usually costly) change from the status quo. The losses in such a situation represent real dollar losses in the accounting sense. The losses associated with a Type II error, on the other hand, are more often opportunity losses—that is, they are losses in potential profit arising from failure to follow the best available course of action. Opportunity losses are generally considered less painful than real accounting losses.

For the problem at hand, we will proceed as if Deft management had decided that α of .05 is adequate. This means that, if μ_0 of $3600 is correct, only about 5 samples of size $n = 225$ out of 100 will have arithmetic means differing sufficiently from $3600 to indicate (incorrectly) that the population mean has changed. In the future, we will usually make use of two common values for α—namely, .05 and .01. This is an expediency based on our lack of an insider's awareness of a company's problems and the relative seriousness of the two

Table 9.1 Decision Table

Part I: The Structure of a Hypothesis-Testing Problem

Reaction to Null Hypothesis	State of Nature	
	H_0 **Is True**	H_0 **Is False**
Reject H_0	Type I error	Correct conclusion
Accept H_0	Correct conclusion	Type II error

Part II: The Corresponding Probabilities

Reaction to Null Hypothesis	State of Nature	
	H_0 **Is True**	H_0 **Is False**
Reject H_0	α: Usually set along with sample size prior to hypothesis test. It is the probability of *rejecting* a null hypothesis which is really *true*. Also called the *level of significance*.	$1 - \beta$: The probability of rejecting a null hypothesis which is really false. Often called the *power of a test* because it measures a test's ability to identify an incorrect null hypothesis.
Accept H_0	$1 - \alpha$: The probability of accepting a null hypothesis which is really true.	β: The probability of *accepting* a null hypothesis which is really *false*.

kinds of error. Still, in practice, careful thought should be given the question of what α to use and what the implications are for β.

Table 9.1 shows the basic structure of a decision problem in Part I. In Part II the probabilities associated with the two possible ways of being right and the two possible ways of being wrong are shown in symbolic form along with some brief elaboration.

Determine the Appropriate Kind of Sampling Distribution to Rely On. In the Deft Tool example, the parameter of interest is the population mean μ. The corresponding sample mean $\overline{X}$ will be used to test the null hypothesis. Therefore, the sampling distribution of $\overline{X}$ will be employed. Here is where our knowledge of the three key characteristics of this sampling distribution will come to our aid.

Establish a Decision Rule. Figure 9.3 shows the sampling distribution of $\overline{X}$ which will be used to complete the Deft Tool Company example. Notice that the curve has been partitioned into two regions: a rejection region—which is really two regions in a two-tailed test—and an acceptance region. These terms mean exactly what they sound like they mean: If $\overline{X}$ (or its Z-value equivalent) is found to be in the acceptance region, it means we should accept the null hypothesis that the population mean is $3600. Conversely, if $\overline{X}$ (or its Z-value equivalent) is found to be in the rejection region, it means we should reject the null hypothesis and conclude that the true population mean is some value other than $3600.

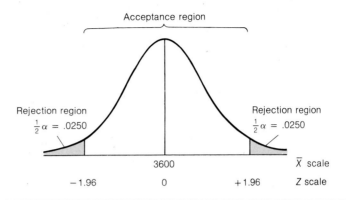

Figure 9.3. Sampling distribution of $\overline{X}$ for a two-tailed test of the null hypothesis H_0: μ = \$3600, with a .05 level of significance.

> The *rejection region* of a test is a set of values of a sample statistic, $\overline{X}$ in the present case, or the Z- or *t*-value equivalents of this set of values, which will lead to rejection of the null hypothesis. The rejection region is also sometimes called the *critical region*.
>
> *Procedural Note:* The specific points where the rejection region begins are called the *action limits*, if the $\overline{X}$ scale is used in the performance of the test, or *critical values* if the Z or *t* scale is used. Although either the $\overline{X}$ scale or the standard-unit scale may be used with equal propriety, we will use the relevant standard-unit scale almost exclusively in this and later chapters dealing with hypothesis testing. (The material in sections 9.5, 9.6, and 9.8 of this chapter call for some modification of this rule.) In two-tailed problems calling for use of the normal curve as a model for the sampling distribution, we will denote the critical values as Z_L and Z_U (L for "lower" and U for "upper"). The absolute values will be the same for both Z_L and Z_U; however, Z_L will take a minus sign and Z_U a plus sign.

You will also observe that in Figure 9.3 the mean of the sampling distribution of $\overline{X}$ has been assumed to be $\mu_{\overline{X}} = \mu_0 = \$3600$. This is tantamount to saying, "Let us assume temporarily that the population mean—and, hence, the mean of the sampling distribution of $\overline{X}$—is exactly what our null hypothesis claims." Although we may ultimately decide that this is an incorrect assumption, beginning in this manner allows us to partition, with relative ease, the sampling distribution of $\overline{X}$ into acceptance and rejection regions.

Because the test is to be two-tailed, the selected value of α, .05 in our example, is divided in half and each half assigned to one tail of the normal curve representing the sampling distribution of $\overline{X}$. The reason is: In a two-tailed test, a sample mean $\overline{X}$ which is either substantially lower than *or* substantially greater than the hypothesized population mean μ_0 will lead to rejection of the null hypothesis. Since the *total* probability of a Type I error is .05, half of this quantity would apply to extremely small values and half to extremely large values of $\overline{X}$.

The remaining steps involved in arriving at a decision rule are quite mechanical: Because .025 is the area in one tail of the curve, then $.5000 - .0250 = .4750$ is the area associated with a range of Z values running from 0 to the value of, say, Z_U. (We could have chosen Z_L just as easily.) From Appendix Table 2 we find Z_U to be $+1.96$. Therefore, Z_L must be -1.96.

A *decision rule* is simply a device for making explicit what should be concluded if the Z-value equivalent of the observed sample mean, hereafter referred to as *observed* Z, is found to be within a specified range and what should be concluded if it is found to be outside that range.

For our problem:

Decision Rule*

1. If $-1.96 \leq$ observed $Z \leq +1.96$, accept H_0.
2. If observed $Z < -1.96$ or if observed $Z > +1.96$, reject H_0.

Take a Random Sample of Size n*, Calculate the Statistic of Interest, and (If Appropriate)*† *Convert It into Its Standard Value.* The sample mean (which we have known right along to be \$3582 but which ordinarily will not be known so early in the analysis) is now converted into its corresponding observed-Z value:

$$\text{Observed Z} = \frac{\overline{X} - \mu_0}{\dfrac{\sigma}{\sqrt{n}}} = \frac{\$3582 - \$3600}{\dfrac{\$200}{\sqrt{225}}} = \frac{-\$18}{\$13.33} = -1.35$$

Apply the Decision Rule to the Observed Sample Statistic and Either Accept or Reject the Null Hypothesis. Since observed Z of -1.35 is within the acceptance region, according to our decision rule, we accept the null hypothesis and conclude that no change in average size of new orders has occurred from last year's \$3600.

Take the Appropriate Action. We have completed the test of the null hypothesis. This subsection is included merely as a reminder that hypothesis testing in a business context is an approach to decision making. There should have been from the start a definite awareness of what action is called for if the null hypothesis is accepted and what action is called for if it is rejected. In the present case, the action is really inaction: Make no change in inventory policy.

---

*Use of the equal sign in connection with the acceptance, rather than the rejection, part of the decision rule is a convention of long standing.
†In this chapter and the next, you can count on this conversion to be appropriate.

"You've got to learn what not to do and when not to do it!"

9.4 Two-Tailed Test for a Population Mean When the Population Standard Deviation Is Unknown

When working through the Deft Tool case, we assumed for convenience that the population standard deviation was known to be $\sigma = \$200$. In practice we will seldom know σ. Therefore, we must now consider ways of modifying the above procedure to make it more useful in connection with practical problems.

Large Sample Size

If the sample size is large (n in excess of about 30 or 40), the modification required is slight: Simply substitute sample s for unknown population σ and then proceed as described above. For example, let us say that all the information about the Deft case is exactly the same as assumed above except that we do not have direct knowledge of σ. Let us further suppose that we take a sample of size $n = 225$ and find the sample standard deviation s to be $\$194$. Clearly, the probability of a Type I error, which up until now we have confidently called .05, is now really *approximately* .05. Otherwise, things are effectively unchanged. The Z-value equivalent of $\overline{X} = \$3582$ is

$$\text{Observed } Z = \frac{\$3582 - \$3600}{\dfrac{\$194}{\sqrt{225}}} = -\frac{\$18}{\$12.93} = -1.39$$

Just as before, the critical Z value associated with $\alpha = .05$ is 1.96 (with plus and minus signs) and, since -1.39 is within the acceptance region (between -1.96 and $+1.96$), we accept the null hypothesis that μ is still $\$3600$.

Small Sample Size

When the population standard deviation σ is unknown and the sample size n is small, the procedural change required is more substantial. Let us work through an example.

ILLUSTRATIVE CASE 9.2

In the past the mean number of man-hours required by the Southwest Sprinkler System Company to install a sprinkler system in a 5000-square-foot lawn had been 20. However, this mean was determined during a time when many people with the necessary skills were seeking work. More recently, tighter labor market conditions have made it necessary for Southwest to hire unskilled workers and give them on-the-job training. Management now wishes to know whether any *change* has occurred in the mean number of man-hours required to install a sprinkler system in a lawn of the indicated size. Examination of a random sample of $n = 9$ time sheets reveals that the number of man-hours required to complete the nine jobs were as follows:

$$25 \quad 29 \quad 23 \quad 23 \quad 31 \quad 21 \quad 27 \quad 25 \quad 33$$

We will follow the steps listed in section 9.2.

1. State the null and alternative hypotheses.

$$H_0: \ \mu = 20 \text{ hours}$$
$$H_a: \ \mu \neq 20 \text{ hours}$$

2. Establish the level of significance.
   Let us say that the probability of a Type I error is set at $\alpha = .01$.

3. Determine the appropriate kind of sampling distribution to rely on.
   Since the null hypothesis has to do with the value of μ and since $\overline{X}$ will be used to test the null hypothesis, it seems reasonable to make use of a normally distributed sampling distribution of $\overline{X}$. However, this would not be quite right. The combination of unknown σ and small n necessitates use of the t distribution, provided that the population does not depart substantially from the normal shape.

4. Establish a decision rule.
   Figure 9.4 shows the t distribution associated with $\alpha = .01$ and $n - 1 = 8$. Notice that the critical values are -3.355 and $+3.355$. Therefore,

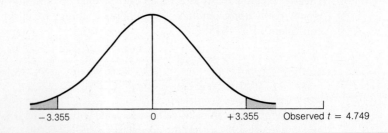

Figure 9.4. The t distribution for $n - 1 = 8$ and critical values of t for the Southwest Company problem.

Decision Rule

1. If $-3.355 \leq$ observed $t \leq +3.355$, accept H_0.
2. If observed $t < -3.355$ or if observed $t > +3.355$, reject H_0.

5. Take a sample of size n, calculate the statistic of interest, and (if appropriate) convert it into its standard value.

   The nine observations listed at the end of the illustrative case are used to determine the sample mean and standard deviation:

   $$\overline{X} = \frac{\Sigma X}{n} = 26.33 \qquad s = \sqrt{\frac{\Sigma (X - \overline{X})^2}{n - 1}} = 4.00$$

   Therefore,

   $$\text{Observed } t = \frac{\overline{X} - \mu_0}{\dfrac{s}{\sqrt{n}}} = \frac{26.33 - 20.00}{\dfrac{4}{\sqrt{9}}} = 4.748$$

6. Apply the decision rule to the observed sample statistic and either accept or reject the null hypothesis.

   Since 4.749 is outside the range of -3.355 to $+3.355$, we reject the null hypothesis and conclude that the average length of time required to install a sprinkler system in a lawn of the indicated size has changed. The large positive value of observed t indicates that the average time has lengthened.

7. Take the appropriate action.

   The appropriate action associated with this rejected hypothesis is not self-evident. Presumably, however, if nothing can be done to reduce the time required, at the very least the longer time should be taken into account when setting work schedules and estimating costs.

---

*9.5 Computing Probabilities of Type II Errors

Let us now return to the Deft Tool case and assume that all information is the same as earlier—namely, $\sigma = \$200$, $n = 225$, $\mu_0 = \$3600$, and $\alpha = .05$.

Because in this problem we accepted the null hypothesis, there is no possible way for us to commit a Type I error. A Type I error can occur only in association with a *rejected null hypothesis*. However, we *can* commit a Type II error. Whenever we accept the null hypothesis, there are two possible explanations: (1) The true population mean is really as hypothesized and $\overline{X}$ is simply attesting to that condition or (2) the true population mean is really not as hypothesized but the sample mean $\overline{X}$ is indicating that it is as hypothesized—hence, a Type II error. The probability of a Type II error cannot be controlled in a situation like this one where α and n are controlled. Still, if one is to control the probability of a Type I error sensibly, he or she should know what effect setting the value of α at a certain level will have on β. Moreover, in a problem like this one where we have already accepted H_0, it is helpful, sometimes downright comforting, to know the risk associated with this other kind of error.

Unfortunately, the value of β cannot be stated in terms having general applicability. It simply has to be worked out on an ad hoc basis. The reason is that there will be a different β value for each different *possible* population mean except for μ_0.

To determine the probability of a Type II error, we begin by converting the critical Z values obtained earlier into action limits.

Action limits, denoted $\overline{X}_L$ and $\overline{X}_U$, play the same role on the $\overline{X}$ scale as Z_L and Z_U play on the Z scale. These are the values at the exact points where the critical region begins. The formulas for $\overline{X}_L$ and $\overline{X}_U$, obtained from rearranging algebraically the formulas for Z_L and Z_U, are

$$\overline{X}_L = \mu_0 - Z_L \frac{\sigma}{\sqrt{n}} \qquad (9.5.1)$$

$$\overline{X}_U = \mu_0 + Z_U \frac{\sigma}{\sqrt{n}}$$

Applying these formulas, we get

$$\overline{X}_L = \$3600 - (1.96)\left(\frac{\$200}{\sqrt{225}}\right) = \$3573.87$$

$$\overline{X}_U = \$3600 + (1.96)\left(\frac{\$200}{\sqrt{225}}\right) = \$3626.13$$

Now let us suppose that the business economists in this example believe that a decrease in average size of new orders of as little as $20 may indicate a potentially serious slump in

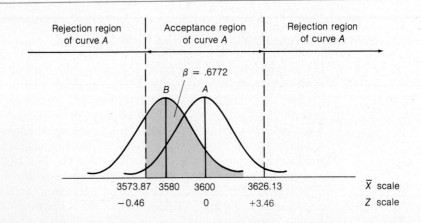

Figure 9.5. Diagram showing the probability of a Type II error when the true population mean, $\mu_a = \mu$, is $3580, $\mu_0 = \$3600$, $\alpha = .05$, and $\sigma_{\overline{x}}$ is \$13.33. (Shaded area indicates the Type II error probability.)

9   *Introduction to Statistical Hypothesis Testing*

future sales. We will rewrite the hypotheses:

$$H_0: \mu = \$3600$$
$$H_a: \mu = \$3580$$

Notice that H_0 and H_a are no longer exact opposites. Although it is still true that if one is right the other must be wrong, it is also now possible that both are wrong.

Figure 9.5 shows how we arrive at β when $n = 225$, $\alpha = .05$, $\mu_0 = \$3600$, and $\mu_a = \$3580$, where μ_a represents the population mean *according to the alternative hypothesis*. (Since μ_a may or may not be right, it may or may not equal the true μ.) We begin by setting up a normal curve representing the sampling distribution of $\overline{X}$ on the assumption that the true population mean is $3600. The normal curve labeled A in this figure represents that sampling distribution. Next, we place the action limits computed above in the appropriate places on the horizontal axis of this curve. Then, we proceed to set up another curve representing the sampling distribution of $\overline{X}$ on the assumption that the true population mean is $3580. The normal curve labeled B in Figure 9.5 represents this sampling distribution.

We must now ask: Where in Figure 9.5 do we find an area representing the probability of a Type II error given that the true population mean is $3580?

> The answer is: The area under curve B that falls between the action limits shown on curve A.

That is, if $3580 is really the value of the population mean, an $\overline{X}$ value leading us to accept the null hypothesis that the population mean is really $3600 will, by definition, result in a Type II error. To determine how much of curve B lies within the acceptance region of curve A, we begin by converting the action limits, as found on curve B, into their Z values:

$$Z_1 = \frac{\overline{X}_L - \mu_a}{\sigma_{\overline{X}}} = \frac{\$3573.87 - \$3580.00}{\$13.33} = -0.46$$

$$Z_2 = \frac{\overline{X}_U - \mu_a}{\sigma_{\overline{X}}} = \frac{\$3626.13 - \$3580.00}{\$13.33} = +3.46$$

According to Appendix Table 2, the area under curve B between $Z_1 = -0.46$ and 0 is .1772. The area under curve B between 0 and $Z_2 = +3.46$ is so large that it is not shown in the table, but we may surmise that it is virtually .5000. Therefore, the probability of accepting a false hypothesis if $\mu_a = \mu = \$3580$ is just slightly less than .6772 (i.e., .1772 + .5000).

Of course, if we vary our assumption about the value of μ_a, we will get a different value for β. For example, suppose that μ_a were (alternatively) hypothesized to be $3560, as indicated in Figure 9.6.

We calculate

$$Z_1 = \frac{\overline{X}_L - \mu_a}{\sigma_{\overline{X}}} = \frac{\$3573.87 - \$3560.00}{\$13.33} = 1.04$$

$$Z_2 = \frac{\overline{X}_U - \mu_a}{\sigma_{\overline{X}}} = \frac{\$3626.13 - \$3560.00}{\$13.33} = 4.96$$

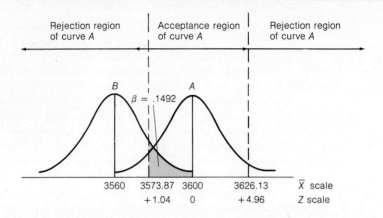

Figure 9.6. Diagram showing the probability of a Type II error when the true population mean, $\mu_a = \mu$, is \$3560, $\mu_0 = \$3600$, $\alpha = .05$, and $\sigma_{\overline{x}} = \$13.33$. (Shaded area indicates the Type II error probability.)

The area under curve B to the right of \$3573.87 is found to be $.5000 - .3508 = .1492$. The area under curve B to the right of \$3626.13 is negligible and may be ignored. Therefore, the probability sought is .1492.

We see that, for a specified sample size n, the greater the difference between μ_a and μ_0, the smaller will be the probability of a Type II error.

Also, for given values of α, μ_0, and μ_a, the larger the sample size n, the smaller will be the probability of a Type II error since the standard error of the mean $\sigma_{\overline{X}}$ will be smaller. A smaller $\sigma_{\overline{X}}$ means that less of curve B will be found within the acceptance region of curve A.

---

*9.6 The Operating Characteristic Curve and the Power Curve

Operating Characteristic Curve

The procedure described in the last section can be extended. That is, we could compute a series of conditional probabilities of the type $P(\overline{X}_L \leq \overline{X} \leq \overline{X}_U | \mu_{a_1} = \mu)$, $P(\overline{X}_L \leq \overline{X} \leq \overline{X}_U | \mu_{a_2} = \mu)$, and so forth. These are read: "The probability that the mean of a random sample will be between $\overline{X}_L$ and $\overline{X}_U$ given that μ_{a_i} is the true population mean." By letting the μ_{a_i} value vary and performing several such conditional probability calculations, one can study and evaluate the probability of committing a Type II error (as well as the probability of accepting the null hypothesis correctly) under each of various assumed conditions even though one cannot control that probability. A useful graphic device for facilitating the evaluation of such probabilities is the *operating characteristic curve*.

> The *operating characteristic curve*, or OC curve, shows a considerable range of *possible values* of the (one and only) true but unknown population mean on the horizontal axis and the corresponding probabilities of *accepting the null hypothesis* on the vertical axis.

To determine the probability of accepting H_0 for each specified possible value of the population mean, we do the same kind of calculation as described in section 9.5 for each value of μ_{a_i} and for the single value of μ_0 from the null hypothesis, as illustrated in Figure 9.7. In Table 9.2 we see displayed a series of possible values for the population mean for the Deft Tool Company case, the corresponding probabilities of accepting the null hypothesis when $\alpha = .05$, and an indication of the type of error, if any, the probability is related to. These values are charted as an operating characteristic curve in Figure 9.8.

The way the entries in Table 9.2 are interpreted will be explained using $\mu_a = \mu = \$3520$: We say, "If $3520 is the true population mean, the probability of accepting the null hypothesis (i.e., concluding that $3600 is the true population mean) is just slightly greater than .0000." Such a reaction to the null hypothesis would, in this instance, constitute a Type II error. Fortunately, we see that the probability of getting such grossly misleading sample information is virtually nil.

Notice that the only situation depicted in Table 9.2 which would not result in a Type II error is the one where the true population mean is assumed to be $3600, the value used in the null hypothesis. Since in that one case the null hypothesis would be correct, the proba-

Table 9.2 Data for Construction of Operating Characteristic Curve for Two-Tailed Test of a Single Mean

(1) Possible Values of the True but Unknown Population Mean (μ_{a_i} and μ_0 Values)	(2) Probability of Accepting the Null Hypothesis H_0. That is, $P(\overline{X}_L \leq \overline{X} \leq \overline{X}_U)$ (where $\overline{X}_L = \$3573.87$; $\overline{X}_U = \$3626.13$).	(3) Type of Error Committed If H_0 Is Accepted
$3520	.0000	Type II
3540	.0057	Type II
3560	.1492	Type II
3570	.3589	Type II
3580	.6772	Type II
3590	.8835	Type II
3600	.9500	No error
3610	.8835	Type II
3620	.6772	Type II
3630	.3589	Type II
3640	.1492	Type II
3660	.0057	Type II
3680	.0000	Type II

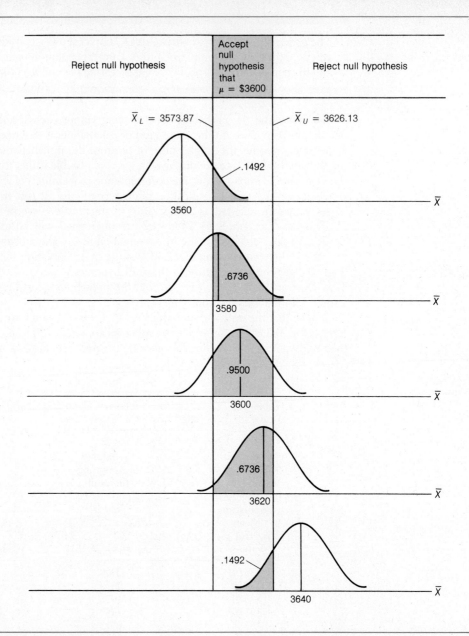

Figure 9.7. Calculations of the probability of accepting the null hypothesis that $\mu = \$3600$ for various assumed values of μ.

bility of .9500 would be that of accepting the null hypothesis *correctly*—not that of committing a Type II error.

The ideal operating characteristic curve for the two-tailed test we have been discussing would look like the one shown in Figure 9.9. The "curve" in this figure is a vertical line (having no width) extending from a probability of 0 to a probability of 1 and situated directly

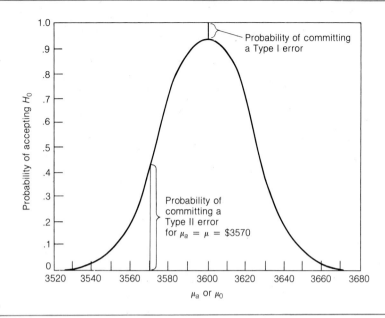

Figure 9.8. Operating characteristic curve for a two-tailed test of a single arithmetic mean.

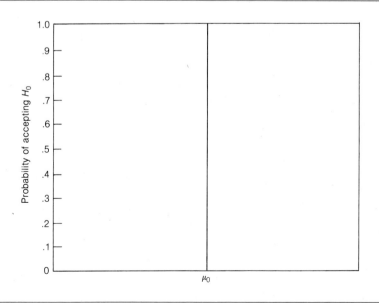

Figure 9.9. The ideal operating characteristic curve. (Technically, it would really be an empty box.)

above $\mu_0 = \mu = \$3600$. Such a "curve" presents in graphic form the idea that the probability of committing a Type I error is 0 and the probability of committing a Type II error, for any specified value of $\mu_a = \mu$, is also 0. Of course, such a "curve" would never occur in practice when sampling is used. However, this ideal curve serves as a standard against which to compare OC curves for actual sampling problems. The narrower the curve the better, because that condition indicates low probabilities of committing a Type II error. Also, the smaller the distance between the tip of the curve and the top of the box enclosing the curve the better, because that condition indicates a small probability of a Type I error.

The specific shape of an OC curve will be affected by (1) the α value and (2) the sample size. Figure 9.10 provides a visual comparison of the operating characteristic curve obtained in the Deft Tool Company example, where $\alpha = .05$ and $n = 225$, and the operating characteristic curve which would have been obtained if the economists had used $\alpha = .01$ and $n = 225$. Figure 9.11 compares the operating characteristic curve for $\alpha = .05$ and $n = 225$ and the one which would be associated with $\alpha = .05$ and $n = 64$, rather than 225.

Power Curve

Some analysts prefer the power curve to the operating characteristic curve.

> The power curve is simply the complement of the operating curve and shows a range of *possible values* for the true but unknown population mean on the horizontal axis and the corresponding probabilities of *rejecting* the null hypothesis on the vertical axis.

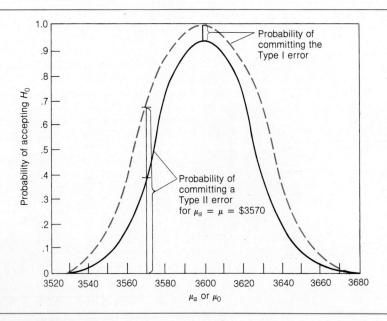

Figure 9.10. Comparison of an operating characteristic curve for $n = 225$ and $\alpha = .05$ (solid line) and an operating characteristic curve for $n = 225$ and $\alpha = .01$ (dashed line).

9 *Introduction to Statistical Hypothesis Testing*

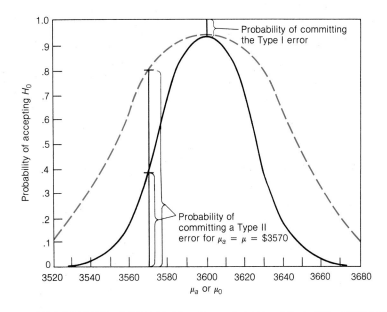

Labels in figure:
1.0
.9
.8
.7
.6
.5
.4
.3
.2
.1
0

Probability of accepting H_0

Probability of committing the Type I error

Probability of committing a Type II error for $\mu_a = \mu = \$3570$

3520   3540   3560   3580   3600   3620   3640   3660   3680

μ_a or μ_0

Figure 9.11. Comparison of an operating characteristic curve for $n = 225$ and $\alpha = .05$ (solid line) and an operating characteristic curve for $n = 64$ and $\alpha = .05$ (dashed line).

The power curve derives its name from the term *power of a test*.

> The *power of a test* of significance is measured by $1 - \beta$, which is the probability of *rejecting* the null hypothesis when it is *incorrect*.

9.7 One-Tailed Test for the Arithmetic Mean of a Single Population

Thus far in our treatment of the Deft Tool Company problem, we have discussed the procedure for performing a two-tailed test only. Implicit in the use of such a test is the assumption that a difference of a given size between the true but unknown mean μ and the hypothesized mean μ_0 has just as much importance if μ_0 exceeds μ as it has if μ_0 is less than μ. For example, up to now we have acted as if the business economists in our illustrative case would view evidence that the true average size of new orders is higher now than a year ago as just as serious in its implications as evidence suggesting that the true average size of new orders is now lower than a year ago. That could conceivably be true. However, sometimes evidence suggesting that the true population mean is, say, less than some value is considered more meaningful than evidence that the true population mean is greater than that same value.

If the Deft Tool economists were to discover that the true average size of new orders has apparently decreased from last year, they might view this as a portent of a future decline in the company's sales and conclude that a switch to a more conservative inventory policy was called for. On the other hand, if tool parts are abundantly available and if prices and interest rates are reasonably stable, evidence of an *increase* in the average size of new orders would not necessarily lead them to recommend a more aggressive inventory policy. Under such a scenario, we would surmise that the economists were much more concerned about the question "Has the average size of new orders *decreased* from a year ago?" than with the question of whether the average size of new orders has changed—never mind the direction of change.

Words like *decreased, increased, declined, deteriorated, improved, expanded,* and many others like them, including *at least* and *at most,* are not neutral with respect to direction. Taken in context, they provide definite clues about which tail of the standard normal curve or the *t* distribution should be used.

We can now explore the hypothesis-testing procedures for the situation just described. You will notice that the statement of the null hypothesis in words is a negation of the thing the economists are concerned about. That is, if the question they wish to answer is "Has the average size of new orders decreased from a year ago?" a helpful and fairly dependable trick is to answer (albeit tentatively): "No, the average size of new orders has not decreased from a year ago." The alternative hypothesis is the exact opposite of this. That is, the alternative tentative answer is: "Yes, the average size of new orders has decreased from a year ago." (*Note:* In words, the null hypothesis can also be read: "The true population mean is *at least* $3600." If the question had been phrased in such a way that the words *at least* were used, the null hypothesis could be set up as shown without your having to resort to the helpful, but rather convoluted, "trick" mentioned immediately above.)

1. State the null and alternative hypotheses.
   In symbols, we have

   $$H_0: \mu \geq \$3600$$
   $$H_a: \mu < \$3600$$

2. Establish the level of significance.
   We will continue to use a level of significance of $\alpha = .05$.

3. Determine the appropriate kind of sampling distribution to rely on.
   The appropriate sampling distribution is the sampling distribution of $\overline{X}$.

4. Establish the decision rule.

   Figure 9.12 shows the correct manner of partitioning the normal sampling distribution of $\overline{X}$. Notice that the mean of the sampling distribution of $\overline{X}$, $\mu_{\overline{X}}$, is $3600, the lowest value in the range of possible population mean values compatible with the null hypothesis. Since a possible population mean of $3650, $3700, and so on, would be just as compatible with the null hypothesis that $\mu \geq \$3600$, we see that use of $3600, while convenient, is rather arbitrary. Therefore, a small change in the way we interpret α is in order.

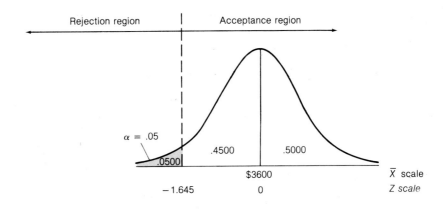

Figure 9.12. Sampling distribution of $\overline{X}$ associated with H_0: $\mu \geq \$3600$ and H_a: $\mu < \$3600$.

In a one-tailed test for a mean of a single population, the α value is not interpreted in quite the same way as with two-tailed tests: Whereas we previously called $\alpha = .05$ *the probability of a Type I error* (or, when s was substituted for σ, the *approximate* probability of a Type I error), we must now view it as the *maximum probability of a Type I error.* *

Notice that the entire value of α has been placed under the lower tail of the distribution. The reason for this is to be found in the statement of the null hypothesis: H_0: $\mu \geq \$3600$. Clearly, a Type I error will be committed only when the null hypothesis is true and the sample mean $\overline{X}$ is appreciably *lower than* the hypothesized population mean μ_0. If the sample mean $\overline{X}$ is found to be larger than the hypothesized population mean μ_0, acceptance of the null hypothesis is a foregone conclusion not dependent on the size of the difference between $\overline{X}$ and μ_0.

If the area in the left-hand tail is .05, the area under the curve between critical Z and 0 must be .4500. By consulting Appendix Table 2, we find that critical Z is either -1.64 or -1.65. We will arbitrarily split the difference and call it -1.645. Therefore, the decision rule is

Decision Rule
1. If observed $Z \geq -1.645$, accept H_0.
2. If observed $Z < -1.645$, reject H_0.

---

*Warning: Although it may be a source of some slight comfort to know that the α value is somewhat overstated in a one-tailed test, it *does* imply two less comforting corollaries: First, values obtained for β may be somewhat understated. Second, the power of the test, as measured by $1 - \beta$, may be somewhat overstated. Consequently, when one employs a one-tailed test, he or she should interpret the probability values with a shade more caution than is called for in connection with two-tailed tests.

5. Take a random sample of size n, calculate the statistic of interest, and (if appropriate) convert it into its standard value.

   To determine observed Z, we compute

$$\text{Observed } Z = \frac{\overline{X} - \mu_0}{\sigma_{\overline{X}}} = \frac{\$3582 - \$3600}{\$13.33} = -1.35$$

6. Apply the decision rule to the observed sample statistic and either accept or reject the null hypothesis.

   Since -1.35 is larger than -1.645, we accept the null hypothesis and conclude that the true but unknown population mean μ is at least \$3600.

7. Take the appropriate action.

   Do not make any change in inventory policy at this time.

Naturally, if the economists' concern was of the opposite kind, the inequality signs in both the null hypothesis and the alternative hypothesis would be turned around. The hypotheses in such a case would be

H_0: The average size of new orders has not increased from the \$3600 of last year, or
   H_0: $\mu \leq \$3600$.

H_a: The average size of new orders has increased from the \$3600 of last year, or
   H_a: $\mu > \$3600$.

The single action limit would be on the right-hand side of the normal sampling distribution of $\overline{X}$, as shown in Figure 9.13.

Let us work some additional one-tailed problems.

ILLUSTRATIVE CASE 9.3

The mean and standard deviation of the breaking strengths of wires used in the construction of retractable clotheslines are $\mu = 110$ pounds of pressure and $\sigma = 20$ pounds of pressure. A

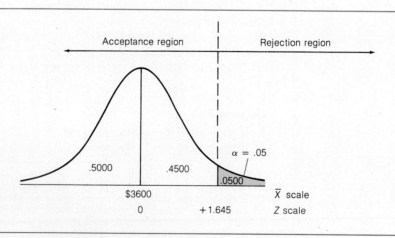

Figure 9.13. Sampling distribution of $\overline{X}$ associated with H_0: $\mu \leq \$3600$ and H_a: $\mu > \$3600$.

manufacturer believes that by using a new process developed by one of the company's engineers, the strength of the wires can be increased without necessitating an increase in their diameters. A random sample of $n = 81$ clothesline wires produced by the new process was examined and the mean breaking strength found to be $\overline{X} = 120$ pounds.

Assume that the $\sigma = 20$ value still applies and that $\alpha = .01$ is used. Has the new process resulted in an increase in average strength?

Let us begin by asserting tentatively: "No, the new process did not result in an increase in average strength." In symbols, H_0: $\mu \leq 110$ pounds. If that is H_0, then H_a will be: "Yes, the new process did result in an increase in average strength," or H_a: $\mu > 110$ pounds. (*Note:* The null hypothesis could be read: "The true population mean is *at most* 110." If the problem were phrased in such a way that the words *at most*—or what amounts to the same thing, "110 or less"—were used, one could immediately set up the null hypothesis in the manner shown.) The normal sampling distribution of $\overline{X}$ for this problem is shown in Figure 9.14.

Decision Rule
1. If observed $Z \leq +2.33$, accept H_0.
2. If observed $Z > +2.33$, reject H_0.

$$\text{Observed } Z = \frac{\overline{X} - \mu_0}{\dfrac{\sigma}{\sqrt{n}}} = \frac{120 - 110}{2.22} = 4.50$$

Since $+4.50$ is greater than $+2.33$, we reject the null hypothesis and conclude that the new process has led to an increase in average strength. Presumably, these results indicate that the new process should be used in the future.

ILLUSTRATIVE CASE 9.4

The manager of a large grocery store is concerned about the waiting time required by customers who use the express checkout lanes; these are lanes which serve customers who pay

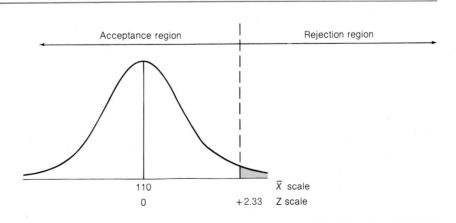

Figure 9.14. Normally distributed sampling distribution of $\overline{X}$ for Illustrative Case 9.3.

cash and who have eight or fewer items. During a past period the mean waiting time was found to be $\mu = 8$ minutes. More recently, a laser-beam scanning method of reading coded prices has been instituted which is expected to reduce the mean waiting time. The waiting times were observed for a random sample of $n = 100$ customers with the following results: $\overline{X} = 7$ minutes; $s = 4$ minutes.

Using a .05 level of significance, has the mean waiting time been reduced? Let us say tentatively that the mean waiting time has not been reduced. That is, H_0: $\mu \geq 8$ minutes. The alternative hypothesis will be H_a: $\mu < 8$ minutes. Figure 9.15 shows the normal sampling distribution of $\overline{X}$ for this problem.

Decision Rule
1. If observed $Z \geq -1.645$, accept H_0.
2. If observed $Z < -1.645$, reject H_0.

In this problem we do not know the population standard deviation σ and must make use of the sample standard deviation s. However, because the sample size is large, the standard normal curve may still be used.

$$\text{Observed } Z = \frac{\overline{X} - \mu_0}{\dfrac{s}{\sqrt{n}}} = \frac{7 - 8}{\dfrac{4}{\sqrt{100}}} = -2.50$$

Since -2.50 is smaller than -1.645, we reject the null hypothesis and conclude that the mean waiting time has decreased.

ILLUSTRATIVE CASE 9.5

A manufacturer claimed that the microwave ovens produced by his company had an average defect-free life of at least two years. Four randomly selected people were given free microwave ovens manufactured by this company and asked to use them in their homes in a normal manner and to note when the first defect appeared. The years of defect-free service

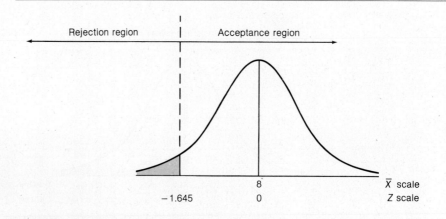

Figure 9.15. Normally distributed sampling distribution of $\overline{X}$ for Illustrative Case 9.4.

9 *Introduction to Statistical Hypothesis Testing*

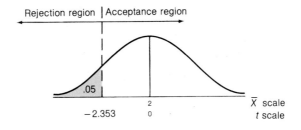

Figure 9.16. The *t* distributed sampling distribution of $\overline{X}$ for Illustrative Case 9.5.

provided by this small sample of microwave ovens were found to be 1.8, 1.9, 2.2, and 1.9 years. The sample mean $\overline{X}$ and standard deviation s were found to be 1.95 and .17 years, respectively. The α value is set at .05.

The words *at least* imply a null hypothesis of the type H_0: $\mu \geq 2$ and an alternative hypothesis of the type H_a: $\mu < 2$. Since we do not know σ, we must make use of s in its place. This fact coupled with the very small sample size of $n = 4$ tells us that the t, rather than the Z, table must be used. The distribution of the t statistic for this problem is shown in Figure 9.16. You will notice that the critical t value shown in this figure and in the decision rule below is not the one associated with $\alpha = .05$ and $n - 1 = 3$ degrees of freedom. It is the one associated with $\alpha = .10$ and $n - 1 = 3$ degrees of freedom. Do you see why? Because this is a one-tailed test, the entire α value must be placed under one tail of the curve, the lower tail in the present case. However, Appendix Table 3 is set up in such a way that the column headings represent the total area in both tails. Therefore, in order to get an area of .05 under the right-hand tail, we must use the column associated with $\alpha = .10$ under both tails.

Decision Rule
1. If observed $t \geq -2.353$, accept H_0.
2. If observed $t < -2.353$, reject H_0.

$$\text{Observed } t = \frac{\overline{X} - \mu_0}{\dfrac{s}{\sqrt{n}}} = \frac{1.95 - 2.00}{\dfrac{.17}{\sqrt{4}}} = -0.588$$

Since -0.588 is larger than -2.353, we accept the null hypothesis and conclude that the population mean defect-free life is at least two years.

*9.8 Determining the Required Sample Size for Specified Values of α and β

Thus far we have assumed that sample size n was predetermined by some means external to the hypothesis-testing task. However, if the sample size is not predetermined, we can find out what it should be if we are able to be quite precise about the risk we are willing to take of committing the Type I error, on the one hand, and the Type II error on the other.

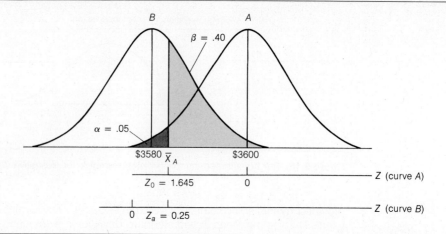

Figure 9.17. Situation implied by the computational steps involved in determining sample size n when values for α and β are given.

Let us assume that management of the Deft Tool Company wishes, as before, to run a risk of no more than .05 of rejecting the null hypothesis when it should be accepted—that is, they set $\alpha = .05$. Let us further suppose that they wish to hold the probability of a Type II error β to .40 *if the true population mean is 3580.* This situation is depicted in Figure 9.17.

Figure 9.17 shows the sampling distribution of $\overline{X}$ when the true population mean is assumed to be 3600 as normal curve A. It shows the sampling distribution of $\overline{X}$ when the population mean is assumed to be 3580 as normal curve B. The shaded area in the lower tail of curve A represents the probability of .05 of a Type I error; the shaded area in the upper tail of curve B represents the probability of .40 of a Type II error if the true population mean is 3580. The Z value, denoted Z_0 for action limit $\overline{X}_A$ when μ_0 is assumed to be true, is $+1.645$ from Appendix Table 2. The Z value, denoted Z_a for action limit $\overline{X}_A$ when $\mu_a = 3580$ is assumed true, is $+0.25$. Notice that Z_0 and Z_a are taken as positive.

We can write the following simultaneous equations:

$$\overline{X}_A = \mu_0 + Z_0 \frac{\sigma}{\sqrt{n}}$$

$$\overline{X}_A = \mu_a + Z_a \frac{\sigma}{\sqrt{n}}$$

Subtracting the second equation from the first, we obtain

$$0 = (\mu_0 - \mu_a) + \frac{\sigma}{\sqrt{n}}(Z_0 - Z_a) \qquad (9.8.1)$$

Rearranging, we get

$$n = \frac{(Z_0 + Z_a)^2 \sigma^2}{(\mu_0 - \mu_a)^2} \qquad (9.8.2)$$

We still need an advance estimate of the population standard deviation σ. Let us continue to use $\sigma = \$200$. We have

$$\mu_0 = \$3600 \qquad Z_0 = 1.645$$
$$\mu_a = \$3580 \qquad Z_a = 0.25 \qquad \sigma = \$200$$

and, by Equation (9.8.2),

$$n = \frac{(1.645 + 0.25)^2(200)^2}{(\$3600 - \$3580)^2} \cong 360$$

Thus, a sample size of 360 or larger would be required if (1) α is to be no greater than .05 and (2) β is to be no greater than .40 when the true population mean is $3580.

9.9 Tests Concerning the Proportion Associated with a Single Population

Many business problems involve attribute data and are, therefore, amenable to analysis through significance tests applied to the proportion. In this section we consider how the procedures described above may be appropriately modified to allow us to test hypotheses concerning the value of a single population proportion p.

Let us begin by considering the following illustrative case.

ILLUSTRATIVE CASE 9.6

The T. P. Thumm Company sells a wide variety of merchandise by mail order. Over the years the company has conducted regular marketing research on the relative merits of alternative types of advertising layout and design. One such study concerned the sale of lawn and patio furniture, a line of merchandise which had in the past been allotted relatively little space in the company's catalog. Sales records showed that, in a typical year, approximately 1% of the households receiving a copy of the catalog responded by ordering at least one article of lawn and patio furniture. More recently, however, management has come to wonder whether more advertising space should be devoted to this line of merchandise. Consequently, they printed 2000 special copies of the catalog. In the special issue, a full page of multicolored sales material was devoted exclusively to lawn and patio furniture. These special copies were mailed to a random selection of households on the company's mailing list.

A later examination of results revealed that the response rate was 2.5%. Management wishes to know whether 2.5% ($\bar{p} = .025$) represents a significant *improvement* over the historical 1% ($p = .01$).

In general outline, the same steps would be taken to test a null hypothesis about a population proportion as were taken in connection with tests of a population mean. They would be as follows:

1. State the null and alternative hypotheses.
   The null and alternative hypotheses implied by the word *improvement* (meaning an increase in response rate) in this illustrative case are

   H_0: The change in the ad size and design did not lead to an increase in the response rate, or H_0: $p \leq .01$.

H_a: The change in the ad size and design did lead to an increase in the response rate, or
H_a: $p > .01$.

2. Establish the level of significance.

Let us suppose that management decides to set the level of significance at $\alpha = .05$.

3. Determine the appropriate kind of sampling distribution to rely on.

Because the hypothesis has to do with population p and is to be tested using sample $\bar{p}$, we will make use of the sampling distribution of $\bar{p}$ known to have the following characteristics (from Chapter 8):

$$\mu_{\bar{p}} = p$$

$$\sigma_{\bar{p}} = \sqrt{\frac{pq}{n}} \quad \text{or} \quad \sqrt{\frac{pq}{n}} \cdot \sqrt{\frac{N-n}{N-1}}$$

Normal shape if $np > 5$ and $nq > 5$.

4. Establish a decision rule.

The decision rule is developed in essentially the same manner as in previous problems. First, since $np = (2000)(.01) = 20$ and $(2000)(.99) = 1980$, we may safely assume that the relevant sampling distribution of $\bar{p}$ will have a normal shape, as shown in Figure 9.18. Since the population proportion is hypothesized to be equal to or less than .01, we take the mean of the sampling distribution to be $p = \mu_{\bar{p}} = .01$ with the realization that the α value of .05 represents the maximum possible probability of a Type I error. Because this is an upper-tail test, we place the entire α value under the right-hand tail. Now, we seek the critical Z value associated with $.5000 - .0500 = .4500$. We previously found this Z value to be 1.645. Therefore:

Decision Rule

1. If observed $Z \leq +1.645$, accept H_0.
2. If observed $Z > +1.645$, reject H_0.

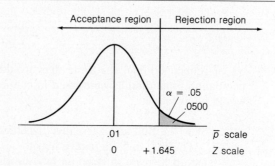

Figure 9.18. Sampling distribution of $\bar{p}$ for Illustrative Case 9.6.

9 *Introduction to Statistical Hypothesis Testing*

5. Take a random sample of size n, compute the sample statistic, and (if appropriate) convert it into its standard value.

We already know that $\bar{p} = .025$. Converting this into its Z value, we get

$$\text{Observed } Z = \frac{\bar{p} - p_0}{\sqrt{\dfrac{p_0 q_0}{n}}} = \frac{.025 - .010}{\sqrt{\dfrac{(.01)(.99)}{2000}}} = \frac{0.15}{.00223} = 6.73$$

where $p_0 =$ population proportion according to the null hypothesis
$q_0 = 1 - p_0$

(*Note*: The relevant population in this case is infinite—all the people who could possibly be exposed to the new version of the ad. Hence, there is no finite correction factor.)

Therefore, even though .010 and .025 do not seem greatly different, in terms of standard units $\bar{p}$ of .025 is seen to be just a little short of seven standard deviations above the hypothesized population proportion of .01!

6. Apply the decision rule to the observed sample statistic and either accept or reject the null hypothesis.

According to our decision rule, observed Z is most emphatically in the rejection region. Therefore, we reject the null hypothesis and conclude that the response rate has increased.

You should be aware, however, that rejecting the null hypothesis does not, in this case, necessarily establish that the new version of the ad was responsible for the significant increase in the response rate. All we have established for certain is that it is highly unlikely that the true population proportion is still .01. Other factors may have contributed to the apparent increase in p. The "experiment" is simply too crude to offer anything but circumstantial evidence that the new type of ad is more effective than the old type.

7. Take the appropriate action.

Even though a tighter experiment would have been desirable, on the basis of the information we have, it would seem sensible to stick with the new ad.

Because the procedures involved in testing a null hypothesis about a single population proportion are so similar to those demonstrated in earlier sections of this chapter, we will not belabor the subject. However, here are a few points you should be aware of:

1. The same three categories of hypotheses described in connection with the mean are also applicable to the proportion. This point is elaborated on somewhat in Figure 9.19.
2. If either np or nq is less than 5, the normal curve is not a particularly good model of the sampling distribution of $\bar{p}$. The method for dealing with this kind of situation is quite different from any methods described in this chapter.
3. The probability of a Type II error for a specified (alternatively) hypothesized population proportion p_a is determined in an analogous manner to that for problems involving a mean. That is, determine the action limits by

$$\bar{p}_A = p_0 - Z\sqrt{\frac{p_0 q_0}{n}} \qquad \text{in a lower-tail test}$$

$$\bar{p}_A = p_0 + Z\sqrt{\frac{p_0 q_0}{n}} \qquad \text{in an upper-tail test}$$

or

$$\left.\begin{array}{l} \bar{p}_L = p_0 - Z_L\sqrt{\dfrac{p_0 q_0}{n}} \\[4mm] \bar{p}_U = p_0 + Z_U\sqrt{\dfrac{p_0 q_0}{n}} \end{array}\right\} \quad \text{in a two-tailed test}$$

However, when converting the action limit into a Z value by $Z = (\bar{p}_A - p_a)/\sigma_{\bar{p}}$, remember that $\sigma_{\bar{p}}$ should be determined by $\sigma_{\bar{p}} = \sqrt{(p_a q_a)/n}$. Therefore, the denominator will change somewhat as different (alternatively) hypothesized proportions, p_a values, are specified.

4. An operating characteristic curve (or power curve, if preferred) can be constructed by extending the operations referred to earlier provided that $\sigma_{\bar{p}} = \sqrt{(p_a q_a)/n}$ is computed anew with each change in specified p_a.

*9.10 Using Observed Significance Levels

The seven-step procedure for testing a null hypothesis set forth in section 9.3 and demonstrated several times in this chapter will be followed, with appropriate minor modifications in some later chapters, throughout the hypothesis-testing parts of this text. This approach, which we might call the *conventional approach,* has as a central feature the establishment of a level of significance α very early in the analysis. An early commitment to a specific α value has certain advantages.

SOME ADVANTAGES OF THE CONVENTIONAL APPROACH TO HYPOTHESIS TESTING

1. It forces, or at least encourages, the analyst to think seriously about the relative adverse consequences of Type I versus Type II errors.
2. It reduces the temptation to reject a null hypothesis at a less rigorous level (that is, at a level associated with a higher α value) than the analyst might originally have thought appropriate.
3. The idea that one can control the α value permits determination of a desirable sample size; it also permits determination of Type II error probabilities independently of the sample results.

You should be aware, however, that a somewhat different approach to statistical hypothesis testing does exist and has enjoyed increasingly widespread popularity in recent years. This approach makes use of an *observed level,* as distinguished from a *controlled level,* of significance. Relating it to our seven-step procedure, we may say that, with the observed-

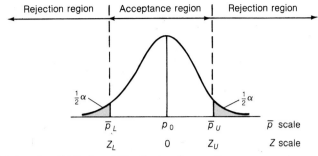

(a) Sampling distribution of $\bar{p}$ for the hypotheses: $H_0: p =$ some value
$H_a: p \neq$ to the value in H_0

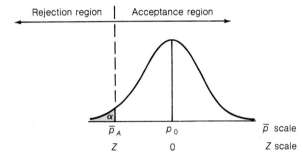

(b) Sampling distribution of $\bar{p}$ for the hypotheses: $H_0: p \geq$ some value
$H_a: p <$ the value in H_0

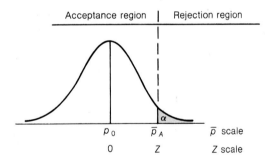

(c) Sampling distribution of $\bar{p}$ for the hypotheses: $H_0: p \leq$ some value
$H_a: p >$ the value in H_0

Figure 9.19. Partitioning of the sampling distribution of $\bar{p}$ for three categories of hypothesis-testing problems.

significance-level approach, Step 2 (setting the value of α) is eliminated and Step 6 (accepting or rejecting the null hypothesis) is modified. More specifically, the investigator works the problem through to the point where he has obtained the test statistic, the observed Z or observed t value (whichever is appropriate); he then determines at what level that observed Z or t value is significant. In other words,

An *observed significance level* is the probability of obtaining (when the null hypothesis is true) a value of the test statistic *at least* as extreme, in the appropriate direction, as the one actually obtained.

SOME ADVANTAGES OF THE OBSERVED-SIGNIFICANCE-LEVEL APPROACH TO HYPOTHESIS TESTING

1. When the statistical analyst and the user of the statistical results are different people, the user need not be constrained by the analyst's preferred level of significance; the user may make his or her own judgment about what kind of result is significant.
2. More precise information is usually conveyed when this approach is used. For example, with the conventional approach one can know that a value of the test statistic is significant at the .05 level but not know whether it is just barely significant at that level or whether it is also significant at the .01 level, the .001 level, and so forth. The observed-significance-level approach reveals precisely how significant a result is.
3. Most computer programs for testing hypotheses take the calculations only to the point where the value of the test statistic is determined. Discovering whether the result is significant is left up to the person using the program. Hence, users of such programs just about have to know how to use this method.

Let us conclude this chapter by briefly reviewing each of the illustrative cases, but this time use the observed-significance-level approach. Sketches pertaining to most of these problems are presented in Figure 9.20.

1. *Illustrative Case 9.1:* The Deft Tool Manufacturing Company (two-tailed version with σ assumed known)
   Pertinent facts:

$$H_0: \mu = \$3600$$
$$H_a: \mu \neq \$3600 \qquad n = 225$$
$$\text{Observed } Z = -1.35$$

To determine at what level the observed Z of -1.35 is significant, we go to the table of areas under the standard normal curve (Appendix Table 2) and find the area associated with $Z = 1.35$. That area is .4115. However, it is the tail area we use to measure significance. Thus, $.5000 - .4115 = .0885$. The .0885 value represents the area in one tail only. The other tail will have the same area. Therefore, $2(.0885) = .1770$. We say that the Z value of -1.35 is significant at about the 18% level. That is, about 18 times out of 100 when $H_0: \mu = \$3600$ is true, samples of size $n = 225$ drawn at random from this same population will have $\overline{X}$ values 1.35 standard deviations *or more* away from $\mu_0 = \$3600$, a situa-

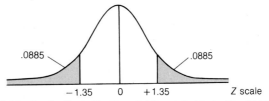

(a) Normal sampling distribution for Illustrative Case 9.1 (two–tailed test with σ known). Significance at the .18 level.

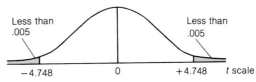

(b) t–distributed sampling distribution for Illustrative Case 9.2. Significance beyond the .01 level.

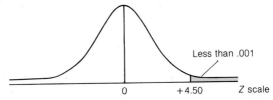

(c) Normal sampling distribution for Illustrative Case 9.3. Significance at well beyond the .001 level.

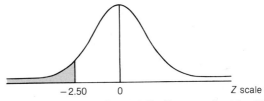

(d) Normal sampling distribution for Illustrative Case 9.4. Significance at about the .006 level.

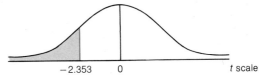

(e) t–distributed sampling distribution for Illustrative Case 9.5. Not significant at the .05 level.

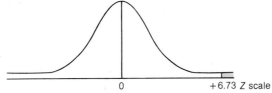

(f) Normal sampling distribution for Illustrative Case 9.6. Significance well beyond the .001 level.

Figure 9.20. Sketches associated with the application of the observed-significance-level method to six illustrative cases.

tion leading to the (incorrect) rejection of the null hypothesis. (See part (a) of Figure 9.20.)

2. *Illustrative Case 9.1:* The Deft Tool Manufacturing Company (two-tailed test with s substituted for unknown σ)
Pertinent facts:

$$H_0: \mu = \$3600$$
$$H_a: \mu \neq \$3600 \qquad n = 225$$
$$\text{Observed } Z = -1.39$$

Following the same procedure as above, we get $.5000 - .4177 = .0823$. And $2(.0823) = .1646$, a result indicating significance at about the 17% level.

3. *Illustrative Case 9.2:* Southwest Sprinkler System Company
Pertinent facts:

$$H_0: \mu = 20 \text{ hours}$$
$$H_a: \mu \neq 20 \text{ hours} \qquad n = 9$$
$$\text{Observed } t = 4.748$$

To determine the observed significance level for this problem, we go to the table of t values (Appendix Table 3) and read across from degrees of freedom = $9 - 1 = 8$ until we find a critical value close to, but not in excess of, 4.748. Regrettably, with this somewhat condensed t table, we can do no better than we did when α was set in advance. That is, the critical t value associated with 8 degrees of freedom and $\alpha = .01$ is found to be 3.355. Our observed t value of 4.748 is obviously significant at a more rigorous level than .01; however, it would require a more detailed table to provide us with the correct observed significance level. (See part (b) of Figure 9.20.)

4. *Illustrative Case 9.1:* The Deft Tool Manufacturing Company (lower-tail version)
Pertinent facts:

$$H_0: \mu \geq \$3600$$
$$H_a: \mu < \$3600 \qquad n = 225$$
$$\text{Observed } Z = -1.35$$

The procedure for obtaining the observed significance level for this problem is exactly the same as in 1 above except that, this time, because the entire α value is under the single relevant tail, a doubling of the area found in the normal curve table is not called for. Thus, the appropriate area is $.0885$, which is to say we have significance at about the 9% level.

5. *Illustrative Case 9.3:* Retractable clothesline wires
Pertinent facts:

$$H_0: \mu \leq 110$$
$$H_a: \mu > 110 \qquad n = 81$$
$$\text{Observed } Z = 4.50$$

Our normal curve table does not go this high. The highest critical Z value in the table is 3.09. The area associated with this value is $.4990$. Thus, $.5000 -$

.4990 = .001. Therefore, we have significance well beyond the .001 level. (See part (c) of Figure 9.20.)

6. *Illustrative Case 9.4:* Checkout lanes
Pertinent facts:

$$H_0: \mu \geq 8 \text{ minutes}$$
$$H_a: \mu < 8 \text{ minutes}$$
$$n = 100$$

$$\text{Observed } Z = -2.50$$

Appendix Table 2 shows the observed significance level to be .5000 − .4938 = .0062. (See part (d) of Figure 9.20.)

7. *Illustrative Case 9.5:* Microwave ovens
Pertinent facts:

$$H_0: \mu \geq 2$$
$$H_a: \mu < 2$$
$$n = 4$$

$$\text{Observed } t = -0.588$$

Reading across from degrees of freedom = 3 in Appendix Table 3, we find that critical t for $\alpha = .10$ (corresponding to $\alpha = .05$ for a one-tailed test) is −2.353. Therefore, we do not have significance at the .05 level, but that is all we can learn from this particular t table. (See part (e) of Figure 9.20.)

8. *Illustrative Case 9.6:* The T. P. Thumm Company
Pertinent facts:

$$H_0: p \leq .01$$
$$H_a: p > .01$$
$$n = 2000$$

$$\text{Observed } Z = 6.73$$

For reasons cited in connection with 5 above, we can only say that $Z = 6.73$ is significant at a level a good deal more rigorous (that is, at a level having a lower α value) than .001. (See part (f) of Figure 9.20.)

9.11 Summary

In statistical hypothesis testing, as in the construction of confidence intervals, the theory of sampling distributions plays a key role. It is our knowledge of the sampling distributions associated with important statistics ($\overline{X}$ and $\overline{p}$ in this chapter) which provides us with a firm theoretical basis for saying that a test statistic of interest is significant at some specified level.

With hypothesis testing, one begins by making an assertion about the true but unknown value of a population parameter. This assertion is called the *null hypothesis*. Usually an alternative hypothesis is also stated. It is this *alternative hypothesis* that, ordinarily, one hopes to "prove" correct. In this chapter we dealt with only two kinds of parameters, μ and p. Their sample counterparts, $\overline{X}$ and $\overline{p}$, were used to test the validity, or lack of it, of the relevant null hypotheses. Because a sample represents only partial information about a situation under study, it may lead us either to a correct or incorrect conclusion about the validity of the null hypothesis. If our sample information leads us to an incorrect conclusion, it may do so in one of two ways: (1) It can indicate that the null hypothesis is false even though it is really true; this is called a *Type I error*. Or (2) it can indicate that the null hypothesis is true even though it is really false; this is called the *Type II error*.

Tests of hypotheses of the kinds described in this chapter may call for use of either or both tails of a normal or *t* distribution. What is used is dependent on the nature of the question being addressed.

Throughout this chapter a seven-part procedure for testing a null hypothesis was used:

1. State the null and alternative hypotheses.
2. Establish the level of significance.
3. Determine the appropriate kind of sampling distribution to rely on.
4. Establish a decision rule.
5. Take a random sample of size *n*, calculate the statistic of interest, and convert the value of that statistic into its standard-value equivalent.
6. Apply the decision rule to the observed test statistic and either accept or reject the null hypothesis.
7. Take the appropriate action.

An alternative approach to hypothesis testing was also touched on. This is the observed-significance-level approach. When using this approach, the investigator does not set the risk of committing a Type I error at the outset. Instead, one calculates the test statistic and determines from it at what level the null hypothesis could be rejected.

In addition to treating hypothesis-testing procedures, this chapter dealt with the related topics of (1) determining probabilities of Type II errors when the sample size and the probability of a Type I error are known, (2) construction of the operating characteristic curve and power curve, and (3) determining the desired sample size when the probability of a Type I error and the probability of a Type II error are known.

Terms Introduced in This Chapter

acceptance region (p. 382)
action limit(s) (p. 388)
alpha, α (p. 381)
alternative hypothesis (p. 374)
beta, β (p. 381)
compound hypothesis (p. 378)
critical region (p. 383)
critical value(s) (p. 383)

decision rule (p. 384)
level of significance (p. 381)
lower-tail test (p. 377)
null hypothesis (p. 374)
observed significance level (p. 408)
one-tailed test (p. 377)
operating characteristic curve (p. 391)
power curve (p. 394)

power of a test, $1 - \beta$ (p. 395)
rejection region (p. 383)
simple hypothesis (p. 378)
two-tailed test (p. 377)
Type I error (p. 376)
Type II error (p. 376)
upper-tail test (p. 377)

Formulas Introduced in This Chapter

Formula for converting an observed sample mean into its Z-value equivalent when σ is known:

$$\text{Observed } Z = \frac{\overline{X} - \mu_0}{\dfrac{\sigma}{\sqrt{n}}}$$

Formula for converting an observed sample mean into its Z-value equivalent when σ is unknown but n is large:

$$\text{Observed } Z = \frac{\overline{X} - \mu_0}{\dfrac{s}{\sqrt{n}}}$$

Formula for converting an observed sample mean into its t-value equivalent when σ is unknown, n is small and the population has a normal shape:

$$\text{Observed } t = \frac{\overline{X} - \mu_0}{\dfrac{s}{\sqrt{n}}}$$

Formulas for converting critical Z values into action limits on the $\overline{X}$ scale when σ is known (if σ is unknown, s may be substituted for σ if n is large; if σ is unknown and n is small, the same substitution may be made, but t_L and t_U will replace Z_L and Z_U in the formulas):

$$\overline{X}_L = \mu_0 - Z_L \frac{\sigma}{\sqrt{n}} \quad \text{or} \quad \overline{X}_U = \mu_0 + Z_U \frac{\sigma}{\sqrt{n}}$$

Formulas for converting action limits on the $\overline{X}$ scale into critical Z values for a sampling distribution of $\overline{X}$ with an alternatively hypothesized mean. Used in the calculation of probabilities of Type II errors:

$$Z_1 = \frac{\overline{X}_L - \mu_a}{\dfrac{\sigma}{\sqrt{n}}} \quad \text{or} \quad Z_2 = \frac{\overline{X} - \mu_a}{\dfrac{\sigma}{\sqrt{n}}}$$

Formula for determining the necessary sample size for specific controlled probabilities of Type I and Type II errors:

$$n \geq \frac{(Z_0 + Z_a)^2 \sigma^2}{(\mu_0 - \mu_a)^2}$$

Formula for converting an observed sample proportion into its Z-value equivalent:

$$\text{Observed } Z = \frac{\overline{p} - p_0}{\sqrt{\dfrac{p_0 q_0}{n}}}$$

Formulas for converting critical Z values into action limits on the $\overline{p}$ scale:

$$\overline{p}_L = p_0 - Z_L \sqrt{\frac{p_0 q_0}{n}} \quad \text{or} \quad \overline{p}_U = p_0 + Z_U \sqrt{\frac{p_0 q_0}{n}}$$

Formulas for converting action limits on the $\overline{X}$ scale into critical Z values for a sampling distribution of $\overline{X}$ whose mean is some alternatively hypothesized p value. Used in the calculation of probabilities of Type II errors:

$$Z_1 = \frac{p_L - p_a}{\sqrt{\dfrac{p_a q_a}{n}}} \quad \text{or} \quad Z_2 = \frac{p_U - p_a}{\sqrt{\dfrac{p_a q_a}{n}}}$$

Questions and Problems

9.1 Explain the meaning of each of the following:
  a. Rejection region   **e.** Power of a test
  b. Critical value of Z   **f.** Operating characteristic curve
  c. Decision rule   **g.** p_a
  d. μ_0

9.2 Distinguish between:
  a. Type I error and Type II error
  b. One-tailed test and two-tailed test
  c. α and β
  d. Type I error and α
  e. Hypothesis test and a confidence interval
  f. Observed significance level and controlled significance level

9.3 Distinguish between:
  a. Operating characteristic curve and power curve
  b. Critical value and action limit
  c. α in a two-tailed test and in a one-tailed test
  d. Acceptance region and rejection region
  e. Power curve and power of a test

9.4 Distinguish between (or among):
  a. Null hypothesis and alternative hypothesis
  b. μ, μ_0, and μ_a
  c. p, p_0, and p_a
  d. β and $1 - \beta$
  e. Simple hypothesis and compound hypothesis

9.5 Indicate which of the following statements you agree with and which you disagree with, and defend your opinions:
  a. $\alpha + \beta = 1$
  b. Since one can make an error of inference either by (1) accepting a false null hypothesis or (2) rejecting a true null hypothesis, when sampling is used there is no way to avoid making an error of inference.
  c. The level of significance selected always represents the maximum probability of a Type I error.
  d. In a statistical hypothesis-testing problem, an investigator who rejects a null hypothesis which is really true commits a Type II error.

9.6 Indicate which of the following statements you agree with and which you disagree with, and defend your opinions:
  a. In a statistical hypothesis-testing problem, an investigator who rejects a null hypothesis which is really true commits neither a Type I nor a Type II error.
  b. When an analyst states in the null hypothesis that a population parameter has a value equal to or in excess of some number, an upper-tail test is called for.
  c. In practice, the level of significance should be established only after careful consideration of the relative seriousness of the two kinds of error of inference, Type I and Type II.
  d. An investigator who avoids committing a Type I error must necessarily commit a Type II error.

9.7 Indicate which of the following statements you agree with and which you disagree with, and defend your opinions:

a. Because when we test a statistical hypothesis we ordinarily take only one random sample, the concept of sampling distribution plays no roll.

b. The null and alternative hypotheses are usually worded in such a way that if one is correct the other must be incorrect.

c. The null and alternative hypotheses are always worded in such a way that if one is incorrect the other must be correct.

d. If the hypotheses are of the form

H_0: $\theta \geq$ some specified value

H_a: $\theta <$ value specified in the null hypothesis

a lower-tail test is called for.

9.8 Indicate which of the following statements you agree with and which you disagree with, and defend your opinions:

a. If the hypotheses are of the form

H_0: $\theta =$ some specified value

H_a: $\theta \neq$ value specified in the null hypothesis

a two-tailed test is called for.

b. If in the situation described in *a* a level of significance of $\alpha = .05$ is used, .05 will be placed under each tail of the standard normal curve or the t distribution, whichever is appropriate.

c. If the true population proportion p is .4, we would expect the $\bar{p}$ values of random samples of size n from this population to be .4 as well.

d. If the true population proportion p is .4, we would expect the $\bar{p}$ values of random samples of size n from this population to cluster around .4.

9.9 Indicate which of the following statements you agree with and which you disagree with, and defend your opinions:

a. If a null hypothesis is rejected at a specified level of significance, that can mean only one thing: The null hypothesis is incorrect.

b. Words like *change*, *difference*, *departure from*, and so forth, in a hypothesis-testing problem alert the analyst to the need for a two-tailed test.

c. One cannot control α and β at the same time.

d. If α and n are controlled, one can calculate values for β but cannot control them.

9.10 Indicate which of the following statements you agree with and which you disagree with, and defend your opinions:

a. One can determine the necessary sample size if α and β are precisely controlled.

b. In business applications, a Type II error is usually considered more serious than a Type I error.

c. The specific points where the rejection region begins are called critical values or action limits, depending on which scale, the $\overline{X}$ scale or the Z scale, is being used.

d. When beginning a two-tailed test of a statistical hypothesis, we assume temporarily that the null hypothesis is true.

9.11 Indicate which of the following statements you agree with and which you disagree with, and defend your opinions:

a. If in a hypothesis-testing problem σ is unknown, and n is, say, 10, critical value(s) of t, rather than Z, will be used.

b. Once we know α and n, we have all the information we need to compute β.

c. In a two-tailed test situation, α can have only one value, whereas β may have many values.

d. In a one-tailed test situation, both α and β may have many values.

9.12 Indicate which of the following statements you agree with and which you disagree with, and defend your opinions:

a. When computing the probability of a Type II error, the null hypothesis may or may not be expressed as a simple hypothesis (as contrasted with a compound hypothesis) but the alternative hypothesis must be expressed as a simple hypothesis.

b. For a specified sample size n the greater the difference between μ_a and μ_0, the greater will be the probability of a Type II error.

c. For given values of α, μ_0, and μ_a, the larger the sample size n, the larger will be the probability of a Type II error.

d. An operating characteristic curve is a graphic device that allows one to get a visual impression of the trade-off between the risks of Type I and Type II errors.

9.13 Indicate which of the following statements you agree with and which you disagree with, and defend your opinions:

a. Whereas an operating characteristic curve shows probabilities of accepting the null hypothesis on the vertical axis, a power curve shows the probabilities of a Type I error on the vertical axis.

b. An ideal operating characteristic curve would have no width whatever.

c. Words like *decreased, increased, declined, improved*, and so on, in a hypothesis-testing problem alert the analyst to the need for a two-tailed test.

d. In a one-tailed test, α is interpreted somewhat differently from the α in a two-tailed test.

9.14 Indicate which of the following statements you agree with and which you disagree with, and defend your opinions:

a. In a one-tailed test such that $\mu_0 \neq \mu_a$, acceptance of the null hypothesis will necessarily lead to a Type II error.

b. In Figure 9.21, if sampling distribution A is erected around the mean stated in the null hypothesis μ_0, and sampling distribution B is erected around the mean of the alternative hypothesis μ_a, the probability of a Type II error is represented by the shaded area.

c. In Figure 9.22, if sampling distribution A is erected around the mean stated in the null hypothesis μ_0, and sampling distribution B is erected around the mean stated in the alternative hypothesis μ_a, the probability of a Type II error is represented by the shaded area.

d. Unlike the standard error of the mean $\sigma_{\bar{X}}$, which will be the same for different values of μ_a, the standard error of the proportion $\sigma_{\bar{p}}$ will change as p_a changes.

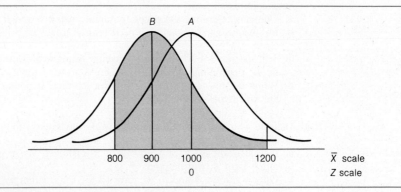

Figure 9.21. Sketch for problem 9.14, part *b*.

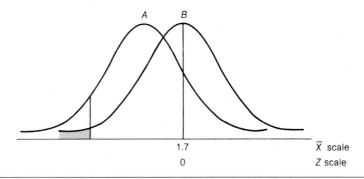

Figure 9.22. Sketch for problem 9.14, part *c*.

9.15 Describe a realistic hypothetical business situation in which
  a. Committing a Type I error would probably be considered more serious than committing a Type II error.
  b. Committing a Type II error would probably be considered more serious than committing a Type I error.

9.16 What assumption is made when the *t* statistic is used to test a null hypothesis concerning the value of a population mean which need not be made when the Z statistic can be appropriately used? (*Hint:* You may wish to review section 8.22 of Chapter 8.)

 9.17 A machine produces twine with an average breaking strength of 40 pounds, when properly adjusted. After the machine has been in use for several working days, a sample of 64 pieces is examined. The sample mean is found to be 38 pounds and the sample standard deviation 5 pounds. The supervisor wishes to know whether a change has occurred in mean breaking strength.
  a. Does the supervisor's desire suggest the need for a one-tailed or a two-tailed test? Why?
  b. Develop the decision rule for this problem assuming that the supervisor is willing to accept a risk of .05 of concluding that the population mean has changed if it really has not changed.
  c. Should the null hypothesis be accepted or rejected? Explain how you arrived at your conclusion.
  d. Determine the probability of concluding that the average strength has not changed if it has really changed to 39 pounds.

 9.18 Refer to problem 9.17. Suppose the supervisor had stated the null and alternative hypotheses to be

H_0: $\mu \geq 40$ minutes

H_a: $\mu < 40$ minutes

Suppose further that the probability of rejecting the null hypothesis incorrectly is set at $\alpha =$ .05, the sample size is $n = 64$, and the observed sample statistics are $\overline{X} = 38$ pounds and $s = 5$ pounds.
  a. What kind of hypothesis-testing procedure is called for now? Why?
  b. Construct the appropriate decision rule.
  c. Tell whether the supervisor should accept or reject the null hypothesis. Justify your answer.

9.19 Refer to the *c* parts of problems 9.17 and 9.18. Determine for each version of this problem at what level the test statistic would be significant if the observed-significance-level method were used.

 9.20 The credit manager of a large department store claims that the average account balance of the store's credit customers is at most $125. An auditor from Haskin and Anderson selects 100 accounts at random and observes that the sample mean is $\overline{X} = \$132$ and that the sample standard deviation is $s = \$50$. Assume that the population size is sufficiently large that the finite correction factor need not be used.
   a. State the null and alternative hypotheses in symbols.
   b. Develop the decision rule for this problem if α is set at .01.
   c. Test the null hypothesis. Should the null hypothesis be accepted or rejected?

 9.21 Refer to part *c* of problem 9.20. At what level would the test statistic be significant if the observed-significance-level method were used?

 9.22 A manufacturer of electric automobiles claims that her company's products will travel at least 100 miles, on the average, before the batteries have to be recharged. A consumer product-rating service tested this claim using $n = 400$ randomly selected new cars produced by this manufacturer and a level of significance of $\alpha = .05$.
   a. State the null and alternative hypotheses that the product-rating service presumably used. Explain why you think it would have stated the two hypotheses in this manner.
   b. Develop the decision rule for this problem.
   c. Suppose that the consumer product-rating service had found the sample mean to be $\overline{X} = 98.5$ miles and the sample standard deviation to be $s = 36$ miles. Should the null hypothesis by accepted or rejected? Justify your conclusion.
   d. Would the answer to part *c* have been any different if the sample size *n* had been 100 rather than 400? Demonstrate that the answer either would or would not have been different.

 9.23 Refer to problem 9.22, part *c*. At what level would the test statistic have been significant if the observed-significance-level method had been used?

 9.24 A paramedical service that covers 40 blocks of a certain large city is concerned about the length of time required to reach the victim(s) after a call has been received. In the past, the mean travel time has been 8 minutes with a standard deviation of 2 minutes. Two new emergency routes have been established whereby the traffic lights are stopped on green until the paramedical unit passes. Sampling will now be used to determine whether the establishment of the emergency routes has been effective. A sample of $n = 9$ emergencies is observed and the sample mean is found to be 6.5 minutes. Assume that the standard deviation of $\sigma = 2$ minutes still applies and that $\alpha = .01$.
   a. State the null and alternative hypotheses.
   b. Develop the decision rule.
   c. Should the null hypothesis be accepted or rejected?

 9.25 Refer to problem 9.24. Suppose that the population standard deviation was not known but that the sample standard deviation for the nine emergency trips was found to be 1.7 minutes. Assume that all other information is the same as in problem 9.24.
   a. State the null and alternative hypotheses.
   b. Develop the decision rule.
   c. Should the null hypothesis be accepted or rejected?
   d. In what way or ways was the hypothesis-testing procedure altered by the fact that σ was unknown?
   e. What, if any, assumption about the population did you accept implicitly when you worked parts *b* and *c* of this problem?

9.26 Refer to the *c* parts of problems 9.24 and 9.25. Determine for each version of this problem at what level the test statistic would be significant if the observed-significance-level method were used.

 9.27 An anti-business political candidate claimed in a fiery speech that the average profit margin measured for the largest corporations in the United States is at least 12%. His pro-business opponent, believing the figure too high and seeing an opportunity to characterize his rival as one who plays fast and loose with the facts, selected a random sample of 49 companies from *Fortune* magazine's list of the country's 500 largest corporations and found that the average profit margin for companies in his sample was only $\overline{X} = 8\%$, with a standard deviation of $s = 2.5\%$.
 a. Indicate the manner in which the pro-business candidate would presumably state his null and alternative hypotheses. Explain why you believe he would state his hypotheses in this particular way.
 b. Develop the decision rule assuming that $\alpha = .10$.
 c. Test the null hypothesis. Should the null hypothesis be accepted or rejected?
 d. When working part *c* of this problem, did you make use of the finite correction factor? Why or why not?

 9.28 Refer to part *c* of problem 9.27. At what level would the test statistic be significant if the observed-significance-level method were used?

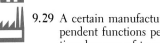 **9.29** A certain manufacturing plant utilizes 200 assembly-line workers. Because of the interdependent functions performed by these workers, some may be involuntarily idle at any given time because of temporary bottlenecks at some prior stage of the assembly process.
   A supervisor conducts a study involving 16 randomly spaced one-minute observations to determine how many of the 200 assembly workers were productively employed at those times. The 16 counts obtained were as follows:

198	192	188	189	200	192	180	200
175	195	190	196	190	190	180	181

   It is known that, during a past period, when the assembly operations were organized along somewhat different lines, the mean number of workers who were productively employed during one-minute observational periods was $\mu = 191$, a number which the supervisor considered probably an irreducible minimum.
 a. State the null and alternative hypotheses assuming the supervisor wishes to know whether a change has occurred since the earlier period.
 b. Develop the appropriate decision rule if the level of significance is set at $\alpha = .01$.
 c. Test the null hypothesis. Should the null hypothesis be accepted or rejected?
 d. Describe in what way or ways the procedure for determining the decision rule would have been different if the sample size had been 81 rather than 16.

 9.30 Refer to problem 9.29. Still assuming a sample size of $n = 16$, a level of significance of $\alpha = .01$, and a past population mean of 191 persons:
 a. State the null and alternative hypotheses assuming the supervisor wishes to know whether a decrease had occurred from the earlier period.
 b. Develop the decision rule.
 c. Test the null hypothesis. Should the null hypothesis be accepted or rejected?
 d. Describe in what way or ways the procedure for determining the decision rule would have been different if the sample size had been 81 rather than 16.

 9.31 A shipment of bolts of polyester-cotton cloth is acceptable to the Sew and Save Company if no more than 3% of the bolts contain flaws. From a shipment of $N = 200$ bolts, the owner of Sew and Save selected $n = 36$ bolts at random for careful examination and found that 2 of these contained flaws.

a. State the null and alternative hypotheses.
b. Develop the appropriate decision rule if the level of significance is set at $\alpha = .05$.
c. Test the null hypothesis. Should the null hypothesis be accepted or rejected?
d. Did you utilize the finite correction factor in connection with part c? Why or why not?

 9.32 Refer to problem 9.31. At what level would the test statistic be significant if the observed-significance-level method were used?

 9.33 A real estate salesperson claims that 60% of all homeowners move to another house within five years of the time of purchase of their first house. Test this null hypothesis against the alternative hypothesis H_a: $p \neq .6$ if, in a random sample of 144 homeowners, 72 were found to have moved within five years after buying their first house. Use a .05 level of significance.

 9.34 Refer to problem 9.33. At what level would the test statistic be significant if the observed-significance-level method were used?

 9.35 Refer to problem 9.33. In what way or ways would the hypothesis-testing procedure be changed if the real estate salesperson had claimed that at least 60% of all homeowners move to another house within five years of the time of purchase of their first house? Still assuming a .05 level of significance, should the null hypothesis be accepted or rejected?

 9.36 A TV network president claimed that, despite what appeared to be widespread public concern over the problem of violence on television, not more than 20% of the viewing audience really believed the amount of violence on television to be excessive. A sample survey of 400 randomly selected viewers revealed that 100 respondents found TV programming too violent. Test the network president's null hypothesis against the alternative hypothesis H_a: $p > .2$. Use a .10 level of significance.

 9.37 Refer to problem 9.36. At what level would the test statistic be significant if the observed-significance-level method were used?

 9.38 In the past 35% of shoppers entering one of Ableson's Grocery stores purchased at least one item from the store's bakery. For a one-week trial period, a device was used to spread the smell of the bakery goods over a larger area of the store. A random sample of $n = 1000$ shoppers was observed during the trial period; it was found that 400 of the sample members had bought at least one bakery item. The manager of the store wishes to know whether an increase has occurred in the proportion of customers who patronize the bakery. If he can show that an increase has indeed occurred, he will recommend the use of the odor-spreading device in other store's in the Ableson's chain.
a. State the null and alternative hypotheses.
b. Develop the decision rule assuming that the probability of concluding that an increase has occurred when no increase actually took place is $\alpha = .05$.
c. Test the null hypothesis. Should the null hypothesis be accepted or rejected?
d. If the null hypothesis had been concerned with a population mean μ, an estimate of the true standard error of the mean, $\sigma_{\bar{x}} = \sigma/\sqrt{n}$, would have to be obtained from the sample data—that is, $s_{\bar{x}} = s/\sqrt{n}$. However, in the present problem it was not necessary to obtain from the sample data an estimate of the true standard error of the proportion $\sigma_{\bar{p}}$. Explain the reason behind this seeming inconsistency.
e. In what way or ways would the hypothesis-testing procedure be altered if the null and alternative hypotheses had been stated

H_0: $p = .35$

H_a: $p \neq .35$

and the level of significance was kept at $\alpha = .05$? Should the null hypothesis be accepted or rejected in light of the sample results cited above?

9.39 Given:

H_0: $\mu \le 9$ minutes

H_a: $\mu > 9$ minutes

$\sigma = 2$ minutes

$\alpha = .05$

What sample size would be required if
a. $\beta = .4$ if the true population mean is really 9.5 minutes?
b. $\beta = .1$ if the true population mean is really 11 minutes?
c. $\beta = .5$ if the true population mean is really 9.2 minutes?

9.40 Given:

H_0: $\mu \ge 100$ miles

H_a: $\mu < 100$ miles

$\sigma = 10$ miles

$\alpha = .01$

What sample size would be required if
a. $\beta = .5$ if the true population mean is 99.5 miles?
b. $\beta = .4$ if the true population mean is 99 miles?
c. $\beta = .2$ if the true population mean is 98 miles?

9.41 Refer to the information given in Figure 9.23. If the null hypothesis is H_0: $\mu = 1000$, what probability is shown by
a. Line AB? Does it represent an error? Explain.
b. Line CD? Does it represent an error? Explain.
c. Line EF? Does it represent an error? Explain.

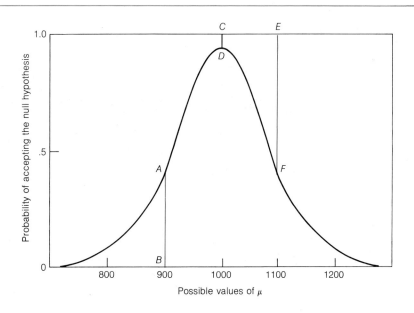

Figure 9.23. Operating characteristic curve associated with problem 9.41.

9.42 In the past a Branahan Company Zephyr-6 power lawnmower could be assembled at the factory in an average time of 18 minutes with a standard deviation of 4 minutes. Recently, the company has moved from the old factory into a more modern facility. The company president now wishes to know whether the move has affected the average assembly time in any way. Accordingly, he issued the following instructions to the supervisor in charge of Zephyr-6 production: Take a random sample of 16 completed Zephyr-6 mowers and determine how many minutes were required, on the average, for assembly. The president decided that (1) if the sample mean turned out to be between 15 and 21 minutes, he would conclude that no change had occurred in average assembly time and (2) if the sample mean turned out to be lower than 15 or higher than 21, he would conclude that a change had occurred.

    a. Assuming that the standard deviation of $\sigma = 4$ minutes is still correct, calculate the probability of accepting the null hypothesis of no change if $\mu = 13$, 14, 15, 16, 17, 18, 19, 20, 21, 22, and 23 minutes.

    b. Is the president's decision rule a good one if it is considered important to detect a change in average assembly time of as little as one minute? Justify your answer.

    c. Under what circumstances would a Type I error be committed if the president's decision rule were followed? Under what circumstances would a Type II error be committed if the president's decision rule were followed?

    d. Present your operating characteristic curve in graphic form.

Special-Challenge Problem

9.43 In the past the proportion of overdue charge accounts with the Darvis Department Store has been .4. The store has recently increased the interest rate charged on overdue accounts in the hope of discouraging slow payment. The store manager now wishes to determine whether the interest rate increase has had the desired result. Her hypotheses are

$H_0: p \geq .4$

$H_a: p < .4$

where p stands for the proportion of accounts that are overdue. The manager asks the credit department personnel to take a random sample of 64 accounts and determine what proportion are overdue. She decides that, if the sample proportion is less than .25 she will conclude that the increase in the interest rate has worked. Otherwise, she will conclude that it has been a failure.

    a. Calculate the probability of accepting the null hypothesis if $p = .2$, .3, .4, .5, and .6.

    b. Is the manager's decision rule a good one if she considers it important to detect a decrease in the proportion of late payers of as little as .1? Justify your answer.

    c. Under what circumstances would a Type I error be committed if the manager's decision rule were followed? Under what circumstances would a Type II error be committed if the manager's decision rule were followed?

Special-Challenge Problem

9.44 In the past the sales staff of the Ark Art Supply Company has achieved average daily sales per salesperson of $2500 with a standard deviation of $800. Recently, the sales staff has engaged in some extensive sensitivity training designed to increase their self-confidence and selling motivation. The sales manager now wishes to get a reading on how much good, if any, this special training has done. His hypotheses are

$H_0: \mu \leq \$2500$

$H_a: \mu > \$2500$

He selects 36 selling days at random and calculates average daily sales per salesperson. He decides that if the sample mean is in excess of $2800, he will conclude that the sensitivity training has done some good. Otherwise, he will conclude that it was a waste of time and company money.

a. Assuming that the standard deviation of $800 is still correct, calculate the probability of accepting the null hypothesis if μ = $2200, $2300, $2400, $2500, $2600, $2700, $2800, and $2900.

b. Is the sales manager's decision rule a good one if he considers it important to detect an increase in average daily sales per salesperson of as little as $200? Justify your answer.

c. Under what circumstances would a Type I error be committed if the sales manager's decision rule were followed? Under what circumstances would a Type II error be committed if the sales manager's decision rule were followed?

d. Present your operating characteristic curve in graphic form.

 9.45 A company is going to add a new men's cologne to its product line and wishes to know which of two package designs, a red-and-white one or a blue-and-white one, would have more consumer appeal. Accordingly, they selected a random sample of 100 potential consumers and asked which package design each prefers. Fifty-six of those questioned expressed a preference for the red-and-white package. Assuming that a respondent was required to express a preference for one of the two proposed package designs, test at the .10 level of significance the null hypothesis that among the population of potential consumers no preference exists for one package design over the other.

 9.46 Refer to problem 9.45. At what level would the test statistic be significant if the observed-significance-level method were used?

Special-Challenge Problem

 9.47 A metal-bending machine is to be observed at random times in order for the company's manager to determine whether the proportion of times it is "down" is at most .1. The maximum risk of concluding that the proportion of down times for this machine exceeds the standard of p = .1 when it does not is set at .05. In addition, if the proportion of down times for this machine is really p = .15, the risk of concluding that it is performing according to standard is also set at .05.

What sample size would be required?

TAKE CHARGE

9.48 Pose a question to yourself pertaining to your area of specialization in college. Your question should be of the kind that would be answerable via the population mean if the relevant population were accessible. For example, a finance major might ask, "Was the average profit per share of common stock earned last year by all the companies making up the airline industry at least $3?" A marketing major might ask, "Was the average amount of money spent by households last year for frozen gourmet dinners at least $25?" And so forth.

When you have posed the question, do the following:

a. Define the target population.

b. Explain briefly how you would probably go about selecting a random sample.

c. State precisely the null and alternative hypotheses you would want to test.

d. Describe the steps you would take to test your null hypothesis.

TAKE CHARGE

9.49 Refer to problem 9.48. Obtain a rough point estimate of the population standard deviation by conducting a small pilot study (about 20 or 30 people or other elementary units, preferably under conditions ensuring independence). From this estimate and information you provide regarding the risk of Type I and Type II errors, determine what size sample you would need to carry out the real study.

TAKE CHARGE

9.50 Pose a question to yourself pertaining to your area of specialization in college. The question should be of the kind that would be answerable via the population proportion if the relevant population were accessible. For example, a real estate major might ask, "Is the proportion of houses up for sale at this moment in the southwest part of my city at least .1?" An advertising major might ask, "Can more than 10% of the people who watched the 1984 Olympics on television recall at least three sponsors of the broadcast?"

When you have posed the question, do the following:
a. Define the target population.
b. Explain briefly how you would probably go about selecting a random sample.
c. State precisely the null and alternative hypotheses you would want to test.
d. Describe the steps you would take to test your null hypothesis.

Cases for Analysis and Discussion

9.1 "MID-CITY BANK OF BIG TOWN": PART II

Review Part I of this case (case 8.2 at the end of Chapter 8).

Prior to the beginning of the 18-month period in which the sample study was in progress, management gathered data for the entire population of "Mid-City's" revolving credit-card customers for the preceding 12-month period; the results are shown in Table 9.3. (Hereafter, this 12-month period will be referred to as period A and the final twelve months of the 18 months will be referred to as period B.) Using the information in Table 9.3 and any relevant information from Part I of this case, answer the following questions:

1. Does the sample evidence suggest that any increase occurred in the average outstanding balance per customer per month from period A to period B with respect to (1) minimum-payback customers, (2) medial-payback customers, (3) full-payback customers, and (4) all customers? Test at the .05 level of significance the null hypotheses that no such increases occurred.

2. Does the sample evidence suggest that any decrease occurred in average payback per customer per month from period A to period B with respect to (1) minimum-payback customers, (2) medial-payback customers, (3) full-payback customers, and (4) all customers? Test at the .05 level of significance the hypotheses that no such decreases occurred.

3. Does the sample evidence suggest that any *increase* occurred in the average interest payment per customer per month from period A to period B with respect to (1) minimum-payback customers, (2) medial-payback customers, (3) full-payback customers, and (4) all customers? Test at the .05 level of significance the null hypotheses that no such increases occurred.

4. Write an overall evaluation (approximately 200 to 400 words) of the desirability of Mid-City's plan to go to the lower minimum-payback requirements.

This case is continued at the end of Chapter 10.

9.2 THE "STAFFORD COMPANY"

The "Stafford Company" has been making use of employee-attitude surveys for almost three decades. In the following report, one of the company's branch managers describes the nature of the surveys and some specific results obtained from a somewhat recent use of the survey approach.

Table 9.3 Summary Table for "Mid-City Bank's" Population Study During Time Period A

Type of Customer	(1) Average Outstanding Balance per Customer per Month	(2) Average Payback per Customer per Month	(3) Average Outstanding Balance per Customer per Month after Payback (1) − (2)
Minimum payback	$273.72	$37.12	$236.60
Medial payback	255.09	43.10	211.99
Full payback	98.00	97.75	0.25
All	$192.49	$65.03	$127.46

Type of Customer	(4) Average Interest Payment per Customer per Month[a] (3) × .015	(5) Population Size N
Minimum payback	$3.55 ($0.64)	25,181
Medial payback	3.18 ($1.50)	22,606
Full payback	0.0015 ($0.20)	36,628
All	$1.9115 ($0.95)	84,415

[a] Population standard deviations for this measure are shown in parentheses.

Many progressive corporations, particularly those with large field organizations, are most eager to determine the attitudes of their employees as they relate to corporate policies and the specific field organization of which they are a part. As a result, attitude surveys are rather widely used.

Within the Stafford Company, an employee-attitude survey is conducted for each individual field organization approximately every three or four years. The principal problems faced by companies utilizing such surveys have to do with determining the key issues which should be explored. Once these issues are established, it is then necessary to construct a list of questions relating to them that will provide the information required to determine employee attitudes. It then becomes necessary to compile this information, analyze it statistically, and present it in a concise way to make it intelligible to the employees who took part. This last step is essential if the surveys are to be meaningful in subsequent years. To raise employee interest and curiosity regarding results and then not give them feedback and an action plan will ultimately destroy the usefulness of the survey system itself.

In June of 1982 the Stafford Company conducted a costly population study of employee attitudes.

In October of 1985 the relatively young Metropolis City Branch, my branch, conducted a sample survey using 100 randomly selected employees. Some of the questions asked were uniquely relevant to the Metropolis City Branch (Branch M hereafter); however, the data to be presented shortly are limited to questions that were identical to those used in the 1982 overall company survey. Although we did not check the matter out with the care we probably should have, it was assumed that, because of Branch M's short time in existence, no employees who took the branch survey in 1985 had also taken the general company survey in 1982. Branch M, by the way, is a marketing organization exclusively,

Table 9.4 Summary Table of Results of the 1982 "Stafford Company" and the 1985 Branch M Employee-Attitude Surveys

| | Proportion Responding Favorably | |
Subject	Company (p)	Branch M ($\bar{p}$)
1. Work organization	.610	.719
2. Work efficiency	.540	.659
3. Management effectiveness	.530	.751
4. Job training and information	.560	.600
5. Work associates	.650	.751
6. Supervisory leadership practices	.670	.778
7. Fringe benefits	.730	.768
8. Job satisfaction	.710	.838
9. Pay	.590	.556

carrying out sales, service, and marketing administration functions. Some branches handle manufacturing functions exclusively, and some carry out both marketing and manufacturing tasks. Exhibit A [Table 9.4] compares some results of the 1982 company survey and the 1985 Branch M survey. All questions consisted of a more or less positive statement, such as "I believe the company's incentive pay system is adequate. Agree ____ Disagree ____."

1. With respect to which subjects, if any, do the Branch M employee responses differ significantly at the .05 level from the overall company employee responses? With respect to these subjects, are the Branch M results more favorable or less favorable than the overall company results? (*Hint:* Use the company-wide p values as the basis for your null hypotheses.)

2. Can the significant differences, if any, that you found in connection with question 1 necessarily be attributable solely to a difference in attitude between Branch M employees and overall company employees? Tell why or why not.

9.3 THE "QUALITY CONTAINER COMPANY"

One of the truly important applications of statistical hypothesis testing within industry is that of quality control, which is a systematic procedure for detecting trouble in a manufacturing process.

Statistical quality-control methods permit the overall variation of some characteristic of a product to be partitioned into a chance variation component and an assignable variation component. *Chance variation* is variation resulting from many, usually unknown, causes, none of which is particularly important. Chance variation is inevitable and not a cause for concern. *Assignable variation* is variation attributable to one, or very few, rather important causes, such as a new operator, wear on parts, improper machine setting, and so forth. When assignable variation is observed, steps are usually taken to eliminate its source or to correct the problem in some manner.

The basic tool of quality control is the control chart. The construction of a control chart is based on one of the two following premises, depending on whether the relevant data are measurement or attribute:

1. The average level of a measurement variable has not changed.
2. The proportion of items having a specified characteristic is unchanged.

If only chance variation is present, 99.7% of the sample means (assuming that the population distribution is normal or that the sample size is sufficiently large that the central limit theorem will apply) will fall within the interval $\mu_0 \pm 3\sigma_{\bar{X}}$. Similarly, 99.7% of the sample proportions will fall within the limits of $p_0 \pm 3\sigma_{\bar{p}}$. If a sample mean or proportion falls within these limits, the hypothesis of no change is accepted and the process is allowed to continue. But if a sample $\bar{X}$ or $\bar{p}$ value (whichever is relevant) falls outside the 3-sigma (3σ) limits, as they are referred to generically, assignable variation is suspected and the no-change hypothesis is rejected. Efforts are then made to determine the source of the problem and to correct it.

A quality-control chart is constructed in such a way that the quality-control personnel can plot the results of periodic monitoring. Usually a relatively small sample size is used in connection with each monitoring.

This very brief summary of quality-control methods should suffice to help you answer the questions posed below about the "Quality Container Company." The following information is conveyed in the words of the quality-control supervisor of "Quality Container."

Aluminum end manufacturing is an extremely complex field which requires a great multitude of individual factors to come together with extreme precision. One of the major components in this process is the quality of the metal. Problems with metal quality must be detected prior to delivery to the marketing area. The primary problem with metal quality centers around the formability features of the metal as determined by the "Longitudinal Yield Strength" test.

Samples are cut from random coils in each lot number and are tested for formability. Yield strength is defined as the strength of the metal divided by the stress. The measure of longitudinal yield strength is the total pounds per square inch (psi) required to pull the metal beyond the point of elasticity.

Sampling is based on a random selection of coils within a given manufacturer's lot size. For the study to be described, our total population size consisted of 500 coils of aluminum end stock from the same manufacturing lot, totaling about 1 million pounds of aluminum. The sample consisted of 24 coils which were selected with the aid of a random number table.

A 12-inch square was cut from each of the 24 coils. Each of the sample squares were, in turn, cut into 3 test strips to assure that the sample would adequately represent the coil. Each selected strip was then tested by means of a tensile test machine, which pulls the

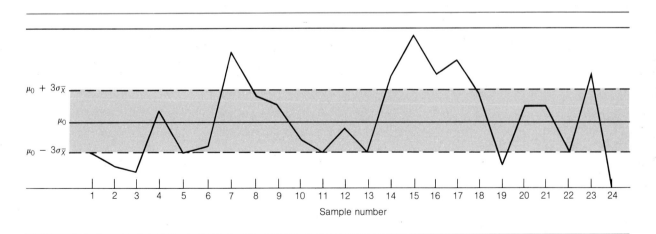

Figure 9.24. Control chart for the "Quality Container Company" case.

Cases for Analysis and Discussion

metal and records the point at which the yield strength limits are reached. This process was continued until all 72 test strips had been tested and the results recorded.

The results of our 24 sample tests were recorded on the control chart shown in Exhibit A [Figure 9.24, p. 427]. Calculations were then completed and the control chart constructed, including the upper and lower control limits of the sample mean.

An analysis of the control chart indicates that something is out of control.

1. Do you agree that something is out of control? On what do you base your conclusion?

2. If your answer to question 1 was Yes, what are some possible factors that might warrant investigation? (Don't be timid; your guesses are as good as anybody's.)

10 Tests of Hypotheses Involving Two Populations

Nothing is good or bad but by comparison.—Thomas Fuller

The great tragedy of science—the slaying of a beautiful hypothesis by an ugly fact.
—Thomas Huxley

What You Should Learn from This Chapter

This chapter builds on the foundation put in place in Chapter 9. Whereas in Chapter 9 we dealt only with hypothesis-testing problems involving a single random sample from a single population, in this chapter we consider how to analyze the results of two samples drawn from two different populations. When you have completed this chapter, you should be able to

1. Explain what is meant by a sampling distribution of (a) $\overline{X}_1 - \overline{X}_2$, (b) $\overline{p}_1 - \overline{p}_2$, (c) s_1^2/s_2^2, (d) $\overline{d}$.
2. Recite the important properties of the above sampling distributions.
3. Perform Z and t tests to determine whether two population means are equal.
4. Perform Z tests to determine whether two population proportions are equal.
5. Perform F tests to determine whether two population variances are equal.
* 6. Perform t tests on paired observations to determine whether an average population difference is equal to 0.

Before beginning this chapter, you may wish to review the following sections. In Chapter 8: sections 8.1–8.6, 8.8, 8.12–8.15, 8.17, 8.22. In Chapter 9: sections 9.1–9.4, 9.7, 9.9.

10.1 Introduction

Frequently, statistical investigations conducted with a decision-making end in mind are concerned with questions like: "Is the mean of one population the same as the mean of another population?" "Is the proportion associated with one population less than the proportion associated with another population?" "Is there more variation in one population than there is in another population?" These are the kinds of questions we address in this chapter. Like the methods presented in Chapter 9, most of those introduced in this chapter will pertain to three different classes of tests: two-tailed, lower-tailed, and upper-tailed. For the most part (section 10.4 is an exception), they will involve use of the normal curve or the t distribution. In very general symbols, the kinds of hypotheses we will discuss can be expressed as

$$H_0: \theta_1 - \theta_2 = 0$$
$$H_a: \theta_1 - \theta_2 \neq 0 \qquad \text{Two-tailed}$$

429

$$H_0: \theta_1 - \theta_2 \geq 0$$
$$H_a: \theta_1 - \theta_2 < 0 \qquad \text{Lower-tailed}$$

$$H_0: \theta_1 - \theta_2 \leq 0$$
$$H_a: \theta_1 - \theta_2 > 0 \qquad \text{Upper-tailed}$$

where θ_1 is the value of a parameter of interest for Population 1 and θ_2 is the value of that same parameter for Population 2.

Notice that with hypotheses such as these, we are not required to hypothesize a specific value for either θ_1 or θ_2. Instead, we hypothesize that the value of θ_1 and θ_2, whatever that value may be, is the same for the two populations.

We will begin with tests pertaining to the equality, or lack of it, of means associated with two independent populations. Keep in mind that to claim $\mu_1 = \mu_2$ is the same as claiming $\mu_1 - \mu_2 = 0$.

10.2 Test for the Difference Between Means of Two Independent Populations

General Nature of the Test

Let us analyze the following illustrative case.

ILLUSTRATIVE CASE 10.1

The Gurney Company holds a four-week orientation class for new management trainees. At the end of four weeks, a written examination is given for the dual purpose of (1) determining how much individual trainees have learned from the program and (2) determining the overall effectiveness of the teaching method used. With respect to the second goal, it is important to note that two rather different methods of instruction have been used in the past and that records have been kept on the test scores of trainees subjected to each method. The two approaches have been (1) a somewhat passive one in which trainees listen to lectures and watch films and demonstrations, and (2) a more involvement-oriented one consisting principally of a variety of role-playing experiences. Over the past ten years, some 300 new employees have taken the orientation classes with results as shown in Table 10.1. The instructor would now like to find out whether there is any difference in the effectiveness of the two approaches. He decides to use a level of significance of $\alpha = .01$.

Table 10.1 Test Results for Management Training Classes Conducted by the Gurney Company

Type of Approach	Mean Test Score Achieved	Standard Deviation of Test Scores	Size of Sample
Passive	$\overline{X}_1 = 89.73$	$s_1 = 12.04$	$n_1 = 200$
Active	$\overline{X}_2 = 85.94$	$s_2 = 10.22$	$n_2 = 100$

You should note that we are dealing in this case with infinite populations that do not actually exist. That is, the population associated with either training approach consists of all the potential management trainees for this company who could conceivably ever be exposed to this approach. Despite the ethereal nature of the populations, just as long as the samples can be viewed as independent, we may still think in terms of there being an arithmetic mean and a standard deviation associated with each population. By comparing $\overline{X}_1$ and $\overline{X}_2$, we may test whether the means of the two populations would be found to be equal were it possible to compute them.

Let us refer to the population means associated with the passive and active approaches as μ_1 and μ_2, respectively. Similarly, σ_1 and σ_2 will be used to denote the corresponding population standard deviations. The hypotheses to be tested are

$$H_0: \ \mu_1 - \mu_2 = 0$$
$$H_a: \ \mu_1 - \mu_2 \neq 0$$

> To perform the desired test, we must compare the sample means to determine whether H_0 should be accepted or rejected. If $\overline{X}_1 - \overline{X}_2$ is found to differ significantly from 0, the hypothesized value of $\mu_1 - \mu_2$, we will reject the null hypothesis. Of course, if $\overline{X}_1 - \overline{X}_2$ is found not to differ significantly from 0, we conclude in favor of the proposition that $\mu_1 - \mu_2$ is 0.

Nature of the Sampling Distribution of $\overline{X}_1 - \overline{X}_2$

Just as the tests in Chapter 9 were carried out with the aid of a sampling distribution of $\overline{X}$ or a sampling distribution of $\overline{p}$, the present test is dependent on the concept of "sampling distribution of $\overline{X}_1 - \overline{X}_2$." Figure 10.1 illustrates what is meant by this term. In brief summary: We begin with two populations labeled Population 1 and Population 2. These are of sizes N_1 and N_2, respectively. Now, if (1) every different sample of size n_1 which could be selected from Population 1 were determined, (2) $\overline{X}_1$ obtained from each such sample, and (3) the resulting $\overline{X}_1$ values organized into probability-distribution form, we would have, as you learned in Chapter 8, a sampling distribution of $\overline{X}$ for this particular population. By the same token, if (1) every different sample of size n_2 which could be selected from Population 2 were determined, (2) $\overline{X}_2$ obtained for each such sample, and (3) the resulting $\overline{X}_2$ values organized into probability-distribution form, we would have a sampling distribution of $\overline{X}$ for the second population. Then if all the $\overline{X}_1$ and $\overline{X}_2$ values were differenced in a consistent way—let us say that $\overline{X}_2$ is always subtracted from $\overline{X}_1$—the resulting differences, when organized into probability-distribution form, would constitute a sampling distribution of $\overline{X}_1 - \overline{X}_2$.

Important Properties of a Sampling Distribution of $\overline{X}_1 - \overline{X}_2$

Keeping in mind that we assume $\overline{X}_1$ and $\overline{X}_2$ to be independent random variables, let us survey three useful properties of a sampling distribution of $\overline{X}_1 - \overline{X}_2$.

1. The mean of a sampling distribution of $\overline{X}_1 - \overline{X}_2$, symbolized $\mu_{\overline{x}_1 - \overline{x}_2}$, is equal to $\mu_1 - \mu_2$, the difference between the means of Populations 1 and 2.
2. The standard deviation of a sampling distribution of $\overline{X}_1 - \overline{X}_2$ (called the standard error of the difference between two means) is computed by

$$\sigma_{\overline{x}_1 - \overline{x}_2} = \sqrt{\frac{\sigma_1^2}{n_1} + \frac{\sigma_2^2}{n_2}} \qquad (10.2.1)$$

if both populations are infinite or finite but at least 20 times the size of the respective samples.
3. For large independent samples regardless of their shapes, the central limit theorem again applies. That is, the sampling distribution of $\overline{X}_1 - \overline{X}_2$ will be approximately normal.

Armed with this information, we are ready to solve the Gurney Company problem—ready, that is, except for one technical detail: Equation (10.2.1), the appropriate equation for the standard error of $\overline{X}_1 - \overline{X}_2$, calls for use of the variances of the two populations. Because

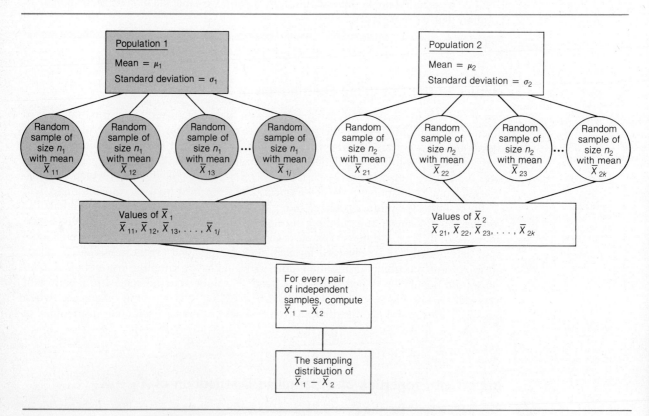

Figure 10.1. Schematic presentation of how a sampling distribution of $\overline{X}_1 - \overline{X}_2$ would be constructed.

10 Tests of Hypotheses Involving Two Populations

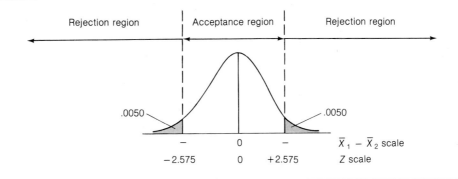

Figure 10.2. Sampling distribution of $\overline{X}_1 - \overline{X}_2$ for the Gurney Company problem (Illustrative Case 10.1).

we will not usually know σ_1^2 and σ_2^2, we must substitute sample variances for the unknown population variances. This substitution is permissible without necessitating any additional changes in procedure when the sample sizes are greater than about 30. The resulting formula for the (estimated) standard error of $\overline{X}_1 - \overline{X}_2$ becomes

$$s_{\bar{x}_1 - \bar{x}_2} = \sqrt{\frac{s_1^2}{n_1} + \frac{s_2^2}{n_2}} \qquad (10.2.2)$$

Since the null hypothesis is H_0: $\mu_1 - \mu_2 = 0$, the test called for is clearly two-tailed because the hypothesis of equal population means would be rejected if $\overline{X}_1 - \overline{X}_2$ for a single pair of independent samples were quite different from 0 *in either direction*. The sampling distribution of $\overline{X}_1 - \overline{X}_2$ for this example is shown in Figure 10.2.

Appendix Table 2 reveals that .0050 (one-half of $\alpha = .01$) of the area under the normal curve lies below $Z = -2.575$ and .0050 of the area lies above $Z = +2.575$. Therefore, we should reject the null hypothesis if our observed $\overline{X}_1 - \overline{X}_2$ is more than 2.575 standard deviations above or below the hypothesized value of 0. Thus, the decision rule is

Decision Rule
1. If $-2.575 \leq$ observed Z $\leq +2.575$, accept H_0.
2. If observed Z < -2.575 or if observed Z $> +2.575$, reject H_0.

Using Equation (10.2.2), we find that the (estimated) standard error of $\overline{X}_1 - \overline{X}_2$ is

$$s_{\bar{x}_1 - \bar{x}_2} = \sqrt{\frac{s_1^2}{n_1} + \frac{s_2^2}{n_2}} = \sqrt{\frac{(12.04)^2}{200} + \frac{(10.22)^2}{100}} = 1.33$$

I hope you have noticed by now a pattern that runs through the methodology of calculating observed Z: The procedure is always

$$\frac{\text{Observed value of the statistic} - \text{Hypothesized parameter value}}{\text{Standard deviation of the sampling distribution of the statistic}}$$

For example, when testing a hypothesis about the mean of a single population, we computed

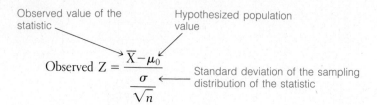

Similarly, when testing a hypothesis about the proportion associated with a single population, we computed

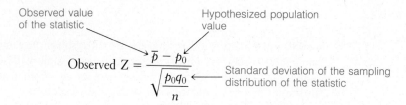

The pattern continues when we turn to problems like the present one:

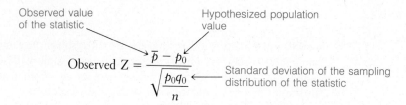

or, because $\mu_1 - \mu_2$ is hypothesized to be 0 (and, of course, because the population variances are unknown), we may simply write

$$\text{Observed } Z = \frac{\overline{X}_1 - \overline{X}_2}{s_{\overline{x}_1 - \overline{x}_2}} \qquad (10.2.3)$$

We calculate the difference between the means of the two independent samples actually observed: $\overline{X}_1 - \overline{X}_2 = 89.73 - 85.94 = 3.79$. We just found $s_{\overline{x}_1 - \overline{x}_2}$ to be 1.33. Therefore, using Equation (10.2.3), we find that observed Z is equal to $3.79/1.33 = +2.85$.

Since $+2.85$ exceeds $+2.575$, we reject the null hypothesis on the grounds that it is unlikely that the two samples came from populations having the same mean. In more practical terms, we conclude that the two approaches to teaching the management orientation class are not equally effective. Curiously, the more passive approach appears better according to this criterion, as evidenced by the larger sample mean associated with it.

Small Sample Sizes

As we worked through the Gurney Company problem, we used sample sizes of $n_1 = 200$ and $n_2 = 100$. Let us change the problem in only one respect: Let us suppose that $n_1 = 16$ and $n_2 = 9$—not, perhaps, terribly unrealistic considering that the problem is concerned with a management training class.

Our hypotheses would be unchanged:

$$H_0: \; \mu_1 - \mu_2 = 0$$
$$H_a: \; \mu_1 - \mu_2 \neq 0$$

Because the sample sizes are small and the population variances are unknown, we must make use of the t distribution. The formula for observed t is

$$\text{Observed } t = \frac{\overline{X}_1 - \overline{X}_2}{s_{\overline{x}_1 - \overline{x}_2}} \qquad (10.2.4)$$

This, of course, is simply Equation (10.2.3) all over again. However, the denominator must now be computed somewhat differently than when the sample sizes were assumed to be large.

When t must be used in a problem such as this, one must make two assumptions about the populations—assumptions which are not necessary when σ_1^2 and σ_2^2 are known or when σ_1^2 and σ_2^2 are unknown but n_1 and n_2 are large. The assumptions are:

1. The populations are both normally distributed.
2. The variances of the populations are equal, that is, $\sigma_1^2 = \sigma_2^2$.

How realistic are these assumptions? No absolutely fail-safe answer is possible. However, there are two sources of reassurance: First, it is known that the test we are discussing is quite robust. What this means is that one or both assumptions can be violated somewhat without our having to fear that the t test is giving a misleading result. Second, we have some, usually adequate, ways of checking these assumptions for validity. The assumption of normal population shapes is the more difficult of the two to "test" when the sample sizes are small. However, one can gain some reassurance (or the opposite) by preparing a box-and-whiskers chart for each sample like the one presented in Figure 10.3. The assumption of equal population variances is discussed in some detail in section 10.4 of this chapter.

Getting back to $s_{\overline{x}_1 - \overline{x}_2}$: The second assumption noted above requires that we use the sample variances in combination as a pooled estimate of the common population variance.

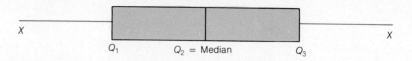

Figure 10.3. Idealized box-and-whiskers chart for a sample from a normally distributed population. Look for: (1) symmetry around the median, (2) 50 percent of the observations within the rectangle, (3) 25 percent of the observations outside of the rectangle on the left-hand side, (4) 25 percent of the observations outside the rectangle on the right-hand side, (5) 95 percent of the observations within a range of $Q_2 \pm (Q_3 - Q_1)$, as indicated by the two X's. If neither sample departs conspicuously from this ideal configuration, one can apply the t test under discussion with confidence.

This we do by using a weighted average of the two sample variances. More specifically, we use degrees of freedom as weights and calculate

$$s_w^2 = \frac{(n_1 - 1)s_1^2 + (n_2 - 1)s_2^2}{n_1 + n_2 - 2} \tag{10.2.5}$$

We may now complete this version of the Gurney Company problem. Substituting values into Equation (10.2.5), we get

$$s_w^2 = \frac{(15)(12.04)^2 + (8)(10.22)^2}{16 + 9 - 2} = 130.8701 \qquad s_w = \sqrt{130.8701} = 11.44$$

The standard error of $\overline{X}_1 - \overline{X}_2$ is given by

$$s_{\bar{x}_1 - \bar{x}_2} = \sqrt{\frac{s_w^2}{n_1} + \frac{s_w^2}{n_2}}$$

which is the same thing as

$$s_{\bar{x}_1 - \bar{x}_2} = s_w \sqrt{\frac{1}{n_1} + \frac{1}{n_2}} \tag{10.2.6}$$

Therefore, the standard error of $\overline{X}_1 - \overline{X}_2$ for the present problem is

$$s_{\bar{x}_1 - \bar{x}_2} = (11.44) \left(\sqrt{\frac{1}{16} + \frac{1}{9}} \right) = 4.77$$

and, using Equation (10.2.4), we find the observed t value to be

$$\text{Observed } t = \frac{89.73 - 85.94}{4.77} = 0.79$$

The number of degrees of freedom for this kind of problem is determined by $n_1 + n_2 - 2$. For our specific problem: $16 + 9 - 2 = 23$. (In the one-sample case where the standard deviation was a point estimate of the true but unknown population standard deviation, there was a loss of one degree of freedom. In the two-sample case, each of the sample standard deviations was used, in squared form, in obtaining the estimate of the population variance. Therefore, the combined degrees of freedom are $[n_1 - 1] + [n_2 - 1] = n_1 + n_2 - 2$.)

The critical t value in Appendix Table 3 for 23 degrees of freedom and $\alpha = .01$ is ± 2.807. The observed t value of 0.79 is safely between the critical values of -2.807 and $+2.807$, and we are led to accept the null hypothesis of no difference in effectiveness between the two approaches, a conclusion quite the opposite of the one drawn when sample sizes were assumed to be larger.

One-Tailed Tests

A test of the kind just described requiring use of only one tail of the sampling distribution of $\overline{X}_1 - \overline{X}_2$ would be indicated when one wishes to know whether one population mean is greater than the other population mean. For example, let us suppose that in the Gurney case the passive approach had been used for quite a long time. Then a switch to a more active approach was tried, let us say, in the expectation of better test scores. In such a situation, the expected relationship between μ_1 and μ_2 is $\mu_1 - \mu_2 < 0$—that is, the more passive approach is expected to be found less effective than the more active approach. The null hypothesis would be a negation of this expectation—that is, H_0: The passive approach is not less effective than the active approach, or H_0: $\mu_1 - \mu_2 \geq 0$. The alternative hypothesis would be an affirmation of the expectation—that is, H_a: The passive approach *is* less effective than the active approach, or H_a: $\mu_1 - \mu_2 < 0$. Then the entire $\alpha = .01$ value would be placed under the left-hand tail of the normal curve or the t distribution, whichever is proper.

10.3 Test for the Difference Between Proportions of Two Independent Populations

General Nature of the Test

Another kind of two-sample, hypothesis-testing problem involves the use of proportions. Let us suppose that a consumer-polling service asks a random sample of adult residents of Des Moines, Iowa, "Do you intend to buy a new automobile anytime within the next 12 months?" Eleven percent, that is, $\overline{p}_1 = .11$, answer affirmatively. The polling service then asks the same question of a random sample of adult residents of Newark, New Jersey. Nine percent, $\overline{p}_2 = .09$, answer affirmatively. The question to be addressed in the test of significance is: "Does the proportion of adult residents intending to buy a new automobile within the next 12 months differ between the two cities?" (Alternatively, the question might be framed in terms of: "Is the proportion of adult residents of Des Moines who intend to buy a new automobile during the next 12 months higher than (lower than) the proportion for Newark?" In such a case, a one-tailed case, the procedures described below could be used with only minor modifications: The entire $\alpha = .10$ value would be placed under the appropriate single tail of the normal curve and the decision rule would be modified appropriately.) Let us say, as alluded to above, that the pollsters feel that they can accept a risk of making a Type I error of $\alpha = .10$. Information to be used in answering the question is shown in Table 10.2.

Table 10.2 Information Resulting from a Two-City Survey of Automobile-Buying Intentions

Des Moines (Population 1)	Newark (Population 2)
$\bar{p}_1 = .11$	$\bar{p}_2 = .09$
$\bar{q}_1 = 1 - \bar{p}_1 = .89$	$\bar{q}_2 = 1 - \bar{p}_2 = .91$
$n_1 = 2000$	$n_2 = 3000$

In symbols, the null and alternative hypotheses would be stated as follows:

$$H_0: \; p_1 - p_2 = 0$$
$$H_a: \; p_1 - p_2 \neq 0$$

where p_1 is the true population proportion of those intending to buy in Des Moines and p_2 is the true population proportion of those intending to buy in Newark.

In order for you to understand the theory underlying this test, you must be conversant with the properties of yet another new sampling distribution, the sampling distribution of $\bar{p}_1 - \bar{p}_2$.

Important Properties of a Sampling Distribution of $\bar{p}_1 - \bar{p}_2$

The following properties of a sampling distribution of $\bar{p}_1 - \bar{p}_2$ are similar to those cited in the preceding section for the sampling distribution of $\bar{X}_1 - \bar{X}_2$:

1. The mean of a sampling distribution of $\bar{p}_1 - \bar{p}_2$, which will be written $\mu_{\bar{p}_1 - \bar{p}_2}$, is equal to the true population difference, $p_1 - p_2$.
2. The standard deviation of a sampling distribution of $\bar{p}_1 - \bar{p}_2$ (called the standard error of the difference between two proportions) is (ignoring the finite correction factor) equal to

$$\sigma_{\bar{p}_1 - \bar{p}_2} = \sqrt{\frac{p_1 q_1}{n_1} + \frac{p_2 q_2}{n_2}} \qquad (10.3.1)$$

where $\quad q_1 = 1 - p_1$
$\qquad\quad q_2 = 1 - p_2$

Since the null hypothesis is $p_1 = p_2$, we may simply use p to represent the assumed common value of p_1 and p_2 and write Equation (10.3.1) as

$$\sigma_{p_1 - p_2} = \sqrt{pq\left(\frac{1}{n_1} + \frac{1}{n_2}\right)} \qquad (10.3.2)$$

3. The sampling distribution of $\bar{p}_1 - \bar{p}_2$ is approximately normal if n_1 and n_2 are sufficiently large. (If $n_1 p_1$, $n_1 q_1$, $n_2 p_2$, and $n_2 q_2$ are all greater than 5, a normal-shaped sampling distribution of $\bar{p}_1 - \bar{p}_2$ may be assumed.)

Testing the Null Hypothesis That Two Population Proportions Are Equal

We wish to test the null hypothesis that $p_1 - p_2 = 0$ by using the following Z formula:

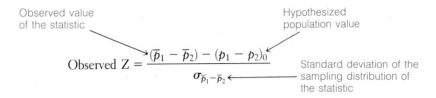

or, because $p_1 - p_2$ is hypothesized to be 0, simply

$$\text{Observed } Z = \frac{\bar{p}_1 - \bar{p}_2}{\sigma_{\bar{p}_1 - \bar{p}_2}} \qquad (10.3.3)$$

Unfortunately, the assumed common value of p in Equation (10.3.3) will not be known; it must be estimated using observed sample data. The estimate needed is obtained by computing a weighted average of the observed $\bar{p}$ values in the following manner:

$$\bar{p}_w = \frac{n_1 \bar{p}_1 + n_2 \bar{p}_2}{n_1 + n_2} \qquad (10.3.4)$$

where $\bar{p}_w$ represents the weighted average of the two sample proportions.

The observed Z value is determined by

$$\text{Observed } Z = \frac{\bar{p}_1 - \bar{p}_2}{s_{\bar{p}_1 - \bar{p}_2}} \qquad (10.3.5)$$

where

$$s_{\bar{p}_1 - \bar{p}_1} = \sqrt{\bar{p}_w \bar{q}_w \left(\frac{1}{n_1} + \frac{1}{n_2} \right)} \qquad (10.3.6)$$

where $\bar{q}_w$ is $1 - \bar{p}_w$

To summarize: We may think of the sampling distribution of $\bar{p}_1 - \bar{p}_2$ as approximately normal with mean, $\mu_{\bar{p}_1 - \bar{p}_2} = 0$ and with a standard deviation adequately approximated by $s_{\bar{p}_1 - \bar{p}_2} = \sqrt{\bar{p}_w \bar{q}_w (1/n_1 + 1/n_2)}$. Such a sampling distribution is shown in Figure 10.4.

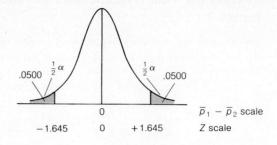

Figure 10.4. Sampling distribution of $\bar{p}_1 - \bar{p}_2$ for the example concerning automobile-buying intentions.

According to Appendix Table 2, the critical Z value when $\alpha = .10$ and the test is two-tailed is ± 1.645. Therefore, the decision rule is

Decision Rule
1. If $-1.645 \leq$ observed $Z \leq +1.645$, accept H_0.
2. If observed $Z < -1.645$ or observed $Z > +1.645$, reject H_0.

We saw above that $\bar{p}_1 - \bar{p}_2$ for our specific random samples is $.11 - .09 = .02$. To apply the decision rule, we must first determine $\bar{p}_w$. Using Equation (10.3.4), we obtain $\bar{p}_w = [(2000)(.11) + (3000)(.09)]/(2000 + 3000) = .098$. The value .098 will be used in Equation (10.3.6) for the estimated standard error of $\bar{p}_1 - \bar{p}_2$.

$$s_{\bar{p}_1-\bar{p}_2} = \sqrt{(.098)(.902)(1/2000 + 1/3000)} = .0086$$

The observed Z value is obtained from Equation (10.3.5):

$$\text{Observed } Z = \frac{.11 - .09}{.0086} = +2.33$$

Because $+2.33$ is greater than $+1.645$, the null hypothesis that $p_1 - p_2 = 0$ must be rejected at the .10 level of significance. It does appear that a larger proportion of the residents of Des Moines is intending to buy new cars during the upcoming 12 months compared to residents of Newark.

10.4 Test for the Difference Between Variances of Two Independent Normally Distributed Populations

Sometimes an analyst will be principally concerned with whether two populations differ in variability. Let's look at another specific case.

ILLUSTRATIVE CASE 10.2

The Rickey Car Company buys original-equipment tires from two companies, Goodgo and Goodgrip, but wishes to use only one of these suppliers in the future. The tires of the two companies cost about the same amount for any specified quantity and are roughly equal in terms of average useful life. However, little is known about variation in useful life for Goodgo tires, on the one hand, and Goodgrip tires on the other. With other relevant considerations so similar, Rickey management decides that the supplier whose tires exhibit less variation in length of useful life will be selected as the exclusive supplier of original-equipment tires.

Ten tires of each supplier are selected and subjected to continuous simulated driving until the treads measure $\frac{1}{4}$ inch or less. The results follow.

	Supplier	
	Goodgo	Goodgrip
n	10	10
$\overline{X}$	6020	6052
s	1700	900

Which supplier should be used?

The hypotheses for a problem of this kind are

$$H_0: \sigma_1^2 - \sigma_2^2 = 0$$
$$H_a: \sigma_1^2 - \sigma_2^2 \neq 0$$

Because neither sample variances nor differences between two sample variances are normally distributed, we must find a way of conducting this test without the use of the normal or t distributions.

The test statistic used in this kind of problem is called F, defined as

$$F = \frac{s_1^2}{s_2^2}$$

In other words, F is simply a ratio obtained by dividing the variance of one sample into the variance of the other sample. Intuitively, if the variation is the same in the two populations, we would expect $F = s_1^2/s_2^2$ to be approximately 1. On the other hand, if the population variances are really different, we would expect $F = s_1^2/s_2^2$ to be substantially larger than or substantially smaller than 1.

Before we solve this problem, it is necessary for you to become familiar with the theoretical sampling distribution we call the F *distribution*. For the moment, project yourself into the following unlikely situation: One night you have a bad dream in which you find yourself in a slave camp where the hapless prisoners are being forced to construct sampling distributions manually. Let us further suppose that you notice two boxes sitting on a table immediately in front of you. The two boxes, you find, are filled with chips and each chip has a number on it. You are informed by a large ugly man with a whip, hereafter known as the bad guy, that the amount of variation, as measured by the variance of the numbers in a specified box, is the same for the two boxes and that the numbers in each box are normally distributed.

The bad guy then instructs you to (1) stir up the chips in each box thoroughly, (2) take n_1 chips at random from the box near your right hand and n_2 chips at random from the box near your left hand, (3) compute the variance for the numbers on the chips taken from the right-hand box, making certain to divide by $n_1 - 1$, (4) compute the variance for the numbers on the chips taken from the left-hand box, making certain to divide by $n_2 - 1$, (5) express the result obtained from Step 3 as a ratio of the result obtained from Step 4 and record the resulting ratio, (6) toss the chips back into their respective boxes, (7) repeat the exact same procedure, say, 100 million times or so, and (8) organize the resulting F ratios into probability-distribution form.

The probability distribution so obtained would be an empirically determined sampling distribution of F. Needless to say, this would not be a very pleasant activity, and if you were lucky you would awaken long before completing 100 million replications of this tedious procedure. Nevertheless, this farfetched scenario does serve to provide a few basic points about the theoretical F distribution.

Worth emphasizing at this juncture is the fact that the bad guy, despite some annoying personal traits, did a few good things: He

1. Insisted on random selection of chips from the obviously independent populations of numbered chips.
2. Informed you that the populations of numbered chips have equal variances.
3. Informed you that each of the populations of numbered chips is normally distributed.

This is all very important because in the absence of these conditions, the sampling distribution you constructed would not be an F distribution. It would be a distribution of something, but not F.

Fortunately for us, the F distribution has been carefully defined mathematically and critical values of F made available in prepared tables. Before turning to the table of critical F values, however, we should note that, like other theoretical probability distributions we have encountered previously, the F distribution is really a large family of similar-appearing distributions. The specific shape of an F distribution is dependent on the degrees of freedom of the numerator (df_1) and the degrees of freedom of the denominator (df_2) of the observed F statistic. Clearly, the smallest possible value for F will be 0 because a negative variance is an impossibility. The largest possible value of F is undefined. The F random variable is continuous and, consequently, probabilities are represented by areas under the curve. Figure 10.5 shows one particular theoretical F distribution.

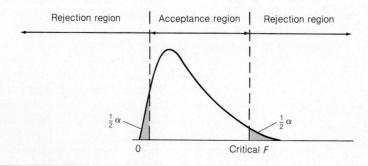

Figure 10.5. An example of an F distribution.

10 Tests of Hypotheses Involving Two Populations

Appendix Table 6, part of which is reproduced here as Table 10.3, presents the critical values of the F distribution for the two most widely used levels of significance; $\alpha = .05$ and $\alpha = .01$. More extensive tables are available if needed. Notice that in Table 10.3 there are two entries associated with each pair of degrees of freedom. The value in lightface type is the critical F ratio associated with an area of .05 under the right-hand tail of the theoretical F distribution for df_1 and df_2 degrees of freedom. This critical F value is the value we expect to be exceeded only 5 times out of 100 in repeated sampling *if the null hypothesis that $\sigma_1^2 - \sigma_2^2 = 0$ is true*. The other critical F value, the one in boldface type, is the one associated with a .01 level of significance.

You may have noted an inconsistency between Figure 10.5, which shows that the null hypothesis may be rejected as a result of a very small, as well as a very large, observed F ratio, and Table 10.3, which gives you critical values only for the right-hand tail of the distribution. Table 10.3 is constructed this way for very good reasons, as will be demonstrated in Chapters 11 and 12. In the meantime, this inconsistency admittedly is potentially troublesome. Since we cannot get from the table the lower-tail critical value of F, we may avoid having to try by *always placing the larger sample variance over the smaller sample variance*

Table 10.3 5% and 1% (Boldface Type) Points for the Distribution of F (From Appendix Table 6)

df_2	1	2	3	4	5	6	7	8	9	10	11	12	14	16	20	24	30	40	50	75	100	200	500	∞	df_2
1	161	200	216	225	230	234	237	239	241	242	243	244	245	246	248	249	250	251	252	253	253	254	254	254	1
	4,052	**4,999**	**5,403**	**5,625**	**5,764**	**5,859**	**5,928**	**5,981**	**6,022**	**6,056**	**6,082**	**6,106**	**6,142**	**6,169**	**6,208**	**6,234**	**6,261**	**6,286**	**6,302**	**6,323**	**6,334**	**6,352**	**6,361**	**6,366**	
2	18.51	19.00	19.16	19.25	19.30	19.33	19.36	19.37	19.38	19.39	19.40	19.41	19.42	19.43	19.44	19.45	19.46	19.47	19.47	19.48	19.49	19.49	19.50	19.50	2
	98.49	**99.00**	**99.17**	**99.25**	**99.30**	**99.33**	**99.36**	**99.37**	**99.39**	**99.40**	**99.41**	**99.42**	**99.43**	**99.44**	**99.45**	**99.46**	**99.47**	**99.48**	**99.48**	**99.49**	**99.49**	**99.49**	**99.50**	**99.50**	
3	10.13	9.55	9.28	9.12	9.01	8.94	8.88	8.84	8.81	8.78	8.76	8.74	8.71	8.69	8.66	8.64	8.62	8.60	8.58	8.57	8.56	8.54	8.54	8.53	3
	34.12	**30.82**	**29.46**	**28.71**	**28.24**	**27.91**	**27.67**	**27.49**	**27.34**	**27.23**	**27.13**	**27.05**	**26.92**	**26.83**	**26.69**	**26.60**	**26.50**	**26.41**	**26.35**	**26.27**	**26.23**	**26.18**	**26.14**	**26.12**	
4	7.71	6.94	6.59	6.39	6.26	6.16	6.09	6.04	6.00	5.96	5.93	5.91	5.87	5.84	5.80	5.77	5.74	5.71	5.70	5.68	5.66	5.65	5.64	5.63	4
	21.20	**18.00**	**16.69**	**15.98**	**15.52**	**15.21**	**14.98**	**14.80**	**14.66**	**14.54**	**14.45**	**14.37**	**14.24**	**14.15**	**14.02**	**13.93**	**13.83**	**13.74**	**13.69**	**13.61**	**13.57**	**13.52**	**13.48**	**13.46**	
5	6.61	5.79	5.41	5.19	5.05	4.95	4.88	4.82	4.78	4.74	4.70	4.68	4.64	4.60	4.56	4.53	4.50	4.46	4.44	4.42	4.40	4.38	4.37	4.36	5
	16.26	**13.27**	**12.06**	**11.39**	**10.97**	**10.67**	**10.45**	**10.29**	**10.15**	**10.05**	**9.96**	**9.89**	**9.77**	**9.68**	**9.55**	**9.47**	**9.38**	**9.29**	**9.24**	**9.17**	**9.13**	**9.07**	**9.04**	**9.02**	
6	5.99	5.14	4.76	4.53	4.39	4.28	4.21	4.15	4.10	4.06	4.03	4.00	3.96	3.92	3.87	3.84	3.81	3.77	3.75	3.72	3.71	3.69	3.68	3.67	6
	13.74	**10.92**	**9.78**	**9.15**	**8.75**	**8.47**	**8.26**	**8.10**	**7.98**	**7.87**	**7.79**	**7.72**	**7.60**	**7.52**	**7.39**	**7.31**	**7.23**	**7.14**	**7.09**	**7.02**	**6.99**	**6.94**	**6.90**	**6.88**	
7	5.59	4.74	4.35	4.12	3.97	3.87	3.79	3.73	3.68	3.63	3.60	3.57	3.52	3.49	3.44	3.41	3.38	3.34	3.32	3.29	3.28	3.25	3.24	3.23	7
	12.25	**9.55**	**8.45**	**7.85**	**7.46**	**7.19**	**7.00**	**6.84**	**6.71**	**6.62**	**6.54**	**6.47**	**6.35**	**6.27**	**6.15**	**6.07**	**5.98**	**5.90**	**5.85**	**5.78**	**5.75**	**5.70**	**5.67**	**5.65**	
8	5.32	4.46	4.07	3.84	3.69	3.58	3.50	3.44	3.39	3.34	3.31	3.28	3.23	3.20	3.15	3.12	3.08	3.05	3.03	3.00	2.98	2.96	2.94	2.93	8
	11.26	**8.65**	**7.59**	**7.01**	**6.63**	**6.37**	**6.19**	**6.03**	**5.91**	**5.82**	**5.74**	**5.67**	**5.56**	**5.48**	**5.36**	**5.28**	**5.20**	**5.11**	**5.06**	**5.00**	**4.96**	**4.91**	**4.88**	**4.86**	
9	5.12	4.26	3.86	3.63	3.48	3.37	3.29	3.23	3.18	3.13	3.10	3.07	3.02	2.98	2.93	2.90	2.86	2.82	2.80	2.77	2.76	2.73	2.72	2.71	9
	10.56	**8.02**	**6.99**	**6.42**	**6.06**	**5.80**	**5.62**	**5.47**	**5.35**	**5.26**	**5.18**	**5.11**	**5.00**	**4.92**	**4.80**	**4.73**	**4.64**	**4.56**	**4.51**	**4.45**	**4.41**	**4.36**	**4.33**	**4.31**	
10	4.96	4.10	3.71	3.48	3.33	3.22	3.14	3.07	3.02	2.97	2.94	2.91	2.86	2.82	2.77	2.74	2.70	2.67	2.64	2.61	2.59	2.56	2.55	2.54	10
	10.04	**7.56**	**6.55**	**5.99**	**5.64**	**5.39**	**5.21**	**5.06**	**4.95**	**4.85**	**4.78**	**4.71**	**4.60**	**4.52**	**4.41**	**4.33**	**4.25**	**4.17**	**4.12**	**4.05**	**4.01**	**3.96**	**3.93**	**3.91**	
11	4.84	3.98	3.59	3.36	3.20	3.09	3.01	2.95	2.90	2.86	2.82	2.79	2.74	2.70	2.65	2.61	2.57	2.53	2.50	2.47	2.45	2.42	2.41	2.40	11
	9.65	**7.20**	**6.22**	**5.67**	**5.32**	**5.07**	**4.88**	**4.74**	**4.63**	**4.54**	**4.46**	**4.40**	**4.29**	**4.21**	**4.10**	**4.02**	**3.94**	**3.86**	**3.80**	**3.74**	**3.70**	**3.66**	**3.62**	**3.60**	
12	4.75	3.88	3.49	3.26	3.11	3.00	2.92	2.85	2.80	2.76	2.72	2.69	2.64	2.60	2.54	2.50	2.46	2.42	2.40	2.36	2.35	2.32	2.31	2.30	12
	9.33	**6.93**	**5.95**	**5.41**	**5.06**	**4.82**	**4.65**	**4.50**	**4.39**	**4.30**	**4.22**	**4.16**	**4.05**	**3.98**	**3.86**	**3.78**	**3.70**	**3.61**	**3.56**	**3.49**	**3.46**	**3.41**	**3.38**	**3.36**	
13	4.67	3.80	3.41	3.18	3.02	2.92	2.84	2.77	2.72	2.67	2.63	2.60	2.55	2.51	2.46	2.42	2.38	2.34	2.32	2.28	2.26	2.24	2.22	2.21	13
	9.07	**6.70**	**5.74**	**5.20**	**4.86**	**4.62**	**4.44**	**4.30**	**4.19**	**4.10**	**4.02**	**3.96**	**3.85**	**3.78**	**3.67**	**3.59**	**3.51**	**3.42**	**3.37**	**3.30**	**3.27**	**3.21**	**3.18**	**3.16**	
14	4.60	3.74	3.34	3.11	2.96	2.85	2.77	2.70	2.65	2.60	2.56	2.53	2.48	2.44	2.39	2.35	2.31	2.27	2.24	2.21	2.19	2.16	2.14	2.13	14
	8.86	**6.51**	**5.56**	**5.03**	**4.69**	**4.46**	**4.28**	**4.14**	**4.03**	**3.94**	**3.86**	**3.80**	**3.70**	**3.62**	**3.51**	**3.43**	**3.34**	**3.26**	**3.21**	**3.14**	**3.11**	**3.06**	**3.02**	**3.00**	
15	4.54	3.68	3.29	3.06	2.90	2.79	2.70	2.64	2.59	2.55	2.51	2.48	2.43	2.39	2.33	2.29	2.25	2.21	2.18	2.15	2.12	2.10	2.08	2.07	15
	8.68	**6.36**	**5.42**	**4.89**	**4.56**	**4.32**	**4.14**	**4.00**	**3.89**	**3.80**	**3.73**	**3.67**	**3.56**	**3.48**	**3.36**	**3.29**	**3.20**	**3.12**	**3.07**	**3.00**	**2.97**	**2.92**	**2.89**	**2.87**	

when computing observed F. That is, $F = s_1^2/s_2^2$, where s_1^2 is, by definition, the larger of the sample variances and s_2^2 is the smaller of the sample variances.

Then, because the area in the right-hand tail represents only $\frac{1}{2}\alpha$, we double this value to obtain the correct probability of a Type I error α. Thus, if we use a critical F ratio presented in lightface type and reject the null hypothesis, we reject at the .10 level of significance; if we use a critical F ratio presented in boldface type and reject the null hypothesis, we reject at the .02 level of significance.

We are now in a position to work the Rickey Car problem (Illustrative Case 10.2):

1. State the null and alternative hypotheses.
   As noted above:
   $$H_0:\ \sigma_1^2 - \sigma_2^2 = 0$$
   $$H_a:\ \sigma_1^2 - \sigma_2^2 \neq 0$$

2. Establish the level of significance.
   Let us say we set the level of significance at $\alpha = .10$. This means that we would use a .05 critical value, one of those in lightface type, from Table 10.3.

3. Determine the appropriate kind of sampling distribution to rely on.
   The appropriate sampling distribution is the F distribution where $df_1 = n_1 - 1 = 10 - 1 = 9$ and $df_2 = n_2 - 1 = 10 - 1 = 9$.

4. Establish a decision rule.
   Figure 10.6 shows the theoretical F distribution divided into appropriate acceptance and rejection regions. The critical value of F, 3.18 in this case, is, of course, from Table 10.3. The decision rule would be:

Decision Rule
1. If observed $F \leq 3.18$, accept H_0.
2. If observed $F > 3.18$, reject H_0.

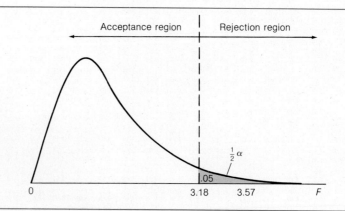

Figure 10.6. *F* distribution for Illustrative Case 10.2.

10 Tests of Hypotheses Involving Two Populations

5. Compute observed F from the two independent random samples.

$$\text{Observed } F = \frac{s_1^2}{s_2^2} = \frac{(1700)^2}{(900)^2} = 3.57$$

6. Apply the decision rule to the sample statistic and either accept or reject the null hypothesis.

Since 3.57 exceeds the critical value of 3.18, we reject at the .10 level of significance the null hypothesis that the two population variances are equal.

7. Take the appropriate action.

Buy all future original-equipment tires from Goodgrip.

In this illustrative case, the population variances were of interest in their own right. The F test we are discussing can serve another useful purpose as well. Recall that when we worked with the small sample version of the Gurney Company case in section 10.2 we noted that the method requires implicit acceptance of two assumptions—namely, that the populations are normally distributed and that the populations have equal variances. This F test can help us to evaluate the validity of the second assumption. The relevant facts (from section 10.2) are $n_1 = 16$, $n_2 = 9$, $s_1 = 12.04$, and $s_2 = 10.22$.

1. State the null and alternative hypotheses.

$$H_0: \sigma_1^2 - \sigma_2^2 = 0$$
$$H_a: \sigma_1^2 - \sigma_2^2 \neq 0$$

2. Establish the level of significance.

Let us use $\alpha = .10$ again. (Since with this application of the F test we wish to *accept* the null hypotheses, α should be made relatively large to guard against too high a probability of accepting the null hypothesis incorrectly.)

3. Determine the appropriate kind of sampling distribution to rely on.

The appropriate sampling distribution is the F distribution for $df_1 = 15$ and $df_2 = 8$ degrees of freedom.

4. Establish a decision rule.

From Table 10.3, we find the critical F value to be 3.22. Therefore,

Decision Rule
1. If observed $F \leq 3.22$, accept H_0.
2. If observed $F > 3.22$, reject H_0.

5. Compute the observed F statistic for the two independent samples.

$$\text{Observed } F = \frac{s_1^2}{s_2^2} = \frac{(12.04)^2}{(10.22)^2} = 1.39$$

6. Apply the decision rule to the test statistic and either accept or reject the null hypothesis.

Since 1.39 is considerably less than 3.22, we accept H_0 and conclude that the two population variances are equal.

7. Take the appropriate action.

Insofar as this assumption is concerned, we can stop worrying about whether the t test used in connection with the Gurney Company problem is valid.

Notice that this F test is an exception to the rule set forth in Chapter 9 to the effect that it is the alternative hypothesis which one usually wishes to "prove." Clearly, in this case, we hope the *null* hypothesis is the correct one.

*10.5 t Test for Paired Observations

A special and most useful kind of t test is called the *t test for paired observations*. When we employ the kind of t test described in section 10.2, we are concerned with determining whether the means of two independent populations are equal.

> However, when we use the t test for paired observations, we are interested in determining whether the true average difference between several sets of paired—and, therefore, probably dependent—observations is equal to 0.

For example, some experimental studies make use of before-and-after measurements taken on the same subjects to determine whether some *treatment variable*, as it is called, has an effect on the *after* measure. Use of the t test described in section 10.2 would be improper for the obvious reason that the *after* values are not independent of the *before* values. For further clarification, let us consider another illustrative case.

ILLUSTRATIVE CASE 10.3

The production superviser of the River City Manufacturing Company, with the cooperation of the company's employees and their union, has conducted a series of experiments having to do with working conditions in the plant. His goal is to develop a set of conditions under which productivity will be maximized and worker discontent minimized. The part of the project on which we focus concerned the determination of the "best" temperature in the plant from the standpoint of getting the most production out of the workers without appearing to push them to produce more. At the outset of the experiment, the production supervisor used only two temperatures—namely, *cool* (defined as 55°F) and *warm* (75°F). If output were found to differ as a result of this large temperature difference, he reasoned, then further experimentation might eventually pinpoint just the right temperature, or range of temperatures, for maximum worker output. Important to note is that all workers included in the experiment were involved in assembly operations which differed very little from worker to worker.

This particular phase of the experiment was run over two working days. The results are shown in Table 10.4. In this table and the discussion that follows, the outputs achieved during the warm day will be designated the X values and the outputs during the cool day the Y values. The differences, calculated by $Y - X$, will be referred to as the d values.

The steps involved in conducting this test are similar to those used in tests already discussed.

Table 10.4 Table of Warm-Day Outputs, Cool-Day Outputs, and Differences for the River City Manufacturing Company Case

Worker Identification Number	Output (Units)		$d = Y - X$	d^2
	Warm Day X	Cool Day Y		
1	27	30	+ 3	9
2	33	37	+ 4	16
3	20	32	+12	144
4	19	29	+10	100
5	26	22	− 4	16
6	29	29	0	0
7	31	29	− 2	4
8	31	32	+ 1	1
9	35	31	− 4	16
10	27	33	+ 6	36
11	23	26	+ 3	9
12	34	30	− 4	16
13	14	23	+ 9	81
14	20	29	+ 9	81
15	26	24	− 2	4
16	18	26	+ 8	64
17	30	30	0	0
18	27	32	+ 5	25
19	33	31	− 2	4
20	18	24	+ 6	36
21	37	35	− 2	4
22	32	30	− 2	4
23	35	40	+ 5	25
24	26	31	+ 5	25
25	18	15	− 3	9
26	24	27	+ 3	9
27	27	32	+ 5	25
28	29	29	0	0
29	32	29	− 3	9
30	18	27	+ 9	81
Total			75	853

1. State the null and alternative hypothesis.

   The statement of the null and alternative hypotheses must be tied to the specific question asked. For example, if the production supervisor strongly suspected that cooler temperatures are more conducive to high output than warmer temperatures, he might elect to perform a one-tailed test having as a null hypothesis—H_0: $\mu_d \leq 0$, where μ_d is the true average difference for the population. (The direction of the inequality sign is based on the assumption that d values were obtained by taking $Y - X$, rather than $X - Y$, an arbitrary decision.) On the other hand, if he is completely open-minded on the question of whether cooler or warmer is better, he might decide that a two-tailed test is more appropriate.

We will complete the operations using a two-tailed test. The hypotheses would be

$$H_0: \mu_d = 0$$
$$H_a: \mu_d \neq 0$$

2. Establish the level of significance.
   Let us suppose that the level of significance is set at $\alpha = .01$.

3. Determine the appropriate kind of sampling distribution to rely on.
   The appropriate sampling distribution is the sampling distribution of $\bar{d}$, where $\bar{d}$ is the average difference for a specific sample and is calculated by $\bar{d} = (\Sigma d)/n$. This sampling distribution has the following important properties:

---

1. The mean of the sampling distribution of $\bar{d}$ is equal to the average difference for the population, μ_d. That is, $\mu_{\bar{d}} = \mu_d$.
2. The standard deviation of the sampling distribution of $\bar{d}$, symbolized $\sigma_{\bar{d}}$, is, for infinite populations or relatively large finite populations, equal to the population standard deviation σ_d divided by the square root of the sample size.
3. The sampling distribution of $\bar{d}$ is approximately normal provided the sample size n is sufficiently large.

---

4. Establish a decision rule.
   Since the population standard deviation σ_d is unknown and must be estimated from the sample standard deviation s_d, we use t values in formulating our decision rule. We begin by finding the critical t value associated with $\alpha = .01$ and $n - 1 = 30 - 1 = 29$ degrees of freedom. Consulting Appendix Table 3, we find the critical t value to be ± 2.756. Consequently, the decision rule is

Decision Rule
1. If $-2.756 \leq$ observed $t \leq +2.756$, accept H_0.
2. If observed $t < -2.756$ or observed $t > +2.756$, reject H_0.

The observed t in this decision rule is the t-value equivalent of $\bar{d}$ and is obtained by

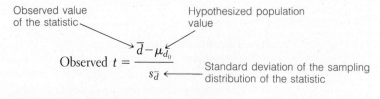

$$\text{Observed } t = \frac{\bar{d} - \mu_{d_0}}{s_{\bar{d}}}$$

or, since μ_d is hypothesized to be 0, simply

$$\text{observed } t = \frac{\bar{d}}{s_{\bar{d}}} \tag{10.5.1}$$

---

5. Calculate the appropriate sample statistic.

   The statistic of primary interest in this problem is the sample average paired difference $\bar{d}$. For the example under discussion, we find

$$\bar{d} = \frac{\Sigma d}{n} = \frac{3 + 4 + 12 + 10 - 4 \cdots - 3 + 9}{30} = 2.50$$

   The $s_{\bar{d}}$ value by which $\bar{d}$ is to be divided to obtain the needed observed t is equal to

$$s_{\bar{d}} = \frac{s_d}{\sqrt{n}} \tag{10.5.2}$$

   We know that $n = 30$ and, therefore, $\sqrt{n} = 5.48$; however, we do not yet know s_d, the standard deviation of the differences for the specific sample under analysis. Since s_d is literally a sample standard deviation, it can be computed by

$$s_d = \sqrt{\frac{\Sigma(d - \bar{d})^2}{n - 1}} \tag{10.5.3}$$

or

$$s_d = \sqrt{\frac{\Sigma d^2 - \dfrac{(\Sigma d)^2}{n}}{n - 1}} \tag{10.5.4}$$

or other standard deviation formula modified to show that we are obtaining the standard deviation of a series of differences, d values, and not of the original X or Y values.

   Using Equation (10.5.4), we calculate

$$s_d = \sqrt{\frac{853 - \dfrac{(75)^2}{30}}{29}} = 4.79$$

   In turn, we find $s_{\bar{d}} = s_d/\sqrt{n}$ to be $4.79/\sqrt{30} = 0.87$. Therefore, $t = \bar{d}/s_{\bar{d}} = 2.50/.87 = 2.87$.

6. Apply the decision rule to the observed sample statistic and either accept or reject the null hypothesis.

   Since 2.87 exceeds the upper critical value of +2.756, we reject the null hypothesis. Apparently, temperature does affect worker output.

7. Take the appropriate action.

   Because the null hypothesis is rejected at the upper tail of the t distribution (and because d values were computed from $Y - X$), it appears that cooler temperatures are more conducive to high worker output than warmer temperatures. The production supervisor might decide to use a 55°F temperature immediately. However, in view of the information given in the case, it seems likely that he will very soon experiment with other temperatures in the vicinity of 55°F in an effort to pinpoint an optimum temperature if there is such a thing.

We noted at the outset of this section that this t test is especially useful in cases involving two samples that are not independent. Moreover, the test is more flexible than the single example used here might suggest in that each pair of measurements need not be on the same elementary unit. The measurements may be taken on two elementary units having some relevant characteristic in common—years of work experience in the present case, for example. As a general rule, such a pairing procedure, though not always practical, is advisable when one has a choice. Because pairing eliminates much extraneous variation, the standard deviation of the sampling distribution of $\overline{d}$ will be smaller than that of the sampling distribution of $\overline{X}_1 - \overline{X}_2$ under circumstances where $n_1 = n_2$ for two independent samples is the same as n for paired observations. Hence, the paired-observations test will produce a significant result in some cases where the other test will not.

On the other hand, the paired-observations test is not completely free of restrictions. First, a normally distributed population of d values is assumed. This is not often a troublesome assumption; but it could be. For example, if a few people in a sample respond to the treatment variable in a similar, extreme way while most respond in a much more subtle way or not at all, this may mean that the assumption of a normally distributed population of d values is being violated. (Of course, if the sample size is large, this assumption will be less important than when the sample size is small. For a large sample size, the central limit theorem comes into play and assures us that the sampling distribution of $\overline{d}$ values will be approximately normal almost regardless of the shape of the population.) Second, the phenomenon or activity of interest must be the same from pair to pair. For example, in our illustrative case, it was stipulated that the workers were all engaged in the same kind of assembly operation. Had the nature of the work performed by the 30 sample members been notably different, the use of this t test would be improper.

10.6 Summary

This chapter has described four tests of hypotheses pertaining to parameters of *two* populations. Table 10.5 presents a summary of pertinent facts for each such test. In this table it is assumed, realistically, that population parameters necessary for use of the mathematically correct standard error formulas are unknown. For example, in the upper left-hand part of the table, we have

$$Z = \frac{\overline{X}_1 - \overline{X}_2}{\sqrt{\dfrac{s_1^2}{n_1} + \dfrac{s_2^2}{n_2}}}$$

rather than the more accurate, but generally unusable, formula involving σ_1^2 and σ_2^2. Moreover, although cutoff points are always somewhat arbitrary, the following rules of thumb will help you interpret this table: "Large sample sizes" will be understood to mean (1) sample

Table 10.5   Summary Information About Four Tests of Hypotheses Involving Two Populations

Purpose of Test	Relationship Between Samples	Large Sample Sizes			Small Sample Sizes		
		Test Statistic	Degrees of Freedom	Assumption About Population Shape	Test Statistic	Degrees of Freedom	Assumption About Population Shape
1. To determine whether the means of two populations are equal	Independent	$Z = \dfrac{\bar{X}_1 - \bar{X}_2}{\sqrt{\dfrac{s_1^2}{n_1} + \dfrac{s_2^2}{n_2}}}$	Not necessary to determine	None	$t = \dfrac{\bar{X}_1 - \bar{X}_2}{s_w \sqrt{\dfrac{1}{n_1} + \dfrac{1}{n_2}}}$ where: $s_w^2 = \dfrac{(n_1 - 1)s_1^2 + (n_2 - 1)s_2^2}{n_1 + n_2 - 2}$	$n_1 + n_2 - 2$	Both populations are normal (Also: Population variances are equal)
2. To determine whether the proportions of two populations are equal	Independent	$Z = \dfrac{\bar{p}_1 - \bar{p}_2}{\sqrt{\bar{p}_w \bar{q}_w \left(\dfrac{1}{n_1} + \dfrac{1}{n_2}\right)}}$ where: $\bar{p}_w = \dfrac{n_1 \bar{p}_1 + n_2 \bar{p}_2}{n_1 + n_2}$ and $\bar{q}_w = 1 - \bar{p}_w$	Not necessary to determine	None	Not covered in chapter	—	—
3. To determine whether the variances of two populations are equal	Independent	$F = \dfrac{s_1^2}{s_2^2}$	$df_1 = n_1 - 1$ and $df_2 = n_2 - 1$	Both populations are normal	$F = \dfrac{s_1^2}{s_2^2}$	$df_1 = n_1 - 1$ and $df_2 = n_2 - 1$	Both populations are normal
4. To determine whether the population $\mu_d = 0$	Dependent	$Z = \dfrac{\bar{d}}{s_d/\sqrt{n}}$	Not necessary to determine	None	$t = \dfrac{\bar{d}}{s_d/\sqrt{n}}$	$n - 1$	Population of d values is normal

sizes ensuring that $n_1 p_1$, $n_1 q_1$, $n_2 p_2$, and $n_2 q_2$ are all greater than 5 in the case of the test for differences between two proportions, or (2) $n > 30$ for the other three tests. "Small sample sizes" for these three tests will mean $n \leq 30$.

Terms Introduced in This Chapter

F distribution (p. 441)
F test (p. 445)
F statistic or F ratio (p. 441)
Paired observations (p. 446)
sampling distribution of $\bar{d}$ (p. 448)

sampling distribution of $\bar{p}_1 - \bar{p}_2$ (p. 438)
sampling distribution of $\bar{X}_1 - \bar{X}_2$ (p. 431)
standard error of $\bar{d}$ (p. 448)

standard error of the difference between two means (p. 432)
standard error of the difference between two proportions (p. 438)

Formulas Introduced in This Chapter

See Table 10.5.

Questions and Problems

10.1 Explain the meaning of each of the following:
   a. Sampling distribution of $\overline{X}_1 - \overline{X}_2$
   b. $\mu_{\overline{x}_1 - \overline{x}_2}$
   c. Standard error of the difference between two means
   d. Pooled estimate of population variance
   e. Observed F
   f. Paired observations

10.2 Explain the meaning of each of the following:
   a. Sampling distribution of $\overline{p}_1 - \overline{p}_2$
   b. Standard error of the difference between two proportions
   c. $\overline{p}_w$
   d. $\overline{d}$
   e. Critical t

10.3 Distinguish between (or among):
   a. θ_1, μ_1, and p_1
   b. $\overline{X}_1 - \overline{X}_2$ and $\mu_1 - \mu_2$
   c. Sample standard deviation and sample variance
   d. $\mu_{\overline{x}_1 - \overline{x}_2}$ and $\mu_1 - \mu_2$
   e. $\sigma_{\overline{x}_1 - \overline{x}_2}$ and $s_{\overline{x}_1 - \overline{x}_2}$

10.4 Distinguish between:
   a. $\sigma_{\overline{p}_1 - \overline{p}_2}$ and $s_{\overline{p}_1 - \overline{p}_2}$
   b. $\overline{p}_1$ and $\overline{p}_w$
   c. Critical t and observed t
   d. df_1 and df_2 in an F test for equality of two population variances
   e. Critical F and observed F

10.5 Distinguish between:
   a. Critical Z and observed Z
   b. σ_1^2 and σ_2^2
   c. σ_1^2 and s_1^2
   d. s_1^2 and s_w^2
   e. $\mu_{\overline{p}_1 - \overline{p}_2}$ and $p_1 - p_2$

10.6 Distinguish between:
   a. Independent populations and populations that are not independent
   b. d and $\mu_1 - \mu_2$
   c. $\sigma_{\overline{d}}$ and σ_d
   d. $s_{\overline{d}}$ and s_d
   e. d and $\overline{d}$

10.7 Indicate which of the following statements you agree with and which you disagree with, and defend your opinions:
   a. When testing a null hypothesis like H_0: $\mu_1 - \mu_2$ or H_0: $p_1 - p_2$, it is necessary to begin by hypothesizing values for μ_1 and μ_2 or for p_1 and p_2.
   b. In a test of the null hypothesis H_0: $\mu_1 - \mu_2 = 0$, we have to contend with fewer restrictive assumptions when Z can be used than when t must be used.
   c. When testing the null hypothesis H_0: $\mu_1 - \mu_2$ by Z or t, we are not concerned at all about the amount of variation in the two populations.
   d. A "robust" test of significance is one which can be applied even under adverse environmental conditions.

10 Tests of Hypotheses Involving Two Populations

10.8 Indicate which of the following statements you agree with and which you disagree with, and defend your opinions:
  a. If the samples are independent, $\bar{p}_1 - \bar{p}_2$ is an unbiased estimator of the true difference between the population proportions $p_1 - p_2$.
  b. When testing the null hypothesis H_0: $\sigma_1^2 - \sigma_2^2 = 0$, we implicitly assume that the population shapes are normal.
  c. The theoretical F distribution associated with $df_1 = 5$ and $df_2 = 10$ degrees of freedom will have exactly the same shape as the theoretical F distribution associated with $df_1 = 10$ and $df_2 = 5$ degrees of freedom.
  d. When comparing observed $F = s_1^2/s_2^2$ with critical F from our prepared table, the probability of a Type I error is really double that indicated in the table.

10.9 Indicate which of the following statements you agree with and which you disagree with, and defend your opinions:
  a. The F test described in this chapter can be used to determine whether one of the assumptions underlying the t test for the difference between two means is valid but it cannot properly be used when one is concerned with population variances in their own right.
  b. The t test for the difference between two independent population means and the t test used to determine whether $\mu_d = 0$ are effectively interchangeable.
  c. The t test for paired observations is applicable only in situations where two measures are taken on the same elementary units.
  d. The t test for paired observations is always applicable when two measures are taken on the same elementary units regardless of how similar or dissimilar the elementary units are in relevant respects.

10.10 Explain in your own words what Figure 10.1 is attempting to convey.

10.11 Given:

Population 1	Population 2
Observations are: 20, 22, 22, 24	Observations are: 18, 20, 20

  a. Compute the mean of Population 1. Of Population 2.
  b. From part a, determine the value of $\mu_1 - \mu_2$.
  c. Compute the standard deviation of Population 1. Of Population 2.
  d. Using $n_1 = 3$ and $n_2 = 2$, construct the sampling distribution of $\overline{X}_1 - \overline{X}_2$.
  e. Compute the mean of the sampling distribution of $\overline{X}_1 - \overline{X}_2$. Is it the same as the difference you obtained in connection with part b?
  f. Using the formula

$$\sigma_{\bar{x}_1 - \bar{x}_2} = \sqrt{\frac{\sigma_1^2}{n_1} \cdot \frac{N_1 - n_1}{N_1 - 1} + \frac{\sigma_2^2}{n_2} \cdot \frac{N_2 - n_2}{N_2 - 1}}$$

obtain the standard error of the difference between two means.
  g. Using the formula

$$\sigma_{\bar{x}_1 - \bar{x}_2} = \sqrt{\frac{\Sigma[\overline{X}_1 - \overline{X}_2) - (\mu_1 - \mu_2)]^2}{\text{Number of combinations of } \overline{X}_1 \text{ and } \overline{X}_2}}$$

obtain the standard error of the difference between two means directly from the sample $\overline{X}_1 - \overline{X}_2$ values. Is your result the same as the result you obtained in connection with part f?
  h. Is your sampling distribution of $\overline{X}_1 - \overline{X}_2$ approximately normal? Elaborate.

Special-Challenge Problem

10.12 Review section 9.5 of Chapter 9. Then explain completely in your own words how you would go about determining the probability of a Type II error in a situation where the hypotheses are H_0: $\mu_1 - \mu_2 = 0$ and H_a: $\mu_1 - \mu_2 = 2$.

10.13 A company with two plants producing the same product, but with two slightly different manufacturing processes, conducted a sample study at each of the two plants with results as follows:

	Sample Drawn From	
	Population 1	**Population 2**
Average time required for employees to complete a specified number of assemblages	$\overline{X}_1 = 8$ minutes	$\overline{X}_2 = 12$ minutes
Standard deviation of times required for employees to complete a specified number of assemblages	$s_1 = 2$ minutes	$s_2 = 1.5$ minutes
Number of observations	$n_1 = 100$	$n_2 = 150$

The null and alternative hypotheses are stated to be

$H_0: \mu_1 - \mu_2 = 0$
$H_a: \mu_1 - \mu_2 \neq 0$

a. Determine the appropriate decision rule if the level of significance is set at $\alpha = .05$.
b. Test the null hypothesis. Should the null hypothesis be accepted or rejected?
c. State precisely what you understand is meant by the conclusion you arrived at in connection with part *b*.
d. Would the conclusion you arrived at in connection with part *b* have been different if the level of significance had been set at $\alpha = .10$? Explain.
e. Would the conclusion you arrived at in connection with part *b* have been different if the sample sizes had been $n_1 = 200$ and $n_2 = 250$? Explain.
f. In what way or ways, if any, would the hypothesis-testing procedure be altered if the null and alternative hypotheses had been stated as

$H_0: \mu_1 - \mu_2 \leq 0$
$H_a: \mu_1 - \mu_2 > 0$?

Would the null hypothesis be accepted or rejected?

10.14 Refer to problem 10.13. Assume that the facts concerning this problem are the same as previously stated:

$H_0: \mu_1 - \mu_2 = 0$
$H_a: \mu_1 - \mu_2 \neq 0$

and that $\alpha = .05$, $\overline{X}_1 = 8$ minutes, $\overline{X}_2 = 12$ minutes, $s_1 = 2$ minutes, $s_2 = 1.5$ minutes. However, assume now that the sample sizes were $n_1 = 10$ and $n_2 = 15$.
a. Determine the decision rule.
b. In arriving at the decision rule, what change or changes in procedure, if any, from the procedure followed in connection with problem 10.13 was (or were) necessitated by the smaller sample sizes? Explain.
c. Test the null hypothesis. Should the null hypothesis be accepted or rejected?

10.15 Refer to problem 10.13, part *b*, and problem 10.14, part *c*. For each version of this problem, determine at what level the test statistic would be significant if the observed-significance-level method of hypothesis testing were used. (*Hint*: See section 9.10 of Chapter 9.)

10.16 Refer to problem 10.14. Is the assumption of equal population variances valid? Use the *F* test and a .10 level of significance to find out.

10.17 Assume that the following data have been obtained from a sample survey conducted with a view to determining whether residents of County 1 and residents of County 2 spend the same mean amount of money on durable goods during a typical year:

County 1	County 2
$\overline{X}_1 = \$3000$	$\overline{X}_2 = \$2700$
$s_1 = \$500$	$s_2 = \$525$
$n_1 = 2000$	$n_2 = 2000$

The null and alternative hypotheses are

$H_0: \mu_1 - \mu_2 = 0$

$H_a: \mu_1 - \mu_2 \neq 0$

a. Determine the decision rule if the level of significance is set at $\alpha = .10$.
b. Test the null hypothesis. Should the null hypothesis be accepted or rejected?
c. State precisely what you understand is meant by the conclusion you arrived at in connection with part b.
d. Would the conclusion you arrived at in connection with part b have been different if the level of significance had been $\alpha = .05$?
e. Would the conclusion you arrived at in connection with part b have been different if the sample sizes had been $n_1 = 1000$ and $n_2 = 1000$?
f. In what way or ways, if any, would the hypothesis-testing procedure have been different if the null and alternative hypotheses had been

$H_0: \mu_1 - \mu_2 \geq 0$

$H_a: \mu_1 - \mu_2 < 0$?

Would the null hypothesis be accepted or rejected?

 10.18 Refer to problem 10.17. Assume that the facts concerning this problem are the same as previously stated:

$H_0: \mu_1 - \mu_2 = 0$

$H_a: \mu_1 - \mu_2 \neq 0$

And that $\alpha = .10$, $\overline{X}_1 = \$3000$, $\overline{X}_2 = \$2700$, $s_1 = \$500$, and $s_2 = \$525$. However, assume now that the sample sizes were $n_1 = 20$ and $n_2 = 10$.
a. Determine the decision rule.
b. In arriving at the decision rule, what change or changes in procedure, if any, from the procedure followed in connection with problem 10.17 was (or were) necessitated by the smaller sample sizes? Explain.
c. Test the null hypothesis. Should the null hypothesis be accepted or rejected?

 10.19 Refer to problem 10.17, part b and problem 10.18, part c. For each version of this problem, determine at what level the test statistic would be significant if the observed-significance-level method of hypothesis testing were used. (*Hint:* See section 9.10 of Chapter 9.)

 10.20 Refer to problem 10.18. Is the assumption of equal population variances valid? Use the appropriate test and a .10 level of significance to find out.

 10.21 It is thought that Appraiser 1 at an automobile body and fender shop tends to arrive at higher estimates for costs of repairs, on the average, than Appraiser 2 at the same shop. The recent records of the two appraisers are examined and the following results found:

Appraiser 1	Appraiser 2
$\overline{X}_1 = \$575$	$\overline{X}_2 = \$498$
$s_1 = \$22$	$s_2 = \$25$
$n_1 = 8$	$n_2 = 6$

a. Assuming independence and a level of significance of .05, develop the decision rule.

b. Test the null hypothesis. Should the null hypothesis be accepted or rejected?

c. What change, if any, in the hypothesis-testing procedures would have been called for if the sample sizes had been $n_1 = 80$ and $n_2 = 60$? Explain.

10.22 Refer to problem 10.21, part *b*. At what level would the test statistic be significant if the observed-significance-level method of hypothesis testing were used? (*Hint:* See section 9.10 of Chapter 9.)

10.23 Refer to problem 10.21. Is the assumption of equal population variances valid? Use the appropriate test and a .10 level of significance to find out.

10.24 It is thought that Appraiser 1 at an automobile body and fender shop tends to arrive at higher estimates for costs of repairs, on the average, than Appraiser 2 at the same shop. The recent records of the two appraisers are examined and the following results found:

Automobile	Appraiser 1	Appraiser 2
1975 Chevrolet	$400	$365
1981 Dodge	$525	$535
1976 Ford	$310	$330
1985 Lincoln	$980	$755
1977 Toyota	$310	$290
1980 Volkswagen	$185	$170

a. Test the appropriate null hypothesis at the .05 level of significance. What is your conclusion?

b. Refer to problem 10.21. In what strategic way or ways does the present problem differ from problem 10.21?

10.25 Refer to problem 10.24. Is the population of *d* values normally distributed? Arrive at an intuitive judgment about this with the aid of a box-and-whiskers plot of the sample *d* values.

10.26 An investor is thinking of buying into one of two mutual funds, A and B. Over the past ten years both funds have averaged a 15% return to shareholders. However, the variance for Fund A over this time has been 25%, whereas it has been only 9% for Fund B. Taking the variance as a measure of risk, test at the .10 level of significance the null hypothesis that the two funds are equal with respect to risk against the alternative hypothesis that they are not equal. (*Note:* $n_1 = n_2 = 10$.)

Special-Challenge Problem

10.27 Given the following data:

Population 1	Population 2
Elementary unit 1 has the characteristic of interest	Elementary unit A has the characteristic of interest
Elementary unit 2 has the characteristic of interest	Elementary unit B does not have the characteristic of interest
Elementary unit 3 does not have the characteristic of interest	Elementary unit C has the characteristic of interest
Elementary unit 4 does not have the characteristic of interest	

a. Determine the population proportion for Population 1. For Population 2.

b. Compute the difference $p_1 - p_2$.

c. Construct the sampling distribution of $\bar{p}_1 - \bar{p}_2$ if $n_1 = 3$ and $n_2 = 2$.
d. Compute the mean of your sampling distribution of $\bar{p}_1 - \bar{p}_2$. Is the figure you obtained the same as that obtained for part *b* of this problem?
e. Is your sampling distribution of $\bar{p}_1 - \bar{p}_2$ approximately normal? Elaborate.

 10.28 A sample study was conducted to determine whether the proportion of television viewers who believe that television shows too much violence differs between Salt Lake City, Utah (hereafter referred to as Population 1), and Detroit, Michigan (Population 2). The following results were obtained:

Sample Obtained From

Population 1	Population 2
$\bar{p}_1 = .7$	$\bar{p}_2 = .6$
$n_1 = 2000$	$n_2 = 4000$

a. Develop the decision rule if the level of significance is set at .05.
b. Test the null hypothesis. Should the null hypothesis be accepted or rejected?
c. State precisely in words what you understand is meant by the conclusion you arrived at in connection with part *b*.
d. Would the conclusion you arrived at in connection with part *b* have been different if the level of significance had been set at $\alpha = .10$?
e. Would the conclusion you arrived at in connection with part *b* have been different if the sample size had been $n_1 = 500$ and $n_2 = 1000$?

 10.29 A sample study was conducted to determine whether a higher proportion of women prefer a new type of magazine advertisement over the old type compared to men. The results are as follows:

Population 1 (Men)	Population 2 (Women)
$\bar{p}_1 = .6$	$\bar{p}_2 = .7$
$n_1 = 200$	$n_2 = 210$

where the $\bar{p}$ values represent sample proportions preferring the new type of magazine advertisement over the old type.
a. Determine the decision rule if the level of significance is set at $\alpha = .01$.
b. Test the null hypothesis. Should the null hypothesis be accepted or rejected?
c. State precisely in words what you understand is meant by the conclusion you arrived at in connection with part *b*.
d. Would the conclusion you arrived at in connection with part *b* have been different if the level of significance had been set at $\alpha = .05$?
e. Would the conclusion you arrived at in connection with part *b* have been different if the sample sizes had been $n_1 = n_2 = 500$?

 10.30 A top manager of a large retail chain of household lighting fixtures kept track of customer complaints related to errors made by specific salespersons in supplying lamp accessories, giving faulty advice, and so forth. The result of one month of such observations at one particular store are shown in the column labeled "Before" in the following table. Feeling that such complaints were too numerous—at this particular store, anyway—the manager instituted an intensive training program aimed at getting the store's salespeople better acquainted with the intricacies of household lighting. After the period of training was completed, careful records were kept of customer complaints for a period of one month. The results are shown in the column labeled "After."

	Number of Complaints	
Salesperson	**Before**	**After**
A	6	4
B	20	6
C	3	2
D	0	0
E	4	0

a. Are the two sets of data independent? Tell why you do or do not think so. What methodological implications, if any, are suggested by your conclusion? (That is, what kind of statistical test regarding the relative sizes of two population means should be performed?)

b. State the null and alternative hypotheses if the manager wishes to know whether the training led to improvement.

c. Develop the decision rule if the level of significance is set at $\alpha = .01$.

d. Test the null hypothesis. Should the null hypothesis be accepted or rejected?

 10.31 Refer to problem 10.30. Is the population of d values normally distributed? Arrive at an intuitive judgment about this with the aid of a box-and-whiskers plot of the sample d values.

 10.32 An automobile manufacturer wished to compare the working lives of two different brands of headlight lamps, H_1 and H_2. The lamps were tested in the following manner: (1) Ten test automobiles were equipped with one of each brand of lamp—the specific brand of lamp being assigned to either the left side or the right side of an automobile as determined from a table of random numbers; (2) the headlights were then run continuously until the lamps burned out and the number of hours of continuous use recorded for each brand. The resulting measurements, in thousands of hours, are shown here.

	Brand of Lamp	
Automobile	H_1	H_2
1	8.7	8.9
2	10.9	10.8
3	7.6	9.4
4	10.1	10.3
5	10.7	10.7
6	10.9	11.0
7	9.2	9.5
8	9.1	9.5
9	7.8	8.2
10	8.5	8.5

a. Are the two sets of data independent? Tell why you do or do not think so. What methodological implications, if any, are suggested by your conclusion? (That is, what kind of statistical test regarding the equality of two population means should be performed?)

b. State the null and alternative hypotheses if the automobile manufacturer wishes to know whether there is any difference between the working lives of the two brands of lamps.

c. Develop the decision rule if the level of significance is set at $\alpha = .05$.

d. Test the null hypothesis. Should the null hypothesis be accepted or rejected?

 10.33 Refer to problem 10.32. At what level would the test statistic be significant if the observed-significance-level method of hypothesis testing were used? (*Hint:* See section 9.10 of Chapter 9.)

10.34 Refer to problem 10.32. Is the population of *d* values normally distributed? Arrive at an intuitive judgment about this with the aid of a box-and-whiskers plot of the sample *d* values.

TAKE CHARGE*

10.35 One long-standing piece of Wall Street folk wisdom is that a stock split will lead to an increase in the market price of a stock (after adjustment for the split itself). A stock split represents a management decision to increase the number of shares outstanding. In theory, a two-for-one split, for example, would result in twice as many shares outstanding, each of which is worth only one-half of its value before the split. However, some people contend that the resulting lower price per share will entice greater numbers of investors to buy the shares. Since investors like to buy in round lots and since the cost of 100 shares of a high-priced stock is too great for many people, a split, so the argument goes, has the effect of "broadening the market." Hence, the announcement of a split should be followed shortly by a rally in the market price.

On the other hand, a substantial body of research has suggested that if a stock split is beneficial to a stock's market price it is because it reflects a recent favorable performance by the company and the expectation of a dividend increase. Moreover, the stock's price tends to rise before, rather than after, the split is announced.

Perform an analysis on the effect, if any, of stock splits on market price by doing the following:

a. Ask your librarian to direct you toward a financial publication that includes information about (1) whether a company's stock has been split at some fairly recent time (you decide what "fairly recent" means) and, if it has been split, (2) on what date the split went into effect, and (3) what was the nature of the split (i.e., two-for-one, three-for-one, three-for-two, etc.).

b. Develop a list of ten stocks that have undergone splits, making certain to record the dates and nature of the splits as well as the names of the stocks.

c. Using appropriate issues of the *Wall Street Journal*, or another newspaper showing stock quotations, find the market price of each stock on your list for the tenth trading day *before* and the tenth trading day *after* the split went into effect.

d. Prepare a list of *before* prices and a list of *after* prices. The *after* prices should be multiplied by two in the case of a two-for-one split, by three in the case of a three-for-one split, and so forth. Also, make certain that each pair of *before* and *after* prices pertains to a single, specific stock. For example, your list will look something like the following:

	Price	
Name of Company	Before Split	After Split
A-to-Z Services	$20\frac{1}{8}$	$26\frac{1}{4}$
Burgermaster	$36\frac{1}{2}$	31

And so forth.

e. Perform the *t* test for paired observations. Let the hypotheses be H_0: $\mu_d = 0$ and H_a: $\mu_d \neq 0$.

f. Do your results appear to support the Wall Street folk wisdom referred to above? The research findings referred to above? Neither? Elaborate.

g. For what population does your conclusion regarding the null hypothesis apply?

h. Is the population normal in shape? Make an intuitive judgment based on a box-and-whiskers chart.

*Group participation is suggested.

TAKE CHARGE*

10.36 Refer to problem 10.35.
   a. Follow the procedures described in problem 10.35, parts *a* through *e*, except that this time list market prices for the thirtieth trading day *before* and the thirtieth trading day *after* the stock split went into effect.
   b. If you did problem 10.35, are your two sets of research findings consistent? If not, how do you explain the inconsistency?

Computer Exercise

 10.37 Refer to Table 11.21.
Have the computer determine observed *t* for the hypotheses

$H_0: \mu_2 = \mu_5$
$H_a: \mu_2 \neq \mu_5$

Computer Exercise

 10.38 Refer to problem 10.30. The manager felt satisfied that the training program had done some good at the one store. Consequently, he decided to put it into effect on a company-wide basis. Table 10.6 shows the results for a random sample of 60 salespeople.

Have the computer do the computational work for a test of paired observations. Test the hypotheses $H_0: \mu_d \geq 0$ and $H_a: \mu_d < 0$, at the .01 level.

Cases for Analysis and Discussion

 ### 10.1 "MID-CITY BANK OF BIG TOWN": PART III

Review Part II of this case (case 9.1 at the end of Chapter 9).

Part I of this case presented results for an experimental group only; this group was made up of credit-card customers who were permitted the lower payback terms during the experimental period. However, management also made use of a control group, a group required to adhere to the existing payback rules. Actually, all credit-card customers who were not selected for membership in the experimental group were automatically members of a very large control group. However, to simplify data gathering and analysis, management selected a random sample of 295 customers from this large group for analysis. Of the 295 customers selected for the control group, there were 88 minimum-payback customers, 79 medial-payback customers, and 128 full-payback customers. The sample results for this group were analyzed along with those of the experimental group for the final 12 months of the experimental period. The results for the control-group sample are presented in Table 10.7. Using information contained within this table and in Tables 8.14 and 8.15, answer the following questions:

1. Why was the use of a control group advisable in this case?

2. Does the evidence obtained from the two samples (i.e., members of the experimental group, on the one hand, and members of the control group on the other) suggest that the population average outstanding balance per customer per month for the experimental group differs from the population average outstanding balance per customer per month for the control group with

*Group participation is suggested.

Table 10.6 Data for Problem 10.38

Salesperson	Number of Complaints		Salesperson	Number of Complaints	
	Before	After		Before	After
1	5	3	31	3	0
2	4	4	32	0	0
3	4	2	33	4	1
4	3	2	34	0	0
5	7	3	35	0	0
6	2	1	36	5	1
7	0	0	37	0	0
8	0	0	38	0	0
9	4	4	39	4	0
10	0	1	40	0	2
11	14	8	41	6	0
12	0	0	42	7	2
13	0	0	43	9	4
14	2	1	44	0	1
15	0	0	45	0	0
16	0	0	46	3	0
17	0	0	47	0	0
18	12	4	48	4	2
19	6	4	49	6	1
20	2	3	50	2	2
21	0	0	51	0	1
22	3	2	52	3	0
23	0	0	53	2	2
24	0	0	54	1	1
25	5	3	55	4	2
26	6	1	56	1	0
27	7	0	57	2	0
28	9	5	58	6	2
29	1	3	59	4	3
30	0	1	60	5	1

respect to (a) minimum-payback customers, (b) medial-payback customers, (c) full-payback customers, and (d) all customers? Test at the .05 level of significance the null hypotheses that no such differences exist (against the alternative hypotheses that such differences do exist).

3. Does the evidence obtained from the two samples suggest that the population average payback per customer per month for the experimental group (Tables 8.14 and 8.15) differs from the population average payback per customer per month for the control group (Table 10.7) with respect to (a) minimum-payback customers, (b) medial-payback customers, (c) full-payback customers, and (d) all customers? Test at the .05 level of significance the null hypotheses that no such differences exist.

4. Does the evidence obtained from the two samples suggest that the population average interest payment per customer per month for the experimental group (Tables 8.14 and 8.15) differs from the population average interest payment per customer per month for the control group (Table 10.7) with respect to (a) minimum-payback customers, (b) medial-payback customers, (c) full-payback customers, and (d) all customers? Test at the .05 level of significance the null hypotheses that no such differences exist.

Table 10.7 Summary Table for "Mid-City Bank's" Control Group

Type of Customer	(1) Average Outstanding Balance per Customer per Month[a]		(2) Average Payback per Customer per Month[a]		(3) (1) − (2) Average Outstanding Balance per Customer per Month After Payback
Minimum payback	$235.23	($212.14)	$35.00	($6.50)	$247.40
Medial payback	275.03	($178.93)	40.67	($32.75)	234.36
Full payback	95.75	($109.50)	95.16	($122.75)	0.59
All	$185.37	($192.00)	$51.73	($88.32)	$ 74.06

Type of Customer	(4) Average Interest Payment per Customer per Month (3) × .015[a]		(5) Population Size N	(6) Sample Size n
Minimum payback	$3.71	($0.65)	25,181	88
Medial payback	3.52	($1.55)	22,606	79
Full payback	0.01	($0.21)	36,628	128
All	$2.05	($0.98)	84,415	295

[a]Sample standard deviations for this measure are shown in parentheses.

 10.2 "MID-CITY BANK OF BIG TOWN": PART IV

Review Part I (case 8.2 at the end of Chapter 8) and Part III (the case preceding this one). Make use of Tables 8.14, 8.15, and 10.7 when answering the following questions:

1. Does the sample evidence suggest that the average outstanding balance per customer per month is higher for the experimental group (Tables 8.14 and 8.15) than for the control group (Table 10.7) with respect to (a) minimum-payback customers, (b) medial-payback customers, (c) full-payback customers, and (d) all customers? Test at the .05 level of significance the null hypotheses that no such "superiority" exists in experimental group over control group.

2. Does the sample evidence suggest that average payback per customer per month is higher for the control group (Table 10.7) than for the experimental group (Tables 8.14 and 8.15) with respect to (1) minimum-payback customers, (2) medial-payback customers, (3) full-payback customers, and (4) all customers? Test at the .05 level of significance the null hypotheses that no such "superiority" exists in control group over experimental group.

3. Does the sample evidence suggest that the average interest payment per customer per month is higher for the experimental group (Tables 8.14 and 8.15) than for the control group (Table 10.7) with respect to (1) minimum-payback customers, (2) medial-payback customers, (3) full-payback customers, and (4) all customers. Test at the .05 level of significance the null hypotheses that no such "superiority" exists in experimental group over control group.

4. Write a general assessment of the effectiveness, or lack of it, of the change to the lower payback requirements insofar as the experimental period is concerned.

This case is continued at the end of Chapter 12.

10.3 "DIXON AND DOXEY ENGINEERING, INC.": PART I

In February of 1975 "Dixon and Doxey Engineering, Inc." ("D&D"), an engineering consulting firm specializing in utilities, was asked by one of its clients to investigate the relative merits of the boiling water reactor (BWR) versus the pressurized water reactor (PWR) for types of nuclear and fossil steam-generating plants. One important aspect of the kind of comparison requested had to do with the problem of construction delay, defined as the difference between the time in which the unit was originally scheduled to begin commercial operation and the time when it did in fact actually begin, or was firmly expected to begin, commercial operations. Expected completions were considered "firm" when at least 70% of the construction had been completed. Data used in this part of the study are presented in Table 10.8.

As shown in Table 10.8, in addition to the PWR-versus-BWR comparison, these categories of steam-generating plants were subcategorized into (1) plants producing 700 megawatts (MW) or less and (2) plants producing more than 700 MW. The reason for this kind of subcategorization system, according to one of D&D's managers who oversaw the research work and who was quoted prior to the outset of the study, was that

> Most, if not all, of the nuclear plants constructed after 1972 have had capacities in excess of 700 MW, whereas prior to 1972 virtually all had capacities of less than 700 MW. So by keeping track of plant size, a time reference is also built into the study. This could be a matter of some importance because the required procedural steps and government regulations were less stringent prior to 1972 than they have been since that time. Even if statistical significance is found, however, determining whether it results from plant size or the licensing procedures in force will be impossible without follow-up research. On the other hand, the absence of statistical significance could be a rather important finding in its own right.

Some background on why construction delays sometimes occur, and why they are matters of considerable concern, is in order. As a result of the public-service nature of their operations and the number of licenses and permits needed before construction of a plant can begin, electric utilities are extremely vulnerable to schedule delays. Such delays can, of course, result from unanticipated external hindrances such as hesitancy on the part of the regulatory agency to grant permits and licenses, as well as from construction difficulties. Because the complexities of building the BWR- and PWR-type plants differ considerably both with respect to licensing requirements and construction techniques, an investigation into the time-delay

Table 10.8 Construction Delay in Months by Type of Reactor

	Pressurized Water Reactor		Boiling Water Reactor	
	700 MW or Less	Over 700 MW	700 MW or Less	Over 700 MW
	3	12	13	1
	36	36	18	18
	5	51	23	19
	6	47	1	39
	16	36	0	22
	16	14	34	46
	8	7	6	38
	26	36	14	39
Total	116	239	109	222

differential was considered a matter of great potential importance. Because accurate cost estimates and cash flows are vital to the public utility's planning of a new plant system, realistic schedules are essential; hence, this analysis.

1. Why do you suppose that in the last sentence of the quoted material the manager said: "On the other hand, the absence of statistical significance could be a rather important finding in its own right?" In what way or ways could such a finding prove important in the present context?

2. For PWRs only, is there a significant difference between average delay for 700 MG or less plants, on the one hand, and over 700 MG plants on the other? Find out using a .10 level of significance.

3. Refer to question 2. Are the population variances equal? Find out using a .10 level of significance.

4. For BWRs only, is there a significant difference between average delay for 700 MG or less plants, on the one hand, and over 700 MG plants on the other? Find out using a .10 level of significance.

5. Refer to question 4. Are the population variances equal? Find out using a .10 level of significance.

6. Ignoring the distinction between 700 MG or less plants and over 700 MG plants, is there a significant difference between average delay for PWRs, on the one hand, and BWRs on the other? Find out using a .10 level of significance.

7. Refer to question 6. Are the population variances equal? Find out using a .10 level of significance.

This case is continued at the end of Chapter 11.

10.4 "GLAMOUR BURGER SYSTEMS": PART III

Refer to Table 3.23, Part V, in case 3.3 at the end of Chapter 3.

1. Does the proportion of respondents giving "good food" as the reason for patronizing a certain restaurant differ significantly between "Glamour Burger" and McDonald's? Glamour Burger and Burger King? Find out using a .01 level of significance.

2. Does the proportion of respondents giving "convenience" as the reason for patronizing a certain restaurant differ significantly between Glamour Burger and McDonald's? Glamour Burger and Burger King? Find out using a .01 level of significance.

3. Does the proportion of respondents giving "fast service" as the reason for patronizing a certain restaurant differ significantly between Glamour Burger and McDonald's? Glamour Burger and Burger King? Find out using a .01 level of significance.

4. Does the proportion of respondents giving "price" as the reason for patronizing a certain restaurant differ significantly between Glamour Burger and McDonald's? Glamour Burger and Burger King? Find out using a .01 level of significance.

5. Write a summary of the findings you arrived at from answering the preceding four questions. Then indicate what implications, if any, your results have with respect to Glamour Burger's efforts to become more competitive with McDonald's and Burger King.

This case is continued at the end of Chapter 13.

11 Introduction to Analysis of Variance

Business may not be the noblest pursuit, but it is true that men are bringing to it some of the qualities that actuate the explorer, scientist, artist: the zest, the open-mindedness, even the disinterestedness, with which the scientific investigator explores some field of pure research.
—Earnest Elmo Calkins

There is no merit in equality unless it be equality with the best.—John Lancaster Spalding

---

What You Should Learn from This Chapter

In sections 9.3, 9.4, and 9.7 of Chapter 9 you learned how to test a null hypothesis about a *single population mean*; for example, H_0: μ = some specified value. In section 10.2 of Chapter 10 you learned how to test a null hypothesis concerned with the *means of two independent populations*; for example, H_0: $\mu_1 - \mu_2 = 0$. This chapter may be viewed as an extension of these earlier sections because analysis of variance is aimed at determining whether the *means of two or more independent populations are equal*. That is, analysis of variance allows us to test a null hypothesis of the following kind—H_0: $\mu_1 = \mu_2 = \mu_3 = \cdots = \mu_c$. As will be demonstrated, the ability to test this kind of null hypothesis is most helpful in connection with both *complex survey-sampling problems* and *designed experiments*.

When you have finished this chapter, you should be able to

1. Work through a standard one-way analysis of variance, that is, an analysis of variance involving a single quantitative or qualitative independent variable.
2. Explain the rationale underlying the F test used in connection with a one-way analysis of variance.
* 3. Apply the method of orthogonal comparisons in connection with a one-way analysis of variance and interpret the results.
* 4. Apply the Duncan multiple-range test in connection with a one-way analysis of variance (where sample sizes are equal) and interpret the results.

Unless you feel quite comfortable with the hypothesis-testing topics treated in Chapters 9 and 10, a quick review of the following sections is advised: 9.1, 9.2, 9.3, 9.4, and 9.7. In addition, a serious review of sections 10.1, 10.2, and 10.4 should prove helpful.

11.1 Introduction

Since both Chapters 11 and 12 are devoted to the subject of analysis of variance, we should waste no time identifying the principal difference between the two chapters: Recall that in Chapter 1 we distinguished between *survey sampling* and *designed experiments*. At that time we stated that with survey-sampling studies the investigator attempts to determine characteristics of a population, or populations, without affecting the elementary units involved. The investigator, in other words, is a spectator, not an active participant. He is a little like a person in the bleachers at a sporting event: He may be wishing for a certain team to win, he may be cheering for that team, he may even be keeping track of earned run averages, yards per carry, or whatever, depending on the kind of sport being witnessed. However, because he is not an actual player, he can do nothing to affect the outcome of the contest.

By contrast, in a designed experiment the investigator is more of a participator—more like a player in a sporting event. In a designed experiment, the investigator varies certain variables systematically with the possible effect of altering the observations taken on the elementary units. If varying a certain variable leads to changes in the observed values, he knows that the variable is important and will presumably put that fact to profitable use at a future time.

In this chapter we treat analysis of variance as a "spectator sport"—one in which the investigator simply observes matters passively. This emphasis in no way makes the present chapter less important than the next one. In business, most statistical analyses are directed toward data obtained from survey-sampling studies. Primarily because of its relatively high cost, actual experimentation is rather rare—though there is some evidence suggesting that its use is increasing, especially in such fields as marketing research and management science. In Chapter 12 our attention shifts to designed experiments. At that time, we treat the subject of analysis of variance as more of a "participation sport."

The following illustrative case is a real one to which one-way analysis of variance was applied. Only the company's name has been changed.

ILLUSTRATIVE CASE 11.1

During a prolonged housing slump, management of Imperial Construction Company, a major builder of single-family residential units with operations throughout the United States, became especially concerned about keeping costs under tight control. One factor among several believed to affect costs importantly was length of time required to sell a completed house. They wondered, for example, whether more time was required to sell a typical high-priced house than a typical low-priced house or whether just the reverse might be the case. In an effort to answer questions such as these, management categorized the company's completed, but unsold, houses into four price classes: (1) low (under $75,000), (2) medium-low ($75,000 but under $90,000), (3) medium-high ($90,000 but under $105,000), and (4) high (over $105,000). (Notice that management *did not assign prices* with a view to seeing what the effect would be; the houses were already built and already sporting price tags, so to speak. Management merely identified which completed houses belonged to each of the price categories.) *A random sample of six houses was selected from each of the populations associated with specified price categories.* The length of time required to sell each house was recorded. The results are shown in Table 11.1.

Before attempting to solve the problem posed by the management of Imperial Construction, we must be clear about what the problem is. Management wishes to answer the question: Does the price of a house have any effect on the length of time required to sell it? But stated in such a manner, the question is much too imprecise for formal statistical testing. Keep in mind that the data to be analyzed represent four separate and independent sets of sample data. Therefore, one way of expressing what management wishes to learn is: Does the number of weeks required to sell a completed house differ from sample to sample (i.e., from column to column in Table 11.1)? If we were to compute, say, the arithmetic mean of each column, we would have four representative values that could be readily compared in order to answer the question. However, although we are gradually becoming more precise in stating

Table 11.1   Data for the Imperial Construction Company Problem

(Figures Represent Number of Weeks Required to Sell House)

	Price Category of House			
	(1) Low	(2) Medium-Low	(3) Medium-High	(4) High
	4	3	3	6
	2	6	8	12
	2	1	6	4
	1	5	5	8
	3	2	4	5
	3	7	7	13
Total	15	24	33	48
Sample mean $\overline{X}_j = 2.5$		4.0	5.5	8.0
Grand mean $\overline{\overline{X}} = 5.0$				

the problem, we still have some distance to go. Management is not really concerned about whether the four sample means per se differ from one another; instead, they are interested in knowing whether the *population means being estimated by the sample means* differ from one another. Therefore, the question being addressed is really: Does $\mu_1 = \mu_2 = \mu_3 = \mu_4$ (where μ_1 is the mean of the population of low-priced houses, μ_2 is the mean of the population of medium-low priced houses, and so forth)? Or, in terms of the null and alternative hypotheses:

$$H_0: \mu_1 = \mu_2 = \mu_3 = \mu_4$$
$$H_a: \mu_1, \mu_2, \mu_3, \text{ and } \mu_4 \text{ are not all equal}$$

An important advantage of analysis of variance over methods introduced in Chapter 10 is that analysis of variance can be employed to evaluate differences among any number of sample means *via a single test*; it does not require several tests on pairs of sample means as would be the case if Z or t tests were used.

11.2 Theoretical Basis of Analysis of Variance

Recall that when discussing the t test for the difference between two population means (section 10.2 of Chapter 10), we noted that we implicitly accept two assumptions when performing the test. These are (1) the populations are both normally distributed and (2) the two populations have equal variances. In analysis of variance, we are dependent on the same two assumptions *but as applied to all relevant populations*. For example, in our illustrative case, we assume that Populations 1, 2, 3, and 4 are all normal in shape and that $\sigma_1^2 = \sigma_2^2 = \sigma_3^2 = \sigma_4^2$. (You should be aware that for analysis of variance these are minimal assumptions. Actually, two additional assumptions are implied in our frequent references in this chapter to random samples from independent populations. More is said about this in section 11.9. Moreover, some advanced kinds of analysis require additional underlying assumptions.)

Let us suppose we are convinced that the assumptions of normal populations and equal population variances are met in the Imperial Construction case. What we now wish to determine is whether $\mu_1 = \mu_2 = \mu_3 = \mu_4$. Suppose that the population means *are* equal; what does that imply about the four populations from which the samples were selected? It implies that they are really exactly the same population, as illustrated in Figure 11.1. In such a case, we would expect each sample to exhibit an amount of variation at least roughly comparable to that of the population. That is, $s_1^2 \cong s_2^2 \cong s_3^2 \cong s_4^2 \cong \sigma^2$. The means of the four samples would be random observations from a normal sampling distribution of sample means for which $\mu_{\bar{X}} = \mu$ and $\sigma_{\bar{X}}^2 = \sigma^2/6$ (the square of the, already familiar, standard error of the mean, $\sigma_{\bar{X}} = \sigma/\sqrt{6}$). Under these circumstances, the unknown population variance σ^2 could properly be estimated from the (1) variance in all 24 sample observations, (2) variance among the sample means, or (3) variance among number of weeks before sale within price categories. Since there would be no between-population variation, any of these approaches to estimating unknown σ^2 would be equally justifiable and would be expected to result in similar numerical values.

Let us now suppose that μ_1, μ_2, μ_3, and μ_4 are all different, a situation depicted in Figure 11.2. (This is just one possibility compatible with the alternative hypothesis. Other possibilities are discussed in section 11.10.) Still assuming normal population shapes and equal population variances, we now have four distinctly different populations and some between-population variation. Because $s_1^2 \cong s_2^2 \cong s_3^2 \cong s_4^2 \cong \sigma_1^2 = \sigma_2^2 = \sigma_3^2 = \sigma_4^2$ still holds, we may still estimate the common population variance σ^2 using the variance within samples. However, we may no longer estimate σ^2 using the variance among sample means or, for that matter, the variance among the 24 sample observations. The presence of between-population variation precludes this. We return to this subject shortly. The main point to remember is that whether (1) the null hypothesis is true or (2) the alternative hypothesis is true makes a great difference to the various ways at our disposal for estimating unknown population variance σ^2 from sample information: If the null hypothesis is true, we may estimate σ^2 in any of three different ways. However, if the alternative hypothesis is true, we may estimate σ^2 in only one way; the other two ways will reflect between-population variation.

Of course, in practice we will not know which hypothesis—the null or the alternative—is true. The test of significance we will use is based on the argument that (1) if variation

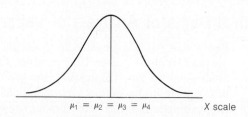

$$\mu_1 = \mu_2 = \mu_3 = \mu_4 \qquad X \text{ scale}$$

Figure 11.1. The single population when the assumptions of (1) normal population shapes and (2) equal population variances are met and, in addition, the population means μ_1, μ_2, μ_3, μ_4 are all equal.

11 Introduction to Analysis of Variance

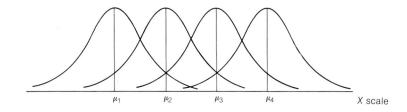

μ_1 μ_2 μ_3 μ_4 X scale

Figure 11.2. The four populations when the assumptions of (1) normal population shapes and (2) equal population variances are met, but the population means are all different.

among sample means is significantly greater than variation within samples, we may infer the presence of between-population variation—which is to say, different population means; (2) if variation among sample means is about on a par with variation within samples, we may infer no between-population variation—that is, no difference among population means.

11.3 Notations to Be Used in a One-Way c-Population Test of Arithmetic Means

We observe that Table 11.1 has four columns. In this chapter we use c to indicate the number of columns; thus, $c = 4$. The number of observations in column (1) will be denoted n_1, the number in column (2), n_2, and so forth. In the present problem, there are $n = n_1 + n_2 + n_3 + n_4 = 6 + 6 + 6 + 6 = 24$ sample observations altogether. Clearly,

$$n = \sum_{\text{all } j} n_j$$

We will let X_{ij} be the number of weeks required to sell the ith house in the jth price class, where $i = 1, 2, 3, 4, 5, 6$ and $j = 1, 2, 3, 4$. For example, (from Table 11.1) the X_{ij} value denoted X_{34} would be 4 because the third house ($i = 3$) in the fourth price class ($j = 4$) required four weeks to sell. We may express the observations in a particular sample (or, equivalently, a particular column) as X_{i1}, X_{i2}, X_{i3}, or X_{i4}. The corresponding sample (column) totals will be denoted

$$\sum_{\text{all } i} X_{i1}, \sum_{\text{all } i} X_{i2}$$

and so forth. The total of all n sample observations will be expressed as

$$\sum_{\text{all } j} \sum_{\text{all } i} X_{ij}$$

When we wish to convey the idea that some sample (column) mean has been divided by n_j to obtain the arithmetic mean of that sample (column), we will simply write $\overline{X}_1$, $\overline{X}_2$, and so forth. Another kind of arithmetic mean of importance to our future work is the grand mean which will be denoted $\overline{\overline{X}}$ (pronounced "X bar bar"); $\overline{\overline{X}}$ is defined as the arithmetic mean of all n sample observations.

11.4 The Meaning of Variance

In statistics the measure we refer to as the *variance* is usually understood to be (for sample data) $s^2 = \Sigma(X - \overline{X})^2/(n - 1)$. In this formula, the $n - 1$ denominator is called *degrees of freedom*, a term conveying the idea that only $n - 1$ of the sample observations are free to vary independently. Because we must know $\overline{X}$ to compute s^2, one of the sample observations is "fixed." (This point is discussed at somewhat greater length in section 8.22 of Chapter 8.)

Although this definition of variance is accurate enough, for analysis-of-variance applications we must broaden the concept: We will define a *sample variance* as

$$\text{Sample variance} = \frac{\text{Sum of squares}}{\text{Degrees of freedom}}$$

"Sum of squares" is simply shorthand for "sum of squared deviations of a set of values from its arithmetic mean" or for the "sum of the summed squared deviations of several sets of values from their respective arithmetic means."

The virtue of this more general definition of sample variance is that it permits a good deal of flexibility in the computational procedures without allowing us to stray too far from the basic notion of variance. In this chapter *sum of squares* will always refer to some kind of sum of squared deviations but it will not always be $\Sigma(X - \overline{X})^2$. Similarly, *degrees of freedom* will always refer to the number of observations constituting the random variable of interest that are free to vary independently, but it will not always be expressed as $n - 1$.

The variance is also frequently referred to as the *mean square*, short for "average squared deviation from the arithmetic mean." Henceforth, we will follow the convention of referring to a sample variance as "mean square." Just think of "sample variance" and "sample mean square" as synonymous.

11.5 Three Ways of Measuring the Variance, or Mean Square, of a Set of Sample Data

Since the term *mean square* refers to the ratio of some kind of sum of squares to some number of degrees of freedom, we may conveniently split the following discussion into two parts: one part pertaining to the measurement of sum of squares and one part pertaining to the measurement of degrees of freedom. We will consider sum of squares first.

Total Sum of Squares

Total sum of squares, denoted $\underset{=}{\text{SSTO}}$, is defined as the sum of the squared deviations from the sample grand mean, $\overline{\overline{X}}$—that is,

$$\text{SSTO} = \sum_{\text{all } j} \sum_{\text{all } i} (X_{ij} - \overline{\overline{X}})^2 \qquad (11.5.1)$$

According to Table 11.2, where the computational steps are demonstrated, the total sum of squares for the Imperial Construction data is 220.

Table 11.2 Calculation of Total Sum of Squares, SSTO, for the Imperial Construction Company Case (Basic Method)

X_{i1}	$\overline{\overline{X}}$	$X_{i1} - \overline{\overline{X}}$	$(X_{i1} - \overline{\overline{X}})^2$	X_{i2}	$\overline{\overline{X}}$	$X_{i2} - \overline{\overline{X}}$	$(X_{i2} - \overline{\overline{X}})^2$
4	5	-1	1	3	5	-2	4
2	5	-3	9	6	5	1	1
2	5	-3	9	1	5	-4	16
1	5	-4	16	5	5	0	0
3	5	-2	4	2	5	-3	9
3	5	-2	4	7	5	2	4
		$\sum_{\text{all } i} (X_{i1} - \overline{\overline{X}})^2 = 43$				$\sum_{\text{all } i} (X_{i2} - \overline{\overline{X}})^2 = 34$	

X_{i3}	$\overline{\overline{X}}$	$X_{i3} - \overline{\overline{X}}$	$(X_{i3} - \overline{\overline{X}})^2$	X_{i4}	$\overline{\overline{X}}$	$X_{i4} - \overline{\overline{X}}$	$(X_{i4} - \overline{\overline{X}})^2$
3	5	-2	4	6	5	1	1
8	5	3	9	12	5	7	49
6	5	1	1	4	5	-1	1
5	5	0	0	8	5	3	9
4	5	-1	1	5	5	0	0
7	5	2	4	13	5	8	64
		$\sum_{\text{all } i} (X_{i3} - \overline{\overline{X}})^2 = 19$				$\sum_{\text{all } i} (X_{i4} - \overline{\overline{X}})^2 = 124$	

$$\text{SSTO} = \sum_{\text{all } j} \sum_{\text{all } i} (X_{ij} - \overline{\overline{X}})^2 = 43 + 34 + 19 + 124 = 220$$

Between-Columns Sum of Squares

> The *between-columns sum of squares*, denoted SSBC, is a measure of variation among sample means. The measure results from the determination of the weighted squared deviations between sample means and the grand mean. The formula is
>
> $$\text{SSBC} = \sum_{\text{all } j} n_j(\overline{X}_j - \overline{\overline{X}})^2 \qquad (11.5.2)$$

The computational steps for obtaining SSBC are demonstrated for our illustrative case in Table 11.3, where SSBC is found to be 99.

Within-Columns Sum of Squares

> The *within-columns sum of squares*, denoted SSWC, is interpreted as a measure of variation resulting from the combined effects of all factors influencing the X_{ij} values except for the independent variable under study (prices of completed houses in our example). The formula is
>
> $$\text{SSWC} = \sum_{\text{all } j} \sum_{\text{all } i} (X_{ij} - \overline{X}_j)^2 \qquad (11.5.3)$$

The computational steps for this measure are demonstrated in Table 11.4, where SSWC is found to be 121.

If our computational work is accurate, the sum of SSBC and SSWC should be exactly equal to SSTO. We have seen that SSBC = 99 and that SSWC = 121. Reassuringly, these results sum to 220, the same value obtained in Table 11.2 for the total sum of squares. Figure 11.3 shows the manner in which we break total sum of squares—part (*a*) of the figure—and total degrees of freedom—part (*b*) of the figure—into two parts: one part pertaining to between columns and the other to within columns.

Table 11.3 Calculation of Between-Columns Sum of Squares, SSBC, for the Imperial Construction Company Case (Basic Method)

(1) n_j	(2) $\overline{X}_j$	(3) $\overline{\overline{X}}$	(4) $\overline{X}_j - \overline{\overline{X}}$	(5) $(\overline{X}_j - \overline{\overline{X}})^2$	(6) $n_j(\overline{X}_j - \overline{\overline{X}})^2$
6	2.5	5.0	−2.5	6.25	37.50
6	4.0	5.0	−1.0	1.00	6.00
6	5.5	5.0	0.5	0.25	1.50
6	8.0	5.0	3.0	9.00	54.00
				Total	99.00 = SSBC

Table 11.4 Calculation of Within-Columns Sum of Squares, SSWC, for the Imperial Construction Company Case (Basic Method)

X_{i1}	$\overline{X}_1$	$X_{i1} - \overline{X}_1$	$(X_{i1} - \overline{X}_1)^2$	X_{i2}	$\overline{X}_2$	$X_{i2} - \overline{X}_2$	$(X_{i2} - \overline{X}_2)^2$
4	2.5	1.5	2.25	3	4.0	−1.0	1.00
2	2.5	−0.5	0.25	6	4.0	2.0	4.00
2	2.5	−0.5	0.25	1	4.0	−3.0	9.00
1	2.5	−1.5	2.25	5	4.0	1.0	1.00
3	2.5	0.5	0.25	2	4.0	−2.0	4.00
3	2.5	0.5	0.25	7	4.0	3.0	9.00
			5.50				28.00

X_{i3}	$\overline{X}_3$	$X_{i3} - \overline{X}_3$	$(X_{i3} - \overline{X}_3)^2$	X_{i4}	$\overline{X}_4$	$X_{i4} - \overline{X}_4$	$(X_{i4} - \overline{X}_4)^2$
3	5.5	−2.5	6.25	6	8.0	−2.0	4.00
8	5.5	2.5	6.25	12	8.0	4.0	16.00
6	5.5	0.5	0.25	4	8.0	−4.0	16.00
5	5.5	−0.5	0.25	8	8.0	0.0	0.00
4	5.5	−1.5	2.25	5	8.0	−3.0	9.00
7	5.5	1.5	2.25	13	8.0	5.0	25.00
			17.50				70.00

$$\text{SSWC} = \sum_{\text{all } j} \sum_{\text{all } i} (X_{ij} - \overline{X}_j)^2 = 5.50 + 28.00 + 17.50 + 70.00 = 121$$

Degrees of Freedom

Because each sum of squares will have associated with it some number of degrees of freedom, the logical next step in our analysis is to determine how many degrees of freedom are associated with our various SS values.

Total Degrees of Freedom. We just saw that the equation for computing the total sum of squares is

$$\text{SSTO} = \sum_{\text{all } j} \sum_{\text{all } i} (X_{ij} - \overline{\overline{X}})^2$$

We noted in Chapter 4 (section 4.9) that a peculiar quality of the arithmetic mean is that the sum of the deviations from it is always 0—that is, $\Sigma(X - \overline{X}) = 0$. This quality of the mean tells us, albeit subtly, that if we know $n - 1$ of the X values and the sample arithmetic mean of all the X values, we know immediately what the nth X value must be (see section 8.22). Thus, we say that $n - 1$ of the X values are free to vary independently and one is fixed or constrained.

Clearly, since the grand mean $\overline{\overline{X}}$ in the total sum of squares equation is simply $\overline{X}$ dressed up a little so that it can be distinguished from a column mean, the number of degrees of freedom associated with SSTO must be $n - 1$. In our example, $df_T = n - 1 = 24 - 1 = 23$. In more general terms:

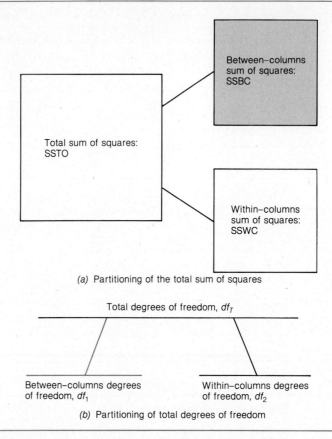

(a) Partitioning of the total sum of squares

Total degrees of freedom, df_T

Between–columns degrees of freedom, df_1

Within–columns degrees of freedom, df_2

(b) Partitioning of total degrees of freedom

Figure 11.3. In one-way analysis of variance, both the total sum of squares and the total degrees of freedom are broken down into parts—one part for between columns and one part for within columns.

In analysis of variance, we determine the appropriate number of degrees of freedom by noting the number of observations, whether they are X values or $\overline{X}$ values, making up the random variable of interest and then subtracting the number of sample means from which deviations are obtained.

Between-Columns Degrees of Freedom. The number of degrees of freedom associated with the between-columns sum of squares, SSBC, is $c - 1$, where c is the number of columns (different independent random samples). The reason for this may readily be seen from a reexamination of Equation (11.5.2). We see that

$$\text{SSBC} = \sum_{\text{all } j} n_j(\overline{X}_j - \overline{\overline{X}})^2$$

The column means are used as representatives of their respective columns and do themselves constitute the random variable of interest. The grand mean $\overline{\overline{X}}$ is subtracted from each column mean $\overline{X}_j$. We conclude, therefore, that there are c column means *potentially* free to vary independently, but because

$$\sum_{\text{all } j} n_j(\overline{X}_j - \overline{\overline{X}}) = 0$$

only $c - 1$ column means are *really* free to vary independently. The act of computing deviations from the grand mean $\overline{\overline{X}}$ "fixes" one of the column means. For our example, the degrees of freedom between columns df_1, as we will denote it, is $df_1 = c - 1 = 4 - 1 = 3$.

Within-Columns Degrees of Freedom.   The number of degrees of freedom associated with within-columns sum of squares, SSWC, is equal to $(n_1 - 1) + (n_2 - 1) + \cdots + (n_c - 1) = n - c$. To see why this is so, let us take a close look at Equation (11.5.3). This equation tells us that

$$\text{SSWC} = \sum_{\text{all } j} \sum_{\text{all } i} (X_{ij} - \overline{X}_j)^2$$

Notice that deviations are obtained around the arithmetic mean of each column. Therefore, each column mean "fixes" one X value in its column because

$$\sum_{\text{all } i} (X_{ij} - \overline{X}_j) = 0$$

Therefore, if we sum the $n_j - 1$ values for all c columns, we get $n - c$ degrees of freedom. In our example, df_2, as we will call it, is found to be $df_2 = n - c = 24 - 4 = 20$.

As we will see shortly, total mean square, between-columns mean square, and within-columns mean square are obtained by dividing total sum of squares by $df_T = n - 1$, between-columns sum of squares by $df_1 = c - 1$, and within-columns sum of squares by $df_2 = n - c$, respectively.

11.6 Shortcut Formulas for Sums of Squares

The sums of squares formulas introduced in the preceding section have the important advantages of (1) keeping the underlying logic in plain view and (2) providing a basis for reasoning through the determination of degrees of freedom. Unfortunately, they are somewhat inconvenient to work with, especially when the number of observations is very large. Frequently, the following, more convenient, formulas are used instead:

$$\text{SSTO} = \sum_{\text{all } j} \sum_{\text{all } i} X_{ij}^2 - \frac{\left(\sum_{\text{all } j} \sum_{\text{all } i} X_{ij} \right)^2}{n} \qquad (11.6.1)$$

$$SSBC = \sum_{\text{all } j} \frac{\left(\sum_{\text{all } i} X_{ij} \right)^2}{n_j} - \frac{\left(\sum_{\text{all } j} \sum_{\text{all } i} X_{ij} \right)^2}{n} \qquad (11.6.2)$$

$$SSWC = \sum_{\text{all } j} \sum_{\text{all } i} X_{ij}^2 - \sum_{\text{all } j} \frac{\left(\sum_{\text{all } i} X_{ij} \right)^2}{n_j} \qquad (11.6.3)$$

Application of these alternative formulas is shown in Tables 11.5, 11.6, and 11.7. (These formulas are derived algebraically from the basic ones by methods similar to those demonstrated in section 4.10 of Chapter 4.)

11.7 The ANOVA Table

A helpful organizing device is the analysis-of-variance table—or ANOVA table for short. An ANOVA table is shown in very general form in Part I of Table 11.8. An ANOVA table for the Imperial Construction problem in particular is shown in Part II of this same table. In

Table 11.5 Calculation of Total Sum of Squares, SSTO, for the Imperial Construction Company Case (Shortcut Method)

X_{i1}	X_{i1}^2	X_{i2}	X_{i2}^2	X_{i3}	X_{i3}^2	X_{i4}	X_{i4}^2
4	16	3	9	3	9	6	36
2	4	6	36	8	64	12	144
2	4	1	1	6	36	4	16
1	1	5	25	5	25	8	64
3	9	2	4	4	16	5	25
3	9	7	49	7	49	13	169
15 $\sum_{\text{all } i} X_{i1}^2 = 43$		24 $\sum_{\text{all } i} X_{i2}^2 = 124$		33 $\sum_{\text{all } i} X_{i3}^2 = 199$		48 $\sum_{\text{all } i} X_{i4}^2 = 454$	

$$\sum_{\text{all } j} \sum_{\text{all } i} X_{ij} = 15 + 24 + 33 + 48 = 120$$

$$\frac{\left(\sum_{\text{all } j} \sum_{\text{all } i} X_{ij} \right)^2}{n} = \frac{(120)^2}{24} = 600$$

$$\sum_{\text{all } j} \sum_{\text{all } i} X_{ij}^2 = 43 + 124 + 199 + 454 = 820$$

Therefore, SSTO = 820 − 600 = 220.

Table 11.6 Calculation of Between-Columns Sum of Squares, SSBC, for the Imperial Construction Company Case (Shortcut Method)

	$\sum\limits_{\text{all } i} X_{ij}$	$\left(\sum\limits_{\text{all } i} X_{ij}\right)^2$	$\dfrac{\left(\sum\limits_{\text{all } i} X_{ij}\right)^2}{n_j}$
	15	225	37.5
	24	576	96.0
	33	1089	181.5
	48	2304	384.0
Total	120		699.0

$$\text{SSBC} = \sum\limits_{\text{all } j} \frac{\left(\sum\limits_{\text{all } i} X_{ij}\right)^2}{n_j} - \frac{\left(\sum\limits_{\text{all } j}\sum\limits_{\text{all } i} X_{ij}\right)^2}{n}$$

$$= 699.0 - \frac{(120)^2}{24} = 699.0 - 600.0 = 99$$

column (1) of both parts, we find listed the sources of variation. For a one-way problem like the one we are now addressing, this list includes only between columns, within columns, and total. In column (2) are listed the sums of squares associated with each source of variation. In column (3) are shown the corresponding degrees of freedom. When the sums of squares in column (2) are divided by the corresponding degrees of freedom in column (3), the results are the between-columns and within-columns mean squares. (Total mean square is

Table 11.7 Calculation of Within-Columns Sum of Squares, SSWC, for the Imperial Construction Company Case (Shortcut Method)

	X_{i1}	X_{i1}^2	X_{i2}	X_{i2}^2	X_{i3}	X_{i3}^2	X_{i4}	X_{i4}^2
	4	16	3	9	3	9	6	36
	2	4	6	36	8	64	12	144
	2	4	1	1	6	36	4	16
	1	1	5	25	5	25	8	64
	3	9	2	4	4	16	5	25
	3	9	7	49	7	49	13	169
Total	15	43	24	124	33	199	48	454

$$\text{SSWC} = \sum\limits_{\text{all } j}\sum\limits_{\text{all } i} X_{ij}^2 - \sum\limits_{\text{all } j} \frac{\left(\sum\limits_{\text{all } i} X_{ij}\right)^2}{n_j}$$

$$= 43 + 124 + 199 + 454 - 699 = 121$$

Table 11.8 General and Specific ANOVA Formats for a One-Way Analysis

Part I: General Structure

(1) Source of Variation	(2) Sum of Squares	(3) Degrees of Freedom	(4) Mean Square	(5) Observed F	(6) Critical F
Between columns	$SSBC = \sum\limits_{all\ j} \dfrac{\left(\sum\limits_{all\ i} X_{ij}\right)^2}{n_j} - \dfrac{\left(\sum\limits_{all\ j}\sum\limits_{all\ i} X_{ij}\right)^2}{n}$	$df_1 = c - 1$	$\dfrac{SSBC}{df_1}$	$\dfrac{MSBC}{MSWC}$	From Appendix Table 6 —
Within columns	$SSWC = \sum\limits_{all\ j}\sum\limits_{all\ i} X_{ij}^2 - \sum\limits_{all\ j} \dfrac{\left(\sum\limits_{all\ i} X_{ij}\right)^2}{n_j}$	$df_2 = n - c$	$\dfrac{SSWC}{df_2}$		
Total	$SSTO = \sum\limits_{all\ j}\sum\limits_{all\ i} X_{ij}^2 - \dfrac{\left(\sum\limits_{all\ j}\sum\limits_{all\ i} X_{ij}\right)^2}{n}$	$df_T = n - 1$	—	—	—

Part II: ANOVA Table for the Imperial Construction Company Problem

(1) Source of Variation	(2) Sum of Squares	(3) Degrees of Freedom	(4) Mean Square	(5) Observed F	(6) .05 Level Critical F
Between columns	99	3	33.00	5.45[a]	3.10
Within columns	121	20	6.05	—	—
Total	220	23	—	—	—

[a] Indicates significant F ratio.

not normally computed.) A glance at columns (4) and (5) of Table 11.8 will provide you with a preview of how the analysis of variance is completed.

11.8 The F Statistic and Test of Significance

The manner of calculating the F statistic and subjecting it to a test of significance flows logically from points made in section 11.2:

> If the null hypothesis regarding the equality of several population means is true, the between-columns mean square (a measure of variation between samples) and within-columns mean square (a measure of variation within samples) would be expected to be approximately equal because both reflect the same thing—chance errors of sampling.

On the other hand,

> If the null hypothesis is false and the population means are not equal, the between-columns mean square would have to be viewed as reflecting two sources of variation: (1) chance sampling error and (2) the effect of prices charged for the houses (more generally, the effect of the independent variable under study). The within-columns mean square, on the other hand, would still reflect only one source of variation—chance sampling error.

It follows from the above that a way to compare MSBC and MSWC (between-columns mean square and within-columns mean square, respectively) is by computing their ratio *using MSBC as the numerator*. This is an F statistic much like the one we worked with in section 10.4 of the preceding chapter. Hence, observed F = MSBC/MSWC. If observed F exceeds 1 by a great enough margin, we are led to reject the null hypothesis on the grounds that MSBC is reflecting some between-population variation. But how much greater than 1 must the observed F ratio be? To answer this question, we rely on the same theoretical sampling distribution of F introduced in Chapter 10. More specifically, we find $df_1 = c - 1 = 3$ along the top margin of Appendix Table 6 and $df_2 = n - c = 20$ along the left-hand margin. Simultaneously projecting downward from 3 and across from 20, we find the numbers 3.10 (light type) and 4.94 (bold type). The 3.10 is the critical value associated with $\alpha = .05$ and 3 and 20 degrees of freedom, whereas 4.94 is the critical value associated with $\alpha = .01$ and 3 and 20 degrees of freedom. If we take our α value to be, say, .05, the associated decision rule is

Decision Rule
1. If observed $F \leq 3.10$, accept H_0.
2. If observed $F > 3.10$, reject H_0.

Since observed F is found to be 5.45 (Table 11.8, Part II) and critical $F_{.05(3,20)}$ is 3.10, we conclude against the null hypothesis—that is, we conclude that the population means are not equal. The price charged for the houses does have a bearing on the speed with which they sell.

Notice that this test is performed at the right-hand tail of the theoretical F distribution only. This will be the case throughout this chapter and the next one; therefore, the α value associated with a critical value in our prepared F table need not be doubled as was true of the test described in Chapter 10.

11.9 Note on Test Assumptions

Although only two underlying assumptions have been emphasized to this point, as noted in section 11.2, one-way analysis of variance is really dependent on these four assumptions:

1. Samples are random samples from their respective populations.
2. The populations are independent.
3. Populations have normal shapes.
4. Population variances are equal.

Let us consider these assumptions in turn as they pertain to the Imperial Construction problem:

1 and 2. These assumptions are usually under the control of the investigator and have to do with the manner in which the populations are defined and the elementary sample units are selected. In the Imperial Construction problem, four random samples were drawn separately from four nonoverlapping populations. There seems to be no need to worry about these assumptions.

3. Although examination of sample data appearing in Table 11.1 gives no compelling reason for doubting that the population shapes are normal, in truth one cannot really establish normality from such small sample sizes. (For larger sample sizes, visual inspection of the sample shapes, aided by box-and-whiskers charts, if desired, will usually suffice. If sample sizes are quite large, formal statistical testing methods may be used—specifically, the chi-square or Kolmogorov-Smirnov tests of goodness of fit to a normal curve. One or both of these methods are covered in many basic statistics textbooks.) When sample sizes are small and one has reason to question the validity of this third assumption, use of the Kruskal-Wallis test described in Chapter 13 in lieu of an F test is advised. This is a weaker test than F in that its power to reject an incorrect null hypothesis is lower. On the other hand, it is also dependent on weaker, and more easily met, assumptions.

4. Of the four assumptions, the one having to do with equal population variances is the one most likely violated in the Imperial Construction case. A quick way to size up the validity, or lack of it, of this assumption is to construct charts of residual values like those shown in Figure 11.4. These charts show the $X_{ij} - \bar{X}_1$ values, the $X_{ij} - \bar{X}_2$ values, and so

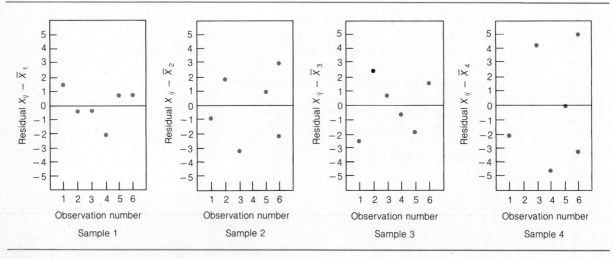

Figure 11.4. Residual plots for the Imperial Construction case.

forth. When the basic procedure for computing SSWC is used, this information will be already available. For example, in Table 11.4 these deviations from the sample means are seen to be an integral part of the computational work. When the shortcut method is used, the residual values must be computed separately.

Notice that, roughly speaking, the variation in the residuals increases as we scan Figure 11.4 from left to right.

Fortunately, this assumption of equal population variances can be violated without serious risk to the validity of the F test *provided that the sample sizes are equal*, as they are in our illustrative case. When the n_j values are different, violation of this assumption can have very serious consequences on the trust we place in the F test.

> For this reason and others yet to be cited, the moral is: If you have a choice, use equal-size samples.

If the sample sizes must be different, it is usually a good idea to perform a formal test of the hypothesis H_0: $\sigma_1^2 = \sigma_2^2 = \sigma_3^2 = \cdots = \sigma_c^2$. Several ways of testing this hypothesis are available—one good one being Bartlett's test for homogeneity of variances—and are covered in virtually all specialized books on experimental designs. If the n_j's must be different and serious doubt exists about the validity of the fourth assumption, then, again, the Kruskal-Wallis test should probably be substituted for the F test.

Under the present circumstances, we conclude that we probably need not be concerned about the validity of the F test in the Imperial Construction case.

＊11.10 Identifying Sources of Significance: Multiple-Comparison Methods

The discovery of significant between-columns variation tells us that the population means are not all equal. But this is often only the beginning of a complete evaluation of the data. The population means may be unequal in a rather wide variety of ways each of which might have unique implications from a decision-making standpoint. Some possible ways that the population means may differ are illustrated in Figure 11.5.

If the investigator (and the decision maker, if they are different people) is to enjoy the full benefits that analysis of variance has to offer, he must ordinarily seek more detailed information regarding differences between pairs, or groupings, of specific means. In recent years quite a number of formal methods, called *multiple-comparison* methods, for doing that have been developed. In this section we will deal with two of these—namely, orthogonal comparisons and the Duncan multiple-range test. The choice of these two methods was motivated by the desire to expose you to one more-or-less representative *a priori* method and one more-or-less representative *a posteriori* method. It should not be construed as a commentary on (or as an entry in the ongoing debate about) the merits of these methods vis-à-vis alternative available methods.[1] The method of orthogonal comparisons is called an *a priori multiple-comparison method* because it requires that the specific ways of following up on a

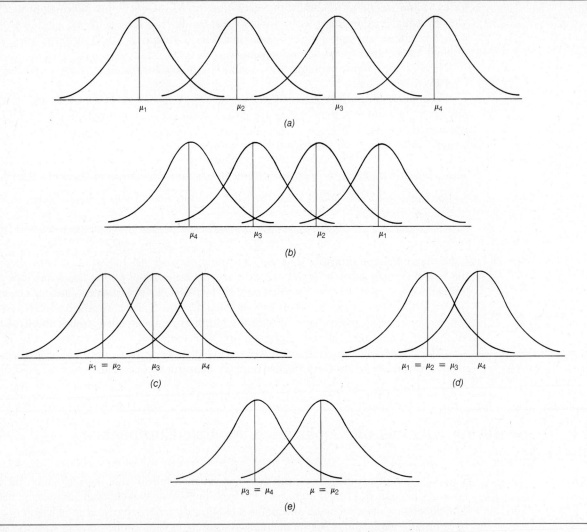

Figure 11.5. Just a few population situations that could lead to rejection of the null hypothesis H_0: $\mu_1 = \mu_2 = \mu_3 = \mu_4$. (a) All population means are different and $\mu_1 < \mu_2 < \mu_3 < \mu_4$. (b) All population means are different and $\mu_1 > \mu_2 > \mu_3 > \mu_4$. (c) Population means μ_1 and μ_2 are equal and both are less than μ_3 which, in turn, is less than μ_4. (d) Population means μ_1, μ_2, μ_3 are equal and less than μ_4. (e) Population means μ_3 and μ_4 are equal and less than μ_1 and μ_2, which are themselves equal.

rejected null hypothesis be spelled out before data are gathered and organized. This kind of prior planning serves to ensure that the probability of a Type I error, α, set earlier will still be valid. The Duncan multiple-range test is called an *a posteriori method* because it permits considerable relatively unrestricted "data snooping" after the main F test has been performed. However, with this approach the α value is altered unpredictably because the comparisons were not identified at random but are instead based on observed results.

Orthogonal Comparisons

The method of orthogonal comparisons attempts to break SSBC and df_1 into parts associated with specific subhypotheses, as we will call them. Let us begin by defining the central term *comparison*. A comparison between the average of two sample means, $\overline{X}_1$ and $\overline{X}_2$, may be written $\overline{X}_1 - \overline{X}_2$; a comparison between the average of two sample means, $\overline{X}_1$ and $\overline{X}_2$, and a third sample mean, $\overline{X}_3$, may be written $[(\overline{X}_1 + \overline{X}_2)/2] - \overline{X}_3$; a comparison between the sum of two sample means, $\overline{X}_1$ and $\overline{X}_2$, and the sum of two other sample means, $\overline{X}_3$ and $\overline{X}_4$, may be written $(\overline{X}_1 + \overline{X}_2) - (\overline{X}_3 + \overline{X}_4)$. Thus, a comparison is a *difference*—a difference between two sample means or two groupings of sample means.

The general expression for any comparison among a set of sample means can be written

$$\text{Com } i = m_{i1}\overline{X}_1 + m_{i2}\overline{X}_2 + m_{i3}\overline{X}_3 + \cdots + m_{ic}\overline{X}_c \qquad (11.10.1)$$

where m_{i1} is read "multiplier 1 of Comparison i," m_{i2} is read "multiplier 2 of Comparison i," and so forth. It is also necessary for

$$\sum_{\text{all } j} m_{ij} = 0$$

The m_{ij} values are simply coefficients or multipliers expressing what is done to specific sample means when comparing them with other sample means. Their role will be demonstrated shortly. As for the term *orthogonal comparisons*:

We say that two comparisons

$$\text{Com } 1 = m_{11}\overline{X}_1 + m_{12}\overline{X}_2 + m_{13}\overline{X}_3 + \cdots + m_{1c}\overline{X}_c$$

and

$$\text{Com } 2 = m_{21}\overline{X}_1 + m_{22}\overline{X}_2 + m_{23}\overline{X}_3 + \cdots + m_{2c}\overline{X}_c$$

are orthogonal if

$$\frac{m_{11}m_{21}}{n_1} + \frac{m_{12}m_{22}}{n_2} + \frac{m_{13}m_{23}}{n_3} + \cdots + \frac{m_{1c}m_{2c}}{n_c} = 0 \qquad (11.10.2)$$

If there are more than two comparisons, all pairs of comparisons must meet the above requirement. This condition is known as *mutual orthogonality*.

Let us suppose that the management of Imperial Construction *does* require more detailed information than that provided by the main F test demonstrated in section 11.8. More specifically, they wish to test three subhypotheses. (The term *subhypotheses* is used here simply to convey the idea that the null hypothesis H_0: $\mu_1 = \mu_2 = \mu_3 = \mu_4$, which was rejected when the main F test was performed, may be viewed as containing several null hypotheses of more limited scope.) The three null subhypotheses that Imperial management wishes to explore are the following:

$H_{0_{\text{sub}\,1}}$: Low-priced and medium-low-priced houses are sold with equal speed, or $\mu_1 - \mu_2 = 0$.

$H_{0_{\text{sub}\,2}}$: High-priced and medium-high-priced houses are sold with equal speed, or $\mu_3 - \mu_4 = 0$.

$H_{0_{\text{sub}\,3}}$: Lower-priced houses (including both low-priced and medium-low-priced) and higher-priced houses (including both high-priced and medium-high-priced) are sold with equal speed, or $(\mu_1 + \mu_2) - (\mu_3 + \mu_4) = 0$.

Of course, the alternative hypotheses associated with these null subhypotheses could be formed simply by substituting "not equal" for "equal" in the symbolic versions.

Notice that there is in this example one null subhypothesis for each degree of freedom between columns. That is, $df_1 = 3$ and we have three subhypotheses. The method of orthogonal comparisons requires this kind of correspondence. *Each between-columns degree of freedom must have one and only one subhypothesis associated with it.* Application of this method will, if all rules are followed, result in a clean breakdown of SSBC into parts—one part for each subhypothesis.

The comparisons implied by the three subhypotheses listed above are summarized in Table 11.9.

Individual lines in Table 11.9 are read as follows:

For Comparison 1: $(+1)(\overline{X}_1) + (-1)(\overline{X}_2) + (0)(\overline{X}_3) + (0)(\overline{X}_4) = \overline{X}_1 - \overline{X}_2$

For Comparison 2: $(0)(\overline{X}_1) + (0)(\overline{X}_2) + (+1)(\overline{X}_3) + (-1)(\overline{X}_4) = \overline{X}_3 - \overline{X}_4$

For Comparison 3: $(+1)(\overline{X}_1) + (+1)(\overline{X}_2) + (-1)(\overline{X}_3) + (-1)(\overline{X}_4) = (\overline{X}_1 + \overline{X}_2) - (\overline{X}_3 + \overline{X}_4)$

Are the three comparisons orthogonal? We can find out by employing Equation (11.10.2) three times:

$$\text{For Com 1 and Com 2: } \frac{(1)(0)}{6} + \frac{(-1)(0)}{6} + \frac{(0)(1)}{6} + \frac{(0)(-1)}{6} = 0$$

Table 11.9 Coefficients (m_{ij} Values) for the Three Comparisons Suggested by $H_{0_{\text{sub}\,1}}$, $H_{0_{\text{sub}\,2}}$, and $H_{0_{\text{sub}\,3}}$

	Sample Mean			
Comparison	$\overline{X}_1$	$\overline{X}_2$	$\overline{X}_3$	$\overline{X}_4$
1	+1	−1	0	0
2	0	0	+1	−1
3	+1	+1	−1	−1

$$\text{For Com 1 and Com 3: } \frac{(1)(1)}{6} + \frac{(-1)(1)}{6} + \frac{(0)(-1)}{6} + \frac{(0)(-1)}{6} = 0$$

$$\text{For Com 2 and Com 3: } \frac{(0)(1)}{6} + \frac{(0)(1)}{6} + \frac{(1)(-1)}{6} + \frac{(-1)(-1)}{6} = 0$$

We see by the zeroes that we have mutual orthogonality. That is our ticket to testing the three null subhypotheses. Absence of mutual orthogonality would have meant that all that follows would be improper.

The sum of squares associated with orthogonal comparisons may be determined by the following equation:

$$SS_{\text{Com } i} = \frac{(\text{Com } i)^2}{\dfrac{(m_{i1})^2}{n_1} + \dfrac{(m_{i2})^2}{n_2} + \cdots + \dfrac{(m_{ic})^2}{n_c}} \qquad (11.10.3)$$

Using Equation (11.10.3) and the sample means first shown in Table 11.1, the sums of squares for comparisons are

$$SS_{\text{Com } 1} = \frac{[(1)(2.5) + (-1)(4.0)]^2}{\dfrac{(1)^2}{6} + \dfrac{(-1)^2}{6}} = 6.75$$

$$SS_{\text{Com } 2} = \frac{[(1)(5.5) + (-1)(8.0)]^2}{\dfrac{(1)^2}{6} + \dfrac{(-1)^2}{6}} = 18.75$$

$$SS_{\text{Com } 3} = \frac{[(1)(2.5) + (1)(4.0) + (-1)(5.5) + (-1)(8.0)]^2}{\dfrac{(1)^2}{6} + \dfrac{(1)^2}{6} + \dfrac{(-1)^2}{6} + \dfrac{(-1)^2}{6}} = 73.50$$

We may check the accuracy of our calculations by summing the resulting sums of squares for comparisons: $6.75 + 18.75 + 73.50 = 99.00$. The between-columns sum of squares was earlier found to be 99. When orthogonality exists, the sum of the comparison sums of squares will be exactly equal to the between-columns sum of squares. That is essentially what orthogonality means: When the total is broken down into parts, the parts sum to the total—not to more than nor less than the original total. If our checks for orthogonality—the three applications of Equation (11.10.2) (see Figure 11.6)—had failed, or if only one or two of them had failed, the comparison sums of squares would not sum to the value of SSBC and the F tests demonstrated below could not be employed.

To test each comparison for significance, we need merely to divide its mean square by the within-columns mean square, SSWC, and compare the resulting F ratio with critical $F_{.05(1, 20)}$ (assuming continued use of a .05 level of significance). This has been done in Table 11.10.

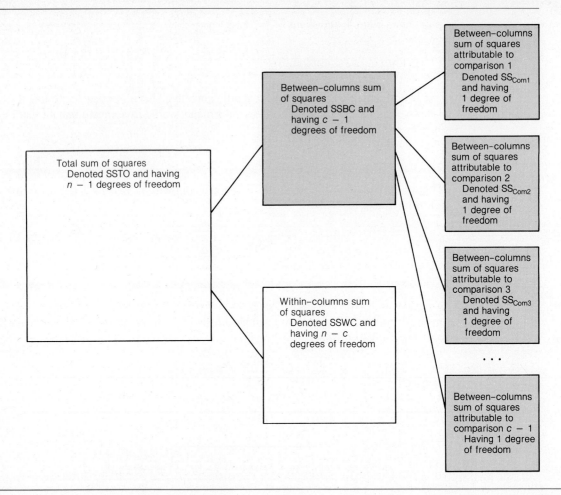

Figure 11.6. Hypothetical decomposition of the total sum of squares into (1) between-columns sum of squares and (2) within-columns sum of squares, and the decomposition of between-columns sum of squares into $c - 1$ parts—one part attributable to comparison 1, another part to comparison 2, and so forth, until all between-columns degrees of freedom have been employed.

Table 11.10 reveals that (1) selling a low-priced house requires the same length of time, on the average, as selling a medium-low-priced house, (2) selling a high-priced house requires the same length of time, on the average, as selling a medium-high-priced house, and (3) a higher-priced house, as defined in $H_{0_{\text{sub }3}}$, requires a different length of time, on the average, to sell than a lower-priced house, as defined in $H_{0_{\text{sub }3}}$. According to the sample means listed in Table 11.1, higher-priced houses take longer to sell than lower-priced houses. Figure 11.7 portrays what appears to be the population situation.

The method of orthogonal comparisons is an elegant and sometimes highly useful analytical procedure. However, it does possess some limitations. First, when the number of degrees of freedom between columns—and, hence, the number of subhypotheses—is large,

Table 11.10 ANOVA Table for the One-Way Problem Including Tests of Three Orthogonal Comparisons

(1) Source of Variation	(2) Sum of Squares	(3) Degrees of Freedom	(4) Mean Square	(5) F	(6) .05 Level Critical F
Between columns	99.00	3	33.00	5.45[a]	3.10
Com 1	(6.75)	(1)	6.75	1.12	4.35
Com 2	(18.75)	(1)	18.75	3.10	4.35
Com 3	(73.50)	(1)	73.50	12.15[a]	4.35
Within columns	121.00	20	6.05	—	—
Total	220.00	23	—	—	—

[a]Indicates significant F ratio.

the investigator may find that she cannot set up the multipliers, the m_{ij} values, in such a way that they are simultaneously (1) compatible with *all* the stated subhypotheses and (2) contributory to a mutually orthogonal system of comparisons. Unfortunately, when such is the case, it is the subhypotheses that must be modified; there can be no compromising the requirement of mutual orthogonality. Sometimes resourceful analysts find that they can rank the subhypotheses according to their relative importance. They will then leave the most important of these alone and change some or all of the others for the baldly pragmatic purpose of achieving a mutually orthogonal system.

Second, if sample sizes are unequal, the task of achieving a mutually orthogonal system of comparisons may be difficult or impossible even for a small number of comparisons. Here we have another good reason why the sample sizes should be equal, if at all possible.

Third, as noted earlier, the method of orthogonal comparisons is strictly legitimate in situations where comparisons have been planned prior to data collection and organization.

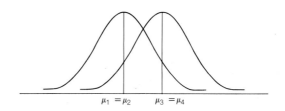

$\mu_1 = \mu_2$ $\mu_3 = \mu_4$

Figure 11.7. The population situation for the Imperial Construction Company problem, as suggested by F tests or orthogonal comparisons. Population means μ_1 and μ_2 are equal and lower than population means μ_3 and μ_4, which are themselves equal.

The Duncan Multiple-Range Test

Frequently, investigators will want to evaluate the differences among sample means after the main analysis of variance has been completed even though they did not commit themselves to any such specific tests beforehand. Until somewhat recently we had no really satisfactory method available for doing this. Today, several good methods are available. The one we will focus on here is called the (new) *Duncan multiple-range test*. Use of this procedure is limited to one-way analyses such as the Imperial Construction Company problem. We will limit our discussion of the technique to situations where $n_1 = n_2 = n_3 = \cdots = n_c$ although a similar methodology does exist for use in cases where the sample sizes are unequal.[2]

The steps involved in performing the Duncan multiple-range test are as follows:

---

1. Arrange all sample means in order from smallest to largest.
2. Compute the standard error of the sample means $s_{\overline{X}_j}$.
3. Obtain critical values from Appendix Table 11.
4. Multiply the standard error of the sample means $s_{\overline{X}_j}$ obtained in Step 2 by the critical values found in Step 3 to get the *least significant ranges*, or *LSR*.
5. Set up a table of differences between pairs of sample means.
6. Test the differences between all pairs of sample means for significance.

Let us apply these steps to the Imperial Construction problem. The data to be used are from Tables 11.1 and 11.8.

$$\overline{X}_1 = 2.5 \qquad \overline{X}_3 = 5.5 \qquad \text{MSWC} = 6.05 \qquad df_2 = 20$$
$$\overline{X}_2 = 4.0 \qquad \overline{X}_4 = 8.0 \qquad n_j = 6 \qquad c = 4$$

We proceed as follows:

1. Arrange all sample means in order from smallest to largest.
   As it happens, the column means in our example are already in ascending order. When this is not the case, they should be so arranged.

2. Compute the standard error of the sample means $s_{\overline{X}_j}$.
   The standard error is computed using the following formula:

$$s_{\overline{X}_j} = \sqrt{\frac{\text{MSWC}}{n_j}} \qquad\qquad (11.10.4)$$

For our example, we have $s_{\overline{X}_j} = \sqrt{6.05/6} = 1.0042$.

3. Obtain critical values from Appendix Table 11. Appendix Table 11, reproduced here as Table 11.11, is made up of two parts, the first part being employed when the level

of significance has been set at .05 and the second part when $\alpha = .01$. We will continue to assume a level of significance of .05 and consequently make use of the first part.

In examining Table 11.11, we find above the leftmost column the symbol df_2. This refers to the within-columns degrees of freedom from the main analysis of variance—20 in our example. Reading across the row associated with $df_2 = 20$, we record the value in that row found in the column under "Range of Comparison = 4" (since there were four samples involved in the main analysis of variance), namely 3.190, as well as all the entries to the left of that one in the same row—3.097 in the "3" column and 2.950 in the "2" column. (More generally, the longest "Range of Comparison" value will be equal to c, the number of samples involved in the main analysis of variance. Therefore, we record the table entries across from the appropriate df_2 value and down from c, $c - 1$, $c - 2$, and so forth, until we reach the "2" column.)

4. Multiply the standard error of the sample means $s_{\bar{x}_j}$ obtained in Step 2 by the critical values found in Step 3 to obtain the *least significant ranges*, or *LSR* values.

By following this instruction, we obtain the LSR values shown in column (4) of Table 11.12.

Table 11.11 Critical Values for Duncan's Multiple-Range Test (From Appendix Table 11)

Part I: Significance Level $\alpha = .05$

Range of Comparison

df_2	2	3	4	5	6	7	8	9	10	11	12	13	14	15
1	17.970	17.970	17.970	17.970	17.970	17.970	17.970	17.970	17.970	17.970	17.970	17.970	17.970	17.970
2	6.085	6.085	6.085	6.085	6.085	6.085	6.085	6.085	6.085	6.085	6.085	6.085	6.085	6.085
3	4.501	4.516	4.516	4.516	4.516	4.516	4.516	4.516	4.516	4.516	4.516	4.516	4.516	4.516
4	3.927	4.013	4.033	4.033	4.033	4.033	4.033	4.033	4.033	4.033	4.033	4.033	4.033	4.033
5	3.635	3.749	3.797	3.814	3.814	3.814	3.814	3.814	3.814	3.814	3.814	3.814	3.814	3.814
6	3.461	3.587	3.649	3.680	3.694	3.697	3.697	3.697	3.697	3.697	3.697	3.697	3.697	3.697
7	3.344	3.477	3.548	3.588	3.611	3.622	3.626	3.626	3.626	3.626	3.626	3.626	3.626	3.626
8	3.261	3.399	3.475	3.521	3.549	3.566	3.575	3.579	3.579	3.579	3.579	3.579	3.579	3.579
9	3.199	3.339	3.420	3.470	3.502	3.523	3.536	3.544	3.547	3.547	3.547	3.547	3.547	3.547
10	3.151	3.293	3.376	3.430	3.465	3.489	3.505	3.516	3.522	3.525	3.526	3.526	3.526	3.526
11	3.113	3.256	3.342	3.397	3.435	3.462	3.480	3.493	3.501	3.506	3.509	3.510	3.510	3.510
12	3.082	3.225	3.313	3.370	3.410	3.439	3.459	3.474	3.484	3.491	3.496	3.498	3.499	3.499
13	3.055	3.200	3.289	3.348	3.389	3.419	3.442	3.458	3.470	3.478	3.484	3.488	3.490	3.490
14	3.033	3.178	3.268	3.329	3.372	3.403	3.426	3.444	3.457	3.467	3.474	3.479	3.482	3.484
15	3.014	3.160	3.250	3.312	3.356	3.389	3.413	3.432	3.446	3.457	3.465	3.471	3.476	3.478
16	2.998	3.144	3.235	3.298	3.343	3.376	3.402	3.422	3.437	3.449	3.458	3.465	3.470	3.473
17	2.984	3.130	3.222	3.285	3.331	3.366	3.392	3.412	3.429	3.441	3.451	3.459	3.465	3.469
18	2.971	3.118	3.210	3.274	3.321	3.356	3.383	3.405	3.421	3.435	3.445	3.454	3.460	3.465
19	2.960	3.107	3.199	3.264	3.311	3.347	3.375	3.397	3.415	3.429	3.440	3.449	3.456	3.462
20	2.950	3.097	3.190	3.255	3.303	3.339	3.368	3.391	3.409	3.424	3.436	3.445	3.453	3.459
24	2.919	3.066	3.160	3.226	3.276	3.315	3.345	3.370	3.390	3.406	3.420	3.432	3.441	3.449
30	2.888	3.035	3.131	3.199	3.250	3.290	3.322	3.349	3.371	3.389	3.405	3.418	3.430	3.439
40	2.858	3.006	3.102	3.171	3.224	3.266	3.300	3.328	3.352	3.373	3.390	3.405	3.418	3.429
60	2.829	2.976	3.073	3.143	3.198	3.241	3.277	3.307	3.333	3.355	3.374	3.391	3.406	3.419
120	2.800	2.947	3.045	3.116	3.172	3.217	3.254	3.287	3.314	3.337	3.359	3.377	3.394	3.409
∞	2.772	2.918	3.017	3.089	3.146	3.193	3.232	3.265	3.294	3.320	3.343	3.363	3.382	3.399

(continued)

Table 11.11 *(Continued)*

Part II: Significance Level $\alpha = .01$

Range of Comparison

df_2	2	3	4	5	6	7	8	9	10	11	12	13	14	15
1	90.030	90.030	90.030	90.030	90.030	90.030	90.030	90.030	90.030	90.030	90.030	90.030	90.030	90.030
2	14.040	14.040	14.040	14.040	14.040	14.040	14.040	14.040	14.040	14.040	14.040	14.040	14.040	14.040
3	8.261	8.321	8.321	8.321	8.321	8.321	8.321	8.321	8.321	8.321	8.321	8.321	8.321	8.321
4	6.512	6.677	6.740	6.756	6.756	6.756	6.756	6.756	6.756	6.756	6.756	6.756	6.756	6.756
5	5.702	5.893	5.989	6.040	6.065	6.074	6.074	6.074	6.074	6.074	6.074	6.074	6.074	6.074
6	5.243	5.439	5.549	5.614	5.655	5.680	5.694	5.701	5.703	5.703	5.703	5.703	5.703	5.703
7	4.949	5.145	5.260	5.334	5.383	5.416	5.439	5.454	5.464	5.470	5.472	5.472	5.472	5.472
8	4.746	4.939	5.057	5.135	5.189	5.227	5.256	5.276	5.291	5.302	5.309	5.314	5.316	5.317
9	4.596	4.787	4.906	4.986	5.043	5.086	5.118	5.142	5.160	5.174	5.185	5.193	5.199	5.203
10	4.482	4.671	4.790	4.871	4.931	4.975	5.010	5.037	5.058	5.074	5.088	5.098	5.106	5.112
11	4.392	4.579	4.697	4.780	4.841	4.887	4.924	4.952	4.975	4.994	5.009	5.021	5.031	5.039
12	4.320	4.504	4.622	4.706	4.767	4.815	4.852	4.883	4.907	4.927	4.944	4.958	4.969	4.978
13	4.260	4.442	4.560	4.644	4.706	4.755	4.793	4.824	4.850	4.872	4.889	4.904	4.917	4.928
14	4.210	4.391	4.508	4.591	4.654	4.704	4.743	4.775	4.802	4.824	4.843	4.859	4.872	4.884
15	4.168	4.347	4.463	4.547	4.610	4.660	4.700	4.733	4.760	4.783	4.803	4.820	4.834	4.846
16	4.131	4.309	4.425	4.509	4.572	4.622	4.663	4.696	4.724	4.748	4.768	4.786	4.800	4.813
17	4.099	4.275	4.391	4.475	4.539	4.589	4.630	4.664	4.693	4.717	4.738	4.756	4.771	4.785
18	4.071	4.246	4.362	4.445	4.509	4.560	4.601	4.635	4.664	4.689	4.711	4.729	4.745	4.759
19	4.046	4.220	4.335	4.419	4.483	4.534	4.575	4.610	4.639	4.665	4.686	4.705	4.722	4.736
20	4.024	4.197	4.312	4.395	4.459	4.510	4.552	4.587	4.617	4.642	4.664	4.684	4.701	4.716
24	3.956	4.126	4.239	4.322	4.386	4.437	4.480	4.516	4.546	4.573	4.596	4.616	4.634	4.651
30	3.889	4.056	4.168	4.250	4.314	4.366	4.409	4.445	4.477	4.504	4.528	4.550	4.569	4.586
40	3.825	3.988	4.098	4.180	4.244	4.296	4.339	4.376	4.408	4.436	4.461	4.483	4.503	4.521
60	3.762	3.922	4.031	4.111	4.174	4.226	4.270	4.307	4.340	4.368	4.394	4.417	4.438	4.456
120	3.702	3.858	3.965	4.044	4.107	4.158	4.202	4.239	4.272	4.301	4.327	4.351	4.372	4.392
∞	3.643	3.796	3.900	3.978	4.040	4.091	4.135	4.172	4.205	4.235	4.261	4.285	4.307	4.327

Source: Adapted from H. L. Harter, "Critical Values for Duncan's New Multiple Range Test," *Biometrics 16* (1960), pp. 671–685. With permission from the Biometric Society.

5. Set up a table of differences between pairs of sample means.

   Table 11.13 is such a table. Ignoring the LSR column for now, we can describe the method of constructing this table in terms of five relatively simple steps: (1) List, in order of magnitude, all sample means except the smallest across the top of the table, using these means as column headings. (2) List, in order of magnitude, all sample means except the largest down the left-hand side, using

Table 11.12 Calculation of Least Significant Ranges (LSR Values)

(1) Range of Comparison	(2) Critical Values	(3) $s_{\bar{x}_j}$	(4) (2) × (3) LSR
2	2.950	1.0042	$LSR_2 = 2.962$
3	3.097	1.0042	$LSR_3 = 3.110$
4	3.190	1.0042	$LSR_4 = 3.203$

Table 11.13 Differences Between All Pairs of Sample Means

	$\overline{X}_2 = 4.0$	$\overline{X}_3 = 5.5$	$\overline{X}_4 = 8.0$	LSR
$\overline{X}_1 = 2.5$	1.5	3.0^a	5.5^a	$LSR_2 = 2.962$
$\overline{X}_2 = 4.0$	—	1.5	4.0^a	$LSR_3 = 3.110$
$\overline{X}_3 = 5.5$	—	—	2.5	$LSR_4 = 3.203$

^{a}Indicates significance at the .05 level.

these means as row headings. (3) Subtract each of the sample means used for row headings from the sample mean serving as the heading of the right-most column of differences. (4) Repeat this procedure for each of the other columns in the emerging table. (5) Enter the resulting differences in the appropriate cells of the table.

The table so constructed shows a top row containing differences between means said to be *over a range of two means*, a second row containing differences *over a range of three means*, and a third row containing differences *over a range of four means*. If there were c sample means in the analysis, there would be $c - 1$ rows in this table and the $c - 1$th row would contain differences *over a range of c means*.

6. Test the differences between all pairs of sample means for significance.

The test of significance can now be performed rather quickly. We simply examine the top row of Table 11.13 and compare the entries with $LSR_2 = 2.962$ which was determined earlier and which we have placed in the extreme right-hand column of Table 11.13. If a difference exceeds LSR_2, it is declared significant; if less than LSR_2, it is declared not significant. We then conduct the same kind of comparison with the second row, but, this time, we compare the differences with $LSR_3 = 3.110$. Finally, we repeat this procedure with the third row using $LSR_4 = 3.203$ as the appropriate standard for determining significance or lack of it.

A simple graphic device sometimes adds clarity to one's interpretation of the findings of this multiple-range test. The sample means are arrayed in ascending order, as shown here, and a line is drawn under those means that *do not* differ significantly:

2.5	4.0	5.5	8.0

For our example, it appears from the positioning of the lines that price is not an important factor in determining the speed with which a house can be sold *when two adjacent price categories are compared*. This is true whether we are examining the two categories of lower-priced houses, the two categories of higher-priced houses, or the two price categories in the middle. However, when the price categories are not adjacent, price is indeed an important independent variable.

```
PROBLEM CODE HOCKEY
NUMBER OF TREATMENT GROUPS        3
NUMBER OF VARIABLE FORMAT CARDS   1
DATA INPUT TAPE                   5

THE VARIABLE FORMAT

(80F1.0)


TREATMENT GROUP          1          2          3

SAMPLE SIZE             35         35         35
MEAN                1.4286     2.4286     1.8000
STANDARD DEVIATION  1.1450     1.9895     1.3016


GRAND MEAN = 1.8857          STANDARD DEVIATION = 1.5647

                ANALYSIS OF VARIANCE

              SUM OF SQUARES      DF      MEAN SQUARE      F RATIO

BETWEEN GROUPS    17.8857         2          8.9429        3.8530
WITHIN GROUPS    236.7429       102          2.3210

TOTAL            254.6286       104
```

Figure 11.8. Computer printout of the hockey problem.

11.11 Computer-Assisted One-Way Analysis of Variance

With only four samples and six observations per sample, the calculations in the Imperial Construction case can be carried out rather quickly and easily with a hand calculator. Even the old-fashioned method of pushing a pencil would not put unreasonable demands on the investigator. However, when many more sample observations are involved, use of one of several available packaged computer programs can save a considerable amount of time. Figure 11.8 shows the result of one such computer-assisted study. The study was conducted by a member of the physical education department of a university and had to do with the question of whether a certain period—the first, second, or third—in hockey games was more productive of points than other periods. As the computer printout shows, the answer is Yes at the .05 level but No at the .01 level of significance. With 105 observations altogether, the computational work could have been considerable. However, the computer turned out the information in Figure 11.8 in just a few seconds after the data had been entered.

11.12 Summary

If one were to remember only one thing about analysis of variance, it should be that, for random sample data obtained from independent populations, *analysis of variance permits the testing of any number of sample means for significance.*

If this sounds like a somewhat austere virtue, it shouldn't because its reach is potentially vast. You have seen in this chapter that analysis of variance can be viewed as an extension of previously treated tests involving one or two sample means. As such, analysis of variance brings to the analysis of *survey-sampling data* a flexibility not offered by the regular *t* or Z tests. In the next chapter, you will see how analysis of variance methods can be applied to quite a variety of *designed experiments* as well.

Notes

1. Many specialized books deal with this subject more comprehensively. One good source is Chapter 3 of Roger E. Kirk, *Experimental Design: Procedures for the Behavioral Sciences* (Belmont, Calif.: Brooks/Cole, 1968).
2. See Clyde Young Kramer, "Extension of Multiple Range Tests to Group Means with Unequal Numbers of Replications," *Biometrics* (September 1956). pp. 307–310.

Terms Introduced in This Chapter

analysis of variance (p. 465)
one-way analysis of variance (p. 465)
ANOVA table (p. 476)
a posteriori multiple-comparison
  method (p. 482)
a priori multiple-comparison method
  (p. 481)

comparison (p. 483)
Duncan multiple-range test (p. 488)
least significant range (LSR) (p. 489)
mean square (p. 470)
orthogonal comparisons (p. 483)

orthogonality (p. 483)
mutual orthogonality (p. 483)
residual plots (p. 480)
subhypothesis (p. 483)
sum of squares (p. 470)

Formulas Introduced in This Chapter

Basic formula for computing total sum of squares:

$$\text{SSTO} = \sum_{\text{all } j} \sum_{\text{all } i} (X_{ij} - \overline{\overline{X}})^2$$

Basic formula for computing between-columns sum of squares:

$$\text{SSBC} = \sum_{\text{all } j} n_j (\overline{X}_j - \overline{\overline{X}})^2$$

Basic formula for computing within-columns sum of squares:

$$\text{SSWC} = \sum_{\text{all } j} \sum_{\text{all } i} (X_{ij} - \overline{X}_j)^2$$

Total degrees of freedom:

$$df_T = n - 1$$

Between-columns degrees of freedom:

$$df_1 = c - 1$$

Within-columns degrees of freedom:

$$df_2 = n - c$$

Shortcut formula for computing total sum of squares:

$$\text{SSTO} = \sum_{\text{all } j} \sum_{\text{all } i} X_{ij}^2 - \frac{\left(\sum_{\text{all } j} \sum_{\text{all } i} X_{ij}\right)^2}{n}$$

Shortcut formula for computing between-columns sum of squares:

$$\text{SSBC} = \sum_{\text{all } j} \frac{\left(\sum_{\text{all } i} X_{ij}\right)^2}{n_j} - \frac{\left(\sum_{\text{all } j} \sum_{\text{all } i} X_{ij}\right)^2}{n}$$

Shortcut formula for computing within-columns sum of squares:

$$\text{SSWC} = \sum_{\text{all } j} \sum_{\text{all } i} X_{ij}^2 - \sum_{\text{all } j} \frac{\left(\sum_{\text{all } i} X_{ij}\right)^2}{n_j}$$

Formal way of expressing the meaning of "comparison":

$$\text{Com}_i = m_{i1}\overline{X}_1 + m_{i2}\overline{X}_2 + \cdots + m_{ic}\overline{X}_c$$

where $\sum_{\text{all } j} m_{ij} = 0$.

Formula for computing the sum of squares for a particular comparison:

$$\text{SS}_{\text{Com } i} = \frac{(\text{Com } i)^2}{\dfrac{(m_{i1})^2}{n_1} + \dfrac{(m_{i2})^2}{n_2} + \cdots + \dfrac{(m_{ic})^2}{n_c}}$$

Formula for the standard error of the sample means:

$$s_{\overline{X}_j} = \sqrt{\frac{\text{MSWC}}{n_j}}$$

Questions and Problems

11.1   Explain the meaning of each of the following:
   a. Mean square
   b. Degrees of freedom
   c. Sum of squares
   d. Total sum of squares
   e. ANOVA table
   f. Comparison
   g. Residual plots

11.2   Explain the meaning of each of the following:
   a. Between-columns sum of squares
   b. Within-columns sum of squares
   c. Observed F
   d. Subhypothesis
   e. Sum of squares for a comparison
   f. Least significant range
   g. $\overline{\overline{X}}$
   h. $s_{\overline{X}_j}$

11.3   Distinguish between:
   a. Use of analysis of variance in connection with a survey-sampling problem and in connection with a designed experiment

b. Between-columns sum of squares and within-columns sum of squares

c. A priori method and a posteriori method of following up on results of an analysis of variance

d. Method of orthogonal comparisons and the Duncan multiple-range test

e. SSBC and MSBC

11.4 Distinguish between:

a. The statement "$\mu_1 \neq \mu_2 \neq \mu_3 \neq \mu_4$" and the statement "$\mu_1, \mu_2, \mu_3, \mu_4$" are not all equal

b. μ_1 and $\overline{X}_1$

c. s_j^2 and $s_{\overline{X}_j}$

d. Sum of squares and mean square

e. df_1 and df_2

11.5 Distinguish between:

a. Variance and mean square

b. $\displaystyle\sum_{\text{all } i} X_{ij}$ and $\displaystyle\sum_{\text{all } j}\sum_{\text{all } i} X_{ij}$

c. Orthogonal comparisons and comparisons that are not orthogonal

d. n_j and n

e. $\overline{X}_j$ and $\overline{\overline{X}}$

11.6 Indicate which of the following statements you agree with and which you disagree with, and defend your opinions:

a. Analysis of variance was designed to determine whether the variances of several independent populations are equal.

b. In a one-way analysis of variance involving 5 samples and 10 observations per sample, the within-columns degrees of freedom, df_2, will be 49.

c. The assumption in analysis of variance regarding normally distributed populations need not concern us very much if the sample sizes are large, the reason being that the central limit theorem assures us that a population distribution will be approximately normal if the sample size is sufficiently large.

d. If, in carrying out a one-way analysis of variance, an analyst finds that the between-columns mean square does not exceed significantly the within-columns mean square, she will be unlikely to follow up her main analysis of variance with an orthogonal comparisons analysis.

11.7 Indicate which of the following statements you agree with and which you disagree with, and defend your opinions:

a. In a one-way analysis of variance, the analyst wishes to know whether the sample means are equal.

b. The presence of between-population variation should have no bearing on how we estimate the unknown population variance σ^2 from sample data.

c. When performing an F test, it makes little difference whether MSBC is used as the numerator or as the denominator.

d. As far as the assumption of equal population variances is concerned, it makes little difference whether the sample sizes are equal or unequal.

11.8 Indicate which of the following statements you agree with and which you disagree with and defend your opinions:

a. In a situation where $df_1 = 3$, an investigator could elect to work with three but not four orthogonal comparisons.

b. In a one-way analysis of variance, there is only one way in which the population means can be consistent with the null hypothesis.

c. In a one-way analysis of variance, there are several ways in which the population means can be consistent with the alternative hypothesis.

d. If one has completed a one-way analysis of variance, he or she already has all the information required to compute the standard error of the sample means $s_{\bar{X}_j}$, for use in the Duncan multiple-range test.

11.9 The value of MSWC is an unbiased estimator of the unknown population variance σ^2 regardless of whether the null hypothesis is true, whereas the value of MSBC is an unbiased estimator of unknown σ^2 only if the null hypothesis is true.

a. Explain how the truth or falsity of the null hypothesis affects the unbiasedness of MSBC as an estimator of σ^2.

b. Explain how a comparison between the values of MSBC and MSWC can help one draw a conclusion about whether the null hypothesis is true or false.

11.10 Someone wishes to test the following subhypotheses:

$H_{0_{\text{sub }1}}$: $\mu_1 - \mu_2 = 0$

$H_{0_{\text{sub }2}}$: $\mu_2 - \mu_3 = 0$

$H_{0_{\text{sub }3}}$: $(\mu_1 + \mu_2 + \mu_3) - 3\mu_4 = 0$

a. Develop a table of multipliers, m_{ij} values, like Table 11.9.

b. Justify the placement and values of the multipliers in your table.

c. Assuming that the sample sizes are the same, are the comparisons mutually orthogonal? Explain.

11.11 Someone wishes to test the following subhypotheses:

$H_{0_{\text{sub }1}}$: $\mu_1 - \mu_2 = 0$

$H_{0_{\text{sub }2}}$: $\mu_3 - \mu_4 = 0$

$H_{0_{\text{sub }3}}$: $(\mu_1 + \mu_2) - (\mu_3 + \mu_4) = 0$

$H_{0_{\text{sub }4}}$: $(\mu_1 + \mu_2 + \mu_3 + \mu_4) - 4\mu_5 = 0$

a. Develop a table of multipliers, m_{ij} values, like Table 11.9.

b. Justify the placement and values of the multipliers in your table.

c. Assuming that the sample sizes are the same, are the comparisons mutually orthogonal? Explain.

11.12 Examine Table 11.14. Are the comparisons orthogonal? How can you tell?

11.13 Examine Table 11.15. Are the comparisons orthogonal? How can you tell?

Table 11.14 Hypothetical ANOVA Table

Source of Variation	Sum of Squares	Degrees of Freedom	Mean Square	Observed F
Between columns	200	2	100.00	16.00
Com 1	25	1	25.00	4.00
Com 2	175	1	175.00	28.00
Within columns	125	20	6.25	—
Total	325	22	—	—

Table 11.15 Hypothetical ANOVA Table

Source of Variation	Sum of Squares	Degrees of Freedom	Mean Square	Observed F
Between columns	30	3	10.00	5.00
Com 1	10	1	10.00	5.00
Com 2	20	1	20.00	10.00
Com 3	10	1	10.00	5.00
Within columns	10	5	2.00	—
Total	40	8	—	—

11.14 The Duncan multiple-range test was applied to a four-sample analysis-of-variance problem with the following results:

<u>5.4   6.8   7.6</u>   10.4

Tell which differences between two sample means are significant and which are not significant.

11.15 The Duncan multiple-range test was applied to a six-sample analysis-of-variance problem with the following results:

<u>12.4   13.7   17.4</u>   25.9   <u>35.5   38.9</u>

Tell which differences between two sample means are significant and which are not significant.

11.16 The number of units of product turned out per hour by two workers in a manufacturing plant was observed for randomly selected hours with the following results:

Worker 1	Worker 2
4	3
3	4
4	2
2	2
5	3
5	3

The null and alternative hypotheses are

H_0: $\mu_1 - \mu_2 = 0$
H_a: $\mu_1 - \mu_2 \neq 0$

a. Test the null hypothesis at the .05 level of significance using the appropriate t test.
b. Test the null hypothesis at the .05 level of significance using analysis of variance.
c. Compare the conclusions you reached in parts a and b.
d. Square observed t and compare the result with observed F. What is your finding?
e. Square critical t and compare the result with critical F. What is your finding?
(*Note*: The results you should have obtained in connection with parts d and e will hold only for situations where two samples are involved—not for more than two samples.)

11.17 Earnings-per-share data for random samples of five companies from two industries were obtained with the following results:

Industry 1	Industry 2
$2.25	$3.00
0.89	1.25
1.17	1.95
1.94	1.18
1.57	2.59

The null and alternative hypotheses are

H_0: $\mu_1 - \mu_2 = 0$
H_a: $\mu_1 - \mu_2 \neq 0$

a. Test the null hypothesis at the .01 level of significance using the appropriate t test.
b. Test the null hypothesis at the .01 level of significance using analysis of variance.
c. Compare the conclusions you reached in parts a and b.
d. Square observed t and compare the result with observed F. What is your finding?
e. Square critical t and compare the result with critical F. What is your finding?
(Note: The results you should have obtained in connection with parts d and e will hold only for situations where two samples are involved—not for more than two samples.)

11.18 a. List the four assumptions underlying a one-way analysis of variance.
b. Which of these are usually under the control of the investigator?
c. What must the investigator do, if possible, to ensure that the assumptions under her control are met?
d. Which of these assumptions are not under the control of the investigator?
e. What are some things the investigator should do to determine whether the assumptions not under her control are met?

11.19 At various places throughout this chapter, advantages have been cited for using equal, rather than unequal, sample sizes whenever possible. What are three such advantages?

11.20 Complete the ANOVA table presented here as Table 11.16.
a. Should the null hypothesis be accepted or rejected at the .05 level of significance?
b. How many random samples were involved?
c. Were the sample sizes even or uneven? How do you know?

 11.21 Prices charged for Kodachrome 64 slide film in random samples of department stores, photo shops, and pharmacies were collected by an amateur photographer. The resulting data are shown in Table 11.17.

Table 11.16 Hypothetical ANOVA Table

Source of Variation	Sum of Squares	Degrees of Freedom	Mean Square	Observed F	.05 Level Critical F
Between columns	259.8	5			
Within columns					
Total	400.0	84			

Table 11.17 Data Related to Problems 11.21 Through 11.24

	Kind of Store	
Department	Photo Shop	Pharmacy
$2.77	$3.49	$2.39
2.69	4.35	2.80
3.29	3.75	3.79
2.76	4.40	2.89
3.29	3.50	3.79
2.99	3.80	4.50
2.89	3.84	3.55
2.76	4.50	3.70
3.15	3.85	4.00

    a. State the null and alternative hypotheses.
    b. How many degrees of freedom are associated with the total sum of squares? With the between-columns sum of squares? With the within-columns sum of squares?
    c. What values do you get for MSBC and MSWC?
    d. State the decision rule if the level of significance is .05.
    e. Perform the F test and state your conclusion.

11.22 Refer to problem 11.21.
    a. Array all sets of sample data and evaluate judgmentally whether you think the assumption of normally distributed populations has been violated. Justify your conclusion.
    b. Develop charts of residual plots and tell whether you think the assumption of equal population variances has been violated. Justify your conclusion.

11.23 Refer to problem 11.21. Suppose that the following subhypotheses are to be tested:

$$H_{0_{\text{sub }1}}: \mu_1 - \mu_2 = 0$$
$$H_{0_{\text{sub }2}}: (\mu_2 + \mu_3) - 2\mu_1 = 0$$

    a. Develop a table of multipliers, m_{ij} values, like Table 11.9.
    b. Perform the check for orthogonality. What is your conclusion?
    c. If the comparisons are orthogonal, test the two subhypotheses at the .05 level of significance and interpret your results.

11.24 Refer to problem 11.21. Perform Duncan's multiple-range test, using a .05 level of significance, and interpret your results.

11.25 Does size of borrowing firm have any effect on the interest rate banks charge? A study was conducted to find out. Borrowing firms were organized into size categories, as shown in Table 11.18, and a random sample of size five was selected from each category.
    a. State the null and alternative hypotheses.
    b. How many degrees of freedom are associated with the total sum of squares? With the between-columns sum of squares? With the within-columns sum of squares?
    c. What values do you get for MSBC and MSWC?
    d. State the decision rule if the level of significance is .01.
    e. Perform the F test and state your conclusion.

Table 11.18 Data Related to Problems 11.25 Through 11.28

(Numbers Represent Interest Rates Charged)

| | Size of Borrowing Firm | | |
Small	Medium-Small	Medium-Large	Large
9.0	9.0	6.0	6.0
9.5	9.0	8.0	6.0
8.5	8.0	8.0	7.0
9.0	8.0	7.0	7.0
9.0	7.0	7.0	6.5

 11.26 Refer to problem 11.25.
  a. Do you have any reason for doubting the validity of the assumption about normally distributed populations? Explain.
  b. Develop charts of residual plots and tell whether you think the assumption about equal population variances has been violated. Justify your conclusion.

 11.27 Refer to problem 11.25. Suppose the following subhypotheses are to be tested:

$$H_{0_{sub\ 1}}: \mu_1 - \mu_2 = 0$$
$$H_{0_{sub\ 2}}: \mu_3 - \mu_4 = 0$$
$$H_{0_{sub\ 3}}: (\mu_1 + \mu_2) - (\mu_3 + \mu_4) = 0$$

  a. Develop a table of multipliers, m_{ij} values, like Table 11.9.
  b. Perform the checks for mutual orthogonality. What is your conclusion?
  c. If the comparisons are mutually orthogonal, test the three subhypotheses at the .01 level of significance and interpret your results.

 11.28 Refer to problem 11.26. Perform Duncan's multiple-range test, using a .01 level of significance, and interpret your results.

 11.29 A random sample of six companies was selected from each of five different industrial categories and their earnings per share determined. The results are shown in Table 11.19.

Table 11.19 Data Related to Problems 11.29 Through 11.32

| | Industry Category | | | |
Petroleum Refiners	Motor Vehicle	Electrical Appliances	Chemical	Food
$6.83	$7.51	$2.93	$8.50	$2.90
3.27	8.52	3.85	3.38	3.13
5.65	4.27	2.24	2.07	1.23
3.20	8.34	4.12	9.30	4.49
5.90	9.13	4.00	7.15	4.86
9.08	4.80	2.45	3.30	1.86

a. State the null and alternative hypotheses.
b. How many degrees of freedom are associated with the total sum of squares? With the between-columns sum of squares? With the within-columns sum of squares?
c. What values do you get for MSBC and MSWC?
d. State the decision rule if the level of significance is .05.
e. Perform the F test and state your conclusion.

 11.30 Refer to problem 11.29.
a. Do you have any reason for doubting the validity of the assumption about normally distributed populations? Explain.
b. Develop charts of residual plots and tell whether you think the assumption about equal population variances has been violated. Justify your answer.

 11.31 Refer to problem 11.29. Suppose that the following subhypotheses are to be tested:

$H_{0_{\text{sub } 1}}$: $\mu_1 - \mu_2 = 0$

$H_{0_{\text{sub } 2}}$: $\mu_3 - \mu_5 = 0$

$H_{0_{\text{sub } 3}}$: $(\mu_1 + \mu_2) - (\mu_3 + \mu_5) = 0$

$H_{0_{\text{sub } 4}}$: Make up your own, paying due respect to the requirement of mutual orthogonality

a. Develop a table of multipliers, m_{ij} values, like Table 11.9.
b. Perform the checks for mutual orthogonality. What is your conclusion?
c. If the comparisons are mutually orthogonal, test the four subhypotheses and interpret your results.

 11.32 Refer to problem 11.29. Perform Duncan's multiple-range test, using a .05 level of significance, and interpret your results.

 11.33 Are some work areas safer than others? A manufacturing company attempted to find out by drawing a random sample of five employees from six ostensibly identical work areas. The results are shown in Table 11.20.
a. State the null and alternative hypotheses.
b. How many degrees of freedom are associated with total sum of squares? With between-columns sum of squares? With the within-columns sum of squares?
c. What values do you get for MSBC and MSWC?

Table 11.20 Data Related to Problems 11.33 Through 11.36

(Entries Represent Number of Work-Related Injuries During the Most Recent Three-Month Period)

		Work Area			
1	2	3	4	5	6
0	1	0	4	0	1
1	1	0	6	1	1
1	0	0	5	0	0
2	1	2	2	1	0
1	1	1	2	0	1

d. State the decision rule if the level of significance is .01.
e. Perform the F test and state your conclusion.

 11.34 Refer to problem 11.33.
   a. Do you have any reason for doubting the validity of the assumption about normally distributed populations? Explain.
   b. Develop charts of residual plots and tell whether you think the assumption about equal population variances has been violated. Justify your answer.

 11.35 Refer to problem 11.33. Can you think of any way to employ the method of orthogonal comparisons in connection with this problem? If so, explain fully and then test your five subhypotheses at the .01 level of significance. Interpret your results.

 11.36 Refer to problem 11.33. Perform Duncan's multiple-range test, using a .01 level of significance, and interpret your results.

TAKE CHARGE

 11.37 Over-the-counter drug prices tend to vary rather widely from store to store. Visit three drug stores, three grocery stores, and three general merchandise discount stores such as K-mart or Target in your city and record the prices charged for a 100-tablet bottle of Bayer aspirin. Place the prices in a table like the following one:

	Price	
Drug Store	**Grocery Store**	**Discount Store**
X_{11}	X_{12}	X_{13}
X_{21}	X_{22}	X_{23}
X_{31}	X_{32}	X_{33}

a. Test, using $\alpha = .05$, the hypotheses

H_0: Type of store has no bearing on price charged, or
$\quad \mu_1 = \mu_2 = \mu_3$
H_a: Type of store does have a bearing on price, or
$\quad \mu_1, \mu_2, \mu_3$ are not all equal

State your conclusion.
b. If part a yields a significant observed F ratio, apply Duncan's multiple-range test to the data. State your conclusions.
c. Did you select your stores randomly? If not, do you still trust the results of your F and Duncan tests? Why or why not?

Computer Exercise

 11.38 Table 11.21 shows average percent return on total capital over a recent five-year period by a random sample of twelve companies belonging to each of six industries. Have the computer determine the observed F associated with the hypotheses

H_0: $\mu_1 = \mu_2 = \mu_3 = \mu_4 = \mu_5 = \mu_6$
H_a: $\mu_1, \mu_2, \mu_3, \mu_4, \mu_5, \mu_6$ are not all equal

Table 11.21 Data for Problem 11.38

(Average Percent Return on Total Capital)

(1) Automotive Parts	(2) Electric Utilities	(3) Forest Products	(4) Oil and Gas	(5) Regional Banks	(6) Steel
3.2	6.4	5.5	7.5	7.7	0.0
4.3	6.5	6.6	8.6	8.7	0.0
4.7	6.6	6.7	9.4	10.6	1.1
5.0	6.6	6.9	9.6	10.8	1.6
5.1	6.7	7.6	9.7	11.0	3.8
5.8	7.0	7.6	10.7	11.8	5.5
9.6	7.2	8.1	12.4	12.3	5.7
10.4	7.4	9.0	13.2	12.9	6.9
11.5	7.6	9.3	13.4	14.1	11.6
12.3	7.7	10.9	14.0	15.3	12.1
14.6	8.3	13.9	20.7	19.3	14.0
15.1	8.8	16.7	20.9	23.9	17.9

Source: Forbes, January 2, 1984.

Cases for Analysis and Discussion

 ## 11.1 "DIXON AND DOXEY ENGINEERING, INC.": PART II

Reread Part I of this case (case 10.3 at the end of Chapter 10) and examine Table 10.8.

1. If you have not previously answered questions 2, 4, and 6 of Part I of this case, do so now.

2. Can you think of any way of testing the three null hypotheses implied in questions 2, 4, and 6 of Part I *using a single one-way analysis of variance and relevant follow-up procedures?* If so, explain how this could be accomplished. Then do the analysis.

3. Comment on the similarities or dissimilarities between the results you obtained in connection with questions 1 and 2 above.

11.2 "DENTAL INTELLIGENCE ASSOCIATES"

The following is a close paraphrase of a report prepared by a consultant working for the dental profession:

> In recent years, the availability of continuing education courses for dentists has increased throughout the United States. Some regions of the country have experienced greater growth than others and could be considered innovators of new techniques and philosophy. The goal of my research was to determine whether the mean numbers of continuing dental education courses offered by dental schools differ in six specific regions. The results could have implications with respect to identifying the most desirable places for new dentists to locate.
>
> The data used in the analysis were collected from data on each U.S. dental school that appeared in the December 1979 *Journal of the American Dental Association.*

To answer the question of whether the number of continuing dental education courses differs from one region to another, a one-way analysis of variance was performed. The areas examined were categorized as: Eastern, Southeastern, Great Lakes, North Central, South Central, and Western. Table 11.22 lists the number of courses by subject and region. The hypotheses were:

H_0: $\mu_1 = \mu_2 = \mu_3 = \mu_4 = \mu_5 = \mu_6$

H_a: μ_1, μ_2, μ_3, μ_4, μ_5, and μ_6 are not all equal

Table 11.22 Number of Continuing Education Courses for Dentistry, by Region

Type of Course	Eastern	South-eastern	Great Lakes	North Central	South Central	Western
Anesthesia and pain control	10	7	5	0	1	12
Auxiliary utilization	2	5	3	1	1	2
Basic sciences	4	6	8	1	2	13
Dental materials	2	5	0	1	1	5
Endodontics	11	10	6	0	8	4
General dentistry	14	16	7	3	6	9
Hospital dentistry	1	0	0	1	0	3
Occlusion	11	4	8	8	3	16
Operative dentistry	8	5	4	0	1	2
Oral diagnosis and radiology	5	7	5	0	4	10
Oral medicine	0	8	2	4	5	2
Oral pathology	3	3	2	0	3	2
Oral rehabilitation	7	0	0	0	0	4
Oral surgery	7	4	2	2	10	4
Orthodontics	12	10	8	0	2	8
Pedodontics	5	11	9	1	1	2
Periodontics	24	14	7	2	7	15
Practice administration	11	8	5	2	5	18
Preventive dentistry	5	8	1	0	0	4
Prosthodontics	16	13	19	2	10	19
Public health dentistry	2	0	1	0	0	2
Restorative dentistry	21	13	4	1	2	5
Courses for dental auxiliary	27	35	18	9	1	16

Table 11.23 ANOVA Table for the "Dental Intelligence Associates" Case

Source of Variation	Sum of Squares	Degrees of Freedom	Mean Square	Observed F	.01 Level Critical F
Between columns	5809.8	5	1161.96	37.98[a]	3.02
Within columns	4038.3	132	30.59	—	—
Total	9848.1	137	—	—	—

[a] Indicates significance at the .01 level.

Table 11.23 shows the completed ANOVA results. The observed F value is much greater than the critical F value. Therefore, we were led to reject the null hypothesis, at the .01 level, that region of the country has no bearing on the number of continuing dental education courses offered by dental schools.

In an effort to "get more mileage" out of my analysis of variance, I made use of the method of orthogonal comparisons. The subhypotheses were

$H_{0_{sub\ 1}}$: $\mu_1 - \mu_2 = 0$

$H_{0_{sub\ 2}}$: $\mu_3 - \mu_4 = 0$

$H_{0_{sub\ 3}}$: $\mu_5 - \mu_6 = 0$

$H_{0_{sub\ 4}}$: $(\mu_1 + \mu_2) - (\mu_3 + \mu_4) = 0$

$H_{0_{sub\ 5}}$: $(\mu_1 + \mu_2 + \mu_3 + \mu_4) - (\mu_5 + \mu_6) = 0$

The m_{ij} values are presented in Table 11.24, and the extended ANOVA table appears in Table 11.25.

1. Write a critical review of this study.

2. Are the five comparisons mutually orthogonal? Elaborate.

11.3 THE "HANDI LAUNDRY COMPANY": PART I

The "Handi Laundry Company" is a factory-authorized route operator/distributor of Speed Queen commercial laundry products with responsibility for distribution to the multi-family industry in a large northern state.

Revenues come from three kinds of operation: (1) outright sales of coin-operated washers and dryers to apartment owners, (2) repair service for those outright sales accounts, and (3)

Table 11.24 m_{ij} Values for the "Dental Intelligence Associates" Case

Comparison	$\overline{X}_1$	$\overline{X}_2$	$\overline{X}_3$	$\overline{X}_4$	$\overline{X}_5$	$\overline{X}_6$
1	+1	−1	0	0	0	0
2	0	0	+1	−1	0	0
3	0	0	0	0	+1	−1
4	+1	+1	−1	−1	0	0
5	+1	+1	+1	+1	−1	−1

Table 11.25 Extended ANOVA Table for the "Dental Intelligence Associates" Case

Source of Variation	Sum of Squares	Degrees of Freedom	Mean Square	Observed F	.01 Level Critical F
Between columns	5809.80	5	1161.96	37.98[a]	3.02
Com 1	4.14	1	4.14	.14	6.63
Com 2	157.36	1	157.36	5.14[a]	6.63
Com 3	232.76	1	232.76	7.61[a]	6.63
Com 4	609.71	1	609.71	19.93[a]	6.63
Com 5	708.66	1	708.66	23.17[a]	6.63
Within columns	4038.30	132	30.59	—	—
Total	9848.10	137	—	—	—

[a]Indicates significance at the .01 level.

lease operations. Lease operations represent the main thrust of the company's efforts. The company owns and operates in excess of 1500 coin-operated washers and dryers serving over 10,000 apartment units in the two-state region.

This case has to do with the question of whether there are significant differences in profits realized from coin-operated washers and dryers installed in apartment buildings in six specific geographical sectors. The dependent variable is "gross revenue per apartment unit" (GPU). It is determined by the formula GPU = AGR/Apts, where AGR represents average gross revenue. GPU is the value that must be predicted by company salespeople when proposing (normally, under competitive bid conditions) a commission percentage to an apartment owner as rent for use of the coin-operated laundry space in his building. The more accurate the forecasted revenue from a prospective installation, the more precise and less risky will be the commission proposal terms. Greater precision and lower uncertainty seem to be associated with greater profitability.

Factors normally considered important in the forecasting process are (1) geographical location, (2) age of the building, and (3) adult, family, or senior users. Forecasts are usually carried out by studying "comparable" buildings from among the buildings served by Handi.

Each month Handi Laundry collects the coins from each machine it operates at the various installations. An accounting is made to each apartment owner/manager of the collection revenue for his particular building for the purpose of calculating the rent the company must pay for use of laundry space in that building (normally a percentage of these receipts).

A random sample of 15 apartment buildings was selected from each of six geographical areas: (1) the downtown metro, (2) the northeast metro, (3) the southeast metro, (4) the northwest metro, (5) the northern part of the state, and (6) the southwest metro.

According to the president of Handi Laundry, two factors are difficult to assess:

a. Prices have been fairly standard at 50 cents per wash, 25 cents per dry. Certain of the areas, such as downtown, have lagged in the complete conversion from 35 cents to 50 cents per wash. In our study, we assumed that no distortion was created by this difference.

b. New construction of apartments has been heaviest in the southeast metro area during the company's tenure. As a result, a concentration of larger, more modern, apartment buildings exists in this part of town. We assumed that size and age of apartment building has no important bearing on gross revenue per apartment unit.

The data are shown in Table 11.26.

Table 11.26 Data for the "Handi Laundry Company" Case

Gross Revenue per Apartment Unit, by Geographical Area

Downtown Metro	Northeast Metro	Southeast Metro	Northwest Metro	Northern Part of State	Southwest Metro
$3.87	$5.88	$5.62	$5.00	$4.59	$6.80
3.99	3.54	4.46	6.45	1.81	6.92
3.97	1.56	3.77	2.82	5.38	4.58
2.08	1.81	3.31	6.11	5.12	3.33
2.26	4.89	4.25	4.33	3.82	6.50
2.11	2.23	3.27	5.45	5.83	3.12
1.36	3.46	3.08	3.16	2.68	9.17
2.74	4.56	3.21	6.65	8.38	6.17
3.61	4.81	2.63	3.53	7.97	5.66
4.19	2.54	4.69	3.88	5.92	5.60
4.18	1.98	4.88	3.83	5.91	2.67
3.24	5.79	2.23	5.57	7.43	6.35
4.34	4.66	5.92	2.85	6.32	5.95
1.75	5.41	5.25	3.79	7.66	4.10
3.09	5.82	4.15	9.73	3.85	4.65

1. In your opinion, is the operational purpose of the study clearly spelled out? Elaborate. State in your own words what you think Handi Laundry hoped to accomplish from this study.

2. Array each set of sample data and evaluate judgmentally whether the assumption about normally distributed populations has been violated. Defend your conclusion.

3. Develop charts of residual plots and indicate whether the assumption about equal population variances has been violated. Defend your conclusion.

4. Refer to the president's remarks about the two factors whose effects are difficult to evaluate. Do you feel that his assumption about their not mattering very much is justified? Give an intuitive answer.

5. With respect to question 4, what kind of statistical information already in hand would be expected to reflect the effects, if any, of these two imponderables?

6. Using SPSSX, SAS, SYSTAT, any other available packaged computer program, or more old-fashioned means, perform the standard one-way analysis of variance on the data in Table 11.26 and state your result. Use $\alpha = .05$.

7. Perform Duncan's multiple-range test, using a .05 level of significance, and interpret your results.

This case is continued at the end of Chapter 13.

12 Experimentation in Business

That which we call sin in others is experiment for us.—Ralph Waldo Emerson

The trick is to make your mistakes on a small scale.—William Feather

What You Should Learn from This Chapter

As promised, in this chapter we treat analysis of variance as a "participation sport." That is, we now assume that at least one independent variable is under the control of the investigator. Aside from this change in viewpoint, this chapter is a continuation of Chapter 11. When you have finished this chapter, you should be able to

   1. Distinguish between a treatment variable and a blocking variable.

* 2. Distinguish among fixed-effects, random-effects, and mixed statistical models.

   3. Work through an analysis of variance for a completely randomized experimental design.

   4. Work through an analysis of variance for a randomized block design with a single observation per cell.

* 5. Obtain the mean squares for a randomized block design having more than one observation per cell (but equal cell frequencies) and state what implications the statistical model holds for the calculation of observed F ratios and the determination of critical F ratios.

* 6. Obtain the mean squares for a two-factor factorial design having more than one observation per cell (but equal cell frequencies) and state what implications the statistical model holds for the calculation of observed F ratios and the determination of critical F ratios.

* 7. Determine whether interaction is present, identify its specific nature, and discuss its theoretical and practical implications.

   8. State the assumptions one implicitly accepts when applying analysis-of-variance methods to each of the above-mentioned experimental designs.

Before beginning this chapter, you may wish to review sections 11.1 through 11.9 of Chapter 11.

12.1 Introduction

In Chapter 11 we studied in some detail the problem of the Imperial Construction Company. The problem, as posed there, was a nonexperimental one;* it was a survey-sampling

---

*We use a stricter delineation between experimental and nonexperimental investigations than that used in some textbooks but one compatible with our earlier distinction between survey-sampling studies, on the one hand, and designed experiments on the other.

problem which happened to lend itself to solution by one-way analysis of variance. Management merely wanted to know whether the number of weeks required to sell a completed house is affected by the *already established* price category to which the house belongs.

We will continue to focus on Imperial's problem for a time, but from a more experimental perspective. Remember that in experimentation, as we are using the term, the investigator plays an active, rather than a passive, role by varying one or more factors whose effects he wishes to study. With respect to the ANOVA aspects of the task, what he seeks is a breakdown of total sum of squares, SSTO, and total degrees of freedom, df_T, into parts—all but one of these parts being attributable to the independent variables under study and the other part being attributable to a collection of many other unknown factors. This goal is illustrated in Figure 12.1.

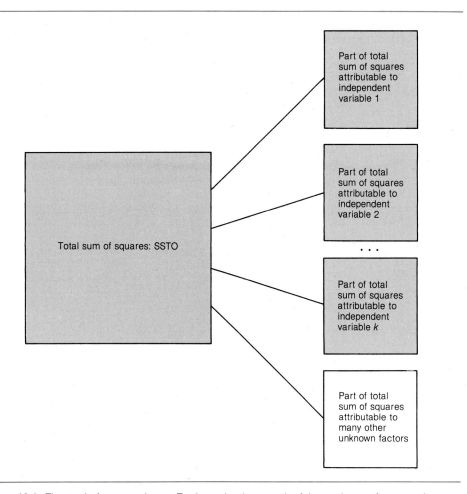

Figure 12.1. The goal of an experiment: To determine how much of the total sum of squares is attributable to specific independent variables and how much to other unknown factors. Total degrees of freedom would be broken down in the same manner.

12.2 The Completely Randomized Design

Actually, we have already dealt with the completely randomized design, according to one commonly used definition, which states that a completely randomized design is one in which an independent random sample is obtained from each of c populations. Therefore, the version of the Imperial Construction Company case presented in Chapter 11 would qualify.

However, in this chapter we will use an alternative, and more proper, definition—one with an unmistakably experimental slant:

> A *completely randomized design* is a kind of experimental design involving use of only one independent variable (a treatment variable) and the random assignment of test units to treatment levels.

For example, let us suppose that Imperial's management decides to act aggressively to sell their houses by offering some special inducements. However, before putting any new program into practice on a large scale, they wish to determine on a small scale which specific kind of special inducement is most effective. For purposes of illustration, we will say that management elects to experiment with the following inducements: (1) Company pays buyer's interest expenses for the first three months, (2) company provides free carpeting, and (3) no special inducement (a control useful in distinguishing between changes really attributable to the *actual* special inducements and changes which would have occurred even in their absence).

Before we proceed, some terms used above require definition:

> The three special inducements, including "no special inducement," are referred to collectively as the *treatment variable*. This is simply an independent variable—but one under the control of the investigator.
>
> Each special inducement—or, more generally, each category of the treatment variable—will be called a *treatment level*. Therefore, the three-months-of-no-interest inducement would be one treatment level among three comprising the treatment variable.
>
> The specific elementary units (houses, people, stores, Canadian rats, etc.) on which measurements are taken are called *test units* or *experimental units*.

Achieving Random Allocation

Achieving a random assignment of test units to treatment levels is usually a relatively simple task. Special tables of random permutations are often used, but a regular table of random digits will suffice. The main thing, as usual with this kind of table, is to determine in advance a systematic way of using it. For example, Imperial's management might begin by labeling the sample houses 1, 2, 3, . . . , 24. Presumably, for reasons mentioned in Chapter 11, they would want an even allocation of test units to treatment levels—that is, with a total sample size of $n = 24$, they would want 8 houses associated with each special inducement. They might decide to assign one house at a time. If the first random number selected (in a predetermined manner) is, say, 0, 1, or 2, sample house 1 is assigned to the three-months-of-no-interest treatment level; if the random number is 3, 4, or 5, sample house 1 is assigned to the free-carpeting treatment level; if the random number is 6, 7, or 8, sample house 1 is assigned to the no-special-inducement treatment level; finally, if the random number is 9, it is ignored. The same procedure would be applied to all 24 sample houses. If, say, the no-special-inducement column has 8 houses associated with it before the other columns do and a random digit of 6, 7, or 8 is obtained, that random digit will be ignored. Random assignment need not preclude an insistence on equal sample sizes.

The preceding, of course, is just one way among many possible ways in which the requirement of random assignment of test units to treatment levels could be met.

Let us suppose that the randomization process resulted in an allocation like that shown in Table 12.1.

Statistical Methodology

The null and alternative hypotheses for this version of the Imperial Construction problem are

H_0: Type of special inducement used has no bearing on the speed of sale of new houses, or, in symbols: $\mu_1 = \mu_2 = \mu_3$

H_a: Type of special inducement does have an effect on the speed of sale of new houses, or, in symbols: μ_1, μ_2, μ_3 are not all equal

Table 12.1 New Data for the Imperial Construction Company Problem

Type of Inducement ←——Treatment variable

Three Months of Interest-Free Living	Free Carpeting	No Special Inducement	Treatment levels
2	1	3	
2	3	4	Observations made
1	2	6	on test units (number
3	5	7	of weeks required
3	4	6	for sale of house)
5	7	8	
4	6	8	
5	12	13	
Total 25	40	55	
Mean 3.125	5.000	6.875	

$$\overline{\overline{X}} = 5.000$$

From this point on, the methodology is identical with that used for an ordinary one-way analysis of variance. The assumptions are also effectively the same. Assumptions 1 and 2 (section 11.9 of Chapter 11) are presumed to be met if random assignment of test units to treatment levels has been used. The assumptions about normally distributed populations and equal population variances are outside the control of the investigator but can usually be evaluated adequately by methods touched on in section 11.9.

Because the computational procedures for a completely randomized design are identical with those described in sections 11.5 through 11.8 of the preceding chapter, they will not be repeated here. Part I of Table 12.2 provides a review of the methodology; Part II of the table gives the results of this experimental version of the Imperial Construction problem.

If desired, the same follow-up, multiple-comparison procedures as discussed in Chapter 11—namely, the method of orthogonal comparisons or the Duncan multiple-range test—may be employed. These are demonstrated in Figures 12.2 and 12.3, respectively. (These figures pertain to material designated "optional" in Chapter 11; if you did not cover this material, you may wish to bypass Figures 12.2 and 12.3 completely.)

Table 12.2 ANOVA Table for Completely Randomized Experiment

Part I: General Structure

Source of Variation	Sum of Squares	Degrees of Freedom	Mean Square	Observed F	Critical F
Between columns (treatment levels)	$SSBC = \sum_{\text{all } j} \dfrac{\left(\sum_{\text{all } i} X_{ij}\right)^2}{n_j} - \dfrac{\left(\sum_{\text{all } j}\sum_{\text{all } i} X_{ij}\right)^2}{n}$	$df_1 = c - 1$	$\dfrac{SSBC}{df_1}$	$\dfrac{MSBC}{MSWC}$	From Appendix Table 6
Within columns	$SSWC = \sum_{\text{all } j}\sum_{\text{all } i} X_{ij}^2 - \sum_{\text{all } j} \dfrac{\left(\sum_{\text{all } i} X_{ij}\right)^2}{n_j}$	$df_2 = n - c$	$\dfrac{SSWC}{df_2}$	—	
Total	$SSTO = \sum_{\text{all } j}\sum_{\text{all } i} X_{ij}^2 - \dfrac{\left(\sum_{\text{all } i}\sum_{\text{all } j} X_{ij}\right)^2}{n}$	$df_T = n - 1$	—	—	—

Part II: ANOVA Table for Completely Randomized Experimental Version of Imperial Construction Company Problem

Source of Variation	Sum of Squares	Degrees of Freedom	Mean Square	Observed F	.05 Level Critical F
Between columns (treatment levels)	56.25	2	28.12	3.61[a]	3.47
Within columns	163.75	21	7.80	—	—
Total	220.00	23	—	—	—

[a]Significant at the .05 level.

$H_{0_{sub1}}$: Special inducements 1 and 2 do not differ in their effects on speed of sales of new houses, or

$$\mu_1 - \mu_2 = 0$$

$H_{a_{sub2}}$: It makes no difference to speed of sales of new houses whether actual special inducements are used or no special inducement is used, or

$$(\mu_1 + \mu_2) - 2\mu_3 = 0$$

Comparison	$\bar{X}_1$	$\bar{X}_2$	$\bar{X}_3$
1	+1	−1	0
2	+1	+1	−2

$$SS_{Comi} = \frac{(Com\ i)^2}{\dfrac{(m_{i1})^2}{n_1} + \dfrac{(m_{i2})^2}{n_2} + \cdots + \dfrac{(m_{iC})^2}{n_C}}$$

$$SS_{Com1} = \frac{[(1)\,(3.125) + (-1)\,(5.000)]^2}{\dfrac{(1)^2}{8} + \dfrac{(-1)^2}{8}} = 14.06$$

$$SS_{Com2} = \frac{[(1)\,(3.125) + (1)\,(5.000) + (-2)\,(6.875)]^2}{\dfrac{(1)^2}{8} + \dfrac{(1)^2}{8} + \dfrac{(-2)^2}{8}} = 42.19$$

Source of Variation	Sum of Squares	Degrees of Freedom	Mean Square	Observed F	.05 Level Critical F
Between columns (treatment levels)	56.25	2	28.12	3.61*	3.47
Comparison 1	14.06	1	14.06	1.80	4.32
Comparison 2	42.19	1	42.19	5.41*	4.32
Within columns	163.75	21	7.80	—	—
Total	220.00	23	—	—	—

*Significant at .05 level.

Figure 12.2. The method of orthogonal comparisons applied to the completely randomized experimental version of the Imperial Construction Company problem.

Conclusions

Part II of Table 12.2 reveals that the treatment variable is significant at the .05 level—but barely so.

In Figure 12.2 two subhypotheses are tested. The first states that there is no difference in effectiveness between the three-months-of-no-interest inducement, on the one hand, and

Sample means arranged in ascending order:

$$3.125 \qquad 5.000 \qquad 6.875$$

- -

Standard error of the sample means:

$$s_{\bar{x}_j} = \sqrt{\frac{MSWC}{n_j}} = \sqrt{\frac{7.80}{8}} = .9874$$

- -

Critical values:

$$2.94 \qquad 3.56$$

- -

Calculation of least significant ranges (LSR):

(1) Value of	(2) Critical Value	(3) $s_{\bar{x}_j}$	(4) (2) × (3) LSR
2	2.94	.9874	2.903
3	3.56	.9874	3.515

- -

Differences between all pairs of sample means:

	$\bar{X}_2 = 5.000$	$\bar{X}_3 = 6.875$	
$\bar{X}_1 = 3.125$	1.875	3.750	LSR 2 = 2.903
$\bar{X}_2 = 5.000$	----	1.875	LSR 3 = 3.515

- -

Conclusions:

$$\underline{3.125 \qquad 5.000} \qquad 6.875$$

Difference between $\bar{X}_1$ and $\bar{X}_2$ is not significant
Difference between $\bar{X}_2$ and $\bar{X}_3$ is not significant
Difference between $\bar{X}_1$ and $\bar{X}_3$ is significant

Figure 12.3. Application of Duncan's multiple-range test to the completely randomized experimental version of the Imperial Construction Company problem.

the free-carpeting inducement on the other. The second subhypothesis states that there is no difference in effectiveness between actual special inducements (considered together), on the one hand, and no special inducement on the other. These subhypotheses were picked somewhat arbitrarily, though it does seem reasonable that management would be most interested in these particular ones. We see in Figure 12.2 that there is no difference in effectiveness between the three-months-of-no-interest and the free-carpeting inducements (subhypothesis $H_{0_{\text{sub }1}}$ is accepted) but that some kind of special inducement, never mind which, is better than no special inducement at all (subhypothesis $H_{0_{\text{sub }2}}$ is rejected).

In Figure 12.3, where the Duncan multiple-range test is applied, the results pretty much confirm those obtained from the method of orthogonal comparisons. However, the Duncan test does reveal a piece of information the other missed: If use of some kind of special inducement is to be instituted, it probably should be the three-months-of-no-interest rather than the free-carpeting inducement. Do you see why?

Advantages and Limitations of the Completely Randomized Design

Despite its relative simplicity, the completely randomized design boasts a number of advantages. These include

1. *Flexibility.* This design can accommodate any number of treatment levels and any number of test units per treatment level.
2. *Unequal sample sizes are allowed.* This design permits use of different sample sizes for different treatment levels. Although, as we have seen, certain advantages accompany use of equal n_j's, in experimentation it is not uncommon for certain test units to "drop out," especially if they are people, or for some observations to be of suspicious quality and thus discardable. Therefore, when the time comes to apply analysis of variance, the sample sizes may differ. Moreover, there are times when samples of unequal size are used intentionally. For example, in a case like ours where actual treatments and a control are used, comparisons can be made more precisely by assigning more test units to the control.
3. *Large number of degrees of freedom for the error mean square.* The number of degrees of freedom associated with the error mean square (that is, the denominator of the F ratio, SSWC in the above example) is larger than for a restricted randomization design with equal n.
4. *Few assumptions.* As mentioned in section 11.9, four underlying assumptions is a minimum for analysis of variance. Some more complex methods require more than the four assumptions associated with use of a completely randomized design.

The most serious disadvantage of a completely randomized design is its relative inefficiency when the test units are quite disparate in size or responsiveness. Restricted randomization designs like the randomized block (section 12.4 and the last part of section 12.5) can utilize knowledge about extraneous sources of variation to reduce the size of the error mean square. A smaller denominator in the F ratio, resulting from elimination of the effects of some extraneous factor, produces a sharper test of the treatment variable.

*12.3 Introduction to Statistical Modeling in ANOVA Designs

In Chapter 11 and to this point in the present chapter, we have focused primarily on the underlying rationale of one-way ANOVA problems and the computational procedures used to solve them. The underlying assumptions have been listed and elaborated on briefly; however, their implications for helping us to determine what kind of F ratio to compute have received little mention. It is time to look more closely at these assumptions and, in so doing, introduce some pertinent points about statistical modeling.

As you will see as you proceed through this chapter, experiments intended to shed light on the possible influence of a treatment variable on the dependent variable can be designed in a great many ways besides the one-way design; analysis of variance methods have proved inestimably helpful in connection with many of these ways. A concise, explicit statement of the conditions assumed to exist in generating the data to fulfill a specific kind of design is often useful. Such a statement is called a *statistical model*. Specification of the model is essential for understanding the form of analysis of variance to be used because the analysis is a direct outgrowth of the assumptions of the model.

At the outset, let us distinguish between two kinds of treatment variables and the kinds of statistical models they imply:

> A treatment variable is said to be *fixed* if the treatment levels actually used in the experiment constitute an *entire population* of treatment levels of interest.
>
> A treatment variable is said to be *random* if the treatment levels used in the experiment constitute a *random sample* from a larger population of treatment levels of interest.

If the investigator actually selects a random sample of treatment levels, there is no mistaking what kind of treatment variable she is using; it will be a random one. However, the treatment levels used in the experiment need not actually be selected at random as long as they are reasonably representative of the large population of levels of interest. Consequently, some treatment variables are inherently random in nature. For example, if someone wished to test the productivity of three brands of similar machines—call them Brands A, B, and C—and used one machine of each brand in the experiment, chances are she would not be particularly interested in the performance of, say, the Brand A machine per se. It seems more reasonable to think that she would care about the performance of this Brand A machine as a *representative of all comparable machines of the same brand*. On the other hand, a marketing researcher testing three kinds of package design for a new product would most likely be interested in the sales effectiveness of, say, Design A *for its own sake*. If so, package design would be a fixed treatment variable. To be sure, if the marketing researcher actually selected three package designs at random from a larger set of designs of interest, then this treatment variable would be random. However, unless we know such to be the case, we are well advised to view the three package designs used in the experiment as a *complete population* of package designs.

> With a completely randomized design, an experiment employing a fixed treatment variable is referred to as having a *fixed-effects model*. This kind of model is also called *Model I*.
>
> With a completely randomized design, an experiment employing a random treatment variable is referred to as having a *random-effects* model. This kind of model is also called *Model II*.

(Though perhaps unnecessary, this writer feels a compulsion to remind you not to confuse treatment levels and test units. In analysis of variance, a set of test units is always viewed as a random sample from some population of possible test units. What we are talking about now is whether a set of treatment levels should itself be viewed as a random sample or as a complete population.)

Do the treatment levels (1) three months of no interest, (2) free carpeting, and (3) no special inducement constitute an entire population of treatment levels of interest? If so, our treatment variable is said to be fixed and our statistical model is a fixed-effects model. Do these treatment levels constitute only a random sample from a larger population of treatment levels? If so, our treatment variable is said to be random and our statistical model is a random-effects model.

Although we have not as yet actually said what kind of treatment variable we are working with, we have spoken as if our three treatment levels are the only ones with which management has any concern. For this reason, we will proceed on the assumption that our model is a fixed-effects one.

We may summarize the assumptions listed in section 11.9 of Chapter 11, along with relevant additional assumptions, with the following statement for the fixed-effects, completely randomized design problem described above (*NID* stands for "normally and independently distributed"):

FIXED-EFFECTS MODEL, OR MODEL I

$$X_{ij} = \mu + T_j + \epsilon_{ij} \qquad (12.3.1)$$

where ϵ_{ij}: NID$(0, \sigma_e^2)$

$$\sum_{\text{all } j} T_j = 0$$

This model states that each observation X_{ij} is composed of three linear and additive components. The first one, μ, is the true population mean effect for which $\overline{\overline{X}} = 5$ (Table 12.1) is a point estimate. The second component, T_j, is the true effect of the jth treatment level. For example, if we focus on, say, the three-months-of-no-interest inducement, we may call the true treatment-level effect T_1, where $T_1 = \mu_1 - \mu$; $\overline{X}_1 - \overline{\overline{X}} = 3.125 - 5.000 = -1.875$ is a point estimate of T_1. The point estimate of T_2 is $\overline{X}_2 - \overline{\overline{X}} = 5 - 5 = 0$. The point estimate of T_3 is $\overline{X}_3 - \overline{\overline{X}} = 6.875 - 5.000 = 1.875$. The third component of the model, ϵ_{ij}, is the residual error associated with the ith test unit and the jth treatment level.

The model also states that the ϵ_{ij} values are assumed to be independently and normally distributed about a mean of 0. The point estimate for the variance of the distribution of ϵ_{ij} values, σ_e^2, for our example, is the sample error mean square—that is, MSWC = SSWC/$(n - c) = 163.75/21 = 7.80$. (This information comes from Table 12.2.)

Finally, the model states that the sum of the true effects of the c treatment levels (three for our example) is zero:

$$\sum_{\text{all } j} T_j = 0$$

This assumption establishes our model as a fixed-effects one.

Given all these assumptions, one can derive expressions for the components of the true population mean squares. If the design dictates that one observation is to be made on each test unit and that there are to be equal numbers of observations associated with each treatment level, it can be shown that the structures of the population mean squares will be as presented in the right-hand column of Part I in Table 12.3.

We may now examine what would happen if we were to conclude from our F test that the null hypothesis is true—that is, if we conclude that type of special inducement does indeed have no effect on speed of sale of new houses. This accepted null hypothesis, together with the assumptions cited above, ensures that not only is the sum of the treatment effects equal to zero but that each treatment-level effect is itself zero. In symbols:

$$\sum_{\text{all } j} T_j = 0$$

and $T_1 = T_2 = T_3 = 0$. From Part I of Table 12.3, we see that the true mean square for which MSBC = SSBC/$(c - 1)$ is a point estimate will be just σ_e^2. Of course, the population mean square for which MSWC = SSWC/$(n - c)$ is a point estimate will also be just σ_e^2. What this says is that SSBC/$(c - 1)$ and SSWC/$(n - c)$ are independent estimates of σ_e^2 from equivalent normal populations and that their ratio will be distributed as F when the null hypothesis is true.

To be sure, if the null hypothesis checks out as not true, that is, if special inducements do have an effect on speed of sale of new houses, then the population mean square for which

Table 12.3 Components of Variance for a Completely Randomized Design

Part I: Fixed-Effects Model

Source of Variation	Sum of Squares	Degrees of Freedom	Sample Mean Square	Components of True Population Mean Square[a]
Between columns	SSBC	$c - 1$	$\dfrac{\text{SSBC}}{c - 1}$	$\sigma_e^2 + \dfrac{\sum\limits_{\text{all } j} n_j T_j^2}{c - 1}$ [b]
Within columns	SSWC	$n - c$	$\dfrac{\text{SSWC}}{n - c}$	σ_e^2

Part II: Random-Effects Model

Source of Variation	Sum of Squares	Degrees of Freedom	Sample Mean Square	Components of True Population Mean Square
Between columns	SSBC	$c - 1$	$\dfrac{\text{SSBC}}{c - 1}$	$\sigma_e^2 + n_j \sigma_T^2$
Within columns	SSWC	$n - c$	$\dfrac{\text{SSWC}}{n - c}$	σ_e^2

[a]In the literature of experimental designs, entries in this column are frequently referred to as *expected mean squares*, abbreviated EMS.
[b]It is common and convenient to write this merely as $\sigma_e^2 + n_j \sigma_T^2$.

SSBC/$(c - 1)$ is a point estimate will be greater than σ_e^2 by an amount equal to

$$\frac{\displaystyle\sum_{all\ j} n_j(\mu_j - \mu)^2}{c - 1} = \frac{\displaystyle\sum_{all\ j} n_j T_j^2}{c - 1}$$

and the sample F ratio, reflecting this population situation, would be expected to exceed 1.

Had we assumed a random-effects model, we would express it in symbols as follows (again, *NID* stands for "normally and independently distributed"):

RANDOM-EFFECTS MODEL, OR MODEL II

$$X_{ij} = \mu + T_j + \epsilon_{ij} \qquad (12.3.2)$$

where ϵ_{ij}: NID$(0, \sigma_e^2)$
T_j: NID$(0, \sigma_T^2)$

In this model the three treatment levels in our example would be presumed to be a random sample from a larger population of special inducements of interest. Hence, we would not assume that $\displaystyle\sum_{all\ j} T_j = 0$, but rather that the T_j values comprise a population that is independently and normally distributed with a mean of zero and a variance of σ_T^2, not necessarily equal to the variance σ_e^2 for the ϵ_{ij} values.

Under such a model, the structures of the true population mean squares would be as shown in Part II of Table 12.3.

Despite the fact that Model II differs from Model I, the analysis of variance is conducted in an identical fashion because in each case $F = $ MSBC/MSWC is a point estimate of a ratio having two components in the numerator and only one in the denominator. In words:

$$\frac{\text{Error variance} + \text{Variation attributable to treatment levels}}{\text{Error variance}}$$

where the error variance, denoted σ_e^2, is a measure of variation in the population data attributable to many, largely unknown, extraneous factors. As we will see, different assumed models will not always lead to the same procedure for testing a null hypothesis. That is why explicit consideration of the statistical model is so very important.

12.4 The Randomized Block Design: Single Observation per Cell

In the experimental version of the Imperial Construction Company problem, we sought to determine whether different kinds of special inducements have different effects on the length of time required to sell a completed house. The special inducements were listed as column headings in Table 12.1 and we were concerned about whether there were significant be-tween-columns effects. In Chapter 11 we worked through the same kind of procedure using

Table 12.4 Data for Simple Randomized Block Experiment, Imperial Construction Company

(Figures Represent Number of Weeks Required to Sell House)

Type of Special Inducement	Low	Medium-Low	Medium-High	High	Total
	Price Categories ← Classification or blocking variable				
Three months of no-interest payments	1	2	3	4	10
Free carpeting	3	1	4	6	14
No special inducement	3	6	8	8	25
Total	7	9	15	18	49

Treatment variable

Treatment levels

price categories for the independent variable. To carry this example a step further, let us assume, for the sake of easing our way into the subject of two-way analysis of variance, that there were only 12 houses—rather than 24—included in the experiment. (The randomized block design utilizing all 24 sample observations will be treated in section 12.5.) The 12 sample houses, let us say, are classified according to the four price categories listed earlier. Moreover, each kind of special inducement—three months of no interest, free carpeting, and no special inducement—was tried in connection with one house in each price category, as shown in Table 12.4. We will give management credit for having assigned the test units (houses) to treatment levels (inducements) in a random manner; it is this precaution, in addition to the presence of a classification variable (price categories), that makes this kind of experiment a *randomized block design*.

Definition and Requirements

A *randomized block design* is one which utilizes two independent variables, one of which is a treatment variable and the other a classification variable (or blocking variable). Moreover, in a randomized block experiment, the test units within each specific category, or *block* as it is called, of the classification variable are assigned to treatment levels in a random manner.

More specifically, a randomized block experiment has three characteristics:

1. For a single-observation-per-cell problem, there must be exactly as many test units within each block as there are treatment levels.
2. The treatment levels must be assigned at random to the test units within each block.
3. For a randomized block design to represent an improvement over a completely randomized design, the test units within a block must be relatively homogeneous with respect to the extraneous factor which prompted the use of blocking. Moreover, the test units should be relatively heterogeneous between blocks.

For the Imperial Construction case, we may find that, by utilizing price categories as a blocking variable, the differences in effectiveness among the kinds of special inducements can be more clearly discerned. That is, if in the completely randomized version of this problem, price had, unknown to us, been inflating the denominator of the sample F ratio, the F ratio would be lower than it would have been if the effect of the price variable had been removed. By measuring the effect of the price categories and removing it from the denominator of the sample F ratio, we may find that the treatment variable is really more important than a completely randomized analysis with the same n would have revealed. Tables 12.5, Part II, and 12.6 illustrate this point.

Figure 12.4 shows in schematic form how we attempt to partition the total sum of squares when employing a randomized block design (with a single observation per cell). It also illustrates the potential advantage of the randomized block design over the completely randomized design.

Achieving Random Allocation

Recall that one requirement of a randomized block design is that the treatment levels must be assigned at random to the test units within each block. When there is only one observa-

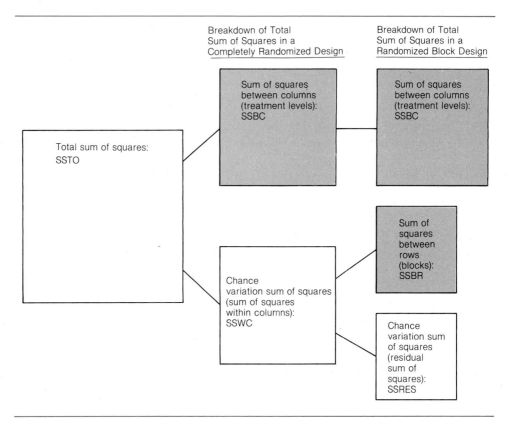

Figure 12.4. The breakdown of the total sum of squares in (1) a completely randomized design and (2) a randomized block design with one observation per cell. If the classification variable in a randomized block design is a meaningful one, the denominator of the F ratio will be reduced.

tion per cell, this task is relatively simple—though extremely important. For simplicity, let us focus on the three houses represented in Table 12.4 which have been classified as low in price. We will call these Houses A, B, and C. The kind of question which must now be answered is: "Should House A be used in connection with the three-months-of-no-interest treatment level, the free-carpeting treatment level, or the no-special-inducement treatment level?" Such questions can be handled adequately through use of a table of random digits. For example, we might decide that if the first random number selected (using a predetermined method) is 0, 1, or 2, House A will be assigned the three-months-of-no-interest treatment level; if the random number is 3, 4, or 5, House A will be assigned to the free-carpeting treatment level; if the random number is 6, 7, or 8, House A will be assigned to the no-special-inducement treatment level; and, finally, if the random number is 9, it will be ignored. When House A has been assigned a treatment level, House B in the low-price category will be assigned to *another* treatment level in exactly the same way. Once Houses A and B have been assigned to treatment levels, House C in the low-price category will be assigned to the remaining treatment level. Then we would move to the medium-low-price category, which includes, let us say, Houses D, E, and F, and repeat the same procedure. Then we would do the same with the medium-high category (made up of, say, Houses G, H, and I) and the high-price category (Houses J, K, and L).

The Assumptions

When using a randomized block design having a single observation per cell of the organizing table (Table 12.4 in the present example), we make the following five assumptions:

1. Each cell entry, that is, each X_{ij} value, is a random sample of size 1 from one of the *r*-times-*c* populations represented, where *r* represents the number of treatment levels and *c* the number of blocks used in the experiment. (*Note:* $rc = n$.)
2. The *n* random samples of size 1 are independent.
3. Each of the populations represented is normally distributed.
4. All of the populations have the same variance.
5. The block (columns) and treatment-level (rows) effects are additive.

An alternative way of expressing assumption 5 is: There is no *interaction* between blocks and treatment levels. Loosely speaking, when we say there is no interaction, we mean that no combination of blocks and treatment levels produces an effect that is greater than or smaller than the sum of their individual effects. Interaction will be discussed in much greater detail in sections 12.6 and 12.7. It is mentioned here only for the purpose of pointing out that interaction cannot be measured when there is only one observation per cell of the organizing table (Table 12.4). Therefore, we are obliged to proceed as if it does not exist. (Of course, to say that something cannot be measured is not the same as saying that it doesn't exist. One should make some effort to evaluate this assumption before proceeding. If an argument can be made for the presence of interaction, one would be well advised to use a more complex randomized block design, for example, the kind treated at the end of section 12.5.) When we make use of a two-way experiment *having more than one observation per cell*, we can measure interaction and can thus drop the fifth assumption.

✳ Possible Models

With any two-way randomized block experiment, three possible models exist and the model specified can sometimes affect importantly the way we obtain the sample F ratios.

> In a two-way analysis, if both independent variables are fixed (as defined in section 12.3), the model is said to be a *fixed-effects model*. This model is also called *Model I*.
>
> In a two-way analysis, if both independent variables are random (as defined in section 12.3), the model is said to be a *random-effects model*. This model is also called *Model II*.
>
> In a two-way analysis, if one independent variable is fixed and the other is random, the model is said to be *mixed*. This model is also called *Model III*. Moreover, since either of the two variables could be the fixed one, there are two possible kinds of mixed models.

Details of these models will be presented in the order listed here. (Remember that interaction is assumed to be nonexistent.)

FIXED-EFFECTS MODEL, OR MODEL I

$$X_{ij} = \mu + B_j + T_i + \epsilon_{ij} \qquad (12.4.1)$$

where ϵ_{ij}: NID$(0, \sigma_e^2)$

$$\sum_{\text{all } j} B_j = 0$$

$$\sum_{\text{all } i} T_i = 0$$

Thus, any observed X_{ij} value is viewed as being composed of the following linear and additive components: (1) the true population mean effect μ, (2) the effect of the blocks B_j, (3) the effect of the treatments T_i, and (4) the residual error associated with the ith treatment level and the jth block ϵ_{ij}.

Moreover, the ϵ_{ij} values are assumed to be normally and independently distributed about a mean of 0 and to have a variance equal to σ_e^2. With respect to the blocking variable, $B_j = \mu_j - \mu$ is the effect of the jth block and $\overline{X}_j - \overline{\overline{X}}$ is a point estimate of B_j. The sum of the B_j values is 0. Similarly, $T_i = \mu_i - \mu$ is the effect of the ith treatment level and $\overline{X}_i - \overline{\overline{X}}$ is a point estimate of T_i. The sum of the T_i values is 0.

$$X_{ij} = \mu + B_j + T_i + \epsilon_{ij} \qquad (12.4.2)$$

where ϵ_{ij}: $NID(0,\ \sigma_e^2)$

B_j: $NID(0,\ \sigma_B^2)$

T_i: $NID(0,\ \sigma_T^2)$

MIXED MODEL, OR MODEL III

$$X_{ij} = \mu + B_j + T_i + \epsilon_{ij} \qquad (12.4.3)$$

where ϵ_{ij}: $NID(0,\ \sigma_e^2)$

$$\sum_{\text{all } j} B_j = 0$$

T_i: $NID(0,\ \sigma_T^2)$

if blocks are fixed and treatment levels are random. Or

$$X_{ij} = \mu + B_j + T_i + \epsilon_{ij} \qquad (12.4.4)$$

where ϵ_{ij}: $NID(0,\ \sigma_e^2)$

B_j: $NID(0,\ \sigma_T^2)$

$$\sum_{\text{all } i} T_i = 0$$

if treatment levels are fixed and blocks are random.

(*Note:* This last-mentioned model is the kind ordinarily associated with a randomized block design: The test units usually represent random samples from the separate blocks whereas the treatment levels, in the absence of information to the contrary, are viewed as the only treatment levels of interest.)

If we were to examine in a manner similar to that used in Table 12.3 the components of variance associated with MSBC, MSBR, and MSRES, respectively, we would find that

1. The population value for which MSBC is a point estimate contains two components of variance, the error variance σ_e^2 and a measure of variation associated with blocks.
2. The population value for which MSBR is a point estimate also contains two components of variance, σ_e^2 and a measure of the variation associated with treatments.
3. The population value for which MSRES (residual mean square, the denominator of the F ratios in this kind of analysis) is a point estimate contains only one component of variance, σ_e^2.

Thus, the model assumed makes no difference to the way in which we calculate the sample F ratios. The MSRES will be the denominator when the treatment-level hypothesis is tested and when (if desired) the blocks hypothesis is tested—regardless of the model.

By now you may be wondering why such a fuss is being made over the matter of model specification. When, if ever, will the model make a difference? Actually, the model specified can make a great deal of difference in problems having measurable interactions, as will be demonstrated in section 12.5.

Notations and Formulas

In a two-way analysis of the kind under discussion, we can decompose SSTO and df_T into three parts each. For the sums of squares, we have

$$SSTO = SSBR + SSBC + SSRES$$

where   SSTO = Total sum of squares

SSBR = Between-rows sum of squares

SSBC = Between-columns sum of squares

SSRES = Residual sum of squares

The residual sum of squares, SSRES, measures the variation in the X_{ij} values not accounted for by either the variable represented by the columns or the variable represented by the rows. Therefore, SSRES divided by the appropriate number of degrees of freedom will serve as our measure of chance variation—that is, the error mean square to be used as the denominator of the F ratios.

For degrees of freedom, the breakdown is

$$df_T = df_{11} + df_{12} + df_2$$

where df_T = Total degrees of freedom

df_{11} = Between-rows degrees of freedom

df_{12} = Between-columns degrees of freedom

df_2 = Residual degrees of freedom

The subscripts assigned the between-columns and between-rows degrees of freedom, df_{11} and df_{12}, respectively, are used to indicate that there may be two F tests and, consequently, two F-ratio numerators. Therefore, insofar as Appendix Table 6 is concerned, there will be two df_1's and only one df_2.

We will also modify somewhat the notations used for summation and means. We observe that Table 12.4 has four columns and three rows. We will continue to use c to indicate the number of columns; thus, $c = 4$. We will indicate the number of rows by r; thus, $r = 3$. Since in the kind of design under discussion the number of observations associated with any given column will necessarily be the same as that associated with any other column, we will dispense with n_j and use r instead. Similarly, the number of observations associated with any specified row must be c. In our illustrative problem, we have $n = rc = (3)(4) = 12$ sample observations altogether.

We will let X_{ij} represent the number of weeks required to sell a house in the jth price class using the ith special inducement, where $j = 1, 2, 3, 4$ and $i = 1, 2, 3$. For example, the X_{ij} value denoted X_{34} would be 8 because the house in the fourth price class ($j = 4$) sold with the help of the third special inducement ($i = 3$) is seen to have required 8 weeks to sell. We will express the values in a specific column as $X_{i1}, X_{i2}, X_{i3},$ or X_{i4} and in a specific row as $X_{1j}, X_{2j},$ or X_{3j}. Column totals will be written

$$\sum_{\text{all } i} X_{\bullet 1}, \ \sum_{\text{all } i} X_{\bullet 2}, \ \sum_{\text{all } i} X_{\bullet 3}, \ \text{or} \ \sum_{\text{all } i} X_{\bullet 4}$$

Row totals will be written

$$\sum_{\text{all } j} X_{1 \bullet}, \ \sum_{\text{all } j} X_{2 \bullet}, \ \text{or} \ \sum_{\text{all } j} X_{3 \bullet}$$

The total of all n observations will be expressed

$$\sum_{\text{all } i} \sum_{\text{all } j} X_{\bullet\bullet}$$

When we wish to express the idea that some column total has been divided by r to obtain the arithmetic mean of that column, we will write $\overline{X}_{\bullet 1}, \overline{X}_{\bullet 2}, \overline{X}_{\bullet 3},$ or $\overline{X}_{\bullet 4}$. When we wish to do the same thing for rows, we will write $\overline{X}_{1 \bullet}, \overline{X}_{2 \bullet}, \overline{X}_{3 \bullet}$. As before, the grand mean will be denoted $\overline{\overline{X}}$.

This notational system implies the following formulas for SSTO, SSBR, and SSBC, respectively:

Basic Formula   **Shortcut Formula**

$$\text{SSTO} = \sum_{\text{all } i} \sum_{\text{all } j} (X_{ij} - \overline{\overline{X}})^2 = \sum_{\text{all } i} \sum_{\text{all } j} X_{ij}^2 - \frac{\left(\sum\limits_{\text{all } i} \sum\limits_{\text{all } j} X_{\bullet\bullet} \right)^2}{n} \qquad (12.4.5)$$

$$\text{SSBR} = \sum_{\text{all } i} \sum_{\text{all } j} (\overline{X}_{i\bullet} - \overline{\overline{X}})^2 = \sum_{\text{all } i} \frac{\left(\sum\limits_{\text{all } j} X_{i\bullet} \right)^2}{c} - \frac{\left(\sum\limits_{\text{all } i} \sum\limits_{\text{all } j} X_{\bullet\bullet} \right)^2}{n} \qquad (12.4.6)$$

$$\text{SSBC} = \sum_{\text{all } j} \sum_{\text{all } i} (\overline{X}_{\bullet j} - \overline{\overline{X}})^2 = \sum_{\text{all } j} \frac{\left(\sum\limits_{\text{all } i} X_{\bullet j} \right)^2}{r} - \frac{\left(\sum\limits_{\text{all } i} \sum\limits_{\text{all } j} X_{\bullet\bullet} \right)^2}{n} \qquad (12.4.7)$$

Notice that two of these equations, (12.4.5) and (12.4.7), are identical, aside from very slight changes in notations, with those used in connection with the one-way analysis of variance. The only new formula in the above list is that for SSBR, and it is so closely analogous to the one for SSBC that it does not represent any noteworthy departure from methods already used.

Once SSTO, SSBR, and SSBC have been determined, SSRES can be computed readily by

$$SSRES = SSTO - (SSBR + SSBC) \qquad (12.4.8)$$

The degrees of freedom are determined by the following formulas:

$$df_T = n - 1 \qquad (12.4.9)$$
$$df_{11} = r - 1 \qquad (12.4.10)$$
$$df_{12} = c - 1 \qquad (12.4.11)$$
$$df_2 = (r - 1)(c - 1) \qquad (12.4.12)$$

Let us now proceed in step-by-step fashion to solve our illustrative problem.

1. State the null and alternative hypotheses.
   There can be two null hypotheses now:

   For Rows (Treatment Levels)

   H_{0_1}: $\mu_{1\bullet} = \mu_{2\bullet} = \mu_{3\bullet}$

   H_{a_1}: $\mu_{1\bullet}, \mu_{2\bullet}, \mu_{3\bullet}$ are not all equal

   For Columns (Blocks)

   H_{0_2}: $\mu_{\bullet 1} = \mu_{\bullet 2} = \mu_{\bullet 3} = \mu_{\bullet 4}$

   H_{a_2}: $\mu_{\bullet 1}, \mu_{\bullet 2}, \mu_{\bullet 3}, \mu_{\bullet 4}$ are not all equal

2. State the level of significance to be used.
   Let us say that α is set at .05.

3. Using the shortcut versions of Equations (12.4.5), (12.4.6), and (12.4.7), compute total sum of squares, between-rows sum of squares, and between-columns sum of squares.

$$SSTO = \sum_{\text{all } i}\sum_{\text{all } j} X_{ij}^2 - \frac{\left(\sum_{\text{all } i}\sum_{\text{all } j} X_{\bullet\bullet}\right)^2}{n} = (1)^2 + (3)^2 + \cdots + (8)^2 - \frac{(49)^2}{12}$$

$$= 265.00 - 200.08 = \underline{64.92}$$

$$SSBR = \sum_{all\ i} \frac{\left(\sum\limits_{all\ j} X_{i\bullet}\right)^2}{c} - \frac{\left(\sum\limits_{all\ i}\sum\limits_{all\ j} X_{\bullet\bullet}\right)^2}{n} = \frac{(10)^2 + (14)^2 + (25)^2}{4} - \frac{(49)^2}{12}$$

$$= 230.25 - 200.08 = \underline{30.17}$$

$$SSBC = \sum_{all\ j} \frac{\left(\sum\limits_{all\ i} X_{\bullet j}\right)^2}{n} - \frac{\left(\sum\limits_{all\ i}\sum\limits_{all\ j} X_{\bullet\bullet}\right)^2}{n} = \frac{(7)^2 + (9)^2 + (15)^2 + (18)^2}{3}$$

$$- \frac{(49)^2}{12} = 226.33 - 200.08 = \underline{26.25}$$

4. Compute the residual sum of squares, SSRES.

   SSRES = SSTO − (SSBR + SSBC) = 64.92 − (30.17 + 26.25) = $\underline{8.50}$

5. Using Equations (12.4.9) through (12.4.12), compute the degrees of freedom.

$$df_T = n - 1 = 12 - 1 = 11$$
$$df_{11} = r - 1 = 3 - 1 = 2$$
$$df_{12} = c - 1 = 4 - 1 = 3$$
$$df_2 = (r - 1)(c - 1) = (4 - 1)(3 - 1) = 6$$

6. Place all sums of squares and all degrees-of-freedom values into an ANOVA table and test the two null hypotheses.

Part I of Table 12.5 shows a general ANOVA table for a randomized block design with a single observation per cell. Part II shows the ANOVA table for the Imperial Construction randomized block design. We see that the special-inducements variable *does* appear to influence importantly the speed with which the company's houses can be sold.

The advantage of using a blocking variable, provided it is a relevant one—which is to say, one capable of producing homogeneity among test units within blocks and heterogeneity between blocks—is demonstrated in Table 12.6. In this table, results are shown for a one-way analysis involving $n = 12$ observations and the treatment variable only. In other words, everything related to Table 12.6 is the same as for Part II of Table 12.5 except that in Table 12.6 a completely randomized design (with the price variable ignored) is assumed. As a result, the effect of special inducements seems not to be significant. Judging from the results of our randomized block experiment, the importance of the treatment variable is masked in the simpler experimental design by the presence of an unacknowledged, but influential, extraneous factor.

Comparison with the *t* Test for Paired Observations

In section 10.5 of Chapter 10, we performed a *t* test for paired observations on some experimental data for the River City Manufacturing Company. In this problem, the treatment variable was temperature within the plant. There were only two treatment levels, 55°F and 75°F. The purpose of the experiment was to determine whether temperature affects worker output. Data for this study were presented in Table 10.4. The analysis resulted in an ob-

Table 12.5 ANOVA Table for a Randomized Block Experiment: Single Observation per Cell

Part I: General Structure

Source of Variation	Sum of Squares	Degrees of Freedom	Mean Square	Observed F
Between rows (treatment levels)	SSBR	$r - 1$	$MSBR = \dfrac{SSBR}{r - 1}$	$\dfrac{MSBR}{MSRES}$
Between columns (blocks)	SSBC	$c - 1$	$MSBC = \dfrac{SSBC}{c - 1}$	$\dfrac{MSBC}{MSRES}$
Residual	SSRES	$(r - 1)(c - 1)$	$MSRES = \dfrac{SSRES}{(c - 1)(r - 1)}$	—
Total	SSTO	$n - 1$	—	—

Part II: ANOVA Table for the Randomized-Block Experimental Version of the Imperial Construction Company Problem

Source of Variation	Sum of Squares	Degrees of Freedom	Mean Square	Observed F	.05 Level Critical F
Between rows (treatment levels)	30.17	2	15.08	10.62[a]	5.14
Between columns (blocks)	26.25	3	8.75	6.16[a]	4.76
Residual	8.50	6	1.42	—	—
Total	64.92	11	—	—	—

[a]Significant at the .05 level.

Table 12.6 Data from Table 12.5 Analyzed as a Completely Randomized Design with Price Categories (Columns) Ignored

(1) Source of Variation	(2) Sum of Squares	(3) Degrees of Freedom	(4) Mean Square	(5) F	(6) .05 Level Critical F
Between rows	30.17	2	15.08	3.91	4.26
Within rows	34.75	9	3.86	—	—
Total	64.92	11	—	—	—

served t value of 2.87 which was compared with a critical t of 2.756 and found significant at the .01 level.

Table 12.7 presents the results of a two-way analysis of variance applied to the same data. The importance of the temperature variable, originally brought to light by the t test for paired observations, is confirmed by the F test. It is no accident that the results of these tests are in accord. Notice that

Table 12.7 ANOVA Table for a Two-Way Analysis of Variance Applied to the River City Manufacturing Company Data

Source of Variation	Sum of Squares	Degrees of Freedom	Mean Square	Observed F	.01 Level Critical F
Between rows (workers)	1439.68	29	49.64	—	—
Between columns (temperatures)	93.75	1	93.75	8.17^a	7.60
Residual	332.75	29	11.47	—	—
Total	1866.18	59	—	—	—

^{a}Significant at the .01 level.

$$\sqrt{\text{Observed } F} = \sqrt{8.17} = 2.86 \quad \text{and} \quad \sqrt{\text{Critical } F} = \sqrt{7.60} = 2.756$$

$$\text{Observed } t = 2.86 \qquad\qquad \text{Critical } t = 2.756$$

We see, then, that these are perfectly equivalent tests when (1) the t test has $n - 1$ degrees of freedom associated with it and (2) the F test has one degree of freedom associated with the numerator and the degrees of freedom associated with the denominator $[(r - 1)(c - 1)]$ is the same as the $n - 1$ value for the t test. Consequently, an experiment which can properly be analyzed by the t test for paired observations may also properly be viewed as a particular kind of randomized block experiment.

12.5 Two-Way Factorial Arrangement of Treatment Levels: More Than One Observation per Cell

In this section you will have an opportunity to learn why knowledge of the statistical model underlying an experiment is so very important to proper analysis. You will also see why the ability to measure interaction is an advantage—not only from the standpoint of the statistical analysis but also from the standpoint of sensible business decision making.

We will limit our attention to experiments involving the same number of observations per cell (but a number greater than 1). With the same number of observations per cell, the kind of experimental design we will be discussing is just a shade more difficult than the randomized block design discussed in section 12.4. When cell frequencies are not all equal, the analysis becomes considerably more complicated. Although it would be most unusual to find equal cell frequencies resulting from a survey-sampling study, it is not nearly so unusual in experimental studies. In the latter case, the investigator controls the choice of treatment levels and the manner of allocating test units to the treatment levels. Two closely related advantages of using more than one observation per cell are that (1) it permits the investigator to measure sample interaction and to test it for significance and (2) it provides a "pure" estimate of the population error variance σ_e^2. This estimate in the present kind of design is the mean square within cells, MSWCL. MSWCL is a "pure" estimate of σ_e^2 in that it is unaffected by the variable represented by the rows, the variable represented by the columns, and any interaction that may exist between the two variables.

> The phrase *factorial arrangement of treatment levels* refers to an arrangement of experimental data such that each level of one independent variable is paired with each level of every other independent variable.

We have already dealt with one such factorial design—namely, the randomized block design. At the end of this section we will return to the randomized block design approach to analyzing the Imperial Construction Company case—only this time making use of two houses per cell.

Before that, however, we will analyze the following illustrative case involving two *treatment variables*. This characteristic of the case makes the example our first "two-factor design." Previous designs have been single-factor designs either because (1) only one independent variable was considered or (2) only one treatment variable was used even though the analysis benefitted from the presence of a blocking variable as well.

ILLUSTRATIVE CASE 12.1

Management of Specialty Foods, Inc., plans to introduce an imported cheese product for sale in supermarkets. However, this product has never been mass-marketed before in any country, and management is concerned about a number of elements of the marketing mix. For example, the packaging department has developed three quite different kinds of containers for the product: (1) an old-fashioned-looking package with a Scandinavian motif, (2) a subdued package—rather plain and neither obviously old-fashioned nor obviously modern, and (3) a modern-looking package. We will refer to these simply as (1) old-fashioned, (2) neutral, and (3) modern. Management feels that each type of package has some obvious merits and perhaps some drawbacks. For convenience, we will refer to the type-of-package treatment variable as *Factor A*.

Price is also a puzzler. The product could be sold relatively inexpensively and, ordinarily, one would expect more sales at a lower price than at a higher price. However, this is a gourmet item; sometimes a higher price will lend an image of class to a product. The price treatment variable, made up of two treatment levels, $1 and $4, will be referred to as *Factor B*.

Management decides to engage in some test-marketing to help them determine how best to market this product. Toward this end, they obtain cooperation from 18 grocery stores thought to be roughly alike in size and type of customer served. In three stores, determined at random, the old-fashioned package and the $1 price were used; in three others, determined at random, the old-fashioned package and the $4 price were used; and so forth.

The distribution of grocery stores involved in this experiment is shown in Table 12.8. The sales data we will be analyzing are presented in Table 12.9.

The Assumptions and Possible Models

When employing the kind of experimental design treated in this section, we make the following assumptions:

1. The n_{ij} entries in each cell together constitute a random sample drawn from the population defined by the particular combination of levels for the two factors.
2. The samples are independent.
3. Each of the corresponding populations is normally distributed.
4. All populations have the same variance.

Table 12.8 Tabular Setup for a Two-Way Analysis of Variance with Three Observations per Cell

	Price Category		
Type of Package	Low	High	Total
Old-fashioned	3 stores	3 stores	6 stores
Neutral	3 stores	3 stores	6 stores
Modern	3 stores	3 stores	6 stores
Total	9 stores	9 stores	18 stores

Table 12.9 Data Used in a Two-Way Analysis of Variance Where both Independent Variables Are Treatment Variables (Three Observations per Cell)

	Price Category		
Type of Package	Low	High	Total
Old-fashioned	10	12	
	20	5	76
	25	4	
Cell total	55	21	
Neutral	7	11	
	6	8	41
	4	5	
Cell total	17	24	
Modern	8	6	
	5	4	31
	6	2	
Cell total	19	12	
Total	91	57	148

Notice that we are not required to assume that column and row effects are additive—which is to say that there is no interaction present. As previously mentioned, by using more than once observation per cell, we are able to tell whether there is or is not any interaction and, if there is, to separate its effects from the other effects.

The statistical model for a factorial arrangement of treatment levels (more than one observation per cell, equal frequencies per cell) can be described in the following possible ways:

$$X_{ijk} = \mu + A_i + B_j + (AB)_{ij} + \epsilon_{ijk} \qquad (12.5.1)$$

where ϵ_{ijk}: NID$(0, \sigma_e^2)$

and

		Mixed Models	*(Model III)*	
(1)	**(2)**	**(3)**	**(4)**	
Fixed-Effects Model (Model I)	**Factor A Fixed, Factor B Random**	**Factor A Random Factor B Fixed**	**Random-Effects Model (Model II)**	
$\displaystyle\sum_{\text{all } i} A_i = 0$	$\displaystyle\sum_{\text{all } i} A_i = 0$	A_i: NID$(0, \sigma_A^2)$	A_i: NID$(0, \sigma_A^2)$	
$\displaystyle\sum_{\text{all } j} B_j = 0$	B_j: NID$(0, \sigma_B^2)$	$\displaystyle\sum_{\text{all } j} B_j = 0$	B_j: NID$(0, \sigma_B^2)$	
$\displaystyle\sum_{\text{all } i} (AB)_{ij} = 0$	$(AB)_{ij}$: NID$(0, \sigma_{AB}^2)$	$(AB)_{ij}$: NID$(0, \sigma_{AB}^2)$	$(AB)_{ij}$: NID$(0, \sigma_{AB}^2)$	
	But $\displaystyle\sum_{\text{all } i} (AB)_{ij} = 0$	But $\displaystyle\sum_{\text{all } i} (AB)_{ij} \neq 0$		
	$\displaystyle\sum_{\text{all } j} (AB)_{ij} \neq 0$	$\displaystyle\sum_{\text{all } j} (AB)_{ij} = 0$		

We see from the above table that each X_{ijk} value is the sum of five linear components: (1) a mean effect μ, (2) a main effect of Factor A, A_i, (3) a main effect of Factor B, B_j, (4) an effect from the interaction of each level of Factor A in combination with each level of Factor B, $(AB)_{ij}$, and (5) a residual error assumed to be normally and independently distributed with a true mean of 0 and variance σ_e^2.

For a fixed-effects model—column (1) in the above table—where, recall, Factors A and B are both fixed and each, therefore, is regarded as an entire population of factor levels of interest, the sums of the Factor A effects, the Factor B effects, and the AB interaction effects are all 0.

We see from columns (2) and (3) of the above table that for a mixed model (one factor fixed while the other is viewed as a random sample from a larger population of factor levels), the relevant assumptions will depend on which factor is fixed and which is random. If we take Factor A to be fixed and Factor B to be random, then (1) the sum of the Factor A effects is 0, (2) the Factor B effects are assumed to be normally and independently distributed with a true mean of 0 and variance σ_B^2, (3) the interaction effects are also assumed to be normally and independently distributed with a mean of 0 and variance σ_{AB}^2, and (4) the sum of the interaction effects is 0 insofar as Factor A (the fixed factor) is concerned but not insofar as Factor B (the random factor) is concerned.

For the random effects model shown in column (4) of the above table, all components—A_i, B_j, $(AB)_{ij}$, and, of course ϵ_{ijk}—are assumed to be normally and independently distributed with a mean of 0 and a variance unique to the component being considered (that is, σ_A^2, σ_B^2, σ_{AB}^2, or σ_e^2).

Diagnostic Warmup: A Table of Means

An easy but useful way of getting a quick preview of what is likely to result from the formal analysis of variance is to prepare a table of cell, row, and column means. Such a table, constructed from the data presented in Table 12.9, is shown in Table 12.10 and suggests the following tentative conclusions:

1. A comparison of row means reveals that the old-fashioned design makes the best package for stimulating sales.
2. A comparison of column means reveals that the low price ($1) is better than the high price ($4) for stimulating sales.
3. But both of the above conclusions occur largely as a result of the spectacular effects of combining the old-fashioned package with the $1 price. In brief, we have strong reasons to believe that interaction is present.
4. In view of tentative conclusion 3, it seems likely that interaction will check out as significant. The main effects tests (that is, the test to determine whether type of package, on the one hand, and price category, on the other, make a difference) may or may not reveal significance. As we will see, that depends to a considerable degree on the kind of statistical model assumed.

Notations and Formulas

The statistical model specified will not affect the computational steps through the calculation of the mean squares. Therefore, we will describe these steps with an attempt at brevity. But then we must contemplate quite seriously the implications of the model before we can compute the F ratios correctly.

Table 12.10 The Means for the Specialty Foods Problem

Type of Package	Price Category		Row Means
	Low	High	
Old-fashioned	18.33	7.00	12.67
Neutral	5.67	8.00	6.83
Modern	6.33	4.00	5.17
Column means	10.11	6.33	8.22

In the type of problem under discussion, the total sum of squares is decomposed into the following parts:

$$SSTO = SSBR + SSBC + SS(rc) + SSWCL$$

where   SSTO = Total sum of squares

SSBR = Between-rows sum of squares

SSBC = Between-columns sum of squares

SS(rc) = Interaction sum of squares

SSWCL = Within-cells sum of squares

This breakdown is illustrated in Figure 12.5. The figure also shows why the present kind of two-way analysis is potentially superior to the kind involving only one observation per cell.

Worth noting is that, when SSBR, SSBC, and SS(rc) are summed, the result is the between-cells sum of squares, SSBCL. This measure is not ordinarily used in a test of significance, but it does serve as a facilitator of other computational steps.

Total degrees of freedom will be decomposed as follows:

$$df_T = df_{11} + df_{12} + df_{13} + df_2$$

where df_T = Total degrees of freedom

df_{11} = Between-rows degrees of freedom

df_{12} = Between-columns degrees of freedom

df_{13} = Interaction degrees of freedom

df_2 = Within-cells degrees of freedom

We observe that Table 12.9 has three rows, two columns, and six cells. In symbols: $r = 3$, $c = 2$, and $CL = 6$. The number of observations associated with any given cell can be denoted n_{ij}. The number of observations for any specified row is equal to cn_{ij} and for any specified column rn_{ij}. Therefore, in the Specialty Foods problem we have $n_{ij} = 3$ entries per cell, $cn_{ij} = (2)(3) = 6$ entries per row, and $r\,n_{ij} = (3)(3) = 9$ entries per column.

We will let X_{ijk} represent sales achieved by the kth store subjected to the ith package type and the jth price, where $k = 1, 2, 3$; $j = 1, 2$; and $i = 1, 2, 3$. For example, the X value denoted X_{321} would be 6 because the first store ($k = 1$) subjected to the third package type ($i = 3$) and the second price ($j = 2$) in Table 12.9 sold six dozen units of the product during the experimental period. We will express a value in a specific column as X_{i1k} or X_{i2k}, in a specific row as X_{1jk}, X_{2jk}, or X_{3jk}, and in a specific cell as X_{11k}, X_{21k}, X_{31k}, X_{12k}, X_{22k}, or X_{32k}. Row totals will be expressed

$$\sum_{\text{all } i}\sum_{\text{all } k} X_{1\bullet\bullet} \quad \text{or} \quad \sum_{\text{all } i}\sum_{\text{all } k} X_{2\bullet\bullet}$$

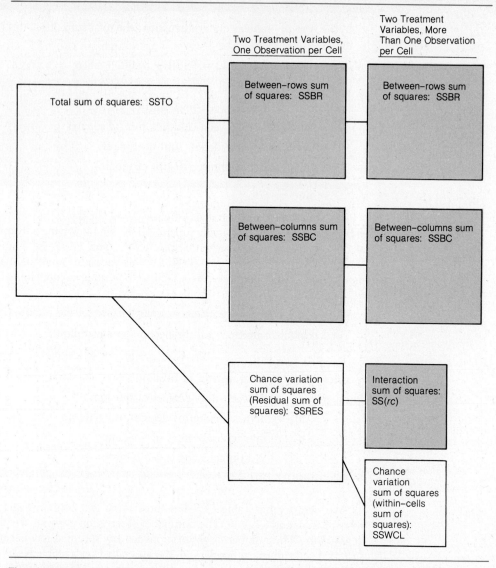

Figure 12.5. Decomposition of total sum of squares, SSTO, for two kinds of two-way analysis-of-variance problems.

Column totals will be expressed

$$\sum_{\text{all } j} \sum_{\text{all } k} X_{\bullet 1 \bullet}, \ \sum_{\text{all } j} \sum_{\text{all } k} X_{\bullet 2 \bullet}, \ \text{or} \ \sum_{\text{all } j} \sum_{\text{all } k} X_{\bullet 3 \bullet}$$

Cell totals will be written

$$\sum_{\text{all } k} X_{11\bullet}, \sum_{\text{all } k} X_{21\bullet}, \ldots, \sum_{\text{all } k} X_{32\bullet}.$$

The total of all n observations will be expressed

$$\sum_{\text{all } i} \sum_{\text{all } j} \sum_{\text{all } k} X_{\ldots}$$

When we wish to express the idea that (1) some column total has been divided by rn_{ij} to obtain the arithmetic mean of that column, we will write $\overline{X}_{\bullet 1\bullet}$ or $\overline{X}_{\bullet 2\bullet}$, (2) some row total has been divided by cn_{ij} to obtain the arithmetic mean of that row, we will write $\overline{X}_{1\bullet\bullet}$, $\overline{X}_{2\bullet\bullet}$, or $\overline{X}_{3\bullet\bullet}$, (3) some cell total has been divided by n_{ij} to obtain the mean of that cell, we will write $\overline{X}_{11\bullet}$, $\overline{X}_{21\bullet}$, . . . , $\overline{X}_{32\bullet}$. As usual, the grand mean will be denoted $\overline{\overline{X}}$. The basic and shortcut formulas for SSTO, SSBR, SSBC, and SSBCL are as follows:

Basic Formulas **Shortcut Formulas**

$$\text{SSTO} = \sum_{\text{all } i} \sum_{\text{all } j} \sum_{\text{all } k} (X_{ijk} - \overline{\overline{X}})^2$$

$$= \sum_{\text{all } i} \sum_{\text{all } j} \sum_{\text{all } k} X_{ijk}^2 - \frac{\left(\sum_{\text{all } i} \sum_{\text{all } j} \sum_{\text{all } k} X_{\ldots}\right)^2}{n} \qquad (12.5.2)$$

$$\text{SSBR} = \sum_{\text{all } i} \sum_{\text{all } j} \sum_{\text{all } k} (\overline{X}_{i\bullet\bullet} - \overline{\overline{X}})^2$$

$$= \sum_{\text{all } i} \frac{\left(\sum_{\text{all } j} \sum_{\text{all } k} X_{i\bullet\bullet}\right)^2}{cn_{ij}} - \frac{\left(\sum_{\text{all } i} \sum_{\text{all } j} \sum_{\text{all } k} X_{\ldots}\right)^2}{n} \qquad (12.5.3)$$

$$\text{SSBC} = \sum_{\text{all } i} \sum_{\text{all } j} \sum_{\text{all } k} (\overline{X}_{\bullet j\bullet} - \overline{\overline{X}})^2$$

$$= \sum_{\text{all } j} \frac{\left(\sum_{\text{all } i} \sum_{\text{all } k} X_{\bullet j\bullet}\right)^2}{rn_{ij}} - \frac{\left(\sum_{\text{all } i} \sum_{\text{all } j} \sum_{\text{all } k} X_{\ldots}\right)^2}{n} \qquad (12.5.4)$$

$$\text{SSBCL} = \sum_{\text{all } i} \sum_{\text{all } j} \sum_{\text{all } k} (X_{ij\bullet} - \overline{\overline{X}})^2$$

$$= \sum_{\text{all } i} \sum_{\text{all } j} \frac{\left(\sum_{\text{all } k} X_{ij\bullet}\right)^2}{n_{ij}} - \frac{\left(\sum_{\text{all } i} \sum_{\text{all } j} \sum_{\text{all } k} X_{\ldots}\right)^2}{n} \qquad (12.5.5)$$

Once SSTO, SSBR, SSBC, and SSBCL have been determined, SS(rc) and SSWCL can be computed readily by

$$SS(rc) = SSBCL - (SSBR + SSBC) \qquad (12.5.6)$$
$$SSWCL = SSTO - SSBCL \qquad (12.5.7)$$

The needed degrees of freedom formulas are

$$df_T = n - 1 \qquad (12.5.8)$$
$$df_{11} = r - 1 \qquad (12.5.9)$$
$$df_{12} = c - 1 \qquad (12.5.10)$$
$$df_{13} = (r - 1)(c - 1) \qquad (12.5.11)$$
$$df_2 = n - CL \text{ or } rc(n_{ij} - 1) \qquad (12.5.12)$$

Procedural Steps

The steps required to do the analysis of the type of problem under discussion are as follows:

1. State the *three* null and alternative hypotheses.

   For Rows (Package Types)

   H_{0_1}: $\mu_{1\bullet\bullet} = \mu_{2\bullet\bullet} = \mu_{3\bullet\bullet}$

   H_{a_1}: $\mu_{1\bullet\bullet}, \mu_{2\bullet\bullet}, \mu_{3\bullet\bullet}$ are not all equal

   For Columns (Price)

   H_{0_2}: $\mu_{\bullet1\bullet} = \mu_{\bullet2\bullet}$

   H_{a_2}: $\mu_{\bullet1\bullet}, \mu_{\bullet2\bullet}$ are not equal

   For Interaction

   H_{0_3}: INT(rc) = 0 for all cells

   H_{a_3}: INT(rc) $\neq$ 0 for at least one cell

The first two hypotheses are referred to as main effects hypotheses and the third as the first-order interaction hypothesis.

2. Determine the level of significance to be used.
   Let us continue to use $\alpha = .05$.

3. Compute CF (for "correction factor").

$$CF = \frac{\left(\sum_{\text{all } i} \sum_{\text{all } j} \sum_{\text{all } k} X_{\cdots} \right)^2}{n} = \frac{(148)^2}{18} = 1216.89$$

4. Using the shortcut versions of Equations (12.5.2) through (12.5.5), compute SSTO, SSBR, SSBC, and SSBCL.

$$\text{SSTO} = \sum_{\text{all } i} \sum_{\text{all } j} \sum_{\text{all } k} X_{ijk}^2 - \text{CF} = (10)^2 + (20)^2 + \ldots + (2)^2 - 1216.89$$

$$= 1802.00 - 1216.89 = \underline{585.11}$$

$$\text{SSBR} = \sum_{\text{all } i} \frac{\left(\sum_{\text{all } j} \sum_{\text{all } k} X_{i\bullet\bullet}\right)^2}{cn_{ij}} - \text{CF} = \frac{(76)^2 + (41)^2 + (31)^2}{6} - 1216.89$$

$$= 1403.00 - 1216.89 = \underline{186.11}$$

$$\text{SSBC} = \sum_{\text{all } j} \frac{\left(\sum_{\text{all } i} \sum_{\text{all } k} X_{\bullet j\bullet}\right)^2}{rn_{ij}} - \text{CF} = \frac{(91)^2 + (57)^2}{9} - 1216.89$$

$$= 1281.11 - 1216.89 = \underline{64.22}$$

$$\text{SSBCL} = \sum_{\text{all } i} \sum_{\text{all } j} \frac{\left(\sum_{\text{all } k} X_{ij\bullet}\right)^2}{n_{ij}} - \text{CF} = \frac{(55)^2 + (17)^2 + \cdots + (12)^2}{3} - 1216.89$$

$$= 1612.00 - 1216.89 = \underline{395.11}$$

5. Using Equations (12.5.6) and (12.5.7), compute SS(rc) and SSWCL.

$$\text{SS}(rc) = \text{SSBCL} - (\text{SSBR} + \text{SSBC}) = 395.11 - (64.22 + 196.11) = \underline{144.78}$$
$$\text{SSWCL} = \text{SSTO} - \text{SSBCL} = 585.11 - 395.11 = \underline{190.00}$$

6. Using Equations (12.5.8) through (12.5.12), determine df_T, df_{11}, df_{12}, df_{13}, and df_2.

$$df_T = n - 1 = 18 - 1 = 17$$
$$df_{11} = r - 1 = 3 - 1 = 2$$
$$df_{12} = c - 1 = 2 - 1 = 1$$
$$df_{13} = (r - 1)(c - 1) = (1)(2) = 2$$
$$df_2 = n - \text{CL} = 18 - 6 = 12$$

7. Place the sums of squares and degrees of freedom obtained above (excluding SSBCL) into an ANOVA table and calculate the mean squares.

Table 12.11 shows this, partially completed, ANOVA table (partially completed because we cannot finish the analysis without specifying the model we are using).

F Tests: Random-Effects Model Assumed

We will examine the implications of the random-effects model first. It seems rather unlikely that this model would really have been the one used in practice—but it could have been. Besides, beginning with the assumption of a random-effects model will permit you to (1) see why sometimes it is not the error mean square (MSWCL in the present case) that is used as the denominator of an F ratio and (2) become familiar with the most complex type of components-of-variance structure of the possible models. Once you become familiar with this structure, it is an easy matter to modify it to suit either a fixed-effects model or one of the mixed models.

Table 12.11 ANOVA Table for the Specialty Foods Company Case

Source of Variation	Sum of Squares	Degrees of Freedom	Mean Square	Observed F	.05 Level Critical F
Between rows (packages)	186.11	2	93.06		
Between columns (prices)	64.22	1	64.22		
Interaction	144.78	2	72.39		
Within cells	190.00	12	15.83		
Total	585.11	17	—	—	—

With a two-way random-effects factorial model, the r levels of Factor A actually used in the experiment constitute a random sample from a population of R levels of that factor, where $R > r$. Similarly, the c levels of Factor B actually used in the experiment constitute a random sample from a population of C levels of that factor, where $C > c$.

It can be shown that, by assuming a random-effects model, the population mean squares for which MSBR, MSBC, MS(rc), and MSWCL are point estimates possess the following sets of variance components:

$$\text{For Factor A (rows):} \quad \sigma_e^2 + n\left(1 - \frac{c}{C}\right)\sigma_{AB}^2 + nc\sigma_A^2 \quad (12.5.13)$$

$$\text{For Factor B (columns):} \quad \sigma_e^2 + n\left(1 - \frac{r}{R}\right)\sigma_{AB}^2 + nr\sigma_B^2 \quad (12.5.14)$$

$$\text{For interaction:} \quad \sigma_e^2 + n\,\sigma_{AB}^2 \quad (12.5.15)$$

$$\text{For within cells:} \quad \sigma_e^2 \quad (12.5.16)$$

Notice that:

1. The population mean square for which MSWCL is a point estimate has only one component of variance, namely, σ_e^2.
2. The population mean square for which SS(rc) is a point estimate has two components of variance, σ_e^2 and σ_{AB}^2.
3. The population mean square for which MSBR is a point estimate has three components of variance, σ_e^2, σ_{AB}^2, and σ_A^2.
4. The population means square for which MSBC is a point estimate also has three components of variance, σ_e^2, σ_{AB}^2, and σ_B^2.

A valid F test requires that the following structural conditions be met: (1) The numerator of the F ratio must have only one more component of variance associated with it than the denominator has, and (2) all components of variance except for the one whose effect is to be tested must be the same for both numerator and denominator.

To illustrate, we could not do this: $F = \text{MSBR/MSWCL}$ because the population measure associated with the numerator has three components of variance whereas that associated with the denominator has only one. But we could do: $F = \text{MS}(rc)/\text{MSWCL}$, $F = \text{MSBR/MS}(rc)$, and $F = \text{MSBC/MS}(rc)$. In all three of these F calculations, the two structural requirements cited above are met. We see, then, that in order to complete the analysis of a two-way random-effects model we must compute the F ratios in the following order and manner:

1. Compute the F ratio associated with interaction—that is, $F = \text{MS}(rc)/\text{MSWCL}$.
2. Compute the F ratios associated with Factors A and B using $F = \text{MSBR/MS}(rc)$ and $F = \text{MSBC/MS}(rc)$, respectively.

Part I of Table 12.12 shows the completed analysis of variance, begun in Table 12.11, for the random-effects model.

F Tests: Mixed Model Assumed

Let us now assume a mixed model with Factor A (package designs, represented by rows) random and Factor B (prices, represented by columns) fixed.

A convenient way to determine how the F ratios should be calculated is by looking once more at the components-of-variance structures for the random-effects model in (12.5.13) through (12.5.16). The rule is: Eliminate from either (12.5.13) or (12.5.14) the interaction component if the factor being considered is random. If the factor is fixed, leave the interaction component in. The rationale behind this rule can be understood readily by our taking a close look at the components-of-variance structure of the random factor, Factor A in our present example. We have $\sigma_e^2 + n[1 - (c/C)]\,\sigma_{AB}^2 + nc\sigma_A^2$. We are assuming the columns factor, Factor B, to be fixed. If it is fixed, then $c/C = 1$ and $1 - (c/C) = 0$. It follows that $n(0)\sigma_{AB}^2 = 0$. Therefore, the interaction component drops out of the components-of-variance structure for the *random factor*. This conclusion is quite the opposite of what armchair theorizing might suggest. Just remember that the $n[1 - (c/C)]\sigma_{AB}^2$ term is associated with the *rows variable* and the $n[1 - (r/R)]\sigma_{AB}^2$ term with the *columns variable*. With respect to the columns variable: Since rows are random, then $r/R < 1$ and $1 - (r/R)$ will be a nonzero positive value. Consequently, $n[1 - (r/R)]\sigma_{AB}^2$ must be greater than zero. The conclusion is that the interaction term will remain part of the components-of-variance structure of the fixed factor. Accordingly,

$$\text{For Factor A (rows):} \qquad \sigma_e^2 + nc\sigma_A^2 \tag{12.5.17}$$

$$\text{For Factor B (columns):} \quad \sigma_e^2 + n\left(1 - \frac{r}{R}\right)\sigma_{AB}^2 + nc\sigma_B^2 \tag{12.5.18}$$

$$\text{For interaction:} \qquad \sigma_e^2 + n\sigma_{AB}^2 \tag{12.5.19}$$

$$\text{For within cells:} \qquad \sigma_e^2 \tag{12.5.20}$$

This components-of-variance structure implies:

1. Compute $F = \dfrac{MS(rc)}{MSWCL}$ and $F = \dfrac{MSBR}{MSWCL}$

2. Compute $F = \dfrac{MSBC}{MS(rc)}$

Part II of Table 12.12 shows the completed analysis of variance, begun in Table 12.11, for this kind of mixed model.

F Tests: Fixed-Effects Model Assumed

Now let us return to (12.5.13) through (12.5.16) and realize that, because both factors are now assumed to be fixed, we have $c/C = 1$ and $1 - (c/C) = 0$, and $r/R = 1$ and $1 - (r/R) = 0$. Consequently, the interaction term is eliminated from the components-of-variance structures of both factors. That is,

For Factor A (rows): $\sigma_e^2 + nc\sigma_A^2$ (12.5.21)

For Factor B (columns): $\sigma_e^2 + nr\sigma_B^2$ (12.5.22)

For interaction: $\sigma_e^2 + n\sigma_{AB}^2$ (12.5.23)

For within cells: σ_e^2 (12.5.24)

This components-of-variance structure implies that all sample F ratios may be computed with a denominator of MSWCL. That is, $F = MS(rc)/MSWCL$, $F = MSBR/MSWCL$, and $F = MSBC/MSWCL$

Part III of Table 12.12 shows for the fixed-effects model the completed analysis of variance begun in Table 12.11. Notice that the .05 level critical F values were determined using the following:

1. For the random-effects model: 2 and 2 degrees of freedom for between rows, 1 and 2 degrees of freedom for between columns, and 2 and 12 degrees of freedom for interaction.
2. For the mixed model: 2 and 12 degrees of freedom for between rows, 1 and 2 degrees of freedom for between columns, and 2 and 12 degrees of freedom for interaction.
3. For the fixed-effects model: 2 and 12 degrees of freedom for between rows and interaction, and 1 and 12 degrees of freedom for between columns.

Also notice that it is more difficult to establish significance for the main effects (the column or row effects) when some factors are random. Our tests on the fixed-effects model, for example, resulted in significance at the .05 level for two of the tests and was close for the third. In the mixed model, two tests also resulted in significance but the third was nowhere close. In the random-effects model, only the test for interaction resulted in significance, and

Table 12.12 Table 12.11 Completed, Assuming Various Statistical Models

Part I: Random-Effects Model

Source of Variation	Sum of Squares	Degrees of Freedom	Mean Square	Observed F	.05 Level Critical F
Between rows (packages)	186.11	2	93.06	1.29	19.00
Between columns (prices)	64.22	1	64.22	0.89	18.51
Interaction	144.78	2	72.39	4.57[a]	3.89
Within cells	190.00	12	15.83	—	—
Total	585.11	17	—	—	—

[a]Significant at the .05 level.

Part II: Mixed Model, Rows Random and Columns Fixed

Source of Variation	Sum of Squares	Degrees of Freedom	Mean Square	Observed F	.05 Level Critical F
Between rows (packages)	186.11	2	93.06	5.88[a]	3.89
Between columns (prices)	64.22	1	64.22	0.89	18.51
Interaction	144.78	2	72.39	4.57[a]	3.89
Within cells	190.00	12	15.83	—	—
Total	585.11	17	—	—	—

[a]Significant at the .05 level.

Part III: Fixed-Effects Model

Source of Variation	Sum of Squares	Degrees of Freedom	Mean Square	Observed T	.05 Level Critical F
Between rows (packages)	186.11	2	93.06	5.88[a]	3.89
Between columns (prices)	64.22	1	64.22	4.06	4.75
Interaction	144.78	2	72.39	4.57[a]	3.89
Within cells	190.00	12	15.83	—	—
Total	585.11	17	—	—	—

[a]Significant at the .05 level.

the other two were nowhere close. When either or both factors are random, not only is the denominator of some of the tests usually larger than for a fixed-effects model (MS(rc) versus MSWCL) but, because of small values for degrees of freedom, the critical values tend to be large.

Randomized Block Design: More Than One Observation per Cell

When we introduced the randomized block design for the Imperial Construction Company in section 12.4, we made use of only 12 of the 24 sample observations available. This was a sensible strategy at the time because it allowed us to sidestep the issue of model type. However, now that we have been introduced to the various possible models and shown how to

Table 12.13 Data for the Randomized Block Version of the Imperial Construction Company Problem, Two Observations per Cell

Type of Special Inducement	Price Category				Row Total
	Low	Medium-Low	Medium-High	High	
Three months of no-interest payments	1, 2	2, 3	3, 5	4, 5	25
Free carpeting	3, 2	1, 5	4, 7	6, 12	40
No special inducement	3, 4	6, 7	8, 6	8, 13	55
Column total	15	24	33	48	120

complete an analysis of variance for a specified model type, we can now redo the problem utilizing all 24 sample observations and test for interaction as well as for main effects. Table 12.13 organizes all the necessary data. The steps involved in calculating the mean squares are the same as those above and will not be repeated here. However, do recall that earlier we said that with a randomized block design the blocking variable is usually regarded as random and the treatment variable as fixed. This seems reasonable for the Imperial Construction problem since test units associated with a specific block constitute a sample from the corresponding subpopulation, whereas the special inducements (the treatment levels) are presumably the only ones with which management has any concern. Table 12.14 shows the completed ANOVA table for this illustrative case, and Figure 12.6 outlines an approach for achieving random allocation in a randomized block experiment of the kind being discussed here.

---

*12.6 A Closer Look at Interaction

Although we will attempt a more rigorous explanation of interaction in short order, a simple analogy might get us off to a meaningful start. Let us consider four highly skilled tennis

Table 12.14 ANOVA Table Developed from Data in Table 12.13, Using Randomized Block Design, Two Observations per Cell, Mixed Model with Rows Fixed and Columns Random

Source of Variation	Sum of Squares	Degrees of Freedom	Mean Square	Observed F	.05 Level Critical F
Between rows (inducements)	56.25	2	28.12^a	11.43^a	5.14
Between columns (price categories)	99.00	3	33.00^a	7.91^a	3.49
Interaction	14.75	6	2.46	0.59	3.00
Within cells	50.00	12	4.17	—	—
Total	220.00	23	—	—	—

^{a}Significant at the .05 level.

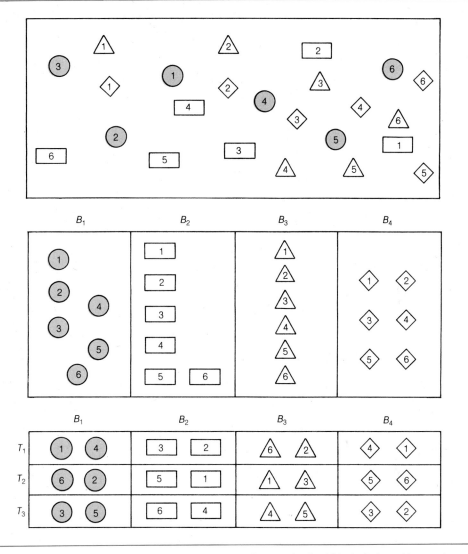

Figure 12.6. Example of a random allocation process in a randomized block design with two observations per cell. Begin with heterogeneous test units. Organize them into blocks. Assign test units within blocks to treatment levels at random.

players: Adams, Baker, Caldwell, and Dunne. Individually they have attained equally impressive records. Moreover, they have over the years played one another many times, with none emerging as clearly superior or inferior; from all indications, they are equally capable players. Now, we will imagine them pairing up to play doubles. Able and Baker become a team and Caldwell and Dunne become a team. The two teams compete against each other with predictable results: Neither team emerges as clearly superior. We may liken this situation to an experiment where no interaction is present. But now let us say that Adams and

Caldwell become a team and Baker and Dunne another team. This time something unexpected does happen: Adams and Caldwell beat Baker and Dunne time after time. For some inexplicable reason or reasons (they might even be identifiable by one with a trained eye for the game), the Adams-Caldwell pair make an especially effective team and/or the Baker-Dunne pair an especially ineffective one. Apparently, Adams and Caldwell together perform better than one would be lead to expect just taking into account their individual abilities and/or Baker and Dunne together perform worse than one would be lead to expect on the same basis.

But, mathematically speaking, what do we mean by interaction? Actually, it is easier to say what we mean by no interaction; so let's start there.

> When there is no interaction, the relationship among the column effects is the same regardless of the row being considered, and the relationship among the row effects is the same regardless of the column being considered. In symbols:
>
> $$\mu_{ij} - \mu_{i\bullet} - \mu_{\bullet j} + \mu_{\bullet\bullet} = 0$$

To illustrate: Let us consider a table of means similar to that shown in Table 12.10 earlier in the chapter. However, to keep things reasonably simple, let us say that (1) the relevant experiment has only two levels associated with rows and only two levels associated with columns and (2) the table is composed completely of *known population means*. Such a situation is shown in Figure 12.7. Notice that $\mu_{11} - \mu_{1\bullet} - \mu_{\bullet 1} + \mu_{\bullet\bullet} = 12 - 11 - 8.5 + 7.5 = 0$. Therefore, according to the above definition, this situation is devoid of interaction. By way of contrast, Figure 12.8 shows the same values for the cell means in Figure 12.7, but in a rearranged order. The results are three examples of situations in which interaction *would be* present.

12.7 The Interaction Graph: An Aid to Determining Sources of Significant Interaction

In the preceding section, we concluded that significant interaction was present in the Specialty Foods Company data. However, interaction can occur in many different ways. From

	C_1	C_2	
R_1 R_2	$\mu_{11} = 12$ $\mu_{21} = 5$	$\mu_{12} = 10$ $\mu_{22} = 3$	$\mu_{1\bullet} = 11$ $\mu_{2\bullet} = 4$
	$\mu_{\bullet 1} = 8.5$	$\mu_{\bullet 2} = 6.5$	$\mu_{\bullet\bullet} = 7.5$

Note: $\mu_{11} - \mu_{1\bullet} - \mu_{\bullet 1} + \mu_{\bullet\bullet} = 12 - 11 - 8.5 + 7.5 = 0$

Figure 12.7. Table of population means for a situation having no interaction.

	C_1	C_2	
R_1	$\mu_{11} = 12$	$\mu_{12} = 3$	$\mu_{1.} = 7.5$
R_2	$\mu_{21} = 5$	$\mu_{22} = 10$	$\mu_{2.} = 7.5$
	$\mu_{.1} = 8.5$	$\mu_{.2} = 6.5$	$\mu_{..} = 7.5$

Note: $\mu_{11} - \mu_{1.} - \mu_{.1} + \mu_{..} = 12 - 7.5 - 8.5 + 7.5 = 3.5 \neq 0$

	C_1	C_2	
R_1	$\mu_{11} = 12$	$\mu_{12} = 10$	$\mu_{1.} = 11$
R_2	$\mu_{21} = 3$	$\mu_{22} = 5$	$\mu_{2.} = 4$
	$\mu_{.1} = 7.5$	$\mu_{.2} = 7.5$	$\mu_{..} = 7.5$

Note: $\mu_{11} - \mu_{1.} - \mu_{.1} + \mu_{..} = 12 - 11 - 7.5 + 7.5 = 1 \neq 0$

	C_1	C_2	
R_1	$\mu_{11} = 12$	$\mu_{12} = 5$	$\mu_{1.} = 8.5$
R_2	$\mu_{21} = 3$	$\mu_{22} = 10$	$\mu_{2.} = 6.5$
	$\mu_{.1} = 7.5$	$\mu_{.2} = 7.5$	$\mu_{..} = 7.5$

Note: $\mu_{11} - \mu_{1.} - \mu_{.1} + \mu_{..} = 12 - 8.5 - 7.5 + 7.5 = 3.5 \neq 0$

Figure 12.8. Tables of population means (taken from Figure 12.7 but rearranged), showing three situations in which interaction would be present.

the standpoint of developing a successful marketing mix, knowledge of the sources of interaction might be extremely important. In this section we consider a helpful device for making sense out of the interaction—namely, the interaction graph.

In Figure 12.9 the vertical axis is used to accommodate the measurements being analyzed (sales in our example). On the horizontal axis, the levels of one or the other treatment variables are listed. One dot is placed on the chart for each cell average. Each such dot relates a specified level of the treatment variable shown on the horizontal axis to (1) the cell average associated with that level and (2) a particular level of the other treatment variable. For example, the lowest price is \$1. The cell means corresponding to this price level are $55/3 = 18.33$ for the old-fashioned package, 5.67 for the neutral package, and 6.33 for the modern package. Therefore, dots are placed on the graph straight up from \$1 and straight across from 18.33, 5.67, and 6.33. This same procedure is followed with the \$4 price. Next, a line is drawn that connects all dots associated with the second treatment variable whose levels are not shown on the horizontal axis of the graph. We see in Figure 12.9, *by the lack of parallelism* of the three lines, that interaction is present. More specifically, the combination of the \$1 price and the old-fashioned package again stands out as an especially favorable one. Figure 12.10 gives some examples of what the interaction graph might look like if interaction were not present.

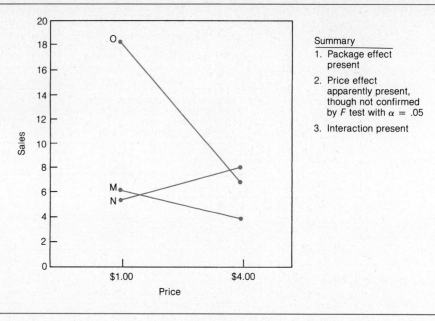

Summary
1. Package effect present
2. Price effect apparently present, though not confirmed by F test with $\alpha = .05$
3. Interaction present

Figure 12.9. Interaction graph for the Specialty Foods Company problem (Illustrative Case 12.1). O = old-fashioned, N = neutral, and M = modern.

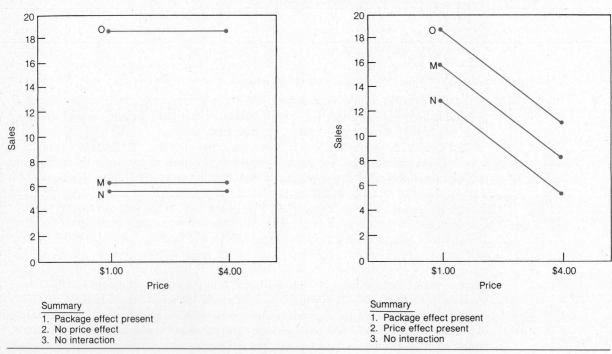

Summary
1. Package effect present
2. No price effect
3. No interaction

Summary
1. Package effect present
2. Price effect present
3. No interaction

Figure 12.10. Two examples of how the interaction graph for the Speciality Foods Company problem (Illustrative Case 12.1) *might have* looked in the absence of interaction. O = old-fashioned, N = neutral, and M = modern.

12 Experimentation in Business

12.8 Summary

In this chapter we have dealt with three kinds of experimental designs for which analysis-of-variance methods are extremely helpful. These were

1. Completely randomized
2. Randomized block (one observation per cell)
3. Two-way factorial arrangement of treatment levels (more than one observation per cell), including a randomized block design as a special case

The first two of these are referred to as single-factor designs—that is, each makes use of only one treatment variable. With the completely randomized design, the single treatment variable is the only independent variable used. The ANOVA procedure for this design may be summarized as follows:

Source of Variation	Sum of Squares	Degrees of Freedom	Mean Square	Observed F	Critical F
Between columns	SSBC	$c - 1$	MSBC	MSBC/MSWC	Table value for $df_1 = c - 1$ and $df_2 = n - c$
Within columns	SSWC	$n - c$	MSWC	—	—

The type of statistical model used may affect the interpretation given a significant F value but will not affect the manner in which F is computed in the case of a completely randomized design.

The randomized block design is called a restricted randomization, single-factor design. It is a two-way analysis, but only one treatment variable is really of interest. The other variable is a blocking variable employed to control a specific source of extraneous variation. For a one-observation-per-cell problem, the ANOVA procedures may be summarized as follows:

Source of Variation	Sum of Squares	Degrees of Freedom	Mean Square	Observed F	Critical F
Between rows	SSBR	$r - 1$	MSBR	MSBR/MSRES	Table values for (1) $df_{11} = r - 1$ and $df_2 = (r - 1)(c - 1)$ and (2) $df_{12} = c - 1$ and $df_2 = (r - 1)(c - 1)$
Between columns	SSBC	$c - 1$	MSBC	MSBC/MSRES	
Residual	SSRES	$(r - 1)(c - 1)$	MSRES	—	—

Again, the statistical model does not affect the manner of obtaining and testing sample F ratios.

The third experimental design we examined differed from the preceding ones in two respects: (1) Both independent variables were treatment variables and (2) more than one observation per cell was used (for simplicity, we used equal cell frequencies). The method of applying analysis of variance to this kind of design begins in the following manner:

Source of Variation	Sum of Squares	Degrees of Freedom	Mean Square
Between rows (Factor A)	SSBR	$r - 1$	MSBR
Between columns (Factor B)	SSBC	$c - 1$	MSBC
Interaction	SS(rc)	$(r - 1)(c - 1)$	MS(rc)
Within cells	SSWCL	$n - CL$ or $rc(n_{ij} - 1)$	MSWCL

What we do next depends on the statistical model. With a fixed-effects model, the sample F ratios are determined by dividing MSBR, MSBC and MS(rc) by MSWCL. With a random-effects model, MS(rc) is divided by MSWCL, whereas both MSBR and MSBC are divided by MS(rc). With a mixed model, the manner of computing observed F will depend on which factor is fixed and which is random. MSWCL is divided into MS(rc) and the mean square associated with the random factor. The mean square associated with the fixed factor is divided by MS(rc).

If the F test for interaction is rejected, indicating that interaction is present, a fairly simple diagnostic tool, the interaction graph, can serve to help clarify the specific nature of the interaction.

Terms Introduced in This Chapter

block (p. 520)
blocking variable (p. 520)
classification variable (p. 520)
completely randomized design (p. 510)
components-of-variance structure (p. 519)
correction factor, CF (p. 538)
error mean square (p. 517)
error variance, σ_e^2 (p. 517)
extraneous source of variation (p. 520)
factorial arrangement of treatment levels (p. 531)

fixed-effects model (p. 516)
fixed factor (p. 516)
interaction (p. 546)
main effects (p. 534)
mixed model (p. 523)
Model I (p. 516)
Model II (p. 516)
Model III (p. 523)
random-effects model (p. 516)
random factor (p. 516)
randomized block design (p. 520)
residual error (p. 517)

single-factor experiment (p. 531)
statistical model (p. 516)
table of means (p. 534)
test units (p. 510)
treatment level (p. 510)
treatment variable (p. 510)
two-factor experiment (p. 531)

Formulas Introduced in This Chapter

Formula for finding the correction factor in a randomized block design (two-way, one observation per cell):

$$CF = \frac{\left(\sum_{\text{all } i} \sum_{\text{all } j} X_{\bullet\bullet}\right)^2}{n}$$

Formulas for finding the *sum of squares* in a randomized block design (two-way, one observation per cell):

$$SSTO = \sum_{\text{all } i} \sum_{\text{all } j} (X_{ij} - \overline{\overline{X}})^2 = \sum_{\text{all } i} \sum_{\text{all } j} X^2_{ij} - CF \qquad \text{Total}$$

$$SSBR = \sum_{\text{all } i} \sum_{\text{all } j} (\overline{X}_{i\bullet} - \overline{\overline{X}})^2 = \sum_{\text{all } i} \frac{\left(\sum_{\text{all } j} X_{i\bullet}\right)^2}{c} - CF \qquad \text{Between rows}$$

$$SSBC = \sum_{\text{all } i} \sum_{\text{all } j} (\overline{X}_{\bullet j} - \overline{\overline{X}})^2 = \sum_{\text{all } j} \frac{\left(\sum_{\text{all } i} X_{\bullet j}\right)^2}{r} - CF \qquad \text{Between columns}$$

$$SSRES = SSTO = (SSBR + SSBC) \qquad \text{Residual}$$

Formulas for finding degrees of freedom in a randomized block design (two-way, one observation per cell):

$df_T = n - 1$ Total
$df_{11} = r - 1$ Between rows
$df_{12} = c - 1$ Between columns
$df_2 = (r - 1)(c - 1)$ Residual

Formula for finding the correction factor in a two-way factorial arrangement of treatment levels (more than one observation per cell):

$$CF = \frac{\left(\sum_{\text{all } i} \sum_{\text{all } j} \sum_{\text{all } k} X_{\bullet\bullet\bullet}\right)^2}{n}$$

Formulas for finding the sum of squares in a two-way factorial arrangement of treatment levels:

$$SSTO = \sum_{\text{all } i} \sum_{\text{all } j} \sum_{\text{all } k} (X_{ijk} - \overline{\overline{X}})^2 = \sum_{\text{all } i} \sum_{\text{all } j} \sum_{\text{all } k} X^2_{ijk} - CF \qquad \text{Total}$$

$$SSBR = \sum_{\text{all } i} \sum_{\text{all } j} \sum_{\text{all } k} (\overline{X}_{1\bullet\bullet} - \overline{\overline{X}})^2 = \sum_{\text{all } i} \frac{\left(\sum_{\text{all } j} \sum_{\text{all } k} X_{i\bullet\bullet}\right)^2}{cn_{ij}} - CF \qquad \text{Between rows}$$

$$SSBC = \sum_{\text{all } i} \sum_{\text{all } j} \sum_{\text{all } k} (\overline{X}_{\bullet j\bullet} - \overline{\overline{X}})^2 = \sum_{\text{all } j} \frac{\left(\sum_{\text{all } i} \sum_{\text{all } k} X_{\bullet j\bullet}\right)^2}{rn_{ij}} - CF \qquad \text{Between columns}$$

$$SSBCL = \sum_{\text{all } i} \sum_{\text{all } j} \sum_{\text{all } k} (X_{ij\bullet} - \overline{\overline{X}})^2 = \sum_{\text{all } i} \sum_{\text{all } j} \frac{\left(\sum_{\text{all } k} X_{ij\bullet}\right)^2}{n_{ij}} - CF \qquad \text{Between cells}$$

$$SS(rc) = SSBCL - (SSBR + SSBC) \qquad \text{Interaction}$$
$$SSWCL = SSTO - SSBCL \qquad \text{Within cells}$$

Formulas for finding degrees of freedom in a two-way factorial arrangement of treatment levels:

$$df_T = n - 1 \qquad\qquad \text{Total}$$
$$df_{11} = r - 1 \qquad\qquad \text{Between rows}$$
$$df_{12} = c - 1 \qquad\qquad \text{Between columns}$$
$$df_{13} = (r - 1)(c - 1) \qquad \text{Interaction}$$
$$df_2 = n - CL \text{ or } rc(n_{ij} - 1) \qquad \text{Within cells}$$

Formulas for finding components of variance in a two-way factorial arrangement of treatment levels, using statistical models:

Random-effects model:

$$\sigma_e^2 + n\left(1 - \frac{c}{C}\right)\sigma_{AB}^2 + nc\sigma_A^2 \qquad \text{Factor A (rows)}$$

$$\sigma_e^2 + n\left(1 - \frac{r}{R}\right)\sigma_{AB}^2 + nr\sigma_B^2 \qquad \text{Factor B (columns)}$$

$$\sigma_e^2 + n\sigma_{AB}^2 \qquad\qquad\qquad\qquad \text{Interaction}$$

$$\sigma_e^2 \qquad\qquad\qquad\qquad\qquad\quad \text{Error}$$

Mixed model with rows fixed and columns random:

$$\sigma_e^2 + n\left(1 - \frac{c}{C}\right)\sigma_{AB}^2 + nc\sigma_A^2 \qquad \text{Factor A (rows)}$$

$$\sigma_e^2 + nr\sigma_B^2 \qquad\qquad\qquad\qquad \text{Factor B (columns)}$$

$$\sigma_e^2 + n\sigma_{AB}^2 \qquad\qquad\qquad\qquad \text{Interaction}$$

$$\sigma_e^2 \qquad\qquad\qquad\qquad\qquad\quad \text{Error}$$

Mixed model with rows random and columns fixed:

$$\sigma_e^2 + nc\sigma_A^2 \qquad\qquad\qquad\qquad \text{Factor A (rows)}$$

$$\sigma_e^2 + n\left(1 - \frac{r}{R}\right)\sigma_{AB}^2 + nr\sigma_B^2 \qquad \text{Factor B (columns)}$$

$$\sigma_e^2 + n\sigma_{AB}^2 \qquad\qquad\qquad\qquad \text{Interaction}$$

$$\sigma_e^2 \qquad\qquad\qquad\qquad\qquad\quad \text{Error}$$

Fixed-effects model

$$\sigma_e^2 + nc\sigma_A^2 \qquad \text{Factor A (rows)}$$

$$\sigma_e^2 + nr\sigma_B^2 \qquad \text{Factor B (columns)}$$

$$\sigma_e^2 + n\sigma_{AB}^2 \qquad \text{Interaction}$$

$$\sigma_e^2 \qquad\qquad\qquad \text{Error}$$

Questions and Problems

12.1 Explain the meaning of each of the following:

    a. Test unit
    b. Interaction
    c. Treatment variable
    d. Experiment

    e. Completely randomized design
    f. Components-of-variance structure
    g. Fixed-effects model

12 Experimentation in Business

12.2 Explain the meaning of each of the following:

    a. Mixed model

    b. Factorial arrangement of treatment levels

    c. Treatment level

    d. Block

    e. Statistical model in analysis of variance

12.3 Explain the meaning of each of the following:

    a. Error mean square

    b. Random-effects model

    c. Interaction graph

    d. Blocking variable

    e. SSBR

    f. MS(rc)

    g. MSRES

12.4 Distinguish between:

    a. Use of analysis of variance in connection with an experiment and use of analysis of variance in connection with a nonexperimental investigation

    b. Treatment variable and blocking variable

    c. Completely randomized design and randomized block design

    d. Fixed-effects model and random-effects model

12.5 Distinguish between:

    a. df_{11} and df_{12}

    b. SSRES and SSWCL

    c. Two-way analysis involving two treatment variables and having one observation per cell and two-way analysis involving two treatment variables and having more than one observation per cell

    d. $\sum\limits_{\text{all } i}\sum\limits_{\text{all } j} (X_{ij} - \bar{\bar{X}})^2$ and $\sum\limits_{\text{all } i}\sum\limits_{\text{all } j} X_{ij}^2 - \dfrac{\left(\sum\limits_{\text{all } i}\sum\limits_{\text{all } j} X_{..}\right)^2}{n}$

    e. $\sum\limits_{\text{all } i}\sum\limits_{\text{all } j} X_{..}$ and $\sum\limits_{\text{all } i}\sum\limits_{\text{all } j}\sum\limits_{\text{all } k} X_{...}$

12.6 Distinguish between:

    a. Fixed-effects model and mixed model

    b. Model I and Model II

    c. Model II and Model III

    d. Fixed factor and random factor

    e. $\sum\limits_{\text{all } j} T_j = 0$ and T_j: NID$(0, \sigma_T^2)$

12.7 Indicate which of the following statements you agree with and which you disagree with, and defend your opinions:

    a. In a two-way analysis of variance having a fixed-effects model and 6 columns, 4 rows, and 5 observations per cell, the number of degrees of freedom associated with the sample measure of chance variation (that is, the denominator of the F ratio) will be $(6 - 1)(4 - 1) = 15$.

    b. In the situation described in part a, the number of degrees of freedom associated with SS(rc) will be $rc(n_{ij} - 1) = 96$.

    c. In a two-way analysis of variance having more than one observation per cell, if the columns-and-rows interaction is found to be significant at some specified level, we may be pretty sure that either the F ratio associated with columns or the F ratio associated with rows, or both, will also be significant at the same level.

    d. If, in carrying out a two-way analysis of variance with more than one observation per cell, an analyst finds that the interaction F ratio is not significant, he or she will be unlikely to follow up on the main analysis by utilizing an interaction graph.

12.8 Indicate which of the following statements you agree with and which you disagree with, and defend your opinions:

a. Determining the statistical model underlying an analysis of variance is not too important because often the type of model has no effect on the way the computational procedures are carried out.

b. In a two-way factorial arrangement of treatment levels with more than one observation per cell, the within-cells mean square will always be used as the denominator of the sample F ratios.

c. An experiment making use of paired observations would be a special randomized block experiment.

d. A randomized block experiment, as we dealt with it in this chapter, is a special case of a factorial design.

12.9 Three brands of automobile tires are tested to determine whether they are equally durable. Sixty tires were tested on 15 randomly assigned taxicabs in a certain city (each taxicab having four tires of the same brand), and the number of miles that a taxicab traveled before the first tire in a set of four failed to meet state inspection standards was recorded for each vehicle. The results are presented in Table 12.15.

Indicate the value in Table 12.15 corresponding to each of the following symbolic descriptions:

a. n_3 **d.** $\overline{\overline{X}}_1$

b. n **e.** $\overline{\overline{X}}$

c. $\displaystyle\sum_{\text{all } i} X_{i2}$ **f.** $\displaystyle\sum_{\text{all } i}\sum_{\text{all } j} X_{ij}$

12.10 Refer to problem 12.9. Assume a fixed-effects model and

a. State the null and alternative hypotheses.

b. Set up an ANOVA table and test the null hypothesis. Should the null hypothesis be accepted or rejected?

c. State briefly some practical implications of the conclusion you arrived at in part *b*.

Table 12.15 Data for the Tire Durability Experiment

(Numbers Represent Thousands of Miles)

	Brand of Tire		
	Goodgo	Goodwill	Goodbody
	30	22	32
	36	28	40
	23	25	40
	40	33	38
	37	28	42
Total	166	136	192
Mean	33.2	27.2	38.4
Sample size	5	5	5

Grand total = 494

Grand mean = 32.93

Overall sample size = 15

12.11 Refer to problem 12.9. What assumptions regarding the experimental design and the statistical model does one implicitly make when carrying out this kind of analysis?

 12.12 Refer to problem 12.9. Would your procedures be any different if you were told to assume a random-effects model? Tell why or why not.

12.13 Soapy Sales, Inc., tested their calla lily–scented bar soap in four different-colored wrappers—red, blue, green, and yellow—for an experimental period extending over eight weeks. The red wrapper was assigned at random to four cooperating retail stores; the blue wrapper to four stores; the green wrapper to six stores; and the yellow wrapper to six stores. Sales achieved during the experimental period are shown in Table 12.16.

Assume a random-effects model.
a. State the null and alternative hypotheses.
b. Set up an ANOVA table and test the null hypothesis at the .01 level. State your conclusion.
c. State briefly in words the practical implications, if any, of the conclusions you arrived at in part *b*.

 12.14 Refer to problem 12.13. What assumptions regarding the experimental design and the statistical model does one implicitly make when carrying out this kind of analysis?

 12.15 Refer to problem 12.13. Would your procedures be any different if you were told to assume a fixed-effects model? Tell why or why not.

12.16 A manufacturer of television sets wished to identify the best location within retail stores for a new, high-priced videodisk appliance. He viewed the possible locations as (1) left rear of store, (2) right rear of store, (3) left front of store, (4) right front of store, and (5) center of store. He gained the cooperation of 30 retail television store managers and, for a six-month period, the manufacturer's research staff oversaw the use of the left rear location in six ran-

Table 12.16 Data for the Soap Wrapper Color Experiment

(Numbers Represent Sales in Dozens of Bars)

	Color of Wrapper			
	Red	Blue	Green	Yellow
	40	50	32	21
	30	45	30	27
	45	55	28	19
	40	45	31	22
			28	17
			18	20
Total	155	195	167	126
Mean	38.75	48.75	27.83	21.00
Sample size	4	4	6	6

Grand total = 643
Grand mean = 32.15
Overall sample size = 20

Table 12.17 Data for the Location-in-Store Experiment

(Figures Represent Number of Videodisks Sold)

	Location Within Store				
	Left Rear	**Right Rear**	**Left Front**	**Right Front**	**Center**
	3	2	3	5	7
	2	1	4	7	9
	1	2	2	3	7
	2	2	3	2	10
	2	4	3	4	8
	3	1	3	7	8
Total	13	12	18	28	49
Mean	2.17	2.00	3.00	4.67	8.17
Sample size	6	6	6	6	6

Grand total = 120

Grand mean = 4.00

Overall sample size = 30

domly assigned stores, the use of the right rear location in six randomly assigned stores, and so forth. The resulting sales, in units, are presented in Table 12.17. Assume a fixed-effects model.
a. State the null and alternative hypotheses.
b. Set up an ANOVA table and test the null hypothesis at the .05 level of significance. Should the null hypothesis be accepted or rejected?

12.17 Refer to problem 12.16. What assumptions regarding the experimental design and the statistical model does one implicitly make when carrying out this kind of analysis?

12.18 Refer to problem 12.16. Would your procedures be any different if you were told to assume a random-effects model? Tell why or why not.

12.19 Refer to problem 12.9. Suppose that the entire experiment was repeated (replicated) using 15 different randomly assigned taxicabs. Suppose further that the following sample information resulted:

Goodgo	Goodwill	Goodbody
$\overline{X}_1 = 34.5$	$\overline{X}_2 = 26.0$	$\overline{X}_3 = 40.0$
$s_1^2 = 6.8$	$s_2^2 = 4.1$	$s_3^2 = 3.8$
$n_1 = 5$	$n_2 = 5$	$n_3 = 5$

Can you conclude that the mean useful lives of the three brands of tires are equal? Use a .05 level of significance.

12.20 Refer to problem 12.13. Suppose that the entire experiment was repeated (replicated) using 20 different randomly assigned stores. Suppose further that the following sample information resulted:

Red	Blue	Green	Yellow
$\overline{X}_1 = 40.0$	$\overline{X}_2 = 50.0$	$\overline{X}_3 = 30.0$	$\overline{X}_4 = 20.0$
$s_1^2 = 6.3$	$s_2^2 = 5.0$	$s_3^2 = 5.0$	$s_4^2 = 3.4$
$n_1 = 4$	$n_2 = 4$	$n_3 = 6$	$n_4 = 6$

Can you conclude that the mean number of sales for the four soap-wrapper colors are equal? Use a .01 level of significance.

 12.21 Refer to problem 12.16. Suppose that the entire experiment was repeated (replicated) using 30 different randomly assigned stores. Suppose further that the following sample information resulted:

Left Rear	Right Rear	Left Front	Right Front	Center
$\overline{X}_1 = 2.0$	$\overline{X}_2 = 2.0$	$\overline{X}_3 = 3.0$	$\overline{X}_4 = 5.0$	$\overline{X}_5 = 7.5$
$s_1^2 = 0.75$	$s_2^2 = 1.00$	$s_3^2 = 0.63$	$s_4^2 = 2.56$	$s_5^2 = 1.30$
$n_1 = 6$	$n_2 = 6$	$n_3 = 6$	$n_4 = 6$	$n_5 = 6$

Can you conclude that the mean number of sales for the five in-store locations are equal? Use a .05 level of significance.

 12.22 A manufacturing plant utilizes a conveyor belt and human labor in combination. The speed of the conveyor belt can be increased or decreased, depending on how fast the employees are able to complete their assembling tasks. Management decided to experiment with various factors that might increase worker speed without making the work crew feel pushed or badgered in any way. Factors to be considered included lighting fixtures, temperatures in the work area, presence or absence of background music, and so forth. One factor studied, which will be our sole concern in this problem, is the side of the conveyor belt on which the workers are situated. Eight employees were used in a four-week experiment. Four of the workers selected were assigned at random to the right-hand side of the conveyor belt and the other four to the left-hand side. These locational assignments were maintained for two weeks and then all workers changed sides for the remaining two weeks. The resulting output values, expressed as "percent of standard," are presented in Table 12.18.

Management now wishes to know whether the side of the conveyor belt on which a worker is situated has any effect on output. Assume a mixed model with columns fixed and rows random.

a. Would you describe this as a completely randomized experiment? Tell why or why not.
b. Should there be one or two null hypotheses? Explain.
c. State the one or two null hypotheses and alternative hypotheses.
d. Set up the ANOVA table and test the null hypothesis or hypotheses. State your conclusions.

Table 12.18 Data for the Conveyer Belt Problem

(Numbers Represent "Percent of Standard")

Worker	Side of Conveyer Belt	
	Left Side	Right Side
A	75	79
B	82	83
C	80	83
D	85	86
E	85	89
F	84	88
G	90	95
H	92	94

12.23 Refer to problem 12.22. What assumptions regarding the experimental design and the statistical model does one implicitly make when carrying out this kind of analysis?

 12.24 Refer to problem 12.22. Would your procedures be any different if you were to assume a random-effects model? Tell why or why not. Does it seem reasonable to think that this experiment could have had a random-effects model associated with it? Tell why or why not.

 12.25 Refer to problem 12.22. Can you think of any hypothesis-testing procedure treated in an earlier chapter that would be applicable to this experimental situation? If so, what is it?

 12.26 A merchandising experiment involving a new Christmas mix of dried fruits was performed in 15 stores in the following manner: The treatment variables were (1) price charged and (2) type of display used. The price variable had associated with it five treatment levels—namely, $1.00, $1.10, $1.20, $1.50, and $2.00. The display variable had associated with it three treatment levels—namely, checkout stand display, end-aisle display, and rear-of-store standing display. The results are presented in Table 12.19.

According to Table 12.19, what value is associated with each of the following symbolic descriptions:

a. c

b. r

c. n_{ij}

d. X_{14}

e. X_{35}

f. $\sum_{\text{all } i} X_{\bullet 1}$

g. $\sum_{\text{all } j} X_{2 \bullet}$

h. $\overline{X}_{\bullet 2}$

i. $\sum_{\text{all } i} \sum_{\text{all } j} X_{\bullet\bullet}$

 12.27 Refer to problem 12.26. Assume a fixed-effects model.

a. State the two relevant null hypotheses and the two corresponding alternative hypotheses.

b. Set up an ANOVA table and test the two null hypotheses. State your conclusions. Use a .05 level of significance.

c. Might interaction conceivably exist between price charged and type of display used? Explain. Is it possible to apply an F test for interaction in this specific problem? Why or why not?

d. What assumptions about the experimental design and the statistical model did you implicitly accept while carrying out the above analysis?

Table 12.19 Data for the Dried Fruit Problem

(Numbers Represent Dozens of Packages Sold)

	Prices				
Type of Display	$1.00	$1.10	$1.20	$1.50	$2.00
Checkout stand	17	15	13	10	5
End-aisle	20	18	16	10	5
Standing, rear of store	10	10	9	4	1

12.28 Refer to problem 12.26. Would your procedures be any different if you were to assume a random-effects model? Tell why or why not? Is it reasonable to think that this experiment might have had a random-effects model associated with it? Explain.

12.29 Table 12.20 shows the results, in miles per gallon, of an experiment designed for the purpose of comparing the effectiveness of four brands of gasoline additives. The experiment also involved the use of four different makes of automobile. Assume a mixed model, with the automobile factor random and the additive factor fixed.
  a. State the two relevant null hypotheses and the corresponding alternative hypotheses.
  b. Set up an ANOVA table and test the two null hypotheses at the .05 level of significance. State your conclusions.
  c. Might interaction conceivably exist between brand of gasoline additive and make of automobile driven? Explain. Is it possible to apply an F test for interaction in this specific problem? Why or why not?
  d. What assumption(s) did you implicitly accept when carrying out this analysis?

12.30 A study was conducted to determine the amount of time vacationers at three different resort areas spend at these vacation spots during the following three-month periods: (1) January, February, and March (hereafter referred to as Quarter 1), (2) April, May, and June (Quarter 2), (3) July, August, and September (Quarter 3), and (4) October, November, and December (Quarter 4). A random sample of four registered guests was selected at each of the resort areas during each of the indicated quarters and the number of days that each of the sample members stayed was recorded. The results are shown in Table 12.21. Assume a fixed-effects model.
  a. In your opinion, is this an experimental problem? Explain why you think it is or is not.
  b. State the three relevant null hypotheses and the corresponding alternative hypotheses.
  c. What would be the measure of "chance variation" (i.e., the denominator of the F ratios) in this analysis? Why? How many degrees of freedom would the measure of chance variation have associated with it?
  d. How many degrees of freedom are associated with the between-columns sum of squares? With the between-rows sum of squares? With the interaction sum of squares? With the total sum of squares?
  e. Determine the decision rule for (1) the between-columns F ratio, (2) the between-rows F ratio, and (3) the interaction F ratio if the level of significance is set at $\alpha = .05$.
  f. Set up an ANOVA table and test the null hypotheses. State your conclusions clearly.
  g. What assumptions did you implicitly accept regarding the experimental design and the statistical model when carrying out the above analysis?

12.31 Refer to problem 12.30. Work all parts of this problem assuming a mixed model, with the quarters factor fixed and the resorts factor random.

Table 12.20 Data for the Gasoline Additive Problem

(Numbers Represent Miles per Gallon)

Make of Car	Brand of Additive			
	I	II	III	IV
A	18	20	15	22
B	12	15	10	18
C	20	22	18	25
D	17	20	15	21

Table 12.21 Data for the Vacation Resort Problem

(Numbers Represent Length of Stay in Days)

Quarter	Resort A	B	C
I	2	3	13
	4	2	25
	3	3	28
	2	2	35
II	7	7	6
	10	10	4
	12	14	4
	15	14	6
III	14	10	7
	16	14	3
	20	20	4
	30	12	5
IV	3	2	14
	4	3	21
	5	6	21
	4	2	28

 12.32 Refer to Table 12.21. Set up an interaction graph and interpret your results.

12.33 An experiment is conducted on three brands of power lawn mowers to determine (1) whether there is any difference in useful life among the three brands, (2) whether the speed (high, medium, or low) at which they are consistently operated affects length of useful life, and (3) whether there is any interaction between speed and brand of mower. The results of the experiment are shown in Table 12.22.

Assume a mixed model, with the type-of-mower factor random and the speed factor fixed.

a. State the three relevant null hypotheses and the corresponding alternative hypotheses.

b. Set up an ANOVA table and test the three null hypotheses at the .05 level of significance.

c. Indicate some possible practical implications of the conclusions you reached in part *b*.

d. What assumptions did you implicitly make about the experimental design and the statistical model when carrying out this analysis?

12.34 Refer to problem 12.33.

a. Is it possible that the speed factor could have been random as well as the brand-of-mower factor? Explain.

b. Work all parts of problem 12.33 assuming a random-effects model.

12.35 Refer to Table 12.22. Set up an interaction graph and interpret your results.

Table 12.22 Data for the Power Lawn Mower Problem

(Numbers Represent Hundreds of Hours of Continuous Use)

Speed	Type of Mower		
	A	B	C
High	10	11	18
	6	9	16
	12	5	8
Medium	20	24	26
	18	18	22
	16	20	28
Low	16	24	18
	12	30	16
	10	27	22

12.36 What kind of experimental design and what kind of statistical model are described below? Explain how you know.

$$X_{ij} = \mu + T_j + \epsilon_{ij}$$

where ϵ_{ij}: NID$(0, \sigma_e^2)$
   T_j: NID$(0, \sigma_T^2)$

12.37 What kind of experimental design and what kind of statistical model are described below? Explain how you know.

$$X_{ij} = \mu + B_j + T_i + \epsilon_{ij}$$

where ϵ_{ij}: NID$(0, \sigma_e^2)$

$$\sum_{\text{all } j} B_j = 0$$

   T_i: NID$(0, \sigma_T^2)$

12.38 What kind of experimental design and what kind of statistical model are described below? Explain how you know.

$$X_{ijk} = \mu + A_i + B_j + (AB)_{ij} + \epsilon_{ijk}$$

where ϵ_{ijk}: NID$(0, \sigma_e^2)$

$$\sum_{\text{all } i} A_i = 0 \qquad \sum_{\text{all } j} B_j = 0 \qquad \sum_{\text{all } i} (AB)_{ij} = 0$$

12.39 What kind of experimental design and what kind of statistical model are described below? Explain how you know.

$$X_{ijk} = \mu + A_i + B_j + (AB)_{ij} + \epsilon_{ijk}$$

where ϵ_{ijk}: NID$(0, \sigma_e^2)$
   A_i: NID$(0, \sigma_A^2)$
   B_j: NID$(0, \sigma_B^2)$
   $(AB)_{ij}$: NID$(0, \sigma_{AB}^2)$

12.40 What kind of experimental design and what kind of statistical model are described below? Explain how you know.

$$X_{ijk} = \mu + A_i + B_j + (AB)_{ij} + \epsilon_{ijk}$$

where $\quad \epsilon_{ijk}$: NID$(0, \sigma_e^2)$

$\qquad\qquad A_i$: NID$(0, \sigma_A^2)$

$$\sum_{\text{all } j} B_j = 0$$

$\qquad\qquad (AB)_{ij}$: NID$(0, \sigma_{AB}^2)$

But $\qquad \displaystyle\sum_{\text{all } i} (AB)_{ij} \neq 0$

$$\sum_{\text{all } j} (AB)_{ij} = 0$$

TAKE CHARGE

12.41 Only a few experimental designs useful for experimentation in business situations have been treated in this chapter. Some others are

1. Latin square design
2. Graeco-Latin square design
3. Double changeover design
4. Split-plot design

a. Think of a realistic problem related to your major field of study in which experimentation might be employed.
b. Get a book on experimental designs from the library and study the above-mentioned designs just enough to understand under what circumstances they are typically used and what their principal strengths and weaknesses are.
c. Tell briefly in writing how each of the above-mentioned designs might be usefully employed to help solve the problem you posed in part *a*. If you believe that certain of these designs would not be useful in connection with the problem posed, tell why.

 Computer Exercise

12.42 An experiment aimed at determining the best way to merchandise a new brand of men's cologne was conducted over a three-week period and 108 retail stores were involved—12 stores for each combination of two treatment levels. The treatment variables, Factors A and B, had to do with package design and product name.

Factor A: Kind of image projected by the picture on the package

A_1: Macho
A_2: Suave
A_3: Wise

Factor B: Name

B_1: Zap
B_2: Debonair
B_3: Insight

Table 12.23 Data for Problem 12.42

Nature of Picture	Name		
	B₁: Zap	B₂: Debonair	B₃: Insight
A₁: Macho	20, 18, 24, 20, 18, 14, 12, 15, 10, 8, 8, 11	18, 15, 15, 17, 15, 10, 12, 12, 7, 8, 6, 5	16, 13, 12, 13 13, 12, 12, 11, 5, 6, 7, 8
A₂: Suave	7, 6, 4, 5 9, 7, 6, 7, 6, 5, 4, 6	10, 8, 9, 6, 25, 22, 18, 27, 18, 16, 14, 22	8, 7, 7, 8, 22, 20, 18, 19, 16, 14, 13, 16,
A₃: Wise	7, 6, 4, 7 9, 8, 7, 6, 10, 14, 11, 9	9, 8, 8, 5, 15, 14, 14, 11, 16, 15, 13, 12	6, 5, 7, 8, 13, 14, 12, 14, 15, 14, 16, 14

The results are shown in Table 12.23. Entries represent number of bottles sold during the experimental period.

a. Have the computer calculate all mean squares.

b. Test the three relevant null hypotheses at the .05 level of significance. Assume first a fixed-effects model and then a random-effects model. Interpret both sets of results.

c. Which model considered in connection with part *b* seems more realistic to you? Why?

Cases for Analysis and Discussion

12.1 "MID-CITY BANK OF BIG TOWN": PART V

Review briefly the first four parts of this case (end of Chapters 8, 9, and 10).

What follows is a hypothetical, though sensible, variation on the research methods actually used by "Mid-City Bank."

Assume that management had decided to include three separate samples, or categories of customers, in its sample study. The three sample groups were (1) an experimental group whose members were required to pay only 3% of the outstanding balance per month or $5, whichever is greater, (2) a second experimental group whose members were required to pay 4.5% of the outstanding balance per month or $7.50, whichever is greater, and (3) a control group whose members were required to pay 6% of the outstanding balance per month or $10, whichever is greater. Some results of the sample study are presented in Table 12.24.

1. Does the sample evidence for minimum-payback customers shown in Table 12.24 suggest that the population average interest payment per customer per month differs among the three minimum-payback-requirement categories? Test at the .05 level of significance the hypothesis that no such differences exist.

2. Does the sample evidence for medial-payback customers shown in Table 12.24 suggest that the population average interest payment per customer per month differs among the three minimum-payback-requirement categories? Test at the .05 level of significance the hypothesis that no such differences exist.

Table 12.24 Sample Information Pertaining to Interest Payment per Customer per Month for Three Categories of Minimum-Payback-Requirement Customers

Part I: Minimum-Payback Customers

	Minimum-Payback Requirement		
Sample Information	(1) 3% or $5 (Experimental Group Number 1)	(2) 4.5% or $7.50 (Experimental Group Number 2)	(3) 6% or $10 (Control Group)
Sample mean	$4.67	$4.50	$3.71
Sample standard deviation	$0.80	$0.75	$0.65
Sample size	88	88	88

Part II: Medial-Payback Customers

	Minimum-Payback Requirement		
Sample Information	(1) 3% or $5 (Experimental Group Number 1)	(2) 4.5% or $7.50 (Experimental Group Number 2)	(3) 6% or $10 (Control Group)
Sample mean	$3.54	$3.53	$3.52
Sample standard deviation	$1.40	$1.37	$1.55
Sample size	79	79	79

Part III: Full-Payback Customers

	Minimum-Payback Requirement		
Sample Information	(1) 3% or $5 (Experimental Group Number 1)	(2) 4.5% or $7.50 (Experimental Group Number 2)	(3) 6% or $10 (Control Group)
Sample mean	$0.0012	$0.0015	$0.01
Sample standard deviation	$0.20	$0.18	$0.21
Sample size	128	128	128

Part IV: All Customers

	Minimum-Payback Requirement		
Sample Information	(1) 3% or $5 (Experimental Group Number 1)	(2) 4.5% or $7.50 (Experimental Group Number 2)	(3) 6% or $10 (Control Group)
Sample mean	$2.34	$2.28	$2.05
Sample standard deviation	$0.98	$1.00	$0.98
Sample size	295	295	295

3. Does the sample evidence for full-payback customers shown in Table 12.24 suggest that the population average interest payment per customer per month differs among the three minimum-payback-requirement categories? Test at the .05 level of significance that no such differences exist.

4. Does the sample evidence for all customers shown in Table 12.24 suggest that the population average interest payment per customer per month differs among the three minimum-payback-requirement categories? Test at the .05 level of significance the hypothesis that no such differences exist.

5. What assumptions, if any, did you accept when you applied the statistical test you selected in questions 1 through 4?

6. (Pertains to material labeled "optional" in Chapter 11.) Refer to questions 1 through 4 above. For each of these questions, test at the .05 level of significance the following two subhypotheses using the method of orthogonal comparisons:

$H_{0_{sub\ 1}}$: Population mean associated with the experimental group paying 3% or $5 does not differ from the population mean associated with the experimental group paying 4.5% or $7.50

$H_{0_{sub\ 2}}$: Population mean associated with the two experimental groups (considered together) does not differ from the population mean associated with the control group

7. Do the test results suggest that 3% (or $5) minimum-payback requirement would be preferable to the 4.5% (or $7.50) minimum-payback requirement? Justify your answer.

 ## 12.2 "MODERN COMPUTER DYNAMICS, INC."*

"Modern Computer Dynamics, Inc." (MCD) is a medium-sized electronic equipment manufacturer. The company designs, develops, manufactures, and markets computer peripherals. Even though the company is less than 15 years old, it has experienced such rapid growth that it is now one of the *Fortune* 500 companies. This brisk growth has led to many changes, frustrations, and pressures as well as rewards and benefits for the employees at MCD. One of the problem areas has been a high employee turnover rate. Management considered this to be quite a serious problem and enlisted a member of the technical staff to conduct a study of turnover rates. The study had many facets. The one we will focus on here has to do with the question of whether employee turnover was higher for certain job classifications than for others.

The measure of employee turnover used was the separation ratio defined as

$$\text{Separation ratio} = \frac{\text{Number of terminations during 1979}}{\text{Number of employees during 1979}} \cdot 100$$

Table 12.25 shows the basic data. Results of the analysis of variance are presented in Table 12.26. The Duncan multiple-range test was then performed; these results are shown in Table 12.27.

1. In your opinion, was this study an experiment? Tell why you do or do not think so.

2. In your opinion, is the separation ratio, as defined in the case, an adequate measure of employee turnover? Tell why or why not.

*Partly dependent on material labeled "Optional" in Chapter 11.

Table 12.25 Separation Ratios for "Modern Computer Dynamics, Inc.," Organized by Job Classification

	Job Classification				
	Managers	**Professionals**	**Technicals**	**Clericals**	**Direct Labor**
	5.5	5.8	11.6	24.5	19.7
	14.9	9.6	14.1	43.4	8.6
	8.4	19.6	6.9	30.7	29.8
	12.4	14.0	20.1	25.3	39.9
Total	41.2	49.0	52.7	123.9	98.0
Mean	10.30	12.25	13.18	30.98	24.50

Table 12.26 ANOVA Table for the "Modern Computer Dynamics" Case

Source of Variation	Sum of Squares	Degrees of Freedom	Mean Square	Observed F	.05 Level Critical F
Between columns	1303.78	4	325.94	4.80	3.06
Within columns	1018.29	15	67.89	—	—
Total	2322.07	19	—	—	—

Table 12.27 Results of Duncan Multiple-Range Test Applied to Column Means in Table 12.25

	Managers	Professionals	Technicals	Direct Labor	Clerical
Mean	10.30	12.25	13.18	24.50	30.98

3. Prepare a brief written summary of the results contained in Tables 12.26 and 12.27. (*Hint:* You may wish to refer to section 11.10.)

4. What would you guess are some of the reasons for the results you commented on in question 2? Is there any hard evidence in this case tying rate of employee turnover to rate of growth of the company? Explain.

 ## 12.3 "STRANGWAYS TEXTILES, INC."*

As a consequence of the growing concern over consumer protection, the federal government has focused much recent attention on the problem of textile flammability. The Federal Trade

*Partly dependent on material labeled "Optional" in Chapter 11.

Commission (F.T.C.) has set stringent safety standards aimed at protecting the public against the hazards of "rapid flash burning," "continuous slow burning," and "smoldering" attributable to textiles. To determine whether a product meets the safety standards established by the F.T.C., a set of procedures has been outlined for use by the manufacturing company. Products that pass this battery of tests are marketable; those that do not must either be discarded or adjusted so that they can pass the test. If the F.T.C., in one of its frequent spot checks of products already on the shelves, finds a product that does not pass, the manufacturer is then required to recall the entire lot and is subject to a fine.

In its attempts to improve its products with respect to the flammabilty criterion, the research and development department of "Strangways Textiles" continually experiments with new chemical additives that are claimed to enhance the flame-retardant quality of the textile. In doing so, it employs the F.T.C. testing procedures in evaluating both new products and modifications to old ones. In accordance with F.T.C. specifications, four items must be selected at random from each production unit. Each of these items must then be cut to provide eight specimens of uniform size. The actual testing procedure is composed of a battery of four tests, each requiring eight trials. If two to three specimens fail in any of the four tests, the sample is considered marginal; if four or more of any of the eight trials fail, the sample is considered flammable.

The specific problem we will consider here is concerned with testing the effects of two additives mixed into the latex normally used as backing for certain rugs produced by Strangways. To determine whether either of the additives was helpful as a flame retardant, a sample lot was run using each of the additives. Specimens made with each additive were then cut according to F.T.C. rules. Additional specimens were taken from this lot, which were made with the latex backing normally used. In all, three hypotheses were tested:

H_0: There is no difference among the flame-retardant qualities of the three versions of the latex backing

H_{0_1}: There is no difference in the flame-retardant qualities of Additives A and B

H_{0_2}: There is no difference between the flame-retardant qualities of Additives A and B (considered together), on the one hand, and no additive on the other hand

The basic data for a single test run are shown in Table 12.28. The measurements used are rating scores—the higher the score, the greater the flame-retardant capability.

Table 12.28 Data for the "Strangways Textile" Case

	Construction of Backing	
	Additive	
No Additive	A	B
4	8	5
6	7	8
5	7	5
5	7	6
6	8	7
6	8	7
7	8	8
6	8	7

Table 12.29 ANOVA Table for the "Strangways Textile" Case

Source of Variation	Sum of Squares	Degrees of Freedom	Mean Square	Observed F	.05 Level Critical F
Between columns					
Com$_1$					
Com$_2$					
Within columns				—	—
Total			—	—	—

1. Fill in all parts of Table 12.29.

2. Interpret the results of your analysis of variance.

13 Selected Nonparametric Tests

We must never assume that which is incapable of proof.—George Henry Lewes

If the only tool you have is a hammer, you tend to see every problem as a nail.
—Abraham Maslow

What You Should Learn from This Chapter

In this chapter we continue with tests of hypotheses. However, unlike most tests we have studied thus far, those presented here require only mild—and, hence, easily met—assumptions. When you have finished this chapter, you should be able to

1. Define nonparametric test.
2. Distinguish between parametric and nonparametric tests of hypotheses.
3. Recognize when chi-square analysis would be called for.
4. Work through chi-square problems involving (1) a single classification variable, (2) two classification variables, and (3) several sample proportions.
* 5. Recognize when each of the following kinds of tests might be used to advantage: (1) the sign test for matched pairs, (2) The Wilcoxon matched-pairs signed-rank test, (3) the Mann-Whitney U test, (4) the Kruskal-Wallis one-way "analysis of variance," and (5) the Friedman two-way "analysis of variance."
* 6. Work through problems calling for each of the five kinds of nonparametric tests listed in 5 above.

No specific review material is recommended before beginning this chapter.

13.1 Introduction

Some unpleasant person once offered the following definition of a statistician: "A statistician is a person who draws a mathematically precise line from an unwarranted assumption to a foregone conclusion." Naturally, your author finds this assessment a little harsh. Still, in calling attention to the possibility of unwarranted assumptions, this anonymous cynic has partly redeemed himself. He has reminded us that we should not blindly apply, say, a t or F test without attempting to judge the validity of the underlying assumptions. And what are these assumptions? They vary somewhat; however, you can hardly help recalling two which have already been referred to quite often. When applying t or F tests to hypotheses concerned with the equality of two or more population means, we assume that the relevant populations have normal shapes and equal variances. Often, we will suspect that one or both of these assumptions have been violated. Probably more often, we simply will not have an adequate basis for judging their validity. Yet these assumptions are inseparable from the tests themselves. If they are violated in the extreme, the validity of the test of significance will be undermined.

In this chapter we concern ourselves with a collection of more cooperative hypothesis-testing methods—methods which require no, or very mild, assumptions. These are called *nonparametric methods*.

Some authors have distinguished between nonparametric methods and distribution-free methods, arguing that a nonparametric method is one that involves no hypothesis or assumptions about the values of parameters, whereas a distribution-free method is one which requires no assumption about the specific shape of the population(s) of interest. More commonly, this distinction is played down and both categories of tests are lumped together under "nonparametric."

For our purposes, we will understand a *nonparametric test* to be one that satisfies at least one of the following criteria:

1. The data being analyzed are frequency data.
2. The data being analyzed represent ranks.
3. The null hypothesis is not concerned with the value of a parameter, such as μ or σ^2, which requires that each elementary unit in the population has an identifiable value associated with it.
4. The hypothesis-testing procedure is not dependent on assumptions about parameters such as μ or σ^2.
5. The probability distribution of the statistic used is not dependent on information or assumptions about the specific shape of the population(s) from which the sample(s) came.

It should be obvious from what has been said thus far that nonparametric methods represent useful additions to the statistician's tool kit. You should be aware, however, that in statistics, just as in any other aspect of life, there is no such thing as a free lunch. We pay a price for the increased versatility and freedom from stringent assumptions offered by these methods: In a situation where either a parametric or a nonparametric procedure could be used and the assumptions associated with the parametric procedures are met, the parametric procedure will be more powerful. That is, it will lead to rejection of an incorrect null hypothesis more surely than will its nonparametric counterpart.

13.2 Introduction to Chi-Square Analysis

We will begin our discussion of nonparametric methods with one of the most versatile hypothesis-testing tools available. Chi-square analysis is so versatile, in fact, that it has parametric as well as nonparametric uses. However, we will confine ourselves to its nonparametric applications. Important to note at the outset is the fact that

---

All of the chi-square applications discussed here will be concerned with the analysis of *frequency distributions*.

GRIN AND BEAR IT   by Lichty & Wagner

"Give me a yes or no in 25 words or less."

Research in business often generates frequency data. This is certainly the case in most opinion surveys in which the person interviewed is asked to respond to a question by marking, say, "Agree," "Not Sure," or "Disagree" or some other such collection of categories. In a case like this, the investigator might be concerned with determining what proportion of respondents marked each of the choices or whether there is any relationship between the opinion marked and the sex, age, or occupation of the respondent.

Chi-square methods make possible the meaningful analysis of frequency data by permitting the comparison of *frequencies actually observed* with *frequencies which would be expected if the null hypothesis were true*. Let us consider, in a general way for now, this idea of "expected frequency."

The Basis of Expected Frequencies

Picture yourself playing a video game having the following rules: The Sky Monster zooms onto the screen in his flying saucer from a random direction and following a random course. You have a split second to get a shot off. If you zap him with that shot, you win. Otherwise, he zonks you and he wins. One or the other thing must happen: Either he is zapped or you are zonked. The probability of your hitting him with your one shot is p_1, which means that the probability of your being "hit" is $p_2 = 1 - p_1$. Let us say that your ability to hit the Sky Monster in the limited time allowed is a function of your reflex action, something which cannot be improved upon to any meaningful degree. In other words, we are assuming that probabilities p_1 and p_2 remain constant from trial to trial. The number of times you play the game—the number of trials—may be denoted n.

The situation described is suggestive of a binomial probability problem: We have (1) an either-or situation, (2) unchanging probabilities, and (3) trials that are identical in nature. Clearly, if we knew the value of p_1 we could readily determine E_1, the expected frequency of games won by you, by determining the value of np_1. Similarly, E_2, the expected frequency of games won by the Sky Monster, could be determined either by $n(1 - p_1)$ or $n - E_1$. Of course, $E_1 + E_2 = n$.

Let us push this fantasy a step further by assuming that you are allowed two shots at the Sky Monster. You may hit him (1) not at all, (2) one time, or (3) two times. We now have three p_i values to consider, two of which have to do with your hitting the Sky Monster. These we will call p_1 and p_2; they represent the probability of your hitting him on the first shot and on the second shot, respectively. (p_1 and p_2 are presumably, though not necessarily equal; what matters is that they be independent within trials and constant from trial to trial.) The remaining probability is p_3, the probability that the Sky Monster will win the battle. Of course, $p_3 = 1 - (p_1 + p_2)$. The corresponding expected frequencies are $E_1 = np_1$, $E_2 = np_2$, and $E_3 = np_3$. Necessarily, $E_1 + E_2 + E_3 = n$. We could go on like this, increasing the number of shots you are allowed by increments of 1. However, we soon realize that the situation described, if stated in very general terms, is one in which $p_k = 1 - (p_1 + p_2 + p_3 + \cdots + p_{k-1})$ and $E_k = n - (E_1 + E_2 + E_3 + \cdots + E_{k-1})$ or, rearranging, $n = E_1 + E_2 + E_3 + \cdots + E_k$.

Now, suppose that prior to playing any games you hypothesize values for p_1, p_2, $p_3, \ldots, p_k$. You do not know what these probabilities really are but you bravely hypothesize that $p_1 =$ some stated value between 0 and 1, $p_2 =$ some stated value between 0 and 1, and so forth, making certain that

$$\sum_{\text{all } i} p_i = 1$$

Clearly, you could, on the basis of your hypothesized p_i values, obtain a series of expected frequencies, or $E_i = np_i$ values. A reasonable proposition would be: If the hypothesized p_i values are all true, the corresponding O_i values (the observed frequencies) would not differ greatly from the hypothesized E_i values. On the other hand, if the hypothesized p_i values are not all correct, you would expect at least some observed frequencies to differ somewhat—and maybe substantially—from the corresponding expected frequencies. (See Figure 13.1.)

A sensible conclusion extractable from all this is: If you wish to test a set of hypothesized p_i values, you should make use of a test statistic that utilizes all k deviations of the type $O_i - E_i$. Fortunately, we have available such a statistic, the χ^2 statistic, computed by

$$\chi^2 = \sum_{\text{all } i} \left[\frac{(O_i - E_i)^2}{E_i} \right]$$

or, more generally,

$$\chi^2 = \sum_{\text{all cells}} \left[\frac{(O - E)^2}{E} \right] \tag{13.2.1}$$

The video game example—and the implications drawn from it—admittedly oversimplifies the idea of expected frequency because it is strictly relevant only to univariate-data

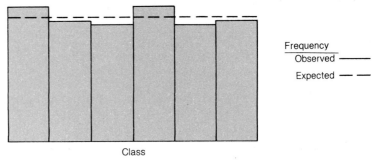

(a) Example of a situation where observed and expected frequencies differ relatively little. Hypothesized p_i values may be correct.

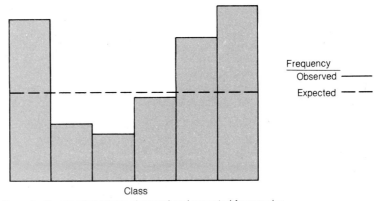

(b) Example of a situation where observed and expected frequencies differ greatly. Hypothesized p_i values are probably wrong.

Figure 13.1. Two sets of comparisons between observed and expected frequencies. (For simplicity, it has been assumed that $H_0: p_1 = p_2 = p_3 = \cdots = p_6$.)

problems. Some of the most interesting and useful applications of chi-square analysis involve bivariate frequency distributions. Nevertheless, the example should give you some feel for "expected frequency" as well as conveying the idea that, in the chi-square applications that follow, what counts is the sizes of the differences between observed frequencies and the corresponding expected frequencies.

General Procedure for Testing a Null Hypothesis Using Chi-Square Analysis

In very general terms, the steps involved in a chi-square analysis are as follows:

1. State the null and alternative hypotheses.
2. Decide on the level of significance to use.
3. Obtain the sample observed frequencies—the O values.

4. Proceeding on the assumption that the null hypothesis is true, compute the expected frequencies—the E values—implied by it.
5. Compare the two kinds of frequencies obtained in Steps 3 and 4 by applying Equation (13.2.1) to the data.
6. Determine the number of degrees of freedom, *df*.
7. Obtain from Appendix Table 7 the critical value of χ^2 associated with the α value set in Step 2 and the degrees of freedom determined in Step 6.
8. Compare observed χ^2 and critical χ^2 and either accept or reject the null hypothesis.

This list of steps implies that the user of chi-square analysis must know how to (1) determine the correct number of degrees of freedom and (2) use the table of critical χ^2 values. These topics are discussed in the next two subsections.

Determining Degrees of Freedom

The method of determining degrees of freedom in chi-square analysis will differ from one type of application to another.

> As a general rule, however, we determine the number of degrees of freedom by (1) counting the number of expected frequencies altogether and (2) subtracting the number of *expected frequencies* that are not free to vary independently.

For example, in the simplest of the chi-square problems, such as the kind implied by our video game example, one expected frequency is not free to vary independently of the others. Why? Because the sum of the expected frequencies must be $np_1 + np_2 + \cdots + np_k = n$, or stated more simply: *The sum of the expected frequencies must be exactly equal to the sum of the observed frequencies, n.* This being so, we must conclude that to know (1) the sample size n and (2) values of all but one of the expected frequencies is equivalent to knowing *all* expected frequencies. Therefore, one expected frequency is said to be "fixed."

When the problem under analysis involves a bivariate frequency distribution, the number of degrees of freedom will be determined somewhat differently, though still in a manner consistent with the general guideline given above. This subject is discussed in Section 13.4.

The χ^2 Distribution

Technically, the chi-square distribution is the sampling distribution which would result if we (1) drew n values at random from a normally distributed population of X values, (2) converted each sample X into its Z-value equivalent, (3) squared the resulting Z's, and (4) summed these squared values. That is,

$$\chi^2 = \sum_{\text{all } i} Z_i^2 = \sum_{\text{all } i} [(X_i - \mu)/\sigma]^2$$

If this procedure were repeated a great many times with samples of size n *and a probability distribution of the results constructed, that probability distribution would be a chi-square sampling distribution.*

And just how is this information beneficial? With respect to the applications described in the next several pages, this background information on the χ^2 distribution is of greater *indirect* than direct importance. We rely heavily on the fact that

> For a sample of fairly large size, we are assured that the statistic
>
> $$\sum_{\text{all cells}} \left[\frac{(O - E)^2}{E} \right]$$
>
> is distributed in approximately a χ^2 manner.

We should note two other things about the theoretical χ^2 sampling distribution: First, because any χ^2 value is obtained by summing a series of squared values, as Equation (13.2.1) makes clear, that sum can never be negative. It could conceivably be zero; a zero would occur in a situation where perfect correspondence existed between each observed frequency and the related expected frequency. This assures us that our chi-square tests will inevitably be upper-tail tests, as illustrated in Figure 13.2. It is only when large discrepancies between O and E values—and, hence, a large observed χ^2 value—occur that we are led to doubt the null hypothesis. If these discrepancies—and, hence, observed χ^2—are small, we have no reason to reject the null hypothesis.

Second, the theoretical χ^2 distribution is not really a single distribution but rather a family of similar distributions. There exists a different χ^2 distribution for each number of degrees of freedom, as shown in Figure 13.3.

Hereafter, we will distinguish between observed χ^2 and critical χ^2 in the following way: If we wish to refer to the observed value, that is, the value obtained using Equation (13.2.1),

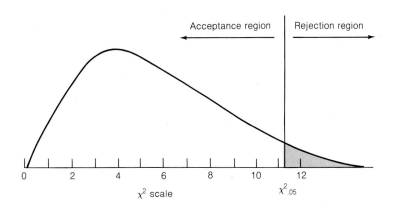

Figure 13.2. The χ^2 distribution for *df* = 5 with a .05 rejection region indicated.

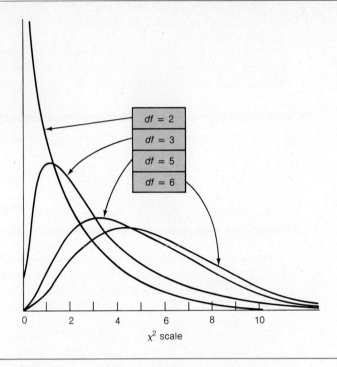

Figure 13.3. The χ^2 distributions associated with $df = 2$, 3, 5, and 6.

we will do so by writing χ^2. If we wish to indicate a critical value, we will write $\chi^2_{\alpha(df)}$. Thus, for a problem in which $\alpha = .05$ and $df = 4$, we write $\chi^2_{.05(4)}$.

Critical values of the χ^2 statistic are shown in Appendix Table 7. This table is reproduced in part here as Table 13.1. To find the critical value to be compared with observed χ^2, we locate the value of α—.05, .01, and so forth—across the top of the table. We then read

Table 13.1   Critical Values of χ^2

Degrees of Freedom (*df*)	Area in Right Tail			
	.10	.05	.02	.01
1	2.706	3.841	5.412	6.635
2	4.605	5.991	7.824	9.210
3	6.251	7.815	9.837	11.345
4	7.779	9.488	11.668	13.277
5	9.236	11.070	13.388	15.086
6	10.645	12.592	15.033	16.812

down this column until we locate the entry directly across from the appropriate number of degrees of freedom. That number is $\chi^2_{\alpha(df)}$. If observed χ^2 is less than this value, we accept the null hypothesis; if larger, we reject H_0.

Let us now survey some ways that the chi-square analysis can be applied to business problems.

13.3 Chi-Square Single-Classification Problems

In a single-classification problem, the n observations are hypothesized to be assignable to classes of a single variable according to probabilities stated in advance of any examination of the data. Suppose, for example, that 600 consumers were questioned regarding their reactions to a new food product. The categories to be checked on a simple questionnaire were (1) I don't like it, (2) I neither like nor dislike it, and (3) I like it. Each respondent was limited to *one choice*. Since the concerned company, let us say, had no prior experience with consumer reactions to the product, it used a null hypothesis requiring an absolute minimum of preconception—namely, that responding consumers would be evenly distributed over the three categories of opinion. Let us work through this problem following the steps presented in section 13.2.

1. State the null and alternative hypotheses.
   In chi-square analysis, the null hypothesis is stated in such a way that, if it is true, a specific kind of frequency pattern would be expected. The null and alternative hypotheses for the present problem can be stated as follows:

   $$H_0:\ p_1 = p_2 = p_3 = 1/3$$
   $$H_a:\ p_1,\ p_2,\ p_3\ \text{are not all equal to } 1/3$$

2. Decide on the level of significance to use.
   Let us say that the level of significance was set at $\alpha = .05$.

3. Obtain the observed frequencies—the O values.
   The observed frequencies are shown in column (1) of Table 13.2.

Table 13.2 Chi-Square Worksheet for a Single-Classification Problem
($H_0:\ p_1 = p_2 = p_3 = 1/3$)

Consumer Reaction	(1) O_i	(2) E_i	(3) $O_i - E_i$	(4) $(O_i - E_i)^2$	(5) $\dfrac{(O_i - E_i)^2}{E_i}$
1. I don't like it	200	200	0	0	0
2. I neither like nor dislike it	100	200	-100	10,000	50
3. I like it	300	200	100	10,000	50
Total	600	600			$100 = \chi^2$

4. Proceeding on the assumption that the null hypothesis is true, compute the expected frequencies—the E values—implied by it.

   The expected frequencies are obtained by multiplying the sample size by p_i. We have $n = 600$ times $1/3 = 200$. The value 200 would then be assigned to each of the three response categories, as shown in column (2) of Table 13.2.

5. Compare the two kinds of frequencies obtained in Steps 3 and 4 by applying Equation (13.2.1) to the data.

   This step is also demonstrated in Table 13.2. Observed χ^2, shown in the lower right-hand corner of this table, was found to be 100.

6. Determine the number of degrees of freedom, df.

   We calculate: $df = k - 1 = 3 - 1 = 2$.

7. Obtain from Table 13.1 the critical value of χ^2 associated with the α value set in Step 2 and the degrees of freedom determined in Step 6.

   Table 13.1 reveals that the critical value $\chi^2_{.05(2)}$ is 5.991.

8. Compare observed χ^2 and critical χ^2 and either accept or reject the null hypothesis.

   Since observed χ^2 was found to be 100, a considerably larger value than 5.991, we resolutely reject the null hypothesis and conclude that the response-category probabilities are not all 1/3. However,

---

In business applications of chi-square analysis, one is always wise to scrutinize the O and E values closely before basing a decision on a rejected null hypothesis.

Consider, for example, the following two explanations for the significant χ^2 statistic: (1) More people say they dislike the product than would be expected under the null hypothesis $(O_1 > E_1)$ and fewer say they like it $(O_3 < E_3)$, and (2) fewer people say they dislike the product than would be expected $(O_1 < E_1)$ and more than expected say they like the product $(O_3 > E_3)$. Clearly, either of these possibilities could lead to a rejected null hypothesis, but for reasons that are polar opposites. The decision flowing from such an analysis could be a sound one or a disaster, depending on the reason for the large observed χ^2 value.

Actually, the truth is less clear-cut in its decision-making implications than either of the possibilities recited above, though closer to the second, more favorable, possibility than to the first. Table 13.2 reveals that exactly the same number of respondents claimed to dislike the product as expected under the null hypothesis. However, only half as many respondents were indifferent, and more claimed that they like it than would be the case if the null hypothesis were true.

The kind of problem that hypothesizes the equal distribution of the n sample observations over the k classes is probably the most frequently used type of single-classification problem because it calls for a minimum amount of knowledge and few preconceived ideas about the situation being analyzed. Nevertheless, unequal hypothesized p_i values—and, hence, unequal E values—can be used by following exactly the same procedure as demonstrated above.

13.4 Chi-Square Tests of Independence Between Two Principles of Classification

Some Fundamentals

The rationale behind the chi-square test of independence is tied in with the concept of independence in probability theory. We noted in Chapter 6 that two events, E_1 and E_2, are independent if the conditional and unconditional probabilities for each event are equal. At that time we illustrated the point using data shown in Table 6.4, which pertained to a population of 200 salespeople with the Exmont Company. This table is reproduced for convenience of reference as Table 13.3.

If we were told that one salesperson was to be picked at random from among the 200, we could determine the probability that the salesperson picked enjoys a highly successful "Sales Success Status" (A_1) by expressing the number of salespeople classified as highly successful—namely, 60—to the total number of salespeople altogether (200), giving us $P(A_1) = 60/200 = .3$.

On the other hand, if we wished to determine $P(A_1|C_1)$, where C_1 indicates that consumer goods is the type of product sold, we would express the number of highly successful consumer-goods salespeople—namely, 45—as a ratio to the total number of consumer-goods salespeople (150). Doing so, we obtain $P(A_1|C_1) = 45/150 = .3$. Thus, we see that $P(A_1) = P(A_1|C_1)$. Moreover, we also see that $P(C_1) = P(C_1|A_1) = .75$. Clearly, A_1 and C_1 in this example are independent events. If this quality holds for all possible A_i and C_j combinations, we say that the variables A and C are independent variables.

> Whether two variables are independent is what the present chi-square test is designed to determine. The test is performed on a random sample from a bivariate population. The test result indicates whether the sample data are consistent with or inconsistent with the hypothesis of independence.

Table 13.3 Bivariate Frequency Distribution Showing the Relationship Between Sales Success Status and Type of Product Sold for 200 Salespeople Employed by the Exmont Company

Sales Success Status		Type of Product		Total
		Consumer (C_1)	Industrial (C_2)	
Highly successful	(A_1)	45	15	60
Moderately successful	(A_2)	90	30	120
Unsuccessful	(A_3)	15	5	20
Total		150	50	200

2 × 2 Contingency Tables

The organizing device used for the chi-square test of independence is the *contingency table*, which is really nothing more than a bivariate frequency distribution. A convenient kind of contingency-table problem with which to begin our discussion of this test is the kind involving two qualitative variables each of which has only two categories and is, therefore, appropriately described as a 2 × 2 contingency table (two rows and two columns). One of these is presented in Table 13.4, Part I. All entries within this part of the table represent observed frequencies and, as always, the cell frequencies sum to the grand total which is the sample size n. The manner of subjecting data in a 2 × 2 contingency table to chi-square analysis will be demonstrated by referring to the following illustrative case:

ILLUSTRATIVE CASE 13.1

The Pry Brothers Company was planning the introduction of a new toothpaste which was to be accompanied by an extensive magazine advertising campaign. The company's advertising staff had developed two rather different appeals which we will call (1) Appeal F, so-called because it was thought to be of greater interest to women than men, and (2) Appeal M, because of its presumed special interest to men. Because of the high cost of the planned advertising campaign, management considered it quite important to know whether Appeal F did in fact interest women more than men and whether Appeal M did in fact interest men more than women. They reasoned that if such were the case, a wise strategy would be to run Appeal F in women's magazines and Appeal M in men's magazines despite the somewhat higher cost of this more selective approach relative to that of running the same appeal in both categories of magazines.

A random sample of 800 adults was selected and each sample member asked to indicate whether he or she preferred Appeal F or Appeal M. They could base their choice on any criteria they thought relevant and take whatever time was needed to decide. But they were required to state a preference: Ties of any nature were not permitted.

The null hypothesis, to be tested at the .01 level of significance, was H_0: The appeal preferred is independent of the sex of the respondent. Of course, the alternative hypothesis was H_a: The appeal preferred is not independent of the sex of the respondent.

We will perform the chi-square analysis by following the same sequence of steps used in the previous example. Since the first two steps are contained within the case description itself, we will begin with Step 3.

3. Obtain the sample observed frequencies—the O values.
   These are presented in Part I of Table 13.4.

4. Proceeding on the assumption that the null hypothesis is true, compute the expected frequencies—the E values—implied by it.
   Carrying out this step for a contingency table problem is somewhat more involved than the analogous step in a single-classification problem.

> The procedure for getting the expected frequency for the upper left-hand cell of the table, for example, involves (1) dividing the total of the leftmost column of observed frequencies by the overall sample size n, and (2) multiplying the resulting quotient by the total of the top row of observed frequencies.

Table 13.4 Observed and Expected Frequencies for the
Pry Brothers Problem

Part I: Observed Frequencies

	Appeal Preferred		
Sex of Respondent	F	M	Total
Male	125	300	425
Female	200	175	375
Total	325	475	800

Part II: Expected Frequencies

	Appeal Preferred		
Sex of Respondent	F	M	Total
Male	172.66	252.34	425.00
Female	152.34	222.66	375.00
Total	325.00	475.00	800.00

To illustrate this procedure, look for the total of the leftmost column of observed frequencies in Part I of Table 13.4. We find it is 325. When this total is divided by $n = 800$, the result is .40625. This quotient represents the proportion of people, regardless of sex, in the sample preferring Appeal F. Or, in the language of probability, it is the unconditional probability that a randomly selected member of this sample would prefer Appeal F.

Now, if the appeal preferred is independent of the sex of the respondent, as the null hypothesis asserts, the proportion of males preferring Appeal F (that is, the probability that a person picked at random will be male given that the person picked prefers Appeal F) should also be .40625. Therefore, granting for the moment that the null hypothesis is true, we multiply 425, the total number of males in the sample, by .40625, getting 172.66. This result is the *expected frequency* associated with the upper left-hand cell.

This procedure is followed for all cells in the contingency table. For our example:*

$$(325/800)(425) = 172.66$$
$$(325/800)(375) = 152.34$$
$$(475/800)(425) = 252.34$$
$$(475/800)(375) = 222.66$$

5. Compare the two kinds of frequencies obtained in Steps 3 and 4 by applying Equation (13.2.1) to the data.

*Actually, only one such calculation need be performed when working with a 2 × 2 contingency table because, when the investigator knows all marginal totals and *one* expected frequency, he can determine the remaining expected frequencies by subtracting the one which is known from the row and column totals.

Ignoring for the moment a technical adjustment called for in this kind of problem, we calculate χ^2 as follows:

$$\chi^2 = \frac{(125 - 172.66)^2}{172.66} + \frac{(200 - 152.34)^2}{152.34} + \frac{(300 - 252.34)^2}{252.34}$$
$$+ \frac{(175 - 222.66)^2}{222.60} = 47.27$$

6. Determine the number of degrees of freedom, *df*.

The procedure used to obtain the expected frequencies in this problem ensures that row and column totals will be the same for the expected frequencies as for the corresponding observed frequencies. This fact is clearly evident in Table 13.4. Consequently, we find that just as soon as the expected frequency associated with one cell has been determined, the expected frequencies associated with the remaining three cells are "fixed." We conclude, therefore, that any 2 × 2 contingency table will have but one degree of freedom.

More generally, we observe that

The number of degrees of freedom for any contingency table can be determined by $df = (r - 1)(c - 1)$, where r is the number of rows and c is the number of columns.

7. Obtain from Table 13.1 the critical value of χ^2 associated with the α value set in Step 2 and the degrees of freedom determined in Step 6.

Table 13.1 reveals that critical $\chi^2_{.01(1)} = 6.635$.

8. Compare observed χ^2 and critical χ^2, and either accept or reject the null hypothesis.

Since observed χ^2 of 47.27 is greater than critical $\chi^2_{.01(1)}$ of 6.635, we reject the null hypothesis and conclude that preference is not independent of sex. Careful comparison of observed and expected frequencies reveals that female respondents tend to favor Appeal F and male respondents, Appeal M.

Yates' Correction. For chi-square problems involving only one degree of freedom, the accuracy of the test of significance can be improved somewhat if the χ^2 equation is modified slightly. The recommended procedure is

$$\chi^2 = \sum_{\text{all cells}} \left[\frac{(|O - E| - .5)^2}{E} \right] \tag{13.4.1}$$

In this modified equation, the absolute value of the difference between O and E for each cell is reduced by .5 before the squaring is done. This refinement, known as Yates' correction for continuity, has the effect of reducing a bias in the direction of rejecting too many null hypotheses.* For our advertising problem, use of Yates' correction resulted in observed χ^2 of 46.28, only slightly less than the 47.27 obtained without its use. Clearly, this refinement is important only in connection with borderline cases.

*Some critics of this refinement argue that it overcorrects. Nevertheless, the prevailing view is that its use is desirable in one-degree-of-freedom problems and unnecessary in problems having more than one degree of freedom.

The $r \times c$ Contingency Table

A contingency table may be of any size. For example, Table 13.5 shows a more extensive one. The observed frequencies are shown in Part I of this table, the expected frequencies in Part II, and the calculation of the χ^2 statistic in Part III. The expected frequencies were obtained in the same manner as described above for the Pry Brothers problem. Part I of Table

Table 13.5 Chi-Square Test of Whether Price-Earnings Ratios Are Independent of Earnings per Share

Part I: Observed Frequencies

	Earnings per Share				
Price-Earnings Ratio	0 and Under $1	$1 and Under $2	$2 and Under $3	$3 and Over	Total
3 and under 9	6	27	32	19	84
9 and under 12	10	13	13	7	43
12 and over	15	33	12	13	73
Total	31	73	57	39	200

Part II: Expected Frequencies

	Earnings per Share				
Price-Earnings Ratio	0 and Under $1	$1 and Under $2	$2 and Under $3	$3 and Over	Total
3 and under 9	13.02	30.66	23.94	16.38	84.00
9 and under 12	6.67	15.70	12.26	8.39	43.00
12 and over	11.32	26.65	20.81	14.24	73.01
Total	31.01	73.01	57.01	39.01	200.00

Part III: Calculation of the χ^2 Statistic

Observed Frequency O	Expected Frequency E	$O - E$	$(O - E)^2$	$\dfrac{(O - E)^2}{E}$
6.00	13.02	−7.02	49.2804	3.7850
10.00	6.67	3.33	11.0889	1.6625
15.00	11.32	3.68	13.5424	1.1963
27.00	30.66	−3.66	13.3956	.4369
13.00	15.70	−2.70	7.2900	.4643
33.00	26.65	6.35	40.3225	1.5130
32.00	23.94	8.06	64.9636	2.7136
13.00	12.26	.74	.5476	.0447
12.00	20.81	−8.81	77.6161	3.7298
19.00	16.38	2.62	6.8644	.4191
7.00	8.39	−1.39	1.9321	.2303
13.00	14.24	−1.24	1.5376	.1080
Total				$16.304 = \chi^2$

13.5 is called a 3×4 contingency table because it has $r = 3$ rows and $c = 4$ columns. The number of degrees of freedom, therefore, would be $df = (r - 1)(c - 1) = (3 - 1)(4 - 1) = 6$.

The question addressed in this problem is: Is the price-earnings ratio independent of earnings? The null hypothesis is H_0: Price-earnings ratios are independent of earnings per share. Since $\chi^2 = 16.304$ (Part III of Table 13.5) is greater than $\chi^2_{.05(6)} = 12.592$ (Table 13.1), we reject the null hypothesis at the .05 level of significance.

13.5 Chi-Square Test for Equality Among Several Population Proportions

Problems sometimes arise which require the comparison of several sample proportions with a view to determining whether the corresponding population proportions are equal. Recall that in section 9.9 of Chapter 9 we dealt with hypotheses about a single population proportion. In section 10.3 of Chapter 10, we considered tests pertaining to the equality, or lack of it, of two population proportions. The present section may be viewed as an extension of these earlier ones. You may recall that in section 10.3 we tackled a problem concerned with a consumer-polling service which asked a random sample of residents of each of two cities, Des Moines, Iowa, and Newark, New Jersey, the question "Do you intend to buy a new automobile anytime within the next twelve months?" Let us now suppose that this consumer-polling service also asked the same question of a random sample of 2500 residents of San Jose, California, and 1000 residents of Helena, Montana. Part I of Table 13.6 summarizes the results obtained in the four cities.

The polling service now seeks an answer to the question: "Is the proportion of residents intending to buy a new automobile during the next year the same for the *four* cities?" We can answer this question for them by following the now-familiar list of procedural steps:

1. State the null and alternative hypotheses.

$$H_0: \ p_1 = p_2 = p_3 = p_4$$
$$H_a: \ p_1, p_2, p_3, \text{ and } p_4 \text{ are not all equal}$$

2. Decide on the level of significance to use.
    Let us assume that α is set at .10 as was the case in Chapter 10 when only two cities were considered.

3. Obtain the sample observed frequencies—the O values.

WARNING

Chi-square methods cannot properly be used on relative frequencies—whether they are called proportions, percents, or something else. Therefore, the sample proportions must be converted to observed frequencies through multiplication by the corresponding sample sizes.

Table 13.6 Chi-Square Test to Determine Whether Four Population Proportions Are Equal

Part I: Consumer Poll Results

		City	
Des Moines	Newark	San Jose	Helena
$\bar{p}_1 = .11$	$\bar{p}_2 = .09$	$\bar{p}_3 = .12$	$\bar{p}_4 = .13$
$n_1 = 2000$	$n_2 = 3000$	$n_3 = 2500$	$n_4 = 1000$
$O_1 = 220$	$O_2 = 270$	$O_3 = 300$	$O_4 = 130$
$n_1 - O_1 = 1780$	$n_2 - O_2 = 2730$	$n_3 - O_3 = 2200$	$n_4 - O_4 = 870$

Part II: Observed and Expected Frequencies (Expected in Parentheses)

Nature of Response	Des Moines	Newark	San Jose	Helena	Total
Answered Yes	220 (216.47)	270 (324.71)	300 (270.59)	130 (108.24)	920
Did not answer Yes	1780 (1783.53)	2730 (2675.29)	2200 (2229.41)	870 (891.76)	7580
Total	2000	3000	2500	1000	8500

Part III: Calculation of χ^2 Statistic

Observed Frequency O	Expected Frequency E	$(O - E)$	$(O - E)^2$	$\dfrac{(O - E)^2}{E}$
220	216.47	3.53	12.4609	.0576
1780	1783.53	−3.53	12.4609	.0070
270	324.71	−54.71	2993.1841	9.2180
2730	2675.29	54.71	2993.1841	1.1188
300	270.59	29.41	864.9481	3.1965
2200	2229.41	−29.41	864.9481	.3880
130	108.24	21.76	473.4976	4.3745
870	891.76	−21.76	473.4976	.5310
Total				$18.8910 = \chi^2$

For example, the observed frequency associated with Des Moines is $O_1 = n_1\bar{p}_1 = (2000)(.11) = 220$; the observed frequency for Newark is $O_2 = n_2\bar{p}_2 = (3000)(.09) = 270$; and so forth. The resulting observed frequencies are the *actual number of persons* who answered Yes to the question.

WARNING

In order to achieve a valid chi-square test for this kind of problem, we must secure another set of observed frequencies—namely, the number of people who *did not* answer Yes. More generally, we must obtain the *complements* of the observed frequencies computed by the $O_i = n_i\bar{p}_i$ process.

Such complements for the problem under discussion are shown in the bottom line of Part I of Table 13.6. In Part II of this same table, all necessary observed frequencies are displayed. (These are the numbers not within parentheses.)

4. Proceeding on the assumption that the null hypothesis is true, compute the expected frequencies—the E values—implied by it.

   From here on the procedural steps are carried out in the same way as the tests of independence. The expected frequencies for our problem are displayed within parentheses in Part II of Table 13.6.

5. Compare the two kinds of frequencies obtained in Steps 3 and 4 by applying Equation (13.2.1) to the data.

   The χ^2 calculations are carried out in the usual way. Computational details and results for the present problem are displayed in Part III of Table 13.6. Observed χ^2 is 18.89.

6. Determine the number of degrees of freedom, df.

   We determine that $df = (r - 1)(c - 1) = (2 - 1)(4 - 1) = 3$.

7. Obtain from Table 13.1 the critical values of χ^2 associated with the α value set in Step 2 and the degrees of freedom determined in Step 6.

   Table 13.1 informs us that the critical value $\chi^2_{.10(3)}$ is 6.251.

8. Compare observed χ^2 and critical χ^2, and either accept or reject the null hypothesis.

   Since χ^2 of 18.89 is greater than $\chi^2_{.10(3)}$ of 6.251, we are led to reject H_0, concluding, therefore, that the population proportions are not the same for the four cities.

The chi-square tests for equality of population proportions and for independence appear very similar and, indeed, most of the computational procedures are identical for both. However, the two chi-square tests do differ in a most fundamental way: The test of independence is applied to a *single sample* whose elementary units can be categorized on the basis of two sets of attributes. The analyst applying this test is concerned about whether one set of attributes is independent of the other set. The present test is applied to proportions associated with *several independent samples* with a view to answering the question: Are the corresponding population proportions equal?

13.6 Some Precautions in the Use of Chi-Square Analysis

Although the applications of chi-square analysis described above are all nonparametric, this very versatile family of testing procedures is not entirely free of restrictions. Here are some rules that should be observed when performing a chi-square analysis:

1. As mentioned, if the number of degrees of freedom, df, is 1, Yates' correction for continuity should be used. (See the example in section 13.4.)
2. The observed frequencies must be absolute, not relative. (See the example in section 13.5.)
3. The separate observations making up the random sample(s) should be independent.

In particular, this means that caution must be exercised in situations where repeated observations are made on the same individuals. In such situations, it is not inevitable that independence will be lacking, but the presumption is strong that it will be.

4. Small expected frequencies are to be avoided. A good rule is that no E values should be less than 5. Some authorities allow that, for a contingency-table problem, some cells—not to exceed 20% of the total number of cells—may have expected frequencies of less than 5.

5. The sample size n should not be too small. Usually $n = 50$ is considered an absolute minimum.

In the remainder of this chapter, we describe ways of applying several other kinds of nonparametric tests.

✻ 13.7 The Sign Test for Matched Pairs

Application of the sign test for matched pairs involves the matching of pairs of test units according to some characteristic or combination of characteristics. It has particular usefulness in situations where two samples are obviously not independent, such as in an experiment in which two readings, like "before-and-after" measures, must be made on each elementary unit. Also, it may be used when the shapes of the parent populations are thought to be nonnormal or when successive pairs of observations do not measure exactly the same thing. It can even be used on nonquantifiable pairs of observations provided that there is some way of recognizing which member of a pair is "better," "brighter," "more beautiful," or whatever. In short, you are about to become acquainted with a very flexible test of significance. Alas, as will be shown, it is not a very powerful one.

The analysis utilizes the differences between measurements (when the study involves something measurable) obtained from each member of the pair—but not the differences themselves; *only the signs of the differences are used*. The only restriction on this test is that successive pairs of observations are assumed to be independent.

The sign test can be demonstrated most conveniently by referring to Illustrative Case 10.3 in Chapter 10. You may recall that in section 10.5 we discussed the t test for matched pairs using information obtained from an experiment conducted at the River City Manufacturing Company. The production supervisor wondered whether worker output was different when the temperature within the plant was "cool" (defined as 55°F) compared to when the temperature was "warm" (defined as 75°F). According to the t test, a parametric test, output was found to differ significantly between the cool and the warm temperatures, with the cool temperature being better.

Procedural Steps for the Sign Test for Matched Pairs

The steps we will follow when applying the sign test for matched pairs on the data from Illustrative Case 10.3 can be summarized as follows:

1. State the null and alternative hypotheses.
2. Decide on the level of significance to use.

3. Calculate the $Y - X$ values.
4. When the $Y - X$ values have been calculated, eliminate all 0 values (all ties) from the analysis.
5. Count the number of plus signs and the number of minus signs associated with the remaining differences.
6. Calculate the probability of obtaining, in a random sample of size n, the number of plus signs (or the number of minus signs) which were actually obtained if the probability of a plus (or minus) were .5.
7. Determine whether the null hypothesis should be accepted or rejected.

Application of the Procedural Steps

We will consider each of the above steps in turn.

1. State the null and alternative hypotheses.
   The null hypothesis can be stated

   $$H_0: P(Y > X) = P(Y < X) = .5$$

   That is, the probability of any given Y value's exceeding the corresponding X value is the same as the X value's exceeding the Y value—namely, .5.
   For a two-sided test, the alternative hypothesis is

   $$H_a: P(Y > X) \neq P(Y < X) \neq .5$$

2. Decide on the level of significance to be used.
   Let us suppose the production supervisor chooses $\alpha = .05$.

3. Calculate the $Y - X$ values.
   The $Y - X$ values are displayed in Tables 10.4 and 13.7.

4. When $Y - X$ values have been calculated, eliminate all 0 values (all ties) from the analysis.
   Three ties were found among the $Y - X$ values; these are eliminated from the analysis and the sample size correspondingly reduced from $n = 30$ to $n = 27$.

5. Count the number of plus signs and the number of minus signs associated with the remaining differences.
   The count reveals 17 pluses and 10 minuses.

6. Calculate the probability of obtaining, in a random sample of size n, the number of plus signs (or the number of minus signs) which were actually obtained if the probability of a plus (or minus) were .5.
   Since a difference must have either a plus sign or a minus sign and since the probability of a "success" in a single trial is stated in the null hypothesis to be .5, we have a problem that suggests the use of binomial probabilities. Let us define the number of times the *least frequently occurring* sign is present in the $Y - X$ values as S. That is, S = the number of minus signs; in this case, 10. Our task, then, would be to calculate the probability $P(S \leq 10)$, where this probability is equal to the sum of several binomial probabilities—namely,

$$P(0) + P(1) + \cdots + P(10)$$

$$= \frac{27!}{0!27!}(.5)^0(.5)^{27} + \frac{27!}{1!26!}(.5)^1(.5)^{26} + \cdots + \frac{27!}{10!17!}(.5)^{10}(.5)^{17}$$

For very small samples ($5 \leq n \leq 10$) the above procedure should be used. If the probability obtained is less than .025—for $\alpha = .05$ and a two-tailed test—the null hypothesis is rejected; otherwise, H_0 is accepted. Samples of size $n < 6$ should not be used at all for tests having $\alpha \leq .05$ because it will be impossible to reject the null hypothesis even if all the signs in the sample are identical.

Fortunately, for larger sample sizes the direct use of the binomial formula is not really necessary; satisfactory approximations to cumulative binomial probabilities can be secured through use of the normal curve. This procedure, like all procedures involving use of the normal curve, calls for the determination of a Z value where, in this type of problem,

$$Z = \frac{S - \mu \pm .5}{\sigma} \qquad (13.7.1)$$

This expression for Z is essentially the same as the one introduced in Chapter 7. The only differences are (1) the use of S in place of X and (2) the $\pm.5$ in the numerator. This latter difference is the continuity correction introduced toward the end of section 7.6.

The arithmetic mean appears in the Z formula, as usual—but, this time, it is the arithmetic mean of a binomial probability distribution. Hence, by Equation (7.4.2), $\mu = np = (27)(.5) = 13.5$.

The standard deviation in this Z formula is determined by applying Equation (7.4.3) and is $\sigma = \sqrt{npq} = \sqrt{(27)(.5)(.5)} = 2.60$. Therefore, the Z value sought is

$$Z = \frac{S - \mu + .5}{\sigma} = \frac{10.5 - 13.5}{2.60} = -1.15$$

7. Determine whether the null hypothesis should be accepted or rejected.
   The decision rule for this problem is

Decision Rule
1. If observed $Z \geq -1.96$, accept H_0.
2. If observed $Z < -1.96$, reject H_0.

Notice that, even though we are conducting a two-tailed test and even though the critical value of -1.96 is the appropriate lower-tail critical value for a two-tailed test with $\alpha = .05$, the decision rule suggests a one-tailed test. Keep in mind that we defined S to be the number of times the *least frequently occurring* sign was observed. This being so, we force observed Z to be a negative number (or 0), a situation requiring that the comparison between observed Z and critical Z be made in the left-hand tail of the normal curve. If a one-tailed test were called for, regardless of the tail, the entire α value would be placed under the left-hand tail, as illustrated in Figure 13.4.

In this problem the observed value of Z, -1.15, is within the acceptance region, so we are led to accept the null hypothesis that $P(Y > X) = P(Y < X) = .5$.

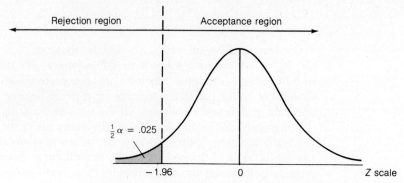

(a) Acceptance and rejection regions for a two–tailed matched–pairs sign test. The null hypothesis is H_0: $P(Y > X) = P(Y < X) = .5$.

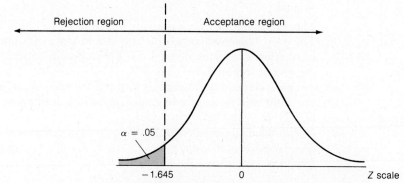

(b) Acceptance and rejection regions for a one-tailed (regardless of the tail), matched-pairs sign test. The null hypothesis may be H_0: $P(Y > X)$ is greater than .5 or H_0: $P(Y > X)$ is less than .5.

Figure 13.4. How the sign test for matched pairs is applied in the case of a two-tailed test and in the case of a one-tailed test, where $\alpha = .05$ for both kinds of test.

Or, expressing the same idea in more practical terms, we conclude that temperature has no effect on worker output. This, of course, is contrary to the conclusion reached through application of the t test for matched pairs in Chapter 10. These conflicting results illustrate rather well the (usually) greater power of a parametric test relative to a "comparable" nonparametric test.

*13.8 The Wilcoxon Matched-Pairs Signed-Rank Test

The Wilcoxon test is similar to the sign test in that it also makes use of differences between pairs of observations. However, it is a more powerful test because it takes into account not only the signs, but also the magnitudes, of the differences.

In our discussion of this useful nonparametric tool, we will use the symbol d_a to represent the *absolute* value of a $Y - X$ difference. That is, if we look at the output of worker 5 in Table 10.4, for example, we note that the corresponding values of X and Y are, respectively, 26 and 22. Therefore, $Y - X = 22 - 26 = -4$; but $d_a = |Y - X| = 4$.

Procedural Steps for the Wilcoxan Matched-Pairs Signed-Rank Test

The Wilcoxan test calls for the following sequence of steps:

1. State the null and alternative hypotheses.
2. Decide on the level of significance to be used.
3. Calculate the d_a values, that is, the $|Y - X|$ values.
4. When d_a values have been calculated, eliminate all zeros (all ties) from the analysis.
5. Rank the d_a values in order of magnitude from smallest to largest.
6. When the ranking of the d_a values has been completed, the signs originally associated with the $Y - X$ differences are attached to the ranks.
7. Determine the value of the test statistic T.
8. Compute μ_T, the mean of the sampling distribution of T.
9. Compute σ_T, the standard deviation of the sampling distribution of T.
10. Determine the Z-value equivalent of computed T.
11. Either accept or reject the null hypothesis.

Application of the Procedural Steps

Let us now apply these general steps to a specific problem. As a means of demonstrating the greater power of this test over the similar sign test, we will again use data from the River City Manufacturing Company case.

1. State the null and alternative hypotheses.

   H_0: For the population, the sum of ranks of positive sign is the same as the sum of ranks of negative sign

   H_a: For the population, the sum of ranks of positive sign is not the same as the sum of ranks of negative sign

2. Decide on the level of significance to be used.
   Let us continue to assume that $\alpha = .05$.

3. Calculate the d_a values.
   This step is shown in column (2) of Table 13.7.

4. When d_a values have been calculated, eliminate all zeros (all ties) from the analysis.
   As before, three ties were found and eliminated from the analysis. Consequently, the sample size is reduced from $n = 30$ to $n = 27$.

5. Rank the d_a values in order of magnitude from smallest to largest.
   This step is demonstrated in columns (3) and (4) of Table 13.7. In column (3) the d_a values appearing in column (2) are presented in an array. In column (4) the ranks of these arrayed d_a values are listed.

Table 13.7 Worksheet for Calculating the T Statistic

(1) $Y - X$	(2) $\lvert Y - X \rvert = d_a$	(3) Arrayed d_a Values	(4) Rank of d_a Values	(5) Ranks with Signs Attached
+3	3	1	1	+1
+4	4	2	4	−4
+12	12	2	4	−4
+10	10	2	4	−4
−4	4	2	4	−4
0	Eliminated	2	4	−4
−2	2	3	9	+9
+1	1	3	9	+9
−4	4	3	9	−9
+6	6	3	9	+9
+3	3	3	9	−9
−4	4	4	13.5	+13.5
+9	9	4	13.5	−13.5
+9	9	4	13.5	−13.5
−2	2	4	13.5	−13.5
+8	8	5	17.5	+17.5
0	Eliminated	5	17.5	+17.5
+5	5	5	17.5	+17.5
−2	2	5	17.5	+17.5
+6	6	6	20.5	+20.5
−2	2	6	20.5	+20.5
−2	2	8	22	+22
+5	5	9	24	+24
+5	5	9	24	+24
−3	3	9	24	+24
+3	3	10	26	+26
+5	5	12	27	+27
0	Eliminated			
−3	3		*Sums*	
+9	9			

Minus Ranks	**Plus Ranks**
78.5	299.5

$$\therefore T = 78.5$$

Ties sometimes occur at this stage of the analysis. If some tying ranks are observed, all tying values are given the same rank—a value representing the *average* of what the ranks would have been in the absence of a tie. For example, if $Y - X$ values of +6 and −6—that is, 6 and 6 after the elimination of signs—are tied for the rank of 4, they would each be assigned a rank of 4.5 because, if they had not been tied, they would have received ranks of 4 and 5. Of course, $(4 + 5)/2 = 4.5$. The d_a value just greater than these would be assigned the rank of 6.

6. When the ranking of the d_a values has been completed, the signs originally associated with the $Y - X$ differences are attached to the ranks.

This step is shown in column (5) of Table 13.7.

13 *Selected Nonparametric Tests*

7. Determine the value of the test statistic T.

The test statistic to be used will be symbolized T and defined as the sum of either the positive ranks or the negative ranks, *whichever is smaller.*

For our problem, $T = 78.5$, as indicated at the bottom of Table 13.7.

8. Compute μ_T, the mean of the sampling distribution of T.

The mean is determined by

$$\mu_T = \frac{n(n+1)}{4} \qquad (13.8.1)$$

For the present problem,

$$\mu_T = \frac{(27)(28)}{4} = 189$$

9. Compute σ_T, the standard deviation of the sampling distribution of T.

This standard error is determined by

$$\sigma_T = \sqrt{\frac{n(n+1)(2n+1)}{24}} \qquad (13.8.2)$$

Therefore,

$$\sigma_T = \sqrt{\frac{(27)(28)(55)}{24}} = 41.62$$

10. Determine the Z-value equivalent of T.

For large samples ($n > 16$ as a rule of thumb), the theoretical sampling distribution of T is approximately normal.

This being the case, we can convert T to its Z-value equivalent by

$$Z = \frac{T - \mu_T}{\sigma_T} = \frac{78.50 - 189.00}{41.62} = -2.65.$$

11. Either accept or reject the null hypothesis.
    The decision rule is the same as it was for the sign test—namely,

Decision Rule

1. If $Z \geq -1.96$, accept H_0.
2. If $Z < -1.96$, reject H_0.

Since -2.65 is less than -1.96 and outside the acceptance region of the standard normal curve, we reject the null hypothesis. Evidently the cooler temperature is more conducive to high worker output than is the warmer temperature.

This conclusion is contrary to the one reached when applying the sign test to the same data, the reason being that in the case of the sign test neither sign was dominant enough to suggest convincingly that temperature matters. However, when the magnitudes of the differences as well as their signs were taken into account, the superiority of the cooler temperature became evident.

A disadvantage of the Wilcoxon test relative to the sign test is that its underlying assumptions are more restrictive. In addition to the assumption of independence among pairs of observations, an assumption which the two tests have in common, the Wilcoxon test when used in connection with the kind of null hypothesis stated above is dependent on a second, more stringent, assumption. This second assumption is that the population associated with the X values and the population associated with the Y values have identical frequency distributions and, by implication, equal variances.

Greater comparability between the two tests can be achieved by stating the null hypothesis for the Wilcoxon test as H_0: The two populations are identical. Testing this null hypothesis requires only the independence assumption. However, a rejected null hypothesis is now more difficult to interpret. One would know that the two populations are not identical, but there is more than one way in which they could differ. The explanation, if it could be known, might bear little relationship to the question which prompted use of the Wilcoxon test in the first place.

When $n < 16$, a special probability table should be used. This table is Appendix Table 8, part of which is reproduced here.

One-Tailed	Two-Tailed	$n = 5$	$n = 6$	$n = 7$	$n = 8$	$n = 9$	$n = 10$
$\alpha = .05$	$\alpha = .10$	1	2	4	6	8	11
$\alpha = .025$	$\alpha = .05$		1	2	4	6	8
$\alpha = .01$	$\alpha = .02$			0	2	3	5
$\alpha = .005$	$\alpha = .01$				0	2	3

Integers within this table represent critical values of T. For example, if $n = 9$ and $\alpha = .05$ for a two-tailed test, we would reject H_0 if $T \leq 6$. If $n = 9$ and $\alpha = .01$ for a two-tailed test, we would reject H_0 if $T \leq 2$. (Recall that the smaller the value of T, the greater the evidence that the two distributions differ.)

✷13.9 The Mann-Whitney U Test

In Chapter 10 we indicated how the Z and t statistics could be used to test the difference between the means of two independent populations for significance. The Mann-Whitney U test serves a similar function and may be used to test whether two independent samples were drawn from the same population (or from two different populations having the same mean). It is a useful substitute for the *t* test when either the assumption of normally distributed populations or the assumption of equal population variances appears to be violated. It also has the virtue of being relatively easy to use.

Procedural Steps for the Mann-Whitney U Test

The steps involved in employing the Mann-Whitney U test are as follows:

1. State the null and alternative hypotheses.
2. Decide on the level of significance.
3. Combine the observations for the two samples and rank them from smallest to largest; however, keep track of the sample to which each such observation originally belonged.
4. Determine W, that is, the sum of the ranks for the sample designated 1.
5. Compute the test statistic U.
6. Compute μ_T, the mean of the sampling distribution of U.
7. Compute σ_T, the standard deviation of the sampling distribution of U.
8. Determine the Z-value equivalent of U.
9. Either accept or reject the null hypothesis.

Application of the Procedural Steps

We will demonstrate the procedural steps listed above by referring to the following illustrative case:

ILLUSTRATIVE CASE 13.2

The Faraday Company had been selling its Earthquake brand of men's after-shave lotion in a relatively plain bottle for over 30 years. They were now contemplating a change to a more ornate bottle. In order to determine which, if either, bottle would have a more beneficial effect on sales, they decided to conduct an experiment in which (1) sales for a randomly selected sample of 15 retail distributors presently selling the product in the plain bottle would be carefully monitored for a period of six months and (2) 11 randomly selected retailers would sell the product in the new, more ornate bottle for the same period of time. At the end of the six-month period, average weekly sales information would be compared.

Data resulting from the experiment are displayed in Table 13.8.

Let us now apply the list of procedural steps to these results.

1. State the null and alternative hypotheses.

H_0: The two bottles are equally effective in influencing sales

H_a: The two bottles influence sales differently

2. Decide on the level of significance.
   Let us assume that α is set at .05.

Table 13.8 Average Weekly
Sales of Earthquake Brand
After-Shave Lotion for a Six-
Month Experimental Period

*(Numbers Represent Average
Number of Bottles Sold per
Week by Individual Stores)*

Sample 1: Old Bottle	Sample 2: New Bottle
50	31
35	37
43	59
30	67
62	44
52	49
43	54
57	62
33	34
70	42
64	40
58	
53	
65	
39	

3. Combine the observations for the two samples into a single group and rank them from smallest to largest; however, keep track of the sample to which each observation originally belonged.

   This step is shown in columns (1) through (3) of Table 13.9. Column (1) displays the 26 sample observations in an array. Column (2) simply identifies the sample to which each observation in column (1) originally belonged. Column (3) shows ranks assigned the entries in column (1). Notice that, in the case of ties, the ranks which would have been assigned the relevant values had they not been tied are averaged as in the Wilcoxon test.

4. Determine W.

   For this test, the value to be denoted W is simply the sum of the ranks associated with the sample designated 1, as shown in column (4) of Table 13.9. The value for W is 213.5.

5. Compute the test statistic U.

   The statistic to be evaluated using the Mann-Whitney test will be denoted U and computed by

$$U = n_1 n_2 + \frac{n_1(n_1 + 1)}{2} - W \qquad (13.9.1)$$

Table 13.9 Worksheet for Calculating the U Statistic

(1) Array of Sample Values	(2) Sample from Which Value Was Obtained	(3) Rank	(4) Ranks Assigned to Values of Sample 1
30	1	1	1
31	2	2	
33	1	3	3
34	2	4	
35	1	5	5
37	2	6	
39	1	7	7
40	2	8	
42	2	9	
43	1	10.5	10.5
43	1	10.5	10.5
44	2	12	
49	2	13	
50	1	14	14
52	1	15	15
53	1	16	16
54	2	17	
57	1	18	18
58	1	19	19
59	2	20	
62	1	21.5	21.5
62	2	21.5	
64	1	23	23
65	1	24	24
67	2	25	
70	1	26	26
			$\overline{213.5 = W}$

For the present problem, we get

$$U = (15)(11) + \frac{(15)(16)}{2} - 213.5 = 71.5$$

6. Compute μ_U, the mean of the sampling distribution of U.

This mean is obtained by

$$\mu_U = \frac{n_1 n_2}{2} \qquad (13.9.2)$$

Therefore,

$$\mu_U = \frac{(15)(11)}{2} = 82.5$$

7. Compute σ_U, the standard deviation of the sampling distribution of U.

The formula for this standard error is

$$\sigma_U = \sqrt{\frac{n_1 n_2(n_1 + n_2 + 1)}{12}} \qquad (13.9.3)$$

For the present problem:

$$\sigma_U = \sqrt{\frac{(15)(11)(15 + 11 + 1)}{12}} = 19.27$$

8. Determine the Z-value equivalent of U.

If n_1 and n_2 are each greater than 10, the theoretical sampling distribution of U is approximately normal. Taking advantage of this fact, we proceed to determine the Z value associated with our observed U of 71.5 as follows:

$$Z = \frac{U - \mu_U}{\sigma_U} = \frac{71.5 - 82.5}{19.27} = -0.57$$

9. Either accept or reject the null hypothesis.

U of -0.57 is safely within the acceptance region running from -1.96 to $+1.96$. Therefore, we conclude in favor of the null hypothesis.

If either n_1 or n_2, or both, had been less than 10, a more valid test could have been performed by using a special table of critical U values such as Appendix Table 9, a table similar to the brief one shown here. This abbreviated table shows critical values of U associated with $\alpha = .025$ for a one-tail test or $\alpha = .05$ for a two-tailed test.

n_2 ↓	$n_1 \to$ 2	3	4	5	6	7	8	9	10
3				0	1	1	2	2	3
4			0	1	2	3	4	4	5
5			1	2	3	5	6	7	8
6		1	2	3	5	6	8	10	11
7		1	3	5	6	8	10	12	14
8	0	2	4	6	8	10	13	15	17
9	0	2	4	7	10	12	15	17	20
10	0	3	5	8	11	14	17	20	23

For example, if $n_1 = n_2 = 5$ and $\alpha = .05$ for a two-tailed test, we would learn from this table that the critical value of U is 2. Therefore, the null hypothesis would be rejected if the observed U value were 2 or less.

*13.10 The Kruskal-Wallis One-Way "Analysis of Variance"

The Kruskal-Wallis test is similar to the Mann-Whitney test in that (1) it utilizes a statistic computed from ranks of pooled sample observations and (2) it is concerned with determining whether different samples came from identical populations. On the other hand, the Kruskal-Wallis test differs from—or really is an extension of—the Mann-Whitney test in that it can be applied to several independent samples rather than only two. This last point suggests an analogy with one-way analysis of variance. The analogy is quite appropriate. However, one of the advantages of the Kruskal-Wallis test is that it does not require us to assume that the populations are normally distributed with equal variances.

Procedural Steps for Applying the Kruskal-Wallis Test

The procedural steps used when applying this new test are as follows:

1. State the null and alternative hypotheses.
2. Decide on the level of significance.
3. Combine the observations of the c samples and rank them from smallest to largest; however, keep track of the sample to which each observation originally belonged.
4. Assign each resulting *rank* to the sample in which the corresponding actual observation was found.
5. Compute the test statistic H described below.
6. Compute the test statistic H' (H corrected for ties in ranks).
7. Either accept or reject the null hypothesis.

Application of the Procedural Steps

Let us again refer to the Faraday Company case (Illustrative Case 13.2) but extend one dimension of the problem somewhat: Whereas we previously examined the effects of two styles of bottle, we will now assume that there are three different new-style bottles, as well as the same old-style bottle, to be compared. Sales data for all four kinds of bottles are displayed in Table 13.10.

To apply the Kruskal-Wallis test, we proceed as follows:

1. State the null and alternative hypotheses.
   Technically, the null hypothesis is: The rank assigned to any specified observation has an equal chance of being any number between 1 and n, regardless of the sample in which it occurred.
   A more operationally oriented way of expressing the null hypothesis is

   H_0: Type of bottle used has no effect on sales of Earthquake
   Brand after-shave lotion

   Of course, the alternative hypothesis would be

   H_a: Type of bottle used does have an effect on sales of
   Earthquake Brand after-shave lotion

Table 13.10 Average Weekly Sales of Earthquake Brand After-Shave Lotion for a Six-Month Experimental Period

(Figures Represent Average Number of Bottles Sold per Week by Individual Stores)

Sample 1: Old Bottle	Sample 2: New Bottle 1	Sample 3: New Bottle 2	Sample 4: New Bottle 3
50	31	27	35
35	37	19	39
43	59	32	37
30	67	20	38
62	44	18	28
52	49	23	33
43	54		
57	62		
33	34		
70	42		
64	40		
58			
53			
65			
39			

Table 13.11 Sales Data from Table 13.10 Arrayed and Ranked

(1) Array of Sample Values	(2) Sample from Which the Value Was Obtained	(3) Rank	(1) Array of Sample Values	(2) Sample from Which the Value Was Obtained	(3) Rank
18	3	1	40	2	20
19	3	2	42	2	21
20	3	3	43	1	22.5
23	3	4	43	1	22.5
27	3	5	44	2	24
28	4	6	49	2	25
30	1	7	50	1	26
31	2	8	52	1	27
32	3	9	53	1	28
33	1	10.5	54	2	29
33	4	10.5	57	1	30
34	2	12	58	1	31
35	1	13.5	59	2	32
35	4	13.5	62	1	33.5
37	2	15.5	62	2	33.5
37	4	15.5	64	1	35
38	4	17	65	1	36
39	1	18.5	67	2	37
39	4	18.5	70	1	38

2. Decide on the level of significance.
   Let us continue to use $\alpha = .05$.

3. Combine the observations in the c samples and rank them from smallest to largest; however, keep track of the sample to which each observation originally belonged.
   This step is demonstrated in Table 13.11. Notice that in column (3) ties are handled in the same way as in the Wilcoxon and Mann-Whitney tests.

4. Assign each resulting *rank* to the sample in which the corresponding actual observation was found.
   This step is demonstrated in Table 13.12. For example, in Table 13.11 the value 18 is seen to have (1) come from Sample 3 and (2) been assigned the rank of 1. Therefore, in Table 13.12 the rank of 1 replaces the measurement of 18 under the heading "Sample 3." The same procedure is followed for all 38 (that is, all n) ranks.
   Once all ranks have been assigned to the appropriate samples, several calculations are made, as shown at the bottom of Table 13.12. The ranks within samples are summed, giving what we will refer to as T_j values. Other calculations are also shown at the bottom of this table; these will be used in connection with the determination of the H and H' statistics to be discussed next.

Table 13.12 Ranks Associated with Sales Data Appearing in Table 13.10

Sample 1: Old Bottle	Sample 2: New Bottle 1	Sample 3: New Bottle 2	Sample 4: New Bottle 3
7	8	1	6
10.5	12	2	10.5
13.5	15.5	3	13.5
18.5	20	4	15.5
22.5	21	5	17
22.5	24	9	18.5
26	25		
27	29		
28	32		
30	33.5		
31	37		
33.5			
35			
36			
38			
$T_1 = 379$	$T_2 = 257$	$T_3 = 24$	$T_4 = 81$
$T_1^2 = 143,641$	$T_2^2 = 66,049$	$T_3^2 = 576$	$T_4^2 = 6,561$
$n_1 = 15$	$n_2 = 11$	$n_3 = 6$	$n_4 = 6$

$$\frac{T_1^2}{n_1} = 9,576.07 \qquad \frac{T_2^2}{n_2} = 6,004.45 \qquad \frac{T_3^2}{n_3} = 96 \qquad \frac{T_4^2}{n_4} = 1,093.50$$

5. Compute H.

In the absence of any ties, our test statistic would be H, determined by

$$H = \left[\frac{12}{n(n+1)} \right] \left[\frac{\sum\limits_{\text{all } j} T_j^2}{n_j} \right] - 3(n+1) \qquad (13.10.1)$$

Using the information displayed at the bottom of Table 13.12, we obtain the following for our illustrative problem:

$$H = \left[\frac{12}{(38)(39)} \right] \left[\frac{(379)^2}{15} + \frac{(257)^2}{11} + \frac{(24)^2}{6} + \frac{(81)^2}{6} \right] - (3)(39) = 18.79$$

6. Compute the test statistic H' (H corrected for ties in ranks).

Usually little harm is done if one simply tests H for significance. However, a somewhat more accurate test is achieved if H is corrected for the effect of ties by dividing by a value found from

$$C_t = 1 - \left[\frac{\sum\limits_{i} t_i(i^3 - i)}{n^3 - n} \right] \qquad (13.10.2)$$

where C_t stands for "correction factor for ties" and t_i is the number of ties of order i, $i = 2, 3, 4, \ldots$.

For example, we find in Table 13.11 six sets of two-way ties, 0 sets of three-way ties, 0 sets of four-way ties, and so forth. Therefore,

$$C_t = 1 - \left\{ \frac{6[(2)^3 - 2] + 0[(3)^3 - 3] + 0[(4)^3 - 4] + \cdots}{(38)^3 - 38} \right\}$$

$$= 1 - \left\{ \frac{6[(2)^3 - 2]}{(38)^3 - 38} \right\} = .999344$$

We determine H' by $(H/C_t) = 18.79/.999344 = 18.80$.

Unless the number of tied observations is quite large relative to the sample size n, use of C_t will make little difference—as in our example. Moreover, if each set of tied observations is found within the same sample, the correction becomes unnecessary.

7. Either accept or reject the null hypothesis.

The statistic H' is distributed in a manner approximating the χ^2 distribution with $c - 1$ degrees of freedom. In our illustrative problem, α was set at .05 and there were $c = 4$ columns or samples. Therefore, we seek the critical value $\chi^2_{.05(3)}$ which, from either Table 13.1 or Appendix Table 7, is found to be 7.815. Our decision rule is

Decision Rule
1. If $H' \leq 7.815$, accept H_0.
2. If $H' > 7.815$, reject H_0.

Since our H' value of 18.80 is greater than 7.815, we reject the null hypothesis that type of bottle used has no effect on sales.

This conclusion is contrary to the one arrived at in the preceding section where only two types of bottles were considered. However, the significant H' value in the present case seems to reflect the extremely poor sales ranks associated with New Bottles 2 and 3. The management of Faraday Company is still left with the conclusion that the old-style bottle and New Bottle 1 are about equally effective. The choice must be made between them and not on the grounds of sales effectiveness. New Bottles 2 and 3 would presumably not be seriously considered.

---

*13.11 The Friedman Two-Way "Analysis of Variance"

When we discussed the sign test (section 13.7), the Wilcoxon test (section 13.8), and the t test for matched pairs (section 10.5), we were focusing on methods for analyzing *two* interdependent samples. These tests are very useful when the investigation concerns a clear dichotomy, such as scores before and after a treatment variable has been applied to the test units. However, not all problems involving interdependent samples are concerned with only two such samples. For example, a treatment variable might be suspected of having cumulative effects, in which case the investigator would presumably desire, in addition to a "before" score, two or more "after" scores as well. Moreover, some problems involving two or more interdependent samples may not be concerned with quantifiable properties. Often ranks, rather than actual measurements, must be used. Fortunately, we have in the Friedman two-way "analysis of variance" a tool for analyzing problems involving three or more interdependent samples and rank data.

Procedural Steps for Applying the Friedman Test

The steps we will follow for the Friedman two-way analysis are as follows:

1. State the null and alternative hypotheses.
2. Decide on a level of significance.
3. Arrange data in tabular form, with treatment levels serving as column headings and names or identification numbers of the test units serving as row headings.
4. Compute the test statistic χ^2_r.
5. Either accept or reject the null hypothesis.

Application of the Procedural Steps

The procedural steps will be demonstrated using the following illustrative case:

ILLUSTRATIVE CASE 13.3

Management of the Biggelow Bottling Company planned to add a grape soda to the company's growing line of soft drink flavors. The ingredients of the drink had been adequately worked out by Biggelow's chemists except for the relative amount of grape flavoring to use. This, the chemists felt, was more of a marketing research problem than a pure chemistry problem. Management agreed and asked the marketing research staff to run an experiment with five different strengths of grape flavoring to determine which, if any, the experimental subjects preferred. Strength 1 was the name assigned the mix with the smallest amount of grape flavoring; Strength 2 was the name assigned the mix having a somewhat greater amount of grape flavoring; and so forth, up to Strength 5, the most concentrated mix.

Twenty people, both male and female, of varying ages, consented to participate in the experiment. The experiment itself was conceptually quite simple: The five different versions of the beverage (all ingredients being identical except for the amount of grape flavoring) were placed before each (isolated) subject. Each subject was instructed to take as many tastes as he or she wished from each cup in any desired order until a ranking from 1 (best) to 5 (worst) could be made. Subjects were not permitted to report ties (though this is not a restriction of the statistical test under consideration—only one of the experimental procedure in this specific case). The results are shown in Table 13.13.

Before proceeding with the Friedman test, we should note two things about the work thus far. First, when assigning ranks, 1 was declared to mean "best" and 5 "worst." The test results would not have been affected if the meanings of the ranks had been just the reverse. Second, the fact that the five treatment levels are themselves, in this example, amenable to ranking on the basis of strength of grape flavoring is immaterial to this test.

Let us now look at the specific steps involved in applying the Friedman test.

1. State the null and alternative hypotheses.

    Basically, the null hypothesis states that, for a given treatment level (a particular column in Table 13.13), any specified sequence of ranks 1 through c is just as probable as any other sequence of ranks of like span. Therefore, the column rank sums, $T_1, T_2, \ldots, T_c$, are all equal.

    In more down-to-earth terms, we state

    > H_0: The five versions of the grape drink are liked equally well
    >
    > H_a: Some versions of the grape drink are preferred over others

2. Decide on a level of significance.
    Let us say that the level of significance was set at $\alpha = .10$.

3. Arrange data in tabular form, with treatment levels serving as column headings and names or identification numbers of test units serving as row headings.

    Although this instruction is self-explanatory, it does merit some emphasis. Within each *row* are found the ranks assigned *by* a specific subject *to* each listed treatment level. That is, the ranks 1, 2, 3, . . ., c are listed along rows, though seldom in that order—not down columns. The subjects are not being ranked.

4. Compute the test statistic, χ_r^2.

Table 13.13 Rank Data Obtained from the Biggelow Bottling Company Experiment

Subject Identification Number	Strength of the Grape Flavoring				
	1	2	3	4	5
1	5	4	3	1	2
2	5	3	4	2	1
3	4	5	2	1	3
4	5	4	1	3	2
5	4	5	3	2	1
6	1	2	3	4	5
7	5	2	4	1	3
8	1	3	5	2	4
9	4	5	2	1	3
10	5	2	4	1	3
11	4	5	1	2	3
12	2	1	4	3	5
13	1	2	3	4	5
14	5	3	4	2	1
15	5	4	2	1	3
16	5	4	1	2	3
17	4	5	1	2	3
18	5	3	4	1	2
19	1	2	4	3	5
20	2	1	3	4	5
Total T_j	73	65	58	42	62

$$T = 73 + 65 + 58 + 42 + 62 = 300$$

The test statistic we will use is denoted χ_r^2 and determined by the following formula:

$$\chi_r^2 = \frac{12}{rc(c+1)}(T_1^2 + T_2^2 + T_3^2 + \cdots + T_c^2] - 3r(c+1) \qquad (13.11.1)$$

We compute

$$\chi_r^2 = \frac{12}{(20)(5)(6)}[(73)^2 + (65)^2 + (58)^2 + (42)^2 + (62)^2] - (3)(20)(6)$$

$$= \underline{10.52}$$

5. Either accept or reject the null hypothesis.

If $c \geq 4$ and $r \geq 10$, the test statistic χ_r^2 is distributed in approximately a χ^2 manner with degrees of freedom equal to $c - 1$; $c - 1$ in our present example equals $5 - 1 = 4$. This fact, plus our knowledge of the α value, leads us to the following decision rule:

Decision Rule

1. If $\chi_r^2 \leq 7.779$, accept H_0.
2. If $\chi_r^2 > 7.779$, reject H_0.

Since χ_r^2 is 10.52 and $\chi_{.10(4)}^2$ was found to be a smaller 7.779, we conclude that the null hypothesis must be rejected at the .10 level of significance. Apparently, consumers are not indifferent to the concentration of the grape flavoring. A glance at the T_j values in Table 13.13 reveals that Strength 4 is the one most preferred.

When (1) c is 3 and r is less than 10 or (2) c is 4 and r is less than 5, the test should be carried out using a special table. This is Appendix Table 10, part of which is shown below as Table 13.14. A value in a column labeled P represents the probability of getting the corresponding χ_r^2 value or a higher one when the null hypothesis is true. For example, a null hypothesis would be rejected at the .05 level of significance if χ_r^2 were 6.0 or higher and $r = 2$, or if χ_r^2 were 7.4 or higher and $r = 3$.

Table 13.14 Critical χ^2 Values

			$c = 4$		
$r = 2$				$r = 3$	
χ_r^2	P	χ_r^2	P	χ_r^2	P
4.8	.208	4.2	.300	6.6	.075
5.4	.167	5.0	.207	7.0	.054
6.0	.042	5.4	.175	7.4	.033
		5.8	.148	8.2	.017
				9.0	.0017

13.12 Summary

In this chapter we have discussed several useful nonparametric tests. A "nonparametric test" has been understood to mean a hypothesis-testing procedure which satisfies *at least one* of the following criteria:

1. The data being analyzed are frequencies.
2. The data being analyzed are ranks.
3. The null hypothesis is not concerned with the value of a parameter such as μ or σ^2, which requires that each elementary unit in the population has an identifiable value associated with it.
4. The hypothesis-testing procedure is not dependent on assumptions about parameters such as μ or σ^2.
5. The probability distribution of the statistic used is not dependent on information or assumptions about the specific shape of the population(s) from which the sample(s) came.

The principal advantage of nonparametric methods is their general applicability. They can be applied in situations where actual measurements would be difficult or impossible to obtain. Also—and not unimportant—they ease the mind of the analyst who may be harboring serious misgivings about the assumptions required by a parametric test under consideration. The principal shortcoming of nonparametric methods is that they are less powerful than parametric tests. That is, for a situation where either a parametric or a nonparametric test could be used and the assumptions associated with the parametric test are met, the latter will lead to rejection of a false null hypothesis more surely than will the nonparametric test.

Chi-square tests are applied to frequency data. In this chapter, we looked at three somewhat different applications of this versatile tool: single-classification tests, tests concerned with whether two principles of classification are independent, and tests concerned with whether the proportions of several independent populations are equal.

The *sign test* and the *Wilcoxon matched-pairs signed-rank test* may be viewed as nonparametric counterparts of the *t* test for paired observations. These are useful in connection with problems involving two populations that are obviously not independent, such as when "before-and-after" readings are taken on the same elementary units. Of the two, the Wilcoxon test is more powerful but also requires somewhat more stringent assumptions.

The *Mann-Whitney* U *test* is the nonparametric counterpart of the Z or *t* test for the difference between the means of two independent populations. Its principal virtue is that it does not require that one assume populations of normal shape and having equal variances.

The *Kruskal-Wallis test* may be viewed as an extension of the Mann-Whitney test. It permits comparison of data obtained from several independent populations. Thus, it is the nonparametric counterpart of a one-way analysis of variance. However, its application is dependent on fewer and weaker assumptions than is true of ANOVA.

The *Friedman test* may be thought of as an extension of the sign test or the Wilcoxon test in that it can be applied to data from several different nonindependent populations. It is the nonparametric counterpart of a randomized-block experimental design.

Terms Introduced in This Chapter*

contingency table (p. 580)
expected frequencies (p. 571)
observed frequencies (p. 571)

Yates' correction for continuity (p. 582)

Formulas Introduced in This Chapter

For all applications of chi-square analysis treated in this chapter, the chi-square statistic is computed by

$$\chi^2 = \sum_{\text{all cells}} \left[\frac{(O - E)^2}{E} \right]$$

if the number of degrees of freedom is greater than 1, or by

$$\chi^2 = \sum_{\text{all cells}} \left[\frac{(|O - E| - .5)^2}{E} \right]$$

if the number of degrees of freedom is 1, where O stands for observed frequency and E stands for expected frequency.

*Several terms introduced in this chapter have already been presented in the Summary (section 13.12).

The formula used in *the sign test for matched pairs*, in cases where sample sizes are greater than about 10, is

$$Z = \frac{S - \mu \pm .5}{\sigma}$$

where S is the number of plus signs or the number of minus signs (whichever occurs less frequently), and where Z is S converted to standard form, $\mu = np$, $\sigma = \sqrt{npq}$, and $\pm.5$ is a correction for continuity. For smaller sample sizes use the binomial probability formula.

The formulas used in the *Wilcoxon matched-pairs signed-rank test* are

$$\mu_T = \frac{n(n + 1)}{4} \qquad \sigma_T = \sqrt{\frac{n(n + 1)(2n + 1)}{24}}$$

$$Z = \frac{T - \mu_T}{\sigma_T}$$

where T is the sum of the positive ranks or the sum of the negative ranks, whichever is smaller; μ_T is the mean of the sampling distribution of T; σ_T is the standard deviation of the sampling distribution of T; n is the sample size after elimination of ties; and Z is T converted to standard form.

The formulas used in the *Mann-Whitney U Test* are

$$U = n_1 n_2 + \frac{n_1(n_1 + 1)}{2} - W \qquad \mu_U = \frac{n_1 n_2}{2}$$

$$\sigma_U = \sqrt{\frac{n_1 n_2(n_1 + n_2 + 1)}{12}}$$

$$Z = \frac{U - \mu_U}{\sigma_U}$$

where S is the number of plus signs or the number of minus signs (whichever occurs less frequently), and where Z is S converted to standard form, $\mu = np$, $\sigma = \sqrt{npq}$, and $\pm.5$ is a correction for continuity. For smaller sample sizes use the binomial probability formula.

The formulas used in the *Kruskal-Wallis one-way "analysis of variance"* are

$$H = \frac{12}{n(n + 1)} \left[\frac{\sum_{all\ i} T_j^2}{n_j} \right] - 3(n + 1)$$

$$C_t = 1 - \left[\frac{\sum_i t_i(i^3 - i)}{n^3 - n} \right] \qquad H' = \frac{H}{C_t}$$

where H is as defined above; T_j is the sum of the ranks associated with sample j; n_j is the number of observations associated with sample j; n is the total number of observations in all the samples; C_t is a correction factor for ties; and H' is H after the correction for ties.

The formula for *the Friedman two-way "analysis of variance"* is

$$\chi_r^2 = \frac{12}{rc(c + 1)} [T_1^2 + T_2^2 + T_3^2 + \cdots + T_c^2] - 3r(c + 1)$$

where χ_r^2 is as defined above; r is the number of rows and c is the number of columns; and T's are the sums of ranks associated with specific samples.

Questions and Problems

13.1 Explain the meaning of each of the following:
    a. Nonparametric test   e. Contingency table
    b. Chi-square distribution   f. $\chi^2_{.05(4)}$
    c. Expected frequency   g. Yates' correction
    d. Independence

13.2 Explain the meaning of each of the following:
    a. T d. W
    b. μ_T e. H'
    c. σ_T f. χ^2_r

13.3 Distinguish between:
    a. Observed χ^2 and critical χ^2
    b. Observed frequency and expected frequency in chi-square analysis
    c. Chi-square test of two principles of classification and chi-square test for equality among several population proportions
    d. H and H'
    e. H' and U

13.4 Distinguish between:
    a. Chi-square methodology used in connection with a 2×2 contingency table and chi-square methodology used in connection with a larger contingency table
    b. Observed χ^2 and observed χ^2_r
    c. The Mann-Whitney U test and the Kruskal-Wallis one-way "analysis of variance"
    d. The Kruskal-Wallis one-way "analysis of variance" and the Friedman two-way "analysis of variance"

13.5 Distinguish between:
    a. The kind of data used in chi-square analysis and the kind(s) of data used in the other nonparametric tests presented in this chapter
    b. Assumptions underlying the sign test for matched pairs, on the one hand, and assumptions underlying the Wilcoxon matched-pairs signed-rank test on the other
    c. d and d_a
    d. Chi-square single-classification test and chi-square test for equality among several population proportions

13.6 Indicate which of the following statements you agree with and which you disagree with, and defend your opinions:
    a. Yates' correction should be used in a one-degree-of-freedom chi-square analysis to offset a bias in the direction of accepting too many null hypotheses.
    b. Chi-square analysis may be used in connection with any kind of frequency data regardless of whether the data represent actual frequencies or relative frequencies.
    c. Although the Kruskal-Wallis one-way "analysis of variance" is similar in certain respects to the Mann-Whitney U test, the former can be applied to a greater variety of problems.
    d. The chi-square test of two principles of classification and the chi-square test for equality among several population proportions are identical.

13.7 Indicate which of the following statements you agree with and which you disagree with, and defend your opinions:
    a. Like the standard normal distribution, the χ^2 distribution has the same shape regardless of the number of degrees of freedom.

b. In a chi-square problem, all one really cares about is whether the null hypothesis should be accepted or rejected—not in the nature of the discrepancies between the *O* and *E* values.

c. Even though the sign test for matched pairs and the Wilcoxon matched-pairs signed-rank test attempt to accomplish the same thing, the latter test is usually better at identifying a false null hypothesis.

13.8 Indicate which of the following statements you agree with and which you disagree with, and defend your opinions:

a. A contingency table having 7 columns and 4 rows would have $(7)(4) - 4 = 24$ degrees of freedom associated with it.

b. In order to find the critical values for the sign test for matched pairs and the Wilcoxon matched-pairs signed-rank test, it is necessary to know the degrees of freedom.

c. In a problem involving the Kruskal-Wallis one-way "analysis of variance" and five independent samples, the number of degrees of freedom will be $5 - 2 = 3$.

13.9 The manager of the Piedmont Grocery Store counted the number of customers using the store's five checkout lanes during Friday and Saturday of a certain week. The results follow:

	Checkout Lane				
	1	2	3	4	5
Number of customers	160	200	300	120	100

a. Does this problem lend itself to analysis by chi-square procedures? Defend your answer.

b. If this problem does lend itself to analysis by chi-square methods, is it a problem involving (1) a test applied to a single-classification system or (2) a test to determine whether five population proportions are equal? Defend your choice.

c. State the null and alternative hypotheses assuming that the manager harbors no preconceptions about certain checkout lanes being used more than others.

d. Determine the decision rule if the level of significance is set at $\alpha = .05$.

e. Test the null hypothesis. Should the null hypothesis be accepted or rejected?

f. Indicate some possible practical implications of the conclusion you arrived at in part *e*.

13.10 Refer to problem 13.9. Suppose that checkout lane 5 is closed much of the time because it is serviced by the assistant manager during busy times only. Consequently, the manager suspects, prior to taking the actual count, that this checkout lane will be used only half as often as checkout lanes 1, 2, and 4. Moreover, checkout lane 3 is an express lane and the manager suspects, prior to taking the actual count, that this lane might be used by twice as many people as lanes 1, 2, and 4. Assuming that the empirical results of his count are as shown in problem 13.9, do the following:

a. State the null and alternative hypotheses.

b. Determine the decision rule if the level of significance is set at $\alpha = .05$.

c. Test the null hypothesis. Should the null hypothesis be accepted or rejected?

d. Indicate some possible practical implications of the conclusion you arrived at in part *c*.

13.11 An experiment was conducted within a manufacturing plant to determine whether the number of hours workers were required to perform uninterrupted labor had any bearing on the number of accidents that occurred. For the experimental period, some employees were assigned at random to one-hour shifts, others to two-hour shifts, still others to three-hour shifts, and so forth, up to shifts of eight-hour duration. Care was taken to ensure that the number of man-hours of exposure time was equated among the groups. That is, (1) the number of workers assigned to each length shift was the same and (2) the one-hour shift group was required to work eight times as many well-separated shifts as the eight-hour shift

Table 13.15 Data for the Plant Accident Problem

Length of Shift (Hours)	Number of Accidents
1	20
2	18
3	15
4	12
5	14
6	14
7	21
8	22
Total	136

group; the two-hour shift group was required to work four times as many well-separated shifts as the eight-hour shift group; and so forth. Furthermore, considerable care was taken to define exactly what constitutes an "accident." The results are shown in Table 13.15.

Management now wishes to test in a formal manner whether length of shift affects occurrence of accidents.

a. Does this problem lend itself to analysis by chi-square procedures? Defend your answer.
b. If this problem does lend itself to analysis by chi-square methods, is it a problem involving (1) a test applied to a single-classification system or (2) a test to determine whether eight population proportions are equal? Defend your choice.
c. State the null and alternative hypotheses assuming that management harbors no preconceptions about whether shifts of certain duration are more or less conducive to accidents than shifts of other duration.
d. Determine the decision rule if the level of significance is set at $\alpha = .01$.
e. Test the null hypothesis. Should the null hypothesis be accepted or rejected?
f. Indicate some possible practical implications of the conclusion you arrived at in part *e*.

 13.12 Refer to problem 13.11. Suppose that, prior to the experiment, a supervisor suspected that the most dangerous shift lengths were the shortest ones and the longest ones. More specifically, he believed that only about half as many accidents would be associated with shifts of 3, 4, 5, and 6 hours as would be associated with shifts of 1, 2, 7, or 8 hours. Assuming that the results of the experiment are the same as those shown in Table 13.15, do the following:
a. State the null and alternative hypotheses.
b. Determine the decision rule if the level of significance is set at $\alpha = .01$.
c. Test the null hypothesis. Should the null hypothesis be accepted or rejected?
d. Indicate some possible practical implications of the conclusion you arrived at in part *c*.

 13.13 Refer to problems 13.11 and 13.12. Suppose that, instead of counting the number of accidents associated with shifts of various lengths, management counted the number of workers involved in at least one accident, as indicated in Table 13.16. Suppose further that exactly 100 workers had been assigned to each of the shift-duration categories.
a. What change or changes, if any, in methodology from that used in problems 13.11 and 13.12 would be called for? Why?
b. State the null and alternative hypotheses.
c. Determine the decision rule if the level of significance is set at $\alpha = .01$.
d. Test the null hypothesis. Should the null hypothesis be accepted or rejected?
e. Indicate some possible practical implications of the conclusion you reached in part *d*.

Table 13.16 Data for the Revised Version of
the Plant Accident Problem

Length of Shift (Hours)	Number of Workers Involved in at Least One Accident
1	20
2	18
3	15
4	12
5	14
6	14
7	21
8	22
Total	136

 13.14 A sample of 800 stockholders of the Martin Balsam Company, a manufacturer of wood products, was surveyed to determine stockholder attitudes toward a threatened takeover of the company by Engulf and Devour, Inc. The people surveyed were classified according to the length of time they had held their shares. The results are presented in Table 13.17.

    a. Does this problem lend itself to analysis by chi-square procedures? Defend your answer.

    b. If this problem does lend itself to chi-square analysis, is the purpose of the test of significance to determine (1) whether two principles of classification are independent or (2) whether three population proportions are equal? Defend your choice.

    c. State the null and alternative hypothesis.

    d. Determine the decision rule if the level of significance is set at $\alpha = .05$. How many degrees of freedom did you use in setting up your decision rule? Why?

    e. Test the null hypothesis. Should the null hypothesis be accepted or rejected?

    f. Indicate some possible practical implications of the conclusion you arrived at in part *e*.

    g. In your opinion, is it better to include or exclude the "indifferent" category of stockholder? Why?

 13.15 In a small doll-manufacturing plant, two employees, Dobbs and Smeers, are responsible for applying the final paint touches to the finished dolls' faces. A manager wishes to compare the two employees to determine whether their work is of comparable quality. A sample of 350 dolls is selected and the paint jobs rated on a quality scale of A (best), B, C, and D (unacceptable). The results are shown in Table 13.18.

Table 13.17 Data for the Stockholder-Attitude Problem

(Figures Represent Number of Respondents)

Length of Holding	Attitude			Total
	Approve	Disapprove	Indifferent	
Less than 1 year	40	30	30	100
1 to 3 years	60	100	40	200
More than 3 years	100	300	100	500
Total	200	430	170	800

Table 13.18 Data for the Painting-Quality Problem

(Figures Represent Number of Dolls)

Quality	Employee Dobbs	Smeers	Total
A	50	45	95
B	100	75	175
C	40	20	60
D	10	10	20
Total	200	150	350

a. Does this problem lend itself to analysis by chi-square procedures? Defend your answer.
b. If this problem does lend itself to chi-square analysis, is the purpose of the test of significance to determine (1) whether two principles of classification are independent, (2) whether two population proportions are equal, or (3) whether four population proportions are equal? Defend your choice.
c. State the null and alternative hypotheses.
d. Determine the decision rule if the level of significance is set at $\alpha = .10$. How many degrees of freedom did you use in setting up your decision rule? Why?
e. Test the null hypothesis. Should the null hypothesis be accepted or rejected?
f. Indicate some possible practical implications of the conclusion you reached in part *e*.

✓ 13.16 A marketing research organization sent questionnaires to test attitudes concerning a certain brand of laundry product. In City A, 65% of the returns indicated a favorable attitude toward the product. In Cities B and C, the percentage responding favorably were 70% and 75%, respectively. The marketing research organization now wishes to determine whether these results suggest that attitudes toward the product differ among the three cities.
a. Does this problem lend itself to analysis by chi-square procedures? Defend your answer.
b. If this problem does lend itself to a chi-square analysis, is the purpose of the test of significance to determine (1) whether two principles of classification are independent or (2) whether three population proportions are equal? Defend your choice.
c. State the null and alternative hypotheses.
d. Determine the decision rule if the level of significance is set at $\alpha = .05$. How many degrees of freedom did you use when setting up the decision rule? Why?
e. Test the null hypothesis. Should the null hypothesis be accepted or rejected?
f. Indicate some possible practical implications of the conclusion you reached in part *e*.

13.17 None of the frequency distributions in Table 13.19 can properly be analyzed by chi-square analysis. In each case, explain what requirement of chi-square analysis is not met.

13.18 Refer to problem 10.30. (If you did not previously work this problem using the *t* test for matched pairs, do so now.) Suppose that some question exists in the mind of the company's president regarding the normality of the population of differences. As a result, he decides to apply the sign tests for matched pairs.
a. State the null and alternative hypotheses.
b. Determine the appropriate decision rule if the level of significance is set at $\alpha = .10$.
c. Test the null hypothesis. Should the null hypothesis be accepted or rejected?
d. Is the conclusion you reached in part *c* of this problem the same as the conclusion you reached in part *d* of problem 10.30? If not, how do you explain the difference in results?

Table 13.19 Five Frequency Distributions That Cannot Properly Be Analyzed Using Chi-Square Analysis

I. Fabric-Color Preference	Number of Respondents
Red	10
Blue	15
Other	10
Total	35

	Length of Time with Company			
II. Attitude Toward Company Promotion Policies	Less Than 1 Year	1 to 5 Years	Over 5 Years	Total
Approve	300	5	5	310
Disapprove	12	8	5	25
Indifferent	5	5	5	15
Total	317	18	15	350

	Sex of Respondent	
III. Automobile Preference	Male	Female
Domestic	70%	80%
Foreign	30	20
Total	100%	100%

IV. Number of complaints received by a vending machine company during the month of June: 395

	Flavor of Ice Cream Preferred		
V. Age of Respondent	Vanilla	Strawberry	Chocolate
1 to 3 years	200	500	250
Under 5 years	350	750	300
5 years and over	500	1500	750

13.19 Refer to problem 10.32. (If you did not previously work this problem using the t test for matched pairs, do so now.) Suppose that the automobile-manufacturing researchers are not entirely satisfied that the population of differences is normally distributed. As a result they decide to apply the sign test for matched pairs.
   a. State the null and alternative hypotheses.
   b. Determine the decision rule if the level of significance is set at $\alpha = .05$.
   c. Test the null hypothesis. Should the null hypothesis be accepted or rejected?
   d. Is the conclusion you reached in part c of this problem the same as the conclusion you reached in part d of problem 10.32? If not, how do you explain the difference in results?

Table 13.20 Data for the Absenteeism Problem

Worker	Number of Days Absent Old Conditions	Number of Days Absent New Conditions	Worker	Number of Days Absent Old Conditions	Number of Days Absent New Conditions
1	14	14	11	12	14
2	16	10	12	11	7
3	2	2	13	14	13
4	5	0	14	6	8
5	12	11	15	20	8
6	30	10	16	10	10
7	5	4	17	11	10
8	5	5	18	14	15
9	7	5	19	0	0
10	8	3	20	18	10

13.20 A supervisor in a manufacturing plant wished to determine whether several specific changes in the working environment would lead to a reduction in absenteeism by workers in her charge. For each such worker, number of working days missed was determined from employee records for the past year; Table 13.20 lists this information. The contemplated changes in working conditions were then put into effect. Subsequently, the working days missed for the same workers during the following year were gathered. Workers who had not been on the job for the entire two-year period were excluded from the study.) Suppose the supervisor elected to use the Wilcoxon matched-pairs signed-rank test on these data. Suppose further that the supervisor wished to determine whether the changes in working conditions have led to a *decrease* in absenteeism.
a. State the null and alternative hypotheses.
b. Determine the decision rule if the level of significance is set at $\alpha = .05$. Is the test a one-tailed or a two-tailed test? How do you know? How will the method of applying the Wilcoxon test be affected by your answer to this question?
c. Compute the test statistic T.
d. Test the null hypothesis. Should the null hypothesis be accepted or rejected?
e. Apply the sign test for matched pairs to the data in Table 13.20. Does this test lead you to the same conclusion you reached in part d of this problem? If not, how do you explain the difference?

13.21 Table 13.21 presents sales achieved by a sample of stores selling Serene brand crystal ware during a six-month period preceding and a six-month period following an expensive nationwide advertising campaign. Suppose that the Wilcoxon matched-pairs signed-rank test is to be used and that the question of concern is whether the advertising campaign has led to an *increase* in sales.
a. State the null and alternative hypotheses.
b. Determine the decision rule if the level of significance is set at $\alpha = .01$. Is the test a one-tailed or a two-tailed test? How do you know? How will the method of applying the Wilcoxon test be affected by your answer to this question?
c. Compute the test statistic T.
d. Test the null hypothesis. Should the null hypothesis be accepted or rejected?
e. Did (or would) rejection of the null hypothesis necessarily prove that the advertising campaign had a beneficial effect on sales? Explain why or why not.
f. Apply the sign test for matched pairs to the data in Table 13.21. Does this test lead you to the same conclusion you reached in part d of this problem? If not, how do you explain the difference?

Table 13.21 Data for the Serene Crystal Ware Problem

| Store | Dozens of Sets Sold | |
	Before Campaign	After Campaign
1	10	12
2	6	6
3	8	16
4	18	22
5	3	9
6	12	13
7	15	17
8	6	8
9	7	7
10	4	9
11	6	7
12	7	7
13	8	4
14	10	8
15	4	10
16	6	9
17	12	13
18	35	50
19	6	9
20	8	14
21	40	42
22	6	8
23	5	3
24	12	14
25	15	17

 13.22 Refer to problem 12.22. (If you did not previously work this problem using a randomized block analysis, do so now.) Suppose that management harbors some misgivings about some of the assumptions associated with the use of the F test and, thus, decides to use the Wilcoxon matched-pairs signed-rank test.
  a. Determine the decision rule if the level of significance is set at $\alpha = .05$. Is the test a one-tailed or a two-tailed test? How do you know? How will the method of applying the Wilcoxon test be affected by your answer to this question?
  b. Compute the test statistic T.
  c. Test the null hypothesis. Should the null hypothesis be accepted or rejected?
  d. Is the conclusion you reached in part c of this problem the same as the one you reached when employing the randomized block analysis of variance? If not, how do you explain the difference?

 13.23 Table 13.22 shows the response rates associated with two distinct mail-order promotional pieces. Each type of promotional piece has been sent on a random basis to a sample of households in 10 cities. Suppose that management of the mail-order firm wishes to determine, using the Mann-Whitney U test, whether one promotional piece was more effective than the other.
  a. State the null and alternative hypotheses.
  b. Determine the decision rule if the level of significance is set at $\alpha = .05$. Is the test a one-tailed or a two-tailed test? How do you know?

Table 13.22 Data for the
Direct-Mail Promotion Problem

(Numbers Represent the Proportion of Recipients Responding Favorably)

Promotional Piece A	Promotional Piece B
.08	.10
.10	.10
.12	.15
.07	.22
.05	.06
.01	.08
.08	.18
.05	.06
.04	.07
.09	.11

c. Compute the test statistic U.

d. Test the null hypothesis. Should the null hypothesis be accepted or rejected?

 13.24 A company has in the past made use of two quite different teaching techniques in connection with its management training classes. Test scores achieved under the two teaching methods are shown in Table 13.23 Suppose that management now wishes to determine, using the Mann-Whitney U test, whether one kind of teaching method results in higher test scores than the other kind of teaching method.

a. State the null and alternative hypotheses.

Table 13.23 Data for the
Management-Training Teaching
Method Problem

(Numbers Represent Test Scores)

Method A	Method B
82	95
97	51
54	78
82	71
76	84
90	85
92	80
88	69
75	72
	64
	68
	75
	92

b. Determine the decision rule if the level of significance is set at $\alpha = .05$. Is the test a one-tailed or a two-tailed test? How do you know?

c. Compute the test statistic U.

d. Test the null hypothesis. Should the null hypothesis be accepted or rejected?

 13.25 Refer to Table 11.17 presented in connection with problem 11.21. Suppose that some uncertainty exists regarding the validity of the assumptions underlying the use of the one-way analysis of variance and that the Kruskal-Wallis one-way "analysis of variance" is thought to be a more cautious method of analysis.

a. State the null and alternative hypotheses.

b. Determine the appropriate decision rule if the level of significance is set at $\alpha = .05$.

c. Compute H.

d. If appropriate, convert H to H' by correcting for the presence of ties.

e. Test the null hypothesis. Should the null hypothesis be accepted or rejected?

 13.26 Refer to Table 11.18 presented in connection with problem 11.25. Suppose that some uncertainty exists regarding the validity of the assumptions underlying the use of the one-way analysis of variance and that the Kruskal-Wallis one-way "analysis of variance" is thought to be a more cautious method of analysis.

a. State the null and alternative hypotheses.

b. Determine the decision rule if the level of significance is set at $\alpha = .10$.

c. Compute H.

d. If appropriate, convert H to H' by correcting for the presence of ties.

e. Test the null hypothesis. Should the null hypothesis be accepted or rejected?

 13.27 Refer to Table 12.15 presented in connection with problem 12.9. Suppose that some uncertainty exists regarding the validity of the assumptions underlying a completely randomized block design and that the Kruskal-Wallis one-way "analysis of variance" is thought to be a more cautious method of analysis.

a. State the null and alternative hypotheses.

b. Determine the decision rule if the level of significance is set at $\alpha = .01$.

c. Compute H.

d. If appropriate, convert H to H' by correcting for the presence of ties.

e. Test the null hypothesis. Should the null hypothesis be accepted or rejected?

 13.28 Refer to Table 12.16 presented in connection with problem 12.13. Suppose that some uncertainty exists regarding the validity of the assumptions underlying a completely random-

Table 13.24 Data for the Cake-Mix Rating Problem

Housewife	Brand of Cake Mix			
	A	B	C	D
1	3	2	1	4
2	1	2	4	3
3	3	2	1	4
4	4	1	2	3
5	3	1.5	1.5	4
6	3.5	2	1	3.5
7	4	1	2	3
8	4	2	3	1
9	4	2.5	1	2.5
10	2	3	1	4

Table 13.25 Data for the Employee Rating Problem

Manager	Employee		
	A	B	C
1	1	3	2
2	1	2	3
3	2	3	1
4	1	2.5	2.5
5	2	1	3
6	1.5	1.5	3

ized design and that a Kruskal-Wallis one-way "analysis of variance" is thought to be a more cautious method of analysis.
a. State the null and alternative hypotheses.
b. Determine the decision rule if the level of significance is set at $\alpha = .05$.
c. Compute H.
d. If appropriate, convert H to H' by correcting for the presence of ties.
e. Test the null hypothesis. Should the null hypothesis be accepted or rejected?

13.29 Ten housewives were asked to rank four different brands of cake mix on a scale from 1 (best) to 4 (worst) according to whatever criteria they thought appropriate. The results are shown in Table 13.24. These results are to be tested using the Friedman two-way "analysis-of-variance."
a. State the null and alternative hypotheses.
b. Determine the decision rule if the level of significance is set at $\alpha = .05$.
c. Compute the value of the test statistic χ_r^2.
d. Test the null hypothesis. Should the null hypothesis be accepted or rejected?

13.30 Six managers were asked to rank three employees with respect to their promotability. Each manager was to assign a rank of 1 to the employee he deemed most worthy of promotion and a rank of 3 to the least worthy. The results are shown in Table 13.25. These results are to be tested using the Friedman two-way "analysis of variance."
a. State the null and alternative hypotheses.
b. Determine the appropriate decision rule if the level of significance is set at $\alpha = .05$.
c. Compute the value of the test statistic χ_r^2.
d. Test the null hypothesis. Should the null hypothesis be accepted or rejected?

TAKE CHARGE

13.31 Using the Biggelow Bottling Company example (Illustrative Case 13.3) as your model, perform a preference-ranking experiment. Let your experiment include at least four roughly similar things to be ranked and at least ten subjects to do the ranking. (These may be acquaintances of yours or fellow class members. Just make certain that one person's opinion is not allowed to influence anyone else participating in the experiment.)
Then write a short report describing (1) the purpose and nature of the experiment, (2) the way it was conducted, and (3) the conclusion you drew from application of the Friedman test.

Computer Exercise

13.32 Refer to Table 11.21, presented in connection with problem 11.38. Have the computer perform a Kruskal-Wallis one-way "analysis of variance" on these data.

Computer Exercise

13.33 Refer to Table 10.6, presented in connection with problem 10.38. Have the computer perform a Wilcoxon matched-pairs signed-rank test on these data.

Cases for Analysis and Discussion

13.1 "GLAMOUR BURGER SYSTEMS": PART IV

Review Part I (case 1.3 at the end of Chapter 1), Part II (case 3.3 at the end of Chapter 3), and Part III (case 10.4 at the end of Chapter 10) of this case.

1. Using the appropriate kind of chi-square analysis and a .05 level of significance, analyze all parts of Table 3.23. State your null hypotheses and your conclusions clearly.

2. On the basis of the data in Table 3.23 and your chi-square analysis, what do you think is (are) the principle competitive disadvantage(s) Glamour Burger suffers from? Elaborate.

3. Suggest some things that Glamour Burger needs to do to compete more effectively with McDonald's and Burger King.

13.2 THE JANUARY STOCK MARKET INDICATOR

The following is an excerpt from "Up and Down Wall Street" by Alan Ableson which appeared in the January 11, 1982 issue of *Barron's* magazine:*

> As a die-hard fundamentalist (our bible is Graham and Dodd), we view voodoo market analysis with grave misgivings and unwaivering skepticism. Yet, we must admit, for reasons too deep for mere mortals to comprehend, sometimes the stuff works. The magic market potion we have in mind is the January Barometer. This particular divining rod uses the action of the market in the first month of the year—up or down—as a harbinger of what the market will do over the entire year.
>
> The January Barometer was invented by Yale Hirsch, a certified stock market pundit. We queried Yale last Friday on possible rational explanations for the barometer's effectiveness—in 28 out of the past 32 years, if the market finished higher in January it finished higher for the full year and the reverse was true as well. In a nutshell, Yale thought that it was tied to the fact that January is so freighted with political and economic events—budget, tax measures, Congress coming to life—that the month often sets the tone for the economy and the market for the entire year.
>
> Well, here we are in January and so far, so bad.

On January 25, 1982, the Letter to the Editor column ran the following letter from a reader, John R. McGinley Jr.:*

> At first glance, the indicator created by Yale Hirsch ("Up and Down Wall Street," January 11) from the market action of the first five days of January seems to have a good record: right 27 [sic] times out of 32 years (for chi^2 of 15, highly significant: 0.01% chance the result could have been by accident).

On closer inspection, however, one finds the indicator is not as accurate when the first five days are down—as [they were] this year. There have been 11 down years in the last 32, but in three the indicator was wrong! Eight right and three wrong produce a chi^2 of 2.27, not significant; and a high 13% chance the result was random.

The first-five-days indicator is great, but only if it's up.

1. Write a critical evaluation of the magazine's evidence.

2. Write a critical evaluation of the reader's evidence.

13.3 THE "HANDI LAUNDRY COMPANY": PART II

Review Part I of this case (case 11.3 at the end of Chapter 11) and examine Table 11.26.

1. Can you think of any method described in this chapter that might be useful in determining whether significant differences exist in gross revenues per apartment unit among the six geographical areas?

2. If you believe that one of our nonparametric methods could be applied, do so using $\alpha = .05$.

3. In what way or ways might the nonparametric method you picked be preferable to a one-way analysis of variance?

4. In what way or ways might the nonparametric method you picked be inferior to a one-way analysis of variance?

5. If you previously performed the exercise called for in question 6 of case 11.3, compare these results with the results obtained from your nonparametric test. Do the results differ? If so, how do you account for the difference?

14 Simple Linear Regression and Correlation Analysis

Whatever phenomenon varies in any manner whenever another phenomenon varies in some particular manner is either a cause of or an effect of that phenomenon, or is connected with it through some fact of causation.—John Stuart Mill

Whether the stone hits the pitcher or the pitcher hits the stone, it's going to be bad for the pitcher.
—Miguel de Cervantes

What You Should Learn from This Chapter

In this chapter we study statistical relationships, a subject not entirely new to you: cross-classification tables (Chapter 3), analysis of variance (Chapters 11 and 12), and some of the chi-square applications and other nonparametric techniques (Chapter 13) all addressed the question of whether two or more variables are related.

Two things distinguish the subject matter of the present chapter from this earlier material: First, the contents of this chapter span both description and induction; several useful descriptive measures will be introduced before we attempt to set up confidence intervals or to test hypotheses about population parameters. Second, in this chapter emphasis is placed on how to describe a relationship by means of a *precise estimating equation*. When you have completed this chapter, you should be able to

1. Distinguish between regression and correlation; between simple and multiple regression; between linear and nonlinear relationships; and between mathematical and statistical relationships.
2. Construct and interpret a scatter diagram.
3. Obtain the following descriptive measures: (*a*) Y intercept, (*b*) slope, (*c*) standard error of the estimate, and (*d*) coefficients of determination and correlation.
4. Recite the assumptions underlying inductive methods in (*a*) simple regression analysis and (*b*) simple correlation analysis, and check their validity where possible.
5. Construct confidence intervals for (*a*) the population slope, (*b*) the subpopulation mean of Y given X, and (*c*) a subpopulation Y_{NEW} given X.
6. Test hypotheses about the population slope and the population coefficient of determination.
7. Recite some hazards associated with simple regression and correlation analysis.

No specific review material is suggested. However, if you are not now on friendly terms with the contents of Chapter 8, a serious effort to become so is advised. Also, a review of one-way analysis of variance (Chapter 11) should prove helpful in some parts of this chapter.

14.1 Introduction

Central to the art of business decision making is the analysis of relationships among variables. Whether developing next year's budget, setting sales quotas, deciding whether to declare a dividend, or contemplating whether to bring a new product to market, the sensible decision maker will take into account, in some manner, the probable effects of a number of variables thought to be important.

In this and the next chapter we consider some useful *formal* techniques for arriving at precise information about the relationship between two or more variables. These techniques are primarily concerned with the determination of estimates for a variable of interest using known or assumed values of other variables.

14.2 Basic Terminology

The title of this chapter—"Simple Linear Regression and Correlation Analysis"—tells us a great deal about both its scope and its limitations. Let us reflect on the terms *simple*, *linear*, *regression*, and *correlation*.

> When we speak of *simple* regression and *simple* correlation analyses, we refer to analyses involving only *two* variables: a dependent variable, denoted Y, and an independent variable, denoted X.
>
> The term *linear* indicates that we will be limiting our attention to situations in which the variables of interest, Y and X, are related by means of a straight-line function.

Thus, in this chapter we limit ourselves to analyzing straight-line relationships between only two variables. Analysis of relationships involving more than two variables will be the province of Chapter 15.

> *Regression analysis* is concerned with the *form* of the relationship between two or more variables. This division of the subject addresses such questions as: Is the relationship linear? If so, what is the specific linear function relating the variables? If the relationship is not linear, just how should the form of the relationship be described?
>
> *Correlation analysis* is concerned with the *strength* of the relationship between two or more variables—that is, with how dependably the variables vary together.

The distinction between regression and correlation is important because in the inductive aspects of this subject the assumptions the investigator implicitly accepts when performing a regression analysis are not exactly the same as those associated with a correlation analysis. The differences in the assumptions are spelled out in section 14.7.

You should be aware, however, that regression and correlation analyses are tightly intertwined. For example, one cannot safely obtain a measure of correlation without first specifying the nature of the regression relationship. Moreover, one will have little faith in estimates obtained from a regression equation if a measure of correlation fails to show that the variables are related in a reasonably strong way.

In this chapter we will introduce *descriptive* regression and correlation measures without being overly concerned with distinguishing between them. However, beginning with section 14.7, where we enter the inductive realm, we will take pains to distinguish between the two kinds of analyses and the assumptions underlying the proper use of each.

Let us begin our study of regression and correlation analyses by focusing on the following illustrative case:

ILLUSTRATIVE CASE 14.1

The sales manager of the Granby Specialties Company, manufacturer of a line of relatively low-priced specialty products, wished to find a statistical procedure by which to estimate sales by territory. Such a procedure, if sufficiently accurate, would be useful to her efforts to set sales quotas for individual territories and to evaluate sales force performance fairly. Such information in the aggregate would also be most helpful to the production supervisor and the inventory-control personnel.

The Granby Specialties Company serves a total of 200 clearly defined geographical territories throughout the United States and Canada, and each territory is covered by a single salesperson. In an effort to make her investigation manageable, the executive decided against examining all 200 territories. Instead, she selected, at random, 10 of the territories for careful analysis.

The dependent variable was sales; individual observations represented number of units of sales achieved in the individual territories during the preceding year. One of the independent variables selected—one whose values are easily determined—was number of households. Data for the sales and household variables are presented in Table 14.1.

Two other independent variables, a price variable and an advertising expenditures variable, were also used in the overall study; however, our discussion of these additional variables will be postponed until Chapter 15.

Table 14.1 Sales and Household Data for the Granby Specialties Company Problem

Sales Territory	Y Sales in Units Achieved Last Year (Tens of Thousands)	X Number of Households in Territory Last Year (Thousands)
1	20	21
2	1	7
3	14	19
4	11	12
5	17	18
6	4	3
7	23	17
8	10	9
9	12	7
10	6	12

14.3 The Scatter Diagram

A simple graphic device, called the scatter diagram, can provide a remarkable amount of information about the relationship between two variables.

> A *scatter diagram* is a graphic tool for presenting information about the existence, form, and closeness of the relationship between two variables. As illustrated in Figure 14.1, the dependent variable Y is shown on the vertical axis and the independent variable X on the horizontal axis. For each value of X, the corresponding Y value is located and a dot made at the point of intersection of imaginary lines projected from these values.
>
> A scatter diagram can provide four important kinds of information: (1) whether a statistical relationship really exists between two variables, and, if so, whether it is (2) direct or inverse, (3) linear or nonlinear, and (4) "strong" or "weak."

Let us look closely at how one interprets a scatter diagram.

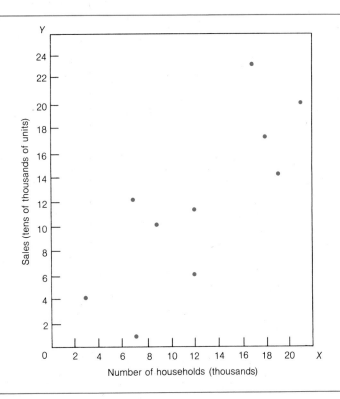

Figure 14.1. Scatter diagram for the Granby Specialties Company case.

Presence of a Statistical Relationship

> If the arrangement of dots in a scatter diagram is such that a straight line drawn through their center would not be parallel to the horizontal axis, this condition suggests the presence of some kind of *statistical relationship*.

Parts (*a*) and (*b*) of Figure 14.2 show two arrangements of dots suggesting the total absence of a relationship between the X and Y variables. All other parts of this figure show arrangements of dots suggesting that the two variables are related in some manner.

The scatter diagram associated with our illustrative case (Figure 14.1) reveals a tendency for sales to increase as the number of households increases. Therefore, the two variables do appear to be statistically related.

Direct and Inverse Relationships

> A statistical relationship is said to be (1) *direct* or *positive* if the dots in a scatter diagram tend to rise as we scan the chart from left to right and (2) *inverse* or *negative* if the dots tend to decline.

Examples of hypothetical direct (linear) relationships are shown in parts (*c*) and (*d*) of Figure 14.2. An inverse (linear) relationship is shown in part (*e*) of the same figure.

For the illustrative case under study, the presence of a *direct* relationship between sales and number of households is evident in Figure 14.1.

Linear and Nonlinear Statistical Relationships

> A statistical relationship between two variables is said to be *linear* if the pattern of dots in a scatter diagram can be adequately described by a *straight line* passing through its center. A statistical relationship is said to be *nonlinear* or *curvilinear* if something other than a straight line is required to describe the pattern of the dots.

Parts (*c*), (*d*), and (*e*) of Figure 14.2 show three kinds of linear relationships, whereas parts (*f*), (*g*), and (*h*) show different types of nonlinear relationships.

Determining whether a statistical relationship is linear or nonlinear is a task usually carried out simply by "eyeballing" the pattern of dots in the scatter diagram—admittedly a

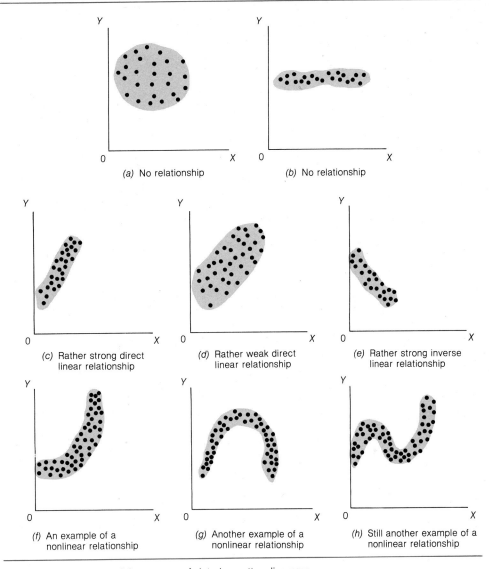

Figure 14.2. Some possible patterns of dots in scatter diagrams.

pretty unscientific approach. Fortunately, if the path of the dots looks linear, we can usually asume it to be linear. It is the clear-cut departures from linearity which concern us most because the unjustified assumption of linearity can lead to a serious loss of information. Figure 14.3 illustrates this point.

The scatter diagram in Figure 14.1 strongly suggests that Granby sales and the number of households in the sample sales territories are related in a linear manner.

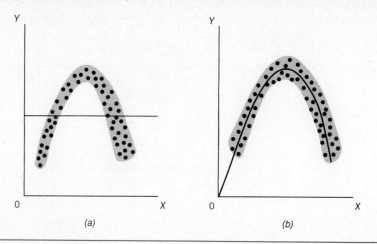

Figure 14.3. The way information is lost when a straight line is incorrectly assumed to be the appropriate way of expressing the form of the relationship between two variables. (a) A linear relationship is incorrectly assumed, with the result that no correlation is registered, whereas in (b) a rather strong correlation is discernible when a more appropriate function is used.

Weak and Strong Statistical Relationships

> A linear statistical relationship is said to be *strong* if there is relatively *little variation* in the dots around a line passing through their center. Conversely, a statistical relationship is said to be *weak* if there is *much variation* around the line.

Parts (c) and (d) of Figure 14.2 show the difference between an apparently strong linear relationship and a weaker linear relationship.

The scatter diagram for the Granby case (Figure 14.1) suggests a relationship between sales and number of households which is neither particularly strong nor weak—just "so-so."

The principal advantage of the scatter diagram is that it provides the investigator with a considerable amount of information quickly and easily. The principal disadvantage is that all the information the scatter diagram conveys—however useful it may be—is conveyed through visual impression only. Usually we prefer to be more precise. In the following three sections, we deal with ways of quantifying the kinds of information gleaned from examination of a scatter diagram.

14.4 The Estimating Equation

Mathematical and Statistical Relationships

We have spoken loosely of a line running through the center of a set of dots in a scatter diagram. In this section we present a method for actually *fitting* a straight line to the dots so as to capture the nature of the relationship in the form of a mathematical equation.

The general equation for a linear function is $Y = a + bX$. In this equation, a is the Y intercept (the value of Y when $X = 0$) and b is the slope of the line, interpreted as the change in Y associated with a one-unit increase in the value of X.

Figure 14.4 shows the line for the mathematical equation $Y = 10 + 3X$. To find the value of Y corresponding to, say, $X = 3$, we calculate $Y = 10 + (3)(3) = 19$. This same value may be obtained directly from Figure 14.4 by reading straight up from $X = 3$ to the line and then reading directly across to the Y axis.

The relationship between X and Y, as described by the equation $Y = 10 + 3X$, is the kind we refer to as a *mathematical*, as distinguished from a *statistical*, relationship.

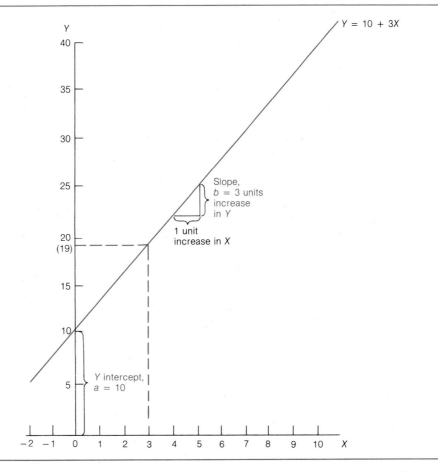

Figure 14.4. The mathematical relationship $Y = 10 + 3X$, with Y intercept, slope, and one point on the line identified.

> In a *mathematical relationship*, there is no opportunity for actual Y values to depart from the values which the equation dictates.

For example, if you enjoy eating popcorn when watching a movie and you know that the movie costs $5 and that popcorn costs $1 a bag, you can readily develop a perfectly accurate equation that will tell you your overall cost. If the prices of the movie ticket and the popcorn consumed are the only relevant costs, the equation would be Y = $5 + $1X, where Y is the total cost of going to the movie and X is the number of bags of popcorn purchased while there. In such a situation, you would know without a doubt that if you purchased, say, five bags of popcorn, the overall cost of the movie would be Y = $5 + ($1)(5) = $10.

Alas, few variables with decision-making implications are related by such unerring mathematical expressions.

> Most dependent variables of interest consist of observations that result from the interplay of a great many other variables. Even when it is possible to express the relationship between two variables in, say, linear-equation form, that relationship is always imperfect in the sense that some variation in Y remains "unexplained" even after the X variable has "explained" as much as it can. In such a case, we speak of the variables X and Y as having a *statistical relationship*.

Consequently, instead of saying, "Given a value of X, what would be the corresponding value of Y?" as we did in the movie example above, we say, "Given a value of X, what would be the corresponding *estimated* value of Y?"

> The equation relating variables X and Y, therefore, is in the form
>
> $$\hat{Y} = a + bX \qquad (14.4.1)$$
>
> *where* $\hat{Y}$ is the estimated sample value of Y associated with a specific known value of X
>
> a is the Y intercept of the sample regression line, that is, the estimated value of Y when X = 0
>
> b is the slope of the sample regression line; it tells by how much $\hat{Y}$, rather than Y per se, changes when X increases by 1

Obtaining the Estimating Equation

There are a number of ways of arriving at a linear estimating equation relating two variables. However, the most frequently used method by far is the method of *least squares*.

A *least squares regression line* is one which meets the following pair of requirements:

$$\Sigma\,(Y - \hat{Y}) = 0$$

and

$$\Sigma\,(Y - \hat{Y})^2 \text{ is a minimum for a specified type of}$$
functional relationship (such as linear)

Our task with respect to the Granby problem is to develop an equation of the type $\hat{Y} = a + bX$ which meets these requirements. More precisely, we seek values for a and b that will minimize the quantity $\Sigma\,(Y - \hat{Y})^2$. Such values can be obtained through a calculus approach. Fortunately for those not on intimate terms with the calculus, one can use the least squares method readily and accurately with no knowledge of the calculus whatever. A pair of *normal equations*, as they are called, a by-product of the calculus approach alluded to above, can be manipulated algebraically to obtain the a and b constants. These normal equations are

$$\Sigma\,Y = na + b\,\Sigma\,X \qquad\qquad (14.4.2)$$
$$\Sigma\,XY = a\,\Sigma\,X + b\,\Sigma\,X^2 \qquad\qquad (14.4.3)$$

To use these equations we would set up a worksheet, like the one shown in Table 14.2, with column headings of Y, X, XY, and X^2. (Table 14.2 also shows a column headed Y^2; this column is used in later calculations.) Column totals are obtained and substituted into Equations (14.4.2) and (14.4.3):

$$118 = 10a + 125b$$
$$1780 = 125a + 1891b$$

We next solve for a and b simultaneously. In so doing, we obtain a b value of .928463 and an a value of .194213. Thus, the equation for the least squares straight line for our illustrative case is $\hat{Y} = .194213 + .928463X$. (It is a good idea in regression and correlation analysis to carry your results out to five or six decimal places. The measures are very interdependent. For example, if b had been rounded to two decimal places, making it .93, the resulting a value would have been .175 rather than the correct .194213. Of course, the $\hat{Y}$ values are dependent on both a and b and may be quite wrong if one is too liberal with his rounding. Once the analysis has been completed, it is permissible to round to one or two decimal places.)

In practice, the two normal equations are usually used in the following algebraically rearranged form:

$$b = \frac{\Sigma\,XY - \dfrac{(\Sigma\,X)(\Sigma\,Y)}{n}}{\Sigma\,X^2 - \dfrac{(\Sigma\,X)^2}{n}} \qquad (14.4.4)$$

$$a = \frac{\Sigma\,Y}{n} - b\frac{\Sigma\,X}{n} \qquad (14.4.5)$$

Using data from Table 14.2 again, we calculate, from Equation (14.4.4),

$$b = \frac{1780 - \dfrac{(125)(118)}{10}}{1891 - \dfrac{(125)^2}{10}} = .928463$$

and, from Equation (14.4.5),

$$a = \frac{118}{10} - (.928463)\left(\frac{125}{10}\right) = .194213$$

Of course, the estimating equation is $\hat{Y} = .194213 + .928463X$, the same as that obtained from the original form of the normal equations.

According to this estimating equation, the estimated Y value associated with X = 0 is .194213 (i.e., 1942 units). However, you are well advised to interpret the a value with caution for reasons to be discussed in section 14.14. The estimating equation also tells us, through the b value, that if the number of households increases by 1000 we can expect an increase in sales of about 9285 (i.e., .928463 times 10,000) units.

Table 14.2 Worksheet Used to Compute the Least Squares a and b Values and Other Regression and Correlation Measures for the Granby Specialties Company Problem

Y	X	XY	X^2	Y^2
20	21	420	441	400
1	7	7	49	1
14	19	266	361	196
11	12	132	144	121
17	18	306	324	289
4	3	12	9	16
23	17	391	289	529
10	9	90	81	100
12	7	84	49	144
6	12	72	144	36
118	125	1780	1891	1832

Now, if each observed X value were substituted into the estimating equation, the corresponding $\hat{Y}$ values could be obtained. For example,

$$\hat{Y} = .194213 + (.928463)(21) = 19.69$$
$$\hat{Y} = .194213 + (.928463)(7) \ = \ 6.69$$

and so forth.

If a $\hat{Y}$ value were obtained for, and plotted against, each observed X value, the result would indeed be a straight line, as illustrated in Figure 14.5. The actual Y values and the corresponding $\hat{Y}$ values are also presented in columns (1) and (2) of Table 14.3.

14.5 Measuring the Adequacy of the Estimating Equation

When we speak of measuring the adequacy of the estimating equation, we are really alluding to the question of how good our chosen independent variable is in providing estimates of values for the dependent variable. If our independent variable is a good one for this purpose, we would expect to find relatively little variation around the line of regression (the line

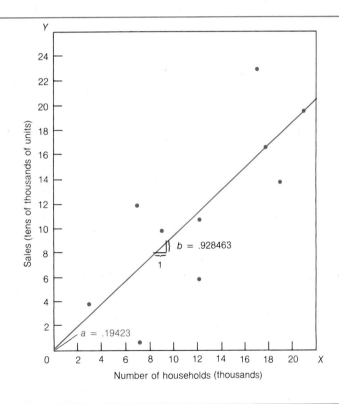

Figure 14.5. Scatter diagram from Figure 14.1, with the least squares line of regression for the linear function $\hat{Y} = .19423 + .928463X$ shown.

Table 14.3 Comparison of Y and $\hat{Y}$ Values and Presentation of Their Differences and Squared Differences for the Granby Specialties Company Problem

(1) Y	(2) $\hat{Y}$	(3) $Y - \hat{Y}$	(4) $(Y - \hat{Y})^2$
20	19.69	.31	.0961
1	6.69	−5.69	32.3761
14	17.84	−3.84	14.7456
11	11.34	−.34	.1156
17	16.91	.09	.0081
4	2.98	1.02	1.0404
23	15.98	7.02	49.2804
10	8.55	1.45	2.1025
12	6.69	5.31	28.1961
6	11.34	−5.34	28.5156
118	118.01	−0.01	156.4765

showing the $\hat{Y}$ values). Conversely, if the independent variable is a poor one from the standpoint of accuracy of estimation, we would expect to see a considerable amount of scatter around the line of regression.

> The *standard error of the estimate*, denoted s_e for sample data, serves as a summary measure of the amount of variation around the line of regression. It does so in much the same way that the ordinary standard deviation serves as a summary measure of the amount of variation around the arithmetic mean of a single variable.

Because of their very close similarity, we may best approach a discussion of the standard error of the estimate by reviewing briefly the regular standard deviation. Recall that in Chapter 4 we presented the standard deviation for a set of sample data as $s = \sqrt{\Sigma (X - \overline{X})^2/(n - 1)}$. Naturally, if the variable under study were designated Y instead of X, the standard deviation formula would be $s = \sqrt{\Sigma (Y - \overline{Y})^2/(n - 1)}$.

We may now ask, "If we knew the value of $\overline{Y}$ and nothing else, how accurately might we estimate the actual Y values? That is, if we were to assert for all n of the sample values that $\overline{Y} = Y$, how accurate would we be overall?" The answer, of course, would depend on how much variation there was around $\overline{Y}$—or, more specifically, on whether the value of the sample standard deviation was large or small.

The standard error of the estimate works in much the same way except that it is a measure of variation around the $\hat{Y}$ values rather than around $\overline{Y}$. The basic formula for the standard error of the estimate for sample data is

$$s_e = \sqrt{\frac{\Sigma (Y - \hat{Y})^2}{n - 2}} \qquad (14.5.1)$$

Equation (14.5.1) differs in two respects from the basic equation for the standard deviation s: First, as noted above, it is the $\hat{Y}$ values, rather than $\overline{Y}$, which are subtracted from each of the Y values. Second, the denominator is $n - 2$ rather than $n - 1$.

> In general, the number of degrees of freedom to be used in the denominator of the standard error of the estimate formula is $n - \lambda$, where λ represents the number of constants in the estimating equation.

In simple linear regression we lose $\lambda = 2$ degrees of freedom because, when determining estimated values for Y associated with specified X values, (1) instead of using the true population Y-intercept value A, we are obliged to estimate it using our known sample a value, and (2) instead of using the true population slope value B, we are required to estimate it using b, the sample slope value.[*] Hence, $n - \lambda = n - 2$ sample observations are said to vary independently and two sample observations are said to be "fixed." An implication of this is that, if we happened to be working with population data and simply wished to use the population standard error of the estimate, denoted σ_e, as a descriptive measure of the variation around the population line of regression,[†] we would not be required to subtract 2 from the number of observations N.

A somewhat different way of showing the importance of taking the loss of 2 degrees of freedom into account when calculating the sample standard error of the estimate is as follows: If the sample size were merely $n = 2$, there would be only two dots in the scatter diagram. If we were to fit a least squares line to the dots, that line would pass exactly through them, regardless of their placement in the chart. Therefore, no variation around the line would be possible.

Applying Equation (14.5.1) to the sample data given in our illustrative case (column (4) of Table 14.3), we find $s_e = \sqrt{156.4765/(10 - 2)} = 4.42$ (in 10,000 of units). Of course, the ideal result would have been $s_e = 0$; the worst possible result would have been $s_e \cong s$.

Equation (14.5.1) has the desirable property of keeping the underlying logic of the standard error of the estimate in plain view. Unfortunately, it is a somewhat difficult equation to work with when the number of observations is large. However, it is possible to replace $\hat{Y}$ in Equation (14.5.1) with $a + bX$ and then, by rearranging the parts of the expression

---

[*] Beginning with Section 14.7, we write the true population regression equation as $\mu_{Y|X} = A + BX$.
[†] This is not exactly the same measure denoted σ_e in Chapter 12. In the present context σ_e is used as a measure of variation around the *population line of regression*.

algebraically, to produce a formula which is often easier to use in practice. This alternative formula is

$$s_e = \sqrt{\frac{\Sigma Y^2 - a \Sigma Y - b \Sigma XY}{n-2}}$$

(14.5.2)

All information required by Equation (14.5.2) can be obtained for our illustrative case from the worksheet (Table 14.2) and from the estimating equation already determined— namely, $\hat{Y} = .194213 + .928463X$. The value resulting from this alternative formula will be identical with that resulting from Equation (14.5.1):

$$s_e = \sqrt{\frac{1832 - (.194213)(118) - (.928463)(1780)}{10 - 2}} = 4.42$$

14.6 Measuring Strength of Association

Total, Explained, and Unexplained Variation

Recall that in our discussions of analysis of variance in Chapters 11 and 12, we always began the computational work by measuring the total sum of squares, SSTO, for the dependent variable. We would then proceed to break the total sum of squares into parts, each part being related either to a specific independent variable or to a collection of unidentified variables. We can follow an analogous procedure with our present subject: We can begin by measuring the *total variation* (the total sum of squares, SSTO) in the Y variable. We can then break this total into two parts, one part representing the variation in Y said to be *explained* by variation in the X variable and the other the variation in Y said to be *unexplained* by variation in the X variable. We will call these parts *regression sum of squares* (SSR) and *error sum of squares* (SSE), respectively. Figure 14.6 illustrates the breakdown of SSTO into its SSR and SSE components for two quite different situations.

Information about explained and unexplained variation could be of interest in its own right. However, these measures will be discussed here more as a means of showing the rationale behind our measures of strength of association.

> *Total variation, SSTO,* is the total amount of variation in the dependent variable Y as measured by the sum of the squared deviations from the arithmetic mean $\overline{Y}$. More specifically, if one were to (1) determine the arithmetic mean of the Y variable, (2) calculate the deviations from the arithmetic mean by subtracting $\overline{Y}$ from each of the Y values, (3) square the resulting deviations, and (4) sum the squared deviations, the resulting value would be the total sum of squares, or, in the present context, *total variation, SSTO.* In symbols: SSTO = $\Sigma (Y - \overline{Y})^2$.

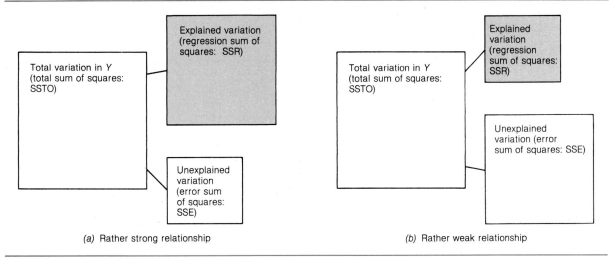

(a) Rather strong relationship   (b) Rather weak relationship

Figure 14.6. The concepts of *total, explained,* and *unexplained variation* for two very different situations.

The computational work for determining total variation for the Granby Specialties Company case is presented in part (*a*) of Figure 14.7. The nature of the deviations is illustrated graphically in part (*b*) of the same figure.

Explained variation, SSR, is the amount of variation in the Y variable that is "explained" by introducing the X variable into the analysis. This measure, like the previous one, calls for the determination of the sum of squared deviations from the arithmetic mean of the Y variable. However, rather than subtracting $\overline{Y}$ from each actual Y value, we subtract $\overline{Y}$ from each $\hat{Y}$ value. That is, SSR $= \Sigma (\hat{Y} - \overline{Y})^2$.

The computational procedures required to secure a value for explained variation for the Granby problem are shown in part (*a*) of Figure 14.8; the nature of the deviations is illustrated in part (*b*) of the same figure.

Unexplained variation is a measure of the amount of variation in Y not explained by variation in the X variable. It is the sum of the squared deviations from the Y values and is obtained by subtracting the appropriate $\hat{Y}$ value from each Y value. In symbols: SSE $= \Sigma (Y - \hat{Y})^2$.

Sales District	Sales Y	$\bar{Y}$	$Y - \bar{Y}$	$(Y - \bar{Y})^2$
1	20	11.8	8.2	67.24
2	1	11.8	-10.8	116.64
3	14	11.8	2.2	4.84
4	11	11.8	-0.8	0.64
5	17	11.8	5.2	27.04
6	4	11.8	-7.8	60.84
7	23	11.8	11.2	125.44
8	10	11.8	-1.8	3.24
9	12	11.8	0.2	0.04
10	6	11.8	-5.8	33.64
Total	118	118.0	0.0	439.60 = SSTO

(a) Computational work for determining total variation: SSTO

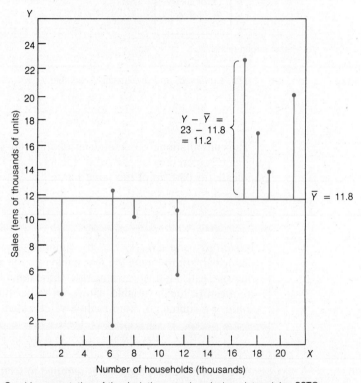

(b) Graphic presentation of the deviations employed when determining SSTO

Figure 14.7. Tabular and graphic presentation of how total variation, SSTO, is obtained.

The computational procedures for determining unexplained variation for the Granby problem were presented in Table 14.3. The nature of the deviations is illustrated in Figure 14.9.

If the calculations have been carried out correctly, the explained variation and the unexplained variation will sum to exactly the same value as the one computed for total variation.

Sales District	$\hat{Y}$	$\overline{Y}$	$\hat{Y} - \overline{Y}$	$(\hat{Y} - \overline{Y})^2$
1	19.69	11.8	7.89	62.2521
2	6.69	11.8	−5.11	26.1121
3	17.83	11.8	6.03	36.3609
4	11.34	11.8	−0.46	0.2116
5	16.91	11.8	5.11	26.1121
6	2.98	11.8	−8.82	77.7924
7	15.98	11.8	4.18	17.4724
8	8.55	11.8	−3.25	10.5625
9	6.69	11.8	−5.11	26.1121
10	11.34	11.8	−0.46	0.2116
Total	118.00	118.0	0.00	283.1998 = SSR

(a) Computational work for determining explained variation: SSR

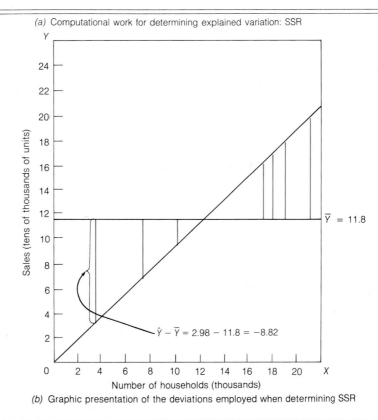

(b) Graphic presentation of the deviations employed when determining SSR

Figure 14.8. Tabular and graphic presentation of how explained variation, SSR, is obtained.

We see below that such is the case with our example:

Total variation, SSTO = Explained variation, SSR + Unexplained variation, SSE

$$\Sigma\,(Y - \overline{Y})^2 \quad = \quad \Sigma\,(\hat{Y} - \overline{Y})^2 \quad + \quad \Sigma\,(Y - \hat{Y})^2$$

$$439.60 \quad = \quad 283.1998 \quad + \quad 156.4765$$

(from Figure 14.7)   (from Figure 14.8)   (from Table 14.3)

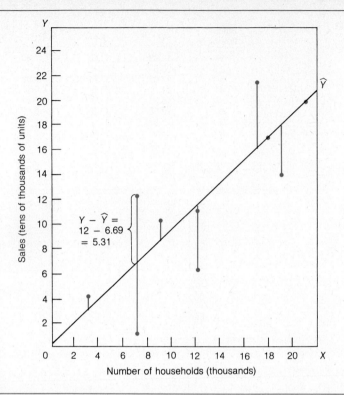

Figure 14.9. The nature of the deviations obtained when computing unexplained variation, SSE.

The Coefficient of Determination

Measures of strength of relationship are correlation, as distinguished from regression, measures. The one we will emphasize in this chapter is the *coefficient of determination*.

> The *coefficient of determination* is the proportion of the variation in the dependent variable Y (the total variation) which is "explained" by variation in the independent variable X. In short, this measure, denoted r^2, is simply the ratio of explained variation to total variation.

Clearly, as little as 0% of the variation in Y might be explained by variation in X—a situation indicating complete absence of a linear relationship. At the opposite extreme, r^2 could conceivably be as great as 1. In reality, r^2 will be some number between 0 and 1. For example, if we apply this measure to the Granby Specialties Company data, we find

$$r^2 = \frac{\Sigma\,(\hat{Y} - \overline{Y})^2}{\Sigma\,(Y - \overline{Y})^2} = \frac{283.1998}{439.60} = .644222$$

Thus we say, "About 64% of the variation in unit sales is 'explained by' variation in number of households," a result which seemingly confirms the suspicion expressed earlier that the linear relationship under investigation is neither especially strong nor especially weak.

Although the above approach to determining the coefficient of determination r^2 is one which keeps the logic of the measure in view, it leaves something to be desired from the standpoint of computational efficiency. An alternative formula that is often more convenient to apply is called the *Pearson product-moment formula*:

$$r^2 = \frac{\left[\Sigma\, XY - \dfrac{(\Sigma\, X)(\Sigma\, Y)}{n} \right]^2}{\left[\Sigma\, X^2 - \dfrac{(\Sigma\, X)^2}{n} \right]\left[\Sigma\, Y^2 - \dfrac{(\Sigma\, Y)^2}{n} \right]} \qquad (14.6.1)$$

Applying Equation (14.6.1) to our illustrative case by utilizing the column totals in Table 14.2, we get

$$r^2 = \frac{\left[1780 - \dfrac{(125)(118)}{10} \right]^2}{\left[1891 - \dfrac{(125)^2}{10} \right]\left[1832 - \dfrac{(118)^2}{10} \right]} = .644179$$

the same result, aside from the effects of rounding errors, determined from use of the other method. Formidable though Equation (14.6.1) may appear at first, a little reflection will confirm that much of the necessary work will already have been completed in the process of determining the slope coefficient b.

Adjusting r^2 for Degrees of Freedom Lost. Neither way described for determining r^2 takes into account degrees of freedom lost. As a result, the value of r^2 is undoubtedly somewhat exaggerated. For example, ask two of your friends to tell you (1) the amount of change they are carrying (the Y variable) and (2) the last digit of their telephone number (the X variable) and run the results through Equation (14.6.1). What do you think r^2 will be? Logically, it should be zero (or, anyway, pretty small) because there is no reason to think that any relationship exists between these two variables. However, you will find that $r^2 = 1$—a perfect linear correlation! Why? When there are only two points, a least squares straight line will pass exactly through those points. In the present context, this means that *all variation in Y is explained variation*. There is no possibility for any unexplained variation. If you were to ask *three* friends for the same information and work r^2, the result would probably not be 1. We cannot say for certain what it would be. But we can say for certain that the r^2 value will be inflated somewhat if an adjustment is not made for the number of degrees of freedom lost. The smaller the number of degrees of freedom associated with s_e and the smaller the value of r^2, the greater will be the impact of the adjustment. The adjustment referred to is carried out as follows:

$$\bar{r}^2 = 1 - (1 - r^2)\left(\frac{n-1}{n-2}\right) \tag{14.6.2}$$

where $\bar{r}^2$ is the coefficient of determination after adjustment for degrees of freedom lost.

For the Granby case, we get $\bar{r}^2 = 1 - (1 - .644179)(\frac{9}{8}) = .599701$. Hence, a more conservative, and more valid, statement about the strength of association in this case is "About 60% (rather than about 64%) of the variation in Y is 'explained by' variation in X." In simple linear correlation, this adjustment will seldom make much difference. However, in complex nonlinear cases and cases involving several independent variables, it can make the difference between a correct and a dangerously incorrect assessment of strength of association.

The Coefficient of Correlation

If we simply compute the square root of the coefficient of determination, we obtain a second widely used measure of strength of association, called the *coefficient of correlation* and denoted r. Unlike the coefficient of determination, the correlation coefficient does assume a sign—the same sign as the regression coefficient b. Therefore, the value of r may be anything between -1 and $+1$ inclusive—with -1 indicating a perfect inverse relationship and $+1$ a perfect direct relationship. Of course, a value of 0 indicates no relationship at all. For the specific problem under discussion (using the result of the product-moment approach), we find $r = \sqrt{r^2} = \sqrt{.644179} = +.802608$.

The *coefficient of correlation* r is defined simply as an abstract measure of strength of linear association between two variables.

Although the coefficient of correlation is widely used, it is somewhat more difficult to interpret than its square. Moreover, because the coefficient of correlation is the square root of a decimal fraction, it will almost always (the exceptions being when $r^2 = 1$ or 0) suggest that more correlation is present than the corresponding coefficient of determination will indicate. However, in terms of information conveyed about the absolute strength of the relationship, neither measure contains any information not possessed by the other.

Also worth noting is that neither r^2 nor r is dependent on the units of measurement used for either variable X or Y. Nor is either r or r^2 dependent on which variable is selected to serve as the dependent variable and which the independent variable.

The r value can be adjusted for degrees of freedom lost by squaring it and using the squared value in Equation (14.6.2). When r^2 has been so adjusted, the square root will be taken to obtain $\bar{r}$.

14.7 Introduction to the Inductive Aspects of Regression and Correlation Analysis

So far in our discussion of regression and correlation analysis, most of our attention has been directed toward the important descriptive tools. To be sure, we have suggested that s_e be obtained using an $n - 2$ denominator, thereby improving its adequacy as an estimator of the population standard error of the estimate σ_e. For the most part, however, the discussion has been limited to description as distinguished from induction. Nevertheless, you should be aware that all the descriptive measures we have considered are, in the context of a sampling problem, merely point estimates of their population counterparts—and it is these unknown population values about which one is usually concerned.

Accordingly, it becomes necessary at this juncture to introduce some population symbols:

The *population regression equation* will be written

$$\mu_{Y|X} = A + BX$$

where $\mu_{Y|X}$ is interpreted as the population mean Y value associated with a specified X value—that is, it is the population counterpart of $\hat{Y}$. A is the population Y-intercept value and B is the population slope.

The *population standard error of the estimate* will be written σ_e and interpreted as a measure of the amount of variation around the population line of regression; ρ^2 will be used to denote the *population coefficient of* (linear) *determination*.

Clearly, a $\hat{Y}$ value obtained from sample data will seldom be the same as the population $\mu_{Y|X}$ value for the same specified value of X. By the same token, a will seldom be equal to A; b will seldom be equal to B; s_e will seldom be equal to σ_e; and r^2 will seldom be equal to ρ^2. Hence, the need for inductive techniques.

In the next three sections we will examine ways of utilizing sample results to draw inferential conclusions about population regression and correlation measures.

Assumptions Underlying Inductive Procedures in Simple Linear Regression

It is important to be clear about the assumptions we tacitly accept when we undertake a simple linear regression analysis with an inductive goal in mind. There are five such assumptions:

---

1. For each value of X there is a subpopulation of Y values.
2. The means of the subpopulations of Y values, the $\mu_{Y|X}$ values, all lie on the same straight line.

---

3. The Y values are statistically independent. This means that the values of Y associated with one X value are in no way dependent on the values of Y associated with any other X value.
4. Each subpopulation of Y values is normally distributed around $\mu_{Y|X}$.
5. The variances of all subpopulations are equal.

These assumptions collectively suggest that each population Y value can be expressed in the following way:

$$Y_i = A + BX_i + \epsilon_i \qquad \epsilon_i: \text{NID}(0, \sigma_e) \qquad (14.7.1)$$

Here the dependent variable Y is assumed to be composed of (1) a linear function, $A + BX$, of the independent variable X and (2) a random variable that is normally and independently distributed about a mean of 0 and that has a standard deviation of σ_e.

An implication of this model is that values of the independent variable X are known exactly, while the dependent variable Y varies in accordance with the linear regression model and the effects of a number of other unidentified variables.

In practice, it is convenient to express the population relationship as $\mu_{Y|X} = A + BX$, thereby interpreting the $\mu_{Y|X}$ values as subpopulation means. (Refer to the second assumption in the list.) The population situation for the Granby case as suggested by this assumption is shown in Figure 14.10.

Assumptions Underlying Inductive Procedures in Simple Linear Correlation

In the list of five assumptions, no mention is made of the X variable's need to be either fixed or random. By *fixed* we mean that the X values are assigned by the investigator at the outset and, hence, are not allowed to vary freely. By *random* we mean that no prior restrictions are imposed on the X values. The five assumptions related to regression analysis are appropriate whether X is a fixed or a random variable. Unfortunately, the assumptions underlying inductive procedures in correlation analysis (analysis involving r^2 or r) are more restrictive and often, alas, more unrealistic than those pertaining to regression.

In addition to the assumptions listed already for regression analysis, in correlation analysis we also assume that the population X and Y values are distributed in a *bivariate normal manner*, as illustrated in Figure 14.11.

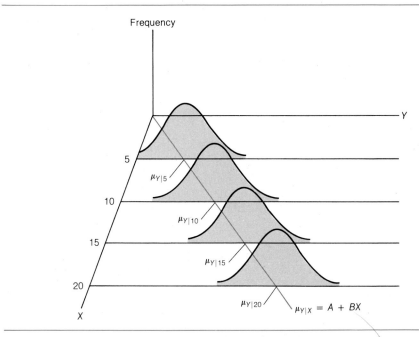

Figure 14.10. Simple linear regression model for the Granby Specialties Company example.

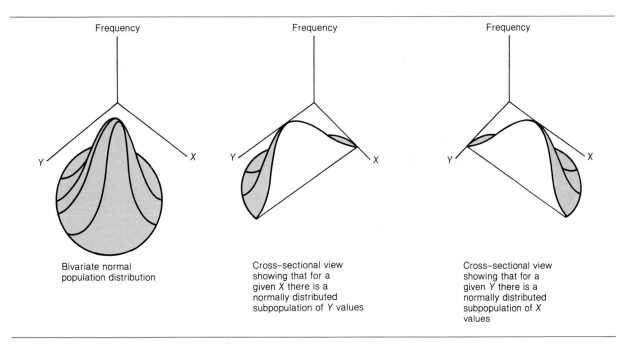

Bivariate normal population distribution

Cross-sectional view showing that for a given X there is a normally distributed subpopulation of Y values

Cross-sectional view showing that for a given Y there is a normally distributed subpopulation of X values

Figure 14.11. A bivariate normal population.

This assumption of bivariate normality is quite restrictive even when the investigation involves a random X variable. For those involving a fixed X variable, it is devastating because it precludes any use of inductive correlation procedures.*

Let us now explore some of the inductive techniques within the realms of regression and correlation analysis. We will begin with regression methods.

14.8 Inferences About the Population Slope

If we could know it, the true population slope B would tell us exactly by how much Y changes, on the average, given a one-unit increase in X. Of course, we will not ordinarily know the value of B. At best, we can guess at it by making use of its sample counterpart b and our knowledge of the nature of the sampling distribution of b.

Drawing correct conclusions about B based on b can be a matter of considerable importance. For example, let us consider the case of a horizontal line of regression—that is, the case of B = 0. In such a situation there is no relationship whatever between variables X and Y. The b, our point estimate of B, however, is subject to sampling variation and may assume a nonzero value—either positive or negative. (This condition is depicted in Figure 14.12.) Such a situation is rife with dangers. For example, earlier we found the b value for the sample of ten Granby sales territories to be .928463. This b value comes very close to saying, "Another household, another nine or ten sales." Let us suppose that an analysis of population movements reveals that a flood of people will probably be moving into, say, Territory 175 during the coming year. This information could be used as grounds for increasing the sales quota for this territory rather dramatically—or even splitting the territory. Such an action might be appropriate if the true population B value were not too different from the sample b value of .928463. However, if the true B value were 0, any action along these lines would be a mistake. It is this kind of possibility that, in practice, leads us to insist that b differ *significantly* from zero before we accept it as evidence supporting the proposition that a real statistical relationship exists between variables X and Y.

Moreover, the sensitivity of Y to a change in X, as measured by B, is frequently of interest in its own right, aside from the question of whether B is 0 or some nonzero value. If the Granby sales manager knew for certain that the true population slope was, say, B = .5, she would presumably react differently to news about an expected large increase in number of households in Territory 175 than she would if she knew that B was, say, 3. And if B were actually −2, that would call for still a different kind of reaction. It is useful, therefore, to have a procedure for setting up a confidence interval around b which, with a specified degree of confidence, can be expected to include the true but unknown population B value.

---

*In a way, it is not so much this assumption as the meaning of r^2 (or r) that is troublesome in such cases. In a truly bivariate population, ρ^2 (or ρ) is a valid parameter which does measure the dependability with which Y and X co-vary. The statistic r^2 (or r) serves as a point estimate of its population counterpart. However, when the X values are assigned by the investigator, as they often are in experimental studies, r^2 will be partly determined by the X values assigned. In other words, investigators themselves create some of the correlation they discover just by deciding by how much the X values will differ from one another. In such situations, r^2 is more properly viewed as a measure (akin to s_e) indicating the amount of variation around the regression line—not the extent to which Y and X co-vary.

In this section we show how (1) a confidence interval for B may be constructed and (2) a hypothesis concerning the true value of B can be tested. First, however, we must become familiar with the important properties of the sampling distribution of the b statistic.

Important Properties of the Sampling Distribution of b

For the model implied by the assumptions listed in section 14.7, it can be shown that the sampling distribution of b has the following properties:

1. The mean of the sampling distribution of b is equal to the true B value for the population. That is, $\mu_b = B$.
2. The standard deviation of the sampling distribution of b, called the standard error of b, is

$$\sigma_b = \frac{\sigma_e}{\sqrt{\Sigma X^2 - \dfrac{(\Sigma X)^2}{n}}} \qquad (14.8.1)$$

3. The sampling distribution b has a normal shape.

Confidence Interval for the Population Slope

On the basis of the three properties of the sampling distribution of b, we can construct a basic formula for a confidence interval for B as follows:

$$b - Z_{\alpha/2}\sigma_b \le B \le b + Z_{\alpha/2}\sigma_b \qquad (14.8.2)$$

where $Z_{\alpha/2}$ is the tabled Z value associated with the confidence coefficient selected.

Equation (14.8.2) is mathematically correct but does not often lend itself to practical use because it requires knowledge of the population standard error of the estimate.

Therefore, instead of using σ_b in the confidence interval formula, we must use an estimate of σ_b. We may proceed, as we have on several previous occasions, by using a known sample value in place of the desired but unavailable population value. We substitute for

$$\sigma_b = \frac{\sigma_e}{\sqrt{\Sigma X^2 - \dfrac{(\Sigma X)^2}{n}}}$$

the following:

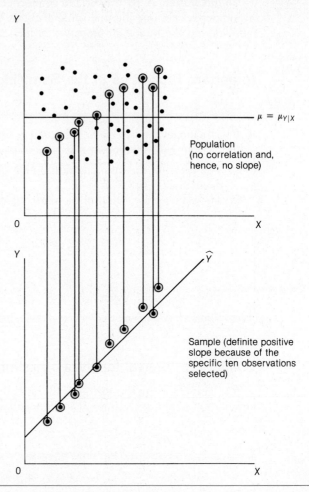

Figure 14.12. The case where the population values have no slope, but the sample values suggest that a slope exists in the population.

$$s_b = \frac{s_e}{\sqrt{\Sigma\, X^2 - \dfrac{(\Sigma\, X)^2}{n}}}$$

(14.8.3)

where s_b is read "the estimated standard error of b."

For reasons similar to those cited earlier in connection with univariate analysis, substitution of s_b for σ_b makes the t distribution a more accurate model than the normal distribution for the sampling distribution of b. Throughout this chapter, the t value used in connection with confidence-interval construction or hypothesis testing will be the one associated with the α value selected and $n - 2$ degrees of freedom.

$$b - t_{\alpha/2} s_b \leq B \leq b + t_{\alpha/2} s_b \qquad (14.8.4)$$

Let us apply Equation (14.8.4) to the Granby Specialties Company problem using a confidence coefficient of .95 ($\alpha = 1 - .95 = .05$). Since $n = 10$, the number of degrees of freedom would be $n - 2 = 8$. The critical t value from Appendix Table 3 is 2.306.

The value of s_e was earlier found to be 4.42. This value must now be divided by $\sqrt{\Sigma X^2 - [(\Sigma X)^2/n]}$ which is simply the square root of the denominator obtained when computing the b value itself. Earlier, this denominator was found to be 328.5 (from section 14.4). Therefore, $\sqrt{\Sigma X^2 - [(\Sigma X)^2/n]} = \sqrt{328.5} = 18.124569$. Thus,

$$s_b = \frac{s_e}{\sqrt{\Sigma X^2 - \dfrac{(\Sigma X)^2}{n}}} = \frac{4.42}{18.124569} = .243868$$

The b value itself we know to be .928463. Substituting the known values into Equation (14.8.4), we get

$$.928463 - (2.306)(.243868) \leq B \leq .928463 + (2.306)(.243868)$$
$$= .366103 \leq B \leq 1.490823$$

We can say with 95% confidence that the range .366103 to 1.490823 includes the true but unknown value of the population slope. This is a rather large interval; whether it would serve as a sufficiently precise estimate of the true B value is questionable. However, the lower limit of $+.366103$ *does* allow us to make two firm conclusions: the population B value *is not* zero and it *is* some positive value. Therefore, some kind of direct relationship almost certainly exists between the two variables.

Testing the Null Hypothesis That the Population Slope Is Zero

As mentioned, in the situation where $B = 0$, the relationship between the two variables is nil regardless of what the sample slope b might suggest. We can determine whether $B = 0$ by following essentially the same steps presented in Chapter 9—namely, we

1. State the null and alternative hypotheses.
   The two hypotheses would be

$$H_0: B = 0$$
$$H_a: B \neq 0$$

2. Decide on a level of significance.
   Let us say that the level of significance is set at $\alpha = .05$.

3. Determine the kind of sampling distribution to rely on.

The appropriate kind of sampling distribution would be the normally distributed sampling distribution of b were it not for the fact that σ_e is unknown. As it is, we use the t distribution as the appropriate model.

4. State the decision rule.

   Decision Rule

   1. If $-2.306 \leq$ observed $t \leq +2.306$, accept H_0.
   2. If observed $t < -2.306$ or observed $t > +2.306$, reject H_0.

5. Calculate the value of the relevant sample statistic.

   The relevant sample statistic, of course, is b, which was found in our example to be .928463. This value converted to its t-value equivalent is $t = (b - B_0)/s_b = b/s_b = .928463/.243868 = 3.807$.

6. Apply the decision rule stated in Step 4 and either accept or reject the null hypothesis.

   Since the observed t of $+3.807$ is greater than the critical t value of $+2.306$, we are led to reject the null hypothesis that $B = 0$ and to accept the alternative hypothesis.

This test has been presented simply to show how it could be used in the absence of a known confidence interval. Actually, when the confidence interval is already known, it is not really necessary to perform this test of significance. If the confidence interval does not include 0, we would simply reject the hypothesis that $B = 0$ at a level of significance equal to the complement of the confidence coefficient.

14.9 Confidence Intervals Using Sample Estimated Y

When discussing the estimating equation determined from sample data, we have thus far usually spoken as if the purpose of such an equation is to produce $\hat{Y}$ values which can be used to estimate, in the descriptive sense of the term, the Y values making up the related subsamples. However, the $\hat{Y}$ values can serve some useful inductive purposes as well. Figure 14.13 illustrates this point.

Two kinds of interval estimates can be developed for values of the dependent variable associated with a specified value of the independent variable: (1) the subpopulation *mean* value of Y and (2) what we will call the subpopulation *new* value of Y. Sample values of $\hat{Y}$ are utilized for both, the only difference in procedure being found in the determination of the standard errors.

To illustrate the difference between the things being estimated—$\mu_{Y|X}$ or Y_{NEW}—we will refer again to the Granby Specialties Company problem. The sales manager might be interested in the *average* sales (the $\mu_{Y|X}$ value) for territories with a specified number of households. But she will almost certainly also need to estimate sales for specific territories with that same number of households (a Y_{NEW} value). By following the steps spelled out here, she can make use of her $\hat{Y}$ values and other sample information to secure both kinds of estimates.

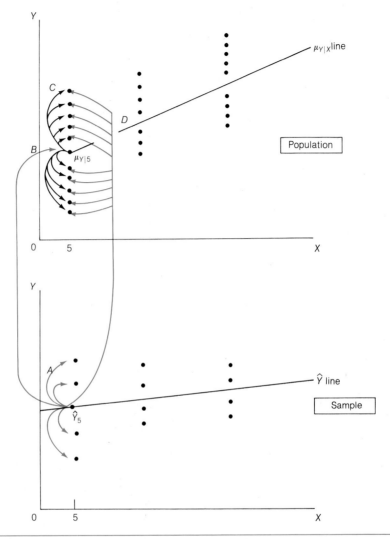

Figure 14.13. Various estimation roles played by a specific $\hat{Y}$ value in regression analysis. *Explanation:* $\hat{Y}$ for $X = 5$ may be used to estimate, in the descriptive sense, all sample Y values associated with $X = .5$. (See arrows at location A.) At the same time, $\hat{Y}$ is a point estimate, in the inductive sense, of its population counterpart $\mu_{Y|5}$. (See arrow at location B.) But since $\mu_{Y|5}$ may be viewed as an estimate, in the descriptive sense, of each of the population Y values associated with $X = 5$ (see arrows at location C), $\hat{Y}$ is also a point estimate, in the inductive sense, of each of the population Y values associated with $X = 5$. (See arrows at location D.)

Constructing Confidence Intervals for Subpopulation Means

Let us suppose that the sales manager wishes to estimate *average* sales associated with *all* territories having 12,000 households ($X = 12$). The point estimate for the $\mu_{Y|12}$ value would be the sample $\hat{Y}$ value associated with $X = 12$. That point estimate is

$$\hat{Y} = .194213 + (.928463)(12) = 11.34$$

But will the true $\mu_{Y|12}$ value be exactly 11.34 as implied by the point estimate? As usual, because of sampling variation, it is highly unlikely.

Because the line obtained using the sample estimating equation was defined in terms of the Y intercept a and the slope b, and because both a and b are subject to sampling variation, we may describe the standard error of $\hat{Y}$, denoted $\sigma_{\hat{Y}}$, as a combination of the sampling error associated with a *and* the sampling error associated with b. Figure 14.14 shows the sample regression line obtained above and a line which, for purposes of illustration, is assumed to be the true population line of regression. Clearly, the two lines have different Y intercepts and different slopes. Especially noteworthy is the tendency for the $\hat{Y}$ values to differ increasingly from the corresponding $\mu_{Y|X}$ values as X gets farther from $\overline{X}$ in either direction.

In Figure 14.15, a line labeled $A + .928463$ is shown running parallel to the sample line of regression. This parallel line conveys the fact that the Y-intercept error may be measured

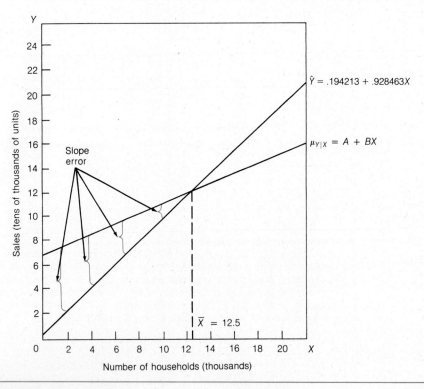

Figure 14.14. Slope error and the reason confidence intervals for $\mu_{Y|X}$ and Y_{next} are wider when X is some value far away from $\overline{X}$ than when X is some value close to X.

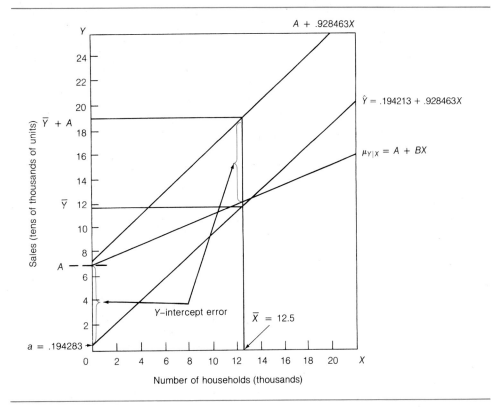

Figure 14.15. The Y-intercept error and the two ways of measuring it.

in either of two ways: (1) by the difference between a and A and (2) by an equivalent vertical distance from $\overline{Y}$. Because a and b are not independent statistics whereas $\overline{Y}$ and b *are* independent, it is more convenient to focus on variation in $\overline{Y}$, rather than on variation in a, when attempting to measure the Y-intercept error.

The two sources of sampling variation in the sample line of regression, when combined, provide the following standard error formula for $\hat{Y}$:

$$\sigma_{\hat{Y}} = \sigma_e \sqrt{\frac{1}{n} + \frac{(X - \overline{X})^2}{\Sigma X^2 - \dfrac{(\Sigma X)^2}{n}}} \qquad (14.9.1)$$

In past discussions, we have noted that the mathematically correct standard error expression may not be usable without some modification because often some kind of population standard deviation must be known. Equation (14.9.1) is no exception since its use would require knowledge of the population standard error of the estimate σ_e, a quantity not ordinarily known in situations where $\mu_{Y|X}$ values are themselves unknown. Fortunately, we can get around this problem reasonably well by substituting sample s_e for σ_e and making use of a t rather than a Z multiplier. Therefore, the (estimated) standard error of $\hat{Y}$ would be obtained by

$$s_{\hat{Y}} = s_e \sqrt{\frac{1}{n} + \frac{(X - \bar{X})^2}{\Sigma X^2 - \dfrac{(\Sigma X)^2}{n}}} \qquad (14.9.2)$$

and the confidence interval formula would be

$$\hat{Y} - t_{\alpha/2} s_{\hat{Y}} \leq \mu_{Y|X} \leq \hat{Y} + t_{\alpha/2} s_{\hat{Y}} \qquad (14.9.3)$$

Using Equation (14.9.2), we determine the estimated standard error of $\hat{Y}$ to be

$$s_{\hat{Y}} = 4.42 \sqrt{\frac{1}{10} + \frac{(12 - 12.5)^2}{328.5}} = 1.403$$

We now use Equation (14.9.3) to help us construct a 95% confidence interval for $\mu_{Y|12}$ for the Granby Specialties Company case. From Appendix Table 3 for $10 - 2 = 8$ degrees of freedom and an area under both tails of $1 - .95 = .05$, we obtain the t value of 2.306. Substituting this and the value of $s_{\hat{Y}} = 1.403$ into this equation, we estimate $\mu_{Y|12}$ to be $11.34 - (2.306)(1.403) \leq \mu_{Y|12} \leq 11.34 + (2.306)(1.403)$, or

$$8.10 \leq \mu_{Y|12} \leq 14.58$$

The sales manager could calculate 95% confidence intervals around the other $\hat{Y}$ values as well. These are shown for the Granby case in Figure 14.16.

The interval of 8.10 to 14.58 for our illustrative use of Equation (14.9.3) is somewhat large—perhaps too large for the purpose for which it is to be used. When it is possible to increase the sample size, precision can be improved without a reduction in confidence. Also, a search for a better (or additional) independent variable should help, provided the search is successful, since that would reduce the sample standard error of the estimate s_e.

When n is quite large, the appropriate Z value from Appendix Table 2 can be used in Equation (14.9.3) in place of t. In fact, for large n a further simplification is also often made: Since the ratio

$$\frac{(X - \bar{X})^2}{\Sigma X^2 - \dfrac{(\Sigma X)^2}{n}}$$

tends to become very small when the sample size is increased, it is often simply eliminated. Hence the abbreviated expression

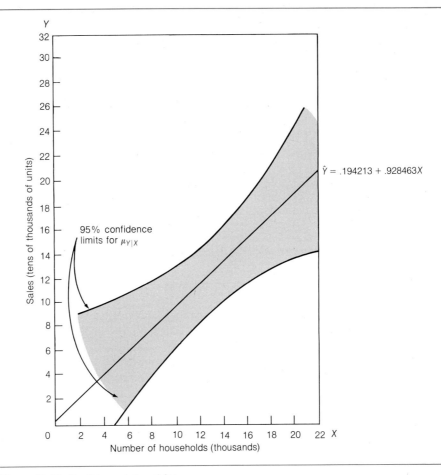

$\hat{Y} = .194213 + .928463X$

95% confidence limits for $\mu_{Y|X}$

Figure 14.16. Confidence interval for $\mu_{Y|X}$.

$$\hat{Y} - Z_{\alpha/2}\frac{s_e}{\sqrt{n}} \leq \mu_{Y|X} \leq \hat{Y} + Z_{\alpha/2}\frac{s_e}{\sqrt{n}} \qquad (14.9.4)$$

Estimation of Y_{NEW}

The procedure for developing a confidence interval for a new value of Y given a specific value of X is similar to that just described—only the standard error formula is altered. As noted in Figure 14.13, $\hat{Y}$ can be viewed as a point estimate of a subpopulation Y value as well as a point estimate of a subpopulation $\mu_{Y|X}$ value. For example, if the sales manager of the Granby Specialties Company wished to estimate sales for a territory not included in the sample study but having 12,000 households, then the same point estimate would be used as was used in connection with the estimation of $\mu_{Y|12}$—namely, $\hat{Y} = 11.34$. To distinguish $\hat{Y}$ in its present role from $\hat{Y}$ in its previous role, we will symbolize the population Y value to be estimated as Y_{NEW}. The difference between the point estimate for Y_{NEW} of 11.34 and the point estimate of $\mu_{Y|12}$ of 11.34 is that Y_{NEW} represents a single value for Y while $\mu_{Y|12}$ is the population *average* of all Y values associated with $X = 12$. From our previous discussion of confidence intervals, we would expect $\sigma_{Y_{NEW}}$, the standard error of Y_{NEW}, to be greater than $\sigma_{\hat{Y}}$, the standard error of $\hat{Y}$. Such is indeed the case, as Figure 14.17 illustrates.

To obtain the standard error of Y_{NEW}, we use the following formula:

$$\sigma_{Y_{NEW}} = \sigma_e \sqrt{1 + \frac{1}{n} + \frac{(X - \overline{X})^2}{\Sigma X^2 - \dfrac{(\Sigma X)^2}{n}}} \qquad (14.9.5)$$

This equation is the same as Equation (14.9.1) except for the presence of the extra 1, an addition used to account for the variability of Y around the true population regression line. Thus, the equation for the standard error of Y_{NEW} acknowledges the influence of three sources of variation: (1) variation in a from sample to sample, (2) variation in b from sample to sample, and (3) variation in the Y values around the population line of regression.

To obtain an estimate of $\sigma_{Y_{NEW}}$, we make the usual substitution of s_e for σ_e in the standard error equation. Thus, the estimated standard error of Y_{NEW} is

$$s_{Y_{NEW}} = s_e \sqrt{1 + \frac{1}{n} + \frac{(X - \overline{X})^2}{\Sigma X^2 - \dfrac{(\Sigma X)^2}{n}}} \qquad (14.9.6)$$

and the equation for the confidence interval is

$$\hat{Y} - t_{\alpha/2} s_{Y_{NEW}} \leq Y_{NEW} \leq \hat{Y} + t_{\alpha/2} s_{Y_{NEW}} \qquad (14.9.7)$$

We may now construct the 95% confidence interval for Y_{NEW} for the Granby Specialties Company when $X = 12$. First, substituting into Equation (14.9.6), we obtain for $s_{Y_{NEW}}$

$$s_{Y_{NEW}} = 4.42 \sqrt{1 + \frac{1}{10} + \frac{(12 - 12.5)^2}{328.5}} = 4.637$$

Then, substituting into Equation (14.9.7), we get

$$11.34 - (2.306)(4.637) \leq Y_{NEW} \leq 11.34 + (2.306)(4.637) = .65 \leq Y_{NEW} \leq 22.03$$

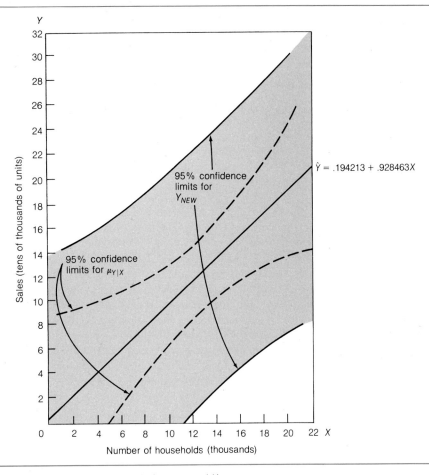

Figure 14.17. Confidence intervals for $\mu_{Y|X}$ and Y_{new}.

You will note that this interval is considerably wider than the one obtained previously for $\mu_{Y|12}$ ($8.10 \leq \mu_{Y|12} \leq 14.58$). This should not be surprising since we are now attempting to estimate sales for a *particular territory* having within it 12,000 households, not the *mean* of all territories having 12,000 households.

When the sample size is large, the normal curve and the following simplified confidence interval equation may be used:

$$\hat{Y} - Z_{\alpha/2}s_e \leq Y_{\text{NEW}} \leq \hat{Y} + Z_{\alpha/2}s_e \qquad (14.9.8)$$

However, expert opinion holds that n should exceed 100 before Equation (14.9.7) is replaced by Equation (14.9.8), a serious limitation of this more convenient formula.

14.10 *F* Test for Population Coefficient of Determination

So far, all of our inductive efforts have been related to regression analysis. By making use of the measures explained variation and unexplained variation, we can put together a useful test of significance for correlation analysis. Keep in mind that it is possible to obtain a high value for the sample coefficient of determination r^2 even though a complete absence of correlation exists in the population. We insist, therefore, that r^2 differ significantly from 0 before we conclude $\rho^2 \neq 0$.

We found in section 14.6 that

$$\text{Explained variation SSR} = \Sigma\,(\hat{Y} - \overline{Y})^2 = 283.1998$$
$$\text{Unexplained variation SSE} = \Sigma\,(Y - \hat{Y})^2 = 156.4765$$

Recall that (1) explained variation is the variation in Y "explained by" variation in the specific independent variable X and (2) unexplained variation is the variation in Y not "explained by" variation in that specific independent variable. Thus, unexplained variation may be likened to *chance variation*, as this term was used in Chapters 11 and 12. Viewed in this light, these two measures of variation are much like between-columns sum of squares and within-columns sum of squares in one-way analysis of variance. This being so, why could we not divide unexplained variation into explained variation and have a legitimate *F* test? The answer is that we can—just as long as (1) the model is a random one and (2) we can determine how many degrees of freedom are associated with the numerator and how many with the denominator of the *F* ratio. In the Granby case, we know that the model is random because the X values were not assigned by an investigator. With respect to degrees of freedom, let us start with the denominator of the *F* ratio.

Because the standard error of the estimate is in effect

$$s_e = \sqrt{\dfrac{\text{Unexplained variation}}{n - 2}}$$

it would seem that $n - 2$ would be the appropriate number of degrees of freedom; and so it is. As we have seen, if there are not at least three observations in the sample, there will be no leeway whatever for any unexplained variation.

The problem is a little more subtle with explained variation. We reason: Explained variation has to do with the $\hat{Y}$ values. The $\hat{Y}$ values can vary, but only in a restricted way—as the *b* value dictates. Thus, the degrees of freedom must be 1 for the one slope constant in the estimating equation. (More generally, it will be $\lambda - 1$, where λ is the number of constants in the estimating equation.)

Therefore, our *F* ratio is

$$\text{Observed } F = \dfrac{\text{Explained variation}/1}{\text{Unexplained variation}/n - 2} \qquad (14.10.1)$$

For our specific problem, we have

$$\text{Observed } F = \dfrac{283.1998/1}{156.4765/(10 - 2)} = \dfrac{283.1998}{19.5596} = 14.48$$

The critical F value, $F_{.05(1,8)}$, is 5.32. Therefore, since observed $F > F_{.05(1,8)}$, we reject the null hypothesis that the population coefficient of determination ρ^2 is 0.

What would happen if we were to divide both explained variation and unexplained variation, as they appear in Equation (14.10.1), by total variation before dividing by the degrees of freedom? The answer is that nothing would happen to the value of the F ratio because the same quantity, total variation, is being divided into both numerator and denominator. Therefore, the F ratio could be expressed as

$$\text{Observed } F = \frac{r^2/1}{(1 - r^2)/(n - 2)} \qquad (14.10.2)$$

This is so because the definition of r^2 is Explained variation/Total variation. Thus, we calculate F as follows:

$$\text{Observed } F = \frac{.644179/1}{.355821/8} = \frac{.644179}{.044478} = 14.48$$

This is the same value as was obtained using Equation (14.10.1). This latter F formula is especially convenient when the Pearson product-moment method, a method which bypasses the calculation of the three kinds of variation, has been used to compute r^2.

Postscript: We knew that r^2 would be significant before we even applied this test. Although the present test was billed as a test of the null hypothesis H_0: $\rho^2 = 0$, it is simultaneously a test of H_0: $B = 0$, a hypothesis we tested and rejected in section 14.8 using t. Clearly, both this F test and the earlier t test need not be applied in the simple linear analysis. In *multiple* regression and correlation analysis, on the other hand, the results of a t test and an F test may convey quite different kinds of information, as will be demonstrated in the next chapter.

14.11 Some Caveats

Although regression and correlation methods can be extremely useful, they do possess great potential for misleading the decision maker. Here are some suggestions that should be heeded whenever regression and correlation analysis is undertaken:

WARNING

1. It is very important to give careful consideration to the functional form of the relationship and to do all statistical calculations and analysis in strict accordance with that form.

All formulas used in regression and correlation analyses are selected on the assumption that the correct functional form of the relationship (i.e., linear, second degree, etc.) has been identified. However, as Figure 14.3 demonstrated, using a simple estimating equation when a more complex one would be better not only means losing information that is there for the taking but also may be dangerously misleading. On the other hand, using a complex estimating equation in a situation where a simpler one would suffice complicates the analysis unnecessarily.

WARNING

2. Regardless of the strength of the sample relationship, as measured by, say, $\bar{r}^2$, one should not interpret such results as proof of a cause-and-effect link between the variables. Any such interpretation must come from outside the statistical analysis.

Let us consider the following hypothetical situation with a view to contemplating the various ways that a sample coefficient of determination, r^2, might be very high. Suppose you wish to know whether your firm's sales would increase if it were to spend more money on advertising than it customarily has. Let us say that you decide to see what the experience of other companies has been and, toward this end, you select a random sample of businesses and run a correlation analysis between sales and advertising expenditures using the latter as the independent variable and presumed cause. The result is an $\bar{r}^2$ close to 1. The following are all possible (and plausible) explanations:

1. The two variables correlate because X is the cause of Y.
   Perhaps it is as you suspected all along: Spend more for advertising and get more sales.

2. The two variables correlate because Y is the cause of X.
   Maybe the presumed cause is really the effect, and vice versa. Perhaps it is the businesses with the highest volumes of sales that are most inclined, and best able, to spend lavishly on advertising.

3. The two variables correlate because they interact with each other.
   Perhaps, rather than there being a one-directional relationship between the two variables, X is sometimes cause and sometimes effect, and the same with Y. That is, maybe advertising helps to bolster sales of small firms which, as their sales increase, can afford to spend more liberally on advertising. The additional advertising helps to boost sales even more, a condition making possible even more lavish spending on advertising. And so it goes.

4. The two variables correlate by chance.
   Whenever correlation analysis is applied to sample data, the risk exists that the sample data will correlate even though there is absolutely no correlation in the population data. Figure 14.12, which was presented earlier to illustrate a slightly different point, illustrates the present one as well. This kind of mischief is not likely to occur very often, especially if larger sample sizes are used, but the possibility is ever present.

"Stop dancing to acid rock . . . it's causing acid rain."

5. The two variables correlate because they are both effects of a third variable outside the analysis.

 The story is told about a man who wrote a letter to an airline requesting that its pilots cease turning on the little light that says fasten seat belts, because every time they do, the ride gets bumpy. Needless to say, leaving the light off would not make the ride any smoother. Both the light and the bumpiness are the results of a third variable—namely, turbulence. Similarly, both sales volume and amount spent on advertising for your sample of companies might be quite dependent on, say, type of industry, or stage of the business cycle, or who knows what other kinds of external factors.

WARNING

3. The sample estimating equation should not be used to estimate Y values associated with X values that are outside the range of the X values used in the sample analysis.

The form of the relationship outside the range of values actually observed will usually be unknown. Therefore, extrapololating is a risky thing to do. Moreover, as our discussion of confidence intervals for $\mu_{Y|X}$ and Y_{NEW} revealed, even if the form of the statistical relationship were to remain the same outside the range of the observed data, the precision of estimates based on sample $\hat{Y}$ values falls off rapidly as X departs more and more from $\overline{X}$.

Along this same line, you will be wise to remember that the Y-intercept value in the sample estimating equation is itself usually an extrapolation. When we interpret the *a* value as the value of $\hat{Y}$ when $X = 0$, we are implicitly assuming that the same form of relationship holds outside the range of the observed X values as exists within that range. Sometimes this assumption is patently untrue. The *a* value is a necessary part of most estimating equations, but it should not ordinarily be interpreted literally as the value of estimated Y in a nonexistent world, so to speak, where $X = 0$.

WARNING

4. If the analysis is conducted with a formal inductive end in mind, the data should be scrutinized carefully to determine how well the underlying assumptions, discussed in section 14.7, are met.

Chances are that the assumptions in any given analysis will not be met perfectly well; this unfortunate fact does not necessarily mean that the results of the investigation will be worthless. The important thing for the investigator to have some feeling for is the approximate size of the discrepancy between the "ideal" and the "real"—that is, between the perfect situation as described by the assumptions and the real situation as represented by his observed data—and, hence, the degree of confidence he may have in his findings. Alas, occasions do sometimes arise when some basic assumptions are violated so grossly that the analysis should be abandoned or continued only for the *descriptive information* it may hold.

A useful diagnostic aid when attempting to evaluate the assumptions is one with which you became familiar in Chapter 11—namely, *residual plots*. The residuals (the $Y - \hat{Y}$ values) for the linear fit to the Granby Specialties Company data (Illustrative Case 14.1) are plotted against the X values in Figure 14.18. We observe that

1. A relatively large amount of dispersion about the 0 line is clearly evident. This could indicate that one or more important independent variables have been omitted from the analysis. This does not pertain to any assumptions listed in section 14.7, but it does suggest the advisability of performing a multiple regression analysis—the subject of the next chapter.
2. No pattern in the dots is evident as we scan the chart from left to right—that is, we do not find, for example, several negative residuals, followed by several positive residuals, followed by more negative residuals. Apparently a straight line fits the data quite well.
3. As we scan the chart from left to right, no tendency is evident suggesting that the amount of variation around the 0 line becomes larger or smaller as the X values become larger.

In addition to these points, because of the nature of the study and the manner in which the sample data were gathered, there is seemingly no reason to believe that the assumption of independence has been violated.

Generally speaking, it is difficult to tell from the residual plots whether the assumption of normally distributed subpopulations has been violated. However, we do have available several fairly good diagnostic methods for evaluating this assumption. One approach involves standardizing the residuals—that is, computing $e = (Y - \hat{Y})/s_e$ values. The standardized residuals for the Granby case follow:

$$e: \quad .07 \quad -1.29 \quad -.87 \quad -.08 \quad .02 \quad .23 \quad 1.59 \quad .33 \quad 1.20 \quad -1.21$$

(*Note*: The regular residuals were obtained from Table 14.3; the standard error of the estimate s_e was found earlier to be 4.42.)

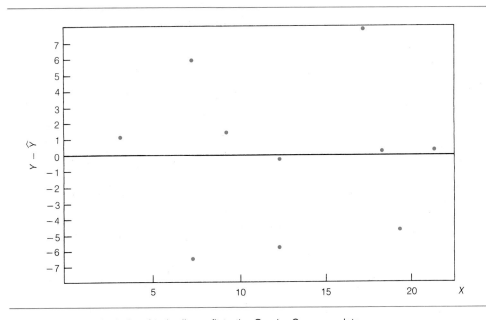

Figure 14.18. Residual plots for the linear fit to the Granby Company data.

Having computed the standardized residuals, we now consult the t table (Appendix Table 3) to determine the critical values associated with the middle, say, 50% of the theoretical t distribution, the middle 70%, and so forth. Once these critical values have been found, we determine the proportion of *observed* standardized residuals between each set of critical t values and compare the results. This is done here:

Critical t Values	Proportion of t Distribution Falling Between the Indicated Critical Values	Proportion of Observed Standardized Residuals Falling Between the Indicated Critical Values
$-.706$ to $+.706$	.50	$5/10 = .50$
-1.108 to $+1.108$	.70	$6/10 = .60$
-1.397 to $+1.397$	.80	$9/10 = .90$
-2.306 to $+2.306$	.95	$10/10 = 1.00$

The correspondence between the two lists of proportions is obviously imperfect. However, the discrepancies are probably not sufficiently great to lead us to throw out our analysis on the grounds that the normality assumption has been violated.

Another easy check on the normality assumption can be made through use of a box-and-whiskers chart, a subject introduced in Chapter 3. Such a chart is shown in Figure 14.19. Recall that the left-hand edge of the box represents the first quartile, the vertical line within the box the second quartile, or median, and the right-hand edge of the box the third quartile. In our problem (using regular residuals rather than standardized residuals), the corresponding values are -3.83, $.20$, and $+1.45$, respectively. The interquartile range is the difference between the third and first quartiles: $1.45 - (-3.83) = 5.28$. In Figure 14.19, an X has been placed a distance of 5.28 below the left-hand edge of the box and a distance 5.28 above the right-hand edge. If the residuals were strictly normally distributed, we would expect to find (1) symmetry around the median, (2) 50% of the residuals within the box, (3) 25% of the residuals outside the box on the left-hand side, (4) 25% of the residuals outside the box on the right-hand side, and (5) approximately 95% of the residuals between the two X's.

Our residuals are clearly not symmetrical. However, 50% of them are found inside the box (counting those on the edges as half in) and another 50% evenly distributed between the two ends outside the box. Between the two X's, we find 90% of our residuals. The results pertaining to the normality assumption are admittedly ambiguous—but probably not devastating.

The assumption of *bivariate normality* required in connection with tests of correlation is difficult to evaluate. One possibility is to apply one of the checks for normality to both the Y values and the X values (a requirement implying use of an equation of the form $\hat{X} = a + bY$ as well as $\hat{Y} = a + bX$). Acceptance of normality for both variables suggests the possibility of bivariate normality—but does not guarantee it. Still, such positive findings should be sufficient to assure us that the results of the F tests can be believed. On the other hand, rejection of this assumption with respect to either variable is strong evidence that the assumption of bivariate normality has not been met.

Most packaged computer programs for regression and correlation analysis will provide a list of residuals and will also show the residuals converted into standardized form. They will also permit use of nonlinear relationships and a switching of roles by the X and Y variables. Figure 14.20 shows a fairly typical computer printout for this kind of analysis.

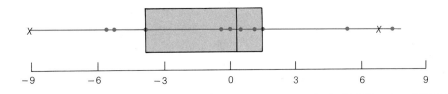

Figure 14.19. Box-and-whiskers chart of residuals for the Granby Company problem.

```
THE REGRESSION EQUATION IS
Y = - 168931 + 67.9 X1

                         ST. DEV.    T-RATIO =
COLUMN      COEFFICIENT  OF COEF.    COEF/S.D.
            -168931      6263        -26.97
X1          67.877       1.994       34.05

STANDARD ERROR OF THE ESTIMATE IS:
S = 4731

R-SQUARED = 98.3 PERCENT
R-SQUARED = 98.2 PERCENT, ADJUSTED FOR D.F.

ANALYSIS OF VARIANCE

DUE TO       DF      SS           MS=SS/DF
REGRESSION   1       25950717952  25950717952
RESIDUAL     20      447713760    22385688
TOTAL        21      26398431232

          X1      Y       PRED. Y  ST.DEV.
ROW       C1      C2      VALUE    PRED. Y  RESIDUAL  ST.RES.
 1        2491    3719    151      1579     3568      0.80
 2        2476    5321    -867     1602     6188      1.39
 3        2577    6716    5988     1451     728       0.16
 4        2613    8066    8432     1401     -366      -0.08
 5        2650    9487    10943    1351     -1456     -0.32
 6        2636    10045   9993     1369     52        0.01
 7        2696    12274   14066    1291     -1792     -0.39
 8        2697    14307   14134    1290     173       0.04
 9        2725    17081   16034    1256     1047      0.23
10        2796    20873   20853    1177     20        0.00
11        2849    24596   24451    1126     145       0.03
12        3009    29970   35311    1025     -5341     -1.16
13        3152    36507   45018    1014     -8511     -1.84
14        3274    43425   53299    1066     -9874     -2.14R
15        3371    56656   59883    1144     -3227     -0.70
16        3464    66558   66195    1242     363       0.08
17        3515    80005   69657    1304     10348     2.28R
18        3619    81899   76716    1444     5183      1.15
19        3714    85632   83164    1585     2468      0.55
20        3837    95056   91513    1781     3543      0.81
21        4068    102197  107193   2177     -4996     -1.19
22        3981    103023  101288   2025     1736      0.41

R DENOTES AN OBS. WITH A LARGE ST. RES.

X DENOTES AN OBS. WHOSE X VALUE GIVES IT LARGE INFLUENCE
```

The regression equation is $Y = -169886 + 68.2X_1$, where X_1 is the same as the X symbol used in this chapter.

The "COEFFICIENT" column simply lists the sample a and b values shown in the estimating equation above.

The "ST. DEV. OF COEF." column shows the s_a (not discussed in this chapter) and the s_b values. The latter is the estimated standard deviation of the sampling distribution of b.

The "T-RATIO = COEF./S.D." column shows the result of dividing the Y intercept, a, by its standard error value and the slope, b, by its standard error value. The results are observed t ratios which would be compared with tabled t values for α and 20 degrees of freedom to test the null hypotheses:

$$H_0: A = 0 \qquad\qquad H_0: B = 0$$
$$\text{and}$$
$$H_a: A \neq 0 \qquad\qquad H_a: B \neq 0$$

The "STANDARD ERROR OF THE ESTIMATE" is simply our standard error of the estimate, s_e.

"R-SQUARED" is our coefficient of simple linear determination, r^2.

The other "R-SQUARED" value is our $\bar{r}^2$, that is, r^2 adjusted for loss of degrees of freedom.

The analysis of variance information may be used to test either

$$H_0: B = 0 \qquad\qquad H_0: \rho^2 = 0$$
$$\text{or}$$
$$H_a: B \neq 0 \qquad\qquad H_a: \rho^2 \neq 0$$

Observed F would have to be determined by the analyst and compared with critical F for α and 1 and 20 degrees of freedom.

The columns labeled "X1" and "Y" are simply the original observations.

"PRED. Y VALUES" are what we have been calling *estimated Y values*, that is, $\hat{Y}$ values.

"ST. DEV. PRED. Y" values are the $s_{Y\text{new}}$ values used for estimating Y_{new}. The values shown here would have to be multiplied by $t_{\alpha/2}$ and the resulting product added to and subtracted from the related Y value.

"RESIDUAL" refers to the $Y - \hat{Y}$ values.

"ST. RES." (short for *standardized residuals*) are the $Y - \hat{Y}$ values divided by the standard error of the estimate, $S_e = 4731$.

Figure 14.20. Computer printout of regression and correlation results.

14.11 Some Caveats

14.12 Summary

In this chapter you have been introduced to an extremely useful and diverse family of techniques for describing and making use of the relationship between two variables. These techniques have run the gamut from pure description to statistical induction.

The descriptive methods we examined included (1) construction and interpretation of a scatter diagram, (2) determination of the sample least squares linear equation, (3) calculation of the sample standard error of the estimate, and (4) calculation of sample strength-of-relationship measures.

We then turned to subjects pertaining to induction. We considered how to construct confidence intervals for the population slope, the subpopulation mean of Y values associated with a specified X value, and the value of a particular population Y value associated with a specified X value. We also considered how to test null hypotheses concerning the population slope and the population coefficient of simple linear determination.

I hope that by now you are convinced that regression and correlation methods are capable of providing a rich assortment of information about a pair of variables. And the fun is just beginning. In the next chapter you will learn how to employ variations on the methods described here in order to analyze situations involving more than one independent variable. Most variables of any interest to business and economic analysts are influenced importantly by more than one independent variable. Consequently, the potential usefulness of regression and correlation methods is greatly expanded when one can deal with two or more independent variables at a time.

Terms Introduced in This Chapter

bivariate normal population distribution (p. 645)
coefficient of correlation (p. 642)
coefficient of determination (p. 640)
confidence interval for B (p. 647)
confidence interval for $\mu_{Y|X}$ (p. 652)
confidence interval for Y_{NEW} (p. 655)
correlation analysis (p. 623)
direct relationship (p. 626)
error sum of squares, SSE (p. 626)
estimated Y (p. 630)
estimating equation (p. 630)
explained variation (p. 637)
inverse relationship (p. 626)

least squares criteria (p. 631)
least squares estimating equation (p. 631)
line of regression (p. 633)
linear relationship (p. 623)
mathematical relationship (p. 630)
multiple regression and correlation analysis (p. 623)
nonlinear relationship (p. 626)
normal equations (p. 631)
regression analysis (p. 623)
regression sum of squares, SSR (p. 637)

sampling distribution of b (p. 647)
scatter diagram (p. 625)
simple regression and correlation analysis (p. 623)
standard error of b (p. 647)
standard error of the estimate (p. 634)
standard error of $\hat{Y}$ (p. 653)
statistical relationship (p. 626)
total sum of squares, SSTO (p. 636)
unexplained variation (p. 637)
Y intercept (p. 629)
Y_{NEW} (p. 655)

Formulas Introduced in This Chapter

General formula for a simple linear estimating equation. The $\hat{Y}$, a, and b indicate that the equation was obtained from sample, rather than population, data:

$$\hat{Y} = a + bX$$

The normal equations for deriving a simple linear *least squares* estimating equation from sample data:

$$\Sigma\ Y = na + b\ \Sigma\ X$$
$$\Sigma\ XY = a\ \Sigma\ X + b\ \Sigma\ X^2$$

Alternative form of the normal equations:

$$b = \frac{\Sigma\ XY - \dfrac{(\Sigma\ X)(\Sigma\ Y)}{n}}{\Sigma\ X^2 - \dfrac{(\Sigma\ X)^2}{n}}$$

$$a = \frac{\Sigma\ Y}{n} - b\frac{\Sigma\ Y}{n}$$

Basic formula for obtaining the sample standard error of the estimate in a simple linear analysis:

$$s_e = \sqrt{\frac{\Sigma\ (Y - \hat{Y})^2}{n - 2}}$$

Shortcut formula for the sample standard error of the estimate in a simple linear analysis:

$$s_e = \sqrt{\frac{\Sigma\ Y^2 - a\ \Sigma\ Y - b\ \Sigma\ XY}{n - 2}}$$

Total variation in the dependent variable:

$$\text{SSTO} = \Sigma\ (Y - \overline{Y})^2$$

Variation in the dependent variable *explained by* variation in the independent variable:

$$\text{SSR} = \Sigma\ (\hat{Y} - \overline{Y})^2$$

Variation in the dependent variable *not explained* by variation in the independent variable:

$$\text{SSE} = \Sigma\ (Y - \hat{Y})^2$$

Basic formula for the sample coefficient of simple linear determination:

$$r^2 = \frac{\Sigma\ (\hat{Y} - \overline{Y})^2}{\Sigma\ (Y - \overline{Y})^2}$$

Pearson product-moment formula for the sample coefficient of simple linear determination:

$$r^2 = \frac{\left[\Sigma\ XY - \dfrac{(\Sigma\ X)(\Sigma\ Y)}{n}\right]^2}{\left[\Sigma\ X^2 - \dfrac{(\Sigma\ X)^2}{n}\right]\left[\Sigma\ Y^2 - \dfrac{(\Sigma\ Y)^2}{n}\right]}$$

The sample coefficient of simple linear correlation can be obtained by

$$r = \sqrt{r^2}$$

and attaching the same sign as possessed by *b*.

Sample coefficient of simple linear determination adjusted for degrees of freedom lost:

$$\overline{r}^2 = 1 - (1 - r^2)\left(\frac{n - 1}{n - 2}\right)$$

General formula for a simple linear estimating equation. The $\mu_{Y|X}$, A, and B indicate that the equation was obtained from population, rather than sample, data:

$$\mu_{Y|X} = A + BX$$

True standard error of b:

$$\sigma_b = \frac{\sigma_e}{\sqrt{\Sigma X^2 - \dfrac{(\Sigma X)^2}{n}}}$$

Estimated standard error of b:

$$s_b = \frac{s_e}{\sqrt{\Sigma X^2 - \dfrac{(\Sigma X)^2}{n}}}$$

Correct formula for a confidence interval for the population slope:

$$b - Z_{\alpha/2}\sigma_b \leq B \leq b + Z_{\alpha/2}\sigma_b$$

Formula for a confidence interval for the population slope when σ_e is unknown:

$$b - t_{\alpha/2}\,s_b \leq B \leq b + t_{\alpha/2}\,s_b$$

Observed t for testing the null hypothesis that B = 0:

$$t = \frac{b - B_0}{s_b} = \frac{b}{s_b}$$

True standard error of $\hat{Y}$:

$$\sigma_{\hat{Y}} = \sigma_e \sqrt{\frac{1}{n} + \frac{(X - \overline{X})^2}{\Sigma X^2 - \dfrac{(\Sigma X)^2}{n}}}$$

Estimated standard error of $\hat{Y}$:

$$s_{\hat{Y}} = s_e \sqrt{\frac{1}{n} + \frac{(X - \overline{X})^2}{\Sigma X^2 - \dfrac{(\Sigma X)^2}{n}}}$$

Formula for a confidence interval for a subpopulation $\mu_{Y|X}$ value when σ_e is unknown:

$$\hat{Y} - t_{\alpha/2}\,s_{\hat{Y}} \leq \mu_{Y|X} \leq Y + t_{\alpha/2}\,s_{\hat{Y}}$$

Simpler alternative to the above formula (limited to large sample sizes):

$$\hat{Y} - Z_{\alpha/2}\frac{s_e}{\sqrt{n}} \leq \mu_{Y|X} \leq \hat{Y} + Z_{\alpha/2}\frac{s_e}{\sqrt{n}}$$

True standard error of Y_{NEW}:

$$\sigma_{Y_{NEW}} = \sigma_e \sqrt{1 + \frac{1}{n} + \frac{(X - \overline{X})^2}{\Sigma X^2 - \dfrac{(\Sigma X)^2}{n}}}$$

Estimated standard error of Y_{NEW}:

$$s_{Y_{\text{NEW}}} = s_e \sqrt{1 + \frac{1}{n} + \frac{(X - \overline{X})^2}{\Sigma X^2 - \dfrac{(\Sigma X)^2}{n}}}$$

Formula for a confidence interval for a subpopulation Y_{NEW} value when σ_e is unknown:

$$\hat{Y} - t_{\alpha/2}\, s_{Y_{\text{NEW}}} \leq Y_{\text{NEW}} \leq Y + t_{\alpha/2}\, s_{Y_{\text{NEW}}}$$

Simple alternative to the above formula (limited to large sample sizes):

$$\hat{Y} - Z_{\alpha/2}\, s_e \leq Y_{\text{NEW}} \leq \hat{Y} + Z_{\alpha/2}\, s_e$$

The F ratio formula for testing the null hypothesis that $H_0: \rho^2 = 0$:

$$\text{Observed } F = \frac{r^2/1}{(1 - r^2)/(n - 2)}$$

Questions and Problems

14.1 Explain the meaning of each of the following:
   a. Scatter diagram
   b. Explained variation
   c. Simple regression
   d. Inverse relationship
   e. $\mu_{Y|X}$
   f. $\sigma_{\hat{Y}}$
   g. r^2

14.2 Explain the meaning of each of the following:
   a. Least squares estimating equation
   b. Total variation
   c. Standard error of the estimate
   d. Bivariate normal distribution
   e. $\sigma_{Y_{\text{NEW}}}$
   f. ρ^2
   g. B

14.3 Explain the meaning of each of the following:
   a. Pearson product-moment formula
   b. Linear regression
   c. Unexplained variation
   d. Coefficient of correlation
   e. b
   f. σ_e
   g. Standard error of b

14.4 Explain the meaning of each of the following:
   a. Standardized residual
   b. s_e
   c. r^2
   d. A
   e. Extrapolation in regression analysis

14.5 Distinguish between:
   a. Regression analysis and correlation analysis
   b. Simple regression analysis and multiple regression analysis
   c. Direct and inverse relationships
   d. Linear and nonlinear relationships
   e. Mathematical and statistical relationships

14.6 Distinguish between:
   a. Strong and weak relationships
   b. a and b

c. b and B
d. Standard deviation and standard error of the estimate
e. Explained variation and unexplained variation

14.7 Distinguish between:
   a. Coefficient of determination and coefficient of correlation
   b. Fixed X variable and random X variable
   c. r^2 and $\bar{r}^2$
   d. $\hat{Y}$ and $\mu_{Y|X}$

14.8 Distinguish between:
   a. Assumptions underlying induction in simple linear regression analysis and assumptions underlying induction in simple linear correlation analysis
   b. σ_b and $\sigma_{\hat{Y}}$ and $\sigma_{Y_{NEW}}$
   c. Confidence interval for $\mu_{Y|X}$ and confidence interval for Y_{NEW}
   d. Residuals and standardized residuals

14.9 Indicate which of the following statements you agree with and which you disagree with, and defend your opinions:
   a. The coefficient of determination r^2 may be as low as 0 or as high as 1.
   b. The coefficient of determination r^2 is a correlation, as distinguished from a regression, measure.
   c. If in a simple linear correlation analysis one finds that the sample total variation in Y is 5 and that the sample explained variation is 3, the value of the sample coefficient of determination will be .75.
   d. If a simple linear regression analysis yields an estimating equation of $\hat{Y} = 5 + .7X$, this means that, given an increase of one unit in X, $\hat{Y}$ will increase by .7.

14.10 Indicate which of the following statements you agree with and which you disagree with, and defend your opinions:
   a. Imagine the following experiment: Four one-digit numbers are selected from a table of random digits and separated in a random manner into two groups of two numbers each and the coefficient of determination r^2 (unadjusted for the loss of degrees of freedom) is computed. In such a situation, we might reasonably expect r^2 to be as high as, say, .2 or .3 by accident, but could not reasonably expect r^2 to be in excess of about .8.
   b. If in a simple linear correlation analysis sample total variation were found to be 250 and the sample standard error of the estimate (computed with a denominator of n) were 10, the sample coefficient of determination r^2 would be .6.
   c. If, according to an F test of the null hypothesis H_0: $\rho^2 = 0$, the null hypothesis is rejected at the specified level of significance, nothing can be gained by employing a nonlinear regression equation.
   d. The assumptions underlying the use of tests of significance for correlation measures are more restrictive than the assumptions underlying tests of significance for regression measures.

14.11 Indicate which of the following statements you agree with and which you disagree with, and defend your opinions:
   a. If all of the observed Y and X values in a simple linear regression analysis are positive, it is impossible for the slope regression coefficient b to be negative.
   b. Except for the minor detail that 1 is included under the radical sign in one case and not in the other, there is no practical difference between a .95 confidence interval for $\mu_{Y|X}$ and a .95 confidence interval for Y_{NEW}.
   c. If a sample coefficient of determination r^2 is tested and found to be statistically significant, it means that, for the sample at least, the relationship between Y and X is strong.
   d. If a sample coefficient of determination r^2 has a value close to 1, it necessarily follows that it must be statistically significant at the .01 level.

14.12 Indicate which of the following statements you agree with and which you disagree with, and defend your opinions:

a. A test of the null hypothesis H_0: $\rho^2 = 0$ is simultaneously a test of the null hypothesis H_0: $B = 0$ at the same level of significance.

b. A confidence interval for population B can serve as a test of the null hypothesis H_0: $B = 0$.

c. If a test performed on the sample slope b leads to acceptance of the null hypothesis H_0: $B = 0$, this result suggests an absence of any kind of relationship whatever between the X and Y values making up the population.

d. The larger the sample size n, the more nearly equal will be the values of r and r^2.

14.13 If examination of a scatter diagram revealed that all of the original sample points fell on the line of regression, what would be the value of:

a. The coefficient of determination? (If it is impossible to say from the information given, explain why.)

b. The standard error of the estimate? (If it is impossible to say from the information given, explain why.)

c. The slope coefficient b? (If it is impossible to say from the information given, explain why.)

 14.14 Management of a life insurance company wished to determine whether dollar sales by the company's salespeople are related to the number of years of insurance selling experience.

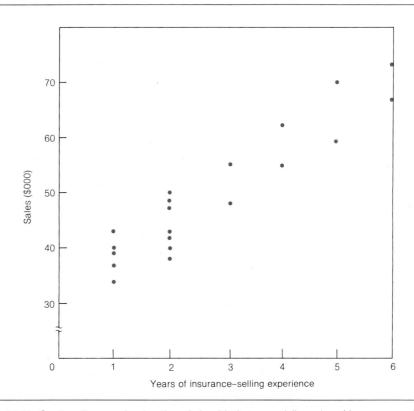

Figure 14.21. Scatter diagram showing the relationship between dollar sales of insurance and number of years of selling experience for 20 salespeople.

Toward this end, the managers drew a random sample from the population consisting of all the salespeople employed by the company and recorded the sales (Y) of each for the most recent calendar year and the number of years of selling experience of each (X). The scatter diagram obtained from the data is shown in Figure 14.21 (p. 671).

a. Does a statistical relationship exist between dollar sales and the number of years of experience? Tell why you believe that Figure 14.21 indicates the presence or absence of a statistical relationship.

If a relationship does exist between the two variables:

b. Is it linear or nonlinear? Briefly defend your conclusion.
c. Is it direct or inverse? Briefly defend your conclusion.
d. Is it weak or strong? Briefly defend your conclusion.

14.15 Refer to problem 14.14. The scatter diagram shown in Figure 14.21 was obtained from the data presented in Table 14.4.

a. Derive the linear least squares estimating equation relating these two variables.
b. Compute the $\hat{Y}$ value associated with each of the X values. What are your impressions of the adequacy of the estimating equation as a tool for estimating sales?
c. Compute and interpret the standard error of the estimate s_e.
d. Interpret the b value you obtained in connection with part a. Test this b value for significance at the .05 level and state your conclusion.
e. Compute and interpret the coefficient of correlation r and the coefficient of determination r^2.
f. Test r^2 for significance at the .05 level and state your conclusion.
g. For the entire population of salespeople employed by this company, what is your estimate of the average dollar sales, $\mu_{Y|X}$, of those salespeople with four years' experience? Use a .95 confidence interval.

Table 14.4 Data from Which the Scatter Diagram in Figure 14.21 Was Constructed

Salesperson	Y Sales (Thousands)	X Years of Experience
A	55	3
B	50	2
C	40	1
D	35	1
E	47	2
F	39	1
G	37	1
H	38	2
I	59	5
J	70	5
K	73	6
L	67	6
M	62	4
N	55	4
O	43	1
P	43	2
Q	48	3
R	40	2
S	49	2
T	42	2

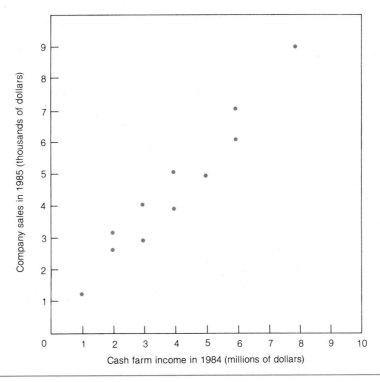

14.16 A manager with the Phyllis Tiller Company, manufacturer and distributor of farm equipment, suspects that there exists a strong positive relationship between sales made to farmers in several specified counties of the Midwestern United States in a given year and cash farm income for the same counties in the preceding year. To determine whether her suspicion is correct, she examines sales in a random sample of these counties for the year 1985 and cash farm income for the same sample of counties for the year 1984. The resulting data are presented in scatter diagram form in Figure 14.22.
  a. Does a statistical relationship exist between company sales in these counties in 1985 and cash farm income in these same counties in 1984? Tell why you believe Figure 14.22 reveals the presence or absence of a statistical relationship.
  If a relationship does exist between the two variables:
  b. Is it linear or nonlinear? Defend your choice.
  c. Is it direct or inverse? Defend your choice.
  d. Is it rather weak or reasonably strong? Defend your choice.

14.17 Refer to problem 14.16. The values of the X and Y variables used to construct the scatter diagram in Figure 14.22 are presented in Table 14.5.
  a. Derive the linear least squares estimating equation relating these two variables.
  b. Compute the $\hat{Y}$ value associated with each of the X values. What is your impression of the adequacy of your equation as a tool for estimating sales?
  c. Compute and interpret the standard error of the estimate s_e.
  d. Interpret the meaning of the b value you obtained in connection with part a. Test the b value for significance at the .05 level.

Figure 14.22. Scatter diagram showing the relationship between sales of the Phyllis Tiller Company in 1985 and cash farm income in 1984 for 11 randomly selected Midwestern counties.

Table 14.5 Data from Which the Scatter Diagram in Figure 14.22 Was Constructed

County	Company Sales, 1985 (Thousands of Dollars)	Cash Farm Income, 1984 (Millions of Dollars)
A	3.0	3
B	5.0	5
C	6.0	6
D	9.0	8
E	5.0	4
F	3.5	2
G	4.0	3
H	2.5	2
I	7.0	6
J	4.0	4
K	1.2	1

e. Compute and interpret the coefficient of correlation r and the coefficient of determination r^2.
f. Test r^2 for significance at the .01 level.
g. What is your estimate of the average 1985 dollar sales, $\mu_{Y|X}$, to those counties that received a cash farm income in 1984 of $3 million? Use a .99 confidence interval.
h. What is your estimate of the company's 1985 dollar sales to Crosby County, a county not included in the sample but which had $3 million of cash farm income in 1984? Use a .99 confidence interval.

 14.18 The Witchita Control Corporation, a large manufacturer of custom-built control valves used by various public utilities, has utilized a rather haphazard pricing policy based on management's assessment of similar past experiences. In an effort to improve on its pricing methods, management sought to develop an estimating equation for determining the marginal cost of the last control valve produced for each order and to do so by using size of order as the independent variable. The ensuing investigation involved the use of the 20 most recent observations on size of order in units, X, and the marginal cost of the last unit produced for that order in dollars, Y. The worksheet results were as follows:

$$\Sigma Y = 608 \quad \Sigma Y^2 = 21{,}464 \quad \Sigma XY = 14{,}550$$
$$\Sigma X = 681 \quad \Sigma X^2 = 37{,}371 \quad n = 20$$

Assuming a linear relationship, do the following:
a. Derive the least squares estimating equation.
b. Interpret the meaning of the b value you obtained in part a.
c. Compute and interpret the standard error of the estimate s_e.
d. Compute and interpret the coefficient of correlation r and the coefficient of determination r^2.
e. Test r^2 for significance at the .01 level.
f. What is your estimate of the average marginal cost of the last unit, $\mu_{Y|X}$, associated with all orders of size 30? Use a .95 confidence interval.
g. What is your estimate of the marginal cost of the last unit associated with an order from Southwest Services, a company not included in the sample but which bought 30 valves? Use a .95 confidence interval.
h. Cite some possible practical implications of the results you obtained in parts a through g of this problem.

14.19 Management of Tot's Toyland, Inc., a large discount toy supermarket, conducted a study to determine the relationship between the number of people entering the store on each of ten randomly selected days, X, and the dollar value of sales, expressed in thousands of dollars, for the same ten days, Y. The worksheet results were as follows:

$$\Sigma Y = 157 \qquad \Sigma Y^2 = 2727 \qquad \Sigma XY = 19,700$$
$$\Sigma X = 1150 \qquad \Sigma X^2 = 143,500 \qquad n = 10$$

Assuming a linear relationship, do the following:
a. Derive the least squares estimating equation.
b. Interpret the meaning of the b value you obtained in part a.
c. Compute and interpret the standard error of the estimate s_e.
d. Compute and interpret the coefficient of correlation r and the coefficient of determination r^2.
e. Test r^2 for significance at the .05 level.
f. What is your estimate of the average amount of dollar sales $\mu_{Y|X}$ associated with days when 100 people entered the store? Use a .95 confidence interval.
g. What is your estimate of the amount of dollar sales on a Tuesday, two weeks ago, a day not included in the sample but which saw 100 people enter the store? Use of .95 confidence interval.
h. Cite some possible practical implications of the results you arrived at in parts a through g of this problem.

TAKE CHARGE

14.20 What is the best single predictor of a university student's grade-point average (GPA)? IQ? Grade-point average in high school? Performance in high school English or mathematics? SAT score? Motivation? Something else?

Select a single quantitative variable that you believe would serve as a good predictor of university GPA. Make certain that your predictor variable is one whose value will be known to the students you question (or can easily be obtained by them).

Select a random sample of 10 to 20 students from your statistics class—or other class with the instructor's permission—and secure from each (1) his or her GPA and (2) the value of the other variable you chose as a predictor of GPA.

Subject the results to a simple regression and correlation analysis and write a brief report on your findings. Don't fail to define in your report what you believe to be the population to which your *inductive conclusions* apply.

Computer Exercise

14.21 Many investors believe that the rate of growth of a company's earnings per share of common stock is a most important influencing factor on the rate of growth in that stock's market price.

The data set shown in Table 14.6 displays observations on a random sample of companies from Forbes' 36th Annual Report on American Industry (*Forbes Magazine*, January 2, 1984, pp. 265ff.) The company name is shown in the leftmost column. Values for the dependent variable are shown in the center column. This dependent variable is percent change in market price over the five-year period from 1979 to 1983. Values for the independent variable are shown in the rightmost column. These values represent average annual percent change in earnings per share over the same five-year period. Is average annual percent change in earnings per share a good predictor variable for percent change in market

Table 14.6 Data for Computer Exercise (Problem 14.21)

Company Name	Percent Change in Market Price (Y)	Average Annual Percent Change in Earnings per Share (X)
1. Roses' Stores	800.1	21.9
2. E. F. Hutton	641.5	30.3
3. Avnet	512.4	11.3
4. Sherwin-Williams	422.5	19.8
5. E.G. & G.	350.0	25.0
6. Associated Drygoods	312.2	7.5
7. Seagrams	302.7	21.6
8. Nordstrom	283.1	16.8
9. General Dynamics	267.1	25.5
10. Curtice-Burns	255.3	10.4
11. Borg-Warner	241.8	12.5
12. Albertson's	228.1	16.9
13. Masco	217.0	13.5
14. West Point-Pepperell	206.5	9.1
15. Ceco	201.1	16.8
16. Westinghouse Electric	196.2	15.5
17. Morrison-Knudsen	189.7	16.4
18. Weis Markets	176.3	15.6
19. Wyman-Gordon	170.0	15.2
20. Melville	162.6	16.3
21. Monfort of Colorado	158.7	19.0
22. Quaker Oats	154.6	12.2
23. BanCal Tri State	148.0	6.4
24. Clorox	140.7	9.7
25. Signal Cos.	137.2	6.2
26. LTV	132.1	7.1
27. May Department Stores	126.5	7.8
28. Handy and Harmon	123.7	6.8
29. Gordon Jewelry	120.6	12.0
30. Crown Central Petroleum	118.4	10.3
31. Union Pacific	107.8	13.5
32. Greyhound	105.7	9.8
33. Sun Company	102.9	12.2
34. ASARCO	98.2	15.1
35. U.S. Gypsum	92.3	10.1
36. Parker-Hannifin	91.3	6.8
37. American National Insurance	88.9	12.4
38. Bank of New York	87.2	9.8
39. W. W. Grainger	81.5	12.6
40. McDonnell Douglas	79.7	10.4
41. Interlake	77.7	−4.8
42. Consolidated Paper	73.6	12.9
43. Texas Instruments	67.0	−2.1
44. A. O. Smith	65.4	−17.7
45. Federal Paper Board	63.6	4.6
46. Louisiana Pacific	60.8	−5.2
47. Joy Manufacturing	58.8	6.2
48. Gates Learjet	57.4	6.5

(continued)

Table 14.6 *(Continued)*

Company Name	Percent Change in Market Price (*Y*)	Average Annual Percent Change in Earnings per Share (*X*)
49. Ogden	55.8	3.4
50. Stewart-Warner	53.6	−5.0
51. United Energy Resources	52.6	18.9
52. Amerace	50.0	−13.3
53. Cameron Iron Works	48.1	14.3
54. Washington National	46.8	6.0
55. Carter Hawley Hale	44.0	−2.2
56. Texas Utilities	41.7	8.0
57. Edison Brothers Stores	39.7	6.0
58. Copperweld	37.7	−11.9
59. Anderson-Clayton	35.8	6.0
60. Utah Power and Light	32.9	3.4
61. Dayco	31.5	−10.3
62. Columbia Gas Systems	28.4	6.4
63. Hawaii Electric	25.0	2.3
64. Oklahoma Gas and Electric	23.1	1.4
65. Diamond Shamrock	18.2	−6.1
66. Phelps-Dodge	16.1	−11.3
67. United Illuminating	12.1	4.7
68. UGI	8.8	2.7
69. Deere	5.8	−6.7
70. Southwest Forest Industries	0.0	−26.4
71. Amsted Industries	−3.5	3.0
72. Westmoreland Coal	−9.4	−55.5
73. Amdahl	−19.3	−14.1
74. Caterpillar Tractor	−27.2	−14.3
75. Avon Products	−54.7	3.4

Source: Adapted from *Forbes*, January 2, 1984, pp. 265ff.

price for the same period of time? Using a prepared computer program, subject these data to a simple linear regression and correlation analysis. Then write a brief report on your findings. If your findings are affirmative, tell how this information might prove useful to a long-term investor in common stocks.

Cases for Analysis and Discussion

14.1 "OCTAGONAL INDUSTRIES, INC.": PART I

The compensation officer of "Octagonal Industries" (OI) reported the following instructive case:

Approximately six months ago, a manufacturing department director in charge of some 393 salaried employees registered a complaint with the company's general manager. He argued that the company's failure to grant sufficiently high annual pay increases to outstanding performers was causing him to lose some of his most proficient people. At the time of the complaint, this manager offered little evidence to support his assertion. However, it was true that during recent months a dozen above-average performers had left the company, and the exit interview forms had turned up little useful information regarding their reasons for leaving.

The general manager established a study group consisting of the complainant, the director of industrial relations, the chief of finance, and myself. This study group was charged with the responsibility of determining whether the company had indeed been negligent in rewarding its better performers adequately.

Important to note is the fact that several years ago OI had established precise guidelines, applicable to every department, for rewarding salaried employees. These guidelines follow:

1. Award up to 10% annual salary increases as a performance reward for persons considered outstanding.
2. Compensation granted salaried employees should reflect information contained within the annual performance report.
3. An annual salary increase is not required; should performance records indicate the advisability of lengthening the time period between increases, this would be acceptable.
4. In cases of outstanding performance, the annual pay increase may be granted somewhat earlier than a time exactly 12 months after the previous increase.
5. Certain unusual levels of compensation require approval from corporation headquarters.

Before proceeding further, it is important to introduce and define some basic terms:

Performance appraisal: This is a written evaluation of an individual's performance prepared, usually on an annual basis, by that individual's immediate superior.

Rank list: This is a composite grouping of all salaried employees in a quality grade, the possible grades being (1) Outstanding, (2) Above Average, (3) Average, (4) Minimal Performance for Disposition, and (5) Below Average. The numbers 1 through 5 are used to indicate the quality grade of an employee. In addition, letter ranks of A, B, C, and D are attached to each numerical quality rank to indicate degrees of job difficulty, the most difficult jobs being assigned a rank of A. Thus, the rating suggesting the largest annual percent salary increase would be 1A and the rating suggesting the smallest annual percent salary increase would be 5D.

Position grade: This is a term applied to the *level* of salary grade to which a salaried employee is assigned for compensation and control. This particular case study is concerned with position grades 10, 9, 8, 7, 6, and 5. The 5 salary grade is the highest grade held under a manager in any specified department.

A sample of 96 salaried employees (roughly 24% of employees in the manufacturing division under discussion) was selected from the population of 393 salaried persons. I must admit to the possible presence of bias in our sampling technique. The time constraint imposed by higher management was terribly stringent; only two days were allowed for the preliminary investigation. As a result, "key areas" were weighted heavily in the sampling procedure. By "key areas" I mean those units of salaried employees that had experienced relatively high rates of attrition.

The following scatter diagrams [Figures 14.23, 14.24, and 14.25] show some of the results of the preliminary investigation. The first one [Figure 14.23] relates percent pay increases (the dependent variable) and performance rating. The second scatter diagram [Figure 14.24] relates percent pay increases and time since last pay increase. The third scatter diagram [Figure 14.25] relates percent pay increases to the position grades 5 through 10 mentioned above. (Our sample was devoid of people in both the lowest and the highest position grades.)

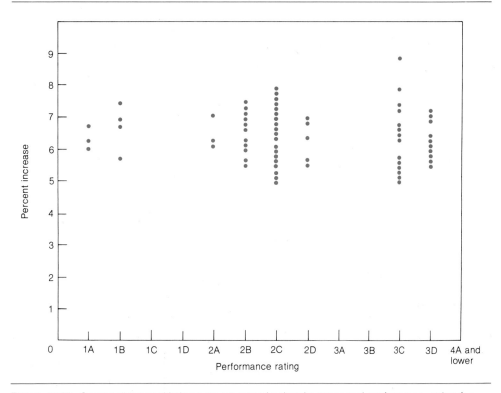

Figure 14.23. Scatter diagram relating percent annual salary increase and performance rating for Octagonal Industries, Inc.

1. Do you approve of the manner in which the scatter diagrams appearing in Figures 14.23, 14.24, and 14.25 are constructed? Why or why not?

2. Do the scatter diagrams appearing in Figures 14.23 through 14.25 suggest that the company guidelines for salaried-employee compensation are being carefully adhered to? Justify your answer.

3. The compensation officer concluded: "The exodus of 12 gifted employees during the past few months might very possibly be related to inadequate compensation." Judging from the evidence presented, do you agree with this conclusion? Why or why not?

This case is continued at the end of Chapter 15.

14.2 "UNIVERSAL DEVICES, INC."

A manager of the Industrial Engineering Division of "Universal Devices'" Sun City plant wrote the following case report concerning indirect workload analysis:

Within the various divisions of Universal Devices, different definitions and methods of measuring workload are used. The purpose of the present report is to illustrate a statistical technique which has been used by my company for several years to analyze the indirect workload within a plant and to provide a basis for comparing the indirect workload of a

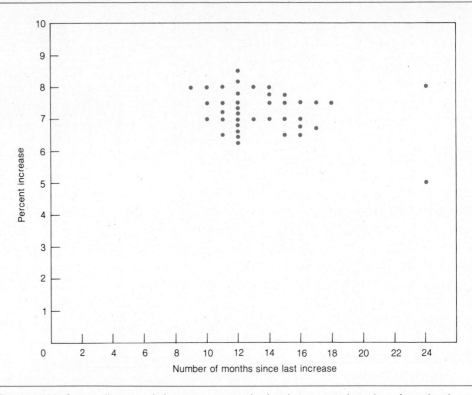

Figure 14.24. Scatter diagram relating percent annual salary increase and number of months since last increase for Octagonal Industries, Inc.

plant with that of other plants within the company. This analysis, if used as a management tool, enables plant and corporate management to evaluate the distribution and potential effectiveness of its indirect labor work force. This technique is called *Indirect Workload Analysis.* To simplify the description of this analytical technique, I will use my own division, the Industrial Engineering Division, as an example. However, the same basic technique is used in connection with plant administration, personnel, finance, plant engineering, information systems, production control, procurement, manufacturing engineering, and quality evaluation.

To be most useful, Indirect Workload Analysis must permit comparisons of workloads of indirect areas, like Industrial Engineering, among similar plant locations. Conceptually, only if the Industrial Engineering groups are identical can differences among workloads be attributed to plant size, special conditions, and differences in efficiencies.

To establish a commonality of functional definition, a series of what has been called *model activities* has been defined. Each of these activities is a fundamental package of work that can be given a definition that is commonly and consistently applicable to all locations. Each plant is required to define its work in terms of an appropriate model-activity category.

For each model activity, there may be several factors that control the amount of workload the plant must devote to the accomplishment of the activity. However, at any given location, usually one factor, or weighted combination of factors, is the major cause of work.

This same factor can be found at other sites and is generally the major work-causing factor at all of them. This is an independent variable which we call a *modifier*. Each model activity has a corresponding work-causing modifier. An example of the model activity and its modifier for Industrial Engineering is shown below [Table 14.7].

The combination of the model activity and the modifier is a model of that activity as it is performed at a specific location. Because this model of the activity is commonly defined for all locations, the activities at each of several locations can legitimately be compared. Since the most significant cost for these activities has been found to be the salaries of the personnel involved and since material costs are already well controlled, the cost that is measured is that of workload. However, before any comparison can be made, the nature of the effect of the modifier must be determined.

Ideally, the relationship between the activity and the work-causing factor (the modifier) would be expressed as

Workload at a given plant = Constant × Modifier for plant

In actual practice, if a graph of the workload activity were plotted as a function of its modifier, it would generally appear as in the following graph [Figure 14.26], where each of the plotted points represents the workload and modifier factor at a specified plant. Thus, the line that best fits the data must be found. The least squares straight line is shown [in this same figure].

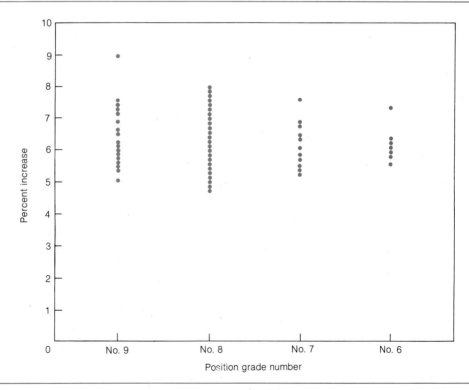

Figure 14.25. Scatter diagram relating percent annual salary increase and position grade number for Octagonal Industries, Inc.

Table 14.7 Examples of Model Activities and Modifiers for the Industrial Engineering Function of "Universal Devices, Inc."

Model Activity	Modifier
1. Cost estimating—total product cost	Dollar value added per million
2. Procurement cost estimating	Dollar value of purchased parts per million
3. Direct labor controls	EOP inplant direct manufacturing workload per million
4. Direct manpower planning	EOP total direct manufacturing workload per million
5. Indirect manpower planning	Inplant, indirect manpower per million
6. Space and facility planning	Gross outside square feet per 100 million
7. Facility layout	Gross outside square feet per 100 million
8. Material handling engineering	Inventory dollars per million
9. Special studies and advertising	Total-plant regular manpower per million

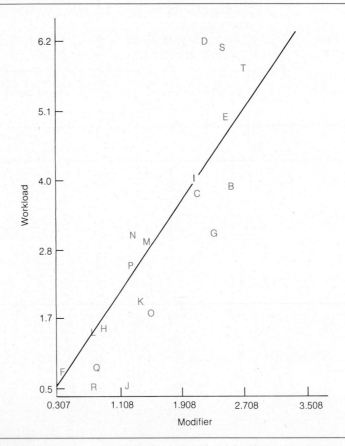

Figure 14.26. Scatter diagram for model activity, indirect manpower planning, and modifier, inplant indirect manpower per thousand for the industrial engineering function at 19 Universal Devices plants.

14 Simple Linear Regression and Correlation Analysis

At Universal Devices we have developed acceptable levels of correlation for various sample sizes. A sample size is the number of plants reporting on workload activity. These guidelines are displayed in the following table [Table 14.8]. It can be seen that the smaller the sample size, the higher the coefficient of correlation must be for acceptance.

From the linear regression function, a workload index is calculated for each plant location. The workload index is the ratio of the measured workload and the workload estimated by the least squares estimating equation. It is the basic element used for comparison tracking and assessment. An index of 1.0 indicates that the subject plant has an average workload for that activity at that modifier quantity. If the plant's index is, say, .8, this means that the plant is performing the activity with 20% less manpower than the average or "model" plant. Naturally, if a plant has an index of 1.50 for an activity, it should be seeking ways of improving its effectiveness in that activity.

How are these data used by management at Universal Devices? The division or functional manager of each group receives a summary of all his workload activities. In the Industrial Engineering case, the manager would receive a graph and a report listing the group's index for each activity and the group's mean overall effectiveness. If the overall effectiveness is equal to or less than 1 for all workload activities, then this is an indication that the Industrial Engineering function has allocated its manpower reasonably well as compared to the model Industrial Engineering function as calculated by the regression technique. If the overall index is greater than 1, then the manager would use graphs to indicate which locations are more effective than his. He usually makes contact with his counterparts at the more efficient plants. It is this analysis-and-communications exchange that often brings the potential benefits of the Indirect Workload Analysis to fruition.

The graphs and indexes can also be used for future projection of workload. For example, if several new buildings are planned in the future, the manager can use the planned gross outside square footage of these buildings to determine, at a minimum, how many industrial engineers are required for the space-planning and facilities-layout activities.

Table 14.8 Schedule of Minimum Acceptable Coefficients of Correlation for Specified Sample Sizes

Sample Size	Good	Fair
3	.997	.988
4	.950	.900
5	.878	.805
6	.811	.729
7	.755	.669
8	.707	.622
9	.666	.582
10	.632	.549
11	.602	.521
12	.576	.497
13	.553	.476
14	.531	.458
15	.514	.441
16	.497	.426
17	.482	.412
18	.468	.400
19	.456	.388
20	.444	.378

1. Some of the advantages of Universal Devices' Indirect-Workload-Analysis procedures were mentioned in this case. What do you see as some of the possible disadvantages, or limitations, if any, of this approach? Be specific.

2. Can you think of any management uses for this kind of analysis in addition to those already cited? What are they?

3. Explain in your own words your understanding of how coefficients of correlation are used in the kind of analysis described in this case.

15 Multiple Linear Regression and Correlation Analysis

The real trouble with this world of ours is not that it is an unreasonable one. The trouble is that it is nearly reasonable but not quite.—G. K. Chesterton

In any series of elements to be controlled, a selected small fraction, in terms of numbers of elements, always account for a large fraction in terms of effect.—Pareto

---

What You Should Learn from This Chapter

In this chapter, we push the methods introduced in Chapter 14 several steps further by showing how to apply regression and correlation methods to situations involving more than one independent variable. When you have finished this chapter, you should be able to

1. Derive the estimating equation for an analysis involving three variables and linear relationships.
2. Compute the standard error of the estimate and the coefficient of determination for an analysis involving three variables and linear relationships.
3. Compute and interpret coefficients of partial determination for an analysis involving three variables and linear relationships.
4. Recite the assumptions underlying inductive methods in multiple regression and correlation analysis.
5. Construct confidence intervals for (1) the population slopes, (2) the subpopulation mean of Y given X_1 and X_2, and (3) a subpopulation Y_{NEW} given X_1 and X_2.
6. Test hypotheses about (1) the population slopes, (2) the population coefficient of multiple determination, and (3) the population coefficients of partial determination.
7. Recite the hazards uniquely associated with multiple regression and correlation analysis.
8. Interpret a computer printout for a problem utilizing one dependent variable and more than two independent variables.
9. Demonstrate an awareness of how to deal with situations calling for use of a qualitative independent variable.
* 10. Demonstrate a familiarity with computer methods commonly used to select the "best" set of independent variables.

This chapter presupposes that you are familiar with the contents of Chapter 14.

15.1 Introduction

When examining the Granby Specialties Company problem (Illustrative Case 14.1), we studied the effect of only one independent variable—namely, number of households. However, we were told that two other variables, average price and advertising expenditures, were also thought to exert an influence on Granby's sales. It seems reasonable to assume that three independent variables used in combination could lead to more accurate estimates of sales than could a single independent variable. Granted, such a result may not be inevitable, but we will find it is common. After all, we live in a world of interrelated parts.

Table 15.1 presents the data we will be analyzing.

15.2 Some Fundamentals of Three-Variable Regression and Correlation Analysis

We will begin our treatment of multiple regression and correlation analysis by ignoring variable X_3, advertising expenditures, and showing how the analysis would be conducted if the effects of number of households, X_1, and average price charged, X_2, were the only ones to be analyzed. Later we will make use of all three independent variables.

> The regression equation we seek will be of the form
> $$\hat{Y} = a + b_1X_1 + b_2X_2 \qquad (15.2.1)$$

Just as we previously relied on the least squares method to provide us with the best fitting two-variable linear regression equation, we now rely on this same method to provide us with the best fitting three-variable linear regression equation of the form $\hat{Y} = a + b_1X_1 + b_2X_2$.

Table 15.1 Granby Company Sales Y (Dependent Variable) and Number of Households X_1, Average Price X_2, and Advertising Expenditures X_3 (Independent Variables) Organized by Sales Territory

Sales Territory	Y Sales Achieved Last Year (Tens of Thousands of Units)	X_1 Number of Households in Territory Last Year (Thousands)	X_2 Weighted Average Retail Price Charged for Company's Products Last Year (Dollars per Unit)	X_3 Advertising Expenditures Last Year (Thousands of Dollars)
1	20	21	1.92	4
2	1	7	2.75	2
3	14	19	2.00	4
4	11	12	1.94	7
5	17	18	2.02	6·
6	4	3	2.50	5
7	23	17	1.60	7
8	10	9	1.95	4
9	12	7	1.76	5
10	6	12	2.25	2
Total	118	125	20.69	46

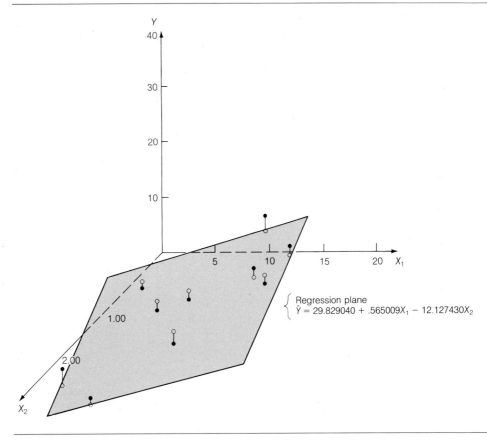

Figure 15.1. Multiple regression surface for the three-variable version of the Granby Company problem. Actual Y values are shown with black circles and $\hat{Y}$ values with white circles.

In multiple regression analysis, the least squares method gives the investigator an equation describing the best fitting *plane* rather than the best fitting line for the data under analysis. In a three-variable regression problem, the points can be plotted in three dimensions, as shown in Figure 15.1. The best fitting plane would pass through the points in such a way that $\Sigma (Y - \hat{Y}) = 0$ and $\Sigma (Y - \hat{Y})^2$ is a minimum.

The a value in the equation $\hat{Y} = a + b_1 X_1 + b_2 X_2$ represents the value of $\hat{Y}$ where the regression plane cuts the Y axis so that we may still refer to it as the Y intercept—in theory, the value of $\hat{Y}$ when $X_1 = 0$ and $X_2 = 0$.

The b_1 and b_2 values are called *net regression coefficients* or *partial regression coefficients* and are interpreted somewhat differently from the b value in simple linear regression analysis. Net regression coefficient b_1 represents the change in $\hat{Y}$ associated with a one-unit increase in X_1, holding X_2 constant. Net regression coefficient b_2

represents the change in $\hat{Y}$ associated with a one-unit increase in X_2, holding X_1 constant.

When we speak of *holding* an independent variable constant in the present context, we mean that we are eliminating mathematically from Y any variation associated with changes in that particular independent variable. This idea is illustrated in Figure 15.2.

To understand the meaning of "holding a variable constant," we can draw an analogy with a chemist who performs an experiment but is handicapped at first by the presence of impurities in his test tube. Naturally, he is not certain whether the results he has been getting are identical with those he would achieve in the absence of the impurities. However, if he is able to remove one of the impurities (hold it constant), he has some reason to believe that the next result he achieves will be closer to the truth than the previous one. The more impurities

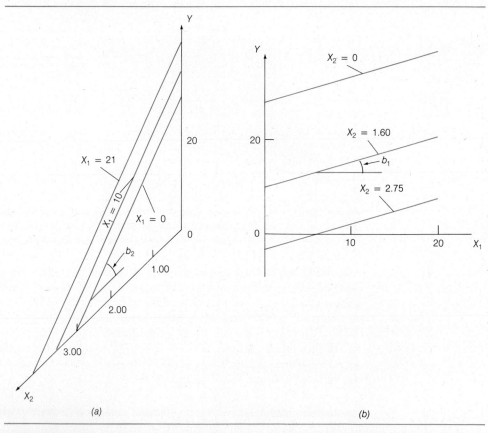

(a)   (b)

Figure 15.2. (a) Concept of net or partial regression. Variable X_1 is held constant at the indicated levels, while variables Y and X_2 vary freely. The slope b_2 is the same regardless of the level of X_1. (b) Another illustration of the concept of net or partial regression. X_2 is held constant at the indicated levels, while variables Y and X_1 vary freely. The slope b_1 is the same regardless of the level of X_2.

he removes (the more variables he holds constant), the better he feels about the probable validity of his findings. To be sure, business researchers and social scientists will always suffer from dirty test tubes, so to speak; the phenomena they study are influenced by a great many, largely uncontrollable, variables. Nevertheless, our ability to hold variables constant *statistically* represents an important step in the direction of removing impurities from our investigations.

15.3 Determining the Least Squares Estimating Equation for a Three-Variable Problem

As with simple regression analysis, the constants in the least squares estimating equation can be determined through use of a set of normal equations. When three variables are included in the analysis, the appropriate normal equations are

$$\Sigma Y = na + b_1 \Sigma X_1 \qquad + b_2 \Sigma X_2 \qquad (15.3.1)$$

$$\Sigma X_1 Y = a \Sigma X_1 + b_1 \Sigma X_1^2 \qquad + b_2 \Sigma X_1 X_2 \qquad (15.3.2)$$

$$\Sigma X_2 Y = a \Sigma X_2 + b_1 \Sigma X_1 X_2 + b_2 \Sigma X_2^2 \qquad (15.3.3)$$

Note: The rectangle constructed of dashed lines serves to call attention to the fact that the normal equations for simple linear regression are part of the larger system of normal equations for three-variable linear regression.

To apply these equations, we begin with a worksheet like the one shown in Table 15.2 where the Y, X_1, and X_2 values are shown along with their squares and the products obtained by multiplying each Y value by the corresponding X_1 value, each Y value by the corresponding X_2 value, and each X_1 value by the corresponding X_2 value. The $X_1 Y$, $X_2 Y$, and $X_1 X_2$ values are called *cross products*.

The three normal equations, with appropriate column totals from Table 15.2 included, are as follows:

$$118 = 10a + 125b_1 + 20.69b_2$$
$$1780 = 125a + 1891b_1 + 248.78b_2$$
$$225.75 = 20.69a + 248.78b_1 + 43.8655b_2$$

Solving these equations simultaneously for a, b_1, and b_2, we get

$$a = 29.829040$$
$$b_1 = .565009$$
$$b_2 = -12.127430$$

Therefore, the estimating equation is $\hat{Y} = 29.829040 + .565009X_1 - 12.127430X_2$.

Table 15.2 Worksheet Used to Compute the Least Squares a, b_1, and b_2 Values, and Multiple and Partial Regression and Correlation Measures for the Granby Specialties Company (Three-Variable) Problem

Sales Territory	Y Sales	X_1 Households	X_2 Price	X_1Y	X_2Y	X_1X_2	X_1^2	X_2^2	Y^2
1	20	21	1.92	420	38.40	40.32	441	3.6864	400
2	1	7	2.75	7	2.75	19.25	49	7.5625	1
3	14	19	2.00	266	28.00	38.00	361	4.0000	196
4	11	12	1.94	132	21.34	23.28	144	3.7636	121
5	17	18	2.02	306	34.34	36.36	324	4.0804	289
6	4	3	2.50	12	10.00	7.50	9	6.2500	16
7	23	17	1.60	391	36.80	27.20	289	2.5600	529
8	10	9	1.95	90	19.50	17.55	81	3.8025	100
9	12	7	1.76	84	21.12	12.32	49	3.0976	144
10	6	12	2.25	72	13.50	27.00	144	5.0625	36
Total	118	125	20.69	1780	225.75	248.78	1891	43.8655	1832

$$\Sigma Y = 118 \qquad \Sigma X_1Y = 1780 \qquad \Sigma X_1^2 = 1891$$
$$\Sigma X_1 = 125 \qquad \Sigma X_2Y = 225.75 \qquad \Sigma X_2^2 = 43.8655$$
$$\Sigma X_2 = 20.69 \qquad \Sigma X_1X_2 = 248.78 \qquad \Sigma Y^2 = 1832$$
$$n = 10$$

Shortcut Method

When calculation of a, b_1, and b_2 must be done manually, investigators can spare themselves a substantial amount of time and effort (and, usually, much of the anguish that comes from making minor errors along the way and not discovering that unfortunate fact until near the end of the work) by using a shortcut method that calls for the substitution of y, x_1, and x_2 for Y, X_1 and X_2, respectively, where (1) y is defined as $Y - \overline{Y}$, (2) x_1 is defined as $X_1 - \overline{X}_1$, and (3) x_2 is defined as $X_2 - \overline{X}_2$. That is, instead of using the original values for the variables under study, we use, in effect, numbers representing the differences between these values and their respective arithmetic means.

The advantage of this approach is that, because the sum of the deviations of a variable from its arithmetic mean will always be 0, one entire normal equation can be eliminated as well as parts of the remaining two. The normal equations with which we work using this approach are

$$\Sigma x_1y = b_1 \Sigma x_1^2 + b_2 \Sigma x_1x_2 \qquad (15.3.4)$$
$$\Sigma x_2y = b_1 \Sigma x_1x_2 + b_2 \Sigma x_2^2 \qquad (15.3.5)$$

Solving the two equations will provide values for the b_1 and b_2 constants but not for the a value. This constant can be obtained readily by using the following equation:

$$a = \overline{Y} - b_1\overline{X}_1 - b_2\overline{X}_2 \qquad (15.3.6)$$

Another nice feature of this approach is that we need not actually figure out the y, x_1, and x_2 values. We can determine all the sums needed for Equations (15.3.4) and (15.3.5), without calculating a single deviation, by making use of the following equations:

$$\sum y^2 = \sum Y^2 - n\overline{Y}^2$$
$$\sum x_1^2 = \sum X_1^2 - n\overline{X}_1^2$$
$$\sum x_2^2 = \sum X_2^2 - n\overline{X}_2^2$$
$$\sum x_1 y = \sum X_1 Y - n\overline{X}_1\overline{Y} \qquad (15.3.7)$$
$$\sum x_2 y = \sum X_2 Y - n\overline{X}_2\overline{Y}$$
$$\sum x_1 x_2 = \sum X_1 X_2 - n\overline{X}_1\overline{X}_2$$

Substituting the appropriate values from Table 15.2 into these expressions, we obtain the following results:

$$\sum y^2 = \sum Y^2 - n\overline{Y}^2 = 1832 - 1392.4 = 439.6$$
$$\sum x_1^2 = \sum X_1^2 - n\overline{X}_1^2 = 1891 - 1562.5 = 328.5$$
$$\sum x_2^2 = \sum X_2^2 - n\overline{X}_2^2 = 43.8655 - 42.80761 = 1.05789$$
$$\sum x_1 y = \sum X_1 Y - n\overline{X}_1\overline{Y} = 1780 - 1475 = 305$$
$$\sum x_2 y = \sum X_2 Y - n\overline{X}_2\overline{Y} = 225.75 - 244.142 = -18.392$$
$$\sum x_1 x_2 = \sum X_1 X_2 - n\overline{X}_1\overline{X}_2 = 248.78 - 258.625 = -9.845$$

Placing these results into the transformed normal equations, Equations (15.3.4) and (15.3.5), we have

$$305 = 328.5b_1 - 9.845b_2$$
$$-18.392 = -9.845b_1 + 1.05789b_2$$

By solving the equations, we obtain

$$b_1 = .565009$$
$$b_2 = -12.127430$$

Using Equation (15.3.6), we determine a to be

$$a = 11.8 - (.565009)(12.5) - (-12.127430)(2.069) = 29.829040$$

Therefore, for this problem, the least squares estimating equation of the form $\hat{Y} = a + b_1 X_1 + b_2 X_2$ is

$$\hat{Y} = 29.829040 + .565009X_1 - 12.127430X_2$$

which is exactly the same as that obtained using the longer method.

If the sales manager wished to estimate sales for a territory with, say, 12,000 ($X_1 = 12$) households and with an average price charged for Granby's products of, say, \$1.94 per unit, she could simply substitute the indicated values into the estimating equation in the following manner:

$$\hat{Y} = 29.829040 + (.565009)(12) - (12.127430)(1.94) = 13.082, \text{ or } 130,820 \text{ units}$$

We may interpret the value of $b_1 = .565009$ as follows: For each increase of 1000 households, there will be, on the average, an increase in sales of 5650 units when the price charged per unit is held constant. The value of $b_2 = -12.127430$ is interpreted similarly. For each *increase* of \$1 in average price charged per unit, there will be, on the average, a *decrease* in sales of 121,274.3 units when the number of households is held constant. At first glance, price appears to exert a considerable influence on sales. However, we must be aware that a \$1 per unit increase in price would amount to an increase in the neighborhood of 50% per unit, a rather large percent change for items which, we are informed, have traditionally been low-priced. Therefore, the large b_2 value, while still somewhat surprising, is not necessarily a persuasive reason for doubting either the accuracy of the calculations or the plausibility of the findings. Reassuring is the fact that the sign associated with b_1 is positive, indicating that the greater the number of households, the greater the quantity of unit sales, and that the sign of b_2 is negative, indicating that higher prices are detrimental and lower prices beneficial to sales. Both signs are in accord with what an intuitive analysis would suggest.

A question which might reasonably be raised at this juncture is: How accurate are the $\hat{Y}$ values in their role of providing estimates of the actual sample Y values? More specifically, do the $\hat{Y}$ values resulting from this three-variable analysis provide more accurate estimates of Y than were obtained from the two-variable equation? Let us seek the answer by comparing standard errors of the estimate.

15.4 The Standard Error of the Estimate

We saw in connection with two-variable regression analysis that a convenient measure of the amount of variation around the line of regression is the standard error of the estimate s_e. This measure can also be used, with slight modification in the method of computing it, to summarize the amount of variation around the $\hat{Y}$ values in multiple regression analysis.

For a three-variable regression analysis, the standard error of the estimate of the sample data will be denoted S_e. It can be calculated by

$$S_e = \sqrt{\frac{\Sigma (Y - \hat{Y})^2}{n - 3}} \qquad (15.4.1)$$

where it is understood that the $\hat{Y}$ values were obtained from an equation of the form $\hat{Y} = a + b_1X_1 + b_2X_2$.

(Notice the use of S in place of s in Equation (15.4.1). We will use a lowercase s in connection with simple regression analysis and a capital S in connection with multiple regression analysis.)

The denominator $n - 3$ is used to comply with the rule, presented in section 14.5 of the preceding chapter, that in regression analysis the degrees of freedom are equal to $n - \lambda$, where n is the sample size and λ is the number of constants in the estimating equation. For the problem under discussion, we compute (using data shown in Table 15.3):

$$S_e = \sqrt{\frac{\Sigma\,(Y - \hat{Y})^2}{n - 3}} = \sqrt{\frac{44.1994}{7}} = 2.51, \text{ or } 25,100 \text{ units}$$

The value 2.51 is encouragingly smaller than the 4.42 obtained in section 14.5 of the preceding chapter.

In practice, the standard error of the estimate, S_e, can be determined with less computational time and effort when the following equation is used in lieu of Equation (15.4.1):

$$S_e = \sqrt{\frac{\Sigma\, y^2 - b_1 \Sigma\, x_1 y - b_2 \Sigma\, x_2 y}{n - 3}} \qquad (15.4.2)$$

The virtue of this equation is that it does not require the calculation of the $\hat{Y}$ value associated with every pair of X_1 and X_2 values. Substituting known values into this equation, we get

$$S_e = \sqrt{\frac{439.6 - (.565009)(305) - (-12.127430)(-18.392)}{10 - 3}} = 2.51$$

the same result as was obtained from use of Equation (15.4.1).

Table 15.3 Comparison of Y and $\hat{Y}$ Values and Presentation of Their Differences and Squared Differences for the Three-Variable Version of the Granby Specialties Company Problem

	(1) Y	(2) $\hat{Y}$	(3) $Y - \hat{Y}$	(4) $(Y - \hat{Y})^2$
	20	18.41	1.59	2.5281
	1	.43	.57	.3249
	14	16.31	-2.31	5.3361
	11	13.08	-2.08	4.3264
	17	15.50	1.50	2.2500
	4	1.21	2.79	7.7841
	23	20.03	2.97	8.8209
	10	11.27	-1.27	1.6129
	12	12.44	-.44	.1936
	6	9.32	-3.32	11.0224
Total	118	118.00	0.00	44.1994

15.5 Multiple Determination

You will recall that in Chapter 14 we measured the strength of linear association between two variables by means of the coefficient of (simple) determination, denoted r^2 for sample data. The r^2 value was defined as the ratio of the amount of variation in Y "explained by" variation in X to the total variation in Y. That is, $r^2 = \Sigma (\hat{Y} - \overline{Y})^2 / \Sigma (Y - \overline{Y})^2$. The more variation explained, that is, the closer this ratio is to 1, the stronger the relationship is.

We may use a similar measure when the analysis involves a second independent variable. This measure for a problem involving variables Y, X_1, and X_2 is denoted $R^2_{Y.12}$ and called the *coefficient of multiple determination*. The subscript Y.12 indicates which and how many variables are being used in the analysis. The coefficient of multiple determination, $R^2_{Y.12}$, then, is defined as the ratio of the amount of variation in Y "explained by" variation in variables X_1 *and* X_2 *together* to the total amount of variation in Y. That is,

$$R^2_{Y.12} = \frac{\text{Explained variation}}{\text{Total variation}} = \frac{\Sigma (\hat{Y} - \overline{Y})^2}{\Sigma (Y - \overline{Y})^2} \qquad (15.5.1)$$

The calculations could be done in literal accord with Equation (15.5.1) but with an investment of more computational time than is really necessary. The same results can be achieved more easily from the following alternative equation:

$$R^2_{Y.12} = \frac{b_1 \Sigma x_1 y + b_2 \Sigma x_2 y}{\Sigma y^2} \qquad (15.5.2)$$

The coefficient of multiple determination for the Granby Specialties Company problem is

$$R^2_{Y.12} = \frac{(.565009)(305) + (-12.127430)(-18.392)}{439.6} = .899399$$

Therefore, for the observed sample data, approximately 90% of the variation in sales is "explained by" variation in number of households and price in combination, a value noticeably greater than the 64% obtained earlier using number of households as the sole independent variable.

After allowing for degrees of freedom lost, this measure becomes

$$\overline{R}^2_{Y.12} = 1 - (1 - R^2_{Y.12})\left(\frac{n-1}{n-3}\right) = 1 - (1 - .899399)(1.285714) = .870656$$

For a given sample size n, this adjustment becomes increasingly important as more independent variables are added to the analysis.

15.6 Coefficients of Partial Determination

With the exception of the net regression coefficients, thus far in this chapter we have treated only measures that describe certain aspects of the relationship among variables Y, X_1, and X_2 in an *overall* manner—measures which tell us how much influence X_1 and X_2 *in combination* have upon Y. In this section we continue with the subject of measuring strength of association, but our focus now shifts to the task of *measuring the strength of association between the dependent variable and a specific independent variable when the other independent variable is left out of consideration (i.e., is held constant).*

A convenient place to start is with the calculation of the coefficients of simple determination r_{Y1}^2 and r_{Y2}^2, where r_{Y1}^2 represents the coefficient of simple determination between variables Y and X_1, and r_{Y2}^2 represents the same thing for variables Y and X_2. This can be accomplished rather quickly by utilizing the following slightly modified versions of the product-moment formula:

$$r_{Y1}^2 = \frac{(\Sigma\, x_1 y)^2}{(\Sigma\, x_1^2)(\Sigma\, y^2)} \qquad (15.6.1)$$

$$r_{Y2}^2 = \frac{(\Sigma\, x_2 y)^2}{(\Sigma\, x_2^2)(\Sigma\, y^2)} \qquad (15.6.2)$$

Applying these equations, we get

$$r_{Y1}^2 = \frac{(305)^2}{(328.5)(439.6)} = .644179$$

(This result should look familiar; it was computed in Chapter 14 and denoted simply r^2.)

$$r_{Y2}^2 = \frac{(-18.392)^2}{(1.05789)(439.6)} = .727377$$

The coefficient of partial determination for variables Y and X_2 is denoted $r_{Y2.1}^2$ and interpreted as follows: It is the proportion of variation in Y which was left unexplained after variable X_1 has done all it can do to explain the variation in Y but which is now explained by variation in variable X_2 (see Figure 15.3). Less cumbersome, though also less precise, is the following interpretation: $r_{Y2.1}^2$ is the proportion of variation in Y "explained by" variation in X_2 when X_1 is held constant. This measure is computed by

$$r_{Y2.1}^2 = \frac{R_{Y.12}^2 - r_{Y1}^2}{1 - r_{Y1}^2} \qquad (15.6.3)$$

Similarly, the coefficient of partial determination for variables Y and X_1 is denoted $r_{Y1.2}^2$ and interpreted as follows: The proportion of variation in Y "explained by" variation in X_1 when X_2 is held constant. The equation for this measure is

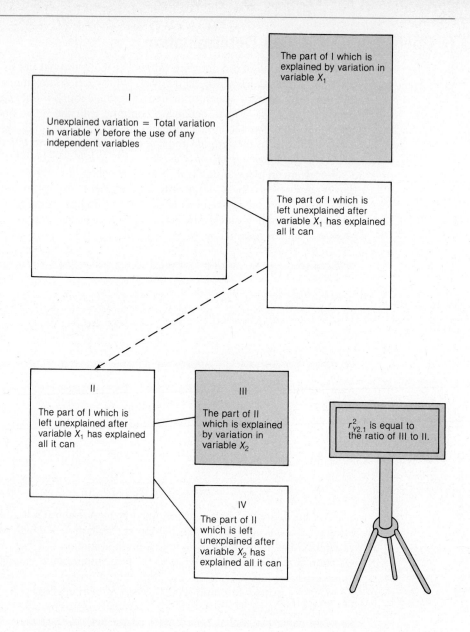

Figure 15.3. Rationale behind coefficient of partial determination, $r^2_{Y2.1}$.

$$r_{Y1.2}^2 = \frac{R_{Y.12}^2 - r_{Y2}^2}{1 - r_{Y2}^2} \qquad (15.6.4)$$

Applying these equations to data from our illustrative case, we get

$$r_{Y2.1}^2 = \frac{.899399 - .644179}{1 - .644179} = .717271$$

$$r_{Y1.2}^2 = \frac{.899299 - .727377}{1 - .727377} = .630622$$

Thus, we say (1) about 71.7% of the variation in sales is "explained by" variation in price when number of households is held constant and (2) about 63.1% of the variation in sales is "explained by" variation in number of households when the price charged is held constant.

It appears that variable X_2, average price charged, is the more important independent variable according to the criterion of strength of association. This variable explains somewhat more of the original variation in Y than does variable X_1 ($r_{Y2}^2 = .727377$ versus $r_{Y1}^2 = .644179$) and somewhat more of the residual variation as well ($r_{Y1.2}^2 = .630989$ versus $r_{Y2.1}^2 = .717271$). However, the apparent superiority of X_2 should not be overemphasized in the present case since both variables make substantial contributions to the explanation of the variation in sales.

Of course, if the coefficients of partial *correlation* were desired, we would simply compute the square roots of the coefficients determined above and attach the same sign as that possessed by the corresponding *b* value. Hence,

$$r_{Y2.1} = \sqrt{.717271} = -.846919 \qquad r_{Y1.2} = \sqrt{.630989} = +.794117$$

15.7 Introduction to Inductive Procedures in Three-Variable Multiple Regression and Correlation Analysis

For three-variable problems, the population estimating equation will be written

$$\mu_{Y|12} = A + B_1X_1 + B_2X_2 \qquad (15.7.1)$$

where $\mu_{Y|12}$ is interpreted as the subpopulation mean Y value associated with specified values of variables X_1 and X_2, that is, it is the population counterpart of $\hat{Y}$ ($\mu_{Y|12}$ is read "mu sub Y given X_1 and X_2"—not "given twelve"; X's will not hereafter be used in subscripts)

A is the population Y intercept

B_1 is the population slope relating Y and X_1 (X_2 held constant)

B_2 is the population slope relating Y and X_2 (X_1 held constant)

σ_e is the population standard error of the estimate

$\rho_{Y|12}^2$ will be used to denote the population coefficient of multiple determination

The assumptions one tacitly—though, one hopes, not uncritically—accepts when using inductive methods in multiple linear regression and correlation analysis are similar to those of two-variable analysis.

Multiple Regression

The assumptions underlying inductive procedures in three-variable linear regression are as follows:

1. Associated with each combination of X_1 and X_2 values there is a subpopulation of Y values.
2. The means of the subpopulations of Y values are related to the X_1 and X_2 values by a regression plane of the form $\mu_{Y|12} = A + B_1X_1 + B_2X_2$.
3. The Y values are statistically independent.
4. Each subpopulation of Y values is normally distributed around $\mu_{Y|12}$.
5. The variances of the subpopulations of Y values are all equal regardless of the X_1 and X_2 values specified.

Thus, each population Y value is expressible as

$$Y_i = A + B_1X_1 + B_2X_2 + \epsilon_i \qquad (15.7.2)$$

where ϵ_i: NID(0, σ_e^2)

Multiple Correlation

In multiple correlation analysis, the underlying assumptions are painfully restrictive. The model assumed in three-variable multiple correlation analysis is the same as that for three-variable regression analysis, but with the proviso that variables X_1 and X_2, as well as Y, must be random (not assigned by the investigator). Moreover, the *joint distribution* of the variables is assumed to be normal. Such a distribution is called a *multivariate normal distribution* and is like the bivariate normal distribution mentioned in Chapter 14 but with one more dimension in the three-variable case.

15.8 Gauss Multipliers

To calculate the standard errors we use in the next three sections, we will need the use of measures called *Gauss multipliers*. Gauss multipliers have their basis in matrix inversion but can also be obtained through use of some simple algebra. The latter approach is the one shown here. To determine Gauss multipliers for the present example, we begin by creating two pairs of equations as follows:

$$c_{11} \, \Sigma \, x_1^2 + c_{12} \, \Sigma \, x_1 x_2 = 1$$
$$c_{11} \, \Sigma \, x_1 x_2 + c_{12} \, \Sigma \, x_2^2 = 0$$

$$(15.8.1)$$

$$c_{21} \, \Sigma \, x_1^2 + c_{22} \, \Sigma \, x_1 x_2 = 0$$
$$c_{21} \, \Sigma \, x_1 x_2 + c_{22} \, \Sigma \, x_2^2 = 1$$

$$(15.8.2)$$

These are merely the two normal equations set up in accordance with the shortcut method but with (1) c_{ij} values substituted for b values and (2) 1 and 0 substituted for the $\Sigma \, x_1 y$ and $\Sigma \, x_2 y$ terms. The c_{ij} values in the first pair of equations are those with a subscript beginning with "1." The second digits of the subscripts run from small to large as we scan the equations from left to right. The c_{ij} values in the second pair of equations are those with a subscript beginning with "2." Again, the second digits in the subscripts run from small on the left-hand side to large on the right-hand side.

Determining where to place the 1 and 0 within a pair of equations like (15.8.1) or (15.8.2) is a matter of noting whether the equation contains a term including (1) $\Sigma \, x_i^2$ and (2) a subscript with identical digits, that is, 11 or 22. For example, in the first pair of equations, (15.8.1), $\Sigma \, x_1^2$ and c_{11} appear together in the first equation of this pair. Therefore, that equation is set equal to 1. In the second pair of equations, (15.8.2), $\Sigma \, x_2^2$ and c_{22} appear together in the second equation of this pair, and that equation is therefore set equal to 1. The other equations are set equal to 0.

Let us now solve for Gauss multipliers c_{11}, c_{12}, c_{21}, and c_{22} for our specific problem. Using Equation (15.8.1) first, we have

$$328.5c_{11} - 9.845c_{12} = 1$$
$$-9.845c_{11} + 1.05789c_{12} = 0$$

Solving these simultaneously, we get

$$c_{11} = .004222$$
$$c_{12} = .039287$$

Turning to Equations (15.8.2), we note

$$328.5c_{21} - 9.845c_{22} = 0$$
$$-9.845c_{21} + 1.05789c_{22} = 1$$

However, in a multiple regression problem, c_{21} will always equal c_{12} so the work becomes easier as we go along. We can write

$$(328.5)(.039287) - 9.845c_{22} = 0$$
$$(-9.845)(.039287) + 1.05789c_{22} = 1$$

Therefore, we can solve for c_{22} using either equation. In doing so, we get $c_{22} = 1.310892$. The Gauss multipliers to be used in the next three sections are summarized below.

	c_{i1}	c_{i2}
c_{1j}	.004222	.039287
c_{2j}	.039287	1.310892

15.9 Inferences Concerning B_1 and B_2

As you are well aware by now, the net regression coefficient denoted b_1 is, in a sampling problem, merely an estimate of the true B_1 value for the population and, similarly, b_2 is merely an estimate of the true population B_2 value. That being so, both b_1 and b_2 will vary from sample to sample. We can measure and make use of the sampling errors of these statistics both in hypothesis-testing and confidence interval problems.

Testing Hypotheses of Zero Population Slopes

A frequently used test of significance in multiple regression analysis is one which asserts that the population net regression coefficient relating Y and a specific independent variable is 0. This test is often used to determine whether a particular independent variable should continue to be represented in the estimating equation or whether it should be dropped on the grounds that it is probably not contributing anything to the accuracy of estimates of the Y values.[*] We will demonstrate the steps involved in testing this hypothesis by using the b_1 value, found earlier to be .565009 for the linear, three-variable version of the Granby problem. We proceed as follows:

1. State the null and alternative hypotheses.
$$H_0\!: B_1 = 0$$
$$H_a\!: B_1 \neq 0$$

2. Decide on the level of significance to use.
    Let us say that the level of significance is set at .05.

3. Determine the kind of sampling distribution to rely on.
    The appropriate kind of sampling distribution is the sampling distribution of b_1 which, under the model and assumptions specified, is normal with a mean equal to B_1 and a standard deviation, σ_{b_1}, which can be approximated by

$$S_{b_1} = S_e\sqrt{c_{11}} \tag{15.9.1}$$

4. Develop the decision rule.
    Because of the small sample size and the fact that the population standard error of the estimate σ_e is unknown, we must make use of the t-value equivalent of b_1 as the test statistic. Accordingly,

Decision Rule
1. If $-2.365 \leq$ observed $t \leq +2.365$, accept H_0.
2. If observed $t < -2.365$ or observed $t > +2.365$, reject H_0.

---

[*] Although widely used and reasonably sound, this method is not free of shortcomings, as will be shown in section 15.16.

The critical t value used above was determined from Appendix Table 3 for $\alpha = .05$ and $df = n - 3 = 10 - 3 = 7$.

5. Calculate the value of the relevant test statistic.

The relevant test statistic is the t-value equivalent of $b_1 = .565009$. Observed t is found by

$$\text{Observed } t = \frac{b_1}{S_{b_1}} \qquad (15.9.2)$$

Thus, for our problem observed t is $.565009/.163092 = 3.46$, where the denominator was obtained by applying Equation (15.9.1): $S_{b_1} = S_e\sqrt{c_{11}} = (2.51)(\sqrt{.004222}) = .163092$.

6. Apply the decision rule stated in Step 4 and either accept or reject the null hypothesis.

Since observed t of 3.46 exceeds the critical t value of 2.365, we are led to reject at the .05 level of significance the null hypothesis that $B_1 = 0$. We conclude that on the population level number of households does exert an influence on the company's sales.

Usually this test is applied to all sample net regression coefficients. In our example, b_2 was found to be -12.127430. Subjecting this value to the same kind of analysis, we first note that

$$S_{b_2} = S_e\sqrt{c_{22}} \qquad (15.9.3)$$

We know that S_e is 2.51 and that c_{22} is 1.310892. Therefore, $S_{b_2} = (2.51)(\sqrt{1.310892}) = 2.873804$. We determine observed t using

$$\text{Observed } t = \frac{b_2}{S_{b_2}} \qquad (15.9.4)$$

Substituting known values into Equation (15.9.4), we find $t = -12.127430/2.873804 = -4.22$. This observed t value is also outside the acceptance region of the decision rule stated above. Therefore, the null hypothesis $H_0: B_2 = 0$ is rejected at the .05 level of significance. We conclude that on the population level price charged for this company's products also exerts an influence on sales.

An alternative procedure for computing S_{b_1} and S_{b_2} permits one to bypass the use of the Gauss multipliers. If the Gauss multipliers are not to be used in any other part of the analysis, this alternative approach is probably more efficient in terms of computational time invested. The S_{b_1} and S_{b_2} are computed by

$$S_{b_1} = \frac{S_e}{\sqrt{\Sigma \, x_1^2(1 - r_{12}^2)}} \qquad (15.9.5)$$

$$S_{b_2} = \frac{S_e}{\sqrt{\Sigma \, x_2^2(1 - r_{12}^2)}} \qquad (15.9.6)$$

We know from Equations (15.3.7) that $\Sigma \, x_1^2 = 328.5$ and $\Sigma \, x_2^2 = 1.05789$. We find the coefficient of simple determination between the two independent variables r_{12}^2 as follows:

$$r_{12}^2 = \frac{(\Sigma \, x_1 x_2)^2}{(\Sigma \, x_1^2)(\Sigma \, x_2^2)} = \frac{(-9.845)^2}{(328.5)(1.05789)} = .278905$$

Therefore,

$$S_{b_1} = \frac{2.51}{\sqrt{(328.5)(1 - .278905)}} = .163083$$

and

$$S_{b_2} = \frac{2.51}{\sqrt{(1.05789)(1 - .278905)}} = 2.873803$$

These are the same S_{b_1} and S_{b_2} values, aside from the effects of minor rounding errors, as obtained through the use of the other method.

Confidence Intervals for Population Slopes

The standard errors of the net regression coefficients can also be used for constructing confidence intervals by calculating

$$b_1 - t_{\alpha/2}S_{b_1} \leq B_1 \leq b_1 + t_{\alpha/2}S_{b_1} \qquad (15.9.7)$$
$$b_2 - t_{\alpha/2}S_{b_2} \leq B_2 \leq b_2 + t_{\alpha/2}S_{b_2} \qquad (15.9.8)$$

For a .95 confidence coefficient we would have

$$.565009 - (2.365)(.163092) \leq B_1 \leq .565009 + (2.365)(.163092)$$
$$= .179296 \leq B_1 \leq .950722$$

and

$$-12.127430 - (2.365)(2.873804) \leq B_2 \leq -12.127430 + (2.365)(2.873804) =$$
$$-18.923976 \leq B_2 \leq -5.330884.$$

15 Multiple Linear Regression and Correlation Analysis

15.10 Confidence Interval for a Subpopulation Mean

The method of computing confidence intervals for the means of subpopulations is very nearly the same in multiple linear regression as in simple linear regression except that the standard error is more complicated. For a three-variable problem, the equation for the (estimated) standard error of $\hat{Y}$ is

$$S_{\hat{Y}} = S_e\sqrt{1/n + c_{11}x_1^2 + c_{22}x_2^2 + 2c_{12}x_1x_2} \qquad (15.10.1)$$

Notice that the x's in this expression are lowercase ones (not caps), indicating that when we specify values for the two independent variables we express them in terms of deviations from their respective means.

Let us suppose that $\mu_{Y|12}$ for all territories having 12,000 households and an average price of $1.94 is to be estimated using a .95 confidence interval. We saw earlier that $\hat{Y}$ associated with $X_1 = 12$ and $X_2 = 1.94$ is 13.08. From Appendix Table 3 we find the critical t for $\alpha = .05$ and $df = 7$ is 2.365. We have known for some time that S_e is 2.51. Now all we need is the standard error of $\hat{Y}$. To find the value for this measure we draw upon our knowledge of S_e, n, and the c_{ij} values determined earlier. We must also express $X_1 = 12$ as $x_1 = 12.0 - 12.5 = -.5$ and $X_2 = 1.94$ as $x_2 = 1.94 - 2.069 = -.129$. Therefore,

$$S_{\hat{Y}} = (2.51)(\sqrt{1/10 + (.004222)(.25) + (1.310892)(.016641) + (2)(.039287)(.0645)}$$
$$= .897788$$

The confidence interval itself is obtained by

$$\hat{Y} - t_{\alpha/2}S_{\hat{Y}} \leq \mu_{Y|12} \leq \hat{Y} + t_{\alpha/2}S_{\hat{Y}} \qquad (15.10.2)$$

Therefore,

$$13.08 - (2.365)(.897788) \leq \mu_{Y|12} \leq 13.08 + (2.365)(.897788) = 10.96 \leq \mu_{Y|12} \leq 15.20$$

This .95 confidence interval is notably smaller than the one obtained in Chapter 14 when $X_1 = 12$ and number of households was the only independent variable.

Of course, the precision could be increased if use of a larger sample size were feasible or additional relevant independent variables were used. Estimates are also more precise when X_1 and X_2 values are close to their respective means.

15.11 Confidence Interval for Y_{NEW}

As pointed out in Chapter 14, the sales manager of the Granby Specialties Company might find an interval estimate for $\mu_{Y|12}$ most helpful, but chances are she would also desire similar information about individual subpopulation Y values, which we have been denoting Y_{NEW}.

The procedure for computing a confidence interval for Y_{NEW} is similar to that for $\mu_{Y|12}$, the equation for the confidence interval being

$$\hat{Y} - t_{\alpha/2}S_{Y_{\text{NEW}}} \leq Y_{\text{NEW}} \leq \hat{Y} + t_{\alpha/2}S_{Y_{\text{NEW}}} \qquad (15.11.1)$$

The equation for the (estimated) standard error of Y_{NEW} is

$$S_{Y_{\text{NEW}}} = S_e\sqrt{1 + \frac{1}{n} + c_{11}x_1^2 + c_{22}x_2^2 + 2c_{12}x_1x_2} \qquad (15.11.2)$$

Again, we will use $X_1 = 12$ and $X_2 = 1.94$, which is to say, $x_1 = -.5$ and $x_2 = -.129$. (Remember that X values must be expressed as deviations from their means.) Substituting known values into Equation (15.11.2), we get

$$S_{Y_{\text{NEW}}} = 2.51\sqrt{1 + \frac{1}{10} + (.004222)(.25) + (1.310892)(.016641) + (2)(.039287)(.0645)}$$

$$= 2.665731$$

Therefore, the .95 confidence interval, from Equation (15.11.1), is

$$13.08 - (2.365)(2.666) \leq Y_{\text{NEW}} \leq 13.08 + (2.365)(2.666) = 6.77 \leq Y_{\text{NEW}} \leq 19.39$$

15.12 Tests of Significance for Coefficients of Determination

In section 14.10 of the preceding chapter we introduced an F ratio for testing the null hypothesis H_0: $\rho^2 = 0$. This F ratio, presented as Equation (14.10.2), is

$$\text{Observed } F = \frac{\dfrac{r^2}{1}}{\dfrac{1 - r^2}{n - 2}}$$

By altering this equation slightly, we can test null hypotheses regarding the significance of the sample coefficient of multiple determination, $R_{Y.12}^2$, and the sample coefficients of partial determination, $r_{Y1.2}^2$ and $r_{Y2.1}^2$.

Beginning with the test for the coefficient of multiple determination, we note that the value obtained in our illustrative case for $R_{Y.12}^2$ was .899399. The question this test of significance attempts to answer is: Is .899399 sufficiently close to 0 to be consistent with the notion that the coefficient of multiple determination for the population, $\rho_{Y.12}^2$, is equal to 0?

The null and alternative hypotheses are

$$H_0: \rho_{Y.12}^2 = 0$$

$$H_a: \rho_{Y.12}^2 \neq 0$$

or, equivalently, $H_0: B_1 = B_2 = 0$ and $H_a: B_1, B_2$ are not both zero.*

Let us say that the level of significance to be used is $\alpha = .05$. The test statistic is computed by

$$\text{Observed } F = \frac{\dfrac{R_{Y.12}^2}{\lambda - 1}}{\dfrac{1 - R_{Y.12}^2}{n - \lambda}} \qquad (15.12.1)$$

where λ is the number of constants in the estimating equation.

Substituting, we find

$$\text{Observed } F = \frac{\dfrac{.899399}{2}}{\dfrac{.100601}{7}} = 31.29$$

Critical $F_{.05(2,7)}$ is, from Appendix Table 6, 4.74. Since $F > F_{.05(2,7)}$, we reject the null hypothesis that the population coefficient of multiple determination is 0.

The two sample coefficients of partial determination for our example were found to be $r_{Y1.2}^2 = .630989$ and $r_{Y2.1}^2 = .717271$. These can be tested for significance in a manner similar to that just described. The only difference is that one degree of freedom, rather than $\lambda - 1$ degrees of freedom, is assigned the numerators of both observed F ratios. Therefore, the appropriate formulas are

$$\text{Observed } F_1 = \frac{\dfrac{r_{Y1.2}^2}{1}}{\dfrac{1 - r_{Y1.2}^2}{n - \lambda}} \qquad (15.12.2)$$

$$\text{Observed } F_2 = \frac{\dfrac{r_{Y2.1}^2}{1}}{\dfrac{1 - r_{Y2.1}^2}{n - \lambda}} \qquad (15.12.3)$$

*If the null hypothesis $H_0: B_1 = B_2 = 0$ were accepted, that would imply that both population slopes were zero, and consequently that the Y, X_1, and X_2 variables were completely uncorrelated on the population level.

Substituting, we get

$$\text{Observed } F_1 = \frac{\dfrac{.630622}{1}}{\dfrac{.369011}{7}} = 11.96$$

and

$$\text{Observed } F_2 = \frac{\dfrac{.717271}{1}}{\dfrac{.282729}{7}} = 17.76$$

Critical $F_{.05(1,7)}$ is found to be 5.59. Since (1) $F_1 > F_{.05(2,7)}$, we reject the null hypothesis that the population coefficient of partial determination $\rho_{Y1.2}^2$ is 0; and since (2) $F_2 > F_{.05(1,7)}$, we also reject the null hypothesis that the population coefficient of partial determination $\rho_{Y2.1}^2$ is 0.

An alternative way of performing the tests on the partials involves the use of coefficients of partial *correlation*, as well as their squares, and the following t formulas:

$$\text{Observed } t_1 = \frac{r_{Y1.2}\sqrt{n - \lambda}}{\sqrt{1 - r_{Y1.2}^2}} \tag{15.12.4}$$

$$\text{Observed } t_2 = \frac{r_{Y2.1}\sqrt{n - \lambda}}{\sqrt{1 - r_{Y2.1}^2}} \tag{15.12.5}$$

Actually, if the net regression coefficients, b_1 and b_2, have been tested for significance, testing the coefficients of partial determination (or correlation) is unnecessary since the tests of b_1 and b_2 will provide the same information.

15.13 Utilizing a Qualitative Variable

As a general rule, our regression and correlation procedures are used in connection with quantitative variables only. However, a qualitative independent variable can sometimes be used to advantage if the observations can be viewed as belonging to two clearly distinguishable categories. For the sake of illustration, let us change the three-variable version of the Granby Specialties Company problem and assume that the two independent variables are (1) number of households, X_1, and (2) whether or not a particular sales territory has within it a Granby distribution center, X_2. Clearly, X_2 this time is a "yes-or-no" type of variable. Nevertheless, as Figure 15.4 suggests, this variable might be rather important in helping to "explain" variation in sales. In this figure, X_1 and Y points for sales territories not having distribution centers are shown with dots, whereas the X_1 and Y points associated with territo-

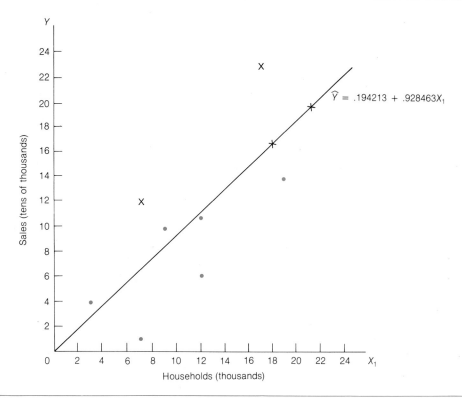

Figure 15.4. Scatter diagram showing the relationship between sales and number of households, but with territories segregated according to whether they have a distribution center (crosses) or do not have a distribution center (dots). The straight line is the regression line obtained in Chapter 14, where variables Y and X_1 were analyzed using simple regression methodology.

ries having distribution centers are shown with crosses. The two sets of points look as if they would be better described by two different straight lines rather than by only one.

Let us assign to each sample sales territory that has a distribution center a value of 1 and each sample territory without a distribution center a value of 0. A qualitative variable whose observations are assigned values of 1, indicating that a specific attribute is present, or 0, indicating that the attribute is absent, is called a *dummy variable*. For our specific problem, we will assume that Sales Territories 1, 5, 7, and 9 happen to have distribution centers, as indicated in column (4) of Table 15.4, where a standard multiple regression worksheet is presented.

Using column totals from Table 15.4 (or their transformed values if the shortcut method is used) and solving for a, b_1, and b_2 in the same way as with the standard three-variable problem just discussed, we find

$$a = .833389$$
$$b_1 = .661285$$
$$b_2 = 6.751372$$

Table 15.4 Worksheet Used to Compute the Least Squares a, b_1, and b_2 Values for a Three-Variable Multiple Regression Analysis with Two Quantitative Variables and One Qualitative Variable

(1) Sales Territory	(2) Sales Y	(3) Households X_1	(4) Distribution Center or Not X_2	(5) X_1Y	(6) X_2Y	(7) X_1X_2	(8) X_1^2	(9) X_2^2	(10) Y^2
1	20	21	1	420	20	21	441	1	400
2	1	7	0	7	0	0	49	0	1
3	14	19	0	266	0	0	361	0	196
4	11	12	0	132	0	0	144	0	121
5	17	18	1	306	17	18	324	1	289
6	4	3	0	12	0	0	9	0	16
7	23	17	1	391	23	17	289	1	529
8	10	9	0	90	0	0	81	0	100
9	12	7	1	84	12	7	49	1	144
10	6	12	0	72	0	0	144	0	36
Total	118	125	4	1780	72	63	1891	4	1832

$$\Sigma Y = 118 \qquad \Sigma X_1Y = 1780 \qquad \Sigma X_1^2 = 1891$$
$$\Sigma X_1 = 125 \qquad \Sigma X_2Y = 72 \qquad \Sigma X_2^2 = 4$$
$$\Sigma X_2 = 4 \qquad \Sigma X_1X_2 = 63 \qquad \Sigma Y^2 = 1832$$

Therefore, the estimating equation, expressed in the usual way, is

$$\hat{Y} = .833389 + .661285X_1 + 6.751372X_2$$

However, since variable X_2 can assume only two values, 0 and 1, we can view this estimating equation as really two equations in disguise—one such equation being appropriate for sales territories having distribution centers and the other appropriate for those lacking distribution centers. For example, for territories lacking distribution centers, X_2 will be 0 and the right-most term in the estimating equation vanishes. We are left with

$$\hat{Y} = .833389 + .661285X_1$$

For territories with distribution centers, X_2 will be 1 and the estimating equation can be written

$$\hat{Y} = .833389 + .661285X_1 + 6.751372$$

or, summing the a and b_2 values, we obtain

$$\hat{Y} = 7.584761 + .661285X_1$$

Judging from Figure 15.5, which shows the two regression lines obtained by manipulating the multiple regression equation in the manner described above, use of variable X_2 appears to have done some good. The two solid regression lines in this figure fit their respective sets of points better than does the single dashed line, which portrays the overall relationship between X_1 and Y.

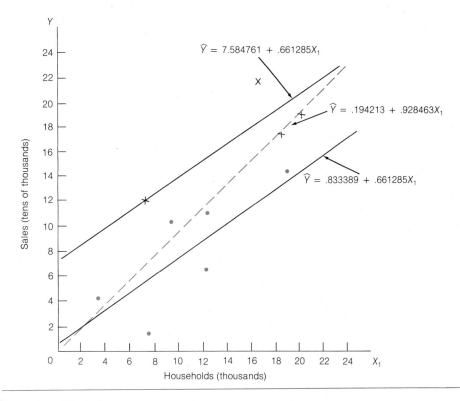

Figure 15.5. This is Figure 15.4, but with the two regression lines obtained from the dummy variable procedure added.

15.14 Computer Assistance

Up to now, we have presented the nuts and bolts of three-variable linear regression and correlation analysis in considerable detail. Moreover, emphasis has been placed on methods promising computational efficiency. This is an appropriate kind of emphasis for one just learning the subject. The sample size and the number of variables has been kept small intentionally so that the thought of performing the calculations manually would not seem totally outlandish.

In practice, however, samples sizes are usually larger than $n = 10$ and there may be a great many independent variables to grapple with. Consequently, the prodigious amount of computational work required will preclude—or at least render highly distasteful—carrying out the analysis manually. Fortunately, many excellent computer programs are available that can greatly reduce the computational drudgery of such an analysis. Figure 15.6 presents some of the results of the Granby problem—the original version—as they would be shown in a computer printout. (Notice that the standard error of the estimate is here called "the standard deviation of Y about the regression line.")

```
THE REGRESSION EQUATION IS
Y = 29.83 + .5650 X1 - 12.13 X2


                          ST. DEV.    T-RATIO =
COLUMN     COEFFICIENT    OF COEF.    COEF/S.D.
            29.830         7.290       4.09
X1           0.565         0.163       3.46
X2         -12.130         2.880      -4.21

STANDARD ERROR OF THE ESTIMATE IS:
S = 13.19

R-SQUARED = 89.9 PERCENT
R-SQUARED = 87.1 PERCENT, ADJUSTED FOR D.F.

ANALYSIS OF VARIANCE

DUE TO        DF     SS        MS=SS/DF        F
REGRESSION     2   395.38       197.69       31.28
RESIDUAL       7    44.22         6.32
TOTAL          9   439.60




CORRELATION COEFFICIENTS

              Y          X1
X1        .802608
X2       -.852864    -.528115
```

The estimating equation is $Y = 29.83 + .5650X_1 - 12.13X_2$

Entries in the column labeled "COEFFICIENT" are simply the sample a, b_1, and b_2 values (in that order). Entries in the column labeled "ST. DEV. OF COEF." are the standard error of a (not treated in the chapter), the standard error of b_1, and the standard error of b_2. Entries in the column labeled "T-RATIO COEF/S.D." are observed t ratios for testing hypotheses:

$$H_0: A = 0 \qquad H_0: B_1 = 0 \qquad H_0: B_2 = 0$$

$$H_a: A \neq 0 \qquad H_a: B_1 \neq 0 \qquad H_a: B_2 \neq 0$$

Observed t values would be compared with critical t for a specified level of significance and degrees of freedom of $n - 3$.

$R^2_{Y.12}$ is 89.9 percent, that is, 89.9 percent of the variation in Y is explained by variation in X_1 and X_2 together.

This same measure, adjusted for degrees of freedom lost, is $\bar{R}_{Y.12} = 87.1$ percent.

The ANOVA data enables one to test the hypotheses

$$H_0: \rho^2_{Y.12}$$

$$H_a: \rho^2_{Y.12}$$

or what amounts to the same thing:

$$H_0: B_1 = B_2 = 0$$

$$H_a: B_1, B_2 \text{ not both } 0$$

At the bottom of the printout is a correlation matrix, showing all zero-order (i.e., *simple*) coefficients of correlation in an organized manner.

Figure 15.6. Computer output for the three-variable version of the Granby Company case.

Similar computer printout material is presented in Figure 15.7. This time, however, three independent variables have been used. In addition to the households variable X_1 and the price variable X_2, the advertising variable X_3 (see the right-most column of Table 15.1) has been included. A comparison of Figures 15.6 and 15.7 reveals that the advertising variable contributes little or nothing to the analysis. Can you find at least two pieces of evidence in these figures to support this negative assessment?

15.15 Two Special Problems

The cautionary notes sounded in section 14.11 of the preceding chapter with respect to two-variable relationships also apply to relationships involving more than two variables. In addition, two dangers accompany multiple regression that are not considerations in simple regression—namely, (1) hidden extrapolation and (2) multicollinearity.

Hidden Extrapolation

In section 14.11 you were warned about the dangers of extrapolation—that is, of using the sample estimating equation to estimate Y values associated with X values outside the range of those used in obtaining that sample estimating equation. At that time, of course, the kind of extrapolation referred to was explicit extrapolation, the only kind possible in simple regression analysis.

With multiple regression analysis, a second, more subtle, kind of extrapolation is possible, which we will call *hidden extrapolation*.

> *Hidden extrapolation* occurs when, even though numbers in a set of values for variables X_1, X_2, . . . , $X_{\lambda-1}$ are individually within the realm of experience, they are *collectively* outside that realm but are nevertheless substituted into the estimating equation together.

For example, in our four-variable problem, data for which were presented in Table 15.1, the largest value for X_1 is 21; for X_2, 2.75; and for X_3, 7. Since this is a set of observed values, the values are necessarily within the realm of experience. However, their appearance together would *not* be within the realm of experience. The X_1 of 21 was associated with Territory 1; X_2 of \$2.75, with Territory 2; and X_3 of 7, with Territories 4 and 7. If these values were to be used together in the sample estimating equation, their user would be extrapolating implicitly. Certainly this is not a heinous felony, but in view of the sometimes serious problems that extrapolation can invite, it is good to be aware of this potential problem.

Multicollinearity

A common and, for some applications, most troublesome problem in multiple regression and correlation analysis is *multicollinearity*.

```
THE REGRESSION EQUATION IS
Y = 18.88 + 0.6091 X1 + 8.959 X2 + 0.8350 X3


                                   ST. DEV.      T-RATIO =
       COLUMN      COEFFICIENT     OF COEF.      COEF/S.D.
                     18.900         10.9000        1.85
       X1             0.609          0.1550        3.92
       X2            -8.960          3.4700       -2.58
       X3             0.835          0.5820        1.43

       STANDARD ERROR OF THE ESTIMATE IS:
       S = 2.34

       R-SQUARED = 92.5 PERCENT
       R-SQUARED = 88.8 PERCENT, ADJUSTED FOR D.F.

       ANALYSIS OF VARIANCE

          DUE TO      DF        SS       MS=SS/DF        F
       REGRESSION      3      406.67      135.56       24.69
       RESIDUAL        6       32.93        5.49
       TOTAL           9      439.60




       CORRELATION COEFFICIENTS

                      Y           X1          X2
       X1          .802608
       X2         -.852864    -.528115
       X3          .601425     .207063    -.637446
```

Figure 15.7. Computer output for the four-variable version of the Granby Company case. (Explanations are essentially the same as those given in Figure 15.6; see Explanatory Notes.)

> *Multicollinearity* occurs when the independent variables are strongly correlated among themselves.

When multicollinearity exists, the separate effects of the independent variables on the dependent variable are difficult to distinguish and, for reasons to be demonstrated, the sample net regression coefficients may be quite unreliable. Descriptively, these coefficients may possess values or signs quite different from what a plausible body of theory or just good old common sense would lead one to expect. They may also be highly sensitive to additions or deletions in the set of independent variables used. Finally, they may vary greatly from one sample to another. Inductively, the standard errors may be much too large for precise estimation of the population B values. The reason for this latter problem can be seen readily enough by inspecting the alternative equations for the S_b values. In the three-variable case, for example, the alternative formula, Equation (15.9.5), for S_{b_1} was shown to be

$$S_{b_1} = \frac{S_e}{\sqrt{\Sigma x_1^2(1 - r_{12}^2)}}$$

For the four-variable version, it would be

$$S_{b_1} = \frac{S_e}{\sqrt{\Sigma x_1^2(1 - R_{1.23}^2)}}$$

These equations both contain coefficients of determination that measure the strength of association *among all independent variables.*

We may surmise that when r_{12}^2 (or $R_{1.23}^2$) approaches 1, the denominator approaches 0 and S_{b_1} becomes very large. Consequently, the estimation of population B_1 with any useful degree of precision may be rendered impossible even with a large sample size.

One of the surer ways of determining whether multicollinearity is present and causing trouble is to compare results of t tests associated with the individual net regression coefficients with the F test for determining whether there is any (linear) correlation at all in the population. If, for example, the F test of H_0: $B_1 = B_2 = \cdots = B_{\lambda-1} = 0$ indicates that not all population B values are zero but the t tests indicate that, individually, $B_1 = 0$, $B_2 = 0, \ldots, B_{\lambda-1} = 0$, one is usually justified in pinning the blame on multicollinearity.

The usual solution to the multicollinearity problem is to eliminate one or more independent variables from the analysis. This solution is admittedly not very satisfying. Nevertheless, at the present time there is little else we can do without abandoning the least squares approach.* Some of the computer-assisted methods for selecting the "best" set of independent variables (discussed in the next section) have built-in safeguards against selecting two or more highly correlated independent variables.

---

*A relatively new, and still controversial, method for dealing with multicollinearity is *ridge regression.* In a nutshell, this method substitutes for the unbiased least-squares estimators of B_1, B_2, etc., biased estimators having smaller standard errors than those associated with the least squares method when multicollinearity is present. The motivated reader is referred to Norman Draper and Harry Smith, *Applied Regression Analysis,* second edition (New York: Wiley, 1981), pp. 313–325.

*15.16 Computer-Assisted Approaches to Selecting the "Best" Set of Independent Variables

Up to now we have spoken as if the set of independent variables to be used in a regression analysis was a given. Sometimes it will be. Often, however, the selection of the independent variables from the pool of *possible* independent variables is the most challenging—and most perplexing—aspect of the entire analysis. What is needed is a screening device—a quality control procedure, so to speak.

Despite the considerable importance of the variable-selection puzzle and the fact that many first-rate statistical minds have attempted to solve it, there is at present no generally accepted "best" way to proceed. However, this disclaimer should not be interpreted as a cry of despair. We *do* have available several good selection procedures, each of which offers a systematic, objective way of distinguishing between the winners and the losers, the starting line up and the bench, among the independent variables "aspiring to be" part of the regression equation. Moreover, it is not uncommon for two or more of these methods to select the same set of independent variables. Such close agreement is not necessarily to be expected. Still, it does occur rather often.

Five of these selection procedures are described in this section. They are

1. Simultaneous backward elimination
2. All possible regressions
3. Backward elimination by stages
4. Forward selection
5. Stepwise regression

ILLUSTRATIVE CASE 15.1

An Officer Candidate Training School instructor wished to develop a method of predicting the performance of future army officers as measured by Officer Evaluation Report (OER) scores. These OER scores are performance ratings assigned by superior officers and can have values between 0 and 200. The scores used in this study represent the average of four separate scoring efforts by relevant superior officers—hence, the label "AVG. OER" appearing on the computer output.

The OER scores are the dependent variable. The pool of possible independent variables was made up of

X_1: Scholastic Aptitude Test Score (SAT)
X_2: College grade point average (GPA COLL)
X_3: ROTC grade point average (GPA ROTC)
X_4: Leadership Potential Index (LPI)
X_5: Army Adaptation Inventory (AAI)
X_6: Basic Officer Course Score (BAS. OFF)

A random sample of 40 officers was used. The data are presented in Figure 15.8.

Simultaneous Backward Elimination

This method is introduced first because it is the most frequently used—not because it is the best.

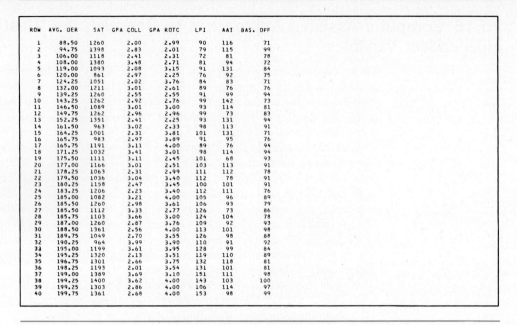

ROW	AVG. OER	SAT	GPA COLL	GPA ROTC	LPI	AAI	BAS. OFF
1	88.50	1260	2.00	2.99	90	116	71
2	94.75	1398	2.83	2.01	79	115	99
3	106.00	1118	2.41	2.31	72	81	78
4	108.00	1380	3.48	2.71	81	94	72
5	119.00	1093	2.08	3.15	91	131	84
6	120.00	861	2.97	2.25	76	92	75
7	124.25	1051	2.02	3.76	84	83	71
8	132.00	1211	3.01	2.61	89	76	76
9	139.25	1260	2.55	2.55	91	99	94
10	143.25	1262	2.92	2.76	99	142	73
11	146.50	1089	3.01	3.00	93	114	81
12	149.75	1262	2.96	2.96	99	73	83
13	152.25	1351	2.41	2.25	93	131	94
14	161.50	963	3.02	2.33	98	113	91
15	164.25	1001	2.31	3.81	101	131	71
16	165.75	983	2.97	3.89	91	95	76
17	165.75	1191	3.11	4.00	89	76	94
18	171.25	1032	3.41	3.01	98	114	94
19	175.50	1111	3.11	2.45	101	68	93
20	177.00	1166	3.01	2.51	103	113	91
21	178.25	1063	2.31	2.99	111	112	78
22	179.50	1036	3.04	3.40	112	78	91
23	180.25	1158	2.47	3.45	100	101	91
24	183.25	1206	2.23	3.40	112	111	76
25	185.00	1082	3.21	4.00	105	96	89
26	185.50	1260	2.98	3.61	106	93	79
27	185.50	1112	3.33	2.77	126	73	86
28	185.75	1103	3.66	3.00	124	104	78
29	187.00	1260	2.87	3.76	109	92	93
30	188.50	1361	2.56	4.00	113	101	98
31	189.75	1049	2.70	3.55	126	98	88
32	190.25	964	3.99	3.90	110	91	92
33	195.00	1199	3.61	3.95	128	99	84
34	195.25	1320	2.13	3.51	119	110	89
35	196.75	1301	2.66	3.75	132	118	81
36	198.25	1193	2.01	3.54	131	101	81
37	199.00	1389	3.69	3.10	151	111	98
38	199.25	1400	3.62	4.00	143	103	100
39	199.25	1303	2.86	4.00	106	114	97
40	199.75	1361	2.68	4.00	153	98	99

Figure 15.8. Basic data for the Officer Evaluation Report regression problem.

Simultaneous backward elimination employs the following steps:

1. Develop the multiple linear regression equation for all potential independent variables.
2. Examine the *t* ratios associated with the individual net regression coefficients; retain those variables whose *t* ratios are significant at a predetermined level and eliminate the others.
3. (*a*) Develop the new multiple regression equation utilizing all independent variables retained in Step 2.
   (*b*) Check to make certain all retained independent variables have significant *t* ratios; if these *t* ratios are all significant, use the resulting regression equation.
   (*c*) If some *t* ratios are still not significant, eliminate the corresponding variable or variables and develop the new, simpler, multiple regression equation.

Parts *a* and *c* of Figure 15.11 (page 721) show how this procedure would be applied to the present problem.

From Part *a* we see that the regression equation is

$$\text{AVG. OER} = -7.5 - .0522 \text{ SAT} + 2.46 \text{ GPA COLL} + 14.2 \text{ GPA ROTC} + 1.07 \text{ LPI}$$
$$(-2.56) \qquad (0.44) \qquad\qquad (2.94) \qquad\qquad (6.31)$$

$$- .014 \text{ AAI} + .809 \text{ BAS. OFF}$$
$$(-0.09) \qquad\quad (2.50)$$

Observed t values are shown directly below the corresponding net regression coefficients. We see that GPA COLL and AAI do not have significant t ratios. (The critical values of t for $\alpha = .05$ and $n - 7 = 33$ degrees of freedom are about ± 2.03.) Therefore, according to the above list of steps, we eliminate these two variables and recompute the regression equation using only the four remaining variables. In doing so, we get (as shown in Part c of this same figure):

$$\text{AVG. OER} = -4.0 - \underset{(-2.82)}{.0544 \text{ SAT}} + \underset{(3.02)}{14.0 \text{ GPA ROTC}} + \underset{(6.79)}{1.09 \text{ LPI}} + \underset{(2.81)}{.853 \text{ BAS. OFF}}$$

All four independent variables have significant t ratios. Therefore, the process stops.

Despite its popularity and relative simplicity, this method can be dangerously misleading when multicollinearity is present and pumping up the standard errors of the slope coefficients. For example, the two "best" independent variables could be strongly correlated with each other and both thrown out quite unceremoniously.

A preferred elimination method of this same basic type would be one which eliminates independent variables one at a time and, therefore, makes possible the retention of one of the intercorrelated "best" independent variables. Backward elimination by stages, to be described shortly, is one such preferred method.

Despite the potential hazards of the present method, the five-variable equation shown here, which includes X_1, X_3, X_4, and X_6, turns out to be the same multiple regression equation selected by all methods described in this section.

Notice the inverse relationship between the OER and the SAT scores. The reason for this is not obvious. Multicollinearity, the usual suspect, is not clearly implicated this time.

All Possible Regressions

> The "*all possible regressions*" approach is for analysts who attach a great deal of importance to thoroughness. This selection procedure requires one to investigate all possible regression equations having $0, 1, 2, \ldots, \lambda - 1$ independent variables and to select the best equation on the basis of some criterion.

Various criteria are used in connection with the "all possible regressions" approach. We will limit our treatment to one criterion, MSE_λ. (For the present example, other frequently used criteria resulted in the same set of "best" independent variables.)

Notice that for our seven-variable problem there will be $2^{\lambda-1} = 2^6 = 64$ different regression equations to determine and study. If there were, say, ten potential independent variables under scrutiny, the number of regression equations to be analyzed would jump to 1024! Clearly, access to a computer is a prerequisite for this approach to independent variable selection. Even with a computer, the size of the task can be staggering if very many potential independent variables are to be screened.

Table 15.5 presents the complete results for the problem under discussion.

Table 15.5 Descriptive Data for the "All Possible Regressions" Selection Procedure Applied to the OER Score Problem

	(1) Independent Variables Represented in Regression Equation	(2) Number of Independent Variables $\lambda - 1$	(3) Degrees of Freedom	(4) SSE	(5) Selection Criterion MSE_λ
Set 0	None	0	39	40135.0	1029.1$\sqrt{}$
Set 1	X_1	1	38	39963.0	1052.0
	X_2	1	38	36948.5	972.3
	X_3	1	38	25164.0	662.0
	X_4	1	38	13656.0	359.0$\sqrt{}$
	X_5	1	38	40048.0	1054.0
	X_6	1	38	32457.1	854.1
Set 2	X_1, X_2	2	37	36709.4	992.1
	X_1, X_3	2	37	25130.4	679.2
	X_1, X_4	2	37	12179.0	329.0
	X_1, X_5	2	37	39824.0	1076.0
	X_1, X_6	2	37	32104.4	867.7
	X_2, X_3	2	37	22368.3	604.5
	X_2, X_4	2	37	13320.0	360.0
	X_2, X_5	2	37	36908.2	997.5
	X_2, X_6	2	37	31437.6	849.7
	X_3, X_4	2	37	11326.0	306.0$\sqrt{}$
	X_3, X_5	2	37	25159.9	680.0
	X_3, X_6	2	37	19750.0	534.0
	X_4, X_5	2	37	13289.0	359.0
	X_4, X_6	2	37	12805.0	346.0
	X_5, X_6	2	37	32434.1	876.6
Set 3	X_1, X_2, X_3	3	36	22303.5	619.5
	X_1, X_2, X_4	3	36	11990.6	331.1
	X_1, X_2, X_5	3	36	36695.0	1019.0
	X_1, X_2, X_6	3	36	31247.5	868.0
	X_1, X_3, X_4	3	36	10213.8	283.7$\sqrt{}$
	X_1, X_3, X_5	3	36	25121.4	697.8
	X_1, X_3, X_6	3	36	19307.4	536.3
	X_1, X_4, X_5	3	36	12005.8	333.5
	X_1, X_4, X_6	3	36	10499.6	291.7
	X_1, X_5, X_6	3	36	32103.4	891.8
	X_2, X_3, X_4	3	36	10774.2	299.3
	X_2, X_3, X_5	3	36	22200.4	616.7
	X_2, X_3, X_6	3	36	18685.4	519.0
	X_2, X_4, X_5	3	36	13113.3	364.0
	X_2, X_4, X_6	3	36	12662.4	351.7
	X_3, X_5, X_6	3	36	31421.5	872.8
	X_3, X_4, X_5	3	36	11123.4	309.0
	X_3, X_4, X_6	3	36	10226.6	284.1
	X_3, X_5, X_6	3	36	19747.7	548.5
	X_4, X_5, X_6	3	36	12519.9	347.8

(continued)

Table 15.5 (Continued)

	(1) Independent Variables Represented in Regression Equation	(2) Number of Independent Variables $\lambda - 1$	(3) Degrees of Freedom	(4) SSE	(5) Selection Criterion MSE_λ
	X_1, X_2, X_3, X_4	4	35	9847.6	281.4
	X_1, X_2, X_3, X_5	4	35	22165.3	633.3
	X_1, X_2, X_3, X_6	4	35	18430.8	526.6
	X_1, X_2, X_4, X_5	4	35	11897.2	339.9
	X_1, X_2, X_4, X_6	4	35	10493.2	299.8
	X_1, X_2, X_5, X_6	4	35	31206.5	891.6
	X_1, X_3, X_4, X_5	4	35	10127.5	289.4
Set 4	X_1, X_3, X_4, X_6	4	35	8330.8	238.0 ✓
	X_1, X_3, X_5, X_6	4	35	19273.7	550.7
	X_1, X_4, X_5, X_6	4	35	10437.3	298.2
	X_2, X_3, X_4, X_5	4	35	10721.2	306.3
	X_2, X_3, X_4, X_6	4	35	9955.0	284.4
	X_2, X_4, X_5, X_6	4	35	12465.2	356.1
	X_2, X_3, X_5, X_6	4	35	18571.9	530.6
	X_3, X_4, X_5, X_6	4	35	10098.7	288.5
	X_1, X_2, X_3, X_4, X_5	5	34	9833.3	289.2
	X_1, X_2, X_3, X_4, X_6	5	34	8272.5	243.3 ✓
Set 5	X_1, X_2, X_3, X_5, X_6	5	34	18247.5	536.7
	X_1, X_2, X_4, X_5, X_6	5	34	10437.1	307.0
	X_1, X_3, X_4, X_5, X_6	5	34	8318.9	244.7
	X_2, X_3, X_4, X_5, X_6	5	34	9910.5	291.5
Set 6	$X_1, X_2, X_3, X_4, X_5, X_6$	6	33	8270.5	250.6

✓ Best value in a set.
Optimum values overall are circled.

MSE_{all}

MSE_λ, obtained by

$$\frac{\Sigma \, (Y - \hat{Y})^2}{n - \lambda}$$

will decrease at first as the number of independent variables increases. However, at some point, the favorable effect of adding one more independent variable is insufficiently great to overcome the adverse effect of one less degree of freedom, and, consequently, MSE_λ begins to rise. Users of this criterion seek to find the set of independent variables which minimizes MSE_λ or a set for which MSE_λ is so close to the minimum that adding more independent variables is not worthwhile.

Number of Independent Variables $\lambda - 1$	New Independent Variable Associated with Lowest MSE_λ in Set		Lowest MSE_λ in Set	Amount of Reduction in MSE_λ
0	None		1029.1	——
1	Leadership Potential Index	(X_4)	359.0	670.1
2	Grade point average—ROTC	(X_3)	306.0	53.0
3	Scholastic Aptitude Test	(X_1)	283.7	22.3
4	Basic Officer Course Score	(X_6)	238.0	45.7
5	Grade point average—College	(X_2)	243.3	−5.3
6	Army Adaptation Index	(X_5)	250.6	−7.3

Minimum MSE_λ values for each set of regression equations represented in Table 15.5, column (5), follow; they are also shown in chart form in Figure 15.9.

Thus we are led to select the same four independent variables—X_1, X_3, X_4, and X_6.

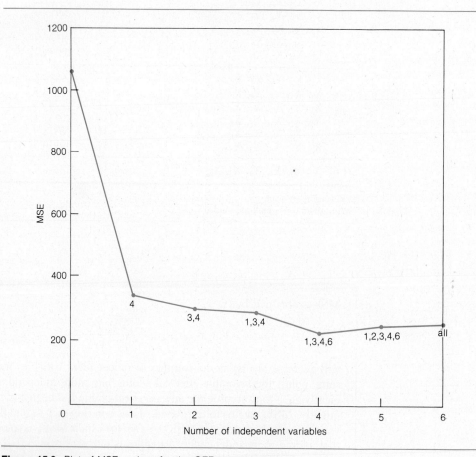

Figure 15.9. Plot of MSE_λ values for the OER scores problem.

Backward Elimination by Stages

The selection procedure here called *backward elimination by stages* is a refinement of, and improvement on, the first method described in this section.

> With the process of *backward elimination by stages*, the analyst begins by developing the multiple regression equation, utilizing all variables in the pool of potential independent variables. The single least promising independent variable is eliminated (assuming its t ratio is not significant at a predetermined level*) and a new, simpler multiple regression equation is obtained. The least promising independent variable in the new, smaller set is eliminated and a new multiple regression equation obtained. This process is continued until the only independent variables remaining are all significant at the chosen level.

This method is illustrated using a flowchart in Figure 15.10. Computer results for the problem under discussion are presented in Figure 15.11.

Forward Selection

As its name suggests, forward selection is just the reverse of backward elimination. With forward selection, one variable is *entered* at a time. The first independent variable selected will usually be the one correlating most strongly with the dependent variable. The simple linear regression equation is developed and the t ratio checked for significance. If the t ratio is not significant, this implies that no independent variable or combination of them is likely to be very useful. However, if the first independent variable has a significant t ratio (at some predetermined level), then all possible two-independent-variable equations are calculated. The *new* variable with the largest observed t value will be the second one entered (provided, of course, that the largest t value is significant at the chosen level). This process continues until there is no standby independent variable that, when entered into the regression along with the previously picked ones, will produce a significant t value.

This method is illustrated in Figures 15.12 and 15.13 (pp. 722 and 723).

Stepwise Regression

Stepwise regression is much like the forward selection method but also has something in common with the backward elimination method. The forward selection procedure is followed; however, as new variables are entered, previously picked variables are examined to determine whether they all still have significant t ratios. If a variable that had been significant is found not to be significant now, it is eliminated. And then on with the forward selection process.

---

* Many packaged computer programs use F tests applied to the net regression coefficients for this and other methods described in this section. They are equivalent tests.

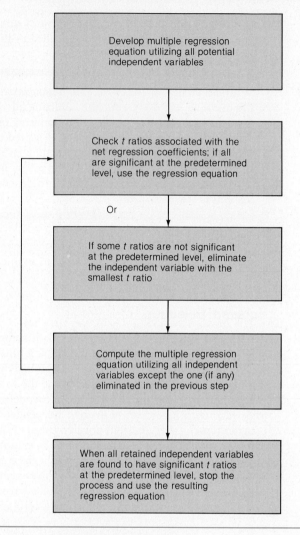

Figure 15.10. Flowchart for backward elimination by stages.

For example, we have seen that LPI is the independent variable most strongly correlated with OER scores. Using the forward selection procedure, GPA ROTC was the next independent variable included. The third independent variable entered by the forward selection method was SAT. Now suppose, contrary to fact, that when SAT is included the t ratio for GPA ROTC is no longer significant. In such a case GPA ROTC would be eliminated. Then the forward selection search would begin for a third independent variable to accompany LPI and SAT in the multiple regression equation.

For our example, no independent variables selected in the early stages were eliminated. Therefore, the stepwise and forward selection methods turned out to be identical. Nevertheless, this backward elimination feature of stepwise regression is an attractive, and sometimes quite important, improvement upon the regular forward selection method.

In practice, the user of stepwise regression will specify two levels of significance α_1 and α_2. The value of α_1 is used to screen variables for inclusion in the model and α_2 to identify variables entered at an earlier stage which should now be dropped. Usually, $\alpha_2 > \alpha_1$.

Despite the brief treatment afforded it here, stepwise regression is for many purposes the most desirable of the independent-variable selection methods. As noted, it makes use of features from both backward elimination by stages and forward selection. Consequently, it is a somewhat more refined method than either. Moreover, it does not require, as does the "all possible regressions" approach, the development of a great many regression equations, most of which could have been rejected at an earlier stage by some kind of simple "eyeballing" approach. Finally, because it involves the addition or deletion of one variable at a time, stepwise regression is much less likely than simultaneous backward elimination to reject potentially useful independent variables simply because multicollinearity is present.

```
THE REGRESSION EQUATION IS
AVG. OER = - 7.5 - 0.0522 SAT + 2.46 GPA COLL + 14.2 GPA ROTC + 1.07 LPI - 0.014 AAI
           + 0.809 BAS. OFF
Part (a)
                                  ST. DEV.     T-RATIO =
COLUMN        COEFFICIENT         OF COEF.     COEF/S.D.
                    -7.55            36.30         -0.21
SAT              -0.05218          0.02040         -2.56
GPA COLL            2.456            5.585          0.44
GPA ROTC           14.241            4.843          2.94
LPI                1.0705           0.1697          6.31
AAI               -0.0138           0.1551         -0.09   Worst - Eliminate
BAS. OFF           0.8088           0.3239          2.50

STANDARD ERROR OF THE ESTIMATE IS:
S = 15.83

R-SQUARED = 79.4 PERCENT
R-SQUARED = 75.6 PERCENT, ADJUSTED FOR D.F


Part (b)
THE REGRESSION EQUATION IS
AVG. OER = - 9.2 - 0.0524 SAT + 2.59 GPA COLL + 14.3 GPA ROTC
           + 0.811 BAS. OFF

                                  ST. DEV.     T-RATIO =
COLUMN        COEFFICIENT         OF COEF.     COEF/S.D.
                    -9.16            31.05         -0.29
SAT              -0.05241          0.01993         -2.63
GPA COLL            2.592            5.294          0.49   Worst - Eliminate
GPA ROTC           14.297            4.732          3.02
LPI                1.0683           0.1653          6.46
BAS. OFF           0.8105           0.3186          2.54

STANDARD ERROR OF THE ESTIMATE IS:
S = 15.60

R-SQUARED = 79.4 PERCENT
R-SQUARED = 76.4 PERCENT, ADJUSTED FOR D.F.


THE REGRESSION EQUATION IS
AVG. OER = - 4.0 - 0.0544 SAT + 14.0 GPA ROTC + 1.09 LPI + 0.853 BAS. OFF
Part (c)
                                  ST. DEV.     T-RATIO =
COLUMN        COEFFICIENT         OF COEF.     COEF/S.D.
                    -4.00            28.90         -0.14   All significant -
SAT              -0.05443          0.01929         -2.82
GPA ROTC           14.045            4.653          3.02     Stop the process
LPI                1.0854           0.1598          6.79
BAS. OFF           0.8529           0.3032          2.81

STANDARD ERROR OF THE ESTIMATE IS:
S = 15.43

R-SQUARED = 79.2 PERCENT
R-SQUARED = 76.9 PERCENT, ADJUSTED FOR D.F.
```

Figure 15.11. Computer output for OER scores problem, using backward elimination by stages.

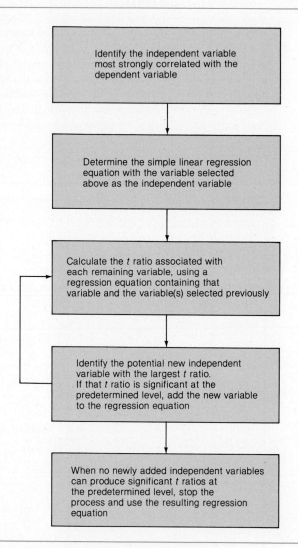

Figure 15.12. Flowchart of forward selection.

```
(1)                                            Simple Correlation Coefficients
                  AVG. OER    SATGPA COLLGPA ROTC      LPI      AAI
SAT               0.066
GPA COLL          0.282  -0.041
GPA ROTC          0.611   0.060    0.029
LPI              (0.812)  0.306    0.237   0.494
AAI              -0.047   0.172   -0.274  -0.061   0.060
BAS. OFF          0.437   0.351    0.296   0.119   0.372  -0.052

                 Best


THE REGRESSION EQUATION IS
AVG. OER = 24.1 + 1.33 LPI

                                 ST. DEV.    T-RATIO =
COLUMN         COEFFICIENT       OF COEF.    COEF/S.D.
                  24.07            16.71        1.44
LPI                1.3324          0.1552       8.58

S = 18.96

R-SQUARED = 66.0 PERCENT
R-SQUARED = 65.1 PERCENT, ADJUSTED FOR D.F.


(2-a)
THE REGRESSION EQUATION IS
AVG. OER = 67.7 + 1.43 LPI - 0.0461 SAT

                                 ST. DEV.    T-RATIO =
COLUMN         COEFFICIENT       OF COEF.    COEF/S.D.
                  67.73            26.08        2.60
LPI                1.4335          0.1560       9.19
SAT               -0.04610         0.02176     -2.12

STANDARD ERROR OF THE ESTIMATE IS:
S = 18.14

R-SQUARED = 69.7 PERCENT
R-SQUARED = 68.0 PERCENT, ADJUSTED FOR D.F.


(2-b)

THE REGRESSION EQUATION IS
AVG. OER = 11.1 + 1.30 LPI + 5.90 GPA COLL

                                 ST. DEV.    T-RATIO =
COLUMN         COEFFICIENT       OF COEF.    COEF/S.D.
                  11.10            21.44        0.52
LPI                1.2957          0.1599       8.10
GPA COLL           5.904           6.106        0.97

STANDARD ERROR OF THE ESTIMATE IS:
S = 18.97

R-SQUARED = 66.8 PERCENT
R-SQUARED = 65.0 PERCENT, ADJUSTED FOR D.F.


(2-c)

THE REGRESSION EQUATION IS
AVG. OER = 1.2 + 1.11 LPI + 14.5 GPA ROTC

                                 ST. DEV.    T-RATIO =
COLUMN         COEFFICIENT       OF COEF.    COEF/S.D.
                   1.25            17.50        0.07
LPI                1.1078          0.1648       6.72
GPA ROTC          14.450           5.237       (2.76)- Best

STANDARD ERROR OF THE ESTIMATE IS:
S = 17.50

R-SQUARED = 71.8 PERCENT
R-SQUARED = 70.3 PERCENT, ADJUSTED FOR D.F.


(2-d)
THE REGRESSION EQUATION IS
AVG. OER = 40.9 + 1.34 LPI - 0.176 AAI

                                 ST. DEV.    T-RATIO =
COLUMN         COEFFICIENT       OF COEF.    COEF/S.D.
                  40.93            23.60        1.73
LPI                1.3419          0.1555       8.63
AAI               -0.1760          0.1740      -1.01
STANDARD ERROR OF THE ESTIMATE IS:
S = 18.95

R-SQUARED = 66.9 PERCENT
R-SQUARED = 65.1 PERCENT, ADJUSTED FOR D.F.
```

Figure 15.13. Computer output for OER scores problem, using forward selection.

(2-e)

```
THE REGRESSION EQUATION IS
AVG. OER = - 13.1 + 1.24 LPI + 0.552 BAS. OFF

                        ST. DEV.    T-RATIO =
COLUMN     COEFFICIENT  OF COEF.    COEF/S.D.
             -13.09       28.81       -0.45
LPI           1.2367      0.1641       7.54
BAS. OFF      0.5518      0.3518       1.57

STANDARD ERROR OF THE ESTIMATE IS:
S = 18.60

R-SQUARED = 68.1 PERCENT
R-SQUARED = 66.4 PERCENT, ADJUSTED FOR D.F.
```

(3-a)

```
THE REGRESSION EQUATION IS
AVG. OER = 41.1 + 1.21 LPI + 13.3 GPA ROTC - 0.0402 SAT

                        ST. DEV.    T-RATIO =
COLUMN     COEFFICIENT  OF COEF.    COEF/S.D.
              41.10       26.24        1.57
LPI            1.2131      0.1673       7.25
GPA ROTC      13.350       5.073        2.63
SAT           -0.04024     0.02032     (-1.98)  ── Best

STANDARD ERROR OF THE ESTIMATE IS:
S = 16.84

R-SQUARED = 74.6 PERCENT
R-SQUARED = 72.4 PERCENT, ADJUSTED FOR D.F.
```

(3-b)

```
THE REGRESSION EQUATION IS
AVG. OER = - 16.6 + 1.05 LPI + 15.2 GPA ROTC + 7.60 GPA COLL

                        ST. DEV.    T-RATIO =
COLUMN     COEFFICIENT  OF COEF.    COEF/S.D.
             -16.61       21.73       -0.76
LPI            1.0491      0.1686       6.22
GPA ROTC      15.186       5.207        2.92
GPA COLL       7.602       5.598        1.36

STANDARD ERROR OF THE ESTIMATE IS:
S = 17.30

R-SQUARED = 73.2 PERCENT
R-SQUARED = 70.9 PERCENT, ADJUSTED FOR D.F.
```

(3-c)

```
THE REGRESSION EQUATION IS
AVG. OER = 14.5 + 1.12 LPI + 14.0 GPA ROTC - 0.131 AAI

                        ST. DEV.    T-RATIO =
COLUMN     COEFFICIENT  OF COEF.    COEF/S.D.
              14.54       24.05        0.60
LPI            1.1218      0.1664       6.74
GPA ROTC      14.005       5.290        2.65
AAI           -0.1314      0.1622      -0.81

STANDARD ERROR OF THE ESTIMATE IS:
S = 17.58

R-SQUARED = 72.3 PERCENT
R-SQUARED = 70.0 PERCENT, ADJUSTED FOR D.F.
```

(3-d)

```
THE REGRESSION EQUATION IS
AVG. OER = - 42.4 + 0.986 LPI + 15.2 GPA ROTC + 0.629 BAS. OFF

                        ST. DEV.    T-RATIO =
COLUMN     COEFFICIENT  OF COEF.    COEF/S.D.
             -42.38       27.85       -1.52
LPI            0.9863      0.1703       5.79
GPA ROTC      15.249       5.061        3.01
BAS. OFF       0.6290      0.3197       1.97

STANDARD ERROR OF THE ESTIMATE IS:
S = 16.85

R-SQUARED = 74.5 PERCENT
R-SQUARED = 72.4 PERCENT, ADJUSTED FOR D.F.
```

(4-a)

```
THE REGRESSION EQUATION IS
AVG. OER = 23.3 + 1.16 LPI + 14.0 GPA ROTC - 0.0371 SAT + 6.25 GPA COLL

                        ST. DEV.    T-RATIO =
COLUMN     COEFFICIENT  OF COEF.    COEF/S.D.
              23.27       30.45        0.76
LPI            1.1566      0.1738       6.65
GPA ROTC      14.041       5.088        2.76
SAT           -0.03707     0.02043     -1.81
GPA COLL       6.250       5.479        1.14

STANDARD ERROR OF THE ESTIMATE IS:
S = 16.77

R-SQUARED = 75.5 PERCENT
R-SQUARED = 72.7 PERCENT, ADJUSTED FOR D.F.
```

Figure 15.13. (continued)

15 Multiple Linear Regression and Correlation Analysis

```
(4-b)
THE REGRESSION EQUATION IS
AVG. OER = 48.2 + 1.22 LPI + 13.1 GPA ROTC - 0.0385 SAT - 0.087 AAI

                                ST. DEV.      T-RATIO =
COLUMN       COEFFICIENT        OF COEF.      COEF/S.D.
                48.17            29.50           1.63
LPI              1.2179           0.1692         7.20
GPA ROTC        13.103            5.143          2.55
SAT             -0.03852          0.02076       -1.86
AAI             -0.0867           0.1588        -0.55

STANDARD ERROR OF THE ESTIMATE IS:
S = 17.01

R-SQUARED = 74.8 PERCENT
R-SQUARED = 71.9 PERCENT, ADJUSTED FOR D.F.

(4-c)
THE REGRESSION EQUATION IS
AVG. OER = - 4.0 + 1.09 LPI + 14.0 GPA ROTC - 0.0544 SAT + 0.853 BAS. OFF

                                ST. DEV.      T-RATIO =
COLUMN       COEFFICIENT        OF COEF.      COEF/S.D.
                -4.00            28.90          -0.14
LPI              1.0854           0.1598         6.79
GPA ROTC        14.045            4.653          3.02
SAT             -0.05443          0.01929       -2.82
BAS. OFF         0.8529           0.3032         2.81    — Best

STANDARD ERROR OF THE ESTIMATE IS:
S = 15.43

R-SQUARED = 79.2 PERCENT
R-SQUARED = 76.9 PERCENT, ADJUSTED FOR D.F.

(5-a)
THE REGRESSION EQUATION IS
AVG. OER = - 9.2 + 1.07 LPI + 14.3 GPA ROTC - 0.0524 SAT + 0.811 BAS. OFF
           + 2.59 GPA COLL

                                ST. DEV.      T-RATIO =
COLUMN       COEFFICIENT        OF COEF.      COEF/S.D.
                -9.16            31.05          -0.29       GPA Coll ⎱ Do not result in
LPI              1.0683           0.1653         6.46       AAI      ⎰ significant improvement
GPA ROTC        14.297            4.732          3.02
SAT             -0.05241          0.01993       -2.63
BAS. OFF         0.8105           0.3186         2.54
GPA COLL         2.592            5.294          0.49

STANDARD ERROR OF THE ESTIMATE:
S = 15.60

R-SQUARED = 79.4 PERCENT
R-SQUARED = 76.4 PERCENT, ADJUSTED FOR D.F.

(5-b)
THE REGRESSION EQUATION IS
AVG. OER = - 0.9 + 1.09 LPI + 13.9 GPA ROTC - 0.0536 SAT + 0.844 BAS. OFF
           - 0.032 AAI

                                ST. DEV.      T-RATIO =
COLUMN       COEFFICIENT        OF COEF.      COEF/S.D.
                -0.87            32.58          -0.03
LPI              1.0885           0.1627         6.69
GPA ROTC        13.945            4.739          2.94
SAT             -0.05363          0.01989       -2.70
BAS. OFF         0.8436           0.3103         2.72
AAI             -0.0325           0.1474        -0.22

STANDARD ERROR OF THE ESTIMATE IS:
S = 15.64

R-SQUARED = 79.3 PERCENT
R-SQUARED = 76.2 PERCENT, ADJUSTED FOR D.F.
```

Figure 15.13. (concluded)

15.17 Summary

Often analysts using only one independent variable will not achieve the strength of association or the precision of estimation desired. Therefore, they must search for and use additional relevant independent variables. This chapter has shown how to apply regression and correlation methods when more than one independent variable is employed.

At the outset, we dealt exclusively with the three-variable linear model of the form $Y_i = A + B_1X_1 + B_2X_2 + \epsilon_i$, where ϵ_i: $NID(0, \sigma_e^2)$. We divided the subject into a descriptive part and an inductive part. In the descriptive part, we showed how to compute the sample:

1. Least squares estimating equation, $\hat{Y} = a + b_1 X_1 + b_2 X_2$
2. Standard error of the estimate
3. Coefficient of multiple determination
4. Coefficients of partial determination or partial correlation

Then we crossed the threshold into the inductive realm, pausing briefly to examine the underlying assumptions and to get acquainted with some ancillary measures called Gauss multipliers.

We then looked at the following inductive topics:

1. Testing a hypothesis about a population net regression coefficient
2. Developing a confidence interval for a population net regression coefficient
3. Developing a confidence interval for a subpopulation mean
4. Developing a confidence interval for a subpopulation Y_{NEW} value
5. Testing a hypothesis about the population coefficient of multiple determination
6. Testing a hypothesis about a population coefficient of partial determination

In an effort to convey something about the versatility of multiple regression methods, we showed how to (1) utilize a qualitative independent variable, and (2) how to interpret a computer printout for a problem involving one dependent variable and *three* independent variables.

We then sounded a warning about two potential problems encountered in multiple regression analysis which are of no concern in simple analysis. These are hidden extrapolation and multicollinearity.

To this point, we had spoken as if the analyst had known from the start exactly what independent variables were to be used. Because this will not necessarily be the case in practice, we concluded the chapter with a survey of ways used to select the "best" independent variables from a pool of potential independent variables under consideration. The methods treated in this final section were

1. Simultaneous backward elimination
2. All possible regressions
3. Backward elimination by stages
4. Forward selection
5. Stepwise regression

Although this chapter and the previous one have been rather extensive in their coverage of the more commonly used regression and correlation methods, you should be aware that you have really only seen the tip of the iceberg. For one thing, we have assumed throughout that all variables are related in a linear manner even though, in reality, better results can often be achieved by departing from the linear form. In addition, a great many other kinds of variations on and extensions of the methods presented here do exist, and their numbers have been growing very rapidly during recent years.

Finally, keep in mind that the methods described in this and the preceding chapter have been used in connection with problems in which a random sample is quite literally selected from a stable population. Regression and correlation methods can also be used to advantage within the context of business forecasting. Hence, we will return to this subject in Chapter 17.

Terms Introduced in This Chapter

all possible regressions (p. 715)
backward elimination by stages
   (p. 719)
coefficient of multiple determination
   (p. 694)
coefficient of partial determination
   (p. 695)

cross products (p. 689)
forward selection (p. 719)
Gauss multipliers (p. 698)
hidden extrapolation (p. 710)
multicollinearity (p. 711)

net regression coefficient (p. 607)
simultaneous backward elimination
   (p. 713)
stepwise regression (p. 719)

Formulas Introduced in This Chapter

Note: Most of the equations presented in this chapter pertained to the three-variable linear model. This is the model assumed in this glossary.

Estimating equation or regression equation:

$$\hat{Y} = a + b_1 X_1 + b_2 X_2$$

Normal equations for determining a, b_1, and b_2 by the least squares method:

$$\Sigma\, Y = na + b_1\, \Sigma\, X_1 + b_2\, \Sigma\, X_2$$
$$\Sigma\, X_1 Y = a\, \Sigma\, X_1 + b_1\, \Sigma\, X_1 + b_2\, \Sigma\, X_1 X_2$$
$$\Sigma\, X_2 Y = a\, \Sigma\, X_2 + b_1\, \Sigma\, X_1 X_2 + b_2\, \Sigma\, X_2^2$$

Shortcut version of the above normal equations:

$$\Sigma\, x_1 y = b_1\, \Sigma\, x_1^2 + b_2\, \Sigma\, x_1 x_2$$
$$\Sigma\, x_2 y = b_1\, \Sigma\, x_1 x_2 + b_2\, \Sigma\, x_2^2$$

Formulas for obtaining the sums required by the shortcut method:

$$\Sigma\, y^2 = \Sigma\, Y^2 - n\bar{Y}^2$$
$$\Sigma\, x_1^2 = \Sigma\, X_1^2 - n\bar{X}_1^2$$
$$\Sigma\, x_2^2 = \Sigma\, X_2^2 - n\bar{X}_2^2$$
$$\Sigma\, x_1 y = \Sigma\, X_1 Y - n\bar{X}_1 \bar{Y}$$
$$\Sigma\, x_2 y = \Sigma\, X_2 Y - n\bar{X}_2 \bar{Y}$$
$$\Sigma\, x_1 x_2 = \Sigma\, X_1 X_2 - n\bar{X}_1 \bar{X}_2$$

Y intercept when the shortcut method is used:

$$a = \bar{Y} - b_1 \bar{X}_1 - b_2 \bar{X}_2$$

Sample standard error of the estimate (where it is understood that the Y values were determined from the equation $\hat{Y} = a + b_1 X_1 + b_2 X_2$):

$$S_e = \sqrt{\frac{\Sigma\,(Y - \hat{Y})^2}{n - 3}}$$

Shortcut version of the above formula:

$$S_e = \sqrt{\frac{\Sigma\, y^2 - b_1\, \Sigma\, x_1 y - b_2\, \Sigma\, x_2 y}{n - 3}}$$

Basic formula for the sample coefficient of determination:

$$R_{Y.12}^2 = \frac{\Sigma \, (\hat{Y} - \overline{Y})^2}{\Sigma \, (Y - \overline{Y})^2}$$

Shortcut version of the above formula:

$$R_{Y.12}^2 = \frac{b_1 \, \Sigma \, x_1 y + b_2 \, \Sigma \, x_2 y}{\Sigma \, y^2}$$

Formula for adjusting $R_{Y.12}^2$ for degrees of freedom lost:

$$\overline{R}_{Y.12} = 1 - (1 - R_{Y.12}^2)\left(\frac{n-1}{n-3}\right)$$

Coefficient of partial determination between Y and X_2 when X_1 is held constant:

$$r_{Y2.1}^2 = \frac{R_{Y.12}^2 - r_{Y1}^2}{1 - r_{Y1}^2}$$

Coefficient of partial determination between Y and X_1 when X_2 is held constant:

$$r_{Y1.2}^2 = \frac{R_{Y.12}^2 - r_{Y2}^2}{1 - r_{Y2}^2}$$

General form of a three-variable linear regression equation for population data:

$$\mu_{Y|12} = A + B_1 X_1 + B_2 X_2$$

The model for a three-variable linear regression equation:

$$Y_i = A + B_1 X_1 + B_2 X_2 + \epsilon_i$$
where ϵ_i: NID$(0, \sigma_e^2)$

Formulas for obtaining Gauss multipliers c_{11} and c_{12}:

$$c_{11} \, \Sigma \, x_1^2 + c_{12} \, \Sigma \, x_1 x_2 = 1$$
$$c_{11} \, \Sigma \, x_1 x_2 + c_{12} \, \Sigma \, x_2^2 = 0$$

Formulas for obtaining Gauss multipliers c_{21} and c_{22}:

$$c_{21} \, \Sigma \, x_1^2 + c_{22} \, \Sigma \, x_1 x_2 = 0$$
$$c_{21} \, \Sigma \, x_1 x_2 + c_{22} \, \Sigma \, x_2^2 = 1$$

The estimated standard error of b_1:

$$S_{b_1} = S_e \sqrt{c_{11}}$$

The estimated standard error of b_2:

$$S_{b_2} = S_e \sqrt{c_{22}}$$

Alternative formulas for the standard errors of b_1 and b_2:

$$S_{b_1} = \frac{S_e}{\sqrt{\Sigma \, x_1^2 (1 - r_{12}^2)}}$$

$$S_{b_2} = \frac{S_e}{\sqrt{\Sigma \, x_2^2 (1 - r_{12}^2)}}$$

Formulas for obtaining observed t ratios for testing the null hypotheses $B_1 = 0$ or $B_2 = 0$ against the alternative hypotheses $B_1 \neq 0$ or $B_2 \neq 0$:

$$t = \frac{b_1}{S_{b_1}} \quad \text{and} \quad t = \frac{b_2}{S_{b_2}}$$

Formulas for obtaining confidence intervals for population B_1 and B_2 values:

$$b_1 - t_{\alpha/2}S_{b_1} \leq B_1 \leq b_1 + t_{\alpha/2}S_{b_2}$$
$$b_2 - t_{\alpha/2}S_{b_2} \leq B_2 \leq b_2 + t_{\alpha/2}S_{b_2}$$

Estimated standard error of $\hat{Y}$:

$$S_{\hat{Y}} = S_e \sqrt{\frac{1}{n} + c_{11}x_1^2 + c_{22}x_2^2 + 2c_{12}x_1x_2}$$

Formula for obtaining a confidence interval for a subpopulation mean:

$$\hat{Y} - t_{\alpha/2}S_{\hat{Y}} \leq \mu_{Y|12} \leq \hat{Y} + t_{\alpha/2}S_{\hat{Y}}$$

Formula for obtaining a confidence interval for a subpopulation individual Y value when values of X_1 and X_2 are specified:

$$\hat{Y} - t_{\alpha/2}S_{Y_{\text{NEW}}} \leq Y_{\text{NEW}} \leq \hat{Y} + t_{\alpha/2}S_{Y_{\text{NEW}}}$$

Estimated standard error of Y_{NEW}:

$$S_{Y_{\text{NEW}}} = S_e \sqrt{1 + \frac{1}{n} + c_{11}x_1^2 + c_{22}x_2^2 + 2c_{12}x_1x_2}$$

F ratio for testing the null hypothesis that $\rho_{Y.12}^2 = 0$ against the alternative hypothesis that $\rho_{Y.12}^2 \neq 0$ (equivalent to testing H_0: $B_1 = B_2 = 0$ against H_a: B_1, B_2 are not both 0):

$$F = \frac{\dfrac{R_{Y1.2}^2}{3 - 1}}{\dfrac{1 - R_{Y.12}^2}{n - 3}}$$

F ratio for testing the null hypothesis that $\rho_{Y1.2}^2 = 0$ against the alternative hypothesis that $\rho_{Y1.2}^2 \neq 0$ (equivalent to t test of H_0: $B_1 = 0$ against H_a: $B_1 \neq 0$):

$$F = \frac{\dfrac{r_{Y1.2}^2}{1}}{\dfrac{1 - r_{Y1.2}^2}{n - 3}}$$

F ratio for testing the null hypothesis that $\rho_{Y2.1}^2 = 0$ against the alternative hypothesis that $\rho_{Y2.1}^2 \neq 0$ (equivalent of t test of H_0: $B_2 = 0$ against H_a: $B_2 \neq 0$):

$$F = \frac{\dfrac{r_{Y2.1}^2}{1}}{\dfrac{1 - r_{Y2.1}^2}{n - 3}}$$

The t ratio for testing the null hypothesis that $\rho_{Y1.2} = 0$ against the alternative hypothesis that

$\rho_{Y1.2} \neq 0$ (equivalent to t test for H_0: $B_1 = 0$ versus H_a: $B_1 \neq 0$ and F test for H_0: $\rho^2_{Y1.2} = 0$ versus H_a: $\rho^2_{Y1.2} \neq 0$):

$$t = \frac{r_{Y1.2}\sqrt{n-3}}{\sqrt{1-r^2_{Y1.2}}}$$

The t ratio for testing the null hypothesis that $\rho_{Y2.1} = 0$ against the alternative hypothesis that $\rho_{Y2.1} \neq 0$ (equivalent to t test for H_0: $B_2 = 0$ versus H_a: $B_2 \neq 0$ and F test for H_0: $\rho^2_{Y2.1} = 0$ versus H_a: $\rho^2_{Y2.1} \neq 0$):

$$t = \frac{r_{Y2.1}\sqrt{n-3}}{\sqrt{1-r^2_{Y2.1}}}$$

Questions and Problems

15.1 Explain the meaning of each of the following:
  a. Net regression coefficient
  b. Shortcut method of obtaining the least squares multiple regression equation
  c. Standard error of the estimate
  d. Coefficient of multiple determination
  e. Coefficient of partial determination
  f. Gauss multiplier

15.2 Explain the meaning of each of the following:
  a. S_{b_1}
  b. $S_{\hat{Y}}$
  c. $S_{Y_{NEW}}$
  d. Stepwise regression
  e. $R^2_{Y.12}$
  f. Dummy variable
  g. Multicollinearity
  h. Hidden extrapolation

15.3 Distinguish between:
  a. Y value and y value
  b. b_1 and b_2
  c. S_e and s_e
  d. b_1 and B_1
  e. $S_{\hat{Y}}$ and $S_{Y_{NEW}}$

15.4 Distinguish between:
  a. S_e and $S_{\hat{Y}}$
  b. S_{b_1} and S_{b_2}
  c. $r^2_{Y1.2}$ and $r^2_{Y2.1}$
  d. $r^2_{Y1.2}$ and $r^2_{Y1.23}$
  e. $F = \dfrac{\dfrac{R^2_{Y.12}}{\lambda-1}}{\dfrac{1-R^2_{Y.12}}{n-\lambda}}$ and $F = \dfrac{\dfrac{r^2_{Y2.1}}{1}}{\dfrac{1-r^2_{Y2.1}}{n-\lambda}}$

15.5 Distinguish between the methods for selecting the "best" set of independent variables referred to in each of the following pairs of methods:
  a. Simultaneous backward elimination and backward elimination by stages
  b. Backward elimination by stages and forward selection
  c. Forward selection and stepwise regression
  d. Stepwise regression and all possible regressions

15.6 Indicate which of the following statements you agree with and which you disagree with, and defend your opinions:
  a. Of the five methods described in this chapter for selecting the "best" set of independent variables, all possible regressions is the most economical of computer time.
  b. Since simultaneous backward elimination and backward elimination by stages represent essentially the same process, it matters little which of the two methods is used.
  c. Different methods for selecting the "best" set of independent variables will always result in the same multiple regression equation.

15.7 Indicate which of the following statements you agree with and which you disagree with, and defend your opinions:
  a. If a multiple regression analysis results in the estimating equation $\hat{Y} = 5 + .7X_1 + .8X_2$, the .7 figure tells us that, given an increase of 1 unit in the X_1 variable, there will be an increase in Y of .7.
  b. Imagine the following experiment: 25 one-digit numbers are selected at random from a table of random digits and separated in a random manner into five groups of five numbers each (that is, $n = 5$) and the coefficient of determination R^2 (unadjusted for degrees of freedom lost) is computed. In such a situation, we might reasonably expect R^2 to be as high as, say, .2 or .3 by accident but could not reasonably expect it to be in excess of about .8.
  c. If (1) total variation is 100 and (2) the variation remaining unexplained after independent variables X_1, X_2, and X_3 have explained as much of the variation in Y as they can is 20, the coefficient of multiple determination $R^2_{Y.123}$ will be .8.

15.8 Indicate which of the following statements you agree with and which you disagree with, and defend your opinions:
  a. If $R^2_{Y.12}$ is .8 and r^2_{Y1} is .7, then $r^2_{Y2.1}$ will be .33.
  b. Multicollinearity means that the coefficient of multiple determination $R^2_{Y.12}$ is very high.
  c. Even if the net regression coefficients have been tested for significance, additional information can be obtained by testing the coefficients of partial determination for significance.

15.9 a. Using Equations (15.3.1), (15.3.2), and (15.3.3) as a guide, write the normal equations for a problem involving one dependent variable and three independent variables.
  b. Show how the normal equations you obtained in connection with part a would be written if the shortcut method were used. Also, show the equation for obtaining the a constant.
  c. Refer to part a. Do the same thing for a problem involving one dependent variable and four independent variables.
  d. Show how the normal equations you obtained in connection with part c would be written if the shortcut method were used. Also, show the equation for obtaining the a constant.

15.10 a. Using Equations (14.5.2) and (15.4.2) as a guide, write the equation for the standard error of the estimate for a problem involving one dependent variable and three independent variables.
  b. Do the same thing for a problem involving one dependent variable and four independent variables.

15.11 a. Using Equation (15.5.2) as a guide, write the equation for the coefficient of multiple determination for a problem involving one dependent variable and three independent variables.
  b. Do the same thing for a problem involving one dependent variable and four independent variables.

15.12 a. Using Equations (15.6.3) and (15.6.4) as a guide, write the equation for the coefficient of partial determination $r^2_{Y1.23}$, for a problem involving one dependent variable and three independent variables.
  b. Do the same thing for $r^2_{Y1.234}$ for a problem involving one dependent variable and four independent variables.

15.13 a. Using Equations (15.9.5) and (15.9.6) as a guide, write the equation for the standard error of b_1, S_{b_1}, for a problem involving one dependent variable and three independent variables.
  b. Do the same thing for a problem involving one dependent variable and four independent variables.

15.14 Given: $\hat{Y} = 4 + 1.3X_1 + 5X_2$. The X_2 is a dummy variable having a value of 0 when an attribute of interest is absent and a value of 1 when that attribute is present. Convert this multiple regression equation into two simple regression equations—one appropriate for situations in which the attribute is absent and the other appropriate for situations in which that attribute is present.

 15.15 The management of a large toy store wished to determine how important annual family income and number of children in family are in determining total dollar expenditures for toy items per year by families patronizing this particular establishment. Consequently, management obtained a random sample of family heads for close study. The results for expenditures during a recent year (Y), family income during the same year (X_1), and number of children in family prior to June 30 of the same year (X_2) are presented in Table 15.6.

Table 15.6 Data for Multiple Regression and Correlation Analysis Regarding Annual Toy Expenditures

Family	Y Amount Spent on Toys During a Recent Year (Hundreds of Dollars)	X_1 Family Income (Thousands of Dollars)	X_2 Number of Children
A	0.8	15	1
B	1.0	20	2
C	1.0	25	1
D	0.5	12	1
E	1.5	35	2
F	1.0	15	3
G	1.5	30	1
H	2.5	50	6
I	2.0	25	4
J	1.5	27	2
K	1.5	22	4
L	1.0	18	2
M	2.0	55	3
N	2.0	32	3
O	1.0	21	1

Also given:
$\Sigma Y = 20.80 \qquad \Sigma X_1 = 402.00 \qquad \Sigma X_2 = 36.00$
$\Sigma X_1 Y = 637.00 \qquad \Sigma X_2 Y = 58.80 \qquad \Sigma X_1 X_2 = 1097.00$
$\Sigma Y^2 = 33.14 \qquad \Sigma X_1^2 = 12{,}896.00 \qquad \Sigma X_2^2 = 116.00$

a. Derive the least squares multiple linear estimating equation.
b. Interpret the b_1 and b_2 values obtained in connection with part a.
c. Compute the standard error of the estimate making certain to allow for the loss of three degrees of freedom.
d. Using the appropriate t test, test the net regression coefficients, b_1 and b_2, for significance at the .05 level. What are your conclusions? What assumption(s) do you implicitly accept whenever you employ this test?
e. Compute and interpret the coefficient of multiple determination $R^2_{Y.12}$.
f. Using the appropriate F test, test the coefficient of multiple determination $R^2_{Y.12}$ for significance at the .05 level. What is your conclusion?
g. Judging from the results you obtained in connection with parts d and e of this problem, do you believe that the two independent variables, X_1 and X_2, acting together serve as useful aids for estimating toy expenditures? Why or why not?

 15.16 Refer to problem 15.15. (You must have worked problem 15.15 in order to work this one.)
a. Compute and interpret the coefficients of partial determination, $r^2_{Y1.2}$ and $r^2_{Y2.1}$. According to these measures of strength of relationship, which independent variable, X_1 or X_2, is more important?
b. Using the appropriate F test, test the coefficients of partial determination, $r^2_{Y1.2}$ and $r^2_{Y2.1}$, for significance at the .05 level. What are your conclusions?

 15.17 Refer to problem 15.15. (You must have worked problem 15.15 in order to work this one.)
a. Set up a .95 confidence interval around b_1. Around b_2.
b. Set up a .95 confidence interval which you would expect to include the true but unknown subpopulation $\mu_{y|12}$value if X_1 is 20 and X_2 is 4.
c. Set up a .95 confidence interval which you would expect to include the true but unknown population value of Y_{NEW} if $X_1 = 20$ and $X_2 = 4$.
d. Explain the difference between the confidence interval you obtained in connection with part b of this problem and the confidence interval you obtained in connection with part c.

15.18 A sales manager suspects that sales performance might be related to IQ and to years of sales experience. To evaluate the validity of this suspicion, the sales manager selects a random sample of ten salespeople and records their dollar sales for a recent year (Y), their IQs (X_1), and the number of years of sales experience each had had prior to the year under study (X_2). The resulting data are shown in Table 15.7.
a. Derive the least squares multiple linear estimating equation.
b. Interpret the b_1 and b_2 values obtained in connection with part a.
c. Compute the standard error of the estimate making certain to allow for the loss of three degrees of freedom.
d. Using the appropriate t test, test the net regression coefficients for significance at the .01 level. What are your conclusions? What assumption(s) did you implicitly accept when applying this t test?
e. Compute and interpret the coefficient of multiple determination $R^2_{Y.12}$.
f. Using the appropriate F test, test the coefficient of multiple determination for significance at the .01 level. What is your conclusion?
g. Judging from the results you obtained in connection with parts d and e of this problem, do you believe that the two independent variables, X_1 and X_2, acting together serve as useful aids for estimating sales? Why or why not?

15.19 Refer to problem 15.18. (You must have worked problem 15.18 in order to work this one.)
a. Compute and interpret the coefficients of partial determination, $r^2_{Y1.2}$ and $r^2_{Y2.1}$. According to these measures of strength of relationship, which independent variable, X_1 or X_2, is more important?
b. Using the appropriate F test, test the coefficients of partial determination, $r^2_{Y1.2}$ and $r^2_{Y2.1}$, for significance at the .01 level. What are your conclusions?

Table 15.7 Data for Multiple Regression and Correlation Analysis Regarding Salesperson Performance

Salesperson	Y Sales During a Recent Year (Thousands of Dollars)	X_1 IQ	X_2 Years of Sales Experience
A	50	110	10
B	35	104	9
C	12	108	4
D	45	120	10
E	55	143	11
F	47	118	10
G	52	133	10
H	33	112	7
I	30	119	6
J	12	117	4

Also given:

$$\Sigma Y = 371 \qquad \Sigma X_1 = 1184 \qquad \Sigma X_2 = 81$$
$$\Sigma Y^2 = 15{,}965 \qquad \Sigma X_1^2 = 141{,}436 \qquad \Sigma X_2^2 = 719$$
$$\Sigma X_1 Y = 44{,}833 \qquad \Sigma X_2 Y = 3367 \qquad \Sigma X_1 X_2 = 9717$$

15.20 Refer to problem 15.18. (You must have worked problem 15.18 in order to work this one.)
   a. Set up a .99 confidence interval around b_1. Around b_2.
   b. Set up a .99 confidence interval which you would expect to include the true but unknown subpopulation $\mu_{Y|12}$ value if $X_1 = 120$ and $X_2 = 10$.
   c. Estimate sales for a salesperson not included in the sample study but having an IQ of 120 and ten years of selling experience. Use a .99 confidence coefficient.
   d. Explain the difference between the confidence interval you obtained in connection with part *b* of this problem and the confidence interval you obtained in connection with part *c*.

15.21 A sales manager suspected that sales performance might be related to years of sales experience and to whether a particular salesperson has a university education. In an effort to test this suspicion, the sales manager selected a random sample of 20 salespersons and recorded for each sample member dollar sales for a recent year (Y), number of years of sales experience prior to the year under analysis (X_1), and whether or not a particular salesperson has a university education (X_2). In the case of variable X_2, a value of 1 was assigned to salespeople who possess a university degree and a value of 0 to those who do not possess a university degree. The results are shown in Table 15.8.
   a. Derive the least squares multiple linear estimating equation.
   b. Interpret the b_1 and b_2 values obtained in connection with part *a*.
   c. Compute the standard error of the estimate making certain to allow for the loss of three degrees of freedom.
   d. Using the appropriate *t* test, test the net regression coefficients, b_1 and b_2, for significance at the .05 level. What are your conclusions? What assumption(s) did you implicitly accept when applying this test?
   e. Compute and interpret the coefficient of multiple determination $R^2_{Y.12}$.
   f. Using the appropriate *F* test, test the coefficient of multiple determination for significance at the .05 level. What is your conclusion?

Table 15.8 Data for Multiple Regression and Correlation Analysis Regarding Sales Performance

Salesperson	Y Sales During a Recent Year (Thousands of Dollars)	X_1 Years of Sales Experience	X_2 University Education or Not?
A	45	10	1
B	20	7	0
C	35	9	1
D	42	11	0
E	17	4	0
F	32	7	0
G	50	10	1
H	29	7	0
I	55	11	1
J	30	6	0
K	50	10	1
L	25	5	1
M	52	10	0
N	22	3	0
O	33	7	0
P	27	5	1
Q	30	6	1
R	22	7	0
S	28	4	0
T	20	4	0

Also given:

$$\Sigma Y = 664 \qquad \Sigma X_1 = 143 \qquad \Sigma X_2 = 8$$
$$\Sigma Y^2 = 24{,}692 \qquad \Sigma X_1^2 = 1147 \qquad \Sigma X_2^2 = 8$$
$$\Sigma X_1Y = 5250 \qquad \Sigma X_2Y = 317 \qquad \Sigma X_1X_2 = 66$$

    g. Judging from the results you obtained in connection with parts *d* and *e* of this problem, do you believe that the two independent variables, X_1 and X_2, acting together serve as useful aids for estimating sales? Why or why not?

15.22 Refer to problem 15.21. (You must have worked problem 15.21 in order to work this one.)
    a. Compute and interpret the coefficients of partial determination, $r^2_{Y1.2}$ and $r^2_{Y2.1}$. According to these measures of strength of relationship, which independent variable, X_1 or X_2, is more important?
    b. Using the appropriate F test, test the coefficients of partial determination for significance at the .05 level. What are your conclusions?

15.23 Refer to problem 15.21. (You must have worked problem 15.21 in order to work this one.)
    a. Set up a .95 confidence interval around b_1. Around b_2.
    b. Set up a .95 confidence interval which you would expect to include the true but unknown subpopulation $\mu_{Y|12}$ value if $X_1 = 5$ and $X_2 = 1$.
    c. Estimate sales for a salesperson not included in the sample study who has five years of sales experience and a college degree. Use a .95 confidence coefficient.

d. Explain the difference between the confidence interval you obtained in connection with part *b* of this problem and the confidence interval you obtained in connection with part *c*.

 15.24 Refer to problem 15.21. (You must have worked problem 15.21 in order to work this one.)

a. Convert the multiple regression equation obtained in connection with part *a* of problem 15.21 into two separate simple linear regression equations—one equation pertaining to the presence of a university degree and one equation pertaining to the absence of a university degree.

b. Obtain the simple linear regression equation relating variable Y to variable X_1 only.

c. Obtain estimated Y values using (1) the two equations you derived in connection with part *a* above and (2) the one equation you obtained in connection with part *b* above, and tell which approach to obtaining Y values produces better results.

d. Support the conclusion you arrived at in part *c* of this problem by presenting two scatter diagrams constructed according to the following instructions: One scatter diagram should have sales on the vertical axis and number of years of sales experience on the horizontal axis. The line of regression should be based on the use of the estimating equation you obtained in part *b*. The second scatter diagram should also show sales on the vertical axis and number of years of selling experience on the horizontal axis; however, this time there should be two lines of regression—one associated with the presence of a university degree and one associated with the absence of a university degree.

✔ **15.25** It is believed by some that the number of miles per gallon achieved by an automobile is influenced by (1) weight of the automobile, and (2) age of the automobile. To test this belief, an experiment was performed in the following manner: Eighteen automobiles were used each of which belonged to one of three weight categories, and one of the three age categories, as shown in Table 15.9.

a. Derive the least squares multiple linear estimating equation.

b. Interpret the b_1 and b_2 values obtained in part *a*.

c. Using the appropriate *t* test, test the net regression coefficients, b_1 and b_2, for significance at the .05 level. What are your conclusions? What assumption(s) did you implicitly accept when applying this *t* test?

d. Compute the standard error of the estimate making certain to allow for the loss of three degrees of freedom.

e. Compute and interpret the coefficient of multiple determination, $R^2_{Y.12}$.

f. Using the "appropriate" *F* test, test the coefficient of multiple determination for significance at the .05 level. What is your conclusion? What assumption(s) did you implicitly accept when applying this *F* test? Given the nature of this problem as described above, is the use of *F* *really* appropriate? Justify your conclusion. (*Hint*: You may wish to consult the footnote on page 646.)

g. Judging from the results you obtained in parts *d* and *e* of this problem, do you believe that the two independent variables, X_1 and X_2, acting together serve as useful aids for estimating miles per gallon achieved? Why or why not?

✔ **15.26** Refer to problem 15.25. (You must have worked problem 15.25 in order to work this one.)

a. Compute and interpret the coefficients of partial determination, $r^2_{Y1.2}$ and $r^2_{Y2.1}$. According to these measures of strength of relationship, which independent variable, X_1 or X_2, is most important?

b. Using the "appropriate" *F* test, test the coefficients of partial determination for significance at the .05 level. What are your conclusions? What assumption(s) did you implicitly accept when applying this *F* test? Given the nature of this problem as described in problem 15.25, is the use of this *F* test *really* appropriate? Justify your conclusion. (*Hint*: You may wish to consult the footnote on page 646.)

Table 15.9 Data for Multiple Regression and Correlation Analysis Pertaining to Miles per Gallon Achieved by 18 Experimental Automobiles

Automobile	Y Miles per Gallon	X_1 Weight (Thousands of Pounds)	X_2 Age (Years)
A	25	2	1
B	15	2	5
C	15	2	9
D	20	2	1
E	18	2	5
F	12	2	9
G	23	3	1
H	11	3	5
I	13	3	9
J	16	3	1
K	14	3	5
L	10	3	9
M	15	4	1
N	9	4	5
O	10	4	9
P	12	4	1
Q	15	4	5
R	7	4	9

Also given:

$$\Sigma Y = 260 \qquad \Sigma X_1 = 54 \qquad \Sigma X_2 = 90$$
$$\Sigma Y^2 = 4138 \qquad \Sigma X_1^2 = 174 \qquad \Sigma X_2^2 = 642$$
$$\Sigma X_1 Y = 743 \qquad \Sigma X_2 Y = 1124$$

 15.27 Refer to problem 15.25. (You must have worked problem 15.25 in order to work this one.)
   a. Set up a .95 confidence interval around b_1 and around b_2.
   b. Set up a .95 confidence interval which you would expect to include the true but unknown subpopulation $\mu_{Y|12}$ value if $X_1 = 3$ and $X_2 = 1$.
   c. Estimate miles per gallon for a two-year-old car not included in the experiment but weighing 3000 pounds. Use a .95 confidence coefficient.
   d. Explain the difference between the confidence interval you obtained in part *b* of this problem and the confidence interval you obtained in part *c*.

 15.28 A homebuilder believes that the number of weeks required to sell a new house (Y) is influenced by the price of the house (X_1), and proximity to a major freeway (X_2). Variable X_2 is quantified in the following manner: If the house is located within two miles of a major freeway, a value of 1 is assigned; if the house is located more than two miles away from a major freeway, a value of 0 is assigned. For a random sample of ten recently completed houses, the builder obtained the data shown in Table 15.10.
   a. Derive the least squares multiple linear estimating equation.
   b. Interpret the b_1 and b_2 values obtained in part *a*.

Table 15.10 Data for Multiple Regression and Correlation Analysis Regarding Number of Weeks Required to Sell a New House

House	Y Number of Weeks	X_1 Price (Thousands of Dollars)	X_2 Proximity to Major Freeway
A	1	45	0
B	2	50	1
C	2	52	1
D	3	53	1
E	4	55	0
F	5	58	0
G	5	67	0
H	5	65	0
I	7	70	1
J	10	100	0

Also given:

$$\Sigma Y = 44 \qquad \Sigma X_1 = 615 \qquad \Sigma X_2^2 = 4$$
$$\Sigma Y^2 = 258 \qquad \Sigma X_1^2 = 40{,}041 \qquad \Sigma X_2 = 4$$
$$\Sigma X_1Y = 3068 \qquad \Sigma X_2Y = 14 \qquad \Sigma X_1X_2 = 225$$

c. Using the appropriate t test, test the net regression coefficients, b_1 and b_2, for significance at the .05 level. What are your conclusions? What assumption(s) did you implicitly accept when applying this t test?

d. Compute the standard error of the estimate making certain to allow for the loss of three degrees of freedom.

e. Compute and interpret the coefficient of multiple determination, $R_{Y.12}^2$.

f. Using the appropriate F test, test the coefficient of multiple determination for significance at the .05 level. What is your conclusion?

g. Judging from the results you obtained in parts d and e of this problem, do you believe that the two independent variables, X_1 and X_2, acting together serve as useful aids in estimating sales? Why or why not?

h. Convert the multiple regression equation obtained in part a into two separate two-variable regression equations—one equation pertaining to new houses in close proximity to a major freeway and one equation pertaining to new houses not in close proximity to a major freeway.

15.29 The Bon Ton Clothing Company conducted a questionnaire survey in order to identify factors which influence amount of money spent on women's clothing. The variables were: amount of money spent (in hundreds of dollars, Y), number of children in respondent's family (X_1), age of oldest child (X_2), and age of youngest child (X_3). Part of the computer printout is shown in Figure 15.14.

a. Discuss the advisability of using variables X_1, X_2, and X_3 in the same analysis. (*Hint:* What if a woman had no children? What if she had only one child?)

b. Test the null hypothesis that $\rho_{Y.123}^2 = 0$ at the .05 level of significance. What is your conclusion?

```
THE REGRESSION EQUATION IS
Y = 35.87 + 0.4311 X1 + 2.2355 X2 - 1.0472 X3


                                ST. DEV.      T-RATIO =
COLUMN      COEFFICIENT         OF COEF.      COEF/S.D.
              35.8700            18.221          1.97
X1             0.4311             6.027          0.07
X2             2.2355             0.914          2.45
X3            -1.0473             0.806         -1.30

STANDARD ERROR OF THE ESTIMATE IS:
S = 75.40

R-SQUARED =   7.2 PERCENT
R-SQUARED =   2.6 PERCENT, ADJUSTED FOR D.F.

ANALYSIS OF VARIANCE

  DUE TO     DF          SS        MS=SS/DF
REGRESSION    3       38944.65     12981.55
RESIDUAL     60      341109.60      5685.16
TOTAL        63      380054.25




CORRELATION COEFFICIENTS

             Y          X1         X2
X1        .07775
X2        .08181      .22390
X3        .53606     -.25348     -.0056
```

Figure 15.14. Computer output for the Bon Ton Clothing Company problem.

c. Using a .05 level of significance, test the null hypothesis that
   1. $B_1 = 0$
   2. $B_2 = 0$
   3. $B_3 = 0$
d. Write a brief overall evaluation of the results of this analysis from the standpoint of possible practical applications.

 15.30 The Christopher Plumbing Company wished to predict the shipping cost in dollars (Y) for different mixes of plumbing items when the height of the crate in feet (X_1) and the estimated amount of dead space (in square inches) inside the crate (X_2) are known. Altogether they conducted four regression and correlation analyses. These included (1) a multiple analysis with X_1 and X_2 as independent variables, (2) a simple analysis with X_1 as the sole independent variable, (3) a simple analysis with X_2 as the sole independent variable, and (4) a simple analysis between X_1 and X_2. The computer results for (4) are shown in Figure

```
THE REGRESSION EQUATION IS
X1 = -1.02 + .3012 X2


                                ST. DEV.      T-RATIO =
COLUMN      COEFFICIENT         OF COEF.      COEF/S.D.
X2             .3012             .0367           8.21

STANDARD ERROR OF THE ESTIMATE IS:
S = 24.02

R-SQUARED = 93.1 PERCENT
R-SQUARED = 91.7 PERCENT, ADJUSTED FOR D.F.
```

Figure 15.15. Computer output for the Christopher Plumbing Company problem.

15.15 (p. 739). Also given: $\hat{Y} = 42.19 + 9.3321X_1 - 1.288395X_2$, $\overline{R}^2_{Y.12} = .566$; $\hat{Y} = 36.40 + 5.3493X_1$, $\overline{r}^2_{Y1} = .608$; $\hat{Y} = 29.22 + 1.7X_2$, $\overline{r}^2_{Y2} = .638$.

On the basis of this limited information, which kind of regression equation would you recommend that management of this company use: (1) the multiple equation, (2) the simple equation with X_1 as the only independent variable, or (3) the simple equation with X_2 as the only independent variable? Justify your choice.

TAKE CHARGE

15.31 Without question, in business and economic research, multiple regression and correlation analysis is the most frequently used kind of formal statistical method. The intent of this exercise is twofold: To encourage you to (1) become acquainted with a specific use of multiple regression and correlation in connection with research related to your major and (2) familiarize yourself with some of the professional journals related to your major.

Go to the library and find an article in a professional journal pertaining to your major. (For example, if you are a marketing major, you will want to examine *The Journal of Marketing, The Journal of Marketing Research, The Journal of Advertising Research*, and so forth.) Make certain the article you pick describes research conducted with the aid of multiple regression and correlation methods.

Write a brief summary of the article. In your summary, treat such matters as (1) purpose of the research, (2) general research procedures, (3) general conclusions, and (4) interpretations of as many multiple regression and correlation measures as you feel comfortable discussing.

Computer Exercise

15.32 This problem builds on problem 14.21 of the preceding chapter.

Conduct a multiple regression and correlation analysis aimed at determining what factors are most influential in determining percent change in a company's stock over the five-year period from 1979 to 1983.

Table 15.11 presents data for a sample of 22 companies selected from among those included in Forbes' 36th Annual Report of American Industry (*Forbes Magazine*, January 2, 1984.) The variables are as follows:

Y: Percent change in market price of company's common stock during the 5-year period from 1979 to 1983

X_1: Average annual percent change in earnings per share, 1979 to 1983

X_2: Average return on equity, 1979 to 1983

X_3: Average net profit margin, 1979 to 1983

X_4: Debt/equity ratio, last reported fiscal year

X_5: Average annual percent change in sales, 1979 to 1983

Computer Exercise

15.33 Refer to Table 15.11. Select the "best" set of independent variables using the "all possible regressions" approach, a .10 level of significance (if relevant), and the MSE_λ criterion.

Computer Exercise

15.34 Refer to Table 15.11. Select the "best" set of independent variables using simultaneous backward elimination and a .10 level of significance.

Table 15.11 Data for Computer Exercises 15.32 to 15.37

Company Name	Y	X_1	X_2	X_3	X_4	X_5
1. General Dynamics	267.1	25.5	19.3	3.8	.0	16.3
2. Curtice-Burns	255.3	10.4	18.2	1.6	2.4	18.3
3. Albertson's	228.1	16.9	22.4	1.5	.6	16.9
4. Morrison-Knudsen	189.7	16.4	15.6	1.8	.4	17.7
5. Wyman-Gordon	170.0	15.2	20.0	10.5	.1	12.4
6. Quaker Oats	154.6	12.2	15.5	2.3	.2	11.2
7. BanCal Tri State	148.0	6.4	9.2	6.2	.3	12.9
8. LTV	132.1	7.1	9.8	0.0	1.3	11.9
9. May Department Stores	126.5	7.8	14.5	4.2	.5	9.8
10. Handy and Harmon	123.7	6.8	16.9	1.6	.6	9.6
11. Gordon Jewelry	120.6	12.0	14.2	4.2	.4	16.5
12. Parker Hannifin	91.3	6.8	15.4	3.4	.3	16.5
13. American National Insurance	88.9	12.4	11.8	12.2	.0	7.2
14. Federal Paper Board	63.6	4.6	13.2	1.3	1.4	6.1
15. Amerace	50.0	−13.3	5.5	0.0	.4	3.5
16. Cameron Iron Works	48.1	14.3	19.2	0.0	.4	16.6
17. Washington National	46.8	6.0	10.8	2.9	.1	7.4
18. Edison Brothers Stores	39.7	6.0	17.1	4.0	.2	9.9
19. Copperweld	37.7	−11.9	6.6	0.0	.5	6.0
20. Anderson-Clayton	35.8	6.0	10.0	1.9	.1	4.9
21. Phelps-Dodge	16.1	−11.3	2.9	0.0	.6	5.2
22. Amsted Industries	−3.5	3.0	13.1	0.0	.0	6.5

Source: Adapted from *Forbes*, January 2, 1984, pp. 265ff.

Computer Exercise

15.35 Refer to Table 15.11. Select the "best" set of independent variables using backward elimination by stages and a .10 level of significance.

Computer Exercise

15.36 Refer to Table 15.11. Select the "best" set of independent variables using forward selection and a .10 level of significance.

Computer Exercise

15.37 Refer to Table 15.11. Select the "best" set of independent variables using stepwise regression, a .05 level of significance for entering new variables, and a .10 level of significance for eliminating "old" variables.

Cases for Analysis and Discussion

15.1 "OCTAGONAL INDUSTRIES, INC.": PART II

The compensation officer who described the procedures used in connection with Part I of this case (Case 14.1 at the end of Chapter 14) continued his commentary as follows:

The study is presently being continued on a company-wide level. As I look upon the preliminary analysis now, it impresses me as pretty crude. I believe that the investigation lent itself—and may still lend itself—to multiple regression and correlation analysis using annual percent pay increases as the dependent variable and (1) performance ranking, (2) time since last increase, and (3) position grade as independent variables.

1. Judging from the scatter diagrams presented in Figures 14.23 through 14.25, do you agree that use of multiple regression and correlation analysis might prove beneficial? Why or why not?

2. In general, do you believe that the kind of analysis used is a good way to determine whether the company's guidelines for compensation of salaried employees had been adhered to? Tell why you do or do not think so.

15.2 "THE DUPLIKWIK CORPORATION"

Give this case only a quick reading before answering question 1 on page 746.

In the following case report, a "Duplikwik" sales executive describes the sales analysis program now in use in his company and suggests ways of employing formal statistical techniques to make the program even more useful.

Obtaining maximum results from the Duplikwik sales force is essential to achieving national as well as local marketing goals. From a national standpoint, programs, incentives, and proper pricing of products must be provided if these goals are to be met. From a local standpoint, branch office activities must be planned and evaluated. A key element in determining the quantity and expected effectiveness of these programs and activities is timely and accurate sales-call information.

In order to meet the need for specific field-related information, Duplikwik samples branch sales activity on a monthly basis. During the field surveys, all sales representatives in the surveyed branches are required to participate. Care is taken by local management to ensure that all sales personnel are well informed and properly motivated to report timely and accurate information.

Upon completion of a month's sampling of all sales activities, a branch manager receives a computerized printout of his branch's average sales time and call patterns along with a national comparison of these statistics.

Each month, ten branch offices are sampled from the over eighty branch offices nationwide. During the survey, each sales representative is asked to record all sales activities. The representative is asked to identify the account name, where the work was performed, whether the work was administrative or sales in nature, whether or not the customer already uses Duplikwik equipment, the specific selling action, the targeted equipment involved for placement, and the time spent on each of these activities.

The sales representatives use separate forms for each of the four weeks of the survey period. At the end of the period, all of the forms from all of the salespeople are forwarded to the company's national headquarters. There the information from the ten sample branches is assimilated and summarized. In addition, national as well as local reports are generated and distributed.

Two reports are generated from the sales-call data. The first is the capsule summary. It highlights the main findings pertaining to calls reported and time spent during calls; included in this report are the average national statistics for the same month.

The second report is the total-branch statistical report. This computerized report

(1) gives the details of a branch's reported activities by product type and (2) displays time patterns and call patterns for various categories.

Upon receiving these two reports, branch management can compare the national data with its own specific branch data and evaluate any significant variance. More important, individual branch goals, performance, and problem areas can be evaluated.

Essentially, the monthly sales-call activity report classifies and summarizes input provided by each of the ten sample branches on a national and local basis. The shortcoming of this type of reporting procedure is that it tells headquarters and the surveyed branch offices what has happened in specific areas; it does not indicate what the desired results should have been. Each manager is left to make that decision. The national averages cannot properly be construed to a desired norm. Each is merely an average of ten surveyed branches.

Certainly, these reports are adequate for diagnosing a significant problem or area of needed improvement. However, if maximum time utilization for the sales force is to be attained, a specific call standard must be supplied in addition to the information regarding what actually occurred in each category. In this way, branch management will know exactly what corrective action to take in each area to maximize efficiency and sales of the sales staff. No longer will there be as many interpretations of a report, and as many corrective actions, as there are people reading it.

Recommendation: By using the statistical method of multiple regression and correlation analysis, these sales optimizing standards can be obtained for each category on the sales-call activity report. This can be done on a national, regional, or local basis to compensate for geographical differences.

For example, one of the major goals of the company is that of obtaining new net rental placements. This is accomplished by

1. Preventing losses through cancellation. (Sales calls made for this purpose will be referred to as "cancel-save calls.")
2. Adding to the company's rental population. (Sales calls made for this purpose will be referred to as "new-business calls.")
3. Trading from one renting customer to another. (Sales calls made for this purpose will be referred to as "trade calls.")

Since cancel-save, new-business, and trade calls are important to increasing "net adds," it would be very useful for a branch manager to know exactly how salespeople's time should be spent in these areas. This can be done in three steps.

First, by deriving the multiple regression equation, it is possible for a manger to know what change he can expect in the dependent variable, the "net adds" (y), with a one-unit change in an independent variable—cancel-save calls (X_1), new-business calls (X_2), or trade calls (X_3)—with the other two independent variables held constant. The equation from sample data will be in the form

$$\hat{Y} = a + b_1X_1 + b_2X_2 + b_3X_3$$

Second, it would be helpful to know what percent of the variation in "net adds" can be explained by all three of the independent variables taken together. This can be found by determining the coefficient of multiple determination.

Finally, it would be helpful to know the relative importance of cancel-save calls, new-business calls, and trade calls on the dependent variable, "net adds." This can be accomplished by determining the coefficients of partial determination.

The sales-call activity report shows the average number of hours per week and the average number of calls per week that salespeople spend making cancel-save, new-business, and trade-selling calls. However, it is left to an individual manager's intuition to determine whether adequate time is being spent on these types of calls and whether the amount of time spent on each is in proper proportion.

By taking national and weekly totals of several observations of time spent on these three types of calls and relating them to "net adds" achieved, specific answers to the above questions can be obtained.

Using hypothetical data for illustrative purposes, the following results were obtained:

$a = 73.11491$

$b_1 = .005843$

$b_2 = .554914$

$b_3 = 1.602222$

Thus, the multiple regression equation is

$$\hat{Y} = 73.11 + .0058X_1 + .5549X_2 + 1.6022X_3$$

Interpreting the equation for, say, X_2, it is apparent that, if a manager increases the total weekly number of hours per representative by 1 for making new-business calls, an average increase of .5549 "add points" can be achieved when the effects of X_1 and X_3 are held constant. Similar interpretations can be made for X_1 and X_3.

The coefficient of multiple determination, $R^2_{Y.123}$, is found to be .7755. This figure says that cancel-save, new-business, and trade calls taken together account for approximately 77.5% of the variation in "net add." Therefore, at least 77.5% of the time a salesperson spends in a week on the job should be spent making these types of calls. An examination of an actual weekly sales report revealed that the average percent of time per representative spent working on the job was for a specific week: 6.22 hours in the office, 19.82 hours with customers, and .71 hours at home. These sum to 26.75 hours per week. The actual time spent making these types of calls was 6.24 hours making cancel-save calls, 7.97 hours making new-business calls, and 3.1 hours making trade calls. These sum to 17.31 hours. This total, 17.31 hours, divided by the total hours spent on the job, 26.75, should be equal to 77.5%. In fact, it is equal to 64.7%. Therefore, the branch manager should increase the amount of time salespeople spend making these types of calls by 12.8%. This translates into an additional 20.8 hours per week per representative, rather than the current 17.31 hours.

Once it is known how much time should be spent doing these types of jobs, it is important to know the relative importance of each independent variable in influencing the dependent variable. The partial determination coefficient gives the percent of explained variation in the dependent variable for each independent variable, with the other two independent variables held constant. The resulting coefficients of partial determination follow:

Dependent Variable	Independent Variable	Coefficients of Partial Determination	Relative Importance
Net adds	Cancel-save calls	.053	9.2%
Net adds	New-business calls	.365	63.4
Net adds	Trade calls	.158	17.4
	Total	.576	100.0%

By dividing each coefficient of partial determination by the sum of all three coefficients, the relative importance of each independent variable on the dependent variable becomes apparent. What these figures suggest is that 9.2% of a salesperson's time should be spent on cancel-save calls, 63.4% on new-business calls, and 27.4% on trade calls.

The coefficient of multiple determination shows that 20.8 hours is the optimum total weekly time spent by a representative making these types of calls. By multiplying this optimum by the percent just derived, it is possible to know how the 20.8 hours should be divided between the three selling activities:

20.8 hours $\times$ .092 = 1.9 hours for cancel-save calls

20.8 hours $\times$ .634 = 13.2 hours for new-business calls

20.8 hours $\times$ .274 = 5.7 hours for trade calls

By subtracting the actual weekly figures from the calculated optimum figures, it is possible to know where sales time should be increased or decreased, and by exactly how much. This kind of calculation is presented here:

	Kind of Call		
	Cancel Save	New Business	Trade
Optimum	1.9 hours	13.2 hours	5.7 hours
Actual	−6.24 hours	− 7.97 hours	−3.10 hours
Needed change	−4.34 hours	+ 5.23 hours	+2.6 hours

By dividing the actual number of calls, for each of the three types, into the total observed time for making that type of call, an average time per type of call can be derived:

6.24 hours ÷ 9.6 calls = .65 hours per cancel-save call

7.97 hours ÷ 11.1 calls = .72 hours per new-business call

3.10 hours ÷ 4.1 calls = .76 hours per trade call

Once the average time per type of call is known, this can be divided into the optimum number of hours for each type of call, to determine the number of each type of call that should be made each week:

1.9 hours ÷ .65 hours = 3 cancel-save calls per week

13.2 hours ÷ .72 hours = 18 new-business calls per week

5.7 hours ÷ .76 hours = 8 trade calls per week

By subtracting the observed number of calls from the calculated optimum number of each type of call, the exact change in calling activity can be calculated:

	Kind of Call		
	Cancel Save	New Business	Trade
Optimum	3.0 calls	18.0 calls	8.0 calls
Observed	−9.6 calls	−11.1 calls	−4.1 calls
Net change	−6.6 calls	+ 6.9 calls	+3.9 calls

Projection: By including this kind of information in each sales-call activity report, in an easy-to-understand format, meaningful interpretation of this valuable report will not be left to the individual manager's judgment. Correct, positive decisions will be made immediately, and the entire Duplikwik sales force will operate closer to its maximum efficiency.

The concept of applying multiple regression and correlation analysis to other segments of this same report, and indeed to other areas in the company, can be a key factor in the ability of Duplikwik to maintain its place of leadership in the marketplace.

1. On the basis of your quick reading of this case, write an evaluation of the sales executive's proposal. Include comments on (1) the feasibility of implementing the program and (2) the advantages claimed for the proposed changes.

2. Now read the case with greater care, paying close attention to the executive's interpretations of statistical measures. For the most part, are his interpretations proper or improper? Explain.

3. After having analyzed the interpretations in this report, do you still agree with your answer to question 1?

16 Introduction to Time Series Analysis

A trend is a trend is a trend,
But the question is, will it end?
Will it alter its course
Through some unforeseen force
And come to a premature end?—"Stein Age Forecaster," as quoted by Alec Cairncross

This summer one-third of the nation will be ill-housed, ill-nourished, and ill-clad. Only they call it vacation.—Joseph Salak

What You Should Learn from This Chapter

From time immemorial, people have tried to second-guess the future. The many forms these attempts have taken, such as analyzing dreams or studying the entrails of dead birds, may strike us today as laughably unscientific, but then it is possible that our great grandchildren will feel the same way about our present forecasting methods. For that matter, many modern-day observers feel that the forecasting methods being used right now are no more scientific or any less riddled with superstition than those employed by oracles of old.

This is probably not true. But true or false, the fact remains that *forecasting must be done. It is an absolutely essential part of business planning.* If our methods are crude and imperfect, then we simply must do the best we can with crude and imperfect methods while working to develop better ones.

In this chapter we deal with the granddaddy—but still a very active granddaddy—of statistical forecasting methods—namely, forecasting by means of decomposed time series. When you have completed this chapter, you should be able to

1. Define *time series.*
2. Define *secular trend, cyclical oscillations, seasonal variation,* and *irregular fluctuations.*
3. Distinguish between the additive and multiplicative models.
4. Fit and project linear, exponential, and second-degree parabolic trends.
5. Scale down a trend equation from a yearly basis to a monthly or other short-term basis.
6. Demonstrate an awareness of the steps involved in isolating the seasonal component for situations in which (1) the seasonal pattern is stable and (2) the seasonal pattern is changing.
7. Isolate the cyclical and irregular components of a time series.
8. Show some awareness of how to harness this decomposition method to obtain forecasts.

Because the least squares method is used in this chapter, you may wish to review section 14.4 in Chapter 14.

16.1 Introduction

First, we will need some definitions.

> A *time series* is a set of measurements describing the state of some ongoing activity over time.

> When dealing with a time series in a formal way, we require that:
>
> 1. The time periods at which measurements are observed be of equal duration—that is, years, quarters, months, weeks, and so forth.
> 2. The relevant historical data be free of missing observations.
> 3. The observations be arranged in chronological order from the earliest time period to the most recent.
> 4. The data possess comparability over time—that is, the activity being measured, or the manner of measuring it, does not change abruptly.

Figure 16.1 shows some sample time series. Notice how different their appearances are even though these particular series measure similar economic activities.

16.2 The Components of a Time Series

Just as the human personality, according to the Freudians, is determined by the interplay of id, ego, and superego, the "personality" of a time series is determined by the interplay of (1) the secular trend, (2) cyclical oscillations, (3) seasonal variation, and (4) irregular fluctuations.

Any given time series may be affected greatly, little, or not at all by population growth or the rate of capital goods formation or changes in the weather or the vicissitudes of consumer sentiment, or . . . . A complete list of all relevant influencing variables might be endless; consequently, time series analysts do not ordinarily attempt such a listing. What they do instead is to organize, conceptually, the various influencing factors into broad categories according to the span of time over which their effects are felt. In practice, they do this indirectly by decomposing a time series of interest into parts. One part represents the combined effects of long-term factors (the secular trend); one part, the combined effects of intermediate-term factors (cyclical oscillations); one part, the combined effects of short-term but systematic factors (seasonal variation); and, finally, one part, the combined effects of short-term and unsystematic factors (irregular fluctuations).

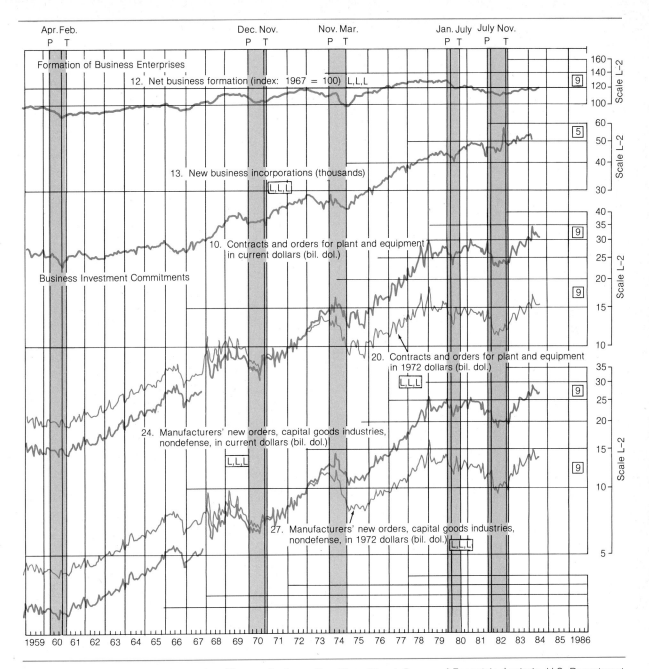

Figure 16.1. Some illustrative time series. (*Source: Business Conditions Digest,* Bureau of Economic Analysis, U.S. Department of Commerce.)

Secular Trend

> The term *secular trend* refers to the smooth upward or downward path of a time series over a period encompassing several years.

The concept of secular trend implies a fundamental continuity. Frequent changes in direction or in rates of increase or decrease are inconsistent with the very essence of the term. The direction and nature of the secular trend tend to be influenced by such things as population growth, stage of industry life cycle, technological developments, and longstanding changes in consumers' tastes or ways of doing things.

Cyclical Oscillations

> *Cyclical oscillations* consist of recurring, but not perfectly periodic, intermediate-term cumulative and reversible movements characterized by alternating periods of expansion and contraction. Most cyclical oscillations in specific time series are intimately linked to similar movements in the general economy.

Cyclical oscillations are in many respects the most vexing time series component. They are systematic—at least in the sense of recurring with some hint of regularity—but their durations vary over a substantial range and their shapes are anything but uniform. No completely dependable mathematical approach to forecasting the cyclical component of a time series has, as yet, been developed. This is not to say that we should not attempt to forecast cyclical movements; their importance to business planning dictates otherwise.

Seasonal Variation

> *Seasonal variation* consists of intrayear movements that recur each year with a more-or-less regular pattern, although in some time series this component changes gradually over the years.

Seasonal variation results from annually recurring changes in the weather and from customs related to the time of the year such as vacation periods, the school-year cycle, Christmas shopping, and so forth.

Irregular Fluctuations

The *irregular fluctuations* are those that remain after the trend, cyclical, and seasonal movements have been identified. These movements are of two kinds: (1) sporadic changes and (2) minor random movements.

Sporadic changes result from single, identifiable, one-time events, such as weather conditions that are highly unusual by historical standards, strikes in major industries, unusual political developments, and so forth. These produce temporary drops or spurts in the underlying trend-cycle pattern.

Minor random movements result from many, generally unknown causes including (1) factors related to the commonsense, though somewhat untidy, possibility that economic activities may indeed unfold in a jerky, jagged manner and (2) statistical inadequacies such as sampling errors, response errors, and errors caused by faulty seasonal adjustment. These give the time series a sawtooth appearance.

Irregular fluctuations, particularly the minor random movements, are considered unpredictable and are therefore ignored when preparing a forecast by the time series decomposition method described in this chapter. However, to say they are ignored is not the same as saying they are unimportant. For example, the analyst who must recognize quickly and with little likelihood of error when a cyclical turning point has occurred will find his ability to do so greatly impaired if large irregular fluctuations are present in the data. This point is illustrated in Figure 16.2.

16.3 How the Components of a Time Series Are Combined

Two basic models have been employed by statisticians and economists to indicate the manner in which the four components of a time series are combined: the additive model and the multiplicative model.

The *additive model* can be expressed as

$$Y = T + C + S + I \tag{16.3.1}$$

where Y represents the value of a time series of interest at a specific point in time and T, C, S, and I represent values of the trend, cyclical, seasonal, and irregular components, respectively, at the same point in time.

The additive model is used in connection with time series that have both negative and positive values. In general, however, it has met with less favor than the multiplicative model

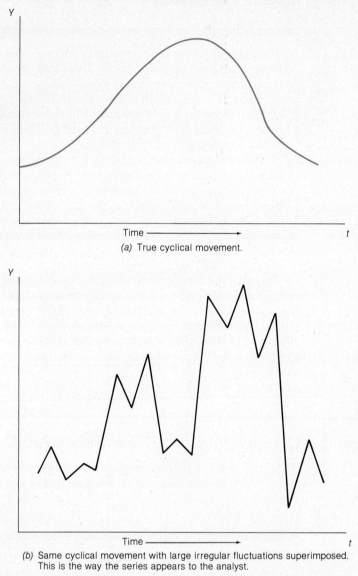

(a) True cyclical movement.

(b) Same cyclical movement with large irregular fluctuations superimposed. This is the way the series appears to the analyst.

Figure 16.2. How irregular fluctuations can interfere with the correct interpretation of the current stage of a cyclical movement.

because of the built-in assumption of independence among the four time series components. The additive model says, in effect, that regardless of how sustained the trend movement may be over time or how great the values of the trend component eventually become, the magnitudes of the cyclical, seasonal, and irregular fluctuations will remain unaffected.

The *multiplicative model* can be expressed as

$$Y = TCSI \qquad (16.3.2)$$

The symbols represent the same things they did in the additive model.

When this model is used, the implicit assumption is that the time series components are interdependent in some manner. We will follow the example of the majority of economists and economic statisticians and deal exclusively with the multiplicative model.

Before proceeding with the measurement of the four time series components, let us consider the following illustrative case:

ILLUSTRATIVE CASE 16.1

Outdoor Activities, Inc., is a small but growing manufacturer of camping and warm weather sports equipment. Ten years of sales data are presented in Table 16.1. These data are also shown graphically in Figure 16.3.

Forrest Hill, the young founder and president of the company, is painfully aware that sales are influenced importantly by both seasonal and cyclical factors. Beyond this awareness, however, he knows nothing about the structure of the company's sales data from the standpoint of trend, cyclical, seasonal, and irregular elements. He asks us to perform a standard time series analysis and to report on our findings.

Table 16.1 Monthly Data and Annual Totals for the Outdoor Activities Case, 1975 to 1984

(Thousands of Current Dollars)

Month	Year									
	1975	1976	1977	1978	1979	1980	1981	1982	1983	1984
January	78	87	131	74	71	126	143	91	87	138
February	82	111	117	88	81	129	141	88	97	156
March	122	157	158	118	125	162	175	104	121	196
April	130	189	177	114	130	173	185	102	141	228
May	146	192	176	109	159	214	195	96	151	242
June	141	173	165	71	149	238	214	85	158	245
July	135	191	170	90	142	216	173	86	162	253
August	155	187	152	101	167	212	149	97	167	260
September	141	186	136	104	141	197	137	101	178	274
October	148	193	130	94	157	205	131	103	180	273
November	135	143	108	81	131	191	114	96	166	240
December	109	125	79	81	126	152	91	87	139	198
Total	1522	1934	1699	1125	1579	2215	1848	1136	1747	2703

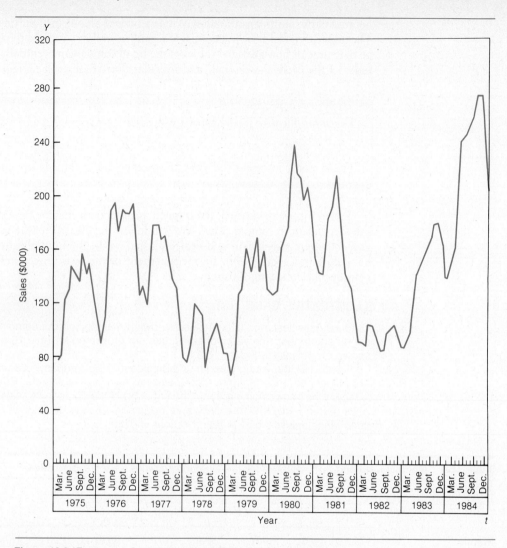

Figure 16.3. Ten years of monthly sales data for Outdoor Activities, Inc.

16.4 Introduction to Trend Fitting

A convenient place to begin analyzing a time series is with the trend component since the methodology involved is much like that of simple regression analysis.

Steps Preparatory to Trend Fitting

Before beginning to fit a trend, certain preliminary steps should be taken. First, annual totals should be obtained from the monthly totals if the former are not already known. The reason is that it makes no difference to the results whether the time series analyst (1) determines the trend equation from monthly data directly or (2) determines the trend equation from annual data and then modifies the equation appropriately to get monthly trend values. In the meantime, he or she can obviously work with, say, 10 or 20 values much more easily than with 12 times as many.

Second, serious thought should be given to whether the data are already in the appropriate form or whether some systematic alteration needs to be made. For example, for some purposes, the data should be divided by a price index so that changes solely attributable to price inflation are removed from the analysis. For other purposes, the values should be divided by population data so that they are expressed on a per capita basis. Although we will not encumber the example by doing so, one could argue that the Outdoor Activities data should be divided by a price index.

Third, some detective work should be undertaken to make certain that the data are comparable over time. If sales have been measured differently or if the nature of the company's activities has changed appreciably during the period represented by the historical data, such noncomparability sabotages the analysis before it is even under way.

Fourth, the data should be presented in chart form, as in Figure 16.3. Not only does this preliminary step give the analyst some idea of whether the data actually contain all four time series components—some series lack one or more of the usual components—but it also provides one with a feel for the appropriate type of trend to fit.

16.5 Fitting a Linear Trend by the Method of Least Squares

By far the most frequently used method of fitting a trend to a set of time series data is the least squares method. The method of least squares is so widely used because it provides a mathematical method, based on a reasonable criterion of goodness of fit, which can not only describe the trend in the historical data but which also permits extrapolation of the trend into the future. However, you should be aware of this:

> The least squares method, in the context of time series analysis, does not have associated with it the probability features that it boasts in regression analysis. Consequently, it does not require acceptance of the relatively restrictive assumptions cited in Chapters 14 and 15, nor does it deliver as great a variety of information as in regression analysis.

Since the data in Figure 16.3 appear to be describable by a straight line, we will proceed to determine the linear trend equation. In accordance with the first of the preparatory steps suggested a few paragraphs back, we will work with annual data and later adapt the trend equation so that it will provide monthly trend estimates. The annual data for the Outdoor Activities problem are shown in Table 16.2. Data in column (2) are the values assigned to time, hereafter called t values (but having nothing whatever to do with the t distribution discussed in earlier chapters).

One can be rather arbitrary in assigning values to time periods provided that two simple rules are followed: First, the smallest numbers must be assigned to years furthest back in time and the largest numbers to the most recent years. Second, the difference between successive pairs of time values, when the smaller member of the pair is subtracted from the larger, must be constant throughout the list of differences.

Accordingly, in Table 16.2, since 1975 is the least recent year, it was assigned the smallest time value, namely, 1; since 1984 is the most recent year, it was assigned the largest value, namely, 10. Moreover, the difference between the values assigned to the years 1975 and 1976 is $2 - 1 = 1$; the same difference holds for the years 1976 and 1977: $3 - 2 = 1$; and so forth, the difference being 1 for all adjacent values for time. Columns (3) and (4) of Table 16.2 are simply the familiar worksheet columns showing the values of tY (corresponding to XY in simple linear regression analysis) and t^2 (corresponding to X^2) whose sums are needed when applying the least squares method.

The linear trend equation will be of the form

$$T_t = a + bt \qquad (16.5.1)$$

where T_t represents the trend value at time period t.

Because we are using the least squares method, we determine a and b by the formulas

$$b = \frac{\Sigma\, tY - \dfrac{(\Sigma\, t)(\Sigma\, Y)}{n}}{\Sigma\, t^2 - \dfrac{(\Sigma\, t)^2}{n}} \qquad (16.5.2)$$

$$a = \frac{\Sigma\, Y}{n} - b\,\frac{\Sigma\, t}{n} \qquad (16.5.3)$$

These, of course, are the same equations used in Chapter 14. Only the symbols have been changed to make them consistent with those used in connection with the time series models introduced earlier.

As indicated at the bottom of Table 16.2, the trend equation was found to be

$$T_t = 1440.47 + 56.4242t$$

$$0 = 1974 \qquad t \text{ unit} = 1 \text{ year}$$

Supporting credentials, like those shown under the trend equation (read "1974 is the year associated with $t = 0$; the slope value shows by how much the trend value increases *per year*"), should always accompany the trend equation.

Table 16.2 Worksheet for Deriving the Linear Trend for the Outdoor Activities Problem

(1) Year	(2) Time Code t	(3) Sales Y	(4) tY	(5) t^2	(6) T_t
1975	1	1522	1522	1	1496.89
1976	2	1934	3868	4	1553.32
1977	3	1699	5097	9	1609.74
1978	4	1125	4500	16	1666.17
1979	5	1579	7895	25	1722.59
1980	6	2215	13290	36	1779.02
1981	7	1848	12936	49	1835.44
1982	8	1136	9088	64	1891.86
1983	9	1747	15723	81	1948.29
1984	10	2703	27030	100	2004.71
Total	55	17,508	100,949	385	17,508.03

$$b = \frac{\Sigma\, tY - \dfrac{(\Sigma\, t)(\Sigma\, Y)}{n}}{\Sigma\, t^2 - \dfrac{(\Sigma\, t)^2}{n}} = \frac{100,949 - \dfrac{(55)(17,508)}{10}}{385 - \dfrac{(55)^2}{10}} = 56.4242$$

$$a = \frac{\Sigma\, Y}{n} - b\left(\frac{\Sigma\, t}{n}\right) = \frac{17,508}{10} - (56.4242)\left(\frac{55}{10}\right) = 1440.47$$

$$T_t = 1440.47 + 56.4242t$$
$$0 = 1974 \qquad t \text{ unit} = 1 \text{ year}$$

To obtain the trend values, we merely substitute t values associated with specified years of interest into the equation and solve. For example, the trend value for the year 1984 is $T_{10} = 1440.47 + (56.4242)(10) = 2004.71$. The T_t values for other years used to determine the trend equation are displayed in column (6) of Table 16.2.

Projecting the Trend

Once the trend equation has been determined, projecting the trend line is simplicity itself. All that is required is the substitution of the appropriate values for t into the trend equation. For example, let us say that the analyst wishes early in 1985 to obtain trend estimates for the years 1989 and 1990. Since the t value assigned the year 1984 is 10 and the constant difference between adjacent t values is 1, he would calculate

$$T_{15} = 1440.47 + (56.4242)(15) = 2286.83$$

$$T_{16} = 1440.47 + (56.4242)(16) = 2343.26$$

A little later, we will list some precautions which should be exercised when projecting a trend.

Changing the Scale of the Trend Equation

We stated earlier that, when fitting the trend equation to data where monthly values are of interest, we could (1) actually fit the trend to the monthly Y values or (2) fit the trend to the yearly Y values and then change the scale of the resulting equation so that the trend values produced represent monthly rather than yearly values. We chose the latter option, which means we are now obliged to make the trend equation better suited to the production of monthly trend values. (Obliged, that is, if we wish to complete the time series decomposition procedure which is the main concern of this chapter. If the trend component were the analyst's sole concern, chances are that he or she would not bother to adapt the trend equation to monthly values.)

Before we proceed, it is important for you to be aware that the procedures we look at will pertain only to situations where the annual values are *sums* of the corresponding 12 monthly values. Such is the case with our Outdoor Activities example. It is not always the case with time series data. For example, annual values for most index number series are *averages*, rather than sums, of the monthly values. For such series the only modification in the trend equation needed to convert trend values to a monthly basis would be $T_{t/12} = a + b(t/12)$. Or better from the standpoint of ease of use: $T_{t/12} = a + (b/12)t$. The rationale behind this kind of arithmetic should become evident as we discuss the more difficult problem of converting a trend equation to a monthly basis when annual Y values represent *totals* of 12 monthly Y values. The scaling down (or "stepping down") process involves two steps: (1) scaling down the Y variable and (2) scaling down the t variable.

Scaling Down the Y Variable. The first step in changing the scale of measurement is to divide both sides of the trend equation by a factor representing the ratio of the length of the period of the old unit to the length of the period of the new unit. That is, because the length of the old period represented by t is 12 (months) and the length of period to be represented by t when the conversion of the trend equation is completed is 1 (month), we divide both sides of the yearly trend equation by 12. Thus, if we begin with the equation of the form

$$T_t = a + bt$$

we would now write it

$$\frac{T_t}{12} = \frac{a}{12} + \frac{b}{12}t \qquad (16.5.4)$$

This is reasonable. Since T_0, the trend value associated with the year for which $t = 0$, represents the sum of 12 monthly values and we now wish it to represent the trend value associated with a single month, dividing a by 12 seems like the commonsense solution. Also, since b originally represented the average *annual* increase in sales and now represents merely the average *monthly* increase over a year's time, dividing by 12 seems like a reasonable way of interpolating. These points are presented graphically in part (*a*) of Figure 16.4.

Scaling Down the t Variable. Equation (16.5.4), called the *annual monthly average* trend equation, is useful for some purposes. For example, it does convey the idea that the trend value this month is higher by $b/12$ than it was 12 months ago. For our purposes, however, the equation requires one more refinement, the reason being that we do not seek *monthly average* trend values but rather monthly trend values per se. The distinction is illustrated in

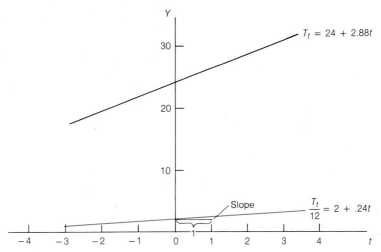

(a) When the Y, but not the t, scale is reduced from accommodating annual totals to accommodating monthly totals, a and b become $\frac{1}{12}$ of their original values. b, however, is a monthly increment associated with a one-year time period, as shown, and is interpreted as the annual monthly average increase (or decrease). It indicates how much higher (or lower) the trend value is compared to the year before.

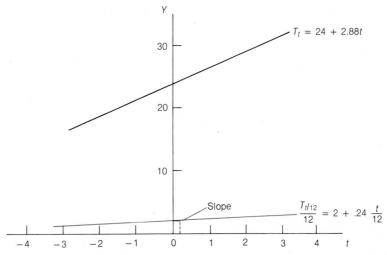

(b) The "scaling down" procedure is completed by, in addition to the divisions performed in (a), dividing t by 12, which has the effect of reducing b to $\frac{1}{144}$ of its original value, as shown above, where the same monthly equation as in (a)—but with a different time increment—is used. The trend equation now produces actual trend values for specified months.

Figure 16.4. What happens (a) when Y is scaled down and (b) when both Y and t are scaled down. Trend equations used are hypothetical.

Figure 16.4. To obtain the equation we seek, we must, in addition to the dividing already done, divide t by 12 giving

$$\frac{T_{t/12}}{12} = \frac{a}{12} + \left(\frac{b}{12} \cdot \frac{t}{12}\right) = \frac{a}{12} + \frac{b}{144} t \qquad (16.5.5)$$

Of course, if we wished to scale the trend equation down to a quarterly basis, we would write it $(T_{t/4}/4) = (a/4) + (b/16)t$. For a weekly basis: $(T_{t/52}/52) = (a/52) + (b/2704)t$. The notations on the left-hand side of the scaled-down equations are becoming rather cumbersome. They have been used here simply to distinguish between trend equations before and after the scaling-down process. The symbol T_t would be quite adequate for the monthly trend equation, provided that an identification statement were appended to the equation indicating that the t unit is now one month.

We will illustrate the application of Equation (16.5.5) by using the Outdoor Activities yearly equation presented earlier. That equation was $T_t = 1440.47 + 56.4242t$ ($0 = 1974$; t unit = 1 year). Converting this equation as indicated by Equation (16.5.5), we get

$$T_t = 120.04 + .3918t$$

$$0 = \text{June–July 1974} \qquad t \text{ unit} = 1 \text{ month}$$

Shifting the Origin of a Trend Equation

A minor problem still remains: When one scales down from yearly to monthly data, the origin is situated between two months rather than on a single month. In the present case, the origin is between June and July of 1974. To shift the origin of the above trend equation to July 1974, we would calculate

$$T_t = 120.04 + .3918(t + .5) =$$

$$T_t = 120.24 + .3918t$$

$$0 = \text{July 1974} \qquad t \text{ unit} = 1 \text{ month}$$

Thus, the trend value for, say, January 1975 would be obtained by $T_t = 120.24 + (.3918)(6) = 122.59$. Other monthly trend values would be obtained in a similar manner.

If desired, the origin can be moved even further: Just (1) count the number of months ahead that the origin is to be shifted, (2) add .5 to that number (but only if the shift from between two months onto a single month has not already been accomplished), (3) multiply the resulting sum by the slope value b, and (4) add the resulting product to the Y-intercept value a.

Figure 16.5 shows the original monthly sales data and the linear trend line using the equation $T_t = 120.24 + .3918t$ ($0 = \text{July 1974}$; t unit = 1 month).

Let us now set the Outdoor Activities problem aside temporarily while we consider some other trend types frequently used in time series analysis.

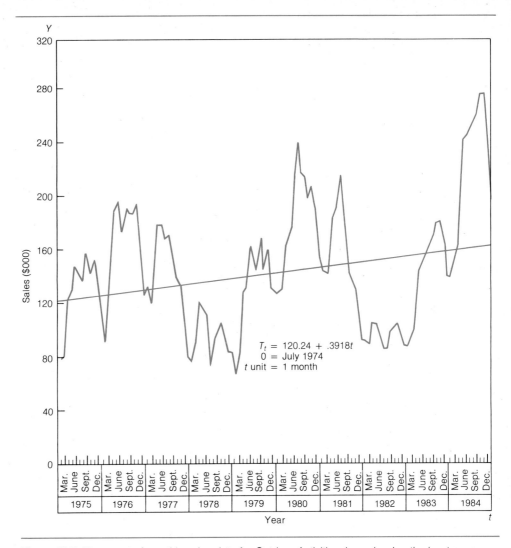

Figure 16.5. Ten years of monthly sales data for Outdoor Activities, Inc., showing the least squares linear trend line.

The chart annotation reads:
$$T_t = 120.24 + .3918t$$
$$0 = \text{July 1974}$$
$$t \text{ unit} = 1 \text{ month}$$

16.6 Fitting an Exponential Trend by the Method of Least Squares

Use of a trend equation of the type $T_t = a + bt$ is based on the assumption that the true trend values increase (or decrease) by a constant amount each year (each time period). This is an apt description of the nature of the trend in many time series. For others, however, the straight line simply does not adequately describe the long-run "drift" of the series. In such a case, use of some kind of nonlinear trend may be called for. One type of nonlinear trend equation which describes the pattern of growth in many time series is

$$T_t = ab^t \qquad\qquad (16.6.1)$$

This is called an *exponential* trend and is simply a thinly disguised version of an equation you may have used to solve compound-interest problems. Use of this kind of trend equation is based on the assumption that the true trend values in the series under analysis increase (or decrease) by a *constant* percent each year (each time period).

Recognizing an Exponential Trend

For reasons to be explained momentarily, one can determine whether an equation of the type $T_t = ab^t$ is called for by plotting the Y values first on arithmetic grid graph paper and then on semilogarithmic graph paper.

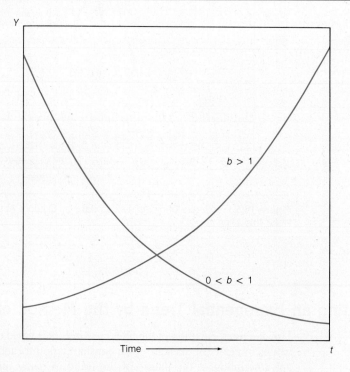

Figure 16.6. Two exponential trend curves.

16 Introduction to Time Series Analysis

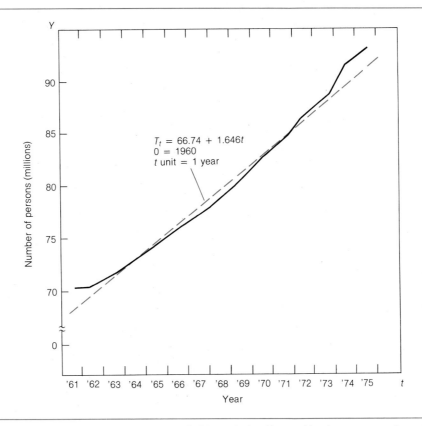

$T_t = 66.74 + 1.646t$
$0 = 1960$
t unit = 1 year

Figure 16.7. Fifteen years of data on the U.S. Civilian Labor Force with a least squares linear trend (dashed line). (*Source:* U.S. Bureau of Labor Statistics.)

> *Semilogarithmic*, or just *semilog*, paper has an arithmetic horizontal grid and a logarithmic vertical grid. That is, the vertical axis is calibrated in such a way that the vertical distances are proportional to logarithms of the numbers rather than to the numbers themselves.

Use of this kind of graph paper is a convenient alternative to plotting t values against log Y values on graph paper having two arithmetic scales. With semilog paper, one can plot t against the original Y values, and the vertical grid will do the transforming.

Here is what one looks for in using the two kinds of graph paper: (1) If the points when plotted on arithmetic grid paper show a nonlinear pattern resembling either of the curves in Figure 16.6 and (2) if the points when plotted on semilog paper appear to have a pattern describable by a straight-line trend, then an equation of the form $T_t = ab^t$ will probably be satisfactory. For example, let us examine Figures 16.7 and 16.8, which show 15 years of data for the civilian labor force of the United States. In Figure 16.7 the data are plotted on regular

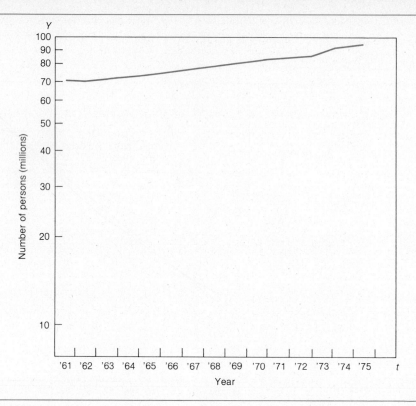

Figure 16.8. Fifteen years of data on the U.S. Civilian Labor Force charted on semilog paper. (*Source:* U.S. Bureau of Labor Statistics.)

arithmetic grid paper. Although a linear trend does not provide a grossly poor fit to the data, the nature of the deviations, $Y - T_t$ values—especially the fact that they are positive for a few consecutive years at the beginning and the end of the 15-year period and negative for several consecutive years in the middle—suggests that a trend with an upward curvature would be better. Figure 16.8 appears to support this conclusion. Notice that when the Y values are presented on semilog paper, the dots form what looks like a linear pattern.

Using Logarithms to Fit an Exponential Trend

To fit an exponential trend to a set of time series data, we could actually work with the equation $T_t = ab^t$. However, because t in the equation is an exponent and could conceivably be very large, it is usually easier to work with this equation in its logarithmic form, which is

$$\log T_t = \log a + t \log b \qquad (16.6.2)$$

Clearly, since a and b are constants, log a and log b are also constants. The equation is now effectively the straight-line equation save for the fact that the left-hand side is log T_t rather than just T_t. To obtain the constants needed for Equation (16.6.2), we make use of the regular a and b equations for a linear trend, except that log Y will appear wherever Y appeared previously. That is,

$$\log b = \frac{\Sigma\, t \log Y - \dfrac{(\Sigma\, t)(\Sigma\, \log Y)}{n}}{\Sigma\, t^2 - \dfrac{(\Sigma\, t)^2}{n}} \qquad (16.6.3)$$

$$\log a = \frac{\Sigma\, \log Y}{n} - \log b\!\left(\frac{\Sigma\, t}{n}\right) \qquad (16.6.4)$$

The computational work required is shown in Table 16.3. Column (1) shows the original Y values for the civilian labor force. The logarithms of these Y values are presented in column (2). Column (3) shows the time values assigned to each of the 15 years. Columns (4) and (5) are simply worksheet columns needed for application of Equations (16.6.3) and (16.6.4). The trend equation turns out to be

$$\log T_t = 1.829658 + .008904t$$
$$0 = 1960 \qquad t \text{ unit} = 1 \text{ year}$$

Column (6) of Table 16.3 shows the trend values in logarithmic form—obtained by substituting t values of 1, 2, 3, . . . , 15 into the trend equation. Because the trend values are usually not very meaningful in logarithmic form, most time series analysts using an exponential trend will make use of antilogs to convert the log T_t values to T_t values. The T_t values for our present problem are displayed in column (7) of Table 16.3.

Projecting the Trend

To project the trend into the future, the same procedure is followed as was demonstrated in connection with linear trends. For example, the trend value for the civilian labor force for 1976 would be

$$\log T_t = 1.829658 + (.008904)(16) = 1.972122$$

Converting 1.972122 to its antilog, we get approximately 93.8.

Reducing the t Scale

Since annual values for this labor force series are expressed as *averages*, rather than totals, of the 12 corresponding monthly values, scaling the equation down to a monthly basis would entail merely reducing the t scale by dividing each t value by 12. That is,

Table 16.3 Least Squares Straight-Line Trend Fitted to Logarithms of Data for the Civilian Labor Force of the United States, 1961 to 1975

(Millions of Persons)

Year	(1) Y	(2) $\log Y$	(3) t	(4) $t \log Y$	(5) t^2	(6) $\log T_t$	(7) T_t
1961	70.5	1.848189	1	1.848189	1	1.837257	68.8
1962	70.6	1.848805	2	3.697610	4	1.847466	70.4
1963	71.8	1.856124	3	5.568372	9	1.856370	71.7
1964	73.1	1.863917	4	7.455668	16	1.865274	73.3
1965	74.5	1.872156	5	9.3607	25	1.874178	74.9
1966	75.8	1.879669	6	11.278014	36	1.883082	76.4
1967	77.4	1.888741	7	13.221187	49	1.891986	78.0
1968	78.7	1.895975	8	15.167800	64	1.900890	79.4
1969	80.7	1.906874	9	17.161866	81	1.909794	81.2
1970	82.7	1.917506	10	19.175060	100	1.918698	82.9
1971	84.1	1.924796	11	21.172756	121	1.927602	84.6
1972	86.5	1.937016	12	23.244192	144	1.936506	86.4
1973	88.7	1.947924	13	25.323012	169	1.945410	88.2
1974	91.1	1.959518	14	27.433252	196	1.954314	90.0
1975	92.5	1.966142	15	29.492130	225	1.963218	91.9
Total		28.513352	120	230.599888	1240	28.513350	

$$\log b = \frac{\Sigma\, t \log Y - \dfrac{(\Sigma\, t)(\Sigma \log Y)}{n}}{\Sigma\, t^2 - \dfrac{(\Sigma\, t)^2}{n}} = \frac{230.599888 - \dfrac{(120)(28.513352)}{15}}{1240 - \dfrac{(120)^2}{15}} = .008904$$

$$\log a = \frac{\Sigma \log Y}{n} - \log b\, \frac{\Sigma\, t}{n} = \frac{28.513352}{15} - (.008904)\left(\frac{120}{15}\right) = 1.829658$$

$$\log T_t = 1.829658 + .008904t$$
$$0 = 1960 \qquad t \text{ unit} = 1 \text{ year}$$

$$\log T_{t/12} = \log a + \left(\frac{t}{12}\right)\log b$$

or, often more conveniently,

$$\log T_{t/12} = \log a + \frac{\log b}{12}\, t \qquad\qquad (16.6.5)$$

For our specific example, the trend equation for yearly data was found to be

$$\log T_t = 1.829658 + .008904t$$
$$0 = 1960 \qquad t \text{ unit} = 1 \text{ year}$$

For monthly data, it would be $\log T_{t/12} = 1.829658 + .000742t$ (0 = June–July 1960; t unit = 1 month).

Shifting the Origin of an Exponential Trend

To center the exponential trend values on a specific month, we can multiply the log of the slope for monthly data by $(t + .5)$ and add the result to log a. That is, if we begin with the equation

$$\log T_{t/12} = 1.829658 + .000742t$$
$$0 = \text{June–July } 1960 \qquad t \text{ unit} = 1 \text{ month}$$

and wish to shift the origin to July 1960, we would calculate

$$\log T_{t/12} = 1.829658 + .000742(t + .5)$$
$$= \log T_{t/12} = 1.830029 + .000742t$$
$$0 = \text{July } 1960 \qquad t \text{ unit} = 1 \text{ month}$$

Reducing the Y Scale

As we noted earlier, the present illustrative problem is simpler to "step down" to a monthly basis than is the kind involving annual values which are *totals*, rather than averages, of monthly values. If the problem had been one of the other kind, we would be required to scale down the Y variable in addition to the t variable. This is accomplished in the following manner: In exponential form, the trend equation using annual data is expressed as $T_t = ab^t$. After the t scale has been reduced by division of t by 12, the equation becomes $T_{t/12} = ab^{t/12}$. To reduce the Y scale, we now must divide through by 12, that is, $(T_{t/12}/12) = (ab^{t/12})/12 = (a/12)b^{t/12}$. In logarithmic form, the yearly equation $\log T_t = \log a + t \log b$ becomes, when similarly "stepped down,"

$$\log T_t = \log a - \log 12 + \frac{\log b}{12} t \qquad (16.6.6)$$

16.7 Fitting Parabolic Trends by the Method of Least Squares

Occasionally the expression $T_t = a + bt + ct^2$, a quadratic, or second-degree, equation, will describe the underlying sweep of the data better than either a linear or an exponential equation. Because the method used to obtain a second-degree estimating equation is a rather straightforward extension of the linear procedures, we will not work through an example here. However, presentation of a few sundry facts, paralleling those cited above for the linear and exponential trends, is in order.

The a, b, and c constants are obtained by solving simultaneously the following normal equations:

$$\Sigma\, Y = na + b\,\Sigma\, t + c\,\Sigma\, t^2$$
$$\Sigma\, tY = a\,\Sigma\, t + b\,\Sigma\, t^2 + c\,\Sigma\, t^3 \qquad (16.7.1)$$
$$\Sigma\, t^2Y = a\,\Sigma\, t^2 + b\,\Sigma\, t^3 + c\,\Sigma\, t^4$$

If Y values represent annual values (the sums of 12 monthly values), the scale of the Y variable can be reduced by dividing through by 12. That is, we begin with a trend equation of the type $T_t = a + bt + ct^2$. Dividing through by 12 gives $T_t/12 = (a/12) + (b/12)t + (c/12)t^2$. To scale down the t variable as well, we divide t by 12 wherever it appears in the yearly trend equation. Thus,

$$\frac{T_{t/12}}{12} = \frac{a}{12} + \left(\frac{b}{12}\right)\left(\frac{t}{12}\right) + \left(\frac{c}{12}\right)\left(\frac{t}{12}\right)^2 = \frac{a}{12} + \frac{b}{144}t + \frac{c}{1728}t^2 \qquad (16.7.2)$$

To shift the origin from a point between two months to the next month, that is, a one-half month change, we would calculate

$$\frac{T_{t/12}}{12} = \frac{a}{12} + \frac{b}{144}(t + .5) + \frac{c}{1728}(t + .5)^2$$

Higher-degree parabolic trends have the virtue of being more flexible than the linear or exponential trends. Some kind of parabolic trend curve can be made to fit historical data whose trend pattern is too complex to be described adequately by either of the simpler trend types. Indeed, if a parabolic equation of high enough degree is used, all the waves and wiggles of the historical data can be described quite well by the trend curve. However, the point is quickly reached where the trend becomes not a trend at all as economic statisticians have traditionally used the term. The trend becomes a very complex curve tending to follow the cyclical and even shorter-term movements. Moreover this "trend" cannot be projected without great risk. Because of the large exponents attached to the t values, the "trend" is likely to go rocketing skyward (or earthward) outside the range of the observed data in a manner completely unrelated to what the realities of the situation would permit. One must, therefore, use wisdom and restraint when employing parabolic trend equations.

16.8 Cautionary Notes on Trend Projection

Trends are sometimes fitted to time series data for purposes other than forecasting. For example, analysts of economic developments will sometimes fit trends to historical data for two or more time series simply to compare descriptively their rates of growth over several past years. Another nonforecasting purpose of fitting trends is that of isolating the cyclical movements so they can be studied free of the (sometimes) obscuring effects of the trend movement. Still, forecasting is certainly the most common motivation behind trend fitting.

When using a trend for forecasting purposes, one must be somewhat more cautious than is necessary when using the other applications just cited. These other applications usually require simply that a reasonably well-fitting trend curve be used. Projecting a trend, on the other hand, is a kind of inductive, as distinguished from a purely descriptive, process. As such, the goal is more heroic and the hazards greater. What follows are some cautionary remarks that should be heeded when fitting a trend for purposes of projection.

1. Keep in mind and, if possible, evaluate critically the underlying assumption of trend projection.

> Projecting a trend, although a task relatively free of restrictive assumptions, does depend on the assumption that the collection of factors influencing a time series of interest during the period used in obtaining the trend equation will continue to influence the series in essentially the same way in the forecast period.

You should be aware that sometimes foreseeable changes in causal factors do promise to occur, in which case modification of the projected trend values will be required.

2. Keep in mind that projected trend values are estimates of only one component of a time series.

   For forecasting just a few periods ahead, the accuracy of forecasts based on trend projection will depend not only on the adequacy of the fit to the historical data and the validity of the assumption mentioned above, but also on the relative magnitudes of cyclical, seasonal, and irregular fluctuations. For example, forecasts based on trend projections might be quite erroneous, in a sense, in any given year because the cyclical component departs substantially from the trend line during that year. Such a situation would not constitute a condemnation of the trend equation for forecasting purposes. Use of trend projections is usually oriented toward long-term prognosis. Over several years the shorter-term movements, including the cyclical ones, can be expected to average out but the trend will, one hopes, continue to move along its historically familiar path.

3. When in doubt about trend type, a straight-line trend should generally be used.
   The linear trend type is the simplest to fit and its equation easiest to interpret. Moreover, the linear trend does not produce explosive projections as do some of the more complex curves.

4. When in doubt about trend type, do not use a higher-degree parabolic trend.
   A higher-degree parabolic trend may provide a beautiful fit to the historical data, but its propensity for generating unrealistic and unattainable projections is great enough to warrant much caution in its use.

5. The amount of historical data used in deriving the trend equation should be such that it reflects as closely as possible the influences of factors that are likely to continue to affect the series in the future.

This is sometimes a difficult requirement to observe because it requires more knowledge than one will usually have in practice. Still, a few guidelines regarding the amount of historical data to use might prove helpful. (1) In general, more recent historical data contain more information regarding the future path of the series than do less recent historical data. (2) However, several years of past data are required to ensure that the effects of cyclical oscillations are neutralized, as illustrated in part (*a*) of Figure 16.9. (3) As a general rule, *10 to 15 years of historical data will suffice.*

16.9 Introduction to the Measurement of Seasonal Variation

Seasonal variation is a second time series component characterized by a reasonably high order of predictability. Seasonal variation may be of interest for either of two reasons: (1) The seasonal movements are important in their own right or (2) the seasonal movements may

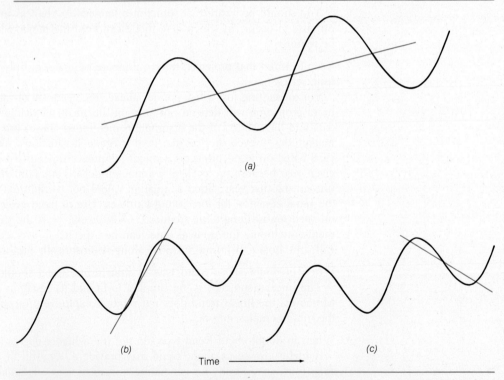

Figure 16.9. Good and bad quantities of historical data for trend projection. (*a*) Very nearly an ideal amount of historical data. Two complete cycles (about 7 to maybe 12 years) are included in the observed data. Notice also that the beginning and end points of the trend line are situated at approximate halfway points on the cycles. Realistic trend projections should be achieved. (*b–c*) Obvious examples of uses of inadequate amounts of historical data for trend fitting. Misleading trend projections will be the inevitable result.

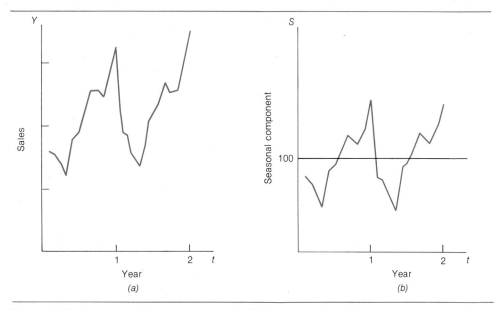

Figure 16.10. A case in which seasonal variation (*b*) dominates the original sales series (*a*) and is of interest in its own right rather than being of interest only as a hindrance to analysis of the cyclical oscillations.

hinder interpretation of the more important cyclical oscillations and must, therefore, be removed from the data. Figure 16.10 illustrates the first purpose of measuring seasonal variation. In part (*a*) the original series for sales of several kinds of hobby items by a retail store is shown. In part (*b*) the seasonal fluctuations have been isolated from the other time series components. Clearly, the seasonal variation in this case dominates the behavior of the overall series. Presumably, therefore, for purposes of determining the quantities of these items to order for intrayear needs, the store owner would wish to pay much attention to the historical seasonal pattern and would be little interested in the cyclical oscillations, which are moderate, the trend movement, which is almost nonexistent, or the irregular fluctuations, which are small.

On the other hand, in a situation like that depicted earlier in Figure 16.2, it could just as easily be the seasonal variation as the irregular fluctuations that obscures the underlying cyclical movement. If the cyclical oscillations are of principal interest, seasonal variation is usually removed statistically because, otherwise, one cannot tell until too late to take advantage of the fact that a cyclical turning point has occurred.

16.10 The Role of Seasonal Indexes

Since the number of calculations is great and the list of steps long for measuring and isolating the seasonal variation, confusion will be minimized by our watching the last act first—that is, by examining the role played by the seasonal indexes before we attempt to derive them ourselves. For simplicity, we will, until further notice, assume that the seasonal pattern of

Table 16.4 Seasonal Indexes
for the Outdoor Activities
Sales Data

Month	Seasonal Index
January	76.8
February	78.9
March	101.8
April	108.8
May	119.4
June	110.6
July	110.1
August	112.5
September	105.7
October	108.7
November	89.2
December	77.7

the Outdoor Activities sales data is one which is repeated in pretty much the same way year after year rather than one which exhibits a gradual change with the passage of time. When monthly data are used, 12 seasonal indexes are computed that are both (1) measures of the relative importance of each of the 12 months in a year and (2) divisors for the original observations which render the latter "free" of seasonal variation. For example, the seasonal indexes for the Outdoor Activities problem are shown in Table 16.4.

We see that during the summer months, sales tend to be higher than average, as indicated by seasonal indexes in excess of 100, and in the winter months, lower than average. This finding is quite in keeping with what one would expect of sales data for a company selling camping and outdoor sports equipment. In part (a) of Figure 16.11 the original sales data for this company are shown. In part (b) we see the *seasonally adjusted* data—that is, data freed of the seasonal influences through division of the original values by the seasonal indexes. More specifically, the January seasonal index of 76.8 (Table 16.4) was divided, as .768, into all original Y values associated with the month of January, the February seasonal index of 78.9, or .789, was divided into all original Y values associated with the month of February, and so forth, through the month of December. The seasonally adjusted series in part (b) of Figure 16.11 continues to exhibit a zigzag pattern because of the presence of irregular fluctuations. However, it is still considerably smoother than the original series shown in part (a).

That, then, is our destination. Let us now turn our attention to the process of getting there.

16.11 Deriving Seasonal Indexes When the Seasonal Pattern Is Stable

To develop a set of monthly seasonal indexes for a series with a stable seasonal pattern, at least five years of monthly data should be used. The procedure to be described is called the *ratio-to-moving-average method* and consists of six steps.

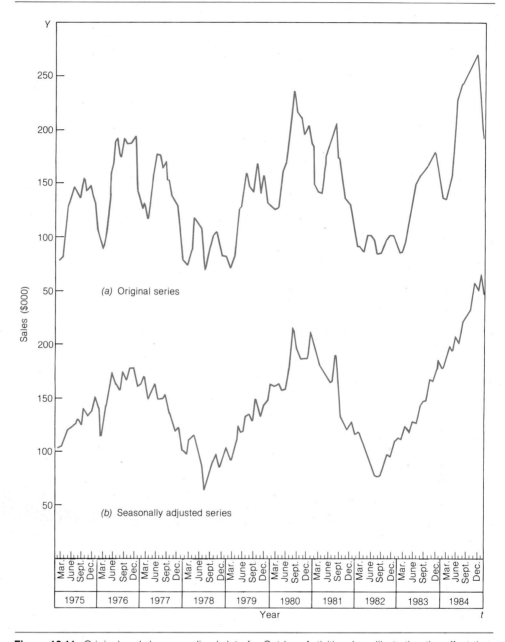

Figure 16.11. Original and deseasonalized data for Outdoor Activities, Inc., illustrating the effect that adjusting data for seasonal variation can have upon the appearance of a time series.

Step 1. Compute a 12-month moving average of the original Y values.

Recall that under our assumed model the Y values represent the effects of trend, cyclical, seasonal, and irregular components combined in a multiplicative manner—that is, $Y = TCSI$. By developing a 12-month moving average, we will obtain relatively smooth data representing the TC values. When the TC values are divided into the corresponding $Y = TCSI$ values, the effect is a set of SI values. If, when we reach this point, we can figure out how to rid ourselves of the I part of the SI values, we will have the seasonal component nicely separated from the other components.

In the meantime, let us consider the mechanics of obtaining a 12-month moving average for our data. We begin by summing the original Y values for January through December of 1975 and dividing the sum by 12. The resulting average would be viewed as applying to the middle month of the 12, which, alas, is not a month at all but rather a point between June and July of 1975. Then we sum the sales values for February of 1975 through January of 1976 and divide the result by 12. This average is assigned the middle month of the new 12-month period (really *between* July and August of 1975). We repeat this procedure until the last average, reflecting the sales values for January through December of 1984, has been calculated.

Step 2. Center the moving-average values *on* months by computing a 2-month moving average of them.

Since the eventual seasonal indexes will apply to specific months—not the cracks between months—we must shift our averages onto the middles of months. This process is known as *centering*. Centering can be achieved rather easily by developing a 2-month moving average of the 12-month moving averages. For example, we saw that the first 12-month moving average is assigned to a point between June and July of 1975; the second such average is assigned to a point between July and August of the same year. Now, if we average the first and second 12-month averages, the result will be a new average which corresponds to the middle of the month of July 1975. Of course, we continue this process until the final two 12-month averages have been averaged.

Steps 1 and 2 are demonstrated, albeit somewhat indirectly, in columns (1), (2), and (3) of Table 16.5. For computational ease, a twelve-month moving *total*, rather than *average*, was computed and entered in column (1) and a 2-month moving *total* of the 12-month totals was determined and entered in column (2). Finally, the totals in column (2) were divided by 24 to get the centered 12-month moving-average values shown in column (3). This way of completing Steps 1 and 2 eliminates one set of divisions. The centered moving-average values are the official TC (trend-cycle) values because (1) averaging over overlapping 12-month periods eliminates the seasonal component and (2) increases and decreases in the Y values resulting from irregular fluctuations are assumed to cancel out over a 12-month period.

Step 3. Divide the original Y values for all months by their corresponding moving-averages, thereby obtaining the ratio-to-moving-average values or, more conveniently, the SI values.

As mentioned, the resulting ratios contain only the seasonal and irregular movements, since the trend and cycle components are eliminated in the divisions by the moving averages. In symbols:

$$\frac{Y}{MA} = \frac{TCSI}{TC} = SI$$

This step is presented in column (4) of Table 16.5.

Table 16.5 Calculation of the Ratio-to-Moving-Average Values (*SI* Values) for Sales Data Presented in Table 16.1

Year and Month		(1) Sales (Thousands of Dollars)	(2) 12-Month Moving Total	(3) 2-Month Moving Total	(4) Centered 12-Month Moving Average	(5) Ratio of Actual to Moving Average
1975	January	78	——	——	——	——
	February	82	——	——	——	——
	March	122	——	——	——	——
	April	130	——	——	——	——
	May	146	——	——	——	——
	June	141	1522	——	——	——
	July	135	1531	3053	127.2	106.1
	August	155	1560	3091	128.8	120.3
	September	141	1595	3155	131.5	107.2
	October	148	1654	3249	135.4	109.3
	November	135	1700	3354	139.8	96.6
	December	109	1732	3432	143.0	76.2
1976	January	87	1788	3520	146.7	59.3
	February	111	1820	3608	150.3	73.9
	March	157	1865	3685	153.5	102.3
	April	189	1910	3775	157.3	120.2
	May	192	1918	3828	159.5	120.4
	June	173	1934	3852	160.5	107.8
	July	191	1978	3912	163.0	117.2
	August	187	1984	3962	165.1	113.3
	September	186	1985	3969	165.4	112.5
	October	193	1973	3958	164.9	117.0
	November	143	1957	3930	163.8	83.3
	December	125	1949	3906	162.8	76.8
	⋮	⋮	⋮	⋮	⋮	⋮
1983	January	87	1401	2726	113.6	76.6
	February	97	1471	2872	119.7	81.0
	March	121	1548	3019	125.8	96.2
	April	141	1625	3173	132.2	106.7
	May	151	1695	3320	138.3	109.2
	June	158	1747	3442	143.3	110.2
	July	162	1798	3545	147.7	109.7
	August	167	1857	3655	152.3	109.7
	September	178	1932	3789	157.9	112.7
	October	180	2019	3951	164.6	109.4
	November	166	2110	4129	172.0	96.5
	December	139	2197	4307	179.5	77.4
1984	January	138	2288	4485	186.9	73.8
	February	156	2381	4669	194.5	80.2
	March	196	2477	4858	202.4	96.8
	April	228	2570	5047	210.3	108.4
	May	242	2644	5214	217.2	111.4
	June	245	2703	5347	222.8	110.0
	July	253	——	——	——	——
	August	260	——	——	——	——
	September	274	——	——	——	——
	October	273	——	——	——	——
	November	240	——	——	——	——
	December	198	——	——	——	——

Step 4. Organize the *SI* values resulting from Step 3 by months and years.
A convenient way of performing this step is to use a table like that shown in Table 16.6.

Step 5. Determine representative *SI* values for all months.
What we now seek is a representative value for each column in Table 16.6. Because of the presence of sometimes large irregular fluctuations, some of which are of a sporadic nature, and because of the marked sensitivity of the regular arithmetic mean to extreme values, use of the arithmetic mean in the present context is considered rather risky. What time series analysts usually do is to obtain the medians or the modified means (arithmetic means for data after extremely high and extremely low values have been eliminated) of the columns. Medians have been used in connection with the Outdoor Activities case.

In the process of determining the median for each column in a table like Table 16.6, the irregular fluctuations are assumed to cancel out. Because of the presence of irregular fluctuations, some *SI* values will be higher than and some lower than what they would be if only seasonal forces were operating. However, the process of averaging (deriving the median in the present example) effectively eliminates the irregular movements. Therefore, once the column medians are adjusted according to the requirements of Step 6, they are full-fledged seasonal indexes and may be viewed as the seasonal component of the original time series.

A most helpful visual impression of the *SI* values—with respect to both level and variability (and, hence, dependability)—can be achieved through use of box-and-whiskers plots, as shown in Figure 16.12. Notice, for example, the relative instability of the *SI* values for the month of June.

Step 6. Adjust the representative *SI* values so that they sum to 1200 (that is, average 100).
In Table 16.6 the column medians are shown immediately below the corresponding *SI* values. These may be viewed as first-approximation seasonal indexes. However, we see that the medians total to 1196.1, which means that they average 99.7, a little less than the

Table 16.6 *SI* Values for the Outdoor Activities Sales Data and Determination of Constant Seasonal Indexes

Year		J	F	M	A	M · J	J	J	A	S	O	N	D
1975		—	—	—	—	—	—	106.1	120.3	107.2	109.3	96.6	76.2
1976		59.3	73.9	102.3	120.2	120.4	107.8	117.2	113.3	112.5	117.0	87.3	76.8
1977		81.1	73.5	101.5	117.2	119.8	115.0	122.1	112.1	102.4	101.2	87.8	67.9
1978		67.8	84.9	117.8	117.2	115.1	75.8	96.2	108.4	111.6	99.8	83.6	79.3
1979		65.9	71.9	106.9	107.4	126.4	114.9	106.0	120.8	99.5	108.3	87.9	81.2
1980		77.8	77.3	94.7	98.6	119.0	129.7	116.6	113.7	105.1	108.7	101.4	81.5
1981		77.8	78.6	100.5	109.7	120.1	136.7	119.0	101.0	96.3	96.4	88.9	76.6
1982		82.9	84.6	103.7	104.4	100.2	89.7	91.0	102.4	105.4	105.0	94.1	81.0
1983		76.6	81.0	96.2	106.7	109.2	110.2	104.7	109.7	112.7	109.4	96.5	77.4
1984		73.8	80.2	96.8	108.4	111.4	110.0	—	—	—	—	—	—
Column median		76.6	78.6	101.5	108.4	119.0	110.2	109.7	112.1	105.4	108.3	88.9	77.4
						Total = 1196.1							
Seasonal index		76.8	78.9	101.8	108.8	119.4	110.6	110.1	112.5	105.7	108.7	89.2	77.7
						Total = 1200.2							

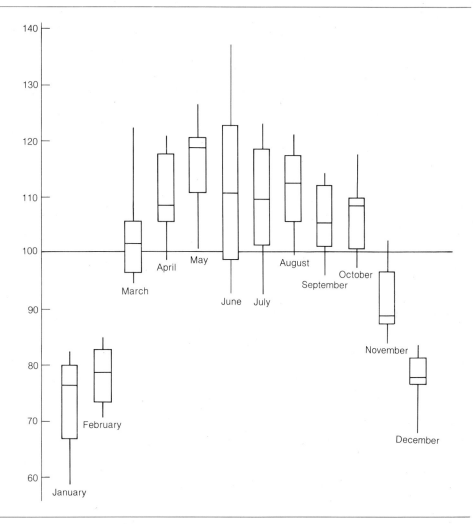

Figure 16.12. Box plots of *SI* values organized by month.

required 100. An average somewhat above or somewhat below 100 is to be expected at this point in the work. Nevertheless, this situation cannot go uncorrected because in the multiplicative model the seasonal indexes must be such that an index value of, say, 110 can be interpreted as 110% of the annual average. Fortunately, this is one of those relatively rare statistical problems with an easy solution: We simply divide the sum of the column medians into 1200 to get a multiplier to be applied to each of the individual medians. That is, $1200 \div 1196.1 = 1.003261$. When all column medians are multiplied by this value, the resulting products will sum to 1200, aside from minor rounding errors.

We have now come full circle. The entries in Table 16.4, discussed earlier, were obtained by the process just described. The series shown in part (*b*) of Figure 16.11 was developed by dividing each original January Y value by the January seasonal index, each original February Y value by the February seasonal index, and so forth.

GRIN & BEAR IT   **BY WAGNER**

"Only 22 more shopping days until the Christmas shopping season begins!"

16.12 Deriving Seasonal Indexes When the Seasonal Pattern Is Changing

In the preceding section we rather casually assumed the seasonal pattern to be stable because the statistical procedures are much simpler to discuss when such is the case. However, for many time series the seasonal pattern changes gradually with the passage of time and seldom will the analyst know at the outset whether the seasonal pattern is a stable one or a changing one. Therefore, the above list of six steps should really be a nine-step list. Since Steps 1 through 4 above were demonstrated in the preceding section, no further comment will be made on them. We start with a new Step 5.

Step 5. Prepare 12 charts of *SI* values, one for each month, using the table obtained in connection with Step 4.

    The table alluded to is the kind shown in Table 16.6. For purposes of illustration, let us chart, say, the March data from this table. This is done in Figure 16.13.

Step 6. Scrutinize each chart resulting from Step 5 to determine whether the *SI* values appear to fluctuate around (1) a trend line with zero slope or (2) a trend line with a nonzero slope.

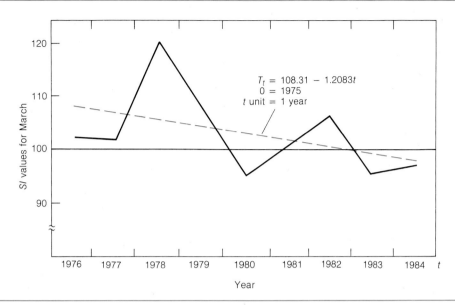

Figure 16.13. *SI* values for the month of March and the least squares linear trend line.

If the *SI* values for a particular month fluctuate around a trend line with slope 0, that is tantamount to saying that they fluctuate around their own arithmetic mean with no tendency either to increase or decrease over time. The reason for our singling out the *SI* values for the month of March is that they do, for whatever reasons, appear to exhibit a downward trend.

Step 7. If all sets of *SI* values appear to fluctuate around a trend line with 0 slope, complete the task of determining seasonal indexes in the manner described in section 16.11; on the other hand, if the *SI* values for some months fluctuate around a trend line having a nonzero slope, complete the task of determining seasonal indexes in the manner described in Steps 8 and 9 below.

We have in the data under study a somewhat ambiguous situation—and, in this sphere of statistical analysis, there exist few guidelines for coping with ambiguous situations. In our example, *SI* values in chart form (not shown for any month but March) indicate that some months hint at a changing seasonal pattern (though not terribly persuasively) and other months do not. Although we have not followed our own advice in future dealings with this case, presumably if *SI* values for a few months call for it, it is best to proceed with the changing-seasonal-pattern methodology. If this decision is wrong, little harm is done because, in a problem where doubt exists, the more complex approach should lead to seasonal indexes quite close to those which would have been derived using the simpler methodology. To complete the discussion of this methodology, we will move to Steps 8 and 9.

Step 8. Derive trend curves for the *SI* values associated with each individual month and use the trend values as first-approximation seasonal indexes.

Whereas when using the methodology described in the preceding section we obtained one seasonal index for each month of the year—12 such indexes altogether—the procedure for a changing seasonal pattern would require a number of seasonal indexes equal to 12 times the number of relevant years. For example, the number of seasonal indexes required in our illustrative case, for the 10 years of historical data only, would be $10 \times 12 = 120$. Of course, if one wishes to use the seasonal indexes to deseasonalize future data, still more such indexes will have to be determined through trend projection.

No inalterable rules exist for fitting trend lines to the *SI* values. Freehand methods, least squares linear or nonlinear trends, and moving averages have all been used by time series analysts. The straight line shown in Figure 16.13 is a least squares trend line. This approach is an objective one which has produced satisfactory results for many analysts and one which allows easy projection of the trend line. Still, it is probably best to take each situation as it comes and decide on the trend fitting procedure, or combination of procedures, after careful inspection of the 12 charts of *SI* values.

Step 9. For each year of historical data, adjust the first-approximation seasonal indexes obtained from the trend lines so that they sum to 1200 (average 100).

You will recall that, in describing the procedures to use when the seasonal pattern is stable, we noted that the column medians of the *SI* values will not necessarily average 100. We further noted that, if the column medians do not average 100, the analyst must divide the actual total of the column medians into 1200 and apply the resulting quotient to each column median. When the seasonal pattern is changing over time, this same procedure must be followed *for each relevant past and future year*. For example, once trend lines were determined for the January, February, . . . , December *SI* values for the Outdoor Activities case and the first-approximation seasonal indexes determined for the year 1975 by using trend lines, the 12 resulting first-approximation indexes would have to be adjusted so that they average 100 in the year 1975. The same argument applies to 1976, 1977, and so on. This requirement usually calls for judgmentally taking from some first-approximation seasonal indexes and giving to others. Consequently, regardless of the degree of objectivity of the original trend-fitting process, chances are excellent that the trend estimates will have to be shaved a little here and padded a little there before the work is completed. When carrying out this process of taking from one month and giving to another, a broad guideline should be heeded: The trend associated with each month, although perhaps nonlinear, should be kept smooth and well fitting over the entire span of relevant years.

16.13 Measuring Cyclical Oscillations

We are now ready to separate out the two remaining time series components: cyclical oscillations, *C*, and irregular fluctuations, *I*. Let us start with the cyclical component. We noted earlier that we can describe the centered 12-month moving average as a *TC* series—which is to say as a time series having only trend and cyclical components. Since we have already determined the trend values for the Outdoor Activities data, we may now isolate the cyclical component by dividing each moving-average value by the corresponding trend value. In symbols: $TC/T = C$. This operation is demonstrated in Table 16.7. The resulting cyclical movements are shown graphically in part (*c*) of Figure 16.14.

With the realization that the small number of complete cycles renders any conclusions about the nature of the cyclical oscillations in the Outdoor Activities data tentative at best, let

Table 16.7 Demonstration of the Way in Which Cyclical Values Are Determined from the Trend and Centered 12-Month Moving Averages

Year and Month		Time Code t	(1) Centered 12-Month Moving Averages (*TC* Values)	(2) Trend values from $T_t = 120.24 + .3918t$ (*T* Values)	(3) (1) ÷ (2) $TC/T \times 100$ (*C* Values)
1975	January	6	——	122.5	——
	February	7	——	123.0	——
	March	8	——	123.4	——
	April	9	——	123.8	——
	May	10	——	124.2	——
	June	11	——	124.5	——
	July	12	127.2	124.9	101.8
	August	13	128.8	125.3	102.8
	September	14	131.5	125.7	104.6
	October	15	135.4	126.1	107.4
	November	16	139.8	126.5	110.5
	December	17	143.0	126.9	112.7
1983	July	108	147.7	162.6	90.8
	August	109	152.3	162.9	93.5
	September	110	157.9	163.3	96.7
	October	111	164.6	163.7	100.5
	November	112	172.0	164.1	104.8
	December	113	179.5	164.5	109.1
1984	January	114	186.9	164.9	113.3
	February	115	194.5	165.3	117.7
	March	116	202.4	165.7	122.1
	April	117	210.3	166.1	126.6
	May	118	217.2	166.5	130.5
	June	119	222.8	166.9	133.5
	July	120	——	167.3	——
	August	121	——	167.6	——
	September	122	——	168.0	——
	October	123	——	168.4	——
	November	124	——	168.8	——
	December	125	——	169.2	——

us attempt to list some characteristics of the cyclical movements. Part (c) of Figure 16.14 suggests that

1. The cyclical component is a most substantial one.
2. The amplitudes of the cyclical swings may be getting more extreme with the passage of time.
3. The cyclical highs, on the one hand, and the cyclical lows, on the other, appear to occur roughly four years apart.

We will make use of these facts in section 16.15 where we show how the components of a decomposed time series can be put to use when obtaining a forecast.

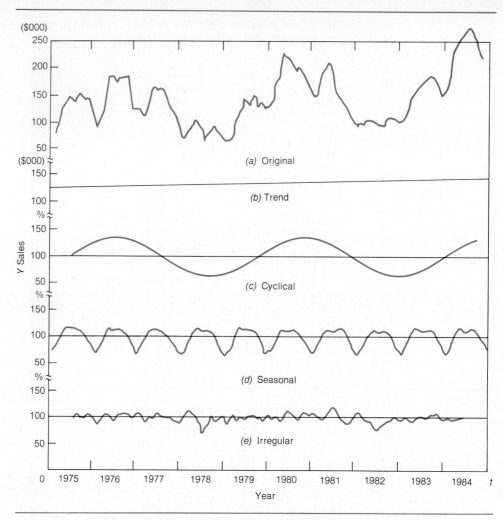

Figure 16.14. Sales data for Outdoor Activities, Inc.—original and the four components.

16.14 Measuring Irregular Fluctuations

Let us review briefly what we have done to this point. We started with the original Y values where $Y = TCSI$. We then proceeded in two rather diverse directions: First, we fitted a linear trend to the Y values, thereby separating them into T and CSI values. Second, we deseasonalized the Y values, thereby separating them into S and TCI values. In the process of determining the seasonal indexes, we obtained the TC and the SI values. Finally, we detrended the TC series, thereby splitting it into a T series and a C series. Collecting all of this information, we find that we know, in an approximate way, the values of T, C, S, TC, SI, TCI, CSI, and, of course, $TCSI$. One more thing we can do is to divide the deseasonal-

782   *16   Introduction to Time Series Analysis*

ized values, TCI, by the corresponding centered 12-month moving-average values, TC, in order to obtain the irregular values, I. That is, $TCI/TC \times 100 = I$. (Alternatively, we could "deseasonalize" the SI values by $SI/S \times 100 = I$.) The results of the TCI/TC calculations are shown graphically in part (e) of Figure 16.14.

Figure 16.14 reveals that the irregular fluctuations are, fortunately, a relatively minor component of this series. From a forecasting standpoint, the irregular fluctuations of any time series are considered a lost cause because of their random pattern. Still, it is good to know how they compare in size with the systematic components we have been discussing. Knowledge of the magnitudes of the swings in this unpredictable component helps one to judge realistically the degree of accuracy that can be achieved by this forecasting approach in the short run.

16.15 Preparing a Forecast from a Decomposed Time Series

Throughout this chapter, we have assumed the multiplicative time series model $Y = TCSI$. Because in preparing a forecast the irregular fluctuations cannot be predicted statistically, we will ignore them and proceed as if the appropriate model were $Y = TCS$. Let us now attempt to develop a forecast of sales for the month of January 1985.

For January 1985 the trend value to be expected is

$$T_{121} = 120.24 + (.3918)(126) = 169.6$$

By referring to Table 16.4 and assuming a constant seasonal pattern, we find the value of the seasonal index for this specific month is 76.8, or .768.

The value of the cyclical component will be estimated using its own past history. It is a little unfortunate that the most recent known cyclical value is the one associated with June 1984, a condition indicating that we must project the cyclical movement to the end of 1984 before even attempting to forecast the cyclical value for January 1985. Moreover, if the past cyclical pattern of turning-point timing continues, 1985 will be a year in which the cyclical movement reverses itself and begins to decline. In certain respects, then, we are not in an enviable position regarding this time series component. Let us, however, take into account the direction of the cyclical movement up until June of 1984 and also the size of the month-to-month changes in C in the first half of 1984. The changes were +4.2, +4.4, +4.4, +4.5, +3.9, +3.0. The gradual reduction in these increases plus the fact that the cyclical component may be about due for a reversal in direction would suggest that increasingly smaller increments to the cyclical values should probably be assumed through the final months of 1984 and January of 1985. Exactly what number should be attached is a matter over which there might be heated disagreement among different analysts. However, differences in estimates obtained might not differ too greatly in view of the rather dependable cyclical pattern in the past. For the sake of illustration, let us say that we estimate that the value of the cyclical component for January 1985 will be 148, or 1.48. The forecast for the month under consideration would therefore be

$$Y = TCS = 169.6 \times .768 \times 1.48 = 192.8$$

Forecast estimates for later months could be developed in a similar manner. Keep in mind, however, that our confidence in forecasting the cyclical component, which might not be too great to begin with, may fall off rapidly as we push further into the future.

Warning: The way of dealing with the cyclical component in this problem was suggested by the apparent regularity of cyclical shape and duration of this series. This discussion should not be interpreted as suggesting that the method will work with the cyclical component of other time series. Indeed, it may even prove faulty with this one. Often with respect to the cyclical component, one should experiment with different forecasting techniques.[1]

16.16 Summary

The time series decomposition forecasting procedure just described is admittedly based on rather crude and approximate methods of disentangling the trend, cyclical, seasonal, and irregular components. Moreover, much personal judgment must sometimes be exercised in forecasting these components, especially the cyclical one. Still, if you look once again at Figure 16.3 to remind yourself of the sorry appearance of the original time series, you will most likely agree that we have come quite a way using this method. First, we have defined the secular trend and, in so doing, have developed an equation for projecting trend values into the future. Second, we have isolated the seasonal variation sufficiently well to find that there is a seasonal rhythm which can be exploited for forecasting purposes. Third, we found that the cyclical swings, though substantial, do appear to unfold with encouraging regularity.

In summary, then, through use of this decomposition method, we have at least succeeded in finding quite a high degree of order amid fluctuations which, upon first examination, looked almost hopelessly chaotic.

Note

1. For an exposure to quite a wide variety of approaches to forecasting, see Chapter 17 of this text and the following, more specialized, books on the subject: J. S. Armstrong, *Long-Range Forecasting: From Crystal Ball to Computer* (New York: Wiley, 1978); William F. Butler, Robert A. Kavesh, and Robert B. Platt, eds., *Methods and Techniques of Business Forecasting* (Englewood Cliffs, N.J.: Prentice-Hall, 1974); James P. Cleary and Hans Levenbach, *The Beginning Forecaster* and *The Professional Forecaster* (Belmont, Calif.: Lifetime Learning Publications, 1982); and Charles W. Gross and Robin T. Peterson, *Business Forecasting*, 2nd ed. (Boston, Mass.: Houghton Mifflin, 1983).

Terms Introduced in This Chapter

additive time series model (p. 751)
cyclical oscillations (p. 750)
exponential trend equation (p. 762)
higher-degree parabolic trend equation (p. 767)
irregular fluctuations (p. 751)

minor random movements (p. 751)
sporadic changes (p. 751)
linear trend equation (p. 756)
multiplicative time series model (p. 753)
seasonal indexes (p. 772)

seasonally adjusted data (p. 772)
seasonal variation (p. 750)
secular trend (p. 750)
time series (p. 748)

Formulas Introduced in This Chapter

The additive time series model:

$$Y = T + C + S + I$$

The multiplicative time series model:

$$Y = TCSI$$

Linear trend equation:

$$T_t = a + bt$$

Least squares equation for obtaining the slope value of a linear trend equation:

$$b = \frac{\Sigma\, tY - \dfrac{(\Sigma\, t)(\Sigma\, Y)}{n}}{\Sigma\, t^2 - \dfrac{(\Sigma\, t)^2}{n}}$$

Least squares equation for obtaining the Y-intercept value of a linear trend equation:

$$a = \frac{\Sigma\, Y}{n} - b\,\frac{\Sigma\, t}{n}$$

Exponential trend equation:

$$T_t = ab^t$$

Least squares equation for obtaining the slope value of an exponential trend equation in log linear form:

$$\log b = \frac{\Sigma\, t \log Y - \dfrac{(\Sigma\, t)(\Sigma \log Y)}{n}}{\Sigma\, t^2 - \dfrac{(\Sigma\, t)^2}{n}}$$

Least squares equation for obtaining the Y-intercept value for an exponential trend equation in log linear form:

$$\log a = \frac{\Sigma \log Y}{n} - \log b\,\frac{\Sigma\, t}{n}$$

Second-degree parabolic trend equation:

$$T_t = a + bt + bt^2$$

Normal equations for obtaining a, b, and c constants for a second-degree parabolic trend equation:

$$\Sigma\, Y \;\;= na \;\;\;\;+ b\,\Sigma\, t \;\;+ c\,\Sigma\, t^2$$
$$\Sigma\, tY \;= a\,\Sigma\, t \;+ b\,\Sigma\, t^2 + c\,\Sigma\, t^3$$
$$\Sigma\, t^2Y = a\,\Sigma\, t^2 + b\,\Sigma\, t^3 + c\,\Sigma\, t^4$$

Manner of isolating the seasonal-irregular component of a time series; TC represents a centered 12-month moving average (or centered 4-quarter moving average, etc.):

$$\frac{Y}{MA} = \frac{TCSI}{TC} = SI$$

Manner of isolating the cyclical component of a time series:

$$C = \frac{TC}{T}$$

Manner of expressing how a time series is adjusted for seasonal variation:

$$TCI = \frac{TCSI}{S}$$

Manner of isolating the irregular component of a time series:

$$I = \frac{SI}{S}, \text{ or } \frac{TCI}{TC}$$

Questions and Problems

16.1 Explain the meaning of each of the following:
    a. Time series   e. Semilog graph paper
    b. Least squares trend   f. Moving average
    c. Seasonal variation   g. Ratio-to-moving-average method
    d. Seasonal index

16.2 Explain the meaning of each of the following:
    a. Cyclical oscillations   d. Exponential trend
    b. Sporadic change   e. Scaling down a trend equation
    c. Minor random movements   f. Centered 12-month moving average

16.3 Explain the meaning of each of the following:
    a. *TCSI*   c. *TC*
    b. *TCI*   d. *SI*

16.4 Distinguish between:
    a. Seasonal variation and irregular fluctuations
    b. Secular trend and cyclical oscillations
    c. Sporadic change and minor random movements
    d. $T + C + S + I$ and $TCSI$
    e. $T_t = a + bt$ and $T_t = ab^t$

16.5 Distinguish between:
    a. $T_t = ab^t$ and $T_t = a + bt + ct^2$
    b. Stable seasonal pattern and changing seasonal pattern
    c. Use of least squares method in trend fitting and use of the least squares method in simple regression analysis
    d. Scaling down the Y variable and scaling down the t variable

16.6 Indicate which of the following statements you agree with and which you disagree with, and defend your opinions:
    a. The term *cyclical oscillation* is appropriate in view of the very dependable periodicities of these movements.
    b. The additive model of a time series has been favored by statisticians and economists primarily because it acknowledges the interdependencies that probably exist among the four time series components.

c. If a time series analyst wishes to determine trend values on an annual basis, he must develop his trend equation from annual data; if he wishes to determine trend values on a quarterly basis, he must develop his trend equation from quarterly data; if he wishes to determine trend values on a monthly basis, he must develop his trend equation from monthly data.

16.7 Indicate which of the following statements you agree with and which you disagree with, and defend your opinions:
  a. Projecting a trend requires no assumptions whatever.
  b. Generally speaking, the cyclical component of a time series is more difficult to predict than either the trend component or the seasonal component.
  c. Generally speaking, the cyclical component of a time series is more difficult to predict than either the seasonal component or the irregular component.
  d. If an analyst who has developed a linear trend equation wishes to scale that equation down so that trend estimates can be determined for each *third of a year*, she will not have to shift the origin by one-half a period.

16.8 a. Cite a realistic hypothetical business problem in which the fitting and projecting of a trend of some kind might be helpful. Be as specific as possible.
  b. Indicate some steps that should be taken *prior to* the fitting of the trend and explain why each would be advisable in the situation you described in part *a* of this problem.
  c. Cite some possible hazards, or limitations, of utilizing a trend in the situation you described in part *a* of this problem.

 16.9 Table 16.8 presents 15 years of time series data for each of three unemployment measures.
  a. Present the three sets of data in chart form.
  b. Do the graphs suggest that any of these employment-related series is (are) not well suited to being forecasted by means of trend projection? Justify your answer.
  c. Which, if any, of the three employment-related series in Table 16.8 appear(s) to have a linear trend?

Table 16.8 Data for Three Employment-Related Time Series

Year	Total Unemployment Rate (Percent)	Employees on Nonagricultural Payrolls (Thousands of Persons)	Employees in Goods-Producing Industries (Thousands of Persons)
1961	5.5	54,042	19,814
1962	5.7	55,596	20,405
1963	5.2	56,702	20,593
1964	4.5	58,331	20,958
1965	3.8	60,815	21,880
1966	3.8	63,955	23,116
1967	3.6	65,857	23,268
1968	3.5	67,951	23,693
1969	4.9	70,442	24,311
1970	5.9	70,920	23,507
1971	5.6	71,222	22,820
1972	4.9	73,714	23,546
1973	5.6	76,896	24,727
1974	8.5	78,413	24,697
1975	8.5	76,985	22,549
1976	7.7	79,437	23,334

d. Which, if any, of the three employment-related series in Table 16.8 appear(s) to have an exponential trend?
e. Could the trend of any of these three series be adequately described and projected by a second-degree parabolic equation? Why or why not?

 16.10 Table 16.9 presents ten years of time series data for each of three inventory series.
a. Present the three sets of data in chart form.
b. Do your graphs suggest that any of these inventory-related series is (are) not well suited to being forecasted by means of a trend projection? Justify your answer.
c. Which, if any, of the three inventory-related series in Table 16.9 appear(s) to have a linear trend?
d. Which, if any, of the three inventory-related series in Table 16.9 appear(s) to have an exponential trend?
e. Could the trend of any of these three series be adequately described and projected by a second-degree parabolic trend equation? Elaborate.

 16.11 In Table 16.10, sales achieved by the Strong Manufacturing Company are categorized according to whether the items sold are consumer goods or industrial goods.
a. Fit a least squares linear trend to each series and present the trend equations. Explain what the b values represent.
b. Determine the linear trend values for each series for the years 1977 through 1990, inclusive.
c. Plot the original data for the two series on arithmetic-grid graph paper along with all linear trend values obtained in part b. Is the long-run tendency of the consumer sales series adequately described by the least squares linear trend? Elaborate. Is the long-run tendency of the industrial sales series adequately described by the least squares linear trend? Elaborate.
d. Compute the cyclical values for consumer sales by dividing the original values by the corresponding linear trend values. Are the cyclical values *really* only cyclical or do they also contain seasonal and irregular elements? Explain.
e. Compute the cyclical values for industrial sales by dividing the original values by the corresponding linear trend values. Are the cyclical values *really* only cyclical or do they also contain seasonal and irregular elements? Explain.

Table 16.9   Data for Three Inventory-Related Time Series

Year	Book Value of Manufacturing and Trade Inventories (Billions of Dollars)	Change in Business Inventories (Billions of Dollars)	Inventories-to-Sales Ratio
1967	136.8	10.1	1.62
1968	145.4	7.7	1.60
1969	155.4	9.4	1.64
1970	168.3	3.8	1.73
1971	175.4	6.4	1.70
1972	184.8	9.4	1.62
1973	198.0	17.9	1.58
1974	237.9	10.7	1.70
1975	278.4	−14.6	1.80
1976	275.5	11.9	1.69

Table 16.10 Sales Achieved by the Strong Manufacturing Company

| | Sales (Millions of Dollars) | |
Year	(1) Consumer Goods	(2) Industrial Goods
1977	10	5
1978	13	5
1979	15	6
1980	19	7
1981	25	6
1982	25	8
1983	28	9
1984	29	9
1985	34	10
1986	37	12

    f. Scale down the trend equation for consumer sales so that it will produce trend values on a monthly, rather than on an annual, basis. Tell in what way or ways you altered the original linear trend equation and why you did so.

    g. Use the modified trend equation obtained in part *f* to develop a trend estimate for July of 1990. Do you have much confidence in this estimate? Why or why not?

    h. Scale down the trend equation for industrial sales so that it will produce trend values on a monthly, rather than on an annual, basis. Tell in what way or ways you altered the original linear trend equation and why you did so.

    i. Use the modified trend equation obtained in part *h* to develop a trend estimate for July of 1990. Do you have much confidence in this estimate? Why or why not?

 16.12 Refer to Table 16.10.

    a. Fit a least squares exponential trend to each series and present the trend equations. Explain what the log *b* values represent.

    b. Determine the exponential trend values for each series for the years 1977 through 1990 inclusive.

    c. Plot the original data for the two series on either arithmetic-grid or semilog graph paper along with all exponential trend values obtained in part *b*. Is the long-run tendency of the consumer sales series adequately described by the least squares exponential trend? Elaborate. Is the long-run tendency of the industrial sales series adequately described by the least squares exponential trend? Elaborate.

    d. Compute the cyclical values for consumer sales by dividing the original values by the corresponding exponential trend values. Are the cyclical values *really* only cyclical or do they also contain seasonal and irregular elements? Explain.

    e. Compute the cyclical values for industrial sales by dividing the original values by the corresponding exponential trend values. Are the cyclical values *really* only cyclical or do they also contain seasonal and irregular elements? Explain.

    f. Scale down the trend equation for consumer sales so that it will produce trend values on a monthly, rather than on an annual, basis. Tell in what way or ways you altered the original exponential trend equation and why you did so.

    g. Use the modified trend equation obtained in part *f* to develop a trend estimate for July of 1990. Do you have much confidence in this estimate? Why or why not?

    h. Scale down the trend equation for industrial sales so that it will produce trend values on a monthly, rather than on an annual, basis. Tell in what way or ways you altered the original exponential trend equation and why you did so.

  i. Use the modified trend equation obtained in part *h* to develop a trend estimate for July of 1990. Do you have much confidence in this estimate? Why or why not?

Special-Challenge Problem

 16.13 Refer to Table 16.10.

  a. Fit a least squares second-degree parabolic trend to each series and present the trend equations.
  b. Determine the second-degree parabolic trend values for each series for the years 1977 through 1990 inclusive.
  c. Plot the original data for the two series on arithmetic-grid graph paper along with all second-degree parabolic trend values obtained in part *b*. Is the long-run tendency of the consumer sales series adequately described by the least squares second-degree parabolic trend? Elaborate. Is the long-run tendency of the industrial sales series adequately described by the least squares second-degree parabolic trend? Elaborate.
  d. Compute the cyclical values for consumer sales by dividing the original values by the corresponding second-degree parabolic trend values. Are the cyclical values *really* only cyclical or do they also contain seasonal and irregular elements? Explain.
  e. Compute the cyclical values for industrial sales by dividing the original values by the corresponding second-degree parabolic trend values. Are the cyclical values *really* only cyclical or do they also contain seasonal and irregular elements? Explain.
  f. Scale down the trend equation for consumer sales so that it will produce trend values on a monthly, rather than on an annual, basis. Tell in what way or ways you altered the original second-degree parabolic trend equation and why you did so.
  g. Use the modified trend equation obtained in part *f* to develop a trend estimate for July of 1990. Do you have much confidence in this estimate? Why or why not?
  h. Scale down the trend equation for industrial sales so that it will produce trend values on a monthly, rather than on an annual, basis. Tell in what way or ways you altered the original second-degree parabolic trend equation and why you did so.
  i. Use the modified trend equation obtained in part *h* to develop a trend estimate for July of 1990. Do you have much confidence in this estimate? Why or why not?

 16.14 Table 16.11 presents nine years of past revenue, cost, and profit data for the Spectre Company.

  a. Fit a least squares linear trend to each series and present the trend equations. Explain what the *b* values represent.
  b. Determine the linear trend values for each series for the years 1977 through 1986 inclusive.
  c. Plot the original data for the three series on arithmetic-grid graph paper along with all linear trend values obtained in part *b*. Is the long-run tendency of the revenues series adequately described by the least squares linear trend? Elaborate. Is the long-run tendency of the costs series adequately described by the least squares linear trend? Elaborate. Is the long-run tendency of the profits series adequately described by the least squares linear trend? Elaborate.
  d. Compute the cyclical values for each series by dividing the original values by the corresponding linear trend values. Are the cyclical values *really* only cyclical or do they also contain seasonal and irregular elements? Explain.
  e. Scale down the trend equation for revenues so that it will produce trend values on a quarterly, rather than on an annual, basis. Tell in what way or ways you altered the original linear trend equation and why you did so.
  f. Use the modified equation obtained in part *e* to develop a trend estimate for September 1986. Do you have much confidence in this estimate? Why or why not?

Table 16.11 Revenue, Cost, and Profit Data for the Spectre Company

(All Figures in Thousands of Dollars)

Year	(1) Revenues	(2) Costs	(3) (1) − (2) Profits
1977	130	67	63
1978	225	85	140
1979	240	110	130
1980	320	140	180
1981	350	190	160
1982	420	240	180
1983	440	320	120
1984	525	400	125
1985	550	533	17

g. Scale down the trend equation for costs so that it will produce trend values on a quarterly, rather than on an annual, basis. Tell in what way or ways you altered the original linear trend equation and why you did so.

h. Use the modified equation obtained in part g to develop a trend estimate for the third quarter of 1986. Do you have much confidence in this estimate? Why or why not?

i. Scale down the trend equation for profits so that it will produce trend values on a quarterly, rather than on an annual, basis. Tell in what way or ways you altered the original linear trend equation and why you did so.

j. Use the modified equation obtained in part i to develop a trend estimate for the third quarter of 1986. Do you have much confidence in this estimate? Why or why not?

 16.15 Refer to Table 16.11.

a. Fit a least squares exponential trend to each series and present the trend equations.

b. Determine the exponential trend values for each series for the years 1977 through 1986, inclusive.

c. Plot the original data for the three series on either arithmetic-grid or semilog graph paper along with all exponential trend values obtained in part b. Is the long-run tendency of the revenues series adequately described by the least squares exponential trend? Elaborate. Is the long-run tendency of the costs series adequately described by the least squares exponential trend? Elaborate. Is the long-run tendency of the profits series adequately described by the least squares exponential trend? Elaborate.

d. Compute the cyclical values for each series by dividing the original values by the corresponding exponential trend values. Are the trend values *really* only cyclical or do they also contain seasonal and irregular elements? Explain.

e. Scale down the trend equation for revenues so that it will produce trend values on a quarterly, rather than on an annual, basis. Tell in what way or ways you altered the original exponential trend equation and why you did so.

f. Use the modified trend equation obtained in part e to develop a trend estimate for the third quarter of 1986. Do you have much confidence in this estimate? Why or why not?

g. Scale down the trend equation for costs so that it will produce trend values on a quarterly, rather than on an annual, basis. Tell in what way or ways you altered the original exponential trend equation and why you did so.

h. Use the modified equation obtained in part g to develop a trend estimate for the third quarter of 1986. Do you have much confidence in this estimate? Why or why not?

i. Step down the trend equation for profits so that it will produce trend values on a quarterly, rather than on an annual, basis. Tell in what way or ways you altered the original exponential trend equation and why you did so.

j. Use the modified trend equation obtained in part *i* to develop a trend estimate for September 1986. Do you have much confidence in this estimate? Why or why not?

 16.16 Table 16.12 presents five years of monthly sales data (in thousands of units) for the Kant Container Company.

    a. Plot the sales data on arithmetic grid paper.

    b. Does your graph reveal any evidence of seasonal variation in these data? Explain why you do or do not believe so.

    c. Does your graph reveal the presence of an upward or downward trend? Elaborate. Does it suggest the presence of cyclical movements? Elaborate. Does it suggest the presence of irregular fluctuations? Elaborate.

 16.17 Refer to Table 16.12.

    a. Obtain the centered 12-month moving-average values for these sales data.

    b. Divide the results obtained in part *a* into the original series in order to obtain the *SI* values.

    c. Present *SI* values for January, February, . . . , December in box-and-whiskers form and tell what this chart reveals about the seasonal pattern.

    d. Using the *SI* values obtained in part *b*, obtain the seasonal index associated with each month, assuming a stable seasonal pattern. (Get the first-approximation seasonal indexes by determining the median *SI* associated with all Januarys, the median *SI* associated with all Februarys, and so forth.)

    e. Explain how you would employ the seasonal indexes obtained in part *b* to adjust the original data for seasonal variation.

    f. Use the seasonal indexes in the manner you described above. Plot the deseasonalized values on the same graph as the original values. Briefly comment on any differences in appearance you can observe between the series of original *Y* values and the series of deseasonalized values.

Table 16.12 Monthly Sales Data for the Kant Container Company

(All Sales Figures in Thousands of Units)

Month	Year				
	1981	1982	1983	1984	1985
January	704	753	855	1026	1238
February	641	688	783	953	1137
March	641	698	800	979	1162
April	642	705	816	996	1173
May	630	689	808	990	1154
June	628	705	822	1020	1192
July	654	737	868	1094	1252
August	649	735	852	1070	1233
September	625	708	835	1040	1185
October	643	742	866	1066	1212
November	652	758	885	1074	1237
December	811	940	1114	1349	1554

16.18 Explain your understanding of the term *deseasonalized data*.

16.19 Given:
   a. $T = 175.4$, $C = 132.2$, $S = 114.3$, $I = 93.6$
   b. $T = 27.4$, $C = 73.5$, $S = 144.2$, $I = 105.4$
   c. $T = 47.8$, $C = 157.3$, $S = 90.7$, $I = 80.5$
   In each case, find Y, assuming a multiplicative model.

 16.20 a. Name three products whose sales probably experience considerable seasonal variation over the course of a year.
   b. Name three products whose sales are probably quite sensitive to changes in general business conditions and, hence, have a substantial cyclical component.

16.21 How can one tell whether a seasonal pattern has been changing? Describe the changes in the seasonal adjustment procedure which would be called for if the seasonal pattern has been observed to have been changing.

16.22 Indicate which time series components are (normally) contained within:
   a. The original series.
   b. The centered 12-month moving-average series.
   c. The series obtained when the centered 12-month moving average values are divided into the corresponding original values.

TAKE CHARGE

16.23 Obtain ten years of quarterly data for a time series of interest to you. Decompose the chosen time series into trend, cyclical, seasonal, and irregular components. Describe your results briefly in writing. Also, tell whether the time series you selected appears predictable by an approach similar to that described in section 16.15.

Computer Exercise

16.24 Refer to Table 16.8.
   a. Make use of a packaged regression analysis program to fit a linear least squares trend equation to the data under the column heading "Employees on Nonagricultural Payrolls."
   b. Gather several years of more current data for the "Employees on Nonagricultural Payrolls" series and determine whether the linear trend equation you obtained in part *a* is still describing the basic path of this time series adequately. What is your conclusion?

Cases for Analysis and Discussion

✎ 16.1 "SUPERIOR TUNNELING COMPANY"

The following report describes the nature of the testimony given by a construction company executive, acting as an expert witness in behalf of "Superior Tunneling Company," in a business impairment damages trial against the Southwest Transportation Company. As you read the report, project yourself into the role of the lawyer for the "Southwest Transportation Company" and see whether you can develop a persuasive rebuttal against the dollar amount proposed by the expert witness—a rebuttal which, if effective, will save your client money.

Superior Tunneling Company (STC) in its role as a subcontractor has suffered business impairment damages on the Delta Tunnel as a direct result of the misrepresentation of factual data in the plans and specifications which prolonged the project and increased its cost above what could have been reasonably expected at the time of bid. The business impairment damages, as quantified herein, are in excess of those damages already claimed for project direct cost, indirect cost, and reasonable profit.

The failure of Southwest Transportation Company to properly recognize these additional costs at the time of performance has severely impacted Superior Tunneling Company's cash flow and ability to finance subsequent projects. The reduction in the firm's net worth by 27% as a result of this project caused the surety company to reduce its bonding limit from $4 million to $2.5 million. This dramatically curtailed Superior's bidding capacity and thus the company's ability to obtain new work.

It should be noted that the Delta Tunnel project was the only major contract Superior Tunneling was performing during the years 1976 and 1977.

Superior Tunneling was incorporated in 1963 with an initial capitalization of $6000. Up to the time of the Delta Tunnel project, STC had successfully performed over 11 projects with an in-place value of over $17 million.

The financial data for the firm have been compiled from the annual statements and are shown in tabular form in Visual Sheet 1. [Not shown here.]

From an examination of the financial data it can be seen that STC has experienced a steady growth, with the exception of 1968 when they sustained their only loss in 12 years of operation prior to the Delta Tunnel. Gross profit on contracts (that is, profit before home office, general, and administrative expenses) has been 20% or better in 10 of the 12 years, and operating income (profit before taxes) has averaged above 7%.

As a measure of management's dedication to growth, it should be noted that Superior Tunneling through March 31, 1976, had built its net worth up to $689,475, an annualized growth of 44%.

Superior Tunneling's annual contract income from the company's inception through 1978 is plotted on the chart [Figure 16.15]. Contract income is revenue received from the sole performance of construction projects and does not include any other sources of income. It could reasonably be expected that the contract income would continue on an upward trend after 1976 based both on historical data as well as management's commitment to continued growth. However, because of the severe delays and cost overruns STC was experiencing on the Delta Tunnel, they were unable to increase their annual contract volume. As mentioned earlier, the curtailment in their cash flow and reduction of bonding capacity were the major contributors to the financial reversals experienced after 1976.

Statistical methods have been developed for forecasting trends. One of the most widely accepted methods is that of least squares trend projection.

Three different least squares trends were developed for the data shown [in Figure 16.15]—one straight line and two exponential. The results showed the straight-line trend to be the best fitting one. The trend equation is

$T_t = -1,515,570 + 340,000t$

$0 = 1963$ t unit = 1 year

This linear trend, also shown on the chart [Figure 16.15], tells us that, on the average, there was a yearly increase in contract income of $340,000 from 1969 to 1976.

Had STC been compensated for the problems they encountered during the performance of their work, there would be no need for a business impairment claim. However, they were not and, as explained earlier, their contract income *did not* continue upward on its historical trend. It is reasonable to project an annual increase in contract income from 1976, without the Delta Tunnel project, since profitable work could have been obtained elsewhere. The period from 1976 to 1978 was one of an expanding economy and considerable work of the nature STC excels in was available for bid.

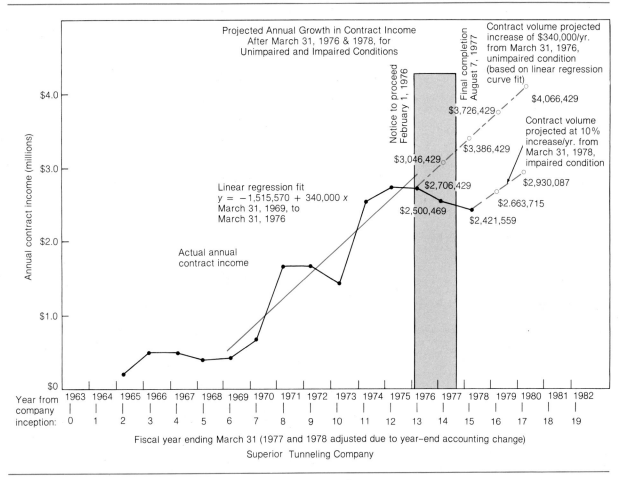

Figure 16.15. Graph of annual contract income for Superior Tunneling Company.

Projecting ahead from 1976, based on past performance, it would be reasonable to expect STC to have an annual income of over $4.066 million by 1980.

By the time of completion of the Delta Tunnel project, STC had been severely impacted in their ability to bid on and obtain new work. However, based on the assumption of their obtaining new work if they were able to maintain their level of activity at that experienced for 1978, there would still be some inflationary factors to contend with. In terms of "real" dollars of work completed, adjustment would therefore have to be made for inflation. The Dodge Building Cost Services report shows the nonresidential building cost index at year ending 1976 = 616 and year ending 1978 = 753—an annualized increase of 10.56%. It would therefore be reasonable to assign a minimum of 10% annual growth in contract income for STC's operations after 1978.

The annual operating income is that income STC is left with after contract revenues have been reduced by project contract costs and home office, general, and administrative expenses. Two methods of computing the average yearly operating income are shown below [Table 16.13].

Table 16.13 Two Ways of Computing Annual Operating Income for "STC"

Year	Contract Income	Operating Income	Operating as a Percent of Contract Income
1969	$ 414,937	$ 50,423	12.15
1970	677,617	66,617	9.83
1971	1,640,935	246,665	15.03
1972	1,645,935	142,816	8.68
1973	1,412,872	43,286	3.06
1974	2,513,261	285,066	11.34
1975	2,704,455	168,118	6.22
1976	2,706,429	77,707	2.87
Total	$13,715,426	$1,080,698	69.18

1. Average percent income based on yearly averages:

$$\frac{69.18}{8} = 8.65\%$$

2. Average percent income based on total operating income versus total contract income for the period 1969–1976:

$$\frac{1,080,698}{13,715,426} = 7.88\%$$

As a conservative estimate of operating income as a percent of contract income, the lower percentage should be used.

The total amount of damages (business impairment and project compensation damages) that STC suffered on the project can now be calculated. The contract income would have continued to increase, excluding the Delta Tunnel project, to a value of approximately $4.088 million by 1980. Operating income on this revenue would have averaged 7.88% based on historical experience. In the impaired condition, the contract income dropped to about $2.420 million by 1978 at the end of the project. In the next two years it could be assumed to increase at 10% a year to $2.930 million by 1980. The actual operating income is known through 1978 under the impaired conditions, and therefore it should amount to 7.88% of the yearly projected contract income. The difference between the total projected operating income for the unimpaired operation less that of the impaired operation is the total amount of lost operating income, or total damages. This is summarized below.

Unimpaired Condition

Year	Projected Contract Income	Percent Operating Income	Projected Operating Income
1977	$3,046,429	7.88	$240,059
1978	3,386,429	7.88	266,851
1979	3,726,429	7.88	293,643
1980	4,066,429	7.88	320,435
Total			$1,120,988

Impaired Condition

1977	—	—	($184,809)
1978	—	—	(270,356)
1979	$2,663,715	7.88	209,901
1980	2,930,087	7.88	230,891
Total			($ 14,373)

Total Loss in Operating Income

$$\$1{,}120{,}988 - (\$-14{,}373) = \$1{,}135{,}361$$

The total damages, or total loss in operating income, is calculated at $1,135,361. Any award of project compensatory damages will go to offset the total damages, since this amount will be taken in as operating income.

1. On what grounds do you suppose this expert witness concluded that the linear trend equation was better than the two exponential trend equations that were considered?

2. Here is your chance to save your client company, Southwest Transporation Company, some money. What inadequacies, if any, do you find in the estimate developed by this expert witness? (By the way, STC was actually awarded a somewhat smaller amount than that suggested by this witness.)

16.2 "UNIVERSAL INSTRUMENTS, INC."

The Environmental Technology Division (ETD) of "Universal Instruments" has grown very rapidly during its brief existence. This division manufactures and sells instruments that measure various facets of our ecological environment as well as equipment designed to help control or improve our environment. In its present state, ETD is made up of three business centers:

Analytical instruments. All relevant functions, including manufacturing and distributing functions, are planned and implemented in this center.

Electro-optical instruments. All relevant functions, except actual production of major instruments, are planned and carried out in this center.

Laslow filter systems. All administrative functions, but no manufacturing functions, are carried out at this center.

The Environmental Technology Division had its beginning in 1971 at which time it employed only ten people. Growth was slow until mid-1972 when the electro-optical systems group started to penetrate the market for smoke-detection and measurement instruments. Since that time, the division has experienced rapid growth, a condition generating the need to expand plant size. Because of its rapid rate of growth, ETD has never been able to stay in a facility for very long. Since 1973, three facility changes have taken place with their attendant costs and inefficiencies. A new 20,300 square foot facility was occupied in September of 1973 and furnished with Herman Miller concept relocatable furniture so that change and growth could be accomplished with a minimum of inconvenience. Nevertheless, four months after the adoption of this new facility, the office areas became crowded, inefficient, and unattractive. The general feeling among members of ETD's management was that, as the office areas became overcrowded, efficiency decreased, communication became more difficult, and organizational patterns degenerated.

One member of management describes an analytical effort that was attempted to improve the division's planning of growth requirements:

ETD has planned a new office building to be occupied after the first of the upcoming calendar year. It is planned to be 40,000 square feet in floor area with 50% of that space devoted to office area and 50% to production. (The 20,300 square foot building has 55% of its area devoted to office space and 45% to production.)

The problem I will be describing has to do with an approach I used to analyze the personnel growth rate during ETD's short history. The analysis utilized a trend projection procedure to estimate growth in personnel assuming that growth during the next two years will be influenced by the same factors as in the past. Using this information along with our qualitative knowledge regarding crowding, discomfort, and inefficiencies in the present facility, it should be possible, I reasoned, to determine approximately when the new facility will be overcrowded. Moreover, the study should also suggest some steps that might be taken to displace this overcrowding point further into the future.

The following table [Table 16.14] presents a summary of personnel growth from Janu-

Table 16.14 Increase in Personnel at ETD Between 1971 and 1974

Month	1971 General Office	Production	Staff	1972 General Office	Production	Staff
January	0	0	1	0	0	0
February	1	0	0	0	0	0
March	1	0	1	0	0	0
April	0	0	1	1	0	0
May	0	0	0	0	0	0
June	0	0	0	0	0	0
July	0	0	0	1	0	0
August	0	0	1	1	0	0
September	0	0	0	1	0	0
October	0	0	0	2	0	0
November	0	0	0	2	1	1
December	0	0	0	2	0	0
Year	2	0	4	10	1	1

Month	1973 General Office	Production	Staff	1974 General Office	Production	Staff
January	6	0	0	3	4	0
February	1	1	0	5	1	1
March	1	1	0	2	1	0
April	2	1	0	4	1	0
May	0	1	0	6	1	0
June	0	0	0	4	1	0
July	2	1	0	9	0	0
August	0	1	0	1	0	0
September	6	1	1	0	0	0
October	6	3	0	0	0	0
November	0	0	0	0	0	0
December	1	0	0	0	0	0
Year	25	10	1	34	9	1

ary of 1971. Between that time and the present, ETD has grown at a rate of approximately four employees per month. This employee growth record is shown pictorially in the following chart [Figure 16.16]. The chart suggests that ETD changed its growth characteristics dramatically in late 1972. For this reason, a linear trend line was fitted to observations 22 through 44 on this graph without regard to earlier points. The least squares linear trend equation was found to be $T_t = -69.656 + 3.9029t$.

The present facility was occupied in September of 1973 (Month 34). Conditions were then ideal for a period of about four months.

If we assume that proportionality exists regarding such things as conference room areas, reception areas, walkways, lavatory space, lunch rooms, and so forth, then the number of employees that can be comfortably housed within the planned new facility would be approximately 140. If space is used in the same way as in the present facility, then overcrowding will become evident approximately eight months after this ideal number of employees has been reached and approximately 12 months after occupancy. The chart [Figure 16.16] indicates that the overcrowding referred to will occur at the planned level of 170 employees. However, it may be that a 50–50 allocation of facility space to production and office employees is not optimal.

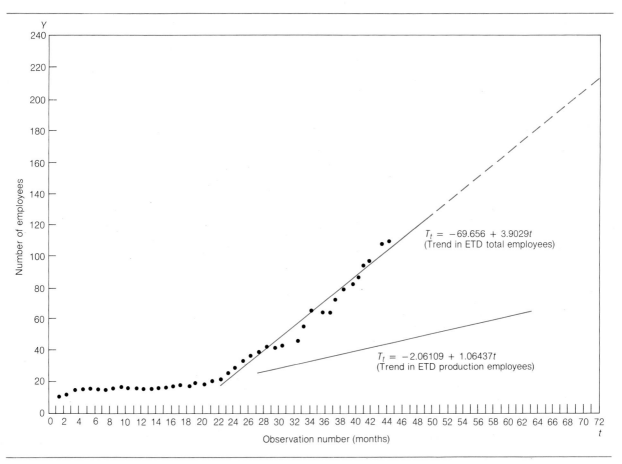

Figure 16.16. Linear trend lines for (1) total employees and (2) production employees in Environmental Technology Division of Universal Instruments, Inc.

Cases for Analysis and Discussion

A linear trend analysis was performed on the subpopulation of production workers with results as shown in the chart. The equation was found to be

$$T_t = -2.06109 + 1.0644t$$

Clearly, the ratio of slopes between the overall population and that of the production population is quite high—3.9029/1.0644 = 3.67. What this suggests is that a 50% allocation for production workers may be too high for the new facility. This conclusion is confirmed by the fact that the office area of the present facility is incontestably overcrowded, whereas the production area does not appear to be overcrowded at this writing. A 60–40 allocation (that is, 60% office personnel and 40% production personnel) would result in the following improvements:

1. It would increase the new facility's office space from 20,000 to 24,000 square feet.
2. This would then increase the ideal employee population from 120 to 165.
3. Adding the eight-month delay to 165 employees gives us a maximum of 200 employees and a period of almost two years between initial occupancy and the onset of overcrowding.

    In designing the new building, ETD should insist on a design which has about 10% to 15% flexibility—that is, it could be used for either production workers or office workers as the need arises.

1. What are some of the principal advantages and principal limitations of the manner of using trend analysis in this case?

2. Do you believe that this manager's emphasis on space allocation goes very far toward reducing this division's recurring problem of outgrowing its physical facilities? Elaborate. What alternative approach(es) to facility planning do you believe should be considered? Why?

3. In this chapter, you were told that about 10 to 15 years of historical data should be used when fitting a trend. In this case there are 22 data points. Has the rule been observed or violated? Explain.

 ## 16.3 "UBIQUITY INSURANCE COMPANY, INC.": PART I

Every insurance company is required by law to have enough money in loss reserves to be able to meet all outstanding claims. These are claims that have been reported to the insurance companies that may not be settled for several years but the insurance company is required to make an estimate of the ultimate losses it will incur. There are also claims that have occurred and that the insurance company is liable for but that have yet to be reported to the insurance company. The insurance company is required to estimate the ultimate loss on these unreported claims and carry enough reserves to cover them.

The accidents that have occurred but that the insurance company is not aware of are *unreported claims*. In order to estimate reserves for unreported claims, the company must estimate the number of unreported claims and the average severity of these claims. Multiplying the estimate of the unreported claims by the estimated severity gives an estimate of the unreported reserves the insurance company needs. At the end of 1976, "Ubiquity Insurance Company" had approximately $190 million in unreported reserves.

Ubiquity estimates the unreported reserves at the end of each month. To estimate how the number of unreported claims outstanding at the end of each month changes, knowledge of fluctuations of claim occurrence from month to month is helpful.

A statistician retained by Ubiquity Insurance analyzed four auto coverages and one homeowners coverage. The four auto coverages and the homeowner's coverage represent some of the largest unreported liabilities Ubiquity has, accounting for approximately 62% of the total unreported reserves. The statistician describes his approach as follows:

I gathered 93 months of data for each line and coverage from January 1977 to September 1984. These figures are from Ubiquity Insurance Company and are based on their claim experience in the United States and Canada.

I used the Census X-11 packaged computer program which has some unique features, including an analysis-of-variance test for the presence of stable seasonality. The test showed that all five lines and coverages do experience stable seasonality. The results are shown in the following chart [Figure 16.17].

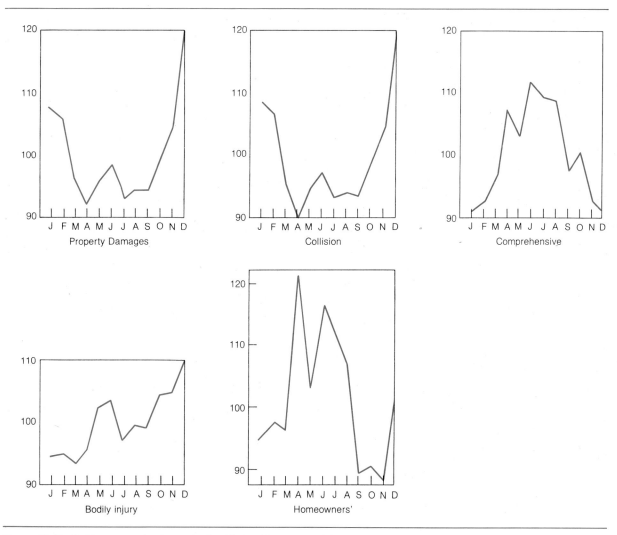

Figure 16.17. Stable seasonal patterns for five kinds of insurance claims.

1. Why do you think the seasonal patterns of the various types of claims are as they are?

2. In what way or ways would a knowledge of the seasonal patterns be helpful to the insurance company? Would knowledge of these seasonal patterns be of any use to society at large? Explain.

 ## 16.4 "UBIQUITY INSURANCE COMPANY, INC.": PART II

Reread Part I of this case. The statistician refers to a useful time series computer program, namely, Census X-11. As an outside project, learn as much as you can about the kinds of information provided by this program.

17 Short-Term Forecasting with Autoprojective Methods

Life can only be understood backward, but it must be lived forward.—Niels Bohr

I know of no way of judging the future but by the past.—Patrick Henry

What You Should Learn from This Chapter

In this chapter we continue the subject of time series forecasting. A great many different forecasting approaches are used in practice, and this chapter makes no attempt to deal with all, or even most, of these. What is attempted is a reasonably comprehensive treatment of a family of methods having one thing in common: Forecasts are obtained through some kind of systematic manipulation of past data only. We will refer to these as *autoprojective methods*.

When we use the term *autoprojective forecasting method*, we refer to a method that utilizes only past data for a particular variable and a formula for obtaining estimates of future values of that same variable. No attempt is made to identify and take advantage of any relationships between the variable to be predicted and other variables.

You may already have surmised that autoprojective methods have something in common with the trend projection methods described in Chapter 16. The analogy is a useful one but one which should not be pushed too far for two reasons: First, trend projection is usually used in connection with long-range forecasting, whereas the methods described in this chapter are almost universally limited to much shorter-term concerns. Second, trend projection attempts to provide estimates for only one component of the time series, whereas the autoprojective methods to be described here attempt to provide estimates for the combined effects of all or most of the systematic components of the series.

When you have completed this chapter, you should be able to

1. Apply the MAD, MSE, $\bar{r}$, and ADR measures of forecast adequacy to a set of predicted time series values.
2. Apply a no-change model to a set of time series data and obtain predicted values.
3. Apply moving-average models, both unweighted and weighted, to a set of time series data and obtain predicted values.
4. Apply exponential smoothing models to a set of time series data and obtain predicted values.
5. Apply autoregressive models, both simple and multiple, to a set of time series data and obtain predicted values.
* 6. Describe in a general way how the Box-Jenkins methodology can be employed to obtain predicted values for a variable.

Before beginning this chapter, you may wish to review the following sections: in Chapter 4, sections 4.3, 4.9, and 4.10; in Chapter 7, section 7.2; in Chapter 14, sections 14.4, 14.5, and 14.6; in Chapter 15, sections 15.2, 15.3, 15.4, 15.5, 15.6, 15.15, and 15.16; and in Chapter 16, sections 16.1, 16.2, and 16.3.

17.1 Introduction

When using most autoprojective forecasting methods, the analyst views the time series of interest as being made up of (1) some kind of systematic pattern and (2) a set of randomly arranged irregular fluctuations (Figure 17.1). The forecaster's job is to identify the systematic pattern and to project that pattern into the future.

An important background assumption is that the underlying systematic pattern will not change between time t, the time when the forecast is prepared, and time $t + \tau$, the future time period to which the predicted value corresponds. (The τ value will be 1, 2, 3, and so on, where 1 indicates that the predicted value is for one period in the future, 2 indicates that it is for two periods in the future, and so forth.) Clearly, the assumption of a nonchanging

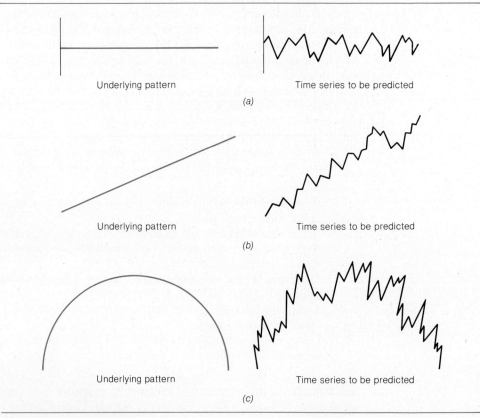

Figure 17.1. Some illustrative examples of how time series are viewed when autoprojective methods are used. Time series data actually observed represent a combination of the underlying systematic pattern and random irregular fluctuations.

systematic pattern will sometimes be invalid. The further into the future one attempts to project, the greater is the likelihood that the pattern will change. For this reason, all of the methods described in this chapter are ordinarily used in situations requiring forecasts for time periods only a short distance—that is, a week, a month, or, at most, a few months—into the future.

Even if the requirement of an unchanging underlying pattern is met, forecasting errors will still be made because the irregular component is random and hence unpredictable. Of course, if the pattern is identified incorrectly or if the pattern changes, the forecasting errors will be even greater. The term *forecasting error* is defined in a commonsense manner:

The *forecasting error* equals the actual value of the variable of interest, denoted Y, for a specific time period t, less the *previously predicted* value of that variable for the same time period. In symbols:

$$E_t = Y_t - \hat{Y}_t$$

where E_t is the forecasting error at time t
Y_t is the actual value of Y at time t
$\hat{Y}_t$ is the predicted value of Y for time t obtained at a previous time

17.2 Ways of Measuring the Adequacy of a Forecasting Model

The question of how to evaluate the adequacy of a forecasting model is potentially complex and one on which little agreement is to be found. Much depends on the forecasting horizon, the purpose for which the forecast is prepared, the relative importance of foreseeing future levels of a time series versus recognizing an upcoming cyclical turning point, and so forth.

Without pretending to present an exhaustive treatment of the subject of forecasting-model evaluation, we will make frequent use of four measures which can be applied quite conveniently to autoprojective methods. These measures are (1) the mean absolute deviation (MAD), (2) the mean squared error (MSE), (3) the coefficient of simple linear correlation ($\bar{r}$), and (4) the average duration of run (ADR).

MAD and MSE

A reasonable way of measuring the overall accuracy of an autoprojective forecasting model would seemingly be to calculate the average value of E_t over a large number of time periods. But, of course, we cannot simply average the E_t's, some of which will be positive and some negative, without getting a meaningless result.* What we *can* do is make use of average absolute or average squared deviations from the predicted values. That is:

---

*Although such an average *is* sometimes used for the purpose of detecting any systematic bias present in forecasts generated by a particular model. Unlike deviations from the arithmetic mean, E_t values will not average zero unless such bias is absent.

The mean absolute deviation, denoted MAD, is determined by

$$MAD = \frac{\sum\limits_{t=1}^{N} |Y_t - \hat{Y}_t|}{N} \qquad t = 1, 2, 3, \ldots, N \qquad (17.2.1)$$

where N represents the number of pairs of Y_t and $\hat{Y}_t$ values.
The mean squared error, denoted MSE, is computed by

$$MSE = \frac{\sum\limits_{t=1}^{N} (Y_t - \hat{Y}_t)^2}{N} \qquad t = 1, 2, 3, \ldots, N \qquad (17.2.2)$$

where, again, N represents the number of pairs of Y_t and $\hat{Y}_t$ values.

Which of these measures is used to compare the forecasting adequacy of several models under consideration will depend importantly on one's attitude toward large errors. The mean absolute deviation, MAD, is the easier measure to interpret because it is literally the average of N forecasting errors. On the other hand, because of the squaring of the forecasting errors, the mean squared error, MSE, penalizes large errors relatively severely. When a forecaster elects to use MSE in preference to MAD, it usually means that he or she would prefer to commit a great many small errors rather than a few very large ones. Thus, a forecasting model which results in a smaller MSE than other models under consideration will be the one chosen.

$\bar{r}$ and ADR

We will also apply two other measures of forecast adequacy in this chapter: $\bar{r}$, between original and predicted values, and average duration of run, ADR. The $\bar{r}$ measure is the now familiar coefficient of simple linear correlation, corrected for the loss of two degrees of freedom.* This measure is frequently used to indicate the degree of closeness of the *patterns* (but not necessarily the values) of the original Y values, on the one hand, and the predicted $\hat{Y}$ values on the other. The $\bar{r}$ measure is computed by

$$\bar{r} = \sqrt{1 - (1 - r^2)\left(\frac{n-1}{n-2}\right)}$$

where r^2 is determined by squaring the correlation coefficient, r, and where r itself is obtained by

$$\frac{\sum Y_t \hat{Y}_t - \dfrac{(\sum Y_t)(\sum \hat{Y}_t)^2}{N}}{\sqrt{\left[\sum Y_t^2 - \dfrac{(\sum Y_t)^2}{N}\right]\left[\sum \hat{Y}_t^2 - \dfrac{(\sum \hat{Y}_t)^2}{N}\right]}}$$

---

*This correction will seldom be necessary when the number of historical observations is fairly large. It is used here in anticipation of comparisons between two time periods having quite different numbers of observations. See section 17.3 for details.

The closer $\bar{r}$ is to $+1$, the closer the patterns of the original Y values and the predicted $\hat{Y}$ values correspond. An $\bar{r}$ of 0 would indicate no relationship between the two patterns, whereas $\bar{r}$ of -1 would indicate that the two sets of Y values, original and predicted, possess completely opposite patterns.

The average duration of run may be defined as

$$\text{ADR} = \frac{\text{Number of change signs in truncated set of such signs for predicted values}}{\text{Number of runs in the same truncated set of change signs}}$$

The numerator of the above formula is obtained after dropping incomplete runs at either end of the set of period-to-period changes in the predicted values. For example, if the sequence of change signs in the predicted values were as shown in the following table, the first two and the final plus signs would be discarded on the grounds that they are probably associated with incomplete runs. A *run* is *a sequence of like signs in the period-to-period changes in the predicted values*. Therefore, ADR would be obtained by dividing the remaining number of signs—namely, 6 in our example—by the number of runs found between the horizontal lines in the right-hand column—3 in this case. Thus, $\text{ADR} = (6/3) = 2$. What this means is that the predicted values proceed in the same direction for two periods, on the average, before reversing direction. Expected average duration of run for a random series of infinite length is 1.5.[1] For a nonrandom series with a substantial cyclical component, ADR could be as large as the average duration of cyclical expansions and contractions in the historical data.

Time t	$\hat{Y}_t$	Sign of Change	
1	5	———	
2	7	+	←— Ignored
3	8	+	
4	6	− }	
5	3	−	
6	7	+ }	Number of runs = 3
7	8	+	Number of signs = 6
8	6	− }	
9	4	−	
10	5	+	←— Ignored

ADR is not really a measure of forecasting *accuracy* because it entails no comparisons with the actual Y values. However, under some circumstances it can be a most useful measure of the *practical adequacy* of a forecasting model. Suppose that the forecasts are to be used for business planning purposes and that it is important that a change in direction in the predicted values *not be acted upon* unless there is good reason to believe that it signals a *true cyclical reversal*. (After all, it could be just a wiggle in the irregular component.) In such a situation, the forecasting model boasting the largest ADR would be considered most desirable—other things being equal—because a minimum of noncyclical reversals in direction will appear in the predicted values.

Clearly, in practice, not all measures of forecasting adequacy will be applied at the same time because they often work at cross purposes. For example, with a highly irregular time series, a forecasting model which results in an $\bar{r}$ close to $+1$ (high) may have an ADR value close to 1.5 (low). Therefore, that model will be considered good if similarity of pattern between original and predicted Y values is important and bad if elimination of false signals of cyclical change is important. The measure(s) of forecasting adequacy selected must always be predicated on the requirements of the user of the forecasts.

Table 17.1 Change in Business Inventories in 1972 Dollars, First Quarter of 1969 Through Second Quarter of 1984

(Annual Rates, Billions of Dollars)

Year and Quarter	Inventory Change	Year and Quarter	Inventory Change
1969.1	$11.7	1977.1	10.5
.2	11.8	.2	13.8
.3	13.7	.3	18.7
.4	7.0	.4	10.1
1970.1	2.1	1978.1	17.3
.2	5.0	.2	18.4
.3	6.5	.3	13.3
.4	1.4	.4	15.2
1971.1	11.2	1979.1	12.9
.2	10.4	.2	13.7
.3	7.0	.3	4.8
.4	3.6	.4	− 2.3
1972.1	6.3	1980.1	− 0.5
.2	12.1	.2	− 2.1
.3	12.8	.3	−10.1
.4	9.7	.4	− 4.7
1973.1	16.0	1981.1	8.1
.2	15.2	.2	12.4
.3	13.8	.3	17.5
.4	23.7	.4	7.2
1974.1	13.2	1982.1	− 6.7
.2	12.6	.2	− 4.0
.3	7.7	.3	− 6.4
.4	12.9	.4	−24.6
1975.1	−14.3	1983.1	−16.5
.2	−11.3	.2	− 6.1
.3	1.0	.3	0.9
.4	− 2.3	.4	7.2
1976.1	10.0	1984.1	31.6
.2	11.3	.2	20.3
.3	7.3		
.4	2.4		

Source: U.S. Department of Commerce, Bureau of Economic Analysis.

17.3 No-Change Models

The data set we will use throughout this chapter to illustrate how to apply several kinds of autoprojective forecasting models is presented in tabular form in Table 17.1 and in graphic form in Figure 17.2.

The data represent inflation-adjusted changes in business inventories for the United States for the first quarter of 1969 through the second quarter of 1984. This series was selected for illustrative purposes for two principal reasons: First, it is a component of real GNP and one which is watched closely by economic forecasters because of its considerable cyclical volatility. Second, it is devoid of a trend and a seasonal component, a fact which will help us keep the treatment of some future topics (for example, Box-Jenkins methods described in sections 17.7 and 17.8) as simple as possible.

When applying the MAD, MSE, $\bar{r}$, and ADR criteria to the predicted values, we will divide the historical data into two periods: (1) the first quarter of 1969 (abbreviated 1969.1) through the fourth quarter of 1981 (1981.4) and (2) the first quarter of 1982 (1982.1) through the second quarter of 1984 (1984.2). The period from 1969.1 to 1981.4 will be used in the development of forecasting models that must be "fitted" in some manner to the historical

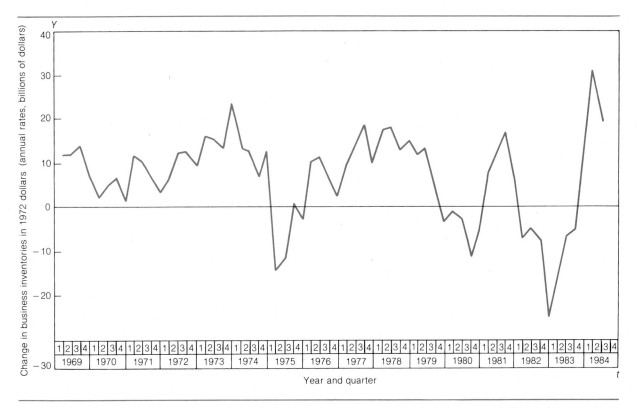

Figure 17.2. Original data for inflation adjusted change in business inventories from the first quarter of 1969 to the second quarter of 1984. (*Source:* U.S. Department of Commerce, Bureau of Economic Analysis.)

YEAR,QT	ACTUAL INVENTR'S	FORECAST	ERROR	SQ. ERR.
1969.1	11.70			
.2	11.80	11.70	0.10	0.01
.3	13.70	11.80	1.90	3.61
.4	7.00	13.70	6.70	44.89
1970.1	2.10	7.00	4.90	24.01
.2	5.00	2.10	2.90	8.41
.3	6.50	5.00	1.50	2.25
.4	1.40	6.50	5.10	26.01
1971.1	11.20	1.40	9.80	96.04
.2	10.40	11.20	0.80	0.64
.3	7.00	10.40	3.40	11.56
.4	3.60	7.00	3.40	11.56
1972.1	6.30	3.60	2.70	7.29
.2	12.10	6.30	5.80	33.64
.3	12.80	12.10	0.70	0.49
.4	9.70	12.80	3.10	9.61
1973.1	16.00	9.70	6.30	39.69
.2	15.20	16.00	0.80	0.64
.3	13.80	15.20	1.40	1.96
.4	23.70	13.80	9.90	98.01
1974.1	13.20	23.70	10.50	110.25
.2	12.60	13.20	0.60	0.36
.3	7.70	12.60	4.90	24.01
.4	12.90	7.70	5.20	27.04
1975.1	(14.30)	12.90	27.20	739.84
.2	(11.30)	(14.30)	3.00	9.00
.3	1.00	11.30	10.30	106.09
.4	(2.30)	1.00	3.30	10.89
1976.1	10.00	(2.30)	12.30	151.29
.2	11.30	10.00	1.30	1.69
.3	7.30	11.30	4.00	16.00
.4	2.40	7.30	4.90	24.01
1977.1	10.50	2.40	8.10	65.61
.2	13.80	10.50	3.30	10.89
.3	18.70	13.80	4.90	24.01
.4	10.10	18.70	8.60	73.96
1978.1	17.30	10.10	7.20	51.84
.2	18.40	17.30	1.10	1.21
.3	13.30	18.40	5.10	26.01
.4	15.20	13.30	1.90	3.61
1979.1	12.90	15.20	2.30	5.29
.2	13.70	12.90	0.80	0.64
.3	4.80	13.70	8.90	79.21
.4	(2.30)	4.80	7.10	50.41
1980.1	(0.50)	(2.30)	1.80	3.24
.2	(2.10)	(0.50)	1.60	2.56
.3	(10.10)	(2.10)	8.00	64.00
.4	(4.70)	(10.10)	5.40	29.16
1981.1	8.10	(4.70)	12.80	163.84
.2	12.40	8.10	4.30	18.49
.3	17.50	12.40	5.10	26.01
.4	7.20	17.50	10.30	106.09
1982.1	(6.70)	7.20	13.90	193.21
.2	(4.00)	(6.70)	2.70	7.29
.3	(6.40)	(4.00)	2.40	5.76
.4	(24.60)	(6.40)	18.20	331.24
1983.1	(16.50)	(24.60)	8.10	65.61
.2	(6.10)	(16.50)	10.40	108.16
.3	0.90	(6.10)	7.00	49.00
.4	7.20	0.90	6.30	39.69
1984.1	31.60	7.20	24.40	595.36
.2	20.30	31.60	11.30	127.69
.3		20.30		

Figure 17.3. Computer printout showing application of the no-change formula $\hat{Y}_{t+1} = Y_t$ to the inventory data in Table 17.1 and the calculation of MAD and MSE.

data.* The period from 1982.1 to 1984.2 will be used to determine how well the models perform on data outside the range used in their development.

When applying any adequacy criteria to a forecasting model, it is a good idea to have a standard for comparison. No-change models can serve this purpose nicely. As the name suggests, no-change models are simple forecasting models based on the assumption that something will remain exactly the same between time period t and time period $t + \tau$, the forecast period. For example:

The simplest no-change model is

$$\hat{Y}_{t+\tau} = Y_t \tag{17.3.1}$$

where $\hat{Y}_{t+\tau}$ is the predicted value for a time τ periods ahead and Y_t is the actual Y value for period t.

Let us suppose that a forecast of real inventory change one quarter ahead is what is desired. In such a case the forecasting equation would be $\hat{Y}_{t+1} = Y_t$. Accordingly, the predicted value for the third quarter of 1984 would be the same as the observed value for the

---

* Since this second period is rather short and since application of the ADR measure results in the waste of some data, ADR will be applied only to the entire 1969.1 to 1984.2 period.

17 *Short-Term Forecasting with Autoprojective Methods*

second quarter of the same year—namely, $20.3 billion. Figure 17.3 shows how this no-change model is applied to our inventory data. The arrows in the figure serve to show that each observed value of Y is shifted forward one quarter each time and used as the predicted value for the quarter coming up. (The observed values of Y would be shifted forward two or more quarters if a longer forecasting horizon were required.) Notice that, even though there are 62 historical values, only 61 pairs of Y_t and $\hat{Y}_t$ values exist. Therefore, in this example $t = 1$ corresponds to the *second quarter of 1969* and $t = N = 61$ corresponds to the second quarter of 1984.

Values for the four measures of forecasting adequacy are presented in the lower left-hand corner of Figure 17.4. Naturally, we hope that use of a more complex forecasting model will result in better numbers for at least some of these adequacy criteria. We reason that, if this ultrasimple no-change model results in more favorable values for MAD, MSE, $\bar{r}$, or ADR than more complex models, why even bother with the latter? As will be shown, beating this no-change model is not always as easy as it may sound.

Because our inventory-change data lack a trend, the no-change model described immediately above, $\hat{Y}_{t+1} = Y_t$, is probably an appropriate one to use as a standard for comparison. However, you should be aware that other no-change models can also be employed. For example, if the data under analysis exhibit a definite upward or downward trend, a no-change-in-the-change model, $\hat{Y}_{t+1} = Y_t + (Y_t - Y_{t-1})$, or even a no-change-in-the-percent-change model, $\hat{Y}_{t+1} = Y_t(Y_t/Y_{t-1})$, will ordinarily track the observed data more closely and serve as a more rigorous standard for comparison than $\hat{Y}_{t+1} = Y_t$.

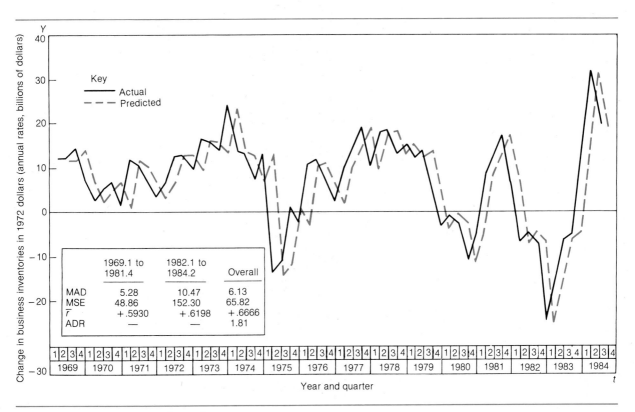

Figure 17.4. No-change model $\hat{Y}_{t+1} = Y_t$ applied to the inventory data.

17.4 Moving Averages

One serious shortcoming of the no-change model described above is that with a highly irregular time series the irregular fluctuations are merely projected into the future. As a result, the predicted values are difficult to interpret and of limited usefulness as aids to decision making. Moreover, it seems reasonable to think that the next observed Y value will be the outgrowth of several past values rather than only one. One way around these problems is to *average* the observed values for several recent time periods and view the average so obtained as the predicted value for time $t + \tau$.

Unweighted Moving Averages

An unweighted moving average is calculated by assigning a weight of 1 to each value in a set of n original values and dividing the sum of the resulting products by n. In symbols:

$$\hat{Y}_{t+\tau} = \frac{(1)Y_t + (1)Y_{t-1} + \cdots + (1)Y_{t-(n-1)}}{n}$$

or simply

$$\hat{Y}_{t+\tau} = \frac{Y_t + Y_{t-1} + \cdots + Y_{t-(n-1)}}{n} \tag{17.4.1}$$

where $\hat{Y}_{t+\tau}$ is the predicted value for τ periods ahead, Y_t is the most recent observed Y value, Y_{t-1} is the second most recent observed Y value, and $Y_{t-(n-1)}$ is the oldest value in a set of n recent observed values. The n is the number of original observations reflected in each average and is often referred to as the *moving-average period* or *span of the moving average.*

This procedure is then repeated with a new set of Y values, n in number, consisting of a new, more recent, observed Y and all Y values used previously, except for the oldest one.

For example, Figure 17.5 shows how a three-quarter unweighted moving average is applied to our inventory data. Notice that, even though the number of historical data points is 62, there are only 59 pairs of Y_t and $\hat{Y}_t$ values. Consequently, the N value used to calculate overall MAD, MSE, and $\bar{r}$ is 59.

Figure 17.6 (p. 814) shows the results of forecasts obtained from application of a three-quarter, a five-quarter, and a seven-quarter unweighted moving average to our inventory data. Table 17.2 (p. 815) presents the measures of forecasting adequacy. Notice that the MAD, MSE, and $\bar{r}$ values are inferior to those generated by the no-change model (Figure 17.4). Only ADR has improved.

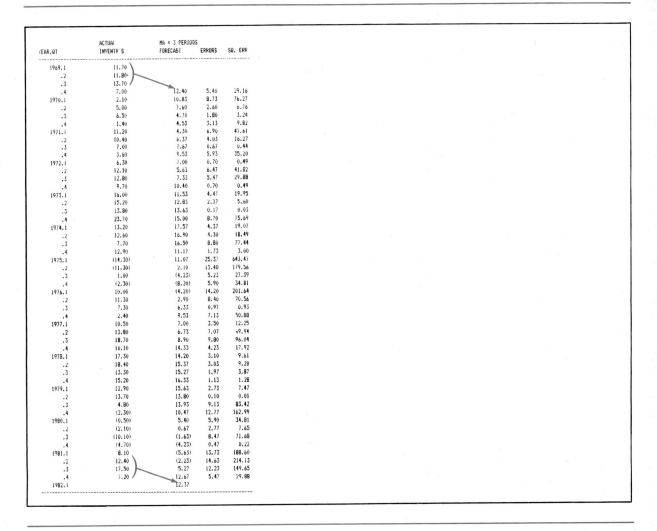

Figure 17.5. Computer printout showing application of a three-period unweighted forecasting model to our inventory data.

Weighted Moving Averages

Although unweighted moving averages are used more often than weighted ones for short-term forecasting, there is no reason why the Y values making up a set could not be assigned different weights. For example, if one believes that the most recent Y value in a set of n values holds greater forecasting implications than older values in the same set, weights of declining magnitudes could be applied. One possibility might be

$$\hat{Y}_{t+1} = \frac{(4)Y_t + (3)Y_{t-1} + (2)Y_{t-2} + (1)Y_{t-3}}{10}$$

Exponential smoothing, described in the next section, is based in part on this point of view.

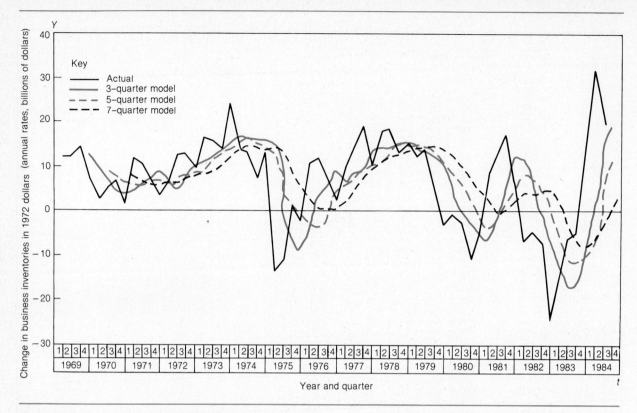

Figure 17.6. Predictions obtained from applying 3-quarter, 5-quarter, and 7-quarter unweighted moving-average forecasting models to the inventory data presented in Table 17.1.

Weighted moving averages are also sometimes employed in the hope of achieving predicted values having certain desirable qualities. For example, inspired by an article by J. Bongard,[2] this writer developed a family of weighted moving-average formulas which promised to minimize, for a specific moving-average period n, the number of noncyclical reversals in direction in the predicted values.* Results of the application of three members of this family of smoothing formulas are presented in Figure 17.7 and Table 17.3. The formulas are

$$\hat{Y}_{t+1} = \frac{(3)Y_t + (4)Y_{t-1} + (3)Y_{t-2}}{10}$$

$$\hat{Y}_{t+1} = \frac{(5)Y_t + (8)Y_{t-1} + (9)Y_{t-2} + (8)Y_{t-3} + (5)Y_{t-4}}{35}$$

$$\hat{Y}_{t+1} = \frac{(7)Y_t + (12)Y_{t-1} + (15)Y_{t-2} + (16)Y_{t-3} + (15)Y_{t-4} + (12)Y_{t-5} + (7)Y_{t-6}}{84}$$

*The manner of deriving these weights is somewhat complicated. However, a simple procedure emerging from the derivations may be summarized as follows: Assign a weight of n to observations Y_t and $Y_{t-(n-1)}$, a weight of $n + (n-2)$ to observations Y_{t-1} and $Y_{t-(n-2)}$, a weight of $n + (n-2) + (n-4)$ to observations Y_{t-2} and $Y_{t-(n-3)}$, and so forth.

Table 17.2 Measures of Forecasting Adequacy Resulting from Three Kinds of Unweighted Moving Averages

	Three-Quarter Moving Average		
	1969.1 to 1981.4	1982.1 to 1984.2	Overall
MAD	5.92	13.68	7.24
MSE	58.59	245.80	90.43
$\bar{r}$	+.4099	+.0762	+.4622
ADR	—	—	3.06

	Five-Quarter Moving Average		
	1969.1 to 1981.4	1982.1 to 1984.2	Overall
MAD	7.12	16.59	8.08
MSE	72.77	360.53	123.25
$\bar{r}$	+.2236	+.0000	+.2042
ADR	—	—	3.00

	Seven-Quarter Moving Average		
	1969.1 to 1981.4	1982.1 to 1984.2	Overall
MAD	7.07	16.08	8.71
MSE	84.47	369.35	136.27
$\bar{r}$	+.0000	+.5068	+.0000
ADR	—	—	4.00

Unfortunately, these weighted moving-average formulas perform worse than—or, at best, no better than—the no-change model and the unweighted moving-average models with respect to MAD, MSE, and $\bar{r}$. However, they are conspicuously superior to the no-change model (Figure 17.4) with respect to ADR. Presumably, one would not wish to use such weighted moving averages except in situations where ADR is considered an important criterion of forecast adequacy. But when ADR *is* regarded as very important, these formulas do well what they were designed to do—namely, keeping the number of false signals of cyclical reversal as small as possible.

17.5 Exponential Smoothing

We now come to an autoprojective forecasting method which has enjoyed wide and growing popularity during recent years—namely, exponential smoothing. A convenient way of introducing this subject will be to return briefly to the topic of unweighted moving averages.

For our inventory-change data, Table 17.2 showed that, when judged by MAD and MSE, the three-quarter unweighted moving average is superior to the five-quarter unweighted moving average. In turn, the five-quarter unweighted moving average is superior to

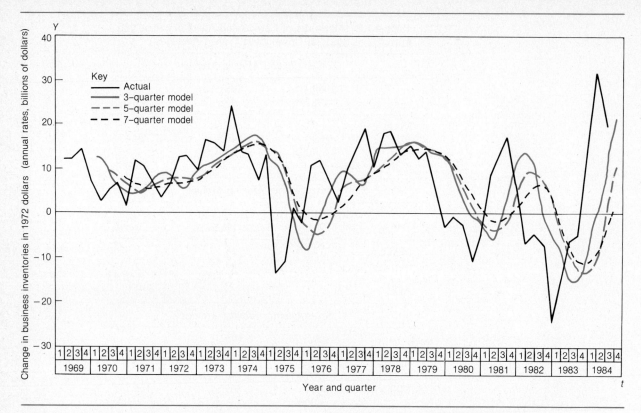

Figure 17.7. Predictions obtained from applying 3-quarter, 5-quarter, and 7-quarter weighted moving-average formulas (with weight sets of 3,4,3; 5,8,9,8,5 and 7,12,15,16,15,12,7, respectively) to the inventory data presented in Table 17.1.

the seven-quarter unweighted moving average. (This same conclusion holds for the weighted moving averages whose results are summarized in Table 17.3.)

With respect to these criteria of forecasting accuracy, it appears that, for our inventory data, the ability of a forecasting method to adjust rapidly to a change in the systematic component of the original time series is more than its ability to smooth the irregular fluctuations. A characteristic of unweighted moving averages worth emphasizing at this time is that the greater the number of original values n, represented in each moving average value, the greater will be the smoothing effect on the resulting predicted values. Thus, if we knew that the historical data contain much randomness, we might be inclined to pick a moving-average formula of longer period on the grounds that it will produce less confusing forecasts than a formula of shorter period. On the other hand, an unweighted moving average of longer period will react to changes in the systematic component in a more sluggish manner than one of shorter period, a fact revealed rather clearly in Figure 17.6.

Some useful insights into the nature of the unweighted moving average as a forecasting tool can be gained through a little algebraic manipulation. We have seen that the predicted value $\hat{Y}_{t+1}$ obtained by computing an unweighted average over the most recent n observations and assigning that average to time period $t + 1$ can be expressed by the formula

$$\hat{Y}_{t+1} = \frac{Y_t + Y_{t-1} + \cdots + Y_{t-(n-1)}}{n}$$

Table 17.3 Measures of Forecasting Adequacy Resulting from Three Kinds of Weighted Moving Averages

Three-Quarter Moving Average (Weights: 3, 4, 3)

	1969.1 to 1981.4	1982.1 to 1984.2	Overall
MAD	6.02	13.90	7.35
MSE	59.87	251.12	92.29
$\bar{r}$	+.4274	+.0000	+.4654
ADR	—	—	4.25

Five-Quarter Moving Average (Weights: 5, 8, 9, 8, 5)

	1969.1 to 1981.4	1982.1 to 1984.2	Overall
MAD	6.48	17.18	8.36
MSE	76.23	382.28	128.68
$\bar{r}$	+.2074	+.0000	+.1873
ADR	—	—	6.25

Seven-Quarter Moving Average (Weights, 7, 12, 15, 16, 15, 12, 7)

	1969.1 to 1981.4	1982.1 to 1984.2	Overall
MAD	7.38	17.36	9.19
MSE	91.36	420.23	151.16
$\bar{r}$	+.1378	+.0000	+.2149
ADR	—	—	8.17

This formula makes clear that, in computing a single unweighted moving-average value, equal weights are given to the most recent n observations, a practice suggesting that these n observations may be viewed as being of equal importance. Note also that observations prior to time period $t - (n - 1)$ receive weights of 0, suggesting that they are completely devoid of importance.

If the above formula tells us how to calculate the unweighted moving-average value to be assigned to time period $t + 1$ and identified as $\hat{Y}_{t+1}$, it follows that the formula for calculating the unweighted moving-average forecast for time period t must have been

$$\hat{Y}_t = \frac{Y_{t-1} + Y_{t-2} + \cdots + Y_{t-n}}{n}$$

The change in these two predicted values can thus be expressed as

$$\hat{Y}_{t+1} - \hat{Y}_t = \left(\frac{Y_t + Y_{t-1} + \cdots + Y_{t-(n-1)}}{n}\right) - \left(\frac{Y_{t-1} + Y_{t-2} + \cdots + Y_{t-n}}{n}\right)$$

Or, after some simplification, $\hat{Y}_{t+1} - \hat{Y}_t = (Y_t - Y_{t-n})/n$. Finally, we can express $\hat{Y}_{t+1}$ in terms of $\hat{Y}_t$ and the change; that is

$$\hat{Y}_{t+1} = \hat{Y}_t + \frac{Y_t - Y_{t-n}}{n} \tag{17.5.1}$$

In this form, the moving-average forecast for time period $t + 1$ reflects (1) the moving-average forecast for the prior period and (2) an adjustment to that prior forecast. Clearly, with a longer moving-average period, a smaller adjustment is made to the prior forecast each time a new forecast is obtained. This is the reason longer-period unweighted moving-average forecasts adjust more slowly to a change in the pattern of the original time series than do shorter-period unweighted moving-average forecasts.

As indicated above, a major limitation of unweighted moving averages is that their use entails applying the same weight to each of the n observations reflected in a moving-average value and no weight whatever to earlier observations. Let us suppose that the following assumptions impress us as reasonable: (1) The most recent observation on a time series contains more information of relevance to the near-term future of that series than do less recent observations and (2) all past observations contain some information relevant to the near-term future of that series. Acceptance of these assumptions implies a need for a moving average such that (1) larger weights are assigned the more recent observations and (2) all past observations are assigned some nonzero weight. Exponential smoothing as a forecasting method meets these requirements.

Recall that Equation (17.5.1) was given as $\hat{Y}_{t+1} = \hat{Y}_t + (Y_t - Y_{t-n})/n$. This equation may also be written

$$\hat{Y}_{t+1} = \hat{Y}_t + \frac{Y_t}{n} - \frac{Y_{t-n}}{n} \tag{17.5.2}$$

We can modify Equation (17.5.2) by replacing the Y_{t-n} value with an *approximate Y* value. Of course, there are any number of candidates for the job, but the one best qualified in terms of getting us to the basic equation for exponential smoothing is $\hat{Y}_t$, the predicted value of Y for period t obtained in period $t - 1$. Thus, we have

$$\hat{Y}_{t+1} = \hat{Y}_t + \frac{Y_t}{n} - \frac{\hat{Y}_t}{n}$$

which can be rewritten

$$\hat{Y}_{t+1} = \left(\frac{1}{n}\right)Y_t + \left(1 - \frac{1}{n}\right)\hat{Y}_t$$

Let us now use the symbol w in place of $1/n$. We have

$$\hat{Y}_{t+1} = wY_t + (1 - w)\hat{Y}_t \tag{17.5.3}$$

where $\hat{Y}_{t+1}$ is the predicted value of Y for period $t + 1$, w is assigned a value between 0 and 1 and is called the *smoothing constant*, and $\hat{Y}_t$ is the predicted value of Y for period t obtained in period $t - 1$.

Equation (17.5.3), then, is the basic formula for exponential smoothing. An alternative way of writing this equation can provide further insights into the nature of the technique:

$$\hat{Y}_{t+1} = \hat{Y}_t + w(Y_t - \hat{Y}_t)$$

Viewed in this light, the new forecast obtained through exponential smoothing is seen to be simply the previous forecast plus an adjustment for the error resulting from that previous forecast. Therefore, if w were close to 1, the new forecast would contain a substantial adjustment component attributable to the previous forecasting error. On the other hand, if w were close to 0, the new forecast would contain only a small adjustment component. We see, then, that a small w in an exponential smoothing formula implies the same thing as a large n in a moving-average formula: rather smooth predicted values, but slow adjustment to meaningful changes in the systematic component of the time series under study.

To apply the method of exponential smoothing to our inventory data, we begin by selecting a value for w. For purposes of illustration, we will use $w = .5$, $w = .3$, and $w = .1$. Thus, the exponential smoothing formulas used in the following demonstration are: $\hat{Y}_{t+1} = .5Y_t + .5\hat{Y}_t$, $\hat{Y}_{t+1} = .3Y_t + .7\hat{Y}_t$, and $\hat{Y}_{t+1} = .1Y_t + .9\hat{Y}_t$. Application of the first of these is demonstrated in Figure 17.8. (Notice that the first Y_t value is simply shifted to the "forecast" column and used as if it were the first $\hat{Y}_t$ value. After that, each Y_t value and each previously predicted $\hat{Y}_t$ value is multiplied by .5 and the resulting products summed.) The results obtained from applying all three exponential smoothing formulas are presented in Figure 17.9. The four measures of forecasting adequacy achieved by these formulas are shown in Table 17.4 (p. 822).

Comparing the information in Table 17.4 with that contained within Figure 17.4 (pertaining to the no-change model), we find that none of the exponential smoothing formulas surpasses the no-change model with respect to MAD and MSE. The exponential smoothing formula with $w = .3$ surpasses the no-change model with respect to $\bar{r}$. (But this "advantage" is not shared by the other two exponential formulas.) All three exponential smoothing formulas surpass the no-change model with respect to ADR.

Although comparing the results of exponential smoothing formulas with those achieved by moving-average formulas can be tricky because of the different n's involved, we will still hazard some tentative conclusions. When we compare the information in Table 17.4 with that of Table 17.2 (pertaining to the unweighted moving-average formulas) and Table 17.3 (pertaining to the weighted moving-average formulas), we find the exponential formulas slightly better. These formulas appear to be superior to the moving-average formulas, both unweighted and weighted, with respect to MAD and MSE. The exponential smoothing formula with $w = .3$ surpasses the moving-average formulas with respect to $\bar{r}$. All of the exponential smoothing formulas except the one with $w = .5$ surpass the unweighted moving-average formulas with respect to ADR but fall way short of the weighted moving-average formulas with respect to this adequacy criterion.

A practical advantage enjoyed by the exponential smoothing approach to short-term forecasting over the moving-average approach—both unweighted and weighted—is that a new average need not be computed each time a new Y value becomes available. One simply has to multiply the new Y value by w and the preceding predicted $\hat{Y}$ value by $(1 - w)$, and sum the two products. For this reason, exponential smoothing is often used in practice in preference to moving averages even though the resulting predicted values may be slightly inferior.

As a postscript to the subject of forecasting by means of smoothing formulas, we must point out that the methods described in this and the preceding section are strictly appropriate

YEAR,QT	ACTUAL INVENTR'S	ALPHA = .5 FORECAST	ERRORS	SQ. ERR.
1969.1	11.70	11.70	0.00	0.00
.2	11.80	11.70	0.10	0.01
.3	13.70	11.75	1.95	3.80
.4	7.00	12.73	5.73	32.78
1970.1	2.10	9.86	7.76	60.26
.2	5.00	5.98	0.98	0.96
.3	6.50	5.49	1.01	1.02
.4	1.40	6.00	4.60	21.12
1971.1	11.20	3.70	7.50	56.29
.2	10.40	7.45	2.95	8.71
.3	7.00	8.92	1.92	3.70
.4	3.60	7.96	4.36	19.03
1972.1	6.30	5.78	0.52	0.27
.2	12.10	6.04	6.06	36.72
.3	12.80	9.07	3.73	13.91
.4	9.70	10.94	1.24	1.53
1973.1	16.00	10.32	5.68	32.29
.2	15.20	13.16	2.04	4.17
.3	13.80	14.18	0.38	0.14
.4	23.70	13.99	9.71	94.29
1974.1	13.20	18.84	5.64	31.86
.2	12.60	16.02	3.42	11.71
.3	7.70	14.31	6.61	43.71
.4	12.90	11.01	1.89	3.59
1975.1	(14.30)	11.95	26.25	689.21
.2	(11.30)	(1.17)	10.13	102.54
.3	1.00	(6.24)	7.24	52.37
.4	(2.30)	(2.62)	0.32	0.10
1976.1	10.00	(2.46)	12.46	155.23
.2	11.30	3.77	7.53	56.69
.3	7.30	7.54	0.24	0.06
.4	2.40	7.42	5.02	25.18
1977.1	10.50	4.91	5.59	31.26
.2	13.80	7.70	6.10	37.16
.3	18.70	10.75	7.95	63.17
.4	10.10	14.73	4.63	21.40
1978.1	17.30	12.41	4.89	23.88
.2	18.40	14.86	3.54	12.56
.3	13.30	16.63	3.33	11.08
.4	15.20	14.96	0.24	0.06
1979.1	12.90	15.08	2.18	4.76
.2	13.70	13.99	0.29	0.08
.3	4.80	13.85	9.05	81.82
.4	(2.30)	9.32	11.62	135.09
1980.1	(0.50)	3.51	4.01	16.09
.2	(2.10)	1.51	3.61	13.00
.3	(10.10)	(0.30)	9.80	96.10
.4	(4.70)	(5.20)	0.50	0.25
1981.1	8.10	(4.95)	13.05	170.28
.2	12.40	1.58	10.82	117.17
.3	17.50	6.99	10.51	110.51
.4	7.20	12.24	5.04	25.44
1982.1		9.72		

Figure 17.8. Application of exponential smoothing formula $\hat{Y}_{t+1} = .5Y_t + .5\hat{Y}_t$ to the inventory data presented in Table 17.1.

only when the time series of interest is *stationary*, a term meaning that the series values tend to exhibit a horizontal pattern. When such a pattern is not present or when the pattern changes frequently, these techniques, if used at all, should be employed in such a way as to require only a modest amount of time displacement. That is, n should be small, or w large, and τ should be small. When this suggestion is observed, one can usually proceed as if the underlying pattern were horizontal—whether or not it actually is—with little harm being done.

Also, be aware that many specialized books on forecasting methods describe how to adapt the moving-average and exponential smoothing methods to accommodate time series data having an upward or downward trend or widely swinging cyclical oscillations.[3]

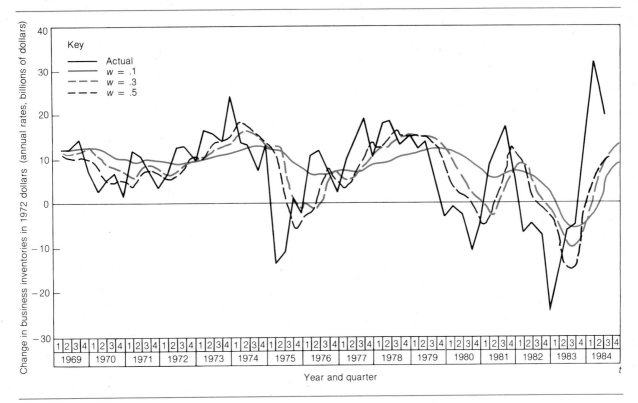

Figure 17.9. Predictions obtained from applying exponential smoothing formulas with *w* values of .5, .3, and .1 to the inventory data presented in Table 17.1.

17.6 Autoregressive Models

We noted earlier that all of the autoprojective forecasting methods described in this chapter presuppose that the time series of interest is composed of some kind of underlying pattern and a set of random irregular fluctuations. The presence of an underlying pattern suggests that Y_t values will be correlated with Y_{t-1} values, Y_{t-2} values, Y_{t-3} values, and so forth. That being so, we should be able to think in terms of a regression equation like the following one:

$$\hat{Y}_{t+\tau} = a + b_1 Y_t + b_2 Y_{t-1} + b_3 Y_{t-2} + \cdots + b_k Y_{t-(k-1)} \qquad (17.6.1)$$

where it is understood that some, but not all, of the net regression coefficients could have values of zero.

In other words, we should be able to generate short-term forecasts for a time series of interest by using some of its own past values in a regression equation.

Table 17.4 Measures of Forecasting Adequacy Resulting from Three Kinds of Exponential Smoothing Formulas

Exponential Smoothing Formula with $w = .5$

	1969.1 to 1981.4	1982.1 to 1984.2	Overall
MAD	5.32	11.76	6.38
MSE	49.69	207.43	75.55
$\bar{r}$	+.4071	+.3165	+.5125
ADR	—	—	2.35

Exponential Smoothing Formula with $w = .3$

	1969.1 to 1981.4	1982.1 to 1984.2	Overall
MAD	5.62	13.62	6.93
MSE	56.60	255.64	89.56
$\bar{r}$	+.6801	+.7440	+.7314
ADR	—	—	3.12

Exponential Smoothing Formula with $w = .1$

	1969.1 to 1981.4	1982.1 to 1984.2	Overall
MAD	6.13	14.52	7.51
MSE	65.01	292.53	102.30
$\bar{r}$	+.0000	+.0000	+.1371
ADR	—	—	4.23

In this section you will be shown how to use the regression and correlation techniques introduced in Chapters 14 and 15 in a new way. Because autoregression models form an important part of the Box-Jenkins methods described in the next section, the subject of autoregressive forecasting will be treated only briefly at this time.

Simple Autoregression

Often the strongest correlation to be found within a set of time series data will be that between the Y_t values and the Y_{t-1} values. With that in mind, we will develop an autoregressive forecasting equation of the form $\hat{Y}_{t+1} = a + bY_t$ for our inventory data. We can do this by deriving the least squares regression equation of the form $\hat{Y}_t = a + bY_{t-1}$. Once the equation has been derived, we can always substitute Y_t values into it and get $\hat{Y}_{t+1}$ values. The slope and Y-intercept values are determined by the same procedures employed in Chapter 14. For slope,

$$b = \frac{\sum Y_t Y_{t-1} - \dfrac{(\sum Y_t)(\sum Y_{t-1})}{N}}{\sum Y_{t-1}^2 - \dfrac{(\sum Y_{t-1})^2}{N}}$$

For Y intercept,

$$a = \frac{\Sigma\,Y_t}{N} - b\,\frac{\Sigma\,Y_{t-1}}{N}$$

Figure 17.10 shows some computer output for the inventory data for the period from 1969.1 to 1981.4. Figure 17.11 compares the original Y_t values with the $\hat{Y}_t$ values predicted one quarter ahead by the simple linear regression equation. This graph covers the period from 1969.1 to 1984.2; it also presents the four measures of forecasting adequacy.

Notice that we have finally come up with a forecasting formula that is generally superior

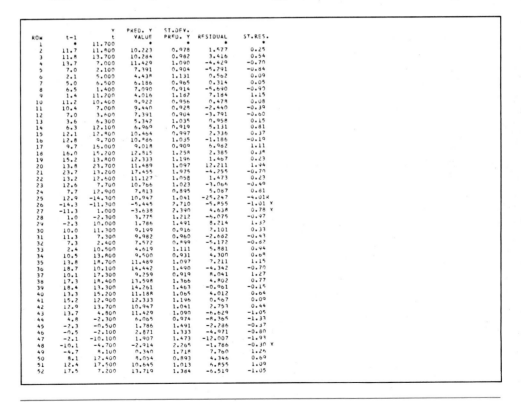

Figure 17.10. Computer output for a simple linear autoregression analysis on the inventory data presented in Table 17.1.

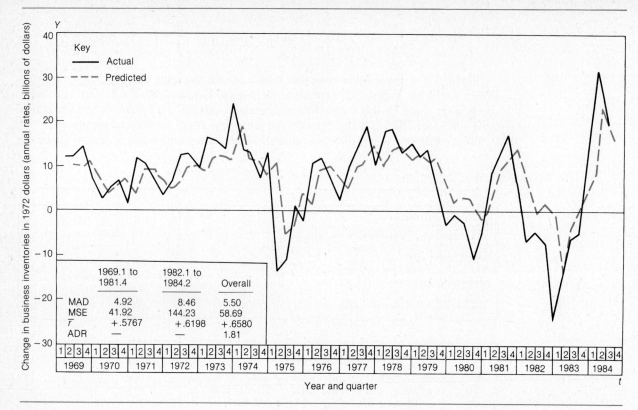

Figure 17.11. Comparison of actual values and values predicted one quarter ahead by the simple linear regression equation $\hat{Y}_{t+1} = 3.17 + .603Y_t$.

to the intuitively unsatisfying, but disarmingly accurate, no-change model. For the period 1969.1 to 1981.4, we have

	No-Change Model	Simple Autoregressive Model
MAD	5.28	4.92
MSE	48.86	41.92
$\bar{r}$	+.5930	+.5767

For the period 1982.1 to 1984.2:

	No-Change Model	Simple Autoregressive Model
MAD	10.47	8.46
MSE	152.30	144.23
$\bar{r}$	+.6198	+.6198

Both the no-change model and the simple autoregressive model lack smoothing power as measured by the average duration of run. Therefore, ADR for both models is the same as that for the original data—namely, 1.81.

Multiple Autoregression

It seems reasonable to suppose that an even better forecasting equation than the simple autoregressive one just described could be developed by utilizing more than a single lag. For example, if an equation of the form $\hat{Y}_t = a + bY_{t-1}$ surpasses the no-change model in most respects, should not a regression equation of the form, say, $\hat{Y}_t = a + b_1Y_{t-1} + b_2Y_{t-2} + b_3Y_{t-3}$ be even better? The answer is an equivocal "maybe and maybe not." The problem is that, when the Y_t and the Y_{t-1} values are rather strongly correlated, the Y_{t-1} values will be rather strongly correlated with the Y_{t-2} values and the Y_{t-2} values with the Y_{t-3} values, and so forth. Consequently, it is possible that no net good will result from the incorporation of more "independent variables." Indeed, a net harm might result via the introduction of multicollinearity into the analysis. Still, multiple autoregression is usually worth a try.

The question of how many "independent variables" to use and which ones they should be will be addressed in greater detail in the next section. For now, the procedures for selecting independent variables demonstrated in section 15.16 of Chapter 15 will usually serve adequately. For example, Figure 17.12 shows how backward elimination by stages was employed in connection with our inventory data. To begin with, the Y_t values were regressed on the Y_{t-1} values, the Y_{t-2} values, and so forth, through the Y_{t-10} values. The t tests* applied to the net regression coefficients for the resulting forecasting equation showed only one lag, $t - 1$, to be significant (as indicated by a t ratio equal to or in excess of 2.00). However, as insignificant "independent variables" were eliminated one at a time, beginning with the one with the lowest t statistic, it was found that three lags, $t - 1$, $t - 5$, and $t - 10$ produced significant results. Some results of the four-variable regression analysis are shown in Figure 17.13. Comparisons of actual and predicted Y_t values, along with measures of forecasting adequacy, are presented in Figure 17.14 (p. 828).

Not surprisingly, the multiple autoregression *does* result in superior adequacy measures than those obtained from simple autoregression—*for the data actually used in the development of the prediction equations.* We find that for 1969.1 to 1981.4:

	Simple Autoregression	Multiple Autoregression
MAD	4.92	4.56
MSE	41.92	37.61
$\bar{r}$	+.5767	+.6763

However, the same conclusion cannot be drawn about the period from 1982.1 to 1984.2—at least with respect to the MAD and MSE criteria:

---

*The t tests referred to are those introduced in section 15.9 of Chapter 15. Do not confuse t used in this way with t used as time-period indicators in the subscripts for Y and $\hat{Y}$ values.

Stage 1   Multiple autoregression equation with ten "independent variables" (t ratios are shown in parentheses beneath the net regression coefficients).

$$\widehat{Y}_t = 11.038 + .4423Y_{t-1} - .0312Y_{t-2} + .0326Y_{t-3} - .0464Y_{t-4} - .1446Y_{t-5}$$
$$\qquad\qquad (2.40) \qquad (-.15) \qquad (.17) \qquad (-.24) \qquad (-.73)$$
$$\qquad - .0824Y_{t-6} + .0631Y_{t-7} - .1952Y_{t-8} - .0764Y_{t-9} - .2517Y_{t-10}$$
$$\qquad (-.41) \qquad (.31) \qquad (-.97) \qquad (-.37) \qquad (-1.29)$$

$\bar{R}^2 = .344$

Stage 2   Variable Y_{t-2} has the smallest t ratio and is eliminated.

$$\widehat{Y}_t = 10.965 + .4291Y_{t-1} + .0199Y_{t-3} - .0462Y_{t-4} - .1473Y_{t-5} - .0834Y_{t-6}$$
$$\qquad\qquad (2.67) \qquad (.11) \qquad (-.24) \qquad (-.75) \qquad (-.42)$$
$$\qquad + .0671Y_{t-7} - .1932Y_{t-8} - .0815Y_{t-9} - .2462Y_{t-10}$$
$$\qquad (.34) \qquad (-.97) \qquad (-.40) \qquad (-1.30)$$

$\bar{R}^2 = .364$

Stage 3   Variable Y_{t-3} has the smallest t ratio and is eliminated.

$$\widehat{Y}_t = 11.002 + .4341Y_{t-1} - .0367Y_{t-4} - .1462Y_{t-5} - .0794Y_{t-6} + .0669Y_{t-7}$$
$$\qquad\qquad (2.85) \qquad (-.21) \qquad (-.76) \qquad (-.41) \qquad (.34)$$
$$\qquad - .1979Y_{t-8} - .0814Y_{t-9} - .2457Y_{t-10}$$
$$\qquad (-1.03) \qquad (-.41) \qquad (-1.32)$$

$\bar{R}^2 = .383$

Stage 4   Variable Y_{t-4} has the smallest t ratio and is eliminated.

$$\widehat{Y}_t = 10.933 + .4279Y_{t-1} - .1658Y_{t-5} - .0836Y_{t-6} + .0606Y_{t-7} - .1962Y_{t-8}$$
$$\qquad\qquad (2.90) \qquad (-1.00) \qquad (-.44) \qquad (.32) \qquad (-1.04)$$
$$\qquad - .0749Y_{t-9} - .2453Y_{t-10}$$
$$\qquad (-.39) \qquad (-1.33)$$

$\bar{R}^2 = .401$

Stage 5   Variable Y_{t-7} has the smallest t ratio and is eliminated.

$$\widehat{Y}_t = 11.042 + .4231Y_{t-1} - .1611Y_{t-5} - .0604Y_{t-6} - .1735Y_{t-8} - .0772Y_{t-9}$$
$$\qquad\qquad (2.92) \qquad (-.99) \qquad (-.35) \qquad (-1.01) \qquad (-.40)$$
$$\qquad - .2414Y_{t-10}$$
$$\qquad (-1.33)$$

$\bar{R}^2 = .416$

Stage 6   Variable Y_{t-6} has the smallest t ratio and is eliminated.

$$\widehat{Y}_t = 10.802 + .4318Y_{t-1} - .1931Y_{t-5} - .1818Y_{t-8} - .0811Y_{t-9} - .2380Y_{t-10}$$
$$\qquad\qquad (3.06) \qquad (-1.44) \qquad (-1.08) \qquad (-.43) \qquad (-1.33)$$

$\bar{R}^2 = .430$

Stage 7   Variable Y_{t-9} has the smallest t ratio and is eliminated.

$$\widehat{Y}_t = 10.458 + .4446Y_{t-1} - .1921Y_{t-5} - .2118Y_{t-8} - .2648Y_{t-10}$$
$$\qquad\qquad (3.26) \qquad (-1.45) \qquad (-1.40) \qquad (-1.60)$$

$\bar{R}^2 = .443$

Stage 8   Variable Y_{t-8} has the smallest t ratio and is eliminated.

$$\widehat{Y}_t = 9.298 + .4843Y_{t-1} - .2534Y_{t-5} - .3245Y_{t-10}$$
$$\qquad\qquad (3.59) \qquad (-2.01) \qquad (-2.00)$$

Variables Y_{t-1}, Y_{t-5}, and Y_{t-10} meet the requirement of having a t ratio equal to or in excess of ± 2.00.

$\bar{R}^2 = .429$

Figure 17.12. The method of backward elimination by stages applied to a multiple autoregression equation.

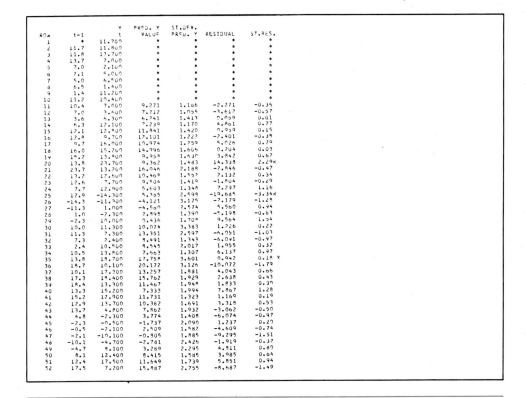

```
THE REGRESSION EQUATION IS
t = 9.30 + 0.484 t-1 - 0.253 t-5 - 0.325 t-10

    42 CASES USED     10 CASES CONTAIN MISSING VALUES

                                 ST. DEV.    T-RATIO =
    COLUMN      COEFFICIENT      OF COEF.    COEF/S.D.
                    9.298          2.884        3.22
    t-1             0.4843         0.1349       3.59
    t-5            -0.2534         0.1262      -2.01
    t-10           -0.3245         0.1619      -2.00

STANDARD ERROR OF THE RESIDUAL IS:
S = 6.447

R-SQUARED = 47.1 PERCENT
R-SQUARED = 42.9 PERCENT, ADJUSTED FOR D.F.
```

```
                      Y    PRED. Y   ST.DEV.
ROW      t-1          t    VALUE     PRED. Y   RESIDUAL    ST.RES.
  1        *       11.700     *         *          *          *
  2      11.7      11.800     *         *          *          *
  3      11.8      13.700     *         *          *          *
  4      13.7       7.000     *         *          *          *
  5       7.0       2.100     *         *          *          *
  6       2.1       5.000     *         *          *          *
  7       5.0       6.500     *         *          *          *
  8       6.5       1.400     *         *          *          *
  9       1.4      11.200     *         *          *          *
 10      11.2      10.400     *         *          *          *
 11      10.4       7.000    9.271     1.166     -2.271      -0.35
 12       7.0       3.600    7.212     1.055     -3.612      -0.57
 13       3.6       6.300    6.241     1.413      0.059       0.01
 14       6.3      12.100    7.239     1.170      4.861       0.77
 15      12.1      12.800   11.841     1.420      0.959       0.15
 16      12.8       9.700   12.101     1.222     -2.401      -0.38
 17       9.7      16.000   10.974     1.259      5.026       0.79
 18      16.0      15.200   14.996     1.605      0.204       0.03
 19      15.2      13.800    9.958     1.630      3.842       0.62
 20      13.8      23.700    9.362     1.483     14.338       2.29R
 21      23.7      13.700   16.046     2.188     -2.846      -0.47
 22      13.7      12.600   10.468     1.552      2.132       0.34
 23      12.6       7.700    9.504     1.419     -1.804      -0.29
 24       7.7      12.900    5.603     1.348      7.297       1.16
 25      12.9     -14.300    5.365     2.598    -19.665      -3.34R
 26     -14.3     -11.300   -4.121     3.175     -7.179      -1.28
 27     -11.3       1.000   -4.560     2.574      5.560       0.94
 28       1.0      -2.300    2.898     1.390     -5.198      -0.63
 29      -2.3      10.000    0.436     1.708      9.564       1.54
 30      10.0      11.300   10.074     3.383      1.226       0.22
 31      11.3       7.300   13.351     2.597     -6.051      -1.03
 32       7.3       2.400    8.491     1.343     -6.091      -0.97
 33       2.4      10.500    8.545     2.017      1.955       0.32
 34      10.5      13.800    7.663     1.307      6.137       0.97
 35      13.8      18.700   17.758     3.601      0.942       0.18 X
 36      18.7      10.100   20.172     3.126    -10.072      -1.79
 37      10.1      17.300   13.257     1.881      4.043       0.66
 38      17.3      18.400   15.762     1.929      2.638       0.43
 39      18.4      13.300   11.467     1.948      1.833       0.30
 40      13.3      15.200    7.333     1.994      7.867       1.28
 41      15.2      12.900   11.731     1.323      1.169       0.19
 42      12.9      13.700   10.362     1.691      3.318       0.53
 43      13.7       4.800    7.862     1.932     -3.062      -0.50
 44       4.8      -2.300    3.774     1.408     -6.074      -0.97
 45      -2.3      -0.500   -1.737     2.090      1.237       0.20
 46      -0.5      -7.100    2.509     1.582     -4.609      -0.74
 47      -2.1     -10.100   -0.805     1.885     -9.295      -1.51
 48     -10.1      -4.700   -2.781     2.426     -1.919      -0.32
 49      -4.7       8.100    3.289     2.295      4.811       0.80
 50       8.1      12.400    8.415     1.585      3.985       0.64
 51      12.4      17.500   11.649     1.739      5.851       0.94
 52      17.5       7.200   15.887     2.755     -8.687      -1.49
```

Figure 17.13. Computer output for the multiple autoregressive forecasting equation.

	Simple Autoregression	Multiple Autoregression
MAD	8.46	12.00
MSE	144.23	202.45
$\bar{r}$	+.6198	+.6873

Unfortunately, results like these are not at all uncommon. Frequently, a simple forecasting model will enjoy a longer and more successful working life than a more complex one even though the latter may fit the past data better. This is why one is well advised to test a forecasting model on data outside the range of that used in its development.

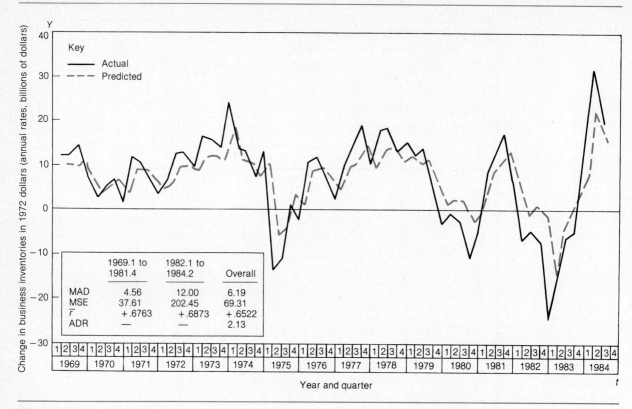

Figure 17.14. Comparison of actual Y_t values and the values predicted by the multiple autoregressive equation $\hat{Y}_t = 9.3 + .484Y_{t-1} - .253Y_{t-5} - .325Y_{t-10}$.

*17.7 A Glimpse at Box-Jenkins Models

One of the truly exciting breakthroughs in forecasting occurred in 1970 with the publication of *Time Series Analysis: Forecasting and Control* by George E. P. Box and Gwilym M. Jenkins.[4] Although some of the techniques comprising the Box-Jenkins approach apparently date back to the 1930s, it was Box and Jenkins who brought the techniques together and developed a *general model-building strategy* for time series forecasting—a strategy quite satisfying from the standpoint of statistical theory.

Be aware, however, that Box-Jenkins forecasting is not nearly so easy to understand as the methods presented earlier in this chapter.[5] Nor can the approach be employed without the aid of a computer. Moreover, a respectable body of research has shown that Box-Jenkins methods often provide forecasts that are not significantly more accurate than those generated by simpler autoprojective models; sometimes the simpler models perform better.[6]

So why study a forecasting approach that is admittedly complicated, that cannot be put to use without a computer, and that will not necessarily produce more accurate forecasts than simpler approaches? There are several good reasons. For one thing, despite some sobering exceptions, the Box-Jenkins methodology has established a credible record of accuracy when compared with alternative forecasting methods. In addition, this approach enjoys some impressive theoretical advantages. Let us look at some.

Theoretical Advantages of the Box-Jenkins Approach

Without intending to impugn the practical usefulness of any of the autoprojective techniques described earlier, we must call to your attention several theoretical limitations of these. Some of the more significant limitations are listed here alongside the Box-Jenkins advantages.

Most Other Auto-Projective Methods	Box-Jenkins Methods
There is a limited number of models—that is, horizontal, trending, moving second-degree parabolic, and so on. When the underlying pattern is quite complex, regular autoprojective techniques are usually incapable of reproducing them.	The Box-Jenkins procedures incorporate a large class of models. These can model a wide variety of time series patterns—including quite complex ones.
There is no systematic way of finding the right model. In practice, much trial and error or just plain luck might be involved in identifying an autoprojective model that works well when applied to a particular time series. It is a little like buying a suit "off the rack."	There is a systematic approach to model identification. The Box-Jenkins approach includes statistical techniques for determining what forecasting model is best for a particular time series. It might be likened to buying a custom-made suit from a skilled tailor.
There is difficulty in verifying the adequacy of the model. There are few, if any, obvious criteria for judging whether the model selected is the "best" one.	The adequacy of the model can be verified. There are several statistical checks compatible with Box-Jenkins methodology that can be used to verify the appropriateness of the model.
There is no prior assurance that the forecasting errors will be normally and independently distributed, a fact which inhibits the use and interpretation of confidence intervals. Consequently, one ordinarily settles for point estimates and proceeds without the aid of measures of reliability.	The methodology ensures that the forecasting errors are random. Thus, the Box-Jenkins approach permits the construction of confidence intervals.

In brief summary, then, the Box-Jenkins methodology is used to aid the time series analyst in developing forecasting equations particularly well suited to a specific series. A Box-Jenkins forecasting equation will be one that is optimal for a particular series in terms of fitting the underlying pattern and minimizing the forecasting error, as measured by MSE. This is so even when the underlying pattern is quite complex.

A Note on the Theoretical Basis of the Box-Jenkins Approach

> Box-Jenkins methods involve forecasting models based on the assumption that the time series has been generated by a *stochastic process*—that is, that each Y_t value making up the series is drawn at random from some kind of probability distribution. When modeling such a process, the forecaster attempts to describe the characteristics of the relevant probability distribution.

When employing the Box-Jenkins methodology, we proceed on the assumption that the series $Y_1, Y_2, Y_3, \ldots, Y_t$ is a set of jointly distributed random variables. There is, so the argument goes, some probability distribution function $P(Y_1, Y_2, Y_3, \ldots, Y_t)$ that assigns probabilities to all possible combinations of values of $Y_1, Y_2, Y_3, \ldots, Y_t$. We would like to be able to specify precisely what that probability distribution function is for a given time series. If that were possible, we could actually determine probabilities associated with specified future outcomes.

Of course, this goal will seldom be fully realized in practice. Nevertheless, it is usually possible to develop a simplified model of the series which explains its randomness sufficiently well to be useful in practical forecasting. For example, let us suppose that the underlying probability distribution is normal in shape with an unchanging mean and variance. Figure 17.15 illustrates such a situation (with $\mu \pm 3\sigma$ bands being used). Notice that the observed $Y_1, Y_2, Y_3, \ldots, Y_t$ values are situated at different points within the normal curves. Still, the curves remain the same with respect to shape, mean, and variance. This ensures that the probability of a departure from the mean by a specified amount remains the same from one point in time to the next.

Stationary Versus Nonstationary Processes

When modeling a time series using the Box-Jenkins approach, it is a matter of considerable importance to know whether the underlying stochastic process can be assumed to be *invariant with respect to time*. If the characteristics of the process change over time (as illustrated in an extreme way in Figure 17.16), we will have difficulty representing the movements of the series by some simple algebraic model. On the other hand, if the stochastic process is fixed in time, that is, if it is *stationary*, then it is possible to model the process using an equation with fixed coefficients determined from past data. Therefore, stationarity must be assumed in connection with the Box-Jenkins procedures. (Later in this section we will address the question of how the Box-Jenkins approach can be adapted to suit nonstationary time series. This is important because most time series one encounters in practice are nonstationary. Still, the stationarity assumption will allow us to introduce Box-Jenkins procedures in the most painless possible way.)

A really rigorous definition of stationarity is quite complicated. However, the following will suffice for our purposes:

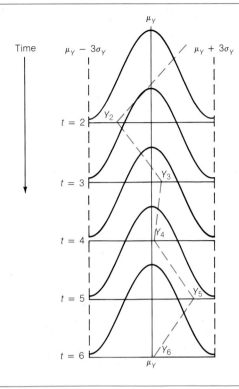

Figure 17.15. A hypothetical stationary process. Shape, mean, and variance remain the same with the passage of time.

> We say that a time series is *stationary* if it tends to fluctuate around an unchanging arithmetic mean and has a constant variance.

Brief Overview of How the Method Is Applied

Box and Jenkins developed the schematic diagram shown in Figure 17.17. This diagram will provide a useful starting point for a discussion of how Box-Jenkins methods are put into practice. It is a three-stage process.

Of the three stages indicated in Figure 17.17, the most difficult by far is Stage 1—identifying a tentative forecasting model. Before we can deal with this stage meaningfully, you must be somewhat familiar with two important diagnostic tools, the *autocorrelogram* (AC) and the *partial autocorrelogram* (PAC).

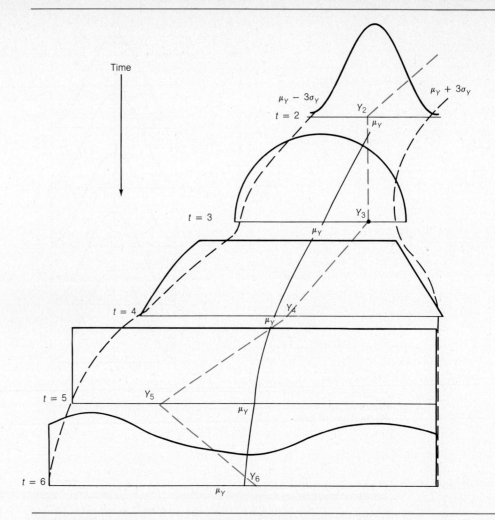

Figure 17.16. A stochastic process that most emphatically is not stationary. The probability distribution evolves rapidly from normal to bi-modal. Also, the mean and variance change.

The Autocorrelogram and the Partial Autocorrelogram

In connection with simple autoregressive forecasting models, we noted that frequently the Y_t and Y_{t-1} values will be relatively strongly correlated. To measure the degree of correlation, we could simply adapt the formula for the coefficient of simple linear correlation introduced in Chapter 14. That is:

$$\rho_1 = \frac{\sum Y_t Y_{t-1} - \dfrac{(\sum Y_t)(\sum Y_{t-1})}{N}}{\sqrt{\left[\sum Y_t^2 - \dfrac{(\sum Y_t)^2}{N}\right]\left[\sum Y_{t-1}^2 - \dfrac{(\sum Y_{t-1})^2}{N}\right]}}$$

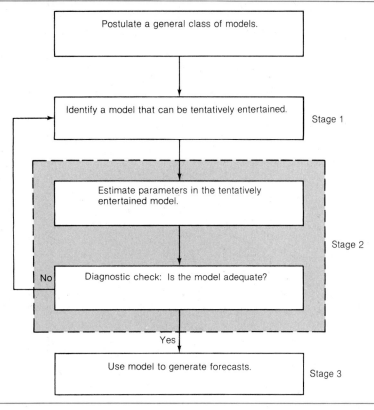

Figure 17.17. Flowchart of the Box-Jenkins method.

where the "1" subscript applied to ρ indicates that the "independent variable" is simply Y lagged one period.

PROCEDURAL NOTE

The use of ρ rather than r is temporary. For a time, we will speak as if the true correlation structure and the true forecasting models were known with certainty. In practice, of course, the best we can do is estimate the true parameter values. Consequently, when the Box-Jenkins methods are employed on actual time series data, we view the observed data as a single random sample drawn from the underlying stochastic process. Sample notations are then appropriate.

Similarly, if we wished to know the strength of association between Y_t values and Y_{t-2} values, we could compute

$$\rho_2 = \frac{\sum Y_t Y_{t-2} - \dfrac{(\sum Y_t)(\sum Y_{t-2})}{N}}{\sqrt{\left[\sum Y_t^2 - \dfrac{(\sum Y_t)^2}{N}\right]\left[\sum Y_{t-2}^2 - \dfrac{(\sum Y_{t-2})^2}{N}\right]}}$$

The Autocorrelogram. We could, of course, apply this same correlational procedure to all pairs of Y_t and Y_{t-3} values, all pairs of Y_t and Y_{t-4} values, and so forth. When the resulting coefficients of correlation—let us call them ρ_k values, where $k = 1, 2, 3, \ldots$—are plotted against the k values themselves, the result is an autocorrelogram, denoted AC. (Many computer programs call it the *autocorrelation function* and abbreviate it *ACF*.)

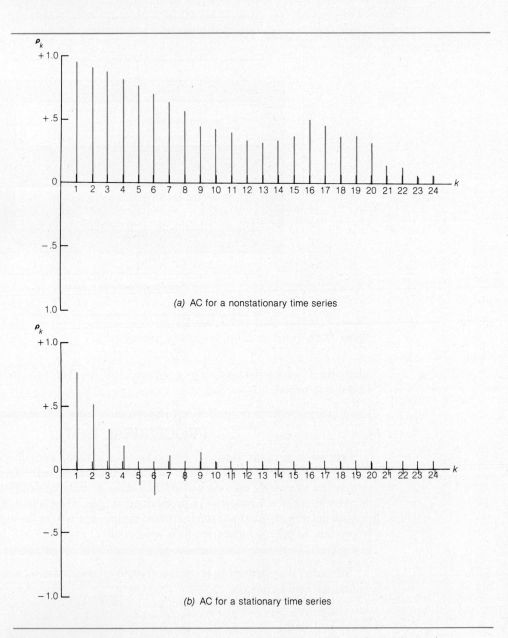

(a) AC for a nonstationary time series

(b) AC for a stationary time series

Figure 17.18. Hypothetical autocorrelograms for a nonstationary and a stationary time series.

17 Short-Term Forecasting with Autoprojective Methods

If, in the original time series, higher-than-average Y values tend to be followed by higher-than-average Y values k periods later, or if lower-than-average Y values tend to be followed by lower-than-average Y values k periods later, the resulting ρ_k value will be positive. On the other hand, if higher-than-average Y values tend to be followed by lower-than-average Y values k periods later, or if lower-than-average Y values tend to be followed by higher-than-average Y values k periods later, ρ_k will be negative. Clearly, the autocorrelation structure as depicted in an autocorrelogram is a kind of description of the repeating patterns, if any, in the time series.

Two hypothetical autocorrelograms (with quarterly time series data assumed) are shown in Figure 17.18. Part (a) of Figure 17.18 attempts to show how a strongly trending (therefore, nonstationary) time series would be depicted in an autocorrelogram. Notice that several periods must pass before the coefficients of simple linear correlation approximate zero—if they ever do.

Part (b) of Figure 17.18 is more typical of a stationary time series. Here the ρ_k values start out rather strongly positive but drop below zero after the passage of only a few time periods. Thus, we see that, in addition to its other diagnostic uses to be described shortly, the autocorrelogram provides a quick, convenient, visual check on the approximate stationarity of the original data. Notice also that the vertical lines at $k = 4$, $k = 8$, $k = 12$, and so forth, are not conspicuously large relative to those associated with the other k values. This observation suggests that seasonality is absent—or virtually absent. If seasonality were strongly present, it would be reflected in large spikes at the above-mentioned k values. (The large spikes would be observed at $k = 12$, $k = 24$, and so forth, in the case of monthly data.)

The Partial Autocorrelogram. The partial autocorrelogram, denoted PAC, is structured in the same way as the regular autocorrelogram. The only difference is that the measures of strength of relationship used are coefficients of partial correlation rather than coefficients of simple correlation. Coefficients of partial correlation (and their squares, the coefficients of partial determination) were described in section 15.6 of Chapter 15. These coefficients are essentially the same here: A particular coefficient of partial autocorrelation measures the amount of correlation between the Y_t and the Y_{t-k} variables when the effects of other lags are eliminated. Figure 17.19 shows a hypothetical partial autocorrelogram that might well accompany the regular autocorrelogram shown in part (b) of Figure 17.18. As will be shown, partial autocorrelograms are complementary to ordinary autocorrelograms with respect to the patterns they display. AC and PAC are used together in the model-identification stage of the Box-Jenkins analysis.

Kinds of Box-Jenkins Models

We now begin a general survey of the three families of possible forecasting models that might be suggested by a pair of AC and PAC graphs:

Autoregressive, or AR, Models. In view of the discussion of autoregressive models in the preceding section, the nature of this kind of Box-Jenkins model is readily apparent. An autoregression, or AR, model is one of the form

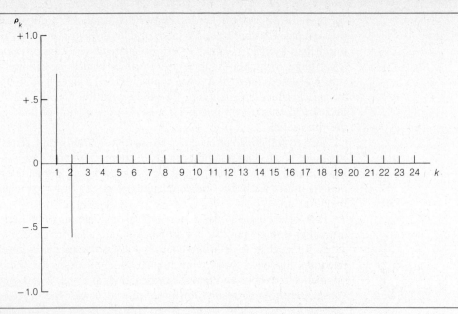

Figure 17.19. Hypothetical partial autocorrelogram which might accompany the regular autocorrelogram shown in (b) of Figure 17.18.

$$Y_t = \gamma + \phi_1 Y_{t-1} + \phi_2 Y_{t-2} + \phi_3 Y_{t-3} + \cdots + \phi_k Y_{t-k} + E_t \qquad (17.7.1)$$

where Y_t values serve as the dependent variable; Y_{t-1}, Y_{t-2}, Y_{t-3}, . . . , Y_{t-k} values serve as independent variables; E_t is the forecasting error term, which, when a correct forecasting model has been identified, will represent only the effects of random factors; γ is the intercept parameter; and ϕ_1, ϕ_2, ϕ_3, . . . , ϕ_k are slope parameters. (These parameters are comparable with A, B_1, B_2, . . . , B_k, in a multiple regression equation.)

An autoregressive equation of the form $Y_t = \gamma + \phi_1 Y_{t-1} + E_t$ is described as a model of order 1 and denoted AR(1) because the largest lag utilized is $k = 1$. An equation of the form $Y_t = \gamma + \phi_1 Y_{t-1} + \phi Y_{t-2} + E_t$ is said to be of order 2 and denoted AR(2) because the largest lag is $k = 2$. And so forth. An equation of the form $Y_t = \gamma + \phi_2 Y_{t-2} + E_t$ is also said to be of order 2 and denoted AR(2) even though the parameter associated with lag $k = 1$ is absent from the equation. In general, an AR model is denoted AR(p), where p represents the largest lag employed.

Moving-Average, or MA, Models. At first glance MA models appear quite similar to AR models. However, they differ in one very important way: Whereas an autoregression model employs lagged values of Y, a moving-average model employs lagged values of E. In other words, Y_t is viewed as a function of past forecasting errors. A moving-average, or MA, model is of the form

$$Y_t = \gamma - \theta_1 E_{t-1} - \theta_2 E_{t-2} - \theta_3 E_{t-3} - \cdots - \theta_k E_{t-k} + E_t \quad (17.7.2)$$

where γ is the intercept parameter; θ_1, θ_2, θ_3, . . . , θ_k are slope parameters; E_{t-1}, E_{t-2}, E_{t-3}, . . . , E_{t-k} are forecasting errors for one, two, three, . . . , k periods earlier; and E_t is the forecasting error associated with the newest predicted value. (Use of negative signs in connection with MA slope parameters is a convention used in most Box-Jenkins literature.)[*]

The notion that Y_t can be expressed as a function of E_{t-k} is a unique feature of the Box-Jenkins approach to forecasting. Implicit within it is the idea that a random error constitutes a small random shock that sets the process in motion initially and keeps it going thereafter.

Notations used in connection with MA models are analogous to those for AR models. For example, a model of the form $Y_t = \gamma - \theta_1 E_{t-1} + E_t$ is said to be of order 1 and denoted MA(1). A model of the form $Y_t = \gamma - \theta_1 E_{t-1} - \theta_2 E_{t-2} - \theta_3 E_{t-3} + E_t$ is said to be of order 3 and denoted MA(3). And so forth. In general, an MA model is denoted MA(q), where q represents the largest lag employed.

Mixed Autoregression and Moving-Average, or ARMA, Models. A forecasting equation could conceivably make use of lagged values of both Y and E. Such an equation is referred to as an ARMA model. Such models are denoted ARMA(1,1), ARMA(2,2), ARMA(1,2), and so forth—in general, ARMA(p,q)—the numbers within parentheses being dependent on the largest lag used in connection with the Y values and in connection with the E values, respectively. An ARMA forecasting model would be of the form

$$Y_t = \underbrace{\gamma + \phi_1 Y_{t-1} + \phi_2 Y_{t-2} + \cdots + \phi_k Y_{t-k}}_{\text{Autoregression}} \underbrace{-\theta_1 E_{t-1} - \theta_2 E_{t-2} - \cdots - \theta_k E_{t-k} + E_t}_{\text{Moving average}} \quad (17.7.3)$$

The AR, MA, and ARMA labels are strictly appropriate in connection with stationary time series. Slight changes in the notations will be necessary when we consider nonstationary series.

Identifying Forecasting Models by AC and PAC

In this subsection we limit our attention to highly idealized situations. As mentioned above, the model-identification stage of a Box-Jenkins analysis is the most difficult one. By considering how an *unmistakably* AR, MA, or ARMA model would be suggested by the AC and PAC patterns, we establish some ideal standards against which to judge the less-than-ideal patterns that life has a tendency to serve up.

---

[*]Clearly, the term *moving average* is used differently here from the way it was used in section 17.4. In the present context, a *moving average* is a weighted average of a fixed number of past random errors which is updated as t increases.

MODEL-IDENTIFICATION RULE 1

If the appropriate forecasting model for a time series is an AR(p) model, then

1. The coefficients of simple autocorrelation appearing in AC will tend to decrease exponentially toward zero as the length of lag k becomes larger.
2. The coefficients of partial autocorrelation appearing in PAC will have a spike at $k = p$ and perhaps at some shorter-period lags as well.

For example, Figures 17.20 and 17.21 show ACs and PACs associated with two different AR(1) models. The alternating pattern in the AC of the second of these figures indicates that the slope parameter has a negative value. Figure 17.22 shows AC and PAC for an AR(2) model.

MODEL-IDENTIFICATION RULE 2

If the appropriate forecasting model for a time series is an MA(q) model, then the patterns in AC and PAC will be much the reverse of those for an AR(p) model. That is,

1. The coefficients of autocorrelation appearing in AC will have a spike at $k = p$ and perhaps at some shorter-period lags as well.
2. The coefficients of partial autocorrelation appearing in PAC will tend to decrease exponentially toward zero as the length of lag k becomes larger.

Figures 17.23, 17.24, and 17.25 (pp. 842–44) show AC and PAC for three different kinds of moving-average models.

MODEL-IDENTIFICATION RULE 3

If the appropriate forecasting model for a time series is an ARMA(p,q) model, then

1. The coefficients of autocorrelation appearing in AC will ordinarily have an irregular pattern between lags 1 and q (where q represents the order of the MA part of the model) and a tendency to decrease exponentially toward zero thereafter as the length of lag k becomes larger.
2. The coefficients of partial autocorrelation appearing in PAC will ordinarily have an irregular pattern between lags 1 and p (where p represents the order of the AR part of the model) and a tendency to decrease exponentially toward zero as the length of lag k becomes larger.

Figure 17.26 (p. 845) shows AC and PAC patterns for an ARMA(1,1) forecasting model.

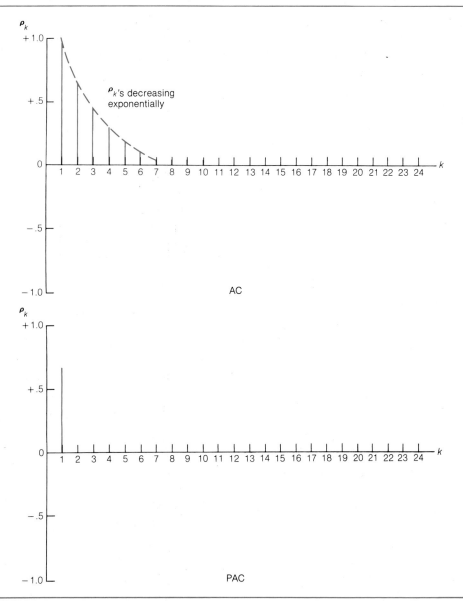

Figure 17.20. Hypothetical AC and PAC for an AR(1) forecasting model. The sign of the slope parameter is positive.

Parameter Estimation

Once a tentative model has been identified by examining AC and PAC, the parameters must be estimated. For example, let us suppose that we identify the desired forecasting model to be one of the ARMA(1,2) kind (that is, one AR parameter with lag 1 and two MA parameters with lags 1 and 2). The forecasting model would be of the form $Y_t = \gamma + \phi_1 Y_{t-1} -$

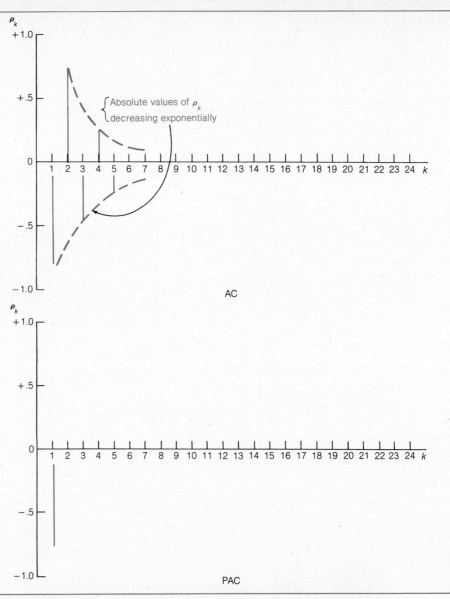

Figure 17.21. Hypothetical AC and PAC for an AR(1) forecasting model. The sign of the slope parameter is negative.

$\theta_1 E_{t-1} - \theta_2 E_{t-2} + E_t$. Clearly, ϕ_1, θ_1, and θ_2 can take on a great many different values. To make matters worse, for each different set of ϕ_1, θ_1, and θ_2 values there will almost certainly be different γ and E_t values. At this point, one ordinarily relies on the computer to perform a search for the optimum parameter values—optimum in the sense of minimizing the mean squared error, MSE. With most packaged programs, the analyst merely has to indicate that

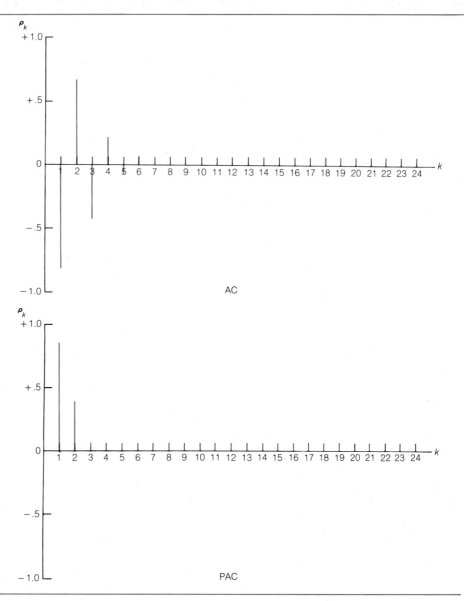

Figure 17.22. Hypothetical AC and PAC for an AR(2) forecasting model.

he or she wishes an autoregressive parameter of order 1 and moving-average parameters of orders 1 and 2. The computer then makes use of a nonlinear multiple regression algorithm to determine what values of ϕ_1, θ_1, and θ_2 will generate predicted values having minimum MSE. The adequacy of the tentative model can then be evaluated using a number of diagnostic checks.

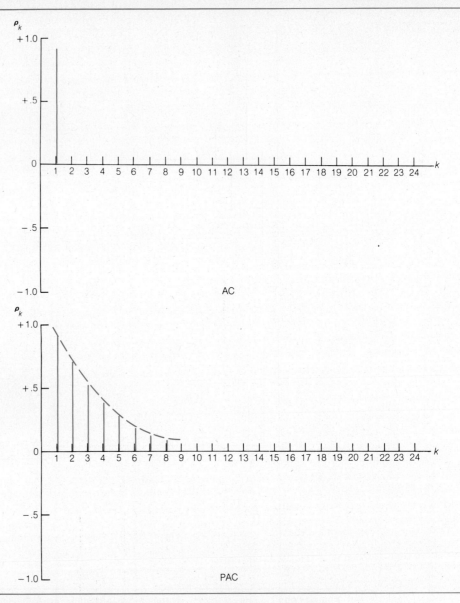

Figure 17.23. Hypothetical AC and PAC for an MA(1) forecasting model.

Diagnostic Checks

Diagnostic measures and tests appearing in the computer printout will vary somewhat from program to program. However, we will discuss briefly some of the methods commonly available in packaged Box-Jenkins computer programs. These diagnostics will be employed with an actual time series in section 17.8. The diagnostic checks described here will be limited to those applied to the residual values (that is, the forecast errors or E_t values). The

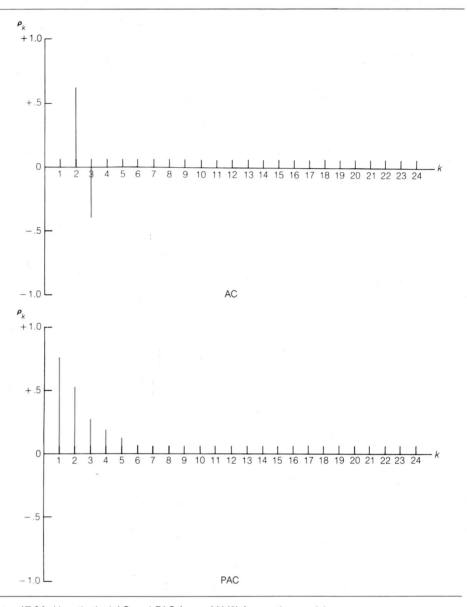

Figure 17.24. Hypothetical AC and PAC for an MA(3) forecasting model.

residual values should be unrelated to one another and fluctuate around a mean of zero. From a practical standpoint, it would be nice if they were small.

*1. **Residual Mean.*** The residual mean is simply the average of the E_t values. That is, $\overline{E}_t = \Sigma\, E_t/N$. Ideally, this average will be zero. In practice, since the data we actually work with are viewed as a sample from the stochastic process, we are content with an $\overline{E}_t$ that does not differ significantly from zero. Finding an average error that *is* significantly different from

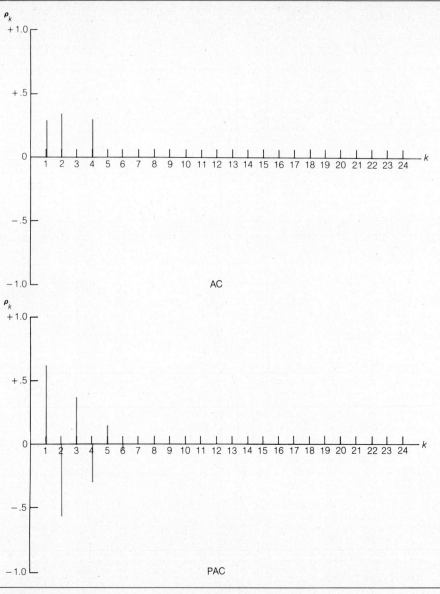

Figure 17.25. Hypothetical AC and PAC for an MA(4) forecasting model. The equation would not include a parameter for lag 3.

zero would be unfortunate: It would indicate that either the predicted Y's tend to be higher or lower than the original series values, or errors in a positive (negative) direction tend to be larger in absolute terms than errors in a negative (positive) direction. Either way, future forecasts will be affected adversely and the theoretical requirement that $\overline{E}_t$ be zero will not have been met.

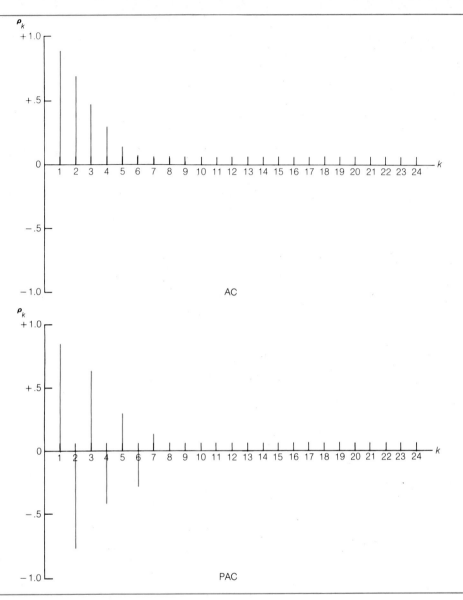

Figure 17.26. Hypothetical AC and PAC for an ARMA forecasting model—presumably an ARMA(1, 1) model, as evidenced by the absence of an irregular pattern before exponential decrease begins.

2. Residual ACs and PACs. An autocorrelogram and partial autocorrelogram can be obtained for the residuals. The computer will develop these for you. Ideally, the AC and PAC would look like those presented in Figure 17.27, in which all correlation coefficients, both simple and partial, are zero. In practice, we make use of a sample and are thus content with results that are compatible with this ideal. More specifically, as long as the correlation

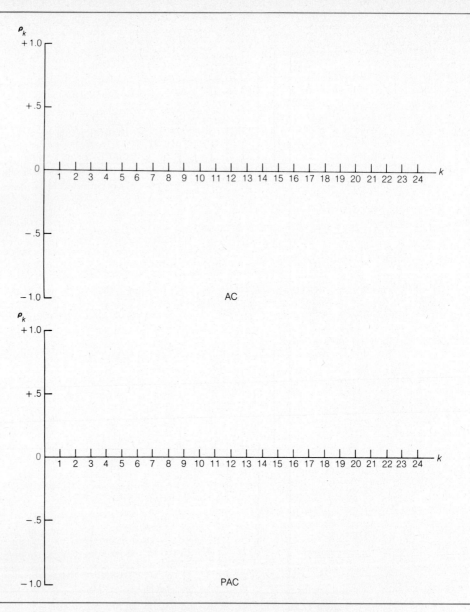

Figure 17.27. Ideal AC and PAC of the residuals. All simple and partial autocorrelations are zero.

coefficients are within a pair of confidence bands (95% confidence bands are commonly used), we accept the assumption of uncorrelated random errors; see Figure 17.28.

The AC and PAC patterns can also be used as guides to improving the forecasting model when necessary. Notice the large spike at lag 5 in the AC of Figure 17.28. This spike might well be suggesting that an MA parameter of order 5 should be added to the model.

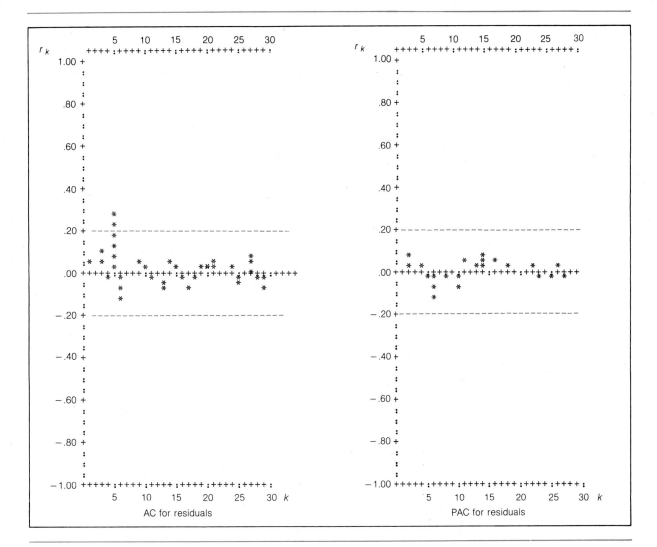

Figure 17.28. AC and PAC for residual values. All but one of the coefficients of correlation are within the .95 confidence bands.

3. The Q Statistic for Residual Autocorrelation. The Q statistic is a kind of chi-square statistic computed from the sample autocorrelogram of the residuals. It can be used to test whether a set of residual autocorrelations, considered *together* (as opposed to individually), is statistically significant. When applying this test, one hopes that any specified set of residual autocorrelations will check out as nonsignificant. A good idea is to apply the Q-statistic test to a relatively small set of autocorrelations (say 6 quarters) and then to a larger set (containing the first, smaller set—say 12 quarters), and so forth. If computed Q's are smaller than the appropriate critical χ^2 values, the model is acceptable according to this criterion.

4. Residual Variance, MSE.   After a valid forecasting model has been found, a descriptive measure of the variance of the residual values provides an indication of the maximum possible accuracy of forecasts generated by the model. Naturally, the smaller the residual variance, the better one likes it because the actual and predicted Y values will be close together.

Step-Ahead Forecasts and Confidence Intervals

Let us suppose that we have the following information:

1. $Y_t = 10$.
2. $Y_{t-1} = 8$.
3. $Y_{t-2} = 5$.
4. The forecasting model obtained by following the procedures described above turns out to be an AR(3) model of the form $Y_t = 0 + .4Y_{t-1} + .3Y_{t-2} + .1Y_{t-3} + E_t$.

Moreover, let us assume that any unknown E_t's will be equal to their expected value—namely, zero.

One can obtain forecasts for one or more future periods quite simply by applying the AR equation to the known Y values. For example, for one period ahead:

$$\hat{Y}_{t+1} = 0 + .4Y_t + .3Y_{t-1} + .1Y_{t-2}$$
$$= 0 + (.4)(10) + (.3)(8) + (.1)(5) = 6.9$$

For two periods ahead, unless the lags represented in the model happen to be quite long, we soon become faced with having to use $\hat{Y}$ values as if they were actual Y values. For example:

$$\hat{Y}_{t+2} = 0 + .4\hat{Y}_{t+1} + .3Y_t + .1Y_{t-1}$$
$$= 0 + (.4)(6.9) + (.3)(10) + (.1)(8) = 6.56$$

For three periods ahead:

$$\hat{Y}_{t+3} = 0 + .4\hat{Y}_{t+2} + .3\hat{Y}_{t+1} + .1Y_t$$
$$= 0 + (.4)(6.56) + (.3)(6.9) + (.1)(10) = 5.69$$

For four periods ahead:

$$\hat{Y}_{t+4} = 0 + .4\hat{Y}_{t+3} + .3\hat{Y}_{t+2} + .1\hat{Y}_{t+1}$$
$$= 0 + (.4)(5.69) + (.3)(6.56) + (.1)(6.9) = 4.93$$

Seldom does it pay to push this procedure too far because the predicted values tend rather rapidly toward the mean of the original series. Similarly, the variance of the predictions tends toward the variance of the original series. Thus, for time periods very far into the future, one can do just as well, or just as poorly, by using the historical mean of the series to predict future values.

The method of obtaining predicted values would have been slightly different had we assumed an MA or an ARMA model. However, because the computer will provide these predicted values, we will rest content with this very brief treatment.

A most attractive Box-Jenkins "extra" is that the underlying theory permits construction of meaningful confidence intervals. Thus, the user of the forecasts is equipped with a range of probable future values rather than just a set of point estimates. As with regular confidence intervals, the degree of confidence one wishes to have in the interval estimate can be set by the forecaster or manager. The confidence intervals can also be used conveniently as a check on the constancy of the presumed stochastic process. When actual Y values begin to fall consistently outside the, say, .95 limits, one has reason to suspect that it is time to redo the analysis and develop a new forecasting model.

As intuition—as well as our discussions in Chapters 14 and 15 of confidence intervals for Y_{NEW}—would suggest, the further out one tries to project, the larger will be the confidence interval associated with any specified confidence coefficient. This tendency for the limits to diverge is illustrated in Figure 17.29.

Dealing with Nonstationarity

We earlier confessed that most time series encountered in practice are nonstationary. The most frequently encountered impediment to stationarity is the presence of a rising trend. Fortunately, most nonstationary time series can be made stationary through the use of some kind of transformation process. Often, the process of differencing—that is, converting Y_1, Y_2, Y_3, . . . , Y_t values into what we might call dY values by $dY_2 = Y_2 - Y_1$, $dY_3 = Y_3 - Y_2$, . . . , $dY_t = Y_t - Y_{t-1}$—will suffice. The effect of this single-differencing process applied to a time series with a linear trend is illustrated in Figure 17.30.

If the trend in the original nonstationary time series is something other than linear, a more complex differencing procedure may have to be employed. For example, for some time series, transforming the original Y values into logarithms and then differencing the logarithms will result in stationarity. For other time series, double differencing of the original series values may be required for stationarity. *Double differencing* means that the differencing process described above is carried one step further. That is, $ddY_3 = dY_3 - dY_2$, $ddY_4 =$

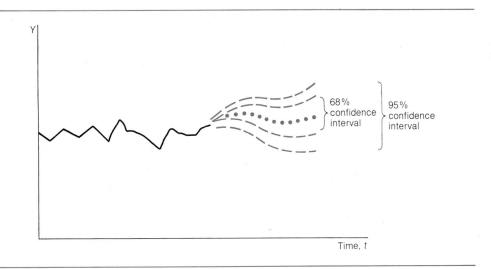

Figure 17.29. How confidence intervals widen as t becomes larger.

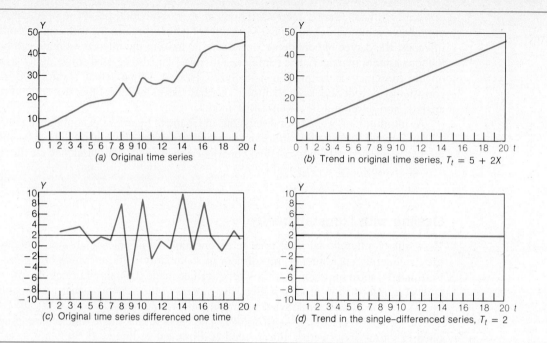

Figure 17.30. How single-differencing converts a (linearly) trending series into a stationary series.

$dY_4 - dY_3, \ldots, ddY_t = dY_t - dY_{t-1}$. Figure 17.31 demonstrates the effect of such double differencing on a time series having a second-degree parabolic trend.

Once approximate stationarity has been achieved, the Box-Jenkins methodology we have sketched out may be employed. Of course, the predicted values will be for the differences rather than for the original values. However, the predicted differences can be converted into predicted $\hat{Y}$'s quite easily.

As with a stationary time series problem, the basic task of the Box-Jenkins forecaster is to determine whether the appropriate model is an AR, an MA, or an ARMA model. These models are denoted in much the same way as those discussed above, except that the symbol I for "integrated" is introduced. This term *integrated*, which essentially means "summed," is used because the differencing process can be reversed to obtain predicted Y's by summing successive values of the differenced series.

Thus, we now distinguish among three other models.

1. *ARI Models (Autoregressive Integrated Models).* For example, what would have been an AR(2) model had the series been stationary to begin with would now be identified as, say, an ARI(2,1) model, where the "1" indicates that single differencing was used to achieve stationarity.

2. *IMA Models (Integrated Moving-Average Models).* What would be an MA(2) model for a naturally stationary series would be an IMA(1,2) model if single differencing had been used.

3. *ARIMA Models (Autoregressive Integrated Moving-Average Models).* An ARMA(2,2) model now becomes an ARIMA(2,1,2) model if single differencing had been used.

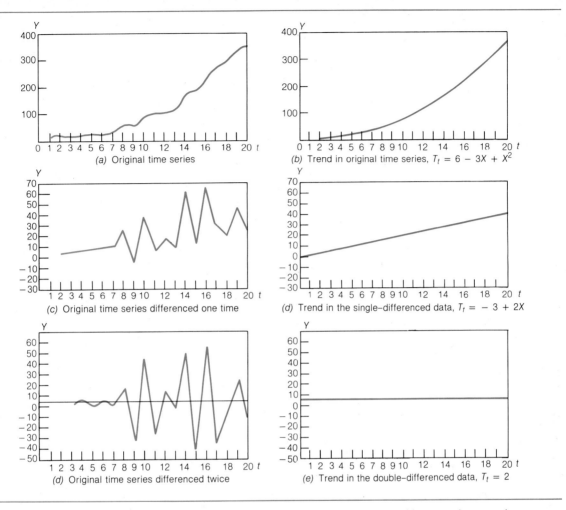

Figure 17.31. How double-differencing converts a series with a second-degree parabolic trend into a stationary series.

In general, we speak of ARIMA(p,d,q) models where, as before, p represents the longest lag used in connection with the Y values and q the longest lag used in connection with the E values. The d represents the number of differencing stages required to achieve approximate stationarity.

Dealing with Seasonality

For simplicity, we have disregarded the possibility of a seasonal pattern in the time series to be predicted. However, for some series the seasonal component is a dominant and most important one from the standpoint of short-term decision making. A nice feature of the Box-Jenkins methodology is that seasonal parameters can be incorporated into the forecasting model. Assuming for the moment that the time series of interest is naturally stationary and that the values correspond to quarters, we observe the following possibilities:

$$\text{AR(4): } Y_t = \gamma + \phi_4 Y_{t-4} + E_t \tag{17.7.4}$$

$$\text{MA(4): } Y_t = \gamma - \theta_4 E_{t-4} + E_t \tag{17.7.5}$$

$$\text{ARMA(4,4): } Y_t = \gamma + \phi_4 Y_{t-4} - \theta_4 E_{t-4} + E_t \tag{17.7.6}$$

Implicit within Equations (17.7.4) to (17.7.6) is the assumption that the only systematic pattern in the original series is a seasonal one.* Since this will seldom be the case, we see that the seasonal parameter(s) associated with these equations can be incorporated into the more general Equations (17.7.1) to (17.7.3) described earlier. Even though the Box-Jenkins approach allows for—and provides a logical justification for—such a merger of equations, the subject can become quite complicated. For one thing, as noted in Chapter 16, the components of a time series are usually combined in a multiplicative, rather than an additive, manner. Thus we will leave this aspect of Box-Jenkins forecasting to the many more specialized books on the subject.[7]

*17.8 An Application of the Box-Jenkins Methodology

We now apply the procedures described in a general way in section 17.7 to our inventory-change data.

Evaluating Stationarity

The first order of business is to determine whether the series is already effectively stationary or whether some kind of transformation of the data is called for. A sensible way to begin addressing this matter is to break the original data into two or more segments and see how the means and the variances of the segments compare. One such comparison applied to the inventory-change data was encouraging: For the 52-quarter period running from 1969.1 through 1981.4 (the period actually used in fitting the autoregressive models), the mean for the first 26 quarters was found to be 8.53 and for the second 26 quarters, 7.84. Similarly, the variance for the first 26 quarters was 61.76 and for the second 26 quarters, 62.56. This comparison speaks well for the likelihood of approximate stationarity.†

Another way of attacking this question is to examine the autocorrelogram of the original series before and after single differencing is performed. A stationary series would be expected to produce an autocorrelation pattern similar to the hypothetical one shown in part (b) of Figure 17.18. Figures 17.32 and 17.33 show the autocorrelograms for the original series and the first-difference series, respectively.

Figure 17.32 reveals a pattern quite consistent with the concept of stationarity. Therefore, we will work with the original inventory-change data rather than with the first differ-

---

*And even then, they have been intentionally oversimplified. It is possible that the Y_{t-8} and E_{t-8} terms should be included and perhaps the Y_{t-12} and E_{t-12} terms, and so forth.
†Though a similar comparison for the entire 62 quarters (1969.1 to 1984.2) was less reassuring.

```
VARIABLE - INVENTRY SERIES LENGTH -   54
DEGREE OF NONSEASONAL DIFFERENCING -  (0) DEGREE OF SEASONAL DIFFERENCING -   0

MEAN VALUE OF THE PROCESS                    Original series
  0.82241E+01

STANDARD DEVIATION OF THE PROCESS
  0.76420E+01

AUTOCORRELATION FUNCTION FOR VARIABLE INVENTRY
AUTOCORRELATIONS *
TWO STANDARD ERROR LIMITS .

     AUTO. STAND.
LAG  CORR.  ERR.  -1  -.75  -.5  -.25   0   .25   .5   .75   1
                  :----:----:----:----:----:----:----:----:
  1   0.603  0.130                      .    :    .   *
  2   0.349  0.129                      .    :    . *
  3   0.176  0.127                      .    :   *.
  4  -0.015  0.126                      .    *    .
  5  -0.195  0.125                      .  *  :    .
  6  -0.267  0.123                      *    :    .
  7  -0.289  0.122                     *.    :    .
  8  -0.377  0.121                   *  .    :    .
  9  -0.383  0.119                   *  .    :    .
 10  -0.365  0.118                   *  .    :    .
 11  -0.262  0.116                     *.    :    .
 12  -0.283  0.115                    *.    :    .
 13  -0.125  0.114                      . *  :    .
 14  -0.047  0.112                      .   *:    .
 15   0.029  0.111                      .   :*   .
 16   0.092  0.109                      .   : *  .
 17   0.192  0.108                      .   :  * .
 18   0.306  0.106                      .   :   .*
 19   0.352  0.104                      .   :   . *
 20   0.440  0.103                      .   :   .   *
 21   0.387  0.101                      .   :   .  *
 22   0.224  0.100                      .   :   *
 23  -0.006  0.098                      . * :   .
 24  -0.153  0.096                     .* :   .
 25  -0.225  0.094                    *. :   .
```

Figure 17.32. AC for original inventory-change data.

```
VARIABLE - INVENTRY SERIES LENGTH -   52
DEGREE OF NONSEASONAL DIFFERENCING -  (1) DEGREE OF SEASONAL DIFFERENCING -   0

MEAN VALUE OF THE PROCESS                    First-difference series
 -0.49057E-01

STANDARD DEVIATION OF THE PROCESS
  0.68721E+01

AUTOCORRELATION FUNCTION FOR VARIABLE INVENTRY
AUTOCORRELATIONS *
TWO STANDARD ERROR LIMITS .

     AUTO. STAND.
LAG  CORR.  ERR.  -1  -.75  -.5  -.25   0   .25   .5   .75   1
                  :----:----:----:----:----:----:----:----:
  1  -0.176  0.131                      .* :    .
  2  -0.105  0.130                      . * :    .
  3   0.021  0.128                      .   *    .
  4  -0.021  0.127                      .   *    .
  5  -0.135  0.126                      . * :    .
  6  -0.056  0.124                      .  *:    .
  7   0.081  0.123                      .   : *  .
  8  -0.104  0.121                      . * :    .
  9  -0.025  0.120                      .  *:    .
 10  -0.108  0.119                      . * :    .
 11   0.155  0.117                      .   :  * .
 12  -0.222  0.116                     .* :    .
 13   0.103  0.114                      .   : *  .
 14   0.003  0.113                      .   *    .
 15   0.012  0.111                      .   *    .
 16  -0.050  0.110                      .  *:    .
 17  -0.012  0.108                      .   *    .
 18   0.078  0.106                      .   : *  .
 19  -0.052  0.105                      .  *:    .
 20   0.173  0.103                      .   :  *.
 21   0.136  0.101                      .   :  *.
 22   0.092  0.100                      .   : * .
 23  -0.106  0.098                      . * :    .
 24  -0.096  0.096                      . * :    .
 25  -0.090  0.094                      . * :    .
```

Figure 17.33. AC for first differences of the inventory-change data.

ences. (We must always try to avoid overdifferencing—that is, differencing a greater number of times than that required to achieve stationarity. Overdifferencing merely complicates the analysis and accomplishes nothing.)

Model Identification

Figure 17.32 shows what appears to be a pattern of exponentially decreasing coefficients of autocorrelation, a pattern suggesting an AR model or, possibly, an ARMA model. A regular MA model is not indicated. For further clarification we consult the partial autocorrelogram of the original inventory-change data. This PAC is shown in Figure 17.34 and reveals a spike of significant size (that is, outside the 95% confidence limits) at lags 1 and 12. The spike at lag 12 is barely outside the confidence limits and could easily be a fluke. With 25 partial autocorrelation coefficients, one could expect significant partial correlation coefficients resulting from chance alone about 1.25 times. On the other hand, there is no mistaking the meaningfulness of the spike at lag 1. AC and PAC appear to be pointing us toward a forecasting model of the AR(1) type.

Diagnostic Checks

Upon employing the AR(1) model, we find that the diagnostic checks confirm the impression that this model is appropriate for our data. More specifically:

1. The mean value of the residual series is .00067—definitely not significant.
2. The autocorrelogram of the residuals (shown in Figure 17.35) shows a significant spike only at lag 20, and it is barely significant. (No attempt was made to incorporate this lag into the forecasting model.)

```
PARTIAL AUTOCORRELATION FUNCTION FOR VARIABLE INVENTRY
PARTIAL AUTOCORRELATIONS *
TWO STANDARD ERROR LIMITS .

     PR-AUT STAND.
LAG  CORR.  ERR.  -1  -.75  -.5 -.25   0   .25  .5   .75   1
                  !----!----!----!----!----!----!----!----!
  1   0.603  0.136                   .        .        *
  2  -0.022  0.136                   .   *    .
  3  -0.041  0.136                   .  *:    .
  4  -0.157  0.136                   . *  :   .
  5  -0.180  0.136                  .*    :   .
  6  -0.064  0.136                   .  *:    .
  7  -0.064  0.136                   .  *:    .
  8  -0.218  0.136                  .* :      .
  9  -0.118  0.136                   . * :    .
 10  -0.155  0.136                   . * :    .
 11  -0.016  0.136                   .   *    .
 12  -0.279  0.136                 *. :       .
 13   0.039  0.136                   . : *    .
 14  -0.186  0.136                  .* :      .
 15  -0.069  0.136                   . *:     .
 16  -0.139  0.136                   . * :    .
 17  -0.019  0.136                   .   *    .
 18   0.039  0.136                   . :*     .
 19   0.052  0.136                   . :*     .
 20   0.137  0.136                   . : *    .
 21   0.024  0.136                   . :*     .
 22  -0.136  0.136                   . * :    .
 23  -0.145  0.136                   . * :    .
 24  -0.168  0.136                   . * :    .
 25   0.057  0.136                   . :*     .

*** BOX-JENKINS REQUIRES      15928 BYTES OF WORKSPACE FOR IDENTIFICATION ***
```

Figure 17.34. PAC for original inventory-change data.

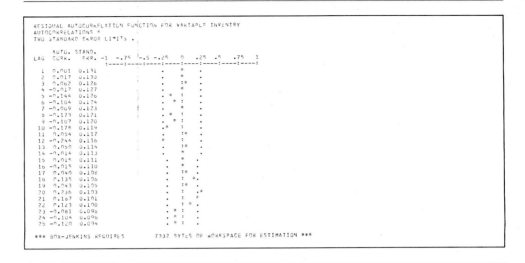

Figure 17.35. Autocorrelogram of the residuals.

3. The Q-statistic results were found to be these:

Lag	Chi-Square (Q)	df	Probability
6	2.23	4	.6933
12	11.13	10	.3476
18	12.89	16	.6810
24	24.74	22	.3096
25	26.35	23	.2847

None of these is significant at a more rigorous level than .28. Nonsignificant results are what we want.

4. The variance of the residual series was found to be 37.84.

For purposes of comparison, let us contrast the above results with those associated with a model clearly not indicated by AC and PAC—namely, the MA(1) model:

1. The mean value of the residual series is $-.0296$ (not significant at $\alpha = .05$).
2. The autocorrelogram of the residuals (not shown) has significant spikes at lags 2, 8, 10, 12, and 20.
3. The Q-statistic results were found to be these:

Lag	Chi-Square (Q)	df	Probability
6	9.57	4	.0484
12	31.88	10	.0004
18	37.56	16	.0017
24	58.14	22	.0000
25	60.05	23	.0000

All are significant at the .05 level.

4. The variance of the residual series was found to be 41.28, higher than the one for AR(1).

Since application of the AR(1) model has already been displayed in Figure 17.11, we will stop the analysis at this point.

Actually, some additional experimentation showed that MSE for model ARMA(2,1) was smaller than that for model AR(1) *for the quarters employed in the fitting of the models.* However, use of AR(1) resulted in an MSE not much larger than that for ARMA(2,1) for these earlier quarters and a smaller MSE for the most recent ten quarters. Recall that these ten quarters were not employed when determining what kind of model to use.

*17.9 Some Limitations of the Box-Jenkins Approach

Early in section 17.7, we took pains to emphasize the superiority of Box-Jenkins methods over other autoprojective techniques with respect to the former's theoretical underpinnings and (usually) greater forecasting accuracy. However, you should also be aware of some of the shortcomings of this more sophisticated approach:

1. The Box-Jenkins approach does not always lead to more accurate forecasts than those produced by simpler autoprojective methods.
2. When Box-Jenkins *does* lead to greater accuracy, it does so at a high cost in terms of the greater computer time and technical expertise required in the development of an appropriate model.
3. At least 50 historical observations are required with the Box-Jenkins approach and 100 would be better yet. Consequently, the method is limited to use on time series having a somewhat long recorded history.
4. At present, there is no convenient way of updating the estimates of model parameters as new observations become available. From time to time the model must be completely refitted.
5. According to some authors who have tried to use the technique in a consulting capacity,[8] sometimes the packaged Box-Jenkins programs do not work properly. This can be a most annoying problem because the procedure is too complicated to permit meaningful judgmental corrections.

17.10 Summary

In this chapter we have surveyed quite a variety of autoprojective forecasting methods. To be sure, this survey does not begin to exhaust the list of methods available to the practicing forecaster. It does not even begin to exhaust the list of forecasting methods that are primarily statistical (as distinguished from primarily judgmental) in nature. Nor are autoprojective methods best for all purposes. Sometimes, for example, forecasts based on a thoroughgoing analysis of underlying causes is desirable. Still, autoprojective methods do occupy a secure space within the forecaster's tool kit. One reason for this is that the system of causes generating values of a time series is sometimes so complex and so little understood that it defies analysis by means of a deeper, causally oriented, analysis. Another reason is that many a manager requires regularly updated forecasts of sales for hundreds, or even thousands, of products. With an autoprojective method—even a fairly complicated one—he or she can quite easily get the computer to generate new forecasts as new data become available.

Two limitations common to all autoprojective techniques are: (1) They are usually not too useful if one must attempt to look very far into the future and (2) they are not designed to signal upcoming cyclical turning points, a most unfortunate limitation because it is in the vicinity of turning points that forecasts can do the most good in guiding the decision maker.

The complexity of autoprojective forecasting methods varies over an enormous range. As has been demonstrated in this chapter, greater complexity does not necessarily beget greater accuracy.

Notes

1. A formal proof of this assertion is presented in Alison M. Grant, "Some Properties of Runs in Smoothed Random Series," *Biometrika*, XXXIX (May 1962), p. 199. However, it will suffice to recognize that once a change in a truly random series has been observed, the expectation is that half the time the next change will be in the same direction and half the time in the opposite direction.
2. J. Bongard, "Some Remarks on Moving Averages," *Seasonal Adjustments on Electronic Computers* (Paris: Organization for Economic Cooperation and Development, 1961), p. 368.
3. A most readable example is Charles W. Gross and Robin T. Peterson, *Business Forecasting*, Second Edition (Boston, Mass.: Houghton Mifflin Company, 1983), Chapter 3. A more advanced treatment appears in Douglas C. Montgomery and Lynwood A. Johnson, *Forecasting and Time Series Analysis* (New York: McGraw-Hill Book Company, 1976), Chapters 2 and 3.
4. George E. P. Box and Gwilym M. Jenkins, *Time Series Analysis: Forecasting and Control* (San Francisco: Holden-Day, 1970). A revised edition was published in 1976.
5. However, the motivated reader is referred to the following book which succeeds in making the Box-Jenkins methodology almost intelligible: John C. Hoff, *A Practical Guide to Box-Jenkins Forecasting* (Belmont, Calif.: Lifetime Learning Publications, 1983).
6. See, for example, Chapter 7 of J. Scott Armstrong, *Long-Range Forecasting from Crystal Ball to Computer* (New York: Wiley, 1978).
7. The Hoff book, noted earlier, treats the subject with unusual clarity; see Chapters 14, 15, and 16. Other good sources include Thomas M. O'Donovan, *Short-Term Forecasting: An Introduction to the Box-Jenkins Approach* (New York: Wiley, 1983), Chapter 6 and James P. Cleary and Hans Levenbach, *The Professional Forecaster* (Belmont, Calif.: Lifetime Learning Publications, 1982), Chapter 21.
8. For example, Gross and Peterson, *Business Forecasting*, 2nd ed. (Boston, Mass.: Houghton Mifflin, 1983), p. 231.

Terms Introduced in This Chapter

AR model (p. 836)
ARI model (p. 850)
ARIMA model (p. 850)
ARMA model (p. 837)
autocorrelogram (p. 834)
autoprojective forecasting model (p. 803)

average duration of run, ADR (p. 807)
autoregressive forecasting model (p. 821)
Box-Jenkins methods (p. 828)
differencing (p. 849)
double differencing (p. 849)

single differencing (p. 849)
exponential smoothing (p. 818)
forecasting error (p. 805)
IMA model (p. 850)
MA model (p. 837)
mean absolute deviation, MAD (p. 806)

mean squared error, MSE (p. 806)
moving-average forecasting model
  (p. 812)
moving-average forecasting model for
  Box-Jenkins methods (p. 837)

no-change forecasting model (p. 810)
nonstationary time series (p. 835)
partial autocorrelogram, PAC (p. 835)
Q statistic (p. 847)
residual AC (p. 845)
residual mean (p. 843)

residual PAC (p. 845)
residual variance (p. 848)
run (p. 807)
stationary time series (p. 831)
step-ahead forecasts (p. 848)

Formulas Introduced in This Chapter

Forecasting error at time t:

$$E_t = Y_t - \hat{Y}_t$$

Mean absolute deviation, a measure of forecast accuracy, where N represents the number of pairs of Y_t and $\hat{Y}_t$ values:

$$\text{MAD} = \frac{\sum_{t=1}^{N} |Y_t - \hat{Y}_t|}{N} \qquad t = 1, 2, 3, \ldots, N$$

Mean squared error—a measure of forecast accuracy which penalizes large errors especially severely:

$$\text{MSE} = \frac{\sum_{t=1}^{N} (Y_t - \hat{Y}_t)^2}{N} \qquad t = 1, 2, 3, \ldots, N$$

Average duration of run, a measure of how well a forecasting model eliminates noncyclical reversals in direction:

$$\text{ADR} = \frac{\text{Number of change signs in a truncated set of such change signs for predicted values}}{\text{Number of runs in the same truncated set of change signs}}$$

No-change (in the level) model:

$$\hat{Y}_{t+\tau} = Y_t$$

No-change (in the change) model:

$$\hat{Y}_{t+\tau} = Y_t + (Y_t - Y_{t-\tau})$$

No-change (in the percent change) model:

$$\hat{Y}_{t+\tau} = Y_t\left(\frac{Y_t}{Y_{t-\tau}}\right)$$

General formula for unweighted moving-average forecasting model:

$$\hat{Y}_{t+\tau} = \frac{Y_t + Y_{t-1} + Y_{t-2} + \cdots + Y_{t-(n-1)}}{n}$$

General formula for an exponential smoothing forecasting model:

$$\hat{Y}_{t+\tau} = wY_t + (1 - w)\hat{Y}_t$$

Alternative formula for an exponential smoothing forecasting model:

$$\hat{Y}_{t+\tau} = \hat{Y}_t + w(Y_t - \hat{Y}_t)$$

General formula for a simple autoregressive forecasting model:

$$\hat{Y}_t = a + bY_{t-k}$$

Slope formula for a simple autoregressive forecasting equation for lag $k = 1$:

$$b = \frac{\Sigma Y_t Y_{t-1} - \dfrac{(\Sigma Y_t)(\Sigma Y_{t-1})}{N}}{\Sigma Y_{t-1}^2 - \dfrac{(\Sigma Y_{t-1})^2}{N}}$$

Y-intercept formula for a simple autoregressive forecasting equation for lag $k = 1$:

$$a = \frac{\Sigma Y_t}{N} - b\frac{\Sigma Y_{t-1}}{N}$$

Formula for determining the (population) coefficient of simple linear correlation between Y_t values and Y_{t-1} values:

$$\rho_k = \frac{\Sigma Y_t Y_{t-1} - \dfrac{(\Sigma Y_t)(\Sigma Y_{t-1})}{N}}{\sqrt{\left[\Sigma Y_t^2 - \dfrac{(\Sigma Y_t)^2}{N}\right]\left[\Sigma Y_{t-1}^2 - \dfrac{(\Sigma Y_{t-1})^2}{N}\right]}}$$

An autoregressive model expressed in Box-Jenkins notations:

$$Y_t = \gamma + \phi_1 Y_{t-1} + \phi_2 Y_{t-2} + \phi_3 Y_{t-3} + \cdots + \phi_k Y_{t-k} + E_t$$

Moving-average model expressed in Box-Jenkins notations:

$$Y_t = \gamma - \theta_1 E_{t-1} - \theta_2 E_{t-2} - \theta_3 E_{t-3} - \cdots - \theta_k E_{t-k} + E_t$$

General formula for an ARMA (or ARIMA) model expressed in Box-Jenkins notations:

$$Y_t = \gamma + \phi_1 Y_{t-1} + \phi_2 Y_{t-2} + \cdots + \phi_k Y_{t-k} - \theta_1 E_{t-1} - \theta_2 E_{t-2} - \cdots - \theta_k E_{t-k} + E_t$$

Average forecasting error:

$$\overline{E}_t = \frac{\Sigma E_t}{N}$$

An AR model for a time series whose only systematic component is seasonality (quarterly data assumed):

AR(4): $Y_t = \gamma + \phi_4 Y_{t-4} + E_t$

An MA model for a time series whose only systematic component is seasonality (quarterly data assumed):

MA(4): $Y_t = \gamma - \theta_4 E_{t-4} + E_t$

An ARMA model for a time series whose only systematic component is seasonality (quarterly data assumed):

ARMA(4,4) $= \gamma + \phi_4 Y_{t-4} - \theta_4 E_{t-4} + E_t$

Questions and Problems

17.1 Explain the meaning of each of the following:
   a. Forecasting error
   b. No-change model
   c. Autoregression
   d. Autoprojective forecasting method
   e. Mean absolute error
   f. Average duration of run
   g. Exponential smoothing

17.2 Explain the meaning of each of the following:
   a. Mean squared error
   b. Lag
   c. Moving-average forecasting model
   d. $\hat{Y}_{t+\tau}$
   e. Smoothing constant
   f. Moving-average period

17.3 Explain the meaning of each of the following:
   a. Autocorrelogram
   b. Box-Jenkins models
   c. Stochastic forecasting model
   d. Stationary time series
   e. AR(3)
   f. ARI(2)
   g. MA(1)

17.4 Explain the meaning of each of the following:
   a. Nonstationary time series
   b. Stochastic process
   c. Partial autocorrelogram
   d. ARMA model
   e. ARIMA(3,1,2)
   f. Autocorrelation
   g. Q statistic
   h. Step-ahead forecasts

17.5 Distinguish between:
   a. Unweighted moving-average forecasting model and weighted moving-average forecasting model
   b. Simple autoregressive forecasting model and multiple autoregressive forecasting model
   c. Mean absolute deviation and mean squared error
   d. Weighted moving-average forecasting model and exponential smoothing forecasting model

17.6 Distinguish between:
   a. No-change-in-the-level forecasting model and no-change-in-the-change forecasting model
   b. $\bar{r}$ and ADR
   c. $\hat{Y}_{t+\tau}$ and $\hat{Y}_{t+1}$
   d. $\hat{Y}_{t+1} = a + bY_t$ and $\hat{Y}_{t+1} = a + b_1 Y_t + b_2 Y_{t-1}$
   e. $\hat{Y}_{t+1} = a + bY_t$ and $\hat{Y}_{t+1} = a + bY_{t-2}$

17.7 Distinguish between:
   a. Single differencing and double differencing
   b. Stationary time series and nonstationary time series
   c. AR(1) model and MA(1) model
   d. ARMA(3,1) model and ARIMA(3,1,1) model
   e. AR and ARI

17.8 Distinguish between:
   a. Autocorrelogram and partial autocorrelogram
   b. AC and PAC
   c. ϕ_1 and θ_1
   d. Autocorrelogram (original series) and residual autocorrelogram
   e. Residual autocorrelogram and the Q statistic.

17.9 Indicate which of the following statements you agree with and which you disagree with, and defend your opinions:
  a. The autoprojective forecasting methods described in this chapter are used primarily for long-range projection.
  b. The autoprojective forecasting methods described in this chapter are especially useful in signaling forthcoming reversals in cyclical direction.
  c. One of the few truisms in forecasting is that, if a forecasting model performs well according to the $\bar{r}$ criterion, it will also perform well according to the ADR criterion.

17.10 Indicate which of the following statements you agree with and which you disagree with, and defend your opinions:
  a. The larger the smoothing constant w in an exponential smoothing forecasting model, the faster the forecasts will adapt to a change in the underlying pattern of the time series.
  b. The longer the moving-average period, the faster the forecasts will adapt to a change in the underlying pattern of the time series.
  c. A no-change forecasting model of the type $\hat{Y}_{t+\tau} = Y_t$ does nothing to smooth the irregular fluctuations of a time series.

17.11 Indicate which of the following statements you agree with and which you disagree with, and defend your opinions:
  a. There is always a close direct relationship between the complexity of a forecasting model and its forecasting accuracy.
  b. According to enthusiasts of exponential smoothing, forecasting from moving-average models leaves something to be desired because only the most recent n observations are reflected in the forecasts.
  c. A multiple autoregressive forecasting model will always perform better than a simple autoregressive forecasting model.
  d. Forecasting errors obtained by $E_t = Y_t - \hat{Y}_t$ will always sum to zero.

17.12 Indicate which of the following statements you agree with and which you disagree with, and defend your opinions:
  a. If a time series has a marked upward trend, the no-change model $\hat{Y}_{t+\tau} = Y_t$ would be expected to track the original values more closely than the no-change model $\hat{Y}_{t+\tau} = Y_t + (Y_t - Y_{t-1})$.
  b. Sometimes an exponential smoothing forecasting model will be used in practice even when an unweighted moving-average forecasting model has performed somewhat better on the historical data.
  c. Ordinarily, the moving-average and exponential smoothing forecasting models do not lend themselves to the construction of meaningful, probabilistically based confidence intervals.

17.13 Indicate which of the following statements you agree with and which you disagree with, and defend your opinions:
  a. Single differencing will sometimes convert a nonstationary time series into an effectively stationary time series.
  b. If a time series really lends itself to prediction by a Box-Jenkins model, AC and PAC for the residuals should display essentially the same patterns as AC and PAC for the original series.
  c. The most difficult stage of a Box-Jenkins analysis is that of applying the diagnostic checks.

17.14 Indicate which of the following statements you agree with and which you disagree with, and defend your opinions:

a. A major advantage of Box-Jenkins methods is that they can model a wide variety of time series patterns.
b. Box-Jenkins methods enable one to construct meaningful confidence intervals to accompany forecasts.
c. The concept of random sampling plays no role in connection with any forecasting methods described in this chapter.
d. If a time series is found to be nonstationary, one must forego use of the Box-Jenkins procedures.

17.15 Indicate which of the following statements you agree with and which you disagree with, and defend your opinions:
a. The notation ARIMA(2,1,1) conveys the idea that the forecasting equation includes estimates of the parameters ϕ_2 (and possibly ϕ_1) and θ_1 and that the original series was stationary to begin with.
b. The AC pattern for an MA(1) model will look much the same as that for an AR(1) model, a condition necessitating an examination of the PAC patterns.
c. If an autocorrelogram displays positive coefficients of simple linear correlation for periods 1 through 25, one has reason to suspect the time series to be nonstationary.
d. If an ARMA model is called for, both AC and PAC will display autocorrelation patterns tending to decrease exponentially.

17.16 Cite two hypothetical, but realistic, business situations requiring forecasting. Make one such situation suggest the need for autoprojective methods and the other the need for a fundamental analysis of causal factors.

 17.17 Ten months of sales data of interest to the inventory manager of the Fairweather Company are shown in Table 17.5.
a. Apply the no-change model $\hat{Y}_{t+1} = Y_t$ to these sales data.
b. Apply a moving-average forecasting model of the form

$$\hat{Y}_{t+1} = \frac{Y_t + Y_{t-1} + Y_{t-2}}{3}$$

to these sales data.
c. Which model is better according to the MAD criterion? Elaborate.

Table 17.5
Ten Months of Sales Data for the Fairweather Company

Month	Sales ($000)
1	225
2	245
3	190
4	182
5	263
6	290
7	210
8	240
9	320
10	250

d. Which model is better according to the MSE criterion? Elaborate.
e. Which model is better according to the $\bar{r}$ criterion? Elaborate.
f. Which model is better according to the ADR criterion? Elaborate.
g. Which model appears better overall? Defend your choice.

17.18 Refer to Table 17.5.

  a. If you did not previously work problem 17.17, apply the no-change model $\hat{Y}_{t+1} = Y_t$ to these sales data.

  b. Apply a moving-average forecasting model of the form

$$\hat{Y}_{t+1} = \frac{Y_t + Y_{t-1} + Y_{t-2} + Y_{t-3} + Y_{t-4}}{5}$$

  to these sales data.

  c. Which model is better according to the MAD criterion? Elaborate.
  d. Which model is better according to the MSE criterion? Elaborate.
  e. Which model is better according to the $\bar{r}$ criterion? Elaborate.
  f. Which model is better according to the ADR criterion? Elaborate.
  g. Which model appears better overall? Defend your choice.

17.19 Refer to Table 17.5.

  a. If you did not previously work either problem 17.17 or 17.18, apply the no-change model $\hat{Y}_{t+1} = Y_t$ to these sales data.

  b. Apply a moving-average forecasting model of the form

$$\hat{Y}_{t+1} = \frac{5Y_t + 8Y_{t-1} + 9Y_{t-2} + 8Y_{t-3} + 5Y_{t-4}}{34}$$

  to these sales data.

  c. Which model is better according to the MAD criterion? Elaborate.
  d. Which model is better according to the MSE criterion? Elaborate.
  e. Which model is better according to the $\bar{r}$ criterion? Elaborate.
  f. Which model is better according to the ADR criterion? Elaborate.
  g. Which model appears better overall? Defend your choice.

17.20 Refer to problems 17.17, 17.18, and 17.19. (The present problem depends on your having already worked problems 17.17, 17.18, and 17.19.) Compare the forecasting adequacy of the three moving-average forecasting models utilized in problems 17.17, 17.18, and 17.19 with respect to

  a. The MAD criterion.
  b. The MSE criterion.
  c. The $\bar{r}$ criterion.
  d. The ADR criterion.
  e. Overall. Tell how you arrived at this judgment.

17.21 Productivity (that is, output per worker-hour) results achieved by assemblers at Trendy Fads Incorporated over the past two years are presented in Table 17.6.

  a. Apply the no-change model $\hat{Y}_{t+2} = Y_t + (Y_t - Y_{t-2})$ to these productivity data.
  b. Apply an exponential smoothing forecasting model of the form $\hat{Y}_{t+2} = .1Y_t + .9\hat{Y}_t$ to these productivity data.
  c. Which model is better according to the MAD criterion? Elaborate.
  d. Which model is better according to the MSE criterion? Elaborate.
  e. Which model is better according to the $\bar{r}$ criterion? Elaborate.
  f. Which model is better according to the ADR criterion? Elaborate.
  g. Which model appears better overall? Defend your choice.

Table 17.6 Twenty-Four Months of Productivity Data for Trendy Fads Incorporated

Month	Productivity	Month	Productivity
1	7	13	11
2	5	14	12
3	4	15	10
4	4	16	7
5	6	17	8
6	6	18	10
7	8	19	11
8	10	20	14
9	12	21	17
10	12	22	24
11	14	23	22
12	13	24	28

 17.22 Refer to Table 17.6.
  a. If you did not previously work problem 17.21, apply the no-change model $\hat{Y}_{t+2} = Y_t + (Y_t - Y_{t-2})$ to these productivity data.
  b. Apply an exponential smoothing forecasting model of the form $\hat{Y}_{t+2} = .5Y_t + .5\hat{Y}_t$ to these productivity data.
  c. Which model is better according to the MAD criterion? Elaborate.
  d. Which model is better according to the MSE criterion? Elaborate.
  e. Which model is better according to the $\bar{r}$ criterion? Elaborate.
  f. Which model is better according to the ADR criterion? Elaborate.
  g. Which model appears better overall? Defend your choice.

 17.23 Refer to problems 17.21 and 17.22. (The present problem depends on your having already worked problems 17.21 and 17.22.) Compare the forecasting adequacy of the two exponential smoothing forecasting models utilized in problems 17.21 and 17.22 with respect to:
  a. The MAD criterion.
  b. The MSE criterion.
  c. The $\bar{r}$ criterion.
  d. The ADR criterion.
  e. Overall. Tell how you arrived at this judgment.

 17.24 Fifteen months of data for the discount rate on U.S. Treasury bills are presented in Table 17.7.
  a. Apply the no-change model $\hat{Y}_{t+1} = Y_t$ to these discount-rate data.
  b. Develop the simple linear least squares autoregressive forecasting model of the form $\hat{Y}_{t+1} = a + bY_t$.
  c. Apply the forecasting equation developed in part b to these discount-rate data.
  d. Which model is better according to the MAD criterion? Elaborate.
  e. Which model is better according to the MSE criterion? Elaborate.
  f. Which model is better according to the $\bar{r}$ criterion? Elaborate.
  g. Which model appears better overall? Defend your choice.

 17.25 Refer to Table 17.7.
  a. If you did not previously work problem 17.24, apply the no-change model $\hat{Y}_{t+1} = Y_t$ to these discount-rate data.

Table 17.7 Fifteen Months of Discount-Rate Data for U.S. Treasury Bills

Year and Month		Discount Rate
1984	October	8.71
	November	8.71
	December	8.96
1985	January	8.93
	February	9.03
	March	9.44
	April	9.69
	May	9.90
	June	9.94
	July	10.13
	August	10.49
	September	10.41
	October	9.97
	November	8.79
	December	8.29

    b. Develop a multiple linear least squares autoregressive forecasting model of the form $\hat{Y}_{t+1} = a + b_1 Y_t + b_2 Y_{t-3}$.

    c. Apply the forecasting equation developed in part *b* to these discount-rate data.

    d. Which model is better according to the MAD criterion? Elaborate.

    e. Which model is better according to the MSE criterion? Elaborate.

    f. Which model is better according to the $\bar{r}$ criterion? Elaborate.

    g. Which model appears better overall? Defend your choice.

 17.26 Refer to problems 17.24 and 17.25. (The present problem depends on your having already worked problems 17.24 and 17.25.) Compare the autoregressive forecasting equations developed in problems 17.24 and 17.25 with respect to

    a. The MAD criterion.

    b. The MSE criterion.

    c. The $\bar{r}$ criterion.

    d. Overall. Tell how you arrived at this judgment.

17.27 Is the time series depicted in the AC in Figure 17.36 stationary or nonstationary? Defend your choice.

17.28 What kind of Box-Jenkins forecasting model is suggested by the AC and PAC shown in Figure 17.37? Tell how you arrived at your conclusion.

17.29 What kind of Box-Jenkins forecasting model is suggested by the AC and PAC shown in Figure 17.38 (p. 868)? Tell how you arrived at your conclusion.

TAKE CHARGE

17.30 Obtain 24 months of (deseasonalized) time series data for an activity of interest to you. Then apply three different autoprojective forecasting models of your own choice to the data and tell which model works best.

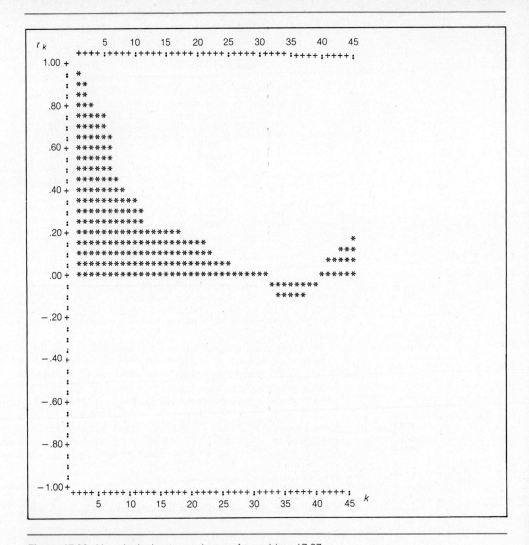

Figure 17.36. Hypothetical autocorrelogram for problem 17.27.

Computer Exercise

17.31 The differential between long-term and short-term interest rates is a most important variable to the residential construction industry. When the differential becomes small or turns negative, as it does from time to time, especially during advanced stages of business cycle expansions, investment funds which might otherwise be invested long term are diverted to shorter-term, less risky, investments. Table 17.8 presents ten years of monthly data for (1) yield on U.S. Treasury bonds (long term), (2) discount rate on U.S. Treasury bills (short term), and (3) difference between bond and bill rates.

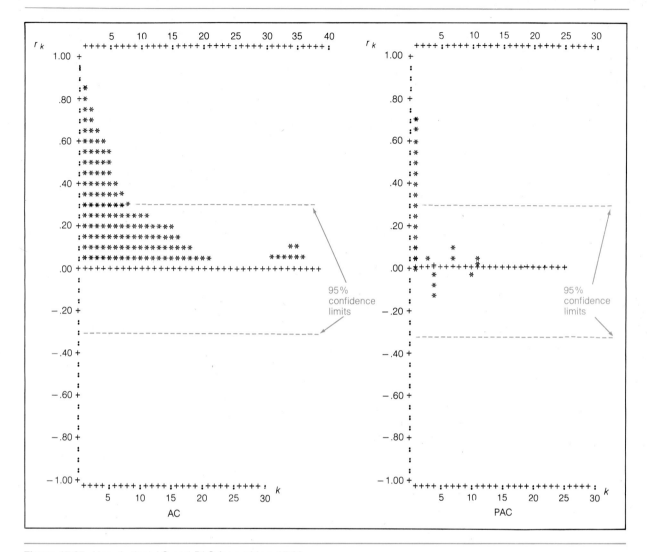

Figure 17.37. Hypothetical AC and PAC for problem 17.28.

Apply three exponential smoothing forecasting equations—one with $w = .1$, one with $w = .3$, and one with $w = .5$—to the difference data and tell which such equation generates the most accurate one-month-ahead forecasts. Explain how you arrived at this conclusion.

Computer Exercise

 17.32 Refer to Table 17.8. Subject the data for U.S. Treasury bonds to Box-Jenkins procedures and determine what kind of forecasting model is called for. (*Hint:* You will have to employ [nonseasonal] differencing of order $d = 1$.)

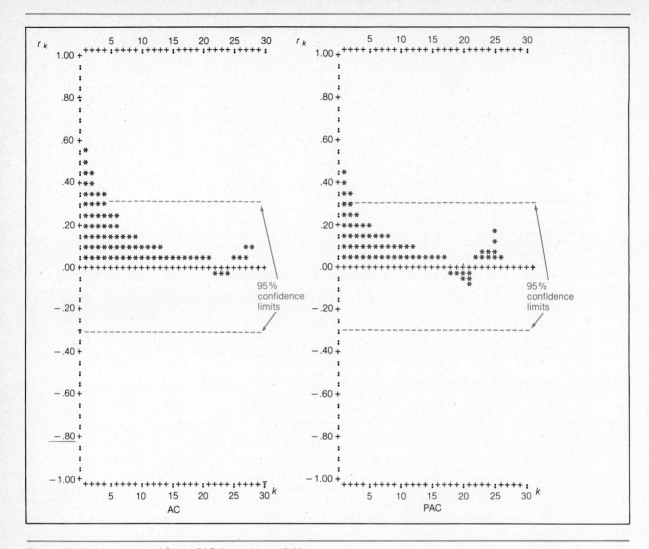

Figure 17.38. Hypothetical AC and PAC for problem 17.29.

Cases for Analysis and Discussion

 ### 17.1 "DATA DYNAMIX CORPORATION"

The "Data Dynamix Corporation" was founded in 1979 by Harry Walters and a group of fellow computer engineers who had gained their experience working for IBM. The company was organized for the purpose of designing, developing, manufacturing, and marketing computer peripheral equipment. Data Dynamix's major effort until 1982 entailed the development of its first products—a line of magnetic tape subsystems. After 1981, the company entered into a

Table 17.8 Ten Years of Monthly Data for Yield on U.S. Treasury Bonds, Discount Rate on U.S. Treasury Bills, and Difference Between Bond and Bill Rates

Year and Month	Bonds	Bills	Difference	Year and Month	Bonds	Bills	Difference
1975 January	6.68	6.49	.19	October	8.07	8.13	− .06
February	6.66	5.58	1.08	November	8.16	8.79	− .63
March	6.77	5.54	1.23	December	8.36	9.12	− .76
April	7.05	5.69	1.36				
May	7.01	5.32	1.69	1979 January	8.43	9.35	− .92
June	6.86	5.19	1.67	February	8.43	9.27	− .84
July	6.89	6.16	.73	March	8.45	9.46	−1.01
August	7.11	6.46	.65	April	8.44	9.49	−1.05
September	7.28	6.38	.90	May	8.55	9.58	−1.03
October	7.29	6.08	1.21	June	8.32	9.05	− .73
November	7.21	5.47	1.74	July	8.35	9.26	− .91
December	7.17	5.50	1.67	August	8.42	9.45	−1.03
				September	8.68	10.18	−1.50
1976 January	6.93	4.96	1.97	October	9.44	11.47	−2.03
February	6.92	4.85	2.07	November	9.80	11.87	−2.07
March	6.88	5.05	1.83	December	9.58	12.07	−2.49
April	6.73	4.88	1.85				
May	7.01	5.18	1.83	1980 January	10.03	12.04	−2.01
June	6.92	5.44	1.48	February	11.55	12.81	−1.26
July	6.85	5.28	1.57	March	11.87	15.53	−3.66
August	6.82	5.15	1.67	April	10.83	14.00	−3.17
September	6.70	5.08	1.62	May	9.82	9.15	.67
October	6.65	4.93	1.72	June	9.40	7.00	2.40
November	6.62	4.81	1.81	July	9.83	8.13	1.70
December	6.38	4.35	2.03	August	10.53	9.26	1.27
				September	10.94	10.32	.62
1977 January	6.68	4.60	2.08	October	11.20	11.58	− .38
February	7.16	4.66	2.50	November	11.83	13.89	−2.06
March	7.20	4.61	2.59	December	11.89	15.66	−3.77
April	7.13	4.54	2.59				
May	7.17	4.94	2.23	1981 January	11.65	14.72	−3.07
June	6.99	5.00	1.99	February	12.23	14.90	−2.67
July	6.98	5.15	1.83	March	12.15	13.48	−1.33
August	7.01	5.50	1.51	April	12.62	13.63	−1.01
September	6.94	5.77	1.17	May	12.96	16.30	−3.34
October	7.08	6.19	.89	June	12.39	14.56	−2.17
November	7.16	6.16	1.00	July	13.05	14.70	−1.65
December	7.24	6.06	1.18	August	13.61	15.61	−2.00
				September	14.14	14.95	− .81
1978 January	7.51	6.45	1.06	October	14.13	13.87	.26
February	7.60	6.46	1.14	November	12.68	11.27	1.41
March	7.63	6.32	1.31	December	12.88	10.93	1.95
April	7.74	6.31	1.43				
May	7.87	6.47	1.40	1982 January	13.73	12.41	1.32
June	7.94	6.71	1.23	February	13.63	13.78	− .15
July	8.10	7.07	1.03	March	12.98	12.49	.49
August	7.88	7.04	.84	April	12.84	12.82	.02
September	7.82	7.84	− .02	May	12.67	12.15	.52
				June	13.32	12.11	1.21

(continued)

Table 17.8 *(Continued)*

Year and Month	Bonds	Bills	Difference	Year and Month	Bonds	Bills	Difference
July	12.97	11.91	1.06	October	11.21	8.71	2.50
August	12.15	9.01	3.14	November	11.32	8.71	2.61
September	11.48	8.20	3.28	December	11.44	8.96	2.48
October	10.51	7.75	2.76				
November	10.18	8.04	2.14	1984 January	11.29	8.93	2.36
December	10.33	8.01	2.32	February	11.44	9.03	2.41
				March	11.90	9.44	2.46
1983 January	10.37	7.81	2.56	April	12.17	9.69	2.48
February	10.60	8.13	2.47	May	12.89	9.90	2.99
March	10.34	8.30	2.04	June	13.00	9.94	3.06
April	10.19	8.25	1.94	July	12.82	10.13	2.69
May	10.21	8.19	2.02	August	12.23	10.49	1.74
June	10.64	8.82	1.82	September	11.97	10.41	1.56
July	11.10	9.12	1.98	October	11.66	9.97	1.69
August	11.42	9.39	2.03	November	11.25	8.79	2.46
September	11.26	9.05	2.21	December	11.23	8.29	2.94

Source: Board of Governors of the Federal Reserve System and U.S. Treasury Department.

marketing agreement to sell under its own name a line of magnetic disk subsystems manufactured by another producer of peripheral equipment. In 1983, the company founded "Disk Development Corporation" for the purpose of developing advanced disk subsystems. Disk Development Corporation was then merged under the name of Data Dynamix Corporation in 1984 after the initial development work had been accomplished.

Data Dynamix now manufactures its own complete lines of magnetic tape and disk subsystems. Tape and disk subsystems are used for the storage and transfer of data to and from computer systems. The company's primary market is the replacement market for IBM equipment both on the domestic scene and in several foreign countries.

In addition, the original-equipment manufacturing (OEM) market has been developed by Data Dynamix. The OEM market involves the sale of equipment to other computer manufacturers to attach to their main computers. Data Dynamix's OEM products are marketed under the names of the purchasing manufacturers.

Competing with IBM and other large computer manufacturers is an extremely challenging task. The products have to be reliable and well supported. Engineering, field service, and spare parts support are essential to the credibility of the firm. Negligent and insufficient support in any of these areas would result in the loss of customers and in a poor reputation for the company.

Engineering and field service depend primarily on the knowledge and experience of the individuals involved. Good hiring and training practices will move the firm in the right direction toward gaining and maintaining credibility in these fields.

Spare parts support, on the other hand, goes well beyond good hiring and training techniques. In order adequately to support the equipment in the field, forecasts of the needed number of spare parts of a wide variety of types must be prepared over short intervals of time. Inventory levels for field warehouses, new areas, and new accounts must be determined. Moreover, future safety stocks must be calculated in order for the company to have a sufficient supply on hand to meet emergencies.

Data Dynamix's spare-parts support department has chosen exponential smoothing as its method for estimating near-term demand for each spare part related to established products.

Exponential smoothing, they reasoned, allows the smoothing out of meaningless irregularities that appear in the data as well as—with appropriate refinements in methodology—the adjusting for differences in number of working days per month.

Exponential smoothing was chosen, according to one company officer, "because it will work well with a small history base. Other techniques require a much larger base for any reasonable degree of confidence. A small history base is an important consideration for Data Dynamix because of the company's short time in operation."

1. Do you agree that exponential smoothing was a wise choice of a short-term forecasting technique for estimating spare-parts demand for this company? Why or why not?

2. Assuming that use of some kind of computerized autoprojective forecasting method seems advisable for this company, can you think of any other autoprojective approach whose possible application should be explored as the company's data base increases? What is it (or what are they)? Why do you believe it (or they) might prove advantageous?

 ## 17.2 UNIVERSAL RESOURCES CORPORATION

A security analyst sought a simple forecasting equation for predicting the market price of Universal Resources common stock one month ahead. She began by developing the autocorrelogram shown in Figure 17.39 based on monthly averages of daily closing prices.

1. Tell as much about the nature of the original price data as you can surmise from the autocorrelogram.

2. Does the autocorrelogram suggest that the stock's price might be fairly predictable? Explain why you do or do not think so.

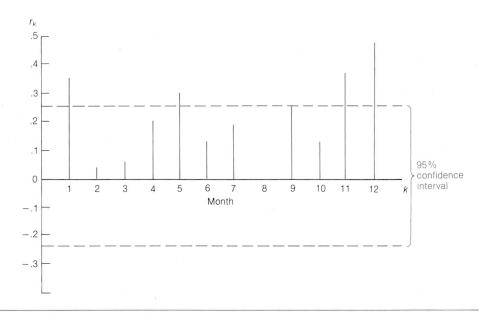

Figure 17.39. Autocorrelogram of Universal Resources common stock prices. (Data represent monthly averages of daily closing prices.)

18 Decision Theory: Prior Analysis

O God, give us serenity to accept what cannot be changed, courage to change what should be changed, and wisdom to distinguish one from the other.—Reinhold Niebuhr

There is only one thing certain and that is that nothing is certain.—G. K. Chesterton

What You Should Learn from This Chapter

This is the first of two chapters on Bayesian decision making. Here we will be concerned exclusively with prior analysis—that is, with methods of choosing the best payoff action using estimated payoffs and *prior probabilities* only. Chapter 19 will show how the prior probabilities and, in turn, the decision procedures can be sharpened by taking additional empirical information into account. When you have completed this chapter, you should be able to

1. Distinguish between classical and Bayesian decision theory.
2. Define *decision maker, action, state of nature,* and *prior probability distribution.*
* 3. Employ the reference-lottery technique to arrive at meaningful subjective prior probabilities, to determine utilities, and to translate between utilities and monetary payoffs.
4. Distinguish among these action-selection criteria: maximin criterion,* maximax criterion,* maximum-likelihood criterion,* insufficient-reason criterion,* and the Bayes decision rule.
5. Tell why use of the Bayes decision rule is usually preferred to the other action-selection criteria listed in (4).
6. Recognize and eliminate inadmissible actions when they exist.
7. Perform prior analysis using either a tabular or a decision-tree format and either utility, monetary, or opportunity loss payoffs.
* 8. Work two-action problems with linear payoff functions.

Before beginning this chapter, you may wish to review section 6.9 of Chapter 6 and section 7.2 of Chapter 7.

18.1 Introduction

We dealt with some aspects of the broad subject of statistical decision theory previously when we studied hypothesis testing. Whenever one tests a null hypothesis statistically, he or she concludes the analysis by either accepting or rejecting that hypothesis. In the context of business problem solving, acceptance of the null hypothesis leads to one course of action (which may be intentional inaction) and rejection of the null hypothesis leads to a different course of action.

*These topics are optional.

> The term *statistical decision theory* refers to those methods of statistical analysis whose application results in a direct answer to the question, "Which one of two or more actions under consideration should I take?"

When the subject of hypothesis testing is viewed as a branch of statistical decision theory, it is customary to refer to it as *classical decision theory*. In this chapter and the next we deal with quite a different approach to arriving at decisions by statistical means. This approach we refer to as *Bayesian decision theory*.

You should be aware at the outset that the label *Bayesian decision theory* is something of a misnomer in many decision-making contexts. This label derives from the fact that often a decision maker will begin his analysis by using subjectively determined probabilities and later *revise them* by making use of additional empirical information and the *Bayes formula* introduced in section 6.9 of Chapter 6. However, Bayesian decision theory has come to be the name applied to a great many decision problems characterized by (1) the possibility of more than two available actions, (2) the possibility of more than two parameter conditions, and (3) the use of subjective prior probabilities. Under this usage, then, it is possible that a Bayesian decision problem will not entail use of the Bayes formula. This broader meaning of "Bayesian decision theory" is the one used in this chapter. In Chapter 19 you will be shown how to inject a truly Bayesian feature into Bayesian decision theory.

Classical and Bayesian Decision Theory Compared

The main differences between classical and Bayesian decision theory have already been alluded to. They can be summarized as follows:

1. Classical decision theory usually permits recognition of only two possible parameter conditions and only two possible alternative actions. *

   You will recall that the first step in applying any of the classical hypothesis-testing procedures is to set up a null and alternative hypothesis. For example, if H_0 is, say, $\mu = 50$ and H_a: $\mu \neq 50$, the true population mean is viewed as being either 50 or not 50. Needless to say, the highly condensed "not 50" alternative contains within it an infinitude of possible population mean values. Similarly, in classical decision theory the actions available are also limited to two—one related to acceptance and the other to rejection of the null hypothesis.

   When using the Bayesian approach, on the other hand, the decision maker is free to contemplate as many possible parameter values, or conditions, and as many alternative actions as he or she considers relevant.

2. Classical analysis does not require explicit statement of the consequences of a bad decision.

   In classical decision methodology, the consequences of a bad decision are supposedly taken into account in the process of determining the level of significance, α. In practice, however, it is common to use a traditional, and thus easily de-

---

*Though, as we saw in section 11.10, some classical methods *do* permit the analyst to pin down the reason(s) behind a rejected null hypothesis—a fact rendering this assertion untrue in some situations.

fended, value of .05 or .01, or to rely on the observed significance level. In the Bayesian approach, a serious effort is made to make the unpleasant consequences of a bad decision an explicit and integral part of the analysis.

3. In classical analysis only objective probabilities, determined independently of the judgment of the decision maker, are used.

   Classical decision methods do not utilize, within the formal analysis, information in hand prior to the sample study. Bayesian theory does. Use of the Bayesian approach allows the decision maker to exploit any or all such prior information which he or she may possess in addition to information obtained from the sample study. This is accomplished through a marriage of subjective prior probabilities and objective conditional probabilities.

Subjective Probability: The Vulnerable Part of Bayesian Methods

As mentioned above, subjective probabilities usually play an important role in Bayesian decision theory, a fact which has made many classical statisticians understandably distrustful of the approach. On the other hand:

> The justification for using *subjective probabilities* is a most persuasive one with respect to many business decision-making situations: Business executives must make judgments regardless of whether they have complete and flawless data in hand; moreover, when they make judgments, they will think in probability terms regardless of whether they actually use numbers to measure the intensity of their convictions. For these reasons, use of subjective probabilities may be justified on pragmatic, if not on purely scientific, grounds.

Fortunately, the Bayesian decision maker does have some safeguards against being too foolish in his assignment of subjective probabilities. For one thing, the subjective probabilities will often actually reflect past experience or empirical information. Moreover, the Bayesian analyst may avail himself of such protective devices as (1) lottery techniques for reducing the vagueness of subjective probability assessments, (2) procedures for revising the initial subjective probabilities using new observed evidence, and (3) methods for evaluating the sensitivity of the decision situation to variation in the probabilities employed. The first two of these protective devices are treated in this and the following chapter. The third is too complicated for problems other than those of very small size, so it will not be covered in this textbook.

18.2 The General Structure of a Bayesian Decision Problem

Let us consider an everyday problem of a kind with which you may be familiar. We will suppose that you are anxiously studying for an important examination to be given first thing tomorrow morning in your accounting course. You are a little nervous because you have as yet done no studying of one large block of assigned reading material. In your favor are the facts that (1) this material was accorded only brief treatment in class and (2) the test is known to consist of only four problems—probably, though not necessarily, equally weighted. Consequently, you feel reasonably sure that the examination will contain no problems on this particular subject matter (to be called *Material Q* henceforth—Q for *questionable*). If you could count on that being so, you could help your score by refraining from studying Material Q and devoting your remaining, and rapidly evaporating, time to more strategic topics. On the other hand, you have learned from harsh experience that some professors delight in asking questions on material receiving little emphasis in class. Can you really afford not to study Material Q?

While you are turning this question over in your mind, let us pause here and define some important terms.

In Bayesian decision theory, the *decision maker* is the person or group of persons charged with determining which of the two or more contemplated actions is best to take.

The *actions* are the various possible avenues of response to the decision situation which are being considered by the decision maker and *under his control*. The actions under consideration will be symbolized A_1, A_2, A_3, . . . , A_n.

The *states of nature* are those conditions generated by forces over which the decision maker *has no control*. Some examples of uncontrollable conditions—that is, possible states of nature—are the state of the general economy, the manner in which consumers react to a new product, the actions of the company's principal competitors, and the decisions of high government officials. The possible states of nature will be symbolized θ_1, θ_2, θ_3, . . . , θ_m.

Notice that the symbol θ (theta) used to identify a specific possible state of nature is the same symbol used previously to indicate an unspecified kind of population parameter. Before now, we have thought of a population of interest as if it had some specific, albeit unknown, parameter value. In Bayesian analysis, we assume a somewhat different viewpoint and regard the relevant descriptive measure—or specific category or set of conditions—denoted by θ_i, as a *variable* to which probabilities may be assigned.

Probabilities assigned the various possible states of nature prior to the gathering of additional information form the *prior probability distribution*. This is quite literally a probability distribution as we have used the term on previous occasions. Thus, it is of utmost importance that the decision maker define the possible states of nature to ensure that they are *mutually exclusive* and that their corresponding probability assessments sum to 1.

In the present hypothetical example, the possible states of nature would be, let us say, (1) θ_1: At least one problem on Material Q will appear on the examination and (2) θ_2: No problems on Material Q will appear on the examination.

For simplicity's sake, we will also assert that Material Q is of such a nature that (1) if you do not study it, you will surely be unable to tackle correctly any problems which might appear on the test, and (2) if you do study for it, you will surely be able to work correctly any problem pertaining to this material. Let us conclude, then, that your available actions are (1) A_1: Do not study Material Q and (2) A_2: Study Material Q.

Except for making allowances for subjective prior probabilities, all information required for dealing with the problem under discussion is displayed in Table 18.1. This kind of table, called a *payoff table* or *payoff matrix*, shows the possible states of nature along one side and the available actions along an adjacent side.

The entries within the cells of this particular payoff table are called *outcomes*, to be symbolized O_{ij}. Each of these represents a *verbal description* of the result of a "joint decision" between the decision maker and nature.

In future discussions, we will often refer to such "joint decisions" as *state-and-action combinations*.

Incidentally, we will not ordinarily begin, as we have here, with a verbal description of the results of a specific state-and-action combination; instead, we will take for granted that O_{ij} can be expressed numerically as $U(O_{ij})$, the utility value associated with Outcome O_{ij}, or as $MP(O_{ij})$, the monetary payoff associated with Outcome O_{ij}. However, one goal of this chapter is to show how outcomes can be converted into utility or monetary values.

Table 18.1 Payoff Table for the Material-Q Decision Problem

State of Nature	Action	
	A_1: Do Not Study Material Q	A_2: Study Material Q
θ_1: At least one problem on Material Q will appear on the examination	O_{11}: Work at least one problem incorrectly which you could have otherwise worked correctly	O_{12}: Work at least one problem correctly which you would have otherwise worked incorrectly
θ_2: No problems on Material Q will appear on the examination	O_{21}: Achieve greater mastery over the "non-Q" material by devoting time to it which would otherwise have been invested in Material Q	O_{22}: Waste some of your limited time which could have been put to better use

θ_i = ith possible state of nature
A_j = jth possible action
O_{ij} = outcome associated with the ith state of nature and the jth action

Table 18.2 Generalized Payoff Table

State of Nature	Action				
	A_1	A_2	A_3	$\cdots$	A_n
θ_1	O_{11}	O_{12}	O_{13}	$\cdots$	O_{1n}
θ_2	O_{21}	O_{22}	O_{23}	$\cdots$	O_{2n}
θ_3	O_{31}	O_{32}	O_{33}	$\cdots$	O_{3n}
$\vdots$	$\vdots$	$\vdots$	$\vdots$	$\vdots$	$\vdots$
θ_m	O_{m1}	O_{m2}	O_{m3}	$\cdots$	O_{mn}

θ_i = ith possible state of nature

A_j = jth possible action

O_{ij} = outcome associated with the ith state of nature and the jth action

m = number of possible states of nature considered in the decision-making effort

n = number of possible actions perceived to be available to the decision maker

Table 18.2 presents a generalized payoff table capable of accommodating any number of states of nature and any number of contemplated actions.

18.3 Determining the Expected Payoff of an Action

Recall that in Chapter 7 (section 7.2) we introduced the measure "expected value of a random variable." We showed that the expected value of a (discrete) random variable, designated X, could be obtained by $P(X_1)X_1 + P(X_2)X_2 + \cdots + P(X_n)X_n = \Sigma_{\text{all } i}[P(X_i)X_i]$. The resulting expected value is interpreted as the long-run average value of X.

By making a few minor changes, we may advantageously use the concept of expected value within decision analysis. We may, for example, think in terms of the "expected payoff" of Action A_1 and thus calculate $EP(A_1) = P(\theta_1) \cdot f(O_{11}) + P(\theta_2) \cdot f(O_{21}) + \cdots + P(\theta_m) \cdot f(O_{m1}) = \Sigma_{\text{all } i}[P(\theta_i) \cdot f(O_{i1})]$, where $EP(A_1)$ is the expected payoff associated with Action A_1, $P(\theta_i)$ is the probability of occurrence of state of nature θ_i, and $f(O_{i1})$ is the payoff associated with Outcome O_{i1}. (In short order, U or MP, or other relevant notation indicating the nature of the payoff, will replace f.)

Since in decision theory the payoffs are frequently stated in terms of either utility values or money values (and may, where appropriate, be stated in terms of, say, efficiency ratings, share-of-market values, and so forth), it will be helpful to employ a symbol system that indicates the kind of units in which the expected payoff is expressed. Therefore,

The *expected utility payoff* of Action A_j, $EU(A_j)$, is the sum of two or more products associated with that action. The products referred to are the utility value corresponding to Outcome O_{1j} times the probability of state of nature θ_1, the utility value corresponding to Outcome O_{2j} times the probability of state of nature θ_2, and so on, through the utility value corresponding to Outcome O_{mj} times the probability of state of nature θ_m. In formula form:

$$EU(A_j) = \sum_{\text{all } i} [P(\theta_i)U(O_{ij})]$$

where it is understood that each verbally expressed outcome has been converted into a number representing its *utility value*, or just *utility*, as indicated by the use of U.

The concept of *utility* is discussed in detail in the following section.

Stated briefly, *utility* is a number expressing the satisfaction or usefulness associated in the decision maker's mind with a specific outcome.

Easier to understand at this stage of the discussion and more often used in practice is the concept of *expected monetary payoff*.

The *expected monetary payoff* of Action A_j, $EMP(A_j)$, is also the sum of a series of probability and payoff-value products but, with this measure, the payoffs are monetary, rather than utility, payoffs. In formula form:

$$EMP(A_j) = \sum_{\text{all } i} [P(\theta_i)MP(O_{ij})]$$

where $MP(O_{ij})$ represents the monetary payoff corresponding to Outcome O_{ij}.

*18.4 The Reference Lottery and Its Uses

The reference-lottery concept may be employed in connection with Bayesian decision problems in various ways. In this section we explore three ways, namely: (1) firming up subjective prior probabilities, (2) measuring utility, and (3) translating between utilities and monetary payoffs.

Firming Up the Subjective Prior Probabilities

Despite the arbitrariness inherent in the act of assigning subjective probabilities to the possible states of nature, the analyst can, using a purely imaginary lottery, at least limit the range of arbitrariness.

Suppose, for example, that you are asked whether you would prefer Lottery 1 or Lottery II, as described here:

Lottery I: Win $0 with probability .5
            Win $1000 with probability .5

Lottery II: Win $0 if a problem on Material Q does appear on the test tomorrow
            Win $1000 if a problem on Material Q does not appear on the test tomorrow

Clearly, since the prize associated with a win is the same for both lotteries ($1000), you would certainly choose the lottery that promises you the higher expected monetary payoff, EMP. The expected monetary payoff of Lottery I is EMP(Lottery I) = (.5)($0) + (.5)($1000) = $500. Therefore, if you were to choose Lottery I, you would be revealing your belief that the probability of your encountering a problem on Material Q, $P(\theta_1)$, is greater than .5 because, if EMP(Lottery II) = (Value slightly higher than .5)($0) + (Value slightly lower than .5)($1000), the result is an expected monetary payoff of less than $500. Conversely, if you were to choose Lottery II, you would be acting as if you thought the probability of your not encountering a problem on Material Q, that is, $P(\theta_2)$, was greater than .5. This must be so because EMP(Lottery II) = (.5 −)($0) + (.5 +)($1000) = an expected monetary payoff greater than $500.* Let us say that you select Lottery II, thereby proclaiming your belief that $P(\theta_2)$ is a number greater than .5.

Next, let us change the probabilities associated with Lottery I so that the choice is now

Lottery I: Win $0 with probability .45
            Win $1000 with probability .55

Lottery II: Win $0 if a problem on Material Q does appear on the test tomorrow
            Win $1000 if a problem on Material Q does not appear on the test tomorrow

Because the expected monetary payoff of Lottery I is now EMP(Lottery I) = (.45)($0) + (.55)($1000) = $550, your continued preference for Lottery II would mean that you believe the probability of not encountering a problem on Material Q, $P(\theta_2)$, is greater than .55 because EMP(Lottery II) = (.45 −)($0) + (.55 +)($1000) = an expected monetary payoff greater than $550.

You could continue to change systematically the probabilities associated with Lottery I until you were indifferent in your choice between the two lotteries. Let us suppose that—after investing quite a lot of time and considerable mental anguish—you are now indifferent and that the probabilities making you indifferent are .3 and .7. That is,

Lottery I: Win $0 with probability .3
            Win $1000 with probability .7

*Small minus and plus signs following values mean "a little less than" and "a little more than," respectively.

Lottery II: Win \$0 if a problem on Material Q does appear on the test tomorrow

Win \$1000 if a problem on Material Q does not appear on the test tomorrow

The expected monetary payoff of Lottery I would now be EMP(Lottery I) = (.3)(\$0) + (.7)(\$1000) = \$700; if you are truly indifferent between the two lotteries, that must mean that you believe EMP(Lottery II) is also \$700.

The above line of reasoning can be generalized to suit virtually any kind of probability-selection situation:

Lottery I: Receive A with probability p
             Receive B with probability $1 - p$
Lottery II: Receive A if θ occurs
             Receive B if θ does not occur

The only stipulation is that A and B cannot be equally desirable prizes. If, for example, you were indifferent between receiving \$1000 and receiving \$0, you would necessarily be indifferent between the two lotteries regardless of the probabilities used. Such global indifference would, of course, render the lottery device unusable.

Measuring Utility

Had the Material-Q problem offered monetary rewards or penalties, we could now complete it rather quickly. As it stands, the problem is more complicated than that because it involves payoffs that are not expressable in money terms. Moreover, the outcomes have not yet been converted into any kind of numerical equivalents whatever. In view of the nature of the problem, it seems appropriate to complete the analysis using utility payoffs.

The first step in the process of obtaining utility values is that of determining which of the listed outcomes you would *most like to have happen* and which you would *least like to have happen*. In making this determination for the present problem, you would presumably wish to be guided in part by the degree of mastery you feel you possess over the "non-Q" material. Therefore, let us assume that you feel reasonably confident toward all the relevant material except for Material Q; however, you do not feel sufficiently confident to welcome the opportunity to turn your attention away from the "non-Q" material and toward Material Q in the absence of evidence strongly suggesting this as a wise course of action.

On the basis of this assumption, you might reason that the *least-preferred outcome* would be O_{22}—waste some of your limited time which could have been put to better use.

What would be the *most-preferred outcome*? Outcome O_{11}, work at least one problem incorrectly which you could otherwise have worked correctly, would presumably have to be ruled out. For the sake of illustration, let us say that of the two remaining outcomes you find Outcome O_{12}, work at least one problem correctly which you would have otherwise worked incorrectly, impresses you as preferable to Outcome O_{21}, achieve greater mastery over the "non-Q" material by devoting more time to it.

Clearly, since O_{12} is considered the *most-preferred outcome* and O_{22}, the *least-preferred outcome* of the four outcomes possible, $U(O_{12})$ must be greater than $U(O_{22})$. Fortunately, you may assign any utility value you like to Outcomes O_{12} and O_{22}, provided that the value assigned the former is greater than that assigned the latter. You should be aware, however, that certain practical advantages do accompany the use of $U(O_{12}) = 1$ and $U(O_{22}) = 0$.

Now let us contemplate Outcome O_{21}. Since this is neither the most-preferred nor the least-preferred outcome, we observe that

$$U(O_{12}) \geq U(O_{21}) \geq U(O_{22})$$

and, because of the arbitrary but wise assignments of 1 to $U(O_{12})$ and 0 to $U(O_{22})$, we also observe that

$$1 \geq U(O_{21}) \geq 0$$

To determine a numerical value for $U(O_{21})$, you could now consider the following choice of lotteries:

Lottery I: Receive $U(O_{21})$ for certain

Lottery II: Receive $U(O_{12})$ with probability p and receive $U(O_{22})$ with probability $1 - p$

How should you decide between the two lotteries? It would seem reasonable to select the lottery promising the *greater expected utility*, EU. The expected utility of Lottery I is obviously $U(O_{21})$; the expected utility of Lottery II is $EU(\text{Lottery II}) = pU(O_{12}) + (1 - p)U(O_{22})$. Because we know $U(O_{12})$ to be 1 and $U(O_{22})$ to be 0, we may simply state $EU(\text{Lottery II}) = p$. Therefore, if $U(O_{21}) > p$, then Lottery I should be chosen; if $U(O_{21}) < p$, then Lottery II should be chosen; and if $U(O_{21}) = p$, you are indifferent between the two lotteries.

This relationship between $U(O_{21})$ and p can be exploited to determine the utility value associated with Outcome O_{21}. By experimenting systematically with various values of p, you will eventually arrive at one which makes you indifferent between the two lotteries—that is, a value such that $EU(\text{Lottery I}) = EU(\text{Lottery II})$ which, as we have seen, is the same thing as $U(O_{21}) = p$. Let us suppose that in due time you come to realize that a value of .7 for p—and, thus, for $U(O_{21})$—would be required to make you indifferent. Hence, $U(O_{21}) = .7$.

The only remaining outcome for which a numerical utility value has not yet been assigned is Outcome O_{11}. As above, we may assert that

$$U(O_{12}) \geq U(O_{11}) \geq U(O_{22})$$

or, equivalently,

$$1 \geq U(O_{11}) \geq 0$$

To determine $U(O_{11})$ you could proceed along the same lines as described above—namely, set up a choice of lotteries like the following:

Lottery I: Receive $U(O_{11})$ for certain

Lottery II: Receive $U(O_{12})$ with probability p and receive $U(O_{22})$ with probability $1 - p$

As before, you would decide between the two lotteries by comparing the respective expected utilities and settling on a value of p such that $EU(\text{Lottery I}) = EU(\text{Lottery II})$ which, after a little simplification, amounts to $U(O_{11}) = p$. Consequently, if $U(O_{11}) > p$,

Table 18.3 Utility-Value Payoff Table for the Material-Q Decision Problem

State of Nature	Action	
	A_1: Do Not Study Material Q	A_2: Study Material Q
θ_1: At least one problem on Material Q will appear on the examination	$U(O_{11}) = .2$	$U(O_{12}) = 1.0$
θ_2: No problems on Material Q will appear on the examination	$U(O_{21}) = .7$	$U(O_{22}) = .0$
	$P(\theta_1) = .3 \qquad P(\theta_2) = .7$	

then Lottery I should be chosen; if $U(O_{11}) < p$, then Lottery II should be chosen; and if $U(O_{11}) = p$, you are indifferent between the two lotteries. Let us say that the value of p which would make you indifferent in this case is found, after some serious experimentation, to be .2. If so, then $U(O_{11})$ must also be .2. You now find yourself in possession of all the utility values needed to construct the payoff table shown in Table 18.3.

Translating Between Utilities and Monetary Payoffs

In the early days of television a show called "The Millionaire" was popular weekly fare for many viewers. The long-standing situational format was that each week a different individual would unexpectedly be handed a check for $1 million by the envoy of an eccentric multimillionaire. Each episode then explored the ways in which the sudden possession of $1 million affected the life of the recipient.

Let us borrow from this program's format but revise the deal somewhat so that the lucky person is permitted a choice between two monetary gifts. These choices are expressed in terms of the following lotteries:

Lottery I: Receive $1 million for certain

Lottery II: Receive $10 million with probability .5 and receive $0 with probability .5.

If we compute the expected monetary payoffs of the two lotteries, we find that EMP(Lottery I) = $1 million and that EMP(Lottery II) = (.5)($10 million) + (.5)($0) = $5 million. Seemingly, the rational choice would be Lottery II. However, it might surprise you to learn that many people faced with this pleasant, but difficult, choice *on a one-shot basis* would choose Lottery I for the very sensible reason that $1 million is a lot of money. The lucky recipient could certainly turn many of his lifelong dreams into pleasant realities and still have enough money left over to ensure a comfortable old age.

To be sure, if he were to choose the second lottery and win, that would be better yet; however, many people would consider $10 million to have a value *to them* of something less than ten times the value *to them* of $1 million. Besides, with the second lottery the recipient faces a 50–50 chance of ending up with no more money than he started with. Consequently, the process of converting utility payoff decision problems into monetary payoff decision problems (or vice versa) must take account of the decision maker's attitude toward taking financial risks.

WARNING

Because essentially it is expected utility which any rational decision maker wishes to maximize, and because monetary payoffs are used in most business decision situations as convenient surrogates for utilities, it is good to know whether the relationship between monetary values and utilities is linear or nonlinear. If it is linear, then substituting monetary payoffs for utilities will not affect the preferred action choice; however, if it is nonlinear, use of such a substitution procedure does carry with it the risk of altering the preferred action choice.

The subject of how to determine the nature of the relationship between utility values and money values via the development of a *utility curve* can be treated most meaningfully by relating it to the situations faced by three quite different decision makers. Let us call them Mr. Walker, Mr. Yale, and Mr. Zellman.

Mr. Walker is entering late middle age. His job of 30 years at the post office has enabled him to accumulate a respectable, though by no means opulent, estate. He is beginning to look forward to retirement in the expectation of doing some limited traveling with Mrs. Walker. Of possible relevance is the fact that many years ago Mr. Walker spent $1 for a chance to win a new Buick in a community July 4th drawing. Since he failed to win the car, he decided against buying another $1 ticket the following year and, indeed, has not bought another one since.

Mr. Yale is also in the latter part of his middle years. As senior vice-president of a broadcasting company, he has accumulated a net worth which would be the envy of most Americans. Although not known to use money recklessly, he is not averse to taking an occasional plunge into the stock market when his instincts tell him the time is right. Accustomed as he is to being where the action is, Mr. Yale plans to remain an active member of the business community until his dying day.

Mr. Zellman is 24 years old and the father of six children. He recently lost $10,000 in the form of IOUs playing craps in an illegal gambling establishment. A representative of the gambling establishment has just informed him that, unless the debt is paid in full by tomorrow noon, come nightfall he will find himself at the bottom of the Atlantic Ocean wearing a pair of cement boots. An acquaintance has euphemistically described Mr. Zellman as "tending toward the impetuous."

Let us now play the modified millionaire game described at the outset of this section with all three. So that the stakes will appear a little less breathtaking to Mr. Walker and Mr. Zellman but still reasonably attractive to Mr. Yale, and so that each person will face the risk of a loss as well as the prospect of a gain, we will revise the "prizes" substantially. Let us say that the choice of lotteries is now as follows:

Lottery I: Receive $0 for certain

Lottery II: Receive $10,000 with probability p and receive $-\$10,000$ with probability $1 - p$

"And how do you plan to pay us back if you don't win the lottery?"

Before allowing anyone a choice of lotteries, we may automatically convert the two monetary payoffs contained within the description of Lottery II into their utility equivalents by following the same procedure we have already used. That is, since the $10,000 monetary payoff indicated in Lottery II is clearly the most attractive payoff of any within the stated range of +$10,000 to −$10,000, we may assign it an arbitrary utility of 1. Similarly, since −$10,000 is the least attractive monetary payoff within this range, we may assign it an arbitrary utility of 0. Thus, $U(\$10,000) = 1$ and $U(-\$10,000) = 0$. These utility values will apply to all three individuals regardless of their subsequent lottery choices and resulting p values.

We will permit Mr. Yale the first opportunity to choose between the two lotteries. Let us begin by determining what, for him, is $U(\$0)$. We will suppose that, after much varying of the p value in Lottery II, Mr. Yale informs us that $p = .5$ would be the probability of winning $10,000 which would make him indifferent about the two lotteries. If $p = .5$, then $U(\$0)$ must also be .5. We now have three points from which to construct a utility curve— namely, (1) $10,000 and 1, (2) $0 and .5, and (3) −$10,000 and 0. If we wished to establish some additional points on the emerging utility curve, we could rewrite the quantity in Lottery I and have it read, for example,

Lottery I: Receive $1000 for certain

Lottery II would be left unchanged.

Let us suppose that Mr. Yale informs us that he would be indifferent about the lotteries if p—and, therefore, $U(\$1000)$—were .55. Similarly, if the choice of lotteries were stated as

Lottery I: Receive $-\$1000$ for certain

Lottery II: Receive $10,000 with probability p and receive $-\$10,000$ with probability $1 - p$

we could again vary the p value in Lottery II and eventually determine the utility corresponding to a loss of $1000. Let us say that Mr. Yale would eventually arrive at a p—and, therefore, a $U(-\$1000)$—value of .45. Part II of Table 18.4 presents the five pairs of values assumed above.

From the schedule in Part II of Table 18.4 we can sketch in Mr. Yale's utility curve. It is roughly as shown in part (b) of Figure 18.1. Clearly, for Mr. Yale, the utility curve is linear, indicating that he derives as much pleasure from winning a stated amount of money as he does displeasure from losing the same amount of money. This symmetry in his attitude toward monetary risk is indicated by the fact that $\Delta_1 = \Delta_2$. With such a utility curve, Mr. Yale could readily proceed to convert a problem involving potentially highly complex utilities into one utilizing conceptually much simpler monetary equivalents. Conversely, if he were to find himself confronted with a personal investment problem, he could arrive at his decision with the assurance that whatever action suits him in terms of money would also suit him in terms of utility.

Let us now construct Mr. Walker's utility curve. Given the choice

Lottery I: Receive $0 for certain

Lottery II: Receive $10,000 with probability p and receive $-\$10,000$ with probability $1 - p$

what value of p is likely to make him indifferent about the two lotteries? Let us say that upon contemplating various possible p values for Lottery II, Mr. Walker indicates that a probability of .8 of receiving $10,000 would be required to make him indifferent. The value .8 would also represent his value for $U(\$0)$. Let us further suppose that, with continued questioning based upon our varying the monetary payoff of Lottery I and leaving Lottery II unchanged, we arrive at the schedule presented in Part I of Table 18.4 and part (a) of Figure 18.1.

Clearly, Mr. Walker is a risk avoider. That is, he derives less utility from receiving a specified amount of money than he derives disutility from losing an identical amount of money. (From Figure 18.1 we note that, for Mr. Walker, $\Delta_1 < \Delta_2$.) According to some scholars, this utility curve probably resembles the utility curves of most people.

Be that as it may, there *are* exceptions, and Mr. Zellman is unquestionably one. Having nothing whatever to lose and a life to keep for a while longer, he would probably present us with a schedule looking something like that shown in Part III of Table 18.4 and part (c) of Figure 18.1.

We see that Mr. Zellman, a bona fide risk taker, would be willing to choose Lottery II despite an expected monetary payoff of much less than $0. He is a decision maker for whom $\Delta_1 > \Delta_2$.

Table 18.4 The Utility Curves of
Three Individuals

Part I: Mr. Walker

Monetary Payoff	Utility
$10,000	1.00
1,000	.84
0	.80
−1,000	.74
−10,000	.00

Part II: Mr. Yale

Monetary Payoff	Utility
$10,000	1.00
1,000	.55
0	.50
−1,000	.45
−10,000	.00

Part III: Mr. Zellman

Monetary Payoff	Utility
$10,000	1.00
1,000	.06
0	.05
−1,000	.04
−10,000	.00

Since different individuals (or different companies) may have different utility curves, we specify that, in any business decision-making problem, the utility curve of interest is the utility curve of the person (or the company) who will receive the payoff.

*18.5 The Fundamental Role of Expected Utility

A point made earlier merits repetition here:

Whether the payoffs associated with specified outcomes are expressed in terms of monetary payoffs, utilities, or something else, we accept as the ultimate selection criterion that the action which leads to the highest *expected utility* is the one which should be chosen.

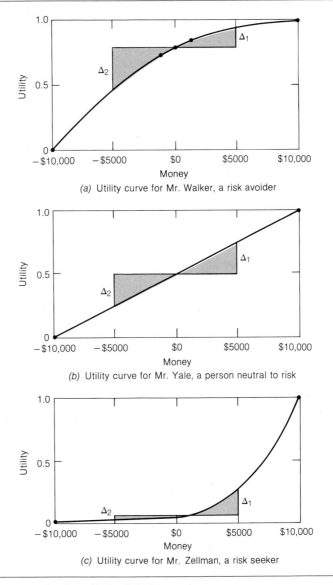

(a) Utility curve for Mr. Walker, a risk avoider

(b) Utility curve for Mr. Yale, a person neutral to risk

(c) Utility curve for Mr. Zellman, a risk seeker

Figure 18.1. Utility curves for three kinds of decision makers.

When the utility associated with each possible outcome is known or readily available, this rule is easy to apply. However, if (1) the utilities are unknown and virtually impossible to determine or (2) the decision maker faces a great many possible outcomes, this rule may be directly applied only with much travail and with little confidence in the superiority of the action selected. Consequently, what is usually done in practice is to assume that the decision maker's utility curve is *linear with respect to money* and to use the more readily accessible and more easily comprehended monetary payoffs as proxies for utilities. In spite of the obvious convenience attending this practice, we must still ask how reasonable it is in the context of business decision making. Fortunately, it seems more generally realistic when applied to businesses than when applied to individuals. Because most business decision problems involve monetary payoffs which are sufficiently small (relative to the overall assets of the company making the decision) that neither a good decision nor a poor decision will noticeably alter the financial health of the firm, we may usually safely assume that the company's utility curve is approximately linear.

Of course, exceptions do exist. For example, even a thriving well-established oil company might find a contemplated drilling venture too rich for its corporate blood almost regardless of the size of the optimal expected monetary payoff. In view of the huge costs involved, management might reason that a failure to find oil would be more disastrous to the company than success in finding oil would be beneficial.

In order to analyze the occasional exception, the analyst would be required to (1) determine the company's utility curve, (2) use the utility curve to convert monetary payoffs into utilities, and (3) select the preferred action on the basis of expected utility. This procedure might be lengthy and arduous; but it could be done.

PROCEDURAL NOTE

Fortunately, situations strongly suggesting nonlinear utility curves are mercifully rare where large business entities are concerned. For this reason, in all examples presented in the remainder of this chapter and in the chapter to follow, when monetary payoffs are used the *linearity assumption* will be presumed valid.

---

*18.6 Some Ways of Selecting the "Best" Action

Table 18.3 presents all the information one needs to make a choice between Actions A_1 and A_2 of the Material-Q problem. All that is lacking is a satisfactory criterion for making the choice. A great many, markedly different, selection criteria have been championed by both scholars and practitioners. A few of these are summarized here.

The Maximin Criterion

When employing the maximin criterion the decision maker proceeds (1) by assuming that nature will be as malevolent as possible and (2) by choosing the action that promises the highest payoff under such undesirable circumstances. In the Material-Q problem, use of this criterion would lead to the selection of Action A_1—do not study Material Q. The reasoning is as follows: Since the best possible payoff ($U(O_{12}) = 1.0$) is associated with state θ_1, θ_1 is the preferred state. However, with this selection criterion, one proceeds as if the less desirable state is inevitable and selects the action with the highest payoff associated with it. Since if state of nature θ_2 were to occur, and the utility values are .7 for Action A_1 and 0 for Action A_2, Action A_1 would be the one selected.

The Maximax Criterion

The maximax selection criterion is similar to the preceding one in some respects. However, whereas a pessimistic bias is built into the use of the maximin criterion, an optimistic bias is built into the use of this one. When employing the maximax approach, the decision maker assumes that nature will be as benevolent as possible and serve up the most desirable state of nature—namely, θ_1—for the problem under discussion. If state of nature θ_1 were to occur, the utility payoffs would be .2 for Action A_1 and 1.0 for Action A_2. Therefore, the better action according to this criterion would be A_2—study Material Q.

The Maximum-Likelihood Criterion

When the maximum-likelihood criterion is used, the decision maker attempts to determine the most probable state of nature and then proceeds to treat its occurrence as inevitable. In the Material-Q example, the reasoning would run something like: Since the state of nature θ_2 is more probable than the state of nature θ_1, then why even worry about the latter? Assume that State θ_2 will occur, that is, that no problems on Material Q will appear on the examination, and devote all remaining study time to the non-Q material. In other words, Action A_1 would be the action selected.

The Insufficient-Reason Criterion

The insufficient-reason criterion can be viewed as a denial of the underpinnings of the maximum-likelihood criterion. The implicit assumption is that anyone who believes he can second guess the true state of nature under conditions of uncertainty is simply fooling himself. Advocates of this selection criterion argue for viewing the possible states as equally likely (for the simple reason that it is futile to do otherwise) and choose the action having the highest payoff. For example, the expected utility for the Material-Q problem would be, according to this reasoning, for Action A_1: $\text{EUP}(A_1) = (.5)(.2) + (.5)(.7) = .45$, and for Action A_2: $\text{EUP}(A_2) = (.5)(1) = .5$. Since $.5 > .45$, Action A_2 (study Material Q) is the better choice.

---

*18.7 Shortcomings of Most Selection Criteria

Each of the selection criteria mentioned (and many others which could have been discussed) boasts its own unique strengths. However, one blanket shortcoming should be emphasized: Each of these selection criteria call for the decision maker to ignore certain information in his possession. With the maximin, maximax, and the maximum-likelihood criteria, the decision maker is asked to proceed as if he knew that a certain state of nature was inevitable. Thus, the decision maker using any of these ignores the fact that there is in reality some nonzero probability that another state might actually occur. Moreover, with these selection criteria, once the inevitable state of nature has been stated, all data on payoffs associated with other possible states are ignored.

The insufficient-reason criterion *does* make use of all payoff values but denies the possibility that the decision maker can realistically claim that one state of nature is more probable than others. Perhaps such cynicism is appropriate in some decision-making situations. However, those situations are relatively rare. Usually the decision maker will have personal knowledge and access to information by which to make reasonable guesses about the relative frequency of occurrence of states of nature. When some such knowledge or external information is available, it should be used—tentatively, perhaps, but used nonetheless. As was pointed out in Chapter 6, we humans appear to have an inherent or acquired bias toward underestimating the value of evidence. The decision maker seldom benefits from indulging this bias.

18.8 The Bayes Decision Rule

Because it is designed to utilize all information available to the decision maker, the selection criterion which has come to be preeminent among statistically oriented analysts is the *Bayes decision rule*. Not only does this approach to decision making utilize all payoff estimates the analyst has prepared and all subjective probabilities he has assigned, but it is also amenable to being refined and sharpened as new information is acquired. We made use of the Bayes decision rule earlier in this chapter without calling it by name:

According to the *Bayes decision rule*, the available action which yields the best expected payoff (that is, the highest expected payoff in a situation where more is better and the lowest in a situation where less is better) when all actions under consideration are subjected to the appropriate form of the equation $EP(A_j) = \sum_{\text{all } i} [P(\theta_i) \cdot f(O_{ij})]$ is the one that should be chosen. That action is called the *Bayes action*.

With respect to the Material-Q data, we find the method of calculating the expected utility of each available action demonstrated in Table 18.5. We learn from this table that Action A_1, do not study Material Q, is the Bayes action.

18.9 A Decision Problem Involving Monetary Payoffs

Business decisions usually involve monetary, rather than utility, payoffs. Let us, therefore, now turn our attention to a monetary payoff problem.

ILLUSTRATIVE CASE 18.1

The Claybourne and Clark Manufacturing Company (CCMC) supplies various kinds of small machine tools to five large companies, all of which have suffered sales slumps during recent months as a result of recession conditions in the general economy. CCMC's reaction to the

Table 18.5 Calculation of the Expected Utilities for the Material-Q Problem

Action A_1: *Do Not Study Material Q*

State of Nature	Probability	Utility	Weighted Utility
θ_1: At least one problem on Material Q will appear on the examination	.3	.2	.06
θ_2: No problems on Material Q will appear on the examination	.7	.7	.49
Total			.55

$$EU(A_1) = .55$$

Action A_2: *Study Material Q*

State of Nature	Probability	Utility	Weighted Utility
θ_1: At least one problem on Material Q will appear on the examination	.3	1.0	.3
θ_2: No problems on Material Q will appear on the examination	.7	.0	.0
Total			.3

$$EU(A_2) = .3$$

recession has been to allow its inventories of finished goods to decline to unusually low levels. The concerned members of management perceive their action choices to be

A_1: Attempt to reduce finished goods inventories even further
A_2: Attempt to maintain finished goods inventories at the present level
A_3: Attempt to increase finished goods inventories
A_4: Follow no particular inventory policy; simply respond to needs as they arise

Despite differences of opinion on some matters, the management group has succeeded in finding certain areas of agreement. First, they have agreed that considering only three possible states of nature would be satisfactory for their purposes. The states designated are

θ_1: Strong demand

θ_2: Typical demand

θ_3: Weak demand

Second, they have agreed on the exact definition of these states of nature—that is, what range of demand values constitutes "strong demand," what range "typical demand," and what range "weak demand." Third, they have developed and agreed upon estimates of profits that would occur if the combination θ_1 and A_1, θ_1 and A_2, . . ., θ_3 and A_4 should occur. The payoff table displaying the 3 indicated states of nature, the 4 available actions, and the 12 profit estimates is shown as Table 18.6 here.

Table 18.6 Payoff Table for the CCMC Inventory Decision Problem

State of Nature	Action			
	A_1: Attempt to Reduce Finished Goods Inventories Even Further	A_2: Attempt to Maintain Finished Goods Inventories at Present Level	A_3: Attempt to Increase Finished Goods Inventories	A_4: Follow No Particular Inventory Policy
θ_1: Strong demand	$900,000	$1,500,000	$2,500,000	$700,000
θ_2: Typical demand	700,000	800,000	500,000	700,000
θ_3: Weak demand	500,000	0	−500,000	300,000

Simplifying the Problem by Eliminating Inadmissible Actions

A peculiarity of the payoff table presented in Table 18.6 should be noted because it suggests that the problem can be simplified at the outset. We refer to the presence of an *inadmissible action*—that is, an action that is "dominated by" some other action. The term *dominated by* may be defined precisely as follows:

> Assume two courses of action, A and B. If for every state of nature appearing in the payoff table the monetary payoff of Action A is (1) at least as great as that of Action B and (2) greater than that of Action B for at least one state of nature, we say that Action B is *dominated by* Action A. In such a case, the decision problem can be simplified by eliminating Action B from the analysis on the grounds that, since Action B could not possibly be chosen as the Bayes action, it contributes nothing to the analysis and its retention makes determining the Bayes' action more difficult than necessary.

Examining the column entries of Table 18.6, we note that each monetary payoff in the column associated with Action A_4 is less than, or at best equal to, the corresponding monetary payoff for Action A_1. Therefore, we have no choice but to conclude that Action A_1 dominates Action A_4. This conclusion, as noted above, allows us to drop Action A_4 from the analysis altogether. The payoff table modified to allow for the ouster of Action A_4 is presented in Table 18.7. Can you find any remaining inadmissible actions in this modified version of the payoff table?

Table 18.7 Revised Payoff Table for the CCMC Inventory Decision Problem

State of Nature	A_1: Attempt to Reduce Finished Goods Inventories Even Further	A_2: Attempt to Maintain Finished Goods Inventories at Present Level	A_3: Attempt to Increase Finished Goods Inventories
θ_1: Strong demand	\$900,000	\$1,500,000	\$2,500,000
θ_2: Typical demand	700,000	800,000	500,000
θ_3: Weak demand	500,000	0	−500,000

Selecting the Bayes Action

Let us suppose that the CCMC management group expects demand for the company's products to be typical. Both weak demand and strong demand are thought to be relative longshots. Moreover, they consider the state of weak demand to be more likely than that of strong demand. Accordingly, they arrive at subjective prior probabilities of $P(\theta_1) = .1$, $P(\theta_2) = .6$, and $P(\theta_3) = .3$.

Determination of the Bayes action is accomplished by computing the expected monetary payoffs for Actions A_1, A_2, and A_3 and by choosing the action with the highest EMP. These calculations are presented in Table 18.8.

Table 18.8 reveals that Action A_1, an attempt to reduce finished goods inventories even further, would generate an estimated \$660,000 of inventory profit. Actions A_2 and A_3, more aggressive actions, would result in expected monetary payoffs of only \$630,000 and \$400,000, respectively. Therefore, Action A_1 would be the profit-maximizing action. Stated a little differently: Action A_1 is the prior-analysis Bayes action.

18.10 Expected Opportunity Loss

The Bayes decision rule can be applied using still a different payoff concept. Have you ever held a good poker hand—a sure winner ordinarily—but been forced to bet it cautiously because an opponent was betting as if he had six of a kind? If this has ever happened to you, you know that, even if you won the pot, you experienced a certain sense of loss. To be sure, you finished the hand with more chips than you started with. So you really didn't lose anything in the accounting sense of "loss." However, you may have failed to finish the hand with as many chips as you would have had with a program of more aggressive betting. The loss you suffered is what we call an *opportunity loss*.

Table 18.8 Calculation of Expected Monetary Payoffs for the CCMC Case

Action A_1: Attempt to Reduce Finished Goods Inventories Even Further

State of Nature	Prior Probability	Monetary Payoff	Weighted Monetary Payoff
θ_1: Strong demand	.1	$900,000	$ 90,000
θ_2: Typical demand	.6	700,000	420,000
θ_3: Weak demand	.3	500,000	150,000
Total			$660,000

$$\text{EMP}(A_1) = \$660,000$$

Action A_2: Attempt to Maintain Finished Goods Inventories at Present Level

State of Nature	Prior Probability	Monetary Payoff	Weighted Monetary Payoff
θ_1: Strong demand	.1	$1,500,000	$150,000
θ_2: Typical demand	.6	800,000	480,000
θ_3: Weak demand	.3	0	0
Total			$630,000

$$\text{EMP}(A_2) = \$630,000$$

Action A_3: Attempt to Increase Finished Goods Inventories

State of Nature	Prior Probability	Monetary Payoff	Weighted Monetary Payoff
θ_1: Strong demand	.1	$2,500,000	$250,000
θ_2: Typical demand	.6	500,000	300,000
θ_3: Weak demand	.3	$-500,000$	$-150,000$
Total			$400,000

$$\text{EMP}(A_3) = \$400,000$$

$$\text{Formula: EMP}(A_j) = \sum_{\text{all } i} [P(\theta_i)\text{MP}(O_{ij})]$$

An *opportunity loss* is a loss one sustains when one fails to take the best possible action available. The opportunity loss of an action is defined as the difference between the payoff of the best action the decision maker might have selected and the payoff realized from the action actually selected.

For example, Table 18.7 shows that if state of nature θ_1 (strong demand) occurs, the best action in the CCMC problem is A_3 with a monetary payoff of $2.5 million. The opportunity loss for that action is thus $2,500,000 − $2,500,000 = $0. The monetary payoff associated with Action A_2 is $1.5 million. Consequently, the opportunity loss incurred from taking this nonoptimal action (for θ_1) is $2,500,000 − $1,500,000 = $1,000,000. Finally, the opportunity loss associated with Action A_1, which generates a monetary payoff of $900,000, is $2,500,000 − $900,000 = $1,600,000. We would, of course, perform the same kind of arithmetic calculations for States θ_2 and θ_3. In fact, this has been done in Table 18.9, which is presented in two parts. Part I is simply a reproduction of the monetary payoffs presented in Table 18.7, except that the most profitable action for each state of nature has been identified with an asterisk. Part II shows the opportunity losses, the OL values, associated with Actions A_1, A_2, and A_3. Notice that the OL values are expressed in positive numbers.

When the OL values associated with a specified action are multiplied by the corresponding prior probabilities and the resulting products are summed, the results represent *expected opportunity losses (EOL values)*. The action promising the *minimum* expected opportunity loss is the Bayes action. With this in mind, let us now calculate the three expected opportunity losses for the CCMC problem. These calculations are shown in Table 18.10.

Again, Action A_1 is the best. *The Bayes action when expected opportunity losses are used will always be the same as the Bayes action when monetary payoffs are used.* Thus, the choice of monetary payoffs or opportunity-loss payoffs will be made on a basis other than the presumed impact on the Bayes action.

18.11 Decision-Tree Diagrams: Schematic Aids to Decision Making

The presentation of a decision situation need not be limited to a tabular format; it may also be accomplished graphically through use of a decision-tree diagram, the subject of this section.

Table 18.9 Monetary Payoffs and Opportunity Losses for the CCMC Inventory Decision Problem

	Part I: Monetary Payoffs		
State of Nature	A_1	A_2	A_3
θ_1: Strong demand	$900,000	$1,500,000	$2,500,000*
θ_2: Typical demand	700,000	800,000*	500,000
θ_3: Weak demand	500,000*	0	−500,000

	Part II: Opportunity Losses		
State of Nature	A_1	A_2	A_3
θ_1: Strong demand	$1,600,000	$1,000,000	$ 0
θ_2: Typical demand	100,000	0	300,000
θ_3: Weak demand	0	500,000	1,000,000

Note: An asterisk denotes the most profitable action for each state of nature.

Table 18.10 Expected Opportunity Loss Calculations for the CCMC Inventory Problem

Action A_1: Attempt to Reduce Finished Goods Inventories Even Further

State of Nature	Prior Probability	Opportunity Loss	Weighted Opportunity Loss
θ_1: Strong demand	.1	$1,600,000	$160,000
θ_2: Typical demand	.6	100,000	60,000
θ_3: Weak demand	.3	0	0
Total			$220,000

$$EOL(A_1) = \$220,000$$

Action A_2: Attempt to Maintain Finished Goods Inventories at Present Level

State of Nature	Prior Probability	Opportunity Loss	Weighted Opportunity Loss
θ_1: Strong demand	.1	$1,000,000	$100,000
θ_2: Typical demand	.6	0	0
θ_3: Weak demand	.3	500,000	150,000
Total			$250,000

$$EOL(A_2) = \$250,000$$

Action A_3: Attempt to Increase Finished Goods Inventories

State of Nature	Prior Probability	Opportunity Loss	Weighted Opportunity Loss
θ_1: Strong demand	.1	$ 0	$ 0
θ_2: Typical demand	.6	300,000	180,000
θ_3: Weak demand	.3	1,000,000	300,000
Total			$480,000

$$EOL(A_3) = \$480,000$$

$$\text{Formula: } EOL(A_j) = \sum_{\text{all } i} [P(\theta_i)OL(O_{ij})]$$

The construction of a decision-tree diagram is a three-stage process, the three stages being as follows:

1. Develop a diagram of all decision forks, chance forks, and branches.
2. Present the payoffs resulting from each joint decision between the decision maker and nature.
3. Introduce the appropriate probability assessments and, utilizing the process of *backward induction*, determine the Bayes action.

These stages will be described and demonstrated using the CCMC profits data as presented in Table 18.7.

1. Develop a diagram of all decision forks, chance forks, and branches.

   This stage of decision-tree construction is illustrated in Figure 18.2. The diagram may be viewed as a visual summary of a kind of game being played between the decision maker and another player (not necessarily an opponent) called *Nature*. The game is played with the decision maker and Nature taking turns; first, the decision maker chooses a course of action and then Nature chooses a state of nature. In a decision-tree diagram, a fork where the decision maker makes a choice is called a *decision fork* and indicated by a rectangle; a fork where Nature makes a choice is called a *chance fork* and indicated by a circle.

   In Figure 18.2 we see that the decision maker, situated at decision fork a, is obliged to choose among Actions A_1, A_2, and A_3. Let us suppose that he elects to follow the branch labeled A_1. Proceeding in a rightward direction along this branch, he soon encounters a chance fork. This means that it is now Nature's

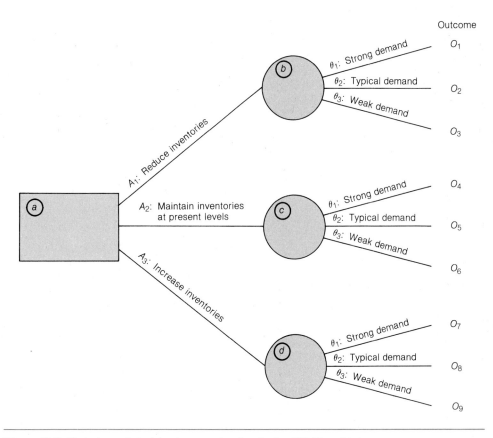

Figure 18.2. First stage of decision-tree construction for the CCMC problem.

turn to make a choice among states θ_1, θ_2, and θ_3. If Nature selects, say, the θ_1 branch, the joint decision process ends up at Outcome O_1 (which will soon be expressed as a monetary payoff as will all other outcomes listed in this figure).

2. Present the payoffs resulting from each joint decision between the decision maker and Nature.

   For a problem as conceptually simple as the CCMC case, this second stage of decision-tree construction consists merely of presenting the monetary payoffs of all outcomes at the ends of the appropriate terminal branches. These payoffs for the CCMC problem are shown in Figure 18.3.

3. Introduce the appropriate probability assessments and, utilizing the process of *backward induction*, determine the Bayes action.

   In Figure 18.4 additional information is included in the decision-tree diagram. This new information consists of (1) the subjective prior probabilities assigned to the three possible states of nature (shown in parentheses above the branches and

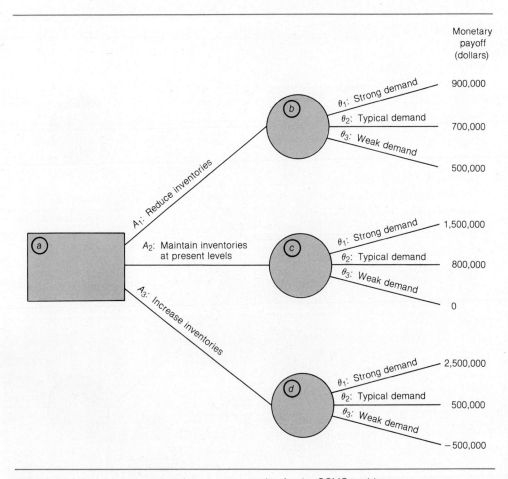

Figure 18.3. Second stage of decision-tree construction for the CCMC problem.

18 Decision Theory: Prior Analysis

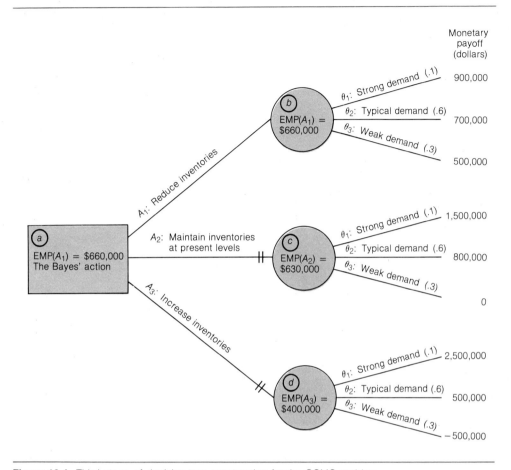

Monetary payoff (dollars)

θ_1: Strong demand (.1)   900,000

EMP(A_1) = $660,000

θ_2: Typical demand (.6)   700,000

θ_3: Weak demand (.3)

500,000

A_1: Reduce inventories

A_2: Maintain inventories at present levels

θ_1: Strong demand (.1)   1,500,000

EMP(A_2) = $630,000

θ_2: Typical demand (.6)   800,000

θ_3: Weak demand (.3)

0

EMP(A_1) = $660,000
The Bayes' action

A_3: Increase inventories

θ_1: Strong demand (.1)   2,500,000

EMP(A_3) = $400,000

θ_2: Typical demand (.6)   500,000

θ_3: Weak demand (.3)

−500,000

Figure 18.4. Third stage of decision-tree construction for the CCMC problem.

on the extreme right of chance forks *b*, *c*, and *d*), (2) the expected monetary payoff of each action (shown inside the circle identifying the corresponding chance fork), and (3) an indication of the best action for the decision maker to take (shown by the *absence of* two short parallel lines through the branch connecting the decision fork and the appropriate chance fork).

The process depicted in Figure 18.4 is called *backward induction*. To interpret this completed decision-tree diagram, we begin at the right-hand side and work backward toward the decision fork. Let us concentrate for the moment on the upper three paths, denoted θ_1, θ_2, and θ_3, emanating from chance fork *b*. To the right of this chance fork and in parentheses are found the corresponding probability assessments .1, .6, and .3. We can thus calculate the expected monetary payoff of being located at chance fork *b* as

EMP(A_1) = (.1)($900,000) + (.6)($700,000) + (.3)($500,000) = $660,000

This result is placed within the circle identifying chance fork *b* and is interpreted as representing the value of standing at that fork after choosing Action A_1 and before Nature selects one of the three state-of-nature paths. The analogous EMP values placed within the circles identifying chance forks c and d are $630,000 and $400,000, respectively. Therefore, if we now imagine the decision maker transported back to the decision fork, and attempting to choose among Actions A_1, A_2, and A_3, we see that the A_1 path yields the highest expected monetary payoff. Accordingly, we block off paths A_2 and A_3 to indicate that these represent nonoptimal courses of action and enter EMP(A_1) = $660,000 in the rectangle identifying the decision fork, a gesture calling attention to the fact that Action A_1 is the Bayes action.

Decision trees are capable of presenting the fundamentals of a decision problem and the necessary lines of reasoning in a manner more readily comprehensible than is sometimes true of payoff tables. That being so, they are marvelous aids to communication between the technical analyst and the manager or among managers trying to come to a decision. Of course, the CCMC problem is sufficiently simple—conceptually, at least—that this advantage here is probably marginal at best. When the problem is more complex, the superiority of decision-tree diagrams over payoff tables is more apparent.

> As a general rule, when decision makers face a problem that (1) calls for the selection of an action sequentially through two or more stages, (2) entails consideration of states of nature in a sequential manner, (3) calls for the use of nonidentical probability distributions in the same stage of the decision process, or (4) involves any combination of two or more of the above, they are well advised to employ a decision-tree diagram rather than a payoff table.

We will look at other uses of decision-tree diagrams in the following section and in Chapter 19.

*18.12 Two-Action Problems with Linear Payoff Functions

A sometimes useful application of decision theory methods arises in connection with problems characterized by (1) only two relevant actions and (2) a linear relationship between the basic random variable and the monetary payoffs. Consider, for example, the following illustrative case wherein management attempts to determine whether it would be better for their company to produce a new product itself or have an outside company carry out the actual manufacturing:

ILLUSTRATIVE CASE 18.2

The Babbett and Byrd Company (B&B) has enjoyed much success as a manufacturer and wholesaler of a line of magic kits for aspiring magicians, a group consisting largely of boys in their preteen and teen years. Management is now considering broadening the company's line into other products which combine, as magic does, the elements of skill and entertainment value. This case is concerned with a proposed juggling set. The juggling set under consideration would sell for a retail price of $20 of which the retailer would receive $5 and B&B $15. If B&B itself were to manufacture the juggling set, the manufacturing costs would be $7.50 per set for variable costs plus $225,000 for fixed costs.

On the other hand, if an outside firm is contracted to produce the proposed new product, all costs would be variable—$13.50 per set. The production superviser has urged the latter course of action until such time as the juggling set proves itself either a success or a failure.

Mr. Babbett, the CEO, preferred that any such decision be tied to an estimate of sales for the first year of the product's life. Consequently, the managers set about developing a judgmental estimate of first-year sales utilizing a probability distribution approach. After much animated discussion, the managers arrived at the distribution shown in Table 18.11.

Let us designate the two actions under consideration as (1) A_1: The company itself will manufacture the juggling set during the one-year trial period and (2) A_2: The company will contract with another manufacturing firm to produce the juggling set during the one-year trial period.

In spite of the obviously discrete nature of the subjective probability distribution shown in Table 18.11, first-year sales could in fact turn out to be any number between 0 and some unknown upper limit. That is, first-year sales is really a continuous variable, and, because of this, we may reason that there must be some break-even level of sales (not necessarily one of the values listed under "θ" in Table 18.11) above which it is more profitable for B&B itself to manufacture the new set and below which it is more profitable for B&B to have another company do the manufacturing. Let us denote the break-even level of θ as θ_e. In a problem of this kind we can determine the value of θ_e by (1) obtaining the two monetary payoff functions associated with the basic random variable, (2) setting these equal to each other, and (3) solving for θ_e. This procedure yields the break-even level of θ in the sense that it gives a value such that the monetary payoffs (in this case, profits) of the alternative actions, A_1 and A_2, are equal.

Table 18.11 Subjective Probability Distribution of First-Year Sales of the Proposed Juggling Set

State of Nature (Number of Sets Sold During the One-Year Trial Period) θ	Probability $P(\theta)$
10,000	.05
20,000	.10
30,000	.15
40,000	.20
50,000	.30
60,000	.15
70,000	.05

GRIN and BEAR IT   by FRED WAGNER

"There's a 30 percent chance it will be in the 50s and a 20 percent chance it will be in the 40s."

Determining the Break-Even Level of Sales

The monetary payoffs for this problem will be profit values derived as follows:

$$MP(A_1) = \text{Income from Action } A_1 - \text{Cost of employing Action } A_1$$
$$= \$15.00\theta - (\$225,000 + \$7.50\theta)$$
$$= \$15.00\theta - \$225,000 - \$7.50\theta$$
$$= -\$225,000 + \$7.50\theta \tag{18.12.1}$$

where $MP(A_1)$ is the monetary payoff, or profit, of Action A_1
θ is a value of the basic random variable, namely, the number of sets sold during the first year
$\$15$ is B&B's income per set sold
The expression $\$225,000 + \$7.50\theta$ is the cost function

Because the θ factor in Equation (18.12.1) may assume many numerical values, we see that this equation expresses a functional relationship, one we refer to as the *monetary payoff function of Action* A_1.

The *monetary payoff function of Action* A_2 is simpler:

$$MP(A_2) = \text{Income from Action } A_2 - \text{Cost of employing Action } A_2$$
$$= \$15.00 - \$13.50\theta$$
$$= \$1.50\theta \qquad (18.12.2)$$

where $MP(A_2)$ is the monetary payoff, or profit, of Action A_2
$13.50 is B&B's cost per set if another company does the manufacturing

To determine θ_e, we equate $MP(A_1)$ and $MP(A_2)$:

$$\$1.50\theta_e = -\$225,000 + \$7.50\theta_e$$
$$= 37,500 \text{ sets}$$

That is, if 37,500 sets were sold during the one-year trial period, the magic company would be indifferent between manufacturing the set itself and paying another company $13.50 per set to do the manufacturing.

Upon further examination, the break-even analysis also reveals that, for first-year sales of less than 37,500, B&B would be better off engaging another company to produce the juggling set; for first-year sales in excess of 37,500, B&B would be better off doing the manufacturing itself. These facts are illustrated in Figure 18.5. The sales and monetary payoff schedules serving as the basis for this chart are presented in Table 18.12.

Table 18.12 Monetary Payoff Schedules for Actions A_1 and A_2

(Illustrative Case 18.2)

(1) State of Nature (Number of Sets Sold During the One-Year Trial Period) θ	(2) Monetary Payoff of Action A_1 $(-\$225,000 + \$7.50\theta)$ $MP(A_1)$	(3) Monetary Payoff of Action A_2 $(\$1.50\theta)$ $MP(A_2)$
$ 0	-\$225,000	$ 0
10,000	-150,000	15,000
20,000	- 75,000	30,000
30,000	0	45,000
37,500[a]	56,250	56,250
40,000	75,000	60,000
50,000	150,000	75,000
60,000	225,000	90,000
70,000	300,000	105,000

[a]Break-even level of sales.

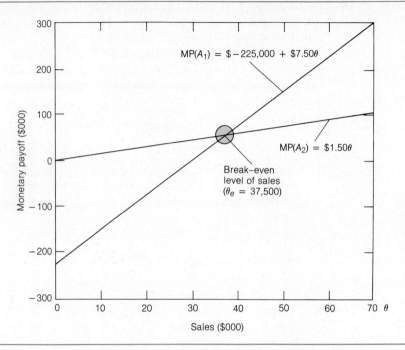

Figure 18.5. Monetary payoff functions for actions A_1 and A_2 of Illustrative Case 18.2.

Determining the Bayes Action

Although the above break-even analysis is of interest in its own right, it tells us neither the Bayes action nor its expected monetary payoff. For this information, we may follow the standard procedure for computing EMP values. For a two-action problem, this means (1) multiplying each monetary payoff associated with Action A_1 by the corresponding $P(\theta)$ value and summing the resulting products, (2) following the same procedure for Action A_2, and (3) choosing the action producing the larger sum, that is, the larger EMP value. Table 18.13, which demonstrates this procedure, reveals that EMP(A_1) = \$93,750 and EMP(A_2) = \$63,750. Thus, Action A_1 is the Bayes action. This conclusion indicates that the Babbett and Byrd Company itself should manufacture the juggling set.

An alternative procedure can be used to secure the same information and, in the process, permit the decision maker to work with an *expected sales* value. Expected sales is a measure of utmost interest to the B&B managers (as it so often is in business decision-making situations) and a major reason behind their developing the subjective probability distribution shown in Table 18.11. The expected sales value is computed by

$$E(\theta) = \Sigma\ [P(\theta)\theta] \qquad\qquad (18.12.3)$$

Table 18.13 Calculation of Expected Monetary Payoffs for the Babbett and Byrd Problem

Action A_1: The Company Itself Will Manufacture the Set During the One-Year Trial Period

(1) State of Nature (First-Year Unit Sales) θ	(2) Probability $P(\theta)$	(3) Monetary Payoff of Action A_1 $MP(A_1) = -\$225,000 + \$7.50\theta$	(4) Weighted Monetary Payoff of Action A_1
10,000	.05	−$150,000	−$ 7,500
20,000	.10	− 75,000	− 7,500
30,000	.15	0	0
40,000	.20	75,000	15,000
50,000	.30	150,000	45,000
60,000	.15	225,000	33,750
70,000	.05	300,000	15,000
Total			$93,750

$$EMP(A_1) = \$93,750$$

Action A_2: The Company Will Contract with Another Manufacturing Concern to Produce the Set During the One-Year Trial Period

(1) State of Nature (First-Year Unit Sales) θ	(2) Probability $P(\theta)$	(3) Monetary Payoff of Action A_2 $MP(A_2) = \$1.50\theta$	(4) Weighted Monetary Payoff of Action A_2
10,000	.05	15,000	$ 750
20,000	.10	30,000	3,000
30,000	.15	45,000	6,750
40,000	.20	60,000	12,000
50,000	.30	75,000	22,500
60,000	.15	90,000	13,500
70,000	.05	105,000	5,250
Total			$63,750

$$EMP(A_2) = \$63,750$$

For our illustrative problem, $E(\theta)$ is found to be

$$E(\theta) = (.05)(10,000) + (.10)(20,000) + (.15)(30,000) + (.20)(40,000)$$
$$+(.30)(50,000) + (.15)(60,000) + (.05)(70,000) = 42,500 \text{ units}$$

To determine the expected monetary payoff of Action A_1, we may now use the following variation on Equation (18.12.1):

$$EMP(A_1) = -\$225,000 + (\$7.50)E(\theta) \qquad (18.12.4)$$

Therefore, $EMP(A_1) = -\$225,000 + (\$7.50)(42,500) = \$93,750.$

To determine the expected monetary payoff of Action A_2, a similar procedure may be used:

$$EMP(A_2) = (\$1.50)E(\theta) \qquad (18.12.5)$$

Thus,

$$EMP(A_1) = (\$1.50)(42,500) = \$63,750$$

Both of the EMP values match the results obtained in Table 18.13.

Some Advantages and Disadvantages of Two-Action Linear Payoff Analysis

One important advantage of this line of analysis is that, if the monetary payoff functions are linear and the true value of the basic random variable (first-year unit sales in our example) is not known with certainty, the analyst can still establish a decision procedure based solely on the *expected value* of the basic random variable. For example, in the problem under analysis, the decision rule would be

Decision Rule

1. Choose Action A_1 if $E(\theta) > 37,500$ units.
2. Choose Action A_2 if $E(\theta) < 37,500$ units.
3. Choose either action if $E(\theta) = 37,500$ units.

A limitation is that this kind of analysis applies strictly to situations where the possible values of the basic random variable are related in a *linear manner* to the monetary payoffs of each of the contemplated actions.

Figure 18.6 shows this same problem analyzed by means of a decision-tree diagram. This decision-tree diagram can be interpreted in the same way as the previous one.

18.13 Summary

This chapter has presented examples of prior analysis using (1) utility payoffs, (2) monetary payoffs, and (3) opportunity-loss payoffs. It has also demonstrated how both payoff tables and decision-tree diagrams can be employed in the search for the Bayes action.

Regardless of type of payoff and organizing format, prior analysis involves (1) obtaining a payoff estimate for each state-and-action combination, (2) preparing a prior probability assessment for each state of nature under consideration, (3) obtaining the expected payoff

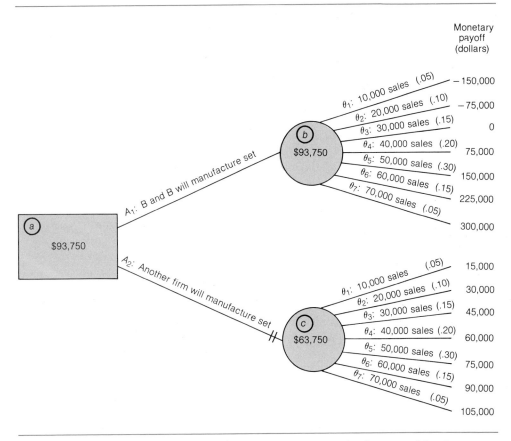

Monetary
payoff
(dollars)

θ_1: 10,000 sales (.05) — -150,000
θ_2: 20,000 sales (.10) — -75,000
θ_3: 30,000 sales (.15) — 0
θ_4: 40,000 sales (.20) — 75,000
θ_5: 50,000 sales (.30) — 150,000
θ_6: 60,000 sales (.15) — 225,000
θ_7: 70,000 sales (.05) — 300,000

A_1: B and B will manufacture set

b
$93,750

a
$93,750

A_2: Another firm will manufacture set

c
$63,750

θ_1: 10,000 sales (.05) — 15,000
θ_2: 20,000 sales (.10) — 30,000
θ_3: 30,000 sales (.15) — 45,000
θ_4: 40,000 sales (.20) — 60,000
θ_5: 50,000 sales (.30) — 75,000
θ_6: 60,000 sales (.15) — 90,000
θ_7: 70,000 sales (.05) — 105,000

Figure 18.6. Decision-tree diagram for the Babbett and Byrd two-action linear payoff function problem.

associated with each available action, and (4) selecting the action with the best expected payoff (highest or lowest, depending on the kind of payoffs used).

A point which should receive some emphasis is that it is expected utility one wishes to maximize. Use of monetary payoffs is based implicitly on the assumption of a linear relationship between utility and monetary payoff.

There are really two levels on which Bayesian decision procedures can be appreciated: (1) When all the input data are carefully estimated and deemed trustworthy, the Bayesian decision rule is a reasonable way of choosing the "best" action from among, perhaps, several available actions. (2) The Bayesian decision rule serves as a useful organizing device for systematically thinking through a decision problem from beginning to end. That being so, the procedure can help guide the decision maker through the process of arriving at a conclusion even if the input data are far from perfect and even if he or she elects not to act upon the conclusion obtained.

Terms Introduced in This Chapter

actions (p. 875)
backward induction (p. 899)
Bayes action (p. 890)
Bayes decision rule (p. 890)
Bayesian decision theory (p. 873)
branch (p. 897)
chance fork (p. 897)
classical decision theory (p. 873)
decision fork (p. 897)
decision maker (p. 875)
decision-tree diagram (p. 896)
expected monetary payoff (p. 878)

expected opportunity loss (p. 895)
expected utility (p. 878)
inadmissible action (p. 892)
insufficient-reason action-selection
  criterion (p. 889)
linear payoff function (p. 902)
maximax action-selection criterion
  (p. 889)
maximin action-selection criterion
  (p. 888)
maximum-likelihood action-selection
  criterion (p. 889)

opportunity loss (p. 894)
outcome (p. 876)
  least-preferred outcome (p. 880)
  most-preferred outcome (p. 880)
payoff table (p. 876)
reference lottery (p. 879)
state-and-action combination (p. 876)
states of nature (p. 875)
statistical decision theory (p. 873)
utility (p. 878)
utility curve (p. 883)

Formulas Introduced in This Chapter

Expected utility for Action A_j:

$$EU(A_j) = \sum_{\text{all } i} [P(\theta_i)U(O_{ij})]$$

Expected monetary payoff for Action A_j:

$$EMP(A_j) = \sum_{\text{all } i} [P(\theta_i)MP(O_{ij})]$$

Expected opportunity loss for Action A_j:

$$EOL(A_j) = \sum_{\text{all } i} [P(\theta_i)OL(O_{ij})]$$

Expected value for the basic random variable in a two-action linear payoff problem:

$$E(\theta) = \sum [P(\theta)\theta]$$

Questions and Problems

18.1   Explain the meaning of each of the following:
  a. Statistical decision theory   **e.** Most-preferred outcome
  b. Decision maker   **f.** Payoff table
  c. Prior probability distribution **g.** Bayes action
  d. Outcome

18.2   Explain the meaning of each of the following:
  a. Utility curve   **e.** O_{32}
  b. Reference lottery   **f.** State-and-action combination
  c. State of nature   **g.** Insufficient-reason criterion
  d. Action

18.3 Explain the meaning of each of the following:
    a. Opportunity loss   e. Maximin criterion
    b. Backward induction   f. Monetary payoff function
    c. Decision-tree diagram   g. Bayes decision rule
    d. Maximum-likelihood criterion

18.4 Distinguish between:
    a. Maximum-likelihood and insufficient-reason action-selection criteria
    b. Bayes decision rule and maximum-likelihood action-selection criteria
    c. Decision fork and chance fork
    d. Admissible action and inadmissible action
    e. $MP(A_1)$ and $MP(A_2)$

18.5 Distinguish between:
    a. θ and θ_e
    b. Classical decision theory and Bayesian decision theory
    c. Action and state of nature
    d. $EU(A_j)$ and $EMP(A_j)$
    e. O_{24} and O_{42}

18.6 Distinguish between:
    a. O_{43} and $MP(O_{43})$
    b. Utility and monetary payoff
    c. Outcome and state-and-action combination
    d. Insufficient-reason criterion and inadmissible action
    e. Monetary payoff and opportunity loss

18.7 Indicate which of the following statements you agree with and which you disagree with, and defend your opinions:
    a. When performing a Bayesian decision analysis, the prior probabilities must inevitably be subjective probabilities.
    b. If Outcome O_1 is preferred to Outcome O_2, then $U(O_1) > U(O_2)$.
    c. If the decision maker is indifferent about having Outcome O_1 occur with certainty, on the one hand, and engaging in a lottery where the probability of occurrence of Outcome O_2 is p and the probability of occurrence of Outcome O_3 is $1 - p$, on the other hand, then $U(O_1) > pU(O_2) + (1 - p)U(O_3)$.
    d. If O_{12} is the least-preferred outcome and O_{21} is the most-preferred outcome, then the utility values assigned the two outcomes must be such that $U(O_{12}) > U(O_{21})$.

18.8 Indicate which of the following statements you agree with and which you disagree with, and defend your opinions:
    a. Since the relationship between utilities and monetary payoffs is always linear, the latter may be freely substituted for the former.
    b. If there are four possible outcomes, O_{11}, O_{12}, O_{21}, and O_{22}, of which O_{12} is the least-preferred outcome and O_{21} is the most-preferred outcome, the utilities assigned Outcome O_{11} and O_{22} must be greater than the utility assigned O_{12} and less than the utility assigned O_{21}.
    c. Whether the payoffs associated with specific outcomes are expressed in terms of monetary payoffs, utilities, market share, or something else, we accept as the ultimate selection criterion that the action chosen will be the one that maximizes expected utility.
    d. The utility curve of a specific large company is more likely to be linear than nonlinear.

18.9 Indicate which of the following statements you agree with and which you disagree with, and defend your opinions:
    a. The world might be devoid of businesses if everyone acted according to the maximin action-selection criterion.

Table 18.14 Payoff Table Based on Specified Outcomes

State of Nature	Action	
	A_1	A_2
θ_1	O_{11}	O_{12}
θ_2	O_{21}	O_{22}

    b. The maximum-likelihood action-selection criterion probably makes more sense in a situation with, say, 2 possible states of nature than it does in a situation with, say, 25 possible states of nature.
    c. The Bayes action will be the same regardless of whether it is selected using expected monetary payoffs or expected utilities.

18.10 Indicate which of the following statements you agree with and which you disagree with, and defend your opinions:
    a. For a given decision situation, the insufficient-reason criterion, on the one hand, and the Bayes decision rule, on the other, will always point toward different actions.
    b. For a given decision situation, the maximum-likelihood criterion, on the one hand, and the Bayes decision rule, on the other, will always point toward different actions.
    c. For a given decision situation, the maximin criterion, on the one hand, and the maximax criterion, on the other, will always point toward different actions.

18.11 Consider the payoff table shown in Table 18.14.
    Suppose that O_{11} is the least-preferred outcome, O_{22} is the most-preferred outcome, and O_{21} and O_{12} are both halfway between O_{11} and O_{22} in desirability.
    a. Assign appropriate values to $U(O_{11})$, $U(O_{12})$, $U(O_{21})$, and $U(O_{22})$.
    b. Determine which action promises the higher expected utility payoff assuming $P(\theta_1) = P(\theta_2)$.
    c. Would the action you selected in part b have been any different if $P(\theta_1) > P(\theta_2)$? Tell why or why not.

18.12 Given the utility schedules shown in Table 18.15 for individuals A, B, and C, determine who is the risk avoider, who is the risk seeker, and who is neutral to risk.

Table 18.15 Monetary and Utility Payoff Data for Three Individuals

A		B		C	
Monetary Payoff	Utility	Monetary Payoff	Utility	Monetary Payoff	Utility
$5000	1.00	$5000	1.00	$5000	1.00
3000	.25	3000	.75	3000	.80
0	.15	0	.70	0	.50
−3000	.10	−3000	.65	−3000	.20
−5000	.00	−5000	.00	−5000	.00

Table 18.16 Data for Problems 18.13 and 18.14

| | Action | | |
State of Nature	A_1	A_2	A_3
θ_1	.40	.60	.50
θ_2	.60	.00	.70
θ_3	.55	.10	.80
θ_4	.45	.30	1.00

18.13 Consider the (utility) payoff table shown in Table 18.16. Suppose you have arrived at the following prior probabilities: $P(\theta_1) = .1$, $P(\theta_2) = .2$, $P(\theta_3) = .4$, and $P(\theta_4) = .3$.
   a. Examine Table 18.16 for inadmissible actions. If you find any inadmissible actions, eliminate it (them).
   b. Compute the expected utility of each remaining action.
   c. Using the results of part *b*, identify the Bayes action.

18.14 Refer to Table 18.16. What would be the best action according to the
   a. Maximin criterion?
   b. Maximax criterion?
   c. Maximum-likelihood criterion?
   d. Insufficient-reason criterion?

18.15 Examine the two monetary payoff tables shown in Table 18.17.
   a. Determine the Bayes action for Payoff Table I.
   b. Determine the Bayes action for Payoff Table II.

Table 18.17 Data for Problem 18.15

Payoff Table I

State of Nature θ_i	Prior Probability $P(\theta_i)$	Actions	
		A_1	A_2
θ_1	.4	\$ 5	\$6
θ_2	.6	10	8

Payoff Table II

State of Nature θ_i	Prior Probability $P(\theta_i)$	Actions	
		A_1	A_2
θ_1	.2	\$ 7	\$10
θ_2	.8	11	2

18.16 The Motherlode Mining Company is contemplating whether to (1) mine a piece of property thought to contain silver ore or (2) sell the property to another mining company and retain a 5% royalty. Table 18.18 shows profit estimates (in millions of dollars) associated with the two possible actions and five possible states of nature.

A team of geoscientists employed by Motherlode hold a conference and arrive at the prior probability assessments shown at the bottom of Table 18.18.
a. Identify and eliminate any inadmissible actions.
b. Compute the expected monetary payoff associated with Action A_1. With Action A_2.
c. Identify the Bayes action.

18.17 Refer to Table 18.18.
a. Convert all of the monetary payoff estimates into opportunity-loss estimates.
b. Compute the expected opportunity losses associated with the two possible actions and identify the Bayes action.

18.18 The president of a manufacturing company assigned the following prior probabilities to three possible sales conditions for the upcoming year: (1) strong sales ($14 million)—.55;

Table 18.18 Monetary Payoff (Profit) Estimates and Prior Probability Assessments for the Motherlode Mining Company

(Profits Are in Millions of Dollars)

State of Nature (Richness of Ore) θ_i	Action	
	A_1: Mine Property	A_2: Sell Property and Retain 5% Royalty
θ_1: Ore is *very rich* in silver	60	12
θ_2: Ore is *rich* in silver	20	9
θ_3: Ore is of *average* richness	5	5
θ_4: Ore has *very little* silver content	− 5	4
θ_5: Ore contains *no silver* whatever	−20	3

$$P(\theta_1) = .1$$
$$P(\theta_2) = .2$$
$$P(\theta_3) = .3$$
$$P(\theta_4) = .2$$
$$P(\theta_5) = .2$$

Table 18.19 Monetary Payoff (Profit) Estimates for a Manufacturing Company

(Profits Are in Millions of Dollars)

State of Nature (Strength of Sales Next Year) θ_i	Action		
	A_1: Build and Plan for Strong Sales	A_2: Build and Plan for Average Sales	A_3: Build and Plan for Weak Sales
θ_1: Strong sales	$980	$600	$300
θ_2: Average sales	500	700	450
θ_3: Weak sales	210	350	290

(2) average sales ($10 million)—.35; and (3) weak sales ($7 million)—.1. The monetary payoff (profit) table he arrived at is shown in Table 18.19.
a. Identify and eliminate any inadmissible actions.
b. Compute the expected monetary payoff associated with all remaining actions.
c. Identify the Bayes action.

 18.19 Refer to Table 18.19.
a. Convert all of the profit estimates to opportunity-loss estimates.
b. Compute the expected opportunity losses associated with the possible actions and identify the Bayes action.

18.20 Using the process of backward induction, determine the Bayes action and the corresponding expected monetary payoff for the decision situation depicted in Figure 18.7.

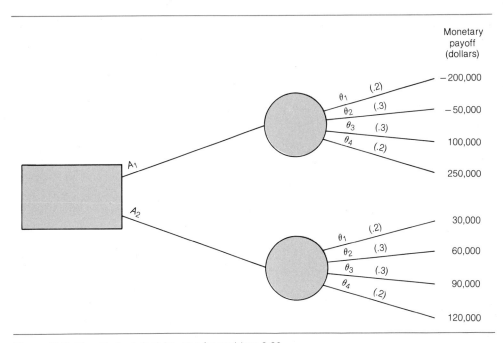

Figure 18.7. Hypothetical decision tree for problem 8.20.

Table 18.20 Partial Payoff Table for a New Kind of Earth-Moving
Machinery ($000)

$P(\theta_i)$	Demand θ_i	Supply A_j					
		0	1	2	3	4	5
.05	0	0					
.15	1						
.30	2						
.25	3						
.15	4						
.10	5						100

Special-Challenge Problem

18.21 Part of a payoff table for a new type of earth-moving machinery is shown in Table 18.20. The product costs $30,000 per unit to manufacture and sells for $50,000 per unit. A contracting company has offered to buy any manufactured but unsold machines for $15,000.
  a. Determine the monetary payoff (profit) estimate for each indicated supply-and-demand combination.
  b. Compute the expected monetary payoff associated with Actions A_1, A_2, A_3, A_4, A_5, and A_6.
  c. Identify the Bayes action.

18.22 A firm plans to begin manufacturing a new product soon. Management is presently considering which of two alternative manufacturing processes, A_1 or A_2, to use. If Process A_1 is employed, fixed costs will be an estimated $100,000 and variable costs an estimated $2 per unit produced. If Process A_2 is employed, fixed costs will be an estimated $150,000 and variable costs an estimated $1 per unit produced. The selling price of the new product will be $5 per unit. The subjective prior probability assessments for first-year sales are shown in Table 18.21.
  a. Determine the linear monetary payoff function for Process A_1.
  b. Determine the linear monetary payoff function for Process A_2.
  c. Determine the break-even level of sales, θ_e.
  d. Using the break-even level of first-year sales determined in part c and the monetary payoff function you determined in parts a and b, indicate which manufacturing process, A_1 or A_2, would be preferred if actual sales turned out to be less than θ_e. If actual sales turned out to be greater than θ_e.

Table 18.21 Prior
Probability Distribution
for a New Product

Sales θ	Probability $P(\theta)$
20,000	.20
30,000	.25
40,000	.30
50,000	.15
60,000	.10

e. Using the information contained within Table 18.21, compute *expected* first-year sales, $E(\theta)$.

f. Using the expected sales value you obtained in part *e* and the monetary payoff functions you determined in parts *a* and *b*, determine the expected monetary payoff of Action A_1, that is, EMP(A_1), and the expected monetary payoff of Action A_2, EMP(A_2), and identify the Bayes action.

 18.23 Refer to problem 18.22. In what important way or ways, if any, would the analytical procedures be altered if variable costs for Action A_1 and A_2 were $2 and $1, respectively, and if fixed costs were $100,000 for both processes?

 18.24 A publishing company is contemplating publication of a new book by a former high political official. The hardcover version of the book would sell for $20 per copy. The cost of publication would be $6 million for fixed costs and $7 per copy for variable costs. The publisher's subjective prior probability distribution for sales during the first year after the book's release is shown in Table 18.22.

a. Determine the linear monetary payoff function assuming the book is published (Action A_1).

b. Determine the linear monetary payoff function assuming the book is not published (Action A_2).

c. Determine the break-even level of sales, θ_e.

d. Using the break-even level of sales, θ_e, you determined in part *c* and the monetary payoff functions you determined in parts *a* and *b*, indicate which action, A_1 or A_2, would be preferred if actual sales turned out to be less than θ_e. If actual sales turned out to be greater than θ_e.

e. Using information contained within Table 18.22, compute expected sales, $E(\theta)$, for the first year.

f. Using the expected sales value obtained in part *e* and the monetary payoff functions determined in parts *a* and *b*, indicate which action, A_1 or A_2, is the Bayes action.

 18.25 Refer to problem 18.24. In what important way or ways, if any, would the analytical procedures be changed if the publishing company had paid $10,000 for the privilege of being the company with first claim to the book should it decide to publish the book before the end of a six-month period? (Assume that, if the company does decide to publish the book, the $10,000 would not have to be paid and would thus not increase the cost of publishing but that, if the company decided not to publish, the $10,000 would be forfeited.)

Table 18.22 Prior Probability Distribution for a Proposed Book

Book Sales θ	Probability $P(\theta)$
200,000	.1
400,000	.3
600,000	.3
800,000	.2
1,000,000	1

18.26 Reread Illustrative Case 18.1 and examine Table 18.6 carefully. Suppose that the monetary payoff associated with state-and-action combination θ_3-A_4 were \$550,000 rather than the \$300,000 shown.
    a. Indicate in what important respect the working of this inventory decision problem would be changed as a result of this higher estimated payoff.
    b. Set up the decision-tree diagram that is appropriate under the change in the θ_3-A_4 payoff figure. (Assume that all other payoff estimates and all prior probability assessments remain unchanged.) Then, using the process of backward induction, compute the expected monetary payoff of each possible action and determine the Bayes action.

18.27 Monotony Movies, Inc., has been offered the opportunity of buying a screenplay. It is estimated that if the screenplay is made into a successful movie, the film company can expect to make \$3 million profit. If, however, the screenplay is made into an unsuccessful movie, the company will lose an estimated \$1 million. After reading the screenplay, the studio chief assesses the probability of success as .2 and the probability of failure as .8.
    a. Keeping in mind that refraining from making the movie is as much an action as making the movie, construct the payoff table for this problem.
    b. Compute the expected monetary payoff of each possible action and identify the Bayes action.

TAKE CHARGE

18.28 Think of a decision situation in your own life, past or present, that lends itself to Bayesian prior analysis. Perform the analysis, determine the Bayes action, and write a brief report (200 to 300 words) telling (1) the nature of the problem, (2) the possible actions, (3) the possible states of nature, (4) the manner in which you measured the payoffs, (5) the manner in which you determined the prior probabilities, and (6) your conclusion. (*Note:* You will be asked to continue this exercise at the end of Chapter 19; try to pick a problem in which objective revision of prior probabilities is possible.)

TAKE CHARGE

18.29 Using the reference-lottery method as described in this chapter, construct your own utility curve for monetary payoffs ranging between +\$10,000 and −\$10,000. Would you characterize yourself as (1) a risk taker, (2) neutral to risk, or (3) a risk avoider? Does this conclusion surprise you or is it what you had expected? Elaborate.

Cases for Analysis and Discussion

18.1 SANDOZ PHARMACEUTICALS*

Research and development (R&D) is an aspect of business planning in which uncertainties abound. Nevertheless, pharmaceutical firms like Sandoz are very dependent on R&D. Ongoing questions R&D managers must ask are (1) What are the most important projects or

---

*This actual case is a greatly condensed version of the account by H. U. Balthasar, R. A. A. Boshi, and M. M. Menke, "Calling the Shots in R and D," *Harvard Business Review*, v. 56 (May–June 1978), 151–160.

potential projects to direct our R&D efforts toward? (2) How should we schedule our R&D projects to ensure a reasonably steady stream of successful new products?

At Sandoz, twice a year a small group of experts is asked to evaluate the probability of technical success for each of the company's R&D projects.

In the early years of the program, subjective probability assessments were elicited "indirectly" by means of a color wheel. The wheel is circular with two colored sections—one orange and the other blue. An important feature of the color wheel is that the relative size of each color can be adjusted. Thus, it could be all orange, all blue, half orange and half blue, 25% orange and 75% blue, and so forth. In the center of the wheel is a pointer that could be spun and that would necessarily stop on one or the other color. An example of the way the wheel was used is as follows:

Let us say that the wheel is 50% orange and 50% blue. An expert is asked which is more likely:

a. The spinner will stop on the orange section.
b. Project X will succeed.

If the answer is (a), the wheel is adjusted to decrease the relative size of the orange section, and the same question is asked again. This procedure is repeated until the expert says that the probabilities of (a) and (b) are equal. The proportion of the wheel that is orange at that point is the expert's subjective probability.

If the answer is (b), the wheel is adjusted to increase the relative size of the orange section, and the same question is asked again. This procedure is repeated until the expert says that the two probabilities are equal. Again, the relative size of the orange section is the expert's subjective probability.

For the first several years of the program, all probabilities were elicited in this "indirect" manner. In more recent years, as the group has become increasingly familiar with the process of assigning probabilities, the individual experts have been asked to give their probability assessments "directly" and then the entire group meets and arrives at a consensus.

Sandoz has been pleased with the results of these efforts to evaluate the probabilities of success of specific R&D projects. A comparison of the expert-consensus probabilities with actual observed relative frequencies of successful projects has revealed that the two are quite close.

1. For a particular expert, what advantages, if any, might there be in using the color wheel over directly asking: "What is the probability of success of project X?"

2. What, if any, advantage might there be in requiring the experts to arrive at a subjective consensus probability?

18.2 THE "DELTA COFFEE COMPANY": PART II

Review case 1.1 of Chapter 1. Using your own subjective probabilities, monetary payoffs, and two possible states of nature, perform the following work:

1. Construct a payoff table.

2. Modify the payoff table, if necessary, by eliminating any inadmissible actions.

3. Determine the expected monetary payoff associated with each of the available actions and identify the Bayes action.

4. Construct a decision-tree diagram for this problem.

This case is continued at the end of Chapter 19.

18.3 THE "HUGHES DOWN COMPANY": PART I

The "Hughes Down Company" designs and markets garments insulated with goose down for use under conditions of extremely cold weather. The company has over the years used expeditions to the Arctic or to very high-altitude zones as testing grounds for new products and has, as a result, acquired the moniker of "Expedition Outfitter." In addition to manufacturing its own products, the company purchases nondown products from other manufacturers to round out its product line of cold-weather outerwear. For years the company marketed its products nationwide by mail order only; then, in 1982, a plan was developed for building a national system of retail outlets for the duo purpose of better serving existing customers and expanding the potential market.

The present case involves an analysis of sales and inventory data for the Mega City store between June 1, 1983, and June 1, 1984. More specifically, the case is concerned with the store manager's attempts to determine (1) the frequency of inventory replenishment needed to achieve the lowest possible processing cost and (2) the optimum level of inventories to be carried in order to satisfy at least 90% of customer demand. What follows is a description of the analysis performed by the manager of the Mega City store. It is expressed in his own words.

Sales in the one-year period amounted to $720,000, an amount representing an estimated 80% of potential demand. The inventory base on hand to support this sales level averaged $337,000 and the inventory turnover was 2.1. The net profit for the period was $69,200.

A simple analysis of cost data revealed that, regardless of the frequency of shipment, the annual handling charges will be approximately the same for a given level of sales. This results from the fact that all charges are calculated on a per-piece basis. The only cost factors that vary with frequency of shipment are the base handling charge of $10 per shipment and the freight allowance for heavier weights. These two factors are not large enough to hold out any real promise of helping us to resolve the problem under discussion.

Finding that a change in the frequency of inventory replenishment would not lead to an annual savings of any consequence, I turned my attention to the other part of the problem, the part concerned with determining the optimum inventory level. There are two parts to this problem: (1) determining the rate at which inventory is turned over annually at a given sales level and (2) determining the dollar sales level at which 90% of potential demand is met. Using $72,000 as the 80% of actual-demand level, I prepared the following table [Table 18.23].

From the past financial statements, the accounting department found that for every turn in a given inventory, 1.5% of gross sales is added to net profit. Using the subjective probabilities of .32 for weak sales, .53 for average sales, and .15 for strong sales, I set up

Table 18.23 Estimated Average Annual Inventories for the "Hughes Down Company" Case

(Thousands of Dollars)

Sales Level	Percent of Estimated Actual Demand	Inventory Turnover			
		2.0	4.0	6.0	8.0
$720,000	80%	360	180	120	90
810,000	90	405	203	135	101
900,000	100	450	225	150	113

Table 18.24 Payoff Table for the "Hughes Down Company" Case

(Net Profit per Inventory Turnover in Dollars)

State of Nature	Action			
	A_1: Turn Inventory Over 2 Times	A_2: Turn Inventory Over 4 Times	A_3: Turn Inventory Over 6 Times	A_4: Turn Inventory Over 8 Times
θ_1: Weak sales ($720,000)	68,400	90,000	111,600	133,200
θ_2: Average sales ($810,000)	76,950	101,250	125,550	149,850
θ_3: Strong sales ($900,000)	85,500	112,500	139,500	166,500

a payoff table. The probabilities used assume no advertising other than that contained within our catalog. The next two tables [Tables 18.24 and 18.25] show in considerable detail the prior analysis I performed.

1. Explain in your own words how the net-profit-per-inventory-turnover values in Table 18.24 were determined.

2. Write an overall evaluation of the prior analysis performed by this store manager.

This case is continued at the end of Chapter 19.

Table 18.25 Calculation of Expected Monetary Payoffs for the "Hughes Down Company" Case

Action A_1: Turn Inventories Over Two Times Annually

State of Nature	Prior Probability	Monetary Payoff	Weighted Monetary Payoff
θ_1: Weak sales ($720,000)	.32	$68,400	$21,888
θ_2: Average sales ($810,000)	.53	76,950	40,784
θ_3: Strong sales ($900,000)	.15	85,500	12,825
Total			$75,497

(continued)

Table 18.25 (*Continued*)

Action A₂: Turn Inventories Over Four Times Annually

State of Nature	Prior Probability	Monetary Payoff	Weighted Monetary Payoff
θ_1: Weak sales ($720,000)	.32	$ 90,000	$28,800
θ_2: Average sales ($810,000)	.53	101,250	53,663
θ_3: Strong sales ($900,000)	.15	112,500	16,875
Total			$99,338

Action A₃: Turn Inventories Over Six Times Annually

State of Nature	Prior Probability	Monetary Payoff	Weighted Monetary Payoff
θ_1: Weak sales ($720,000)	.32	$111,600	$ 35,712
θ_2: Average sales ($810,000)	.53	125,550	66,542
θ_3: Strong sales ($900,000)	.15	139,500	20,925
Total			$123,179

Action A₄: Turn Inventories Over Eight Times Annually

State of Nature	Prior Probability	Monetary Payoff	Weighted Monetary Payoff
θ_1: Weak sales ($720,000)	.32	$133,200	$ 42,624
θ_2: Average sales ($810,000)	.53	149,850	79,421
θ_3: Strong sales ($900,000)	.15	166,500	24,975
Total			$147,020

19 Decision Theory: Posterior and Preposterior Analysis

Take time to deliberate, but when the time for action arrives, stop thinking and go in.
—Andrew Jackson

The open mind never acts: when we have done our utmost to arrive at a reasonable conclusion, we still, when we can reason and investigate no more, must close our minds for the moment with a snap, and act dogmatically on our conclusions. The man who waits to make an entirely reasonable will dies intestate.—George Bernard Shaw

What You Should Learn from This Chapter

In this chapter we continue with the subject of Bayesian decision theory by showing how to base decisions on a combination of prior information and additional empirical evidence. When you have completed this chapter, you should be able to

1. Distinguish among prior analysis, posterior analysis, and preposterior analysis.
2. Compute the expected value of perfect information.
3. Revise prior probabilities and perform posterior analysis.
* 4. Distinguish between extensive-form and normal-form preposterior analysis.
* 5. Perform extensive-form preposterior analysis.
* 6. Compute the expected value of "sample" information and the expected net gain from "sampling."
* 7. Perform normal-form preposterior analysis.

Before beginning this chapter, you may wish to review section 6.9 of Chapter 6 and sections 18.2, 18.3, 18.8, and 18.9 of Chapter 18.

19.1 Introduction

A fascinating feature of probability theory is that the probability of a specified event is dependent on the knowledge of the concerned individual. Consider, for example, the following hypothetical experiment in extrasensory perception (ESP).

Three friends, Adams, Brown, and Clark, decide to test their powers of ESP. They ask a fourth person, Davis, to shuffle an ordinary deck of playing cards and spread the 52 cards face down on a table while the three clairvoyants-in-waiting are out of the room. The plan is: Adams will enter the room and attempt to pick the two of clubs in one selection of a single card. Whatever the card picked, it will be replaced in the deck. Davis will again shuffle the cards and spread them, face down, on the table as before. Then Brown will enter the room and attempt to pick the two of clubs in one try. The card she selects will be replaced and the deck again shuffled and spread. Finally, Clark will enter and make his attempt to pick the two of clubs in one try. The experiment is to involve many replications of the above procedure; however, we will concern ourselves only with the first round.

What is the probability that Adams will be successful? That Brown will be successful? That Clark will be successful? Is it 1/52 in each case? Not really. The reason will become clear after we introduce some additional facts.

Adams has no special knowledge whatever that might help him come up with the target card.

However, Brown happens to know that in the deck there are four cards with slightly bent corners none of which is the two of clubs. Thus, she resolves at the outset, albeit perhaps quite unconsciously, that she will choose a card from among the 48 not having bent corners.

Clark begins with no more knowledge than Adams, but upon entering the room he finds Davis still shuffling the cards. From the doorway he sees Davis inadvertently flash the two of clubs, which just happens to be situated on the bottom of the deck. When the cards are spread, Clark, having made a mental note of the spot where the compact deck was placed prior to being spread, knows the exact whereabouts of the two of clubs.

Even though the three experimenters are performing the same kind of action with the same deck of playing cards, their knowledge, and hence their probabilities of success, differ markedly. Assuming that none of the subjects possesses any ESP talent, the probability that Adams will meet with success is 1/52; the probability for Brown is 1/48; and for Clark, 1—a sure thing.

So what does all this have to do with Bayesian decision theory? Simply this: Up to this point in our discussion of decision theory, we have been concerned with prior analysis only. What this means is that the probabilities employed were determined on the basis of knowledge in hand prior to the acquisition of additional empirical evidence. In the above ESP example, Adams is operating on the level of prior analysis. All Adams knows is that there are 52 cards only one of which is the target card. Brown is also involved in a prior analysis—she just happens to have more prior information than Adams.

Finally, Clark is operating on the level of posterior analysis. Clark began with no special prior knowledge but, because of a fortuitous flaw in the experimental conditions, gained information permitting him to revise his probability of success from 1/52 to 1. We may refer to the 1/52 as his prior probability and to the 1 as his posterior probability, his probability of success after he obtained the additional empirical evidence.

The present chapter is primarily concerned with problems of the Clark type—problems in which the Bayes action is determined through the use of probabilities which reflect not only prior knowledge but additional empirical evidence as well. Such problems call for the use of *posterior analysis*. This chapter also deals with *preposterior analysis* which, conceptually, is much like posterior analysis except that it has the additional virtue of helping the analyst choose between prior analysis and posterior analysis at the outset.

19.2 General Overview of Prior, Posterior, and Preposterior Analysis

Let us attempt to distinguish more precisely among prior, posterior, and preposterior analyst:

Prior Analysis

Prior analysis, as we saw in Chapter 18, entails arriving at a terminal decision, that is, selecting *the* action to be taken, by using the prior probabilities only. In prior analysis the analyst:

1. Calculates or estimates the payoffs associated with all relevant state-and-action combinations and assigns prior probabilities to the possible states of nature.
2. Using data from Step 1, computes the expected payoff of each of the alternative actions under consideration.
3. Selects the action with the best expected payoff.

This three-step procedure is illustrated by the simple flowchart appearing in Figure 19.1.

Under some circumstances prior analysis is the only analysis used. For example, the decision maker may be pressed for a decision and simply cannot afford the luxury of waiting until additional information is collected. On other occasions, he finds that the additional benefits associated with gathering additional information are so slight that they cannot possibly justify the cost of acquiring it. Often, however, when formal decision methodology is used, posterior analysis constitutes an essential part of it.

Posterior Analysis

After the analyst has arrived at his prior probability assessments, he may decide to revise them by making use of additional empirical evidence. If so, the revised state-of-nature probabilities, rather than the prior probabilities, are used as aids in the selection of the terminal action. A decision procedure in which such revised probabilities are used in accordance with the list of steps which follow is called a *posterior analysis*. With posterior analysis, the analyst:

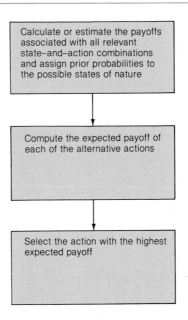

Figure 19.1. Steps involved in a prior analysis.

1. Calculates or estimates the payoffs associated with all state-and-action combinations and assigns prior probabilities to the possible states of nature.
2. Gathers "sample" evidence.
3. Revises the prior probabilities by taking into account the "sample" evidence.
4. Using payoff data from Step 1 and probabilities from Step 3, computes the expected payoff of each of the alternative actions under consideration.
5. Selects the action with the best expected payoff.

The steps are presented pictorially in Figure 19.2.

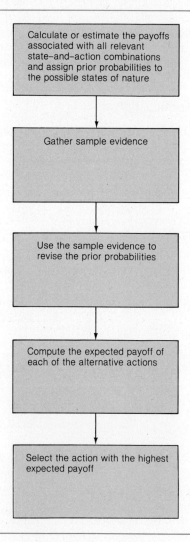

Figure 19.2. Steps involved in a standard posterior analysis.

19   Decision Theory: Posterior and Preposterior Analysis

GRIN and BEAR IT   by FRED WAGNER

"We're at the critical point in Grenada, men . . . We
either need to get out, or build some officers' clubs!"

In this chapter the term *sample* will be used in a most versatile way to refer to any
means of acquiring additional empirical data after the prior probabilities have been
determined. When the word "sample" is used generically, or nonliterally, it will be
shown in quotation marks. "Sample" evidence may result from survey sampling, experimentation, a complete census, or from some other relevant source or activity.

Preposterior Analysis

Often the task of gathering "sample" evidence is a costly undertaking and one which carries
with it no guarantee of improving upon the correctness of action selection. Consequently, an
intervening step may be introduced between the determination of prior probabilities and the
search for meaningful "sample" evidence. In this new step, the analyst grapples with the
question of whether the benefits to be derived from the contemplated "sample" evidence are
likely to exceed the cost of acquiring it. Stated somewhat differently: The new step requires

the analyst to determine prior to collecting sample evidence whether it is likely to prove more profitable (1) to acquire such additional evidence and then employ posterior analysis or (2) to refrain from seeking additional evidence and settle for prior analysis. In brief summary, then, the analyst undertaking a preposterior analysis will follow the procedural steps listed here:

1. Calculate or estimate the payoffs associated with all relevant state-and-action combinations and assign prior probabilities to the possible states of nature.
2. Determine whether it is better to acquire "sample" evidence or to make a terminal decision immediately.
3. Follow Steps 2 and 3 listed in connection with prior analysis if the alternative chosen is "make terminal decision immediately." Or, follow Steps 2 through 5 listed in connection with posterior analysis if the alternative chosen is "obtain 'sample' evidence."

These procedural steps for preposterior analysis are presented pictorially in Figure 19.3. At this juncture, it will be helpful to introduce some subscripts that will be used from now on to distinguish between prior-analysis measures and posterior-analysis measures:

To this point we have had no need to distinguish symbolically between prior and posterior measures. For example, the probability that state of nature θ_1 will occur has been denoted simply $P(\theta_1)$. Hereafter, it will be denoted $P_0(\theta_1)$ if it is a prior probability and $P_1(\theta_1)$ if it is a posterior probability. The "0" and "1" subscripts will be used in this same way in connection with other measures as well.

19.3 Expected Value of Perfect Information

If decision makers can reap no incremental benefit from obtaining information beyond that used in connection with the prior analysis, they should refrain from collecting "sample" evidence. But how can one tell prior to the actual collection of "sample" evidence whether the effort might prove beneficial? The answer is that, in a sense, one cannot tell. That is, unless the data are actually collected, an analyst can hardly know what they would have shown. Fortunately, when we view the matter from a somewhat different vantage point, we see that through a measure called *expected value of perfect information* (EVPI), the analyst *can* tell in advance—assuming reasonably correct prior probability assessments in the first place—when gathering additional information is *not* likely to be a profitable endeavor. How this is possible will be explained after you have become acquainted with this new measure.

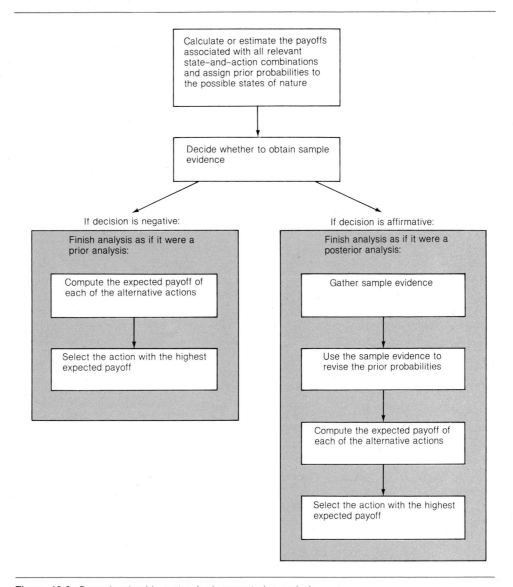

Figure 19.3. Steps involved in a standard preposterior analysis.

The determination of $EVPI_0$ involves the following three steps: (1) calculate a measure we will call *expected value with perfect information* (description to follow) and denote $EVWPI_0$, (2) compute the expected monetary payoff of the Bayes action obtained from prior analysis (which we will denote $EMP_0(A^*)$ when we wish to be very general and

$EMP_0(A_1)$, $EMP_0(A_2)$, . . . , or $EMP_0(A_n)$ when we wish to be more specific), and (3) subtract $EMP_0(A^*)$ from $EVWPI_0$. That is,

$$EVPI_0 = EVWPI_0 - EMP_0(A^*) \qquad (19.3.1)$$

We see from the above list that

The $EVPI_0$ measure defines the *upper limit* of the value of seeking "sample" evidence. That is, it tells the analyst the incremental payoff associated with using a perfectly accurate forecasting device for determining which state of nature will exist during the span of time to which the action pertains.

When we refer to a *perfectly accurate forecasting device*, we will mean exactly that. It is some kind of sampling scheme, experimental procedure, regression model, or something else *having rather definite rules associated with its application*, capable of telling its user *with perfect accuracy* what the true state of nature will be. Such a remarkable forecasting device will not exist in reality, but that unfortunate fact does not negate the usefulness of the concept itself. We will demonstrate how this can be by referring to the CCMC problem introduced in the preceding chapter (Illustrative Case 18.1, p. 890) and computing $EVPI_0$ using Equation (19.3.1).

One need not strain to imagine how the CCMC managers would behave if they possessed a perfectly accurate forecasting device: If, for example, this device informed them that demand would be weak, they could simply run their eyes along the appropriate line of the payoff table presented in Table 18.7 and find the largest monetary payoff—namely, $500,000. Since the $500,000 monetary payoff would result from taking Action A_1, attempt to reduce finished-goods inventories even further, this value would be entered on the "θ_3: Weak demand" line in an emerging table like that shown in Table 19.1. The same procedure would be used in the event of an indication of either typical demand or strong demand.

Notice that the probabilities listed in Table 19.1 are the same as those used previously—namely, $P_0(\theta_1) = .1$, $P_0(\theta_2) = .6$, and $P_0(\theta_3) = .3$. In calculating the $EMP_0(A^*)$, $EVWPI_0$,

Table 19.1 Calculation of Expected Value *with* Perfect Information for the CCMC Inventory Problem (Illustrative Case 18.1)

Predicted State of Nature	Monetary Payoff	Prior Probability	Weighted Monetary Payoff
θ_1: Strong demand	$2,500,000	.1	$250,000
θ_2: Typical demand	800,000	.6	480,000
θ_3: Weak demand	500,000	.3	150,000
		Total	$880,000

$EVWPI_0 = \$880,000$

and EVPI$_0$ measures, these probabilities are interpreted in relative-frequency terms (and are, for now, assumed to be perfectly correct in a relative-frequency sense). That is, they may be interpreted as the proportion of times that each of the indicated states of nature would occur if the present set of circumstances were encountered a great many times.

The expected value with perfect information, EVWPI$_0$, is determined as demonstrated in Table 19.1. Once any inadmissible actions have been eliminated and the highest monetary payoff for each state of nature has been identified, each is entered in a table like this one and then multiplied by the corresponding prior probability. Then, the resulting products are summed; this sum represents EVWPI$_0$.

For the CCMC example, the expected value *with* perfect information is $880,000. We earlier learned that the expected monetary payoff of the prior-analysis Bayes action is EMP$_0$(A*) = EMP$_0$(A$_1$) = $660,000 (computed in Table 18.8). Therefore, following Equation (18.3.1), we determine the expected value *of* perfect information to be

$$EVPI_0 = EVWPI_0 - EMP_0(A_1)$$
$$= \$880,000 - \$660,000$$
$$= \$220,000$$

The $220,000 is interpreted as the expected *increase* in monetary payoff that could be realized if the managers were fortunate enough to have a perfect forecasting device for the future state of nature. It is, therefore, also the amount of money they would be willing to spend for perfect information.

But why concern ourselves with the expected value of perfect information when in the real world we can never be perfectly certain of the future state of nature? The advantage of this concept is that it helps the decision maker to estimate a limit for the worth of less-than-perfect information. For example, in the unlikely event that it costs more than $220,000 to obtain "sample" information, the cost would be too high to make the effort worthwhile. On the other hand, if the cost of obtaining "sample" evidence were considerably less than $220,000, then presumably the rational thing for management to do would be to seek the additional information.

Notice that $220,000 is the same result obtained in connection with the prior determination of the Bayes action when *expected opportunity losses*, rather than expected monetary payoffs, were used in section 18.10. This $220,000 value may also be called the *cost of uncertainty* because it indicates by how much the expected monetary payoff under conditions of uncertainty falls short of the expected monetary payoff under conditions of certainty.

19.4 Determining the Posterior Probabilities

A convenient way of presenting the fundamentals of posterior analysis will be to add an additional feature to the illustrative case we examined in Chapter 18 (Illustrative Case 18.1).

ILLUSTRATIVE CASE 19.1*

Because the Claybourne and Clark Manufacturing Company (CCMC) sells its products to only five large companies, it is in a position to carry out information-gathering efforts of a kind that

---

*This case is a continuation of Illustrative Case 18.1.

might be impracticable and certainly more costly for a company with many irregular customers. One such information-gathering method which CCMC has been using for nearly a quarter of a century is both inexpensive and conceptually simple. Each October the company mails a brief questionnaire to its five industrial customers. The main purpose of this questionnaire is to obtain an estimate of the dollars' worth of merchandise that each recipient company plans to purchase from CCMC between next January 1 and December 31. The five resulting estimates sum to a quantity which qualifies as strong demand, typical demand, or weak demand.

CCMC has found the survey results quite useful over the years. Moreover, the cost of the questionnaire—including materials, postage, handling, and questionnaire construction and interpretation—is generally under $500.

This past October, shortly after the management group completed their determination of prior probabilities, they conducted their purchasing-intentions survey as usual. The resulting estimates, when summed, amounted to a prediction of strong demand.

Before proceeding with the posterior analysis, we must introduce some new symbols. The three possible results of the purchasing-intentions survey will be symbolized by x_k where k may be 1, 2, or 3 and where

x_1 represents "survey indicates strong demand."

x_2 represents "survey indicates typical demand."

x_3 represents "survey indicates weak demand."

An examination of the historical accuracy of the company's survey turned up the results (expressed as conditional probabilities, or likelihoods) shown in Table 19.2.

The manner in which Table 19.2 is read and its relationship to the new symbols we have just introduced can be clarified through a detailed examination of this table. To begin with, we recognize that column (1) is the only column with which we need to be concerned right now. This is the column pertaining to the observed fact that the survey indicated a strong demand. We observe from column (1) that, in the past when strong demand actually occurred, the survey had previously indicated strong demand 70% of the time. However, when the actual level of demand had proved typical, the survey had incorrectly indicated strong

Table 19.2 Relative-Frequency Probabilities of Three Types of Survey Evidence for the CCMC Problem

| State of Nature θ | Conditional Probabilities $P(x_k\|\theta_i)$ | | | Total |
	(1) x_1	(2) x_2	(3) x_3	
θ_1: Strong demand	.7	.2	.1	1.0
θ_2: Typical demand	.3	.6	.1	1.0
θ_3: Weak demand	.2	.1	.7	1.0

x_1 represents "survey indicates strong demand"
x_2 represents "survey indicates typical demand"
x_3 represents "survey indicates weak demand"

demand 30% of the time. Finally, when the actual level of demand had proved weak, the survey had incorrectly signaled strong demand 20% of the time. These percents may be expressed as conditional probabilities, or *likelihoods*, as they are often called, and written

$$P(x_1|\theta_1) = .7$$
$$P(x_1|\theta_2) = .3$$
$$P(x_1|\theta_3) = .2$$

Of course, the entries in columns (2) and (3) of Table 19.2 could also be expressed in terms of similar conditional probability notations. For example, if our "sample" evidence had in fact been "survey indicates typical demand," we would have $P(x_2|\theta_1) = .2$, $P(x_2|\theta_2) = .6$, and $P(x_2|\theta_1) = .1$. Similarly, if the survey had resulted in an indication of weak demand, we would have $P(x_3|\theta_1) = .1$, $P(x_3|\theta_2) = .1$, and $P(x_3|\theta_3) = .7$.

As mentioned earlier, in posterior analysis expected monetary payoffs are determined with the aid of posterior probabilities which are simply prior probabilities revised to reflect information obtained from a "sample." The Bayes theorem, introduced and demonstrated in Chapter 6, is the tool used to determine the posterior probabilities. Recall from Chapter 6 that the formula expressing the Bayes theorem was given as

$$P(A_1|B) = \frac{P(A_1) \cdot P(B|A_1)}{\displaystyle\sum_{\text{all } i} [P(A_i) \cdot P(B|A_i)]}$$

where A_i represents a specified event or state of nature

B represents an observed fact

We stated at that time that the computational work can often be completed most conveniently through use of a table with the states of nature listed along the left-hand side and (1) prior probabilities, (2) likelihoods, (3) joint probabilities, and (4) posterior probabilities appearing in the columns. Since we are here using x_1 in place of B to represent the observed fact and θ_i in place of A_i to represent a specified state of nature, we will make these substitutions and express the Bayes formula as

$$P(\theta_1|x_1) = \frac{P(\theta_1) \cdot P(x_1|\theta_1)}{\displaystyle\sum_{\text{all } i} [P(\theta_i) \cdot P(x_1|\theta_i)]} \qquad (19.4.1)$$

Table 19.3 demonstrates the procedure by which the prior probabilities for the CCMC case may be revised given knowledge that the survey indication is x_1.

Table 19.3 reveals that, as a result of the use of the survey indicator and the related likelihoods, the probability of weak demand has decreased from .3000 to .1935. Consequently, weak demand has moved from the second most probable to the least probable state of nature. This table also indicates a slight reduction in the probability associated with the state of typical demand, decreasing from .6000 to .5806. Finally, the probability associated

Table 19.3 Revision of Prior Probabilities for the CCMC Inventory Problem Based on Survey Indication of Strong Demand

State of Nature θ	(1) Prior Probability $P_0(\theta)$	(2) Likelihood $P(x_1\|\theta)$	(3) Joint Probability $P(\theta) \cdot P(x_1\|\theta)$	(4) Posterior Probability $P_1(\theta\|x_1)$
θ_1: Strong demand	.1	.7	.07	.2258
θ_2: Typical demand	.6	.3	.18	.5806
θ_3: Weak demand	.3	.2	.06	.1935
Total	1.0		.31	.9999[a]

[a] Does not sum to 1.0 because of rounding error.

with the occurrence of strong demand has more than doubled (.2258 up from .1000). The state of strong demand is still only the second most probable state (up from least probable), but the substantial relative increase in its probability of occurrence could result in a marked change in its expected monetary payoff.

19.5 Decision Making Using Posterior Probabilities

Posterior expected monetary payoffs may now be computed using the posterior probabilities (in conjunction, of course, with the monetary payoffs used in the prior analysis). Table 19.4 presents the computational steps required.

We see from Table 19.4 that Action A_2, rather than A_1, is now the Bayes action. Clearly, the optimal action in posterior analysis may be different from the optimal action under prior analysis, as in our present example. But that should not be taken as something to be expected; the Bayes action could have been the same under prior and posterior analyses.

19.6 Posterior Expected Value of Perfect Information

A new measure of expected value of perfect information, $EVPI_1$, may now be obtained, if desired, using procedures identical to those described in section 19.3. Table 19.5 demonstrates the steps involved in determining the revised expected value *with* perfect information, $EVWPI_1$.

Posterior expected value of perfect information is determined by

$$EVPI_1 = EVWPI_1 - EMP_1(A^*)$$

Therefore,

$$EVPI_1 = \$1,125,730 - \$803,180 = \$322,550$$

Notice that the posterior expected value of perfect information is greater than the prior expected value of perfect information ($EVPI_0 = \$220,000$ and $EVPI_1 = \$322,550$). This result may seem nonsensical at first in that the expected value of perfect information is

Table 19.4 Calculation of Expected Monetary Payoffs for the CCMC Problem Using Posterior Probability Weights

Action A$_1$: Attempt to Reduce Finished-Goods Inventories Even Further

State of Nature θ_i	Posterior Probability $P_1(\theta_i)$	Monetary Payoff	Weighted Monetary Payoff
θ_1: Strong demand	.2258	$900,000	$203,220
θ_2: Typical demand	.5806	700,000	406,420
θ_3: Weak demand	.1935	500,000	96,750
		Total	$706,390

$$\text{EMP}_1(A_1) = \$706,390$$

Action A$_2$: Attempt to Maintain Finished-Goods Inventories at Present Levels

State of Nature θ_i	Posterior Probability $P_1(\theta_i)$	Monetary Payoff	Weighted Monetary Payoff
θ_1: Strong demand	.2258	$1,500,000	$338,700
θ_2: Typical demand	.5806	800,000	464,480
θ_3: Weak demand	.1935	0	0
		Total	$803,180

$$\text{EMP}_1(A_2) = \$803,180$$

Action A$_3$: Attempt to Increase Finished-Goods Inventories

State of Nature θ_i	Posterior Probability $P_1(\theta_i)$	Monetary Payoff	Weighted Monetary Payoff
θ_1: Strong demand	.2258	$2,500,000	$564,500
θ_2: Typical demand	.5806	500,000	290,300
θ_3: Weak demand	.1935	−500,000	−96,750
		Total	$758,050

$$\text{EMP}_1(A_3) = \$758,050$$

$$\text{Formula: EMP}_1(A_j) = \sum_{\text{all } i} [P(\theta_i)\text{MP}(O_{ij})]$$

supposed to represent the upper limit of the worth of seeking "sample" evidence. Consequently, we would expect EVPI to become smaller as relevant additional information is introduced. However, occasionally the kind of result we see here *does* occur. When it does, it indicates that the prior probability assessments reflect too much pessimism, or caution, on the part of the decision maker. Remember, in calculating the prior expected value of perfect information, we proceeded as if the prior probabilities were effectively accurate, an assumption which now appears quite wrong.

The value of $322,550 for the posterior expected value of perfect information suggests that if (1) a source of additional "sample" evidence is available, (2) a terminal decision can

Table 19.5 Calculation of Posterior Expected Value *with* Perfect Information

State of Nature θ	(1) Monetary Payoff	(2) Posterior Probability $P_1(\theta)$	(3) Weighted Monetary Payoff (1) × (2)
θ_1: Strong demand	$2,500,000	.2258	$564,500
θ_2: Typical demand	800,000	.5806	464,480
θ_3: Weak demand	500,000	.1935	96,750
		Total	$1,125,730

$$\text{EVPI}_1 = \$1,125,730$$

safely be postponed until still more evidence is amassed, and (3) the new "sample" evidence can be obtained for a cost of substantially less than $322,550, our management group would be well advised to push the posterior analysis into a second stage by revising the state probabilities once more. If a second stage of posterior analysis were to be undertaken, the posterior probabilities—$P_1(\theta_1) = .2258$, $P_1(\theta_2) = .5806$, and $P_1(\theta_3) = .1935$—would be treated as new prior probabilities and denoted by $P_0(\theta_1)$, $P_0(\theta_2)$, and $P_0(\theta_3)$, respectively, and the second-stage revised probabilities would be treated as the new posterior probabilities and denoted $P_1(\theta_1)$, $P_1(\theta_2)$, and $P_1(\theta_3)$. Otherwise, the procedure would be identical to that described above.

Figure 19.4 illustrates the concept of extending posterior analysis beyond the first stage. In this figure the diagram first presented in Figure 19.2 is reproduced and a bent arrow added to indicate that, in theory, the process of revising state-of-nature probabilities could go on and on. The expectation is that each new stage of posterior analysis will bring about a reduction in the value of EVPI. Eventually, the EVPI value will become too small to cover the cost of obtaining still more sample evidence and, at that point, delaying a terminal decision pending new information would be financially irrational.

*19.7 Extensive-Form Preposterior Analysis

In section 19.2 we mentioned that *preposterior analysis* is the term used to describe a decision analysis in which a choice is made at the outset between (1) arriving at a terminal decision immediately and (2) seeking "sample" evidence before arriving at a terminal decision. The choice between these alternatives is based on the expected monetary payoff of each.

The nature of preposterior analysis is clarified in Figure 19.5.

> When the preposterior analysis is carried out using a decision-tree diagram like the one shown in Figure 19.5, it is called *extensive-form preposterior analysis*.

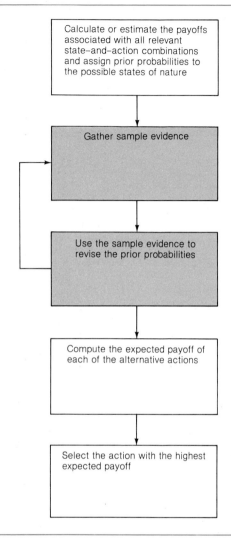

Calculate or estimate the payoffs associated with all relevant state–and–action combinations and assign prior probabilities to the possible states of nature

Gather sample evidence

Use the sample evidence to revise the prior probabilities

Compute the expected payoff of each of the alternative actions

Select the action with the highest expected payoff

Figure 19.4. Steps involved in a posterior analysis involving successive revisions of state-of-nature probabilities.

The choice between prior analysis and posterior analysis is made by comparing EMP(Prior analysis) with EMP(Posterior analysis) − (Cost of obtaining "sample" evidence if posterior analysis were chosen). The decision rule is

Decision Rule
1. If EMP(Prior analysis) > EMP(Posterior analysis) − (Cost of "sample" evidence), choose prior analysis.
2. If EMP(Prior analysis) < EMP(Posterior analysis) − (Cost of "sample" evidence), choose posterior analysis.

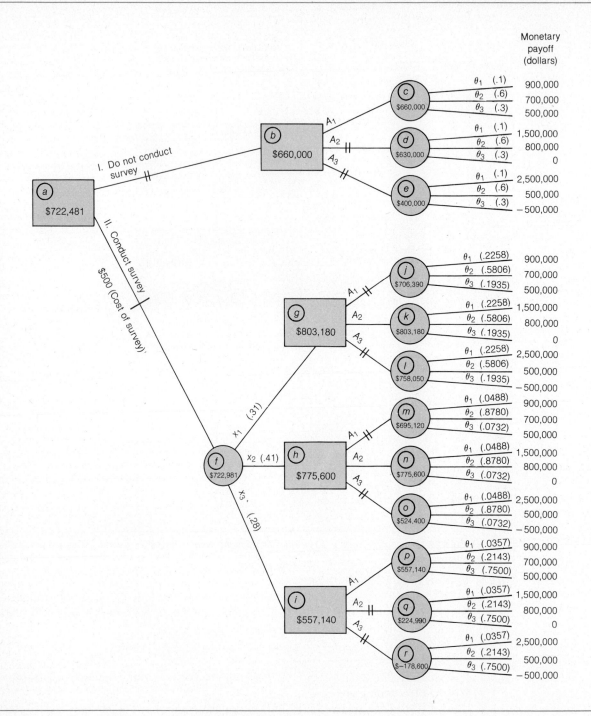

Figure 19.5. Decision-tree diagram of the preposterior analysis of the CCMC inventory problem. (*Symbols:* x_1 = survey indicates *strong* demand; x_2 = survey indicates *typical* demand; x_3 = survey indicates *weak* demand; A_1 = attempt to *decrease* inventories; A_2 = attempt to *maintain* inventories at present levels; A_3 = attempt to *increase* inventories; θ_1 = *strong* demand; θ_2 = *typical* demand; θ_3 = *weak* demand.)

Determining Probabilities for the Posterior Part of Preposterior Analysis

Another aspect of Figure 19.5 calling for special comment concerns the probabilities employed in the lower part of the figure. Recall that when we discussed posterior analysis per se, we made use of a standard table, Table 19.3, for computing posterior probabilities. Whereas at that time we were told that x_1 was the observed survey indication, with preposterior analysis we do not enjoy access to any such "given." Instead, we must consider all possible "givens." That being so, a table for revising prior probabilities, like Table 19.6, is desirable. In this table, as in Table 19.3, the possible states of nature are listed along the left-hand edge and the corresponding prior probabilities are listed in column (1). However, column (2), the likelihood column, differs from its counterpart in Table 19.3 in that it contains *all* the conditional probabilities originally presented in Table 19.2, rather than just those associated with survey indication x_1. Columns (3) and (4) are similarly all-encompassing. In column (3) we have the joint probabilities $P(x_k \text{ and } \theta_i) = P(\theta_i) \cdot P(x_k|\theta_i)$. The sums obtained by adding these joint probabilities both horizontally and vertically are worth noting. These totals, representing unconditional probabilities, are (1) for rows, the prior probabilities $P_0(\theta_1) = .1$, $P_0(\theta_2) = .6$, and $P_0(\theta_3) = .3$, and (2) for columns, the probabilities associated with the three possible survey indications, $P(x_1) = .31$, $P(x_2) = .41$, and $P(x_3) = .28$.

Conditional probabilities for states of nature given the survey evidence—that is, probabilities of the form $P(\theta_i|x_k)$—can be computed by dividing the joint probabilities by the appropriate column totals. For example, the probability of strong demand given a survey indication of strong demand is obtained by

$$\frac{P(\theta_1 \text{ and } x_1)}{P(x_1)} = \frac{.07}{.31} = .2258$$

The probability of strong demand given a survey indication of typical demand is obtained by

$$\frac{P(\theta_1 \text{ and } x_2)}{P(x_2)} = \frac{.02}{.41} = .0488$$

and so forth. All such conditional probabilities for the problem under discussion are presented in column (4) of Table 19.6. Clearly, each of these results is identical to that which would have occurred from application of the Bayes formula, Equation (19.4.1). For example, the probability of θ_1 given x_1 may be written either

$$P(\theta_1|x_1) = \frac{P(\theta_1 \text{ and } x_1)}{P(x_1)} \quad \text{or} \quad P(\theta_1|x_1) = \frac{P(\theta_1) \cdot P(x_1|\theta_1)}{\Sigma[P(\theta_i) \cdot P(x_1|\theta_i)]}$$

The latter form represents a specific application of Equation (19.4.1).

Moving Across the Decision Tree

To demonstrate how these posterior probabilities as well as the probabilities of the form $P(x_k)$ are related to the extensive-form analysis presented in Figure 19.5, we will now describe the process of moving across the decision-tree diagram and performing the required backward induction.

If the decision maker situated at decision fork *a* chooses to perform a prior analysis (branch I), he encounters no intervening chance fork. Instead, he moves directly to decision

Table 19.6 Calculation of All Possible Posterior Probabilities for the CCMC Inventory Decision Problem

State of Nature θ_i	(1) Prior Probability $P_0(\theta_i)$	(2) Likelihood $P(x_k\mid\theta_i)$			(3) Joint Probability $P(x_k \text{ and } \theta_i)$			Total (Unconditional Probability)	(4) Posterior Probability $P_1(\theta_i\mid x_k)$		
		$P(x_1\mid\theta_i)$	$P(x_2\mid\theta_i)$	$P(x_3\mid\theta_i)$	$P(x_1 \text{ and } \theta_i)$	$P(x_2 \text{ and } \theta_i)$	$P(x_3 \text{ and } \theta_i)$		$P_1(\theta_i\mid x_1)$	$P_1(\theta_i\mid x_2)$	$P_1(\theta_i\mid x_3)$
θ_1: Strong demand	.1	.7	.2	.1	.07	.02	.01	.10	.2258	.0488	.0357
θ_2: Typical demand	.6	.3	.6	.1	.18	.36	.06	.60	.5806	.8780	.2143
θ_3: Weak demand	.3	.2	.1	.7	.06	.03	.21	.30	.1935	.0732	.7500
Total (unconditional probability)					.31	.41	.28	1.00	.9999	1.0000	1.0000

fork b, where he must choose among inventory actions A_1, A_2, and A_3. His course from decision fork b (decision forks a in Figures 18.2, 18.3, and 18.4) to the monetary payoffs and back again was detailed in section 18.11 of the preceding chapter and will not be repeated here. Upon finding that Action A_1 (with $EMP(A_1) = \$660,000$) is the optimal action for prior analysis, he places the $\$660,000$ result within the rectangle identifying decision fork b, where it will remain at least as long as is required for the analyst to complete the posterior analysis part of his task.

We will now say that the decision maker (back again at decision fork a) decides to conduct the survey, that is, follow branch II in Figure 19.5. Having so decided, he moves to chance fork f where he finds Nature poised to choose among survey indications x_1, x_2, and x_3. The probabilities $P(x_1) = .31$, $P(x_2) = .41$, and $P(x_3) = .28$ are, as noted, the unconditional probabilities presented at the bottom of Table 19.6, section (3), which has three parts. If Nature selects, say, x_1, the decision maker moves to decision fork g where he must select an action from among A_1, A_2, and A_3. *If he chooses A_1*, for example, he moves to chance fork j where Nature is required to choose either state of nature θ_1, θ_2, or θ_3. The probabilities appearing in parentheses above the terminal branches of the posterior-analysis part of this diagram are the revised probabilities presented in the three parts of section (4), Table 19.6. That is, they are probabilities of the type $P(\theta_i | x_k)$.

Let us suppose that Nature chooses State θ_1, a state having associated with it a monetary payoff of $\$900,000$. The corresponding weighted monetary payoff is obtained by multiplying $\$900,000$ by the posterior probability $.2258$. The same procedure would be followed all the way down the terminal branches.

The weighted monetary payoffs associated with chance fork j, for example, are then summed to get the expected monetary payoff to be placed within the circle identifying this specified chance fork. That is, the weighted monetary payoffs associated with chance fork j are (1) $\$900,000 \times .2258 = \$203,220$ (2) $\$700,000 \times .5806 = \$406,420$, and (3) $\$500,000 \times .1935 = \$96,750$. The sum of these products is $\$706,390$, the value appearing within the circle identifying chance fork j. The same procedure is followed for the remaining chance forks, k through r.

The highest expected monetary payoffs appearing within the circles are as follows: (1) for chance forks j, k, and l—$\$803,180$, (2) for chance forks m, n, and o—$\$775,600$, and (3) for chance forks p, q, and r—$\$557,140$. These preferred expected monetary payoffs are placed within the rectangles identifying decision forks g, h, and i, respectively, and branches associated with nonoptimal decisions are blocked off. The entries within rectangles g, h, and i are then multiplied by the corresponding $p(x_k)$ probabilities, and the resulting products are summed. The result is the expected monetary payoff of $\$722,981$ appearing within the circle identifying chance fork f.

Because the value of $\$722,981$ associated with chance fork f is greater than the one appearing within the rectangle identifying decision fork b ($\$660,000$), the decision maker is predisposed to view posterior analysis as preferable to prior analysis. Before he draws any such conclusion, however, *he must subtract the cost of conducting the purchasing-intentions survey from the expected monetary payoff of the posterior analysis.* We were told earlier that the cost of this survey is a modest $\$500$ or less. When $\$500$ is subtracted from $\$722,981$, the difference is found to be $\$722,481$, a value still well in excess of the $\$660,000$ payoff expected from the use of prior analysis.

Since EMP(Posterior analysis) − (Cost of obtaining "sample" evidence) is greater than EMP(Prior analysis), posterior analysis is chosen. It is chosen, that is, provided it can pass a simple but eminently appropriate test of practical usefulness. We have found that the optimal action with prior analysis was A_1 and the optimal action with posterior analysis given a survey indication of x_1 was A_2. We see, therefore, that *it is possible for the decision maker to arrive at a different Bayes action if "sample" evidence is employed.*

BASIC RULE

If a decision maker's choice of action cannot possibly be affected by the "sample" evidence, regardless of its nature, use of the "sample" procedure is contraindicated on the grounds that it cannot improve upon the decision maker's choice of action determined through use of prior analysis.

Expected Value of "Sample" Information

How much would the CCMC managers be willing to pay to obtain "sample" information? The answer must be the difference between EMP(Posterior analysis) and EMP(Prior analysis).

We found the EMP(Posterior analysis) = \$722,981 and that EMP(Prior analysis) = \$660,000. Therefore, \$722,981 − \$660,000 = \$62,981 is the maximum amount that management would be willing to spend in order to acquire "sample" information, a value conspicuously in excess of the \$500 which they must actually spend. The difference of \$62,981 can be referred to as the *expected value of "sample" information* and denoted EVSI.

EVSI is defined as the expected amount by which the terminal monetary payoff is increased by the information obtained from a "sample" study.

To calculate the *expected net gain from "sampling,"* denoted ENGS, the cost of obtaining "sample" information must be subtracted from the expected value of the "sample" information. Hence, in general,

$$\text{ENGS} = \text{EVSI} - \text{COSI}$$

where COSI stands for "cost of 'sample' information."

✳ 19.8 Normal-Form Preposterior Analysis

In the preceding section we dealt with one of the two widely used approaches to preposterior analysis—namely, extensive-form analysis. In this section we demonstrate the alternative approach, called *normal-form preposterior analysis*, utilizing the same data. First, however, we must explain the meaning of a *strategy* as the term is used in preposterior analysis.

The Role of Strategies in Normal-Form Preposterior Analysis

Perhaps the most meaningful way to approach the concept of strategy is to point out certain services performed by our extensive-form analysis in the preceding section. We note particularly that in the posterior analysis part of the CCMC problem we found the optimal action associated with each possible survey indication. That is, we found (as shown in Figure 19.5) that the decision maker is instructed as follows:

$$x_1 \longrightarrow A_2$$
$$x_2 \longrightarrow A_2$$
$$x_3 \longrightarrow A_1$$

This symbol system merely expresses a decision rule of the kind "If sample outcome x_k occurs, take Action A_j." Thus, for the CCMC problem, the decision maker is advised as follows:

If survey indication x_1 occurs, take Action A_2.

If survey indication x_2 occurs, take Action A_2.

If survey indication x_3 occurs, take Action A_1.

This kind of decision rule is called a *strategy*.

With a *strategy*, the analyst faces two points of uncertainty: first, uncertainty about the outcome of the "sample" study and, second, uncertainty about what the state of nature will be. Therefore, we must consider both sources of uncertainty when choosing a course of action.

We will structure our discussion of normal-form methodology around the following list of steps:

1. List all possible strategies.
2. Using conditional probabilities of "sample" outcomes given specified states of nature, compute all conditional expected monetary payoffs associated with each possible state of nature.
3. Compute the expected monetary payoff of each strategy.
4. Select the strategy with the largest expected monetary payoff.

Each of these steps is demonstrated here using the CCMC data.

1. List all possible strategies.
   To ensure that no possible strategies are overlooked, the analyst should begin by determining the number of different strategies which must be taken into account. This is readily accomplished through use of the expression n^r where n is the number of actions under consideration and r is the number of possible "sample" outcomes. For example, in the CCMC problem, there are altogether 27 possible strategies, a number obtained by recognizing that there are three admissible actions and three possible "sample" indications. Therefore, $n^r = 3^3 = 27$ possible strategies. These are listed in Table 19.7. The strategies, denoted S_1, $S_2, \ldots, S_{27}$, indicate the actions that are taken in response to each specified

Table 19.7 The 27 Possible Strategies for the CCMC Problem

Survey Indication	S_1	S_2	S_3	S_4	S_5	S_6	S_7	S_8	S_9	S_{10}	S_{11}	S_{12}	S_{13}	S_{14}	S_{15}	S_{16}	S_{17}	S_{18}	S_{19}	S_{20}	S_{21}	S_{22}	S_{23}	S_{24}	S_{25}	S_{26}	S_{27}
x_1	A_1	A_2	A_3	A_1	A_1	A_1	A_2	A_2	A_2	A_1	A_1	A_1	A_3	A_3	A_3	A_2	A_2	A_2	A_3	A_3	A_3	A_1	A_1	A_2	A_2	A_3	A_3
x_2	A_1	A_2	A_3	A_1	A_2	A_2	A_1	A_1	A_2	A_1	A_3	A_3	A_1	A_1	A_3	A_2	A_3	A_3	A_2	A_2	A_3	A_2	A_3	A_1	A_3	A_2	A_1
x_3	A_1	A_2	A_3	A_2	A_1	A_2	A_1	A_2	A_1	A_3	A_1	A_3	A_1	A_3	A_1	A_3	A_2	A_3	A_2	A_3	A_2	A_3	A_2	A_3	A_1	A_1	A_2

"sample" indication. For example, Strategy S_{27} in Table 19.7 instructs the decision maker as follows:

$$x_1 \longrightarrow A_3$$
$$x_2 \longrightarrow A_1$$
$$x_3 \longrightarrow A_2$$

We will also sometimes find it helpful to write a strategy—say, Strategy S_{27}—as $S_{27}(A_3, A_1, A_2)$. This notational system tells us that, for Strategy S_{27}, one chooses (1) Action A_3 if sample indication x_1 is observed, (2) Action A_1 if sample indication x_2 is observed, and (3) Action A_2 if sample indication x_3 is observed.

Clearly, some of the strategies listed in Table 19.7 are unlikely candidates for eventual adoption. For example, Strategies $S_1(A_1, A_1, A_1)$, $S_2(A_2, A_2, A_2)$, and $S_3(A_3, A_3, A_3)$ all instruct the decision maker to select the same action regardless of whether the purchasing-intentions survey results in x_1, x_2, or x_3. In other words, once the survey information has been collected, proceed immediately to disregard it.

2. Using conditional probabilities of "sample" outcomes given specified states of nature, compute all conditional expected monetary payoffs associated with each possible state of nature.

Determining the expected monetary payoff of each possible strategy is a two-stage procedure. The first stage is shown here; the second in Step 3, which follows. The first stage consists of calculating for each strategy the expected monetary payoff given (that is, conditional upon) each possible state of nature in turn. We will use the term *conditional expected monetary payoff*, denoted $\text{EMP}(S_l|\theta_i)$ and read "the expected monetary payoff of Strategy S_l given state of nature θ_i," where l goes from 1 to n^r and i goes from 1 to m.

Table 19.8 shows the calculations of the conditional expected monetary payoffs associated with arbitrarily selected Strategies S_6 and S_{22}. For example, the conditional monetary payoffs of Strategy S_6, the $\text{EMP}(S_6|\theta_i)$ values, are shown in Part I of this table. In column (1) the monetary payoffs of Actions A_1, A_2, and A_3 are presented for all three possible states of nature. In column (2) nine conditional probabilities are displayed under the heading "Probability of Action $S_6(A_1, A_2, A_2)$." These nine probabilities—of the form $P(x_k|\theta_i)$—are taken from Table 19.2 and represent the probabilities associated with the decision maker's taking the action specified by the strategy under examination when a particular state of nature and a particular "sample" indication are assumed. For example, the probabilities found in the row of Table 19.8, Part I, labeled "θ_1: Strong demand," are .7, .2, and .1. These are read, respectively, as "the probability of getting 'sample' indication x_1 (and, therefore, taking Action A_1), given state of nature θ_1, is .7," "the probability of getting 'sample' indication x_2 (and, therefore, taking Action A_2), given state of nature θ_1, is .2," and "the probability of getting 'sample' indication x_3 (and taking Action A_2), given state of nature θ_1, is .1."

The conditional expected monetary payoff of Strategy S_6, given state of nature θ_1, denoted EMP$(S_6|\theta_1)$, is computed by (1) multiplying these "probabilities of action" by the respective monetary payoffs generated if these actions are taken and (2) summing the resulting products. Thus, for Strategy S_6, given state of nature θ_1, the expected monetary payoff is determined to be EMP$(S_6|\theta_1)$ = (.7)($900,000) + (.2)($1,500,000) + (.1)($1,500,000) = $1,080,000, as shown near the bottom of Table 19.8, Part I. Similarly, EMP$(S_6|\theta_2)$ is $770,000 and EMP$(S_6|\theta_3)$ is $100,000. For Strategy S_{22} in Part II of Table 19.8, we find that

Table 19.8   Calculation of Conditional Expected Monetary Payoffs for Strategies 6 and 22 from Table 19.7

Part I: Strategy 6

	(1)			(2)	(3)
	Monetary Payoff of Action			Probability of Action	Conditional Expected Monetary
State of Nature θ_i	A_1	A_2	A_2	$S_6(A_1, A_2, A_2)$	Payoff
θ_1: Strong demand	$900,000	$1,500,000	$1,500,000	.7 .2 .1	$1,080,000
θ_2: Typical demand	$700,000	$800,000	$800,000	.3 .6 .1	$770,000
θ_3: Weak demand	$500,000	$0	$0	.2 .1 .7	$100,000

EMP$(S_6|\theta_1)$ = (.7)($900,000) + (.2)($1,500,000) + (.1)($1,500,000) = $1,080,000
EMP$(S_6|\theta_2)$ = (.3)($700,000) + (.6)($800,000) + (.1)($800,000) = $770,000
EMP$(S_6|\theta_3)$ = (.2)($500,000) + (.1)($0) + (.7)($0) = $100,000

Part II: Strategy 22

	(1)			(2)	(3)
	Monetary Payoff of Action			Probability of Action	Conditional Expected Monetary
State of Nature θ_i	A_1	A_2	A_3	$S_{22}(A_1, A_2, A_3)$	Payoff
θ_1: Strong demand	$900,000	$1,500,000	$2,500,000	.7 .2 .1	$1,180,000
θ_2: Typical demand	$700,000	$800,000	$500,000	.3 .6 .1	$740,000
θ_3: Weak demand	$500,000	$0	−$500,000	.2 .1 .7	−$250,000

EMP$(S_{22}|\theta_1$ = (.7)($900,000) + (.2)($1,500,000) + (.1)($2,500,000) = $1,180,000
EMP$(S_{22}|\theta_2)$ = (.3)($700,000) + (.6)($800,000) + (.1)($500,000) = $740,000
EMP$(S_{22}|\theta_3)$ = (.2)($500,000) + (.1)($0) + (.7)(−$500,000) = −$250,000

EMP($S_{22}|\theta_1$) is \$1,180,000, EMP($S_{22}|\theta_2$) is \$740,000, and EMP($S_{22}|\theta_3$) is −\$250,000. Table 19.9 presents the conditional expected monetary payoffs associated with all 27 possible strategies.

3. Compute the expected monetary payoff of each strategy.

The (unconditional) expected monetary payoff of each strategy can now be determined through the use of prior probability weights. More specifically, the three conditional expected monetary payoffs associated with each strategy are multiplied

---

Table 19.9 Complete List of Conditional Expected Monetary Payoffs of the 27 Strategies Appearing in Table 19.7

$S_1(A_1, A_1, A_1)$:   EMP($S_1|\theta_1$) = (.7)\$900,000) + (.2)(\$900,000) + (.1)(\$900,000) = \$900,000
              EMP($S_1|\theta_2$) = (.3)(\$700,000) + (.6)(\$700,000) + (.1)(\$700,000) = \$700,000
              EMP($S_1|\theta_3$) = (.2)(\$500,000) + (.1)(\$500,000) + (.7)(\$500,000) = \$500,000

$S_2(A_2, A_2, A_2)$:   EMP($S_2|\theta_1$) = \$1,500,000; EMP($S_2|\theta_2$) = \$800,000; EMP($S_2|\theta_3$) = \$0
$S_3(A_3, A_3, A_3)$:   EMP($S_3|\theta_1$) = \$2,500,000; EMP($S_3|\theta_2$) = \$500,000; EMP($S_3|\theta_3$) = −\$500,000
$S_4(A_1, A_1, A_2)$:   EMP($S_4|\theta_1$) = \$960,000; EMP($S_4|\theta_2$) = \$710,000; EMP($S_4|\theta_3$) = \$150,000
$S_5(A_1, A_2, A_1)$:   EMP($S_5|\theta_1$) = \$1,020,000; EMP($S_5|\theta_2$) = \$760,000; EMP($S_5|\theta_3$) = \$450,000
$S_6(A_1, A_2, A_2)$:   EMP($S_6|\theta_1$) = \$1,080,000; EMP($S_6|\theta_2$) = \$770,000; EMP($S_6|\theta_3$) = \$100,000
$S_7(A_2, A_1, A_1)$:   EMP($S_7|\theta_1$) = \$1,320,000; EMP($S_7|\theta_2$) = \$730,000; EMP($S_7|\theta_3$) = \$400,000
$S_8(A_2, A_1, A_2)$:   EMP($S_8|\theta_1$) = \$1,380,000; EMP($S_8|\theta_2$) = \$740,000; EMP($S_8|\theta_3$) = \$50,000
$S_9(A_2, A_2, A_1)$:   EMP($S_9|\theta_1$) = \$1,440,000; EMP($S_9|\theta_2$) = \$790,000; EMP($S_9|\theta_3$) = \$350,000
$S_{10}(A_1, A_1, A_3)$: EMP($S_{10}|\theta_1$) = \$1,060,000; EMP($S_{10}|\theta_2$) = \$680,000; EMP($S_{10}|\theta_3$) = −\$200,000
$S_{11}(A_1, A_3, A_1)$: EMP($S_{11}|\theta_1$) = \$1,220,000; EMP($S_{11}|\theta_2$) = \$580,000; EMP($S_{11}|\theta_3$) = \$400,000
$S_{12}(A_1, A_3, A_3)$: EMP($S_{12}|\theta_1$) = \$1,380,000; EMP($S_{12}|\theta_2$) = \$560,000; EMP($S_{12}|\theta_3$) = −\$300,000
$S_{13}(A_3, A_1, A_1)$: EMP($S_{13}|\theta_1$) = \$2,020,000; EMP($S_{13}|\theta_2$) = \$640,000; EMP($S_{13}|\theta_3$) = \$300,000
$S_{14}(A_3, A_1, A_3)$: EMP($S_{14}|\theta_1$) = \$2,180,000; EMP($S_{14}|\theta_2$) = \$620,000; EMP($S_{14}|\theta_3$) = −\$400,000
$S_{15}(A_3, A_3, A_1)$: EMP($S_{15}|\theta_1$) = \$2,340,000; EMP($S_{15}|\theta_2$) = \$520,000; EMP($S_{15}|\theta_3$) = \$200,000
$S_{16}(A_2, A_2, A_3)$: EMP($S_{16}|\theta_1$) = \$1,600,000; EMP($S_{16}|\theta_2$) = \$770,000; EMP($S_{16}|\theta_3$) = −\$350,000
$S_{17}(A_2, A_3, A_2)$: EMP($S_{17}|\theta_1$) = \$1,700,000; EMP($S_{17}|\theta_2$) = \$620,000; EMP($S_{17}|\theta_3$) = −\$50,000
$S_{18}(A_2, A_3, A_3)$: EMP($S_{18}|\theta_1$) = \$1,800,000; EMP($S_{18}|\theta_2$) = \$590,000; EMP($S_{18}|\theta_3$) = −\$400,000
$S_{19}(A_3, A_2, A_2)$: EMP($S_{19}|\theta_1$) = \$2,200,000; EMP($S_{19}|\theta_2$) = \$710,000; EMP($S_{19}|\theta_3$) = −\$100,000
$S_{20}(A_3, A_2, A_3)$: EMP($S_{20}|\theta_1$) = \$2,300,000; EMP($S_{20}|\theta_2$) = \$680,000; EMP($S_{20}|\theta_3$) = −\$450,000
$S_{21}(A_3, A_3, A_2)$: EMP($S_{21}|\theta_1$) = \$2,400,000; EMP($S_{21}|\theta_2$) = \$530,000; EMP($S_{21}|\theta_3$) = −\$150,000
$S_{22}(A_1, A_2, A_3)$: EMP($S_{22}|\theta_1$) = \$1,180,000; EMP($S_{22}|\theta_2$) = \$740,000; EMP($S_{22}|\theta_3$) = −\$250,000
$S_{23}(A_1, A_3, A_2)$: EMP($S_{23}|\theta_1$) = \$1,280,000; EMP($S_{23}|\theta_2$) = \$590,000; EMP($S_{23}|\theta_3$) = \$50,000
$S_{24}(A_2, A_1, A_3)$: EMP($S_{24}|\theta_1$) = \$1,480,000; EMP($S_{24}|\theta_2$) = \$710,000; EMP($S_{24}|\theta_3$) = −\$300,000
$S_{25}(A_2, A_3, A_1)$: EMP($S_{25}|\theta_1$) = \$1,640,000; EMP($S_{25}|\theta_2$) = \$610,000; EMP($S_{25}|\theta_3$) = \$300,000
$S_{26}(A_3, A_2, A_1)$: EMP($S_{26}|\theta_1$) = \$2,140,000; EMP($S_{26}|\theta_2$) = \$700,000; EMP($S_{26}|\theta_3$) = \$250,000
$S_{27}(A_3, A_1, A_2)$: EMP($S_{27}|\theta_1$) = \$2,080,000; EMP($S_{27}|\theta_2$) = \$650,000; EMP($S_{27}|\theta_3$) = −\$50,000

by the corresponding prior probabilities and the resulting products summed. For example, the expected monetary payoff of Strategy S_6 is found to be

$$EMP(S_6) = P_0(\theta_1)EMP(S_6|\theta_1) + P_0(\theta_2)EMP(S_6|\theta_2) + P_0(\theta_3)EMP(S_6|\theta_3)$$
$$= (.1)(\$1,080,000) + (.6)(\$770,000) + (.3)(\$100,000)$$
$$= \$600,000$$

Table 19.10 presents the expected monetary payoffs for all 27 possible strategies ranked from largest to smallest.

4. Select the strategy with the largest expected monetary payoff.

We see from Table 19.10 that the optimal strategy is $S_9(A_2, A_2, A_1)$ with an expected monetary payoff of $722,981—the identical value (aside from the effects of rounding errors) found within the circle identifying chance fork f in Figure 19.5.

*19.9 Extensive-Form and Normal-Form Preposterior Analysis Compared

As we have seen, the extensive-form and the normal-form approaches to preposterior analysis lead the analyst to the same optimal strategy. Despite the inevitability of getting identical results from their use, extensive-form and normal-form methods differ markedly in certain important respects. The analyst attempting to choose between them should be aware of these differences and their implications.

Table 19.10 Expected Monetary Payoffs of the 27 Strategies Listed in Table 19.7 Ranked According to Magnitude

Rank	Strategy	Expected Monetary Payoff	Rank	Strategy	Expected Monetary Payoff
1	$S_9(A_2, A_2, A_1)$	$723,000	15	$S_4(A_1, A_1, A_2)$	567,000
2	$S_{26}(A_3, A_2, A_1)$	709,000	16	$S_{17}(A_2, A_3, A_2)$	527,000
3	$S_5(A_1, A_2, A_1)$	693,000	17	$S_{16}(A_2, A_2, A_3)$	517,000
4	$S_7(A_2, A_1, A_1)$	690,000	18	$S_{21}(A_3, A_3, A_2)$	513,000
5	$S_{13}(A_3, A_1, A_1)$	676,000	19	$S_{20}(A_3, A_2, A_3)$	503,000
6	$S_1(A_1, A_1, A_1)$	660,000	20	$S_{23}(A_1, A_3, A_2)$	497,000
7	$S_2(A_2, A_2, A_2)$	630,000	21	$S_{22}(A_1, A_2, A_3)$	487,000
8	$S_{25}(A_2, A_3, A_1)$	620,000	22	$S_{24}(A_1, A_1, A_3)$	484,000
9	$S_{19}(A_3, A_2, A_2)$	616,000	23	$S_{14}(A_3, A_1, A_2)$	470,000
10	$S_{15}(A_3, A_3, A_1)$	606,000	24	$S_{10}(A_1, A_1, A_3)$	454,000
11	$S_6(A_1, A_2, A_2)$	600,000	25	$S_{18}(A_2, A_3, A_3)$	414,000
12	$S_8(A_2, A_1, A_2)$	597,000	26	$S_3(A_3, A_3, A_3)$	400,000
13	$S_{11}(A_1, A_3, A_1)$	590,000	27	$S_{12}(A_1, A_3, A_3)$	384,000
14	$S_{27}(A_3, A_1, A_2)$	583,000			

The Principal Advantage of Extensive-Form Analysis

When extensive-form preposterior analysis is used, the analyst need not compute the expected monetary payoff of every possible strategy. Consequently, this kind of analysis can be completed with greater speed and ease than can a normal-form analysis applied to the same data. As was demonstrated with the CCMC example, when extensive-form analysis is used, nonoptimal courses of action are blocked off. (See Figure 19.5.) As a result, only the expected monetary payoff of the optimal strategy is computed. This can be a rather important advantage in situations involving a large number of action choices and a large number of possible sample outcomes. For example, in the normal-form version of the CCMC problem, the number of possible strategies was $n^r = 3^3 = 27$, a number implying many computational steps even though the decision system for this problem is relatively simple. Had the number of action choices and the number of possible "sample" outcomes each been 4, the number of possible strategies to be reckoned with would have been almost 9.5 times as large: $n^r = 4^4 = 256$.

The Principal Advantage of Normal-Form Analysis

In our demonstration of both the extensive-form and the normal-form analyses of the CCMC problem, we made use of subjective prior probabilities. For reasons which should be quite understandable by now, many analysts dislike having to rely too heavily on their prior probabilities. This is especially true when the decision situation is thought to be quite sensitive.* Fortunately, in normal-form analysis the prior probabilities are applied toward the end of the computational steps. This feature offers the analyst faced with a sensitive decision situation the convenience of varying prior probability assessments over plausible ranges with a view to determining whether the optimal strategy is altered as a result.

19.10 Summary

Sometimes prior analysis will be the only analysis used. If decision makers are up against time constraints or if they find the expected value of perfect information to be small, they will not extend their analysis to have it include new empirical information.

However, if analysts can afford to do so, or if they have the incentive for doing so, they may decide to revise their prior probability assessments by making use of "sample" evidence. If so, they perform what we call a posterior analysis. A posterior analysis involves: (1) obtaining payoff estimates associated with all state-and-action combinations and assigning prior probabilities, (2) gathering "sample" evidence, (3) revising the prior probabilities by taking into account the "sample" evidence, (4) using payoff data from Step 1 and probabilities from Step 3 and computing the expected payoff of each of the actions under consideration, and (5) selecting the action with the best expected payoff. The resulting Bayes action may be the same as or different from the one arrived at through prior analysis. If it is different, the presumption is that it is also better.

Between prior analysis and posterior analysis, so to speak, is preposterior analysis. Preposterior analysis may be used to help the analyst decide whether he should go to the expense

---

*A "sensitive" decision situation is one in which only small changes in the prior probabilities are required to change the Bayes action. If large changes in the prior probabilities are required to change the Bayes action, the decision situation is said to be "insensitive."

and the bother of performing a posterior analysis or be content with prior analysis. There are two ways of performing preposterior analysis: (1) extensive-form preposterior analysis, a graphic approach, and (2) normal-form preposterior analysis, a tabular approach. Although both approaches will lead to the same Bayes action, one approach may be preferred to the other under some circumstances. With extensive-form preposterior analysis, the computational drudgery is substantially less because the expected monetary payoff of every possible strategy need not be calculated. With normal-form preposterior analysis, the computational requirements are greater, but there is also more leeway for the analyst to vary his prior probabilities over reasonable ranges and to explore the implications of the results.

Terms Introduced in This Chapter

conditional expected monetary payoff
   of a strategy (p. 943)
expected monetary payoff of a strategy
   (p. 944)
cost of sample information, COSI
   (p. 940)
cost of uncertainty (p. 929)
expected net gain from sampling,
   ENGS (p. 940)
expected value of perfect information,
   EVPI (p. 927)

posterior expected value of perfect
   information,
   EVPI (p. 932)
prior expected value of perfect
   information,
   EVPI (p. 927)
expected value of sample information,
   EVSI (p. 940)
expected value with perfect
   information, EVWPI (p. 928)

posterior analysis (p. 923)
preposterior analysis (p. 925)
   extensive-form preposterior analysis
      (p. 934)
   normal-form preposterior analysis
      (p. 940)
"sample" (p. 925)
strategy (p. 941)

Formulas Introduced in This Chapter

Expected value of perfect information based on prior probabilities:

$$EVPI_0 = EVWPI_0 - EMP_0(A^*)$$

Expected value of perfect information based on posterior probabilities:

$$EVPI_1 = EVWPI_1 - EMP_1(A^*)$$

The conditional probability of state of nature θ_1 given sample indication x_1 (this is the Bayes formula introduced in section 6.9 of Chapter 6 with symbols modified to make them compatible with symbols and terms used in this chapter):

$$P(\theta_1|x_1) = \frac{P(\theta_1 \text{ and } x_1)}{P(x_1)} = \frac{P(\theta_1) \cdot P(x_1|\theta_1)}{\sum_{\text{all } i} [P(\theta_i)P(x_1|\theta_i)]}$$

Expected value of "sample" information:

$$EVSI = EMP(\text{Posterior analysis}) - EMP(\text{Prior analysis})$$

Expected net gain from "sampling":

$$ENGS = EVSI - COSI$$

Number of possible strategies (n admissible actions and r possible "sample" indications):

$$n^r$$

Questions and Problems

19.1 Explain what is meant by each of the following:
  a. Posterior analysis
  b. Expected value of perfect information
  c. EMP(A*)
  d. Extensive-form preposterior analysis
  e. Expected value of "sample" information
  f. Expected net gain from "sampling"
  g. Terminal decision
  h. Strategy

19.2 Distinguish between:
  a. Prior analysis and posterior analysis
  b. Preposterior analysis and prior analysis
  c. Posterior analysis and preposterior analysis
  d. Extensive-form preposterior analysis and normal-form preposterior analysis
  e. Expected value of perfect information and expected value of sample information
  f. EVWPI and EVPI
  g. $EMP_0(A_1)$ and $EMP_1(A_1)$

19.3 Distinguish between:
  a. Prior probability and posterior probability
  b. $P(x_1|\theta_1)$ and $P(\theta_1|x_1)$
  c. Prior expected value of perfect information and posterior expected value of perfect information
  d. ENGS and EVSI

19.4 Indicate which of the following statements you agree with and which you disagree with, and defend your opinions:
  a. Whenever EMP(Posterior analysis) less cost of obtaining "sample" evidence is greater than EMP(Prior analysis), one must necessarily employ posterior analysis.
  b. The choice between use of extensive-form preposterior analysis and normal-form preposterior analysis should be made only after serious reflection since the optimal strategy may not be the same under the two methods.
  c. Extensive-form preposterior analysis can usually be completed with greater speed and ease than can normal-form preposterior analysis when applied to the same data.
  d. The sensitivity of the decision situation may be a somewhat less troublesome consideration when normal-form preposterior analysis is used than when extensive-form preposterior analysis is used.

19.5 Indicate which of the following statements you agree with and which you disagree with, and defend your opinions:
  a. The probability of a specific event's occurring is an objective fact quite independent of the knowledge of the decision maker.
  b. Sometimes it is advisable to limit one's analysis of a decision problem to prior analysis.
  c. Since the measure "expected value of perfect information" is dependent on the idea that the analyst possesses a perfectly accurate forecasting device and since no such device exists in reality, the concept of EVPI has no practical importance.
  d. The measure obtained by EVWPI − EMP(A*) is known as the expected value of "sample" information.

 19.6 Refer to problem 18.16
  a. Compute the prior expected value of perfect information.
  b. In your opinion, would it be worthwhile for Motherlode Mining to collect sample information? Why or why not? If so, what kind of sample information would be appropriate?

19.7 Refer to problems 18.16 and 19.6. A small sample of ore is obtained and analyzed. The analysis indicates that the property under consideration is "rich" in silver (state θ_2). The results of a large number of similar past sampling experiences are shown as conditional probabilities in Table 19.11.

   a. Revise the prior probabilities by utilizing relevant information from Table 19.11.

   b. Perform the posterior analysis and identify the Bayes action.

   c. Determine the posterior expected value of perfect information. In your opinion, would it be worthwhile for the Motherlode Mining Company to collect still more sample evidence? Why or why not?

19.8 Refer to problems 18.16, 19.6, and 19.7. Assume that the cost of obtaining and analyzing a sample of ore in the amount used in problem 19.7 is $1 million. Perform an extensive-form preposterior analysis for this problem. State your conclusion.

19.9 Refer to problem 18.18.

   a. Compute the prior expected value of perfect information.

   b. In your opinion, would it be worthwhile for this company to collect "sample" evidence? Why or why not?

19.10 Refer to problems 18.18 and 19.9. The forecast made by the company president was de-rived without reference to a multiple-regression forecasting equation that was prepared sev-

---

Table 19.11 Relative-Frequency Probabilities of Five Types of Sample Evidence for the Motherlode Mining Company

State of Nature θ_i	Conditional Probabilities $P(x_k\|\theta_i)$					Total
	x_1	x_2	x_3	x_4	x_5	
θ_1 Ore is *very rich* in silver	.30	.30	.20	.20	.00	1.00
θ_2: Ore is *rich* in silver	.20	.40	.20	.10	.10	1.00
θ_3: Ore is of *average* richness	.05	.20	.50	.20	.05	1.00
θ_4: Ore has *very little* silver content	.10	.20	.20	.40	.10	1.00
θ_5: Ore contains *no silver* whatever	.10	.20	.20	.20	.30	1.00

x_1 represents "sample indicates State θ_1"
x_2 represents "sample indicates State θ_2"
x_3 represents "sample indicates State θ_3"
x_4 represents "sample indicates State θ_4"
x_5 represents "sample indicates State θ_5"

eral years ago by a consulting firm and that has been used as an ancillary forecasting tool ever since. Table 19.12 presents information regarding the equation's past forecasting accuracy with respect to direction of movement only.

At this time, the forecasting equation is signaling an upward movement.

  a. Revise the prior probabilities by utilizing relevant information from Table 19.12.
  b. Perform the posterior decision analysis and identify the Bayes action.
  c. Determine the posterior expected value of perfect information. In your opinion, would it be worthwhile for the manufacturing company to collect still more sample evidence? Why or why not?

 19.11 Refer to problems 18.18, 19.9, and 19.10. Assume that the cost of keeping the forecasting equation currently operational is estimated to be $500.

  a. Perform an extensive-form preposterior analysis for this problem. State your conclusion.
  b. Perform a normal-form preposterior analysis for this problem. State your conclusion.
  c. How do the results of parts *a* and *b* compare?

Special-Challenge Problem

 19.12 Refer to problem 18.27.

  a. Determine the prior expected value of perfect information.
  b. The studio chief has four critics read the screenplay and offer their independent opinions. The result: Two predict success and two predict failure. If the film should be a failure, the probability that a particular critic would dislike the screenplay is .8. If the film should be a success, the probability that a particular critic would dislike the screenplay is .4. What is the revised probability associated with state of nature θ_1 (success)? What is the revised probability associated with state of nature θ_2 (failure)?
  c. Compute the posterior expected monetary payoff of each possible action and identify the Bayes action.
  d. Determine the posterior expected value of perfect information. Does it seem likely that searching for additional information would be a worthwhile undertaking? Why or why not?

Table 19.12 Data Concerning the Past Direction-of-Change Forecasting Accuracy of the Multiple-Regression Forecasting Equation

I. When actual sales were *strong*, the forecasting equation had previously indicated:

> Up—50% of the time
>
> Flat—33% of the time
>
> Down—17% of the time

II. When actual sales were *average*, the forecasting equation had previously indicated:

> Up—20% of the time
>
> Flat—50% of the time
>
> Down—30% of the time

III. When actual sales were *weak*, the forecasting equation had previously indicated:

> Up—0% of the time
>
> Flat—25% of the time
>
> Down—75% of the time

Special-Challenge Problem

 19.13 Refer to problems 18.27 and 19.12. Perform an extensive-form preposterior analysis using the information provided in problems 18.29 and 19.12 and state your conclusion. (Assume that each critic consulted is paid $500 for his services.)

19.14 a. Examine the decision-tree diagram presented in Figure 19.6 and indicate whether posterior analysis should be employed, assuming that it will cost $50,000 to acquire the needed "sample" information.
   b. Determine the expected monetary payoff of the Bayes choice.
   c. Determine the expected value of "sample" information, if relevant.
   d. Determine the expected net gain from "sampling," if relevant.

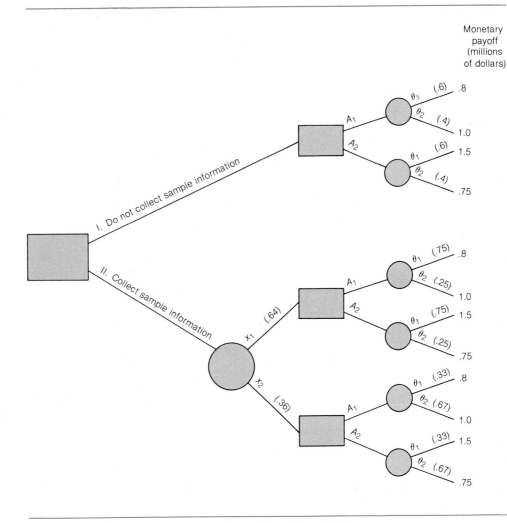

Figure 19.6. Hypothetical decision-tree diagram for problem 19.14.

19.15 a. Examine the decision-tree diagram presented in Figure 19.7 and indicate whether posterior analysis should be utilized, assuming that it will cost $10,000 to acquire the needed "sample" information.

b. Determine the expected monetary payoff of the Bayes choice.

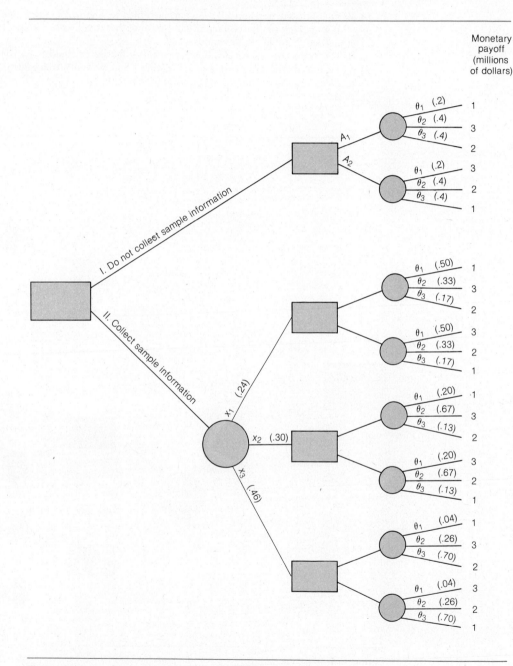

Figure 19.7. Hypothetical decision-tree diagram for problem 19.15.

19 Decision Theory: Posterior and Preposterior Analysis

c. Determine the expected value of "sample" information, if relevant.

d. Determine the expected net gain from "sampling," if relevant.

19.16 The decision problem for a student thinking about opening a surfing equipment store during the summer months is summarized in Table 19.13. Prior probabilities are shown in parentheses in this table.

The student is thinking of conducting a survey among surfers known to frequent the part of the beach where she plans to open her store as well as students of surfing classes which were held nearby the previous summer. She estimates that such a survey would cost about $200. The part of the survey having relevance in this problem asks the respondent to indicate whether the proposed store will be (1) a failure, (2) successful, or (3) very successful. Assume that the reliability of the survey results can be described in conditional probability terms as follows:

P(Survey indicates "failure"|It is a failure) = .7

P(Survey indicates "failure"|It is successful) = .2

P(Survey indicates "failure"|It is very successful) = .2

P(Survey indicates "successful"|It is a failure) = .2

P(Survey indicates "successful"|It is successful) = .5

P(Survey indicates "successful"|It is very successful) = .4

P(Survey indicates "very successful"|It is a failure) = .1

P(Survey indicates "very successful"|It is successful) = .3

P(Survey indicates "very successful"|It is very successful) = .4

a. On the basis of prior analysis, should the student open the surfing equipment store? What is the prior expected monetary payoff of the Bayes action?

b. What is the value of EMP(Posterior analysis) less the cost of obtaining sample information?

c. If relevant, determine the expected value of sample information.

d. If relevant, determine the expected net gain from sampling.

19.17 Refer to problem 19.16. Suppose the student conducted the survey and got an indication of "successful." What is the posterior expected monetary payoff of opening the store?

19.18 An engineer is thinking about quitting his $30,000 a year job and opening up his own consulting firm. His decision situation is shown in Table 19.14. Prior probabilities are shown in parentheses within the table.

Table 19.13 Monetary Payoffs Associated with the Question of Whether or Not to Open a Surfing Equipment Store

(Numbers Represent Profit)

	Action	
State of Nature	A_1: Open Store	A_2: Do Not Open Store
θ_1: Failure (.4)	−$20,000	$0
θ_2: Successful (.5)	$25,000	$0
θ_3: Very successful (.1)	$40,000	$0

Table 19.14 Monetary Payoffs Associated with the Question of Whether or Not to Quit Job and Open an Engineering Consulting Firm

(Numbers Represent Profits Before Taxes and Other Withholdings)

	Action	
State of Nature	A_1: Stay with Present Job	A_2: Quit and Start Consulting Firm
θ_1: Failure (.4)	$30,000	−$50,000
θ_2: Success (.6)	$30,000	−$75,000

He is now thinking about conducting a survey among members of the engineering profession whose opinions he respects. He figures that the cost of such a survey would be about $100, primarily for telephone charges. Assume that the reliability of the survey can be depicted in conditional probability terms as follows:

P(Survey indicates "failure"|Venture proves to be a failure) = .6

P(Survey indicates "failure"|Venture proves to be a success) = .2

P(Survey indicates "success"|Venture proves to be a failure) = .4

P(Survey indicates "success"|Venture proves to be a success) = .8

a. On the basis of prior analysis, should the engineer quit his job and open his own consulting firm? What is the prior expected monetary payoff of the Bayes action?

b. What is the value of EMP(Posterior analysis) less the cost of obtaining sample information?

c. If relevant, determine the expected value of sample information.

d. If relevant, determine the expected net gain from sampling.

 19.19 Refer to problem 19.18. Assume that the engineer conducts the survey and gets an indication of "successful." What is his posterior expected monetary payoff of quitting his present job and opening a consulting firm?

 19.20 A buyer for a chain of retail drugstores is contemplating how many dozen tubes of Preparation F to order. Preparation F is a popular new over-the-counter item for removing freckles. He does not wish to place too large an order because the product has a tendency to evaporate rather quickly on the store shelves. On the other hand, the popularity of Preparation F leads him to fear possible supply shortages if he places too small an order. Each one dozen tubes purchased will cost the drug chain $12 and will sell for $20. Any tubes of the product not sold within one month will be relegated to the one-half price counter where they invariably sell out quickly. The buyer believes that next month the retail chain can sell 1000, 1500, 2000, or 2500 dozen tubes. The buyer's prior probability distribution is as follows:

Sales Next Month θ_i	Probability
1000	.4
1500	.3
2000	.2
2500	.1

Table 19.15 Conditional Probabilities of the Form $P(x_k|\theta_i)$ for the Preparation F Problem

State of Nature (Actual Sales, Dozens)	Forecasted Sales (Dozens)			
	1000	1500	2000	2500
θ_1: 1000	.6	.3	.1	.0
θ_2: 1500	.3	.4	.2	.1
θ_3: 2000	.1	.3	.3	.3
θ_4: 2500	.1	.3	.2	.4

A marketing research firm can be engaged to forecast the demand for Preparation F. Historical records indicate that for similar past studies the reliability of the firm's forecasts can be stated in conditional probability terms as indicated in Table 19.15. How much would the buyer presumably be willing to spend for "sample" information?

TAKE CHARGE

19.21 Refer to problem 18.28. Revise your prior probabilities. Explain clearly how you did this. Then perform the posterior analysis. Has the Bayes action changed?

Cases for Analysis and Discussion

19.1 THE "DELTA COFFEE COMPANY": PART III

Review parts I and II of this case (case 1.1 of Chapter 1 and case 18.2 of Chapter 18). If you have not previously answered the questions pertaining to part II, do so now.

1. Using the hypothetical data you made up for part II of this case, determine the expected value of perfect information.

2. Assume some conditional probabilities, or likelihoods, to be used to revise your subjective prior probabilities. Make sure your likelihoods are clearly defined.

3. Use the likelihoods assumed above and the Bayes theorem to revise your subjective prior probabilities.

4. Determine the posterior expected monetary payoff associated with each of the available actions and identify the Bayes action. Indicate whether the Bayes action based on posterior probabilities is the same as the Bayes action based on prior probabilities.

5. Determine the posterior expected value of perfect information.

19.2 THE "HUGHES DOWN COMPANY": PART II

Review part I of this case (case 18.3 of Chapter 18). If you have not previously written an evaluation of the prior analysis part of this case, do so now.
The store manager continues his account:

The calculation of expected value of perfect information is shown in the following table [Table 19.16]:

Table 19.16 Calculation of Expected Value of Perfect Information for the "Hughes Down Company" Case

State of Nature	Monetary Payoff	Prior Probability	Weighted Monetary Payoff
θ_1: Weak sales ($720,000)	$133,200	.32	$42,624
θ_2: Average sales ($810,000)	149,850	.53	79,421
θ_3: Strong sales ($900,000)	166,500	.15	24,975
		Total = $EVWPI_0$	$147,020

Expected value *with* perfect information: $EVWPI_0 = \$147{,}020$
Expected monetary payoff of the Bayes action: $EMP_0(A_4) = \$147{,}020$
Expected value *of* perfect information: $EVWPI_0 - EMP_0(A_4) = \$147{,}020 - \$147{,}020 = \$0$

I then revised the prior probabilities to reflect the effect of public advertising. The next two tables [Tables 19.17 and 19.18] show the results of my posterior analysis.

This decision analysis shows conclusively that net profit *does* increase with the turn-over rate of inventories. Therefore, to maximize net profit at a given sales level, the base inventory must be lowered to the prescribed level. This will force turnover to be increased. Given that the supply is constant, a daily replenishment frequency will give maximum support to the desired 90% demand level. Cost savings will show through increased net profit.

Table 19.17 Revision of Prior Probabilities Based on the Assumption that Public Advertising is Being Used

State of Nature	(1) Prior Probability $P(\theta_i)$	(2) Likelihood $P(x_2\|\theta_i)$	(3) Joint Probability $P(\theta_i) \cdot P(x_2\|\theta_i)$	(4) Posterior Probability $P(\theta_i\|x_2)$
θ_1: Weak sales ($720,000)	.32	.55	.18	.461
θ_2: Average sales ($810,000)	.53	.35	.19	.487
θ_3: Strong sales ($900,000)	.15	.10	.02	.052
		Total	.39	1.000

19 Decision Theory: Posterior and Preposterior Analysis

Table 19.18 Calculation of Expected Monetary Payoffs for the "Hughes Down Company" Case Using Posterior Probability Weights

Action A_1: *Turn Inventories over Two Times Annually*

State of Nature	Posterior Probability	Monetary Payoff	Weighted Monetary Payoff
θ_1: Weak sales ($720,000)	.462	$68,400	$31,600
θ_2: Average sales ($810,000)	.487	76,950	37,475
θ_3: Strong sales ($900,000)	.051	85,500	4,361
		Total	$73,436

Action A_2: *Turn Inventories over Four Times Annually*

State of Nature	Posterior Probability	Monetary Payoff	Weighted Monetary Payoff
θ_1: Weak sales ($720,000)	.462	$ 90,000	$41,580
θ_2: Average sales ($810,000)	.487	101,250	49,308
θ_3: Strong sales ($900,000)	.051	112,500	5,738
		Total	$96,626

Action A_3: *Turn Inventories over Six Times Annually*

State of Nature	Posterior Probability	Monetary Payoff	Weighted Monetary Payoff
θ_1: Weak sales ($720,000)	.462	$111,600	$ 55,559
θ_2: Average sales ($810,000)	.487	125,550	61,143
θ_3: Strong sales ($900,000)	.051	139,500	7,115
		Total	$123,817

Action A_4: *Turn Inventories over Eight Times Annually*

State of Nature	Posterior Probability	Monetary Payoff	Weighted Monetary Payoff
θ_1: Weak sales ($720,000)	.462	$133,200	$ 61,538
θ_2: Average sales ($810,000)	.487	149,850	72,976
θ_3: Strong sales ($900,000)	.051	166,500	8,492
		Total	$143,006

1. Before looking at questions 2 and 3 (printed upside down to help you avoid temptation), write an evaluation of this store manager's posterior analysis. Be sure to comment on the validity, or lack of it, of his likelihoods.

2. Refer to Table 18.24. Eliminate all inadmissible actions. Does the result in any way affect your opinions about the quality of this store manager's analysis as expressed in question 1 here and question 1 of case 18.37? If so, in what way or ways?

3. Assume that one of the action choices had been A_5: Turn inventories over 365 times annually. How do you suppose the payoff associated with Action A_5 would compare with that associated with Action A_4? Why?

Appendix

Table 1 Individual Terms of the Binomial Distribution

n	X	.05	.10	.15	.20	.25	.30	.35	.40	.45	.50	.55	.60	.65	.70	.75	.80	.85	.90	.95
1	0	.9500	.9000	.8500	.8000	.7500	.7000	.6500	.6000	.5500	.5000	.4500	.4000	.3500	.3000	.2500	.2000	.1500	.1000	.0500
	1	.0500	.1000	.1500	.2000	.2500	.3000	.3500	.4000	.4500	.5000	.5500	.6000	.6500	.7000	.7500	.8000	.8500	.9000	.9500
2	0	.9025	.8100	.7225	.6400	.5625	.4900	.4225	.3600	.3025	.2500	.2025	.1600	.1225	.0900	.0625	.0400	.0225	.0100	.0025
	1	.0950	.1800	.2550	.3200	.3750	.4200	.4550	.4800	.4950	.5000	.4950	.4800	.4550	.4200	.3750	.3200	.2550	.1800	.0950
	2	.0025	.0100	.0225	.0400	.0625	.0900	.1225	.1600	.2025	.2500	.3025	.3600	.4225	.4900	.5625	.6400	.7225	.8100	.9025
3	0	.8574	.7290	.6141	.5120	.4219	.3430	.2746	.2160	.1664	.1250	.0911	.0640	.0429	.0270	.0156	.0080	.0034	.0010	.0001
	1	.1354	.2430	.3251	.3840	.4219	.4410	.4436	.4320	.4084	.3750	.3341	.2880	.2389	.1890	.1406	.0960	.0574	.0270	.0071
	2	.0071	.0270	.0574	.0960	.1406	.1890	.2389	.2880	.3341	.3750	.4084	.4320	.4436	.4410	.4219	.3840	.3251	.2430	.1354
	3	.0001	.0010	.0034	.0080	.0156	.0270	.0429	.0640	.0911	.1250	.1664	.2160	.2746	.3430	.4219	.5120	.6141	.7290	.8574
4	0	.8145	.6561	.5220	.4096	.3164	.2401	.1785	.1296	.0915	.0625	.0410	.0256	.0150	.0081	.0039	.0016	.0005	.0001	.0000
	1	.1715	.2916	.3685	.4096	.4219	.4116	.3845	.3456	.2995	.2500	.2005	.1536	.1115	.0756	.0469	.0256	.0115	.0036	.0005
	2	.0135	.0486	.0975	.1536	.2109	.2646	.3105	.3456	.3675	.3750	.3675	.3456	.3105	.2646	.2109	.1536	.0975	.0486	.0135
	3	.0005	.0036	.0115	.0256	.0469	.0756	.1115	.1536	.2005	.2500	.2995	.3456	.3845	.4116	.4219	.4096	.3685	.2916	.1715
	4	.0000	.0001	.0005	.0016	.0039	.0081	.0150	.0256	.0410	.0625	.0915	.1296	.1785	.2401	.3164	.4096	.5220	.6561	.8145
5	0	.7738	.5905	.4437	.3277	.2373	.1681	.1160	.0778	.0503	.0313	.0185	.0102	.0053	.0024	.0010	.0003	.0001	.0000	.0000
	1	.2036	.3281	.3915	.4096	.3955	.3602	.3124	.2592	.2059	.1563	.1128	.0768	.0488	.0284	.0146	.0064	.0022	.0004	.0000
	2	.0214	.0729	.1382	.2048	.2637	.3087	.3364	.3456	.3369	.3125	.2757	.2304	.1811	.1323	.0879	.0512	.0244	.0081	.0011
	3	.0011	.0081	.0244	.0512	.0879	.1323	.1811	.2304	.2757	.3125	.3369	.3456	.3364	.3087	.2637	.2048	.1382	.0729	.0214
	4	.0000	.0004	.0022	.0064	.0146	.0283	.0488	.0768	.1128	.1562	.2059	.2592	.3124	.3601	.3955	.4096	.3915	.3281	.2036
	5	.0000	.0000	.0001	.0003	.0010	.0024	.0053	.0102	.0185	.0312	.0503	.0778	.1160	.1681	.2373	.3277	.4437	.5905	.7738
6	0	.7351	.5314	.3771	.2621	.1780	.1176	.0754	.0467	.0277	.0156	.0083	.0041	.0018	.0007	.0002	.0001	.0000	.0000	.0000
	1	.2321	.3543	.3993	.3932	.3560	.3025	.2437	.1866	.1359	.0938	.0609	.0369	.0205	.0102	.0044	.0015	.0004	.0001	.0000
	2	.0305	.0984	.1762	.2458	.2966	.3241	.3280	.3110	.2780	.2344	.1861	.1382	.0951	.0595	.0330	.0154	.0055	.0012	.0001
	3	.0021	.0146	.0415	.0819	.1318	.1852	.2355	.2765	.3032	.3125	.3032	.2765	.2355	.1852	.1318	.0819	.0415	.0146	.0021
	4	.0001	.0012	.0055	.0154	.0330	.0595	.0951	.1382	.1861	.2344	.2780	.3110	.3280	.3241	.2966	.2458	.1762	.0984	.0305
	5	.0000	.0001	.0004	.0015	.0044	.0102	.0205	.0369	.0609	.0937	.1359	.1866	.2437	.3025	.3560	.3932	.3993	.3543	.2321
	6	.0000	.0000	.0000	.0001	.0002	.0007	.0018	.0041	.0083	.0156	.0277	.0467	.0754	.1176	.1780	.2621	.3771	.5314	.7351
7	0	.6983	.4783	.3206	.2097	.1335	.0824	.0490	.0280	.0152	.0078	.0037	.0016	.0006	.0002	.0001	.0000	.0000	.0000	.0000
	1	.2573	.3720	.3960	.3670	.3115	.2471	.1848	.1306	.0872	.0547	.0320	.0172	.0084	.0036	.0013	.0004	.0001	.0000	.0000
	2	.0406	.1240	.2097	.2753	.3115	.3177	.2985	.2613	.2140	.1641	.1172	.0774	.0466	.0250	.0115	.0043	.0012	.0002	.0000
	3	.0036	.0230	.0617	.1147	.1730	.2269	.2679	.2903	.2918	.2734	.2388	.1935	.1442	.0972	.0577	.0287	.0109	.0026	.0002
	4	.0002	.0026	.0109	.0287	.0577	.0972	.1442	.1935	.2388	.2734	.2918	.2903	.2679	.2269	.1730	.1147	.0617	.0230	.0036
	5	.0000	.0002	.0012	.0043	.0115	.0250	.0466	.0774	.1172	.1641	.2140	.2613	.2985	.3177	.3115	.2753	.2097	.1240	.0406
	6	.0000	.0000	.0001	.0004	.0013	.0036	.0084	.0172	.0320	.0547	.0872	.1306	.1848	.2471	.3115	.3670	.3960	.3720	.2573
	7	.0000	.0000	.0000	.0000	.0001	.0002	.0006	.0016	.0037	.0078	.0152	.0280	.0490	.0824	.1335	.2097	.3206	.4783	.6983
8	0	.6634	.4305	.2725	.1678	.1001	.0576	.0319	.0168	.0084	.0039	.0017	.0007	.0002	.0001	.0000	.0000	.0000	.0000	.0000
	1	.2793	.3826	.3847	.3355	.2670	.1977	.1373	.0896	.0548	.0313	.0164	.0079	.0033	.0012	.0004	.0001	.0000	.0000	.0000
	2	.0515	.1488	.2376	.2936	.3115	.2965	.2587	.2090	.1569	.1094	.0703	.0413	.0217	.0100	.0038	.0011	.0002	.0000	.0000
	3	.0054	.0331	.0839	.1468	.2076	.2541	.2786	.2787	.2568	.2188	.1719	.1239	.0808	.0467	.0231	.0092	.0026	.0004	.0000
	4	.0004	.0046	.0185	.0459	.0865	.1361	.1875	.2322	.2627	.2734	.2627	.2322	.1875	.1361	.0865	.0459	.0185	.0046	.0004
	5	.0000	.0004	.0026	.0092	.0231	.0467	.0808	.1239	.1719	.2188	.2568	.2787	.2786	.2541	.2076	.1468	.0839	.0331	.0054
	6	.0000	.0000	.0002	.0011	.0038	.0100	.0217	.0413	.0703	.1094	.1569	.2090	.2587	.2965	.3115	.2936	.2376	.1488	.0515
	7	.0000	.0000	.0000	.0001	.0004	.0012	.0033	.0079	.0164	.0312	.0548	.0896	.1373	.1977	.2670	.3355	.3847	.3826	.2793
	8	.0000	.0000	.0000	.0000	.0000	.0001	.0002	.0007	.0017	.0039	.0084	.0168	.0319	.0576	.1001	.1678	.2725	.4305	.6634
9	0	.6302	.3874	.2316	.1342	.0751	.0404	.0207	.0101	.0046	.0020	.0008	.0003	.0001	.0000	.0000	.0000	.0000	.0000	.0000
	1	.2986	.3874	.3679	.3020	.2253	.1556	.1004	.0605	.0339	.0176	.0083	.0035	.0013	.0004	.0001	.0000	.0000	.0000	.0000
	2	.0629	.1722	.2597	.3020	.3003	.2668	.2162	.1612	.1110	.0703	.0407	.0212	.0098	.0039	.0012	.0003	.0000	.0000	.0000
	3	.0077	.0446	.1069	.1762	.2336	.2668	.2716	.2508	.2119	.1641	.1160	.0743	.0424	.0210	.0087	.0028	.0006	.0001	.0000
	4	.0006	.0074	.0283	.0661	.1168	.1715	.2194	.2508	.2600	.2461	.2128	.1672	.1181	.0735	.0389	.0165	.0050	.0008	.0000

(continued)

A-1

Table 1 (*Continued*)

n	X	.05	.10	.15	.20	.25	.30	.35	.40	.45	.50	.55	.60	.65	.70	.75	.80	.85	.90	.95
	5	.0000	.0008	.0050	.0165	.0389	.0735	.1181	.1672	.2128	.2461	.2600	.2508	.2194	.1715	.1168	.0661	.0283	.0074	.0006
	6	.0000	.0001	.0006	.0028	.0087	.0210	.0424	.0743	.1160	.1641	.2119	.2508	.2716	.2668	.2336	.1762	.1069	.0446	.0077
	7	.0000	.0000	.0000	.0003	.0012	.0039	.0098	.0212	.0407	.0703	.1110	.1612	.2162	.2668	.3003	.3020	.2597	.1722	.0629
	8	.0000	.0000	.0000	.0000	.0001	.0004	.0013	.0035	.0083	.0176	.0339	.0605	.1004	.1556	.2253	.3020	.3679	.3874	.2986
	9	.0000	.0000	.0000	.0000	.0000	.0000	.0001	.0003	.0008	.0020	.0046	.0101	.0207	.0404	.0751	.1342	.2316	.3874	.6302
10	0	.5987	.3487	.1969	.1074	.0563	.0282	.0135	.0060	.0025	.0010	.0003	.0001	.0000	.0000	.0000	.0000	.0000	.0000	.0000
	1	.3151	.3874	.3474	.2684	.1877	.1211	.0725	.0403	.0207	.0098	.0042	.0016	.0005	.0001	.0000	.0000	.0000	.0000	.0000
	2	.0746	.1937	.2759	.3020	.2816	.2335	.1757	.1209	.0763	.0439	.0229	.0106	.0043	.0014	.0004	.0001	.0000	.0000	.0000
	3	.0105	.0574	.1298	.2013	.2503	.2668	.2522	.2150	.1665	.1172	.0746	.0425	.0212	.0090	.0031	.0008	.0001	.0000	.0000
	4	.0010	.0112	.0401	.0881	.1460	.2001	.2377	.2508	.2384	.2051	.1596	.1115	.0689	.0368	.0162	.0055	.0012	.0001	.0000
	5	.0001	.0015	.0085	.0264	.0584	.1029	.1536	.2007	.2340	.2461	.2340	.2007	.1536	.1029	.0584	.0264	.0085	.0015	.0001
	6	.0000	.0001	.0012	.0055	.0162	.0368	.0689	.1115	.1596	.2051	.2384	.2508	.2377	.2001	.1460	.0881	.0401	.0112	.0010
	7	.0000	.0000	.0001	.0008	.0031	.0090	.0212	.0425	.0746	.1172	.1665	.2150	.2522	.2668	.2503	.2013	.1298	.0574	.0105
	8	.0000	.0000	.0000	.0001	.0004	.0014	.0043	.0106	.0229	.0439	.0763	.1209	.1757	.2335	.2816	.3020	.2759	.1937	.0746
	9	.0000	.0000	.0000	.0000	.0000	.0001	.0005	.0016	.0042	.0098	.0207	.0403	.0725	.1211	.1877	.2684	.3474	.3874	.3151
	10	.0000	.0000	.0000	.0000	.0000	.0000	.0000	.0001	.0003	.0010	.0025	.0060	.0135	.0282	.0563	.1074	.1969	.3487	.5987
11	0	.5688	.3138	.1673	.0859	.0422	.0198	.0088	.0036	.0014	.0005	.0002	.0000	.0000	.0000	.0000	.0000	.0000	.0000	.0000
	1	.3293	.3835	.3248	.2362	.1549	.0932	.0518	.0266	.0125	.0054	.0021	.0007	.0002	.0000	.0000	.0000	.0000	.0000	.0000
	2	.0867	.2131	.2866	.2953	.2581	.1998	.1395	.0887	.0513	.0269	.0126	.0052	.0018	.0005	.0001	.0000	.0000	.0000	.0000
	3	.0137	.0710	.1517	.2215	.2581	.2568	.2254	.1774	.1259	.0806	.0462	.0234	.0102	.0037	.0011	.0002	.0000	.0000	.0000
	4	.0014	.0158	.0536	.1107	.1721	.2201	.2428	.2365	.2060	.1611	.1128	.0701	.0379	.0173	.0064	.0017	.0003	.0000	.0000
	5	.0001	.0025	.0132	.0388	.0803	.1321	.1830	.2207	.2360	.2256	.1931	.1471	.0985	.0566	.0268	.0097	.0023	.0003	.0000
	6	.0000	.0003	.0023	.0097	.0268	.0566	.0985	.1471	.1931	.2256	.2360	.2207	.1830	.1321	.0803	.0388	.0132	.0025	.0001
	7	.0000	.0000	.0003	.0017	.0064	.0173	.0379	.0701	.1128	.1611	.2060	.2365	.2428	.2201	.1721	.1107	.0536	.0158	.0014
	8	.0000	.0000	.0000	.0002	.0011	.0037	.0102	.0234	.0462	.0806	.1259	.1774	.2254	.2568	.2581	.2215	.1517	.0710	.0137
	9	.0000	.0000	.0000	.0000	.0001	.0005	.0018	.0052	.0126	.0269	.0513	.0887	.1395	.1998	.2581	.2953	.2866	.2131	.0867
	10	.0000	.0000	.0000	.0000	.0000	.0000	.0002	.0007	.0021	.0054	.0125	.0266	.0518	.0932	.1549	.2362	.3248	.3835	.3293
	11	.0000	.0000	.0000	.0000	.0000	.0000	.0000	.0000	.0002	.0005	.0014	.0036	.0088	.0198	.0422	.0859	.1673	.3138	.5688
12	0	.5404	.2824	.1422	.0687	.0317	.0138	.0057	.0022	.0008	.0002	.0001	.0000	.0000	.0000	.0000	.0000	.0000	.0000	.0000
	1	.3413	.3766	.3012	.2062	.1267	.0712	.0368	.0174	.0075	.0029	.0010	.0003	.0001	.0000	.0000	.0000	.0000	.0000	.0000
	2	.0988	.2301	.2924	.2835	.2323	.1678	.1088	.0639	.0339	.0161	.0068	.0025	.0008	.0002	.0000	.0000	.0000	.0000	.0000
	3	.0173	.0852	.1720	.2362	.2581	.2397	.1954	.1419	.0923	.0537	.0277	.0125	.0048	.0015	.0004	.0001	.0000	.0000	.0000
	4	.0021	.0213	.0683	.1329	.1936	.2311	.2367	.2128	.1700	.1208	.0762	.0420	.0199	.0078	.0024	.0005	.0001	.0000	.0000
	5	.0002	.0038	.0193	.0532	.1032	.1585	.2039	.2270	.2225	.1934	.1489	.1009	.0591	.0291	.0115	.0033	.0006	.0000	.0000
	6	.0000	.0005	.0040	.0155	.0401	.0792	.1281	.1766	.2124	.2256	.2124	.1766	.1281	.0792	.0401	.0155	.0040	.0005	.0000
	7	.0000	.0000	.0006	.0033	.0115	.0291	.0591	.1009	.1489	.1934	.2225	.2270	.2039	.1585	.1032	.0532	.0193	.0038	.0002
	8	.0000	.0000	.0001	.0005	.0024	.0078	.0199	.0420	.0762	.1208	.1700	.2128	.2367	.2311	.1936	.1329	.0683	.0213	.0021
	9	.0000	.0000	.0000	.0001	.0004	.0015	.0048	.0125	.0277	.0537	.0923	.1419	.1954	.2397	.2581	.2362	.1720	.0852	.0173
	10	.0000	.0000	.0000	.0000	.0000	.0002	.0008	.0025	.0068	.0161	.0339	.0639	.1088	.1678	.2323	.2835	.2924	.2301	.0988
	11	.0000	.0000	.0000	.0000	.0000	.0000	.0001	.0003	.0010	.0029	.0075	.0174	.0368	.0712	.1267	.2062	.3012	.3766	.3413
	12	.0000	.0000	.0000	.0000	.0000	.0000	.0000	.0000	.0001	.0002	.0008	.0022	.0057	.0138	.0317	.0687	.1422	.2824	.5404

Table 2 Areas for the Standard Normal Probability Distribution

Example: For Z = 2.05, shaded area is .4798 out of the total area of 1.

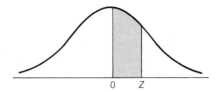

Z	.00	.01	.02	.03	.04	.05	.06	.07	.08	.09
0.0	.0000	.0040	.0080	.0120	.0160	.0199	.0239	.0279	.0319	.0359
0.1	.0398	.0438	.0478	.0517	.0557	.0596	.0636	.0675	.0714	.0753
0.2	.0793	.0832	.0871	.0910	.0948	.0987	.1026	.1064	.1103	.1141
0.3	.1179	.1217	.1255	.1293	.1331	.1368	.1406	.1443	.1480	.1517
0.4	.1554	.1591	.1628	.1664	.1700	.1736	.1772	.1808	.1844	.1879
0.5	.1915	.1950	.1985	.2019	.2054	.2088	.2123	.2157	.2190	.2224
0.6	.2257	.2291	.2324	.2357	.2389	.2422	.2454	.2486	.2518	.2549
0.7	.2580	.2611	.2642	.2673	.2703	.2734	.2764	.2794	.2823	.2852
0.8	.2881	.2910	.2939	.2967	.2995	.3023	.3051	.3078	.3106	.3133
0.9	.3159	.3186	.3212	.3238	.3264	.3289	.3315	.3340	.3365	.3389
1.0	.3413	.3438	.3461	.3485	.3508	.3531	.3554	.3577	.3599	.3621
1.1	.3643	.3665	.3686	.3708	.3729	.3749	.3770	.3790	.3810	.3830
1.2	.3849	.3869	.3888	.3907	.3925	.3944	.3962	.3980	.3997	.4015
1.3	.4032	.4049	.4066	.4082	.4099	.4115	.4131	.4147	.4162	.4177
1.4	.4192	.4207	.4222	.4236	.4251	.4265	.4279	.4292	.4306	.4319
1.5	.4332	.4345	.4357	.4370	.4382	.4394	.4406	.4418	.4429	.4441
1.6	.4452	.4463	.4474	.4484	.4495	.4505	.4515	.4525	.4535	.4545
1.7	.4554	.4564	.4573	.4582	.4591	.4599	.4608	.4616	.4625	.4633
1.8	.4641	.4649	.4656	.4664	.4671	.4678	.4686	.4693	.4699	.4706
1.9	.4713	.4719	.4726	.4732	.4738	.4744	.4750	.4756	.4761	.4767
2.0	.4772	.4778	.4783	.4788	.4793	.4798	.4803	.4808	.4812	.4817
2.1	.4821	.4826	.4830	.4834	.4838	.4842	.4846	.4850	.4854	.4857
2.2	.4861	.4864	.4868	.4871	.4875	.4878	.4881	.4884	.4887	.4890
2.3	.4893	.4896	.4898	.4901	.4904	.4906	.4909	.4911	.4913	.4916
2.4	.4918	.4920	.4922	.4925	.4927	.4929	.4931	.4932	.4934	.4936
2.5	.4938	.4940	.4941	.4943	.4945	.4946	.4948	.4949	.4951	.4952
2.6	.4953	.4955	.4956	.4957	.4959	.4960	.4961	.4962	.4963	.4964
2.7	.4965	.4966	.4967	.4968	.4969	.4970	.4971	.4972	.4973	.4974
2.8	.4974	.4975	.4976	.4977	.4977	.4978	.4979	.4979	.4980	.4981
2.9	.4981	.4982	.4982	.4983	.4984	.4984	.4985	.4985	.4986	.4986
3.0	.4987	.4987	.4987	.4988	.4988	.4989	.4989	.4989	.4990	.4990
3.1	.4990	.4991	.4991	.4991	.4992	.4992	.4992	.4992	.4993	.4993
3.2	.4993	.4993	.4994	.4994	.4994	.4994	.4994	.4995	.4995	.4995
3.3	.4995	.4995	.4995	.4996	.4996	.4996	.4996	.4996	.4996	.4997
3.4	.4997	.4997	.4997	.4997	.4997	.4997	.4997	.4997	.4998	.4998
3.5	.4998	.4998	.4998	.4998	.4998	.4998	.4998	.4998	.4998	.4998
3.6	.4998	.4998	.4999	.4999	.4999	.4999	.4999	.4999	.4999	.4999
3.7	.4999	.4999	.4999	.4999	.4999	.4999	.4999	.4999	.4999	.4999

Table 3   t Distribution

Areas in Both Tails Combined

Example: For 10 degrees of freedom and an area of .05 in the two tails combined, the *t* value is 2.228.

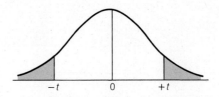

Degrees of Freedom	Area in Both Tails Combined			
	.10	.05	.02	.01
1	6.314	12.706	31.821	63.657
2	2.920	4.303	6.965	9.925
3	2.353	3.182	4.541	5.841
4	2.132	2.776	3.747	4.604
5	2.015	2.571	3.365	4.032
6	1.943	2.447	3.143	3.707
7	1.895	2.365	2.998	3.499
8	1.860	2.306	2.896	3.355
9	1.833	2.262	2.821	3.250
10	1.812	2.228	2.764	3.169
11	1.796	2.201	2.718	3.106
12	1.782	2.179	2.681	3.055
13	1.771	2.160	2.650	3.012
14	1.761	2.145	2.624	2.977
15	1.753	2.131	2.602	2.947
16	1.746	2.120	2.583	2.921
17	1.740	2.110	2.567	2.898
18	1.734	2.101	2.552	2.878
19	1.729	2.093	2.539	2.861
20	1.725	2.086	2.528	2.845
21	1.721	2.080	2.518	2.831
22	1.717	2.074	2.508	2.819
23	1.714	2.069	2.500	2.807
24	1.711	2.064	2.492	2.797
25	1.708	2.060	2.485	2.787
26	1.706	2.056	2.479	2.779
27	1.703	2.052	2.473	2.771
28	1.701	2.048	2.467	2.763
29	1.699	2.045	2.462	2.756
30	1.697	2.042	2.457	2.750
40	1.684	2.021	2.423	2.704
60	1.671	2.000	2.390	2.660
120	1.658	1.980	2.358	2.617
Normal distribution	1.645	1.960	2.326	2.576

Source: From Table III of Fisher and Yates, *Statistical Tables for Biological, Agricultural and Medical Research*, published by Longman Group Ltd., London (previously published by Oliver & Boyd, Edinburgh), by permission of the authors and publishers.

Table 4   Poisson Probabilities

	μ									
X	0.005	0.01	0.02	0.03	0.04	0.05	0.06	0.07	0.08	0.09
0	0.9950	0.9900	0.9802	0.9704	0.9608	0.9512	0.9418	0.9324	0.9231	0.9139
1	0.0050	0.0099	0.0192	0.0291	0.0384	0.0476	0.0565	0.0653	0.0738	0.0823
2	0.0000	0.0000	0.0002	0.0004	0.0008	0.0012	0.0017	0.0023	0.0030	0.0037
3	0.0000	0.0000	0.0000	0.0000	0.0000	0.0000	0.0000	0.0001	0.0001	0.0001

	μ									
X	0.1	0.2	0.3	0.4	0.5	0.6	0.7	0.8	0.9	1.0
0	0.9048	0.8187	0.7408	0.6703	0.6065	0.5488	0.4966	0.4493	0.4066	0.3679
1	0.0905	0.1637	0.2222	0.2681	0.3033	0.3293	0.3476	0.3595	0.3659	0.3679
2	0.0045	0.0164	0.0333	0.0536	0.0758	0.0988	0.1217	0.1438	0.1647	0.1839
3	0.0002	0.0011	0.0033	0.0072	0.0126	0.0198	0.0284	0.0383	0.0494	0.0613
4	0.0000	0.0001	0.0002	0.0007	0.0016	0.0030	0.0050	0.0077	0.0111	0.0153
5	0.0000	0.0000	0.0000	0.0001	0.0002	0.0004	0.0007	0.0012	0.0020	0.0031
6	0.0000	0.0000	0.0000	0.0000	0.0000	0.0000	0.0001	0.0002	0.0003	0.0005
7	0.0000	0.0000	0.0000	0.0000	0.0000	0.0000	0.0000	0.0000	0.0000	0.0001

	μ									
X	1.1	1.2	1.3	1.4	1.5	1.6	1.7	1.8	1.9	2.0
0	0.3329	0.3012	0.2725	0.2466	0.2231	0.2019	0.1827	0.1653	0.1496	0.1353
1	0.3662	0.3614	0.3543	0.3452	0.3347	0.3230	0.3106	0.2975	0.2842	0.2707
2	0.2014	0.2169	0.2303	0.2417	0.2510	0.2584	0.2640	0.2678	0.2700	0.2707
3	0.0738	0.0867	0.0998	0.1128	0.1255	0.1378	0.1496	0.1607	0.1710	0.1804
4	0.0203	0.0260	0.0324	0.0395	0.0471	0.0551	0.0636	0.0723	0.0812	0.0902
5	0.0045	0.0062	0.0084	0.0111	0.0141	0.0176	0.0216	0.0260	0.0309	0.0361
6	0.0008	0.0012	0.0018	0.0026	0.0035	0.0047	0.0061	0.0078	0.0098	0.0120
7	0.0001	0.0002	0.0003	0.0005	0.0008	0.0011	0.0015	0.0020	0.0027	0.0034
8	0.0000	0.0000	0.0001	0.0001	0.0001	0.0002	0.0003	0.0005	0.0006	0.0009
9	0.0000	0.0000	0.0000	0.0000	0.0000	0.0000	0.0001	0.0001	0.0001	0.0002

	μ									
X	2.1	2.2	2.3	2.4	2.5	2.6	2.7	2.8	2.9	3.0
0	0.1225	0.1108	0.1003	0.0907	0.0821	0.0743	0.0672	0.0608	0.0550	0.0498
1	0.2572	0.2438	0.2306	0.2177	0.2052	0.1931	0.1815	0.1703	0.1596	0.1494
2	0.2700	0.2681	0.2652	0.2613	0.2565	0.2510	0.2450	0.2384	0.2314	0.2240
3	0.1890	0.1966	0.2033	0.2090	0.2138	0.2176	0.2205	0.2225	0.2237	0.2240
4	0.0992	0.1082	0.1169	0.1254	0.1336	0.1414	0.1488	0.1557	0.1622	0.1680
5	0.0417	0.0476	0.0538	0.0602	0.0668	0.0735	0.0804	0.0872	0.0940	0.1008
6	0.0146	0.0174	0.0206	0.0241	0.0278	0.0319	0.0362	0.0407	0.0455	0.0504
7	0.0044	0.0055	0.0068	0.0083	0.0099	0.0118	0.0139	0.0163	0.0188	0.0216
8	0.0011	0.0015	0.0019	0.0025	0.0031	0.0038	0.0047	0.0057	0.0068	0.0081
9	0.0003	0.0004	0.0005	0.0007	0.0009	0.0011	0.0014	0.0018	0.0022	0.0027
10	0.0001	0.0001	0.0001	0.0002	0.0002	0.0003	0.0004	0.0005	0.0006	0.0008
11	0.0000	0.0000	0.0000	0.0000	0.0000	0.0001	0.0001	0.0001	0.0002	0.0002
12	0.0000	0.0000	0.0000	0.0000	0.0000	0.0000	0.0000	0.0000	0.0000	0.0001

(continued)

Table 4 (*Continued*)

					μ					
X	3.1	3.2	3.3	3.4	3.5	3.6	3.7	3.8	3.9	4.0
0	0.0450	0.0408	0.0369	0.0334	0.0302	0.0273	0.0247	0.0224	0.0202	0.0183
1	0.1397	0.1304	0.1217	0.1135	0.1057	0.0984	0.0915	0.0850	0.0789	0.0733
2	0.2165	0.2087	0.2008	0.1929	0.1850	0.1771	0.1692	0.1615	0.1539	0.1465
3	0.2237	0.2226	0.2209	0.2186	0.2158	0.2125	0.2087	0.2046	0.2001	0.1954
4	0.1734	0.1781	0.1823	0.1858	0.1888	0.1912	0.1931	0.1944	0.1951	0.1954
5	0.1075	0.1140	0.1203	0.1264	0.1322	0.1377	0.1429	0.1477	0.1522	0.1563
6	0.0555	0.0608	0.0662	0.0716	0.0771	0.0826	0.0881	0.0936	0.0989	0.1042
7	0.0246	0.0278	0.0312	0.0348	0.0385	0.0425	0.0466	0.0508	0.0551	0.0595
8	0.0095	0.0111	0.0129	0.0148	0.0169	0.0191	0.0215	0.0241	0.0269	0.0298
9	0.0033	0.0040	0.0047	0.0056	0.0066	0.0076	0.0089	0.0102	0.0116	0.0132
10	0.0010	0.0013	0.0016	0.0019	0.0023	0.0028	0.0033	0.0039	0.0045	0.0053
11	0.0003	0.0004	0.0005	0.0006	0.0007	0.0009	0.0011	0.0013	0.0016	0.0019
12	0.0001	0.0001	0.0001	0.0002	0.0002	0.0003	0.0003	0.0004	0.0005	0.0006
13	0.0000	0.0000	0.0000	0.0000	0.0001	0.0001	0.0001	0.0001	0.0002	0.0002
14	0.0000	0.0000	0.0000	0.0000	0.0000	0.0000	0.0000	0.0000	0.0000	0.0001

					μ					
X	4.1	4.2	4.3	4.4	4.5	4.6	4.7	4.8	4.9	5.0
0	0.0166	0.0150	0.0136	0.0123	0.0111	0.0101	0.0091	0.0082	0.0074	0.0067
1	0.0679	0.0630	0.0583	0.0540	0.0500	0.0462	0.0427	0.0395	0.0365	0.0337
2	0.1393	0.1323	0.1254	0.1188	0.1125	0.1063	0.1005	0.0948	0.0894	0.0842
3	0.1904	0.1852	0.1798	0.1743	0.1687	0.1631	0.1574	0.1517	0.1460	0.1404
4	0.1951	0.1944	0.1933	0.1917	0.1898	0.1875	0.1849	0.1820	0.1789	0.1755
5	0.1600	0.1633	0.1662	0.1687	0.1708	0.1725	0.1738	0.1747	0.1753	0.1755
6	0.1093	0.1143	0.1191	0.1237	0.1281	0.1323	0.1362	0.1398	0.1432	0.1462
7	0.0640	0.0686	0.0732	0.0778	0.0824	0.0869	0.0914	0.0959	0.1002	0.1044
8	0.0328	0.0360	0.0393	0.0428	0.0463	0.0500	0.0537	0.0575	0.0614	0.0653
9	0.0150	0.0168	0.0188	0.0209	0.0232	0.0255	0.0280	0.0307	0.0334	0.0363
10	0.0061	0.0071	0.0081	0.0092	0.0104	0.0118	0.0132	0.0147	0.0164	0.0181
11	0.0023	0.0027	0.0032	0.0037	0.0043	0.0049	0.0056	0.0064	0.0073	0.0082
12	0.0008	0.0009	0.0011	0.0014	0.0016	0.0019	0.0022	0.0026	0.0030	0.0034
13	0.0002	0.0003	0.0004	0.0005	0.0006	0.0007	0.0008	0.0009	0.0011	0.0013
14	0.0001	0.0001	0.0001	0.0001	0.0002	0.0002	0.0003	0.0003	0.0004	0.0005
15	0.0000	0.0000	0.0000	0.0000	0.0001	0.0001	0.0001	0.0001	0.0001	0.0002

(continued)

Table 4 (*Continued*)

						μ					
X	5.1	5.2	5.3	5.4	5.5	5.6	5.7	5.8	5.9	6.0
0	0.0061	0.0055	0.0050	0.0045	0.0041	0.0037	0.0033	0.0030	0.0027	0.0025
1	0.0311	0.0287	0.0265	0.0244	0.0225	0.0207	0.0191	0.0176	0.0162	0.0149
2	0.0793	0.0746	0.0701	0.0659	0.0618	0.0580	0.0544	0.0509	0.0477	0.0446
3	0.1348	0.1293	0.1239	0.1185	0.1133	0.1082	0.1033	0.0985	0.0938	0.0892
4	0.1719	0.1681	0.1641	0.1600	0.1558	0.1515	0.1472	0.1428	0.1383	0.1339
5	0.1753	0.1748	0.1740	0.1728	0.1714	0.1697	0.1678	0.1656	0.1632	0.1606
6	0.1490	0.1515	0.1537	0.1555	0.1571	0.1584	0.1594	0.1601	0.1605	0.1606
7	0.1086	0.1125	0.1163	0.1200	0.1234	0.1267	0.1298	0.1326	0.1353	0.1377
8	0.0692	0.0731	0.0771	0.0810	0.0849	0.0887	0.0925	0.0962	0.0998	0.1033
9	0.0392	0.0423	0.0454	0.0486	0.0519	0.0552	0.0586	0.0620	0.0654	0.0688
10	0.0200	0.0220	0.0241	0.0262	0.0285	0.0309	0.0334	0.0359	0.0386	0.0413
11	0.0093	0.0104	0.0116	0.0129	0.0143	0.0157	0.0173	0.0190	0.0207	0.0225
12	0.0039	0.0045	0.0051	0.0058	0.0065	0.0073	0.0082	0.0092	0.0102	0.0113
13	0.0015	0.0018	0.0021	0.0024	0.0028	0.0032	0.0036	0.0041	0.0046	0.0052
14	0.0006	0.0007	0.0008	0.0009	0.0011	0.0013	0.0015	0.0017	0.0019	0.0022
15	0.0002	0.0002	0.0003	0.0003	0.0004	0.0005	0.0006	0.0007	0.0008	0.0009
16	0.0001	0.0001	0.0001	0.0001	0.0001	0.0002	0.0002	0.0002	0.0003	0.0003
17	0.0000	0.0000	0.0000	0.0000	0.0000	0.0001	0.0001	0.0001	0.0001	0.0001

						μ					
X	6.1	6.2	6.3	6.4	6.5	6.6	6.7	6.8	6.9	7.0
0	0.0022	0.0020	0.0018	0.0017	0.0015	0.0014	0.0012	0.0011	0.0010	0.0009
1	0.0137	0.0126	0.0116	0.0106	0.0098	0.0090	0.0082	0.0076	0.0070	0.0064
2	0.0417	0.0390	0.0364	0.0340	0.0318	0.0296	0.0276	0.0258	0.0240	0.0223
3	0.0848	0.0806	0.0765	0.0726	0.0688	0.0652	0.0617	0.0584	0.0552	0.0521
4	0.1294	0.1249	0.1205	0.1162	0.1118	0.1076	0.1034	0.0992	0.0952	0.0912
5	0.1579	0.1549	0.1519	0.1487	0.1454	0.1420	0.1385	0.1349	0.1314	0.1277
6	0.1605	0.1601	0.1595	0.1586	0.1575	0.1562	0.1546	0.1529	0.1511	0.1490
7	0.1399	0.1418	0.1435	0.1450	0.1462	0.1472	0.1480	0.1486	0.1489	0.1490
8	0.1066	0.1099	0.1130	0.1160	0.1188	0.1215	0.1240	0.1263	0.1284	0.1304
9	0.0723	0.0757	0.0791	0.0825	0.0858	0.0891	0.0923	0.0954	0.0985	0.1014
10	0.0441	0.0469	0.0498	0.0528	0.0558	0.0588	0.0618	0.0649	0.0679	0.0710
11	0.0245	0.0265	0.0285	0.0307	0.0330	0.0353	0.0377	0.0401	0.0426	0.0452
12	0.0124	0.0137	0.0150	0.0164	0.0179	0.0194	0.0210	0.0227	0.0245	0.0264
13	0.0058	0.0065	0.0073	0.0081	0.0089	0.0098	0.0108	0.0119	0.0130	0.0142
14	0.0025	0.0029	0.0033	0.0037	0.0041	0.0046	0.0052	0.0058	0.0064	0.0071
15	0.0010	0.0012	0.0014	0.0016	0.0018	0.0020	0.0023	0.0026	0.0029	0.0033
16	0.0004	0.0005	0.0005	0.0006	0.0007	0.0008	0.0010	0.0011	0.0013	0.0014
17	0.0001	0.0002	0.0002	0.0002	0.0003	0.0003	0.0004	0.0004	0.0005	0.0006
18	0.0000	0.0001	0.0001	0.0001	0.0001	0.0001	0.0001	0.0002	0.0002	0.0002
19	0.0000	0.0000	0.0000	0.0000	0.0000	0.0000	0.0000	0.0001	0.0001	0.0001

(continued)

Table 4 (Continued)

					μ					
X	7.1	7.2	7.3	7.4	7.5	7.6	7.7	7.8	7.9	8.0
0	0.0008	0.0007	0.0007	0.0006	0.0006	0.0005	0.0005	0.0004	0.0004	0.0003
1	0.0059	0.0054	0.0049	0.0045	0.0041	0.0038	0.0035	0.0032	0.0029	0.0027
2	0.0208	0.0194	0.0180	0.0167	0.0156	0.0145	0.0134	0.0125	0.0116	0.0107
3	0.0492	0.0464	0.0438	0.0413	0.0389	0.0366	0.0345	0.0324	0.0305	0.0286
4	0.0874	0.0836	0.0799	0.0764	0.0729	0.0696	0.0663	0.0632	0.0602	0.0573
5	0.1241	0.1204	0.1167	0.1130	0.1094	0.1057	0.1021	0.0986	0.0951	0.0916
6	0.1468	0.1445	0.1420	0.1394	0.1367	0.1339	0.1311	0.1282	0.1252	0.1221
7	0.1489	0.1486	0.1481	0.1474	0.1465	0.1454	0.1442	0.1428	0.1413	0.1396
8	0.1321	0.1337	0.1351	0.1363	0.1373	0.1382	0.1388	0.1392	0.1395	0.1396
9	0.1042	0.1070	0.1096	0.1121	0.1144	0.1167	0.1187	0.1207	0.1224	0.1241
10	0.0740	0.0770	0.0800	0.0829	0.0858	0.0887	0.0914	0.0941	0.0967	0.0993
11	0.0478	0.0504	0.0531	0.0558	0.0585	0.0613	0.0640	0.0667	0.0695	0.0722
12	0.0283	0.0303	0.0323	0.0344	0.0366	0.0388	0.0411	0.0434	0.0457	0.0481
13	0.0154	0.0168	0.0181	0.0196	0.0211	0.0227	0.0243	0.0260	0.0278	0.0296
14	0.0078	0.0086	0.0095	0.0104	0.0113	0.0123	0.0134	0.0145	0.0157	0.0169
15	0.0037	0.0041	0.0046	0.0051	0.0057	0.0062	0.0069	0.0075	0.0083	0.0090
16	0.0016	0.0019	0.0021	0.0024	0.0026	0.0030	0.0033	0.0037	0.0041	0.0045
17	0.0007	0.0008	0.0009	0.0010	0.0012	0.0013	0.0015	0.0017	0.0019	0.0021
18	0.0003	0.0003	0.0004	0.0004	0.0005	0.0006	0.0006	0.0007	0.0008	0.0009
19	0.0001	0.0001	0.0001	0.0002	0.0002	0.0002	0.0003	0.0003	0.0003	0.0004
20	0.0000	0.0000	0.0001	0.0001	0.0001	0.0001	0.0001	0.0001	0.0001	0.0002
21	0.0000	0.0000	0.0000	0.0000	0.0000	0.0000	0.0000	0.0000	0.0001	0.0001

					μ					
X	8.1	8.2	8.3	8.4	8.5	8.6	8.7	8.8	8.9	9.0
0	0.0003	0.0003	0.0002	0.0002	0.0002	0.0002	0.0002	0.0002	0.0001	0.0001
1	0.0025	0.0023	0.0021	0.0019	0.0017	0.0016	0.0014	0.0013	0.0012	0.0011
2	0.0100	0.0092	0.0086	0.0079	0.0074	0.0068	0.0063	0.0058	0.0054	0.0050
3	0.0269	0.0252	0.0237	0.0222	0.0208	0.0195	0.0183	0.0171	0.0160	0.0150
4	0.0544	0.0517	0.0491	0.0466	0.0443	0.0420	0.0398	0.0377	0.0357	0.0337
5	0.0882	0.0849	0.0816	0.0784	0.0752	0.0722	0.0692	0.0663	0.0635	0.0607
6	0.1191	0.1160	0.1128	0.1097	0.1066	0.1034	0.1003	0.0972	0.0941	0.0911
7	0.1378	0.1358	0.1338	0.1317	0.1294	0.1271	0.1247	0.1222	0.1197	0.1171
8	0.1395	0.1392	0.1388	0.1382	0.1375	0.1366	0.1356	0.1344	0.1332	0.1318
9	0.1256	0.1269	0.1280	0.1290	0.1299	0.1306	0.1311	0.1315	0.1317	0.1318
10	0.1017	0.1040	0.1063	0.1084	0.1104	0.1123	0.1140	0.1157	0.1172	0.1186
11	0.0749	0.0776	0.0802	0.0828	0.0853	0.0878	0.0902	0.0925	0.0948	0.0970
12	0.0505	0.0530	0.0555	0.0579	0.0604	0.0629	0.0654	0.0679	0.0703	0.0728

(continued)

Table 4 (*Continued*)

					μ					
X	8.1	8.2	8.3	8.4	8.5	8.6	8.7	8.8	8.9	9.0
13	0.0315	0.0334	0.0354	0.0374	0.0395	0.0416	0.0438	0.0459	0.0481	0.0504
14	0.0182	0.0196	0.0210	0.0225	0.0240	0.0256	0.0272	0.0289	0.0306	0.0324
15	0.0098	0.0107	0.0116	0.0126	0.0136	0.0147	0.0158	0.0169	0.0182	0.0194
16	0.0050	0.0055	0.0060	0.0066	0.0072	0.0079	0.0086	0.0093	0.0101	0.0109
17	0.0024	0.0026	0.0029	0.0033	0.0036	0.0040	0.0044	0.0048	0.0053	0.0058
18	0.0011	0.0012	0.0014	0.0015	0.0017	0.0019	0.0021	0.0024	0.0026	0.0029
19	0.0005	0.0005	0.0006	0.0007	0.0008	0.0009	0.0010	0.0011	0.0012	0.0014
20	0.0002	0.0002	0.0002	0.0003	0.0003	0.0004	0.0004	0.0005	0.0005	0.0006
21	0.0001	0.0001	0.0001	0.0001	0.0001	0.0002	0.0002	0.0002	0.0002	0.0003
22	0.0000	0.0000	0.0000	0.0000	0.0001	0.0001	0.0001	0.0001	0.0001	0.0001

					μ					
X	9.1	9.2	9.3	9.4	9.5	9.6	9.7	9.8	9.9	10.0
0	0.0001	0.0001	0.0001	0.0001	0.0001	0.0001	0.0001	0.0001	0.0001	0.0000
1	0.0010	0.0009	0.0009	0.0008	0.0007	0.0007	0.0006	0.0005	0.0005	0.0005
2	0.0046	0.0043	0.0040	0.0037	0.0034	0.0031	0.0029	0.0027	0.0025	0.0023
3	0.0140	0.0131	0.0123	0.0115	0.0107	0.0100	0.0093	0.0087	0.0081	0.0076
4	0.0319	0.0302	0.0285	0.0269	0.0254	0.0240	0.0226	0.0213	0.0201	0.0189
5	0.0581	0.0555	0.0530	0.0506	0.0483	0.0460	0.0439	0.0418	0.0398	0.0378
6	0.0881	0.0851	0.0822	0.0793	0.0764	0.0736	0.0709	0.0682	0.0656	0.0631
7	0.1145	0.1118	0.1091	0.1064	0.1037	0.1010	0.0982	0.0955	0.0928	0.0901
8	0.1302	0.1286	0.1269	0.1251	0.1232	0.1212	0.1191	0.1170	0.1148	0.1126
9	0.1317	0.1315	0.1311	0.1306	0.1300	0.1293	0.1284	0.1274	0.1263	0.1251
10	0.1198	0.1210	0.1219	0.1228	0.1235	0.1241	0.1245	0.1249	0.1250	0.1251
11	0.0991	0.1012	0.1031	0.1049	0.1067	0.1083	0.1098	0.1112	0.1125	0.1137
12	0.0752	0.0776	0.0799	0.0822	0.0844	0.0866	0.0888	0.0908	0.0928	0.0948
13	0.0526	0.0549	0.0572	0.0594	0.0617	0.0640	0.0662	0.0685	0.0707	0.0729
14	0.0342	0.0361	0.0380	0.0399	0.0419	0.0439	0.0459	0.0479	0.0500	0.0521
15	0.0208	0.0221	0.0235	0.0250	0.0265	0.0281	0.0297	0.0313	0.0330	0.0347
16	0.0118	0.0127	0.0137	0.0147	0.0157	0.0168	0.0180	0.0192	0.0204	0.0217
17	0.0063	0.0069	0.0075	0.0081	0.0088	0.0095	0.0103	0.0111	0.0119	0.0128
18	0.0032	0.0035	0.0039	0.0042	0.0046	0.0051	0.0055	0.0060	0.0065	0.0071
19	0.0015	0.0017	0.0019	0.0021	0.0023	0.0026	0.0028	0.0031	0.0034	0.0037
20	0.0007	0.0008	0.0009	0.0010	0.0011	0.0012	0.0014	0.0015	0.0017	0.0019
21	0.0003	0.0003	0.0004	0.0004	0.0005	0.0006	0.0006	0.0007	0.0008	0.0009
22	0.0001	0.0001	0.0002	0.0002	0.0002	0.0002	0.0003	0.0003	0.0004	0.0004
23	0.0000	0.0001	0.0001	0.0001	0.0001	0.0001	0.0001	0.0001	0.0002	0.0002
24	0.0000	0.0000	0.0000	0.0000	0.0000	0.0000	0.0000	0.0001	0.0001	0.0001

Table 5 Exponential Distribution

Example: If $\mu = \frac{1}{6}$, the probability of observing a value less than $T = 9$ is found by $F(T)$ for $\mu T = \frac{1}{6}(9) = 1.5$; $P(T < 9) = 0.777$.

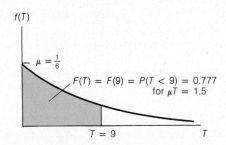

μT	$F(T)$	μT	$F(T)$	μT	$F(T)$	μT	$F(T)$
0.0	0.000	2.5	0.918	5.0	0.9933	7.5	0.99945
0.1	0.095	2.6	0.926	5.1	0.9939	7.6	0.99950
0.2	0.181	2.7	0.933	5.2	0.9945	7.7	0.99955
0.3	0.259	2.8	0.939	5.3	0.9950	7.8	0.99959
0.4	0.330	2.9	0.945	5.4	0.9955	7.9	0.99963
0.5	0.393	3.0	0.950	5.5	0.9959	8.0	0.99966
0.6	0.451	3.1	0.955	5.6	0.9963	8.1	0.99970
0.7	0.503	3.2	0.959	5.7	0.9967	8.2	0.99972
0.8	0.551	3.3	0.963	5.8	0.9970	8.3	0.99975
0.9	0.593	3.4	0.967	5.9	0.9973	8.4	0.99978
1.0	0.632	3.5	0.970	6.0	0.9975	8.5	0.99980
1.1	0.667	3.6	0.973	6.1	0.9978	8.6	0.99982
1.2	0.699	3.7	0.975	6.2	0.9980	8.7	0.99983
1.3	0.727	3.8	0.978	6.3	0.9982	8.8	0.99985
1.4	0.753	3.9	0.980	6.4	0.9983	8.9	0.99986
1.5	0.777	4.0	0.982	6.5	0.9985	9.0	0.99989
1.6	0.798	4.1	0.983	6.6	0.9986	9.1	0.99989
1.7	0.817	4.2	0.985	6.7	0.9988	9.2	0.99990
1.8	0.835	4.3	0.986	6.8	0.9989	9.3	0.99991
1.9	0.850	4.4	0.988	6.9	0.9990	9.4	0.99992
2.0	0.865	4.5	0.989	7.0	0.9991	9.5	0.99992
2.1	0.878	4.6	0.990	7.1	0.9992	9.6	0.99993
2.2	0.889	4.7	0.991	7.2	0.9993	9.7	0.99994
2.3	0.900	4.8	0.992	7.3	0.9993	9.8	0.99994
2.4	0.909	4.9	0.993	7.4	0.9993	9.9	0.99995

Source: Adapted from Harnett and Murphy, *Statistical Analysis for Business and Economics.* © 1985, Addison-Wesley, Reading, Mass., p. A54, Table IV. Reprinted with permission.

Table 6 5% and 1% (Boldface Type) Points for the Distribution of F

df_2	df_1: 1	2	3	4	5	6	7	8	9	10	11	12	14	16	20	24	30	40	50	75	100	200	500	∞	df_2
1	161	200	216	225	230	234	237	239	241	242	243	244	245	246	248	249	250	251	252	253	253	254	254	254	1
	4,052	4,999	5,403	5,625	5,764	5,859	5,928	5,981	6,022	6,056	6,082	6,106	6,142	6,169	6,208	6,234	6,261	6,286	6,302	6,323	6,334	6,352	6,361	6,366	
2	18.51	19.00	19.16	19.25	19.30	19.33	19.36	19.37	19.38	19.39	19.40	19.41	19.42	19.43	19.44	19.45	19.46	19.47	19.47	19.48	19.49	19.49	19.50	19.50	2
	98.49	99.00	99.17	99.25	99.30	99.33	99.36	99.37	99.39	99.40	99.41	99.42	99.43	99.44	99.45	99.46	99.47	99.48	99.48	99.49	99.49	99.49	99.50	99.50	
3	10.13	9.55	9.28	9.12	9.01	8.94	8.88	8.84	8.81	8.78	8.76	8.74	8.71	8.69	8.66	8.64	8.62	8.60	8.58	8.57	8.56	8.54	8.54	8.53	3
	34.12	30.82	29.46	28.71	28.24	27.91	27.67	27.49	27.34	27.23	27.13	27.05	26.92	26.83	26.69	26.60	26.50	26.41	26.35	26.27	26.23	26.18	26.14	26.12	
4	7.71	6.94	6.59	6.39	6.26	6.16	6.09	6.04	6.00	5.96	5.93	5.91	5.87	5.84	5.80	5.77	5.74	5.71	5.70	5.68	5.66	5.65	5.64	5.63	4
	21.20	18.00	16.69	15.98	15.52	15.21	14.98	14.80	14.66	14.54	14.45	14.37	14.24	14.15	14.02	13.93	13.83	13.74	13.69	13.61	13.57	13.52	13.48	13.46	
5	6.61	5.79	5.41	5.19	5.05	4.95	4.88	4.82	4.78	4.74	4.70	4.68	4.64	4.60	4.56	4.53	4.50	4.46	4.44	4.42	4.40	4.38	4.37	4.36	5
	16.26	13.27	12.06	11.39	10.97	10.67	10.45	10.29	10.15	10.05	9.96	9.89	9.77	9.68	9.55	9.47	9.38	9.29	9.24	9.17	9.13	9.07	9.04	9.02	
6	5.99	5.14	4.76	4.53	4.39	4.28	4.21	4.15	4.10	4.06	4.03	4.00	3.96	3.92	3.87	3.84	3.81	3.77	3.75	3.72	3.71	3.69	3.68	3.67	6
	13.74	10.92	9.78	9.15	8.75	8.47	8.26	8.10	7.98	7.87	7.79	7.72	7.60	7.52	7.39	7.31	7.23	7.14	7.09	7.02	6.99	6.94	6.90	6.88	
7	5.59	4.74	4.35	4.12	3.97	3.87	3.79	3.73	3.68	3.63	3.60	3.57	3.52	3.49	3.44	3.41	3.38	3.34	3.32	3.29	3.28	3.25	3.24	3.23	7
	12.25	9.55	8.45	7.85	7.46	7.19	7.00	6.84	6.71	6.62	6.54	6.47	6.35	6.27	6.15	6.07	5.98	5.90	5.85	5.78	5.75	5.70	5.67	5.65	
8	5.32	4.46	4.07	3.84	3.69	3.58	3.50	3.44	3.39	3.34	3.31	3.28	3.23	3.20	3.15	3.12	3.08	3.05	3.03	3.00	2.98	2.96	2.94	2.93	8
	11.26	8.65	7.59	7.01	6.63	6.37	6.19	6.03	5.91	5.82	5.74	5.67	5.56	5.48	5.36	5.28	5.20	5.11	5.06	5.00	4.96	4.91	4.88	4.86	
9	5.12	4.26	3.86	3.63	3.48	3.37	3.29	3.23	3.18	3.13	3.10	3.07	3.02	2.98	2.93	2.90	2.86	2.82	2.80	2.77	2.76	2.73	2.72	2.71	9
	10.56	8.02	6.99	6.42	6.06	5.80	5.62	5.47	5.35	5.26	5.18	5.11	5.00	4.92	4.80	4.73	4.64	4.56	4.51	4.45	4.41	4.36	4.33	4.31	
10	4.96	4.10	3.71	3.48	3.33	3.22	3.14	3.07	3.02	2.97	2.94	2.91	2.86	2.82	2.77	2.74	2.70	2.67	2.64	2.61	2.59	2.56	2.55	2.54	10
	10.04	7.56	6.55	5.99	5.64	5.39	5.21	5.06	4.95	4.85	4.78	4.71	4.60	4.52	4.41	4.33	4.25	4.17	4.12	4.05	4.01	3.96	3.93	3.91	
11	4.84	3.98	3.59	3.36	3.20	3.09	3.01	2.95	2.90	2.86	2.82	2.79	2.74	2.70	2.65	2.61	2.57	2.53	2.50	2.47	2.45	2.42	2.41	2.40	11
	9.65	7.20	6.22	5.67	5.32	5.07	4.88	4.74	4.63	4.54	4.46	4.40	4.29	4.21	4.10	4.02	3.94	3.86	3.80	3.74	3.70	3.66	3.62	3.60	
12	4.75	3.88	3.49	3.26	3.11	3.00	2.92	2.85	2.80	2.76	2.72	2.69	2.64	2.60	2.54	2.50	2.46	2.42	2.40	2.36	2.35	2.32	2.31	2.30	12
	9.33	6.93	5.95	5.41	5.06	4.82	4.65	4.50	4.39	4.30	4.22	4.16	4.05	3.98	3.86	3.78	3.70	3.61	3.56	3.49	3.46	3.41	3.38	3.36	
13	4.67	3.80	3.41	3.18	3.02	2.92	2.84	2.77	2.72	2.67	2.63	2.60	2.55	2.51	2.46	2.42	2.38	2.34	2.32	2.28	2.26	2.24	2.22	2.21	13
	9.07	6.70	5.74	5.20	4.86	4.62	4.44	4.30	4.19	4.10	4.02	3.96	3.85	3.78	3.67	3.59	3.51	3.42	3.37	3.30	3.27	3.21	3.18	3.16	
14	4.60	3.74	3.34	3.11	2.96	2.85	2.77	2.70	2.65	2.60	2.56	2.53	2.48	2.44	2.39	2.35	2.31	2.27	2.24	2.21	2.19	2.16	2.14	2.13	14
	8.86	6.51	5.56	5.03	4.69	4.46	4.28	4.14	4.03	3.94	3.86	3.80	3.70	3.62	3.51	3.43	3.34	3.26	3.21	3.14	3.11	3.06	3.02	3.00	
15	4.54	3.68	3.29	3.06	2.90	2.79	2.70	2.64	2.59	2.55	2.51	2.48	2.43	2.39	2.33	2.29	2.25	2.21	2.18	2.15	2.12	2.10	2.08	2.07	15
	8.68	6.36	5.42	4.89	4.56	4.32	4.14	4.00	3.89	3.80	3.73	3.67	3.56	3.48	3.36	3.29	3.20	3.12	3.07	3.00	2.97	2.92	2.89	2.87	
16	4.49	3.63	3.24	3.01	2.85	2.74	2.66	2.59	2.54	2.49	2.45	2.42	2.37	2.33	2.28	2.24	2.20	2.16	2.13	2.09	2.07	2.04	2.02	2.01	16
	8.53	6.23	5.29	4.77	4.44	4.20	4.03	3.89	3.78	3.69	3.61	3.55	3.45	3.37	3.25	3.18	3.10	3.01	2.96	2.98	2.86	2.80	2.77	2.75	
17	4.45	3.59	3.20	2.96	2.81	2.70	2.62	2.55	2.50	2.45	2.41	2.38	2.33	2.29	2.23	2.19	2.15	2.11	2.08	2.04	2.02	1.99	1.97	1.96	17
	8.40	6.11	5.18	4.67	4.34	4.10	3.93	3.79	3.68	3.59	3.52	3.45	3.35	3.27	3.16	3.08	3.00	2.92	2.86	2.79	2.76	2.70	2.67	2.65	
18	4.41	3.55	3.16	2.93	2.77	3.66	2.58	2.51	2.46	2.41	2.37	2.34	2.29	2.25	2.19	2.15	2.11	2.07	2.04	2.00	1.98	1.95	1.93	1.92	18
	8.28	6.01	5.09	4.58	4.25	4.01	3.85	3.71	3.60	3.51	3.44	3.37	3.27	3.19	3.07	3.00	2.91	2.83	2.78	2.71	2.68	2.62	2.59	2.57	
19	4.38	3.52	3.13	2.90	2.74	2.63	2.55	2.48	2.43	2.38	2.34	2.31	2.26	2.21	2.15	2.11	2.07	2.02	2.00	1.96	1.94	1.91	1.90	1.88	19
	8.18	5.93	5.01	4.50	4.17	3.94	3.77	3.63	3.52	3.43	3.36	3.30	3.19	3.12	3.00	2.92	2.84	2.76	2.70	2.63	2.60	2.54	2.51	2.49	
20	4.35	3.49	3.10	2.87	2.71	2.60	2.52	2.45	2.40	2.35	2.31	2.28	2.23	2.18	2.12	2.08	2.04	1.99	1.96	1.92	1.90	1.87	1.85	1.84	20
	8.10	5.85	4.94	4.43	4.10	3.87	3.71	3.56	3.45	3.37	3.30	3.23	3.13	3.05	2.94	2.86	2.77	2.69	2.63	2.56	2.53	2.47	2.44	2.42	
21	4.32	3.47	3.07	2.84	2.68	2.57	2.49	2.42	2.37	2.32	2.28	2.25	2.20	2.15	2.09	2.05	2.00	1.96	1.93	1.89	1.87	1.84	1.82	1.81	21
	8.02	5.78	4.87	4.37	4.04	3.81	3.65	3.51	3.40	3.31	3.24	3.17	3.07	2.99	2.88	2.80	2.72	2.63	2.58	2.51	2.47	2.42	2.38	2.36	
22	4.30	3.44	3.05	2.82	2.66	2.55	2.47	2.40	2.35	2.30	2.26	2.23	2.18	2.13	2.07	2.03	1.98	1.93	1.91	1.87	1.84	1.81	1.80	1.78	22
	7.94	5.72	4.82	4.31	3.99	3.76	3.59	3.45	3.35	3.26	3.18	3.12	3.02	2.94	2.83	2.75	2.67	2.58	2.53	2.46	2.42	2.37	2.33	2.31	
23	4.28	3.42	3.03	2.80	2.64	2.53	2.45	2.38	2.32	2.28	2.24	2.20	2.14	2.10	2.04	2.00	1.96	1.91	1.88	1.84	1.82	1.79	1.77	1.76	23
	7.88	5.66	4.76	4.26	3.94	3.71	3.54	3.41	3.30	3.21	3.14	3.07	2.97	2.89	2.78	2.70	2.62	2.53	2.48	2.41	2.37	2.32	2.28	2.26	
24	4.26	3.40	3.01	2.78	2.62	2.51	2.43	2.36	2.30	2.26	2.22	2.18	2.13	2.09	2.02	1.98	1.94	1.89	1.86	1.82	1.80	1.76	1.74	1.73	24
	7.82	5.61	4.72	4.22	3.90	3.67	3.50	3.36	3.25	3.17	3.09	3.03	2.93	2.85	2.74	2.66	2.58	2.49	2.44	2.36	2.33	2.27	2.23	2.21	
25	4.24	3.38	2.99	2.76	2.60	2.49	2.41	2.34	2.28	2.24	2.20	2.16	2.11	2.06	2.00	1.96	1.92	1.87	1.84	1.80	1.77	1.74	1.72	1.71	25
	7.77	5.57	4.68	4.18	3.86	3.63	3.46	3.32	3.21	3.13	3.05	2.99	2.89	2.81	2.70	2.62	2.54	2.45	2.40	2.32	2.29	2.23	2.19	2.17	
26	4.22	3.37	2.98	2.74	2.59	2.47	2.39	2.32	2.27	2.22	2.18	2.15	2.10	2.05	1.99	1.95	1.90	1.85	1.82	1.78	1.76	1.72	1.70	1.69	26
	7.72	5.53	4.64	4.14	3.82	3.59	3.42	3.29	3.17	3.09	3.02	2.96	2.86	2.77	2.66	2.58	2.50	2.41	2.36	2.28	2.25	2.19	2.15	2.13	
27	4.21	3.35	2.96	2.73	2.57	2.46	2.37	2.30	2.25	2.20	2.16	2.13	2.08	2.03	1.97	1.93	1.88	1.84	1.80	1.76	1.74	1.71	1.68	1.67	27
	7.68	5.49	4.60	4.11	3.79	3.56	3.39	3.26	3.14	3.06	2.98	2.93	2.83	2.74	2.63	2.55	2.47	2.38	2.33	2.25	2.21	2.16	2.12	2.10	
28	4.20	3.34	2.95	2.71	2.56	2.44	2.36	2.29	2.24	2.19	2.15	2.12	2.06	2.02	1.96	1.91	1.87	1.81	1.78	1.75	1.72	1.69	1.67	1.65	28
	7.64	5.45	4.57	4.07	3.76	3.53	3.36	3.23	3.11	3.03	2.95	2.90	2.80	2.71	2.60	2.52	2.44	2.35	2.30	2.22	2.18	2.13	2.09	2.06	
29	4.18	3.33	2.93	2.70	2.54	2.43	2.35	2.28	2.22	2.18	2.14	2.10	2.05	2.00	1.94	1.90	1.85	1.80	1.77	1.73	1.71	1.68	1.65	1.64	29
	7.60	5.42	4.54	4.04	3.73	3.50	3.33	3.20	3.08	3.00	2.92	2.87	2.77	2.68	2.57	2.49	2.41	2.32	2.27	2.19	2.15	2.10	2.06	2.03	
30	4.17	3.32	2.92	2.69	2.53	2.42	2.34	2.27	2.21	2.16	2.12	2.09	2.04	1.99	1.93	1.89	1.84	1.79	1.76	1.72	1.69	1.66	1.64	1.62	30
	7.56	5.39	4.51	4.02	3.70	3.47	3.30	3.17	3.06	2.98	2.90	2.84	2.74	2.66	2.55	2.47	2.38	2.29	2.24	2.16	2.13	2.07	2.03	2.01	
32	4.15	3.30	2.90	2.67	2.51	2.40	2.32	2.25	2.19	2.14	2.10	2.07	2.02	1.97	1.91	1.86	1.82	1.76	1.74	1.69	1.67	1.64	1.61	1.59	32
	7.50	5.34	4.46	3.97	3.66	3.42	3.25	3.12	3.01	2.94	2.86	2.80	2.70	2.62	2.51	2.42	2.34	2.25	2.20	2.12	2.08	2.02	1.98	1.96	
34	4.13	3.28	2.88	2.65	2.49	2.38	2.30	2.23	2.17	2.12	2.08	2.05	2.00	1.95	1.89	1.84	1.80	1.74	1.71	1.67	1.64	1.61	1.59	1.57	34
	7.44	5.29	4.42	3.93	3.61	3.38	3.21	3.08	2.97	2.89	2.82	2.76	2.66	2.58	2.47	2.38	2.30	2.21	2.15	2.08	2.04	1.98	1.94	1.91	

(continued)

Table 6 (*Continued*)

df_2	1	2	3	4	5	6	7	8	9	10	11	12	14	16	20	24	30	40	50	75	100	200	500	∞	df_2
36	4.11	3.26	2.86	2.63	2.48	2.36	2.28	2.21	2.15	2.10	2.06	2.03	1.98	1.93	1.87	1.82	1.78	1.72	1.69	1.65	1.62	1.59	1.56	1.55	36
	7.39	5.25	4.38	3.89	3.58	3.35	3.18	3.04	2.94	2.86	2.78	2.72	2.62	2.54	2.43	2.35	2.26	2.17	2.12	2.04	2.00	1.94	1.90	1.87	
38	4.10	3.25	2.85	2.62	2.46	2.35	2.26	2.19	2.14	2.09	2.05	2.02	1.96	1.92	1.85	1.80	1.76	1.71	1.67	1.63	1.60	1.57	1.54	1.53	38
	7.35	5.21	4.34	3.86	3.54	3.32	3.15	3.02	2.91	2.82	2.75	2.69	2.59	2.51	2.40	2.32	2.22	2.14	2.08	2.00	1.97	1.90	1.86	1.84	
40	4.08	3.23	2.84	2.61	2.45	2.34	2.25	2.18	2.12	2.07	2.04	2.00	1.95	1.90	1.84	1.79	1.74	1.69	1.66	1.61	1.59	1.55	1.53	1.51	40
	7.31	5.18	4.31	3.83	3.51	3.29	3.12	2.99	2.88	2.80	2.73	2.66	2.56	2.49	2.37	2.29	2.20	2.11	2.05	1.97	1.94	1.88	1.84	1.81	
42	4.07	3.22	2.83	2.59	2.44	2.32	2.24	2.17	2.11	2.06	2.02	1.99	1.94	1.89	1.82	1.78	1.73	1.68	1.64	1.60	1.57	1.54	1.51	1.49	42
	7.27	5.15	4.29	3.80	3.49	3.26	3.10	2.96	2.86	2.77	2.70	2.64	2.54	2.46	2.35	2.26	2.17	2.08	2.02	1.94	1.91	1.85	1.80	1.78	
44	4.06	3.21	2.82	2.58	2.43	2.31	2.23	2.16	2.10	2.05	2.01	1.98	1.92	1.88	1.81	1.76	1.72	1.66	1.63	1.58	1.56	1.52	1.50	1.48	44
	7.24	5.12	4.26	3.78	3.46	3.24	3.07	2.94	2.84	2.75	2.68	2.62	2.52	2.44	2.32	2.24	2.15	2.06	2.00	1.92	1.88	1.82	1.78	1.75	
46	4.05	3.20	2.81	2.57	2.42	2.30	2.22	2.14	2.09	2.04	2.00	1.97	1.91	1.87	1.80	1.75	1.71	1.65	1.62	1.57	1.54	1.51	1.48	1.46	46
	7.21	5.10	4.24	3.76	3.44	3.22	3.05	2.92	2.82	2.73	2.66	2.60	2.50	2.42	2.30	2.22	2.13	2.04	1.98	1.90	1.86	1.80	1.76	1.72	
48	4.04	3.19	2.80	2.56	2.41	2.30	2.21	2.14	2.08	2.03	1.99	1.96	1.90	1.86	1.79	1.74	1.70	1.64	1.61	1.56	1.53	1.50	1.47	1.45	48
	7.19	5.08	4.22	3.74	3.42	3.20	3.04	2.90	2.80	2.71	2.64	2.58	2.48	2.40	2.28	2.20	2.11	2.02	1.96	1.88	1.84	1.78	1.73	1.70	
50	4.03	3.18	2.79	2.56	2.40	2.29	2.20	2.13	2.07	2.02	1.98	1.95	1.90	1.85	1.78	1.74	1.69	1.63	1.60	1.55	1.52	1.48	1.46	1.44	50
	7.17	5.06	4.20	3.72	3.41	3.18	3.02	2.88	2.78	2.70	2.62	2.56	2.46	2.39	2.26	2.18	2.10	2.00	1.94	1.86	1.82	1.76	1.71	1.68	
55	4.02	3.17	2.78	2.54	2.38	2.27	2.18	2.11	2.05	2.00	1.97	1.93	1.88	1.83	1.76	1.72	1.67	1.61	1.58	1.52	1.50	1.46	1.43	1.41	55
	7.12	5.01	4.16	3.68	3.37	3.15	2.98	2.85	2.75	2.66	2.59	2.53	2.43	2.35	2.23	2.15	2.06	1.96	1.90	1.82	1.78	1.71	1.66	1.64	
60	4.00	3.15	2.76	2.52	2.37	2.25	2.17	2.10	2.04	1.99	1.95	1.92	1.86	1.81	1.75	1.70	1.65	1.59	1.56	1.50	1.48	1.44	1.41	1.39	60
	7.08	4.98	4.13	3.65	3.34	3.12	2.95	2.82	2.72	2.63	2.56	2.50	2.40	2.32	2.20	2.12	2.03	1.93	1.87	1.79	1.74	1.68	1.63	1.60	
65	3.99	3.14	2.75	2.51	2.36	2.24	2.15	2.08	2.02	1.98	1.94	1.90	1.85	1.80	1.73	1.68	1.63	1.57	1.54	1.49	1.46	1.42	1.39	1.37	65
	7.04	4.95	4.10	3.62	3.31	3.09	2.93	2.79	2.70	2.61	2.54	2.47	2.37	2.30	2.18	2.09	2.00	1.90	1.84	1.76	1.71	1.64	1.60	1.56	
70	3.98	3.13	2.74	2.50	2.35	2.23	2.14	2.07	2.01	1.97	1.93	1.89	1.84	1.79	1.72	1.67	1.62	1.56	1.53	1.47	1.45	1.40	1.37	1.35	70
	7.01	4.92	4.08	3.60	3.29	3.07	2.91	2.77	2.67	2.59	2.51	2.45	2.35	2.28	2.15	2.07	1.98	1.88	1.82	1.74	1.69	1.62	1.56	1.53	
80	3.96	3.11	2.72	2.48	2.33	2.21	2.12	2.05	1.99	1.95	1.91	1.88	1.82	1.77	1.70	1.65	1.60	1.54	1.51	1.45	1.42	1.38	1.35	1.32	80
	6.96	4.88	4.04	3.56	3.25	3.04	2.87	2.74	2.64	2.55	2.48	2.41	2.32	2.24	2.11	2.03	1.94	1.84	1.78	1.70	1.65	1.57	1.52	1.49	
100	3.94	3.09	2.70	2.46	2.30	2.19	2.10	2.03	1.97	1.92	1.88	1.85	1.79	1.75	1.68	1.63	1.57	1.51	1.48	1.42	1.39	1.34	1.30	1.28	100
	6.90	4.82	3.98	3.51	3.20	2.99	2.82	2.69	2.59	2.51	2.43	2.36	2.26	2.19	2.06	1.98	1.89	1.79	1.73	1.64	1.59	1.51	1.46	1.43	
125	3.92	3.07	2.68	2.44	2.29	2.17	2.08	2.01	1.95	1.90	1.86	1.83	1.77	1.72	1.65	1.60	1.55	1.49	1.45	1.39	1.36	1.31	1.27	1.25	125
	6.84	4.78	3.94	3.47	3.17	2.95	2.79	2.65	2.56	2.47	2.40	2.33	2.23	2.15	2.03	1.94	1.85	1.75	1.68	1.59	1.54	1.46	1.40	1.37	
150	3.91	3.06	2.67	2.43	2.27	2.16	2.07	2.00	1.94	1.89	1.85	1.82	1.76	1.71	1.64	1.59	1.54	1.47	1.44	1.37	1.34	1.29	1.25	1.22	150
	6.81	4.75	3.91	3.44	3.14	2.92	2.76	2.62	2.53	2.44	2.37	2.30	2.20	2.12	2.00	1.91	1.83	1.72	1.66	1.56	1.51	1.43	1.37	1.33	
200	3.89	3.04	2.65	2.41	2.26	2.14	2.05	1.98	1.92	1.87	1.83	1.80	1.74	1.69	1.62	1.57	1.52	1.45	1.42	1.35	1.32	1.26	1.22	1.19	200
	6.76	4.71	3.88	3.41	3.11	2.90	2.73	2.60	2.50	2.41	2.34	2.28	2.17	2.09	1.97	1.88	1.79	1.69	1.62	1.53	1.48	1.39	1.33	1.28	
400	3.86	3.02	2.62	2.39	2.23	2.12	2.03	1.96	1.90	1.85	1.81	1.78	1.72	1.67	1.60	1.54	1.49	1.42	1.38	1.32	1.28	1.22	1.16	1.13	400
	6.70	4.66	3.83	3.36	3.06	2.85	2.69	2.55	2.46	2.37	2.29	2.23	2.12	2.04	1.92	1.84	1.74	1.64	1.57	1.47	1.42	1.32	1.24	1.19	
1000	3.85	3.00	2.61	2.38	2.22	2.10	2.02	1.95	1.89	1.84	1.80	1.76	1.70	1.65	1.58	1.53	1.47	1.41	1.36	1.30	1.26	1.19	1.13	1.08	1000
	6.66	4.62	3.80	3.34	3.04	2.82	2.66	2.53	2.43	2.34	2.26	2.20	2.09	2.01	1.89	1.81	1.71	1.61	1.54	1.44	1.38	1.28	1.19	1.11	
∞	3.84	2.99	2.60	2.37	2.21	2.09	2.01	1.94	1.88	1.83	1.79	1.75	1.69	1.64	1.57	1.52	1.46	1.40	1.35	1.28	1.24	1.17	1.11	1.00	∞
	6.64	4.60	3.78	3.32	3.02	2.80	2.64	2.51	2.41	2.32	2.24	2.18	2.07	1.99	1.87	1.79	1.69	1.59	1.52	1.41	1.36	1.25	1.15	1.00	

Source: Reprinted by permission from *Statistical Methods*, Seventh Edition, by George W. Snedecor and William G. Cochran, © 1980 by The Iowa State University Press, Ames, Iowa 50010.

Table 7 χ^2 Distribution

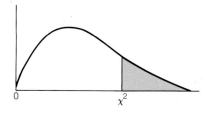

df	.10	.05	.025	.01	.005
1	2.706	3.841	5.024	6.635	7.879
2	4.605	5.991	7.378	9.210	10.597
3	6.251	7.815	9.348	11.345	12.838
4	7.779	9.488	11.143	13.277	14.860
5	9.236	11.070	12.832	15.086	16.750
6	10.645	12.592	14.449	16.812	18.548
7	12.017	14.067	16.013	18.475	20.278
8	13.362	15.507	17.535	20.090	21.955
9	14.684	16.919	19.023	21.666	23.589
10	15.987	18.307	20.483	23.209	25.188
11	17.275	19.675	21.920	24.725	26.757
12	18.549	21.026	23.337	26.217	28.300
13	19.812	22.362	24.736	27.688	29.819
14	21.064	23.685	26.119	29.141	31.319
15	22.307	24.996	27.488	30.578	32.801
16	23.542	26.296	28.845	32.000	34.267
17	24.769	27.587	30.191	33.409	35.718
18	25.989	28.869	31.526	34.805	37.156
19	27.204	30.144	32.852	36.191	38.582
20	28.412	31.410	34.170	37.566	39.997
21	29.615	32.671	35.479	38.932	41.401
22	30.813	33.924	36.781	40.289	42.796
23	32.007	35.172	38.076	41.638	44.181
24	33.196	36.415	39.364	42.980	45.558
25	34.382	37.652	40.646	44.314	46.928
26	35.563	38.885	41.923	45.642	48.290
27	36.741	40.113	43.194	46.963	49.645
28	37.916	41.337	44.461	48.278	50.993
29	39.087	42.557	45.722	49.588	52.336
30	40.256	43.773	46.979	50.892	53.672

Source: Abridged with permission of Macmillan Publishing Company from R. A. Fisher, *Statistical Methods for Research Workers.* Copyright © 1970 University of Adelaide.

Table 8 Critical Values of T_0 in the Wilcoxon Paired-Difference Signed-Rank Test

One-Tailed	Two-Tailed	$n = 5$	$n = 6$	$n = 7$	$n = 8$	$n = 9$	$n = 10$
$\alpha = .05$	$\alpha = .10$	1	2	4	6	8	11
$\alpha = .025$	$\alpha = .05$		1	2	4	6	8
$\alpha = .01$	$\alpha = .02$			0	2	3	5
$\alpha = .005$	$\alpha = .01$				0	2	3
		$n = 11$	$n = 12$	$n = 13$	$n = 14$	$n = 15$	$n = 16$
$\alpha = .05$	$\alpha = .10$	14	17	21	26	30	36
$\alpha = .025$	$\alpha = .05$	11	14	17	21	25	30
$\alpha = .01$	$\alpha = .02$	7	10	13	16	20	24
$\alpha = .005$	$\alpha = .01$	5	7	10	13	16	19
		$n = 17$	$n = 18$	$n = 19$	$n = 20$	$n = 21$	$n = 22$
$\alpha = .05$	$\alpha = .10$	41	47	54	60	68	75
$\alpha = .025$	$\alpha = .05$	35	40	46	52	59	66
$\alpha = .01$	$\alpha = .02$	28	33	38	43	49	56
$\alpha = .005$	$\alpha = .01$	23	28	32	37	43	49
		$n = 23$	$n = 24$	$n = 25$	$n = 26$	$n = 27$	$n = 28$
$\alpha = .05$	$\alpha = .10$	83	92	101	110	120	130
$\alpha = .025$	$\alpha = .05$	73	81	90	98	107	117
$\alpha = .01$	$\alpha = .02$	62	69	77	85	93	102
$\alpha = .005$	$\alpha = .01$	55	61	68	76	84	92
		$n = 29$	$n = 30$	$n = 31$	$n = 32$	$n = 33$	$n = 34$
$\alpha = .05$	$\alpha = .10$	141	152	163	175	188	201
$\alpha = .025$	$\alpha = .05$	127	137	148	159	171	183
$\alpha = .01$	$\alpha = .02$	111	120	130	141	151	162
$\alpha = .005$	$\alpha = .01$	100	109	118	128	138	149
		$n = 35$	$n = 36$	$n = 37$	$n = 38$	$n = 39$	
$\alpha = .05$	$\alpha = .10$	214	228	242	256	271	
$\alpha = .025$	$\alpha = .05$	195	208	222	235	250	
$\alpha = .01$	$\alpha = .02$	174	186	198	211	224	
$\alpha = .005$	$\alpha = .01$	160	171	183	195	208	
		$n = 40$	$n = 41$	$n = 42$	$n = 43$	$n = 44$	$n = 45$
$\alpha = .05$	$\alpha = .10$	287	303	319	336	353	371
$\alpha = .025$	$\alpha = .05$	264	279	295	311	327	344
$\alpha = .01$	$\alpha = .02$	238	252	267	281	297	313
$\alpha = .005$	$\alpha = .01$	221	234	248	262	277	292
		$n = 46$	$n = 47$	$n = 48$	$n = 49$	$n = 50$	
$\alpha = .05$	$\alpha = .10$	389	408	427	446	466	
$\alpha = .025$	$\alpha = .05$	361	379	397	415	434	
$\alpha = .01$	$\alpha = .02$	329	345	362	380	398	
$\alpha = .005$	$\alpha = .01$	307	323	339	356	373	

Source: From F. Wilcoxon and R. A. Wilcox, "Some Rapid Approximate Statistical Procedures," 1964, p. 28. Reproduced with the permission of American Cyanamid Company.

Table 9 The Mann-Whitney Test*

Values of $U_{.01}$ $[n_L; n_S]$

										n_S										
n_L	1	2	3	4	5	6	7	8	9	10	11	12	13	14	15	16	17	18	19	20
1	—																			
2	—	—																		
3	—	—	—																	
4	—	—	—	—																
5	—	—	—	0	1															
6	—	—	—	1	2	3														
7	—	—	0	1	3	4	6													
8	—	—	0	2	4	6	7	9												
9	—	—	1	3	5	7	9	11	14											
10	—	—	1	3	6	8	11	13	16	19										
11	—	—	1	4	7	9	12	15	18	22	25									
12	—	—	2	5	8	11	14	17	21	24	28	31								
13	—	0	2	5	9	12	16	20	23	27	31	35	39							
14	—	0	2	6	10	13	17	22	26	30	34	38	43	47						
15	—	0	3	7	11	15	19	24	28	33	37	42	47	51	56					
16	—	0	3	7	12	16	21	26	31	36	41	46	51	56	61	66				
17	—	0	4	8	13	18	23	28	33	38	44	49	55	60	66	71	77			
18	—	0	4	9	14	19	24	30	36	41	47	53	59	65	70	76	82	88		
19	—	1	4	9	15	20	26	32	38	44	50	56	63	69	75	82	88	94	101	
20	—	1	5	10	16	22	28	34	40	47	53	60	67	73	80	87	93	100	107	114
21	—	1	5	11	17	23	30	36	43	50	57	64	71	78	85	92	99	106	113	121
22	—	1	6	11	18	24	31	38	45	53	60	67	75	82	90	97	105	112	120	127
23	—	1	6	12	19	26	33	40	48	55	63	71	79	87	94	102	110	118	126	134
24	—	1	6	13	20	27	35	42	50	58	66	75	83	91	99	108	116	124	133	141
25	—	1	7	13	21	29	36	45	53	61	70	78	87	95	104	113	122	130	139	148
26	—	1	7	14	22	30	38	47	55	64	73	82	91	100	109	118	127	136	146	155
27	—	2	7	15	23	31	40	49	58	67	76	85	95	104	114	123	133	142	152	162
28	—	2	8	16	24	33	42	51	60	70	79	89	99	109	119	129	139	149	159	169
29	—	2	8	16	25	34	43	53	63	73	83	93	103	113	123	134	144	155	165	176
30	—	2	9	17	26	35	45	55	65	76	86	96	107	118	128	139	150	161	172	182
31	—	2	9	18	27	37	47	57	68	78	89	100	111	122	133	144	156	167	178	189
32	—	2	9	18	28	38	49	59	70	81	92	104	115	127	138	150	161	173	185	196
33	—	2	10	19	29	40	50	61	73	84	96	107	119	131	143	155	167	179	191	203
34	—	3	10	20	30	41	52	64	75	87	99	111	123	135	148	160	173	185	198	210
35	—	3	11	20	31	42	54	66	78	90	102	115	127	140	153	165	178	191	204	217
36	—	3	11	21	32	44	56	68	80	93	106	118	131	144	158	171	184	197	211	224
37	—	3	11	22	33	45	57	70	83	96	109	122	135	149	162	176	190	203	217	231
38	—	3	12	22	34	46	59	72	85	99	112	126	139	153	167	181	195	209	224	238
39	—	3	12	23	35	48	61	74	88	101	115	129	144	158	172	187	201	216	230	245
40	—	3	13	24	36	49	63	76	90	104	119	133	148	162	177	192	207	222	237	252

*In this table n_S is the number of observations associated with the smaller of the two samples and n_L is the number associated with the larger of the two samples.

In the first part, the table entries are the critical values of U for a one-tailed test at the .01 level and a two-tailed test at the .02 level. In the second part, table entries are the critical values of U for a one-tailed test at the .05 level and a two-tailed test at the .10 level.

(continued)

Table 9 (Continued)

Values of $U_{.05}\,[n_L; n_S]$

										n_S										
n_L	1	2	3	4	5	6	7	8	9	10	11	12	13	14	15	16	17	18	19	20
1	—																			
2	—	—																		
3	—	—	0																	
4	—	—	0	1																
5	—	0	1	2	4															
6	—	0	2	3	5	7														
7	—	0	2	4	6	8	11													
8	—	1	3	5	8	10	13	15												
9	—	1	4	6	9	12	15	18	21											
10	—	1	4	7	11	14	17	20	24	27										
11	—	1	5	8	12	16	19	23	27	31	34									
12	—	2	5	9	13	17	21	26	30	34	38	42								
13	—	2	6	10	15	19	24	28	33	37	42	47	51							
14	—	3	7	11	16	21	26	31	36	41	46	51	56	61						
15	—	3	7	12	18	23	28	33	39	44	50	55	61	66	72					
16	—	3	8	14	19	25	30	36	42	48	54	60	65	71	77	83				
17	—	3	9	15	20	26	33	39	45	51	57	64	70	77	83	89	96			
18	—	4	9	16	22	28	35	41	48	55	61	68	75	82	88	95	102	109		
19	0	4	10	17	23	30	37	44	51	58	65	72	80	87	94	101	109	116	123	
20	0	4	11	18	25	32	39	47	54	62	69	77	84	92	100	107	115	123	130	138
21	0	5	11	19	26	34	41	49	57	65	73	81	89	97	105	113	121	130	138	146
22	0	5	12	20	28	36	44	52	60	68	77	85	94	102	111	119	128	136	145	154
23	0	5	13	21	29	37	46	54	63	72	81	90	98	107	116	125	134	143	152	161
24	0	6	13	22	30	39	48	57	66	75	85	94	103	113	122	131	141	150	160	169
25	0	6	14	23	32	41	50	60	69	79	89	98	108	118	128	137	147	157	167	177
26	0	6	15	24	33	43	53	62	72	82	92	103	113	123	133	143	154	164	174	185
27	0	7	15	25	35	45	55	65	75	86	96	107	117	128	139	149	160	171	182	192
28	0	7	16	26	36	46	57	68	78	89	100	111	122	133	144	156	167	178	189	200
29	0	7	17	27	38	48	59	70	82	93	104	116	127	138	150	162	173	185	196	208
30	0	7	17	28	39	50	61	73	85	96	108	120	132	144	156	168	180	192	204	216
31	0	8	18	29	40	52	64	76	88	100	112	124	136	149	161	174	186	199	211	224
32	0	8	19	30	42	54	66	78	91	103	116	128	141	154	167	180	193	206	218	231
33	0	8	19	31	43	56	68	81	94	107	120	133	146	159	172	186	199	212	226	239
34	0	9	20	32	45	57	70	84	97	110	124	137	151	164	178	192	206	219	233	247
35	0	9	21	33	46	59	73	86	100	114	128	141	156	170	184	198	212	226	241	255
36	0	9	21	34	48	61	75	89	103	117	131	146	160	175	189	204	219	233	248	263
37	0	10	22	35	49	63	77	91	106	121	135	150	165	180	195	210	225	240	255	271
38	0	10	23	36	50	65	79	94	109	124	139	154	170	185	201	216	232	247	263	278
39	1	10	23	38	52	67	82	97	112	128	143	159	175	190	206	222	238	254	270	286
40	1	11	24	39	53	68	84	99	115	131	147	163	179	196	212	228	245	261	278	294

Source: Roy C. Milton, "An Extended Table of Critical Values for the Mann-Whitney (Wilcoxon) Two-Sample Statistic," *Journal of the American Statistical Association*, Vol. 22 (1937), pp. 675–701.

Table 10A Exact Distribution of χ_r^2 for Tables with Two to Nine Sets of Three Ranks ($c = 3$; $r = 2, 3, 4, 5, 6, 7, 8, 9$)

p is the probability of obtaining a value of χ_r^2 as great as or greater than the corresponding value of χ_r^2.

$r = 2$		$r = 3$		$r = 4$		$r = 5$	
χ_r^2	p	χ_r^2	p	χ_r^2	p	χ_r^2	p
0	1.000	0.000	1.000	0.0	1.000	0.0	1.000
1	0.833	0.667	0.944	0.5	0.931	0.4	0.954
3	0.500	2.000	0.528	1.5	0.653	1.2	0.691
4	0.167	2.667	0.361	2.0	0.431	1.6	0.522
		4.667	0.194	3.5	0.273	2.8	0.367
		6.000	0.028	4.5	0.125	3.6	0.182
				6.0	0.069	4.8	0.124
				6.5	0.042	5.2	0.093
				8.0	0.0046	6.4	0.039
						7.6	0.024
						8.4	0.0085
						10.0	0.00077

$r = 6$		$r = 7$		$r = 8$		$r = 9$	
χ_r^2	p	χ_r^2	p	χ_r^2	p	χ_r^2	p
0.00	1.000	0.000	1.000	0.00	1.000	0.000	1.000
0.33	0.956	0.286	0.964	0.25	0.967	0.222	0.971
1.00	0.740	0.857	0.768	0.75	0.794	0.667	0.814
1.33	0.570	1.143	0.620	1.00	0.654	0.889	0.865
2.33	0.430	2.000	0.486	1.75	0.531	1.556	0.569
3.00	0.252	2.571	0.305	2.25	0.355	2.000	0.398
4.00	0.184	3.429	0.237	3.00	0.285	2.667	0.328
4.33	0.142	3.714	0.192	3.25	0.236	2.889	0.278
5.33	0.072	4.571	0.112	4.00	0.149	3.556	0.187
6.33	0.052	5.429	0.085	4.75	0.120	4.222	0.154
7.00	0.029	6.000	0.052	5.25	0.079	4.667	0.107
8.33	0.012	7.143	0.027	6.25	0.047	5.556	0.069
9.00	0.0081	7.714	0.021	6.75	0.038	6.000	0.057
9.33	0.0055	8.000	0.016	7.00	0.030	6.222	0.048
10.33	0.0017	8.857	0.0084	7.75	0.018	6.889	0.031
12.00	0.00013	10.286	0.0036	9.00	0.0099	8.000	0.019
		10.571	0.0027	9.25	0.0080	8.222	0.016
		11.143	0.0012	9.75	0.0048	8.667	0.010
		12.286	0.00032	10.75	0.0024	9.556	0.0060
		14.000	0.000021	12.00	0.0011	10.667	0.0035
				12.25	0.00086	10.889	0.0029
				13.00	0.00026	11.556	0.0013
				14.25	0.000061	12.667	0.00066
				16.00	0.0000036	13.556	0.00035
						14.000	0.00020
						14.222	0.000097
						14.889	0.000054
						16.222	0.000011
						18.000	0.0000006

Table 10B   Exact Distribution of χ_r^2 for Tables with Two to Four Sets of Four Ranks ($c = 4$; $r = 2, 3, 4$)

p is the probability of obtaining a value of χ_r^2 as great as or greater than the corresponding value of χ_r^2.

	$r = 2$			$r = 3$			$r = 4$		
χ_r^2	p		χ_r^2	p		χ_r^2	p	χ_r^2	p
0.0	1.000		0.2	1.000		0.0	1.000	5.7	0.141
0.6	0.958		0.6	0.958		0.3	0.992	6.0	0.105
1.2	0.834		1.0	0.910		0.6	0.928	6.3	0.094
1.8	0.792		1.8	0.727		0.9	0.900	6.6	0.077
2.4	0.625		2.2	0.608		1.2	0.800	6.9	0.068
3.0	0.542		2.6	0.524		1.5	0.754	7.2	0.054
3.6	0.458		3.4	0.446		1.8	0.677	7.5	0.052
4.2	0.375		3.8	0.342		2.1	0.649	7.8	0.036
4.8	0.208		4.2	0.300		2.4	0.524	8.1	0.033
5.4	0.167		5.0	0.207		2.7	0.508	8.4	0.019
6.0	0.042		5.4	0.175		3.0	0.432	8.7	0.014
			5.8	0.148		3.3	0.389	9.3	0.012
			6.6	0.075		3.6	0.355	9.6	0.0069
			7.0	0.054		3.9	0.324	9.9	0.0062
			7.4	0.033		4.5	0.242	10.2	0.0027
			8.2	0.017		4.8	0.200	10.8	0.0016
			9.0	0.0017		5.1	0.190	11.1	0.00094
						5.4	0.158	12.0	0.000072

Source: Adapted from table of Friedman's χ_r^2 statistic. Appeared in M. Friedman, "The Use of Ranks to Avoid the Assumption of Normality Implicit in the Analysis of Variance," *Journal of the American Statistical Association*, vol. 32 (1937), pp. 675–701.

Table 11   Critical Values for Duncan's Multiple-Range Test

Significance Level $\alpha = .05$

						Range of Comparison								
df_2	2	3	4	5	6	7	8	9	10	11	12	13	14	15
1	17.970	17.970	17.970	17.970	17.970	17.970	17.970	17.970	17.970	17.970	17.970	17.970	17.970	17.970
2	6.085	6.085	6.085	6.085	6.085	6.085	6.085	6.085	6.085	6.085	6.085	6.085	6.085	6.085
3	4.501	4.516	4.516	4.516	4.516	4.516	4.516	4.516	4.516	4.516	4.516	4.516	4.516	4.516
4	3.927	4.013	4.033	4.033	4.033	4.033	4.033	4.033	4.033	4.033	4.033	4.033	4.033	4.033
5	3.635	3.749	3.797	3.814	3.814	3.814	3.814	3.814	3.814	3.814	3.814	3.814	3.814	3.814
6	3.461	3.587	3.649	3.680	3.694	3.697	3.697	3.697	3.697	3.697	3.697	3.697	3.697	3.697
7	3.344	3.477	3.548	3.588	3.611	3.622	3.626	3.626	3.626	3.626	3.626	3.626	3.626	3.626
8	3.261	3.399	3.475	3.521	3.549	3.566	3.575	3.579	3.579	3.579	3.579	3.579	3.579	3.579
9	3.199	3.339	3.420	3.470	3.502	3.523	3.536	3.544	3.547	3.547	3.547	3.547	3.547	3.547
10	3.151	3.293	3.376	3.430	3.465	3.489	3.505	3.516	3.522	3.525	3.526	3.526	3.526	3.526
11	3.113	3.256	3.342	3.397	3.435	3.462	3.480	3.493	3.501	3.506	3.509	3.510	3.510	3.510
12	3.082	3.225	3.313	3.370	3.410	3.439	3.459	3.474	3.484	3.491	3.496	3.498	3.499	3.499
13	3.055	3.200	3.289	3.348	3.389	3.419	3.442	3.458	3.470	3.478	3.484	3.488	3.490	3.490
14	3.033	3.178	3.268	3.329	3.372	3.403	3.426	3.444	3.457	3.467	3.474	3.479	3.482	3.484
15	3.014	3.160	3.250	3.312	3.356	3.389	3.413	3.432	3.446	3.457	3.465	3.471	3.476	3.478
16	2.998	3.144	3.235	3.298	3.343	3.376	3.402	3.422	3.437	3.449	3.458	3.465	3.470	3.473
17	2.984	3.130	3.222	3.285	3.331	3.366	3.392	3.412	3.429	3.441	3.451	3.459	3.465	3.469

(continued)

Table 11 (*Continued*)

Significance Level α = .05

					Range of Comparison									
df_2	2	3	4	5	6	7	8	9	10	11	12	13	14	15
18	2.971	3.118	3.210	3.274	3.321	3.356	3.383	3.405	3.421	3.435	3.445	3.454	3.460	3.465
19	2.960	3.107	3.199	3.264	3.311	3.347	3.375	3.397	3.415	3.429	3.440	3.449	3.456	3.462
20	2.950	3.097	3.190	3.255	3.303	3.339	3.368	3.391	3.409	3.424	3.436	3.445	3.453	3.459
24	2.919	3.066	3.160	3.226	3.276	3.315	3.345	3.370	3.390	3.406	3.420	3.432	3.441	3.449
30	2.888	3.035	3.131	3.199	3.250	3.290	3.322	3.349	3.371	3.389	3.405	3.418	3.430	3.439
40	2.858	3.006	3.102	3.171	3.224	3.266	3.300	3.328	3.352	3.373	3.390	3.405	3.418	3.429
60	2.829	2.976	3.073	3.143	3.198	3.241	3.277	3.307	3.333	3.355	3.374	3.391	3.406	3.419
120	2.800	2.947	3.045	3.116	3.172	3.217	3.254	3.287	3.314	3.337	3.359	3.377	3.394	3.409
∞	2.772	2.918	3.017	3.089	3.146	3.193	3.232	3.265	3.294	3.320	3.343	3.363	3.382	3.399

Significance Level α = .01

					Range of Comparison									
df_2	2	3	4	5	6	7	8	9	10	11	12	13	14	15
1	90.030	90.030	90.030	90.030	90.030	90.030	90.030	90.030	90.030	90.030	90.030	90.030	90.030	90.030
2	14.040	14.040	14.040	14.040	14.040	14.040	14.040	14.040	14.040	14.040	14.040	14.040	14.040	14.040
3	8.261	8.321	8.321	8.321	8.321	8.321	8.321	8.321	8.321	8.321	8.321	8.321	8.321	8.321
4	6.512	6.677	6.740	6.756	6.756	6.756	6.756	6.756	6.756	6.756	6.756	6.756	6.756	6.756
5	5.702	5.893	5.989	6.040	6.065	6.074	6.074	6.074	6.074	6.074	6.074	6.074	6.074	6.074
6	5.243	5.439	5.549	5.614	5.655	5.680	5.694	5.701	5.703	5.703	5.703	5.703	5.703	5.703
7	4.949	5.145	5.260	5.334	5.383	5.416	5.439	5.454	5.464	5.470	5.472	5.472	5.472	5.472
8	4.746	4.939	5.057	5.135	5.189	5.227	5.256	5.276	5.291	5.302	5.309	5.314	5.316	5.317
9	4.596	4.787	4.906	4.986	5.043	5.086	5.118	5.142	5.160	5.174	5.185	5.193	5.199	5.203
10	4.482	4.671	4.790	4.871	4.931	4.975	5.010	5.037	5.058	5.074	5.088	5.098	5.106	5.112
11	4.392	4.579	4.697	4.780	4.841	4.887	4.924	4.952	4.975	4.994	5.009	5.021	5.031	5.039
12	4.320	4.504	4.622	4.706	4.767	4.815	4.852	4.883	4.907	4.927	4.944	4.958	4.969	4.978
13	4.260	4.442	4.560	4.644	4.706	4.755	4.793	4.824	4.850	4.872	4.889	4.904	4.917	4.928
14	4.210	4.391	4.508	4.591	4.654	4.704	4.743	4.775	4.802	4.824	4.843	4.859	4.872	4.884
15	4.168	4.347	4.463	4.547	4.610	4.660	4.700	4.733	4.760	4.783	4.803	4.820	4.834	4.846
16	4.131	4.309	4.425	4.509	4.572	4.622	4.663	4.696	4.724	4.748	4.768	4.786	4.800	4.813
17	4.099	4.275	4.391	4.475	4.539	4.589	4.630	4.664	4.693	4.717	4.738	4.756	4.771	4.785
18	4.071	4.246	4.362	4.445	4.509	4.560	4.601	4.635	4.664	4.689	4.711	4.729	4.745	4.759
19	4.046	4.220	4.335	4.419	4.483	4.534	4.575	4.610	4.639	4.665	4.686	4.705	4.722	4.736
20	4.024	4.197	4.312	4.395	4.459	4.510	4.552	4.587	4.617	4.642	4.664	4.684	4.701	4.716
24	3.956	4.126	4.239	4.322	4.386	4.437	4.480	4.516	4.546	4.573	4.596	4.616	4.634	4.651
30	3.889	4.056	4.168	4.250	4.314	4.366	4.409	4.445	4.477	4.504	4.528	4.550	4.569	4.586
40	3.825	3.988	4.098	4.180	4.244	4.296	4.339	4.376	4.408	4.436	4.461	4.483	4.503	4.521
60	3.762	3.922	4.031	4.111	4.174	4.226	4.270	4.307	4.340	4.368	4.394	4.417	4.438	4.456
120	3.702	3.858	3.965	4.044	4.107	4.158	4.202	4.239	4.272	4.301	4.327	4.351	4.372	4.392
∞	3.643	3.796	3.900	3.978	4.040	4.091	4.135	4.172	4.205	4.235	4.261	4.285	4.307	4.327

Source: Adapted from H. L. Harler, "Critical Values for Duncan's New Multiple Range Test," *Biometrics 16* (1960), pp. 671–685. With permission from the Biometric Society.

Table 12 Table of Random Digits

26488	65184	35213	08911	41771	78761	34881	71011	38769	24817
03377	88678	27022	84903	84513	47052	14385	22169	12337	20836
67360	11209	04703	39197	93794	87287	50308	34204	64143	29808
07859	05673	77437	56001	96183	50414	77605	50494	86175	30721
21174	79881	31784	29177	80907	52946	67446	20201	60409	14530
14916	68132	65010	74111	93003	60176	43652	38637	24846	17161
81615	80233	87860	88405	03328	09140	30501	39434	67801	36037
94308	19026	51141	52030	91280	38243	06521	09759	37885	79255
18737	84431	10210	19135	62392	33368	65096	21635	52768	00335
91808	11434	84510	09724	44804	75493	23848	07525	51220	32981
98109	97606	32651	03797	75607	71252	53208	11220	12464	56543
57438	59987	03073	88112	13637	38195	83731	35312	90660	36780
39419	40518	60169	67593	77874	81275	71598	17925	79731	59571
45324	63563	99292	39986	24099	98128	46527	40652	16618	04084
41069	15749	28541	31100	25983	21706	09643	07666	01573	52145
03825	02399	75383	34402	35331	47832	39017	53635	74501	68789
82887	42543	92642	70000	63614	14826	49424	89663	85509	14280
28795	19978	07330	87676	27559	19564	29702	24204	97089	69210
59117	31864	20413	01258	48115	99390	66464	10103	13691	82780
81841	85493	06282	32567	42696	58381	46118	83007	08915	36396
26454	28487	32450	58558	16169	18995	77039	12866	40426	18284
07573	05228	92017	16315	85887	84969	37262	95862	40855	07788
26070	07301	70134	60577	49443	92914	51603	39632	09308	18023
78789	17386	65408	33061	61336	76264	31684	07421	74411	94873
89502	63120	42146	28239	80403	14009	58786	99347	38137	71018
49900	89267	81039	37192	86746	13303	33858	41254	05114	18748
04596	00469	84147	41439	86139	46286	85496	98215	14644	54949
03032	90178	97586	58485	14777	86467	30520	61690	34076	13101
60407	73065	52537	22210	07620	55251	26638	73767	84644	13066
20049	85142	71961	44114	43840	67648	06484	11383	42522	86489
81870	83709	97610	79950	42214	31597	87211	89188	87472	62569
63161	59799	41696	37290	02170	57803	12988	91659	20840	41839
99417	04688	84622	36662	69190	58303	10249	46573	41177	77616
56280	67821	02461	86316	53174	01469	18511	48170	54274	35223
49640	27628	66188	83630	42092	55752	37156	21654	73948	39454
40985	49558	10525	84926	12748	94011	26487	94169	94770	79757
83866	16648	90314	32706	31814	08857	86393	14702	96954	81180
99746	49460	24504	08069	64188	23469	15887	35938	57377	50222
25531	37403	28661	31842	83494	64740	75668	91909	48839	71560
58525	46229	46661	77378	62943	84154	55650	85352	05221	74645
59168	77022	38849	48418	37059	81142	43724	69633	48123	83852
45556	78930	17822	64085	97000	58207	03992	00524	33338	63320
95941	09941	26646	79930	08941	71644	28300	07758	88565	99968
69974	07396	30168	81184	35167	14183	93215	20689	95890	85617
27790	47571	99301	08247	98616	45998	62406	07600	31083	04375
32065	65133	03253	34611	90918	60893	85595	52459	21400	40246
50824	88029	37647	93690	97465	39804	11207	28485	48862	11813
13251	27156	23257	95377	30224	88312	51181	98961	30460	40066
19281	67639	61205	85952	06230	58813	79229	03874	00138	55781
22161	31143	48783	53936	65274	50813	06125	59984	42227	94356

Source: The Rand Corporation, A *Million Random Digits with 100,000 Normal Deviates*, The Free Press, 1955, p. 293. Reprinted with permission of the Rand Corporation.

Answers to Quantitative Questions

Chapter 2

2.19 $N = 12$ and $n = 7$ results in 792 sample combinations, whereas $N = 8$ and $n = 3$ results in only 56 combinations.

2.20
a.   84
b.   126
c.   330
d.   462
e.   105

2.21
a.   Ten combinations
b.   AB, AC, AD, AE, BC, BD, BE, CD, CE, DE

2.22
a.   35 combinations
b.   ABC, ABD, ABE, ABF, ABG, ACD, ACE, ACF, ACG, ADE, ADF, ADG, AEF, AEG, AFG, BCD, BCE, BCF, BCG, BDE, BDF, BDG, BEF, BEG, BFG, CDE, CDF, CDG, CEF, CEG, CFG, DEF, DEG, DFG, EFG

2.24
a. $3080
b. $n = 6$

2.26
a. $840
b. $n = 10$

2.28
a. $n = 3$ and 6

2.30
a. $n = 4$, 20, and 60
b. $n = 84$
c. $n = 4$, 20, and 60

Chapter 3

3.9
b.   Eight classes
c.   Class interval of 4 (or 3)
d.

Class	Frequency
35 and under 39	7
39 and under 43	18
43 and under 47	10
47 and under 51	14
51 and under 55	7
55 and under 59	25
59 and under 63	10
63 and under 67	9
Total	100

g. $Q_1 = 41.6$, $Q_2 = 51.4$, $Q_3 = 57.8$

3.10

Class	Relative Frequency
35 and under 39	.07
39 and under 43	.18
43 and under 47	.10
47 and under 51	.14
51 and under 55	.07
55 and under 59	.25
59 and under 63	.10
63 and under 67	.09
Total	1.00

3.11 **a.**

Class Limit	"Less Than" Cumulative Frequency
35	0
39	7
43	25
47	35
51	49
55	56
59	81
63	91
67	100

b.

Class Limit	"Or More" Cumulative Frequency
35	100
39	93
43	75
47	65
51	51
55	44
59	19
63	9
67	0

3.12 **a.**

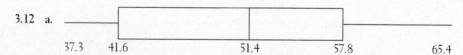

37.3   41.6   51.4   57.8   65.4

b.

3	7 7 8 8 9 9 9 9 9 9 9 9	(12)
4	0 0 0 0 0 0 1 1 1 1 1 2 2 4 4 5 5 5 5 5 5 5 5 5 7 7 7 7 8 8 9 9 9 9 9 9 9	(37)
5	1 1 1 3 3 3 3 5 5 5 5 5 5 5 5 5 5 5 5 6 6 6 6 6 6 8 8 9 9 9 9 9 9 9	(34)
6	0 0 1 1 1 1 1 1 4 4 4 5 5 5 5 5 5	(17)

3.13 **b.** Table indicates seven classes; when the other instructions are followed, eight classes emerge.

c. Class interval is 2850.

d.

Class	Frequency
$ 6,000 and under $8,850	4
8,850 and under 11,700	4
11,700 and under 14,550	6
14,550 and under 17,400	12
17,400 and under 20,250	12

Class	Frequency
20,250 and under 23,100	7
23,100 and under 25,950	4
25,950 and under 28,800	1
Total	50

3.14 g. $Q_1 = \$13,950$; $Q_2 = \$16,600$; $Q_3 = \$19,850$

Class	Relative Frequency
$ 6,000 and under $ 8,850	.08
8,850 and under 11,700	.08
11,700 and under 14,500	.12
14,500 and under 17,400	.24
17,400 and under 20,250	.24
20,250 and under 23,100	.14
23,100 and under 25,950	.08
25,950 and under 28,800	.02
Total	1.00

3.15 a.

Class Limit	"Less Than" Cumulative Frequency
$ 6,000	0
8,850	4
11,700	8
14,500	14
17,400	26
20,250	38
23,100	45
25,950	49
28,800	50

b.

Class Limit	"Or More" Cumulative Frequency
$ 6,000	50
8,850	46
11,700	42
14,500	36
17,400	24
20,250	12
23,100	5
25,950	1
28,800	0

3.16 a.

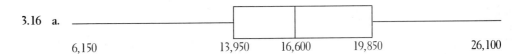

6,150	13,950	16,600	19,850	26,100

b.

```
0 | 6 8 9 9 9                                                  (5)
1 | 0 0 1 2 2 3 4 4 4 5 5 5 5 5 6 6 6 6 7 7 8 8 8 8 9 9 9 9 9 9  (31)
2 | 0 0 0 1 1 1 1 2 3 3 4 5 6 6                                (14)
```

3.17 b. Table indicates five classes; when the other instructions are followed, six classes emerge.

c. Class interval is 10.

d.

Class	Frequency
20 and under 30	3
30 and under 40	8
40 and under 50	6
50 and under 60	3
60 and under 70	0
70 and under 80	1
Total	21

g. $Q_1 = 31$; $Q_2 = 38$; $Q_3 = 46$

3.18 a.

Class	Relative Frequency
20 and under 30	.143
30 and under 40	.381
40 and under 50	.286
50 and under 60	.143
60 and under 70	.000
70 and under 80	.048
Total	1.001

3.19 a.

Class Limit	"Less Than" Cumulative Frequency
20	0
30	3
40	11
50	17
60	20
70	20
80	21

b.

Class Limit	"Or More" Cumulative Frequency
20	21
30	18
40	10
50	4
60	3
70	3
80	1

3.20 a.

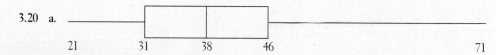

21   31   38   46   71

b.

2	1 7 8	(3)
3	0 0 2 2 4 5 8 8	(8)
4	0 0 1 4 5 7	(6)
5	1 2 4	(3)
6		(0)
7	1	(1)

3.29 a.

	Male	Female
Frequent watcher	.25	.36
Moderate watcher	.44	.43
Infrequent watcher	.31	.21
Total	1.00	1.00

3.30 a.

	Male		Female	
	Single	Married	Single	Married
Frequent watcher	.23	.27	.28	.45
Moderate watcher	.43	.46	.41	.45
Infrequent watcher	.34	.27	.31	.11
Total	1.00	1.00	1.00	1.01

3.31 a.

Perceives Products to Be of:	Eastern U.S.	Western U.S.
High quality	.182	.269
Medium quality	.558	.590
Low quality	.259	.141
Total	.999	1.000

3.32

Perceives Products to Be of:	Eastern U.S.		Western U.S.	
	Male	Female	Male	Female
High quality	.103	.262	.167	.365
Medium quality	.513	.604	.611	.570
Low quality	.385	.134	.222	.065
Total	1.001	1.000	1.000	1.000

3.34

Political Party	Number of Households
Democrat	315
Republican	167
Libertarian	3
Independent	5
Not sure	10
Total	500

Income Class	Number of Households
$ 20,000 and under $ 30,000	103
30,000 and under 40,000	190
40,000 and under 50,000	155
50,000 and under 60,000	0
60,000 and under 70,000	0
70,000 and under 80,000	45
80,000 and under 90,000	0
90,000 and under 100,000	0
100,000 and under 110,000	7
Total	500

Number of Automobiles Owned	Number of Households
1	45
2	295
3	160
Total	500

Chapter 4

4.9 a. $\overline{X} = 46$

b.

X	$\overline{X}$	$X - \overline{X}$
55	46	9
63	46	17
18	46	−28
52	46	6
56	46	10
32	46	−14
Total		0

c. $\Sigma (X - 48) = -12$

d. $\Sigma (X - 55) = -54$

4.10 a. $\Sigma (X - \overline{X})^2 = 1486$

b. $\Sigma (X - 48)^2 = 1510$

c. $\Sigma (X - 55)^2 = 1972$

4.11 a. $\overline{X} = 19.9$

b. $\Sigma (X - 19.9) = 0$

c. $\Sigma (X - 14) = 59$

d. $\Sigma (X - 10) = 99$

4.12 a. $\Sigma (X - \overline{X})^2 = 384.9$

b. $\Sigma (X - 14)^2 = 733$

c. $\Sigma (X - 10)^2 = 1365$

4.13 a. $\mu = 4.8$

b. $\sigma = 1.72$

c. $\mu = 8.8$; $\sigma = 1.72$

4.14 a. $\mu = 2.57$

b. $\sigma = 1.40$

c. $\mu = 7.71$; $\sigma = 4.19$

4.16 a. Mean = $340

b. Median = $300

c. Mode = $300

d. Mean = $533.33

e. Median = $350

f. Mode = $300

g. Mean = $1028.57

h. Median = $400

i. Mode = $300

4.17 a. Ranges: $500; $1400; $3900

b. Mean deviations: $128; $344.44; $983.67

c. Standard deviations: $162.48; $457.04; $1284.76 (N denominators used)

4.18 a. Mean = $38,750

b. Median = $35,000

c. Mode = $35,000

d. Mean = $61,000

e. Median = $35,000

    f. Mode = $35,000

    g. Mean = $217,500

    h. Median = $47,500

    i. Mode = $35,000

4.19 a. Ranges: $35,000; $125,000; $975,000

    b. Mean deviations: $10,625; $35,600; $260,833.33

    c. Standard deviations: $12,930; $45,978.26; $352,452.64 (N denominators used)

4.20 a. $\overline{X} = 51.88$

    b. Median = 55.3

4.21 a. $\Sigma\,(X - 51.88) = 0$

    b. $\Sigma\,(X - 55.3) = -68.4$

4.22 a. 27.6

    b. 7.35

    c. 8.49

4.23 a. $\overline{X} = \$13,062.50$

    b. Median = $13,300

4.24 a. $\Sigma\,(X - 13,062.5) = 0$

    b. $\Sigma\,(X - 13,300) = -4750$

4.25 a. $14,400

    b. $3442.50

    c. $4110.50

4.26 a. 44.44

    b. 45.06

    c. 10.38

4.29 a. 40.87

    b. 38.75

    c. 15.52

4.30 a. (1) $524.74; (2) $638.57

    b. (1) $157.08; (2) $189.92

    d. (1) 29.9%; (2) 29.7%

4.31 a. (1) $21.45; (2) $12.38; (3) $11.93

    b. (1) $4.83; (2) $1.76; (3) $3.66

    d. (1) 22.5%; (2) 14.2%; (3) 30.7%

4.32 a. 2.6   **c.** 2, 2, 2.33, 2.33, 2.67, 2.67, 2.67, 3, 3, 3.33   **e.** 2.6; same

    b. 3.0   **d.** 2, 2, 2, 3, 3, 3, 3, 3, 3, 3   **f.** 2.7; lower

                                                                               g. Mean

4.33 a. .42

    b. .46

4.34 a. $\sigma = 1.02;\ \sigma^2 = 1.04$

    b. Mean of the sample standard deviations = 1.09

    d. Mean of the sample variances = 1.30*

---

*The mean of the sample variances will equal $\sigma^2[N/(N-1)]$. For larger populations, $N/(N-1)$ becomes effectively 1, and the mean of the sample variances is an unbiased estimator of σ^2.

4.35 a. 3
     b. 3.5
     c. 4
     g. 3
     h. 3.5
     i. 3.75

4.36 a. .45
     b. .32

4.37 a. $\sigma = 1.41$
       $\sigma^2 = 2.00$
     c. Mean of sample standard deviations = 1.42
     d. Mean of sample variances = 2.40*

Chapter 5

5.5 1978: 100.0; 1979: 108.8; 1980: 110.5; 10.5%

5.6 1978: 100.0; 1979: 106.9; 1980: 111.5; 11.5%

5.7 1978: 100.0; 1979: 107.3; 1980: 111.0; 11%

5.8 1978: 100.0; 1979: 108.0; 1980: 111.5; 11.5%

5.9 1978: 100.0; 1979: 106.9; 1980: 111.5; 11.5%

5.10 1977: 100.0; 1978: 117.1; 1979: 140.0; 1980: 167.1; 67.1%

5.11 1977: 100.0; 1978: 128.2; 1979: 174.8; 1980: 213.3; 113.3%

5.12 1977: 100.0; 1978: 127.3; 1979: 165.6; 1980: 201.2; 101.2%

5.13 1977: 100.0; 1978: 118.3; 1979: 144.6; 1980: 174.2; 74.2%

5.14 1977: 100.0; 1978: 128.2; 1979: 174.8; 1980: 213.3; 113.3%

5.15 1977: 100.0; 1978: 118.0; 1979: 124.9; 1980: 139.2; 1981: 169.5; 69.5%

5.16 1978: 100.0; 1979: 117.1; 1980: 121.0; 1981: 135.7; 1982: 166.7; 66.7%

5.17 1978: 100.0; 1979: 117.0; 1980: 127.7; 1981: 137.5; 1982: 164.3; 64.3%

5.18 1978: 100.0; 1979: 118.5; 1980: 138.3; 1981: 148.5; 1982: 170.6; 71%

5.19 1978: 100.0; 1979: 117.2; 1980: 121.1; 1981: 135.7; 1982: 166.7; 66.7%

5.20 1967: 37.14; 1968: 42.75; 1969: 43.22; 1970: 40.40; 1971: 52.27; 1972: 62.00; 1973: 59.76; 1974: 45.13; 1975: 38.44; 1976: 47.05

5.21 1967: 81.9; 1968: 85.3; 1969: 89.4; 1970: 93.6; 1971: 96.6; 1972: 100.0; 1973: 107.9; 1974: 124.0; 1975: 133.6; 1976: 138.0

Chapter 6

6.12 b. (1) O_{21}, O_{22}, O_{23}; (2) O_{21}; (3) $O_{11}, O_{21}, O_{31}, O_{22}, O_{23}$; (4) O_{21}, O_{32}; (5) O_{12}, O_{22}; (6) O_{22}

6.13 b. (1) O_{41}, O_{42}, O_{43}; (2) O_{32}; (3) O_{43}; (4) $O_{21}, O_{22}, O_{23}, O_{41}, O_{42}, O_{43}$; (5) O_{31}, O_{42}; (6) O_{31}, O_{32}, O_{33}; (7) O_{12}, O_{32}, O_{42}; (8) O_{11}, O_{12}

6.14 a. .48
      b. .18
      c. .125
      d. .625
      e. .353
      f. .62
      g. .06
      h. .36
      i. .34

6.15 a. $P(A_1|B_1) = .667$; $P(B_1|A_1) = .231$; no
      b. $P(A_1 \text{ and } B_1) = .12$; $P(A_1) \cdot P(B_1) = .094$; not independent
      c. $P(A_1) = .52$; $P(A_1|B_2) = .647$; not independent
      d. $P(A_1) = .52$; $P(A_1|B_3) = .412$; not independent
      e. $P(A_2) = .48$; $P(A_2|B_2) = .353$; not independent

6.16 a. .08
      b. .255
      c. .015
      d. .188
      e. .325
      f. .595
      g. .13

6.17 a. $P(A_3|B_1) = .323$; $P(B_1|A_3) = .233$; no
      b. $P(A_3 \text{ and } B_2) = .065$; $P(A_3) \cdot P(B_2) = .078$; not independent
      c. $P(A_1) = .08$; $P(A_1|B_1) = .032$; not independent
      d. $P(A_5) = .255$; $P(A_5|B_5) = .188$; not independent
      e. No; not mutually exclusive

6.18 a. .6
      b. .267
      c. .2
      d. .8
      e. .4
      f. .12
      g. .72

6.19 a. $P(A_1|B_1) = .267$; $P(B_1|A_1) = .4$; not equal
      b. $P(A_3 \text{ and } B_2) = .12$; $P(A_3) \cdot P(B_2) = .12$; indicates independence
      c. $P(A_1) = .267$; $P(A_1|B_1) = .267$; events are independent
      d. $P(A_2) = .533$; $P(A_2|B_2) = .533$; events are independent

6.20 a. .224
      b. .51
      c. .141
      d. .799
      e. .421

6.21 a. $P(A_1|B_2) = .18$; $P(B_2|A_1) = .41$; no
      b. $P(A_1) \cdot P(B_2|A_1) = .092$; $P(B_2) \cdot P(A_1|B_2) = .092$; yes
      c. $P(A_1) = .224$; $P(A_1|B_2) = .18$; no
      d. $P(A_3) = .202$; $P(A_3|B_2) = .26$; no

6.22   a.   .225
        b.   .489
        c.   .508
        d.   .499
        e.   .314
        f.   .365
        g.   .098
        h.   .839

6.23   a. $P(A_1|C_1) = .133$; $P(C_1|A_1) = .292$; no
        b. $P(A_1|B_1$ and $C_1) = .167$; $P(C_1|A_1$ and $B_1) = .299$; no
        c. $P(A_1) \cdot P(B_1|A_1) P(C_1|A_1$ and $B_1) = .039$; $P(B_1) \cdot P(C_1|B_1) \cdot P(A_1|B_1$ and $C_1) = .039$; yes
        d. $P(A_2) = .574$; $P(A_2|B_2) = .558$; not strictly independent
        e. $P(A_2) = .574$; $P(A_2|C_2) = .587$; not strictly independent
        f. $P(B_2) = .511$; $P(B_2|C_2) = .502$; not strictly independent
        g.   No; $P(A_1$ and $C_1) \neq 0$

6.24   a.   .04
        b.   .667

6.25   a.   .433
        b.   .1
        c.   .35
        d.   .7

6.26   a. $P(A_1) = .333$; $P(A_1|B_1) = .3$; not strictly independent
        b.   No; occurrence of Caucasian does not preclude male; occurrence of male does not preclude Caucasian; or $P(C$ and $F) \neq 0$

6.27   a.   .711
        b.   .967

6.28   a.   .012
        b.   .006
        c.   .04
        d.   .144
        e.   .156

6.29   a. $P(D) = .012$; $P(D|M) = .007$; no
        b.   Not mutually exclusive; $P(D$ and $M) \neq 0$

6.30   a.   .047
        b.   .9995

6.31   a.   .417
        b.   .333
        c.   .583
        d.   .417

6.32   a. $P(\text{Not-}R) = .417$; $P(\text{Not-}R|A) = .5$; no
        b.   Not mutually exclusive; $P(\text{Not-}R$ and $A) \neq 0$

6.33   a.   .929
        b.   .999

6.34   a.   .864
        b.   .132
        c.   .136
        d.   .004

6.36 a. .656
      b. .047

6.39 a. .4096
      b. .5904

6.40 .2

6.41 a. .167
      b. .3

6.45 .0096

6.46 .192

6.47 a. .42
      b. .53
      c. .3

6.48 a. .3
      b. .66
      c. .7

6.49 a. .333
      b. .667

6.50 a. .441
      b. .441
      c. .118

6.51 $P(A \text{ and } R) + P(B \text{ and } R) = P(A) \cdot P(R|A) + P(B) \cdot P(R|B) = (.6)(.8) + (.4)(.5) = .68$

6.52 a. .706
      b. .294

6.53 .903

6.54 .7

6.55 .999

6.56 .279

6.57 .0007

6.58 .6

Chapter 7

7.15 a. .0768
      b. .0230
      c. Approximately .1755
      d. .0000

7.16 a. .1536
      b. .1808
      c. .9728
      d. .1792

7.18 a. .9672
      b. .0473

7.19 a. .1852
      b. .2556
      c. .9294

7.21 a. .0824
      b. .0002
      c. .1222
      d. .1260

7.22 a. .4095
      b. .6723
      c. .9688

7.23 .1681

7.24 a. 2.58
      b. 1.52

7.25 a. Appeal A: mean = 3.50; standard deviation = 1.63
      b. Appeal B: mean = 2.80; standard deviation = 2.06

7.26 a. .4332
      b. .1587
      c. .3085
      d. .6915
      e. 81.45

7.27 a. .0000
      b. .3085
      c. 3.825

7.28 a. 11.495
      b. .675; 5.9%

7.29 a. 961 accounts
      b. 4772 accounts
      c. Between $84.5 and $165.5
      d. $151.4

7.30 28,250 miles

7.31 a. .6554
      b. .3364
      c. .0082

7.32 a. .3859
      b. .4870
      c. .1271

7.34 $5.00

7.35 Supplier A: about .94; Supplier B: about .06

7.36 a. About .33
      b. About .67

7.37 Assuming about equal use of the two services: Jiffy, about .28; Kwick, about .72. A different assumption about historical relative frequency of use would result in different posterior probabilities.

7.38 a. About .34
      b. About .66

Answers to Quantitative Questions

7.39		Poisson Approximation	Normal-Curve Approximation
	a.	.0000	.0009
	b.	.0189	.0209
	c.	.4170	.4364
	d.	.9513	.9599

7.40 a. .5488
      b. .0000
      c. .0000
      d. .9997

7.41 a. .5404
      b. .0000
      c. .0000
      d. .9999

7.42 .7852

7.43 a. .0007
      b. .9279
      c. .7910

7.44 a. .0183
      b. .7852
      c. .0081

7.45 a. .0821
      b. .2423
      c. .4703

7.46 a. .0820
      b. .0067
      c. .9933

7.48 a. .7420
      b. .8495

7.49 a. .3680
      b. .1350
      c. .9500

7.50 a. 3 per hour

7.51 a. .8007
      b. .4232
      c. .0498

7.52 a. .3679
      b. .3679
      c. .9197

7.53 a. .3680
      b. .1350
      c. .3930

7.54

Number of Defectives X		P(X)
0		.3571
1		.5357
2		.1071
	Total	.9999

7.55 At least two: .5758; at most two: .8485

7.56 Crate X: .6774; Crate Y: .3226

Chapter 8

8.15 a. $\mu = 33.4$

b.

$\overline{X}$	$P(\overline{X})$
26.0	.1
27.5	.1
28.5	.1
30.0	.1
31.0	.1
32.5	.1
37.5	.1
38.5	.1
40.0	.1
42.5	.1
Total	1.0

c. (1) .1; (2) .2; (3) .4
d. 0
e. (1) .5; (2) 1.0
f. $\mu_{\overline{X}} = 33.4$; yes

8.16 a. $\sigma = 8.96$
b. $\sigma_{\overline{X}} = 5.49$
c. $\sigma_{\overline{X}} = 5.49$; yes

8.19 a. $\mu = 6.67$

b.

$\overline{X}$	$P(\overline{X})$
5.33	.15
6.00	.20
6.67	.30
7.33	.20
8.00	.15
Total	1.00

c. (1) .3; (2) .35; (3) .35
d. .3
e. (1) .7; (2) .0
f. $\mu_{\overline{X}} = 6.67$; yes

8.20 a. $\sigma = 1.89$
b. $\sigma_{\overline{X}} = .84$
c. $\sigma_{\overline{X}} = .84$; yes

Answers to Quantitative Questions

8.23 a. .7888
b. 116.35 to 118.65

8.24 .9876; 116.92 to 118.08

8.25 .4714; 115.2 to 119.8

8.26 a. .0336
b. .6372
c. .0049; .8030
d. .0031; .8294

8.27 a. .0594
b. .7888
c. .0003; .9940
d. .0188; .9050

8.28 a. $2 million
b. $1.95 million $\leq \mu \leq$ $2.05 million

8.29 $1.94 million $\leq \mu \leq$ $2.06 million

8.30 a. 10%
b. 8.9% $\leq \mu \leq$ 11.1%

8.31 8.69% $\leq \mu \leq$ 11.31%

8.32 a. 3.4
b. 3.297 $\leq \mu \leq$ 3.503
f. 329,700 $\leq$ total $\leq$ 350,300

8.33 a. 3.3216 $\leq \mu \leq$ 3.4784
e. 332,160 $\leq$ total $\leq$ 347,840

8.34 a. $\overline{X}$ = 182.19; s = 42.08; 159.77 $\leq \mu \leq$ 204.61

8.35 a. $\overline{X}$ = 49; s = 7.07
b. 44.17 $\leq \mu \leq$ 53.83
c. $\overline{X}$ = 65; s = 4.10
d. 58.25 $\leq \mu \leq$ 71.75

8.36 Taking σ to be 200, n must be at least 62.

8.37 At least 39

8.38 a. p = .75

b.

$\bar{p}$	$P(\bar{p})$
.0	.0357
.5	.4286
1.0	.5357

c. (1) .5357; (2) .4286; (3) .0357; (4) .0; (5) .4643
d. $\mu_{\bar{p}}$ = .75; yes
e. $\sigma_{\bar{p}}$ = .2835
f. $\sigma_{\bar{p}}$ = .2835; yes

8.40 a. p = .4

b.

$\bar{p}$	$P(\bar{p})$
.00	.1
.33	.6
.67	.3

c. (1) .0; (2) .3; (3) .6; (4) .0; (5) .9

d. $\mu_{\bar{p}} = .4$; yes

e. $\sigma_{\bar{p}} = .2$

f. $\sigma_{\bar{p}} = .2$; yes

8.41 a. .8764

b. .125 to .275

8.42 a. .7242

b. .24 to .36

8.43 a. .3

c. $.26 \le p \le .34$

8.44 $.27 \le p \le .33$

8.45 a. .4

b. $.38 \le p \le .42$

8.46 $.38 \le p \le .42$; slightly larger

8.47 Assuming p to be .5, n would have to be at least 1068.

8.48 Assuming p to be .5, n would have to be at least 377.

Chapter 9

9.17 a. Two-tailed

b. **Decision Rule**
   1. If $-1.96 \le$ observed $Z \le +1.96$, accept H_0.
   2. If observed $Z < -1.96$ or if observed $Z > +1.96$, reject H_0.

c. Observed $Z = -3.20$. Therefore, reject H_0.

d. Prior to selection of the sample and the determination of $\overline{X}$, the probability of accepting H_0 when $\mu = 39$ is .6368.

9.18 a. Lower-tail test

b. **Decision Rule**
   1. If observed $Z \ge -1.645$, accept H_0.
   2. If observed $Z < -1.645$, reject H_0.

c. Observed $Z = -3.20$. Therefore, accept H_0.

9.19 Observed Z is -3.20. Therefore, for the two-tailed test, reject H_0 at the .0014 level. For the lower-tail test, reject at the .0007 level.

9.20 a. $H_0: \mu \le 125$
      $H_a: \mu > 125$

b. **Decision Rule**
   1. If observed $Z \le +2.33$, accept H_0.
   2. If observed $Z > +2.33$, reject H_0.

c. Observed $Z = +1.40$. Therefore, accept H_0.

9.21 Significant at the .0808 level.

9.22   a. $H_0: \mu \geq 100$
           $H_a: \mu < 100$

      b.   **Decision Rule**
          1.   If observed $Z \geq -1.645$, accept H_0.
          2.   If observed $Z < -1.645$, reject H_0.

      c.   Observed $Z = -0.83$. Therefore, accept H_0.
      d.   Observed $Z = -0.42$. Therefore, accept H_0.

9.23   Significant at the .2033 level

9.24   a. $H_0: \mu \geq 8$
           $H_a: \mu < 8$

      b.   **Decision Rule**
          1.   If observed $Z \geq -2.33$, accept H_0.
          2.   If observed $Z < -2.33$, reject H_0.

      c.   Observed $Z = -2.25$. Therefore, accept H_0.

9.25   a. $H_0: \mu \geq 8$
           $H_a: \mu < 8$

      b.   **Decision Rule**
          1.   If observed $t \geq -2.896$, accept H_0.
          2.   If observed $t < -2.896$, reject H_0.

      c.   Observed $t = -2.647$. Therefore, accept H_0.

9.26   Significant at the .0122 level
       Significant at the .025 level (Greater accuracy could be obtained from a more extensive t table.)

9.27   a. $H_0: \mu \geq 12$
           $H_a: \mu < 12$

      b.   **Decision Rule**
          1.   If observed $Z \geq -1.28$, accept H_0.
          2.   If observed $Z < -1.28$, reject H_0.

      c.   Observed $Z = -11.78$. Therefore, reject H_0.

9.28   Observed Z of -3.79 would be significant at the .0001 level. Therefore, observed Z of -11.78 would be significant at an even lower probability level.

9.29 $\overline{X} = 189.75$; $s = 7.50$
      a. $H_0: \mu = 191$
           $H_a: \mu \neq 191$

      b.   **Decision Rule**
          1.   If $-2.947 \leq$ observed $t \leq +2.947$, accept H_0.
          2.   If observed $t < -2.947$ or if observed $t > +2.947$, reject H_0.

      c.   Observe $t = -0.667$. Therefore, accept H_0.

9.30   a. $H_0: \mu \geq 191$
           $H_a: \mu < 191$

      b.   **Decision Rule**
          1.   If observed $t \geq -2.602$, accept H_0.
          2.   If observed $t < -2.602$, reject H_0.

      c.   Observed $t = -0.667$. Therefore, accept H_0.

9.31 a. H_0: $p \le .03$
H_a: $p > .03$

b. **Decision Rule**
1. If observed $Z \le +1.645$, accept H_0.
2. If observed $Z > +1.645$, reject H_0.

c. Observed $Z = +0.99$. Therefore, accept H_0.

9.32 Significant at about the .16 level

9.33 a. H_0: $p = .6$
H_a: $p \ne .6$

b. **Decision Rule**
1. If $-1.96 \le$ observed $Z \le +1.96$, accept H_0.
2. If observed $Z < -1.96$ or if observed $Z > +1.96$, reject H_0.

c. Observed $Z = -2.45$. Therefore, reject H_0.

9.34 Significant at the .0142 level

9.35 a. H_0: $p \ge .6$
H_a: $p < .6$

b. **Decision Rule**
1. If observed $Z \ge -1.645$, accept H_0.
2. If observed $Z < -1.645$, reject H_0.

c. Observed $Z = -2.45$. Therefore, reject H_0.

9.36 a. H_0: $p \le .2$
H_a: $p > .2$

b. **Decision Rule**
1. If observed $Z \le +1.28$, accept H_0.
2. If observed $Z > +1.28$, reject H_0.

c. Observed $Z = +2.50$. Therefore, reject H_0.

9.37 Significant at about the .006 level

9.38 a. H_0: $p \le .35$
H_a: $p > .35$

b. **Decision Rule**
1. If observed $Z \le +1.645$, accept H_0.
2. If observed $Z > +1.645$, reject H_0.

c. Observed $Z = +3.31$. Therefore, reject H_0.

9.39 a. At least 58
b. At least 9
c. At least 271

9.40 a. At least 2172
b. At least 666
c. At least 252

9.41 a. The probability of a Type II error if the true population mean is 900
b. The probability of a Type I error
c. The probability of rejecting the null hypothesis correctly if the true population mean is 1100

Answers to Quantitative Questions

9.42 a.

Possible Population Mean	Probability of Accepting H_0
13	.0228
14	.1587
15	.5000
16	.8413
17	.9772
18	.9974
19	.9772
20	.8413
21	.5000
22	.1587
23	.0228

9.43 a.

Possible Population Proportion	Probability of Accepting H_0
.2	.1587
.3	.8078
.4	.9929
.5	1.0000
.6	1.0000

9.44 a.

Possible Population Mean	Probability of Accepting H_0
2200	1.0000
2300	.9999
2400	.9987
2500	.9878
2600	.9332
2700	.7734
2800	.5000
2900	.2266

9.45 Observed $Z = \dfrac{.56 - .50}{\sqrt{(.5)(.5)/100}} = +1.20$, which is within the range of -1.645 and $+1.645$. Therefore, accept H_0.

9.46 Significant at about the .23 level

9.47 At least 468

Chapter 10

10.11 a. $\mu_1 = 22$; $\mu_2 = 19.33$

b. $\mu_1 - \mu_2 = 2.67$

c. $\sigma_1 = 1.41$; $\sigma_2 = .94$

d.

$\overline{X}_1 - \overline{X}_2$	Probability
1.33	.0833
2.00	.1667
2.33	.1667
2.67	.0833
3.00	.3333
3.67	.1667
Total	1.0000

e. $\mu_{\overline{X}_1 - \overline{X}_2} = 2.67$; yes

f. $\sigma_{\overline{X}_1 - \overline{X}_2} = .67$

g. $\sigma_{\overline{X}_1 - \overline{X}_2} = .67$

10.13 a. Decision Rule
   1. If $-1.96 \leq$ observed $Z \leq +1.96$, accept H_0.
   2. If observed $Z < -1.96$ or if observed $Z > +1.96$, reject H_0.

   b. Observed $Z = -17.09$. Therefore, reject H_0.
   d. No
   e. No
   f. Null hypothesis would be accepted regardless of the level of significance. Critical Z would be positive whereas observed Z is negative.

10.14 a. Decision Rule
   1. If $-2.069 \leq$ observed $t \leq +2.069$, accept H_0.
   2. If observed $t < -2.069$ or if observed $t > +2.069$, reject H_0.

   c. Observed $t = -5.73$. Therefore, reject H_0.

10.15 Problem 10.13, part b: An observed Z of -3.79 would lead to rejection of H_0 at the .0002 level. Therefore, an observed Z of -17.09 would lead to rejection at an even lower probability level. Problem 10.14, part c: An observed Z of -5.73 would lead to rejection of H_0 at a lower probability level than .01.

10.16 Observed $F = 1.78$; critical F for $df_1 = 9$ and $df_2 = 14$ is 2.65. Therefore, accept the null hypothesis that the two population variances are equal.

10.17 a. Decision Rule
   1. If $-1.645 \leq$ observed $Z \leq +1.645$, accept H_0.
   2. If observed $Z < -1.645$ or if observed $Z > +1.645$, reject H_0.

   b. Observed $Z = +18.51$. Therefore, reject H_0.
   f. Null hypothesis would be accepted.

10.18 a. Decision Rule
   1. If $-1.701 \leq$ observed $t \leq +1.701$, accept H_0.
   2. If observed $t < -1.701$ or if observed $t > +1.701$, reject H_0.

   c. Observed $t = 1.52$. Therefore, accept H_0.

10.19 Significant at a lower probability level than .0002; significant at a higher probability level than .10 (a more extensive t table would be required for greater precision).

10.20 Observed $F = 1.10$; critical F for $df_1 = 9$ and $df_2 = 19$ is 2.43. Therefore, accept the null hypothesis of equal population variances.

10.21 a. Decision Rule
   1. If observed $t \leq +1.782$, accept H_0.
   2. If observed $t > +1.782$, reject H_0.

b. Observed $t = 6.12$. Therefore, reject H_0.

10.22 Significant at a lower probability level than .005

10.23 Observed $F = 1.29$; critical F for $df_1 = 5$ and $df_2 = 7$ is 3.97. Therefore, accept the null hypothesis of equal population variances.

10.24 **a.** Observed $t = 1.19$; critical t for 5 degrees of freedom and an upper-tail test is 2.015. Therefore, accept the null hypothesis that $\mu_d \leq 0$.

10.26 Observed $F = 2.78$; critical F for $df_1 = df_2 = 9$ is 3.18. Therefore, accept the null hypothesis of equal population variances.

10.27 **a.** $p_1 = .5; p_2 = .667$
  b. $p_1 - p_2 = -1.67$

  c.

$p_1 - p_2$	Probability
−.667	.1667
−.333	.1667
−.167	.3333
.167	.3333
Total	1.0000

  d. $\mu_{\bar{p}_1 - \bar{p}_2} = -.167$; yes

10.28 **a.** **Decision Rule**
  1. If $-1.96 \leq$ observed $Z \leq +1.96$, accept H_0.
  2. If observed $Z < -1.96$ or if observed $Z > +1.96$, reject H_0.

  b. Observed $Z = 7.58$. Therefore, reject H_0.

10.29 **a.** **Decision Rule**
  1. If observed $Z \geq -2.33$, accept H_0
  2. If observed $Z < -2.33$, reject H_0.

  b. Observed $Z = -2.12$. Therefore, accept H_0.

10.30 **b.** $H_0: \mu_d \geq 0$
  $H_a: \mu_d < 0$

  c. **Decision Rule**
  1. If observed $t \geq -3.747$, accept H_0.
  2. If observed $t < -3.747$, reject H_0.

  d. Observed $t = -1.67$. Therefore, accept H_0.

10.32 **b.** $H_0: \mu_d = 0$
  $H_a: \mu_d \neq 0$

  c. **Decision Rule**
  1. If $-2.262 \leq$ observed $t \leq +2.262$, accept H_0.
  2. If observed $t < -2.262$ or if observed $t > +2.262$, reject H_0.

  d. Observed $t = -1.92$. Therefore, accept H_0.

10.33 Significant at about the .10 level

Chapter 11

11.16 **a.** Observed $t = 1.767$; critical $t = 2.228$. Therefore, accept H_0.
  b. Observed $F = 3.10$; critical $F = 4.96$. Therefore, accept H_0.

    d. $(1.767)^2 = 3.12$

    e. $(2.228)^2 = 4.96$

11.17   **a.** Observed $t = -0.986$; critical $t = 3.355$. Therefore, accept H_0.

    b. Observed $F = 0.97$; critical $F = 11.26$. Therefore, accept H_0.

    d. $(.986)^2 = .972$

    e. $(3.355)^2 = 11.26$

11.20   **a.** SSWC = 140.2, $df_2 = 79$, MSBC = 51.96, MSWC = 1.77, observed F = 29.36, critical F (approximate) = 2.50. Reject the null hypothesis of equal population means.

    b. Six samples

    c. Sample sizes are not all the same.

11.21   **a.** $H_0\colon \mu_1 = \mu_2 = \mu_3$

        $H_a\colon \mu_1, \mu_2, \mu_3$ are not all equal

    b. $df_T = 26$; $df_1 = 2$; $df_2 = 24$

    c. MSBC = 2.20; MSWC = .215

    d. **Decision Rule**

        1. If observed $F \le 3.40$, accept H_0.

        2. If observed $F > 3.40$, reject H_0.

    e. Observed $F = 10.22$. Therefore, reject H_0.

11.23   **a.**

	$\overline{X}_1$	$\overline{X}_2$	$\overline{X}_3$
Com 1	1	−1	0
Com 2	−2	1	1

    b. Comparisons are not orthogonal.

11.24

	$\overline{X}_2 = 3.490$	$\overline{X}_3 = 3.942$	LSR
$\overline{X}_1 = 2.954$	0.536^a	0.998	$LSR_2 = 0.451$
$\overline{X}_2 = 3.490$	—	0.452	$LSR_3 = 0.474$

a denotes a significant difference. Subscripts used here in connection with $\overline{X}$ values were assigned after the values were arranged from small to large. They do not necessarily correspond with the subscripts used in connection with problems 11.21 and 11.23.

11.25   **a.** $H_0\colon \mu_1 = \mu_2 = \mu_3 = \mu_4$

        $H_a\colon \mu_1, \mu_2, \mu_3, \mu_4$ are not all equal

    b. $df_T = 19$; $df_1 = 3$; $df_2 = 16$

    c. MSBC = 6.046; MSWC = 0.444

    d. **Decision Rule**

        1. If observed $F \le 5.29$, accept H_0.

        2. If observed $F > 5.29$, reject H_0.

    e. Observed $F = 13.62$. Therefore, reject H_0.

11.27   **a.**

	$\overline{X}_1$	$\overline{X}_2$	$\overline{X}_3$	$\overline{X}_4$
Com 1	1	−1	0	0
Com 2	0	0	1	−1
Com 3	1	1	−1	−1

    b. Comparisons are mutually orthogonal.

    c. Observed F for $H_{0_{sub1}}$ is 1.60; for $H_{0_{sub2}}$, 1.225; and for $H_{0_{sub3}}$, 15.31.

        Critical F for all three tests is 8.53. Accept the first two subhypotheses and reject the third.

11.28

	$\overline{X}_2 = 7.2$	$\overline{X}_3 = 8.2$	$\overline{X}_4 = 9.0$	LSR
$\overline{X}_1 = 6.5$	0.7	1.7^a	2.5^a	$LSR_2 = 1.231$
$\overline{X}_2 = 7.2$	—	1.0	1.8^a	$LSR_3 = 1.284$
$\overline{X}_3 = 8.2$	—	—	0.8	$LSR_4 = 1.318$

Here a denotes a significant difference. Subscripts used here in connection with $\overline{X}$ values were assigned after the values were arranged from small to large. They do not necessarily correspond with the subscripts used in connection with problems 11.25 and 11.27.

11.29 a. H_0: $\mu_1 = \mu_2 = \mu_3 = \mu_4 = \mu_5$
  H_a: μ_1, μ_2, μ_3, μ_4, and μ_5 are not all equal
  b. $df_T = 29$; $df_1 = 4$; $df_2 = 25$
  c. MSBC = 17.83; MSWC = 4.27

  d. **Decision Rule**
    1. If observed $F \leq 2.76$, accept H_0.
    2. If observed $F > 2.76$, reject H_0.

  e. Observed $F = 4.17$. Therefore, reject H_0.

11.32

	$\overline{X}_2 = 3.265$	$\overline{X}_3 = 5.617$	$\overline{X}_4 = 5.655$	$\overline{X}_5 = 7.095$	LSR
$\overline{X}_1 = 3.078$	.187	2.539^a	2.577^a	4.017^a	$LSR_2 = 2.462$
$\overline{X}_2 = 3.265$	—	2.352	2.390	3.830^a	$LSR_3 = 2.586$
$\overline{X}_3 = 5.617$	—	—	.038	1.478	$LSR_4 = 2.666$
$\overline{X}_4 = 7.095$	—	—	—	1.440	$LSR_5 = 2.721$

Here a denotes a significant difference. Subscripts used here in connection with $\overline{X}$ values were assigned after the values were arranged from small to large. They do not necessarily correspond with the subscripts used in connection with problem 11.29.

11.33 a. H_0: $\mu_1 = \mu_2 = \mu_3 = \mu_4 = \mu_5 = \mu_6$
  H_a: μ_1, μ_2, μ_3, μ_4, μ_5, and μ_6 are not all equal
  b. $df_T = 29$; $df_1 = 5$; $df_2 = 24$
  c. MSBC = 8.32; MSWC = 0.88

  d. **Decision Rule**
    1. If observed $F \leq 3.90$, accept H_0.
    2. If observed $F > 3.90$, reject H_0.

  e. Observed $F = 9.42$. Therefore, reject H_0.

11.36

	$\overline{X}_2 = 0.6$	$\overline{X}_3 = 0.6$	$\overline{X}_4 = 0.8$	$\overline{X}_5 = 1.0$	$\overline{X}_6 = 3.8$	LSR
$\overline{X}_1 = .4$	.2	.2	.4	.6	3.4^a	$LSR_2 = 1.662$
$\overline{X}_2 = .6$	—	.0	.2	.4	3.2^a	$LSR_3 = 1.734$
$\overline{X}_3 = .6$	—	—	.2	.4	3.2^a	$LSR_4 = 1.781$
$\overline{X}_4 = .8$	—	—	—	.2	3.0^a	$LSR_5 = 1.816$
$\overline{X}_5 = 1.0$	—	—	—	—	2.8^a	$LSR_6 = 1.843$

Here a denotes a significant difference. Subscripts used here in connection with $\overline{X}$ values were assigned after the values were arranged from small to large. They do not necessarily correspond with the subscripts used in connection with problem 11.33.

Chapter 12

12.9 a. 5 b. 15 c. 136 d. 33.2 e. 32.93 f. 494

12.10 a. H_0: $\mu_1 = \mu_2 = \mu_3$
H_a: μ_1, μ_2, and μ_3 are not all equal

b.

Source of Variation	Sum of Squares	Degrees of Freedom	Mean Square	Observed F
Between columns	314.1333	2	157.0667	6.10
Within columns	308.8000	12	25.7333	—
Total	622.9333	14	—	—

Conclusion: The .05 level critical F is 3.88; the .01 level critical F is 6.93. Therefore, reject H_0.

12.13 a. H_0: $\mu_1 = \mu_2 = \mu_3 = \mu_4$
H_a: μ_1, μ_2, μ_3, and μ_4 are not all equal

b.

Source of Variation	Sum of Squares	Degrees of Freedom	Mean Square	Observed F	.01 Level Critical F
Between columns	2134.22	3	711.4067	30.41	5.29
Within columns	374.33	16	23.3956	—	—
Total	2508.55	19	—	—	—

Conclusion: Reject H_0.

12.16 a. H_0: $\mu_1 = \mu_2 = \mu_3 = \mu_4 = \mu_5$
H_a: μ_1, μ_2, μ_3, μ_4, and μ_5 are not all equal

b.

Source of Variation	Sum of Squares	Degrees of Freedom	Mean Square	Observed F	.05 Level Critical F
Between columns	157	4	39.25	25.16	2.76
Within columns	39	25	1.56	—	—
Total	196	29	—	—	—

Conclusion: Reject H_0.

12.19 Observed $F = 248.75/4.90 = 50.77$. Therefore, reject H_0 at the .01 level.

12.20 Observed $F = 806.67/4.74 = 170.18$. Therefore, reject H_0.

12.21 Observed $F = 33.300/1.248 = 26.68$. Therefore, reject H_0.

12.22 a. H_{01}: $\mu_1. = \mu_2. = \mu_3. = \mu_4. = \mu_5. = \mu_6. = \mu_7. = \mu_8.$
H_{a1}: $\mu_1.$, $\mu_2.$, $\mu_3.$, $\mu_4.$, $\mu_5.$, $\mu_6.$, $\mu_7.$, and $\mu_8.$ are not all equal
H_{o2}: $\mu_{.1} = \mu_{.2}$
H_{a2}: $\mu_{.1}$ and $\mu_{.2}$ are not equal

b.

Source of Variation	Sum of Squares	Degrees of Freedom	Mean Square	Observed F	.01 Level Critical F
Between rows	409.75	7	58.536	51.21	7.00
Between columns	36.00	1	36.000	31.50	12.25
Residual	8.00	7	1.143	—	—
Total	453.75	15	—	—	—

Conclusion: Reject both null hypotheses at the .01 level.

12.26 a. 5 b. 3 c. 1 d. 10 e. 1 f. 47 g. 69 h. 14.33 i. 163

12.27 **a.** H_{0_1}: $\mu_1. = \mu_2. = \mu_3.$

H_{a_1}: $\mu_1.$, $\mu_2.$, and $\mu_3.$ are not all equal

H_{0_2}: $\mu._1 = \mu._2 = \mu._3 = \mu._4 = \mu._5$

H_{a_2}: $\mu._1$, $\mu._2$, $\mu._3$, $\mu._4$, and $\mu._5$ are not all equal

b.

Source of Variation	Sum of Squares	Degrees of Freedom	Mean Square	Observed F	.05 Level Critical F
Between rows	132.1333	2	66.0667	42.17	4.46
Between columns	295.0665	4	73.7666	47.08	3.84
Residual	12.5335	8	1.5667	—	—
Total	439.7333	14	—	—	—

Conclusion: Reject both null hypotheses.

12.29 **a.** H_{0_1}: $\mu_1. = \mu_2. = \mu_3. = \mu_4.$

H_{a_1}: $\mu_1.$, $\mu_2.$, $\mu_3.$, and $\mu_4.$ are not all equal

H_{0_2}: $\mu._1 = \mu._2 = \mu._3 = \mu._4$

H_{a_2}: $\mu._1$, $\mu._2$, $\mu._3$, and $\mu._4$ are not all equal

b.

Source of Variation	Sum of Squares	Degrees of Freedom	Mean Square	Observed F	.05 Level Critical F
Between rows	117.0	3	39.0000	140.40	3.86
Between columns	110.5	3	36.8333	132.59	3.86
Residual	2.5	9	.2778	—	—
Total	230.0	15	—	—	—

Conclusion: Reject both null hypotheses.

12.30 **a.** H_{0_1}: $\mu_1. = \mu_2. = \mu_3.$

H_{a_1}: $\mu_1.$, $\mu_2.$, and $\mu_3.$ are not all equal

H_{0_2}: $\mu._1 = \mu._2 = \mu._3 = \mu._4$

H_{a_2}: $\mu._1$, $\mu._2$, $\mu._3$, and $\mu._4$ are not all equal

H_{0_3}: INT$(rc) = 0$ for all cells

H_{a_3}: INT$(rc) \neq 0$ for at least one cell

f.

Source of Variation	Sum of Squares	Degrees of Freedom	Mean Square	Observed F	.05 Level Critical F
Between rows	109.0624	3	36.3541	1.99	2.92
Between columns	334.5417	2	167.2708	9.17	3.32
INT(rc)	2409.1251	6	401.5208	22.01	2.42
Within cells	656.7500	36	18.2431	—	—
Total	3509.4792	47	—	—	—

Conclusion: Accept H_{0_1}; reject H_{0_2} and H_{0_3}.

12.31

Source of Variation	Sum of Squares	Degrees of Freedom	Mean Square	Observed F	.05 Level Critical F
Between rows	109.0624	3	36.3541	0.09	4.76
Between columns	334.5417	2	167.2708	9.17	3.32
INT(rc)	2409.1251	6	401.5208	22.01	2.42
Within cells	656.7500	36	18.2431	—	—
Total	3509.4792	47	—	—	—

Conclusion: Accept H_{0_1}; reject H_{0_2} and H_{0_3}.

12.33 a. H_{0_1}: $\mu_{1\cdot} = \mu_{2\cdot} = \mu_{3\cdot}$.
H_{a_1}: $\mu_{1\cdot}$, $\mu_{2\cdot}$, and $\mu_{3\cdot}$ are not all equal
H_{0_2}: $\mu_{\cdot 1} = \mu_{\cdot 2} = \mu_{\cdot 3}$
H_{a_2}: $\mu_{\cdot 1}$, $\mu_{\cdot 2}$, and $\mu_{\cdot 3}$ are not all equal
H_{0_3}: INT(rc) = 0 for all cells
H_{a_3}: INT(rc) $\neq$ 0 for at least one cell

b.

Source of Variation	Sum of Squares	Degrees of Freedom	Mean Square	Observed F	.05 Level Critical F
Between rows	596.2223	2	298.1115	4.70	6.94
Between columns	194.6667	2	97.3333	7.82	3.55
INT(rc)	253.7777	4	63.4444	5.89	2.93
Within cells	194.0000	18	10.7778	—	—
Total	1238.6667	26	—	—	—

Conclusion: Accept H_{0_1}; reject H_{0_2} and H_{0_3}.

12.34 b.

Source of Variation	Sum of Squares	Degrees of Freedom	Mean Square	Observed F	.05 Level Critical F
Between rows	596.2223	2	298.1115	4.70	6.94
Between columns	194.6667	2	97.3333	1.53	6.94
INT(rc)	253.7777	4	63.4444	5.89	2.93
Within cells	194.0000	18	10.7778	—	—
Total	1238.6667	26	—	—	—

Conclusion: Accept H_{0_1} and H_{0_2}; reject H_{0_3}.

Chapter 13

13.9 c. H_0: $p_1 = p_2 = p_3 = p_4 = p_5 = .2$
H_a: p_1, p_2, p_3, p_4, and p_5 are not all .2

d. Decision Rule
1. If observed $\chi^2 \leq 9.488$, accept H_0.
2. If observed $\chi^2 > 9.488$, reject H_0.

e. Observed $\chi^2 = 142.727$. Therefore, reject H_0.

13.10 a. H_0: $p_1 = p_2 = p_4 = .1818$; $p_3 = .3636$; $p_5 = .0909$
H_a: Not all p values are as indicated in the null hypothesis

b. Decision Rule
1. If observed $\chi^2 \leq 9.488$, accept H_0.
2. If observed $\chi^2 > 9.488$, reject H_0.

c. Observed $\chi^2 = 26.25$. Therefore, reject H_0.

13.11 a. H_0: $p_1 = p_2 = p_3 = p_4 = p_5 = p_6 = p_7 = p_8 = .125$
H_a: p_1, p_2, p_3, p_4, p_5, p_6, p_7, and p_8 are not all .125

d. Decision Rule
1. If observed $\chi^2 \leq 18.475$, accept H_0.
2. If observed $\chi^2 > 18.475$, reject H_0.

e. Observed $\chi^2 = 5.765$. Therefore, accept H_0.

13.12 a. $H_0: p_1 = p_2 = p_7 = p_8 = .1667; p_3 = p_4 = p_5 = p_6 = .0833$
H_a: Not all p values are as indicated in the null hypothesis

   b. **Decision Rule**
   1. If observed $\chi^2 \leq 18.475$, accept H_0.
   2. If observed $\chi^2 > 18.475$, reject H_0.

   c. Observed $\chi^2 = 3.906$. Therefore, accept H_0.

13.13 a. In addition to observed and expected frequencies, use their respective complements as well.
   b. $H_0: p_1 = p_2 = p_3 = \cdots = p_8$
$H_a: p_1, p_2, p_3, \ldots,$ and p_8 are not all equal

   c. **Decision Rule**
   1. If observed $\chi^2 \leq 18.475$, accept H_0.
   2. If observed $\chi^2 > 18.475$, reject H_0.

   d. Observed $\chi^2 = 6.945$. Therefore, accept H_0.

13.14 c. H_0: Attitude is independent of length of time shares have been held
H_a: Attitude is not independent of length of time shares have been held

   d. **Decision Rule**
   1. If observed $\chi^2 \leq 9.488$, accept H_0.
   2. If observed $\chi^2 > 9.488$, reject H_0.

   e. Observed $\chi^2 = 34.628$. Therefore, reject H_0.

13.15 c. H_0: Quality of paint job is independent of the painter
H_a: Quality of paint job is not independent of the painter

   d. **Decision Rule**
   1. If observed $\chi^2 \leq 6.251$, accept H_0.
   2. If observed $\chi^2 > 6.251$, reject H_0.

   e. Observed $\chi^2 = 3.432$. Therefore, accept H_0.

13.16 c. $H_0: p_1 = p_2 = p_3$
$H_a: p_1, p_2,$ and p_3 are not all equal

   d. **Decision Rule**
   1. If observed $\chi^2 \leq 5.991$, accept H_0.
   2. If observed $\chi^2 > 5.991$, reject H_0.

   e. Observed $\chi^2 = 23.809$. Therefore, reject H_0.

13.18 a. $H_0: P(Y > X) > .5$
$H_a: P(Y > X) < .5$

   b. **Decision Rule**
   1. If $P(S = 0) \geq .01$, accept H_0.
   2. If $P(S = 0) < .01$, reject H_0.

   c. $P(S = 0) = .0625$. Therefore, accept H_0.

13.19 a. $H_0: P(Y > X) = P(Y < X) = .5$
$H_a: P(Y > X) = P(Y < X) \neq .5$

   b. **Decision Rule**
   1. If $P(S \leq 1) \geq .05$, accept H_0.
   2. If $P(S \leq 1) < .05$, reject H_0.

   c. $P(S \leq 1) = .0352$. Therefore, reject H_0.

13.20 **a.** H_0: For the population, the sum of ranks of negative sign is greater than the sum of ranks of positive sign

H_a: For the population, the sum of ranks of negative sign is not greater than the sum of ranks of positive sign

b. **Decision Rule**
1. If observed $T > 30$, accept H_0.
2. If observed $T \leq 30$, reject H_0.

c. Observed $T = 17$
d. Reject H_0.

13.21 **a.** H_0: For the population, the sum of ranks of positive sign is the same as the sum of ranks of negative sign

H_a: For the population, the sum of ranks of positive sign is not the same as the sum of ranks of positive sign

b. **Decision Rule**
1. If observed $Z \geq -2.575$, accept H_0.
2. If observed $Z < -2.575$, reject H_0.

c. $T = 31.5$
d. Observed $Z = -3.08$. Therefore, reject H_0.

13.22 **a.** H_0: For the population, the sum of ranks of positive sign is the same as the sum of ranks of negative sign.

H_a: For the population, the sum of ranks of positive sign is not the same as the sum of ranks of negative sign.

Decision Rule
1. If $T > 4$, accept H_0.
2. If $T \leq 4$, reject H_0.

b. $T = 0$
c. T is less than 4. Therefore, reject H_0.

13.23 **a.** H_0: The two direct-mail promotional pieces are equally effective

H_a: The two direct-mail promotional pieces are not equally effective

b. **Decision Rule**
1. If $-1.96 \leq$ observed $Z \leq +1.96$, accept H_0.
2. If observed $Z < -1.96$ or if observed $Z > +1.96$, reject H_0.

c. $U = 74.5$
d. Observed $Z = +1.85$. Therefore, accept H_0.

13.24 **a.** H_0: The two teaching methods are equally effective

H_a: The two teaching methods are not equally effective

b. **Decision Rule**
1. If $U > 28$, accept H_0.
2. If $U \leq 28$, reject H_0.

c. $U = 38$
d. U is greater than 33. Therefore, accept H_0.

13.25 **a.** H_0: Price of film does not vary according to type of store where sold

H_a: Type of store does have a bearing on the price of film

b. **Decision Rule**
   1. If $H' \le 5.991$, accept H_0.
   2. If $H' > 5.991$, reject H_0.

c. $H = 12.58$
d. $H' = 12.60$
e. H' is greater than 5.991. Therefore, reject H_0.

13.26 a. H_0: Rate of interest charged is not affected by the size of the borrowing firm
   H_a: Rate of interest charged is affected by the size of the borrowing firm

b. **Decision Rule**
   1. If $H' \le 6.251$, accept H_0.
   2. If $H' > 6.251$, reject H_0.

c. $H = 13.15$
d. $H' = 13.71$
e. H' is greater than 6.251. Therefore, reject H_0.

13.27 a. H_0: The three brands of tires are equally durable
   H_a: The three brands of tires are not equally durable

b. **Decision Rule**
   1. If $H' \le 9.210$, accept H_0.
   2. If $H' > 9.210$, reject H_0.

c. $H = 1.57$
d. $H' = 1.57$
e. H' is less than 9.210. Therefore, accept H_0.

13.28 a. H_0: Color of wrapper has no bearing on quantity of sales
   H_a: Color of wrapper does have a bearing on quantity of sales

b. **Decision Rule**
   1. If $H' \le 7.815$, accept H_0.
   2. If $H' > 7.815$, reject H_0.

c. $H = 15.23$
d. $H' = 15.31$
e. H' is greater than 7.815. Therefore, reject H_0.

13.29 a. H_0: The four brands of cake mix are liked equally well
   H_a: The four brands of cake mix are not liked equally well

b. **Decision Rule**
   1. If $\chi_r^2 \le 7.815$, accept H_0.
   2. If $\chi_r^2 > 7.815$, reject H_0.

c. $\chi_r^2 = 11.01$
d. χ_r^2 is greater than 7.815. Therefore, reject H_0.

13.30 a. H_0: The three employees are considered equally worthy of promotion
   H_a: The three employees are not considered equally worthy of promotion

b. **Decision Rule**
   1. If $\chi_r^2 \le 9.00$, accept H_0.
   2. If $\chi_r^2 > 9.00$, reject H_0.

c. $\chi_r^2 = 3.25$
d. χ_r^2 is less than 9.00. Therefore, accept H_0.

Chapter 14

14.15 a. $\hat{Y} = 31.898 + 6.4372X$

b. $s_e = 4.07$

d. Observed $t = 11.60$; critical $t = 2.101$. Therefore, reject H_0 that $B = 0$.

e. $r = .9392$; $r^2 = .8822$

f. Observed $F = 134.68$; critical $F = 4.41$. Therefore, reject H_0.

g. $55.24 \leq \mu_{Y|X} \leq 60.05$

14.17 a. $\hat{Y} = .527 + 1.0091X$

b. $s_e = .5814$

d. Observed $t = 11.51$; critical $t = 2.262$. Therefore, reject H_0.

e. $r = .9677$; $r^2 = .9364$

f. Observed $F = 132.51$; critical $F = 10.56$. Therefore, reject H_0.

g. $2.92 \leq \mu_{Y|X} \leq 4.19$

h. $1.56 \leq Y_{NEW} \leq 5.55$

14.18 a. $\hat{Y} = 45.17 - 0.4338X$

c. $s_e = 4.16$

d. $r = -.9462$; $r^2 = .8953$

e. Observed $F = 153.92$; critical $F = 8.29$. Therefore, reject H_0.

f. $30.18 \leq \mu_{Y|X} \leq 34.13$

g. $23.19 \leq Y_{NEW} \leq 41.12$

14.19 a. $\hat{Y} = -1.116 + .1462X$

c. $s_e = 1.642$

d. $r = 9580$; $r^2 = .9177$

e. Observed $F = 89.21$; critical $F = 5.32$. Therefore, reject H_0.

f. $12.19 \leq \mu_{Y|X} \leq 14.82$

g. $9.50 \leq Y_{NEW} \leq 17.51$

Chapter 15

15.14 $\hat{Y} = 4 + 1.3X_1$; $\hat{Y} = 9 + 1.3X_1$

15.15 a. $\hat{Y} = .2472 + .026045X_1 + .183678X_2$

c. $s_e = .2225$

d. $c_{11} = .000653$; $c_{12} = c_{21} = -.002915$; $c_{22} = .046804$
$s_{b_1} = .005685$; $s_{b_2} = .048139$
Observed $t_1 = 4.581$; critical $t_1 = 2.179$. Therefore, reject H_{0_1}.
Observed $t_2 = 3.816$; critical $t_2 = 2.179$. Therefore, reject H_{0_1}.

e. $R^2_{Y.12} = .8617$

f. Observed $F = 37.90$; critical $F = 6.93$. Therefore, reject H_0.

15.16 a. $r^2_{Y1.2} = .6362$; $r^2_{Y2.1} = .5483$

b. Observed $F_1 = 20.99$; critical $F_1 = 4.75$. Therefore, reject H_{0_1}.
Observed $F_2 = 14.56$; critical $F_2 = 4.75$. Therefore, reject H_{0_2}.

15.17 a. $0.0137 \leq B_1 \leq 0.0384$; $0.0788 \leq B_2 \leq 0.2886$

b. $1.25 \leq \mu_{Y|12} \leq 1.76$

c. $0.95 \leq Y_{NEW} \leq 2.05$

15.18 a. $\hat{Y} = -27.7780 + .178987X_1 + 5.393328X_2$

c. $s_e = 3.521$

d. $c_{11} = .001004; \quad c_{12} = c_{21} = -.002022; \quad c_{22} = .019967$

$s_{b_1} = .111590; \ s_{b_2} = .497536$

Observed $t_1 = 1.604$; critical $t_1 = 3.499$. Therefore, accept H_{0_1}.

Observed $t_2 = 10.840$; critical $t_2 = 3.499$. Therefore, reject H_{0_2}.

e. $R^2_{Y.12} = .9606$

f. Observed $F = 85.26$; critical $F = 9.55$. Therefore, reject H_0.

15.19 a. $r^2_{Y1.2} = .2687; \ r^2_{Y2.1} = .9438$

b. Observed $F_1 = 1.29$; critical $F_1 = 12.25$. Therefore, accept H_{0_1}.

Observed $F_2 = 58.75$; critical $F_2 = 12.25$. Therefore, reject H_{0_2}.

15.20 a. $-0.2116 \leq B_1 \leq 0.5696; \quad 3.6520 \leq B_2 \leq 7.1347$

b. $46.67 \leq \mu_{Y|12} \leq 52.59$

c. $34.35 \leq Y_{NEW} \leq 60.90$

15.21 a. $\hat{Y} = 4.7592 + 3.764796X_1 + 3.806207X_2$

c. $s_e = 6.454$

d. $c_{11} = .009224; c_{12} = c_{21} = -.016910; c_{22} = .239335$

$s_{b_1} = .619853; \ s_{b_2} = 3.157419$

Observed $t_1 = 6.074$; critical $t_1 = 2.110$. Therefore, accept H_{0_1}.

Observed $t_2 = 1.205$; critical $t_1 = 2.110$. Therefore, reject H_{0_2}.

e. $R^2_{Y.12} = .7884$

f. Observed $F = 31.67$; critical $F = 3.59$. Therefore, reject H_0.

15.22 a. $r^2_{Y1.2} = .7329; \ r^2_{Y2.1} = .0975$

b. Observed $F_1 = 46.64$; critical $F_1 = 4.45$. Therefore, reject H_{0_1}.

Observed $F_2 = 1.84$; critical $F_2 = 4.45$. Therefore, accept H_{0_2}.

15.23 a. $2.4569 \leq B_1 \leq 5.0727; \quad -2.8559 \leq B_2 \leq 10.4684$

b. $20.97 \leq \mu_{Y|12} \leq 33.81$

c. $12.33 \leq Y_{NEW} \leq 42.45$

15.24 a. $\hat{Y} = 4.7592 + 3.764796X_1$ without university education

$\hat{Y} = 8.5654 + 3.764796X_1$ with university education

15.25 a. $\hat{Y} = 28.2778 - 3.083333X_1 - 0.916667X_2$

c. $s_e = 2.671$

d. $c_{11} = .083333; c_{12} = c_{21} = 0; c_{22} = .005208$

$s_{b_1} = .771051; s_{b_2} = .192763$

Observed $t_1 = -3.999$; critical $t_1 = -2.131$. Therefore, reject H_{0_1}.

Observed $t_2 = -4.755$; critical $t_2 = -2.131$. Therefore, reject H_{0_2}.

e. $R^2_{Y.12} = .7201$

f. Observed $F = 19.30$; critical $F = 3.68$. Therefore, reject H_0.

15.26 a. $r^2_{Y1.2} = .5160; \ r^2_{Y2.1} = .6012$

b. Observed $F_1 = 15.99$; critical $F_1 = 4.54$. Therefore, reject H_{0_1}.

Observed $F_2 = 22.61$; critical $F_2 = 4.54$. Therefore, reject H_{0_2}.

15.27 a. $-4.7272 \leq B_1 \leq -1.4395; -1.3276 \leq B_2 \leq -0.5056$

b. $15.92 \leq \mu_{Y|12} \leq 20.30$

c. $12.04 \leq Y_{NEW} \leq 24.08$

15.28 a. $\hat{Y} = -5.5578 + .162428X_1 - .078737X_2$

c. $s_e = .888$

d. $c_{11} = .000491$; $c_{12} = c_{21} = .004300$; $c_{22} = .454294$
$s_{b_1} = .019686$; $s_{b_2} = .598524$
Observed $t_1 = 8.251$; critical $t_1 = 2.365$. Therefore, reject H_{0_1}.
Observed $t_2 = -.132$; critical $t_2 = 2.365$. Therefore, accept H_{0_2}.
e. $R_{Y.12}^2 = .9174$
f. Observed $F = 38.89$; critical $F = 4.74$. Therefore, reject H_0.
h. $\hat{Y} = -5.5578 + .162428X_1$ without close proximity to major freeway
$\hat{Y} = -5.6365 + .162428X_1$ with close proximity to major freeway

15.29 a. Observed $F = 2.28$ or 1.54 (a discrepancy of unknown origin appears in the computer results); critical $F \cong 2.76$. Therefore, accept H_0.
c. Observed $t_1 = 0.07$; critical $t_1 \cong 2.000$. Therefore, accept H_{0_1}.
Observed $t_2 = 2.45$; critical $t_2 \cong 2.000$. Therefore, reject H_{0_2}.
Observed $t_3 = -1.300$; critical $t_3 \cong 2.000$. Therefore, accept H_{0_3}.

Chapter 16

16.11 a. Consumer goods: $T_t = 7.2667 + 2.9515t$ $(0 = 1976; t$ unit $= 1$ year)
Industrial goods: $T_t = 3.6667 + .7333t$ $(0 = 1976; t$ unit $= 1$ year)
b. Consumer goods: 10.2, 13.2, 16.1, 19.1, 22.0, 25.0, 27.9, 30.9, 33.8, 36.8; projected 39.7, 42.7, 45.6, 48.6
Industrial goods: 4.4, 5.1, 5.9, 6.6, 7.3, 8.1, 8.8, 9.5, 10.3, 11.0; projected 11.7, 12.5, 13.2, 13.9
d. Cyclical values for consumer goods: 98.0, 98.5, 93.2, 99.5, 113.6, 100.0, 100.4, 93.9, 100.6, 100.5
e. Cyclical values for industrial goods: 113.6, 98.0, 101.7, 106.1, 82.2, 98.8, 102.3, 94.7, 97.1, 109.1
f. Consumer goods: $T_t = .6158 + .0205t$ $(0 = $ July 1976; t unit $= 1$ month)
g. $T_{168} = .6158 + .0205(168) = 4.06$
h. Industrial goods: $T_t = .3081 + .0051t$ $(0 = $ July 1976; t unit $= 1$ month)
i. $T_{168} = .3081 + (.0051)(168) = 1.16$

16.12 a. Consumer goods: $\log T_t = 1.004947 + .060444t$ $(0 = 1976; t$ unit $= 1$ year)
Industrial goods: $\log T_t = .640278 + .041587t$ $(0 = 1976; t$ unit $= 1$ year)
b. Consumer goods: 11.6, 13.4, 15.4, 17.6, 20.3, 23.3, 26.8, 30.8, 35.4, 40.7; projected 46.8, 53.7, 61.8, 71.0
Industrial goods: 4.8, 5.3, 5.8, 6.4, 7.1, 7.8, 8.5, 9.4, 10.3, 11.4; projected 12.5, 13.8, 15.2, 16.7
d. Cyclical values for consumer goods: 86.2, 97.0, 97.4, 108.0, 123.2, 107.3, 104.5, 94.2, 96.0, 90.9
e. Cyclical values for industrial goods: 104.2, 94.3, 103.4, 109.4, 84.5, 102.6, 105.9, 95.7, 97.1, 105.3
f. Consumer goods: $\log T_t = 1.007466 - 1.079181 + .005037t$ or $\log T_t = -.071715 + .005037t$ $(0 = $ July 1976; t unit $= 1$ month)
g. $\log T_{168} = -.071715 + (.005037)(168) = 0.774502$ in log form; $T_{168} = 5.95$
h. Industrial goods: $\log T_t = .642011 - 1.079181 + .003466t$ or $\log T_t = -.437170 + .003466t$ $(0 = $ July 1976; t unit $= 1$ month)
i. $\log T_{168} = -.437170 + (.003466)(168) = .145118$ in log form; $T_{168} = 1.4$

16.13 a. Consumer goods: $T_t = 6.5167 + 3.3265t - .0341t^2$ $(0 = 1976; t$ unit $= 1$ year)
Industrial goods: $T_t = 4.75 + .1917t + .0492t^2$ $(0 = 1976; t$ unit $= 1$ year)

Answers to Quantitative Questions

b. Consumer goods: 9.81, 13.03, 16.19, 19.28, 22.30, 25.25, 28.13, 30.95, 33.69, 36.37, projected 38.98, 41.52, 44.00, 46.40

Industrial goods: 4.99, 5.33, 5.77, 6.30, 6.94, 7.67, 8.50, 9.43, 10.66, 11.59; projected 12.81, 14.14, 15.56, 17.08

d. Cyclical values for consumer goods: 101.9, 99.8, 92.6, 98.5, 121.1, 99.0, 99.5, 93.7, 100.9, 101.7

e. Cyclical values for industrial goods: 100.2, 93.8, 104.0, 111.1, 86.5, 104.3, 105.9, 95.4, 93.8, 103.5

f. $T_t = .5546 + .023087t - .00002t^2$ (0 = July 1976; t unit = 1 month)

g. $T_{168} = .5546 + (.023087)(168) - (.00002)(168)^2 = 3.87$

h. $T_t = .3965 + .001359t + .000028t^2$ (0 = July 1976; t unit = 1 month)

i. $T_{168} = .3965 + (.001359)(168) + (.000028)(168)^2 = 1.42$

16.14 a. Revenues: $T_t = 98.8889 + 51.3333t$; (0 = 1976; t unit = 1 year)

Costs: $T_t = -45.75 + 55.4833t$; (0 = 1976; t unit = 1 year)

Profits: $T_t = 144.6389 - 4.15t$ (0 = 1976; t unit = 1 year)

b. Trend values for revenues: 150.22, 201.56, 252.89, 304.22, 355.56, 406.89, 458.22, 509.56, 560.89, projection 612.22

Trend values for costs: 9.73, 65.22, 120.70, 176.18, 231.67, 287.15, 342.63, 398.12, 456.60, projection 509.08

Trend values for profits: 140.49, 136.34, 132.19, 128.04, 123.89, 119.74, 115.59, 111.44, 107.29, projection 103.14

d. Cyclical values for revenues: 86.5, 111.6, 94.9, 105.2, 98.4, 103.2, 96.0, 103.0, 98.1

Cyclical values for costs: 688.6, 130.3, 91.1, 79.5, 82.0, 83.6, 93.4, 100.5, 116.7

Cyclical values for profits: 44.8, 102.7, 98.3, 140.6, 129.1, 150.3, 103.8, 112.2, 15.8

e. $T_t = 26.3264 + 3.208331t$ (0 = 3rd quarter 1976; t unit = 1 quarter)

f. $T_{40} = 26.3264 + (3.208331)(40) = 154.66$

g. $T_t = -9.7037 + 3.467706t$ (0 = 3rd quarter 1976; t unit = 1 quarter)

h. $T_{40} = -9.7037 + (3.467706)(40) = 129.00$

i. $T_t = 36.03 - .259375t$ (0 = 3rd quarter 1976; t unit = 1 quarter)

j. $T_{40} = 36.03 - (.259375)(40) = 25.66$

16.15 a. Revenues: $\log T_t = 2.159126 + .070903t$ (0 = 1976; t unit = 1 year)

Costs: $\log T_t = 1.705480 + .113036t$ (0 = 1976; t unit = 1 year)

Profits: $\log T_t = 2.227797 - .041546t$ (0 = 1976; t unit = 1 year)

b. Trend values for revenues: 169.84, 199.96, 235.42, 277.17, 326.32, 384.19, 452.32, 532.54, 626.98, projection 738.17

Trend values for costs: 65.84, 85.42, 110.81, 143.75, 186.49, 241.93, 313.85, 407.16, 528.20, projection 685.23

Trend values for profits: 153.55, 139.54, 126.81, 115.24, 104.73, 95.17, 86.49, 78.60, 71.43, projection 64.91

d. Cyclical values for revenues: 76.5, 112.5, 101.9, 115.5, 107.3, 109.3, 97.3, 98.6, 87.7

Cyclical values for costs: 101.8, 99.5, 99.3, 97.4, 101.9, 99.2, 102.0, 98.2, 100.9

Cyclical values for profits: 41.0, 100.3, 102.5, 156.2, 152.8, 189.1, 138.7, 159.0, 23.8

e. $\log T_t = 1.565929 + .017726t$ (0 = 3rd quarter 1976; t unit 1 quarter)

f. $\log T_{40} = 1.565929 + (.017726)(40) = 2.274959$; $T_{40} = 188.35$

g. $\log T_t = 1.117550 + .028259t$ (0 = 3rd quarter 1976; t unit = 1 quarter)

h. $\log T_{40} = 1.117550 + (.028259)(40) = 2.247910$; $T_{40} = 176.97$

i. $\log T_t = 1.620544 - .010387t$ (0 = 3rd quarter 1976; t unit = 1 quarter)

j. $\log T_{40} = 1.620544 - (.010387)(40) = 1.205064$; $T_{40} = 16.03$

16.17 d. Seasonal indexes:
January: 107.7; February: 97.9; March: 98.9; April: 98.8; May: 96.4; June: 97.4; July: 100.1; August: 98.0; September: 93.6; October: 95.9; November: 96.1; December: 118.9

16.19 a. 248.1
       b. 30.6
       c. 54.9

Chapter 17

17.17 a. Predicted values for no-change model:

Period	Predicted Value
2	225
3	245
4	190
5	182
6	263
7	290
8	210
9	240
10	320
11	250

       b. Predicted values for the three-month moving-average forecasting model:

Period	Predicted Value
4	220.00
5	205.67
6	211.67
7	245.00
8	254.33
9	246.67
10	256.67
11	270.00

       c. MAD for no-change model is 50.11; for 3-month moving-average model, 43.28.
       d. MSE for no-change model is 3264.33; for 3-month moving-average model, 2531.21.
       e. $\bar{r}$ is .0000 for both models.
       f. ADR for the no-change model is 1.78; for 3-month moving-average model, 2.0.

17.18 a. See part *a* of problem 17.17.
       b. Predicted values for the five-month moving-average forecasting model:

Period	Predicted Value
6	221.0
7	234.0
8	227.0
9	237.0
10	264.6
11	262.0

       c. MAD for no-change model is 50.11; for 5-month moving-average model, 40.72.

d. MSE for no-change model is 3264.33; for 5-month moving-average model, 2521.63.

e. $\bar{r}$ is .0000 for both models.

f. ADR for no-change model is 1.75; for 5-month moving-average model, 2.0.

17.19 a. See part *a* of problem 17.17.

b. Predicted values for the five-month weighted moving-average forecasting model:

Period	Predicted Value
6	216.17
7	226.77
8	232.66
9	242.97
10	258.43
11	260.00

c. MAD for no-change model is 50.11; for the 5-month weighted moving-average model, 36.68.

d. MSE for no-change model is 3264.33; for the 5-month weighted moving-average model, 2358.13.

e. $\bar{r}$ = .0000 for both models.

f. ADR for no-change model is 1.75; for the 5-month weighted moving-average model the only run, which may be incomplete, is of 5 months' duration.

17.20 See answers to problems 17.17, 17.18, and 17.19.

17.21 a. No-change-in-the-change model:

Period	Predicted Value	Period	Predicted Value
5	1	16	11
6	3	17	9
7	8	18	2
8	8	19	6
9	10	20	13
10	14	21	14
11	16	22	18
12	14	23	23
13	16	24	34
14	14	25	27
15	8	26	32

b. Exponential smoothing model $\hat{Y}_{t+2} = .1Y_t + .9\hat{Y}_t$:

Period	Predicted Value	Period	Predicted Value
3	7.00	15	9.06
4	6.80	16	9.35
5	6.52	17	9.41
6	6.27	18	9.17
7	6.24	19	9.06
8	6.22	20	9.15
9	6.40	21	9.34
10	6.76	22	9.80
11	7.28	23	10.52
12	7.75	24	11.87
13	8.38	25	12.88
14	8.84	26	14.39

c. No-change-in-the-change model is better: 3.05 versus 4.66.
d. No-change-in-the-change model is better: 13.65 versus 39.62.
e. No-change-in-the-change model is better: .8678 versus .6838.
f. Exponential smoothing model is better: 5.50 versus 2.00.

17.22 a. See answer to part a of problem 17.21.
b. Exponential smoothing formula $\hat{Y}_{t+2} = .5Y_t + .5\hat{Y}_t$:

Period	Predicted Value	Period	Predicted Value	Period	Predicted Value
3	7.00	13	12.55	23	14.53
4	6.00	14	12.78	24	19.27
5	5.00	15	11.89	25	20.63
6	4.50	16	11.94	26	24.32
7	5.25	17	10.97		
8	5.63	18	8.99		
9	6.81	19	8.49		
10	8.41	20	9.25		
11	10.20	21	10.12		
12	11.10	22	12.06		

c. The no-change-in-the-change model is better: 3.05 versus 3.84.
d. The no-change-in-the-change model is better: 13.65 versus 22.38.
e. The no-change-in-the-change model is better: .8678 versus .7329.
f. The exponential smoothing model is better: 3.25 versus 2.00.

17.23 See answers to problems 17.21 and 17.22.

17.24 a. No-change model:

Period	Predicted Value	Period	Predicted Value
2	8.71	10	9.94
3	8.71	11	10.13
4	8.96	12	10.49
5	8.93	13	10.41
6	9.03	14	9.97
7	9.44	15	8.79
8	9.69	16	8.29
9	9.90		

b. $\hat{Y}_{t+1} = 1.2988 + .8602t$

c.

Period	Predicted Value	Period	Predicted Value
2	8.79	10	9.85
3	8.79	11	10.01
4	9.01	12	10.32
5	8.98	13	10.25
6	9.07	14	9.88
7	9.42	15	8.86
8	9.63	16	8.43
9	9.82		

d. The models perform similarly with respect to the MAD criterion: .29 for the no-change model and .30 for the autoregressive model.
e. The models perform similarly with respect to the MSE criterion: .17 for the no-change model and .16 for the autoregressive model.

  f. The models perform similarly with respect to the $\bar{r}$ criterion: .7819 for the no-change model and .7807 for the autoregressive model.

17.25   **a.** See answer to part *a* of problem 17.24.

     b. $\hat{Y}_{t+1} = 4.742 + 1.0622Y_t - .5719Y_{t-3}$

     c.

Period	Predicted Value
6	9.25
7	9.35
8	9.64
9	9.93
10	10.09
11	9.90
12	9.96
13	10.22
14	10.11
15	9.54
16	8.08

     d. No change is better: .29 versus .47.

     e. No change is better: .17 versus .41.

     f. No change is better: .7819 versus .0000.

Chapter 18

18.11   **a.** $U(O_{11}) = 0$; $U(O_{12}) = .5$; $U(O_{21}) = .5$; $U(O_{22}) = 1.0$

     b. $EU(A_1) = .25$; $EU(A_2) = .75$

18.13   **a.** Action A_1 is dominated by Action A_3. Therefore, eliminate A_1.

     b. $EU(A_2) = .19$; $EU(A_3) = .81$

     c. A_3 is the Bayes action.

18.15   **a.** Table I: $EMP(A_1) = 8.0$; $EMP(A_2) = 7.2$; A_1 is the Bayes action.

       Table II: $EMP(A_1) = 10.2$; $EMP(A_2) = 3.6$; A_1 is the Bayes action.

18.16   **a.** No inadmissable actions

     b. $EMP(A_1) = \$6.5$ million; $EMP(A_2) = \$5.9$ million

     c. A_1 is the Bayes action.

18.17   **a.**

	A_1	A_2
θ_1	0	48
θ_2	0	11
θ_3	0	0
θ_4	9	0
θ_4	23	0

     b. $EOL(A_1) = \$6.4$ million; $EOL(A_2) = \$7.0$ million; A_1 is the Bayes action.

18.18   **a.** Action A_3 is dominated by Action A_2. Therefore, eliminate A_3.

     b. $EMP(A_1) = \$735$ million; $EMP(A_2) = \$610$ million

     c. A_1 is the Bayes action.

18.19 a.

	A_1	A_2
θ_1	0	380
θ_2	200	0
θ_3	140	0

b. EOL(A_1) = \$84 million; EOL(A_2) = \$209 million; A_1 is the Bayes action.

18.20 Action A_2 is the Bayes action with EMP of \$75,000 over \$25,000.

18.21 a.

$P(\theta_i)$	Demand θ_i	Supply A_j 0	1	2	3	4	5
.05	0	0	−15	−30	−45	−60	−75
.15	1	0	20	5	−10	−25	−40
.30	2	0	20	40	25	10	−5
.25	3	0	20	40	60	45	30
.15	4	0	20	40	60	80	65
.10	5	0	20	40	60	80	100

b. EMP(A_1) = \$0; EMP(A_2) = \$18.25 thousand; EMP(A_3) = \$31.25 thousand; EMP(A_4) = \$33.75 thousand; EMP(A_5) = \$27.50 thousand; EMP(A_6) = \$16.00 thousand.

c. The Bayes action is A_4 with an expected monetary payoff of \$33.75 thousand.

18.22 a. MP(A_1) = −\$100,000 + \$3.00θ

b. −\$150,000 + \$4.00θ

c. θ_e = 50,000

d. If sales turned out to be less than 50,000, A_1 would be the preferred action; if over 50,000, A_2 would be preferred.

e. E(θ) = 37,000

f. EMP(A_1) = \$11,000; EMP(A_2) = −\$2000; A_1 would be the Bayes action.

18.24 a. MP(A_1) = −\$6,000,000 + \$13.00θ

b. MP(A_2) = \$0

c. θ_e = 461,538

d. If sales are less than 461,538, Action A_2 would be preferred.
If sales are greater than 461,538, Action A_1 would be preferred.

e. E(θ) = 580,000

f. A_1 is the Bayes action.

18.25 θ_e would be reduced to 460,769.

18.26 a. Action A_4 would no longer be inadmissible; EMP(A_4) would have to be determined.

b. EMP(A_1) = \$660,000; EMP(A_2) = \$630,000; EMP(A_3) = \$400,000; EMP(A_4) = \$655,000; A_1 is still the Bayes action.

18.27 a.

	A_1 Make Movie	A_2 Do Not Make Movie
θ_1: Successful	\$3,000,000	\$0
θ_2: Unsuccessful	−\$1,000,000	0
	$P(\theta_1)$ = .2	$P(\theta_2)$ = .8

b. EMP(A_1) = −\$200,000; EMP(A_2) = \$0; Action A_2 is the Bayes action.

Chapter 19

19.6   **a.** $EVPI_0 = \$6.4$ million

19.7   **a.** $P_1(\theta_1) = .12$; $P_1(\theta_2) = .32$; $P_1(\theta_3) = .24$; $P_1(\theta_4) = .16$; $P_1(\theta_5) = .16$
      b. $EMP_1(A_1) = \$10.80$ million; $EMP_1(A_2) = \$6.64$ million; Bayes action is A_1.
      c. $EVPI_1 = \$5.12$ million

19.8   EMP(Prior Analysis) = \$6.5 million
      EMP(Posterior analysis) − cost of sampling = \$8.3 million − \$1 million = \$7.3 million
      Therefore, perform posterior analysis:
      EVSI = \$1.8 million
      ENGS = \$1.8 million − \$1 million = \$.8 million

19.9   **a.** $EVPI_0 = EVWPI_0 − EMP(A_1) = \$819$ million − \$735 million = \$84 million

19.10   **a.** $P_1(\theta_1) = .7971$; $P_1(\theta_2) = .2029$; $P_1(\theta_3) = .0000$
      b. $EMP_1(A_1) = \$882.608$ million; $EMP_1(A_2) = \$620.29$ million; A_1 is the Bayes action.
      c. $EVWPI_1 = \$923.188$ million
         $EVPI_1 = \$923.188 − \$882.608 = \$40.58$ million

19.11   **a.**   EMP(Prior analysis) = \$735 million; EMP(Posterior analysis) = \$735 million
         EMP(Posterior analysis) − Cost of sampling = \$734,999,500. Therefore, do not "sample."
      b.   Payoffs for the eight possible strategies (in millions of dollars):
         $S(A_1,A_1,A_1) = \$735.00$
         $S(A_1,A_2,A_2) = 700.50$
         $S(A_1,A_2,A_1) = 682.66$
         $S(A_1,A_1,A_2) = 657.47$
         $S(A_2,A_1,A_1) = 644.50$
         $S(A_2,A_1,A_2) = 640.47$
         $S(A_2,A_2,A_1) = 614.03$
         $S(A_2,A_2,A_2) = 610.00$
      c.   Since the normal-form analysis says: choose action A_1 regardless of "sample" indication and
         make \$735 million, settle for prior analysis.

19.12   **a.** $EVPI_1 = \$0.6$ million
      b.   Let x_1 represent the fact that two critics out of four like the screenplay and then use the bino-
         mial formula to determine likelihoods. $P_1(\theta_1|x_1) = .36$ and $P_1(\theta_2|x_1) = .64$
      c. $EMP_1(A_1) = \$0.44$ million; $EMP_1(A_2) = \$0$. Therefore, make the film.
      d. $EVPI_1 = \$0.64$ million

19.13   EMP(Prior analysis) = \$0
      EMP(Posterior analysis) − Cost of sampling $\cong \$347,840 − \$2000 \cong \$345,840$

19.14   **a.**   EMP(Prior analysis) = \$1.20 million
         EMP(Posterior analysis) − Cost of sampling = \$1.149 million
         Settle for prior analysis.
      b.   EMP(Prior analysis) = \$1.20 million
      c.   Not relevant
      d.   Not relevant

19.15   **a.**   EMP(Prior analysis) = \$2.20 million
         EMP(Posterior analysis) − Cost of sampling $\cong \$2.31$ million
      b.   EMP(Posterior analysis) $\cong 2.32$ million
      c.   About \$0.12 million
      d.   About \$0.11 million

19.16 a. Yes; $EMP_0(A_1) = \$8500$
b. $10,600
c. EVSI = $2300
d. ENGS = $2100

19.17 a. $EMP_1(A_1) \cong \$16,892$

19.18 a. No; $EMP_0(A_2) = \$25,000$, which is less than his salary of $30,000
b. EMP(Posterior analysis) − Cost of sampling = $38,700
c. $8800
d. $8700

19.19 $43,750; $43,650 when cost of the survey is deducted

19.20 PAYOFF TABLE

Sales Next Month θ_i (Dozens)	Quantity to Purchase (Dozens)			
	A_1: 1000	A_2: 1500	A_3: 2000	A_4: 2500
1000	8000	7,000	6,000	5,000
1500	8000	12,000	11,000	10,000
2000	8000	12,000	16,000	15,000
2500	8000	12,000	16,000	20,000

EMP(Prior analysis) = $10,500; EMP(Posterior analysis) $\cong$ $10,779.99; EVSI $\cong$ $279.99

Index

Note: *Italics* indicate pages with illustrations or tables.

t Distribution

Areas in Both Tails Combined

Example: For 10 degrees of freedom and an area of .05 in the two tails combined, the *t* value is 2.228.

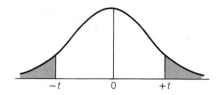

Degrees of Freedom	Area in Both Tails Combined			
	.10	.05	.02	.01
1	6.314	12.706	31.821	63.657
2	2.920	4.303	6.965	9.925
3	2.353	3.182	4.541	5.841
4	2.132	2.776	3.747	4.604
5	2.015	2.571	3.365	4.032
6	1.943	2.447	3.143	3.707
7	1.895	2.365	2.998	3.499
8	1.860	2.306	2.896	3.355
9	1.833	2.262	2.821	3.250
10	1.812	2.228	2.764	3.169
11	1.796	2.201	2.718	3.106
12	1.782	2.179	2.681	3.055
13	1.771	2.160	2.650	3.012
14	1.761	2.145	2.624	2.977
15	1.753	2.131	2.602	2.947
16	1.746	2.120	2.583	2.921
17	1.740	2.110	2.567	2.898
18	1.734	2.101	2.552	2.878
19	1.729	2.093	2.539	2.861
20	1.725	2.086	2.528	2.845
21	1.721	2.080	2.518	2.831
22	1.717	2.074	2.508	2.819
23	1.714	2.069	2.500	2.807
24	1.711	2.064	2.492	2.797
25	1.708	2.060	2.485	2.787
26	1.706	2.056	2.479	2.779
27	1.703	2.052	2.473	2.771
28	1.701	2.048	2.467	2.763
29	1.699	2.045	2.462	2.756
30	1.697	2.042	2.457	2.750
40	1.684	2.021	2.423	2.704
60	1.671	2.000	2.390	2.660
120	1.658	1.980	2.358	2.617
Normal distribution	1.645	1.960	2.326	2.576

Source: From Table III of Fisher and Yates, *Statistical Tables for Biological, Agricultural and Medical Research*, published by Longman Group Ltd., London (previously published by Oliver & Boyd, Edinburgh) and by permission of the authors and publishers.